Landmark Papers in Cell Biology

Landmark Papers in Cell Biology

Selected Research Articles Celebrating Forty Years of The American Society for Cell Biology

EDITORS

Joseph G. Gall
Carnegie Institution of Washington

J. Richard McIntosh
University of Colorado

THE AMERICAN SOCIETY FOR CELL BIOLOGY

COLD SPRING HARBOR LABORATORY PRESS

Landmark Papers in Cell Biology

Published by Cold Spring Harbor Laboratory Press, Cold Spring Harbor, NY, and the American Society for Cell Biology, Bethesda, MD

Printed in the United States of America

Acquisition Editor: John Inglis
Project Coordinator: Mary Cozza
Production Editor: Mala Mazzullo
Desktop Editor: Danny deBruin
Interior Book Designer: Denise Weiss
Cover Designer: Ed Atkeson/Berg Design

Front Cover: Nucleolar genes from an amphibian oocyte. These genes, which code for ribosomal RNA, repeat along the DNA axis and are visualized because approximately 100 enzymes are simultaneously transcribing each gene. The gradient of fibrils extending from each gene contains ribosomal RNA precursor molecules in progressive stages of completion (electron micrograph, x 43,000). See page 101. (Cover figure reprinted, with permission, from Miller O.L., Jr. and Beatty B.R. 1969. *Science* **164:** 955–957 [© American Association for the Advancement of Science].)

Library of Congress Cataloging-in-Publication Data

Landmark papers in cell biology / edited by Joseph G. Gall, J. Richard McIntosh.
p. cm.
Includes bibliographical references and index.
ISBN 0-87969-602-8 (cloth : alk. paper)
1. Cytology. I. Gall, Joseph G. II. McIntosh, J. Richard.
QH574 .L36 2000
571.6--dc21 00-063876

10 9 8 7 6 5 4 3 2 1

Students and members of the American Society for Cell Biology may order discounted copies of this book from the American Society for Cell Biology, 8120 Woodmont Avenue, Suite 750, Bethesda, MD 20814-2755. Phone: (301) 347-9300. Fax: (301) 347-9310. E-mail: ascbinfo@ascb.org. World Wide Web Site http://www.ascb.org

All Cold Spring Harbor Laboratory Press publications may be ordered directly from Cold Spring Harbor Laboratory Press, 10 Skyline Drive, Plainview, New York 11803-2500. Phone: 1-800-843-4388 in Continental U.S. and Canada. All other locations: (516) 349-1930. FAX: (516) 349-1946. E-mail: cshpress@cshl.org. For a complete catalog of Cold Spring Harbor Laboratory Press publications, visit our World Wide Web Site http://www.cshl.org

For orders from Europe, the Middle East, and Africa, British pound pricing is provided. Orders are fulfilled and shipped from Cold Spring Harbor Laboratory Press–Europe c/o Lavis Marketing, 73 Lime Walk, Headington, Oxford OX3 7AD, U.K. Phone: +44 (0) 1865 741541. FAX: +44 (0)1865 750079. E-mail: cshlpress.europe@cshl.org. World Wide Web Site: http://www.cshlpress.co.uk

Contents

Nuclear Envelope and Nuclear Import

Mitosis and Cell Cycle Control

Cell Membrane and Extracellular Matrix

Protein Synthesis and Membrane Traffic

Cytoskeleton

Preface

THE 40 YEARS SINCE THE FOUNDING of the American Society for Cell Biology have brought spectacular advances in our understanding of cells, from the beautiful intricacy of their structure to the incredible complexity of the biochemical reactions and pathways that regulate their behavior. This volume contains reprints of 42 seminal papers that illustrate these advances, along with brief commentaries to place the papers in their historical and intellectual context. The idea for this collection came during a committee meeting in Washington in December 1999 to consider ways of celebrating the Society's 40th anniversary. We agreed that a volume containing reprints of important papers in cell biology would be a good way to highlight advances in the field since the founding of the Society in 1960. The idea seemed simple enough until we actually set about the task. How should the papers be chosen? How many reprints should be included? How, indeed, did one define cell biology, given the blurring that has taken place among the traditional fields of cell, molecular, developmental, and genetic biology? We soon realized that a simple reprinting of papers would be of limited use, especially for students, without some accompanying commentary. At this point we sought outside help.

We started by canvassing members of the editorial board of *Molecular Biology of the Cell*, asking them to nominate the one or two papers they felt should be included in this volume. From nearly 40 responses we received an equal number of suggestions, many of which ended up as selections. We are deeply indebted to the people who wrote us, especially to those who provided critiques that helped to place these papers in proper context. However, as we had expected, the majority of suggestions involved papers that one could classify as "cytoplasmic." Those long-time members of the Society who study the nucleus know how frustrating it has been over the years to find equal time for their favorite part of the cell. So a second appeal went out to about 30 card-carrying nuclear cell biologists, who responded enthusiastically. Again, many of their suggestions ended up in the final volume.

What are the criteria we used in selecting papers? We started with the idea that we should include about 40 papers, in part because we are celebrating the 40th anniversary of the ASCB, but more importantly because this number of papers comes out to about the right number of pages, not so few that we had to skimp on important areas of cell biology and not so many that the volume would be unwieldy. Significance and quality of the papers were, of course, our main criteria. We decided early on that we would arbitrarily limit the selection to studies on eukaryotes. One could easily compile a collection of major advances in cell biology based entirely on *Escherichia coli* and a few other prokaryotes, but this would have a very different flavor from this volume. The hardest decisions came in drawing the line between cell biology and the closely related disciplines of biochemistry, molecular biology, developmental biology, and genetics. In the end, we decided that some purely molecular papers had to be included because of their impact on the field as a whole, but in general we have omitted purely biochemical papers as well as developmental and genetic studies unless they had a strong cell biological bent. Some sub-specialties of cell biology have made stronger gains than others during the past 40 years and some have grown by increments without that one defining experiment everyone knows about. The latter have again been omitted. Finally, for practical reasons, we tended to select shorter papers, even when this meant including the first announcement of a discovery rather than the fully documented follow-up study.

We are painfully aware that not everyone will be happy with our selections, especially with our omissions. Some have expressed the opinion that this sort of compilation gives a distorted picture of how science really progresses—that most fields develop by gradual accretion, not by brilliant insight from single people. We are sympathetic with that viewpoint and hope that this collection of papers will be viewed not as winners in a beauty contest, but as landmarks for students and others who want to see the path that cell biology has followed in the last 40 years and to imagine where it may be going.

Again we thank those who suggested important papers and provided information about them. We are especially grateful to colleagues who read preliminary versions of some of our commentaries: Karen Beemon, Alex Franzusoff, Elizabeth Hay, Kathryn Howell, Joanna Olmsted, Tom Pollard, Andrew Staehelin, Mark Winey, and Yixian Zheng. The beautiful electron micrograph of transcribing genes on the cover was provided by Oscar Miller. We also thank Elizabeth Marincola, Executive Director of the ASCB, Heather Joseph, Managing Editor of *Molecular Biology of the Cell*, and John Inglis, Director of the Cold Spring Harbor Laboratory Press, for their support of this project. Actual production of the volume was overseen by Mary Cozza, Mala Mazzullo, and Danny deBruin, who made certain that the technical aspects of the book met the highest standards.

Joseph G. Gall

October 2000 **J. Richard McIntosh**

Genome Organization and Replication

Taylor J.H., Woods P.S., and Hughes W.L. 1957. The organization and duplication of chromosomes as revealed by autoradiographic studies using tritium-labeled thymidine. *Proc. Natl. Acad. Sci.* **43:** 122–128.

ALTHOUGH BY THE 1950S EVERYONE AGREED THAT DNA was an important constituent of chromosomes, the number and arrangement of DNA molecules in the chromosome was a complete mystery. J. Herbert Taylor and his colleagues labeled chromosomes of the broad bean *Vicia faba* with ^{3}H-thymidine and followed the distribution of label in the progeny chromosomes by autoradiography. The short pathlength of the electrons emitted by ^{3}H was essential for their experiment, which was the first to use ^{3}H-thymidine as a biological tracer. They found that both chromatids of a chromosome were equally labeled at the first mitosis after administration of the isotope, but only one of the two chromatids was labeled at the second mitosis (more accurately, because of sister chromatid exchanges, only one chromatid was labeled at any point along the chromosome). Their demonstration of semi-conservative distribution of label during chromosome replication was published before the better known experiment of Meselson and Stahl (1), which showed semi-conservative distribution of density label during DNA replication in *Escherichia coli*. Although Taylor's experiment is often cited as the first evidence that chromosomes of eukaryotes consist of a single DNA molecule (the unineme hypothesis), the authors themselves did not draw that conclusion. Instead they proposed a complex multi-stranded model, arguing that it was "inconceivable" that the chromosome could be a single supercoiled double helix of DNA. In this respect Taylor and his colleagues reflected the nearly universal opinion of the day that chromosomes were multi-stranded. It was later found that these results could be reproduced without the need for autoradiography by incorporating bromodeoxyuridine into chromosomes followed by staining with Giemsa dye (2). "New" and "old" DNA were differentially stained in this procedure, which had much better spatial resolution than autoradiography and was particularly valuable for studying multiple sister chromatid exchanges. Compelling evidence for the unineme model was later derived from several other types of experiments: kinetics of DNase breakage of chromosomes (3), kinetics of DNA reassociation after denaturation (4), direct visualization of transcription units (5; Miller and Beatty, this volume), and isolation of chromosome-sized lengths of DNA (6).

1. Meselson M. and Stahl F.W. 1958. The replication of DNA in *Escherichia coli*. *Proc. Natl. Acad. Sci.* **44:** 671–682.
2. Wolff S. and Perry P. 1974. Differential Giemsa staining of sister chromatids and the study of sister chromatid exchanges without autoradiography. *Chromosoma* **48:** 341–353.
3. Gall J.G. 1963. Kinetics of deoxyribonuclease action on chromosomes. *Nature* **198:** 36–38.
4. Britten R.J. and Kohne D.E. 1968. Repeated sequences in DNA. *Science* **161:** 529–540.
5. Miller O.L., Beatty B.R., Hamkalo B.A., and Thomas C.A. 1970. Electron microscopic visualization of transcription. *Cold Spring Harbor Symp. Quant. Biol.* **35:** 505–512.
6. Kavenoff R. and Zimm B.H. 1973. Chromosome-sized DNA molecules from *Drosophila*. *Chromosoma* **41:** 1–27.

Reprinted from the Proceedings of the NATIONAL ACADEMY OF SCIENCES,
Vol. 43, No. 1, pp. 122–128. January, 1957.

THE ORGANIZATION AND DUPLICATION OF CHROMOSOMES AS REVEALED BY AUTORADIOGRAPHIC STUDIES USING TRITIUM-LABELED THYMIDINE

BY J. HERBERT TAYLOR,* PHILIP S. WOODS, AND WALTER L. HUGHES

DEPARTMENT OF BOTANY, COLUMBIA UNIVERSITY; BIOLOGY DEPARTMENT AND MEDICAL DEPARTMENT, BROOKHAVEN NATIONAL LABORATORY

Communicated by Franz Schrader, October 26, 1956

Information on the macromolecular organization of chromosomes and their mode of duplication has been difficult to obtain in spite of numerous attempts. One point of attack, long recognized but until recently unattainable, was the selective labeling of some component of the chromosome, the distribution of which could be seen in succeeding cell divisions. Reichard and Estborn[1] demonstrated that N^{15}-labeled thymidine was a precursor of deoxyribonucleic acid (DNA) and that it was not diverted to the synthesis of ribonucleic acid. Recently Friedkin *et al.*[2] and Downing and Schweiger[3] have used C^{14}-labeled thymidine to study DNA synthesis. In chick embryos and *Loctobacillus* there was no appreciable diversion of the tracer to ribonucleic acid. In view of these findings, thymidine appeared to be the intermediate required for the experiment, but the labels so far employed have not been satisfactory for microscopic visualization by autoradiographic means. In order to determine whether an individual chromosome among several in a cell is radioactive, autoradiographs with resolution to chromosomal dimensions must be obtained. Resolution at this level is difficult if not impossible to obtain with most isotopes, since the range of their beta particles is relatively great. Theoretically tritium should provide the highest resolution obtainable, since the beta particles have a maximum energy of only 18 Kev, corresponding to a range of little more than a micron in photographic emulsions. Consequently, identification of this label in particles as small as individual chromosomes should be possible. With this in mind, tritium-labeled thymidine was prepared and used to label chromosomes and to follow their distribution in later divisions by the use of photographic emulsions.

Materials and Methods.—Tritium-labeled thymidine of high specific activity (3×10^3 mc/mM) was prepared by catalytic exchange of tritium from the carboxyl group of acetic acid to a carbon atom in the pyrimidine ring of thymidine (details of the method to be described elsewhere).

Seedlings of *Vicia faba* (English broad bean) were grown in a mineral nutrient solution containing 2–3 μg/ml of the radioactive thymidine. This plant was selected because it has 12 large chromosomes, one pair of which is morphologically distinct, and because the length of the division cycle and the time of DNA synhesisl in the cycle are known.[4] After growth of the seedlings in the isotope solution for the appropriate time, the roots were thoroughly washed with water and the seedlings were transferred to a nonradioactive mineral solution containing colchicine (500 μg/ml) for further growth. At appropriate intervals roots were fixed in ethanol–acetic acid (3:1), hydrolyzed 5 minutes in 1 *N* HCl, stained by the Feulgen reaction, and squashed on microscope slides. Stripping film was applied, and autoradiographs were prepared as described previously.[5]

Experimental Design and Results.—Roots remained in the isotope solution for 8 hours, which is approximately one-third of the division cycle.[4] Since about 8 hours intervene between DNA synthesis in interphase and the next anaphase, few if any nuclei which had incorporated the labeled thymidine should have passed through a division before the roots were transferred to the colchicine solution. In the presence of colchicine, chromosomes contract to the metaphase condition, and the sister chromatids (daughter chromosomes), which ordinarily lie parallel to each other, spread apart. The sister chromatids remain attached at the centromere region for a period of time, but they finally separate completely before transforming into an interphase nucleus. Because colchicine prevents anaphase movement and the formation of daughter cells, but does not prevent chromosomes from duplicating, the number of duplications following exposure to the isotope can be determined for any individual cell by observing the number of chromosomes. Cells without a duplication after transfer to colchicine will have the usual 12 chromosomes at metaphase (c-metaphase), each with the two halves (sister chromatids) spread apart but attached at the centromere. Cells with one intervening duplication will contain 24 chromosomes, and those with two duplications will contain 48 chromosomes.

Two groups of roots were fixed. The first group remained in the colchicine solution 10 hours. The second group remained in the colchicine for 34 hours. In the first group, cells at metaphase had only 12 chromosomes, which indicated that none of these had duplicated more than once during the experiment. The chromosomes in these cells were all labeled, and, furthermore, the two sister chromatids of each chromosome were equally and uniformly labeled (Fig. 1, *a* and *b*). The amount of radioactivity in the chromosomes varied from cell to cell, as would be expected in a nonsynchronized population of cells, but within a given cell the label in different chromosomes was remarkably uniform.

In the second group, cells contained either 12, 24, or 48 chromosomes. Those with 12 chromosomes usually were not labeled, but when labeling occurred, sister chromatids were uniformly labeled as in the first group. In cells with 24 chromosomes, all chromosomes were labeled; however, only one of the two sister chromatids of each was radioactive (Fig. 2, *a* and *b*). Evidently the pool of labeled precursor in the plant had been quickly depleted after the plant was removed from the isotope solution, and these cells with 24 chromosomes had gone through a second duplication in the absence of labeled thymidine.

In the few cells with 48 chromosomes, analysis of all 48 was not possible. However, in several cases where most of the chromosomes were well separated and flattened, approximately one-half of the chromosomes of a complement contained one labeled and one nonlabeled chromatid, while the remainder showed no label in either chromatid. The appearance of cells with 48 chromosomes in a 34-hour period in colchicine also indicates that there was some variation in the predicted 24-hour division cycle.

In cells with 24 and 48 chromosomes a few chromatids were labeled along only a part of their length, but in these cases the sister chromatids were labeled in complementary portions (Fig. 2, *b*, *arrow*). This is the expected situation following sister chromatid exchange and demonstrates that resolution is sufficient to see crossing over in cytological preparations. A careful search of numerous cells with 12 chromosomes failed to yield a decisive case of half-chromatid exchange, which

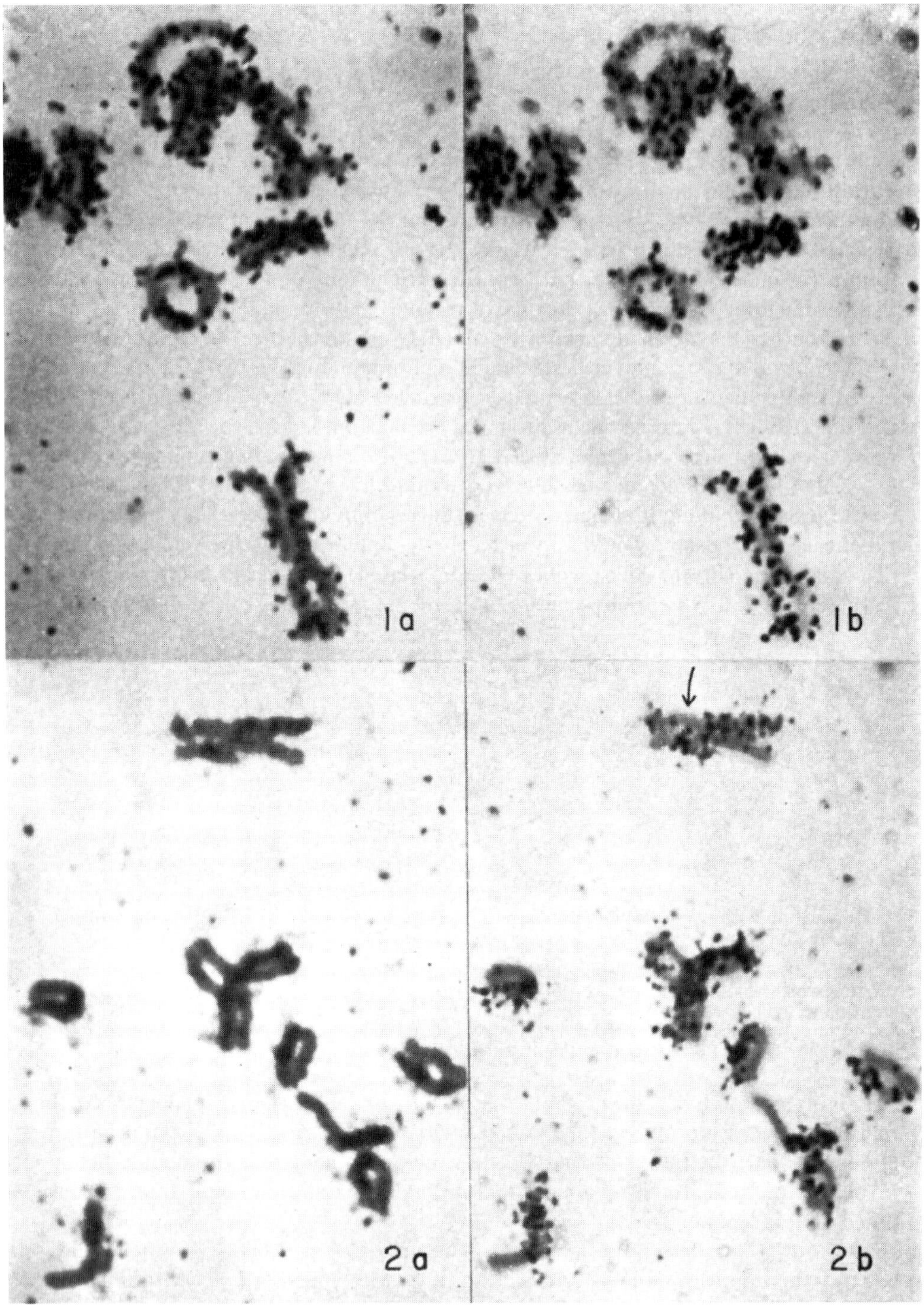

FIG. 1.—Photograph of several chromosomes of a c-metaphase at the first division after labeling occurred; *a*, chromosomes with the chromatids spread apart but still attached at the centromere; *b*, grains in the emulsion above the chromosomes. ×2,200.

FIG. 2.—Photograph of several chromosomes after labeling and one replication in the absence of labeled precursor; *a*, several of the chromosomes from a cell containing 24 chromosomes with chromatids spread but attached at the centromere; *b*, grains in the emulsion above the chromosomes in *a*. ×2,200.

should produce a portion of a chromatid without a label at the first c-metaphase following incorporation of the isotope.

Interpretation and Discussion.—These results indicate (1) that the thymidine built into the DNA of a chromosome is part of a physical entity that remains intact during succeeding replications and nuclear divisions, except for an occasional chromatid exchange; (2) that a chromosome is composed of two such entities probably complementary to each other; and (3) that after replication of each to form a chromosome with four entities, the chromosome divides so that each chromatid (daughter chromosome) regularly receives an "original" and a "new" unit. These conclusions are made clearer by the diagrams in Figure 3. Beginning with two complementary nonlabeled strands in a chromosome, the two strands separate and a complementary labeled strand is produced along each original strand. At the succeeding metaphase each chromatid would appear labeled, although it contains both a labeled and a nonlabeled strand. At a succeeding replication in the absence of labeled precursor, each strand would have a nonlabeled complementary strand produced along its length. At the succeeding metaphase only one chromatid of each chromosome would appear labeled. Following another replication, only one-half of the chromosomes would contain a labeled chromatid, as demonstrated in those cells with 48 chromosomes.

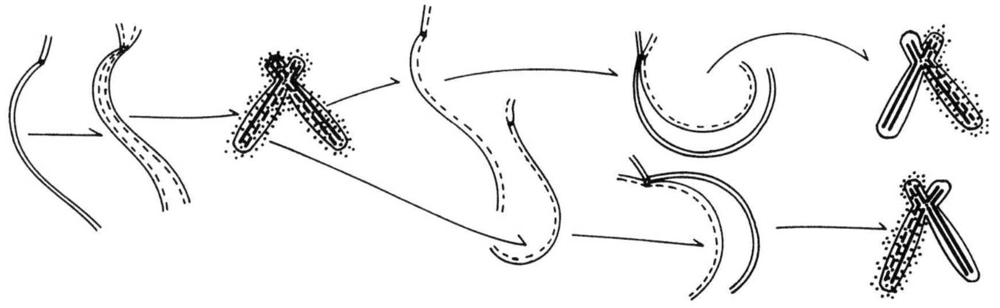

Fig. 3.—Diagrammatic representation of proposed organization and mode of replication which would produce the result seen in the autoradiographs. The two units necessary to explain the results are shown, although these were not resolved by microscopic examination. Solid lines represent nonlabeled units, while those in dashed lines are labeled. The dots represent grains in the autoradiographs.

It is immediately apparent that this pattern of replication is analogous to the replicating scheme proposed for DNA by Watson and Crick.[6] We cannot be sure, of course, that separation of the two polynucleotide chains in the double helix is involved, for the chromosome is several orders of magnitude larger than the proposed double helix of DNA.

We know that these large metaphase chromosomes are coiled into at least one helix at the microscopic level and perhaps are twice coiled, a helix within a helix. That the chromosome could be a single supercoiled double helix of DNA is inconceivable when one considers the amount of DNA in a large chromosome. Chromosomes are much more likely to be composed of multistranded units. To explain their duplication as well as their mechanical properties at the microscopic level, they may be visualized as two complementary multistranded ribbons lying flat upon each other, as shown in Figures 4 and 5. Ribbons of this type with more flexible materials on their outer edges will have a tendency to coil. If the edges contract faster than the central strands when the chromosome begins to shorten, the ribbons fold, one within the other, so as to form a long, trough-shaped cylinder (Fig. 4), and with further contraction they assume the form of a helix. Continued differential contraction would produce a helix within a helix, but the mechanical properties of the model are outside the scope of this discussion.

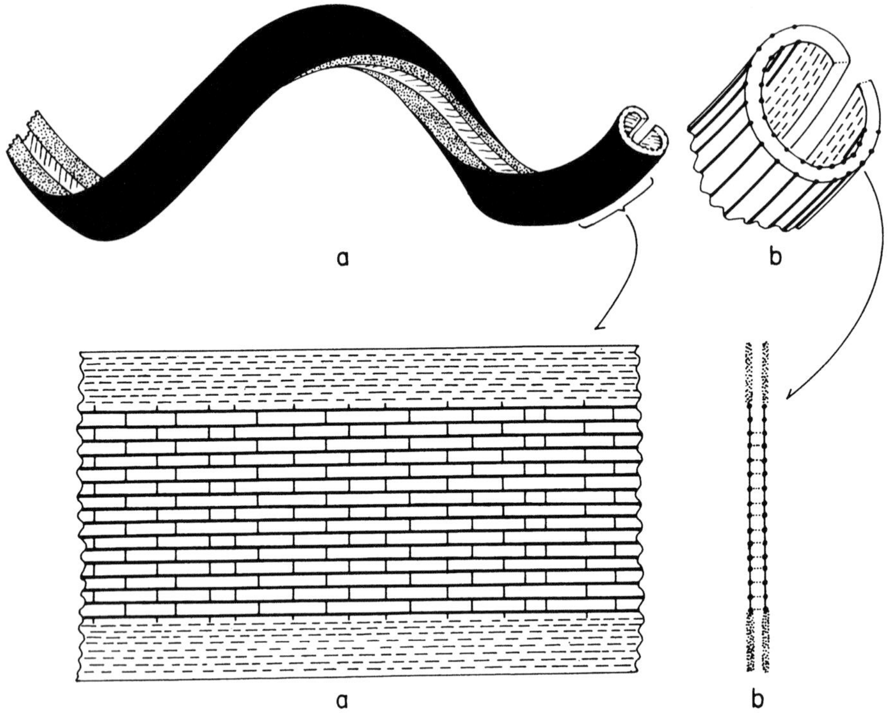

FIG. 4.—Schematic drawing of the proposed ribbon-shaped chromosome with the two multistranded units folded together and coiled; *a*, a single gyre from the coiled chromosome; *b*, detail in cross-section.

FIG. 5.—Diagrammatic sketch of the multistranded units uncoiled and flattened; *a*, short portion in longisection; *b*, cross-section. The number and size of strands shown have no special significance. Although the assumption is made that the strands contain DNA, they do not necessarily correspond to Watson-Crick double helices.

Although the chromosome model is provisional and may require considerable modification and refinement, it has many of the features necessary for duplication and the known stability of genetic materials. The large surface area exposed when the two complementary ribbons are extended would facilitate their rapid duplication. A double-stranded unit with two complementary faces has a high stability,[7] and if the two complementary units are composed of multiple, identical strands cross-bonded, the stability of the larger units should be even greater. Such large units would have a high probability of being transmitted as physical entities. If separation of the complementary faces involves the separation of intertwined double helices of DNA, unwinding presents a problem, but perhaps not an impossible one.[8]

The findings reported here are consistent with those recently reported for the distribution of P^{32}-labeled phage DNA by Levinthal.[9] His data indicated that about 40 per cent of phage DNA is contained in one piece which is divided equally in the formation of two daughter particles, but undergoes no further distribution during the production of about 150 particles that result from the infection of a bacterium. A phage particle would be analogous to a chromosome before duplication, and the first two daughter particles, each of which would be labeled, would be analogous to the two sister chromatids. Since the chromosome is much larger and contains much more DNA than the phage, it is remarkable that they behave in a similar manner in distribution of their DNA during replication.

Our findings are at variance with the report by Mazia and Plaut[10] which was based on the analysis of the anaphase distribution of chromosomes labeled with C^{14}-thymidine. Their data, obtained by the estimation of number of grains over pairs of anaphase or telophase nuclei, indicated a segregation of labeled and non-labeled units at the first division following the incorporation of the isotope. It is entirely possible that in their experiment more than one division occurred between the time of incorporation and the time the telophase nuclei were analyzed. If this had been the case in the present experiment, unequal distribution of activity in sister chromatids would have been observed in diploid cells.

Summary.—Tritium-labeled thymidine was prepared and used for labeling chromosomes during their duplication. Analysis of autoradiographs showed that both daughter chromosomes resulting from duplication in the presence of labeled thymidine appeared equally and uniformly labeled. After an ensuing duplication in the absence of the labeled DNA precursor, the label appeared in only one of each two chromatids (daughter chromosomes). These findings indicate that DNA is synthesized as a unit which extends throughout the length of the chromosome. The units remain intact through succeeding replications and nuclear divisions, except for occasional chromatid exchanges. Each chromosome is composed of two such units, probably complementary to each other. After each replication the four resulting units separate, so that each daughter chromosome always contains an "original" and a "new" unit. To explain the results, a model with two complementary units and a scheme of replication analogous to the Watson-Crick model of DNA is proposed.

* This work was initiated and the original experiments carried out while the senior author was a research collaborator in the Biology Department, Brookhaven National Laboratory. It has been continued at Columbia University under Contract AT (30-1)-1304 with the Atomic Energy Commission.

[1] P. Reichard and B. Estborn, *J. Biol. Chem.*, **188**, 839, 1951.

[2] M. Freidkin, D. Tilson, and D. Roberts, *J. Biol. Chem.*, **220**, 627, 1956.

[3] M. Downing and B. S. Schweigert, *J. Biol. Chem.*, **220**, 521, 1956.

[4] A. Howard and S. R. Pelc, *Exptl. Cell Research*, **2**, 178, 1951.

[5] J. H. Taylor and R. D. McMaster, *Chromosoma*, **6**, 489, 1954.

[6] J. D. Watson and F. H. C. Crick, *Nature*, **171**, 964, 1953; *Cold Spring Harbor Symposia Quant. Biol.*, **18**, 123, 1953.

[7] H. Kacser, *Science*, **124**, 151, 1956.

[8] N. Ardley, *Nature*, **176**, 465, 1955; M. Delbrück, these Proceedings, **40**, 783, 1955; G. Gamow, these Proceedings, **41**, 7, 1955; J. R. Platt, these Proceedings, **41**, 181, 1955; D. P. Bloch, these Proceedings, **41**, 1058, 1955; C. Levinthal and H. R. Crane, these Proceedings, **42**, 436, 1956.

[9] C. Levinthal, these PROCEEDINGS, **42**, 394, 1956.
[10] D. Mazia and W. Plaut, *Biol. Bull.*, **109**, 335, 1955.

Ris H. and Plaut W. 1962. Ultrastructure of DNA-containing areas in the chloroplast of *Chlamydomonas. J. Cell Biol.* **13:** 383–391.

CLASSIC CYTOLOGICAL STUDIES HAD SUGGESTED that both chloroplasts and mitochondria reproduced by fission of the existing organelles, much like the growth and division of micro-organisms. During the 1940s and 1950s genetic studies demonstrated that both the petite phenotype of budding yeast (1) and streptomycin resistance in an alga (2) showed non-Mendelian inheritance, suggestive of a cytoplasmic localization for the relevant genes. The paper reproduced here, and comparable studies on mitochondria that followed by only a year (3), presented convincing cell biological evidence for the presence of DNA in organelles. These discoveries provided a molecular basis for the genetic transmission of organellar traits. The isolation and characterization of organellar DNA from chloroplasts (4) and then from mitochondria (5) put this cytological work on a molecular footing and opened the way for detailed studies of inheritance from multiple genomes within a single organism. We now know that chloroplasts and mitochondria are built from proteins encoded on both nuclear and organellar genomes. Their growth and division depend on a complex interplay between different systems for transcription and translation, so these organelles can be assembled from a proper balance of proteins from two genetic sources.

---

1. Slonimski V. and Ephrussi B. 1949. Action de l'acriflavine sur les levures. V. Le système des cytochromes des mutants "petite colonie." *Ann. Inst. Pasteur, Paris* **77:** 47–63.
2. Sager R. 1954. Mendelian and non-mendelian inheritance of streptomycin resistance in *Chlamydomonas reinhardi. Proc. Natl. Acad. Sci.* **40:** 356–364.
3. Nass M.M.K. and Nass S. 1963. Intramitochondrial fibers with DNA characteristics: I and II. *J. Cell Biol.* **19:** 593–611; 613–629.
4. Sager R. and Ishida M.R. 1963. Chloroplast DNA in *Chlamydomonas. Proc. Natl. Acad. Sci.* **50:** 725–730.
5. Luck D.J.L. and Reich E. 1964. DNA in mitochondria of *Neurospora crassa. Proc. Natl. Acad. Sci.* **52:** 931–938.

ULTRASTRUCTURE OF DNA-CONTAINING AREAS IN THE CHLOROPLAST OF *CHLAMYDOMONAS*

HANS RIS, Ph.D., and WALTER PLAUT, Ph.D.

From the Department of Zoology, University of Wisconsin, Madison

ABSTRACT

The chloroplast of *Chlamydomonas moewusii* was examined by electron microscopic and cytochemical methods for the possible presence of DNA. Both the Feulgen reaction and acridine orange indicated the presence within the chloroplast of one or more irregularly shaped DNA-containing bodies generally in the vicinity of the pyrenoid. Electron micrographs revealed 25 A microfibrils in these areas which correspond to DNA macromolecules with respect to their location, morphology, and sensitivity to deoxyribonuclease digestion. The possibility that this material is the genetic system of the chloroplast and the hypothesis that the chloroplast represents an evolved endosymbiont are discussed.

Except in certain viruses, primary genetic systems appear to be associated with deoxyribonucleic acid (DNA). Electron microscope studies of the DNA-containing cell components have so far revealed two different types of ultrastructure. In bacteria (Kellenberger, 1960), *Streptomyces* (Hopwood and Glauert, 1960 *a*), and blue-gree algae (Hopwood and Glauert, 1960 *b*. Ris and Singh, 1961), one finds areas of low density and variable shape which contain fibrils about 25 to 30 A thick. By spreading protoplasts on the surface of water, these fibrils can be obtained intact for electron microscopy; under these conditions the fibrils are found to be very long, with few free ends visible. They represent DNA macromolecules; the linkage group of a bacterial cell may consist of one or several such macromolecules (Kleinschmidt and Lang, 1960). In most cells of animals and plants the DNA is associated with histone-type protein and more complex proteins to form fibrils 40 and 100 A thick (Ris, 1961). These fibrils, probably as multistranded bundles, are organized into chromosomes which show complicated patterns of replication and structural changes in the mitotic cycle. In those cells the primary genetic system is in the chromosomes, but transmission of hereditary characteristics through cytoplasmic units has been described for a number of cases. One of the best known of these "plasmids" (Lederberg, 1952) is the chloroplast of plant cells (*cf.* Rhoades, 1955). It seemed to us of interest to investigate the possibility that plastids contain structurally organized DNA and thus provide a basis for the supposition that the genetic system of the plastid is basically similar to the known nuclear systems.

The presence of DNA in plastids has been the subject of considerable debate. Cytochemical tests and chemical analyses on isolated chloroplasts (*cf.* Granick, 1955) and incorporation studies with tritiated thymidine (Stocking and Gifford, 1959) have been claimed by some investigators to show the presence of DNA. However, the data are not conclusive at the present time. At least one cytochemical investigation (Littau, 1958) has failed to demonstrate DNA; the chemical analysis of isolated plastids is complicated by the fact that nuclear contamination is

This paper was dedicated to Dr. F. Schrader on the occasion of his 70th birthday. We are saddened by his death in March of this year and dedicate this paper to his memory.

Reprinted from THE JOURNAL OF CELL BIOLOGY, 1962, Vol. 13, No. 3, pp. 383-391
Printed in U.S.A.

difficult to avoid in the isolation of plastids (Jagendorf and Wildman, 1954); lastly, the incorporation of tritiated thymidine is not a sufficient test for the presence of DNA unless the demonstration of incorporation is at least accompanied by a suitable enzymatic control experiment.

In the course of a search for plastids that give a positive Feulgen reaction we found that the chloroplast of *Chlamydomonas* contains several small bodies which are Feulgen-positive; this alga was therefore chosen for a study of the ultrastructure of the plastid.

MATERIAL AND METHODS

Chlamydomonas moewusii Gerloff (Indiana University Algae Collection #96–) were grown in liquid medium (modified Myer medium) or on agar slants. For light microscopy cells were fixed in alcohol-acetic acid for 10 minutes and squashed between slide and coverslip. After freezing in liquid air the coverslips were removed and some slides were subjected to hydrolysis for 10 minutes in N HCl at 60°C and stained with Schiff's reagent for 45 minutes. Control slides were stained without hydrolysis. Other slides, with cells fixed in the same way, were washed in 0.1 M acetate buffer, pH 4.5, stained for 15 to 30 minutes in acridine orange (Harleco) (0.02 mg/ml in 0.02 M acetate buffer, pH 4.5), and then washed with several changes of 0.1 M acetate buffer at the same pH. Fluorescence of acridine orange-stained cells was observed with a 2 mm Zeiss apochromat, NA 1.4, Xenon arc illumination, a BG-12 exciter filter, and a Bausch and Lomb ocular barrier filter, and photographed with High Speed Ektachrome film (Daylight). Some of the preparations stained with acridine orange were treated, before staining, with ribonuclease (Worthington, 0.3 mg/ml in distilled water, pH adjusted to 7.0, 1 hour at 50°C), others with deoxyribonuclease (Worthington, 1X crystallized, 0.3 mg/ml in ¼ strength McIlvaine buffer at pH 7.0, with 4×10^{-3} M $MgSO_4$, 5 hours at 40°C), and still others with both enzymes in sequence.

For electron microscopy, cells growing as a film on liquid medium were fixed in Kellenberger's fixative for 1 hour, post-treated with uranyl acetate or versene (Ryter *et al.*, 1958), and embedded in Epon 812 according to the method of Luft (1958). Similar cells were fixed in 10 per cent formalin in veronal-acetate buffer (final pH 7.8) for 10 minutes, washed in distilled water, and digested with deoxyribonuclease (1 mg of enzyme per ml) at 40° for 5 hours. A control group was incubated in buffer only. The cells were then postfixed in Kellenberger's buffered OsO_4, treated with uranyl acetate, and embedded in Epon 812. One-micron-thick sections were dried on glass slides, hydrolyzed in N HCl at 60°C for 20 minutes, and stained in Schiff's reagent for 45 minutes.

Electron micrographs were taken with a Siemens Elmiskop II b electron microscope at original magnifications of 7500 × and 15,000 × on Gevaert Scientia plates 19D50 and developed in D-19.

RESULTS

1. *Demonstration of DNA in the Chloroplast of Chlamydomonas*

In Feulgen-stained cells the chloroplast contains one or more small bodies of irregular shape which give a reaction of about the same intensity as the nucleus (Fig. 1). In shape and size these bodies vary from spheres 0.5 micron in diameter to oblongs 0.5 by 1.5 micra; they are located mostly around the pyrenoid. Without prior hydrolysis these structures, like the nucleus, do not stain and the only positive reaction is found in the cell walls of some of the cells. After digestion with deoxyribonuclease, the Feulgen reaction of formalin-fixed cells is very faint or absent in both nucleus and chloroplast.

A more sensitive demonstration of DNA is the yellowish-green fluorescence obtained after staining with acridine orange (Rustad, 1958). *Chlamydomonas* cells fixed in alcohol-acetic acid (3:1) and stained with acridine orange show a strong orange-red fluorescence which covers the yellowish-green fluorescence of the nucleus (the latter can be seen with suitable filtration). After ribonuclease treatment, the red fluorescence is absent, and one sees now a moderately strong yellowish-green fluorescence in the nucleus and also in the bodies within the chloroplast which were found to be stained with the Feulgen reaction (Fig. 2). In addition, there is a faint dull green fluorescence in the rest of the cytoplasm. In cells digested with deoxyribonuclease after ribonuclease treatment, the yellowish-green fluorescence of the nucleus and the chloroplast bodies cannot be seen and only the faint green background fluorescence remains (Fig. 3).

We conclude from these observations that in *Chlamydomonas moewusii* small DNA-containing bodies are closely associated with the chloroplast. (DNA in this context is operationally defined as an acid-insoluble substance which is stained by the Feulgen reaction after hydrolysis, gives the characteristic yellowish-green fluorescence after acridine-orange staining, and is sensitive to deoxyribonuclease but not to ribonuclease digestion.) The fact that two other species of *Chlamydomonas*, *rheinhardi* and *eugametos*, were found upon examination of Feulgen-stained preparations to have similar bodies associated with the chloroplasts of every cell suggests that these elements are not restricted to the single species chosen for detailed examination. The specific location of the DNA-containing bodies was studied in the *C. moewusii* cells fixed for electron microscopy, sectioned at one micron thickness and stained with the Feulgen reaction. These preparations showed that the stained bodies are always located within the chloroplast, usually around the pyrenoid, and never outside the chloroplast.

This observation also suggests that the DNA-containing structures of the *Chlamydomonas* chloroplast are not directly attributable to bacterial contaminants which might per chance have been present in our cultures. A subsequent test for bacterial contamination in the cultures from which our experimental cells were taken gave negative results.[1] A Feulgen analysis of *Chlamydomonas* cells derived from a culture known to be infected showed numerous bacteria with typical chromatin bodies outside but never inside the algae. Moreover, the shape of the bacterial chromatin bodies is more regular and clearly different from the DNA-containing structures of the *Chlamydomonas* chloroplast.

If we can thus rule out chance contamination by bacteria as an explanation for the presence of extranuclear DNA in *Chlamydomonas*, we must consider the possibility that a less autonomous but still distinct biological system is associated with

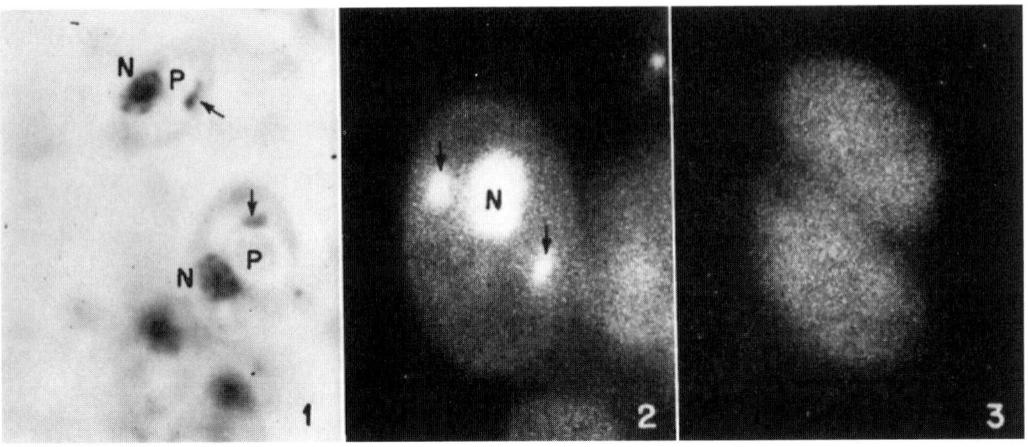

FIGURE 1

Two cells of *Chlamydomonas moewusii* fixed in ethanol-acetic acid 3:1 and stained with the Feulgen reaction. *N* = nucleus, *P* = pyrenoid. The arrows point to the DNA-containing structure in the chloroplast. × 2500.

FIGURE 2

Cell fixed in ethanol-acetic acid 3:1, treated with ribonuclease, stained with acridine orange. Copy of fluorescence micrograph taken on High Speed Ektachrome film. The nucleus (*N*) and the bodies in the chloroplast (arrows) show bright yellowish-green fluorescence. × 3000.

FIGURE 3

Two cells of *Chlamydomonas moewusii* fixed in ethanol-acetic acid 3:1, digested with ribonuclease followed by deoxyribonuclease, stained and photographed as in Fig. 2. The fluorescence in nucleus and chloroplast structures has disappeared. × 3000.

[1] We wish to thank Mr. L. McBride for testing our culture for bacterial contamination.

the chloroplast. As shown below, analysis with the electron microscope suggests a degree of structural integration of the DNA-containing material with the rest of the chloroplast which makes this unlikely.

2. *The Ultrastructure of the DNA-Containing Bodies*

The finely fibrillar nucleoplasm of microorganisms is sensitive to fixation and is only preserved under certain conditions (Ryter *et al.*, 1958). Since we were interested in comparing the DNA-containing regions of the chloroplast with the nucleoplasm of microorganisms, we used fixative and post-treatments developed for bacteria (Ryter *et al.*, 1958).

The identification of the Feulgen-positive regions in the sections proved difficult at first because their small size makes the study of adjacent thin and thick sections almost impossible. However, the general size, shape, and location on the periphery of the pyrenoid led to the recognition of special areas of characteristic appearance located between chloroplast lamellae. Once seen, they are easily identified in every cell. (Figs. 4 to 6). The membranes of chloroplast lamellae are thicker and denser than the outer chloroplast membrane and thus can be distinguished from it (*cm*, Figs. 5 and 6). The Feulgen-positive areas are bounded by chloroplast lamellar membranes and are, therefore, part of the chloroplast and not cytoplasmic pockets extending into the chloroplast. These regions contain granules which resemble the ribosomes in the cytoplasm (*R*, Figs. 5 and 6). Similar granules were found between photosynthetic membranes throughout the chloroplast. In the DNA-containing regions they are associated with areas of low density which contain fibrils about 25 to 30 A thick (arrows, Figs. 5 and 6). There is a striking similarity between these regions and the nucleoplasm in blue-green algae (Hopwood and Glauert, 1960 *b*; Ris and Singh, 1961) and bacteria. In bacteria these fibrils were shown to be DNA (Kleinschmidt and Lang, 1960; Lee, 1960). To test the hypothesis that the 25 A fibrils in the chloroplasts are also DNA, algae fixed in formalin were treated with deoxyribonuclease (controls were treated with buffer without enzyme) and then postfixed with Kellenberger's method.

While the over-all preservation of structure is not so good as in the cells fixed directly in buffered OsO_4, the DNA-containing regions are easily identified in the controls by the ribosome-like granules and the light areas with characteristic 25 A fibrils. After deoxyribonuclease no such fibrils can be found anywhere in the chloroplast, but the ribosome-like granules remain. This suggests that, as in bacteria, the fine fibrils, seen in sections through the Feulgen-positive regions, correspond to the DNA macromolecules which are removed with deoxyribonuclease.

It is of interest to compare this result with the effect of deoxyribonuclease on the nucleus. Thick sections of the same block were stained with the Feulgen reaction and showed staining of nuclei in the control, but not in the enzyme-treated material. Electron micrographs of control and treated nuclei, stained with uranyl acetate, show the same general structure, namely random sections through the characteristic chromosome fibrils. In the enzyme-treated material, however, these fibrils are considerably less dense than in the controls. The nucleolus on the other hand shows no change in density after deoxyribonuclease treatment. If the DNA is combined here with protein as in other chromosomes that have been analyzed chemically, these results could mean that after formalin fixation DNA can be removed from the chromosome fibrils without completely destroying their structure.

DISCUSSION

The involvement of plastids in non-chromosomal, cytoplasmic inheritance has now been well established (*cf.* Rhoades, 1955). In several plants, experiments have shown that plastids are self-dependent (Lederberg, 1952), *i.e.*, they are derived

FIGURE 4

Electron micrograph of section of *Chlamydomonas moewusii*, fixed with Kellenberger's procedure. *CP* = chloroplast; *M* = mitochondrion; *N* = nucleus; *P* = pyrenoid. The arrows point to the DNA-containing structures in the chloroplast. × 21,000.

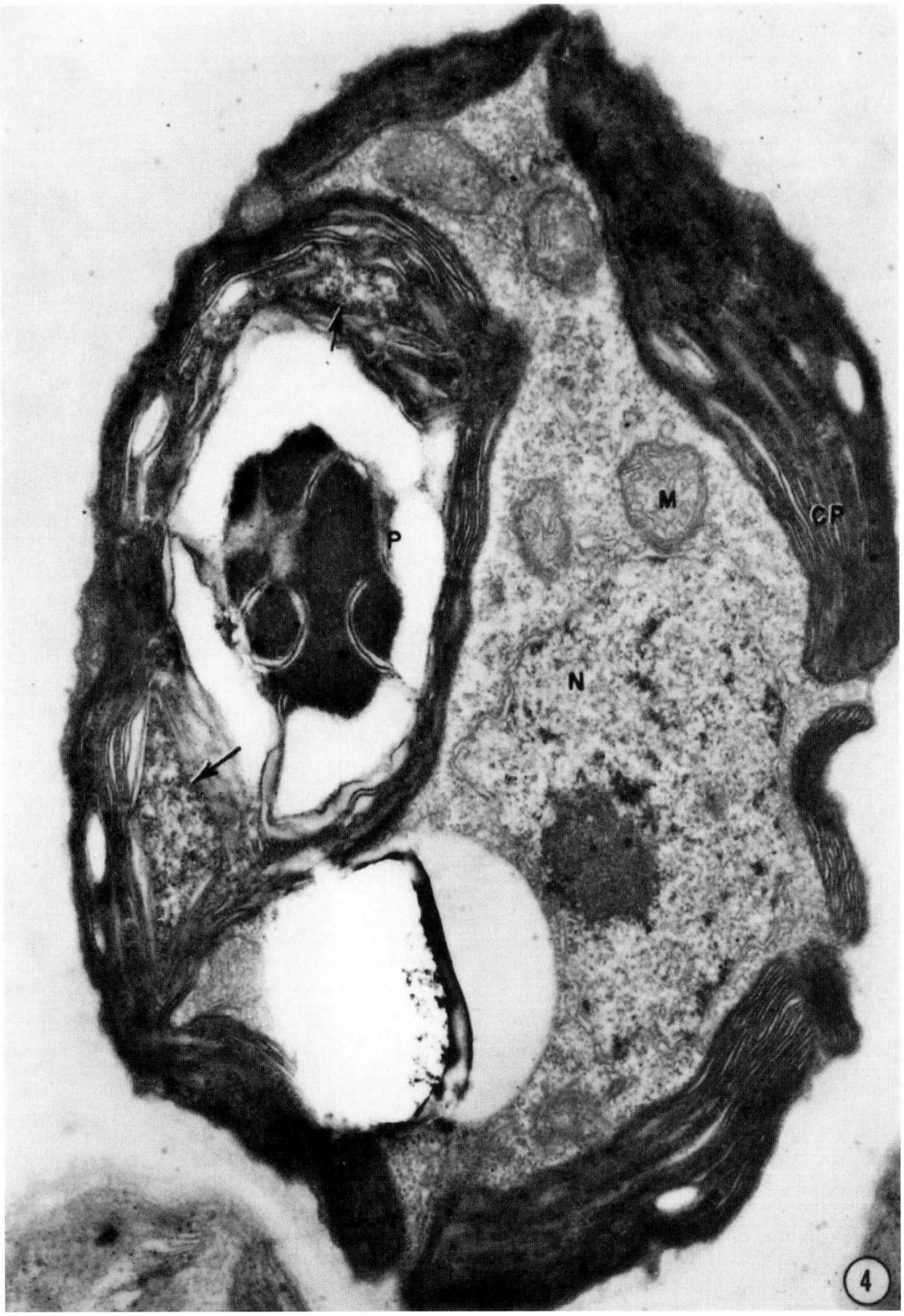
M
CP
P
N
4

from existing plastids or from undifferentiated proplastids and cannot be regenerated by the cell after they have been experimentally eliminated by antibiotics or by abnormal temperatures under conditions which inhibit reproduction of plastids but not of the cell (Pringsheim and Pringsheim, 1952; De Deken-Grenson and Messin, 1958; Provasoli *et al.*, 1951). This partial autonomy of plastids has suggested that they possess some kind of genetic system. In view of the finding that genetic systems in viruses and in nuclear elements of cells are associated with RNA or DNA, many authors have looked for nucleic acids in plastids. The most interesting observations that suggest the presence of DNA in chloroplast are Iwamura's (1960) studies on *Chlorella*.

Chlamydomonas apparently has not been previously investigated for chloroplast DNA. Our cytochemical studies indicate that certain localized areas in *Chlamydomonas* chloroplasts contain DNA in concentrations high enough to give an unmistakably positive Feulgen reaction. This alga was, therefore, especially favorable for investigating the ultrastructure of the DNA-containing material. Sager and Palade (1957) have published an electron microscope study of *Chlamydomonas reinhardi*. They do not mention any special regions in the chloroplast which might correspond to the DNA-containing structures. Since the 25 A fibrils characteristic of these regions show up clearly only with the method of Ryter *et al.* (1958) it is not surprising that they were not visible in the micrographs of Sager and Palade.

The similarity in organization of the DNA-containing regions in the chloroplast and the nucleoplasm of bacteria and blue-green algae (Monera) which we have demonstrated here is of considerable theoretical interest. It is most attractive to assume that the DNA and the corresponding fibrils represent the genetic system of the chloroplast which their genetic properties and autonomy in reproduction imply. This genetic system resembles the nuclear equivalent of Monera (Dougherty's protocaryon, 1957). The "plasmagene" as represented by plastids thus resembles the "chromogene" of the Monera. Is the chloroplast of *Chlamydomonas* unique or are structually similar systems present in plastids of higher plants? In a preliminary study we have found these low density areas containing 25 A fibrils in chloroplasts of a number of plants where the concentration of DNA may not be high enough to give a positive Feulgen reaction (*Elodea canadensis*; *Zea mays*, both normal green chloroplast and albino mutant; *Helianthus tuberosus* and *Anthoceros* sp.). The electron microscope in combination with DNase digestion is a most sensitive tool for the demonstration of such DNA-nucleoplasm and can reveal its presence where direct cytochemical methods fail.

With the demonstration of "nucleoplasm" in chloroplasts, the similarity in ultrastructural organization of a chloroplast and a blue-green algal cell becomes indeed striking. Both are enveloped in a double membrane. Both contain the photosynthetic apparatus in membrane systems of similar organization (*cf.* Mühlethaler, 1960; Ris and Singh, 1961; Lefort, 1960). Both contain particles which look like ribosomes in the electron microscope. Whether they are in fact ribosomes remains to be established by isolation and biochemical analysis. Both contain DNA in the form of a nucleoplasm; *i.e.*, areas of low density which contain fibrils about 25 A thick. We suggest that this similarity in organization is not fortuitous but shows some historical relationship and lends support to the old hypothesis of Famintzin (1907) and Mereschkowski (1905) that chloroplasts originate from endosymbiotic blue-green algae. Endosymbiotic blue-green algae occur today in several types of protozoa. They act as chromatophores and endow these cells with photosynthetic ability (Geitler, 1959 *a*, *b*). This hypothesis also explains why the photosynthetic apparatus is associated

FIGURE 5

Region of the chloroplast of *Chlamydomonas moewusii* which gives a Feulgen-positive reaction. The widened space between the chloroplast membranes contains dense particles, which resemble the ribosomes in the cytoplasm (*R*), and 25 A thick fibrils (arrows) which presumably represent the DNA. *CM* = external chloroplast membrane, *CL* = chloroplast lamellae. × 110,000.

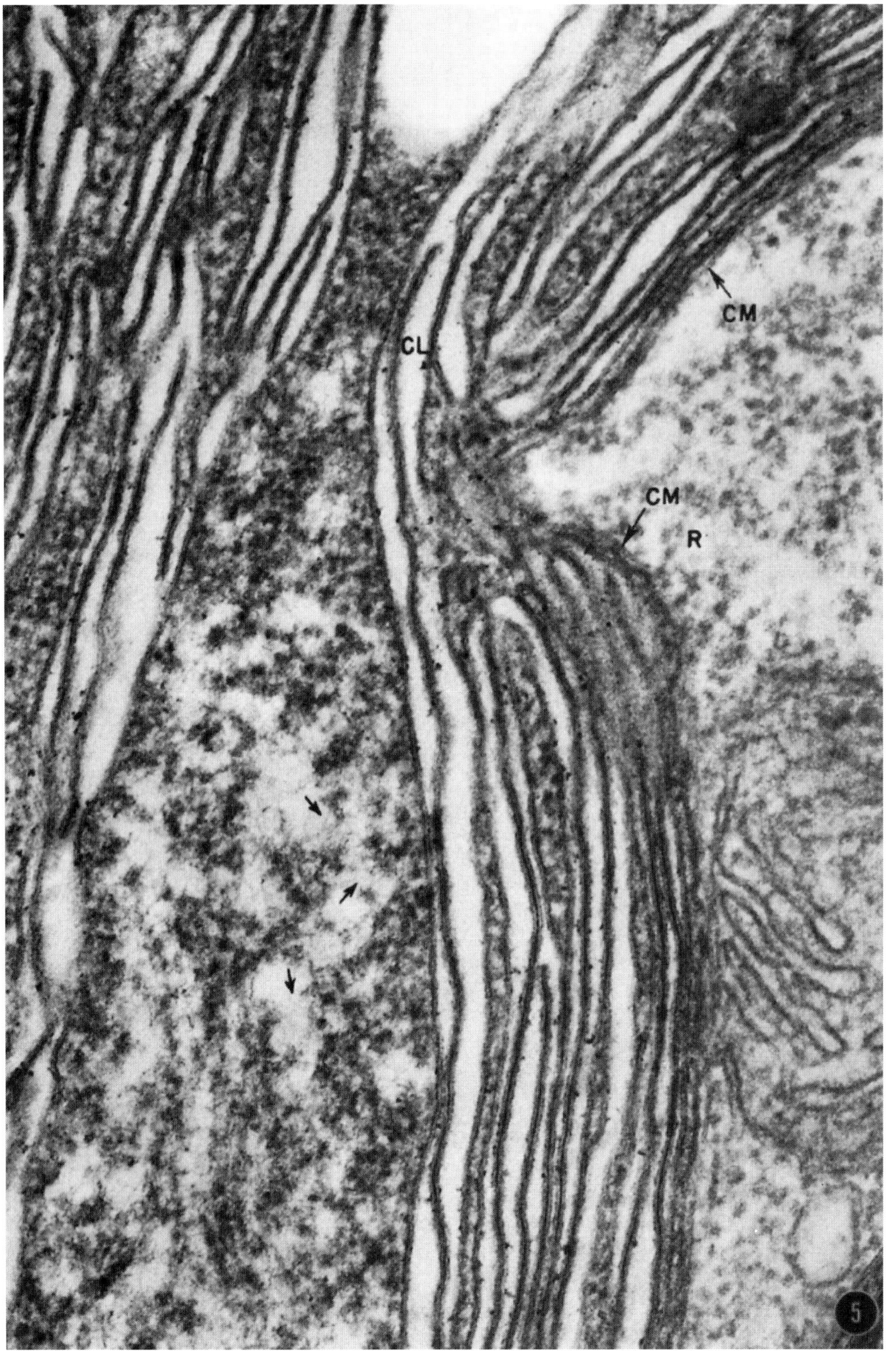
CM
CL
CM
R
5

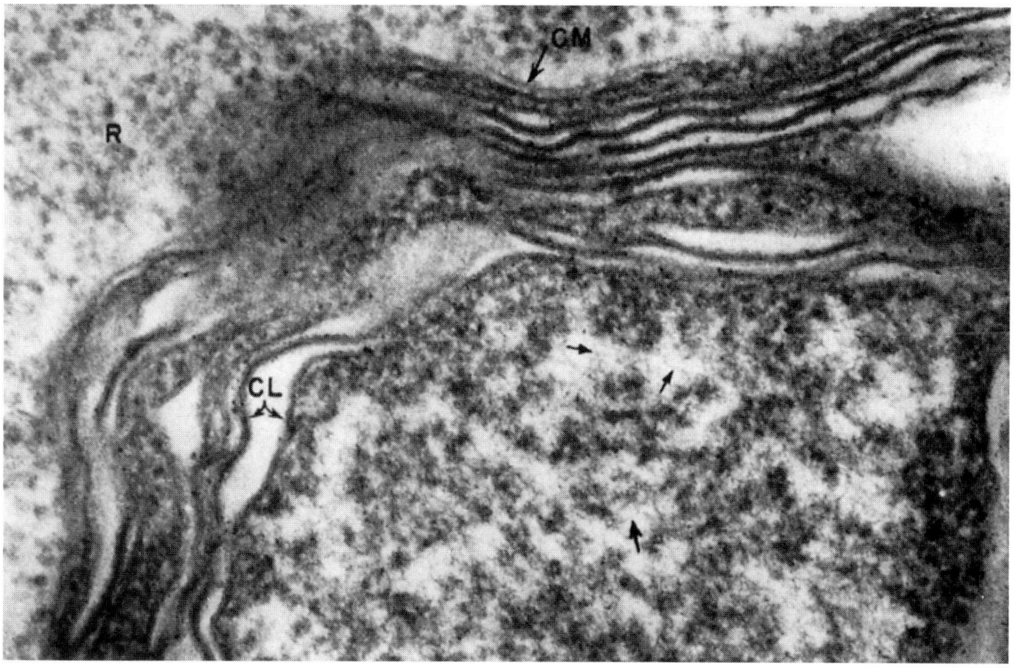

FIGURE 6

Part of the chloroplast of *Chlamydomonas moewusii* with the Feulgen-positive area. *R* = ribosomes in the cytoplasm. *CL* = chloroplast lamellae. The arrows point to the 25 A fibrils which are taken to represent the DNA. *CM* = external chloroplast membrane. × 95,000.

with membrane systems which traverse freely the cytoplasm in blue-green algae but which in higher plants are incorporated into complex cell organelles having a high degree of genetic individuality and containing just about every classified organelle found in free-living blue-green algae. The evolution of the complex cell, with its array of more or less autonomous organelles, from the simpler organization found in Monera is a question that has been neglected. With the demonstration of ultrastructural similarity of a cell organelle and free living organisms, endosymbiosis must again be considered seriously as a possible evolutionary step in the origin of complex cell systems.

Received for publication, September 10, 1961.

BIBLIOGRAPHY

DE DEKEN-GRENSON, M., and MESSIN, S., *Biochim. et Biophysica Acta*, 1958, **27**, 145.

DOUGHERTY, E. C., *J. Protozoology*, 1957, **4**, Suppl., 14.

FAMINTZIN, A., *Biol. Centralblatt*, 1907, **27**, 353.

GEITLER, L., *Oesterreich. Bot. Z.*, 1959 *a*, **106**, 464.

GEITLER, L., *Encyclopedia Plant Physiol.*, 1959 *b*, **11**, 530.

GRANICK, S., *Encyclopedia Plant Physiol.*, 1955, **1**, 507.

HOPWOOD, D. A., and GLAUERT, A. M., *J. Biophysic. and Biochem. Cytol.*, 1960 *a*, **8**, 267.

HOPWOOD, D. A., and GLAUERT, A. M., *J. Biophysic. and Biochem. Cytol.*, 1960 *b*, **8**, 813.

IWAMURA, T., *Biochim. et Biophysica Acta*, 1960, **42**, 161.

JAGENDORF, A. T., and WILDMAN, S. G., *Plant Physiol.*, 1954, **29**, 270.

KELLENBERGER, E., *in* HAYES and CLOWES, Bacterial Genetics, Cambridge University Press, 1960, 39.

KLEINSCHMIDT, A. and LANG, D., *Proc. European Reg. Conf. Electron Micr.*, Delft, 1960, **2**, 690.

LEDERBERG, J., *Physiol. Rev.*, 1952, **32**, 403.

LEE, S., *Exp. Cell Research*, 1960, **21**, 252.

LEFORT, M., *Compt. rend. Acad. sc.*, 1960, **251**, 3046.

LITTAU, V. C., *Am. J. Bot.*, 1958, **45**, 45.

LUFT, J. H., *J. Biophysic. and Biochem. Cytol.*, 1958, **9**, 409.

MERESCHKOWSKI, C., *Biol. Centralblatt*, 1905, **25**, 593.

MÜHLETHALER, K., *Z. Wissensch. Mikr.*, 1960, **64**, 444.

Pringsheim, E. C., and Pringsheim, O., *New Phytol.*, 1952, **51**, 65.

Provasoli, L., Hutner, S. H., and Pintner, I. J., *Cold Spring Harbor Symp. Quant. Biol.*, 1951, **16**, 113.

Rhoades, M. M., *Encyclopedia Plant Physiol.*, 1955, **1**, 19.

Ris, H., *Canad. J. Genetics and Cytol.*, 1961, **3**, 95.

Ris, H., and Singh, R. N., *J. Biophysic. and Biochem. Cytol.*, 1961, **9**, 63.

Rustad, R. C., *Exp. Cell Research*, 1958, **15**, 444.

Ryter, A., Kellenberger, E., Birch-Anderson, A., and Maaløe, O., *Z. Naturforsch.*, 1958, **13b**, 597.

Sager, R., and Palade, G. E., *J. Biophysic. and Biochem. Cytol.*, 1957, **3**, 463.

Stocking, C. R., and Gifford, E. M., *Biochem. Biophys. Res. Communic.*, 1959, **1**, 159.

Brown D.D. and **Gurdon J.B.** 1964. Absence of ribosomal RNA synthesis in the anucleolate mutant of *Xenopus laevis*. *Proc. Natl. Acad. Sci.* **51:** 139–146.

BECAUSE IT WAS READILY VISIBLE IN LIVING CELLS, the nucleolus was recognized in the early 19th century as a nearly universal nuclear organelle. Its association with a specific chromosomal site, the nucleolus organizer, was established by Barbara McClintock in 1934 (1). However, the role of the nucleolus in ribosome biosynthesis became clear only in the 1960s. Perry (2) carried out combined biochemical and autoradiographic studies that established two important points: first, that the 18S and 28S ribosomal RNAs were synthesized as a large precursor, and second, that the nucleolus was the site where this RNA was made. These observations suggested that the genes coding for ribosomal RNA might reside in the nucleolus organizer. Definitive proof for this model came from experiments on mutants of *Xenopus* and *Drosophila* that lacked these genes. In the paper reproduced here Brown and Gurdon showed that anucleolate mutant tadpoles of *Xenopus* failed to produce rRNA and were probably deficient for the rDNA genes themselves (although the homozygous mutant was lethal, the embryos lived to the tadpole stage, making use of ribosomes synthesized during oogenesis). Wallace and Birnstiel (3) subsequently isolated DNA from normal and anucleolate mutants and showed by filter hybridization that the DNA from the mutants did not hybridize to rRNA. *Drosophila* provided an unusually good genetic system for studying the same problem. Ritossa and Spiegelman (4) constructed viable stocks that contained 1, 2, 3, or 4 nucleolus organizers. By filter hybridization they demonstrated that the amount of rDNA in the genome was proportional to the number of nucleolus organizers. Thus, within a few years, the function of the nucleolus as the site of ribosomal RNA synthesis was firmly established.

---

1. McClintock B. 1934. The relation of a particular chromosomal element to the development of the nucleoli in *Zea mays*. *Z. Zellforsch. mikrosk. Anat.* **21:** 294–328.
2. Perry R. P. 1962. The cellular sites of synthesis of ribosomal and 4S RNA. *Proc. Natl. Acad. Sci.* **48:** 2179–2186.
3. Wallace H. and Birnstiel M.L. 1966. Ribosomal cistrons and the nucleolar organizer. *Biochim. Biophys. Acta* **114:** 296–310.
4. Ritossa F.M. and Spiegelman S. 1965. Localization of DNA complementary to ribosomal RNA in the nucleolus organizer region of *Drosophila melanogaster*. *Proc. Natl. Acad. Sci.* **53:** 737–745.

Reprinted from the PROCEEDINGS OF THE NATIONAL ACADEMY OF SCIENCES
Vol. 51, No. 1, pp. 139–146. January, 1964.

ABSENCE OF RIBOSOMAL RNA SYNTHESIS IN THE ANUCLEOLATE MUTANT OF XENOPUS LAEVIS

BY DONALD D. BROWN AND J. B. GURDON

DEPARTMENT OF EMBRYOLOGY, CARNEGIE INSTITUTION OF WASHINGTON, BALTIMORE, AND DEPARTMENT OF ZOÖLOGY, OXFORD, ENGLAND

Communicated by R. B. Roberts, November 18, 1963

Few new ribosomes appear in the cytoplasm of embryos of *Rana pipiens* or *Xenopus laevis* (the South African "clawed toad") before the tail bud stage.[1] At this time the amount of cytoplasmic ribosomes begins to increase; this rise is correlated with an increase of protein in the high speed supernatant fraction as well as with the first appearance or increase of many enzymes. Soon after these events, the embryos develop a requirement for magnesium ions in the medium. Magnesium-starved embryos characteristically stop growing in length and die at early swimming stages (Shumway[2] stages 21–23 for *Rana pipiens*,[1] or Nieuwkoop-Faber[3] stage 40 for *Xenopus laevis*[4]). The magnesium requirement coincides with the onset of intense ribosome synthesis and presumably is based on the important role of magnesium ions in maintaining the integrity of the functional ribosome particle.

The study of ribosome synthesis during amphibian development has been extended utilizing the lethal anucleolate mutant of *Xenopus laevis* first described by Elsdale *et al.*[5] These workers discovered a heterozygote mutant with only one nucleolus (1-*nu*) in each cell, whereas wild-type *Xenopus laevis* have two nucleoli (2-*nu*) in the majority of their diploid cells. The progeny resulting from the mating of two heterozygotes (1-*nu*) fall into three groups having two, one, or zero nucleoli per cell. The ratio of these genotypes is 1:2:1, respectively,[5, 6] as expected of a typical Mendelian factor. The heterozygotes (1-*nu*) lack a secondary constriction ("nucleolar organizer") on one of two homologous chromosomes in diploid cells;[7] the two comparable chromosomes of the anucleolate homozygous mutants (0-*nu*) both lack this secondary constriction. The anucleolate mutant (0-*nu*) has numerous small nucleolar "blobs" instead of typical nucleoli, and both nuclear and cytoplasmic RNA have been shown histochemically to be lower in 0-*nu* embryos after hatching than in controls (1-*nu* and 2-*nu*).[8]

Development of 0-*nu* embryos is first retarded shortly after hatching.[5, 9] The mutant embryos become microcephalic and oedematous and die as swimming tadpoles before feeding. It was apparent that magnesium-starved embryos were to some extent phenocopies of the homozygous mutants (0-*nu*) since retardation of embryogenesis and growth occurred in both groups of embryos at about the same developmental stage (Fig. 1). The above data, as well as recent studies relating nucleolar function to ribosome synthesis, suggested that the anucleolate mutant might be incapable of synthesizing ribosomes and ribosomal RNA.

Material and Methods.—Radioactivity was introduced into developing embryos by incubation with $C^{14}O_2$ at pH 6.0.[10] The methods for measuring ribosome and DNA contents have been described previously.[1] Total RNA was isolated from frozen embryos after homogenization in 0.1 *M* sodium acetate pH 5.0 containing 4 μg/ml polyvinyl sulfate (a ribonuclease inhibitor prepared synthetically by the method of Bernfeld *et al.*)[11] and 0.5% sodium lauryl sulfate (Mann Research Co.). The homogenate was shaken for 5–10 min at 0°C with an equal vol of phenol. Nucleic acids were precipitated from the aqueous phase with 2 vol of ethanol and 0.1 vol of *M*

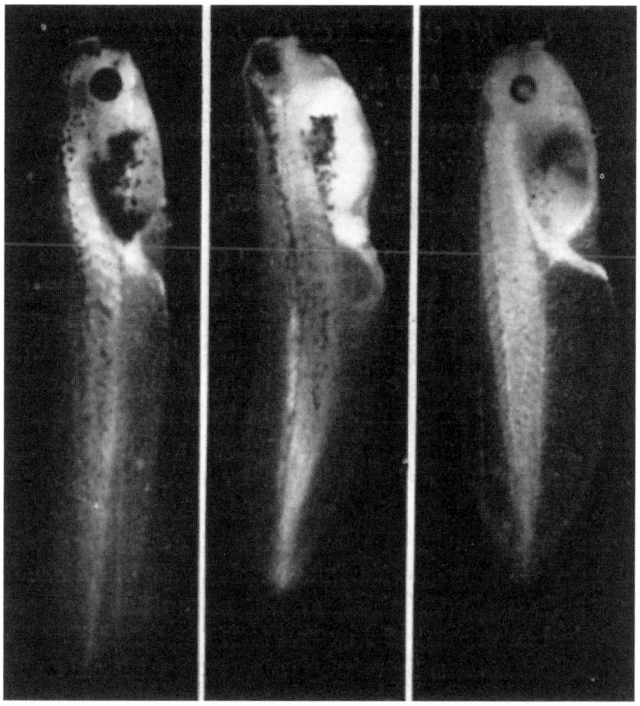

FIG. 1.—Comparison of control (left), anucleolate (middle), and magnesium - deficient (right) embryos of *Xenopus laevis*. These embryos are siblings that have developed for the same length of time. Initial symptoms characterizing the anucleolate mutant and magnesium deficiency syndrome are apparent.

sodium chloride, and the precipitate was dissolved in 0.01 *M* sodium acetate pH 5.0 containing 1 μg/ml polyvinyl sulfate. DNase I (5 μg/ml) and 10^{-3} *M* $MgCl_2$ were added and the solutions incubated for 10 min at 20°C. The RNA was further purified by two subsequent precipitations with NaCl-ethanol and the final precipitate drained of alcohol and dissolved in 1 ml of the 0.01 *M* sodium acetate-polyvinyl sulfate solution. Zonal sucrose gradient centrifugation[12] was performed in the SW-25 rotor of the Spinco Model L centrifuge for $14^1/_2$ hr at 24,000 rpm. The nucleic acid solutions were layered over linear gradients of sucrose which varied from 20% to 5% and which contained 10^{-4} *M* versene and 0.01 *M* sodium acetate pH 5.0. Following centrifugation, the tubes were punctured and fractions collected for optical density measurement at 260 mμ and radioactivity determinations. Nucleic acids precipitated from each fraction by adding trichloroacetic acid to a concentration of 5% were caught on Millipore filters (HA) and dried. After phosphor was added, the filters were counted in a liquid scintillation counter.

Results and Discussion.—Absence of ribosomal RNA synthesis in the anucleolate mutant: Values for the RNA and DNA contents of anucleolate and control embryos are presented in Table 1. The most pronounced difference between the anucleolate *Xenopus* (0-*nu*) and the control mixture of 1-*nu* and 2-*nu* embryos is the reduced quantity of RNA and in particular the small amount of ribosomes in the 0-*nu* mutants.

The relatively small numbers of ribosomes present in the 0-*nu* embryos might have been synthesized entirely during oögenesis before meiotic reduction when the growing oöcytes were heterozygous for

TABLE 1
COMPARISON OF RNA AND DNA CONTENTS OF ANUCLEOLATE AND CONTROL *Xenopus laevis* EMBRYOS

	μg/embryo Homozygous mutant	Control*
DNA†	0.88	0.97
Total RNA†	5.1	11.8
RNA contents of isolated ribosomes‡	3.2	5.4

* Analyses performed on a mixture of heterozygous (1-*nu*) and wild-type (2-*nu*) embryos.
† Control and mutant sibling embryos were at stages 40–41.[3]
‡ Control and mutant sibling embryos were at stages 38–40.[3]

nucleolus formation. Alternatively some ribosome synthesis might have occurred during embryogenesis. To distinguish between these possibilities, radioactive precursor was presented to the developing embryos during neurulation, when ribosomal RNA synthesis is known to have already begun in wild-type embryos;[4, 13] at this stage the 0-*nu* mutants are still developing normally and are morphologically similar to the control embryos. RNA was isolated 48 hr after termination of the radioactive incubation period so that ample time was allowed for complete utilization of the precursor and its incorporation into RNA. The density gradient centrifugation patterns of the total RNA isolated from 90 anucleolate (0-*nu*) mutants and 90 controls (a mixture of 1-*nu* and 2-*nu* embryos) are shown in Figure 2. The mutants contain about one

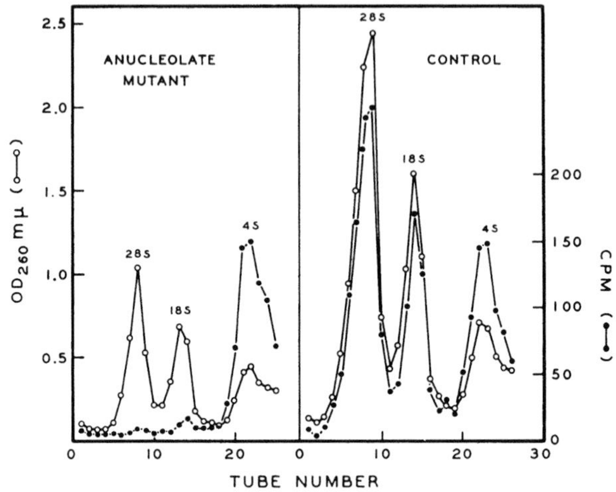

FIG. 2.—Sucrose density gradient centrifugation of total RNA isolated from 0-*nu* and control embryos. Two heterozygote (1-*nu*) adults were mated and the embryos allowed to develop to neurulation (Nieuwkoop-Faber[3] stages 14–18). At this time the embryos were incubated in a closed serum bottle at pH 6.0 with about 0.2 μc $C^{14}O_2$ for 20 hr at 18°C with mild shaking. By the end of this incubation period, development had proceeded to stages 26–28 (muscular response), and the mutant embryos were still indistinguishable grossly from the control embryos. The medium was changed, and the embryos continued development in nonradioactive tap water for 48 hr at 20°C (stages 40–41). The anucleolate mutants were recognized by examination of their tail tips with a phase contrast microscope and separated from the two control genotypes (1 *nu* and 2-*nu*). Both groups were then washed with distilled water and frozen at −20°C. The frozen embryos were packed in dry ice and flown from Oxford to Baltimore for chemical analysis. The bulk of the RNA (O—O) is represented by optical density measurements at 260 mμ. The RNA synthesized between neurula and muscular response stage is represented by the radioactive measurements (●—●).

half as much total RNA as the controls. This quantity (5 μg/embryo) is about the same as that found in the unfertilized egg of *Xenopus laevis*.[4] The control *Xenopus* embryos have synthesized radioactive 28S and 18S ribosomal RNA as well as 4S RNA. However, the 0-*nu* mutants have synthesized less than 5 per cent as much radioactive ribosomal RNA but about the same amount of 4S material as the control embryos. The radioactivity in the 4S region was eluted from the Millipore filters with dilute NH_4OH, made 0.3 *N* with KOH and incubated overnight at 37°C. About 80 per cent of the radioactivity in both mutant and control samples was presumably RNA since it was rendered acid-soluble by this treatment. The remaining alkali-resistant radioactivity was solubilized by hot TCA and probably was partially degraded radioactive DNA.

In these 0-*nu* mutants the ribosomal RNA made during oögenesis has persisted, and the embryos are incapable of synthesizing new ribosomal RNA. Since these embryos develop normally to early swimming stages, it can be concluded that *Xenopus* embryos do not *need* new ribosomal NRA until after this stage of development.

Synthesis of rapidly labeled RNA by the anucleolate mutant: To define the classes of rapidly labeled RNA synthesized by the mutant, RNA was isolated immediately

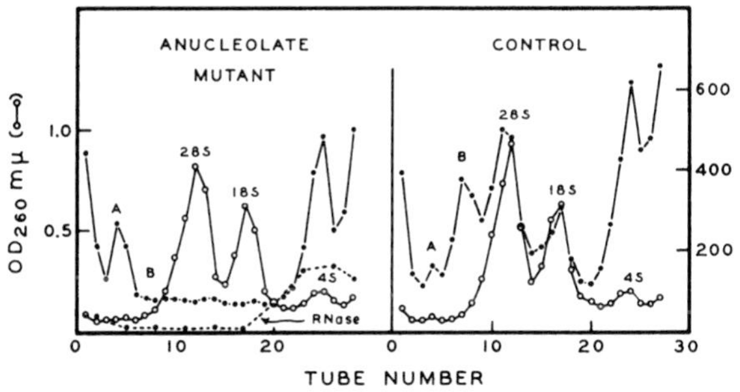

FIG. 3.—Rapidly labeled RNA of mutant (0-*nu*) and control *Xenopus* embryos. Previously separated 0-*nu* mutants and control (1-*nu* and 2-*nu*) embryos (32 embryos each) at stage 27–28 were placed in a 2 ml vial with 1 ml of preboiled Holtfreter-M^{++} medium[1] containing 0.3 *M* sodium phosphate pH 6. After gently blowing CO_2-free air over the medium for 5 min, the vials were sealed with rubber injection caps, and 25 μc of $Na_2C^{14}O_3$ dissolved in 0.01 *M* NaOH was injected into each bottle. The vials were gently shaken for 2 hr at 21°C, then cooled, and the embryos washed with cold distilled H_2O. Nonradioactive RNA (0.8 mg) isolated from *Xenopus* oöcytes was added to each group of embryos and the total RNA purified. The purified RNA was dissolved in 1.5 ml, and 1.0 ml (equivalent to RNA from 20 embryos) was centrifuged. The remaining 0.5 ml of the mutant RNA preparation was made 0.1 *M* with tris buffer pH 7.2 and incubated in a total volume of 1 ml with 20 μg of pancreatic RNase for 10 min at 20°C. Values for the RNase-treated preparation have been corrected for volume and tube number so that they are directly comparable with the untreated RNA.

after a 2-hr incubation with labeled precursor (Fig. 3). Because of the small number of embryos used in this experiment, purified carrier RNA (unlabeled) was added at the beginning of the isolation procedure, so that the optical density peaks of the carrier RNA serve as reference markers for the three classes of bulk RNA, i.e., 28S, 18S, and 4S RNA.

Experiments by Scherrer *et al.*[14] and Perry[15] indicate that the 28S and 18S ribosomal RNA molecules are both derived from larger precursor molecules. Radioactive label appears first in these rapidly sedimenting precursors and only later in 28S and 18S RNA. The results plotted in Figure 3 demonstrate that 2 hr after addition of radioactivity the control already has synthesized 28S and 18S RNA as well as at least two distinct peaks of heavier RNA (labeled *A* and *B*). In contrast, the mutant embryos not only failed to synthesize typical 28S and 18S ribosomal RNA but also lack the heavy precursor RNA that sediments in region *B* of Figure 3 (about 35S). Yet heavier classes of RNA (type A and even more rapidly sedimenting molecules) have been synthesized by the mutant, as well as heterogeneous RNA that sediments throughout the gradient solution. This latter observation is more evident when the sedimentation pattern of rapidly labeled mutant RNA is compared before and after ribonuclease digestion (Fig. 3).

The rapidly labeled RNA synthesized by the mutant is most probably "messenger" RNA. This conclusion is based on the fact that in the mutant all the radioactive RNA sedimenting more rapidly than 4S can be recovered associated with the purified ribosomes of the mutant. When isolated by this technique, the rapidly labeled RNA is degraded to molecules having sedimentation constants between 4 and 18S. The base composition of this heterogeneous RNA labeled with P^{32}, which has been isolated in association with purified ribosomes of normal *Xenopus* embryos, is invariably DNA-like.[4]

The nucleolus as the site of ribosomal RNA synthesis: In control *Xenopus* embryos, although other classes of RNA are synthesized at earlier stages, new *ribosomal* RNA synthesis is not detectable until gastrulation,[4] the same stage that definitive nucleoli first become visible cytologically. (The many small "blobs" seen in blastula nuclei do not seem to be equivalent to "definitive" nucleoli.) Although synthesis of ribosomal RNA begins at gastrulation, the quantity of this newly synthesized RNA remains small when compared to the RNA already present in the unfertilized egg. It is only after hatching that the total RNA content of wild-type *Xenopus* embryos begins to increase significantly.[4] The absence of typical nucleoli at very early stages of development has been reported for other amphibia[16] as well as other developing organisms.[17] Furthermore, Beermann[18] has described developmental arrest in anucleolate recombinants resulting from the mating of two different species of the dipteran, *Chironomus*. The relationship between nucleolar function and ribosome synthesis has been suggested by several observations including electron microscopy,[19, 20] base composition analyses,[20] and radioautographic studies.[15] The close correlation of the time of ribosomal RNA synthesis with the appearance of definitive nucleoli in *Xenopus* and particularly the simultaneous absence of both ribosomal RNA synthesis and normal nucleoli in the 0-*nu* mutant support this relationship.

Difference in nucleotide composition between 28S and 18S RNA: This single mutation prevents the formation of both 28S and 18S RNA molecules (Fig. 2); however, evidence has been presented suggesting that the structure of these two molecules is determined by different gene loci. Yankofsky and Spiegelman[21] have shown that 23S and 16S ribosomal RNA of bacteria hybridize independently with homologous bacterial DNA. Thus, the two molecules must have different nucleotide sequences each complementary to a distinct region of the bacterial DNA. In *Xenopus*, the 28S and 18S ribosomal RNA have different base compositions. Table 2 contains analyses for 28S, 18S, and 4S RNA's separated by density gradient centrifugation

TABLE 2

NUCLEOTIDE COMPOSITION OF 28S AND 18S RIBOSOMAL RNA AND 4S RNA PURIFIED FROM *Xenopus laevis* EGGS AND EMBRYOS

Approximate	Unfertilized Eggs			Stage 45 Embryos		
S Value	28	18	4	28	18	4
AMP	17	22	20	18	22	21
GMP	37	31	28	37	31	31
CMP	30	29	32	28	29	30
UMP	16	18	20	17	18	18
% GC	67	60	60	65	60	61

Following density gradient centrifugation, the RNA was precipitated from the sucrose solutions with cold TCA, washed with ethanol, and hydrolyzed in NHCl at 100°C for 1 hr. Base composition was determined following chromatography in the isopropanol:HCl solvent described by Wyatt.[22]

from RNA purified from ovarian eggs and embryos of *Xenopus laevis*. The 28S RNA has a significantly higher G-C content than the 18S RNA. There is also a difference in base composition between the 23S and 16S ribosomal RNA's of different bacteria[23] as well as the 28S and 18S ribosomal RNA's of chick embryos.[24]

Quantitative regulation of ribosomal RNA gene activity: The rates of ribosomal RNA synthesis in 1-*nu* and 2-*nu* embryos were compared. The results shown in Table 3 demonstrate that all three classes of RNA molecules are synthesized at comparable rates by the heterozygote and homozygous wild-type embryos. Furthermore, the synthesis of 28S and 18S RNA is coordinate since their specific activities are the same. Thus, the haploid complement of ribosomal RNA genes in the heterozygote must produce twice as much ribosomal RNA as do the same genes in the wild-type homozygote. It is of interest to note that the combined

TABLE 3

RNA SYNTHESIS BY 1-*nu* AND 2-*nu* *Xenopus laevis* EMBRYOS

	1-*nu*	2-*nu*
Total RNA μg/embryo	9.1	10.8
	CPM/μg RNA	
28S	0.45	0.38
18S	0.39	0.37
4S	0.95	0.89

The same protocol that is described in the legend to Fig. 2 was followed. Thus, sibling embryos were made radioactive at the same stage and under the same conditions. The radioactive 1-*nu* and 2-*nu* embryos were separated, and the specific activity of 28S, 18S, and 4S RNA was calculated following density gradient centrifugation of the purified RNA.

volume of the 2 nucleoli in 2-*nu* embryos is the same as that of the single nucleolus in 1-*nu* heterozygotes.[6, 25]

Genetic basis of the anucleolate condition: The mutation affecting nucleolar number behaves as a single Mendelian factor and results in a cytologically visible alteration on one chromosome, i.e., in "the nucleolar organizer" region.[7] The defect when homozygous does not alter nucleic acid metabolism generally, but specifically prevents the synthesis of both molecular species of ribosomal RNA. Thus DNA (Table 1), 4S RNA (Fig. 2), and rapidly labeled high molecular weight heterogeneous RNA (Fig. 3) are all synthesized by the mutant 0-*nu* embryos. Furthermore, the relative synthesis of both 28S and 18S RNA is the same (Table 3), even at different developmental stages when ribosomal RNA is formed at widely different rates.[4]

Spiegelman[26] and Scherrer *et al.*[14] have reasoned that closely linked genes (perhaps whole operons[27]) might be transcribed as single large RNA molecules ("polycistronic" RNA[28]) which are subsequently degraded specifically to smaller and, in the case of ribosomal RNA, stable subunits. If adjacent 28S and 18S genes were transcribed together as a single molecule, such a large precursor would be expected to have a molecular weight of about $2–3 \times 10^6$ with a sedimentation constant of approximately 35S. This is about the sedimentation constant of the ribosomal RNA precursor (Region *B*, Fig. 3). This hypothesis accounts for the fact that large precursor molecules give rise to smaller ones as well as providing a molecular basis for coordinate expression of the several genes of an operon. Since the 28S and 18S ribosomal RNA molecules function together as components of a single structure, the ribosome, it is reasonable that their synthesis should be controlled together.

Two general mechanisms adequately account for the characteristics of the anucleolate mutant. The anucleolate condition might be considered as a primary defect of nucleolus formation which secondarily results in the absence of ribosomal RNA synthesis. The alternative hypothesis would have the primary defect of preventing ribosomal RNA synthesis. This latter idea suggests that the nucleolus marks the location of ribosomal RNA and ribosome synthesis in the nucleus, and the presence of the nucleolus is secondary to these synthetic processes. The comments to follow do not distinguish between these two general possibilities but serve to analyze pertinent genetic mechanisms in the light of the data presented in this report.

The consequences of the anucleolate mutation cannot be explained by the alteration of a "repressor or activator" substance[27] that might circulate in the nucleoplasm and inhibit or activate the structural genes for ribosomal RNA. If a substance exists in the nucleoplasm which regulates expression of the ribosomal RNA genes, it would be expected to act equally on *both* nucleolar organizers of a diploid cell (unless it only acts in the immediate vicinity of its own synthesis). Thus an altered repressor such as a "superrepressor"[29] would be *dominant* resulting in an *anucleolate heterozygote* since the altered repressor would inhibit expression of both nucleolar organizers in diploid cells. If, on the other hand, expression of nucleolar organizers required the constant presence of an "activator" substance of endogenous origin, the nonproduction of such a substance would have a *recessive* effect, i.e., the heterozygote embryos (1-*nu*) would contain two nucleoli in each cell just as the control embryos. *In fact, the anucleolate mutation has resulted in nonfunctional gene loci which remain nonfunctional even in the presence of their normal alleles.*

It is highly probable that many genes are involved in the synthesis of ribosomal RNA. Both in bacteria[21] and in mice[30] the ribosomal RNA is complementary to 0.3 per cent of the DNA (0.6% of the nucleotide pairs). It seems likely that a similar proportion will be present in the *Xenopus laevis* genome since ribosomal RNA can constitute such a large fraction of the gene product. If so, there must be thousands of separate gene loci for each class of ribosomal RNA.

If the DNA for ribosomal RNA is distributed amongm any chromosomes, it is difficult to imagine how its products can be concentrated into one nucleolus, while still accounting for the biochemical and cytological properties of the 0-*nu* and 1-*nu* embryos. On the other hand, the entire complement of DNA for ribosomal RNA may be adjacent on a single chromosome. Since there are 18 haploid chromosomes in *Xenopus* the ribosomal region would occupy roughly 10 per cent of one of them, presumably the one containing the nucleolar organizer.

A single deletion would account for the results but would require the loss of an extremely large piece of DNA. Alternatively, an operator mutation[27] would account for the features of the anucleolate mutation. However, this would imply that a single operator locus could control the expression of thousands of genes.

Any mechanism invoked to explain ribosomal RNA synthesis must account for the fact that the activity of the entire complement of genes determining ribosomal RNA structure can be restricted by a single mutation.

Summary.—A mutation in *Xenopus laevis* that prevents the formation of a normal nucleolus at the same time prevents the synthesis of 28S and 18S ribosomal RNA as well as high molecular weight ribosomal RNA precursor molecules. DNA, 4S RNA, and rapidly labeled heterogeneous RNA are synthesized by the anucleolate mutant. Anucleolate mutants survive until the swimming tadpole stage and show normal differentiation of all the main cell types despite their inability to synthesize new ribosomal RNA. Homozygous mutants (0-*nu*) and control embryos conserve the ribosomes made during oögenesis and associate rapidly synthesized RNA with these old ribosomes.

The 28S and 18S ribosomal RNA's differ in base composition and are probably products of different genes; yet their synthesis is coordinate. In the heterozygous (1-*nu*) embryos, the wild-type genes regulate to produce twice as much 28S and 18S ribosomal RNA as do the same genes when present in homozygous wild-type individuals. Since the activity of the entire complement of genes determining ribosomal RNA structure can be curtailed by a single mutation, it is suggested that these genes are under common control and located at the "nucleolar organizer" site of a single chromosome.

The authors are indebted to Miss Elizabeth Littna for her expert technical assistance and to Dr. Igor Dawid for his critical reading of the manuscript.

[1] Brown, D. D., and J. D. Caston, *Develop. Biol.* **5,** 412 (1962).
[2] Shumway, W., *Anat. Record,* **78,** 139 (1940).
[3] Nieuwkoop, P. D., and J. Faber, *Normal Table of Xenopus laevis (Daudin)* (Amsterdam: North-Holland Publishing Company, 1956).
[4] Brown, D. D., manuscript in preparation.
[5] Elsdale, T. R., M. Fischberg, and S. Smith, *Exptl. Cell Res.,* **14,** 642 (1958).
[6] Fischberg, M., and H. Wallace, in *The Cell Nucleus* (London: Butterworth, 1960), p. 30.
[7] Kahn, J., *Quart. J. Microscop. Sci.,* **103,** 407 (1962).
[8] Wallace, H., *Quart. J. Microscop. Sci.,* **103,** 25 (1962).
[9] Wallace, H., *J. Embryol. Exptl. Morphol.,* **8,** 405 (1960).
[10] Cohen, S., *J. Biol. Chem.,* **211,** 337 (1954).
[11] Bernfeld, P., J. Nisselbaum, B. Berkeley, and R. Hanson, *J. Biol. Chem.,* **235,** 2852 (1960).
[12] Britten, R. J., and R. B. Roberts, *Science.* **131,** 32 (1960).
[13] Brown, D. D., and J. D. Caston, *Develop. Biol ,* **5,** 435 (1962).
[14] Scherrer, K., H. Latham, and J. E. Darnell, these PROCEEDINGS, **49,** 240 (1963).
[15] Perry, R. P., these PROCEEDINGS, **48,** 2179 (1962).
[16] Karasaki, S., *Embryologia,* **4,** 273 (1959).
[17] Cowden, R. R., and C. L. Markert, *Acta Embryol. Morphol. Exptl.,* **4,** 142 (1961).
[18] Beermann, W., *Chromosoma,* **11,** 263 (1960).
[19] Birnstiel, M. L., and B. B. Hyde, *J. Cell Biol.,* **18,** 41 (1963). Dr. Elizabeth Hay has found the same particulate structure resembling ribosomes in nucleoli of *Xenopus laevis* embryos (personal communication).
[20] Gall, J. G., in *Cytodifferentiation and Macromolecular Synthesis* (Academic Press, 1963), p. 119.
[21] Yankofsky, S. A., and S. Spiegelman, these PROCEEDINGS, **49,** 538 (1963).
[22] Wyatt, G. R., *Biochem. J.,* **48,** 584 (1951).
[23] Midgley, J. E. M., *Biochim. Biophys. Acta,* **61,** 513 (1962).
[24] Lerner, A. M., E. Bell, and J. E. Darnell, Jr., *Science,* **141,** 1187 (1963).
[25] Barr, H. J., and H. Esper, *Exptl. Cell Res.,* **31,** 211 (1963).
[26] Spiegelman, S., in *Informational Macromolecules* (New York: Academic Press, 1963), p. 27.
[27] Jacob, F., and J. Monod, in *Cellular Regulatory Mechanisms,* Cold Spring Harbor Symposia in Quantitative Biology, vol. 26 (1961), p. 193.
[28] Ohtaka, Y., and S. Spiegelman, *Science,* **142,** 493 (1963).
[29] Willson, C., D. Perrin, F. Jacob, and J. Monod, unpublished observations cited in ref. 27.
[30] Hoyer, B. H., personal communication.

Huberman J.A. and **Riggs A.D.** 1968. On the mechanism of DNA replication in mammalian chromosomes. *J. Mol. Biol.* **32:** 327–341.

TAYLOR ET AL. (THIS VOLUME) INTRODUCED ^{3}H-thymidine to study the distribution of label during the replication of DNA in eukaryotic chromosomes. Cairns (1) realized that he could use the same radioactive tracer to follow the replication of individual DNA molecules by autoradiography. With this technique he showed that the chromosome of *Escherichia coli* consisted of a giant circular molecule that replicated in a bidirectional fashion from a single origin of replication. The more complex issue of DNA replication in chromosomes of higher eukaryotes was addressed by Huberman and Riggs using essentially the same method of fiber autoradiography. Cells were labeled for various times with ^{3}H-thymidine, after which the DNA was carefully extracted and dried onto nitrocellulose filters. The filters were then exposed to photographic emulsion for up to several months, after which time the replicating segments of DNA could be visualized as tracks of silver grains in the developed film. Huberman and Riggs found that replication began at many sites along a single DNA molecule and proceeded bidirectionally away from the initiation site or origin. They made estimates of the rate of fork movement as well as the average size and spacing of replication origins. Subsequent studies with the budding yeast *Saccharomyces cerevisiae* defined sequences at the origins (2, 3) and led the way to eventual isolation and characterization of the macromolecular complex that initiates DNA replication (4).

---

1. Cairns J. 1963. The bacterial chromosome and its manner of replication as seen by autoradiography. *J. Mol. Biol.* **6:** 208–213.
2. Stinchcomb D.T., Struhl K., and Davis R.W. 1979. Isolation and characterization of a yeast chromosomal replicator. *Nature* **282:** 39–43.
3. Brewer B.J. and Fangman W.L. 1987. The localization of replication origins on ARS plasmids in *S. cerevisiae*. *Cell* **51:** 463–471.
4. Bell S.P. and Stillman B. 1992. ATP-dependent recognition of eukaryotic origins of DNA replication by a multiprotein complex. *Nature* **357:** 128–134.

J. Mol. Biol. (1968) **32**, 327–341

On the Mechanism of DNA Replication in Mammalian Chromosomes

Joel A. Huberman and Arthur D. Riggs†

Division of Biology, California Institute of Technology Pasadena, California, U.S.A.

(*Received 7 August 1967, and in revised form 2 November 1967*)

We have combined the techniques of pulse-labeling and DNA autoradiography to investigate the mechanism of DNA replication in the chromosomes of Chinese hamster and HeLa cells. Our results prove that the long fibers of which chromosomal DNA is composed are made up of many tandemly joined sections in each of which DNA is replicated at a fork-like growing point. In Chinese hamster cells most of these sections are probably less than 30 μ long, and the rate of DNA replication per growing point is 2·5 μ per minute or less.

In addition, we have taken advantage of the apparent slowness of equilibration with external thymidine of the internal thymidine triphosphate pool in Chinese hamster cells to determine the direction of DNA synthesis at the conclusion of the pulse in pulse-chase experiments. We have found, unexpectedly, that replication seems to proceed in opposite directions at adjacent growing points. Furthermore, adjacent diverging growing points appear to initiate replication at the same time.

1. Introduction

These studies were undertaken in the hope of increasing our understanding of the mechanism by which the large amount of DNA in the chromosomes of higher organisms is replicated. Previous autoradiographic experiments (Cairns, 1966; Huberman & Riggs, 1966; Sasaki & Norman, 1966) have shown that the DNA of mammalian chromosomes is arranged in the form of long fibers. Maximum fiber lengths of 500 μ (from HeLa cells) and 1800 μ (from Chinese hamster cells) have been reported by Cairns (1966) and Huberman & Riggs (1966), respectively, for DNA from cells lysed with detergent, while Sasaki & Norman (1966) have found DNA fibers more than 2 cm long from nuclei of human lymphocytes lysed without detergent.

In addition, Cairns (1966) has reported the results of pulse experiments which suggest that the long DNA fibers are composed of many separately replicated, tandemly joined sections. One of the aims of our studies was to verify this finding of Cairns. Another aim was to determine whether or not the DNA in each of the separate sections is replicated at a fork-like growing point similar to the growing point for *Escherichia coli* DNA (Cairns, 1963*a*).

Our results do, indeed, show that chromosomal DNA fibers are subdivided into separate sections which are replicated at fork-like growing points. In addition, our results also suggest the unexpected conclusion that DNA is replicated in opposite

† Present address: Salk Institute, La Jolla, Calif., U.S.A.

directions at adjacent growing points. That is, the direction of replication apparently alternates along the length of the long DNA fibers.

In the remainder of this paper we shall use the term replication section to refer to the stretch of DNA replicated by a single growing point. We shall also use the word autoradiogram to refer to any apparently continuous line of grains presumably caused by decay of tritium incorporated into DNA and the word fiber to refer to a single DNA molecule or a single series of DNA molecules joined end-to-end.

2. Materials and Methods

In general the procedures employed were similar to those we have described previously (Huberman & Riggs, 1966). Modifications are given in Figure and Plate legends.

Incubations with pronase were carried out as described previously (Huberman & Riggs, 1966), except that the DNA was dialyzed against a higher concentration of pronase (1 mg/ml.), and that, after the Millipore filters (Millipore Corp.) were glued to glass slides, they were exposed to a formaldehyde-saturated atmosphere for 36 hr and then covered with a thin Parlodion film.

3. Results

(a) *Cold pulse-labeling experiments*

(i) *Significance of tandem arrays*

When Chinese hamster cells were briefly exposed to [^{3}H]thymidine, then immediately lysed and the released DNA subjected to autoradiography, tandem arrays of DNA autoradiograms, similar to those reported by Cairns (1966), were sometimes found (Plate I). In order to prove that these tandem arrays were the result of separate replication sections in single DNA fibers and not the result of side-by-side aggregation of several separate DNA fibers, each containing a single replication section, we performed the following experiments.

We reasoned that if all the DNA fibers were completely labeled by growing the cells in high specific activity [^{3}H]thymidine for 24 hours, then side-by-side aggregates could be distinguished by their higher grain density; we also reasoned that if, during the 24-hour period, we exposed the cells to [^{3}H]thymidine of a lower specific activity for a short interval, we could distinguish the autoradiograms of DNA replicated during that short interval by their lower grain density.

If each DNA fiber contained just one replication section, then after a cold pulse experiment of the type described here, one would expect to find no more than one low grain density (cold) region in any long DNA autoradiogram. On the other hand, if each DNA fiber could contain several replication sections, then after a cold pulse experiment one would expect to find some single long DNA autoradiograms containing several cold regions.

The actual labeling sequences used in these experiments are summarized in Fig. 1. When the stripping films were exposed for sufficient time to bring out the low specific activity regions (three months or more), long DNA autoradiograms containing several regions of low grain density were frequently found for cold-pulsed DNA but not for control DNA (Plate II). Thus Cairns' (1966) hypothesis that the long DNA fibers are composed of many tandemly joined replication sections is proved.

(ii) *Rate of DNA replication during DNA synthesis*

Further information can be obtained from these experiments. The average generation time of the Chinese hamster cells under our growth conditions is about 17 hours.

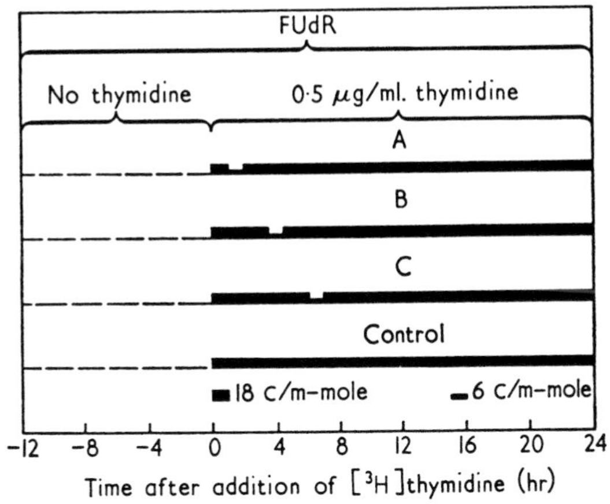

FIG. 1. Labeling schedule for cold pulse experiments.

Chinese hamster cells were grown as described in the legend to Plate I. After a 12-hr pretreatment with FUdR (0·1 μg/ml.), [^{3}H]thymidine (18 c/m-mole; 0·5 μg/ml.) was added to 4 separate Petri plates (A, B, C and control). At various times after the initial addition of [^{3}H]thymidine, as indicated in the Figure, the medium containing [^{3}H]thymidine at 18 c/m-mole was removed from the plates and replaced, for 1 hr, by medium containing [^{3}H]thymidine (0·5 μg/ml.) at 6 c/m-mole. After 24 hr the cells were harvested by trypsinization and diluted to 400 cells/ml. in isotonic saline. Lysis and autoradiography were performed as in the legend to Plate I.

DNA synthesis requires, on the average, about six hours (Taylor, 1960; Hsu, Dewey & Humphrey, 1962). Thus we can calculate that the 12-hour pretreatment with FUdR† (which inhibits thymidine monophosphate biosynthesis) must have blocked nearly two-thirds of the cells at the beginning of DNA synthesis and the remaining cells somewhere in the DNA synthesis period (see Hsu, 1964, for the use of FUdR in synchronizing Chinses hamster cell cultures). Consequently during the cold pulse of Experiment A (Fig. 1), which occurred between one and two hours after the relief of the FUdR block by the addition of [^{3}H]thymidine, the majority of cells were near the beginning of their DNA synthesis period. Similarly, during the cold pulse of experiment B, most cells were in the middle of DNA synthesis, while during the cold pulse of experiment C most cells were either at the end of DNA synthesis or had completed it. If the assumption is made that the size of the regions of low grain density is a measure of the rate of DNA replication per growing point, then our results can provide an answer to the question: Does the rate of DNA replication change during the period of DNA synthesis?

The length distributions of low grain density regions for experiments A, B and C are shown in Fig. 2. Low grain density regions less than 20 μ long were not easily resolved. However, for the low grain density regions of resolvable size there is no large difference among the three distributions. Thus there is apparently no large change in the rate of DNA replication per growing point during DNA synthesis.

(iii) *Distribution of replicating DNA*

In cold pulse experiments A and B, approximately 12% of the total length of autoradiograms was of low grain density; in experiment C, regions of low grain density

† Abbreviations used: FUdR, 5-fluorouracil deoxyriboside; BUdR, 5-bromouracil deoxyriboside.

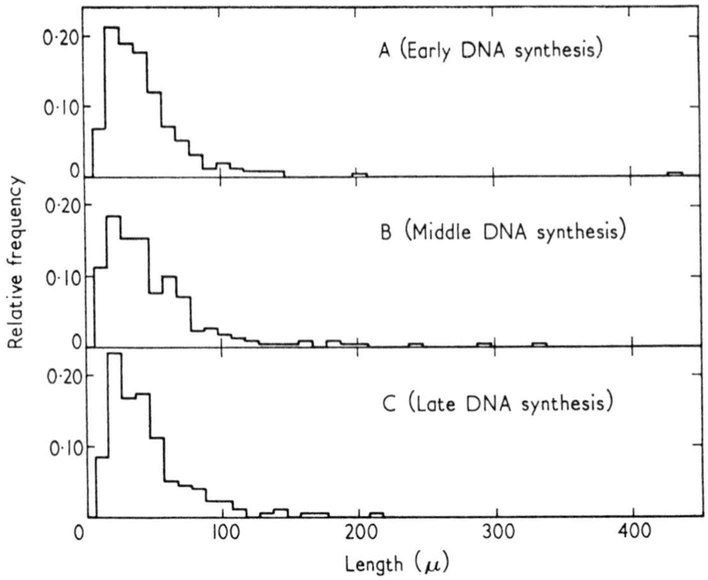

FIG. 2. Distribution of lengths of low grain density regions in cold pulse autoradiograms. Chinese hamster cells were labeled as described in Fig. 1 (A, B and C) and their DNA subjected to autoradiography. The autoradiograms were examined by dark-field microscopy at a magnification of 100. The contours of low grain density regions were traced with a camera lucida, and the lengths of the tracings were then measured and corrected for magnification.

contributed about 7% of the total length. However, in experiments A and B about two-thirds of the autoradiograms contained no regions of low grain density at all; in experiment C four-fifths of the autoradiograms contained no regions of low grain density.

Thus, during any one-hour period, regions of DNA synthesis are not distributed equidistantly along the DNA fibers in chromosomes. The non-uniform distribution of regions of DNA synthesis is apparent in Plate II.

(b) *Effect of pronase on autoradiogram lengths*

We have previously found (Huberman & Riggs, 1966) that low concentrations of the proteolytic enzyme, pronase, acting on the long DNA fibers in solution prior to autoradiography have no effect on autoradiogram lengths. In addition, Macgregor & Callan (1962) have shown that neither of the proteolytic enzymes trypsin nor pepsin is able to break amphibian lampbrush chromosomes when employed at a concentration of 250 μg/ml. for up to two hours at room temperature.

Nevertheless, we felt that the finding that each long DNA fiber may contain many separate replication sections made imperative an investigation with higher concentrations of pronase. In our earlier experiments we were unable to use higher pronase concentrations because some residual pronase, adsorbed to the Millipore filters, digested the stripping film. Therefore, in order to prevent this problem, we inactivated the residual pronase with formaldehyde vapors and further protected the stripping film with a thin Parlodion film placed between it and the Millipore filter. We now report that incubation of the long DNA fibers (from cells exposed to [^{3}H]thymidine for 35 hours) with about 100 μg/ml. of pronase for up to six hours at 34°C has no effect on either the maximum length or general length distribution of the autoradiograms. Thus any linkers connecting the separate replication sections, or connecting other points in the long DNA fibers, must be resistant to pronase. Large protein linkers are therefore extremely unlikely. In this connection it is interesting that, according to Davern (1966), the *E. coli* chromosome also contains no pronase-sensitive linkers.

(c) *Thirty-minute pulse-labeling experiments*

(i) *Does chromosomal DNA replicate at a fork?*

For complete autoradiographic visualization of a replication fork like that of *E. coli* (Cairns, 1963*a*), it is necessary that all three branches of the fork be labeled. This requires that cells complete more than one generation while incorporating [^{3}H]-thymidine. We have found, however, that extensive incorporation of [^{3}H]thymidine at specific activities adequate for DNA autoradiography completely prevents division of Chinese hamster cells. Consequently we have been forced to use a less direct method in our attempt to demonstrate the presence or absence of replication forks in Chinese hamster chromosomal DNA.

An appropriate method was suggested by an experiment of Cairns (1963*a*). He compared the DNA autoradiograms from *E. coli* cells exposed to [^{3}H]thymidine for a short time and then immediately lysed (simple pulse) with the DNA autoradiograms from *E. coli* cells exposed first to [^{3}H]thymidine and then to non-radioactive thymidine and then lysed (pulse–chase). In the simple pulse experiment, the autoradiograms seemed to be the result of *two* DNA fibers lying side-by-side; in contrast, the autoradiograms in the pulse–chase experiment appeared to have been produced by *single* DNA fibers. Cairns suggested that the different appearance of the two types of autoradiograms was the consequence of replication at a fork-like growing point; the labeled regions of DNA in the simple pulse experiment would be held together at the replication fork and might well appear side-by-side in autoradiograms, while the labeled regions of DNA in the pulse–chase experiment would be separated from the replication fork by a considerable length of unlabeled DNA and would be much more likely to appear separated in autoradiograms. These possibilities are shown diagrammatically in Fig. 3.

We reasoned that fork-type replication in Chinese hamster cells, if present, should produce the same phenomenon. To test this possibility we performed an analogous experiment with Chinese hamster cells. We used a pulse time of 30 minutes and a chase time of 45 minutes. In the case of the simple pulse, all the media to which the cells were exposed after removal from the pulse medium contained FUdR to prevent cellular synthesis of unlabeled thymidine monophosphate.

In one respect our results were different from those Cairns (1963*a*) obtained with *E. coli*. Very few of the autoradiograms obtained after simple pulse labeling had an *appearance* suggesting that they might be the result of two DNA fibers lying side-by-side. That is, most such autoradiograms were just single lines of grains with no evidence of separation into two lines of grains. In those few cases where some separation occurred, the separation appeared to be in the middle of the labeled regions rather than at their ends (see below). It is true, however, that many of the autoradiograms obtained after simple pulse labeling had a *grain density* corresponding to that expected for two DNA double helices lying side-by-side if each double helix were labeled in a

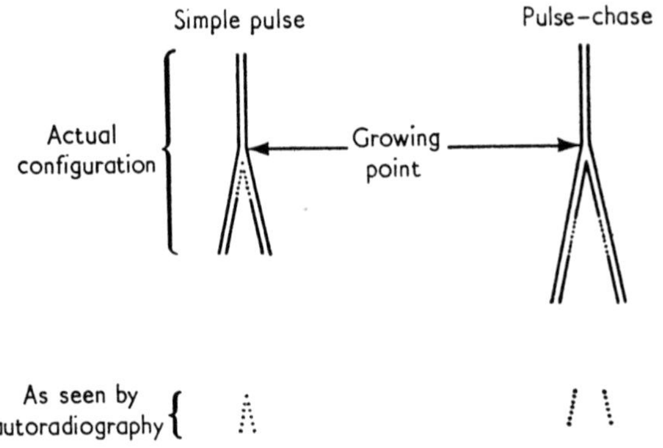

FIG. 3. The effect of replication at a fork-like growing point on the autoradiograms seen after simple pulse and pulse–chase experiments.

single polynucleotide chain. In Fig. 4 are shown histograms of grain-density frequency for both the simple pulse autoradiograms and the pulse–chase autoradiograms. Since a grain density of about 2·5 grains per μ is just slightly less (after correction for specific activity, exposure time, and base composition) than the grain density obtained by Cairns (1963*b*) for a double helix of *E. coli* DNA labeled in a single polynucleotide chain, each autoradiogram having approximately this grain density was probably produced by a double helix containing a single labeled chain. Likewise, autoradiograms having a grain density of about five grains per μ were probably produced by two double helices lying side-by-side and each containing a single labeled chain. The broadness of the distributions can be attributed partly to the statistical error involved in measuring grain density in the shorter autoradiograms and partly to actual errors in grain counting. The higher grain densities may have been underestimated due to the difficulty in distinguishing grains in crowded areas. It is clear, however, that half or more of the simple pulse autoradiograms were probably produced by two labeled chains (the other simple pulse autoradiograms could be explained as the result of breakage). This fact suggests that the two daughter chains are synthesized at about the same time and are held together for a short time after they are synthesized. It is also clear that most of the pulse–chase autoradiograms were produced by single labeled DNA chains (a few pulse–chase autoradiograms produced by two labeled chains might be expected as a result of incomplete chain separation). Thus, although the two daughter chains are held together for a short time after they are synthesized, they later become free to separate from each other. All these properties are consistent with replication at fork-like growing points. Further evidence for the fork-like nature of the growing points will be discussed below.

(ii) *Grain density gradients*

Comparison of typical simple pulse autoradiograms (Plate I) with typical pulse–chase autoradiograms (Plate III) shows another difference besides those mentioned above. Whereas the ends of the simple pulse autoradiograms are always distinct, the ends of the pulse–chase autoradiograms are frequently indistinct because of a gradual decline in grain density from the full grain density expected for one or two labeled chains of DNA in the interior of the autoradiograms to undetectable grain density at the ends.

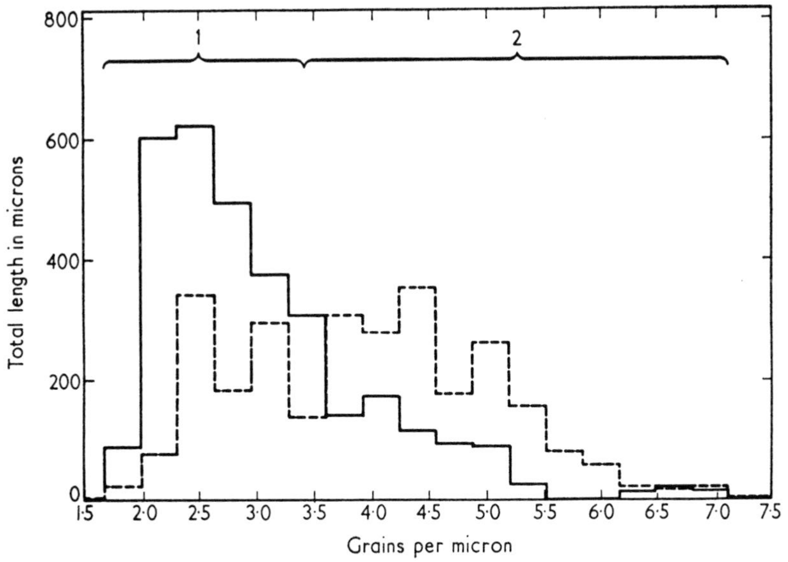

Fig. 4. Frequencies of grain densities in simple pulse and pulse–chase experiments.

Chinese hamster cells were labeled as described in the legend to Plate I (simple pulse) or Plate III (pulse–chase) and their DNA subjected to autoradiography. Exposure time was 4 months. Grain densities were measured with bright-field microscopy at a magnification of 1000 in all the autoradiograms visible over large areas of single Millipore filters. Measurements were made by counting all the grains of each autoradiogram and dividing by the length of the autoradiogram (determined with an eyepiece micrometer). The total length of all autoradiograms in the sample (more than 100 autoradiograms for both distributions shown) having a given grain density is plotted in the Figure as a function of grain density. ———, Pulse–chase; ------, simple pulse. The brackets in the Figure indicate the grain densities presumed to correspond to 1 or 2 labeled polynucleotide chains.

We interpret this reduction in grain density to be the result of gradual change in the specific activity of the intracellular thymidine triphosphate pool upon replacement of the [^{3}H]thymidine in the external medium with non-radioactive thymidine. Accordingly, for the grain-counting experiments above, we counted grains only in internal regions of the pulse–chase autoradiograms where no decline in grain density was evident.

If our interpretation of the grain density gradients is correct, then the direction of decline in grain density is the direction in which DNA was being synthesized at the beginning of the cold chase. Therefore we were surprised to find that many of the autoradiograms had grain densities declining at *both* ends, in *opposite* directions (Plate III). In fact, when tandem arrays of pulse–chase autoradiograms were examined, about 90% of the internal (and therefore unbroken) autoradiograms proved to have declining grain densities at both ends (Plate IV). Thus all these autoradiograms must be the result of at least two growing points proceeding in opposite directions, and most of the pulse–chase autoradiograms which lack gradients at one or both ends (Plate III) must have been produced by broken DNA fibers.

The apparent exceptions to the rule of double gradients in tandem arrays are of two types. The first type is rare (about 3% of the internal autoradiograms) and of uncertain significance. Examples of this type are shown in Plate V(a), (b) and (c). The central structure in Plate V(a) may be an internal autoradiogram without grain density gradients at either end. In Plate V(b) and (c) are examples of internal autoradiograms apparently lacking gradients at *one* end.

The second kind of apparent exception is more frequent (about 7%). Examples are shown in Plates VI(c) and VII(a). Here the distinct autoradiogram ends are paired opposite each other in adjacent autoradiograms, and there are no grains between the paired ends. This configuration suggests that the unlabeled areas between the paired ends are the result of DNA replication completed *before* treatment with FUdR. Since the intracellular thymidine triphosphate pool would be depleted in the presence of FUdR, no time would be required for pool equilibration after introduction of [^{3}H]thymidine and no grain density gradients would be seen in the resulting autoradiograms at the points of resumption of DNA synthesis.

(iii) *Sister double helices*

Notice that all the structures in Plate VI are examples of partial separation of sister double helices. They can be identified as such by the correspondence of grain density patterns in the parallel autoradiograms. This correspondence suggests that corresponding regions of the two labeled polynucleotide chains were synthesized at identical times. Certainly it means that within stretches of DNA longer than the autoradiographic limit of resolution (about 5 μ) DNA was not synthesized first on one parental chain as template and then on the other; rather both parental chains must have acted as template at once. Fork-like growing points provide by far the simplest explanation of this behavior. In fact, the fork-like points of attachment of the sister double helices in Plate VI may be growing points.

Another example of sister double helix separation is shown in Plate V(d). The grain density in the heavily-labeled regions on the top is much less than that expected for a single chain labeled during the pulse (shown in the heavily labeled regions on the bottom). Thus the heavily-labeled regions on the top may represent points where DNA synthesis was initiated after the pulse, during the period of pool equilibration. If this is true then Plate V(d) offers an example of neighboring replication sections beginning replication at different times.

(iv) *The size of replication sections*

We have used the tandem arrays of autoradiograms resulting from the 30-minute pulse and pulse–chase treatments to obtain an estimate of the size of the replication sections in Chinese hamster chromosomal DNA. Separate estimates were made for the simple pulse autoradiograms and for the pulse–chase autoradiograms. To avoid errors resulting from possible breaks in DNA fibers at the ends of tandem arrays, only *internal* autoradiograms of the tandem arrays were used for measurement unless, in the case of pulse–chase autoradiograms, a grain density gradient was present at the end of the array. Due to the fact that most internal autoradiograms in these experiments are the result of at least two growing points (see above), direct measurement of the size of individual replication sections was not possible. Instead, the center-to-center distances between internal or unbroken external autoradiograms of tandem arrays were measured (see Fig. 5). These distances are assumed to correspond to the sum of two adjacent replication sections. Further rationale for this procedure is given in the Discussion section.

Histograms showing the frequency distributions of center-to-center distances are shown in Fig. 6. Note that measurements of simple pulse autoradiograms and of

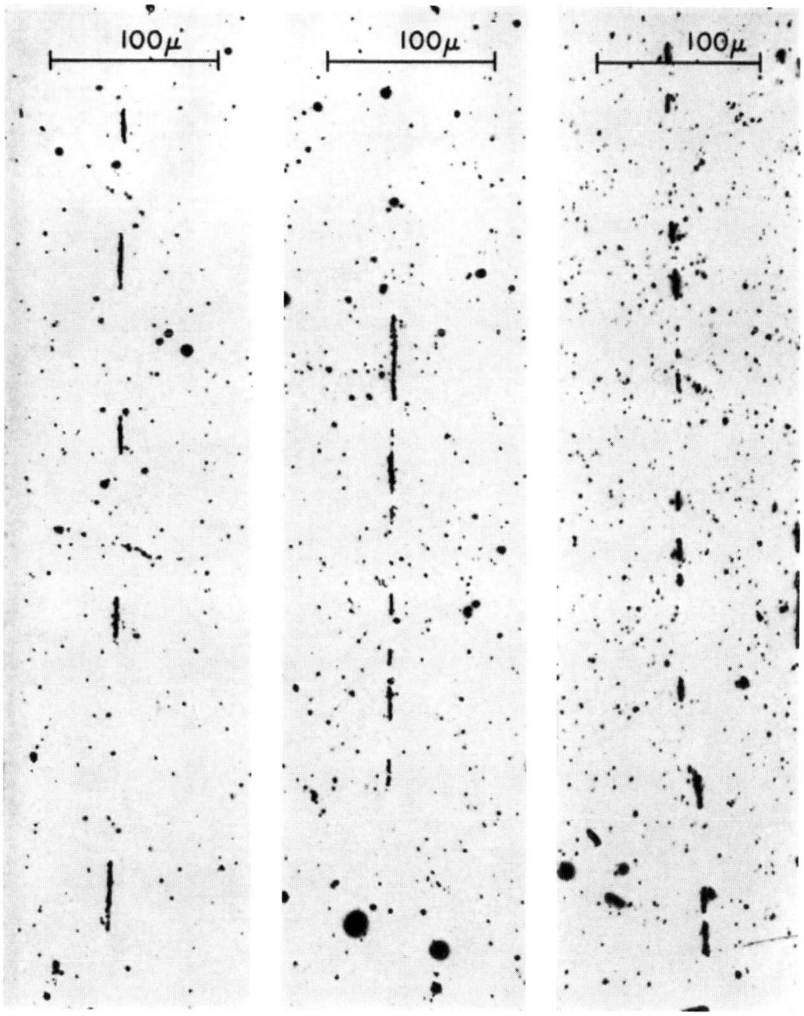

(a) (b) (c)

PLATE I. Tandem arrays of autoradiograms.

Cells of Chinese hamster fibroblast strain B14FAF28 (a gift from Dr T. C. Hsu) were grown as monolayer cultures on plastic Petri dishes in Eagle's medium supplemented with 10% calf serum. After a 12-hr pretreatment with FUdR (0·1 μg/ml.), [^{3}H]thymidine (18 c/m-mole, Nuclear Chicago) was added to 0·5μg/ml. 30 min later the cells were harvested by trypsinization and diluted to 1×10^4 cells/ml. in isotonic saline. The solutions used for trypsinization and dilution contained FUdR (0·1 μg/ml.). The cells were diluted tenfold into lysis medium (1·0 M-sucrose–0·05 M-NaCl–0·01 M-EDTA, pH 8·0), lysed by dialysis against 1% sodium dodecyl sulfate in lysis medium, then dialyzed further against dialysis medium (0·05 M-NaCl–0·005 M-EDTA, pH 8·0). The released DNA was trapped on Millipore VM filters which had served as dialysis membranes. It was then subjected to autoradiography.

Exposure time with Kodak AR10 autoradiographic stripping film (Eastman Kodak Co.) was 4 months. Picture taken by dark-field microscopy.

[facing p. 334

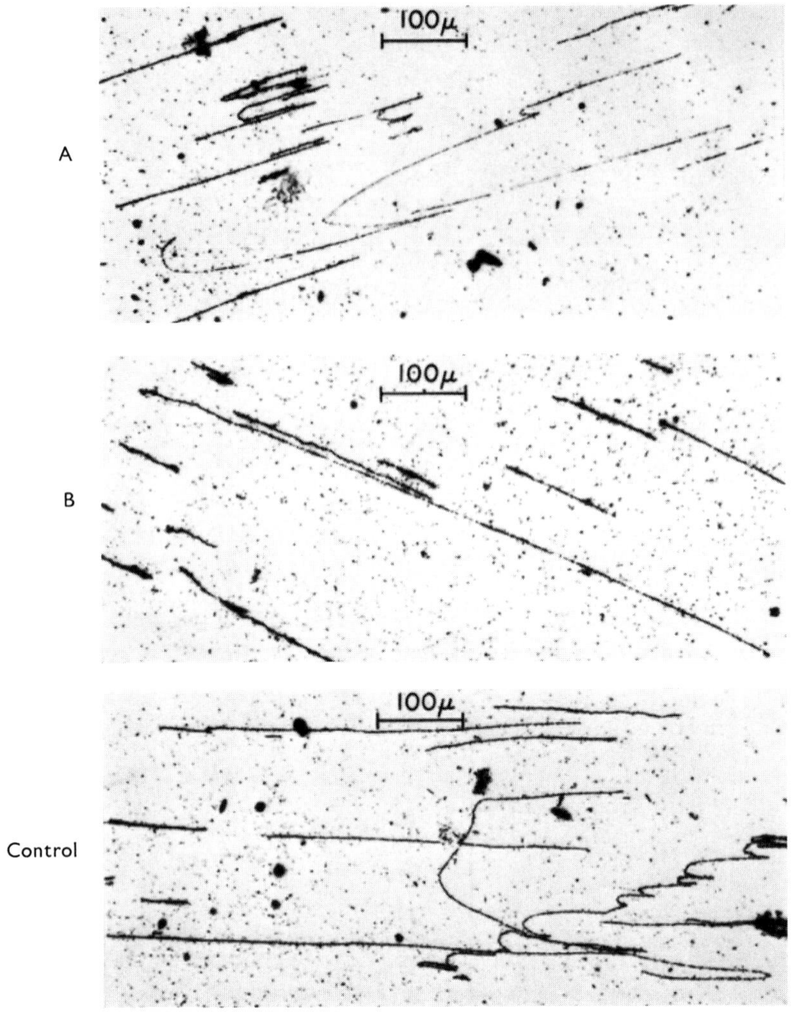

PLATE II. Cold pulse autoradiograms.

Chinese hamster cells were labeled as described in Fig. 1 (A, B and control) and their DNA was subjected to autoradiography. Exposure time was 6 months. Dark field.

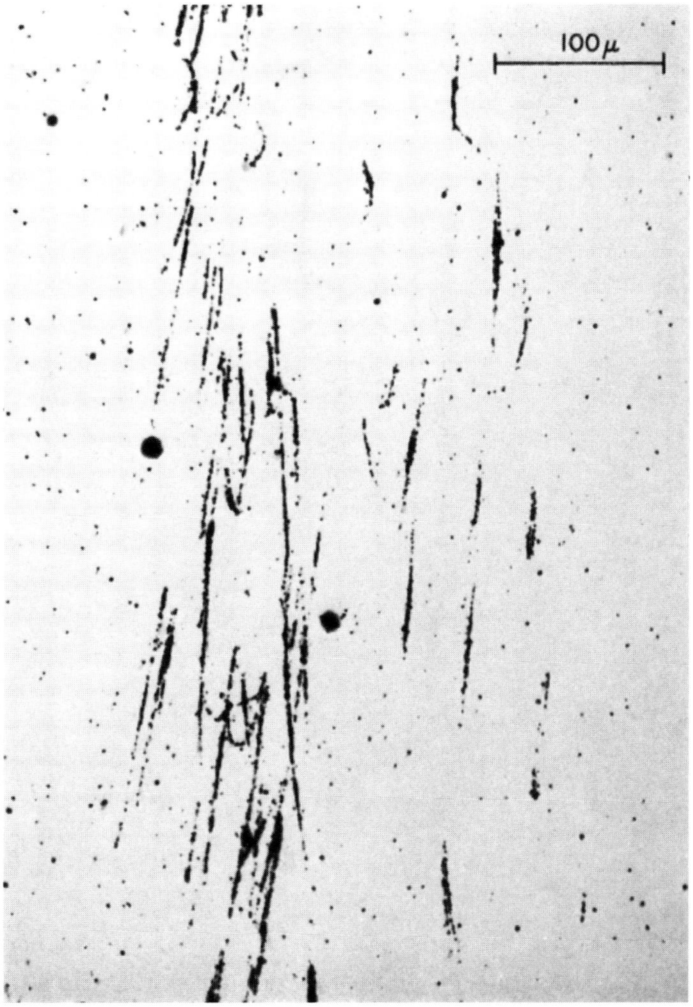

PLATE III. Typical autoradiograms produced in a pulse–chase experiment.

Chinese hamster cells were labeled as described in the legend to Plate I, except that after 30 min of incubation with [^{3}H]thymidine and FUdR, the medium containing [^{3}H]thymidine and FUdR was removed and replaced with medium containing non-radioactive thymidine (5 μg/ml.) and no FUdR. Incubation was continued for 45 min. The cells were then harvested by trypsinization and diluted to 1×10^4 cells/ml. in isotonic saline. Lysis and autoradiography were performed as in the legend to Plate I. Exposure time was 4 months. Dark field.

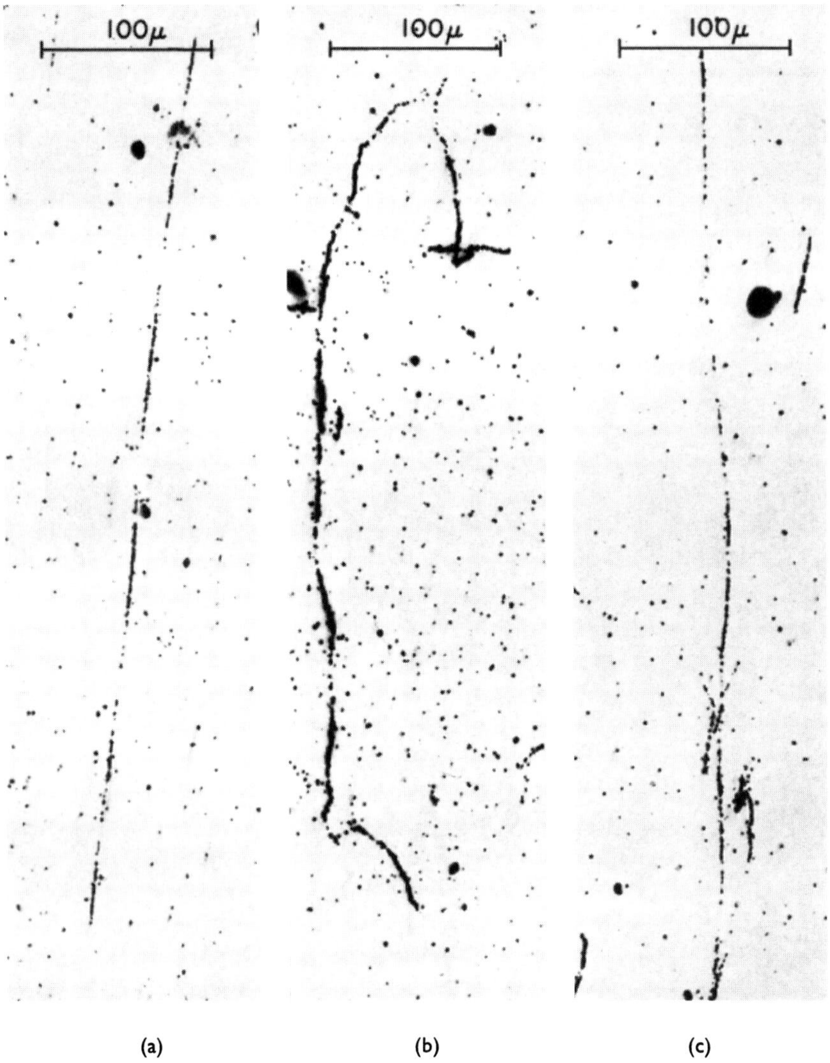

PLATE IV. Tandem arrays of autoradiograms produced in a pulse–chase experiment. Labeling and autoradiography were performed as in the legend to Plate III. Dark field.

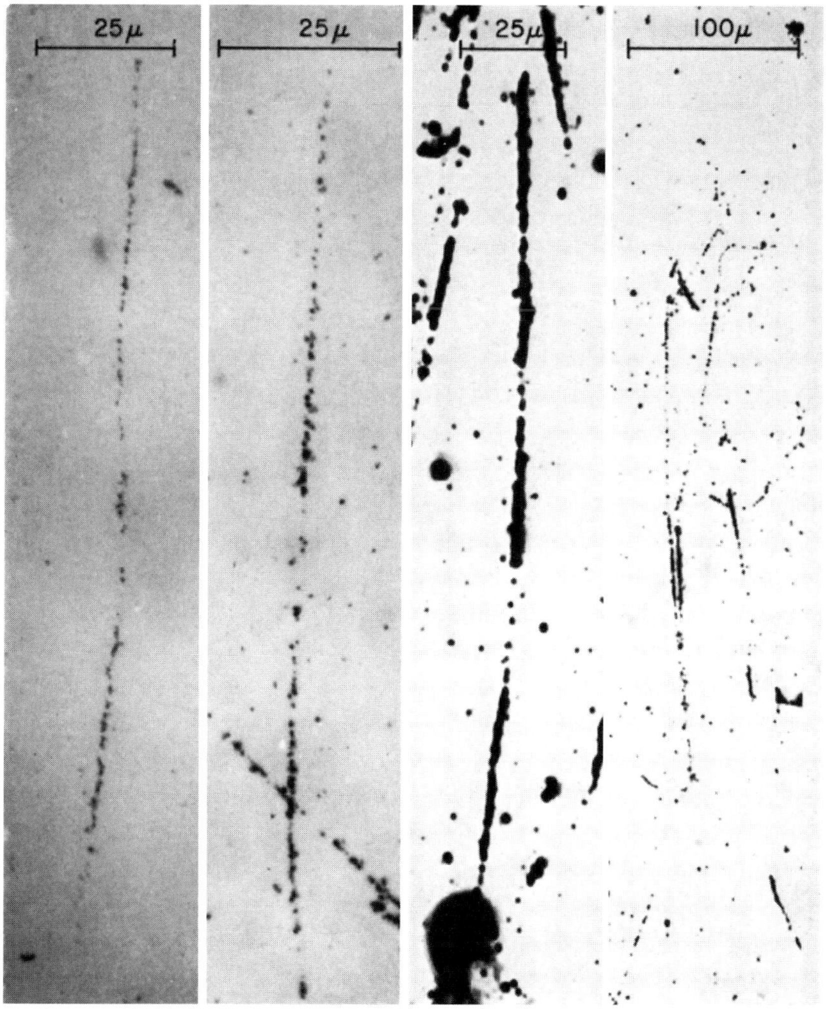

(a) (b) (c) (d)

PLATE V. Examples of autoradiograms produced in a pulse–chase experiment. Labeling and autoradiography were performed as in the legend to Plate III.

(a) Example of autoradiogram apparently without grain density gradients at either end. Picture taken by bright-field microscopy.

(b) Example of autoradiogram apparently lacking a grain density gradient at one end. Bright field.

(c) Example of autoradiogram apparently lacking a grain density gradient at one end. Dark field.

(d) Example of apparent asynchronous initiation of DNA replication in neighboring replication sections. Note that this is also an example of sister double helices. Dark field.

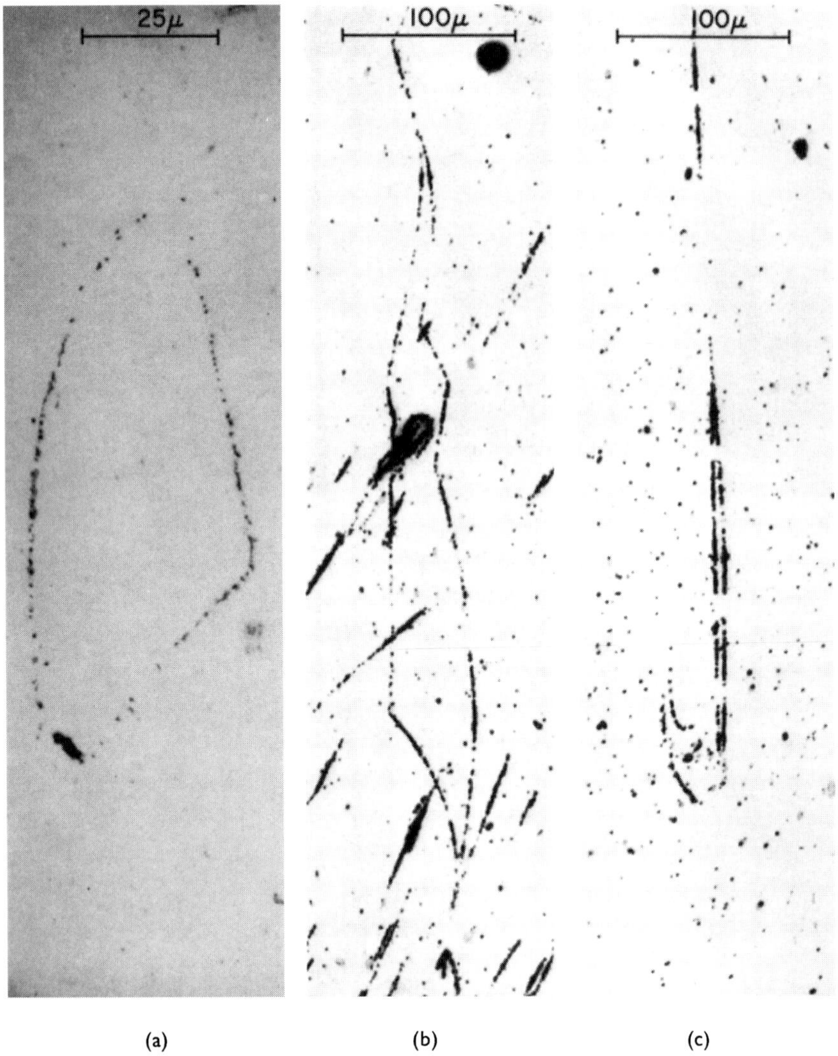

PLATE VI. Examples of sister double helices. Labeling and autoradiography were performed as in the legend to Plate III.
(a) Bright field.
(b) and (c) Dark field.

(a)
(b)
(c)

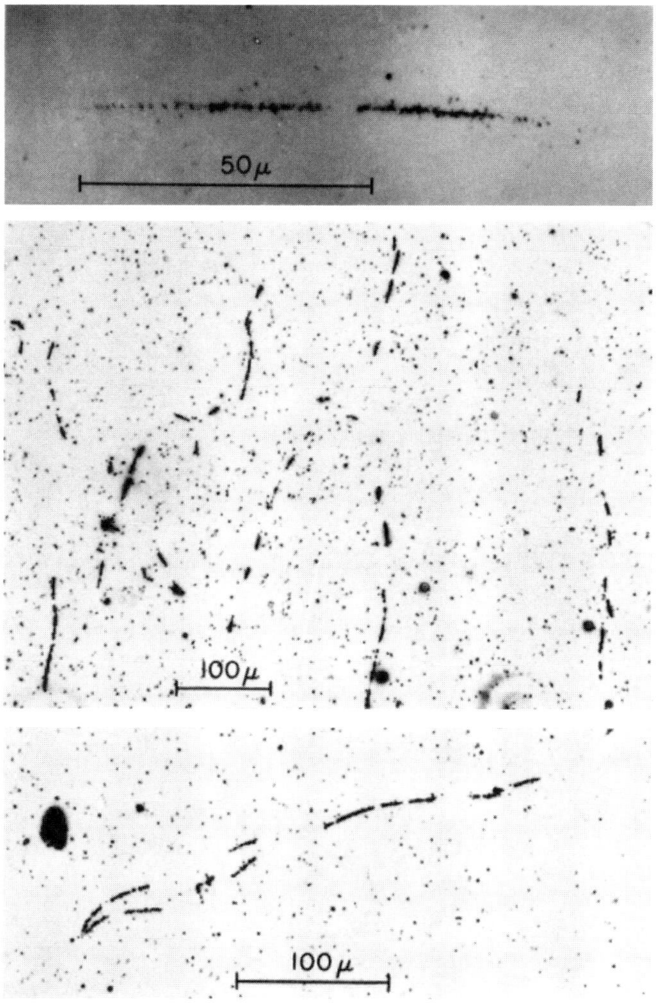

PLATE VII. (a) Example of Chinese hamster cell DNA autoradiogram interrupted by a region where replication presumably took place before addition of FUdR. Conditions were identical to those of Plate III. Bright field.

(b) Example of small, closely spaced autoradiograms of HeLa cell DNA. HeLa S3 cells were grown under conditions identical to those for Chinese hamster cells (legend to Plate I). After a 15-hr pretreatment with FUdR (0·1 μg/ml.), [^{3}H]thymidine (18 c/m-mole) was added to 2 μg/ml. 1 hr later the medium containing [^{3}H]thymidine and FUdR was removed and replaced with medium containing non-radioactive thymidine (5 μg/ml.) and no FUdR. Incubation was continued for 45 min. The cells were then harvested by trypsinization and diluted to 1×10^4 cells/ml. in isotonic saline. Lysis and autoradiography were performed as in the legend to Plate I. Exposure time was 3 months. Dark field.

(c) Example of sister double helices in HeLa cell DNA. Conditions were identical to those of Plate VII(b) except that the cells were exposed to [^{3}H]thymidine for 2 hr. Dark field.

(a)

(b)

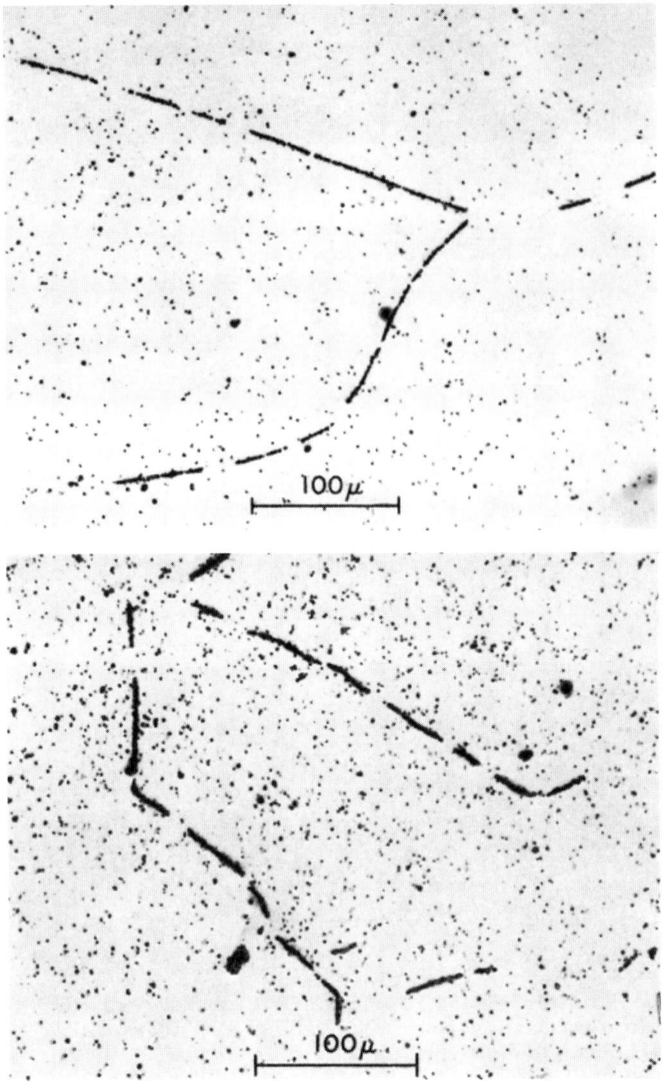

PLATE VIII. Examples of sister double helices in HeLa cell DNA.

(a) Conditions identical to those of Plate VII(c). Dark field.

(b) Conditions identical to those of Plate VII(b) except that there was no pretreatment with FUdR, and FUdR was not present during the pulse. Exposure time was 7·5 months. Dark field.

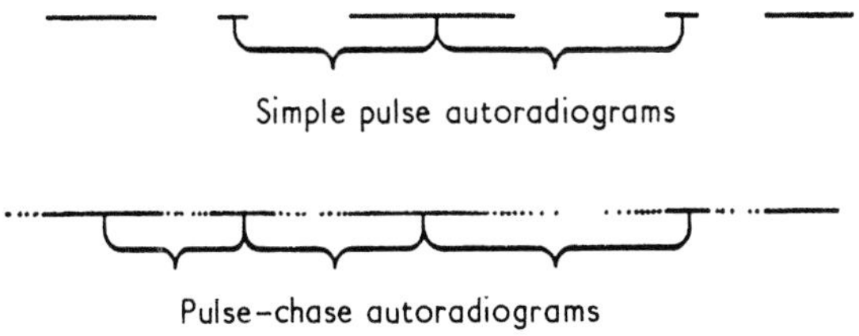

FIG. 5. Method of measurement of the size of pairs of replication sections from tandem arrays of autoradiograms.

The horizontal lines indicate representative autoradiograms. Grain density gradients at the ends of pulse–chase autoradiograms are indicated by dots. The brackets show the lengths which were measured (center-to-center distances between internal or unbroken external autoradiograms).

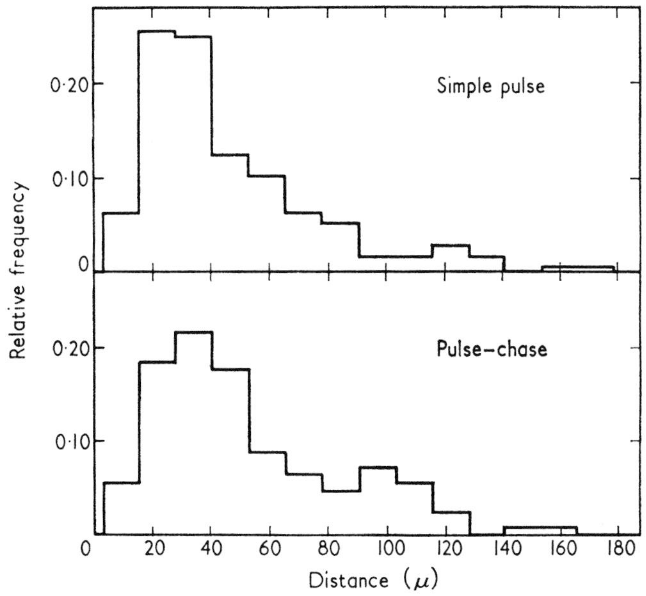

FIG. 6. Distribution of center-to-center distances between autoradiograms of tandem arrays.

Chinese hamster cells were labeled either by the simple pulse method (described in the legend to Plate I) or the pulse–chase method (described in the legend to Plate III), and their DNA subjected to autoradiography. Center-to-center distances between autoradiograms of tandem arrays were measured according to the criteria of Fig. 5. Measurements were made using an eye-piece micrometer and bright-field microscopy at a magnification of 1000.

pulse–chase autoradiograms gave similar results. The most frequently observed distances are between 15 and 60 μ, which would correspond to replication section sizes of about 7 to 30 μ.

The accuracy of these measurements is not certain. The frequency of distances less than 15 μ may have been underestimated due to difficulties in resolution. In addition, the measurements were based on the assumption that each pair of adjacent diverging replication sections in the DNA fibers forming the tandem arrays would be represented by a separate autoradiogram. The possibility that adjacent pairs might have completed replication during the pulse, so that their autoradiograms would be fused, and the possibility that some pairs might not have begun to replicate during the pulse would thus result in an exaggerated estimate of their size.

Since all these uncertainties would cause an over-estimate of pair size, one can conclude that most pairs of adjacent replication sections are at least as small as indicated by the histograms of Fig. 6.

(d) *Estimation of replication rate*

(i) *Estimates from the 30-minute pulse–chase experiment*

We shall use the term replication rate to refer to the rate at which DNA is synthesized at an individual growing point. This rate will be measured in units of μ of movement of the growing point along the parental DNA double helix per minute (μ/min).

The possibility of autoradiogram fusion discussed above and uncertainties in the time of stopping and starting of replication in individual replication units make estimates of replication rate based on the lengths of autoradiograms produced after a 30-minute pulse extremely uncertain. However, the pulse–chase autoradiograms have provided two alternative methods for estimation of replication rate. First, in most cases the grain densities at the ends of the pulse–chase autoradiograms appear to decrease monotonically, suggesting that, within the gradient region, replication was proceeding at a single growing point throughout the equilibration period. Monotonic gradients 20 to 40 μ long were frequently found (Plate III), while monotonic gradients as much as 100 μ long were occasionally found (Plate IV(c)). Since the *maximum* possible equilibration time is 45 minutes (chase time), a *minimum* estimate of replication rate is 0·4 to 0·9 μ per minute for the frequently observed gradients and up to 2 μ per minute for the occasionally observed gradients. The variations found in lengths of monotonic gradients can be explained as the result of heterogeneity between cells in pool equilibration time or as the result of heterogeneity of replication rate or both.

In addition, if our interpretation of the sharp interruptions in the autoradiograms shown in Plates VI(c) and VII(a) is correct, then the ends without grain density gradients represent points where replication started at the beginning of the 30-minute pulse. Likewise the positions where the grain density begins to decline in such autoradiograms represent the points to which replication had proceeded by the end of the pulse. We conclude that each branch of such autoradiograms represents a region synthesized during the entire pulse. We have found samples of such autoradiograms with branches from 15 to 37 μ long. On the assumption that each branch is the result of a single growing point, these lengths correspond to replication rates of 0·5 to 1·2 μ per minute.

(ii) *The maximum possible replication rate*

To measure the maximum possible rate of DNA replication, Chinese hamster cells were exposed to [^{3}H]thymidine for pulse times of 5, 10, 20, 60 and 120 minutes and their DNA subjected to autoradiography. The maximum autoradiogram lengths found after these pulse times increased more or less linearly with pulse time at a rate of 5 to 10 μ per minute.

These measurements set an upper limit to the replication rate of Chinese hamster DNA. Assuming that each autoradiogram is the product of only two growing points, then the maximum possible rate of replication is 2·5 to 5 μ per minute. Although the maximum autoradiogram lengths found after the longer pulse times are probably the result of autoradiogram fusion, the fact that autoradiograms 25 μ long are found even after the shortest pulse time of five minutes suggests that Chinese hamster cell DNA may, indeed, occasionally replicate as rapidly as 2·5 μ per minute.

(e) *Experiments with HeLa cells*

In order to compare DNA replication in HeLa cells with that in Chinese hamster cells, we conducted a series of experiments in which we exposed HeLa cells and Chinese hamster cells to [^{3}H]thymidine for times of 1, 2, 4, 8 and 18 hours. For all these experiments the pulse with [^{3}H]thymidine was followed by a 45-minute chase with non-radioactive thymidine. Because we also wanted to investigate the possible effect of FUdR on DNA replication in these cells, some of the experiments were carried out in the absence of FUdR.

In general, the HeLa cell autoradiograms and the Chinese hamster cell autoradiograms were similar in size and appearance. This was especially true for pulse times of four hours or longer. For pulse times of one or two hours, however, some differences were evident. First, although the HeLa cells were subjected to the same kind of pulse–chase treatment as the Chinese hamster cells, grain density gradients were difficult to observe in the HeLa cell autoradiograms. What might have been short grain density gradients were found only occasionally. We should point out that, even for the Chinese hamster cells, grain density gradients were not nearly so frequent in these longer experiments as they were in the 30-minute pulse–chase experiments. This low frequency of gradients has not been investigated further.

Also, although the absence or presence of FUdR had no detectable effect on the size or type of autoradiograms produced by Chinese hamster DNA, the DNA from HeLa cells exposed to FUdR for 15 hours and then pulse-labeled for one hour produced a much greater proportion of very short autoradiograms, usually close together in tandem arrays, than did DNA from Chinese hamster cells or from HeLa cells not exposed to FUdR. An example of these short HeLa cell autoradiograms is shown in Plate VII(b).

The fact that FUdR did not alter the appearance of Chinese hamster cell autoradiograms suggests that it has no effect on DNA autoradiography in Chinese hamster cells other than the intended ones of preventing thymidine monophosphate biosynthesis and blocking most cells at the beginning of DNA synthesis. If FUdR has only these effects in HeLa cells, then our finding that DNA from HeLa cells treated with FUdR for 15 hours and then pulse-labeled for one hour produces a high proportion of small, close autoradiograms suggests that the DNA which replicates earliest in HeLa cells may be unusually rich in small replication sections. However, the possibilities that the thymidine starvation produced by FUdR may induce replication at an abnormally large number of growing points (Pritchard & Lark, 1964) in HeLa cells, or may allow accumulation of damage which would result in repair replication (Taylor, Haut & Tung, 1962) cannot be excluded.

Like the Chinese hamster DNA, the HeLa DNA produced autoradiograms suggesting the partial separation of sister double helices. Three examples are shown in Plates VII(c) and VIII.

4. Discussion

(a) *Summary of results*

The autoradiographic evidence we have presented above suggests that the process of DNA replication in Chinese hamster chromosomes has the following attributes.

(1) The long fibers of chromosomal DNA are made up of many tandemly joined replication sections, as proposed by Cairns (1966). (See Plates II, V(d) and VI.)
(2) If any linkers connect the replication sections or other points in the long DNA fibers, they must be resistant to pronase.
(3) DNA is replicated at fork-like growing points (Plates V(d) and VI; Fig. 4).
(4) Usually, perhaps always, DNA is synthesized in opposite directions at adjacent growing points with the result that, in our experiments, most unbroken autoradiograms were the result of two or more growing points (Plate IV).
(5) Most replication sections are less than 30 μ long. Many are so short that they are difficult to resolve by autoradiography.
(6) Neighboring replication sections can begin replication at different times (Plate V(d)).
(7) At any one time, regions of DNA synthesis are not distributed equidistantly along the chromosomal DNA fibers (Plate II).
(8) The rate of DNA replication per growing point is 2·5 μ per minute or less. This rate does not vary greatly during the period of DNA synthesis (Fig. 2).

(b) *Bidirectional replication*

(i) *Arrangement of replication sections*

The fact that DNA is usually synthesized in opposite directions at adjacent growing points implies that all or nearly all replication sections must be arranged as shown in Fig. 7(a). In this diagram, each replication section is bounded by an origin (marked O) and a terminus (marked T). Within each section, DNA replication would proceed from the origin to the terminus. Each replication section would share its origin with one adjacent section and share its terminus with the other adjacent section.

(ii) *The unit of control*

The fact that adjacent replication sections share origins or termini suggests the possibility that initiation of replication may be controlled at the level of adjacent *pairs* of sections rather than at the level of individual sections. In fact, at least three models for control of the initiation of DNA replication can be imagined.

(1) Control at the level of the replication section. In this case, each section might initiate replication independently of all its neighbors.
(2) Control at the level of converging pairs of replication sections. In this case adjacent sections sharing a terminus would initiate replication together.
(3) Control at the level of diverging pairs of replication sections. In this case, adjacent sections sharing an origin would initiate replication together.

Each of these models makes certain predictions for the types of internal autoradiograms which might be found in tandem arrays after a pulse–chase experiment, but only the third model predicts that all internal autoradiograms either should have grain density gradients at both ends (Plate IV) or should be arranged in pairs with adjacent distinct ends (Plates VI(c) and VII(a)). With rare exceptions (Plate V(a), (b) and (c), these are the only kinds of internal autoradiograms found. Even the exceptions can be accounted for by the third model if certain assumptions are made about the nature of termini (see below). Furthermore, structures like those in Plate V(d) where neighboring *diverging* pairs of replication sections appear to have started replication at different times are definitely contrary to the predictions of the second model.

The available evidence thus favors the third model above. We propose the term replication unit to mean the basic unit of control in the initiation of replication—presumably an adjacent pair of diverging replication sections. Note that our measurements of center-to-center distances between internal autoradiograms (Fig. 6) are of the order of size of such replication units.

(iii) *The nature of origins and termini*

It is now well established that, at the level of resolution of whole chromosome autoradiography, regions of initiation and termination of DNA replication are reproducibly and heritably controlled (see Huberman, 1967, for review). However, neither whole chromosome autoradiographic studies nor the studies presented in this paper have been able to determine whether sites of initiation (origins) are reproducible at the atomic level. The possibility remains that sites of initiation may occur anywhere within relatively long stretches of the DNA fibers.

Likewise, there are at least two ways in which DNA replication could be terminated at the ends of replication sections. One possibility is that there are specific sites along the DNA fibers where replication, from either direction, is stopped. The other possibility is that replication continues until adjacent converging growing points meet. According to the first possibility one might expect to find, as a result of termination of replication during the pulse in a pulse–chase experiment, occasional internal autoradiograms lacking grain density gradients at one or both ends. Autoradiograms of the type shown in Plate V(a), (b) and (c) may be examples of such termination. However, these structures are so rare (about 3% of the internal autoradiograms) that the existence of specific termini must remain in doubt.

(iv) *The bidirectional model*

Despite the uncertainty noted above about the reproducibility of origins and termini, the fact remains that, in any given replication, each growing point must have both a starting site and a stopping site. Even if one attributes no more to origins and termini than this, one can, by using the results presented in this paper, obtain a better model of the way in which chromosomal DNA fibers are replicated than was previously possible. This model, which we shall call the bidirectional model, is shown in the diagram in Fig. 7; its basic feature is that the replication unit consists of two diverging replication sections. Note that we have assumed in presenting the model that termini are sites where replication stops rather than sites where converging growing points meet. This assumption must be considered tentative. Otherwise the various features of the model seem well supported by the experimental evidence.

(c) *The rate of DNA replication*

(i) *Comparison of results*

The rate of DNA replication in the chromosomes of mammalian cells has previously been measured in two ways. Cairns (1966 and personal communication) measured the lengths of the autoradiograms produced after exposure of HeLa cells to [^{3}H]-thymidine for 45 minutes and 180 minutes, and concluded that the average length increased at a rate of 0·5 μ per minute or less. Taylor (personal communication) measured the BUdR pulse time required before Chinese hamster DNA molecules were fully converted to hybrid density, and estimated from his results that the BUdR-labeled region increased in length at a rate of one to two μ per minute. According to

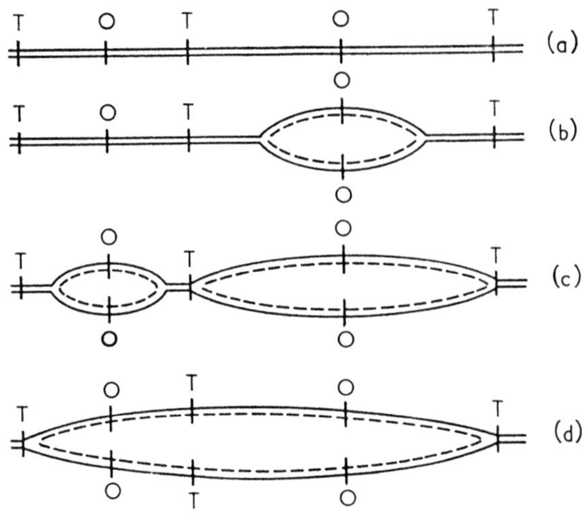

FIG. 7. Summary of the bidirectional model for DNA replication.

Each pair of horizontal lines represents a section of a double helical DNA molecule containing two polynucleotide chains (———, parental chain; ------, newly synthesized chain). The short vertical lines represent positions of origins (O) and termini (T). The diagrams represent different stages in the replication of two adjacent replication units:

(a) Prior to replication; (b) replication started in right-hand unit; (c) replication started in left-hand unit and completed at termini of right-hand unit; (d) replication completed in both units; sister double helices separated at the common terminus.

the bidirectional model, these measurements should be considered over-estimates of the actual rate of DNA replication per growing point, because some of the measured lengths, if not most, probably were the result of *two* growing points.

These measurements are in fairly good agreement with the results we have obtained by a variety of methods. Measurements of the maximum autoradiogram lengths obtained after short pulses suggest that the rate of DNA replication must be less than about 2·5 μ per minute. Measurements of the length of DNA synthesized during equilibration of the thymidine triphosphate pool suggest rates of replication from 0·4 to 2 μ per minute.

Probably the most reliable estimates are provided by measurements of the lengths of the branches in autoradiograms such as those in Plates VI(c) and VII(a) where it is likely that DNA synthesis was going on throughout the 30-minutes pulse. Such measurements suggest replication rates from 0·5 to 1·2 μ per minute.

The uncertainties involved in all these measurements mean that a precise estimate of the replication rate in Chinese hamster cells is still impossible. However, one can draw some general conclusions from the measurements made so far.

All the measurements suggest that there is probably no one replication rate but rather a range of replication rates. Whether or not the replication rate in a single replication unit is always constant is not known.

Also, it is interesting that, despite the fact that mammalian replication sections are much smaller than bacterial chromosomes, the rate of bacterial DNA replication is much larger (30 μ per minute as measured by Cairns, 1963*a*) than the rate of mammalian DNA replication. Perhaps the necessity for synthesizing and organizing the many chromosomal proteins of mammalian cells requires a lower rate of replication.

We gratefully acknowledge the advice and assistance of Dr Giuseppe Attardi in whose laboratory this work was done. This investigation was supported by U.S. Public Health Service grants GM–11726, 5–F1–GM–21,622 and F2–HD–22,991 and by the Arthur McCallum Fund.

REFERENCES

Cairns, J. (1963*a*). *J. Mol. Biol.* **6**, 208.
Cairns, J. (1963*b*). *Cold Spr. Harb. Symp. Quant. Biol.* **28**, 43.
Cairns, J. (1966). *J. Mol. Biol.* **15**, 372.
Davern, C. I. (1966). *Proc. Nat. Acad. Sci., Wash.* **55**, 792.
Huberman, J. A. (1967). In *Some Recent Developments in Biochemistry*. Taipei: Academia Sinica, in the press.
Huberman, J. A. & Riggs, A. D. (1966). *Proc. Nat. Acad. Sci., Wash.* **55**, 599.
Hsu, T. C. (1964). *J. Cell Biol.* **23**, 53.
Hsu, T. C., Dewey, W. C. & Humphrey, R. M. (1962). *Exp. Cell Res.* **27**, 441.
Macgregor, H. C. & Callan, H. G. (1962). *Quart. J. Micro. Sci.* **103**, 173.
Pritchard, R. H. & Lark, K. G. (1964). *J. Mol. Biol.* **9**, 288.
Sasaki, M. S. & Norman, A. (1966). *Exp. Cell Res* **44**, 642.
Taylor, J. H. (1960). *J. Biophys. Biochem. Cytol.* **7**, 455.
Taylor, J. H., Haut, W. F. & Tung, J. (1962). *Proc. Nat. Acad. Sci., Wash.* **48**, 190.

Gall J.G. and **Pardue M.L.** 1969. Formation and detection of RNA-DNA hybrid molecules in cytological preparations. *Proc. Natl. Acad. Sci.* **63:** 378–383.

IT IS A TRUISM OF CELL BIOLOGY THAT IMPORTANT macromolecules are often confined to certain cell types or to specific compartments within the cell, and that the localization of a molecule of unknown function can give insight into its role in cell physiology or development. Fortunately there are highly sensitive and specific methods for localizing both proteins and nucleic acids at the cellular and subcellular level - immunofluorescence for proteins (Lazarides and Weber, this volume) and in situ hybridization for nucleic acids. In situ hybridization takes advantage of the fact that single-stranded nucleic acid molecules, either RNA or DNA, can form stable hybrids with their complementary strands. A suitable technique for in situ hybridization was first developed by Gall and Pardue in 1969. In the original procedure ^{3}H-labeled ribosomal RNA (rRNA) derived from cultured cells was hybridized to squash preparations of immature *Xenopus* oocytes. Oocytes were chosen because they exhibited an extraordinary amplification of the genes coding for rRNA, such that nearly 3/4 of the total DNA in the nucleus was complementary to the probe (1, 2). Without such amplification, it would have been impossible to detect the hybridization, given the low specific activity of the metabolically-labeled probe available at the time. Subsequent experiments demonstrated that some highly repeated sequences, such as the mouse and *Drosophila* satellite DNAs, are localized in the heterochromatin next to the centromeres (3). Eventually probes of higher specific activity permitted localization of single copy sequences. A major improvement in spatial resolution came with the development of fluorescent probes (FISH or fluorescent in situ hybridization), which allowed more precise localization than the original ^{3}H-labeled probes (4). Fluorescent probes have also greatly simplified karyotype analysis by allowing individual chromosomes of a set to be "painted" in different colors (5).

---

1. Gall J.G. 1968. Differential synthesis of the genes for ribosomal RNA during amphibian oogenesis. *Proc. Natl. Acad. Sci.* **60:** 553–560.
2. Brown D.D. and Dawid I.B. 1968. Specific gene amplification in oocytes. *Science* **160:** 272–280.
3. Pardue M.L. and Gall J.G. 1970. Chromosomal localization of mouse satellite DNA. *Science* **168:** 1356–1358.
4. Langer P.R., Waldrop A.A., and Ward D.C. 1981. Enzymatic synthesis of biotin-labeled polynucleotides: novel nucleic acid affinity probes. *Proc. Natl. Acad. Sci.* **78:** 6633–6637.
5. Lichter P., Cremer T., Borden J., Manuelidis L., and Ward D.C. 1988. Delineation of individual human chromosomes in metaphase and interphase cells by in situ suppression hybridization using recombinant DNA libraries. *Hum. Genet.* **80:** 224–234.

Reprinted from the PROCEEDINGS OF THE NATIONAL ACADEMY OF SCIENCES
Vol. 63, No. 2, pp. 378–383. June, 1969.

FORMATION AND DETECTION OF RNA-DNA HYBRID MOLECULES IN CYTOLOGICAL PREPARATIONS*

BY JOSEPH G. GALL AND MARY LOU PARDUE

KLINE BIOLOGY TOWER, YALE UNIVERSITY

Communicated by Norman H. Giles, March 27, 1969

Abstract.—A technique is described for forming molecular hybrids between RNA in solution and the DNA of intact cytological preparations. Cells in a conventional tissue squash are immobilized under a thin layer of agar. Next they are treated with alkali to denature the DNA and then incubated with tritium-labeled RNA. The hybrids are detected by autoradiography. The technique is illustrated by the hybridization of ribosomal RNA to the amplified ribosomal genes in oocytes of the toad *Xenopus*. A low level of gene amplification was also detected in premeiotic nuclei (oogonia).

Several techniques are currently used for annealing RNA molecules to their complementary DNA sequences. For certain purposes both the RNA and DNA can be in solution,[1, 2] but it is often more convenient to have the DNA immobilized in a solid or semisolid matrix,[3, 4] or attached to a nitrocellulose membrane filter.[5] The hybrids are generally detected by scintillation counting of radioactive RNA after treatment with ribonuclease to remove unhybridized RNA.

The hybridization of RNA to the DNA in a cytological preparation should exhibit a high degree of spatial localization, since each RNA species hybridizes only with sequences to which it is complementary. The general principles of a cytological hybridization technique are not difficult to lay down. The chromosomes or nucleus should be fixed in as lifelike a fashion as possible; basic proteins should be removed, since they are known to interfere with the hybridization procedure;[5] the DNA should be denatured in such a way that cytological integrity is not lost; the hybridization should be carried out with radioactive RNA of very high specific activity, since the number of hybridized molecules at a given locus will be small; and detection should be by tritium autoradiography to permit maximal cytological resolution.

This communication describes a cytological hybridization technique applicable to conventional squash preparations. It is illustrated by the hybridization of rRNA to the extrachromosomal rDNA in oocytes of the toad *Xenopus*. A preliminary report on the technique was presented in December 1968 at the International Symposium of Nuclear Physiology and Differentiation, Belo Horizonte, Brazil.[6]

Materials and Methods.—The cytological hybridization technique combines certain features of the agar column[4] and filter methods.[5] It should be generally applicable to any material that can be examined as a squash or smear. The following procedure was used in making the preparation shown in Figure 1.

(1) Ovaries from recently metamorphosed *Xenopus laevis* were fixed for a few minutes in ethanol-acetic acid (3:1).

(2) The tissue was transferred to a drop of 45% acetic acid on a microscope slide and

teased with jewelers' forceps. The larger bits of tissue were removed, a cover slip was added, and the cells were squashed. The slides had been previously subbed by dipping into a solution of 0.1% gelatin in 0.1% chrome alum and draining until dry.

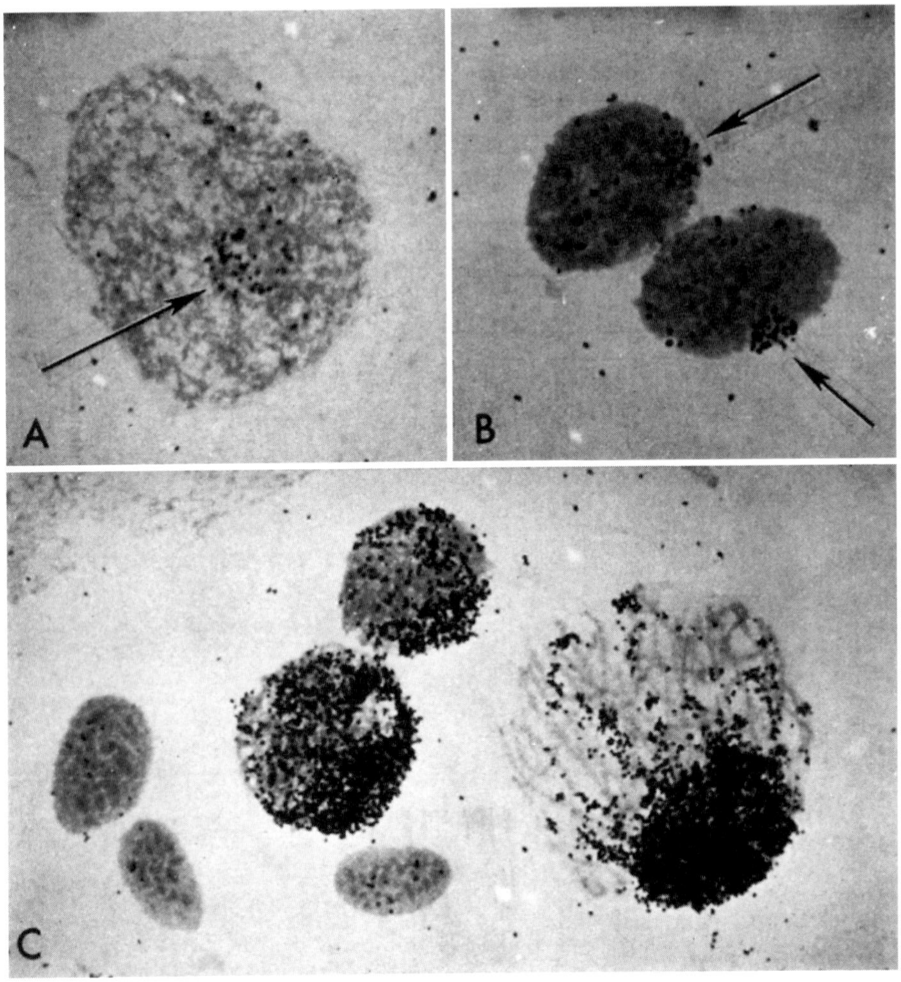

Fig. 1.—Autoradiographs of nuclei from the ovary of the toad *Xenopus*, after *in situ* hybridization with radioactive ribosomal RNA. The preparation was covered with agar, denatured in 0.07 *N* NaOH for 2 min, and hybridized with rRNA having a specific activity of 200,000 cpm/μg. Details are given in *Materials and Methods*. All nuclei from the same slide, exposed for 52 days, stained with Giemsa.

(*A*) Oogonial nucleus showing silver grains above the centrally placed nucleolus (*arrow*). This number of grains indicates that the nucleus contains some 20–40 copies of the nucleolus organizer. ×1500.

(*B*) Two leptotene nuclei with silver grains located over the eccentrically placed nucleolus (*arrows*). ×1500.

(*C*) Three unlabeled follicle nuclei and three labeled pachytene nuclei. The pachytene nuclei illustrate the progressive increase in extrachromosomal rDNA that occurs as the nucleus enlarges. The largest pachytene nucleus contains 25–30 pg of rDNA (approximately 3000 nucleolus organizers). The technique is not sensitive enough to demonstrate the small amount of rDNA in the two nucleolus organizers of the diploid follicle nuclei. ×1200.

(3) The slide was frozen on dry ice, and the cover slip was removed with a razor blade.[7] The slides were transferred to 95% ethanol for a few moments and then dried in air.

(4) The slides were dipped in 0.5% agar held molten at 60°C in a water bath. They were removed and drained vertically at room temperature. In this way a very thin but uniform agar layer covered the slide. The agar was allowed to gel but not to dry completely before the next step.

(5) The slide was placed for 2 min in 0.07 *N* NaOH at room temperature to denature the DNA. It was then transferred to 70 and 95% ethanol for a few minutes each and dried in air. Slides were often stored at this point.

(6) For the hybridization step about 200 μl of rRNA solution was placed directly onto the slide, and a large cover slip was added. The preparation was incubated at 66°C for 12 hr or longer in a moist chamber made from a Petri plate. We used 4-inch plastic plates into which were placed a few sheets of filter paper and enough 6X SSC to moisten thoroughly. The slide was supported above this on two rubber grommets. We used mixed 28*S* and 18*S* rRNA at a concentration of 1–2 μg/ml in 6X SSC. The rRNA was extracted by a detergent-phenol procedure from cultures of *Xenopus* cells[8] and was purified by sucrose density gradient centrifugation. It had a specific activity of 2×10^5 cpm/μg. The specific activity was determined by spotting known amounts of RNA-H^3 on nitrocellulose filters, drying, and counting in toluene-PPO-POPOP in a scintillation counter. The counter had an efficiency of about 40% for *unquenched* H^3 samples.

(7) After incubation the slides were washed in 6X SSC and placed in RNase for 1 hr at room temperature. Pancreatic RNase (1 mg/ml in 0.02 *M* Na acetate, pH 5) was boiled 5 min to remove protease activity and then made up to 20 μg/ml in 2X SSC.

(8) The slides were rinsed in 6X SSC and then in 70 and 95% ethanol before air-drying.

(9) The preparations were covered with Kodak NTB-2 liquid emulsion diluted 1:1 with distilled water. They were developed for 2 min in Kodak D-19, rinsed briefly in 2% acetic acid, and fixed 2 min in Kodak Fixer. After they were rinsed for 10 min in several changes of distilled water, they were stained 10 min with Giemsa, rinsed in distilled water, and air-dried. A drop of Permount medium and a cover slip were added.

Results.—The development of the cytological hybridization technique was facilitated by the use of oocytes of the toad *Xenopus laevis* as a test object. During pachytene of meiosis these cells carry out a differential synthesis of the genes coding for ribosomal RNA.[6, 9–12] Each pachytene nucleus, which contains 12 picograms (pg) of chromosomal DNA, produces about 30 pg of extrachromosomal rDNA. The extra DNA is cytologically detectable as a densely staining cap on one side of the nucleus. During diplotene the rDNA spreads over the inner surface of the nuclear envelope, where it produces the multiple nucleoli that characterize these cells.

The ovaries of recently metamorphosed toads contain many oocytes in the early meiotic stages. Squashes of these ovaries were denatured, hybridized with tritium-labeled rRNA, and subsequently autoradiographed. Heavy label was found in the oocyte nuclei, where it was limited to the extra DNA. Figure 1 (*B* and *C*) shows several stages in the formation of the nuclear cap from leptotene to late pachytene. In each case the label follows the distribution of the extra DNA, the chromosomes being without detectable radioactivity. The nuclei of follicle cells, connective tissue, and red blood cells are unlabeled (Fig. 1*C*), presumably because the technique is not sufficiently sensitive to demonstrate the small amount of rDNA in the normal genome. The unlabeled DNA in these nuclei provides a useful built-in control of the specificity of the hybridization reaction.

Attempts were made to hybridize preparations that had not been denatured with alkali. In autoradiographs exposed for one or two weeks, such control preparations showed no detectable radioactivity in any nuclei. However, most control slides showed weak labeling when exposed for periods of one to two months. In these cases the label displayed the same specific localization seen in alkali-denatured slides, namely, over the rDNA of oocytes. These results suggest that a small amount of DNA is denatured during the fixing and squashing steps.

Two additional tests of specificity have been made. Ovary squashes were treated with DNase (0.3 mg/ml in 0.01 *M* Tris buffer containing 10^{-3} *M* $MgCl_2$, pH 7.2, 37°C, 3 hr). Some were stained with the Feulgen reaction to assess the removal of DNA, and the remainder were covered with agar and hybridized as usual. The preparations showed no detectable Feulgen stain and they gave negative autoradiographs, an indication of the failure to bind rRNA. Some protein remains in such preparations, since cytological details are visible either by phase contrast microscopy or after staining with fast green at pH 2.

We have found that a large excess of heterologous rRNA has no effect on the hybridization reaction (Table 1). Hybridizations were carried out with 2 μg/ml of radioactive *Xenopus* rRNA in the presence of nonradioactive rRNA from *Xenopus* or *Escherichia coli.* The *E. coli* rRNA had no effect on the binding of radioactive *Xenopus* rRNA, even when present at 800 μg/ml. By contrast, nonradioactive *Xenopus* rRNA reduced the binding to low levels. A small fraction of the radioactivity (8%) was not competed by homologous *Xenopus* rRNA. We do not know why competition was incomplete, although it should be noted that the radioactive rRNA was derived from cultured kidney cells, whereas the nonradioactive rRNA came from mature ovaries.

In order to assess the sensitivity of the hybridization procedure, we exposed preparations for periods up to two months. Our aim was to detect the earliest stages of rDNA amplification, which was thought to begin in leptotene or early pachytene.[11, 12] In such preparations the mid- and late pachytene nuclei showed total film blackening above the extra DNA. We were surprised to find easily detectable label not only over all leptotene nuclei (Fig. 1*B*), but also over most oogonial nuclei (Fig. 1*A*). In the leptotene nuclei the label was generally

TABLE 1. *Hybridization of Xenopus rRNA-H^3 in the presence of excess nonradioactive rRNA.*

Competing unlabeled rRNA	Silver grains per nucleus after 28 hr exposure, background subtracted (mean ± SEM)	Nuclei counted (no.)
None	185 ± 20	19 (4 preparations)
50–800 μg/ml *E. coli*	221 ± 10	43 (9 preparations)
50–800 μg/ml *Xenopus*	15 ± 2	50 (10 preparations)

In each case 2 μg/ml of *Xenopus* rRNA-H^3 was present during the hybridization step. *E. coli* rRNA showed no competition even at 800 μg/ml. Unlabeled *Xenopus* rRNA reduced the binding of radioactive rRNA to about 8% of the control value. The small residual binding was unrelated to the quantity of competing rRNA at the levels used here (50–800 μg/ml). The labeled *Xenopus* rRNA was from cultured kidney cells, whereas the unlabeled rRNA came from mature ovaries.

localized at the periphery of the nucleus, whereas in the oogonia it was more often found in one or two clusters within the nucleus. The distribution of silver grains parallels the position of the nucleoli in both cases. These results indicate that a low level of rDNA amplification is present in premeiotic nuclei.

We have begun to examine variables affecting the level of hybridization. The NaOH concentration in the denaturing step has been varied from 0.01 to 0.10 *N*. Concentrations of 0.01–0.02 *N* have given either no hybridization or only a low level. We presume that this is due to inadequate denaturing during the two-minute treatment. Concentrations of 0.05, 0.07, and 0.10 *N* gave roughly comparable levels of hybridization. However, the morphological disruption in preparations treated with 0.10 *N* NaOH was often extensive.

We have had some success in replacing the agar with a collodion (nitrocellulose) layer during the denaturing step. The advantage of collodion is that it can be removed by dipping the slide in ethanol-ether (1:1) before hybridization. If the collodion method proves reliable, it should permit more accurate quantitation, since the nuclei are in direct contact with the autoradiographic emulsion.

Discussion.—The results with *Xenopus* oocytes show that RNA can be hybridized with the DNA of cytological preparations under conditions that preserve the morphological integrity of the nucleus. The following features of the reaction indicate that we are dealing with true hybrid molecules. (1) The DNA must be treated with a denaturing agent to obtain the full reaction. (2) Prior removal of DNA by DNase eliminates the reaction. (3) The complex of RNA with the nucleus is stable to RNase. (4) The reaction is competed by unlabeled *Xenopus* RNA but is unaffected by heterologous *E. coli* RNA. (5) The reaction with rRNA is limited to the nucleoli of oogonia and to the amplified rDNA of oocytes. Its absence from normal diploid nuclei can be explained by the small amount of rDNA in these cells.

In the filter technique of Gillespie and Spiegelman,[5] contaminating basic proteins bind RNA nonspecifically. At the outset, therefore, we were concerned that nuclear histones would interfere with any cytological hybridization procedure. Our first experiments involved pronase digestion after formaldehyde fixation; however, we found such material difficult to denature, even though it retained good morphology. In the method described here, most of the basic proteins are removed by the ethanol-acetic fixative and the 45 per cent acetic acid treatment. This we have demonstrated by acrylamide gel electrophoresis, thus confirming the earlier experiments of Dick and Johns.[13]

Quantitation of our results has been difficult, since we have had no adequate control over self-absorption within the specimen and shielding by the agar layer. The effect of both factors was easy to demonstrate. Shielding was seen when the preparation was covered with 2 per cent agar instead of 0.5 per cent. In this case no autoradiograph was obtained. Self-absorption within the specimen was suggested by a comparison of the silver grains over the large diplotene nuclei and the more compact late pachytenes. Both contain the same amount of rDNA[12] and hence presumably hybridize the same extent; but the number of grains was greater over the larger, more flattened diplotene nuclei.

Keeping these complications in mind, we can make a rough estimate of the sensitivity of the technique. In autoradiographs exposed for one month, the larger diplotene nuclei display approximately 3000 silver grains. These nuclei contain 25–30 pg of extrachromosomal DNA,[9, 12] of which about 18 per cent is complementary to 28*S* and 18*S* rRNA in filter experiments.[6] Thus, these preparations display 15 grains per day per picogram of hybridizable DNA. This can be translated into roughly 0.02–0.03 grain per day per nucleolus organizer.[6] At this level of sensitivity a single nucleolus organizer would be barely detectable after autoradiographic exposure for several months. For this reason the follicle nuclei in our preparations are negative (Fig. 1*C*).

The sensitivity of the hybridization technique can probably be increased by the use of RNA made *in vitro*. We are now preparing RNA enzymatically, using the *Xenopus* rDNA satellite as template. If we can increase the specific activity of the RNA by a factor of 10 or 100 over what is now available, we should be able to demonstrate the locus of rDNA in metaphase chromosomes of most higher organisms.

It should be relatively easy to demonstrate the rDNA in the polytene chromosomes of Diptera, which contain several hundred times as much DNA as a metaphase chromosome. Calculations are somewhat less reliable when dealing with other types of RNA. The genes for transfer RNA and for 5*S* ribosomal RNA can possibly be localized under the favorable conditions afforded by polytene chromosomes. Certain special problems, such as the cytological localization of the mouse satellite DNA[14, 15] and the integrated form of the SV40 viral DNA[16] should be approachable. In the latter case, highly radioactive complementary RNA has already been used effectively in filter hybridization experiments. It is difficult to predict what success may be had with messenger RNA species; hopefully it may be possible to characterize mixtures of messengers from different tissues or from different developmental stages. For such experiments the polytene chromosomes of Diptera offer the best chance of precise cytological localization.

The technical assistance of Mrs. Cherry Barney is gratefully acknowledged.

Abbreviations: rRNA, ribosomal RNA; rDNA, the DNA sequences coding for rRNA; SSC, 0.15 *M* NaCl, 0.015 *M* Na citrate, pH 7.0; toluene-PPO-POPOP, 4 gm of 2,5 diphenyloxazole and 50 mg of 1,4 bis [2-(4-methyl-5-phenyloxazolyl)]-benzene in 1 liter of toluene.

* Aided by USPHS grants GM 12427 and GM 397 from the National Institute of General Medical Sciences.

1 Hall, B. D., and S. Spiegelman, these PROCEEDINGS, **47**, 137 (1961).
2 Nygaard, A. P., and B. D. Hall, *J. Mol. Biol.*, **9**, 125 (1964).
3 Bautz, E. K. F., and B. D. Hall, these PROCEEDINGS, **48**, 400 (1962).
4 Bolton, E. T., and B. J. McCarthy, these PROCEEDINGS, **48**, 1390 (1962).
5 Gillespie, D., and S. Spiegelman, *J. Mol. Biol.*, **12**, 829 (1965).
6 Gall, J. G., *Genetics*, in press.
7 Conger, A. D., and L. M. Fairchild, *Stain Technol.*, **28**, 281 (1953).
8 The culture was kindly furnished by Dr. Keen Rafferty, The Johns Hopkins University School of Medicine, Baltimore.
9 Brown, D. D., and I. B. Dawid, *Science*, **160**, 272 (1968).
10 Evans, D., and M. Birnstiel, *Biochim. Biophys. Acta*, **166**, 274 (1968).
11 Gall, J. G., these PROCEEDINGS, **60**, 553 (1968).
12 Macgregor, H. C., *J. Cell Sci.*, **3**, 437 (1968).
13 Dick, C., and E. W. Johns, *Biochem. J.*, **105**, 46 P (1967).
14 Kit, S., *J. Mol. Biol.*, **3**, 711 (1961).
15 Flamm, W. G., M. McCallum, and P. M. B. Walker, these PROCEEDINGS, **57**, 1729 (1967).
16 Sambrook, J., H. Westphal, P. R. Srinivasan, and R. Dulbecco, these PROCEEDINGS, **60**, 1288 (1968).

Kornberg R.D. 1974. Chromatin structure: A repeating unit of histones and DNA. *Science* **184:** 868–871.

A TYPICAL EUKARYOTIC METAPHASE CHROMOSOME is about 10,000 times shorter than the DNA molecule that it contains. This enormous compaction is achieved in several stages, the first of which involves wrapping the DNA double helix around a histone octamer to form a unit called a nucleosome. A remarkably accurate model of the nucleosome was first proposed by Kornberg in the paper reproduced here. Bringing together data from x-ray diffraction, the behavior of histones in solution, and the cleavage of chromatin into pieces of defined length by nucleases, he suggested that about 200 base pairs of DNA are associated with eight histone molecules in a repeating structure. This model agreed quite well with images of chromatin fibers that had been swollen in water and centrifuged onto a supporting film for observation in the electron microscope (1). Under these conditions chromatin appeared as chains of spheroidal particles about 70Å in diameter connected together by a much thinner fiber. A few years later, it became possible to crystallize the histone core of the nucleosome and thereby obtain x-ray diffraction data showing the arrangement of the individual histone molecules, first at relatively low resolution (2) and most recently at 2.8Å resolution (3).

---

1. Olins A.L. and Olins D.E. 1974. Spheroid chromatin units (ν bodies). *Science* **183:** 330–332.
2. Klug A., Rhodes D., Smith J., Finch J.T., and Thomas J.O. 1980. A low resolution structure for the histone core of the nucleosome. *Nature* **287:** 509–516.
3. Luger K., Mäder A.W., Richmond R.K., Sargent D.F., and Richmond T.J. 1997. Crystal structure of the nucleosome core particle at 2.8 Å resolution. *Nature* **389:** 251–260.

Reprinted from
24 May 1974, Volume 184, pp. 868-871

SCIENCE

Chromatin Structure: A Repeating Unit of Histones and DNA

Chromatin structure is based on a repeating unit of eight histone molecules and about 200 DNA base pairs.

Roger D. Kornberg

Evidence is given in the preceding article (*1*) for oligomers of the histones, both in solution and in chromatin. Here I wish to discuss this and other evidence in relation to the arrangement of histones and DNA in chromatin. In particular, I propose that the structure of chromatin is based on a repeating unit of two each of the four main types of histone and about 200 base pairs of DNA. A chromatin fiber may consist of many such units forming a flexibly jointed chain.

Introduction to Chromatin Structure

Chromatin of eukaryotes contains nearly equal weights of histone and DNA. This corresponds, on the basis of the molecular weights and relative amounts of the five main types of histone, F1, F2A1, F2A2, F2B, and F3, to roughly one of each type of histone per 100 base pairs of DNA with the exception of F1, of which there is half as much. The arrangement of histones and DNA involves repeats of structure. The first evidence of this comes from the work of Wilkins and co-workers (*2*) who obtained x-ray diffraction patterns from whole nuclei of cells showing relatively sharp bands. Chromatin isolated from the nuclei as a nearly pure complex of histone and DNA gives x-ray patterns with the same bands. Further x-ray work (*3–5*) has shown that these bands correspond

The author is a Junior Fellow of the Society of Fellows of Harvard University; he is working at the MRC Laboratory of Molecular Biology, Hills Road, Cambridge CB2 2QH, England.

to structure repeating at intervals of about 100 angstroms along the length of the chromatin fiber. Neither histone nor DNA alone gives x-ray patterns with such bands.

A "super-coil" model has been proposed (*6*) to account for the x-ray data on chromatin. It consists of a DNA double helix with "a coating of histone" coiled into a single larger helix of axial repeat distance 120 Å and diameter 100 Å. There are 340 Å of DNA double helix or 100 DNA base pairs per turn of the larger helix, which is a major drawback of the model in view of the following discussion of the true size of repeating unit in chromatin.

A Repeating Unit

The ratios of histone to DNA and x-ray data mentioned above do not indicate how the five types of histone are distributed in chromatin. The simplest case would be that the histones act together and form a unique structure that gives rise to the x-ray pattern; at the other extreme would be the case of different combinations of histones in different regions of chromatin, some one of which gives rise to the x-ray pattern. Evidence from the preceding article (*1*) helps to distinguish among these and the many possible intermediate cases. It was shown that histones F2A1 and F3 of calf thymus occur entirely as an $(F2A1)_2(F3)_2$ tetramer. It was further shown that a complex of tetramers, F2A2-F2B oligomers, and DNA gives the x-ray pattern of chromatin, and that tetramers and F2A2-F2B oligomers are both required, but F1 is not. The following conclusions may be drawn: F2A1 and F3 form a unique structure; F2A1, F3, F2A2, and F2B act together and form with DNA the repeating structure responsible for the x-ray pattern of chromatin; and F1 is either added on or located elsewhere in chromatin. In sum, four of the histones and DNA form a unique repeating structure.

Now suppose that the $(F2A1)_2(F3)_2$ tetramer defines a repeating unit of this structure and that all the DNA in chromatin is involved in the structure. Then, as chromatin contains roughly one each of F2A1, F3, F2A2, and F2B per 100 base pairs of DNA, the repeating unit may contain two of each of these histones and about 200 base pairs of DNA. This coincides in a rather striking way with the results of digestion of chromatin by certain nucleases, in which most of the DNA is cleaved to pieces of about 200 base pairs. The first such observation was made by Hewish and Burgoyne in work on digestion of chromatin in rat liver nuclei by an endogenous nuclease (*7*). In this digestion more than 80 percent of the DNA is cleaved to multiples of from one to six times 200 base pairs. The occurrence of multiples rather than just 200 base pair pieces is presumably due to some cleavage sites being blocked by nonhistone proteins. A more clear-cut result has come from an extension of the work of Hewish and Burgoyne, in which staphylococcal nuclease has been shown to cleave more than 90 percent of the DNA in rat liver nuclei to pieces of about 200 base pairs (*8*). Both the endogenous and staphylococcal nucleases produce a slight heterogeneity in size of the DNA pieces, the dispersion being about ± 10 percent.

The convergence of work on oligomers of histones and work on cleavage of DNA makes a strong case for a repeating unit containing two each of F2A1, F3, F2A2, and F2B, and about 200 base pairs of DNA. Both kinds of work bear on how much repeating structure there is in chromatin, one kind showing that most of the histone is involved (four of the five types of histone) and the other showing that most of the DNA is involved (more than 90 percent in rat liver). The generality of the results can of course be tested by repeating the work on chromatin from other sources. Short of that, it may be asked whether the relative amounts of the histones and relative amounts of total histone and DNA are independent of source. The relative amounts of the histones have been measured (*9–12*) by extraction from chromatin and fractionation by preparative methods or in polyacrylamide gels. The measurements should be regarded as only approximate, because of possible differential extractability, proteolysis, losses during fractionation, and overlaps of bands in the gels (especially the bands arising from F2A2 and F2B, and minor bands arising from histone modification). The results, expressed as molar ratios of F3, F2A2, and F2B to F2A1, are 0.9, 0.8, and 1.1 in calf thymus (*9*) and nearly the same in other calf tissues (*10*), 0.7, 0.7, and 1.0 in *Drosophila* (*11*), and 0.9, 0.5, and 2.6 in pea bud and other pea tissues (*12*). F2A1 and F3 may in fact be equimolar in all organisms, and F2A2 and F2B roughly equimolar with exceptions.

Despite the approximate nature of these measurements, it may be significant that F2A1 and F3 are more nearly equimolar than F2A2 and F2B. F2A1 and F3, which occur as an $(F2A1)_2(F3)_2$ tetramer in calf thymus, would be expected, on the basis of the conservation of their amino acid sequences during evolution (*13*), to occur as a tetramer in all organisms, and might therefore be expected to occur in equimolar amounts in all organisms. The oligomeric structure of F2A2 and F2B, on the other hand, has not been as well established as for F2A1 and F3, and the amino acid sequences of F2A2 and F2B appear to be less conserved than those of F2A1 and F3 (*1*). The numbers of F2A2 and F2B that I have taken to be in the repeating unit are based on the roughly equimolar amounts of all the histones in calf thymus. These numbers (two each of F2A2 and F2B) may not be exactly right (there may be two of F2A2 and three of F2B in the repeating unit in calf thymus), and they may vary from one organism to another. It is possible to envisage structural roles for F2A2 and F2B compatible with such variation (see below).

Measurements of relative amounts of total histone and DNA are more accurate than measurements of relative amounts of the various histones since amounts of total histone are less sensitive to differential extractability, and so forth. The results, expressed as weight ratios of total histone to DNA, are nearly 1.0 for chromatin from a wide range of sources, for example: 1.15, 0.95, 1.17, 1.08, and 1.10 for rat liver, rat kidney, chicken liver, chicken erythrocytes, and pea bud (*14*); 1.02, 1.04, and 0.86 at three stages in the development of sea urchin embryos (*15*); and 1.05 in the slime mold *Physarum polycephalum* (*16*). This invariance, together with the invariance of amino acid sequences of F2A1 and F3, is the strongest evidence for the generality of a repeating unit of two each of four of the histones and about 200 base pairs of DNA.

F1 is not involved in forming the repeating unit (see above), so it must either be added on to the unit or located in a different region of chromatin. The amount of F1 relative to the other histones suggests that F1 is in fact associated with the unit: the molar ratio of F1 to F2A1 is 0.54 in calf thymus (*9*), 0.40 in *Drosophila* (*11*), and 0.52 in pea bud (*12*); thus there is one F1 for two each of the other histones, or one F1 for every repeating unit.

The repeating structure formed by DNA and all the histones except F1 gives rise to the x-ray pattern of chromatin (see above). It may be asked whether the quantities of histone and DNA in the repeating unit, inferred from biochemical evidence (see above), are compatible with the size of the repeating unit indicated by the x-ray pattern. The answer may be seen by taking the dimensions of the repeating unit from the x-ray pattern and electron microscopy, together with the proportion of chromatin in the repeating unit from additional x-ray data. The x-ray pattern, as mentioned above, shows bands corresponding to structure repeating along the length of the chromatin fiber at intervals of about 100 Å. Electron micrographs generally show fiber diameters of about 100 Å (*17*). This suggests a repeating unit about 100 Å long in the fiber direction and about 100 Å in diameter. The x-ray pattern disappears when the chromatin concentration is raised above about 45 percent by weight (*3, 5*); this suggests that the fibers are packed as closely as the structure permits when the concentration is about 45 percent. A unit 100 Å long and 100 Å in diameter which is 45 percent by weight in chromatin upon close-packing contains 2.8×10^5 daltons of chromatin (*18*). This is equivalent to 2.3 each of F2A1, F3, F2A2, and F2B and 230 base pairs of DNA. Thus, the repeating unit inferred from biochemical evidence and the repeating unit that gives rise to the x-ray pattern may be the same (*19*).

Some indication of the unit of packaging of histones and DNA might be expected in studies of events requiring at least partial unpackaging, such as DNA replication. Kriegstein and Hogness (*20*) have suggested that the rate of movement of DNA replication forks in eukaryotes is limited by a process involving the histones. As discontinuous DNA synthesis in *Drosophila* proceeds in steps of about 200 bases (Kriegstein and Hogness show that the single-stranded gaps at replication forks and the fragments of newly synthesized DNA in *Drosophila* are about 200 and 150 bases), the rate-limiting process could well be unpackaging of units of two each of four of the histones and about 200 base pairs of DNA.

The full significance of the repeating unit of histones and DNA may lie in the relation of the units to base sequences in the DNA. It may be asked, for example, whether there is a specific phase relation between the units and base sequences in the DNA. In other words, do the 200 base pair pieces arising from endonuclease digestion of chromatin form a unique set with respect to base sequence or do they overlap in sequence (*21*)?

A Flexibly Jointed Chain of Repeating Units

My views on the arrangement of histones and DNA in the repeating unit are speculative and meant to be taken as a working hypothesis. The basic idea is that a chromatin fiber is a flexibly jointed chain of repeating units. The point is that a jointed structure may be as flexible as the underlying DNA, whereas a continuous structure, such as a helix, is not. The idea arises from the fact that a chromatin fiber is flexible enough to be extensively coiled or folded. Such coiling or folding must occur, for example, in the bands of polytene chromosomes of *Drosophila*, where the ratio of length of DNA to length of DNA-containing structure is an order of magnitude greater than in a chromatin fiber (*22*).

A possible arrangement of histones and DNA in the repeating unit, leading to a jointed structure, is as follows. The $(F2A1)_2(F3)_2$ tetramer forms the core of the repeating unit [this is suggested by the essentially globular nature of the tetramer (*1*), the conservation in amino acid sequence of F2A1 and F3, and the fact that these histones are the last to be removed from chromatin by mild methods of extraction (*23*)]. F2A2 and F2B determine the spacing of tetramers along the length of the chromatin fiber, perhaps as F2A2-F2B dimers, or as an F2A2-F2B polymer running alongside [suggested by x-ray experiments showing that tetramers and F2A2-F2B oligomers act together to form a structure repeating at regular intervals along the length of the fiber (see above)]. Much of the 200 base pairs of DNA in a repeating unit would follow some path on the tetramer, and the remainder of the DNA would connect tetramers along a path defined by F2A2 and F2B. In brief, I suggest that a chromatin fiber consists of tightly packed DNA and associated protein alternating with more extended DNA and associated protein, rather like beads on a string.

Some evidence for such a structure comes from the nuclease digestion work mentioned above. Endonucleases may produce 200 base pair pieces of DNA by cleaving the connecting strand between tetramers. And recent work (*24*) has shown that the 200 base pair piece and associated protein occurs as a discrete complex in solution.

Electron micrographs of chromatin are also compatible with a jointed structure. Chromatin fibers observed after critical point drying have a generally "knobby" appearance (*25*). Spray-mounted and shadow-cast specimens show "nodules" alternating with thin strands, although the nodules are often widely spaced and are absent from some preparations (*26*). Striking examples of micrographs showing alternate thick and thin regions were published (*27*) while this manuscript was in preparation. In these micrographs, which were obtained by formaldehyde fixation and positive or negative staining, the thick regions are quite closely spaced and have a beadlike appearance. These regions were suggested to contain all five histones, in contrast with the arrangement of histones and DNA suggested above (*28*). Of course electron micrographs alone say nothing of the locations of particular molecules. But it may be possible, for example by selective extraction of histones (*23*) and nuclease digestion, to relate some features of the micrographs to particular histones and to DNA.

Summary

Many lines of evidence on chromatin structure have been discussed. The essential facts are:

1) Chromatin contains roughly one of each type of histone per 100 base pairs of DNA, except for histone F1.

2) X-ray patterns reveal a structure repeating along the length of the chromatin fiber. F2A1, F3, F2A2, and F2B are required in this structure, but F1 is not.

3) Two each of F2A1 and F3 combine to form a tetramer.

4) Certain nucleases cleave almost all the DNA in chromatin to pieces of about 200 base pairs.

5) Chromatin fibers are often extensively coiled or folded.

These facts lead to two proposals:

1) Chromatin structure is based on

a repeating unit of two each of F2A1, F3, F2A2, and F2B and about 200 base pairs of DNA.

2) A chromatin fiber consists of many such units forming a flexibly jointed chain.

References and Notes

1. R. D. Kornberg and J. O. Thomas, *Science* **184**, 865 (1974).
2. M. H. F. Wilkins, *Cold Spring Harbor Symp. Quant. Biol.* **21**, 75 (1956); ———, G. Zubay, H. R. Wilson, *J. Mol. Biol.* **1**, 179 (1959).
3. V. Luzzati and A. Nicolaieff, *J. Mol. Biol.* **7**, 142 (1963).
4. B. M. Richards and J. F. Pardon, *Exp. Cell Res.* **62**, 184 (1970); E. M. Bradbury, H. V. Molgaard, R. M. Stephens, L. A. Bolund, E. W. Johns, *Eur. J. Biochem.* **31**, 474 (1972); C. W. Carter and A. Klug, unpublished.
5. R. D. Kornberg, A. Klug, F. H. C. Crick, in preparation.
6. J. F. Pardon and M. H. F. Wilkins, *J. Mol. Biol.* **68**, 115 (1972).
7. D. R. Hewish and L. A. Burgoyne, *Biochem. Biophys. Res. Commun.* **52**, 504 (1973); the extent of digestion and size of the pieces are from D. R. Hewish, personal communication (the size was determined by velocity sedimentation in alkali and should be regarded as only approximate).
8. Such cleavage (M. Noll, of this laboratory, unpublished) would appear to conflict with the report [R. J. Clark and G. Felsenfeld, *Nat. New Biol.* **229**, 101 (1971)] that about half the DNA in chromatin is converted by staphylococcal nuclease to acid-soluble form while the remaining half is converted to pieces of about 175 base pairs. I suggest the following way of accounting for all the staphylococcal nuclease results. There may be two classes of sites of nuclease action in chromatin: sites between 200 base pair repeating units where nuclease action is rapid, and sites within the repeating units where nuclease action is slow. Brief digestion would be expected to convert most of the chromatin to pieces of about 200 base pairs of DNA with associated protein (the result quoted in the text). Further digestion would be expected to involve breakdown of some of the 200 base pair pieces and binding of the histones that are released to the remaining pieces. The digestion should continue until the binding of extra histone completely protects the pieces that remain. This limit should be reached when about half of the pieces remain (the result of Clark and Felsenfeld) since roughly twice the amount of histone naturally occurring in chromatin is required to neutralize all the negative charge on the DNA (on the basis of the amino acid compositions of the histones and relative amounts in chromatin of histones and DNA).
9. E. W. Johns, *Biochem. J.* **104**, 78 (1967).
10. S. Panyim and R. Chalkley, *Biochemistry* **8**, 3972 (1969).
11. D. Oliver and R. Chalkley, *Exp. Cell Res.* **73**, 295 (1972).
12. D. M. Fambrough, F. Fujimura, J. Bonner, *Biochemistry* **7**, 575 (1968).
13. R. J. DeLange, D. M. Fambrough, E. L. Smith, J. Bonner, *J. Biol. Chem.* **244**, 5669 (1969); L. Patthy, E. L. Smith, J. Johnson, *ibid.* **248**, 6834 (1973).
14. S. C. R. Elgin and J. Bonner, *Biochemistry* **9**, 4440 (1970).
15. R. J. Hill, D. L. Poccia, P. Doty, *J. Mol. Biol.* **61**, 445 (1971).
16. J. Mohberg and H. P. Rusch, *Arch. Biochem. Biophys.* **134**, 577 (1969).
17. H. Ris and D. F. Kubai, *Annu. Rev. Genet.* **4**, 263 (1970).
18. The calculation assumes hexagonal close-packing of chromatin fibers, and densities of F1-depleted chromatin [the material used in recent x-ray work (*5*)] and solvent of 1.50 and 1.00 g/cm^3.
19. The close correspondence of the repeating units inferred from biochemical and x-ray evidence does not necessarily mean they are identical (there may, for example, be 11 of one unit for every 10 of the other), and further work is needed to determine the exact relation between them.
20. H. J. Kriegstein and D. S. Hogness, *Proc. Natl. Acad. Sci. U.S.A.* **71**, 135 (1974).
21. The matter is complicated by the occurrence of repeated sequences in most eukaryote DNA. The most convenient choice for testing would be the genome of a small virus, such as polyoma or SV40. The histone-associated forms of these genomes should be cleaved into about 25 200 base pair pieces by staphylococcal nuclease. It should then be possible—for example, by the use of restriction enzymes—to determine whether these pieces are of only 25 or else a very large number of types with respect to base sequence.
22. The ratio of length of DNA to length of structure in the salivary X chromosome of *Drosophila melanogaster* is about 80 [W. Beermann, in *Results and Problems in Cell Differentiation*, W. Beermann, Ed. (Springer-Verlag, Berlin, 1972), vol. 4, p. 1]. The ratio in a chromatin fiber is about 6.8 (based on 200 base pairs or 680 Å length of DNA in about 100 Å length of fiber).
23. Y. V. Ilyin, A. Ya. Varshavsky, U. N. Mickelsaar, G. P. Georgiev, *Eur. J. Biochem.* **22**, 235 (1971).
24. M. Noll, unpublished result.
25. S. Bram and H. Ris, *J. Mol. Biol.* **55**, 325 (1971).
26. H. S. Slayter, T. Y. Shih, A. J. Adler, G. D. Fasman, *Biochemistry* **11**, 3044 (1972).
27. A. L. Olins and D. E. Olins, *Science* **183**, 330 (1974); similar micrographs have been obtained by Dr. J. T. Finch of this laboratory.
28. The beadlike thick regions are observed to be about 70 Å in diameter. This is compatible with these regions consisting of a globular $(F2A1)_2(F3)_2$ tetramer (diameter 40 to 50 Å) covered by DNA (double helix diameter about 20 Å).
29. I thank Drs. A. Klug, F. H. C. Crick, and M. S. Bretscher for helpful discussions and Drs. F. H. C. Crick, A. Klug, and S. Brenner for criticism of the manuscript.

Stehelin D., Varmus H.E., Bishop J.M., and **Vogt P.K.** 1976. DNA related to the transforming gene(s) of avian sarcoma viruses is present in normal avian DNA. *Nature* **260:** 170–173.

STUDIES ON THE ROUS SARCOMA VIRUS (RSV) have provided many key insights into the relationship between viruses and the cells they infect. In the early decades of the 20th century Peyton Rous of the Rockefeller Institute showed that RSV causes sarcomas in chickens, thereby providing the first identified link between viruses and cancer in any organism. Decades later it was shown that the RSV genome is an RNA molecule that is copied into DNA when the virus infects a cell (1, 2). This DNA copy integrates into the chromosome, where it in turn is transcribed into RNA, some of which is packaged into new virus particles and some of which serves as messenger RNA for the four virus genes (*gag*, *pol*, *env*, and *src*). The *src* gene encodes a protein that is responsible for the neoplastic transformation that accompanies viral infection, and for that reason is called an oncogene (3). In the paper reproduced here it was shown for the first time that the genome of uninfected birds contains sequences homologous to the *src* gene of the virus. Subsequent detailed studies showed that the cellular and viral src genes, referred to as c-*src* and v-*src* respectively, encode a 60 kDa tyrosine kinase (4). This kinase normally resides on the inner surface of the plasma membrane. Because of its relationship to the v-*src* oncogene, the c-*src* gene is called a proto-oncogene. Work on RSV and other RNA tumor viruses led to the award of three Nobel prizes: to Rous in 1966 "for his discovery of tumor inducing viruses," to Baltimore, Dulbecco, and Temin in 1975 "for their discoveries concerning the interaction between tumour viruses and the genetic material of the cell," and to Bishop and Varmus in 1989 "for their discovery of the cellular origin of retroviral oncogenes."

---

1. Baltimore D. 1970. RNA-dependent DNA polymerase in virions of RNA tumour viruses. *Nature* **226:** 1209–1211.
2. Temin H.M. and Mizutani S. 1970. RNA-dependent DNA polymerase in virions of Rous sarcoma virus. *Nature* **226:** 1211–1213.
3. Wang L.-H., Duesberg P., Beemon K., and Vogt P.K. 1975. Mapping RNase T1-resistant oligonucleotides of avian tumor virus RNAs: Sarcoma-specific oligonucleotides are near the poly(A) end and oligonucleotides common to sarcoma and transformation-defective viruses are at the poly(A) end. *J. Virol.* **16:** 1051–1070.
4. Collett M.S., Purchio A.F., and Erikson R.L. 1980. Avian sarcoma virus-transforming protein $pp60^{src}$ shows protein kinase activity specific for tyrosine. *Nature* **285:** 167–169.

DNA related to the transforming gene(s) of avian sarcoma viruses is present in normal avian DNA

INFECTION of fibroblasts by avian sarcoma virus (ASV) leads to neoplastic transformation of the host cell. Genetic analyses have implicated specific viral genes in the transforming process[1-4], and recent results suggest that a single viral gene is responsible[4]. Normal chicken cells contain DNA homologous to part of the ASV genome[5-8]; moreover, embryonic fibroblasts from certain strains of chickens can produce low titres of infectious type C viruses either spontaneously[9] or in response to various inducing agents[10]. None of the viruses obtained from normal chicken cells, however, can transform fibroblasts, and results with molecular hybridisation indicate that the nucleotide sequences responsible for transformation by ASV are not part of the genetic complement of the normal cell[11]. We demonstrate here that the DNA of normal chicken cells contains nucleotide sequences closely related to at least a portion of the transforming gene(s) of ASV; in addition, we have found that similar sequences are widely distributed among DNA of avian species and that they have diverged roughly according to phylogenetic distances among the species. Our data are relevant to current hypotheses of the origin of the genomes of RNA tumour viruses[12] and the potential role of these genomes in oncogenesis[13].

We have prepared radioactive DNA ($cDNA_{sarc}$) complementary to nucleotide sequences which represent most or all of the viral gene(s) required for transformation of fibroblasts by ASV[14]. Our procedure to isolate $cDNA_{sarc}$ exploited the existence of deletion mutants of ASV which lack 10–20% of the viral genome (transformation defective, or td viruses)[11,15-17]; results of genetic analyses indicate that the deleted nucleotide sequences include part or all of the gene(s) responsible for oncogenesis and cellular transformation[4,14]. In our procedure the genome of the Prague-C strain (Pr-C) of ASV was transcribed into complementary DNA by endogenous RNA-directed DNA polymerase activity; we then used molecular hybridisation to select DNA specific for the region missing from the genome of the td deletion mutants. The preparation of $cDNA_{sarc}$ used was a virtually uniform transcript from about 16% of the Pr-C ASV genome[14], a region equivalent in size to the entire deletion in the strain of td virus used in our experiments[11,15-17]. Since the unit genome of ASV contains about 10,000 nucleotides[18,19], the genetic complexity of $cDNA_{sarc}$ is about 1,600 nucleotides, sufficient to represent an entire cistron. Nucleotide sequences homologous to $cDNA_{sarc}$ seem to be ubiquitous in the genomes of ASVs, but are not present in the genomes of avian leukosis viruses (including the endogenous chicken virus, RAV-0) or sarcoma-leukosis viruses from other species[14].

DNAs from several avian species (chicken, turkey, quail, duck and emu) contain nucleotide sequences which can anneal with $cDNA_{sarc}$ (Fig. 1 and Table 1). In contrast, we detected no homology between $cDNA_{sarc}$ and DNA from mammals (mouse and calf thymus; Table 1). The kinetics of the reactions between $cDNA_{sarc}$ and DNAs from chicken, quail and duck (Fig. 1*a*, *c* and *d*) were similar to the kinetics for the reassociation of unique nucleotide sequences in avian DNAs (C_0t about 1,000 mol s^{-1})[20]; thus, nucleotide sequences homologous to $cDNA_{sarc}$ are present as single (or a few) copies in each haploid complement of the avian DNAs tested. This conclusion was substantiated by measuring in a single reaction

Table 1 Homology between $cDNA_{sarc}$ and normal DNAs

Assay	Hybridisation conditions [Na⁺]	Temperature	Chicken	Quail	Turkey	Duck	Emu	Mouse	Calf
			Extent of reaction between $cDNA_{sarc}$ and DNA from						
S1	0.9 M	68°	52%	46%	48%	45%	24%	<2%	<2%
HAP	0.9 M	68°					36%		<5%
HAP	1.5 M	59°					54%		

DNA was extracted from 10–11-d-old embryos of chickens, ducks and quails, 3-d-old mice (strain RIII), livers of adult turkeys, liver and heart of a 22-d-old emu, and calf thymus. Reaction mixtures contining denatured DNA (8 mg ml^{-1}) and ^{3}H-$cDNA_{sarc}$ (0.32 ng ml^{-1}, 7,000 c.p.m. ml^{-1}) in a final volume of 0.3 ml were incubated at either 59 or 68 °C for 48 h. Samples incubated at 59 °C were in 1.5 M NaCl (final C_0t = 40,000), those incubated at 68 °C were in 0.9 M NaCl (final C_0t = 32,000); all reactions also continued 0.001 M EDTA–0.02 M Tris-HCl, *p*H 7.4. Duplex formation was measured by either hydrolysis with S1 nuclease[28] (in 0.3 M NaCl at 50 °C) or fractionation on hydroxyapatite (HAP) (samples adsorbed in 0.14 M sodium phosphate at 50 °C).

mixture the rates of reassociation between chicken DNA and both $cDNA_{sarc}$ (labelled with ^{3}H) and unique nucleotide sequences purified from chicken DNA (labelled with ^{14}C) (Fig. 1*b*); the rates were similar for both labelled DNAs.

The extent of duplex formation with $cDNA_{sarc}$ varied with the amount of chicken DNA used in the reactions

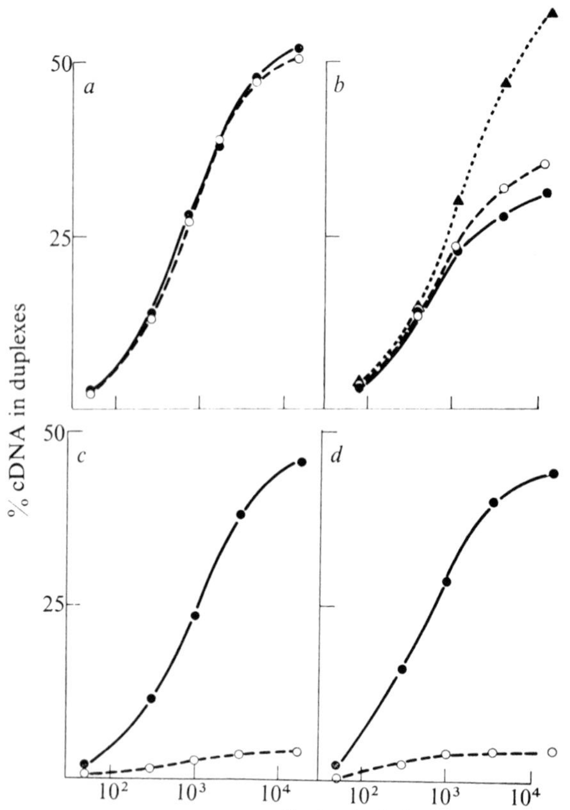

Fig. 1 Annealing of $cDNA_{sarc}$ to normal avian DNAs. DNA was prepared from 10–11-d-old embryos[7], denatured and incubated with ^{3}H-$cDNA_{sarc}$ and ^{32}P-$cDNA_{B77}$ in 0.6 M NaCl–0.001 EDTA–0.02 M Tris-HCl, *p*H 7.4 at 68 °C for 0.1 to 32 h. Samples (0.1–0.2 ml) were withdrawn at different times and the extent of duplex formation with $cDNA_{sarc}$ and $cDNA_{B77}$ was measured by hydrolysis with the single-strand specific nuclease, S1 (ref. 28). The preparation of $cDNA_{sarc}$ and $cDNA_{B77}$ has been described elsewhere[14] and is outlined in the text. Unique sequence chicken DNA was prepared by reannealing denatured ^{14}C-chicken DNA to a C_0t of 500; unique sequence DNA was eluted from a hydroxyapatite column at 60 °C in 0.14 M phosphate buffer[20]. *a*, Chicken DNA (9 mg ml^{-1}) with ^{3}H-$cDNA_{sarc}$ (0.5 ng ml^{-1}, 10,000 c.p.m. ml^{-1}) (●) and ^{32}P-$cDNA_{B77}$ (0.3 ng ml^{-1}, 10,000 c.p.m. ml^{-1}) (○). *b*, Chicken DNA (5 mg ml^{-1}) with ^{3}H-$cDNA_{sarc}$ (0.45 ng ml^{-1}, 9,000 cpm ml^{-1}) (●); ^{32}P-$cDNA_{B77}$ (0.5 ng ml^{-1}, 6,000 c.p.m. ml^{-1}) (○), and ^{14}C-chicken unique sequence DNA (▲). *c*, Quail DNA (9 mg ml^{-1}) with ^{3}H-$cDNA_{sarc}$ and ^{32}P-$cDNA_{B77}$ as in *a*. *d*, Duck DNA (9 mg ml^{-1}) with ^{3}H-$cDNA_{sarc}$ and ^{32}P-$cDNA_{B77}$ as in *a*.

(Fig. 1*a* and *b*), as expected in reactions where identical labelled and unlabelled DNA strands are competing for unlabelled complementary strands, and the reactions were incomplete (about 50%) at the highest value of C_0t (Fig. 1*a*, *c* and *d* and Table 1). Nevertheless, we believe that most or all of the nucleotide sequences of $cDNA_{sarc}$ are present in the DNA of quail since $cDNA_{sarc}$ anneals completely with RNA from certain quail cells (unpublished results of ourselves and C. Moscovici). Moreover, the rates and extents of annealing between $cDNA_{sarc}$ and DNA from chicken, duck and turkey are approximately the same as the rate and extent of annealing between $cDNA_{sarc}$ and quail DNA (Table 1 and Fig. 1). Thus, it is likely that most or all of the nucleotide sequences of $cDNA_{sarc}$ are present in the DNA of all these birds.

The extent of the reaction between $cDNA_{sarc}$ and DNA from emu, a relatively primitive Australian bird, was limited (24%) when tested by hydrolysis with a single strand-specific nuclease (Table 1); the extent of the reaction was greater when analysed on hydroxyapatite, a less stringent procedure than the nuclease test (Table 1), and was further augmented when the annealings were performed in conditions which facilitate pairing of partially matched nucleotide sequences (1.5 M NaCl, 59 °C, ref. 21) (Table 1). These date indicate that the nucleotide sequences homologous to $cDNA_{sarc}$ in emu DNA are substantially diverged from the homologous sequences in the other avian DNAs; we have obtained further data to sustain this conclusion by analysing the thermal stability of the duplexes formed between $cDNA_{sarc}$ and various avian DNAs (see below, Table 2 and Fig. 2).

Avian DNAs were also tested with ^{32}P-labelled single-stranded DNA, complementary to the RNA genome of the B77 strain of ASV ($cDNA_{B77}$), synthesised with detergent-disrupted virions, and purified as described previously[14]. The virus used to prepare $cDNA_{B77}$ consisted mainly of td variants (about 90% of the particles; unpublished observations). Consequently, DNA synthesised with the virus was deficient in nucleotide sequences homologous to $cDNA_{sarc}$ and served principally to detect other portions of the ASV genome. A substantial fraction (about 50%) of $cDNA_{B77}$ reacted with normal chicken DNA, but there was little or no reaction with quail, duck, turkey and emu DNAs (Fig. 1 and unpublished data); these results conform to previous reports[20,22,23]. We conclude that the DNAs from widely divergent avian species all contain nucleotide sequences which are at least partially related to transforming gene(s) of ASV, whereas only chicken DNA has appreciable homology with the remainder of the ASV genome.

The relatedness of DNA sequences homologous to $cDNA_{sarc}$ in different avian species was analysed by denaturing duplexes formed between $cDNA_{sarc}$ and normal avian DNAs (Fig. 2 and Table 2). In addition, we denatured duplexes between $cDNA_{sarc}$ and DNA from XC cells (rat cells transformed by Pr-C ASV) to test completely matched duplexes containing $cDNA_{sarc}$ sequences. The mammalian DNAs we have tested (calf and mouse) contain no nucleotide sequences homologous to $cDNA_{sarc}$ before in-

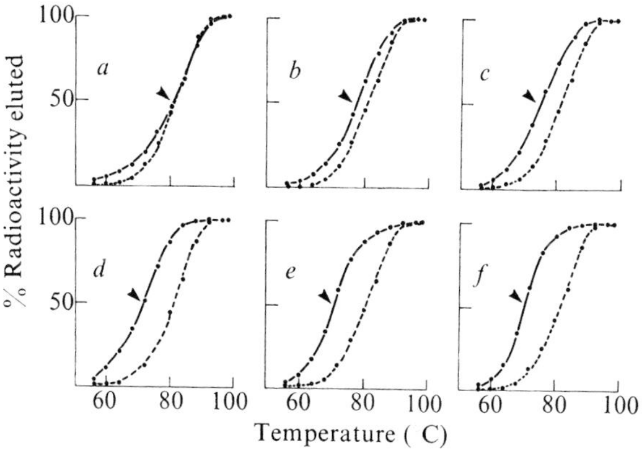

Fig. 2 Thermal denaturation of duplexes formed between cDNA$_{\mathrm{sarc}}$ and cellular DNAs. DNA was extracted from XC cells grown in culture, from 10–11-d-old embryos of chickens, ducks and quails, from the livers of adult turkeys, and from liver and heart of a 22-d-old emu. Denatured DNAs (0.5 mg) were incubated with ^{3}H-cDNA$_{\mathrm{sarc}}$ (0.25 ng, 5,000 c.p.m) in 0.1 ml of 0.9 M NaCl–0.001 M EDTA–0.02 M Tris-HCl, *p*H 7.4, for 100 h at 68 °C, final C_0t = 40,000. In a separate reaction, ^{32}P-cDNA$_{\mathrm{B77}}$ (2.5 ng, 100,000 c.p.m.) was annealed with 7 mg of denatured chicken DNA (final C_0t = 80,000). Samples (0.4 mg of chicken DNA and 6,000 c.p.m. of cDNA$_{\mathrm{B77}}$) of this reaction were added to the DNAs which had been annealed with cDNA$_{\mathrm{sarc}}$. The mixtures were then passed through a column of hydroxyapatite (3 ml packed vol) in 0.12 M sodium phosphate, *p*H 6.8, at 56 °C; in these conditions, stable duplexes bind to the hydroxyapatite, whereas single-stranded DNA does not. The columns were then washed continuously with 0.12 M sodium phosphate (0.4 ml min^{-1}) while the temperature of the column was raised in increments of 4 °C every 10 min. Fractions (4 ml) were collected and analysed for acid-precipitable radioactivity. ●——●, Duplexes with cDNA$_{\mathrm{sarc}}$; ●- - - -●, duplexes with cDNA$_{\mathrm{B77}}$ and chicken DNA (internal standard). *a*, Provirus (XC); *b*, chicken; *c*, quail; *d*, turkey; *e*, duck; *f*, emu.

fection by ASV; consequently, the DNA homologous to cDNA$_{\mathrm{sarc}}$ in rat cells after infection presumably represents nucleotide sequences contained entirely in recently integrated provirus for ASV rather than related sequences present in the DNA of uninfected cells. Denaturation was standardised internally by adding duplexes formed in a separate annealing reaction between normal chicken DNA and ^{32}P-labelled cDNA$_{\mathrm{B77}}$; these duplexes denatured with a T_m of 81±1 °C.

Duplexes between cDNA$_{\mathrm{sarc}}$ and DNA from XC cells had the highest T_m (81 °C; Table 2), as expected for completely matched complementary nucleotide sequences. Duplexes with chicken DNA were slightly less stable (T_m=77 °C); the nucleotide sequences of cDNA$_{\mathrm{sarc}}$ in the ASV genome are apparently diverged from the homologous sequences in the avian host. The reduction in the T_m is consistent with about 3% mismatching of bases[24]. The stability of duplexes with other avian DNAs (T_m=70–74 °C) decreased roughly in accord with phylogenetic distances among the species tested (Table 2); the T_ms suggest 5–8% mismatching of base pairs[24]. We made no effort in these tests to obtain maximum duplex formation with cDNA$_{\mathrm{sarc}}$, and the reactions with quail, turkey, duck and emu were relatively limited in extent (15–37%). In other experiments where at least 50% of cDNA$_{\mathrm{sarc}}$ was annealed into duplexes with chicken, quail and duck DNAs, we observed T_ms similar to those given in Table 2. cDNA$_{\mathrm{sarc}}$ seems to represent nucleotide sequences which arose and diverged during the course of avian speciation.

We have shown previously that the nucleotide sequences of cDNA$_{\mathrm{sarc}}$ include genetic information required for transformation of fibroblasts by ASV[14]; the data reported here indicate that similar or partially related information is present in the genome of the normal chicken cell and widely distributed among the avian species. We suggest that part or all of the transforming gene(s) of ASV was derived from the chicken genome or a species closely related to chicken, either by a process akin to transduction[25] or by other events, including recombination, which are alleged to have generated type C viruses from normal cellular genes[12]. If the entire ASV genome originated from cellular genes[12], then nucleotide sequences in the transforming gene(s) have been conserved relative to other viral genes, because the nucleotide sequences of cDNA$_{\mathrm{sarc}}$ are the only portion of the ASV genome which we can detect in DNA from birds other than chickens. We cannot exclude, however, the possibility that viral genes other than those represented by cDNA$_{\mathrm{sarc}}$ have been introduced into the germ line of chickens after speciation. The sequences homologous to cDNA$_{\mathrm{sarc}}$ in the genome of ASV are slightly diverged from the analogous sequences in chicken genome; this could be the consequence of either the process which generated viral genes from cellular genes[12] or mutations during the course of repeated viral propagation.

Others have reported evidence concerning the genetic divergence during possible spread of unidentified portions of type C viral genomes during evolution[26,27] and the apparent transduction of cellular nucleotide sequences by type C viruses[25], but our study provides the first data on the origins of a genetically identified set of viral nucleotide sequences. It should be possible to carry out similar studies for at least one other gene of ASV (the gene coding for the type-specific glycoprotein), using techniques similar to those reported here.

Neiman and his colleagues have reported that the hybridisation of ASV 70S RNA to normal chicken DNA was competed completely by RNA from non-transforming viruses[11]. They therefore concluded that the genes responsible for transformation by ASV were present in the avian genome only after viral infection. We cannot explain the discrepancy between their results and ours.

We anticipate that cellular DNA homologous to cDNA$_{\mathrm{sarc}}$ serves some function which accounts for its conservation during avian speciation. The nucleotide sequences which anneal with cDNA$_{\mathrm{sarc}}$ are part of the

Table 2 Thermal stabilities of duplexes between cDNA$_{\mathrm{sarc}}$ and normal DNAs

DNA	%cDNA$_{\mathrm{sarc}}$ in duplexes	T_m	ΔT_m	Phylogenetic distance from chicken (Myr)
Provirus (XC)	56	81	0	
Chicken	52	77	− 4	0
Quail	37	74	− 7	35–40
Turkey	30	72	− 9	40
Duck	16	71	−10	80
Emu	15	70	−11	100

Denatured DNAs (0.5 mg) were annealed with ^{3}H-cDNA$_{\mathrm{sarc}}$ (0.25 ng, 5,000 c.p.m.), adsorbed to hydroxyapatite in 0.12 M sodium phosphate at 56 °C, and denatured with a thermal gradient, all as described for Fig. 2. Duplexes between ^{32}P-cDNA$_{\mathrm{B77}}$ and normal chicken DNA, included in each analysis as an internal standard, denatured with T_m = 81 ±1 °C. The estimates of phylogenetic distance, deduced from fossil records and antigenic relationships among proteins[29], were provided by Professor Allan Wilson (personal communication).

Nature Vol. 260 March 11 1976

unique fraction of cellular DNA and could represent either structural or regulatory genes. But the function of those sequences is unknown. We are testing the possibilities that they are involved in the normal regulation of cell growth and development or in the transformation of cell behaviour by physical, chemical or viral agents.

We thank William E. Meeker of Sacramento Zoo for assistance in obtaining the emu, Karen Smith and Jean Jackson for technical assistance, and H. Temin and L. Levintow for editorial advice. This work was supported by grants from the US Public Health Services and the American Cancer Society, and a contract within the Virus Cancer Program of the National Cancer Institute. D.S. was supported by CNRS and H.E.V. is a recipient of a research career development award from the National Cancer Institute.

D. Stehelin*
H. E. Varmus
J. M. Bishop

Department of Microbiology,
University of California
San Francisco, California 94143

P. K. Vogt

Department of Microbiology,
University of California,
Los Angeles, California 90033

Received November 10, 1975; accepted January 2, 1976.

*Present address: IRSC BP8, 94800–Villejuif, France.

1 Martin, G. S., *Nature*, **227**, 1021–1023 (1970).
2 Kawai, S., and Hanafusa, H., *Virology*, **46**, 470–479 (1971).
3 Bader, J. P., *J. Virol.*, **10**, 267–276 (1972).
4 Wyke, J. A., Bell, J. G., and Beamand, J. A., *Cold Spring Harb. Symp. quant. Biol.*, **39**, 897–905 (1974).
5 Baluda, M. A., *Proc. natn. Acad. Sci. U.S.A.*, **69**, 576–580 (1972).
6 Rosenthal, P. N., Robinson, H. L., Robinson, W. S., Hanafusa, T., and Hanafusa, H., *Proc. natn. Acad. Sci. U.S.A.*, **68**, 2336–2340 (1971).
7 Varmus, H. E., Weiss, R. A., Friis, R. R., Levinson, W., and Bishop, J. M., *Proc. natn. Acad. Sci. U.S.A.*, **69**, 20–24 (1972).
8 Wright, S. E., and Neiman, P. E., *Biochemistry*, **13**, 1549–1554 (1974).
9 Vogt, P. K., and Friis, R. R., *Virology*, **43**, 223–234 (1971).
10 Weiss, R. A., Friis, R. R., Katz, E., and Vogt, P. K., *Virology*, **46**, 920–938. (1971).
11 Neiman, P. E., Wright, S. E., McMillin, C., and MacDonnell, D., *J. Virol.*, **13**, 837–846 (1974).
12 Temin, H. M., *Cancer Res.*, **34**, 2835–2841 (1974).
13 Huebner, R. J., and Todaro, G. J., *Proc. natn. Acad. Sci. U.S.A.*, **64**, 1087–1094 (1969).
14 Stehein, D., Guntaka, R. V., Varmus, H. E., and Bishop, J. M., *J. molec. Biol.* (in the press).
15 Duesberg, P. H., and Vogt, P. K., *Proc. natn. Acad. Sci. U.S.A.*, **67**, 1673–1680 (1970).
16 Lai, M. M. C., Duesberg, P. H., Horst, J., and Vogt, P. K., *Proc. natn. Acad. Sci. U.S.A.*, **70**, 2266–2270 (1973).
17 Duesberg, P. H., and Vogt, P. K., *J. Virol.*, **12**, 594–599 (1973).
18 Beemon, K., Duesberg, P., and Vogt, P. K., *Proc. natn. Acad. Sci. U.S.A.*, **71**, 4254–4258 (1974).
19 Billeter, M. A., Parsons, J. T., and Coffin, J. M., *Proc. natn. Acad. Sci. U.S.A.*, **71**, 3560–3564 (1974).
20 Varmus, H. E., Heasley, S., and Bishop, J. M., *J. Virol.*, **14**, 895–903 (1974).
21 Rice, N., and Paul, P., *Yb. Carnegie Instn. Wash.*, **71**, 262–264 (1972).
22 Tereba, A., Skoog, L., and Vogt, P. K., *Virology*, **65**, 524–534 (1975).
23 Kang, C. Y., and Temin, H. M., *J. Virol.*, **14**, 1179–1188 (1974).
24 Ullman, J. S., and McCarthy, B. J., *Biochim. biophys. Acta*, **294**, 405–415 (1973).
25 Scolnick, E. M., Rands, E., Williams, D., and Parks, W. P., *J. Virol.*, **12**, 458–463 (1973).
26 Benveniste, R. E., and Todaro, G. J., *Proc. natn. Acad. Sci. U.S.A.*, **71**, 4513–4518 (1974).
27 Benveniste, R. E., and Todaro, G. J., *Nature*, **252**, 456–459 (1974).
28 Leong, J. A., *et al.*, *J. Virol.*, **9**, 891–902 (1972).
29 Praeger, E. M., Brush, A. H., Nolan, R. A., Nakaniski, M., and Wilson, A. C., *J. molec. Evol.*, **3**, 243 (1974).

Paulson J.R. and **Laemmli U.K.** 1977. The structure of histone-depleted metaphase chromosomes. *Cell* **12:** 817-828.

THE NUCLEOSOME MODEL SHOWED that a chromatin fiber formed by wrapping DNA around histone octamers is some 7 times shorter than the original B-form DNA itself (Kornberg, this volume). However, this chromatin fiber is still 1,000 times longer than the chromosome into which it is packed at metaphase. The study by Paulson and Laemmli suggested a model for the higher order packing of chromatin in which loops of chromatin are attached to a scaffold of non-histone proteins, which is responsible for the basic shape of the metaphase chromosome. Isolated metaphase chromosomes were treated with dextran sulfate and heparin to remove most of the histones, and the histone-depleted chromosomes were then spread on grids for observation in the electron microscope. The dramatic images obtained in this way showed loops of DNA extending laterally from a dense network of fibers. Many loops were 10–30 μm in length and some were nearly 50 μm. Although these observations were the first to show such large loops in metaphase chromosomes, it had been known since the latter part of the 19th century that similar loops occur in the lampbrush chromosomes of amphibian and other vertebrate oocytes (1), and a loop model had been proposed for the structure of puffs and Balbiani Rings of Dipteran polytene chromosomes (2; Skoglund et al., this volume). Whether these different loops have a similar structural basis and just what type of unit they represent remain active areas of investigation and controversy. The scaffold of non-histone proteins appears to be relatively simple in composition. It includes topoisomerase II (3) and members of the SMC (structural maintenance of chromosomes) family of proteins, which play important roles in chromosome condensation, chromatid cohesion, sex chromosome dosage compensation, and DNA recombination repair (reviewed in 4).

---

1. Callan H.G. 1986. *Lampbrush chromosomes*. Springer-Verlag, Berlin.
2. Beermann W. 1962. Riesenchromosomen. *Protoplasmatologia* **6D:** 1–161.
3. Earnshaw W.C. and Heck M.M.S. 1985. Localization of topoisomerase II in mitotic chromosomes. *J. Cell Biol.* **100:** 1716–1725.
4. Cobbe N. and Heck M.M.S. 2000. SMCs in the world of chromosome biology-from prokaryotes to higher eukaryotes. *J. Struct. Biol.* **129:** 123–143.

Cell, Vol. 12, 817-828, November 1977, Copyright © 1977 by MIT

The Structure of Histone-Depleted Metaphase Chromosomes

James R. Paulson and U. K. Laemmli
Department of Biochemical Sciences
Princeton University
Princeton, New Jersey 08540

Summary

We have previously shown that histone-depleted metaphase chromosomes can be isolated by treating purified HeLa chromosomes with dextran sulfate and heparin (Adolph, Cheng and Laemmli, 1977a). The chromosomes form fast-sedimenting complexes which are held together by a few nonhistone proteins.

In this paper, we have studied the histone-depleted chromosomes in the electron microscope. Our results show that: the histone-depleted chromosomes consist of a scaffold or core, which has the shape characteristic of a metaphase chromosome, surrounded by a halo of DNA; the halo consists of many loops of DNA, each anchored in the scaffold at its base; most of the DNA exists in loops at least 10-30 μm long (30-90 kilobases).

We also show that the same results can be obtained when the histones are removed from the chromosomes with 2 M NaCl instead of dextran sulfate. Moreover, the histone-depleted chromosomes are extraordinarily stable in 2 M NaCl, providing further evidence that they are held together by nonhistone proteins.

These results suggest a scaffolding model for metaphase chromosome structure in which a backbone of nonhistone proteins is responsible for the basic shape of metaphase chromosomes, and the scaffold organizes the DNA into loops along its length.

Introduction

Our research efforts are currently directed toward understanding the higher order structure of eucaryotic chromosomes. Considerable progress recently has been made in understanding how histones fold the chromosomal DNA into the basic nucleohistone fiber (see, for example, Oudet, Gross-Bellard and Chambon, 1975; Finch and Klug, 1976). Our work has been motivated, however, by the idea that other components, probably nonhistone proteins, must be responsible for the higher order folding of the basic chromatin fiber. Our approach to the problem of higher order structure, therefore, is to remove the histones from chromosomes and to study the structure and biochemistry of the DNA-protein complexes which remain.

In the accompanying paper (Adolph et al., 1977a), we have shown that histone-depleted chromosomes can be isolated on sucrose gradients following treatment with dextran sulfate and heparin. These complexes are highly sensitive to SDS, urea and proteases, but they are insensitive to high salt or RNAase. SDS-polyacrylamide gel electrophoresis reveals that they contain essentially no histones, and up to 30 major nonhistone proteins. These results suggest that the DNA is maintained in a highly folded, fast-sedimenting structure by nonhistone proteins.

In order to get information on the higher order structure of metaphase chromosomes, we have studied these histone-depleted chromosomes by electron microscopy. The removal of histones should aid in understanding how the nonhistones fold the chromosomal DNA by allowing the DNA to be spread out. When histones are present, the adherence of nucleohistone fibers to one another, and the electron density of the chromosomes, do not permit a view of the underlying structures.

In this paper, we report what we have learned from electron microscope studies of histone-depleted chromosomes spread with cytochrome c.

Results

Preparation of Histone-Depleted Chromosomes for Electron Microscopy

In the preceding paper (Adolph et al., 1977a), we have described the preparation of histone-depleted metaphase chromosomes. When chromosomes are treated with 2 mg/ml dextran sulfate and 0.2 mg/ml heparin, more than 99% of the histones and many of the nonhistone proteins are removed. These histone-depleted chromosomes form compact, fast-sedimenting structures which can be isolated on a sucrose gradient.

When we attempted to prepare this material for electron microscopy by spreading with cytochrome c, two problems arose. The first problem was to obtain histone-depleted chromosomes in sufficient concentration to make routine studies convenient. Since all the DNA exists in only a few very large structures, a very high DNA concentration is needed to see a reasonable number of particles on one electron microscope grid. We solved this problem by sedimenting the chromosomes to a cushion of Metrizamide (see Experimental Procedures). By this method, we can easily obtain a fraction containing 10-50 μg/ml DNA, whereas we can regularly obtain only 1 μg/ml DNA from the peak fraction of a sucrose gradient.

The second problem was that when we used 2 mg/ml dextran sulfate to remove the histones, enough dextran sulfate contaminated the sample to disturb the spreading procedure. Presumably the dextran sulfate binds nearly all the cytochrome c so that a monolayer cannot form. The problem is more serious for the ammonium acetate spreading method than for the carbonate method (see Experimental Procedures), since much more cytochrome c binds to polyanions at pH 6.5 than at pH 10. We solved this problem, however, by using a lower concentration of dextran sulfate to remove the histones.

Thus we routinely prepared histone-depleted chromosomes for electron microscopy by treating chromosomes with 0.2 mg/ml dextran sulfate and 0.02 mg/ml heparin and sedimenting them to a cushion of 0.6 M Metrizamide in gradient buffer. We emphasize that these modifications of the basic procedure of Adolph et al. (1977a) are only a convenience and do not affect the results. We obtained the same results by spreading samples from continuous sucrose gradients. We ran these gradients as described by Adolph et al. (1977a), except that we omitted detergents from the body of the gradient. We also obtained the same results when we used 2 mg/ml dextran sulfate, provided we diluted samples 10–15 times in gradient buffer to reduce the concentration of contaminating dextran sulfate before spreading. As will be shown below, we can obtain the same results without using dextran sulfate at all, if we remove histones by treating chromosomes with 2 M NaCl.

The histone-depleted chromosomes have the same sedimentation properties whether we treat them with 0.2 mg/ml or 2 mg/ml dextran sulfate (Adolph et al., 1977a). Moreover, we have verified, by using the Metrizamide cushion procedure (J. R. Paulson, unpublished data), and also by using continuous sucrose gradients (Adolph et al., 1977a), that 0.2 mg/ml dextran sulfate is sufficient to remove 99% of the histones.

We used two methods of spreading with cytochrome c in these studies (see Experimental Procedures). Both methods gave essentially the same results, but each had special advantages. We preferred the ammonium acetate method because it gave the highest contrast to the DNA strand. The carbonate method was useful in early work, however, because it is rarely affected by contaminating dextran sulfate and because it enables one to spread many different samples in a short period of time. In addition, this method sometimes gives a very informative view of the DNA loops in the histone-depleted chromosomes.

Histone-Depleted Chromosomes Consist of a Protein Scaffold Surrounded by a Halo of DNA

Figure 1 shows a representative micrograph of a histone-depleted chromosome. The chromosome consists of a central darkly staining scaffold or core which is surrounded by a halo of DNA. A magnified view of the central scaffold and part of the DNA halo is shown in Figure 2.

The shape of the scaffold is strikingly similar to the characteristic morphology of metaphase chromosomes, and this fact makes it immediately clear that each chromosome is sedimenting as an intact structure. We find very few free DNA ends around the edges of the DNA halo, and very little free DNA in the background on these grids. This shows that the chromosomes have not been subjected to excessive shear. Complexes such as the one shown in Figure 1 are essentially the only DNA-containing particles we find on these grids. We occasionally find particles which do not contain DNA (presumably cell debris) in these preparations, but they probably do not contribute to the protein pattern of histone-depleted chromosomes purified on sucrose gradients (Adolph et al., 1977a).

It should be noted that in these studies we used only chromosome preparations in which the chromosome morphology was well preserved (as judged by phase-contrast microscopy) for preparation of histone-depleted chromosomes. When chromosomes have been severely stretched or sheared (compare Stubblefield and Wray, 1971), the scaffold is generally found to be pulled apart into fragments and its organization is not discernable.

Two additional examples of chromosome scaffolds are shown in Figure 3. The core is sometimes dense, as in Figure 3a, and sometimes opened up into a fibrous network, as in Figure 3b and Figure 2. In nearly all of the histone-depleted chromosomes, however, the scaffold has the shape of a metaphase chromosome with paired chromatids connected at a centromere. On a typical grid, from which the micrograph in Figure 1 was taken, we examined 194 particles. Of these, 74% (144) had the sister chromatid scaffolds tightly associated, as in Figure 2 and Figure 3b. Another 15% (30) had the sister chromatid scaffolds separated and only held together by a few thin fibers, as in Figure 3a. The remaining 10% (20) appeared to be individual chromatids which had been separated from their sisters. Breaks in a chromatid scaffold are quite rare, occurring in only about 4% (7) of the particles.

When 2 M urea is included in the hypophase, breaks in the scaffold become much more frequent as the scaffold disintegrates. This is in keeping with our finding that the histone-depleted chromosomes are partially unfolded by 2 M urea and completely unfolded by 4 M urea (Adolph et al., 1977a). In the example shown in Figure 4, the scaffold has already begun to dissociate after 3 min in contact with 2 M urea.

Occasionally, we see thick bundles of fibers, similar to those observed by Kavenoff and Bowen

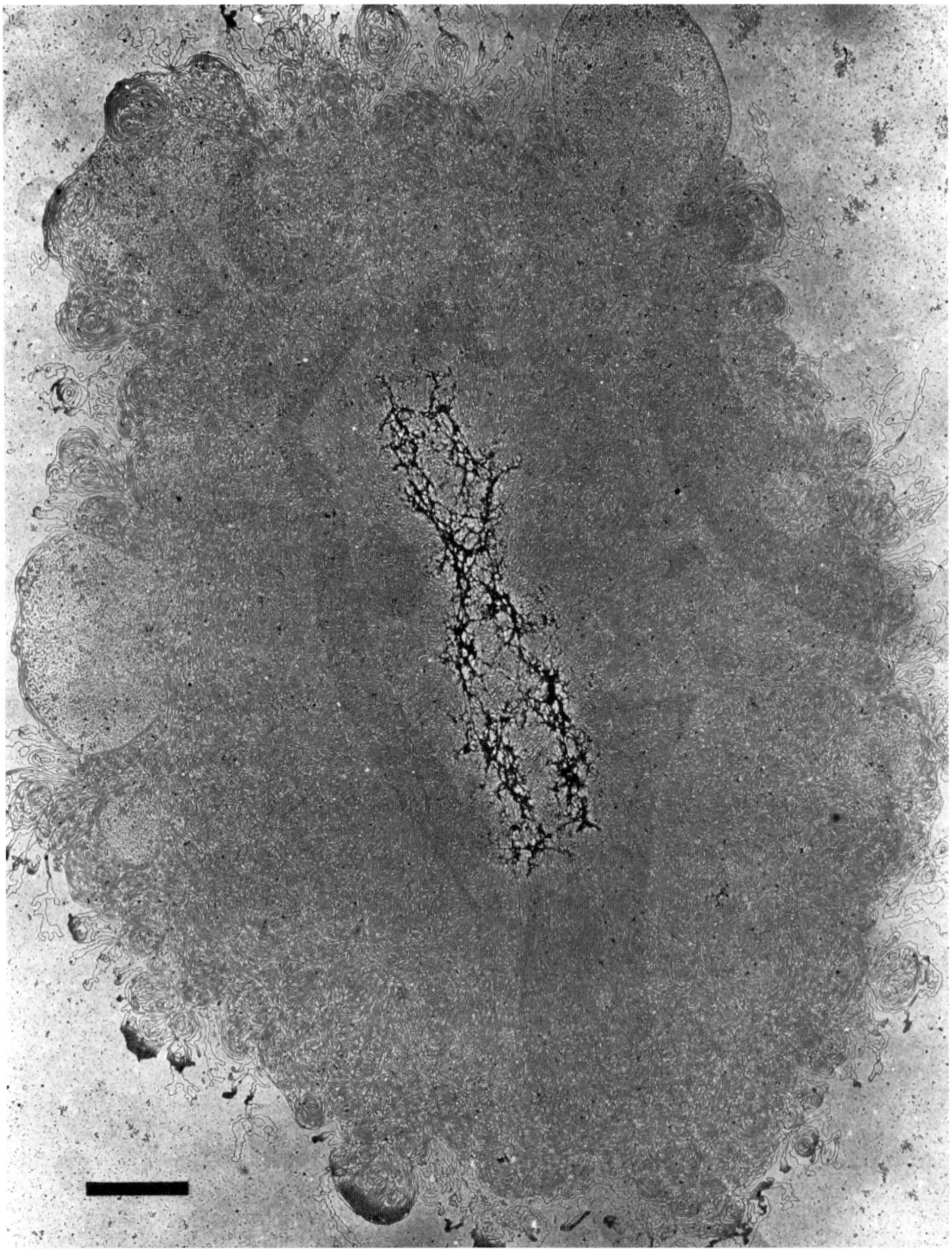

Figure 1. Electron Micrograph of a Histone-Depleted Metaphase Chromosome from Hela

Chromosomes were treated with dextran sulfate and heparin, purified by sedimentation to a cushion of 0.6 M Metrizamide and spread on 0.125 M ammonium acetate (see Experimental Procedures). The cytochrome c monolayer was sampled 1 min after spreading.

The chromosome consists of a central, densely staining scaffold or core surrounded by a halo of DNA extending 6–9 μm outward from the scaffold. The low magnification makes it difficult to see the individual DNA strands except along the edge of the DNA halo. The spider-like flecks in the background (for example, in the upper right corner) are contaminating dextran sulfate.

The bar in this and subsequent micrographs represents 2 μ.

Cell
820

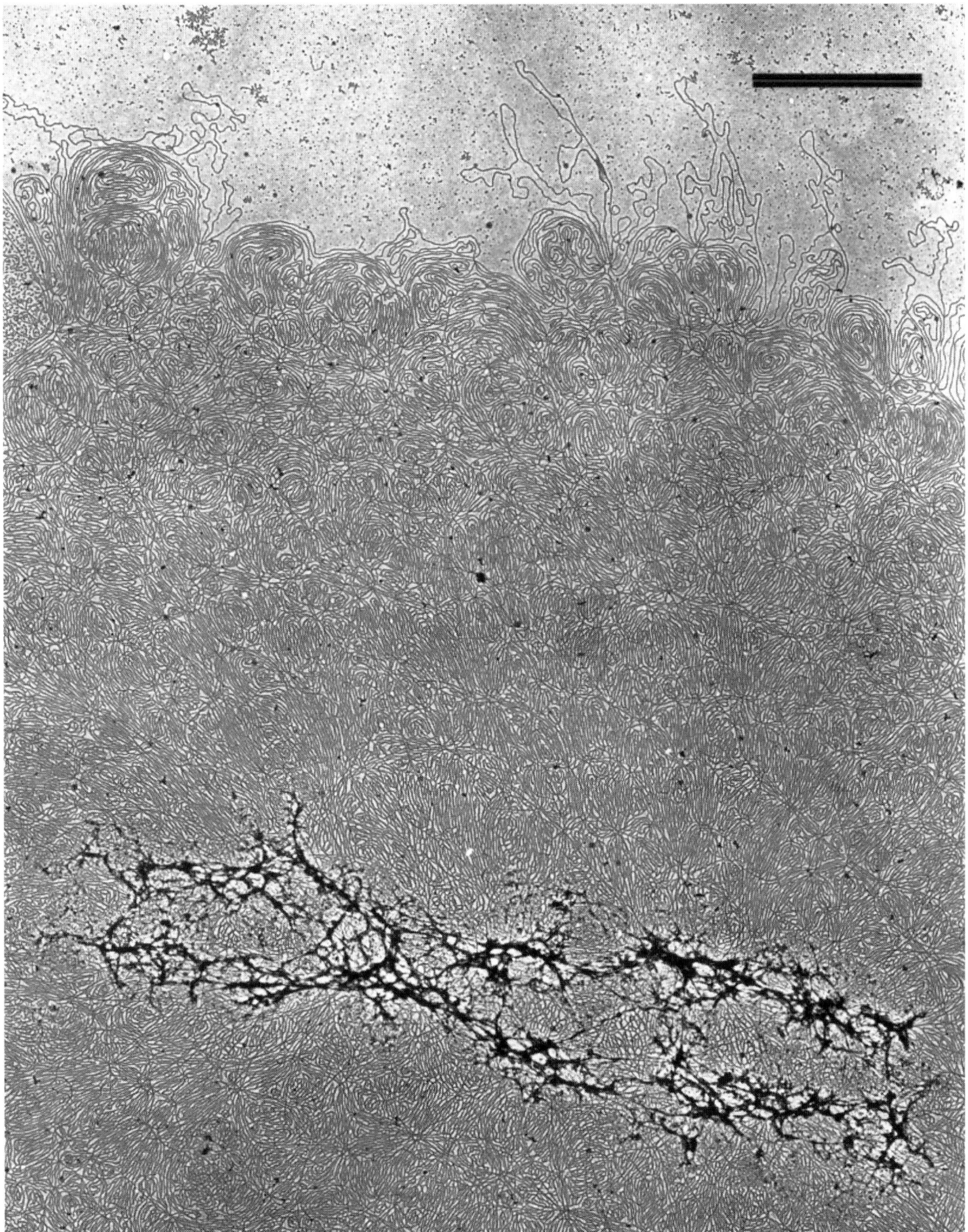

Figure 2. Higher Magnification of the Chromosome Scaffold and Part of the Surrounding DNA of the Chromosome Shown in Figure 1
Note that the scaffold appears to consist of a dense network of fibers. Although individual DNA strands can be easily resolved at this magnification, the large amount of DNA makes it impossible to trace any given strand. Bar = 2 μ.

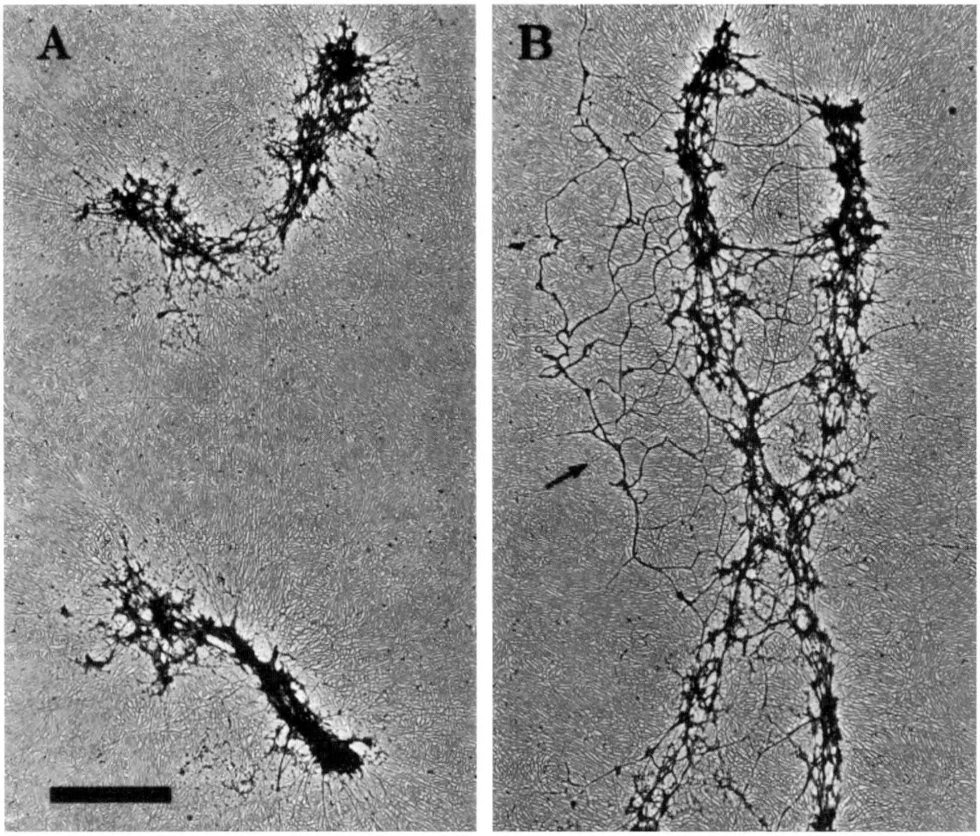

Figure 3. Central Scaffolds of Two Histone-Depleted Chromosomes

Chromosomes were prepared as described in the legend to Figure 1. (A) illustrates a situation seen in about 15% of the chromosomes in which the chromatids are separated and only connected by a few thin fibers. In (B) note the open fibrous appearance of the core. The thick fibrous extensions on the left of the core (arrow) are seen occasionally, and are believed to arise artifactually by collapse and aggregation of DNA strands during dehydration in ethanol (compare Lang, 1969, 1973). Bar = 2 μ.

(1976). An example is seen in Figure 3b (arrow). We believe that these thick fibers are not important structurally, but that they are artifactually produced by ethanol dehydration. Since so much DNA is present in these chromosomes, it is probable that sometimes some of the DNA is sterically prevented from adsorbing to the cytochrome monolayer. As has been shown by Lang (1969, 1973), this DNA could collapse and aggregate into thick fibers when placed in 90% ethanol.

DNA Is Attached to the Scaffold in Loops

When we spread histone-depleted chromosomes on 0.125 M ammonium acetate, the halo of DNA extends outward from the scaffold an average of 10–12 μm radially, although values of 6–20 μm have been observed. In this respect, the chromosome shown in Figure 1 is atypical, since the DNA extends outward only 6–9 μm. In fact, we selected this chromosome for this reason, since larger chromosomes are more difficult to reproduce photographically. In other respects, however, it is typical.

The fairly uniform radius of the DNA halo suggests that the majority of the DNA in these chromosomes exists in loops of approximately the same length, and we would like to determine the lengths of these loops. The radius of the halo tells us that many loops must be at least 20–24 μm long, but there are too many DNA strands piled on top of one another to trace any individual loop, as can be seen from Figure 2.

In some preparations spread by the carbonate method, the DNA loops are well separated (Figure 5). We cannot account for the DNA in these chromosomes in the loops, and indeed the scaffold is surrounded by dense material (Figure 5, bottom)

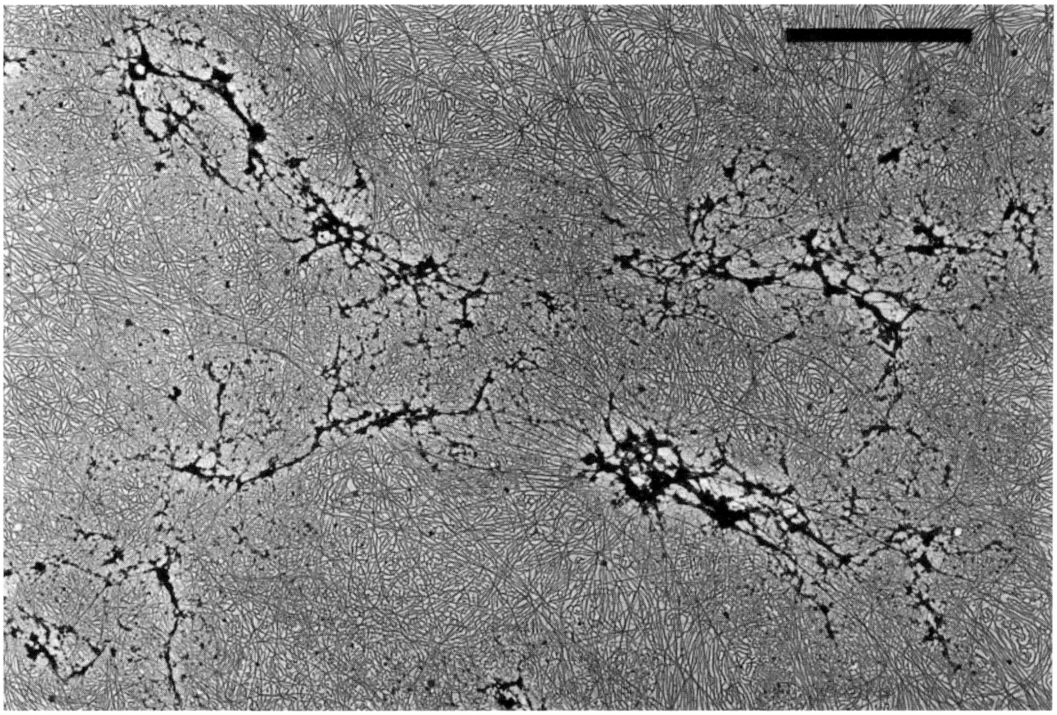

Figure 4. Central Scaffold of a Histone-Depleted Chromosome Spread on a Hypophase Containing Urea

Purified histone-depleted chromosomes were spread by the ammonium acetate method (see Experimental Procedures), except that the hypophase contained 0.125 M ammonium acetate and 2 M urea. The cytochrome c monolayer was sampled after 3 min.

Note that after only brief contact with urea, the scaffold is beginning to disintegrate. This is most easily seen by comparing this scaffold with the one shown in Figure 2. Many other scaffolds on this same grid had dissociated even further. Bar = 2 μ.

which probably consists of incompletely spread DNA. It is possible that in these chromosomes, we have not completely removed the histones, and hence the DNA does not completely unfold.

We took advantage of these unusual preparations to obtain a minimum estimate of the DNA loop size. We measured all the DNA loops along one side of a small chromatid and plotted the results in Figure 6. The number average length of a loop is 14 μm or 42 kilobases [assuming that 1 μm of DNA equals about 3000 base pairs (Chow, Scott and Broker, 1975)]. When the average is weighted according to the amount of DNA in each loop, however, the result is 23 μm (70 kb). This represents the length of the loop in which an average DNA nucleotide finds itself. It is clear from Figure 6b that the majority of the DNA is in loops between 10 and 30 μm long (30–90 kb).

We would like to emphasize that these methods give only a minimum estimate for the DNA loop size. Other micrographs of histone-depleted chromosomes spread by the ammonium acetate methods make clear that some loops in excess of 60 μm do occur. Loops which are very evident, such as those in Figure 5, occur in only a few preparations in which the DNA does not appear to be completely unfolded. Thus we believe 10–30 μm may be an underestimate of the actual loop size, since the DNA loops may not be completely unraveled. In particular, the large number of loops <5 μm long in Figure 6 may be misleading.

An especially important observation is that a loop generally returns to the scaffold adjacent to its point of origin (Figure 5). For the 76 loops measured in Figure 6, 57% (43) have both ends emanating from the core immediately adjacent to each other. Another 28% (21) have six or fewer DNA strands in between.

These results clearly show that the DNA is attached to the scaffold in loops, and that both ends of a DNA loop appear to be anchored at the same place in the scaffold. The majority of the DNA exists in loops that are at least 10–30 μm long (30–90 kb).

We have not directly observed supercoiling in the DNA loops. It is probable, however, that under

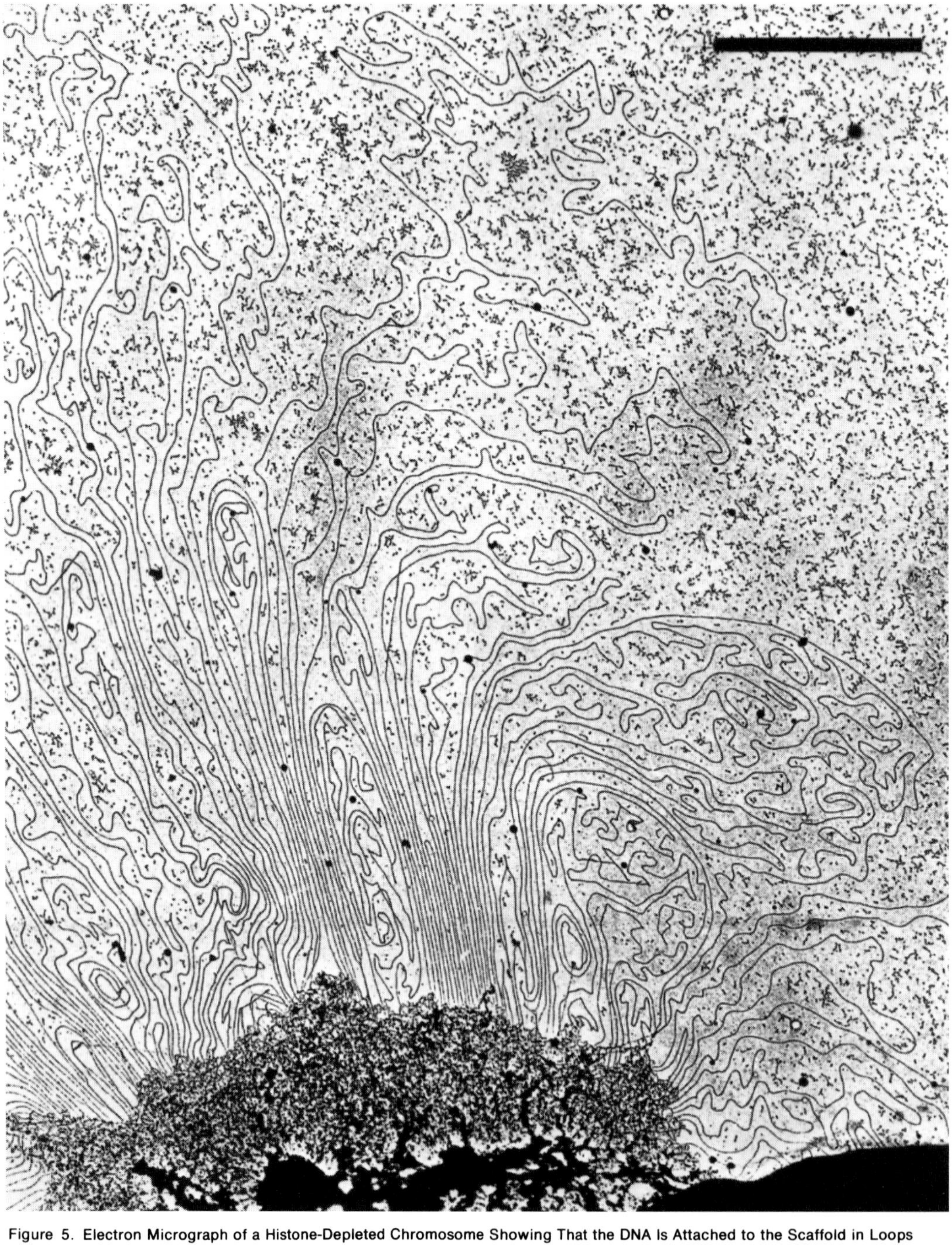

Figure 5. Electron Micrograph of a Histone-Depleted Chromosome Showing That the DNA Is Attached to the Scaffold in Loops

Chromosomes were treated with dextran sulfate and heparin, purified and spread with cytochrome c by the carbonate method (see Experimental Procedures). The contrast of this micrograph was enhanced by copying onto high-contrast plates as described by Chow et al. (1975).

Note that both ends of a DNA loop appear to emanate from adjacent points in the scaffold. A histogram of the loop lengths in this chromosome is shown in Figure. 6. Bar = 2 μ.

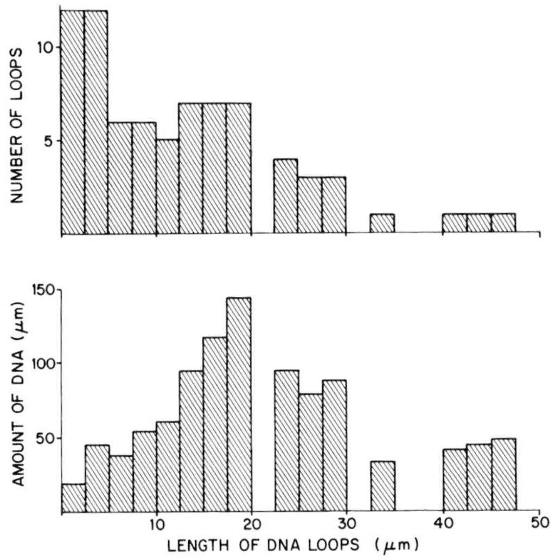

Figure 6. Histogram of Loop Lengths in a Histone-Depleted Metaphase Chromosome

All 76 loops along one side of the chromatid shown in Figure 5 were measured. In the upper panel, the number of loops falling into each size class is plotted. In the lower panel, the total length of DNA occurring in loops of each size class is plotted. This plot makes clear that most of the DNA exists in loops at least 10–30 μm long, although loops up to 47 μm long were observed.

the conditions used to isolate the metaphase chromosomes (pH 6.5, 0.5 mM $CaCl_2$), many nicks are introduced into each loop (Wray, Stubblefield and Humphrey, 1972).

Removing Histones from Chromosomes with 2 M NaCl

It is well known that histones can be removed from DNA with high salt (Spelsberg and Hnilica, 1971). Although we have mainly used the dextran sulfate-heparin procedure to prepare histone-depleted chromosomes, similar results can be obtained by treating chromosomes with 2 M NaCl.

The method described in Experimental Procedures is modified as follows: all step gradient solutions are as before except that they contain 2 M NaCl; chromosomes are added to a solution of 2 M NaCl, 0.1% NP-40, 10 mM EDTA, 10 mM Tris-HCl (pH 9.0) and 1 mM PMSF, gently mixed and layered on top of the step gradient. Centrifugation is begun after a 30 min incubation and the rest of the method is as before. No dextran sulfate or heparin is used.

Figure 7 shows an example of a 2 M NaCl-treated chromosome purified by sedimentation to a Metrizamide cushion. The scaffold is fairly dense and has a somewhat different texture than that of the dextran sulfate-treated chromosomes. Only part of the surrounding DNA is shown. We have not yet characterized the 2 M NaCl-treated chromosomes biochemically, but they give the same profile as dextran sulfate-treated chromosomes when sedimented through sucrose gradients (K. W. Adolph, unpublished observations).

One could argue that metaphase chromosomes are stabilized mainly by side-to-side interactions between the chromatin fibers, and that our scaffolds consist merely of a few histones at the base of the loops which have not been extracted. The following experiment, however, strongly suggests that this is not the case.

Histone-depleted chromosomes prepared by treatment with 2 M NaCl were spread within 2 hr after removal from step gradients, and again 24 hr later. All the chromosomes were intact in both samples, and there were no significant differences between them. In other words, the histone-depleted chromosomes are completely stable for more than 24 hr in 2 M NaCl, as judged by electron microscopy. This strongly suggests that nonhistone proteins, rather than a few remaining histones, stabilize the chromosome scaffold.

Discussion

In this paper, we have examined histone-depleted metaphase chromosomes in the electron microscope by spreading with cytochrome c. Our results show that:

–The histone-depleted chromosomes consist of a central scaffold, which has the basic shape characteristic of metaphase chromosomes, surrounded by a halo of DNA. As we have shown elsewhere, the scaffold consists mainly of a few nonhistone proteins, and few or no proteins are present in the halo (Adolph et al., 1977a, 1977b). The scaffold is extraordinarily stable in 2 M NaCl but disintegrates rapidly when spread on a hypophase containing urea. Since Adolph et al. (1977a) have shown by sedimentation studies that histone-depleted chromosomes fall apart in 2 M or 4 M urea, the rapid dissociation of the scaffold following contact with urea is consistent with the notion that the scaffold is holding the DNA in a compact conformation.

–The chromosomal DNA is organized into many loops which are anchored in the scaffolding. Both ends of a loop appear to be anchored at the same place, since the DNA strand emanates radially from the scaffold and returns to an adjacent point. Most of the DNA exists in loops at least 10–30 μm long, although loops >60 μm in length have been observed. It is interesting to note that similar loop sizes have been observed in E. coli (Kavenoff and Ryder, 1976), and the same range of loop sizes has been inferred from sedimentation studies of eucaryotic interphase nuclei (Cook and Brazell, 1975; Benyajati and Worcel, 1976).

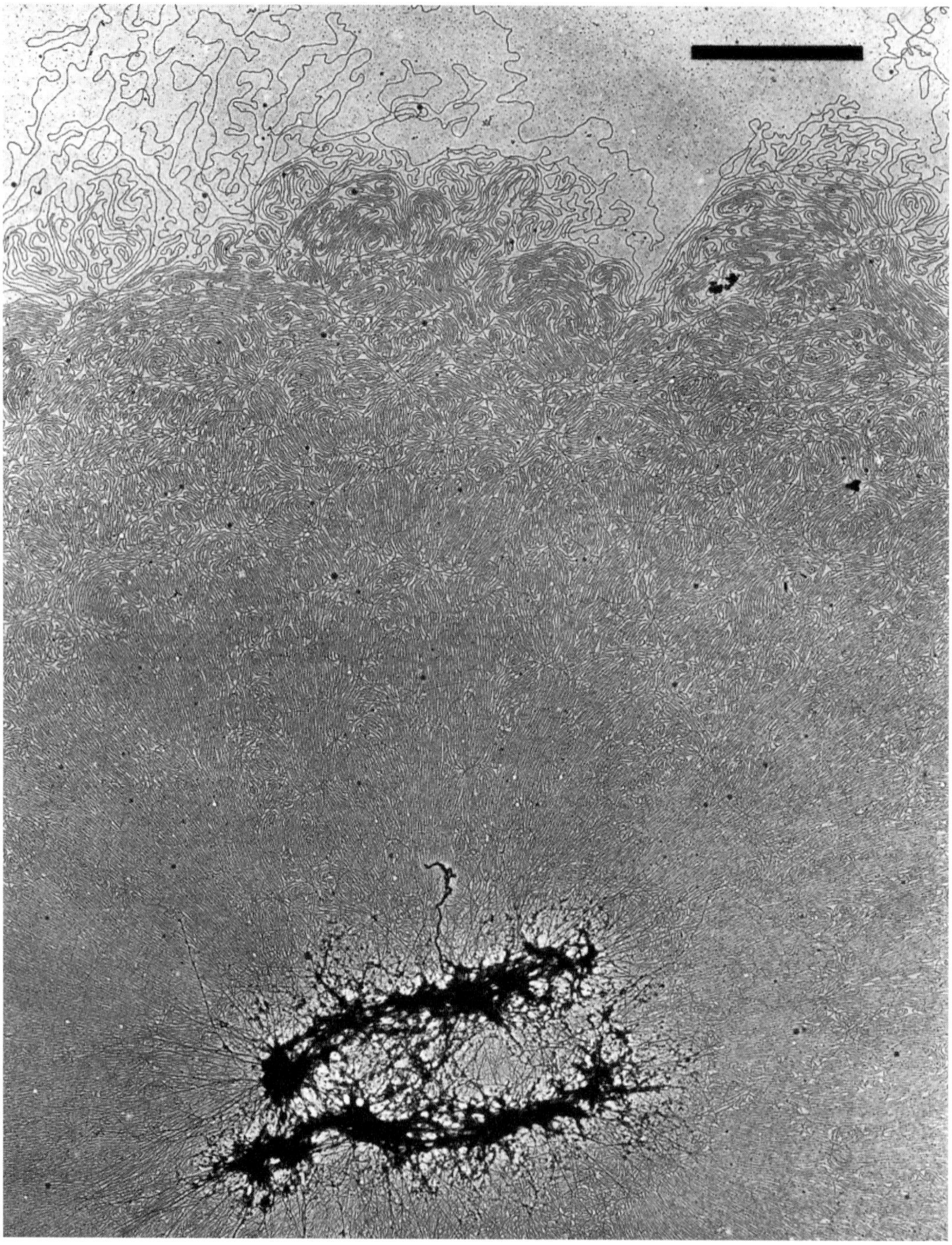

Figure 7. Histone-Depleted Chromosome Prepared by Treating Chromosomes with 2 M NaCl, Showing the Chromosome Scaffold and Part of the Surrounding DNA

Chromosomes were prepared as described in the text and spread on 0.125 M ammonium acetate (see Experimental Procedures). No dextran sulfate or heparin was used.

These chromosomes are virtually indistinguishable from the dextran sulfate-treated chromosomes, either by sedimentation studies (K. W. Adolph, unpublished observations) or in the electron microscope.

—Histone-depleted chromosomes have a similar structure whether the histones are removed with dextran sulfate and heparin or with 2 M NaCl.

These results suggest a scaffolding model for metaphase chromosome structure. In this model, chromosomes receive their basic shape from a scaffolding or backbone of nonhistone proteins which maintains the chromosome shape even when histones are removed, and which organizes the DNA into loops along its length. In the simplest (unineme) model, a single DNA molecule is folded many times, and the folds are maintained by periodic attachments to the scaffold. In this way the DNA molecule could be laid down from one end of the chromatid to the other, without need for any longitudinal fibers.

The DNA loops, of course, do not exist as free DNA but are tightly compacted by histones into 0.5–2 μm loops of 300 Å nucleohistone fiber (see, for example, Rattner, Branch and Hamkalo, 1975; Finch and Klug, 1976; Hozier, Renz and Nells, 1977). It is clear that many such loops attached to a scaffold would give the appearance of a dense bundle of folded chromatin fibers, as is seen in the micrographs of DuPraw (1970). Alternatively, these loops of 300 Å fibers could be further twisted to form microconvules, 500–600 Å in diameter (Daskal et al., 1977). One of the most important features of the scaffolding model, however, is that the compaction of the DNA by the histones into the nucleohistone fiber, and the higher order folding of that fiber by the nonhistone scaffolding proteins, are independent levels of structure. The implications of our data for possible models of chromosome structure have been discussed more fully by Laemmli et al. (1977).

The similarity in electron density between the scaffold and chromatin may explain why the scaffold has not been detected in previous electron microscope studies of intact chromosomes, either in whole mount preparations or in thin sections. There are, of course, suggestions of a chromosome scaffold or core in the literature, particularly in the work of Abuelo and Moore (1969), Stubblefield and Wray (1971), and Sorsa (1973). One of the difficulties in interpreting these electron micrographs, however, is that structures resembling a chromosome core could be an artifact induced by stretching.

Several investigators have proposed chromosome models involving a chromosome core. For example, Stubblefield and Wray (1971) suggested that a few central core chromatin fibers organize the rest of the chromatin into loops of "epichromatin." Sorsa (1974) has reviewed evidence for chromosome models involving an axial DNA filament.

In the scaffolding model, however, the core is not simply composed of a chromatin fiber or fibers, but is an independent structure, made up of a fibrous network of nonhistone proteins. In this respect, the scaffolding model is more like older models such as that of Taylor (1967).

From the electron micrographs presented here, we cannot determine how much DNA is present in the scaffold. We have, however, demonstrated elsewhere that the scaffold may be isolated as a structurally independent entity by treating the chromosomes with micrococcal nuclease before removing the histones (Adolph et al., 1977b). These scaffolds isolated by sucrose gradient sedimentation contain <0.1% of the chromosomal DNA, and they have a similar appearance to the central structure (scaffold) seen in the electron micrographs presented in this paper. The isolated scaffolds also have the same stability properties and the same protein composition as the histone-depleted chromosomes. These experiments show that the scaffold seen in the center of the histone-depleted chromosomes is a structurally independent entity and biochemically distinct from the rest of the chromosomes.

Our results say nothing about possible DNA links between different chromosomes, since the chromosome isolation procedure (Wray and Stubblefield, 1970) would undoubtedly shear such links if they exist. We also have no information on the supposed double nature of the chromosome core (compare Stubblefield and Wray, 1971). The scaffolding model can be easily elaborated to explain uninemy, binemy, chromosome replication and so forth, but we have no experimental data to support such speculations.

In conclusion, the results reported here, together with the results of Adolph et al. (1977a, 1977b), strongly suggest that nonhistone proteins are responsible for the higher order structure of eucaryotic chromosomes, and that these proteins are organized into a structurally independent entity which we call the chromosomal scaffold. Our future work will be directed toward determining which nonhistone proteins are structural elements of the chromosomal scaffold, how they are arranged and what role they play during the rest of the cell cycle.

Experimental Procedures

Cell Culture and Isolation of Metaphase Chromosomes

HeLa S_3 cells were grown in suspension, labeled with ^{3}H-thymidine and blocked in metaphase with colchicine as described by Adolph et al. (1977a). Metaphase chromosomes were then prepared in chromosome isolation buffer [0.1 mM PIPES (piperazine-N,N′-bis-2-ethane sulfonic acid) (pH 6.5), 0.5 mM $CaCl_2$, 1 M hexylene glycol] by the method of Wray and Stubblefield (1970). Chromosomes were used within 1 hr for preparation of histone-depleted chromosomes. Only preparations judged by phase-contrast microscopy to have well preserved chromosome morphologies were used.

Isolation of Histone-Depleted Metaphase Chromosomes
Histone-depleted chromosomes were routinely prepared for electron microscopy as follows. Between 5 and 25 μl of chromosomes containing 2–10 μg of DNA were mixed by gentle rolling with 1 ml of a solution containing 0.2 mg/ml sodium dextran sulfate 500 (Pharmacia, lot number 263), 0.02 mg/ml heparin (Sigma, number H-3125), 10 mM EDTA, 10 mM Tris-HCl (pH 9.0), 0.1% Nonidet P-40 and 1 mM PMSF (phenylmethylsulfonyl fluoride). Immediately after mixing, this solution was carefully layered onto the top of a step gradient using a plastic pipette from which the tip had been cut to have an internal diameter of 2 mm.

The step gradients consisted of three layers beneath the dextran sulfate layer: 1 ml of 2.5% sucrose, 10 mM EDTA, 10 mM Tris-HCl (pH 9.0), 0.1% NP-40 and 1 mM PMSF; 3 ml of 5% sucrose in gradient buffer [10 mM EDTA, 10 mM Tris-HCl (pH 9.0), 0.1 M NaCl], and 0.3 ml of 0.6 M Metrizamide (Nyegaard and Company, Oslo) in gradient buffer.

Following incubation in the dextran sulfate solution for 30 min at 4°C, chromosomes were sedimented to the Metrizamide cushion by centrifugation for 5 min at 500 × g and then 90 min at 3000 X g. Fractions were collected from the top of the tube using a wide bore plastic pipette, and part of each fraction was counted to verify the position of the ^3H-thymidine-labeled DNA. Generally, an 0.2 ml sample was obtained which had a DNA concentration of 10–50 μg/ml.

Preparation of Grids for Electron Microscopy
The following method of making Parlodion-coated grids gives films which are free of holes and strong even on 100 or 200 mesh grids. Parlodion (3.5% in amyl acetate) was stored over Linde molecular sieve (Type 4A) to eliminate moisture. Grids were placed on a stainless steel screen covered by 1 cm of water in a Buchner funnel, and a film was created by touching a drop of Parlodion solution to the water surface. The film was allowed to dry until wrinkles formed at the edges and until the center of the film turned gold and then clear. The film was then lowered onto the grids and screen by draining the water from the funnel. The screen was removed from the funnel, blotted from underneath to remove adhering droplets of water, and immediately placed in a dry oven at 60°C for 3 hr or overnight. Grids were stored in a dessicator and used within 2–3 days. Cytochrome c monolayers were picked up using the side of the Parlodion film which had faced the air.

Spreading Chromosomes with Cytochrome c
Two methods were used, which will be referred to as the ammonium acetate method and the carbonate method. Special care was taken in handling the histone-depleted chromosomes. Mixing operations were performed by gently rolling in a volume of at least 100 μl, and pipetting operations were always done using plastic pipettes from which the tips had been cut to give an internal diameter of 1–2 mm. Considerable care was also taken in the cleanliness of solutions used for spreading. For all purposes relating to electron microscopy, we used only high-purity water (Hydro, Inc., Durham, North Carolina). All containers were thoroughly rinsed with this water before use, and all stock solutions were filtered through 0.2 μm Nalgene filters. The teflon block was cleaned with soap and water and thoroughly rinsed with ethanol, acetone and water before use. Glass slides for ramps were cleaned with acid, soap and water, ethanol and acetone, rinsed with water and stored in water until just before use when they were rinsed with the hypophase solution. In the ammonium acetate method, sterile plastic 35 X 10 mm petri dishes were used for hypophases. We found that cytochrome c solutions could be stored for several months at 4°C, but care had to be taken to avoid contaminating them with any dust. For this reason, pipette tips were thoroughly rinsed with water before inserting into the stock cytochrome solution.

The ammonium acetate method was a modification of the procedure of Kavenoff and Ryder (1976). Chromosomes were diluted as desired with gradient buffer [10 mM EDTA, 10 mM Tris-HCl (pH 9.0), 0.1 M NaCl], and to a 200 μl sample we added 20 μl of cytochrome c solution [1 mg/ml in 5 M ammonium acetate (pH 6.5)]. After gentle mixing, about 50 μl of this sample was spread on a hypophase of 0.125–0.5 M ammonium acetate (pH 6.5) using a clean microscope slide as a ramp. The cytochrome monolayer was sampled beginning 1 min after spreading. Spreading was performed at room temperature.

The carbonate method was a modification of the method of Inman and Schnös (1970) worked out by M. L. Wong in this laboratory. To a 50 μl sample of histone-depleted chromosomes (diluted as desired in gradient buffer), we added consecutively: 50 μl freshly prepared carbonate buffer (0.4 ml H_2O, 40 μl 1 M Na_2CO_3, 32 μl 0.126 M Na_2 EDTA), 100 μl formamide (Matheson, Coleman and Bell), and 30 μl cytochrome c (1 mg/ml in H_2O). After each addition, the sample was thoroughly mixed by gentle rolling. About 10 μl were spread by the droplet method on 0.4 ml water in the teflon tray described by Inman and Schnös (1970). In this method, a 10 μl droplet was formed using a segment of thin-walled teflon tubing attached to a Hamilton syringe, and the droplet was lightly touched to the surface of the hypophase. No ramp was used. Spreading was carried out at 4°C and samples were picked up 1–10 min after spreading.

In both methods, samples were spread within 2 hr after removal from the gradients, unless otherwise noted.

Contrast Enhancement
Immediately after samples had been picked up on Parlodion-coated grids, they were rinsed for 10 sec in 90% ethanol, stained for 20 sec in a fresh 250 fold dilution into 90% ethanol of the stock stain solution, rinsed for 10 sec in 2-methyl butane and air-dried. Finally, samples were rotary-shadowed with platinum from an angle of 8°.

The stock solution of stain was 0.05 F uranyl acetate in 0.05 N HCl (Davis and Davidson, 1968) which can be stored in the dark at 4°C for up to a month. An aliquot of this stock solution was filtered through an 0.3 μ Millipore filter shortly before use.

Electron Microscopy
Samples were viewed in a Phillips EM300, photographed at original magnifications from 2000–10,000X on Kodak Electron Microscope Film (no. 4489) developed in D19.

Acknowledgments

We are grateful to M. L. Wong and R. Kavenoff for technical advice, and to W. R. Baumbach and S. M. Cheng for preparing metaphase chromosomes. This research was supported by grants from the NSF and from the USPHS. J.R.P. is a Fellow in Cancer Research of the Damon Runyon-Walter Winchell Fund.

The costs of publication of this article were defrayed in part by the payment of page charges. This article must therefore be hereby marked "*advertisement*" in accordance with 18 U.S.C. Section 1734 solely to indicate this fact.

Received July 22, 1977; revised August 31, 1977

References

Abuelo, J. G. and Moore, D. E. (1969). J. Cell Biol. *41*, 73–90.

Adolph, K. W., Cheng, S. M. and Laemmli, U. K. (1977a). Cell *12*, 805–816.

Adolph, K. W., Cheng, S. M., Paulson, J. R. and Laemmli, U. K. (1977b). Proc. Nat. Acad. Sci. USA, in press.

Benyajati, C. and Worcel, A. (1976). Cell *9*, 393–407.

Cell
828

Chow, L. T., Scott, J. M. and Broker, T. R. (1975). Electron Microscopy of Nucleic Acids (Cold Spring Harbor, New York: Cold Spring Harbor Laboratory).

Cook, I. and Brazell, P. (1975). J. Cell Sci. *19,* 261-279.

Daskal, Y., Mace, M. L., Wray, W. and Busch, H. (1976). Exp. Cell Res. *100,* 204-212.

Davis, R. W. and Davidson, N. (1968). Proc. Nat. Acad. Sci. USA *60,* 243-250.

DuPraw, E. J. (1970). DNA and Chromosomes (New York: Holt, Rinehart and Winston).

Finch, J. T. and Klug, A. (1976). Proc. Nat. Acad. Sci. USA *73,* 1897-1901.

Hozier, J., Renz, M. and Nells, P. (1977). Cold Spring Harbor Symp. Quant. Biol. *42,* in press.

Inman, R. B. and Schnös, M. (1970). J. Mol. Biol. *49,* 93-98.

Kavenoff, R. and Bowen, B. C. (1976). Chromosoma *59,* 89-101.

Kavenoff, R. and Ryder, O. (1976). Chromosoma *55,* 13-25.

Laemmli, U. K., Cheng, S. M., Adolph, K. W., Paulson, J. R., Brown, J. R. and Baumbach, W. R. (1977). Cold Spring Harbor Symp. Quant. Biol., in press.

Lang, D. (1969). J. Mol. Biol. *46,* 209.

Lang, D. (1973). J. Mol. Biol. *78,* 247-254.

Oudet, P., Gross-Bellard, M. and Chambon, P. (1975). Cell *4,* 281-300.

Rattner, J. B., Branch, A. and Hamkalo, B. A. (1975). Chromosoma *52,* 329-338.

Sorsa, V. (1973). Hereditas *75,* 101-108.

Sorsa, V. (1974). Hereditas *79,* 109-116.

Spelsberg, T. C. and Hnilica, L. S. (1971). Biochim. Biophys. Acta *228,* 202-211.

Stubblefield, E. and Wray, W. (1971). Chromosoma *32,* 262-294.

Taylor, J. H. (1967). In Molecular Genetics, Part II, J. H. Taylor, ed. (New York, London: Academic Press) pp. 95-135.

Wray, W. and Stubblefield, E. (1970). Exp. Cell Res. *59,* 469-478.

Wray, W., Stubblefield, E. and Humphrey, R. (1972). Nature New Biol. *238,* 237-239.

Clarke L. and **Carbon J.** 1980. Isolation of a yeast centromere and construction of functional small circular chromosomes. *Nature* **287:** 504–509.

It has been known since Flemming's original observations in the late 19th century that a specific point on each chromosome makes contact with the spindle during mitosis. This point, the centromere, was later defined genetically as the region of the chromosome that undergoes segregation at the first meiotic division. Using techniques of gene cloning introduced in the 1970s Clarke and Carbon constructed circular plasmids that contained a yeast chromosomal replicator (*ars* sequence) and DNA from the centromere region of yeast chromosome III. One of their plasmids, which contained only 1.6 kb of yeast DNA, was transmitted stably through both mitotic and meiotic divisions, indicating that it retained centromere activity. Later sequence analysis of centromere III and of the comparable region of chromosome XI revealed that both centromeres contained an extremely A+T-rich core segment 87-88 bp in length flanked by two short sequences (14 bp and 11 bp) that were identical in both DNAs (1). The centromeres of *Saccharomyces* seem to represent a minimal condition. Centromeres from other organisms, including the fission yeast *Schizosaccharomyces*, are significantly larger and are flanked by long stretches of highly repetitive DNA. The original centromere plasmids were all circular. With the discovery of telomere sequences, it became possible to construct linear chromosomes that contained a centromere, two telomeres, and additional DNA of any desired sort (2, 3). Such yeast artificial chromosomes or YACs have proved useful for studying the behavior of exogenous sequences in yeast and for propagating pieces of DNA that are too large to maintain in bacteria. Identification of centromere DNA led to the discovery and isolation of centromere-binding proteins, which are important for centromere function (4). Associated with the centromeres of most organisms is a complex structure known as the kinetochore (5; Nicklas and Koch, this volume), whose interaction with spindle microtubules regulates chromosome movement during mitosis.

---

1. Fitzgerald-Hayes M., Clarke L., and Carbon J. 1982. Nucleotide sequence comparisons and functional analysis of yeast centromere DNAs. *Cell* **29:** 235–244.
2. Dani G. M. and Zakian V. A. 1983. Mitotic and meiotic stability of linear plasmids in yeast. *Proc. Natl. Acad. Sci.* **80:** 3406–3410.
3. Murray A.W. and Szostak J.W. 1983. Construction of artificial chromosomes in yeast. *Nature* **305:** 189–193.
4. Lechner J. and Carbon J. 1991. A 240 kd multisubunit protein complex, CBF3, is a major component of the budding yeast centromere. *Cell* **64:** 717–725.
5. Brinkley B.R. and Stubblefield E. 1966. The fine structure of the kinetochore of a mammalian cell in vitro. *Chromosoma* **19:** 28–43.

Reprinted from Nature, Vol 287, No. 5782, pp. 504-509, October 9 1980

Isolation of a yeast centromere and construction of functional small circular chromosomes

Louise Clarke & John Carbon

Department of Biological Sciences, University of California, Santa Barbara, California 93106

The centromeric DNA (CEN3) *from yeast chromosome III has been isolated on a* 1.6 *kilobase-pair segment of DNA located near the centromere-linked* CDC10 *locus of* Saccharomyces cerevisiae. *When present on a plasmid carrying a yeast chromosomal replicator,* CEN3 *enables that plasmid to function as a chromosome both mitotically and meiotically. Minichromosomes containing* CEN3 *are stable in mitosis and segregate as ordinary yeast chromosomes in the first and second meiotic divisions.*

In an attempt to isolate and characterize yeast centromeric DNA, the techniques of overlap hybridization[1], complementation of mutations in yeast through transformation[2,3], immunological screening[4,5], and complementation of auxotrophic mutations in *Escherichia coli*[6] have been used in our laboratory to obtain a number of hybrid plasmids containing DNA segments from around the centromere-linked *leu2, cdc10,* and *pgk* loci on chromosome III of *Saccharomyces cerevisiae.* A contiguous region of about 25 kilobase pairs of DNA that extends from the *LEU2* gene to the *CDC10* gene has been obtained on a set of overlapping plasmids[1,3]. Among these is the plasmid pYe(*CDC10*)*1*, which contains the yeast *CDC10* gene included on an 8 kilobase-pair segment of yeast DNA[3]. This plasmid was selected by complementation of the *trp1* and temperature-sensitive *cdc10* lesions in strain XSB52-23C from a pool of hybrid plasmid DNAs composed of segments of sheared yeast DNA joined by poly (dA · dT) connectors into the *E. coli*-yeast shuttle vector, pLC544 (ref. 3). This vector contains the yeast *TRP1* gene along with the chromosomal replicator, *ars1* (ref. 7).

Unlike other *ars1* bearing plasmids described by Stinchcomb *et al.*[7], Kingsman *et al.*[8], and below, pYe(*CDC10*)1 is relatively stably maintained through mitotic and meiotic cell divisions in yeast. We have investigated the nature of this stability both biochemically by localization and preliminary sizing of the stabilizing segment of DNA, and genetically by examining the behaviour of pYe(*CDC10*)1 and its derivatives in mitosis, meiosis, sporulation and germination. The stabilizing segment is confined to a 1.6-kilobase pair region on the left (*LEU2*) side of the *CDC10* locus. Its presence on plasmids that also carry a yeast chromosomal replicator [*ars1* (ref. 7) or *ars2* (ref. 9)] permits those plasmids to behave mitotically and meiotically as minichromosomes. Genetic markers on a minichromosome act as linked markers. They segregate in the first meiotic division as centromere-linked genes, are thus present in the two sister progeny of the second meiotic division, and are unlinked to genetic markers on other chromosomes. We conclude that the stabilizing DNA segment from pYe(*CDC10*)1 is a functional centromeric DNA sequence, the centromere of chromosome III (*CEN3*).

Plasmid pYe(*CDC10*)1 is mitotically stable

We have reported previously that in cultures of individual XSB52-23C/pYe(*CDC10*)1 transformants growing at 37 °C in the absence of tryptophan (selective conditions) only about 1–2 cells in 500 examined microscopically have lost the plasmid and express the easily distinguished $Cdc10^-$ phenotype[3]. Using overlap hybridization[1], a number of other plasmids have now been identified from our pLC544(*TRP1 ars1*)-yeast DNA genomic library that carry DNA segments from the *cdc10* region of chromosome III (Fig. 1). One of these, pYe98F4T, carries the *CDC10* locus because it complements the *cdc10* mutation in strain XSB52-23C, and it also contains additional DNA in the direction of *PGK*, but lacks about 3.5 kilobase pairs of DNA in the *LEU2* direction that is found on pYe(*CDC10*)1. Microscopic examination of cultures of individual XSB52-23C/pYe98F4T transformants (growing selectively) reveal that approximately one-half of the cells express the mutant Cdc10 phenotype, indicating extensive plasmid loss and instability. The degree of mitotic stability of pYe(*CDC10*)1, pYe98F4T, and two other *TRP1 ars1* plasmids containing DNA from near the *CDC10* locus, pYe65H3T and pYe101C3T (Fig. 1), was further examined by growing individual cultures of XSB52-23C transformants containing each of these plasmids on rich medium at 23 °C (nonselective conditions). After approximately 20–30 generations of nonselective growth, individual clones of the original transformants were scored for loss of plasmid (loss of the Trp^+ phenotype). As seen in Table 1, 97% of the clones derived from transformants harbouring pYe(*CDC10*)1 retained the plasmid after this extensive period of nonselective growth. Plasmids pYe98F4T, pYe65H3T, and pYe101C3T are completely segregated, however, in the same conditions. The parent vector in the above plasmids, pLC544 (ref. 8), other *TRP1 ars1* plasmids such as YRp7 and YRp7–Sc2604 (ref. 7), plasmids containing the chromosomal replicator (*ars2*; ref. 9), and many plasmids containing 2 μm DNA replicators[10] are also mitotically unstable.

Plasmid pYe(*CDC10*)1 is meiotically stable and segregates as a chromosome

Because stability in mitosis is a predicted property of a replicating unit carrying a functional centromere, the unique mitotic stability of pYe(*CDC10*)1 encouraged us to examine the behaviour of this plasmid through meiosis and sporulation. Strain XSB5223Cα/pYe(*CDC10*)1(*cdc10 leu2 trp1*/*CDC10 TRP1*) was crossed with X2928-30-1*Aa*(*trp1 leu1 ade1 met14*), diploids were sporulated and the resulting asci were dissected for genetic analysis. Data from 16 tetrads (cross 1, Table 2) indicate that the plasmid (marked by the wild-type *TRP1* allele) in at least 60% of the asci segregates in the first meiotic division as a chromosome and is thus found in the two sister spores, the products of the second meiotic division. Only parental ditype and nonparental ditype asci were obtained in this cross using as

2

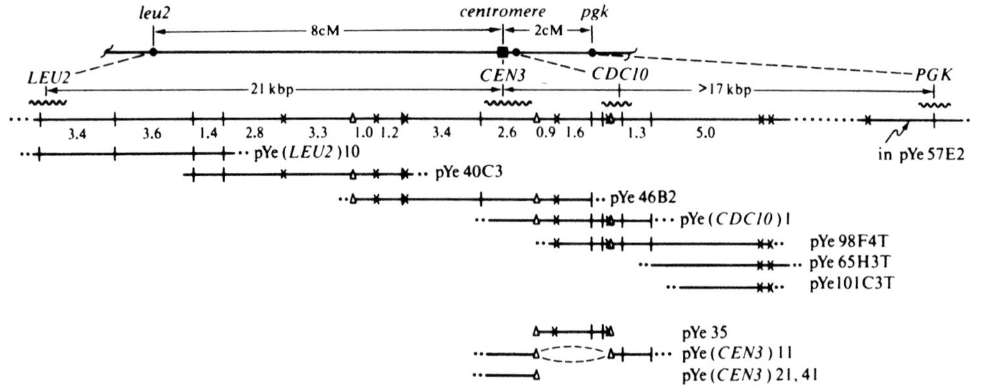

Fig. 1 Genetic and physical maps of the centromere region of *S. cerevisiae* chromosome III. The restriction map gives the location of *Eco*RI (—+—), *Hin*dIII (—✕—), and *Bam*HI (-△-) sites in the DNA. Numbers denote kilobase pairs (kbp). The order of inserts in various overlapping plasmids is indicated below the restriction map. Small dots at the ends of inserts pertain to those sheared segments of yeast DNA that are joined to their respective plasmid vectors by poly (dA · dT) connectors. The isolation and characterization of plasmids pYe(*LEU2*)10 (ref. 6), pYe40C3 (ref. 1), pYe46B2 (ref. 1), pYe(*CDC10*)1 (ref. 3), and pYe57E2 (ref. 5), have been described. Plasmids pYe98F4T, pYe65H3T, and pYe101C3T were obtained by overlap hybridization and characterized as described by Chinault and Carbon[1], with the modification that a single restriction fragment (from pYe(*CDC10*)1) purified by agarose gel electrophoresis was used as probe in the hybridizations[20]. Plasmids pYe(*CDC10*)1, pYe98F4T, pYe65H3T, pYe101C3T, pYe35, and pYe(*CEN3*)11 all contain the *TRP1 ars1* vector, pLC544 (ref. 8). The vector portion of the remaining plasmids is ColE1.

references the centromere markers *leu1*, *leu2*, *met14*, and *ade1* (complete data not shown). Thus the *TRP1* locus on the plasmid behaves as a centromere-linked marker. The *TRP1* gene on pYe(*CDC10*)1 is unlinked to *ade1*, *cdc10*, *leu1*, *leu2*, or *met14* and is therefore not integrated on chromosomes I, III, VII or XI. The plasmid was completely lost in about 30% of the asci and in one ascus was found in all four spores, but did not segregate either 1+ : 3− or 3+ : 1− in any asci analysed in this cross.

A second cross was carried out with one of the progeny of cross 1, SB17*Aa*/pYe(*CDC10*)1(*cdc10 trp1 leu2 ade1*/*CDC10 TRP1*), and strain 6204-18A*α* (*cdc10 leu2*). Data similar to that from Cross 1 were obtained (cross 2, Table 2). The plasmid in cross 2 again segregates 2+ : 2− in the first meiotic division in 92% of the asci and is found in all four spores in one ascus. With *ade1* as the reference centromere marker, again the marker scored on the plasmid (*CDC10* in this cross) is centromere-linked. The plasmid marker, *CDC10*, is unlinked to chromosome IV, because in those tetrads where *CDC10* segregated 2+ : 2−, *TRP1* (wild-type allele on both the minichromosome and one parental chromosome) segregated both 2+ : 2− and 4+ : 0−.

Backcrosses of plasmid-bearing progeny from cross 1 with the original parents XSB52-23C and X2928-3D-1A gave the same pattern of 2+ : 2− segregation of the wild-type plasmid marker that is independent of the segregation of markers on other chromosomes. In all the above crosses, the two markers on pYe(*CDC10*)1, *CDC10* and *TRP1*, were linked in every ascus scored (data not shown). None of the asci obtained consisted of two viable and two nonviable sister spores. Thus the minichromosome does not appear to pair with any of the other yeast chromosomes.

The centromeric DNA (*CEN3*) is located on the left (*LEU2*) side of the *CDC10* locus

The data presented above indicate that plasmid pYe(*CDC10*)1 contains functional centromeric DNA (*CEN3*) from chromosome III. Making use of the set of overlapping plasmids (Fig. 1) and various reclones from the *CDC10* region, we have attempted to determine the location of *CEN3*. In order to locate the *CDC10* gene on pYe(*CDC10*)1 we had previously constructed two reclones of this plasmid[3]. The first, pYe(*CEN3*)11 (see Fig. 1; referred to as pYe(*CDC10*)1-1 in ref. 3) contains all the yeast DNA in pYe(*CDC10*)1 cloned into pLC544, except the 3.5-kilobase pair *Bam*HI fragment from the middle of the insert. The second, pYe35 (referred to as pYe(*CDC10*)1-2 in ref. 3) contains only this 3.5-kilobase pair *Bam*HI restriction fragment and the vector pLC544 (Fig. 1). Neither plasmid complements the *cdc10* mutation in yeast[3], which indicates that one of the two *Bam*HI sites on pYe(*CDC10*)1 is probably close to or within the *CDC10* gene. Both pYe(*CEN3*)11 and pYe35 DNAs transform Trp^- yeast to Trp^+ with high efficiency[3], since

Table 1 Mitotic stability in yeast of various plasmids and minichromosomes containing DNA from the centromere region of chromosome III

	No. of individual transformants tested		Average per cent of Trp^+ or Leu^+ transformants remaining after nonselective growth	
Plasmid	Strain XSB52-23C	Strain RH218	Strain XSB52-23C	Strain RH218
pYe(*CDC10*)1	5	ND	97	ND
pYe(*CEN3*)11	5	3	95	83
pYe(*CEN3*)41	2	ND	91	ND
pYe35	5	5	<1	7
pYe98F4T	5	ND	<1	ND
pYe65H3T	2	ND	<1	ND
pYe101C3T	2	ND	<1	ND
pYe(*CEN3*)21	4	ND	<1	ND
pYe(*ars1–ars2*)1	ND	8	ND	5

Yeast strains XSB52-23C*a*(*cdc10 gal leu2-3 leu2-112 trp1*) and RH218*a* (*CUP1 gal2 mal SVC trp1*) were transformed with the above plasmid DNAs and in each case several individual transformants were isolated and grown nonselectively overnight on YPD (rich) medium. Each culture was subsequently streaked for single colonies on YPD agar, allowed to grow for 2 days at 23 °C (XSB52-23C) or 32 °C(RH218) and replica-plated onto minimal agar medium with or without tryptophan or leucine, depending on the marker carried by the plasmid. Between 40 and 100 individual clones of each original transformant were scored for plasmid loss after growth at 23 °C (XSB52–23C) or 32 °C (RH218). Preparations of media and transformation of yeast were carried out as described by Hsiao and Carbon[9]. ND, not determined.

3

they contain a *TRP1 ars1* vector. However, only one, pYe(*CEN3*)11, yields transformants that are all mitotically stable for the Trp$^+$ phenotype (Table 1). When a mating was carried out between XSB52-23C/pYe(*CEN3*)11(*trp1 cdc10*/*TRP1*) and X2928-3D-1A(*trp1 ade1*), data similar to that for crosses 1–4 were obtained (Table 2). In 60% of the asci 2+:2− segregation was observed for the *TRP1* marker on pYe(*CEN3*)11. The minichromosome was lost entirely in 27% of the asci and in 13% was found in all four spores. Again only parental and non-parental ditype asci were obtained with respect to centromere reference markers, indicating centromere linkage of *TRP1*. No linkage of the minichromosome to chromosomes I or III was observed, since pYe(*CEN3*)11 segregated independently of the *ade1* and *cdc10* markers.

Because *trp1* yeast transformants harbouring pYc98F4T are all unstable for the Cdc10$^+$ (and Trp$^+$) phenotype, and those carrying pYe(*CEN3*)11 are all stable for Trp$^+$ (Table 1), the stabilizing or centromeric DNA (*CEN3*) is most probably contained in pYe(*CEN3*)11 (and pYe(*CDC10*)1) on the 1.6-kilobase pair segment which extends from the *Bam*HI site to the end of the insert in the direction of the *leu2* locus (Fig. 1). To test this hypothesis, we recloned the 1.6-kilobase pair yeast DNA segment into the plasmid, pGT12(*LEU2 ars1*)[11]. The DNA segment was excised from pYe(*CEN3*)11 DNA with the enzymes *Bam*HI and *Hin*dIII. This treatment yields a single *Bam*HI–*Hin*dIII restriction fragment whose *Bam*HI site is derived from the single *Bam*HI site in pYe(*CEN3*)11 and whose *Hin*dIII site is derived from a *Hin*dIII site in the vector (pLC544), located about 400 base pairs from the end of the insert of yeast DNA. The 1.6-kilobase pair segment of interest, which is thus included on a 2.0-kilobase pair *Bam*HI–*Hin*dIII fragment, was then inserted into plasmid pGT12, also cut with a combination of *Bam*HI and *Hin*dIII.

Plasmid pGT12 was constructed by inserting a 1.4-kilobase pair *Eco*RI restriction fragment containing *TRP1* and *ars1* (ref. 7) and a 2.2-kilobase pair *Pst*I fragment containing the *LEU2* gene[11] into the single *Eco*RI and *Pst*I sites of pBR322 (ref. 11). This plasmid also carries resistance to tetracycline. When the 2.0-kilobase pair *CEN3* fragment is inserted into pGT12 cut with a combination of *Bam*HI and *Hin*dIII, both Tet^R and *TRP1* expression are destroyed, but the *ars1* chromosomal replicator is left intact. The resulting plasmid, pYe(*CEN3*)41 (Fig. 2), thus carries both *CEN3* and the replicator *ars1* in addition to the *LEU2* marker.

Plasmid pYe(*CEN3*)41 transforms strain XSB52-23C(*leu2-3 leu2-112*) to Leu$^+$ with high frequency and is stable in the transformants (Table 1). Its behaviour in meiosis (cross 6, Table 2) is essentially the same as that described for pYe(*CDC10*)1 and pYe(*CEN3*)11 in crosses 1–5. Thus the presence of the 1.6-kilobase pair segment (*CEN3*) on a plasmid carrying a yeast chromosomal replicator is all that is required for the plasmid to behave as a chromosome in mitosis and meiosis in the majority of asci analysed.

CEN3 minichromosomes are not integrated into the yeast genome

Genetic data in crosses 1-6 confirm random assortment of *CEN3* minichromosomes with respect to chromosomes I, III, IV, and XI. The possibility that *CEN3*-containing plasmids are integrated near the centromere of one or several other yeast chromosomes was tested both biochemically and genetically. First a mating was performed between two minichromosome-bearing progeny (cross 7, Table 2). In the majority of asci in cross 7, the minichromosomes (both *TRP1*) show the 2+:2− segregation pattern, indicating that the minichromosomes frequently go to the same pole in the first meiotic division and thus do not pair with each other all the time, if at all. This segregation pattern also confirms that pYe(*CDC10*)1 is not integrated into any other single chromosome in the parents in this cross. If it were, all daughter spores should contain plasmid, because the chromosome pair into which the plasmid is integrated would duplicate in meiosis I and go to opposite poles. This point is again confirmed in cross 8 where the minichromosomes are individually marked. In at least two asci both minichromosomes went to the same pole in meiosis I (Tables 2 and 3) and were therefore both found in the same two sister spores.

In addition, biochemical evidence shown in Fig. 3 demonstrates that pYe(*CEN3*)41 is not integrated anywhere in the yeast genome. Crude yeast DNA was prepared from two germinated sister spores that contained the minichromosome

Table 2 Meiotic segregation of the minichromosomes

Genetic cross no.	Minichromosome in cross	Minichromosome marker scored	Distribution in tetrads of genetic marker on minichromosome (%): 4+:0−	3+:1−	2+:2−	1+:3−	0+:4−	Test for centromere linkage of marker on minichromosome: PD	NPD	T	Reference centromere marker (chromosome)
1	pYe(*CDC10*)1	*TRP1*	1 (6%)	0	10 (63%)	0	5 (31%)	2	8	0	*met14*(XI)
2	pYe(*CDC10*)1	*CDC10*	1 (8%)	0	11 (92%)	0	0	2	8	1	*ade1*(I)
3	pYe(*CDC10*)1	*TRP1*	1 (7%)	0	11 (79%)	0	2 (14%)	4	7	0	*met14*(XI)
4	pYe(*CDC10*)1	*TRP1*	4 (21%)	0	11 (58%)	0	4 (21%)	ND	ND	ND	—
5	pYe(*CEN3*)11	*TRP1*	2 (13%)	0	9 (60%)	0	4 (27%)	4	5	0	*ade1*(I)
								5	4	0	*cdc10*(III)
6	pYe(*CEN3*)41	*LEU2*	3 (14%)	3 (14%)	13 (62%)	1 (5%)	1 (5%)	7	6	0	*trp1*(IV)
								7	5	0	*cdc10*(III)
7	pYe(*CDC10*)1 × pYe(*CDC10*)1	*TRP1*	6 (35%)	0	9 (53%)	1 (6%)	1 (6%)	ND	ND	ND	—
8	pYe(*CDC10*)1	*TRP1*	10 (24%)	1 (2%)	31 (74%)	0	0	15	16	0	*met14*(XI)
	pYe(*CEN3*)41	*LEU2*	8 (19%)	2 (5%)	24 (57%)	0	8 (19%)	10	14	0	*met14*(XI)
								17	2	0	*TRP1*(mini)

In all the above crosses the marker used to follow the minichromosome was wild type on the minichromosome and mutant in both parents. The crosses were: 1. XSB52-23Cα/pYe(*CDC10*)1(*cdc10 leu2–3 leu2–112 trp1 gal*/*CDC10 TRP1* with X2928-3D-1A*a*(*ade1 gal1 his2 leu1 met14 trp1 ura3*); 2, SB17A*a*/pYe(*CDC10*)1 (*ade1 cdc10 leu2–3 leu2–112 met14 trp1*/*CDC10 TRP1*) with 6204-18Aα(*cdc10 thr4 leu2–3 leu2–112*); 3, SB1C*a*/pYe(*CDC10*)1(*his2 leu1 met14 trp1*/*CDC10 TRP1*) with XSB52-23Cα (*cdc10 leu2–3 leu2–112 trp1 gal*); 4, SB17Bα/pYe(*CDC10*)1 (*ade1 cdc10 leu2–3 leu2–112 met14 trp1*/*CDC10 TRP1*) with X2928-3D-1A*a*; 5, XSB52-23Cα/pYe(*CEN3*)11 with X2928-3D-1A*a*; 6, XSB52-23Cα/pYe(*CEN3*)41 with X3144-1D*a*(*ade2 arg9 can1 his2 his6 leu2 pet8 trp1*); 7, SB14B*a*/pYe(*CDC10*)1 (*ade1 his2 trp1*/*CDC10 TRP1*) with SB17Bα/pYe(*CDC10*)1(*ade1 cdc10 leu2–3 leu2–112 met14 trp1*/*CDC10 TRP1*); 8, SXB52-23Cα/pYe(*CEN3*)41 with SB17A*a*/pYe(*CDC10*)1. Transformation of yeast in the presence of polyethylene glycol[2] results in a high proportion of diploid transformants (*a*/*a* or α/α)[8]. Diploids are easily recognized by the extremely low spore viability obtained when they are crossed with a haploid, and in this way were eliminated from the group of haploid transformants used in the above crosses. PD, parental ditype; NPD, nonparental ditype; T, tetratype; ND, not determined.

4

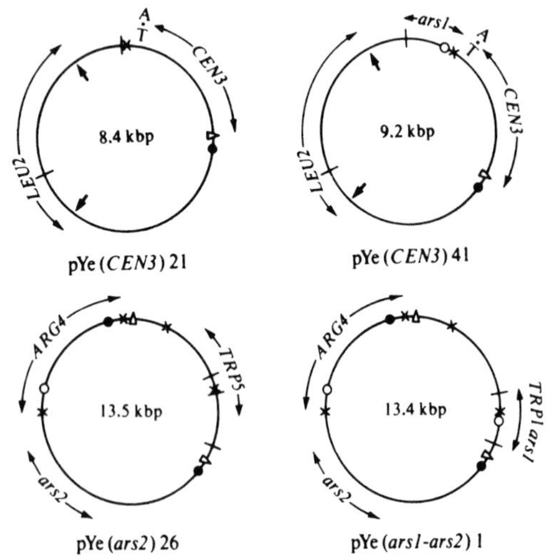

Fig. 2 Physical maps of various plasmid DNAs. The maps show the location of *Eco*RI (—|—), *Hin*dIII (—✕—), *Bam*HI (-△-), *Pst*I (—┬—), *Sal*I (-●-) and *Bgl*II (-○-) sites in the DNAs and indicate approximate locations of pertinent replicators and genes. The construction and use of these plasmids is described in the text. Their sizes are indicated in kilobase pairs (kbp).

(*LEU2*) and from one of the two sisters in the same ascus that was phenotypically Leu$^-$. This tetrad, in which the minichromosome pYe(*CEN3*)41 segregated 2+:2−, was obtained from cross 6 (Table 2). Figure 3 shows an autoradiograph of a Southern blot hybridization[12] in which each of the DNA preparations above was cut to completion with *Sal*I, *Bam*HI, or *Bgl*II. There is a single restriction site for each of these enzymes in the vector portion of pYe(*CEN3*)41 or between the vector and the yeast DNA insert (Fig. 2). In Fig. 3, lanes with DNA from minichromosome-containing progeny (lanes *b* and *c*) show a single band of hybridization to the radioactive probe, [^{32}P]pBR322 DNA. This band has the same mobility regardless of the enzyme used to cut the DNAs, and the mobility is identical to that of linear pYe(*CEN3*)41 generated in the presence of a plasmid-free yeast DNA preparation (lanes *d*) with *Bam*HI or *Bgl*II. The absence of any hybridization bands of sizes other than linear pYe(*CEN3*)41 suggests that this minichromosome is not integrated into yeast genomic DNA. We have not yet attempted to integrate any of the pYe(*CEN3*) plasmids into genomic DNA. Such an event would be expected to lead to dicentric chromosomes and probable mitotic instability[13].

Functional chromosomes require both centromeric DNA and a chromosomal replicator

All the minichromosomes studied and described above contain both *CEN3* and the chromosomal replicator *ars1*. The need for such a replicator with regard to *CEN3* function was tested by inserting the 2.0-kilobase pair *Bam*HI–*Hin*dIII restriction fragment carrying *CEN3* into the plasmid pGT6. Plasmid pGT6 is pBR322 (Tet^r, Amp^r) with a 2.2-kilobase pair fragment carrying the yeast *LEU2* gene inserted into its single *Pst*I site[11]. This plasmid also carries resistance to tetracycline, but not resistance to ampicillin. The 2.0-kilobase pair *Bam*HI–*Hin*dIII *CEN3* fragment was ligated into *Bam*HI–*Hin*dIII cut pGT6 DNA, resulting in the inactivation of Tet^r and yielding the plasmid, pYe(*CEN3*)21, whose structure is shown in Fig. 2.

Somewhat surprisingly, pYe(*CEN3*)21 DNA which contains *CEN3*, but lacks a chromosomal replicator sequence (Fig. 2), also transforms XSB52-23C to Leu+ with a frequency as high as does pYe (*CEN3*)41 DNA. Plasmid pYe(*CEN3*)21 is very unstable mitotically, however (Table 1). When grown under selective conditions in minimal liquid medium without leucine, the doubling time for transformants of pYe(*CEN3*)21 is about 12 hours, as compared to 4 hours for those transformants carrying pYe(*CEN3*)41. In log phase under selective conditions greater than 95% of the cells have segregated pYe(*CEN3*)21, whereas no segregation of pYe(*CEN3*)41 is detected under the same conditions. In general, more than 50% of yeast cells, originally transformed with plasmids carrying chromosomal replicators such as *ars1* (near *TRP1*) or *ars2* (near *ARG4*), lose the plasmid while being grown selectively[7,9]. The high frequency of transformation and high mitotic instability exhibited by pYe(*CEN3*)21 in yeast is perhaps most easily explained by postulating the presence of a weak or partial replication origin on the 1.6-kilobase pair (*CEN3*-containing) segment that permits only enough replication for the plasmid to be transferred to a few daughter cells in the population. It is not known whether this origin plays a role in centromere function.

Plasmids containing two chromosomal replicators are mitotically unstable

Since pYe(*CEN3*)21 DNA does transform yeast with high efficiency in the absence of a chromosomal replicator and seems to have a weak origin of replication of its own, we tested the possibility that any two chromosomal replicators on a plasmid might give mitotic stability to that plasmid in yeast. This notion has been addressed previously by Stinchcomb *et al.*[7], who looked at the behaviour of multimeric forms of *ars1* and found that yeast cultures transformed with monomer, dimer, and trimer *YRp7*(*TRP1 ars1*) were equally unstable mitotically and had identical growth rates.

We have put the *ars1* replicator together on a plasmid with *ars2* (a chromosomal replicator near *ARG4*)[9] in the following way. The plasmid pYe(*ars2*)26 (Fig. 2) contains the yeast *ARG4* and *TRP5* genes and the *ars2* replicator (C. L. Hsiao and J.C., manuscript in preparation). Plasmid pYe(*ars1–ars2*)1 was constructed by treating pYe(*ars2*)26 DNA with *Eco*RI endonuclease and inserting the 1.4-kilobase pair fragment carrying *Trp1* and *ars1* (ref. 7) into the plasmid DNA, thereby partially deleting the *TRP5* gene and replacing it with *TRP1* and ars*1* (Fig. 2). When *trp1* yeast was transformed to Trp$^+$ with pYe(*ars1–ars2*)1 DNA, transformants were as unstable mitotically for the Trp$^+$ phenotype as those transformed with pYe35 which carries only *ars1* (Table 1). In the same experiment, pYe(*CEN3*)11 showed less than 18% loss of the Trp$^+$ phenotype. In addition, the *ars1* replicator segment is in opposite orientations with respect to *CEN3* in pYe(*CDC10*)1 and pYe(*CEN3*)41, and both minichromosomes are equally stable. We conclude that two chromosomal replicators together do not convey mitotic stability upon a plasmid and that *CEN3* is unique

Table 3 Genetic analysis of mating between two haploid parents each carrying an individually marked minichromosome (cross 8)

Distribution in tetrads of markers on minichromosomes: *TRP1*(pYe(*CDC10*)1)	*LEU2*(pYe(*CEN3*)41)	No. of tetrads showing segregation pattern (% of total)
4+:0−	4+:0−	3 (7%)
4+:0−	2+:2−	4(10%)
4+:0−	0+:4−	3 (7%)
3+:1−	2+:2−	1 (2%)
2+:2−	4+:0−	5(12%)
2+:2−	3+:1−	2 (5%)
2+:2−	2+:2−	19(45%)
2+:2−	0+:4−	5(12%)
	Total no. of asci analysed	42(100%)

Strains in cross 8 are described in Table 2, where the meiotic segregation patterns of the minichromosomes are summarized. Among the 19 asci where both minichromosomes segregated 2+:2−, both went to the same pole in the first meiotic division in 2 asci and to opposite poles in 17 asci.

in its stabilizing effect. This conclusion is further verified by the observation that *CEN3*, when isolated on a 5.1-kilobase pair *Eco*RI fragment derived from plasmid pYe46B2 (Fig. 1) and inserted into the *Eco*RI sites at the *TRP5* locus on pYe(*ars2*)26 (Fig. 2), mitotically stabilizes this *ARG4 ars2* plasmid for the Arg^+ phenotype (C. L. Hsiao and J.C., unpublished result).

Minichromosomes do not always pair in the first meiotic division

The absence of asci with two viable and two nonviable sister spores in crosses 1–6 indicates that the minichromosomes do not pair at the first meiotic division with any of the 17 yeast chromosomes. Cross 8 (Tables 2 and 3) was performed with two strains, each harbouring a *CEN3 ars1* plasmid marked with either *LEU2* or *TRP1*. Both parents in cross 8 have mutations at both *leu2* and *trp1*, therefore the behaviour of each minichromosome can easily be followed individually. A detailed analysis of this cross is given in Table 3, from which the following conclusions can be drawn. Each plasmid individually exhibited the 2+ : 2− segregation pattern in more than 60% of the tetrads examined. In the 19 asci where both plasmids segregated 2+ : 2−, the minichromosomes went to opposite poles in the first meiotic division in 17 asci, but they did assort to the same pole in 2 asci. These numbers support the notion that the minichromosomes do not always pair, but there is a strong preference for them to go to opposite poles in the first division. Otherwise, they seem to segregate randomly and independently with respect to one another, and all combinations of 4+ : 0−, 2+ : 2−, and 0+ : 4− are seen (Table 3). For example, both minichromosomes were found in all four spores in three tetrads; in five tetrads pYe(*CEN3*)41 was found in all four spores, but pYe(*CDC10*)1 segregated 2+ : 2−; in three asci pYe(*CDC10*)1 was in all four spores, but pYe(*CEN3*)41 was completely lost. Thus two minichromosomes together in the same sporulating diploid show the same meiotic behaviour as either one exhibits alone.

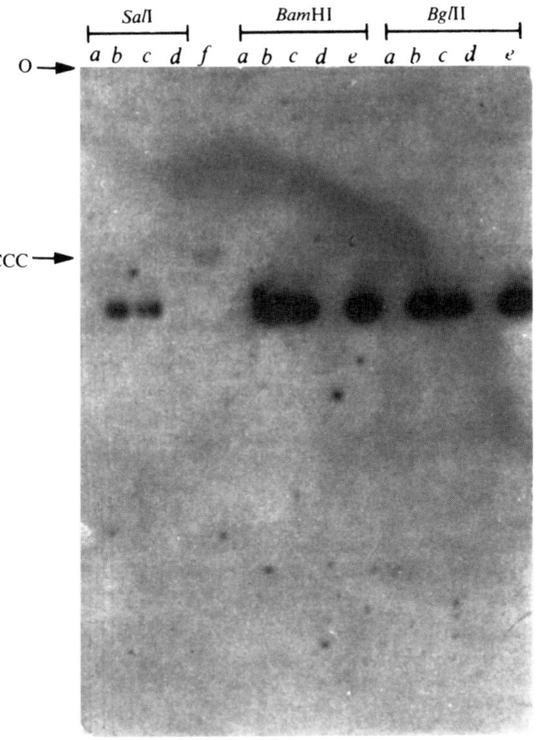

Fig. 3 Southern blot hybridization autoradiograph of total yeast DNAs from pYe(*CEN3*)41-containing strains with [^{32}P]pBR322 DNA as probe. Crude yeast DNA was prepared as described by Stinchcomb *et al.*[7] from one Leu^- progeny (lanes *a*) and two Leu^+ progeny (lanes *b* and *c*) of a single 2+ : 2− ascus in cross 6 and from a Leu^- progeny of a 3+ : 1− ascus in the same cross (lanes *d*). Approximately 2 μg of each DNA preparation was digested with *Bam*HI, *Sal*I, or *Bgl*II and electrophoresed in 0.75% agarose. Other lanes on the gel contained 2 ng of pYe(*CEN3*)41 DNA mixed with 2 μg of crude Leu^- yeast DNA, either undigested (lane *f*) or digested (lanes *e*) with restriction enzyme. Blot hybridization was carried out according to Southern[12], with the following modifications. The fragments were transferred from the gel to nitrocellulose as described by Wahl *et al.*[21] and hybridization conditions were those of Cameron *et al.*[22]. The probe was pBR322 DNA labelled *in vitro* by nick translation[23] as described by Chinault and Carbon[1]. The origin (O) and position of covalently-closed circular pYe(*CEN3*)41 DNA (CCC) are indicated on the autoradiograph.

Discussion

We have identified a small DNA element (*CEN3*) from *S. cerevisiae* chromosome III that when present on a plasmid with a chromosomal replicator (*ars1* or *ars2*) allows the plasmid to behave as a functional chromosome in yeast. Minichromosomes containing *CEN3* are stable in mitosis, and in most cases segregate as chromosomes in the first meiotic division and are thus found in the two sister progeny of the second meiotic division. The *CEN3* element is therefore capable of controlling copy number, thereby ensuring proper meiotic segregation. In approximately 15% of the tetrads analysed in crosses involving *CEN3* plasmids, the minichromosome is lost completely. Generally, in yeast strains that are aneuploid for one or more chromosomes, the extra chromosomes are also somewhat unstable[14]. Tetrads in which a *CEN3* minichromosome segregates 1+ : 3− are rare, therefore the minichromosome is more stable through the second meiotic division than in the first.

In approximately 20% of the tetrads analysed, the minichromosome is found in all four spores after meiosis. This may be a result of selective diploidization of the *CEN3* plasmid or could indicate a loss of control of minichromosome replication. Inability of minichromosomes to pair may cause a certain instability in their attachment or involvement in the meiotic apparatus. Thus control over plasmid replication may be eliminated in some cases resulting in multiple copies that are distributed to all four spores. The 4+ : 0− tetrads are not a separate class, since the same 4+ : 0−, 2+ : 2−, 0+ : 4− patterns are exhibited by the progeny after further matings. Moreover, data from cross 8 (Tables 2 and 3) show that one minichromosome can segregate 2+ :2− while another in the same meiotic event is distributed 4+ : 0−.

The genetic data show that minichromosomes do not pair with any of the normal yeast chromosomes, since asci containing two non-viable sister spores are not seen. However, two minichromosomes in the same diploid exhibit a strong preference to migrate to opposite poles in the first meiotic division. To determine exactly how much DNA, or which regions of DNA are involved in pairing of chromosomes at the first meiotic division, increasingly larger segments of DNA may be added *in vitro* to either or both ends of the *CEN3* segment on a plasmid. Pairing could then be assayed by genetic analysis.

Because plasmid pYe98F4T complements the *cdc10* mutation in yeast and plasmids pYe(*CEN3*)11 and pYe35 do not (Fig. 1), we have concluded that the *CDC10* gene is probably located in the region of the single *Bam*HI site in pYe98F4T. The *cdc10* locus had previously been placed directly next to the centromere of chromosome III (refs 15–17). The data presented here indicate that *cdc10* is located about 3 kilobase pairs from the centromere on the right arm of chromosome III. There are approximately 25 kilobase pairs of DNA between the *LEU2* and *CDC10* genes[3]. The genetic map distance between *leu2* and centromere is about 8 cM (refs 3, 16 and 17). Thus there is an average of about 3 kilobase pairs cM^{-1} from *leu2* across the centromere to *cdc10*. This number is close to the 2.7 kilobase pairs CM^{-1} determined by Strathern *et al.*[18] for very large distances on the circular chromosome III, indicating that the centromere does not appreciably distort recombination frequencies in the *leu2–cdc10* region.

We have shown previously that within the limits of a standard Southern blot hybridization, plasmid pYe(*CDC10*)1 contains only unique DNA[3]. By this same criterion, all the DNA between *leu2* and *cdc10* is unique DNA[1]. Work in progress is expected to define further the size of the centromeric DNA segment and its nucleotide sequence. Whether centromeric DNA sequences are unique for each chromosome may eventually be determined by sequencing other centromeres obtained in the same way.

The small circular chromosomes we have constructed around the *CEN3* element have many properties that mimic those of the much larger parental chromosomes. They are stable both in mitosis and through the first and second divisions of meiosis, probably because *CEN3* is providing an attachment site for spindle structures, a site that may be unique for each chromosome. A chromosomal replicator must accompany the *CEN3* segment to permit proper centromere function, but the replicator alone, or a combination of chromosomal replicators, do not stabilize a plasmid. Therefore the minichromosomes, like the larger chromosomes, require both an attachment site and one or more replicators. Unlike the parental chromosomes, the small circular chromosomes do not always pair, but, as is the case in sporulating aneuploid yeast strains, absence of pairing does not prevent the chromosomes from segregating in meiosis. Although we have no evidence that the weak replicator in the *CEN3* region is involved in centromere function, it is intriguing to speculate that it may be a specialized replicator for that region, perhaps serving to drive sister chromotids apart in a final burst of DNA synthesis, as suggested by DuPraw[19].

The *CEN3* minichromosomes are excellent probes for the study of events in mitosis and meiosis, particularly with regard to the protein–nucleic acid interactions occurring between the centromere and the mitotic and meiotic spindle structures. Minichromosomes with a functional centromere will be useful as well for studies of chromatin structure and the organization of DNA-associated proteins in the centromere region.

Finally, the mitotic stability of *CEN3* plasmids make them especially suitable cloning vehicles. Studies of expression of cloned genes on minichromosomes would not be complicated by the variable mitotic stability displayed by other vectors constructed for cloning in yeast. The functional centromeric sequence, *CEN3*, is not altered by propagation in *E. coli*, since all the *CEN3* minichromosomes were originally isolated in that bacterium before being put into yeast, and the absence of any gross alterations in DNA sequence in cloned segments from around the *CEN3* region has been previously confirmed by Southern blot hybridizations to restriction digests of total yeast genomic DNA[1,3]. It will also be of considerable interest to determine if the yeast centromere will function when introduced into other cell types on the appropriate self-replicating DNA vectors.

We thank Dr Terrance G. Cooper for many helpful discussions about the genetic data, Dr Robert K. Mortimer for introducing us to the intricacies of yeast genetics and for constructing strain XSB52-23C, Robert Gimlich for technical assistance and Drs Craig Chinault, Ronald Hitzeman, and Alan Kingsman for isolating many of the overlapping plasmids around centromere-linked loci on chromosome III. This research was supported by a grant (CA-11034) from the NCI, NIH.

Received 19 May; accepted 1 August 1980.

1. Chinault, A. C. & Carbon, J. *Gene* **5**, 111–126 (1979).
2. Hinnen, A., Hicks, J. B. & Fink, G. B. *Proc. natn. Acad. Sci. U.S.A.* **75**, 1929–1933 (1978).
3. Clarke, L. & Carbon, J. *Proc. natn. Acad. Sci. U.S.A.* **77**, 2173–2177 (1980).
4. Clarke, L., Hitzeman, R. A. & Carbon, J. *Meth. Enzym.* **68**, 436–442 (1979).
5. Hitzeman, R. A., Chinault, A. C., Kingsman, A. J. & Carbon, J., *ICN–UCLA Symposia on Molecular and Cellular Biology* Vol. 14, (eds Maniatis, T. & Fox, C. F.) 57–68 (Academic, New York, 1979).
6. Ratzkin, B. & Carbon, J. *Proc. natn. Acad. Sci. U.S.A.* **74**, 487–491 (1977).
7. Stinchcomb, D. T., Struhl, K. & Davis, R. W. *Nature* **282**, 39–43 (1979).
8. Kingsman, A., Clarke, L., Mortimer, R. K. & Carbon, J. *Gene* **7**, 141–152 (1979).
9. Hsiao, C.-L. & Carbon, J. *Proc. natn. Acad. Sci. U.S.A.* **76**, 3829–3833 (1979).
10. Gerbaud, C. *et al. Gene* **5**, 233–253 (1979).
11. Tschumper, G. & Carbon, J. *Gene* **10**, 157–166 (1980).
12. Southern, E. M. *J. molec. Biol.* **98**, 503–517 (1975).
13. McClintock, B. *Proc. natn. Acad. Sci. U.S.A.* **25**, 405–416 (1939).
14. Parry, E. M. & Cox, B. S. *Genet. Res.* **16**, 333–340 (1971).
15. Hartwell, L. H., Mortimer, R. K., Culotti, J. & Culotti, M. *Genetics* **74**, 267–286 (1973).
16. Mortimer, R. K. & Hawthorne, D. C. *Meth. Cell Biol.* **11**, 221–233 (1975).
17. Culbertson, M. R., Charnas, L., Johnson, M. T. & Fink, G. R. *Genetics* **96**, 745–764 (1977).
18. Strathern, J. N., Newton, C. S., Herskowitz, I. & Hicks, J. B. *Cell* **18**, 309–319 (1979).
19. DuPraw, E. J. *DNA and Chromosomes* (Holt, Rinehart and Winston, New York, 1970).
20. Tabak, H. F. & Flavell, R. A. *Nucleic Acids Res.* **5**, 2321–2332 (1978).
21. Wahl, G. M., Stern, M. & Stark, G. R. *Proc. natn. Acad. Sci. U.S.A.* **76**, 3683–3687 (1979).
22. Cameron, J. R., Loh, E. Y. & Davis, R. W. *Cell* **16**, 739–751 (1979).
23. Rigby, P. W., Dieckmann, M., Rhodes, C. & Berg, P. *J. molec. Biol.* **113**, 237–251 (1977).

Printed in Great Britain by Macmillan Production Ltd., Basingstoke, Hampshire

Greider C.W. and **Blackburn E.H.** 1989. A telomeric sequence in the RNA of *Tetrahymena* telomerase required for telomere repeat synthesis. *Nature* **337**: 331-337.

Cytogenetic experiments on *Drosophila* and maize in the 1930s and 1940s by Hermann Muller and Barbara McClintock showed that the ends of broken chromosomes can fuse with each other, whereas normal ends, which they called telomeres, lack this "sticky"characteristic. Later theoretical considerations suggested that telomeres could not be replicated by the known polymerases, none of which could start at the very end of a DNA molecule. The molecular characterization of telomeres and a solution of the replication problem came in several stages. First, the sequence at the ends of chromosomes was established by Elizabeth Blackburn from an unlikely source, the extrachromosomal genes coding for ribosomal RNA in the ciliated protozoan *Tetrahymena* (1). At first it was not realized that this sequence, the repeated hexanucleotide TTGGGG/AACCCC, had any universal significance, but later experiments showed that most eukaryotic organisms have similar TG/AC rich repeats at their telomeres. A few years later Blackburn and her student Carol Greider discovered an enzymatic activity in extracts of *Tetrahymena* that could add telomeric repeats onto telomeric sequence primers in a non-templated manner (2). They went on to show that this activity, which they named telomerase, was comprised of RNA and protein, both of which were necessary for the addition of telomeric sequences (3). In the study reproduced here, they finally showed that the 159-nucleotide RNA component of telomerase contains the sequence CAACCCCAA , which they suggested was the actual template for the synthesis of TTGGGG repeats. Their study identified a remarkable ribonucleoprotein enzyme and suggested that the synthesis of DNA using an RNA template may be an ancient process.

---

1. Blackburn E.H. and Gall J.G. 1978. A tandemly repeated sequence at the termini of the extrachromosomal ribosomal RNA genes in *Tetrahymena*. *J. Mol. Biol.* **120:** 33–53.
2. Greider C.W. and Blackburn E.H. 1985. Identification of a specific telomere terminal transferase activity in *Tetrahymena* extracts. *Cell* **43:** 405–413.
3. Greider C.W. and Blackburn E.H. 1987. The telomere terminal transferase of *Tetrahymena* is a ribonucleoprotein enzyme with two kinds of primer specificity. *Cell* **51:** 887–898.

Reprinted from Nature, Vol. 337, No. 6205, pp. 331-337, 26th January, 1989

A telomeric sequence in the RNA of *Tetrahymena* telomerase required for telomere repeat synthesis

Carol W. Greider* & Elizabeth H. Blackburn†

* Cold Spring Harbor Laboratory, PO Box 100, Cold Spring Harbor, New York 11724, USA
† Department of Molecular Biology, University of California, Berkeley, California 94720, USA

The telomerase enzyme of Tetrahymena *synthesizes repeats of the telomeric DNA sequence TTGGGG* de novo *in the absence of added template. The essential RNA component of this ribonucleoprotein enzyme has now been cloned and found to contain the sequence CAACCCCAA, which seems to be the template for the synthesis of TTGGGG repeats.*

NUCLEIC acid synthesis reactions are usually template directed. Exceptions include the reactions of terminal deoxynucleotidyl transferase, which adds any deoxynucleotide in a random order onto the 3′ end of a polynucleotide chain[1], poly(A) polymerase which adds rAMP residues to the 3′ end of RNA[2], and tRNA nucleotidyl transferase which adds the sequence CCA to the 3′ end of tRNAs[3]. In addition, we have discovered the enzyme telomerase which synthesizes tandem repeats of the telomeric DNA sequence TTGGGG to the 3′ end of telomeric primers without any added template[4,5].

Telomeres from such diverse organisms as ciliates, yeasts, plants, and mammals are made up of a variable number of tandemly repeated, simple (G+C)-rich sequences[6-12]. In all telomeres examined, there is a strand-specific sequence bias such that the G-rich strand always runs 5′ to 3′ towards the end of the chromosome. As the number of tandem repeats on the end of any given chromosome is variable[13,14], restriction fragments containing telomeres have a characteristic 'fuzzy' appearance in agarose gel electrophoresis. We have proposed that this length heterogeneity arises from a dynamic equilibrium established between the incomplete replication and/or degradation of the chromosome end and the *de novo* synthesis of telomeric sequences by telomerase[4,12-14].

The *Tetrahymena* telomerase extends the free TTGGGG-OH 3′ telomere strand by the addition of further TTGGGG repeats. Both the repeated sequence motif and the orientation of the G strand have been conserved throughout evolution, suggesting that telomerase-like enzymes may be found in all eukaryotes. The presence of telomerase would explain several observations about telomeres. First, when *Tetrahymena* or *Oxytricha* telomeric sequences are introduced into yeast, telomeric sequences characteristic of yeast are added to them[12,15,16]. Second, when trypanosomes or *Tetrahymena* are kept in continuous log-phase growth, there is a net increase in telomere length[13,17]. In *Tetrahymena* this increase involves all of the telomeres in the cell and is due to the addition of telomeric TTGGGG repeats[13,14]. Finally, telomeric sequences from a number of different organisms are maintained as perfect tandem repeats and do not diverge through recombination, unlike most simple repeated sequences in the nuclear genome[18-20] and mitochondrial telomeres[21,22]. This sequence maintenance can be explained by the continual synthesis of new perfect telomeric repeats.

Recently, telomerase activities have been isolated from the nuclei of both the hypotrichous ciliates *Oxytricha* and *Euplotes* (ref. 23; D. Shippen-Lentz and E. Blackburn, submitted). These enzymes synthesize repeats of the telomeric sequences of those organisms, TTTTGGGG, in an analogous way to the *Tetrahymena* telomerase. Together with the conserved structure of telomeres, these results further indicate that telomerase enzymes exist throughout eukaryotes.

We showed previously that the telomerase of *Tetrahymena* is a ribonucleoprotein (RNP) enzyme with essential RNA and protein components[5]. Because the reactions catalysed by many RNPs involve base pairing of the RNA components[24-28], the requirement of an RNA component in the *Tetrahymena* telomerase suggested that the synthesis of TTGGGG repeats might be specified by the RNA[5]. We report here the identification of the essential RNA component of telomerase and show that this RNA contains the sequence CAACCCCAA which could serve as a template for the synthesis of TTGGGG repeats.

Cloning and characterization of 159 RNA

We previously showed that small RNAs of approximately 159 bases and 80 bases co-purified with the telomerase RNP over five column purification steps[5]. Further purification of the telomerase over several other column series indicated that the RNA of about 159 bases (159 RNA) was the only one to reproducibly co-purify with the enzyme activity (data not shown). We directly sequenced the 159 RNA using both sequence-specific RNases and dideoxynucleotide primer extension techniques. Southern blot analysis of *Tetrahymena* macronuclear DNA, using oligonucleotide probes derived from the RNA sequence, indicated that there was a single gene for the 159 RNA in the macronuclear genome. A 2 kilobase (kb) *Hin*dIII fragment containing the gene was cloned from macronuclear DNA. The sequence of a 400-base region of the clone containing the 159 RNA coding sequence is shown in Fig. 1*a*. The most notable feature of the gene is the presence of the sequence 5′ CAACCCCAA 3′ from positions 43 to 51 within the coding region. This sequence could serve as a template for the addition of TTGGGG repeats *in vitro*.

Three ribonucleotide residues at positions 46, 47 and 48 in the 159 RNA were not cut by any of the ribonucleases used in the direct RNA sequencing, indicating that these bases might be modified, although primer extension by reverse transcriptase was unimpeded at these positions. Several different base modifications in tRNAs are known to interfere with ribonuclease cleavage[29]. Ribose 2′-*O* methylation inhibits both RNase and base hydrolysis of RNA by blocking the formation of the cyclic phosphate intermediate[30]. As the base hydrolysis ladder was complete at these positions in the telomerase RNA (data not shown), this type of modification can be ruled out. Both secondary structure modelling of the telomerase RNA (I. Tinoco, personal communication) and data presented below indicate that this region of the RNA is not in a tightly base-paired structure; hence structural interference was probably not responsible for the inability of RNases to cut at these positions. Although the nature of these possible base modifications is not known, it is intriguing that they lie within the CAACCCCAA sequence.

Several features of the 159 RNA gene suggest that it is transcribed by RNA polymerase III. First, the length of the 159 RNA is heterogeneous. Using high-resolution polyacrylamide gel electrophoresis we separated a cluster of four bands and sequenced the two prominent species in the middle. These two species differed only in the number of U residues at the 3′ end. In the sequence of the cloned gene this 3′ end of the RNA falls

2

```
a    Xba I          Pst I         Hind III                    -161
ATCCTCTAGA GTCGACCTGC AGGCATGCAA GCTTATAAAT AATAAATAAT

                                                         -111
AAATATATTT TATATAATTT AATTTAAATA TCTATCTAAA GTGTTGTAAA

                                                          -61
AAAAATATAT ATATATCTTG CTCGTCTTCT ATTATATTTT CTTAAACAGA

                             Ssp I                        -11
CCCATAAAAA TTCGCTTCCA AATATTTTCC ACTTTAAGTA GAAATTATTT

              +1                      Bgl II      Box A    40
TAAGACTTTT ATACCCGCTT AATTCATTCA GATCTGTAAT AGAACTGTCA

                                                           90
TTCAACCCCA AAAATCTAGT GCTGATATAA CCTTCACCAA TTAGGTTCAA

                                                          140
ATAAGTGGTA ATGCGGGACA AAAGACTATC GACATTTGAT ACACTATTTA

      Fok I                                               190
TCAATGGATG TCTTATTTTT TTTATCTATT TCGATTTGTA TTCTATATTT

ATGAGACATT GATCCTATTA AAA
```

```
b                      Oligo 7
AUACCCGCUUAAUUCAUUCAGAUCUGUAAUAGAACUGUCAUU
                       Oligo 8
        Oligo 3
CAACCCCAAAAAUCUAGUGCUGAUAUAACCUUCACCAAUUAG
                       Oligo 6
GUUCAAAUAAGUGGUAAUGCGGGACAAAAGACUAUCGACAUU
              Oligo 10
UGAUACACUAUUUAUCAAUGGAUGUCUUAUUUUU
```

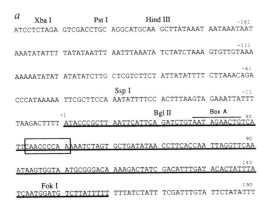

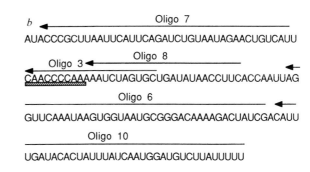

Fig. 1 *a*, Sequence of the telomerase RNA gene. The first 30 nucleotides of the pUC119 polylinker up to the *Hin*dIII site used for cloning are shown for orientation along with 370 bases of the non-template strand of the telomerase RNA gene. The region corresponding to the RNA is underlined. The start site of transcription as determined by dideoxy primer extension sequencing is labelled +1 and the 3′ end of the RNA as determined by direct RNA sequencing is shown within the run of eight T residues near position 160. All of the six-base recognition restriction enzymes which cut in the *Tetrahymena* DNA are shown above their recognition sites. A region of similarity with the consensus Box A for polymerase III genes is shown at positions 28–37. The sequence CAACCCCAA within the RNA gene which may provide the template for the addition of TTGGGG sequences *in vitro* is shown in the box at position 43. *b*, The oligonucleotides (oligos) used for cloning, sequencing and RNase H experiments are shown above the telomerase RNA regions to which they hybridize. The region of the RNA containing the potential template is underlined with a hatched bar. Oligonucleotides 1–6 were made using the sequence information obtained by partial enzymatic RNase sequencing. Oligonucleotides 1, 2, 4 and 5 were shown to contain significant mismatches with the correct sequence obtained by DNA sequence analysis. Oligonucleotides 7, 8 and 10 were synthesized using DNA sequence information. Only the oligonucleotides that were shown to hybridize to the 159 RNA are shown. The sequence and name of all the oligonucleotides used throughout this paper are listed from 5′ to 3′. Oligo 2, TTCTGTTGCAAAATCTGAATGAAT. Oligo 3, GCACTAGATTTTTGGGGTTG. Oligo 5, TTGTTTGAACCTGATT-GTGAAGGTT. Oligo 6, CGATGGTCTTTTGTCCCGCATTGCCACTTGTTTGAACCT. Oligo 7, ATGACAGTTCTATTACAGATCTGAAT-GAATGAATTAAGCGGGT. Oligo 8, GAAGGTTATCAGCACTAGATTT. Oligo 10, AAAAATAAGACATCCATTGATAAATAGT-GTAT-CAAATG.

Methods. Telomerase S100 preparation (60 ml) was fractionated over a sizing column, heparin agarose, and spermine agarose[5]. An aliquot of the active fraction (30 ml) was digested with proteinase K, phenol extracted and ethanol precipitated; half of the RNA was 3′ end-labelled with [^{32}P] pCp using RNA ligase and half was treated with calf intestinal phosphatase, phenol extracted and labelled with [γ-^{32}p]ATP using polynucleotide kinase. The labelled RNA was run on an 80 cm 15% acrylamide/7M urea gel for 21 hours at 3,500 V. The two major bands corresponding to the 159 and 160 species of RNA were cut out, eluted from the acrylamide and partial enzymatic RNA sequencing was done. The sequence obtained from the 5′ and 3′ labelled RNAs overlapped by 50 bases. Thus the entire sequence was obtained except for a few bases where the sequence data from both 5′ and 3′ labelled RNA was unreadable possibly due to RNA base modifications (see text). Using this sequence information, oligonucleotide 6 was synthesized and used as a primer for dideoxy primer extension RNA sequencing. Eighty-one nucleotides of sequence were obtained and oligonucleotides were then synthesized and used for cloning of genomic copy of the RNA. *Tetrahymena* macronuclear DNA was cut with *Hin*dIII and size fractionated on a 0.8% agarose gel. The region around 2.0 kb was cut out and electroeluted. This DNA was ligated into *Hin*dIII-cut pUC119 plasmid and transformed into DH5α F′. 30,000 transformants were obtained: of these 10,000 were screened by colony hybridization. Replica plates were made of the transformation plates, and duplicate colony lifts onto nitrocellulose were made. The duplicate filters were hybridized in 6x SSC, 4x Denhardts and 0.1% SDS with either oligonucleotide 3 or oligonucleotide 6 at 55 °C for 15 hours. The filters were then washed three times at room temperature in 3x SSC 0.1% SDS for 10 min, and then in 3x SSC 0.1% SDS at 55 °C for 10 min. One clone was obtained which hybridized to both oligonucleotides. The clone was sequenced using double stranded dideoxy sequencing (USB-sequenase kit) with the reverse universal primer and an oligonucleotide internal to the RNA gene as primers.

in a stretch of eight T residues. These properties are typical of RNA polymerase III transcription termination[31]. Second, we were able to label the 5′ end of RNA with [γ-^{32}P]ATP and polynucleotide kinase after treatment with phosphatase, indicat- 5′ end of the 159 RNA was not capped (data not shown). Third, there is a region of similarity to the box A consensus RRYNNARYGG[32] beginning 30 nucleotides downstream of the 5′ end of the 159 RNA.

d(TTGGGG)$_4$ elongation interference

We used RNase H inactivation to establish whether the 159 RNA was an essential component of telomerase. This technique has been used to identify essential RNAs for a variety of RNPs[27,28,33–36]. Telomerase was preincubated with antisense oligonucleotides (Fig. 1*b*) in the presence or absence of RNase H. After the preincubation, the primer d(TTGGGG)$_4$ oligonucleotide was added along with [α-^{32}P] dGTP, dTTP and reaction buffer, and the ability of telomerase to synthesize TTGGGG repeats was assayed (Fig. 2). Oligonucleotides 2, 5, 6, 7 and 10, which are complementary to the 159 RNA, had no effect on telomerase activity in either the presence or absence of RNase H. However, oligonucleotide 3 strongly inhibited telomerase activity in either the presence or absence of RNase H, and preincubation with oligonucleotide 8 resulted in an upward shift in the characteristic six-base banding pattern, suggesting this antisense oligonucleotide was itself efficiently priming repeat addition (Fig. 2; see lanes 17–20). In the absence of added RNase H the 159 RNA was intact, indicating that endogenous RNase H was not responsible for the inhibition of telomerase by oligonucleotide 3 (Fig. 4 and data not shown). Although in this initial experiment the effects of RNase H addition were not definitive, the observations that oligonucleotides 3 and 8, which hybridize across, and near to, the CAACCCCAA sequence respectively, specifically affected telomerase activity suggested the 159 RNA is involved in telomerase activity.

Northern analysis of RNA extracted from active telomerase fractions probed with oligonucleotides 3 and 8, showed only one band at 159 nucleotides (data not shown). Thus, these oligonucleotides do not share extensive sequence complementarity with other RNAs in the enzyme fractions used throughout this work.

The inhibition of telomerase activity by preincubation with

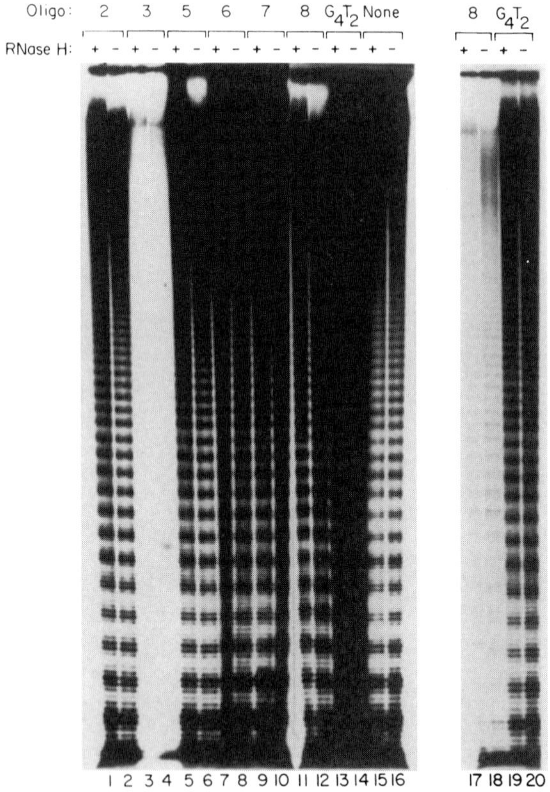

Fig. 2 Treatment of telomerase with antisense oligonucleotides and RNase H. The oligonucleotides indicated were preincubated with telomerase in the presence (+) or absence (−) of RNase H. The ability of telomerase to elongate added d(TTGGGG)$_4$ primer was then assayed. The oligonucleotides added in the preincubation are indicated above the lanes. G_4T_2 is d(TTGGGG)$_4$. Lanes 17–20 show a lighter exposure of lanes 11–14 to indicate better the upward shift in the banding pattern when oligonucleotide 8 is included in the preincubation.

Methods. Telomerase which had been purified over a sizing column and heparin agarose was used[5]. Antisense oligonucleotide (2 μl) at 0.4 μg μl^{-1} was added to 10 μl of telomerase, followed by either 2 μl of RNase H at 2U μl^{-1} (BRL) or 2 μl of TMG + 0.1 M NaCl (TMG, 10 mM Tris *p*H 8.0, 1 mM $MgCl_2$, 10% glycerol) were added. The samples were preincubated at room temperature for 30 min and then 6 μl of TMG + 0.1 M NaCl and 20 μl of 2x reaction mix was added. The 2x reaction mix contained: 0.1 μg d(TTGGGG)$_4$, 200 μM dTTP, 200 mM NaOAc, 100 mM Tris *p*H 8.0, 10 μCi [^{32}P]dGTP at 400 Ci $mmol^{-1}$. Reactions were carried out at 30 °C for 60 min, the mix then phenol extracted, ethanol precipitated and run on a 6% polyacrylamide sequencing gel.

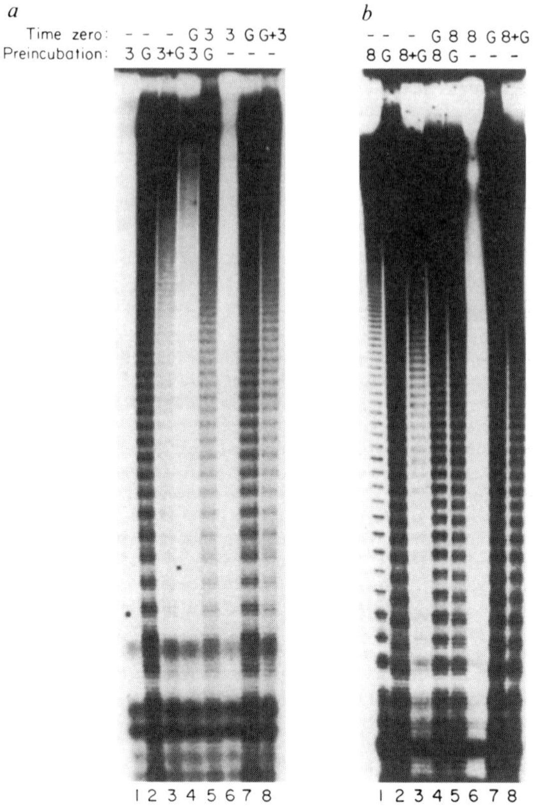

Fig. 3 *a*, Competition between oligonucleotide 3 and d(TTGGGG)$_4$. Telomerase was preincubated at room temperature with the oligonucleotides indicated above the lanes. G is d(TTGGGG)$_4$. At time zero, reaction buffer was added and the reaction was carried out as described in the Fig. 2 legend. The oligonucleotides indicated were added at time zero. *b*, Competition between oligonucleotide 8 and d(TTGGGG)$_4$. The oligonucleotides indicated above the lanes were added in the preincubation. G is d(TTGGGG)$_4$. At time zero the oligonucleotides indicated were added. Darker exposure of the autoradiogram showed a banding pattern in lane 6 similar to that in lane 1.

Methods. 20 μl of telomerase purified as described in the Fig. 2 legend was preincubated with 1 μl of each oligonucleotide at 0.4 μg μl^{-1} or with 1 μl of water as indicated. Preincubation was at room temperature for 30 min. At time zero 20 μl of 2x reaction mix containing 1 μl of the appropriate oligonucleotides or no oligonucleotide was added and the standard activity assay was carried out (see Fig. 2 and ref. 5).

oligonucleotide 3 even in the absence of RNase H suggested that this oligonucleotide might interfere with the access of d(TTGGGG)$_4$ to telomerase. To test for specific inhibition, we reversed the order of addition of the two oligonucleotides. When oligonucleotide 3 was used in the preincubation and the d(TTGGGG)$_4$ was added at the beginning of the elongation reaction (referred to as time zero) there was the same inhibition as in Fig. 2. If d(TTGGGG)$_4$ was added first to the preincubation and then oligonucleotide 3 was added at time zero, there was no inhibition however. When oligonucleotide 3 and d(TTGGGG)$_4$ were added together, an intermediate level of inhibition was seen (Fig. 3*a*). Thus, inhibition by oligonucleotide 3 is a specific effect and its properties are consistent with competition with d(TTGGGG)$_4$.

As oligonucleotide 3 contains the sequence TTGGGGTTG at its 3′ end, its interference with d(TTGGGG)$_4$ elongation could simply be due to its sequence similarity with the d(TTGGGG)$_4$ primer. We also found that oligonucleotide 8, which does not contain any TTGGGG sequence, is itself efficiently elongated by telomerase (Fig. 3*b*). The addition of oligonucleotide 8 alone resulted in the synthesis of a repeat pattern that was shifted upward relative to the pattern primed by d(TTGGGG)$_4$. The repeat pattern visualized by this gel assay is dependent on the sequence and length of the input primer oligonucleotide[5]. Thus, the upward shift in the banding pattern distinguishes the products primed by addition to oligonucleotide 8 from those primed by d(TTGGGG)$_4$. When oligonucleotide 8 was added in the preincubation and d(TTGGGG)$_4$ was added at time zero, the main products had the oligonucleotide 8-primed pattern. In contrast, when the d(TTGGGG)$_4$ oligonucleotide was added

4

first and oligonucleotide 8 was added at time zero, the characteristic d(TTGGGG)$_4$ primed pattern was seen. Simultaneous addition of the two oligonucleotides, either in the preincubation or at time zero, resulted in a mixture of the two patterns indicating that both oligonucleotides were being elongated under these conditions. As oligonucleotide 8 was the only oligonucleotide with a non-telomeric sequence, of 14 tested, that was efficiently elongated by telomerase (see also ref. 5; data not shown), we suggest that the base pairing of oligonucleotide 8 to the 159 RNA adjacent to the CAACCCCAA sequence allows its elongation by telomerase.

Inactivation by cleavage of 159 RNA

The interaction of oligonucleotides 3 and 8 with telomerase *in vitro* suggested that the 159 RNA is a component of telomerase. To investigate further the requirement for this RNA, we sought conditions under which the effects of cleavage of the 159 RNA could be assayed. In RNase H inactivation experiments oligonucleotides 2, 6, 7 and 10, complementary to the 159 RNA, did not inhibit telomerase activity or cause cleavage of the 159 RNA by RNase H (Fig. 2 and data not shown). To assay the effects of oligonucleotide 3 and RNase H, we had first to relieve the inhibition by oligonucleotide 3. We pelleted the telomerase in an ultracentrifuge after preincubation with the test oligonucleotide and RNase H. The enzyme activity of each resuspended pellet was then assayed (Fig. 4*a*) and the RNA examined for evidence of 159 RNA cleavage (Fig. 4*b*). There was no effect on either telomerase activity or the 159 RNA after preincubation with oligonucleotide 7 in either the presence or absence of RNase H. After preincubation and subsequent removal of oligonucleotide 3 in the absence of RNase H, both telomerase activity and the 159 RNA were intact. Preincubation with oligonucleotide 3 in the presence of RNase H however, inhibited telomerase activity and caused cleavage of the 159 RNA. In a separate experiment, we used primer extension to map the sites at which RNase H cuts in the telomerase RNP (Fig. 5). Although in this experiment cutting by RNase H was incomplete, oligonucleotide 3 directed cutting at several positions along its region of complementarity to the 159 RNA, including nucleotides in the CAACCCCAA sequence.

Oligonucleotide 8 also directs RNase H cutting of the 159 RNA; cutting with this oligonucleotide, even when apparently complete, did not however inactivate the telomerase (data not shown). Oligonucleotide 8 cannot base pair in the CAACCCCAA sequence and accordingly all of the cut sites mapped by primer extension are 3′ to the potential template sequence (Fig. 5). Although the ability of telomerase to function after oligonucleotide 8-directed cleavage of the RNA was surprising, it is not unprecedented. The signal recognition particle retains its microsomal translocation activity after micrococcal nuclease digestion which removes up to 150 nucleotides of the 300 nucleotides in the essential 7SL RNA[37,38]. In addition, the 25S rRNA from mature ribosomes of insects is found naturally cleaved into two halves with some sequences missing[39,40]. Furthermore, in *Crithidia*, the 25S RNA is found cleaved into five pieces[41] and in *Chlamydomonas* both the large and small rRNAs are processed into multiple pieces that are held together by intermolecular base pairing[42]. Finally, deletion and replacement analysis of the essential RNA component of the yeast U2 small nuclear RNP has shown that not all regions of this RNA are essential for activity *in vivo*[43]. The specificity of telomerase inactivation by oligonucleotide 3-directed cutting within the CAACCCCAA sequence, compared with the cutting near this sequence directed by oligonucleotide 8, is evidence for the important nature of the CAACCCCAA sequence in telomerase activity.

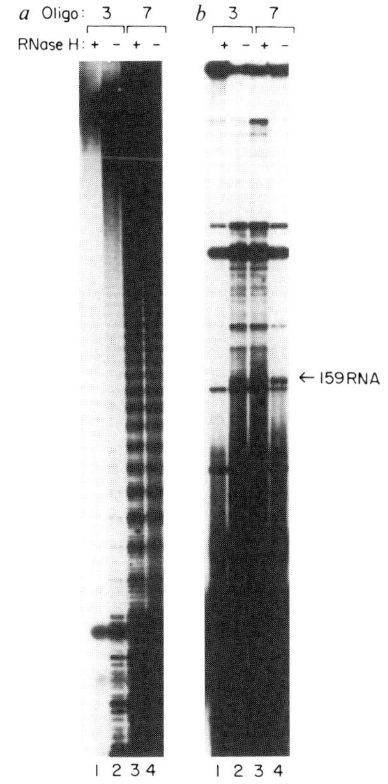

Fig. 4 *a*, The activity of telomerase which had been pretreated with various oligonucleotides in the presence or absence of RNase H and then pelleted in a ultracentrifuge to remove excess oligonucleotide. RNase H was added during the preincubation to the reactions shown in lanes 1 and 3, but not in lanes 2 and 4. The oligonucleotides used in the preincubation are indicated above the lanes. The low level of activity seen in lane 2 relative to lanes 3 and 4 is due to the inhibitory effects of residual oligonucleotide 3 in these fractions. The dark material seen at the top of the gel in lane 1 is due to labelling of high relative molecular mass nucleic acid by some enzyme in the partial purified fraction. More highly purified telomerase fractions do not show this label accumulation. *b*, RNA from the pellets in *a* was extracted, 3′ end-labelled with [^{32}P] pCp and run on a sequencing gel. The lanes are as in *a*. The position of the 159-base telomerase RNA is indicated with an arrow.

Methods. 200 μl of telomerase purified as described in the Fig. 2 legend was preincubated with 6 μl of the appropriate oligonucleotide at 1 μg μl^{-1} in the presence or absence of 5 μl of RNase H at 2U ul^{-1}. Preincubation was at room temperature for 15 min. The extracts were then transferred to Beckman microfuge tubes and spun at 75,000 r.p.m. in a Beckman TLA-100.3 rotor at 4 °C for 3 hours. The supernatant was removed carefully and the pellet was resuspended in 100 μl of TMG+0.1 M NaCl. 20 μl was immediately assayed in the standard activity assay (see Fig. 2 and ref. 5). To the remaining 80 μl, 10 μg of proteinase K was added and the mix was incubated at room temperature for 15 min. The preparation was then phenol-extracted, ethanol-precipitated, 3′ end-labelled with pCp as described[5] and run out on a 6% polyacrylamide/7 M urea sequencing gel.

Discussion

CAACCCCAA sequence may provide a template for TTGGGG repeat synthesis. We have identified and cloned the gene for an essential RNA component of telomerase. Several lines of evidence indicate that the synthesis of TTGGGG repeats is directed by an RNA template. First, the only RNA which reproducibly co-purified over numerous columns contains the sequence CAACCCCAA. Second, oligonucleotide 8 whose 3′ end base pairs just downstream from the CAACCCCAA sequence is efficiently elongated by telomerase, whereas all other non-telomeric sequences are not. Third, oligonucleotide 3, which is complementary to the telomerase RNA sequences up to and including the CAACCCCAA sequence, inhibits telomerase activity. Finally, cleavage by oligonucleotide 3 and RNase H within the CAACCCCAA sequence specifically inactivates telomerase activity. These observations suggest that the CAACCCCAA sequence of the telomerase RNA is in the active site of the RNP and that this sequence provides a template for synthesis of TTGGGG repeats.

Primer elongation by telomerase can result from two modes of primer recognition. Preincubation of telomerase with d(TTGGGG)$_4$ and RNase H did not inhibit enzyme activity (Fig. 2) or result in cleavage of the telomerase RNA, whereas there was RNA cleavage when d(TTGGGG)$_4$ was incubated with naked telomerase RNA and RNase H (data not shown). In contrast, base pairing of both oligonucleotide 3 and oligonucleotide 8 with the telomerase RNA in the intact RNP can direct cleavage by RNase H. Cleavage by RNase H can occur with as few as six base pairs of DNA-RNA duplex[44]. Thus, the base-pairing interaction of oligonucleotides 3 and 8 with the telomerase RNP must be distinct from the way in which d(TTGGGG)$_4$ interacts with the RNP during the preincubation. This difference could be due to an alteration in the RNP structure induced by oligonucleotides 3 and 8, or it may reflect a real distinction between telomerase recognition and elongation. The initial recognition of telomeric primers may not involve the CAACCCCAA sequence of the RNA. This is consistent with our earlier observations that G-rich oligonucleotides with differing primary sequences corresponding to the telomeric sequence of a number of species, as well as synthetic telomere-like sequences, all efficiently prime repeat addition by telomerase[5,45]. Together, these data indicate that a primer can be delivered to the active site in two ways; telomeric G-rich sequence oligonucleotides are specifically recognized, possibly by their unusual secondary structural features[46], whereas some oligonucleotides complementary to the 159 RNA (like 3 and 8) can be positioned in the active site by base hybridization.

The elongation translocation mechanism. When oligonucleotide 8 was preincubated with telomerase and d(TTGGGG)$_4$ was added at time zero, the oligonucleotide 8-primed pattern was seen in the products, indicating that the elongation reaction is processive. This, together with the evidence that hundreds of repeats of TTGGGG are synthesized on the input primer, and that there are one and a half repeats of the template sequence, indicate that an elongation-translocation mechanism operates to synthesize many tandem repeats. Previous work showed that the sequence at the 3′ end of a telomeric primer is recognized by telomerase in the elongation reaction. A primer ending in TTGGGG 3′ is distinguished from those ending in TTGGG 3′, GGGGTT 3′ or TTGG 3′[5,45] and each primer is correctly filled out to complete the last TTGGGG sequence before further repeats are added. These data could be explained by the mechanism outlined in Fig. 6. After primer recognition, the most 3′ nucleotides of the primer are hybridized to the 3′ portion of the CAACCCCAA sequence and the primer is filled out to complete the last full hexanucleotide repeat. Translocation then repositions the primer so that its most 3′ nucleotides are again hybridized ready for elongation.

Such an elongation-translocation mechanism could contribute to the strong six-base periodicity of the *in vitro* addition reaction. Earlier work showed that varying the relative concentrations of dTTP and dGTP in the reaction could change the intensity of the bands in the *in vitro* synthesized banding pattern[4]. This dependance on nucleotide precursor concentrations is also seen for the *Oxytricha* and *Euplotes* enzymes (ref. 23 and D. Shippen-Lentz and E. Blackburn, submitted). Pausing upon the addition of dTTP, as well as primer translocation, may contribute to the six base periodicity in the *in vitro* reaction products. Depending on the nucleotide concentration used, either two or three strong stops are seen in the elongation pattern using (TTGGGG)$_4$ as the primer. The two strong stops correspond to the two T residues in the repeat[4] whereas the three stops correspond to the addition of the nucleotides TTG (data not shown). This pausing pattern is consistent with the model presented in Fig. 6 for elongation and translocation.

Implications of the internal RNA template. The specification of TTGGGG repeat addition by an internal template is analogous to the utilization of the internal binding site as a template in the reactions catalysed by the *Tetrahymena* intron ribozyme. This ribozyme has been shown to catalyse self splicing, ***trans***-splicing, RNA endonuclease and RNA polymerase reactions[47-52]. In all of these reactions a sequence internal to the

5

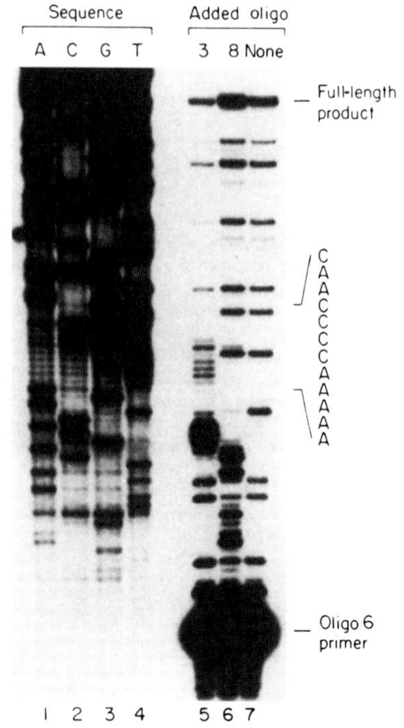

Fig. 5 Primer extension analysis of telomerase RNA cut with RNase H. Lanes 1-4, the sequencing ladder marker using oligonucleotide 6 as a primer. Lanes 5-7, the extension products from [^{32}P] oligonucleotide 6 on RNA extracted from telomerase fractions after treatment with RNase H and the oligonucleotides indicated above the lanes. The bands which appear in all of the lanes including lane 7, where no oligo was added, were probably due to the secondary structure of the 159 RNA. The full-length primer extension product is seen in all of the lanes because the cutting by RNase H in this experiment was not complete. The region corresponding to the CAACCCCAA sequence is indicated.

Methods. 400 μl of telomerase purified over a sizing column and heparin agarose, were incubated with 10 μl of the indicated oligonucleotide and 10 μl of RNase H (2U μl^{-1}) at room temperature for 30 min. 40 μl of a 10% SDS +100 mM EDTA solution and 10 μg of proteinase K were added. The mix was incubated at room temperature for 20 min and then the RNA was phenol extracted and ethanol precipitated after the addition of 5 μg of carrier tRNA. The pellet was resuspended in 6 μl of 125 mM KCl and 5 mM Tris *p*H 8.3. To 3 μl of RNA, 1 μl of [^{32}P] oligo 6 containing 0.05 μg was added and each mix was incubated at 65 °C for 5 min. The annealing reactions were then transferred to 42 °C for 1 hour. To each reaction 6 μl of elongation mix was added. The elongation mix contained 2 μl 5x reverse transcriptase buffer, 1 μl each dNTP at 2.5 mM, 1 μl avian myeloblastosis virus (AMV) reverse transcriptase (Life Sciences), 2 μl DEP-water. The elongation was carried out at 52 °C for 30 min. Sequencing loading dye (4 μl) was then added and 4 μl of each sample was loaded onto a 6% polyacrylamide sequencing gel. The sequencing markers were produced using oligonucleotide 6 as a primer following the procedure for supercoiled DNA sequencing using the Sequenase kit (USB).

6

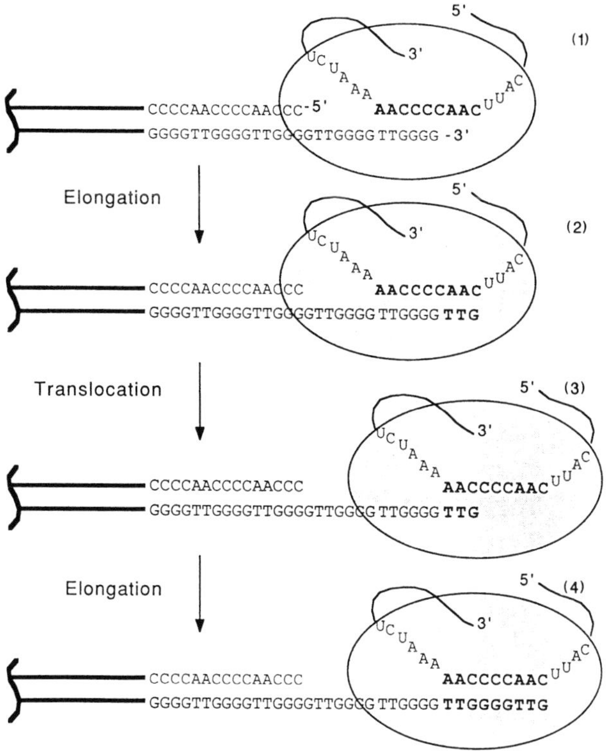

Fig. 6 Model for elongation of telomeres by telomerase. The *Tetrahymena* telomere is shown containing a 13-base overhanging TTGGGG strand[54]. (1) After recognition of the TTGGGG strand by telomerase, the 3′ most nucleotides are hybridized to the CAACCCCAA sequence in the RNA. (2) The sequence TTG is then added one nucleotide at a time[4]. (3) Translocation then repositions the 3′ end of the TTGGGG strand such that the 3′ most TTG nucleotides are hybridized to the RNA component of telomerase. (4) Elongation occurs again, copying the template sequence to complete the TTGGGGTTG sequence. This mechanism explains how oligonucleotides with 3′ ends terminating at any nucleotide within the sequence TTGGGG are correctly elongated to yield perfect tandem repeats of $(TTGGGG)_n$ (ref. 5).

ribozyme interacts with a substrate through Watson–Crick base pairing. Mutations in this internal binding site alter the substrate specificity of the ribozyme[47]. By analogy, it is predicted that changing the sequence of the internal template of telomerase would result in synthesis of telomeric repeats of that altered sequence. Whether the analogy between the telomerase and the intron ribozyme of *Tetrahymena* extends to include catalysis by RNA alone is a question which we can now begin to address.

DNA replication and RNA transcription represent conventional nucleic acid synthesis reactions directed by an external template. Terminal deoxynucleotidyl transferase, poly(A) polymerase and tRNA nucleotidyl transferase represent completely template-independent mechanisms of nucleic acid synthesis. In contrast to both of these types of polymerases, the *Tetrahymena* intron ribozyme and telomerase represent a new class of polymerase which is characterized by containing a template sequence that is an integral component of the polymerase itself. Although many RNPs, as well as the *Tetrahymena* ribozyme, use base pairing of their RNA components with other RNAs to determine reaction specificity[24–28,53], the telomerase enzyme is unique among eukaryotic RNPs in that it uses an RNA template for DNA synthesis.

The synthesis of DNA using an RNA template may be a very ancient process. Telomerase enzymes with RNA components may be relics that have descended from an ancestral type of DNA polymerase whose action was based on an internal RNA template. In this model, the primitive DNA polymerase was replaced in evolution by the current protein-based DNA polymerases and reverse transcriptases which use exogenous templates, whereas telomerase retained the endogenous template and was used by eukaryotes for telomere maintenance.

We thank Bruce Futcher, Bruce Stillman and Kim Arndt for useful discussions and critical reading of the manuscript, and the NIH and Cold Spring Harbor Laboratory for support.

Received 11 November; accepted 8 December 1988.

1. Bollum, F. J. in *The Enzymes* (3rd edn) 145-171 (1974).
2. Jacob, S. T. & Rose, K. M. in *Enzymes of Nucleic Acid Synthesis and Modification* (ed. Jacob, S. T.) 135-157 (CRC, Boca Raton, 1983).
3. Deutscher, M. P. in *Enzymes of Nucleic Acid Synthesis and Modification* (ed. Jacobs, S. T.) 159-177 (CRC, Boca Raton, 1983).
4. Greider, C. W. & Blackburn, E. H. *Cell* **43,** 405-413 (1985).
5. Greider, C. W. & Blackburn, E. H. *Cell* **51,** 887-898 (1987).
6. Blackburn, E. H. *Cell* **37,** 7-8 (1984).
7. Blackburn, E. H. & Szostak, J. W. *A. Rev. Biochem.* **53,** 163-194 (1984).
8. Blackburn, E. H. in *Molecular Developmental Biology* 69-82 (Liss., New York, 1986).
9. Allshire, R. C. *et al. Nature* **332,** 656-659 (1988).
10. Richards, E. R. & Ausubel, F. M. *Cell* **53,** 127-136 (1988).
11. Moyzis, R. K. *et al. Proc. natn. Acad. Sci. U.S.A.* **85,** 6622-6626 (1988).
12. Shampay, J., Szostak, J. W. & Blackburn, E. H. *Nature* **310,** 154-157 (1984).
13. Larson, D. D., Spangler, E. A. & Blackburn, E. H. *Cell* **50,** 477-483 (1987).
14. Henderson, E. *et al.* in *Cancer Cells,* 453-461 (Cold Spring Harbor Laboratory, 1988).
15. Szostak, J. W. & Blackburn, E. H. *Cell* **29,** 245-255 (1982).
16. Pluta, A. F., Dani, G. M., Spear, B. B. & Zakian, V. A. *Proc. natn. Acad. Sci. U.S.A.* **81,** 1475-1479 (1984).
17. Bernards, A., Michels, P. A. M., Lincke, C. R. & Borst, P. *Nature* **303,** 592-597 (1983).
18. Southern, E. M. *Nature* **227,** 794-798 (1970).
19. Lohe, A. R. & Brutlag, D. L. *Proc. natn. Acad. Sci. U.S.A.* **83,** 696-700 (1986).
20. Willard, H. F. & Wayne, J. S. *Trends Genet* **3,** 192-198 (1987).
21. Morin, G. B. & Cech, T. R. *Cell* **46,** 873-883 (1986).
22. Morin, G. B. & Cech, T. R. *Cell* **52,** 367-374 (1988).
23. Zahler, A. M. & Prescott, D. M. *Nucleic Acids Res.* **16,** 6953 (1988).
24. Zhuang, Y. & Weiner, A. M. *Cell* **46,** 827-835 (1986).
25. Parker, R., Silicano, P. G. & Guthrie, C. *Cell* **49,** 229-239 (1987).
26. Brow, D. A. & Guthrie, C. *Nature* **334,** 213-218 (1988).
27. Cotton, M., Glik, O., Vasserot, A., Schaffner, G. & Birnstiel, M. L. *EMBO J.* **7,** 801-808 (1988).
28. Mowry, K. & Steitz, J. A. *Science* **238,** 1682-1687 (1987).
29. Simoncsits, A., Brownlee, G. G., Brown, R. S., Rubin, J. R. & Guilly, H. *Nature* **269,** 833-836 (1977).
30. Anfinsen, C. B. & White, F. H. in *The Enzymes* (eds Boyer, P. D., Lardy, H. & Myrback, K.) 90-99 (Academic, New York, 1961).
31. Bogenhagen, D. F. & Brown, D. F. *Cell* **24,** 261-270 (1981).
32. Cillberto, G., Raugei, G., Costanzo, F., Dente, L. & Cortese, R. *Cell* **32,** 725-733 (1983).
33. Krainer, A. R. & Maniatis, T. *Cell* **42,** 725-736 (1985).
34. Black, D. L., Chabot, B. & Steitz, J. A. *Cell* **42,** 737-750 (1985).
35. Black, D. L. & Steitz, J. A. *Cell* **46,** 697-704 (1986).
36. Chang, D. D. & Clayton, D. A. *Science* **235,** 1178-1184 (1987).
37. Gundelfinger, E. D., Krausse, E., Melli, M. & Dobberstein, B. *Nucleic Acids Res.* **11,** 7363-7374 (1983).
38. Seigel, V. & Walter, P. *Nature* **320,** 81-84 (1986).
39. Ware, V. C., Renkawitz, R. & Gerbi, S. A. *Nucleic Acids Res.* **13,** 3581-3597 (1985).
40. Fujiwara, H. & Ishikawa, H. *Nucleic Acids Res.* **14,** 6393-6401 (1986).
41. Spencer, D. F., Collings, J. C., Schnare, M. N. & Gray, M. W. *EMBO J.* **6,** 1063-1071 (1987).

42. Boer, P. H. & Gray, M. W. *Cell* **55**, 399-411 (1988).
43. Shuster, E. O. & Guthrie, C. *Cell* **55**, 41-48 (1988).
44. Donis-Keller, H. *Nucleic Acids Res.* **7**, 179-192 (1979).
45. Blackburn, E. H. *et al. Genome* (in the press).
46. Henderson, E., Hardin, C., Wolk, S., Tinoco, I. & Blackburn, E. H. *Cell* **51**, 899-908 (1987).
47. Been, M. D. & Cech, T. R. *Cell* **47**, 207-216 (1986).
48. Been, M. D. & Cech, T. R. *Science* **239**, 1412-1416 (1988).
49. Cech, T. R. *Science* **236**, 1532-1539 (1987).
50. Cech, T. R. *Proc. natn. Acad. Sci. U.S.A.* **83**, 4360-4363 (1986).
51. Cech, T. R. & Bass, B. L. *A. Rev. Biochem.* **55**, 599-629 (1986).
52. Zaug, A. J. & Cech, T. R. *Science* **231**, 470-475 (1986).
53. Shine, J. & Dalgarno, L. *Nature* **254**, 34 (1975).
54. Henderson, E. & Blackburn, E. H. *Molec. cell Biol.* (in the press).

Transcription

Miller O.L., Jr. and **Beatty B.R.** 1969. Visualization of nucleolar genes. *Science* **164:** 955–957.

Actively transcribing genes can be visualized in the electron microscope by a simple spreading technique developed by Oscar Miller and his collaborators. Miller took advantage of the amphibian oocyte nucleus (also called the germinal vesicle) in which the genes for ribosomal RNA (rDNA) are amplified more than 1000-fold over their number in a typical somatic nucleus. Several years before molecular cloning was introduced, the rDNA genes were isolated in a biochemically pure form, and molecular analysis had shown that *Xenopus* rDNA consisted of tandem arrays of transcribed sequences separated by non-transcribed spacers (1, 2). Miller's images of these genes illustrated the essential features of their transcription in a dramatic fashion. Each transcribed region displayed about 100 thin fibrils extending laterally from the DNA axis. These fibrils were the nascent RNA chains with associated proteins. They increased in length from the start of the transcription unit, where the polymerase first engaged the DNA template, to the end of the unit, where the nascent chains were released. The large number of fibrils indicated that multiple polymerase molecules were simultaneously active on a single transcription unit. The electron micrograph of rDNA "Christmas trees" that appeared on the cover of Science and is reproduced on the cover of this volume has been used many hundreds of times in textbooks and review articles. It is arguably one of the best known biological images produced in the last 40 years, amply demonstrating that "a picture is worth a thousand words."

---

1. Birnstiel M., Speirs J., Purdom I., Jones K., and Loening U.E. 1968. Properties and composition of the isolated ribosomal DNA satellite of *Xenopus laevis*. *Nature* **219:** 454–463.
2. Brown D.D. and Weber C.S. 1968. Gene linkage by RNA-DNA hybridization. II. Arrangement of the redundant gene sequences for 28s and 18s ribosomal RNA. *J. Mol. Biol.* **34:** 681–697.

Reprinted from
23 May 1969, volume 164, pages 955–957

SCIENCE

Visualization of Nucleolar Genes

O. L. Miller, Jr. and Barbara R. Beatty

Visualization of Nucleolar Genes

Abstract. *The presence of extrachromosomal nucleoli in amphibian oocytes has permitted isolation and electron microscopic observation of the genes coding for ribosomal RNA precursor molecules. Visualization of these genes is possible because many precursor molecules are simultaneously synthesized on each gene. Individual genes are separated by stretches of DNA that apparently are not transcribed at the time of synthesis of precursor rRNA in the extrachromosomal nucleoli.*

During early growth of the amphibian oocyte, the chromosomal nucleolus organizer is multiplied to produce about a thousand extrachromosomal nucleoli within each nucleus (*1*). There is convincing evidence that these nucleoli function similarly to chromosomal nucleoli in the synthesis of rRNA precursor molecules (*2*). In thin sections of fixed oocytes, each extrachromosomal nucleolus typically shows a compact fibrous core surrounded by a granular cortex (Fig. 1). Previous studies have shown that only the core region contains DNA, whereas both components contain RNA and protein (*3*). The large size of the amphibian oocyte nucleus (*4*) allows rapid isolation and manipulation of the extrachromosomal nucleoli before extensive denaturation and cross-linking of proteins occurs. If saline of low molarity or deionized water is used as the isolation medium, nucleolar cores and cortices can be separated and the DNA-containing cores dispersed for electron microscopy (*5*).

Each unwound isolated nucleolar core consists of a thin axial fiber, 100 to 300 Å in diameter, that is periodically coated along its length with matrix material (Figs. 2 and 3). The axial fiber of each core forms a circle, and treatment with deoxyribonuclease breaks the core axes. The diameter of trypsin-treated axial fibers (about 30 Å) suggests that the core axis is a single double-helix DNA molecule coated with protein (*6*). The matrix segments along a core axis exhibit thin to thick gradations, and show similar polarity along the axial fiber. Each unit is separated from its neighbors by matrix-free axis segments.

Nucleolar core axes are stretched to variable degrees depending on preparative procedures. For example, drying preparations out of deionized water before staining causes little or no stretching of axial cores (Fig. 2), whereas precipitating preparations with acetone staining solution before drying stretches the core axes to varible degrees over the grid surface (Fig. 3). When regions of core axes appear unstretched or uniformly stretched, the matrix units along a specific region are similar in length; unstretched matrix units are 2 to 2.5 μ long but can be 5 μ long after severe stretching. The matrix-free segments between matrix units also show variations in length due to stretching, but, in addition, exhibit differences in length independent of stretching (Fig. 3). Most matrix-free segments are about one-third the length of adjacent matrix units, but bare regions up to ten times as long as neighboring matrix

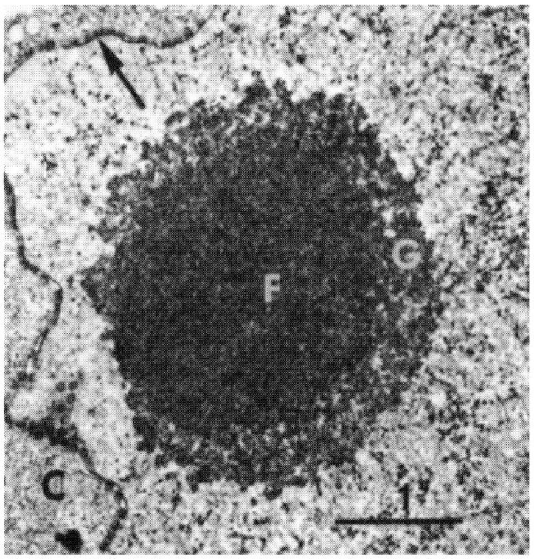

Fig. 1. Thin section of extrachromosomal nucleolus from *Triturus viridescens* oocyte. A granular cortex (*G*) surrounds a compact fibrous core (*F*). Portions of the nuclear envelope (arrow) and cytoplasm (*C*) are visible. Conventional osmium tetroxide fixation, Epon embedding, and uranyl acetate staining. Scale, 1 μ.

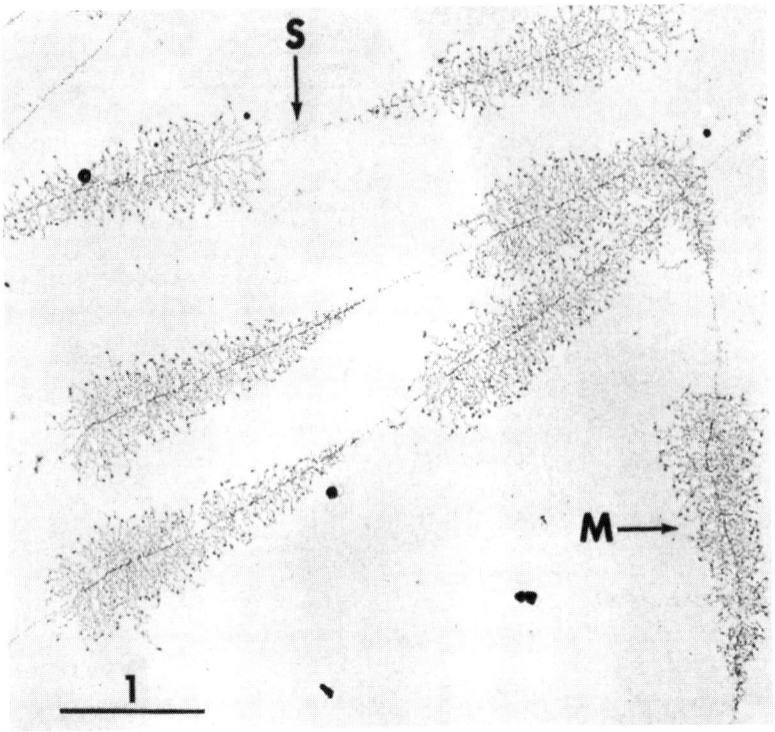

Fig. 2. Portion of a nucleolar core isolated from *Triturus viridescens* oocyte showing matrix units (*M*) separated by matrix-free segments (*S*) of the core axis. The axial fiber can be broken by treatment with deoxyribonuclease, whereas the matrix fibrils can be removed by ribonuclease, trypsin, or pepsin. The specimen was prepared by placing the contents of an oocyte nucleus in deionized water, thus causing dispersal of nucleolar components; the unwound cores were centrifuged through a neutral solution of 0.1*M* sucrose with 10 percent formalin onto a carbon-covered grid; the grid was rinsed in 0.4 percent Kodak Photo-Flo before drying; the preparation was then stained for 1 minute with 1 percent phosphotungstic acid in 50 percent ethanol at *p*H 2.5 (unadjusted) rinsed in 95 percent and then in 100 percent ethanol, and dried with isopentane. The matrix units and intermatrix segments apparently are unstretched by this procedure. Scale, 1 μ.

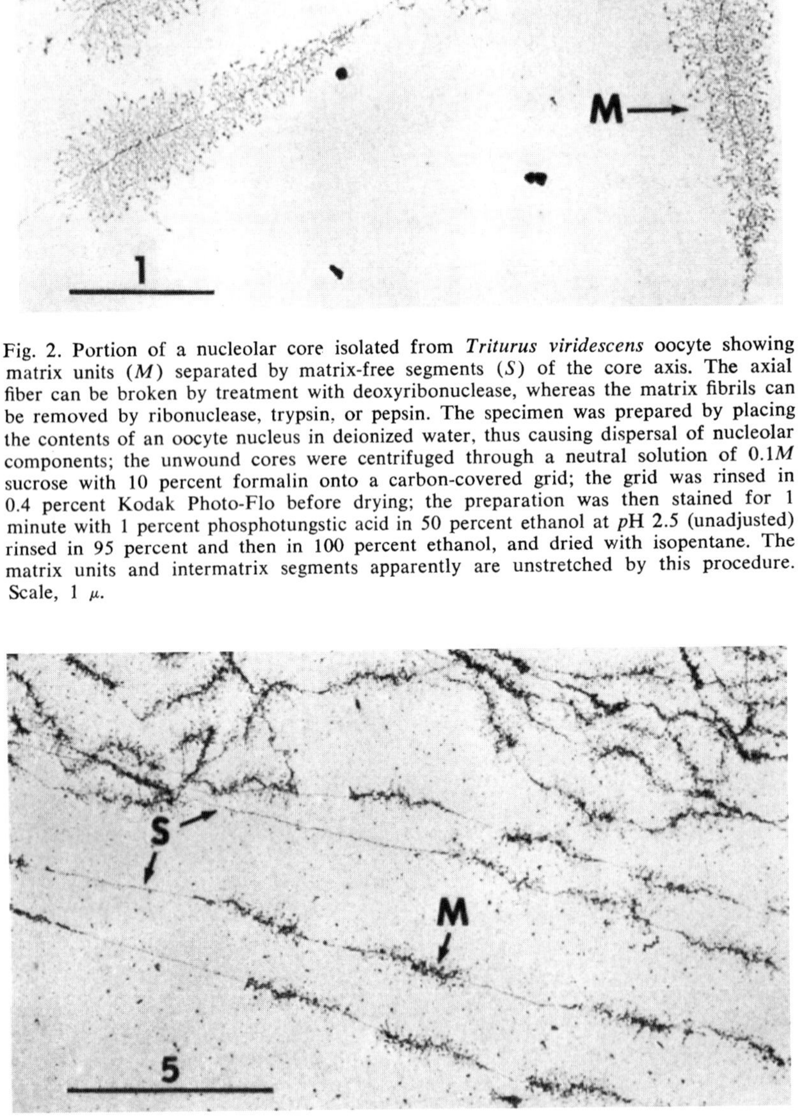

Fig. 3. Portion of nucleolar core isolated from *Triturus viridescens* oocyte, showing matrix units (*M*) separated by matrix-free segments (*S*) of the core axis. Intermatrix segments of various lengths are present. Specimen preparation was similar to that described in Fig. 2, except that the centrifuged specimen was rinsed in water and dipped without drying into 1 percent uranyl acetate in 80 percent acetone and stained for 5 minutes. With this procedure, the matrix units and matrix-free segments appear stretched—slightly to severely—depending on their location on the grid and on how tightly the centrifugation step has compressed them to the grid surface. Scale, 5 μ.

units have been observed in both *Xenopus laevis* and *Triturus viridescens*. There appears to be no pattern to the distribution of the longer matrix-free regions along the core axis.

Detailed examination of matrix units shows that each consists of about 100 thin fibrils connected by one end to the core axis and increasing in length from the thin to the thick end of the unit (Fig. 2). Treatment with ribonuclease, trypsin, or pepsin removes the matrix fibrils from the core axis. After labeling of RNA in intact oocytes for 30 to 60 minutes with tritiated ribonucleosides, electron microscopic autoradiography of unwound cores shows silver grains only over the matrix units. This initial incorporation corresponds in time to the appearance of labeled 40*S* precursor rRNA molecules in nuclear fractions from amphibian oocytes (*7*). Furthermore, the length of unstretched to slightly stretched units (2 to 3 μ) is in close agreement with the length of DNA required to code for the precursor molecule synthesized in amphibians (*8*). Therefore, we believe that each matrix-covered DNA segment is a gene coding for rRNA precursor molecules.

The mechanism of RNA polymerase action in DNA-dependent synthesis of RNA involves a polymerization of monomer ribonucleotides into a polyribonucleotide chain that is immediately dissociated from the template DNA strand (*9*). The structural arrangement of the fibrils in each matrix unit is consistent with a model in which numerous RNA molecules are sequentially initiated before completion of the first. Thus, visualization of the genes coding for rRNA precursors is possible because many molecules are simultaneously synthesized on each gene. In that either ribonuclease or proteases remove matrix fibrils from the core axis, each fibril probably consists of a growing rRNA precursor molecule coated with protein.

If each ribonucleoprotein matrix fibril contains one growing rRNA precursor molecule, the number of fibrils per matrix unit (about 100) and the dimension of the RNA polymerase molecule in the axis of transcription (about 100 Å) (*10*) indicate that about one-third the length of each gene coding for rRNA precursors is covered with polymerase molecules (*8*). A high concentration of RNA polymerase on rRNA genes also has been reported for *Escherichia coli* (*11*).

In equilibrium gradients, the nucleolar DNA of *X. laevis* can be separated from most of the DNA of the cell as a heavier peak (G•C satellite) (*12–14*). Saturation hybridization (*12, 14*) indicates that about 40 percent of the G•C satellite codes for rRNA (*15*). Annealing experiments with fractionated satellite DNA show that the stretches of DNA coding for 18*S* and 28*S* rRNA are alternating, closely adjoining, but separated by stretches of DNA higher in G•C content and not homologous to rRNA (*13, 14*). The latter observations agree with the evidence that each rRNA precursor molecule consists of one 18*S* and one 28*S* rRNA molecule, plus a portion that is degraded during the formation of the two rRNA molecules (*13, 16*).

We propose that the redundant structural arrangement of the rRNA precursor genes and intergene segments seen in isolated nucleolar cores visually confirms the biochemical nature of nucleolus organizer DNA in amphibians. Thus, the DNA axis of the matrix-covered segments corresponds to the satellite portion that is homologous to the entire precursor rRNA molecule (that is, homologous to one 18*S* and one 28*S* rRNA molecule plus the degraded part of the precursor rRNA molecule), and the DNA in the intergene regions corresponds to the remaining portion of the satellite. Measurements of relative lengths of matrix-free and adjacent matrix-covered units in *X. laevis* show that the mean length of intergene segments is about two-thirds the length of a precursor rRNA gene. This indicates that approximately 40 percent of the G•C satellite is inactive nucleolar DNA and about 60 percent consists of genes coding for precursor molecules.

Although the structure of chromosomal loci synthesizing RNA has already been documented, we believe ours are the first observations of the structure of individual genes and associated transcription products whose specific function is known—namely, the extrachromosomal nucleolar genes on which rRNA precursor molecules are synthesized.

O. L. MILLER, JR.
BARBARA R. BEATTY

Biology Division,
Oak Ridge National Laboratory,
Oak Ridge, Tennessee 37830

References and Notes

1. J. G. Gall, *Proc. Nat. Acad. Sci. U.S.* **60**, 553 (1968); H. C. Macgregor, *Quart. J. Microscop. Sci.* **106**, 215 (1968).
2. D. D. Brown, *Nat. Cancer Inst. Monogr.* **23**, 297 (1966); E. H. Davidson and A. E. Mirsky, *Brookhaven Symp. Biol.* **18**, 77 (1965).
3. O. L. Miller, Jr., *Nat. Cancer Inst. Monogr.* **23**, 53 (1966).
4. The nuclei in mature oocytes of some amphibia are near 1 mm in diameter. Mature oocytes of *Xenopus laevis* are about 1.5 mm, and their nuclei are near 0.6 mm in diameter. Mature oocytes of *Triturus viridescens* are near 1.75 mm and their nuclei are about 0.8 mm in diameter.
5. Details of techniques are given in legends for Figs. 2 and 3. Oocytes of *Xenopus laevis*, the African clawed toad, and *Triturus viridescens*, the spotted newt of eastern North America, were used in these studies. Limited examinations in two other genera, *Rana* and *Plethodon*, indicate that these observations probably extend to all amphibians. Earlier reports of these results are found in: O. L. Miller, Jr., and B. R. Beatty, *J. Cell Biol.* **39**, 156a (1968); ———, in *Handbook of Molecular Cytology*, A. Lima-de-Faria, Ed. (North-Holland, Amsterdam, in press); ———, *Genetics*, in press.
6. The diameter of double-helix DNA determined by electron microscopy of shadow-cast molecules [C. E. Hall, *J. Biophys. Biochem. Cytol.* **2**, 625 (1956)] and uranyl acetate–stained molecules [W. Stoeckenius, *J. Biophys. Biochem. Cytol.* **11**, 297 (1961); M. Beer and C. R. Zobel, *J. Mol. Biol.* **3**, 717 (1961)] is approximately 20 Å.
7. J. G. Gall, *Nat. Cancer Inst. Monogr.* **23**, 475 (1966).
8. For double-helix DNA in the B conformation: 2×10^6 daltons = 1 μ; and 1 μ of DNA length codes for 1×10^6 daltons of single-stranded RNA [A. R. Peacocke and R. B. Drysdale, *The Molecular Basis of Heredity* (Butterworths, Washington, D.C., 1965), p. 34]. The molecular weight of the 40*S* precursor rRNA in *X. laevis* has been estimated by acrylamide-gel electrophoresis to be 2.5×10^6 daltons (*14*) and by sedimentation coefficient to be 3.5×10^6 daltons (*13*). These molecules would require, respectively, 2.5 μ and 3.5 μ of double-helix DNA for synthesis.
9. E. K. F. Bautz, in *Molecular Genetics*, J. H. Taylor, Ed. (Academic Press, New York, 1967), pt. 2, p. 213.
10. E. Fuchs, W. Zillig, P. H. Hofschneider, A. Preuss, *J. Mol. Biol.* **10**, 546 (1964); H. S. Slayter and C. E. Hall, *ibid.* **21**, 83 (1966).
11. H. Bremer and D. Yuan, *ibid.* **38**, 163 (1968).
12. J. G. Gall, *Genetics*, in press.
13. D. D. Brown and C. S. Weber, *J. Mol. Biol.* **34**, 681 (1968).
14. M. Birnstiel, J. Spiers, I. Purdom, K. Jones, U. E. Loening, *Nature* **219**, 454 (1968).
15. The saturation hybridization value is about 20 percent. Since only one strand of the double-helix DNA is copied in transcription, the amount of double-helix DNA containing the sequences homologous to rRNA is 40 percent of the total DNA.
16. R. A. Weinberg, U. Loening, M. Willems, S. Penman, *Proc. Nat. Acad. Sci. U.S.* **58**, 1088 (1967).
17. Sponsored by the AEC under contract with Union Carbide Corporation.

25 March 1969

COVER

Nucleolar genes from an amphibian oocyte. These genes, which code for ribosomal RNA, repeat along the DNA axis and are visualized because approximately 100 enzymes are simultaneously transcribing each gene. The gradient of fibrils extending from each gene contains ribosomal RNA precursor molecules in progressive stages of completion (electron micrograph, × 25,000). See page 955. [O. L. Miller, Jr., and Barbara R. Beatty, Biology Division, Oak Ridge National Laboratory]

Fakan S. and **Bernhard W.** 1971. Localisation of rapidly and slowly labelled nuclear RNA as visualized by high resolution autoradiography. *Exp. Cell Res.* **67:** 129–141.

By 1970 biochemical and cytological studies had shown that 18S and 28S ribosomal RNAs were the most abundant RNAs in the cell and that they were synthesized as a large precursor in the nucleolus. The nuclear RNAs that coded for proteins were known to be heterogeneous in size and to label rapidly, but little was known about their intranuclear distribution. Fakan and Bernhard examined the sites of RNA synthesis in nuclei of cultured cells by high resolution ^{3}H-autoradiography. In this technique cells were cultured in the presence of ^{3}H-uridine for up to an hour, fixed and sectioned for electron microscopy, and then covered with an exceedingly thin layer of autoradiographic emulsion. To obtain sufficient numbers of silver grains, Fakan and Bernard held the preparations in the dark for 1–4 months before development. Their technique required great skill and infinite patience. Within the sectioned nuclei, one could distinguish several areas. The prominent nucleoli contained a central clear zone (rDNA) surrounded by two more or less concentric regions—the fibrillar and granular zones. The rest of the nucleus could be divided roughly into chromatin (clear areas in the micrographs), perichromatin fibrils next to the chromatin, and clusters of interchromatin granules. The most important new information was that newly labeled RNA appeared primarily in the area of the perichromatin fibrils, whereas almost no radioactivity was found in the interchromatin granule clusters, even after an extended chase. This study also confirmed earlier observations that nucleolar label appeared within minutes in the fibrillar zone and spread rapidly to the granular zone. It is now possible to follow transcription of RNA by conventional microscopy using fluorescent detection of incorporated BrU (1, 2). Most of the slowly labeled RNA in the interchromatin granules corresponds to snRNA, but may include other species as well (3, 4).

---

1. Jackson D.A., Hassan A.B., Errington R.J., and Cook P.R. 1993. Visualization of focal sites of transcription within human nuclei. *EMBO J.* **12:** 1059–1065.
2. Wansink D.G., Schul W., van der Kraan I., van Steensel B., van Driel R., and de Jong L. 1993. Fluorescent labeling of nascent RNA reveals transcription by RNA polymerase II in domains scattered throughout the nucleus. *J. Cell Biol.* **122:** 283–293.
3. Carmo-Fonseca M., Pepperkok R., Carvalho M. T., and Lamond A. I. 1992. Transcription-dependent colocalization of the U1, U2, U4/U6, and U5 snRNPs in coiled bodies. *J. Cell Biol.* **117:** 1–14.
4. Huang S. and Spector D.L. 1992. U1 and U2 small nuclear RNAs are present in nuclear speckles. *Proc. Natl. Acad. Sci.* **89:** 305–308.

Experimental cell Research 67 (1971) 129–141

LOCALISATION OF RAPIDLY AND SLOWLY LABELLED NUCLEAR RNA AS VISUALIZED BY HIGH RESOLUTION AUTORADIOGRAPHY

S. FAKAN[1] and W. BERNHARD

Institut de Recherches Scientifiques sur le Cancer, 94-Villejuif, France

SUMMARY

Autoradiography combined with electron microscopy is used to visualize the incorporation of ^{3}H-uridine into the nuclei of cultured BSC_1 monkey kidney cells. A preferential RNP stain is applied to differentiate between chromatin and nuclear ribonucleoproteins. It is demonstrated that the incorporation of the radioactive precursor can be localized after pulses of only 2 min., on the one hand at the limit of the chromatin and of fibrillar areas of the nucleolus, and on the other hand in the proximity of condensed chromatin throughout the nucleoplasm where perichromatin fibrils are present. Pulses of 5 to 15 min show a considerable increase of labelling. The number of silver grains over the fibrillar areas of the nucleolus increases. The pattern of extranucleolar incorporation is not changed. If these pulses are followed by a chase varying from 15 min to 3 h, some radioactivity can always be demonstrated throughout the nucleolus, with a considerable amount in the RNP containing interchromatin area as well. However, interchromatin granules are not, or only weakly, labelled even after a labelling of 1 h followed by a chase of up to 3 h.

The results are discussed in view of recent biochemical findings of a rapidly labelled and locally metabolized nuclear RNA. It is suggested that the early-labelled RNA visualized in the vicinity of the condensed chromatin may at least partially correspond to the DNA-like RNA or HnRNA fraction.

There has been an increasing interest in recent years in a biochemically isolated, rapidly labelled nuclear, but extranucleolar RNA fraction which, in addition, has other remarkable properties: (1) its base composition reveals a DNA-like character; (2) most of it does not leave the nucleoplasm and is metabolized in situ. The size variation of its molecules is very wide, ranging between 20s to 100s [2, 6, 10, 30, 31, 34]. This type of RNA has been isolated from HeLa and duck erythroblast cell nuclei but also is present in single puff areas of giant chromosomes of Diptera as polydisperse "chromosomal" RNA [8]. It seemed of interest to find out which type of morphological structure, carrying RNP and being visible in the electron microscope, would correspond to this rapidly labelled RNA species. Possibly good candidates for it were expected to be the perichromatin fibrils, recently described new components of the nucleoplasm which are visualized after bleaching of the chromatin with the EDTA staining technique [4] at the periphery of the condensed chromatin area throughout the nucleoplasm [18].

The most suitable approach for an electron microscopic study of this problem seemed to be a combined application of high resolution

[1] Present address: Institut Suisse de Recheches Expérimentales sur le Cancer, 21 rue de Bugnon, Lausanne, Suisse.

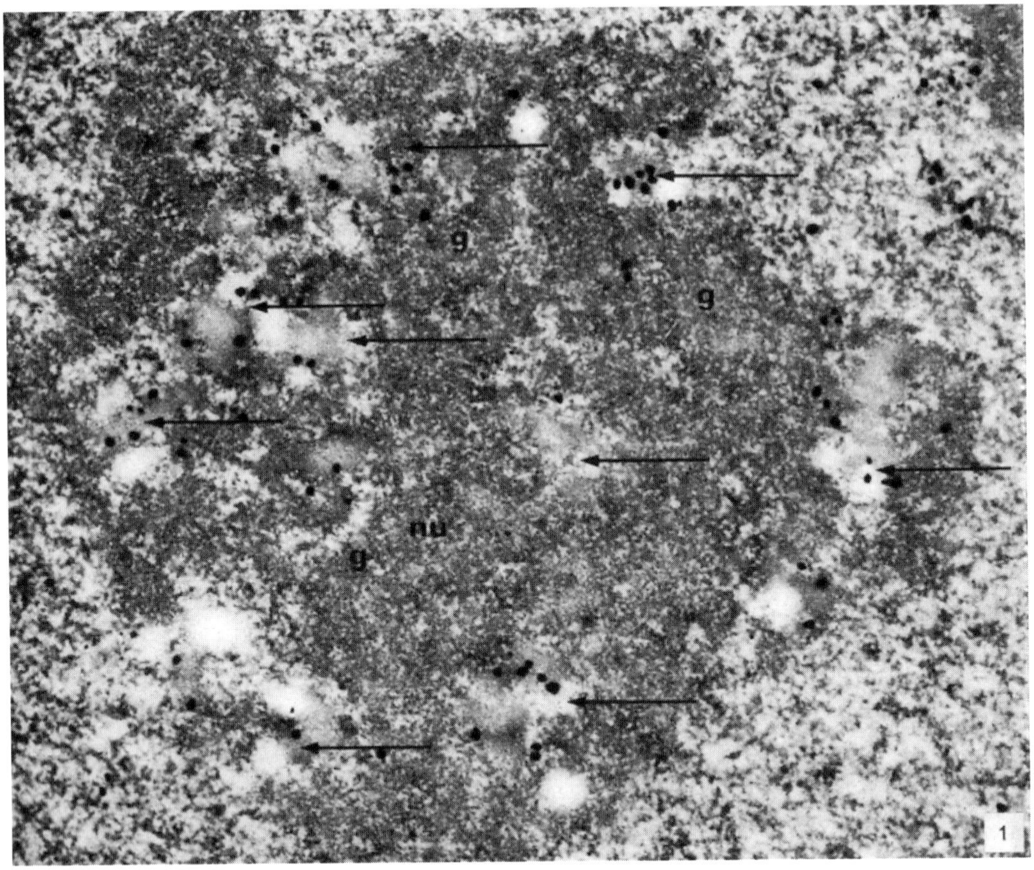

Figs 1–8. BSC_1 cell cultures, labelled with ^{3}H-UdR, fixed 1 h in phosphate-buffered glutraraldehyde 1.6 %. Embedding in Epon. Gold latensification-Elon-ascorbic acid development. EDTA method for preferential RNP stain. *Figs 6–7*, Ilford L4 emulsion; the others, Gevaert NUC 3.07.
Fig. 1. Nucleolus (*nu*) after 2 min of ^{3}H-UdR labelling. The radioactivity is localized at the periphery of the bleached portion of intranucleolar chromatin (→). The granular portion (*g*) is unlabelled. ×23 000.

autoradiography and the preferential RNP-staining method. One of us has already demonstrated that the technical problems linked with the simultaneous use of both can be solved [9]. In a previous paper, Granboulan & Granboulan [12], using short pulses of ^{3}H-uridine and a classical stain, have shown that the precursor was incorporated within 5 min in the fibrillar part of the nucleolus and also in the interchromatin area where euchromatin was supposed to be localized. Extranucleolar label was also found by other workers [1, 17] to be localized outside the condensed chromatin. However, up to the present it was not possible to distinguish the diffuse chromatin fibrils from ribonucleoprotein fibrils as the classical staining techniques did not allow differentiation between RNA- and DNA-carrying structures.

In this study, we have paid particular attention to the extranucleolar incorporation sites of ^{3}H-uridine after short times of labelling, with and without a chase of variable duration with the cold precursor. The behaviour of the nucleolus will be reexamined,

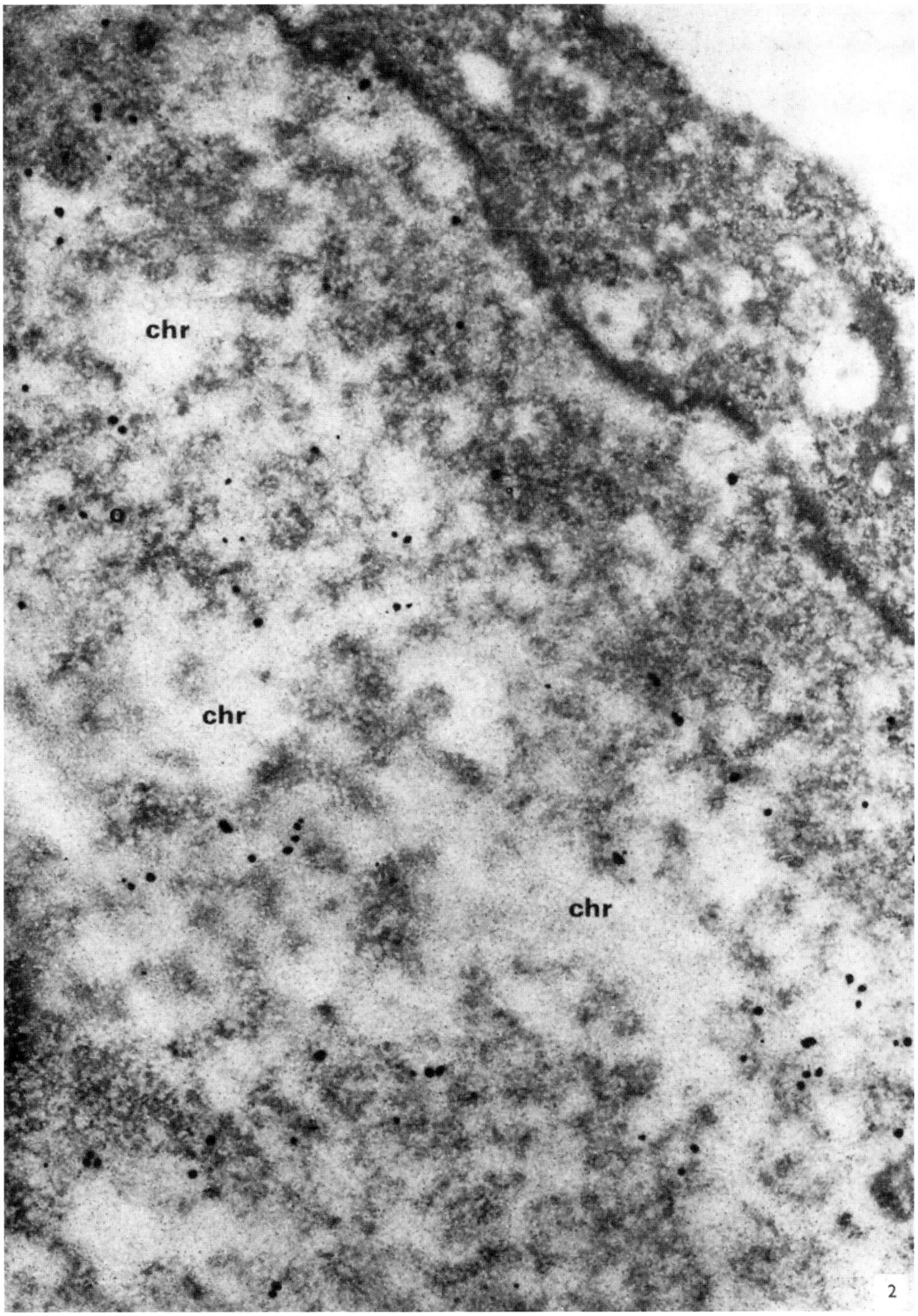

Fig. 2. Extranucleolar portion of a nucleoplasm after 2 min of ^{3}H-UdR labelling. Incorporation unusually high. The majority of the silver grains are localized over the border zone of the bleached chromatin (*chr*) and the RNP of the interchromatin area. No cytoplasmic label. × 23 000.

although this organelle has not been the main object of our investigation.

MATERIAL AND METHODS

A stabilised monkey kidney cell strain, BSC_1 [15] was used in these experiments. The cells were cultivated as monolayers in prescription bottles in Eagle minimal essential medium [7] supplemented with 10 % calf serum.

For labelling of the cells, ^{3}H-5-uridine (spec. act. 18 Ci/mM, CEA) was added to the cultures at the end of exponential growth. The cells were exposed for either 2, 5 or 15 min to 80 μCi/ml, or for 1 h to 60 μCi/ml, of ^{3}H-uridine. After short pulses, some cultures were briefly washed in cold medium containing 1 mg/ml of non-radioactive uridine and immediately fixed. Another group of cultures, after a pulse of 5 min and 15 min, was washed and postincubated in a medium containing 100 μg/ml of non-radioactive uridine for 15 min, 30 min, 1 h and 3 h and fixed. The cells labelled for 1 h with ^{3}H-UdR were fixed after 1 and 3 h of chase.

The fixation was carried out in 1.6 % glutaraldehyde in Sörensen's phosphate buffer for 1 h at 4°C. The cells were then rinsed for about 20 h in frequent changes of cold phosphate buffer with 0.2 M sucrose in order to eliminate the non-incorporated soluble RNA precursor molecules [20]. They were then scraped off the glass with a rubber policeman, centrifuged at low speed for 10 min to form a pellet, dehydrated with acetone and embedded in Epon according to the usual procedure. Ultrathin sections of silver-gold interference colour were cut with an LKB ultramicrotome equipped with a diamond knife. We did not use glycol methacrylate embedding for digestion of the preparations with RNAse, as controls for specific incorporation into RNA have been carried out repeatedly in earlier work by Granboulan & Granboulan [12] published from this laboratory.

The sections were deposited on microscopic glass slides covered with a thin collodion membrane [13]. The slides were dipped in either Gevaert NUC 3.07 or Ilford L4 emulsion, diluted to form a monolayer of silver halide crystals. After exposure for 25 days to 4 months, the majority of preparation were developed by the gold latensification-Elon ascorbic acid method according to Salpeter & Bachmann [27] modified by Wisse & Tates [38]. This technique increases considerably the sensitivity, and therefore, shortens the exposure time. Furthermore, it also increases the resolution, especially when the NUC 3.07 emulsion is used. In order to localize more exactly the radioactive source corresponding to reduced silver grains, we have calculated the error limiting the resolution according to Bachmann & Salpeter [3]. Thus, the radius determining the distance around the developed grain where the radioactive source is located with 50 % probability, or the radius of the circle within which 50 % of developed grains fall is, for an average section thickness of 850 Å and with gold latensification-Elon ascorbic acid development, 1 120 Å for NUC 3.07 emulsion and 1 510 Å for L4 emulsion. Another important advantage of the small grains (300–500 Å) developed by this technique is that the underlying structures remain easily visible. For comparison, the remaining group of preparations was developed in D-19 developer for 4 min at 18–19°C. The background of the autoradiographs was very low for both techniques of development.

After development and photographic fixation, the slides are washed and grids are slipped under the partially detached collodion membrane in the region of the sections, and then dried at 37°C. The staining is carried out on the slides either with uranyl acetate 5 % 10 min, followed by lead citrate 5 min or with the regressive EDTA stain [4]: uranyl acetate 5 % 10 min, rinsing in distilled water, drying 10 min at 37°C, differentiation in 0.2 M EDTA, pH 7, between 10 and 20 min followed by rinsing in distilled water, drying 10 min at 37°C and poststaining with lead citrate 5 min. Only at this stage are the grids detached from the glass slide with a steel needle and examined with the E.M. We used a Philips EM200 at 80 kV with a 50 μ objective aperture.

RESULTS

The combined application of an autoradiographic technique based on physical development to reduce the grain size and to increase the sensitivity, with a staining technique which allows the bleaching of chromatin were indispensable for obtaining the information which we needed. In the controls where classical development with Kodak D-19 was used and where the usual uranyl-lead stain was employed, the limit between chromatin and interchromatin areas could not be determined in most cases and the large grain size obscured the underlying nucleoplasmic structures. The examination of the cultures labelled with H^3-UdR followed or not by chases of various durations gave the following results.

(1) *Two-minute label*

In cells fixed immediately after 2 min labelling with ^{3}H-UdR nearly all radioactivity is clearly localized over the limiting zone between EDTA-resistant RNP structures and the bleached chromatin. This is particularly well seen in the nucleolus where the silver

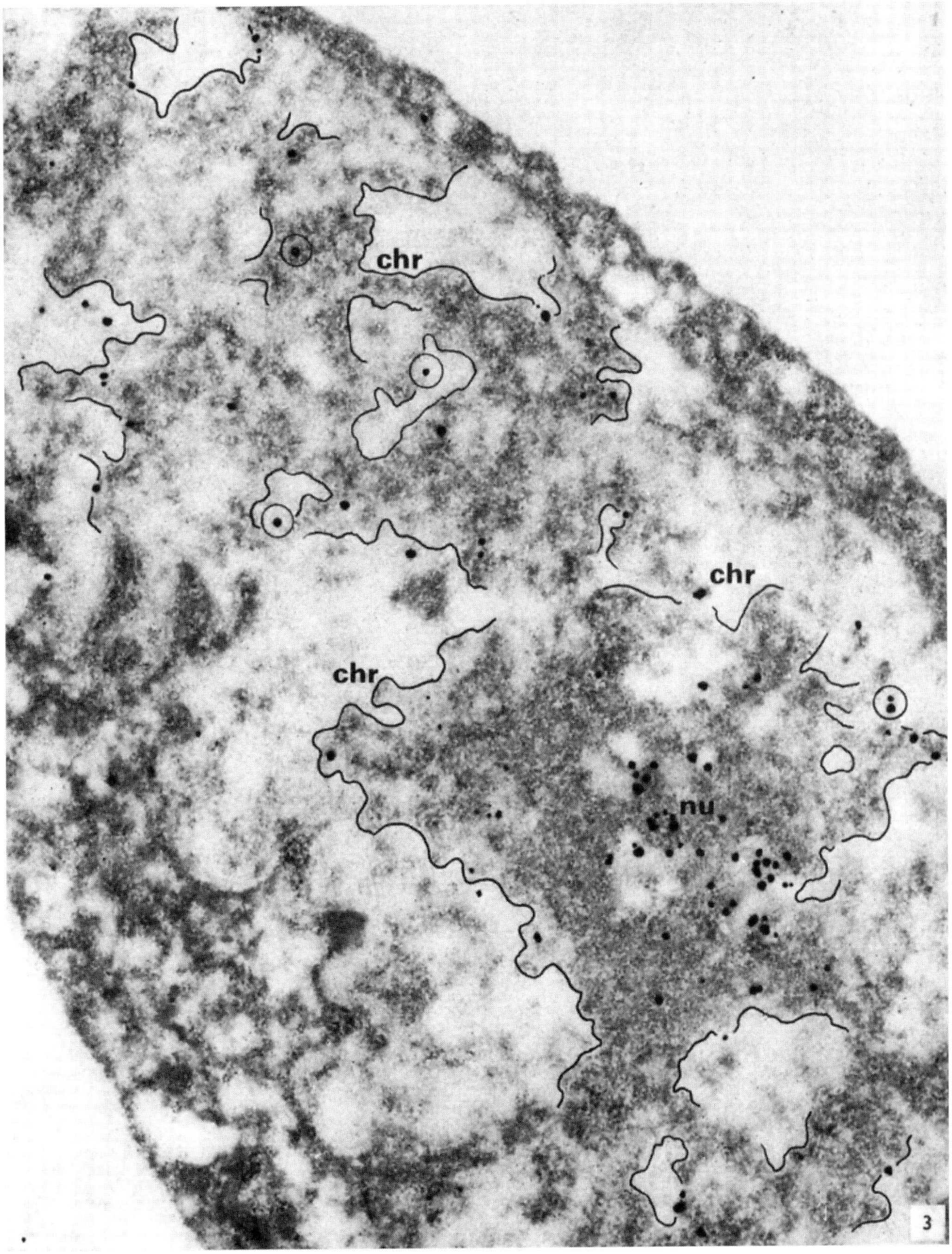

Fig. 3. Portion of a nucleolus (*nu*) and extranucleolar region, after a 5 min pulse of ^{3}H-UdR. The drawing indicates the interface between the bleached chromatin (*chr*) and the RNP-containing interchromatin area. Some silver grains are surrounded by a circle of radius 1 120 Å indicating the resolution of this technique. Most of the label is found over the nucleolus and the border zone between the condensed chromatin and interchromatin which is usually rich in perichromatin fibrils. ×23 000.

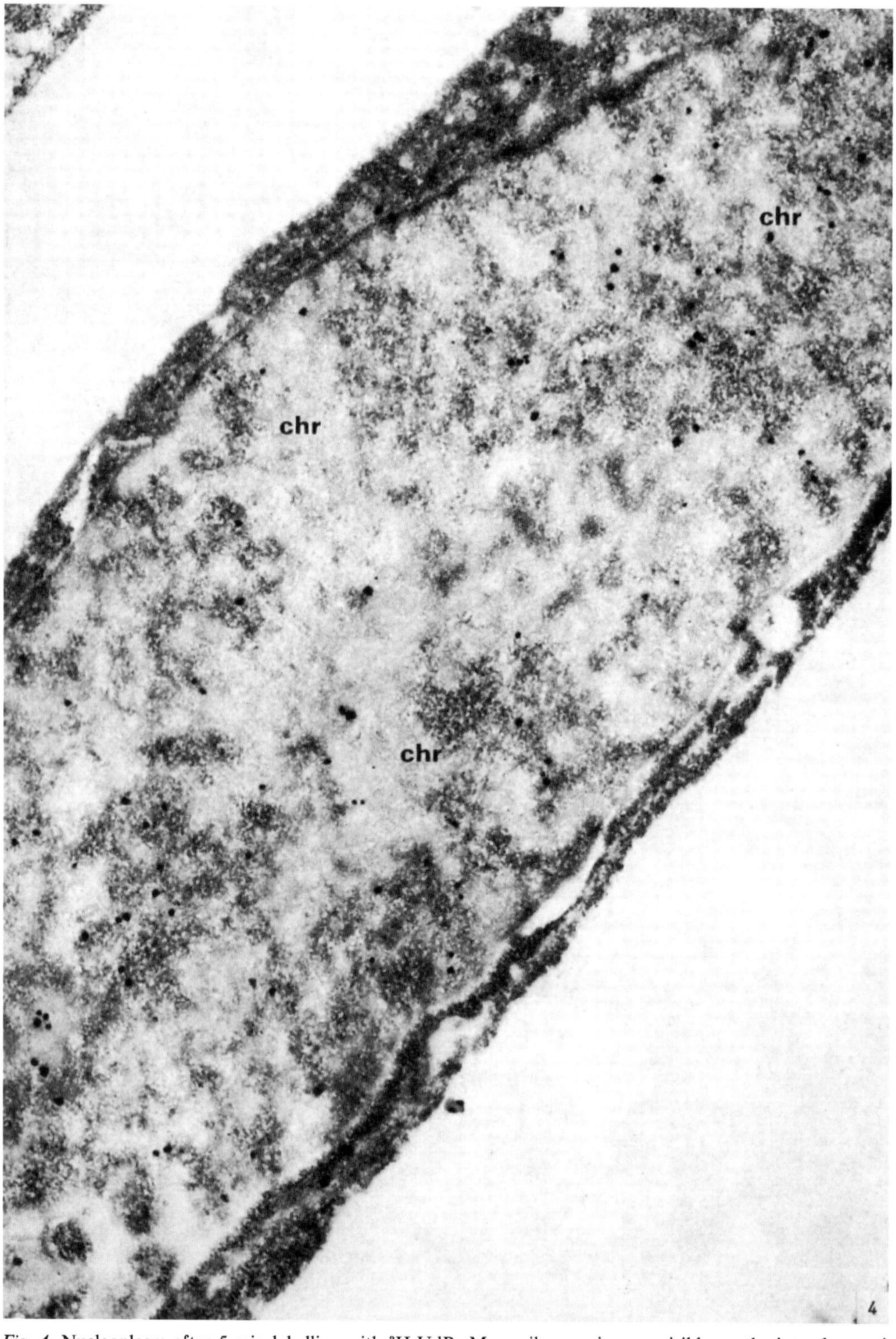

Fig. 4. Nucleoplasm after 5 min labelling with ^{3}H-UdR. More silver grains are visible on the interchromatin zone, and many are still localized in the vicinity of the condensed chromatin (*chr*) which is not labelled. Figs 3 and 4 represent an average intensity of labelling. ×23 000.

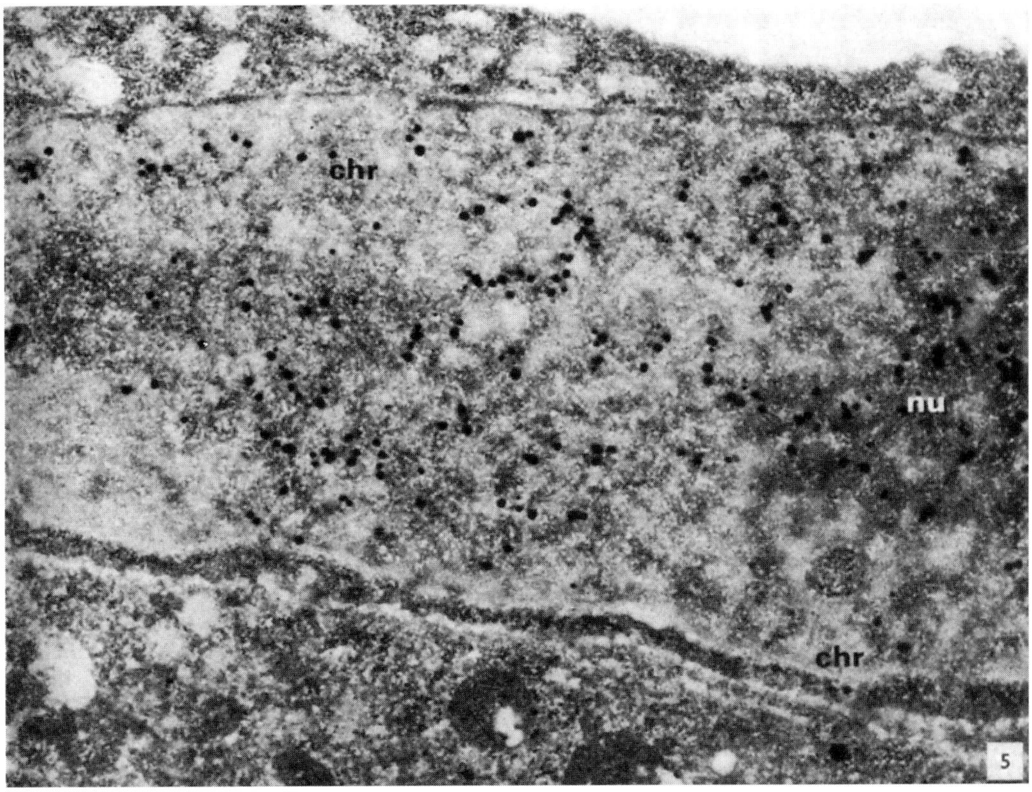

Fig. 5. Intensely labelled nucleoplasm after a 5 min pulse and a chase with cold uridine of 15 min. Practically all radioactivity is found in the RNP-containing interchromatin area and in the nucleolus (*nu*). ×23 000.

grains are localized predominantly at the limit of the fibrillar portions, where intranucleolar nucleohistones are supposed to be present (fig. 1). The same character of labelling is also found in the nucleoplasm, which however, seems relatively less labelled than the nucleolus (fig. 2). Nevertheless it is clearly shown that the large dense chromatin areas are not labelled whereas the majority of the grains are localized over the adjacent zone where RNP carrying structures are visible (fig. 2).

(2) *Five-minute label*

After 5 min of labelling the ^{3}H-UdR incorporated in the nucleolus is seen both on the limiting zone between the bleached intranucleolar chromatin and over the fibrillar part of the nucleolonema as well. The granular portion remains practically unlabelled. The extranucleolar nucleoplasm has the same pattern of labelling as after 2 min of pulse; however, the labelling is more intense and the preferential localization of the radioactivity on the border between dense chromatin and RNP containing interchromatin zone is accentuated (figs 3, 4, table 1). After a 15 min chase, the labelling of the granular portions of the nucleolus has started, but the radioactivity is still predominantly in the fibrillar zones. In the nucleoplasm, there is a relative increase of labelling over the EDTA-resis-

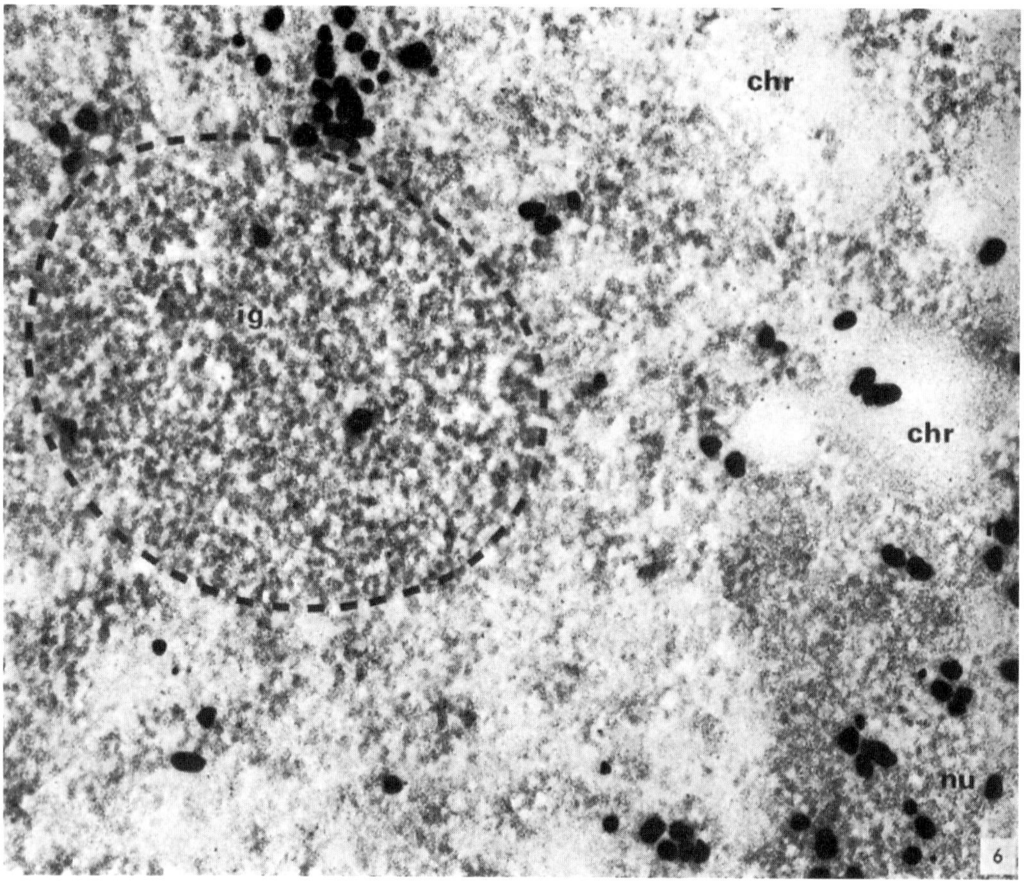

Fig. 6. Nucleoplasm with weakly labelled interchromatin granules (ig) after a 5 min pulse with ^{3}H-UdR and a chase of 1 h. The label is found predominantly at the periphery of the cluster. Compare intense labelling of the nucleolus (*nu*). ×65 000.

tant RNP structures in comparison with the border zones where the labelling decreases (fig. 5). With a prolonged chase the labelling of the interchromatin area increases still further and the distribution of silver grains, over both areas of the nucleolus, becomes more regular. Weak cytoplasmic label has appeared after the 15 min chase and reaches approximately the nuclear level after 3 h. Unlike the intensively labelled perichromatin fibrils, the aggregates of interchromatin granules are weakly labelled. Some radioactivity is observed at the periphery of the clusters, while their interior remains mostly unlabelled even after a 1 h chase (fig. 6).

(3) *Fifteen-minute label*

After a pulse of 15 min, the radioactivity in the nucleolus is distributed more regularly over its components than after shorter periods of labelling. In the nucleoplasm, there is a relatively high amount of silver grains over the limiting region between RNP and chromatin, and the total labelling of the interchromatin area is higher than after shorter pulses. Again, after chases of increasing duration, the

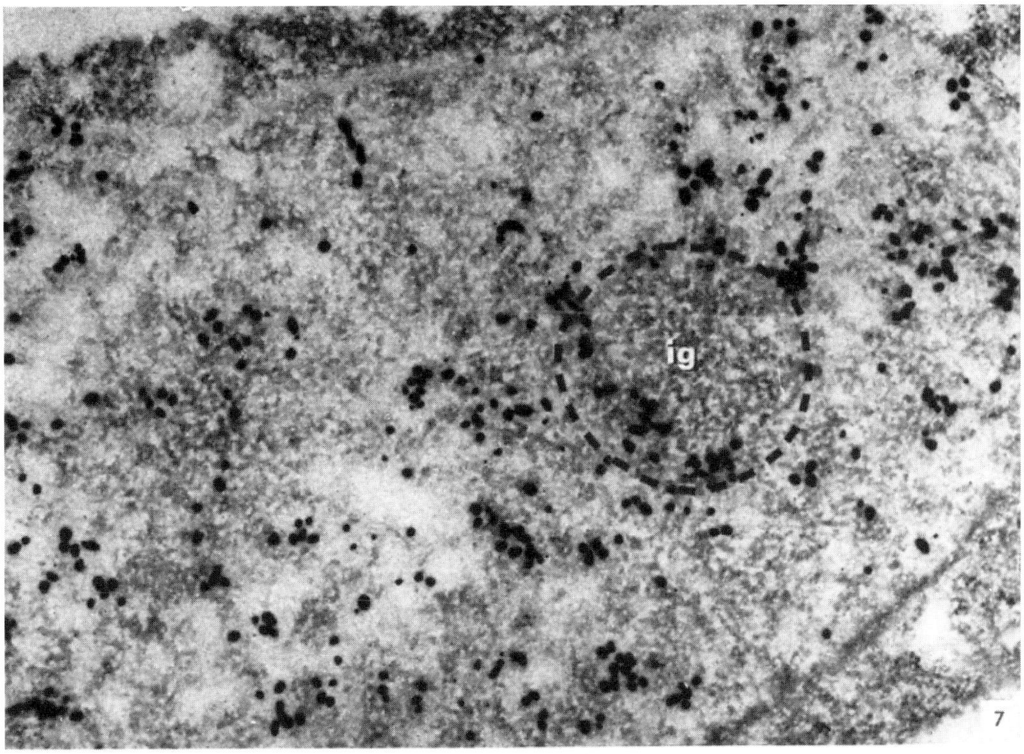

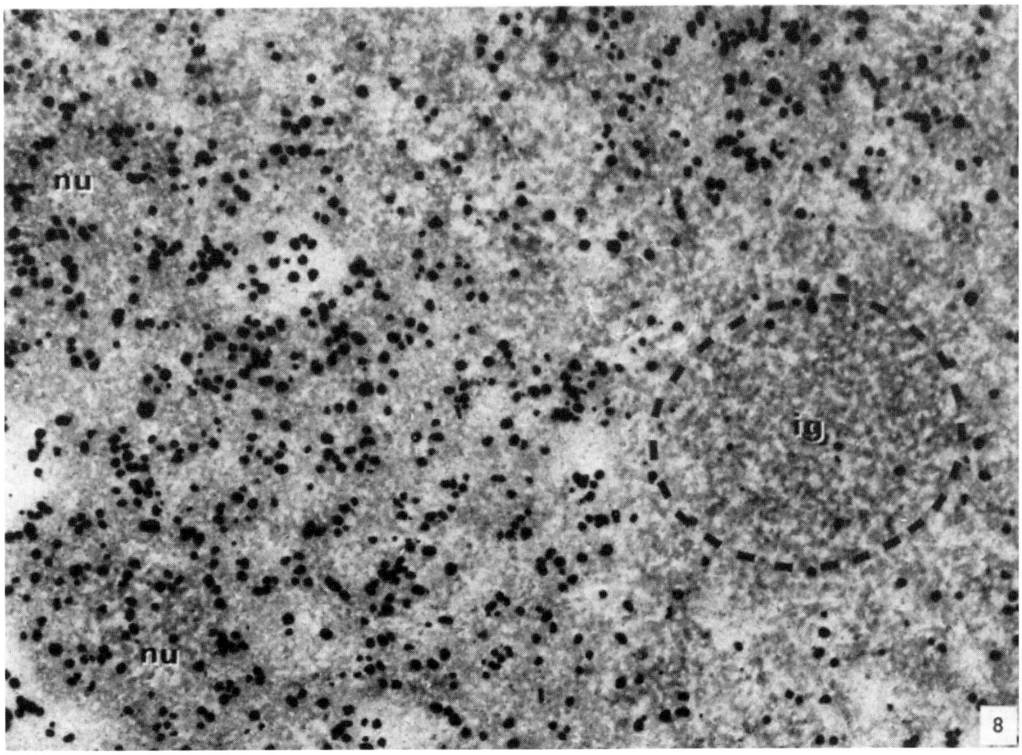

Fig. 7. Nucleoplasm with weakly labelled interchromatin granules (*ig*) after a 15 min pulse with ^{3}H-UdR and a chase of 1 h. All the radioactivity is found at the periphery of the cluster. × 33 000.

Fig. 8. Nucleoplasm with nucleolus (*nu*) and interchromatin granules (*ig*), after labelling of 1 h with ^{3}H-UdR and a chase of 1 h. Compare the very strong nucleolar labelling with the practically absent radioactivity of the cluster of interchromatin granules. × 38 000.

Table 1. *Counts of silver grains over three different regions of the nucleoplasm of 5 cells, fixed after 5 min labelling with ^{3}H-uridine*

Cell no.	Border line of condensed chromatin	Inter-chromatin area	Condensed chromatin
1	57	27	1
2	36	7	3
3	47	9	4
4	19	8	0
5	65	22	6
Total	224	73	14
%	72	23.5	4.5

Emulsion NUC 3.07, Gold latensification Elon ascorbic acid development. The results clearly show that the main radioactivity is found in proximity of the border line between condensed chromatin and the interchromatin area. Very few silver grains are seen over the bleached chromatin. The multiple grains whose centers were closer than 700 Å (mean diameter of silver halide crystals) were counted as a single grain.

total labelling of nuclear RNP structures becomes more intense. However, the clusters of interchromatin granules remain practically unlabelled or show some radioactivity at the periphery (fig. 7).

(4) *One-hour label*

We have examined only cells fixed after 1 and 3 h of chase following 1 h of incubation with ^{3}H-UdR. Our interest was mainly focused on the interchromatin granules. After both chases these constituents showed principally the same character of labelling as mentioned above for short pulses. The aggregates of such granules are labelled mostly over their periphery, while their centre remains practically unlabelled (fig. 8). The rest of the interchromatin area appears very heavily and rather randomly labelled.

DISCUSSION

The results of this investigation clearly show that ^{3}H-uridine incorporation in cells of the BSC_1 strain can be demonstrated in well characterized areas of the nucleoplasm after a pulse of only 2 min. The two sites of radioactivity are:

(1) the fibrillar portion of the nucleolonema at the periphery of intranucleolar chromatin and;

(2) The borderline between the condensed and EDTA-bleached chromatin and the EDTA-resistant interchromatin area.

Concerning the labelling of the nucleolus the original work done by Granboulan & Granboulan [12] is confirmed and, by using the EDTA staining method, it can be shown that the early-appearing silver grains are localized at the peripheral parts of bleached regions of the nucleolonema, where DNA matrixes are believed to be present. The question arises where the DNA used for transcription is localized. We originally assumed that it corresponds to the intranucleolar dense chromatin lamellae [5]. However, according to Recher et al. [26] some DNA might also be present inside the fibrillar RNP zones. After about 15 min of chase, the now well known migration of radioactivity into the granular portions of the nucleolus is demonstrated morphologically. We also noticed the striking fact that there is much residual label in the nucleolus even after 15 min or 1 h of labelling and 3 h of chase. It is interesting that the label, whether nucleolar or extranucleolar, generally still increases after a chase of 15 min up to 1 h. This phenomenon has also been observed by other investigators [24]. One might hypothesize that a certain amount of radioactive precursor is somehow bound to structural elements and cannot be chased.

Concerning the extranucleolar labelling, we can distinguish very rapidly labelled areas of RNP and, on the contrary, very slowly labelled or unlabelled regions believed to contain RNP.

As mentioned above, the very rapidly appearing label is seen at the border of the condensed chromatin, approximately in the region where perichromatin fibrils have been localized. However, in the cell strain we employed these newly described nuclear components are not as clearly found immediately adjacent to the clumped chromatin, and the resolution of our autoradiographic technique is not adequate to localize fibrils whose diameter varies between 50 and 200 Å. Although we have no direct proof that the perichromatin fibrils are indeed the carriers of the radioactivity after very brief labelling, there is indirect evidence for this assumption. Petrov & Bernhard [25] were able to induce perichromatin fibrils in liver within 15 min after cortisone injection, a procedure known to stimulate rapid RNA synthesis. These fibrils were all localized at the periphery of the dense chromatin clumps. In later stages, the fibrils of RNP material are more irregularly dispersed in the interchromatin region. It was concluded that the newly synthesized RNA is visualized in the electron microscope in the form of these fibrillar components. If, in our experimental system, pulses are followed by chases of various intervals, one would expect the labelling to be more diffuse within the interchromatin area, and this is indeed the case.

Earlier findings on localized RNA synthesis in the nucleoplasm are based on light or electron microscopic autoradiography the resolution of which was below that of our present method. Nevertheless, the results obtained can be put in line with our observations. Hsu [16] observed a very weak ^{3}H-UdR incorporation in nuclear (heterochromatin) chromocentres. Littau et al. [17] and Allfrey et al. [1] found after 30 min of uridine labelling that there was considerable radioactivity present in the diffuse chromatin outside the condensed areas in isolated thymus cell nuclei, and Granboulan & Granboulan [12] could demonstrate silver grains in the interchromatin area after a 5 min pulse of ^{3}H-UdR in BSC cells. Unuma et al. [37] also described a predominant labelling of the dispersed chromatin in hepatoma cells labelled in vivo with ^{3}H-UdR.

Recently, Goldstein [11] showed by light microscopic autoradiography, that after 1 and 15 min of incubation of L cells with ^{3}H-UdR no preferential site of labelling over the nuclear membrane can be detected. He thus demonstrated the non-validity for an eukaryote cell of Stent's hypothesis, dealing with a possible coupling of the transcription and translation process [35]. Our results strongly confirm Goldstein's conclusion, as we have never observed any preferential labelling in the proximity of the nuclear membrane.

An important question is to know how the biochemical findings of rapidly labelled extranucleolar RNA can be integrated with our ultrastructural observations. It would seem to us logical to conclude that the heterogeneous RNA fraction isolated in density gradients corresponds to the RNA components which are adjacent to the chromatin, i.e. the perichromatin fibrils. The fibrils are also very heterogeneous from the morphological point of view, as far as their thickness and probably, their length is concerned. It is not yet clear what relationship, if any, exists between mRNA and HnRNA [6]. Both are DNA-like,but whereas the former leaves the nucleus and is associated with polysomes, the majority of the latter (up to 90%) seems to be metabolized in the nucleoplasm [2, 6, 10, 23, 30, 31, 34], thus confirming an observation already mentioned by Harris in 1962 [14]. The rapidly labelled RNA seems to be continuously synthesized throughout the interphase [22]. Scherrer & Marcaud [30] assume that the process of continuous partial disintegration of this "giant nascent messenger-like

RNA" might be linked with regulation on the translational level. Our finding that even after 3 h of chase, there is still a relatively intense labelling of the interchromatin space supports the idea that a good deal of the rapidly labelled RNA species described in this paper is indeed kept in the nucleus and probably to a great extent metabolized in situ. Thus the radioactive precursor may be locally reutilized to resynthesize a very unstable RNA.

The times of labelling in our experiments were generally much shorter than those used in biochemical studies, where the shortest pulses were of the order of 10 min [34], but mostly longer, varying between 30 min and 1 h. We believe that for precise localisation of the sites of incorporation, very short pulses should be used to minimize the effects of subsequent migration of the synthesized product.

Slight cytoplasmic radioactivity was found as early as 15 min after labelling, and was steadily increasing afterwards. After 3 h of chase, it reached about the same level as found in the nucleus.

According to our observations, there also exists a very slowly labelled RNA species, present in the interchromatin granules. The biochemical nature of these components, universally present in all interphase nuclei is still poorly known since their discovery [36] and their function is totally obscure. They are thought to contain proteins and RNA [5, 18, 32, 33] ,but they are very resistant to RNAse digestion even after protease action and can be extracted only with difficulty by cold perchloric acid [18].

In our preparations of ^{3}H-uridine-labelled BSC_1 cells, we have found that after all intervals of pulse or chase most of the interchromatin granules are weakly labelled, if at all, with the radioactivity localized rather at the periphery of the clusters. According to these observations our interchromatin granules do not seem to be the same type of RNP particles as those isolated from rat liver and sedimenting at 40*s* which were found to contain a rapidly labelled RNA [19, 21]. These particles were suggested to represent the interchromatin granules visible in this sections [19]. It is also improbable that the interchromatin granules represent the 30*s* DNA-like RNA containing particles isolated by Samarina et al. [28, 29] which are also said to be labelled rapidly.

Here is an important problem which remains unsolved. Experiments with the same cell strain are now in progress based on long labelling and chase periods in order to determine the metabolic activity of these rather mysterious components.

The authors are grateful to Miss S. Estrade from the Department of Virology (Prof. Tournier) for the preparation of the tissue cultures, and to Mrs J. Fakanova for technical assistance. They wish to thank Dr M. Hill for helpful discussion and Mrs Ch. Taligault for preparing the manuscript.

This study was carried out thanks to a fellowship awarded to one of us (S. Fakan) by the French Government and profited from financial aid from the Centre de Recherches sur les Lymphomes Malins, Lausanne (Prof. F. Cardis).

REFERENCES

1. Allfrey, W G, Pogo, B G T, Pogo, A O, Kleinsmith, L J & Mirsky, A E, Histones (ed A V S de Reuck & J Knight). Ciba Found. Study Group, No. 24, p. 42 Churchill, London (1966).
2. Attardi, G, Parnas, H, Hwang, M I H & Attardi, B, J mol biol 20, (1966) 145.
3. Bachmann, L & Salpeter, M M, Lab invest 14 (1965) 1041.
4. Bernhard, W, J ultrastr res 27 (1969) 250.
5. Bernhard, W & Granboulan, N, Exptl cell res, suppl. 9 (1963) 19.
6. Darnell, J E, Bacteriol rev 32 (1968) 262.
7. Eagle, H, Science 130 (1959) 432.
8. Edström, J E & Daneholt, B, J mol biol 28 (1967) 331.
9. Fakan, S, Proc 7th intern congr electron microscopy, Grenoble, vol. 1, p. 501 (1970).
10. Georgiev, G P, Progress in nucleic acid res, and mol biol (ed J N Davidson & W E Cohn) vol. 6, p. 259. Academic Press, New York and London (1967).
11. Goldstein, L, Exptl cell res 61 (1970) 218.

12. Granboulan, N & Granboulan, Ph, Exptl cell res 38 (1965) 604.
13. Granboulan, Ph, The use of radioautography in investigating protein synthesis (ed C P Leblond & K B Warren) p. 43. Academic Press, New York (1965).
14. Harris, H, Biochem j 84 (1962) 60 p.
15. Hopps, H E, Bernheim, B C, Nisalak, A & Smadel, J E, Fed proc 21 (1962) 454.
16. Hsu, T, Exptl cell res 27 (1962) 332.
17. Littau, V C, Allfrey, V G, Frenster, J H & Mirsky, A E, Proc natl acad sci US 52 (1964) 93.
18. Monneron, A & Bernhard, W, J ultrastruct res 27 (1969) 266.
19. Monneron, A & Moulé, Y, Exptl cell res 51 (1968) 531.
20. — Ibid 56 (1969) 179.
21. Moulé, Y & Chauveau, J, J mol biol 33 (1968) 465.
22. Pagoulatos, G N & Darnell, J E, J cell biol 44 (1970) 476.
23. Penman, S, Vesco, C & Penman, M, J mol biol 34 (1968) 49.
24. Perry, R P, Exptl cell res 29 (1963) 400.
25. Petrov, P & Bernhard, W, J ultrastruct res (1971). In press.
26. Recher, L, Whitescarver, J & Briggs, L, J cell biol 45 (1970) 479.
27. Salpeter, M M & Bachmann, L, J cell biol 22 (1964) 469.
28. Samarina, O P, Krischevskaya, A A & Georgiev, G P, Nature 210 (1966) 1319.
29. Samarina, O P, Lukanidin, E M, Molnar, J, & Georgiev, G P, J mol biol 33 (1968) 251.
30. Scherrer, K & Marcaud, L, J cell physiol 72 suppl. 1 (1968) 181.
31. Scherrer, K, Marcaud, L, Zajdela, F, London, I M & Gros, F, Proc natl acad sci US 56 (1966) 1571.
32. Shankar Narayan, K, Steele, W J, Smetana, K & Busch, H, Exptl cell res 46 (1967) 65.
33. Smetana, K, Steele, W J & Busch, H, Exptl cell res 31 (1963) 198.
34. Soeiro, R, Vaughan, M H, Warner, J R & Darnell, J E, J cell biol 39 (1962) 112.
35. Stent, G S, Science 144 (1964) 816.
36. Swift, H. Interpretation of ultrastructure. Symp. int soc cell biol (ed R J C Harris) vol. 2, p. 21 (1962).
37. Unuma, T, Arendell, J P & Busch, H, Exptl cell res 52 (1968) 429.
38. Wisse, E & Tates, A D, Proc 4th europ reg conf E M, Tipografica Polyglotta Vaticana, Rome (1968) vol. 2, p. 465.

Received January 11, 1971.
Revised version received February 26, 1971

Lerner M.R., Boyle J.A., Mount S.M., Wolin S.L., and **Steitz J.A.** 1980. Are snRNPs involved in splicing? *Nature* **283:** 220–224.

TODAY WE ACCEPT AS ONE of the peculiarities of the molecular world that mature messenger RNA molecules come from longer precursors that must be cut and spliced at precisely defined positions. However, the existence of splicing was originally greeted with astonishment (1, 2), and even now the evolutionary significance of this unlikely phenomenon remains elusive. The involvement of small nuclear ribonucleoproteins (snRNPs or "snurps") in splicing was first suggested in 1980 in a remarkably prescient paper by Joan Steitz and her colleagues. Steitz noted that the primary sequence of U1 snRNA and the proteins with which it was associated were highly conserved among different organisms, and that U1 was particularly abundant in actively transcribing cells, features that might be expected for molecules involved in splicing. Even more suggestive for a role of U1 in splicing was her demonstration that sequences at the 5′ end of U1 were complementary to those spanning the splice junction. Subsequent studies have amply confirmed the involvement of U1, U2, U4/U6 and U5 snRNAs in splicing. Furthermore, a host of other snRNAs has since been identified as active participants in the processing of pre-ribosomal RNA, pre-tRNA, and histone pre-mRNA, suggesting that RNAs have long played a catalytic role in basic cellular processes (reviewed in 3)

---

1. Chow L.T., Gelinas R.E., Broker T.R., and Roberts R.J. 1977. An amazing sequence arrangement at the 5′ ends of adenovirus 2 messenger RNA. *Cell* **12:** 1–8.
2. Berget S.M., Moore C., and Sharp P.A. 1977. Spliced segments at the 5′ terminus of adenovirus 2 late mRNA. *Proc. Natl. Acad. Sci.* **74:** 3171–3175.
3. Gesteland R.F. and Atkins J.F., eds. 1993. *The RNA world.* Cold Spring Harbor Laboratory Press, Cold Spring Harbor, New York.

Reprinted from Nature, Vol. 283, No. 5743, pp. 220-224, January 10 1980

Are snRNPs involved in splicing?

Michael R. Lerner, John A. Boyle, Stephen M. Mount, Sandra L. Wolin & Joan A. Steitz

Department of Molecular Biophysics and Biochemistry, Yale University, New Haven, Connecticut 06510

Discrete, stable small RNA molecules are found in the nuclei of cells[1] from a wide variety of eukaryotic organisms[2]. Many of these small nuclear RNA (snRNA) species, which range in size from about 90 to 220 nucleotides, have been well-characterised biochemically[3–6], and some sequenced[7,8]. However, their function has remained obscure. The most abundant snRNA species exist as a closely related set of RNA–protein complexes called small nuclear ribonucleoproteins (snRNPs)[9]. snRNPs are the antigens recognised by antibodies from some patients with lupus erythematosus (LE), an autoimmune rheumatic disease[10,11]. Anti-RNP antibodies from lupus sera selectively precipitate snRNP species containing Ula[7] and Ulb[9] RNAs from mouse Ehrlich ascites cell nuclei, whereas anti-Sm antibodies bind these snRNPs and four others containing U2 (ref. 8), U4, U5 and U6 (ref. 9) RNAs. Both antibody systems precipitate the same seven prominent nuclear proteins (molecular weight 12,000–32,000). All molecules of the snRNAs U1, U2, U4, U5 and U6 appear to exist in the form of antigenic snRNPs[9]. The particles sediment at about 10S and each probably contains a single snRNA molecule. Indirect immunofluorescence studies (refs 12, 13, and unpublished observations) using anti-RNP and anti-Sm sera confirm the nuclear (but non-nucleolar) location of the antigenic snRNPs. Here we present several lines of evidence that suggest a direct involvement of snRNPs in the splicing of hnRNA. Most intriguing is the observation that the nucleotide sequence at the 5′ end of U1 RNA exhibits extensive complementarity to those across splice junctions in hnRNA molecules.

If snRNPs participate in RNA processing, they might be expected to be strictly conserved across those higher eukaryotic species possessing interchangeable transcription and mRNA processing systems (refs 14–17 and P. Chambon, personal communication). Figure 1 shows that the antigenic snRNPs are indeed highly conserved from man to insects. The anti-RNP lanes reveal that mouse snRNPs containing the closely related Ula and Ulb RNAs[9] are replaced by a single U1-containing snRNP in HeLa and fall armyworm cells. Fingerprints (not shown) indicate that the sequence of HeLa U1 RNA is identical to that of mouse U1a[9] and to that of rat U1a[7]; insect U1 RNA has a quite different fingerprint, although it is comparable in size to the mammalian U1. Likewise, when anti-Sm antibody is used, four additional HeLa or insect cell snRNPs that contain RNAs very close in mobility to mouse U2, U4, U5 and U6 are precipitated; again, fingerprints of the HeLa U2, U4, U5 and U6 RNAs are identical to those of mouse(not shown). Also, human antibodies precipitate snRNPs from the nuclei of frog, chicken and sea urchin cells, indicating high conservation of both RNA and protein components of snRNPs in these species as well. No cross-reacting material is detected in tobacco cell, yeast, *Dictyostelium discoideum* or *Escherichia coli* extracts.

A second prediction concerning snRNP involvement in RNA biogenesis is that snRNAs should be present in highest abundance in metabolically active cell types. Figure 2 compares nuclear extracts from liver and from red blood cells (RBC) of chickens. The sucrose gradient profiles (Fig. 2*a*) showing extracts from equal numbers of the two types of nuclei reveal fewer than one-tenth as many 30S hnRNP particles (the highly conserved ribonucleoprotein complexes that bind hnRNA in the nuclei of higher eukaryotes[18–20]) and much less RNA sedimenting in the 10S region in the case of the RBC. Also, gel analysis of nuclear RNAs reveals snRNAs in both cases, but about 25-fold fewer in liver than in RBC (Fig. 2*b*, compare lane 2 with lanes 3 and 4). Treatment of the chicken with phenylhydrazine, which raised the level of erythroblasts to about 25% of the total RBC, altered the snRNA gel pattern (not shown) to that characteristic of liver cells (Fig. 2*b*, lane 2). Thus, liver cells, which actively synthesise mRNA, have both higher amounts and a slightly different population of snRNAs than the cryptic nuclei of biosynthetically inactive erythrocytes. Were the result otherwise, it would constitute evidence against snRNP participation in RNA processing.

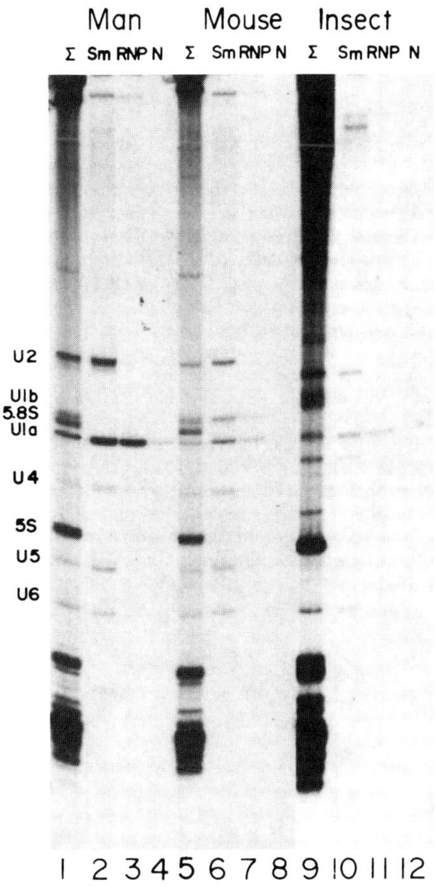

Fig. 1 Comparison of snRNPs from different species. Small RNAs from cell sonicates or immune precipitates from HeLa (lanes 1–4), Ehrlich ascites (lanes 5–8) and fall armyworm (lanes 9–12) cells were prepared as previously described[9] using Pansorbin to obtain the antigen–antibody complexes. The gel contained 10% polyacrylamide, 7 M urea, 1 mM EDTA, 50 mM Tris-borate *p*H 8.3 and was 400 × 200 × 0.5 mm. Lanes 1, 5 and 9 show small RNAs in whole cell extracts; 2, 6 and 10 show typical anti-Sm patterns; while 3, 7 and 11 show typical anti-RNP patterns. Lanes 4, 8 and 12 are controls using normal serum.

Finally, a number of workers[4,21–24] have observed snRNAs co-sedimenting with larger nuclear structures, such as the 30S particles that bind hnRNA. We have argued[9] that the snRNAs U1, U2, U4, U5 and U6 are not structural components of 30S hnRNP. Yet, as shown in Fig. 3*a*, *b*, on sucrose gradient fractionation of a nuclear extract prepared from Ehrlich ascites cells, antigenic snRNPs containing these snRNA species do sediment in the 30S region of the gradient. An RNA-labelled U1a*, however, appears only in the 10S region where the free snRNP particles sediment. Fingerprint analysis (Fig. 3*c*) reveals that

2

U1a* is identical to U1a (Fig. 3*d*) except that it lacks the 5′ trimethylated cap moiety plus six additional nucleotides from the 5′ end of the U1a sequence. Further, U1a* is not present in fresh extracts and the RNA–protein complex containing U1a* is fully antibody precipitable, strongly suggesting that U1a* is an *in vitro* degradation product of U1a. Thus, the absence of a short region at its 5′ terminus seems to prevent the Ula-containing snRNP from co-sedimenting with larger structures containing hnRNA.

It is interesting that several of the residues missing in U1a* fall within a region of U1a which is extensively complementary to sequences across splice junctions in hnRNAs (Fig. 4). DNA sequences corresponding to the ends of 43 introns were analysed (Table 1). Redundant sequences (those which are identical and occur in the same position on homologous RNAs) were discarded, leaving 26 unique sequences at the 5′ ends and 31 unique sequences at the 3′ ends of introns. The two consensus splice junction sequences are presented in Fig. 4*a*; in each position, the frequency of occurrence of the most common base is indicated by underlining as described in the legend. Potential base-pairing interactions between these regions and the nucleotide sequence at the 5′ end of U1a[7] or U1b RNA[9] are also shown. Note that the complementarity of U1 to the consensus sequence at the 5′ end of introns involves eight contiguous residues, plus an adjacent nucleotide (three bases before the splice point) which is generally either C or A. (It seems reasonable to consider that the $^{A}_{C}$ position is also complementary since A·G base pairs are observed where the two strands of an RNA helix diverge in yeast $tRNA^{Phe}$ (ref. 25).) Of the 26 5′ junction sequences, all but one has the dinucleotide GU immediately adjacent to the splice point and at least five (usually six) bases of the consensus sequence. Similarly, at the 3′ end of introns, all but one of the 31 unique sequences has the dinucleotide AG, and all

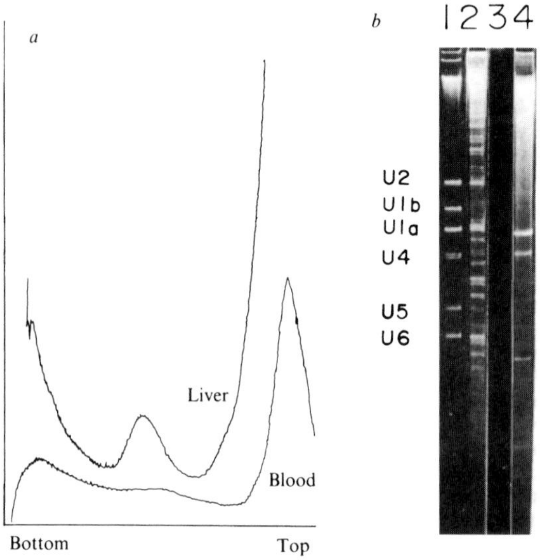

Fig. 2 Sedimentation patterns and gel analysis of small RNAs from nuclear extracts of liver and red blood cells of an adult chicken. *a*, Comparable numbers of liver and RBC nuclei (prepared as in refs 18 and 9, respectively) were extracted with *p*H 8.5 buffer[18]. The extracts were applied to 15–30% sucrose gradients and centrifuged for 17 h at 4 °C and 23,000 r.p.m. in an SW 41 rotor. SDS polyacrylamide gel electrophoresis[9] confirmed that the 30S peak contains primarily the prominent proteins of the 30S hnRNP[18–20]. *b*, RNAs obtained by phenol extraction were analysed by electrophoresis as in Fig. 1. Lane 1, marker Ehrlich ascites snRNAs selected by immunoprecipitation with anti-Sm antibody. Lane 2, total RNA from an extract of chicken liver nuclei. Lane 3, total RNA from an extract of an equivalent number of chicken RBC nuclei. Lane 4, total RNA from an extract of chicken RBC nuclei; the amount of extract loaded represents a 25-fold increase in number of nuclei over that shown in lanes 2 and 3.

Table 1 Intron–exon boundary sequences

Gene	5′Exon|	Intron	|Exon 3′	
Rat insulin	CAGGUAUGU	···	CUAUCUUCCAGG	36†
Rat insulin	AAGGUAAGC	···	CUCCCUGGCAGU	36
Rat insulin	CAGGUAUGU	···	CUAUCUUCCAGG	36
γ1 chain (newborn mouse)		···	UUUUCUUGUAGC	37
γ1 chain (newborn mouse)	UUGGUGAGA	···	UCUCUCCACAGU	37
	CAGGUAAGU		UUCAUCCUUAGU	
γ1 chain (newborn mouse)	AAGGUGAGA	···	CCCACCCACAGG	37
			UUUUCUUGUAGC	
γ1 chain (mouse myeloma)	UUGAGAGGA	···	UCUCUCCACAGU	38
	CAGGUAAGU		UUCAUCCUUAGU	
γ2 chain (mouse myeloma)	AAGGUGAGA	···	CUCACUCACAGG	38
γ1 chain (mouse myeloma)	CAGGUCAGC	···	CCUGUUUGCAGG	38
	CAGGUCAGC		UCUGUUUGCAGG	
γ1 chain (mouse myeloma)	UAGGUGAGU	···	UCAUCCUGCGGC	38
	AACGUAAGU		UCCUUCCUCAGG	
γ2 Chain	AACGUAAGU	···	UCCUUCCUCAGG	39
λ1 Chain	AACCUAAGU	···	UCCUUCCUCAGG	40
λ1 Chain	AACGUAAGU	···	UCCUUCCUCAGG	40
κ Chain	AACGUAAGU	···	UCCUUCCUCAGG	41
κ Chain	AAGGUUAAA	···	UCCACUCCUAGG	41
κ Chain	CAGGUUGGU	···	UCCCUUUUUAGG	41
κ Chain	AGGGUGAGU	···	UAUUCCCACAGC	41
κ Chain	CAGGUUGGU	···	CAUUUUCUCAGG	42
Mouse β-globin	AGGGUGAGU	···	UUUUCCUACAGC	43
Mouse β-globin		···	UCCUCCCACAGC	43
Rabbit β-globin		···	CUUCUCCGCAGC	43
Rabbit β-globin	AAGGUAGGC	···	GUUUGCUCUAGA	43
Human β-globin	AAGGUGAGC	···	UUCAAUUACAGG	43
Human δ-globin	CAGGUACAG	···	UUUCUAUUCAGU	44
Chicken ovalbumin	CCAGUAAGU	···	UUGCUUUACAGG	45
Chicken ovalbumin	AUGGUAAGG	···	CAUUCUUAAAGG	46
Chicken ovalbumin	GAGGUAUAU	···	UGGUUCUCCAGC	46
Chicken ovalbumin	CAGGUAUGG	···	UUUCCUUGCAGC	46
Chicken ovalbumin	AAGGUACCU	···	UUUUAUUUCAGG	46
Chicken ovalbumin	AAGGUUCGU	···	UUUUAUUUCAGG	46
Chicken ovalbumin	CUGGUAAGU	···	UUUUAUUUCAGG	47
SV40 late mRNAs	CUGGUAAGU	···	UUUACUUCUAGG	48, 49
SV40 late mRNAs	AAGGUAAAU	···	GUGUAUUUUAGA	48, 49
SV40 late mRNAs	GAGGUAUUU	···	GUGUAUUUUAGA	48, 49
SV40 late mRNAs				48, 49
SV40 early mRNAs				48, 49
SV40 early mRNAs				48, 49
Polyoma late mRNAs	CAAGUAAGU	···	UAUUUCCCUAGG	*
Polyoma late mRNAs	CAAGUAAGU	···	UUUAAUUCUAGG	*
Polyoma late mRNAs	CAAGUAAGU	···	UCUAUUUUAAGA	*
Silk fibroin	CAGGUGAGU	···	UUUUGUUUCAGU	50
	36 sequences, 26 unique		37 sequences, 31 unique	
Consensus	$^{A}_{C}$AGGUAAGU		UYUYYYU CAGG	

Sequences at splice junctions. The 36 'donor' (5′ end) sequences and 37 'acceptor' (3′ end) sequences represent 43 possible splicing events (each event is depicted on a separate line). Underlined sequences are redundant (either because they are identical to a homologous region or because they represent an alternative splice involving the same region) and were not included in our tabulation. Also not included are sequences at late adenovirus mRNA splice junctions as these are proposed to be recognised by VA RNA[26].

* R. Kamen and B. Griffin, personal communication.

† Refs.

can make at least six (usually eight or nine) of the 11 possible base pairs with U1 RNA.

The complementarity of the U1 sequence to splice junctions in mRNA precursors suggests how hnRNA exons separated by an intervening sequence could be brought into proper orientation for splicing. Because U1 RNA is capable of forming Watson–Crick base pairs with a significant number of residues that lie within the two ends of an intron, a single U1–containing

3

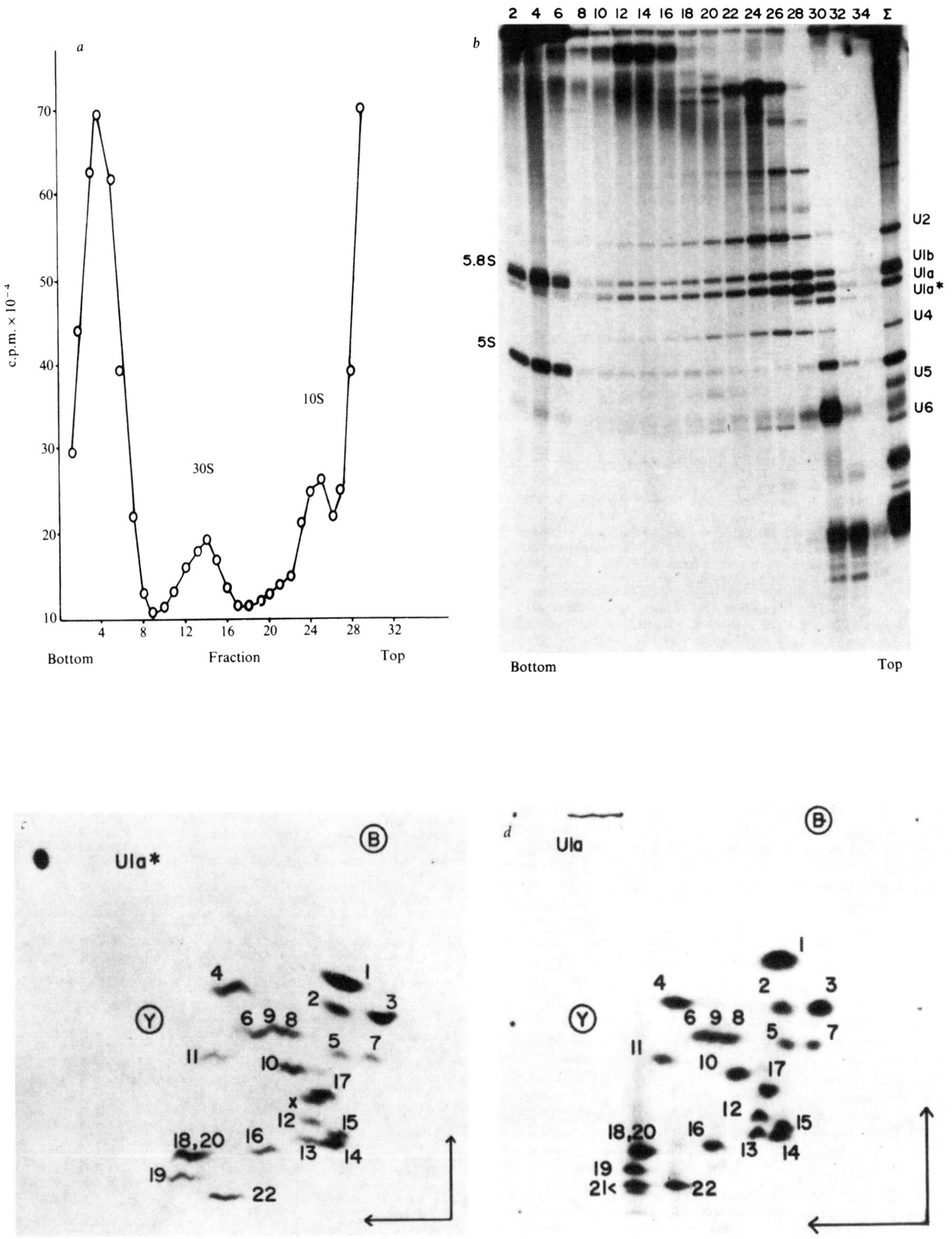

Fig. 3 Location and identification of snRNAs after sucrose gradient fractionation of a nuclear extract. *a*, Nuclei were prepared from ^{32}P-labelled Erlich ascites cells, extracted in high *p*H buffer[9] and the extract applied to a sucrose gradient as in Fig. 2*a*, *b*. The lanes in the gel (run as in Fig. 1) correspond directly to the numbered fractions in the gradient. Direct phenol extraction and prior immunoprecipitation with anti-Sm serum revealed the same distribution of snRNAs. Some contaminating ribosomes appear near the bottom of the gradient, but SDS polyacrylamide gel analysis reveals that the 30,000–40,000 MW 'core proteins'[18-20] of the hnRNP predominate in the 30S region. Indicated RNA species were identified by fingerprint analysis. *c*, T1 RNase fingerprint of U1a. The numbering system corresponds to ref. 7. Ⓑ and Ⓨ indicate the positions of the blue and yellow dyes in first (horizontal) and second (vertical) dimension. Spot 21 is the 5′ end oligonucleotide. *d*, T1 RNase fingerprint of U1a*. Spot X is the remnant of the 5′ oligonucleotide.

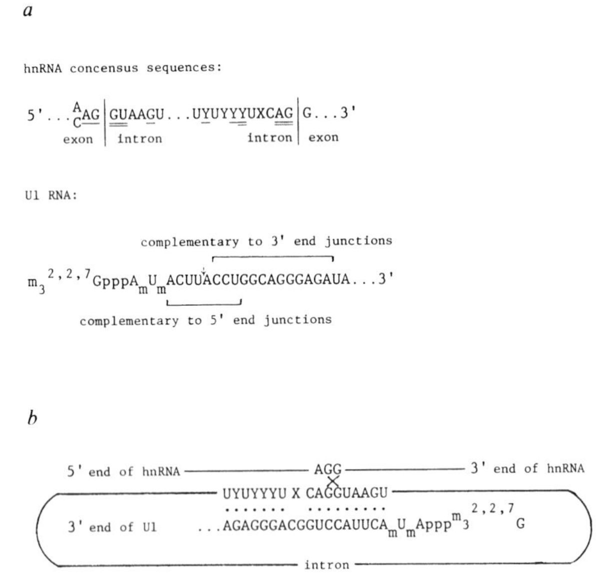

Fig. 4 Possible base-pairing interactions between U1 and splice junction consensus sequences. The U1 sequence is from ref. 7. Sequences used to generate the consensus sequences are presented in Table 1. *a*, To appear in the 5′ or 3′ consensus splice junction sequence, a base must be the most common in that position and occur with a frequency of at least 45%; bases occurring in 75% of the sequences are underlined; those present with 95% or greater frequency are underlined twice. In the position marked $^{A}_{C}$, A occurs 11 times and C 10 times in the 26 unique sequences examined. Y indicates pryimidines. X marks the location of a single nonconserved base in the 3′ consensus sequence. Vertical lines show the most likely location of the splice. The arrow locates the 5′ end of U1a*. All bases occurring with a frequency of 45% or greater are shown. *b*, A possible model for alignment of intron–exon boundaries by base pairing between U1 RNA and sequences at both ends of an intron.

snRNP could interact with both junctions simultaneously (Fig. 4*b*). A similar model has been independently proposed suggesting the involvement of an adenovirus-encoded small RNA known as VA in the processing of late adenovirus messages[26]. Although the complementarity between U1 RNA and splice junction sequences is quite striking, the stability of the predicted helices may be insufficient to achieve the fidelity of splicing observed *in vivo*. Thus, it is attractive to invoke RNA·RNA interactions as only one component of the recognition of intron–exon junctions by a particle containing both U1 RNA and proteins.

In summary, the abundance of U1 RNA in active cells (about 10^6 per nucleus), the high conservation of both its sequence and the proteins with which it associates, and its considerable complementarity to splice junctions are all consistent with the idea that U1-containing snRNPs have a key role in eukaryotic mRNA processing. They could either be the splicing enzyme itself (note the example of *E. coli* RNase P, which contains a small RNA complexed with several small proteins[27]) or an auxiliary factor required to align splice junctions. One specific prediction of our hypothesis is that avian and insect U1 RNAs should possess 5′ end sequences similar to mammalian U1 as the splice junction sequences determined for the chicken ovalbumin and silkworm fibroin genes show the same degree of complementarity to mammalian U1 as do mammalian splice junction sequences (see Table 1). Likewise, it should be possible to demonstrate direct interaction between a U1-containing snRNP and splice junction sequences in a particular hnRNA. Finally, the introduction of antibodies into living cells or *in vitro* splicing systems[28,29] may result in the inhibition of one or more specific steps in the RNA processing pathway. Experiments relating to this are now underway.

The fact that the other closely related snRNPs (containing U2, U4, U5, or U6 snRNAs) possess common antigenic determinants suggests that they may function comparably to the U1-containing snRNP in RNA processing; different RNA sequences at their 'active sites' could facilitate more precise recognition of variant splice junction sequences in hnRNA or in other RNA molecules. Alternatively, the additional related snRNPs could participate in the transport of RNA molecules from the cell nucleus, as is suggested by the finding that splicing and export seem to be inseparably linked for those hnRNAs that undergo splicing[30,31]. Whatever the outcome, it seems reasonable to investigate further the proposal[26,32–35] that RNA·RNA interactions provide specificity for the precise editing of eukaryotic transcripts; this possibility seems especially attractive when one considers the central role of base pairing interactions in other essential processes (for example, transcription and translation) that developed at an early stage in evolution.

We thank John Hardin and J. Pagliaro for gifts of sera, Dennis Knudsen and Dave Ward for cultured cells and Joan Weliky for technical assistance. This work was supported by USPHS training grants to M.R.L., J.A.B. and S.L.W., an NSF predoctoral fellowship to S.M.M. and NSF grant PCM-74-01136 and NIH grant GM26154 to J.A.S.

Received 27 September; accepted 15 November 1979.

1. Weinberg, R. & Penman, S. *J. molec. Biol.* **38**, 289–304 (1968).
2. Hellung-Larsen, P. & Frederiksen, S. *Comp. Biochem. Physiol.* **58B**, 273–281 (1977).
3. Ro-Choi, T. S. & Busch, H. in *The Cell Nucleus* (ed. Busch, H.) 151–208 (Academic, New York, 1974).
4. Zieve, G. & Penman, S. *Cell* **8**, 19–31 (1976).
5. Benecke, B.-J. & Penman, S. *Cell* **12**, 939–946 (1977).
6. Jelinek, W. & Leinwand, L. *Cell* **15**, 205–214 (1978).
7. Reddy, R., Ro-Choi, T. S., Henning, D. & Busch, H. *J. biol. Chem.* **249**, 6486–6494 (1974).
8. Shibata, H. *et al. J. biol. Chem.* **250**, 3909–3920 (1975).
9. Lerner, M. R. & Steitz, J. A. *Proc. natn. Acad. Sci. U.S.A.* **76**, 5495–5499 (1979).
10. Notman, D. D., Kurata, N. & Tan, E. M. *Ann. Int. Med.* **83**, 464–469 (1979).
11. *Provost, T. T. J. Invest. Dermatol.* **72**, 110–113 (1979).
12. Mattioli, M. & Reichlin, M. *J. Immun.* **107**, 1281–1290 (1971).
13. Northway, J. O. & Tan, E. M. *Clin. Immun. Immunopath.* **1**, 140–154 (1972).
14. Weil, P. A., Luse, D. S., Segall, J. & Roeder, R. G. *Cell* **18**, 469–484 (1979).
15. Hamer, D. H. & Leder, P. *Nature* **281**, 35–40 (1979).
16. Mantei, N., Boll, W. & Weissmann, C. *Nature* **281**, 40–46 (1979).
17. Mulligan, R. C., Howard, B. H. & Berg, P. *Nature* **277**, 108–114 (1979).
18. Martin, T. *et al. Cold Spring Harb. Symp. quant. Biol.* **38**, 921–932 (1973).
19. Karn, J., Vidali, G., Boffa, L. C. & Allfrey, V. G. *J. biol. Chem.* **252**, 7307–7322 (1977).
20. Beyer, A. L., Christensen, M. E., Walker, B. W. & L. Stourgeon, W. M. *Cell* **11**, 127–138 (1977).
21. Howard, E. F. *Biochemistry* **17**, 3228–3236 (1978).
22. Deimel, B., Louis, C. & Sekeris, C. E. *FEBS Lett.* **73**, 80–84 (1977).
23. Northemann, W., Scheurlen, M., Gross, V. & Heinrich, P. C. *Biochem. biophys. Res. Commun.* **76**, 1130–1137 (1977).
24. Guimont-Ducamp, C., Sri-Widada, J. & Jeanteur, P. *Biochimie* **59**, 755–758 (1977).
25. Kim, S.-H. in *Transfer RNA* (ed. Altman S.) 248–293 (MIT, Cambridge, Mass., 1979).
26. Murray, V. & Holliday. R. *FEBS Lett.* **106**, 5–7 (1979).
27. Stark, B. C., Kole, R., Bowman, E. J. & Altman, S. *Proc. natn. Acad. Sci. U.S.A.* **75**, 3717–3721 (1978).
28. Blanchard, J.-M., Weber, J., Jelinek, W. & Darnell, J. E. *Proc. natn. Acad. Sci. U.S.A.* **75**, 5344–5348 (1978).
29. Manley, J. L., Sharp, P. A. & Gefter, M. L. *J. molec. Biol.* (in the press).
30. Hamer, D. H., Smith, K. D., Boyer, S. H. & Leder, P. *Cell* **17**, 725–735 (1979).
31. Gruss, P., Lai, C.-J., Dhar, R. & Khoury, G. *Proc. natn. Acad. U.S.A.* **76**, 4317–4321 (1979).
32. Reanney, D. *Nature* **277**, 598–600 (1979).
33. Crick, F. *Science* **204**, 264–271 (1979).
34. Roberts, R. J. *Nature* **274**, 530 (1978).
35. Church, G. M., Slonimski, P. P. & Gilbert, W. *Cell* (in the press).
36. Lomedico, P. *et al. Cell* **18**, 545–558 (1979).
37. Honjo, T. *et al. Cell* **18**, 559–568 (1979).
38. Sakano, H. *et al. Nature* **277**, 627–633 (1979).
39. Tonegawa, S., Maxam, A. M., Tizard, R., Bernard, O. & Gilbert, W. *Proc. natn. Acad. Sci. U.S.A.* **75**, 1485–1489 (1978).
40. Bernard, O., Hozumi, N. & Tonegawa, S. *Cell* **15**, 1133–1144 (1978).
41. Max, E. E., Seidman, J. G. & Leder, P. *Proc. natn. Acad. Sci. U.S.A.* **76**, 3450–3454 (1979).
42. Seidman, J. G., Max, E. E. & Leder, P. *Nature* **280**, 370–375 (1979).
43. van Ooyen, A., van den Berg, J., Mantei, N. & Weissmann, C. *Science* **206**, 337–344 (1979).
44. Lawn, R. M., Fritsch, E. F., Parker, R. C., Blake, G. & Maniatis, T. *Cell* **15**, 1157–1174 (1978).
45. Gannon, F. *et al. Nature* **278**, 428–434 (1979).
46. Catterall, J. F. *et al. Nature* **275**, 510–513 (1978).
47. Breathnach, R., Benoist, C., O'Hare, K., Gannon, F. & Chambon, P. *Proc. natn. Acad. Sci. U.S.A.* **75**, 4853–4857 (1978).
48. Ghosh, P. K., Reddy, V. R., Swinscoe, J., Lebowitz, P. & Weissman, S. M. *J. molec. Biol.* **126**, 813–846 (1978).
49. Fiers, W. *et al. Nature* **273**, 113–120 (1978).
50. Tsujimoto, Y. & Suzuki, Y. *Cell* **18**, 591–600 (1979).

Printed in Great Britain by Macmillan Production Ltd., Basingstoke, Hampshire

Skoglund U., Andersson K., Björkroth B., Lamb M.M., and **Daneholt B.** 1983. Visualization of the formation and transport of a specific hnRNP particle. *Cell* **34:** 847–855.

THE GIANT POLYTENE CHROMOSOMES OF FLY LARVAE and the giant lampbrush chromosomes of amphibian oocytes have clarified numerous issues of nuclear structure and function that are difficult to demonstrate with smaller chromosomes. In the 1950s Wolfgang Beermann made extensive cytological and genetic observations on the midge *Chironomus tentans*, thereby establishing the polytene chromosomes in its salivary glands as a useful model system. Specifically, he showed that the so-called Balbiani Rings (BRs) are products of hyperactive genes that synthesize the mRNAs for the salivary gland proteins (1). Stevens and Swift (2) showed by electron microscopy that BRs produce ribonucleoprotein granules that are released into the nucleoplasm and eventually make their way to the cytoplasm through the nuclear pores. Bertil Daneholt and his collaborators further extended BR studies by showing that individual nascent transcripts could be visualized by electron microscopy. The paper reproduced here describes changes in the structure of the transcripts as they move along the DNA axis, much as Miller and his colleagues described transcripts on the rDNA of amphibian oocytes (Miller and Beatty, this volume). The initial characterization of the BR transcripts permitted later analysis of splicing, assembly of hnRNP complexes, and transport of the completed messenger RNP through the nuclear pore complex (3), one of the few cases in which a specific transcript can be followed from its initial site of synthesis on the chromosome to its final appearance in the cytoplasm.

---

1. Beerman W. 1961. Ein Balbiani-Ring als Locus einer Speicheldrüsen-Mutation. *Chromosoma* **12:** 1–25.
2. Stevens B.J. and Swift H. 1966. RNA transport from nucleus to cytoplasm in *Chironomus* salivary glands. *J. Cell Biol.* **31:** 55–77.
3. Daneholt B. 1997. A look at messenger RNP moving through the nuclear pore. *Cell* **88:** 585–588.

Cell, Vol. 34, 847–855, October 1983, Copyright © 1983 by MIT 0092-8674/83/100847-09 $02.00/0

Visualization of the Formation and Transport of a Specific hnRNP Particle

U. Skoglund,*† K. Andersson,‡ B. Björkroth,‡ M. M. Lamb,‡ and B. Daneholt ‡
*Department of Molecular Biology
Wallenberg Laboratory
University of Uppsala
Box 562
S-75122 Uppsala, Sweden
‡Department of Medical Cell Genetics
Medical Nobel Institute
Karolinska Institutet
S-10401 Stockholm 60, Sweden

Summary

The growth and maturation of the transcription products on the Balbiani ring (BR) genes in Chironomus tentans has been characterized by electron microscopy. The BR transcript is packed into a series of well defined ribonucleoprotein structures of increasing complexity: a 10 nm fiber, a 19 nm fiber, a 26 nm fiber, and a 50 nm granule. The basic 10 nm element was revealed in Miller spreads. The in situ structure of the transcription products and RNA compaction estimates suggested that the 10 nm fiber is packed into the 19 nm fiber as a tight coil. The transition of the 19 nm fiber into the 26 nm fiber is accompanied by a major change of the basic 10 nm fold into a noncoiled structure. Finally, the 26 nm fiber makes a one and one-third left-handed turn forming the final product, the BR granule. Upon translocation through the nuclear pore the BR granule becomes rod-shaped, which most likely corresponds to a relaxation of the highest-order structure into a straight 26 nm fiber.

Introduction

Heterogeneous nuclear RNA (hnRNA) molecules are known to be associated with proteins generating ribonucleoprotein particles (hnRNP; for review, see Van Venrooij and Janssen, 1978). The formation of the particles is intimately coupled to the transcription process, and the growing transcripts appear as RNP fibers on the chromatin template (Miller and Bakken, 1972; Economidis and Pederson, 1983). Early studies suggested a simple and regular organization of the hnRNP particles (Samarina et al., 1968; Martin et al., 1974), but more recent studies emphasize a high complexity in hnRNP with more than 50 protein species involved (Jacob et al., 1981). Since hnRNA is known to be engaged in a series of intranuclear events—e.g., packing, a series of maturation steps and transport (Darnell, 1982)—it is plausible that the complex organization of hnRNP reflects the multitude of processes associated with it. In this context it is interesting to recall that Steitz and Kamen (1981) demonstrated that intervening RNA sequences of polyoma hnRNP are more sensitive to RNAase treatment than premessage segments. Furthermore, Miller and coworkers recently established in chromatin spreads that the arrangement of granules along nascent RNP fibers is nonrandom (Beyer et al., 1980) and when the excision of RNP fiber segments takes place, it usually occurs between two consecutive granules (Beyer et al., 1981). These studies strongly support the idea that the structural organization of hnRNP has functional significance.

As nicely shown by Beyer et al. (1981), ultrastructural studies can contribute to the understanding of specific features of hnRNP. The chromatin-spreading procedure is, however, known to modify or even abolish certain structural features of RNP (and DNP)—in particular the higher-order conformations (Sommerville, 1981). It is therefore most important to characterize the hnRNP structure as it appears within the cell. Perichromatin fibrils and stalked granules have been observed in situ and they are likely to represent growing hnRNP particles (Puvion and Moyne, 1981). More specific studies of the hnRNP structure and the packing process have not been feasible, mainly because it has not been possible to identify and compare consecutive RNP fibers or particles on the template. Only recently, a specific group of active genes, the BR genes in the salivary glands of Chironomus tentans, became amenable for in situ analysis. These genes are known to generate messenger RNA molecules (75S RNA) encoding giant salivary polypeptides (for review see Daneholt, 1982). The gross morphology of proximal transitional, as well as distal, regions of the BR genes were outlined (Andersson et al., 1980).

In the present study we focus our analysis on the formation of the premessenger particles along the BR genes, as well as on the internal structure of the completed and released transcription products, the BR granules. Our study suggests that the hnRNA molecules are complexed with proteins to form a series of discrete, high-order structures: 10, 19, and 26 nm fibers and a 50 nm granule. The separate analysis of the released product is in agreement with the packing of the fiber as observed in its nascent state. When the BR particle passes through the nuclear pore it becomes elongated, which probably implies that the highest-order structure has been relaxed.

Results

The In Situ Structure of the Growing RNP Particles on Balbiani Ring Genes

Balbiani ring genes producing 75S RNA can be visualized in the electron microscope (Andersson et al., 1980). The large size of the genes and the corresponding transcripts (37 kb, according to Case and Daneholt, 1978), have made these active genes particularly suitable for analysis. As shown in Figure 1a, a great number of segments of active BR genes can be observed in a section through a Balbiani ring. Examples of promoter-proximal and pro-

† Present address: Department of Medical Cell Genetics, Medical Nobel Institute, Karolinska Institutet, S-10401 Stockholm 60, Sweden.

moter-distal regions have been encircled in the figure; one proximal region has been marked with p and three distal regions with d1, d2, and d3. Moreover, one example of the transitional region between a proximal and a distal segment has been indicated (t). A schematic figure of the gross morphology of an active BR gene is given in Figure 1b (designations as in Figure 1a). It can be noted how RNP fibers appear in the proximal quarter of the gene and how, further downstream in the gene, the transcription products are visible as more complex RNP structures, often described as stalked granules (Andersson et al., 1980).

In this study we have analyzed the fine morphology of the growing RNP products. A number of proximal transitional, as well as distal, regions of the BR genes have been scrutinized. Figure 2 displays a series of representative RNP structures from the proximal to the most distal portion of the BR genes (illustrative particles have been indicated by arrowheads in the figures). Together these various stages of maturation of the growing RNP product visualize the growth and packing of the transcription product. In the proximal segment, an RNP fiber of roughly even thickness and density is observed (Figure 2a). The diameter has been measured as 19 nm (±3; 30 determinations). In the transitional region, the 5′ end of the RNP fiber becomes packed into a denser structure (Figures 2b and 2c). At this stage of maturation and further downstream in the gene, the transcription product can conveniently be described as an RNP particle with a stalk (the 19 nm fiber) and a peripheral dense portion. The dense portion of the particle will appear as a short, dense fiber with a diameter of 26 nm (±2; 10 determinations). It can also be clearly seen that this thick fiber is somewhat curved (see Figure 2d). The 26 nm fiber will subsequently be bent more and more, until finally the dense portion of the particle attains a globular structure about 50 nm in diameter (Figure 2e–2h). It should be noted that this globular shape is recorded in all the projections of the RNP particle (the attachment of the stalk to the dense part being peripheral in some projections and more or less central in other projections).

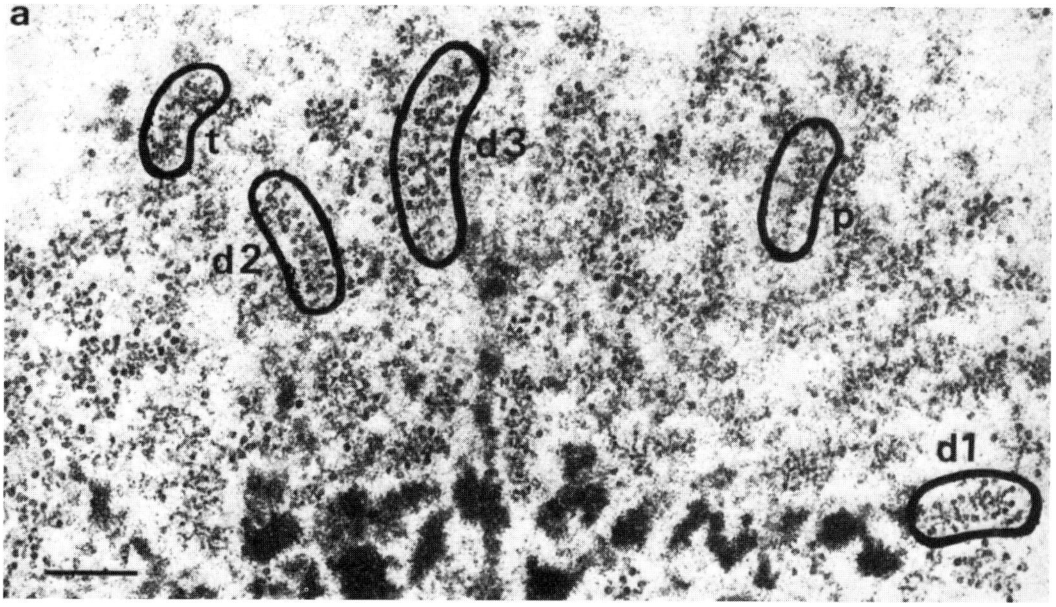

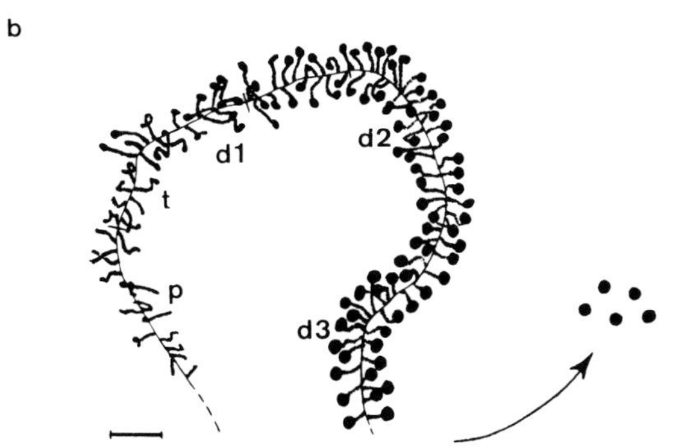

Figure 1. The Ultrastructure of Various Segments of a Balbiani Ring Gene
(a) An electron micrograph of a Balbiani ring on chromosome IV in the salivary glands of Chironomus tentans. Since this micrograph represents a thin section through the puff, the different genes in the polytene bundle have been cut at various stages of transcription. Five representative segments of active BR genes have been encircled and designated: p = proximal, t = transitional, and d1–d3 = distal regions. Bar = 500 nm. (b) Reconstruction of one complete BR gene with the various gene segments indicated. It can be seen how the granule starts in the t region and grows continuously in size throughout the d3 stage, after which it is released. The stalk seems to keep its length constant. Bar = 200 nm.

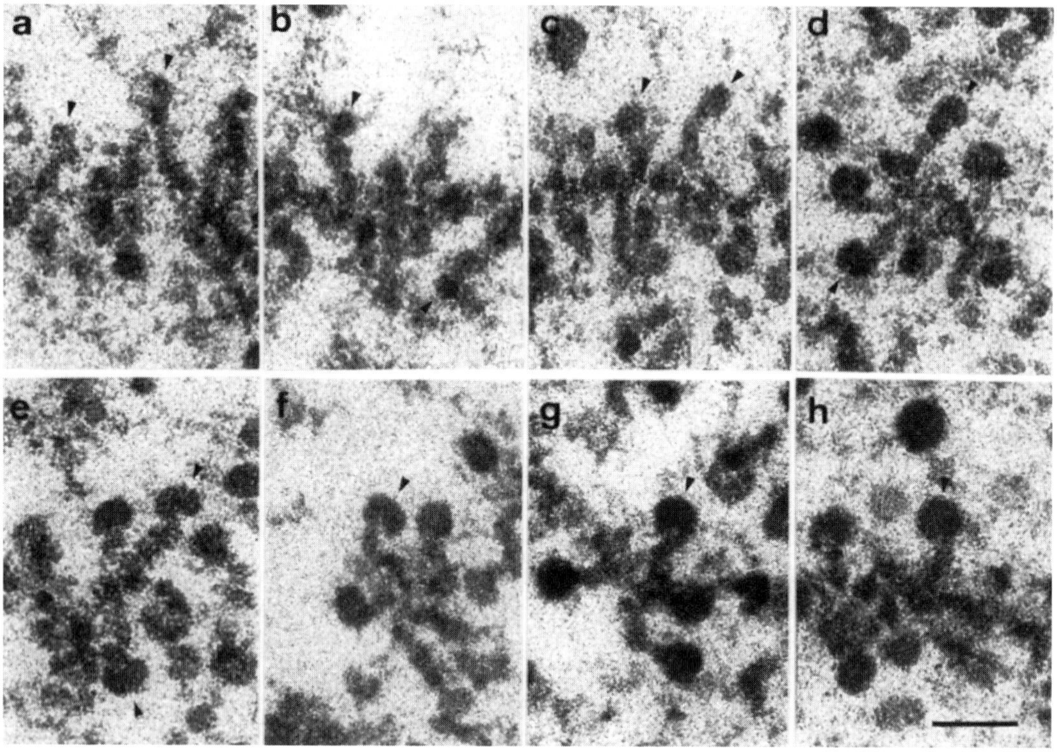

Figure 2. Electron Micrographs of Growing RNP Particles in Various Stages of Maturation

Close-up examples of growing RNP particles from the proximal (a), transitional (b, c), and distal (d–h) portions of the gene. The peripheral dense portion seems to bend continuously into a globular shape. Bar = 100 nm.

Elongated projections, expected to be seen occasionally, resulting from a ring-like configuration, were not recorded. We conclude that the bent 26 nm fiber most likely does not conform to a circular shape, but rather to a spiral-type structure, giving the particle a globular shape. This interpretation is in agreement with the structural analysis of the released product, the Balbiani ring granule (see below). It should also be noted that, while the 26 nm fiber grows in length, the stalk keeps its length constant. Evidently, the newly made transcript is packed into a 19 nm RNP fiber at the same rate that the RNP of the 19 nm fiber is transferred to the 26 nm fiber conformation. As the released granules lack stalks, it seems likely that the stalk is either incorporated into the globular part or discarded after the completion of transcription.

Various portions of the growing RNP particles have been studied in order to reveal further structural elements. Normally, both the 19 and 26 nm fibers display a more or less homogeneous structure. Occasionally we have observed that the stalk portion consists of a 10 nm fiber forming a loose spiral. Three such examples are presented in Figure 3. It should be pointed out that the recorded spirals vary in number and size of the coils. If these structures do reflect the normal structure of the stalk, they suggest that the stalk is built from a 10 nm fiber, tightly packed into a spiral 19 nm in diameter.

The Structure of Growing Balbiani Ring Products in Miller Spreads

Chromosome IVs carrying the BR genes were isolated and spread on an electron microscopic grid according to Miller and Bakken (1972), with the modifications specified in Lamb and Daneholt (1979). Long transcription units corresponding to the BR genes were observed, and we could confirm the observations made by Lamb and Daneholt (1979) that the growing transcripts form a length gradient from the proximal to the distal part of the gene (see Figure 3 in Lamb and Daneholt, 1979).

In the proximal portion of the gene, the fibers are usually uncoiled and form a distinct length gradient (Figure 4a). The thin RNP fibers have a diameter of about 10 nm and exhibit irregularly sized beads with intervening smooth segments. In the more distal parts of the gene, the RNP particles are not as well spread and remnants of the stalked granules are seen. Occasionally, however, it is possible to see how the entire RNP particle has uncoiled into a long, 10 nm RNP fiber (Figure 4b). The long fibers on the distal part of the BR gene display the same structure as the

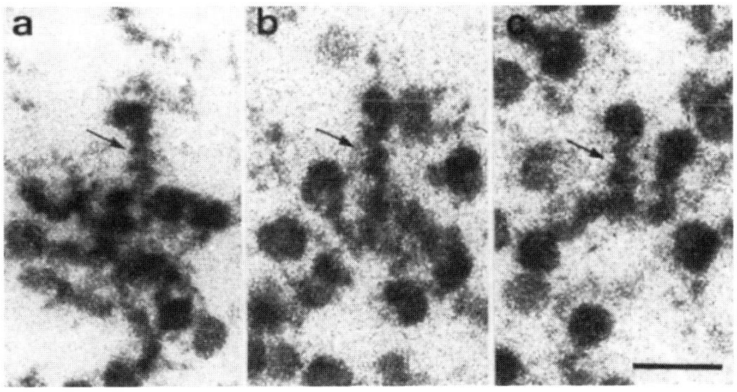

Figure 3. Electron Micrographs of Growing RNP Particles with Partially Uncoiled Stalks

Three examples of growing RNP particles where the 19 nm stalk portion has "loosened up" and displays a 10 nm fiber. Bar = 100 nm.

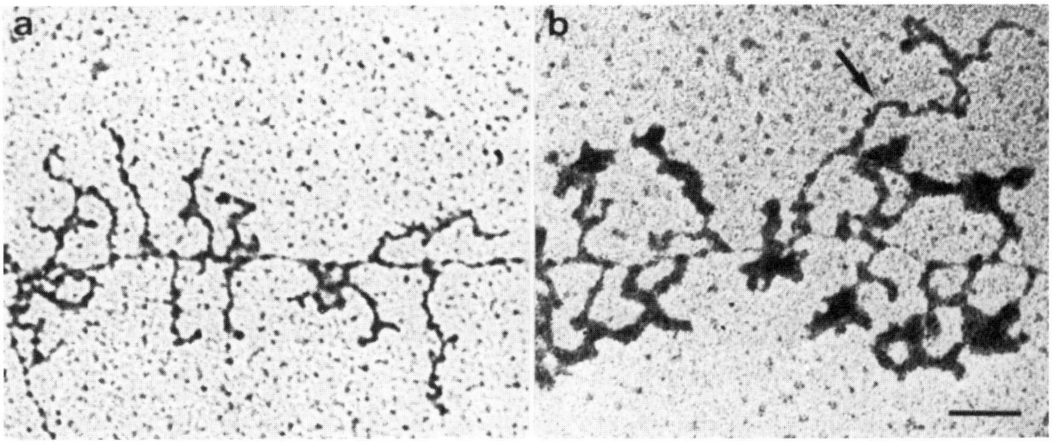

Figure 4. Details from Miller Spreads of Chromosome IV Displaying the Proximal and Distal Parts of the Gene

(a) The spread RNP stalk region has formed a 10 nm fiber with a "knobby" look. (b) The distal region does not spread as readily as the proximal end under mild spreading conditions, but occasionally the entire RNP particle is unfolded and exhibits the 10 nm fiber element (arrow). Bar = 100 nm.

fibers in the proximal part of the gene, and evidently the morphology of the 10 nm fiber is the same all along the gene.

The observed knobby structure of the 10 nm fiber is characteristic of growing RNP fibers of spread nonribosomal transcription units from a large number of sources. (McKnight et al., 1979; Sommerville, 1981). It has been suggested that the beads correspond to the 30–50S monoparticles, which can be released from hnRNP by mild RNAase treatment and are believed to be an essential structural component of hnRNA (Samarina et al., 1968; Martin et al., 1974; Beyer et al., 1977), although not necessarily the only component (e.g., the heterogeneous complexes identified by Jacob et al., 1981). The beads are known to be distributed along the spread RNP fiber in a nonrandom manner (Beyer et al., 1980), and in one way or another they should reflect the in vivo structure. It is important, however, to consider that the disruption of the in vivo structure during the spreading process could cause artifacts, and extrapolation back to the in vivo structure has to be done with great care. In this context, we want to make the point that the BR hnRNP particle contains the same type of basic RNP fiber element as other nonribosomal genes.

Our chromatin spreads suggest that the transcription product is built from an elementary fiber (the 10 nm fiber), which during the transcription process can be packed into the stalk and into other, denser conformations. The observation in situ of a 10 nm fiber as the constituent fiber of the stalks (see above) suggests that the 10 nm fiber revealed in chromatin spreads has its counterpart in vivo.

Determination of RNA Compaction in the 10, 19, and 26 nm RNP Fibers

In order to be able to characterize the RNP fibers recorded in Miller spreads, as well as in the sectioned material, the RNA compaction of the fibers was estimated. Although the figures obtained are only approximate values, they enable us to compare the observed fibers and to evaluate models for folding of the fibers into higher-order structures (see Discussion).

In Miller spreads, entire BR genes are available for

analysis of the RNA compaction of the 10 nm RNP fiber. We have analyzed the proximal parts of four transcription units, since the 10 nm RNP fibers can be satisfactorily uncoiled and measured, and the fibers form a distinct length gradient in this part of the gene. The lengths of the growing RNP fibers were plotted against the distance from the putative initiation site, according to the method devised by Laird and Chooi (1976). The RNA compaction value could be measured directly from the diagram (RNA compaction is defined here as the length of the RNA molecule over the length of the RNP fiber; it is then assumed that the RNA molecule attains the same backbone length as a corresponding B-form DNA molecule). In the four units analyzed we obtained an average figure of about 6 for the compaction (the individual values were: 3.6, 6.3, 6.4, 7.0, average 5.8). This determination of the compaction is in good agreement with similar measurements made on RNP fibers in active genes from other species and cell types (for review, see Sommerville, 1981).

The RNA compaction of the 10 nm fiber, occasionally recorded in situ in sectioned material as the constituent of the stalk, was estimated in the following way. The projected lengths of the 10 nm fibers of the stalks in Figure 3 were measured. The mean projected length (270 nm) was multiplied by $4/\pi$ to compensate for the statistical shortening in the projections (Underwood, 1970), giving the fiber a length of 344 nm. Since the stalk is known to contain 9.3 kb (¼ × 37 kb, Andersson et al., 1980), the compaction was calculated as approximately 9. We conclude that the compaction of the 10 nm fiber in situ is somewhat higher than that of the spread fiber, but the difference can be adequately explained by the assumption that the spread RNP fiber has been slightly extended during the spreading process (for discussion, see Sommerville, 1981).

In order to estimate the RNA compaction in the 19 nm fiber, we measured from the electron micrograph the *projected* length of the 19 nm fiber at a position just upstream of the transitional region between the proximal and the distal portions of the gene. The average length of the fiber in this region amounted to 90 nm (±27; 30 determinations). This figure enabled us to calculate the length of the fibers as 115 nm, using the factor $4/\pi$. Knowing that the fibers correspond to a position about one-quarter of the gene downstream of the start of RNA synthesis, and hence contain an RNA transcript 9.3 kb long, we calculated the RNA compaction as 27.

The compaction of RNA in the 26 nm fiber was estimated in the following way. RNP particles with a long, curved 26 nm fiber were chosen for analysis. A typical example of the stage studied is given in Figure 5a. Particles of this maturation stage were observed at approximately the transition between the third and the fourth quarter of the gene. Upstream, the 26 nm fiber was not as long, and downstream, the fiber was bent into the granular conformation (see Figure 2). The length of the 26 nm fiber was measured as shown in Figure 5b (dotted line), and it amounted to 67 nm (±3; 10 silhouette measurements). Since an RNP particle at this position contains three-fourths of the complete transcript and the stalk harbors one-fourth, the 26 nm fiber should contain half of a complete transcript—that is, 18.5 kb. We then calculated the RNA compaction as 94 for the 26 nm fiber portion of the particle.

It is obvious that the 10 nm fiber, the 19 nm fiber, and the 26 nm fiber constitute a series of increasing degrees of RNA compaction. This result is in good agreement with the concept of a basic element, the 10 nm fiber, which is gradually packed into a number of higher-order structures (see Discussion).

The Structure of Balbiani Ring Granules in the Nuclear Sap

Large granules, 50 nm in diameter, have been observed earlier in nuclear sap and tentatively identified as Balbiani ring products (Beermann 1962; Stevens and Swift, 1966). In the present study, we have further investigated the internal structure of the granules and compared this with the appearance of the transcription products in nascent state. It was noted that the contours of the nuclear sap particles were not always spherical and that many of the particles did not appear as solid granules. When 500 randomly chosen granules were studied, about 10% could be characterized as ring-like and 18% as horseshoe-like in single projections. The less dense portions of the granules clearly indicated an internal organization of the large granules. We therefore proceeded to analyze such granules by tilting the specimen and photographing various projections of the granules (0°; ±15°; ±30°; ±50°).

Stereo electron microscopical analysis of the tilted series has suggested that the granule consists of a fiber, approximately 25 nm in diameter, wound one and one-third turn. We have not measured the exact diameter because of the difficulty of clearly demarcating the fiber. We assume, however, that this fiber in the completed particle is the

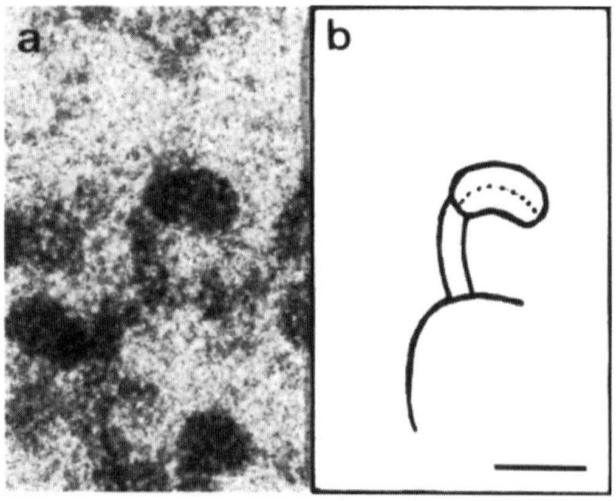

Figure 5. A Representative Growing RNP Particle Chosen for Estimation of the Compaction Value for the 26 nm Fiber

The particles were chosen when three-quarters mature (d2 region in Figure 1). In (a) such a particle is shown and in (b) a dotted line reveals how the length of the 26 nm fiber was measured. Bar = 50 nm.

Cell
852

same as the 26 nm fiber observed during the synthesis of the particle, and we will refer to it as the 26 nm fiber.

A model of the particle was built in both left-handed and right-handed versions and was subsequently compared with eight tilt series of suitable particles. One example of such an analysis is displayed in Figure 6, which presents seven projections from −50° to +50°. In this projection series, we observe examples of the ring-like and the horse-shoe projections often recognized in single electron micrographs of nuclear sap particles. The interpretations drawn from the model are given beneath the electron micrographs. It was found that the projection series of all the eight particles were in agreement with the spiral model. Moreover, in six out of the eight cases, it could be seen clearly that only the left-handed spiral fit all of the projections (see Figure 6). In the remaining two cases it was not possible to decide definitely whether a right- or left-handed spiral fit. Thus large nuclear sap granules probably constitute a left-handed spiral of a 26 nm RNP fiber wound one and one-third turns.

To decide whether the observed structure was the product of a general occurrence, we studied two arbitrarily chosen areas within the nucleus, one containing 15 granules and the other containing 10 granules. The analysis was carried out using a rotation specimen holder and it was possible to investigate the sample so that the tilting axis was in a series of different positions relative to the specimens. With this approach it was possible to investigate thoroughly a large number of different projections of each granule. It was established that 22 out of the 25 granules could be brought into positions showing typical projections exhibiting low-density portions (ring-like or horse-shoe conformations). Moreover, the contouring envelopes of the remaining three granules were in agreement with the spiral arrangement. We infer that the observed structure is representative of all or almost all putative BR granules in the nuclear sap.

From our model of the completed RNP particle we estimated the length of the 26 nm fiber in order to establish the compaction of the 26 nm fiber in its final conformation. Knowing the diameter of the granule (50 nm), we determined the length of the fiber to be 140 nm. A transcript size of 37 kb gives the compaction value of 90 for the fiber in this spiral state. This figure is in good agreement with the compaction calculated for the 26 nm fiber during the formation of the RNP particle (94). Although the spiral model satisfies the various projection pictures, further analysis of the RNP particles by three-dimensional reconstruction procedures is necessary to establish proper dimensions of the particle. It is obvious that the spiral structure observed is in good agreement with the formation process studied at the level of the gene.

Discussion

The Presence of an Ordered Packing Process

In this paper we have followed the gradual packing of a growing transcript into an RNP structure at a specific group of genes—the Balbiani ring genes in C. tentans. It is striking that the process involves a series of discrete structures of increasing complexity: the 10 nm fiber → the 19 nm fiber → the 26 nm fiber → the 50 nm granule (Figure 7). The gradual packing proceeds in a well ordered manner. This is evident when the nascent particles are compared along the gene. There might be some minor structural differences between adjacent particles, but the striking observation is

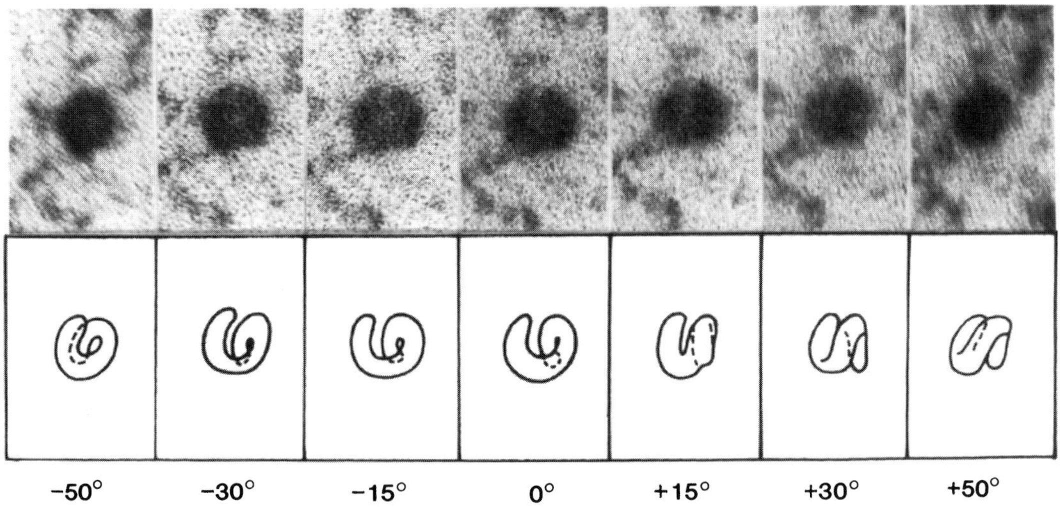

Figure 6. A Tilt Series Example of a Specific, Completed RNP Particle

The tilt axis is roughly vertical. The often observed horse-shoe or ring-like appearing projections of RNP particles are clearly visible at 0° and −30° tilt, respectively. The interpretation in terms of our one and one-third turn left-handed spiral is given below the micrographs.

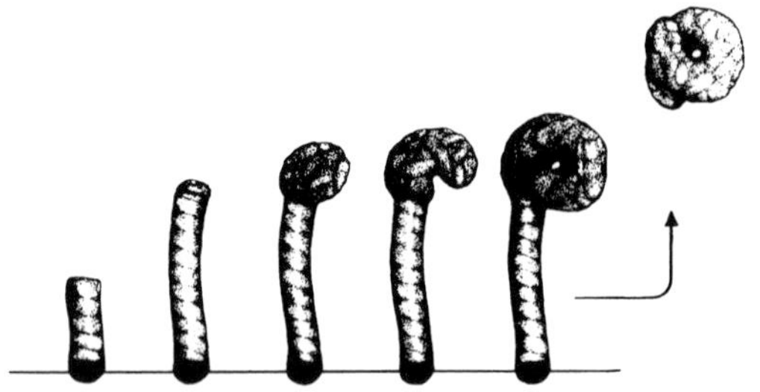

Figure 7. A Model for the Growth and Packing of the BR RNP Particle.

First a right- or left-handed, tightly wound 19 nm RNP stalk is formed from the 10 nm basic RNP fiber. The stalk morphology is kept roughly constant throughout the particle maturation process. At the tip of the stalk a 26 nm fiber is formed, which gradually bends into a globular shape. The 26 nm fiber is made from the 10 nm RNP basic element in a highly convoluted and yet unknown way. The released hnRNP particle consists of the 26 nm fiber curled into a one and one-third turn, left-handed spiral. The particle is transported to the nuclear envelope where, upon nuclear pore passage, it relaxes the spiral and forms a straight 26 nm fiber.

that the maturation scheme is followed by all the consecutive particles with about the same kinetics. For example, it is evident that all the fibers on the proximal segment exhibit a diameter of 19 nm; the 26 nm fiber element appears at the transition from the first to the second quarter of the gene. When the particles are about to be released, all of them appear as stalked granules. It can be concluded that the packing process of the transcription product at a Balbiani ring gene follows a kinetically and structurally well defined maturation pattern.

The type of ordered packing process observed at the BR genes may well be characteristic of several other eucaryotic genes. Perichromatin fibers and granules were found in diploid cells many years ago (Monneron and Bernard, 1969), and subsequent autoradiographic experiments clearly indicated that these structures most likely represent growing hnRNP molecules (Fakan et al., 1976). More recently, it has been shown that most of the perichromatin granules are attached to chromatin fibers with RNP stalks (Puvion-Dutilleul and Puvion, 1981). It is therefore plausible that other genes display the same type of packing as the BR genes (i.e., an RNP fiber is formed initially and subsequently is packed into a globular structure before the transcription process is complete. It seems likely, however, that the dimensions of the structural elements are not equivalent at different genes. For example, it has been shown in Sciara that small granules have thin stalks, while large granules have thick ones (Gabrusewycz-Garcia and Garcia, 1974). Analogous observations have been made in C. tentans when the small granules in BR 3 were compared with the large granules in BR 1 and BR 2 (our unpublished results). In fact, in diploid cells the diameter of the other perichromatin fibrils and perichromatin granules are known to vary over wide ranges (3–20 nm and 30–50 nm, respectively; Puvion and Moyne, 1981). This variation does not, of course, preclude the fact that at a given gene the packing process and the dimensions of the components are as well defined as in the case of the BR genes. Such a conclusion is compatible with the observation in Drosophila that the RNP granules from various puffs often differ considerably, while the granules within a puff attain about the same diameter (Swift, 1963).

The Nature and Significance of the Packing Process

The morphological and RNA compaction data allow us to consider the nature of the packing process in some detail. First, we have observed in Miller spreads that the entire transcription product, including the stalk and the globular portion, is built from a 10 nm fiber. This type of element could, in some cases, be directly observed in situ in partly relaxed stalks. The RNA compaction of the in situ 10 nm fiber turned out to be somewhat higher than that established for the corresponding fiber in Miller spreads (9 vs. 6). If the extending effect of the spreading procedure is taken into consideration, the two estimates are in reasonable agreement. Moreover, the loosely coiled 10 nm fibers suggested to us that the compact 19 nm fiber (and stalk) could be built from a tight coil of the 10 nm fiber. We tested this hypothesis by trying to fit a 10 nm fiber into a cylinder with a diameter of 19 nm. The cylinder was assumed to have a length of 115 nm—i.e., the length of the 19 nm fiber containing 9.3 kb RNA (see Results). A 10 nm fiber with a compaction of 9 and containing the same amount of RNA is 351 nm long. The pitch of the 10 nm fiber in the given cylinder can then be calculated to be about 10 nm; our data satisfy well the hypothesis that the 19 nm fiber (and stalk) is built from a tightly coiled 10 nm fiber.

The next step in the packing process, the formation of the 26 nm fiber, could consist of another regular coiling of a fiber element, or another mode of folding. When the approach mentioned above was used to evaluate the feasibility of fitting 10 or 19 nm fiber elements into a cylinder of 26 nm diameter, assuming a regular helical organization, we found that neither of the two types of fibers permit the required packing of RNP material into the 26 nm structure. Therefore, we assume there is another mode of packing, other than just a regular coiling of the fiber element. If the 19 nm fiber is used as the constituent fiber of the 26 nm structure, it is clear that it has to be substantially deformed or melted, while the 10 nm fiber considered in the same manner could be tightly packed without such major alterations. Thus it seems plausible to postulate that the 26 nm fiber is composed of a 10 nm fiber tightly folded in a still unknown way.

Our electron micrographs show that the 26 nm fiber makes a one and one-third turn, generating the 50 nm granule. The identity of the 26 nm fiber, observed during packing on the template with the corresponding element in the released product, is well substantiated by the RNA compaction data (94 vs. 90). Any major structural change upon the release is therefore not indicated (except for the disappearance of the stalk).

The formation of the 19 nm fiber from the 10 nm fiber,

and the 50 nm granule from the 26 nm fiber, follows a regular coiling theme while the intermediate structure (the 26 nm fiber) is generated in a basically different way. It should be noted that the folding of the 10 nm fiber represents the initiation of the packing of the final product. There is no obvious explanation for the finding that the 10 nm fiber first exists as a helical structure of a given and constant length before it attains its more complex folding in the final product. It is possible that the different modes of packing are a direct consequence of the structure of the 10 nm fiber made rapidly upon transcription and appearing in close proximity to the RNA polymerase itself. An alternative would be that, at the top of the stalk, the 10 nm fiber is changed due to chemical modification of the structure and/or additions or losses of minor protein components. A more specific and speculative possibility would be that initially a "core" 10 nm fiber is made with the few well characterized, abundant hnRNP proteins (Samarina et al., 1968; Beyer et al., 1977), and that on the top of the stalk, minor protein species (maybe sequence- and gene-specific) are added, giving the product a specific folding.

The ordered course of the packing process could be crucial for the proper handling of the transcription product in the nucleus and to accomplish transport of the product to the cytoplasm. Maturation of an hnRNP particle usually involves a series of processing events—like capping, methylation, splicing, and polyadenylation of the RNA molecule (Darnell, 1982)—and possibly other changes of the protein component (in the protein composition and/or modifications of the proteins; Sommerville, 1981). Regarding the processing of the BR transcripts, the available information is still meager. It is known that the BR transcripts become polyadenylated in the nuclear sap (Egyhazi, 1980). Furthermore, when the BR transcripts are translated into the cytoplasm, large portions of the primary transcript are not spliced out, but minor segments might still be (Case and Daneholt, 1978). In further exploring of the significance of the packing process in the BRs, it will be crucial to better characterize the maturation of the BR products, both on the RNA and the RNP level.

After the release from the chromatin template, the BR particles appear attached to the nuclear matrix and are translocated to and through the nuclear pores (Beermann, 1962; Stevens and Swift, 1966; Case and Daneholt, 1977; Olins et al., 1980). Stevens and Swift (1966) first noted that, when passing the nuclear pores, the BR particles change their conformation into elongated structures. They interpreted the process as a squeezing of the particle through a narrow pore. We have measured the dimensions of the rod-like conformation of the BR RNP particle and have found that it attains an average length of 135 nm, with a diameter of 25–30 nm. Considering the structure of the BR granule, it seems most plausible to us that the granule attains the extended conformation by relaxation of the coiled 26 nm fiber during the translocation event. This indicates that the process might be an ordered event and that it takes place in a polar fashion. After the passage through the nuclear pore, the BR RNA molecule sheds the associated proteins and is loaded into the endoplasmic reticulum, where it exerts its messenger function (for review see Daneholt, 1982).

Experimental Procedures

Material

Chironomus tentans was cultured according to Lambert and Daneholt (1975). Salivary glands were isolated from rapidly growing, 6-week-old, fourth-instar larvae.

Electron Microscopy of Sectioned Balbiani Rings

Salivary glands were fixed in 2% glutaraldehyde and later in 1% osmium tetroxide, dehydrated, and embedded in Epon. The sectioned specimens were stained in uranyl acetate followed by lead citrate and studied in a JEOL TEM-SCAN 100-CX electron microscope. The dimensions of RNP structures were determined from tracings of electron micrographs using a Summagraphics digitizer coupled to a Compucorp 445 Statistician. For further experimental details, consult Andersson et al. (1980).

Stereo-electron microscopic analysis was performed on a number of RNP particles. A side-entry goniometer was used with a standard specimen holder or, when appropriate, a rotation holder. The specimens were photographed at a series of tilt angles (0; ±15°; ±30°; ±50°), and the electron micrographs were analyzed pair-wise in a Wild ST4 Mirror stereoscope. When the structure of the released transcription products was analyzed, models were built (from pipe-cleaners) and compared to the various projections of each RNP particle.

Electron Microscopy of Spread Balbiani Ring Genes

Salivary glands were soaked in a buffered solution containing the detergents Triton X-100 (0.2%) and Nonidet P-42 (1%). The polytene chromosomes were released from the glands and chromosome IVs were isolated and processed for electron microscopy, essentially following the procedure of Miller and Bakken (1972). A single fourth chromosome was allowed to spread in alkaline distilled water. The expanded chromosome was centrifuged in a microchamber through formaldehyde-containing sucrose onto an electron microscopic grid. The sample was rinsed in distilled water containing the detergent Photoflo (Kodak), stained in 1% phosphotungstic acid, and examined in the electron microscope. For further experimental details, consult Lamb and Daneholt (1979).

Acknowledgments

We want to thank Åsa Schullström for drawing Figure 7, and Kerstin Ytterman for typing the manuscript. This research was supported by the Swedish Natural Science Research Council, the Swedish Cancer Society, Gunvor and Josef Anérs Stiftelse, and Karolinska Institutet (Reservationsanslaget).

The costs of publication of this article were defrayed in part by the payment of page charges. This article must therefore be hereby marked "*advertisement*" in accordance with 18 U.S.C. Section 1734 solely to indicate this fact.

Received May 10, 1983; revised August 1, 1983

References

Andersson, K., Björkroth, B., and Daneholt, B. (1980). The in situ structure of the active 75 S RNA genes in Balbiani rings of Chironomus tentans. Exp. Cell Res. *130,* 313–327.

Beermann, W. (1962). Cytologische Aspekte der Informationsübertragung von den Chromosomen in das Cytoplasma. In 13. Colloquium der Gesellschaft für physiologische Chemie am 3.–5. Mai 1962 in Mosbach/Baden. (Berlin: Springer-Verlag), pp. 64–100.

Beyer, A. L., Christensen, M. E., Walker, B. W., and LeStourgeon, W. M. (1977). Identification and characterization of the packaging proteins of core 40S hnRNP particles. Cell *11,* 127–138.

Beyer, A. L., Miller, O. L., Jr., and McKnight, S. L. (1980). Ribonucleoprotein structure in nascent hnRNA is nonrandom and sequence-dependant. Cell *20,* 75–84.

Beyer, A. L., Bouton, A. H., and Miller, O. L., Jr. (1981). Correlation of hnRNP structure and nascent transcript cleavage. Cell *26,* 155–165.

Case, S. T., and Daneholt, B. (1977). Cellular and molecular aspects of genetic expression in Chironomus salivary glands. In International Review of Biochemistry, vol. *15,* J. Paul, ed. (Baltimore: University Park Press), pp. 45–77.

Case, S. T., and Daneholt, B. (1978). The size of the transcription unit in Balbiani ring 2 of Chironomus tentans as derived from analysis of the primary transcript and 75 S RNA. J. Mol. Biol. *124,* 223–241.

Daneholt, B. (1982). Structural and functional analysis of Balbiani ring genes in the salivary glands of Chironomus tentans. In Insect Ultrastructure, vol *1,* R. King and H. Akai, eds. (New York: Plenum Publishing Corporation), pp. 382–401.

Darnell, J. E., Jr. (1982). Variety in the level of gene control in eukaryotic cells. Nature *297,* 365–371.

Economidis, I. V., and Pederson, T. (1983). Structure of nuclear ribonucleoprotein: heterogeneous nuclear RNA is complexed with a major sextet of proteins in vivo. Proc. Nat. Acad. Sci. *80,* 1599–1602.

Egyházi, E. (1980). Post-transcriptional polyadenylation is probably an essential step in selection of Balbiani ring transcripts for a cytoplasmic role. Eur. J. Biochem. *107,* 315–322.

Fakan, S., Puvion, E., and Spohr, G. (1976). Localization and characterization of newly synthesized nuclear RNA in isolated rat hepatocytes. Exp. Cell Res. *99,* 155–164.

Gabrusewycz-Garcia, N., and Garcia, A. M. (1974). Studies on the fine structure of puffs in Sciara coprophila. Chromosoma *47,* 385–401.

Jacob, M., Devilliers, G., Fuchs, J.-P., Gallinaro, H., Gattoni, R., Judes, C., and Stevenin, J. (1981). Isolation and structure of the ribonucleoprotein fibrils containing heterogeneous nuclear RNA. In The Cell Nucleus, vol. *8,* H. Busch, ed. (New York: Academic Press), pp. 193–246.

Laird, C. D., and Chooi, W. Y. (1976). Morphology of transcription units in Drosophila melanogaster. Chromosoma *58,* 193–218.

Lamb, M. M., and Daneholt, B. (1979). Characterization of active transcription units in Balbiani rings of Chironomus tentans. Cell *17,* 835–848.

Lambert, B., and Daneholt, B. (1975). Microanalysis of RNA from defined cellular components. In Methods in Cell Biology, vol. 10, D. M. Prescott, ed. (New York: Academic Press), pp. 17–47.

Martin, T., Billings, P., Levey, A., Ozarslan, S., Quinlan, T., Swift, H., and Urbas, L. (1974). Some properties of RNA: protein complexes from the nucleus of eukaryotic cells. Cold Spring Harbor Symp. Quant. Biol. *38,* 921–932.

McKnight, S. L., Martin, K. A., Beyer, A. L., and Miller, O. L., Jr. (1979). Visualization of functionally active chromatin. In The Cell Nucleus, vol. *7,* H. Busch, ed. (New York: Academic Press), pp. 97–122.

Miller, O. L., Jr., and Bakken, A. H. (1972). Morphological studies of transcription. Acta Endocrinol. *168,* 155–177.

Monneron, A., and Bernhard, W. (1969). Fine structural organization of the interphase nucleus in some mammalian cells. J. Ultrastruct. Res. *27,* 266–288.

Olins, A. L., Olins, D. E., and Franke, W. W. (1980). Stereo-electron microscopy of nucleoli, Balbiani rings and endoplasmic reticulum in Chironomus salivary gland cells. Eur. J. Cell Biol. *22,* 714–723.

Puvion, E., and Moyne, G. (1981). In situ localization of RNA structures. In The Cell Nucleus, vol. *8,* H. Busch, ed. (New York: Academic Press), pp. 59–155.

Puvion-Dutilleul, F., and Puvion, E. (1981). Relationship between chromatin and perichromatin granules in cadmium-treated isolated hepatocytes. J. Ultrastruct. Res. *74,* 341–350.

Samarina, O. P., Lukanidin, E. M., Molnar, J., and Georgiev, G. P. (1968). Structural organization of nuclear complexes containing DNA-like RNA. J. Mol. Biol. *33,* 251–263.

Sommerville, J. (1981). Immunolocalization and structural organization of nascent RNP. In The Cell Nucleus, vol. *8,* H. Busch, ed. (New York: Academic Press), pp. 1–57.

Steitz, J. A., and Kamen, R. (1981). Arrangement of 30 S heterogeneous nuclear ribonucleoprotein on polyoma virus late nuclear transcripts. Mol. Cell. Biol. *1,* 21–34.

Nuclear Envelope and Nuclear Import

Unwin P.N.T. and **Milligan R.A.** 1982. A large particle associated with the perimeter of the nuclear pore complex. *J. Cell Biol.* **93:** 63–75.

IT HAS BEEN KNOWN FOR ABOUT 50 YEARS that the continuity of the double nuclear envelope is interrupted by numerous pores with diameters in the range of 60–70 nm. Intimately associated with each pore is a complicated set of structures with eight-fold symmetry known collectively as the nuclear pore complex (NPC). Much of our knowledge about the NPC has come from electron microscopic analysis of pieces of nuclear envelope stripped from the giant nuclei of amphibian oocytes, particular those of *Xenopus*, as first demonstrated by Mick Callan in 1950 (1). With each improvement in microscopic technique the NPC has been described at higher and higher resolution and with more understanding of its chemical composition. The paper reproduced here was one of the first to apply sophisticated image analysis to the NPC, bringing out structural details that were not evident from direct observations of single complexes. The envelopes in this study were examined after chemical fixation, whose effect on the finest structure is difficult to evaluate. Later observations were made on frozen samples using the technique of cryo-electron microscopy (2), with results that were in general agreement with those from chemical fixation. A fibrous network of intermediate filaments, the nuclear lamina, is usually present on the nuclear side of the envelope, and additional structures connected to NPCs on both the cytoplasmic and nuclear sides have been described (3). NPCs from yeast are somewhat simpler in structure than those of *Xenopus*, although they exhibit the same overall octagonal symmetry. A recent study of isolated yeast NPCs suggests that the total number of proteins in the complex may be no more than ~30, and that most of these are symmetrically located on both the cytoplasmic and nuclear sides of the NPC (4).

---

1. Callan H.G. and Tomlin S.G. 1950. Experimental studies on amphibian oocyte nuclei. I. Investigation of the structure of the nuclear membrane by means of the electron microscope. *Proc. R. Soc. Lond., B* **137:** 367–378.
2. Akey C.W. 1989. Interactions and structure of the nuclear pore complex revealed by cryo-electron microscopy. *J. Cell Biol.* **109:** 955–970.
3. Ris H. 1991. The three-dimensional structure of the nuclear pore complex as seen by high voltage electron microscopy and high resolution low voltage scanning electron microscopy. *Electron Microsc. Soc. Am. Bull.* **21:** 54–56.
4. Rout M.P., Aitchison J.D., Suprapto A., Hjertaas K., Zhao Y., and Chait B.T. 2000. The yeast nuclear pore complex: Composition, architecture, and transport mechanism. *J. Cell Biol.* **148:** 635–651.

A Large Particle Associated with the Perimeter of the Nuclear Pore Complex

P. N. T. UNWIN and R. A. MILLIGAN
Department of Structural Biology, Stanford University Medical School, Stanford, California 94305; and Medical Research Council Laboratory of Molecular Biology, Cambridge, England

ABSTRACT The three-dimensional structure of the nuclear pore complex has been determined to a resolution of ~90 Å by electron microscopy using nuclear envelopes from *Xenopus* oocytes. It is shown to be an assembly of several discrete constituents arranged with octagonal symmetry about a central axis. There are apparent twofold axes perpendicular to the octad axis which suggest that the framework of the pore complex is constructed from two equal but oppositely facing halves. The half facing the cytoplasm is in some instances decorated by large particles, similar in appearance and size to ribosomes.

The nuclear pore complex is an organelle, ubiquitous to eucaryotic cells, which serves as a pathway through the nuclear envelope for a variety of nuclear and cytoplasmic molecules. Microinjection experiments involving substances such as dextrans (24), colloidal gold (10), and proteins (4, 7, 16) suggest that it forms an opening for passive movement of molecules up to ~90 Å in diameter. Biochemical studies suggest that it may be composed of only a few major polypeptides (19) and may contain RNA (28), although no direct identification has yet been achieved of these components within the confines of the structure. Besides this knowledge, very little information is available on the chemical or physical nature of the constituents or of their mechanisms of action.

Electron microscopy shows that the pore complex is a cylindrical assembly spanning the two nuclear membranes and having components arranged with octagonal symmetry around its central axis (6, 9, 12, 14). However, the details described by different authors vary considerably, probably because of real differences associated with the cell cycle and also because of artifactual differences related to quality of preservation.

Here we have investigated the structure of pore complexes from *Xenopus* oocytes using Fourier averaging methods to reveal the details. Additional information obtained by selectively releasing some of the constituents has led to a clearer picture of their organization and of the overall three-dimensional configuration. Our results emphasize the subunit nature of the pore complex and we show that particles similar in appearance and size to ribosomes associate with its cytoplasmic perimeter.

MATERIALS AND METHODS

General

Nuclear envelopes used were from the oocytes of *Xenopus laevis*. The oocytes were kept in small ovary pieces in modified Barth's saline solution (15) at 17°C for periods up to several days. Chemicals were obtained from the following sources: triethanolamine chloride from BDH Chemicals Ltd. (Poole, England); poly-L-Lysine (40,000 mol wt) and gold thio-glucose from Sigma Chemical Co. (St. Louis, MO); glutaraldehyde (ultrapure) from Emscope Laboratories (London, England); Triton X-100 from Bio-Rad Laboratories (Richmond, CA).

Low salt medium was 0.5 mM $MgCl_2$ in 1 mM triethanolamine chloride adjusted to pH 8 with KOH. High salt medium was 400 mM KCl, 5 mM $MgCl_2$, 20 mM triethanolamine chloride, adjusted to pH 8 with KOH. All solutions were prepared with double-distilled water.

Isolation of Nuclear Envelope

Nuclei were isolated from mature oocytes directly into the Barth medium by extruding them through small holes punched in the center of the oocyte black hemispheres. Each nucleus, after isolation, was cleaned by several passes up and down a narrow capillary and immediately transferred to an electron microscope grid.

Isolation of Ribosomes

To compare particles on the nuclear envelope with particles known to be ribosomes, we used as markers ribosomes isolated from hypothermic chick embryos (which have about the same molecular weight as those from *Xenopus* [20]). The ribosomes were isolated as tetramers from a postmitochondrial supernatant following, with modifications, the procedure described in (23), and they were shown to be undegraded in terms of their two-dimensional gel electrophoresis patterns (Milligan and Unwin, manuscript in preparation).

THE JOURNAL OF CELL BIOLOGY · VOLUME 93 APRIL 1982 63-75

Sample Preparation

The pore complex constituents were examined *in situ*, attached to the nuclear membranes, and also in isolation, detached from them. For the first purpose we used 600-mesh grids overlaid with a holey carbon support film, and the capillary was withdrawn to leave the nucleus behind within a small hemispherical droplet on one surface. With two fine, but blunted, glass needles, the envelope was disrupted to release the nuclear contents and spread across the surface, cytoplasmic face in contact. Once firmly adhered, the envelope was washed in low salt medium (~30 s), fixed with 2.5% glutaraldehyde in the same medium (1 min), and postfixed with 1% osmium tetroxide (1 min). These and subsequent steps were carried out with the grid completely immersed so that both sides received the same treatments. In some experiments, we carried out an additional incubation for 1 min in high salt, or in low salt with the $MgCl_2$ replaced by 1 mM ethylenediaminetetraacetic acid (EDTA), between the washing and fixation steps.

For the second purpose we used normal carbon-coated grids which had been rendered hydropilic on one surface by bathing in 0.1% polylysine (34). Each grid was placed on parafilm, hydrophilic side uppermost, and immersed in a droplet of low salt medium (sometimes containing 0.1% Triton X-100) before the nucleus was delivered onto it. After its attachment to the grid, the envelope was disrupted, partially spread as described above, and centrifuged at 500 *g* for 1 min. The envelope was then partially detached from the grid by vigorous swirling in high salt medium (or low salt medium when Triton X-100 was used), leaving behind various constituents. A small proportion of pore complexes were released intact when Triton X-100 was used.

Staining was done in the usual way (8) using 2% gold thio-glucose or 1% uranyl acetate. The former stain gave noticeably better preservation of the nuclear envelope; otherwise, at the resolution of this study, both stains gave equivalent results. Fixation with glutaraldehyde was avoided when uranyl acetate was used.

Electron Microscopy and Image Processing

Micrographs were recorded at × 12,000–20,000 using a Philips EM301 or EM400 electron microscope operating at 80 kv. A goniometer stage was used for the tilting experiments. Nuclear envelopes were viewed with their cytoplasmic side nearest to the electron source.

Images of the pore complex were analyzed by numerical Fourier methods. Those images displaying the most perfect eightfold rotational symmetry were assumed to best represent the true structure. By this criterion, and for pore complexes attached to the nuclear membranes, good preservation was achieved only in regions overlying holes in the carbon support film, and in parts where the stain was sufficiently deep that the space between the pores was of uniform density.

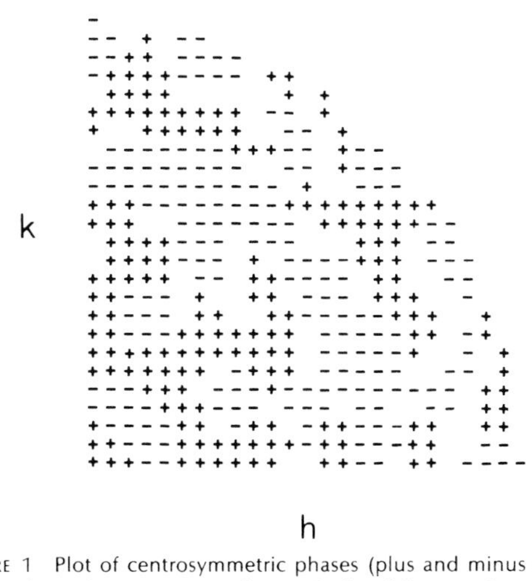

FIGURE 1 Plot of centrosymmetric phases (plus and minus are 0° and 180°) in the Fourier transform calculated from an image of a pore complex attached to the nuclear membranes. The plot was obtained by sampling the continuous transform on a square lattice at intervals of 4.34×10^{-4} Å^{-1} and averaging the complex values at these points, (*h, k*), with additional values collected at the eightfold related positions. The projection computed using these phases and averaged amplitudes is identical to the one in Fig. 9 *a*, derived by harmonic analysis of the same data.

TABLE I

Details of Three-dimensional Data Refinement

Number of images	10
Range in angle of tilt	0–52°
Resolution cut-off	130 Å
Average phase error, based on comparison of individual values along lattice lines	29°

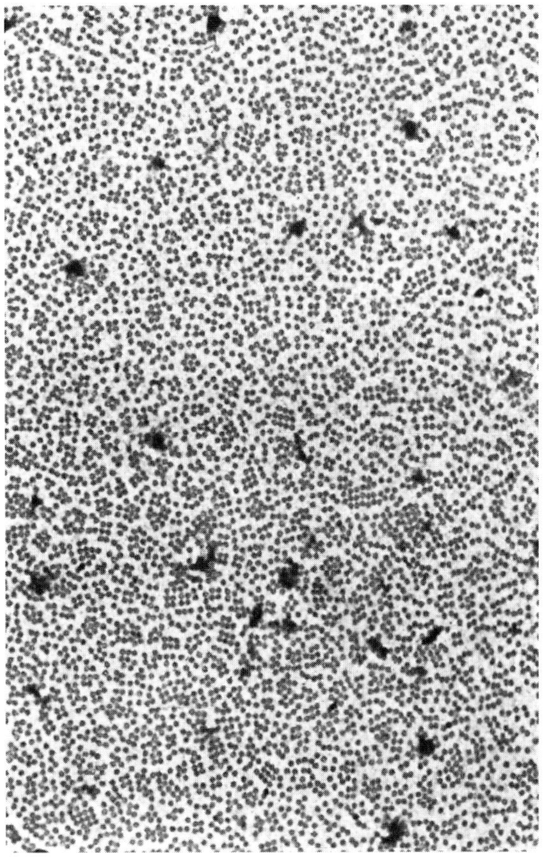

FIGURE 2 Isolated nuclear envelope from *X. laevis* spread over a carbon support film. The pore complexes are usually irregularly organized but sometimes form linear and square arrays. Unstained. × 8,000.

The images were densitometered with a modified Nikon comparator (3) or Perkin-Elmer 1010A automatic microdensitometer (Perkin-Elmer Corp., Instrument Div., Norwalk, CT) to convert the optical densities into numerical arrays. The step size was 30 μm, corresponding to 18 Å at the specimen, and the array size was 256 × 256 points, or, in the case of the three-demensional study, 512 × 512 points. Circular regions in these arrays, enclosing just the pore complex, were boxed off and Fourier transformed, using a program of the type described by DeRosier and Moore (8). The transforms, rather than the images, were used for subsequent manipulations.

Power spectra and rotationally averaged projections were calculated following Crowther and Amos (5). The power spectra are derived by calculating the weight of a specified n-fold rotational harmonic for the best image center consistent with n-fold rotational symmetry, and repeating this calculation over a range of values of n. Displayed are the relative strengths of the rotational harmonics contributing to the image, allowing an objective assessment of the degree of preservation to be made. The projection maps were derived by Fourier-Bessel synthesis of just those harmonics consistent with the observed eightfold rotational symmetry.

A three-dimensional map was calculated from an image of a pore complex attached to the nuclear membranes (Fig. 7) to determine the distribution of matter in the direction perpendicular to the membrane plane. Complex superposition of detail precluded the possibility of obtaining this information directly

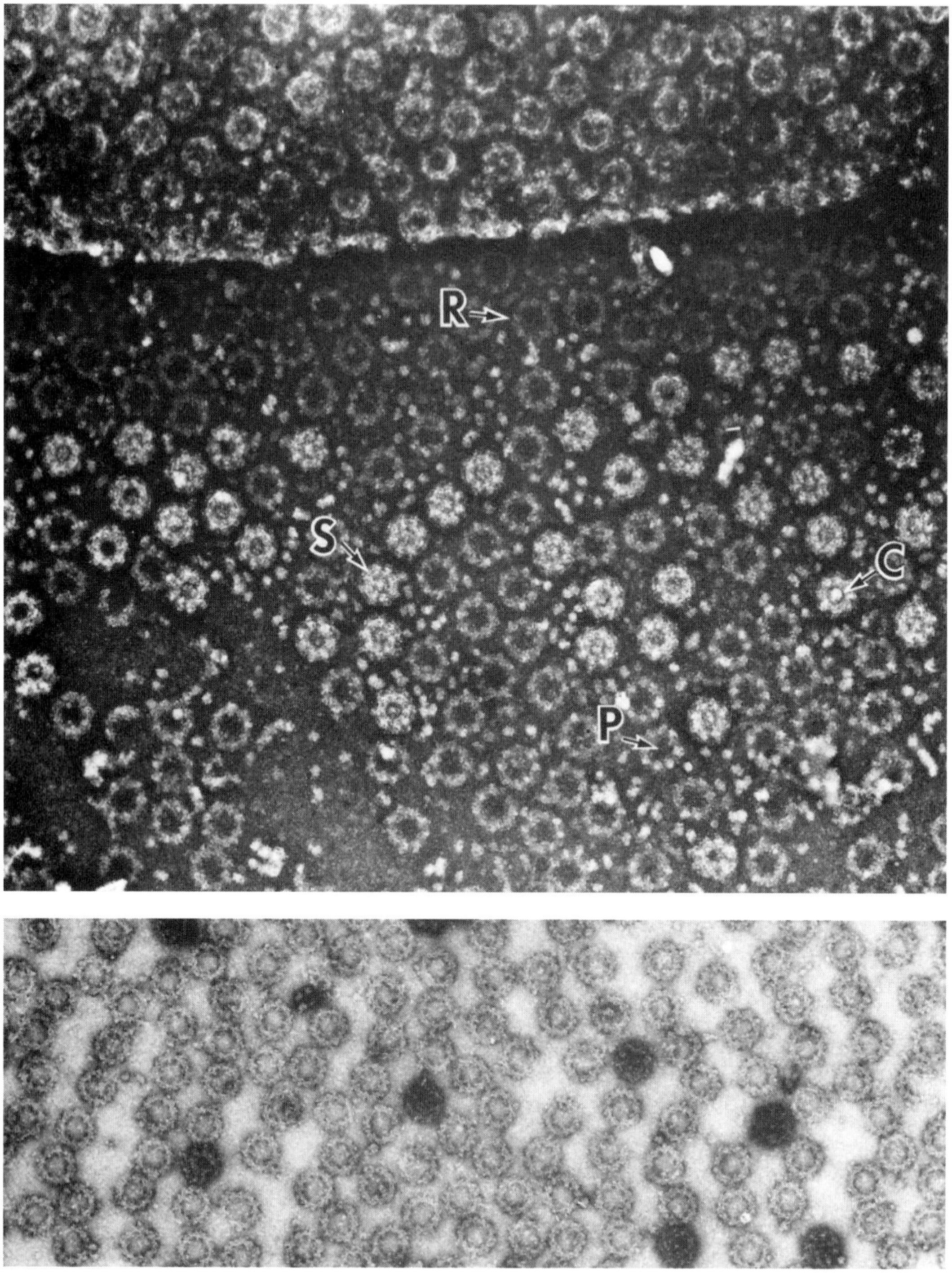

FIGURE 3 Constituents of the nuclear envelope released after its partial detachment from a polylysine-treated carbon film in the presence of 0.1% Triton X-100 (see Methods and Materials). Those constituents most obviously related to the pore complex are: rings (*R*), central plug (*C*), spokes (*S*), and particles (*P*), occasionally observed around the rings. As the lower micrograph shows, the rings are sometimes obtained in large numbers by themselves. Uranyl acetate stain. × 60,000.

from different views. We manipulated the Fourier transforms as if they were derived from a two-dimensional crystal by creating an artificial reciprocal lattice consisting of lines arranged on a square grid and oriented parallel to the octad axis. Images from a tilt series were used to provide many estimates of amplitude and phase at different points along these lines, allowing the continuous variations to be mapped out along them. The continuous curves, sampled at regular intervals, provided the Fourier terms from which to calculate the structure.

We combined the transform measurements starting with the projection data, Fig. 1, and refined the phases, image by image, in order of increasing tilt. Each image, on account of the octagonal symmetry, provided up to four independent

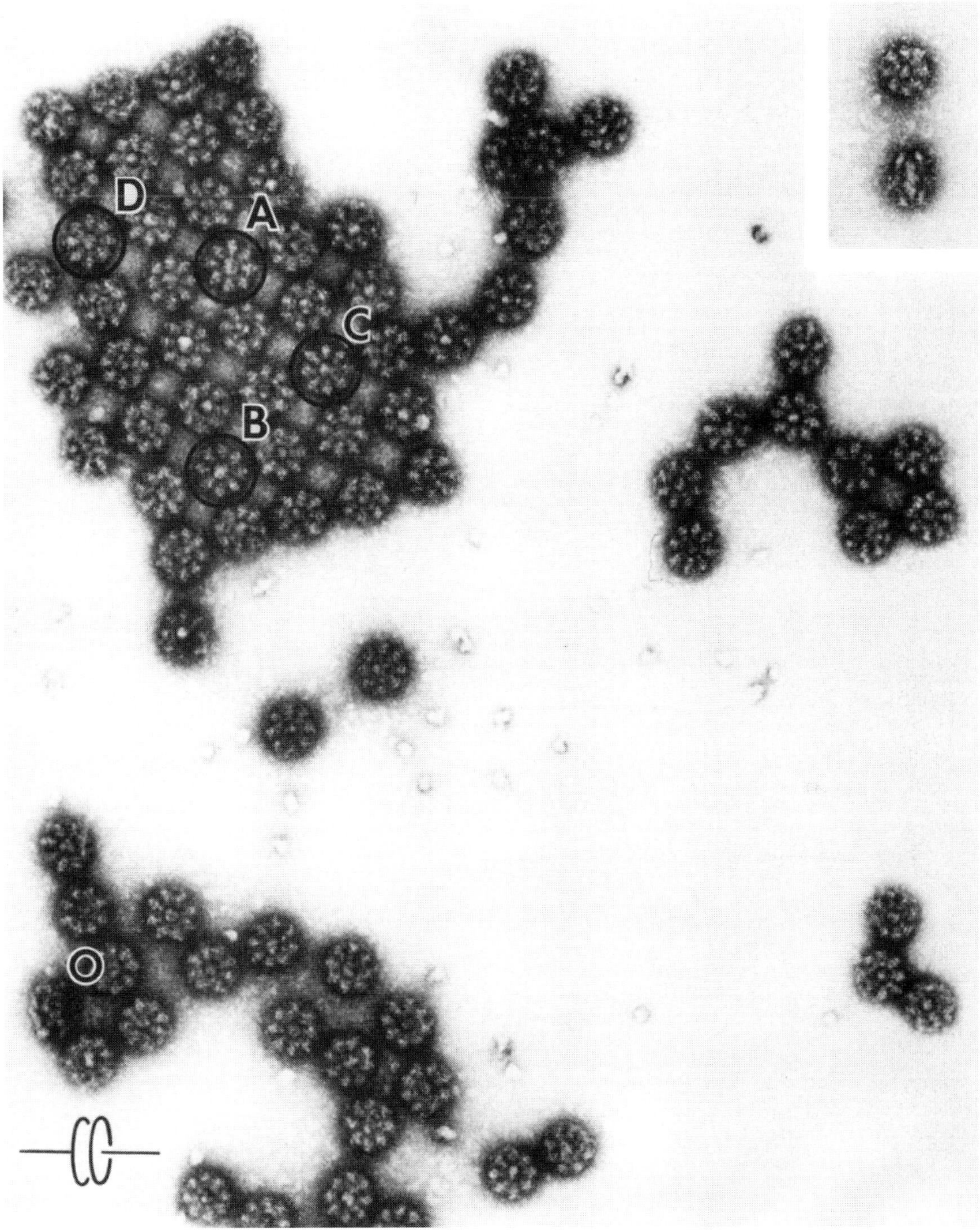

FIGURE 4 Detached pore complexes released onto the microscope grid from a nuclear envelope immersed in low salt medium containing 0.1% Triton X-100. Pore complexes within square arrays are better preserved than others. *A, B, C,* and *D* refer to images from which the results in Fig. 6 were obtained; the contributions from the eightfold harmonics are respectively 51%, 47%, 46%, and 43% of the total power associated with azimuthally varying components. Also shown are oblique, "*O*", and edge-on views (*inset*) of the pore complex. The drawing indicates how the view, "*O*", can be interpreted in three-dimensions in terms of two coaxial rings with matter lying in between them. These rings are not obvious in the *en face* views since they are thin in comparison with the rest of the structure and hence do not contribute much contrast when viewed from this direction. Uranyl acetate stain. × 75,000.

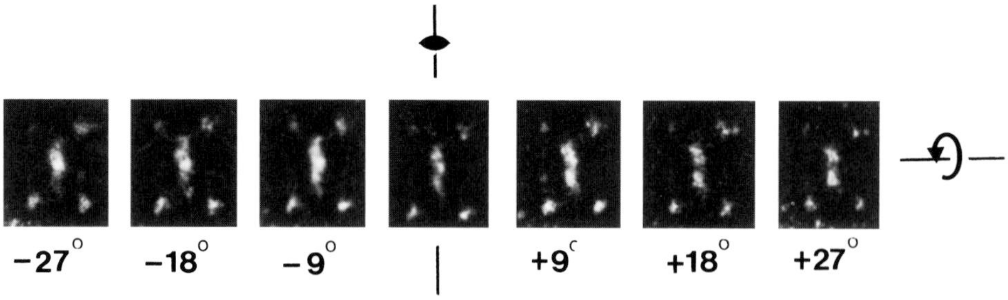

FIGURE 5 Detached nuclear pore complex tilted about the octad axis. The tilt angles are indicated. The position of the putative twofold axis is drawn for the untilted view. Uranyl acetate stain. × 120,000.

estimates of amplitude and phase along each lattice line. The positions of the lattice (and eightfold related) points in the transforms were calculated from the known magnitude of tilt and the direction of the tilt axis on the micrograph. Only those phases for which the corresponding amplitudes were fairly strong (~25% of the total number) and which along any given lattice line lay within 2.17×10^{-4} Å^{-1} of each other were used in the refinement comparisons. The required Fourier terms were collected from the smooth amplitude and phase curves derived from these data by sampling at intervals of 4.34×10^{-4} Å^{-1} along the lattice lines. Errors involved in drawing out these smooth curves caused only small departures from perfect octagonal symmetry. Further details are given in Table I.

RESULTS

Pore complexes in the nuclear envelopes of *Xenopus* oocytes occupy a large fraction of the total membrane surface. Their organization within the membranes is usually rather irregular, but they sometimes form lines or, more rarely, square arrays (Fig. 2). Such motifs are probably a result of their interaction with the thin nuclear lamina (1, 19) and with each other. A complete description of the pore complex, within the envelope, is derived below by bringing together information obtained from several types of experiment.

Plugs, Spokes, Particles, and Rings

We find that the pore complex is constructed from, or related to, several discrete constituents. These are most easily recognized following release with the detergent, Triton X-100 (see Methods and Materials), when both the separated constituents and intact pore complexes appear next to the envelope skeleton (Fig. 3). Clearly visible are: (*a*) "rings" which have an inside diameter close to 800 Å and an outer diameter the same as that of the pore complex itself (~1,200 Å); (*b*) large particles forming "plugs" at the centers of the pore complexes; (*c*) smaller (~220 Å) particles occasionally arranged around the circumference of the rings; and (*d*) "spokes," matter in intact pores extending radially outwards from the plugs towards the periphery. The plugs have a diameter of up to 350 Å, depending possibly on their state of preservation. The rings are composed of globular subunits and their power spectra display weak eightfold components (results not shown).

Detached Pore Complexes

To learn about the three-dimensional arrangement of some of these constituents we investigated further the structure of detached pore complexes. Viewed *en face* (i.e., from a direction perpendicular to the plane in which the membranes would lie) the detached pore complex is divisible into eight parts which are approximately equivalent and symmetrical about lines drawn radially (Fig. 4; see also Fig. 6). It therefore seems to display elements of both octagonal and mirror symmetry. The appearance of true mirror symmetry would suggest that it is composed of two equal but oppositely facing halves (i.e., halves related by twofold axes perpendicular to the octad axis and lying in the central plane). Other views are in accord with this configuration. For example the oblique view, "*O*", in Fig. 4, shows two circular outer rims which are coaxial and equal in diameter and thickness. Views perpendicular to the octad axis (Fig. 5, and *inset* to Fig. 4) show a central zone of matter flanked by two lines (the rims seen edge-on) which are equally prominent at their extremities and symmetrically disposed on either side.

When this structure is tilted about the octad axis (Fig. 5) the rims produce only small variations in contrast and apparent diameter, except at high tilts where flattening effects (11) become most significant. This indicates that the rims are rings of approximately constant thickness rather than, say, circular arrays of the ~220 Å-diameter particles. Moreover the diameter of the rims corresponds with that of the rings in Fig. 3. We thus suppose the pore complex to be framed by two equal rings facing toward the nucleus and the cytoplasm of the cell, respectively. The separation between the rings varies (300–600 Å) probably because of rather flexible, or easily distorted, connections (most obvious in the −18° and −9° tilts) to the rest of the assembly.

In the parts of the pore complex excluding the rings the changes in contrast with angle of tilt are more pronounced. At −27° (Fig. 5), for example, the central zone displays a peak of density over the octad axis, whereas at 0° the density is more evenly distributed. The distribution of matter at the extremities of this zone, where it links up with the rings, changes to a similar degree. These variations are consistent with features varying octagonally around the axis of the pore complex, although poor preservation prevents an exact correspondence of views differing in tilt by 45°.

Power spectra calculated from images of *en face* views which display strong octagonal symmetry indicate that preservation is best in "crystalline" regions, where the pore complexes lie closely apposed. Pore complexes by themselves are apparently more subject to staining distortions, involving differential shrinkage, as described by Moody (22). Best preserved pore complexes show evidence of 16- and 24-fold harmonics in addition to the basic eightfold harmonic (Fig. 6). In general, the stronger the higher order contributions relative to the background the more perfect the mirror symmetry—a phenomenon one would expect if indeed the pore complexes are constructed from two equal and oppositely facing halves.

Consistent features of the projection maps (Fig. 6) are a round central plug, eight prominent spokes emanating from it radially, and a partitioning of each spoke into characteristic regions or domains. We distinguish an inner domain at radii between 220 and 400 Å where the spoke is narrowest and tending to connect up circumferentially with its equivalent

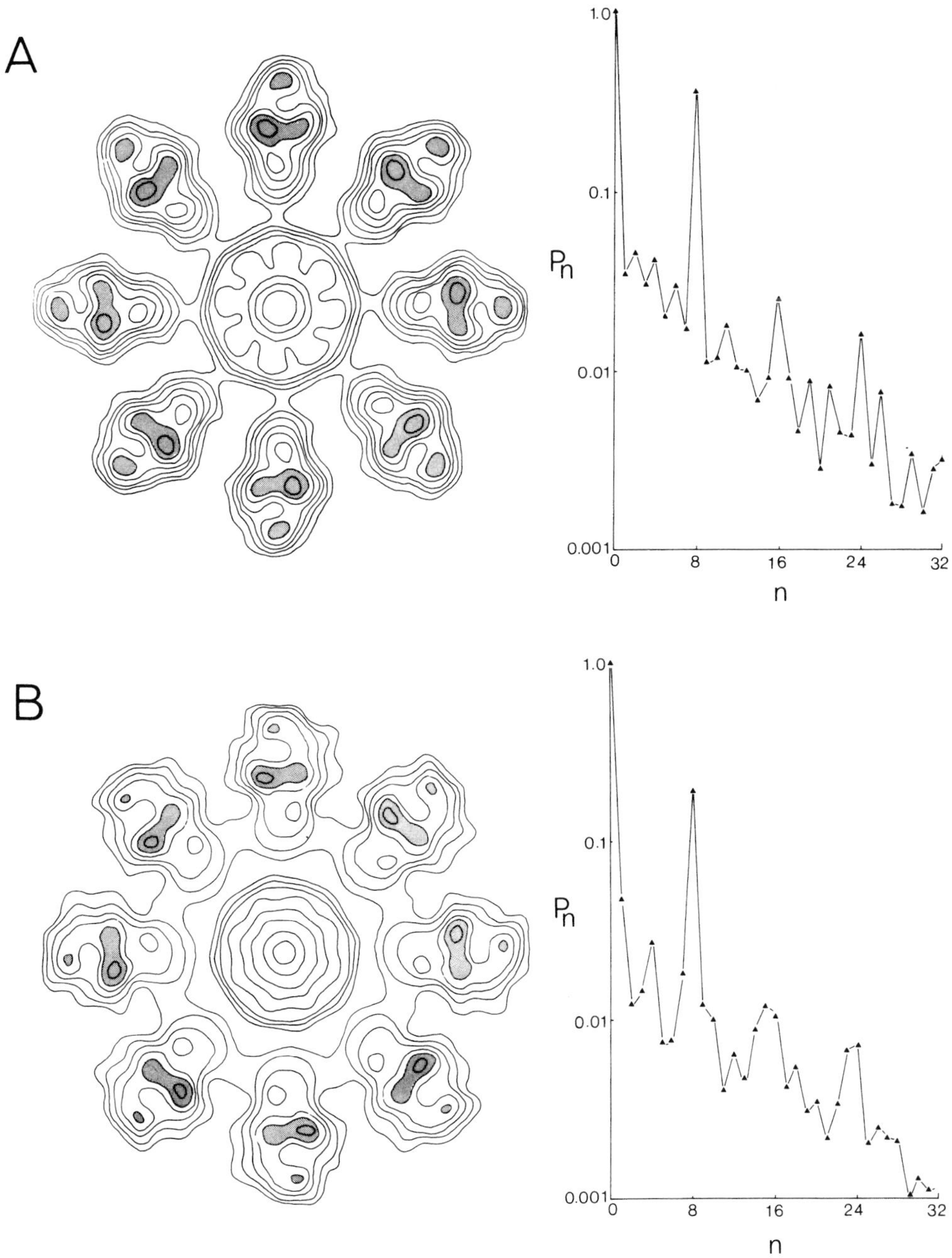

FIGURE 6 Projection maps obtained from the detached pore complexes, *A, B, C,* and *D* in Fig. 4, and their rotational power spectra, plotted on a logarithmic scale. Improvements of the power spectra in terms of enhanced 16-fold (n = 16) and 24-fold (n = 24) harmonics relative to the background level are correlated with a stronger tendency towards mirror symmetry. The resolution is ~90 Å. The consistently observed broad and sharp peaks of density (shaded) are at radii of 450 and 550 Å, respectively. The broken lines drawn in *D* indicate the positions of the membrane border and the particulate matter shown in Fig. 9; the two radial lines correspond to the putative twofold axes which give rise to the appearance of mirror symmetry in projection.

neighbors, and an outer domain composed of two peaks, one rather broad and the other rather sharp at radii of 450 and 550 Å, respectively. Correlation with the edge-on view suggests that the inner domain and plug are located in the central plane of the pore whereas the outer domain encompasses the region where the two oppositely facing halves of the spokes diverge from this plane to link up with the rings. The broad peak (outer domain) appears to be the part of the structure where the rings and the matter in the central plane superimpose. The rings themselves contribute little contrast because, viewed in this direction, they are very thin in comparison to the rest of the structure. The ~220 Å-diameter particles observed in Fig. 3 are not a part of the isolated pore complex.

Pore Complexes Attached to Nuclear Membranes

The pore complex attached to the nuclear membranes (Fig.

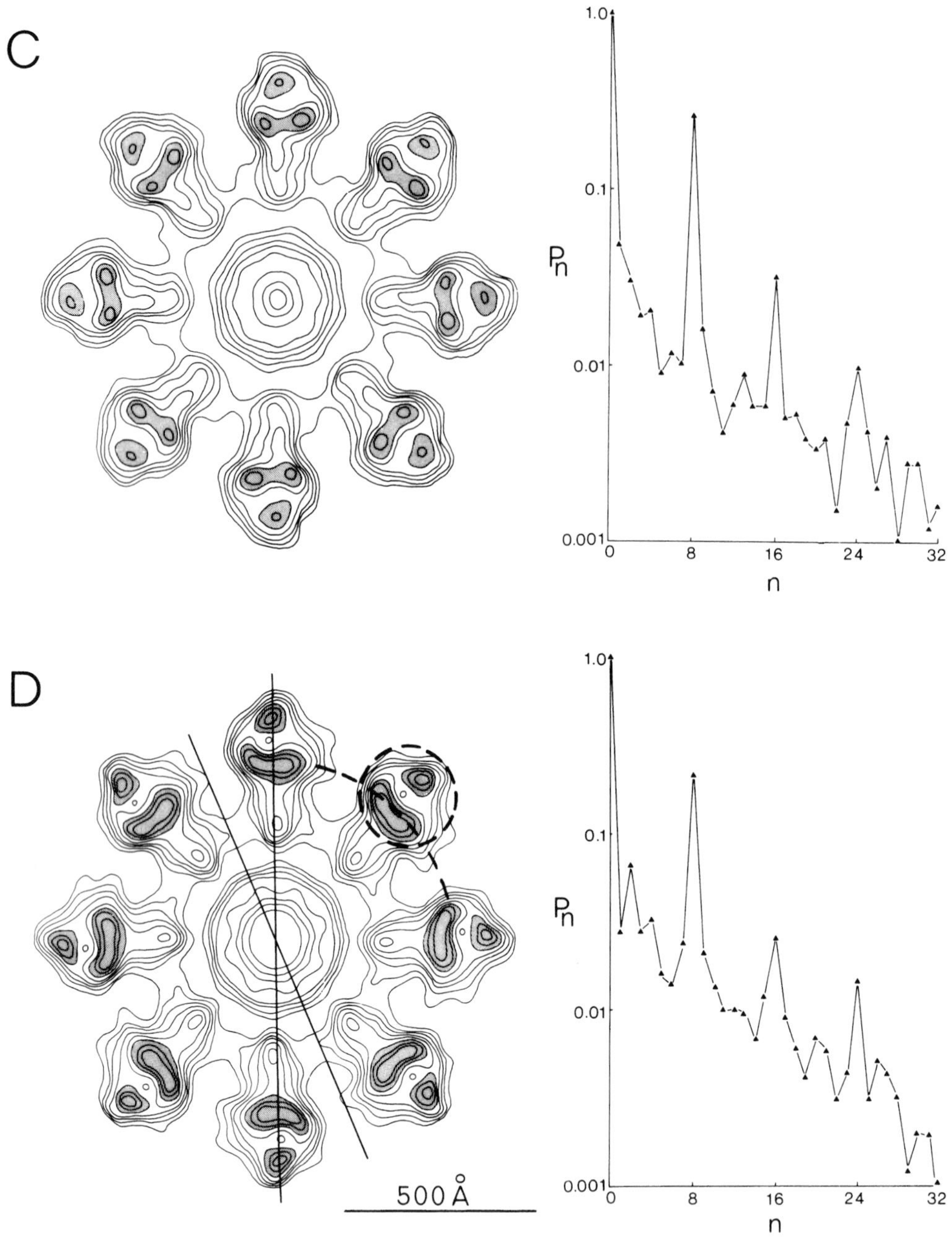

FIGURE 6 *C and D*

7) has additional contrast due to the two membrane layers which are contiguous and come together at the pore periphery (33). We find that the borders of these membranes are made conspicuous by incubating the nuclear envelope either in high salt medium (400 mM KCl) or in low salt medium in which the $MgCl_2$ is replaced by 1 mM EDTA (Fig. 8). The former treatment, although introducing some disorder, largely preserves the integrity of the pore complex. The latter treatment leads to its dissociation, causing also changes in the size of the membrane openings.

Comparison of Figs. 7 and 8 shows that the increased clarity of the membrane borders produced by high salt or EDTA is associated with the detachment of particles from around the perimeters of the pore complex and from intervening spaces.

The effect of the presence of these particles and of the membrane on the appearance of the projection maps is significant only at high radius. In projection maps calculated from pore complexes attached to the nuclear membranes, Fig. 9, the spokes have more matter associated with them in their outer domain than previously (Fig. 6) and appear against a stronger background density (the membrane) beginning abruptly at a radius of ~450 Å.

Particles around Cytoplasmic Perimeter

A low resolution three-dimensional map of a pore complex attached to the nuclear membranes was calculated from a series

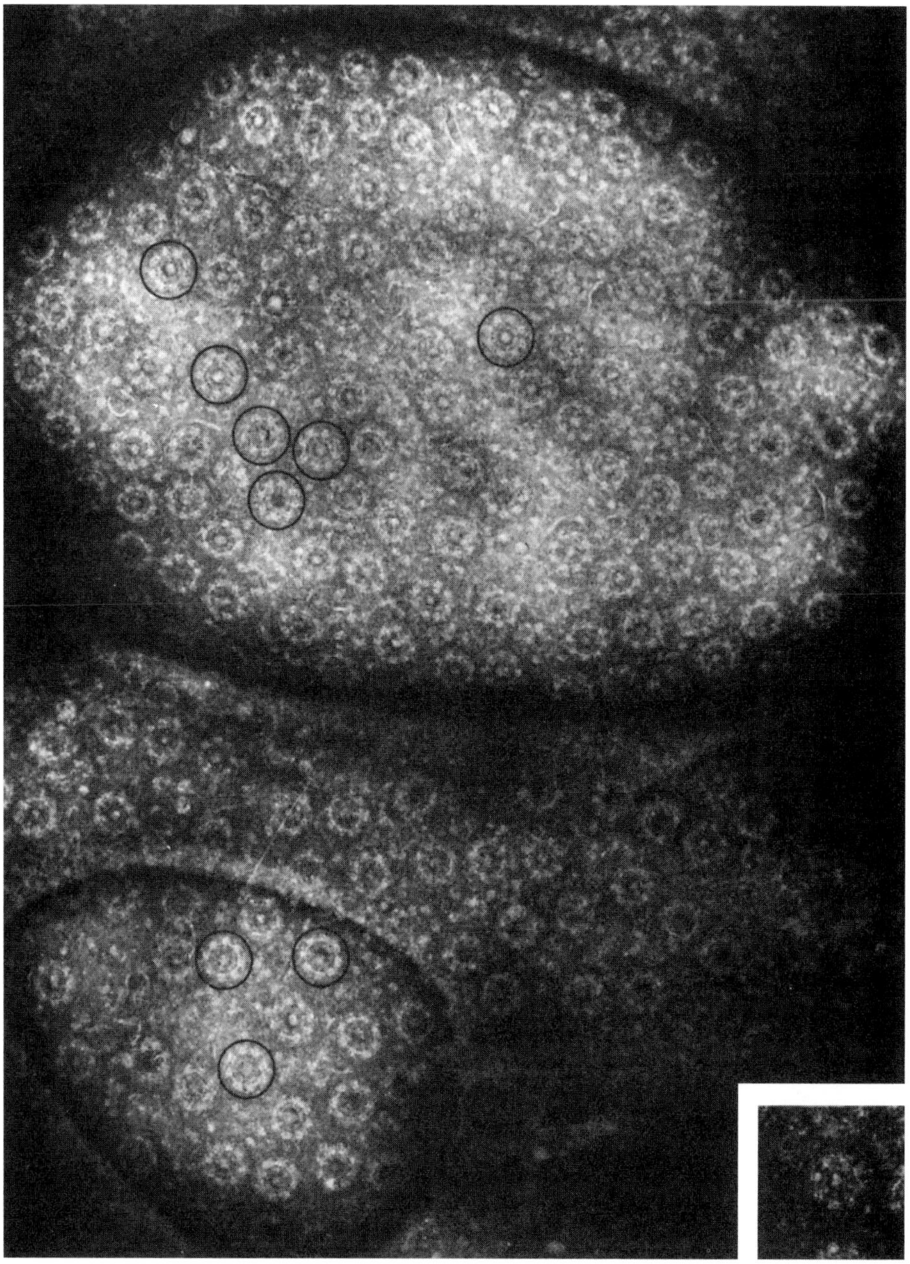

FIGURE 7 Isolated nuclear envelope spanning holes in a carbon support film. For the pore complexes encircled, the contributions from the eightfold harmonics are (top of page to bottom): 23%, 21%, 24%, 27%, 25%, 20%, 15%, 25%, 32%, and 32% of the total power associated with azimuthally varying components. The absence of a central plug in some of the pore complexes may be a consequence of the isolation procedure. The *inset* is of the image from which the results in Fig. 10 were obtained. Gold thioglucose stain. × 58,000. *Inset,* × 73,000.

of images taken with different tilts (see Materials and Methods) to observe how the particulate matter superimposing with the spokes is distributed in the direction of the octad axis.

Some details of this map are given in Fig. 10. Central sections perpendicular to the plane of the membranes (Fig. 10*a*) show a marked departure from the putative twofold relationship described earlier (Fig. 5). The zone contributed by the plug and the inner domain of the spokes is now flanked, at high radius, by matter more heavily weighted towards the cytoplasmic half. This additional matter gives rise to strong eightfold modulations in sections parallel to the membrane plane and a strong asymmetry in terms of the variation in eightfold contrast at this radius with distance through the structure (Fig. 10 *b*). Since both sides of the pore complex were exposed equally to the stain (see Materials and Methods) it is most unlikely that the asymmetry is due to this treatment.

We interpret the details to indicate that the pore complex has, on its cytoplasmic perimeter, particles—presumably the

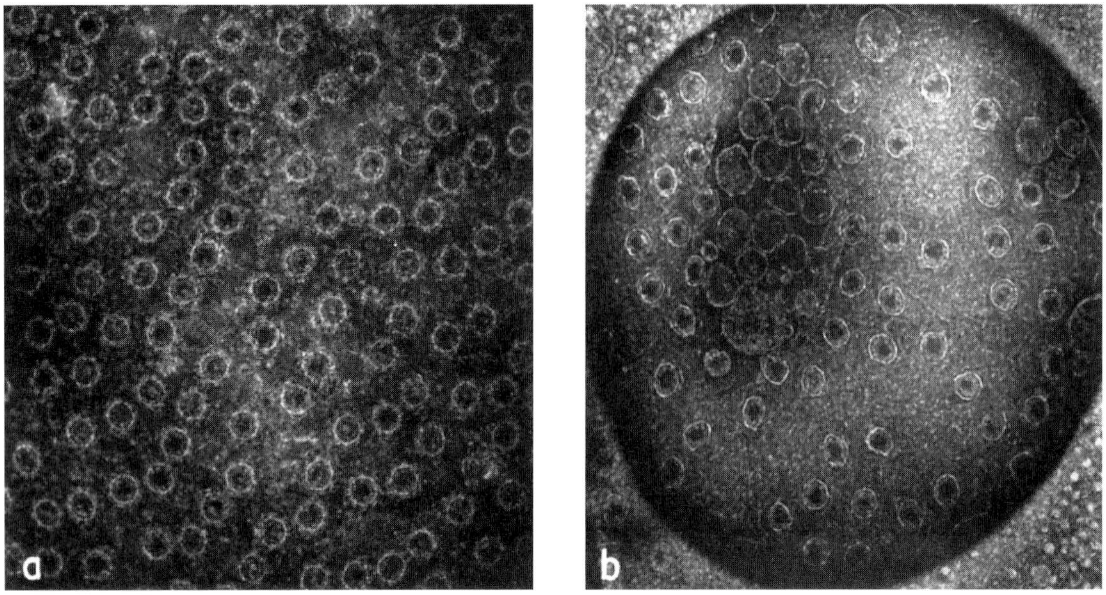

FIGURE 8 Isolated nuclear envelopes after incubation in (*a*) in high salt medium (400 mM KCl) and (*b*) 1 mM EDTA for 1 min (see Materials and Methods). Loss of matter associated with these treatments exposes the membrane borders of the nuclear pores. The diameter of the opening delineated by these borders becomes variable after treatment with EDTA. Gold thio-glucose stain over holes in the carbon support film. × 45,000.

same as those in Fig. 3—which overlay the membrane, the ring and the spokes. It is unlikely that other maps would show other additional features (e.g., particles also on the nuclear side) since, of the 58 pore complexes attached to membranes which we analyzed in projection, the one from which the map was constructed exhibits the best power spectrum in terms of resolution and the strongest eightfold harmonic. On the other hand, inspection of the micrographs suggests that pore complexes often have less than eight particles on their perimeter, and sometimes none.

We confirmed the presence of these particles on the cytoplasmic side, independent of the structure analysis, by releasing them with high salt from envelopes pressed cytoplasmic face downwards onto polylysine-coated carbon films (see Materials and Methods). The "finger-prints" thus obtained show up to eight particles arranged in ~1000 Å diameter circles (Fig. 11). Now isolated from the envelope and the rings, the particles are easily distinguished from the round central plug, on the basis of their smaller size and more angular shape, but seem to be identical to the particles in the spaces between the rings. They also correlate closely with inactive ribosomes given the same treatments (Fig. 12), in terms of their shape and size.

DISCUSSION

The nuclear pore complex is an assembly of several discrete constituents. We have investigated their three-dimensional organization by visualizing them, at a resolution of ~90 Å, both in the presence of and isolated from the nuclear membranes. We used Fourier analysis methods to evaluate and average the images and to derive three-dimensional information from different views. The schematic diagram, Fig. 13, summarizes our results.

We find the pore complex to be a symmetrical structure framed by two widely separated, coaxial rings. The rings attach to the two nuclear membranes so that one faces the nucleus and the other the cytoplasm of the cell. Connected to these rings and extending radially inwards from them, along a central plane, are elongated structures which we call spokes. They approach and appear to contact a central large, approximately spherical particle, the plug. Cytoplasmic particles, also observed decorating the perimeter of many of the pore complexes, are probably not integral components since they are easily detached and are only there when the nuclear membranes are present.

The structural framework of the assembly, i.e., the rings, the spokes, and their connecting links, appears to be arranged with octagonal symmetry about the central axis perpendicular to the plane of the membranes and with twofold symmetry about axes lying in this plane (giving the dihedral point group D_8 [822]). A configuration like this, in which the two halves of the assembly face in opposite directions, would account simply for observations of single pore complexes spanning the two equivalent, but oppositely facing, membranes of ER cisternae (13). An example of another two-layered membrane system incorporating this design principle would be the gap junction (29).

Some earlier models for the pore complex show eight equal "granular subunits" around its perimeter on both the nuclear and cytoplasmic sides (12, 26) and so may be construed as suggesting a twofold symmetry relationship. However, the quality of preservation achieved in the earlier studies was not assessed and alternative interpretations could not therefore be discounted. We suggest that the granular subunits and likewise the proposed tubular subunits (30) or microcylinders (35) should be identified with the detail which we observe in the region where the spokes connect to the rings. It is easy to see how this detail could be interpreted in various ways according to the state of preservation. The features we find in the peripheral region of the pore complex (Figs. 4 and 5) are particularly distinct because we have exposed them by taking away the membranes.

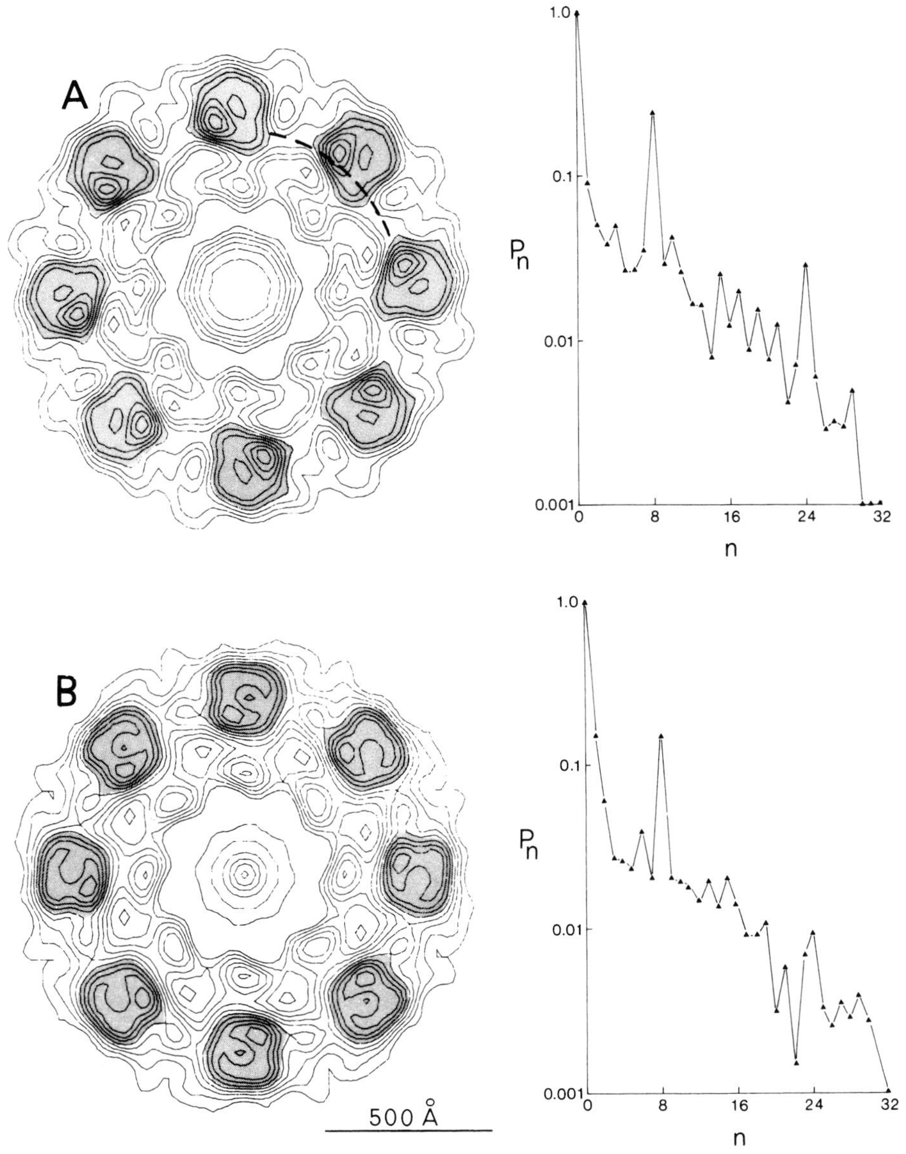

FIGURE 9 Typical projection maps from pore complexes attached to the nuclear membranes, and their rotational power spectra. The resolution is about 90 Å. The shading indicates the position of the particulate matter not present in images of the isolated pore complexes. The broken line in (*A*) indicates the position of the membrane border. The view is from the cytoplasm towards the nucleus. Power spectra of pore complexes attached to the nuclear membranes show 8- and 24-fold harmonics but the 16-fold harmonic is always very weak or absent.

Several independent lines of evidence have led to our conclusion, Fig. 13, that in the cell there are sometimes large (~220 Å-diameter) particles decorating the pore complex around its perimeter on the cytoplasmic side. First, such particles were present on occasions around the rings, following their detachment from the nuclear envelope with Triton X-100 (Fig. 3). Second, these particles were too large to be accommodated as part of the structure of the pore complex itself, yet did give rise to additional density over the rings and spokes of pore complexes which had not been detached from the nuclear envelope. Third, we demonstrated by a three-dimensional analysis that the additional density was concentrated toward the cytoplasmic side rather than the central plane or the side of the nucleus. Fourth, we were able to detach these particles (identified by their appropriate circular configuration) by contact of the cytoplasmic surface of the nuclear envelope against the microscope grid.

At least two reports (6, 17) have clearly demonstrated decoration of the pore perimeter by particles which might be the same as those we observe. In other careful studies (e.g., reference 31), such particles have not been detected, despite conditions being used which should have allowed their retention. Thus the particles appear to be there in vivo on some occasions but not on others. Accordingly, their presence may be related to a process such as the activity of the pore in mediating transfer of molecules between nucleus and cytoplasm.

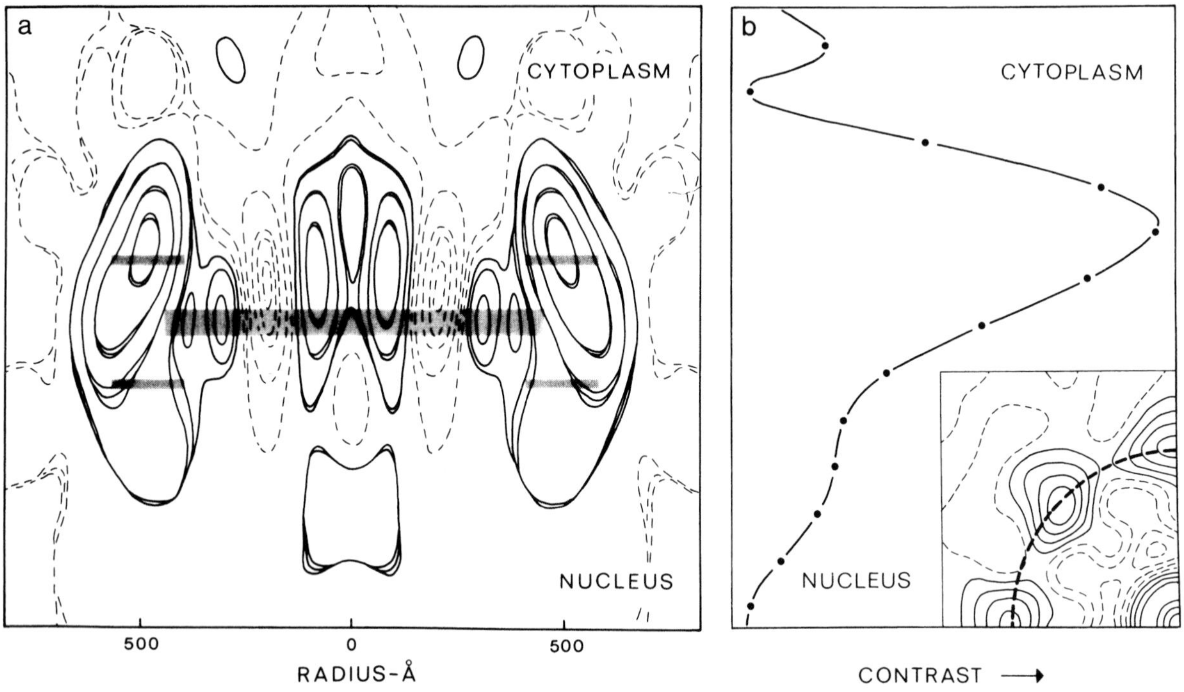

FIGURE 10 Details of a low resolution three-dimensional analysis of a pore complex attached to the nuclear membranes. (*a*) A projection of 25-Å thickness through the three-dimensional map built up from sections perpendicular to the plane of the membranes and intersecting the maxima in the eightfold density modulations. The full lines are the positive contours indicating the regions where the biological matter is concentrated. The broken lines are negative contours. The shading indicates the estimated positions of the twofold related features described earlier (Fig. 5). The two membrane layers are not resolved. Features on the cytoplasmic side give rise to strong eightfold modulations, but those on the nuclear side do not. (*b*) A plot of the variation in contrast associated with the eightfold modulations at a radius of 500 Å. The vertical scale is the same as in *a*. The contrast was estimated as the difference between the maximum and minimum densities in sections perpendicular to the octad axis. The line at the 500-Å radius along which the densities were measured is indicated in the section giving maximum contrast, at the bottom of the figure. There is marked asymmetry in contrast compared to similar views from detached pore complexes. This analysis gives only a qualitative idea of the relative levels and strengths of features. The Fourier terms giving the variation in mean density along the direction of the octad axis are not included. Dimensions of features in this direction are also affected by flattening distortions (19). Because of the variable preservation of the central plug, we attach no significance to details in this region.

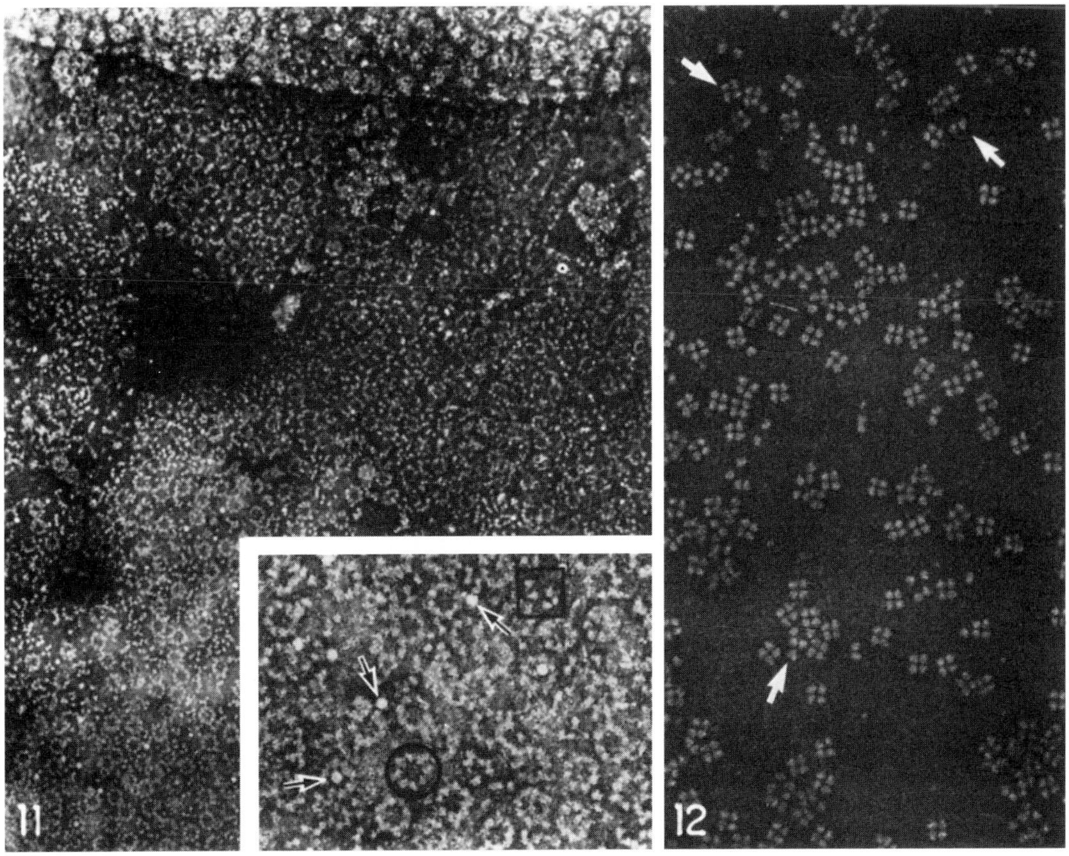

FIGURES 11 and 12 "Finger-print" of material detached from the cytoplasmic surface of the nuclear envelope (top of picture) onto a polylysine-coated carbon support film. The detachment was achieved using high potassium concentration (400 mM; see Methods and Materials). The somewhat angular particles arranged in rings (circle) or randomly (square) can be identified with those in Fig. 7 around the perimeter of the pore complexes and in intervening spaces. These particles are easily distinguished from the larger, round, central plug (arrows). Gold thio-glucose stain. × 25,000. *Inset,* × 55,000. Fig. 12: Ribosomes isolated from hypothermic chick embryos, after incubation in high salt, fixation, and staining as in Fig. 11. Both the tetrameric and single ribosomes appear rather angular (arrows), possibly as a result of detachment of the small subunit. × 55,000.

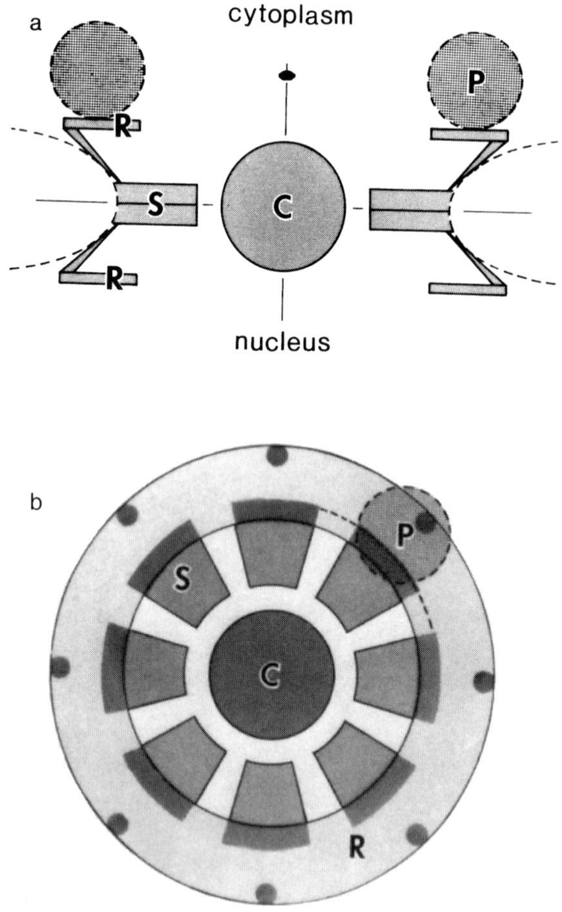

FIGURE 13 Diagram of the nuclear pore complex: (*a*) In central cross-section and (*b*) in projection down the octad axis. The major constituents are the central plug (*C*), the spokes (*S*), and the rings (*R*). The spokes are connected to the rings near to the maximum radius of the assembly (~600 Å). Superposition of detail in projection gives rise to a characteristic hollow or bilobed region of density near the outer extremity of the spokes (see Fig. 6). The broken lines outline the positions of the additional features present with pore complexes embedded within the nuclear envelope: the two membrane layers which come together at the pore complex and octagonally arranged particles (*P*), resembling ribosomes. The position of the membrane border in *a* corresponds with the thin-section view (e.g., reference 6). The particles are easily detached and are not always present.

Our observations on the character of these particles are consistent with the surmise that ribosomes (21, 31), or alternatively ribosomal precursors, are sometimes associated with the nuclear pore complex. The particles resemble inactive ribosomes stained under identical conditions (Figs. 11 and 12). They also resemble the membrane-bound particles in the spaces between the pores, which from studies of nuclear envelopes in other cells one would presume to be ribosomes (25, 31, 32). That they are detached from the periphery of the pore complex by the same biochemical treatments (addition of Triton X-100, EDTA, or a high concentration of potassium ions) that release inactive ribosomes from membranes in secretory cells (2, 27) could reflect the fact that they are attached directly to the nuclear membrane in the immediate vicinity of the pore complex and interact only weakly with the pore complex itself.

We are grateful to John Gurdon and Rick Bram for generously providing us with oocytes, and to Tony Crowther and Linda Amos for the use of their rotational averaging computer program. We thank Roger Kornberg, John Murray, and John Kilmartin for valuable discussions and comments. Guido Zampighi suggested using gold thioglucose as a negative stain.

Received for publication 14 July 1981, and in revised form 21 October 1981.

REFERENCES

1. Aaronson, R. P., and G. Blobel. 1975. Isolation of nuclear pore complexes in association with a lamina. *Proc. Natl. Acad. Sci. U. S. A.* 72:1007–1011.
2. Adelman, M. R., D. D. Sabatini, and G. Blobel. 1973. Ribosome membrane interaction. Nondestructive disassembly of rat liver rough microsomes into ribosomal and membranous components. *J. Cell Biol.* 56:202–229.
3. Arndt, U. W., J. Barrington-Leigh, J. F. W. Mallet, and K. E. Twinn. 1969. A mechanical microdensitometer. *J. Sci. Instrum.* 2:385–387.
4. Bonner, W. M. 1975. Protein migration into nuclei I. Frog oocyte nuclei in vivo accumulate microinjected histones, allow entry to small proteins, and exclude large proteins. *J. Cell Biol.* 64:421–430.
5. Crowther, R. A., and L. A. Amos. 1971. Harmonic analysis of electron microscope images with rotational symmetry. *J. Mol. Biol.* 60:123–130.
6. Daniels, E. W., J. M. McNiff, and D. R. Ekberg. 1969. Nucleopores of the giant amoeba, *Pelomyxa carolinensis*. *Z. Zellforsch. Mikrosk. Anat.* 98:357–368.
7. DeRobertis, E. M., R. F. Longthorne, and J. B. Gurdon. 1978. The intracellular migration of nuclear proteins in *Xenopus* oocytes, *Nature (Lond.)*. 272:254–256.
8. DeRosier, D. J., and P. B. Moore. 1970. Reconstruction of three-dimensional images from electron micrographs of structures with helical symmetry. *J. Mol. Biol.* 52:355–369.
9. Fabergé, A. C. 1973. Direct demonstration of eightfold symmetry in nuclear pores, *Z. Zellforsch. Mikrosk. Anat.* 136:183–190.
10. Feldherr, C. M., and J. M. Marshall. 1962. The use of colloidal gold for studies of intracellular exchanges in ameba *Chaos chaos*. *J. Cell Biol.* 12:641–645.
11. Finch, J. T., and A. Klug. 1965. The structure of viruses of the Papilloma-Polyoma type. III. Structure of rabbit Papilloma virus. *J. Mol. Biol.* 13:12.
12. Franke, W. W., and U. Scheer. 1970. The ultrastructure of the nuclear envelope of amphibian oocytes: a reinvestigation. *J. Ultrastruct. Res.* 30:288–316.
13. Franke, W. W., U. Scheer, and H. Fritsch. 1972. Intranuclear and cytoplasmic annulate lamellae in plant cells. *J. Cell Biol.* 53:823–827.
14. Gall, J. G. 1967. Octagonal nuclear pores. *J. Cell Biol.* 32:391–399.
15. Gurdon, J. B. 1976. Injected nuclei in frog oocytes: fate, enlargement, and chromatin dispersal. *J. Embryol. Exp. Morphol.* 36:523–540.
16. Gurdon, J. B. 1970. Nuclear transplantation and the control of gene activity in animal development. *Proc. R. Soc. Lond. B Biol.* 176:303–314.
17. Hoeijmakers, J. H. J., J. H. N. Schel, and F. Wanka. 1974. Structure of the nuclear pore complex in mammalian cells. Two annular components. *Exp. Cell Res.* 87:195–206.
18. Huxley, H. E., and G. Zubay. 1960. Electron microscope observations on the structure of microsomal particles from *Escherichia coli*. *J. Mol. Biol.* 2:10–18.
19. Krohne, G., W. W. Franke, and U. Scheer. 1978. The major polypeptides of the nuclear pore complex. *Exp. Cell Res.* 116:85–102.
20. Martini, D. H. W., and H. J. Gould. 1975. Molecular weight distribution of ribosomal protein from several vertebrate species. *Mol. Gen. Genet.* 142:317–331.
21. Mepham, R. H., and G. R. Lane. 1969. Nucleopores and polyribosome formation. *Nature (Lond.)*. 221:288–289.
22. Moody, M. F. 1971. Application of optical diffraction to helical structures in the bacteriophage tail. *Philos. Trans. R. Soc. Lond. B Biol. Sci.* 261:181–195.
23. Morimoto, T., G. Blobel, and D. D. Sabatini. 1972. Ribosome crystallization in chicken embryos. I. Isolation, characterization, and in vivo activity of ribosome tetramers. *J. Cell Biol.* 52:338–354.
24. Paine, P. L., L. C. Moore, and S. B. Horowitz. 1975. Nuclear envelope permeability. *Nature (Lond.)*. 254:109–114.
25. Palade, G. E. 1955. A small particulate component of the cytoplasm. *J. Biophys. Biochem. Cytol.* 1:59–68.
26. Roberts, K., and D. H. Northcote. 1970. Structure of the nuclear pore in higher plants. *Nature (Lond.)*. 228:385–386.
27. Sabatini, D. D., Y. Tashiro, and G. E. Palade. 1966. On the attachment of ribosomes to microsomal membranes. *J. Mol. Biol.* 19:503–524.
28. Scheer, U. 1972. The ultrastructure of the nuclear envelope of amphibian oocytes. IV. On the chemical nature of the pore complex material. *Z. Zellforsch. Mikrosk. Anat.* 127:127–148.
29. Unwin, P. N. T., and G. Zampighi. 1980. Structure of the junction between communicating cells. *Nature (Lond.)*. 283:545–549.
30. Vivier, E. 1967. Observations ultrastructurates sur l'envelope nucleaire et ses pores chez des sporozoaires. *J. Microscopie.* 6:371–390.
31. Watson, M. L. 1959. Further observations of the nuclear envelope of the animal cell. *J. Biophys. Biochem. Cytol.* 6:147–156.
32. Watson, M. L. 1955. The nuclear envelope. Its structure and relation to cytoplasmic membranes. *J. Biophys. Biochem. Cytol.* 1:257–270.
33. Watson, M. L. 1954. Pores in the mammalian nuclear membrane. *Biochim. Biophys. Acta.* 15:475–479.
34. Williams, R. C. 1977. Use of polylysine for adsorption of nucleic acids and enzymes to electron microscope specimen films. *Proc. Natl. Acad. Sci. U. S. A.* 74:2311–2315.
35. Wischnitzer, S. 1958. An electron microscope study of the nuclear envelope of amphibian oocytes. *J. Ultrastruct. Res.* 1:201–222.

Lohka M.J. and **Masui Y.** 1984. Roles of cytosol and cytoplasmic particles in nuclear envelope assembly and sperm pronuclear formation in cell-free preparations from amphibian eggs. *J. Cell Biol.* **98:** 1222–1230.

WHEN A SPERM PENETRATES AN EGG AT FERTILIZATION, it undergoes dramatic morphological and biochemical changes, including swelling, formation of a new nuclear envelope, import of various proteins from the egg cytoplasm, DNA synthesis, and eventually mitosis. Remarkably, all these events and more can be observed in a simple egg extract first described by Lohka and Masui. Freshly laid eggs of the frog *Rana pipiens* were centrifuged at 9,000 X g in a minimal volume of buffer. The eggs were crushed by the centrifugal force, lipid floating to the top of the tube and the heavier yolk granules and pigment forming a pellet at the bottom. The clear somewhat viscous intermediate layer of "ooplasm" was collected and used for experiments. When demembranated sperm heads of *Xenopus* were added to the ooplasm, they underwent essentially all the normal events of pronuclear formation. Similar extracts have since been used in numerous studies on formation of the nuclear envelope, import of proteins into the nucleus, microtubule and centrosome assembly, chromosome replication, and regulation of the cell cycle (1-5). Following Lohka and Masui, many workers refer to the extract as "cytosol." The term is technically correct, because the extract is derived from the egg cytoplasm, but the cytoplasm of an unfertilized egg also contains a large store of nuclear proteins derived from breakdown of the germinal vesicle.

---

1. Newmeyer D.D., Finlay D.R., and Forbes D.J. 1986. In vitro transport of a fluorescent nuclear protein and exclusion of non-nuclear proteins. *J. Cell Biol.* **103:** 2091–2102.
2. Mills A.D., Blow J.J., White J.G., Amos W.B., Wilcock D., and Laskey R.A. 1989. Replication occurs at discrete foci spaced throughout nuclei replicating in vitro. *J. Cell Sci.* **94:** 471–477.
3. Newport J.W., Wilson K.L., and Dunphy W.G. 1990. A lamin-independent pathway for nuclear envelope asembly. *J. Cell Biol.* **111:** 2247–2259.
4. Hirano T. and Mitchison T.J. 1994. A heterodimeric coiled-coil protein required for mitotic chromosome condensation in vitro. *Cell* **79:** 449–458.
5. Heald R., Tournebize R., Blank T., Sandaltzopoulos R., Becker P., Hyman A., and Karsenti E. 1996. Self-organization of microtubules into bipolar spindles around artificial chromosomes in *Xenopus* egg extracts. *Nature* **382:** 420–425.

Roles of Cytosol and Cytoplasmic Particles in Nuclear Envelope Assembly and Sperm Pronuclear Formation in Cell-free Preparations from Amphibian Eggs

MANFRED J. LOHKA and YOSHIO MASUI
Department of Zoology, University of Toronto, Toronto, Ontario, Canada M5S 1A1. Dr. Lohka's present address is Department of Pharmacology, University of Colorado School of Medicine, Denver, Colorado 80262.

ABSTRACT A cell-free cytoplasmic preparation from activated *Rana pipiens* eggs could induce in demembranated *Xenopus laevis* sperm nuclei morphological changes similar to those seen during pronuclear formation in intact eggs. The condensed sperm chromatin underwent an initial rapid, but limited, dispersion. A nuclear envelope formed around the dispersed chromatin and the nuclei enlarged. The subcellular distribution of the components required for these changes was examined by separating the preparations into soluble (cytosol) and particulate fractions by centrifugation at 150,000 g for 2 h. Sperm chromatin was incubated with the cytosol or with the particulate material after it had been resuspended in either the cytosol, heat-treated (60 or 100°C) cytosol or buffer. We found that the limited dispersion of chromatin occurred in each of these ooplasmic fractions, but not in the buffer alone. Nuclear envelope assembly required the presence of both untreated cytosol and particulate material. Ultrastructural examination of the sperm chromatin during incubation in the preparations showed that membrane vesicles of ~200 nm in diameter, found in the particulate fraction, flattened and fused together to contribute the membranous components of the nuclear envelope. The enlargement of the sperm nuclei occurred only after the nuclear envelope formed. The pronuclei formed in the cell-free preparations were able to incorporate [^{3}H]dTTP into DNA. This incorporation was inhibited by aphidicolin, suggesting that the DNA synthesis by the pronuclei was dependent on DNA polymerase-α. When sperm chromatin was incubated >3 h, the chromatin of the pronuclei often recondensed to form structures resembling mitotic chromosomes within the nuclear envelope. Therefore, it appeared that these ooplasmic preparations could induce, in vitro, nuclear changes resembling those seen during the first cell cycle in the zygote.

In many species, the cell cycles initiated in the early embryo by fertilization consist of rapidly alternating periods of DNA replication and mitosis (1–3). During the first cell cycle, which is usually longer than succeeding ones, the nucleus of the fertilizing sperm undergoes a transformation into an interphase nucleus (for review, see references 4 and 5). Shortly after the sperm fuses with the plasma membrane of the egg, the nuclear envelope surrounding the sperm chromatin breaks down, the highly condensed sperm chromatin becomes dispersed and a new nuclear envelope is assembled at the periphery of the chromatin to form the male pronucleus. The male pronucleus increases in size, synthesizes DNA, associates with the female pronucleus, and enters mitosis.

The changes in the sperm nucleus following fertilization and during the first cell cycle of zygotes are similar to the changes in the nuclei of other proliferating cells. In both cases, nuclear envelope assembly, chromatin decondensation, DNA synthesis, and chromosome condensation occur in a similar manner. Cell fusion experiments have shown that in proliferating cells these events are controlled by cytoplasmic factors that are active at specific phases in the cell cycle (6). Similarly, the induction of chromatin decondensation, DNA synthesis, and mitosis in nuclei transplanted from embryonic or non-dividing somatic cells into activated eggs, suggests that cytoplasmic factors control nuclear behavior in the zygote (4–9). Therefore, nuclear behavior may be controlled by similar

THE JOURNAL OF CELL BIOLOGY · VOLUME 98 April 1984 1222–1230 cytoplasmic substances in both zygotes and proliferating cells.

The investigation of the molecular basis of the cytoplasmic control of nuclear behavior would be facilitated if the nuclear behavior observed during the cell cycle of intact cells could be reproduced in a cell-free system. Eggs may offer an ideal source of material for such a cell-free system since during oogenesis they accumulate a large store of cellular components required for the rapid cell proliferation that follows fertilization (10). In previous studies, ooplasmic preparations have been shown to induce, in vitro, the decondensation of hen erythrocyte (11) and sea urchin sperm chromatin (12), as well as the initiation of DNA synthesis in isolated somatic cell nuclei (13). We have reported that a cytoplasmic preparation from activated *Rana pipiens* eggs can induce demembranated *Xenopus laevis* sperm nuclei to transform into pronuclei and then mitotic chromosomes (14). In the present experiments, we show that the transformation of the sperm nuclei into pronuclei requires both the soluble and particulate cytoplasmic components found in a heavy ooplasmic fraction and that the particulate material contributes membrane vesicles that form a nuclear envelope in the presence of soluble ooplasmic factors.

MATERIALS AND METHODS

Preparation of Sperm Nuclei: Sperm nuclei were prepared as previously described (14). Testes were dissected from sexually mature *X. laevis* that had been injected with of 100 I.U. human chorionic gonadotropin (Sigma Chemical Co., St. Louis, MO) and kept for 1 h at 22°C. They were washed free of blood and incubated overnight at 18°C in 200% Steinberg's solution containing antibiotics (15) and human chorionic gonadotropin (10 I.U./ml). Sperm were released by gently squeezing the testes, collected by centrifugation at 1,500 *g* for 10 min, and treated for 5 min at 22°C with nuclear isolation medium (16), which contained 0.5% lysolecithin and 1 mg/ml soybean trypsin inhibitor (both from Sigma Chemical Co.). Lysolecithin-treated sperm were washed once with ice-cold nuclear isolation medium + 3% bovine serum albumin (BSA; fraction V, Sigma Chemical Co.), three times with nuclear isolation medium + 0.4% BSA and once with buffer (see below) before use. The suspension contained ~95% sperm nuclei and 5% nuclei from other cells, mostly erythrocytes. For some experiments the isolated sperm nuclei were stored at −80°C in 30% glycerol in nuclear isolation medium. Nuclei stored in this manner were washed extensively with buffer (see below) before use.

Preparation of Cytoplasmic Fractions: Female *R. pipiens* were primed by an injection of 1/6 of a pituitary, kept at 18°C for 24 h and then induced to ovulate by injection of one pituitary and 1 mg progesterone dissolved in corn oil. 40–48 h later, eggs were removed from the ovisac and enzymatically dejellied, in 1% $Na_2HPO_4 7H_2O$, 0.5% crude papain (crude powder, type II; Sigma Chemical Co., St. Louis, MO) and 0.4% cysteine-HCl (Sigma Chemical Co.). The eggs were washed well in 0.1 M NaCl and 200% Steinberg's solution and any damaged or activated eggs were removed. While in 200% Steinberg's solution, the dejellied eggs were activated by an electric shock (80 V, 200 ms), and then incubated in 20% Steinberg's at 19 ± 1°C for 1 h.

Cytoplasmic fractions were prepared as previously described (14). The activated dejellied eggs were washed in ice-cold buffer which consisted of 250 mM sucrose, 200 mM KCl, 1.5 mM $MgCl_2$, 2.0 mM β-mercaptoethanol and 10 mM Tris-HCl at pH 7.5 and transferred to 5 ml centrifuge tubes containing ice-cold buffer. After the excess medium was withdrawn, the eggs were crushed, without homogenization, by centrifugation at 9,000 *g* for 15 min at 2°C. In most experiments, the layer of heavy supernatant above the packed pigment and yolk (Fig. 1*A*) was transferred to an 0.8-ml tube and centrifuged at 9,000 *g* for 30 min to remove most of the pigment. In cytoplasmic preparations made in a similar manner the buffer contributed approximately one-third of the soluble part of the preparation (17). Lysolecithin-treated sperm was incubated at 18°C in 200 μl of the cytoplasmic fraction to give a concentration of 5×10^4 to 1×10^5 sperm/ml.

The soluble and particulate cytoplasmic components were separated by centrifuging the cytoplasmic preparation in a small tube (0.8 ml, Beckman Instruments, Inc., Palo Alto, CA) at 150,000 *g* for 2 h (Fig. 1 *B*). The fluffy part of the pellet, which contains cytoplasmic vesicles, was resuspended, before mixing with the sperm nuclei, in either the supernatant or the supernatant that had been heated at 60°C or 100°C for 10 min and centrifuged at 1,500 *g* for 10 min to remove the precipitated material (heat-treated supernatant), or 1/3 buffer. The final volume of each mixture was adjusted to equal its initial volume.

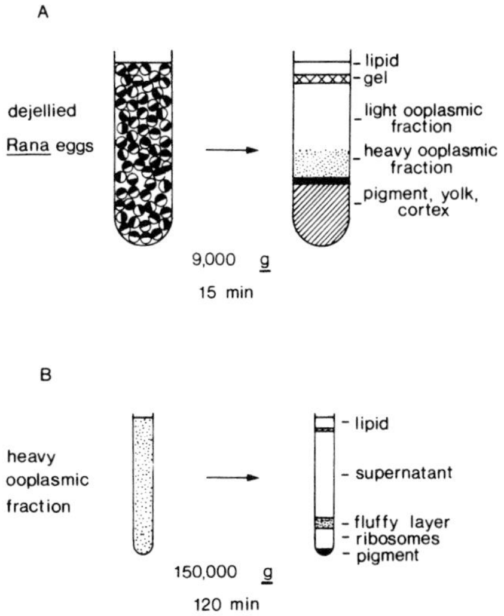

FIGURE 1 Diagram of the ooplasmic fractions obtained following the centrifugation of dejellied *R. pipiens* eggs. The centrifuge used was a Beckmann L3-4 with an SW-50.1 rotor.

Assay for Nuclear DNA Synthesis: To determine whether the sperm nuclei were induced to synthesize DNA, we mixed 100 μl of the ooplasmic preparation with an equal volume of 1/3 buffer containing [methyl-^{3}H]thymidine-5′-triphosphate ([^{3}H]dTTP[1]; 44 Ci/mmol, Amersham) at a concentration of 40 or 80 μCi/ml. Aphidicolin (a gift from Dr. J. Rossant) was dissolved in dimethysulfoxide at a concentration of 5 mg/ml and added to the 1/3 buffer at twice the desired concentration. As a control, dimethylsulfoxide alone was added to the 1/3 buffer. The ooplasmic preparation was then diluted with an equal volume of 1/3 buffer containing both the label and the drug. Freshly prepared sperm nuclei were suspended in the ooplasmic preparations and incubated at 18°C. Samples taken from the incubation mixture at various times were fixed and processed for autoradiography.

Cytological and Histological Procedures: Aliquots of the incubation mixture were fixed in a cold mixture of ethanol and acetic acid (3:1). The fixed material was stained by the Feulgen procedure and portions were transferred to a droplet of 50% acetic acid on a microscope slide and squashed. The squashed preparations were frozen in a mixture of dry ice and ethanol, the coverslips were removed and the slides were air-dried. The specimens were restained with 2% Giemsa in 0.01 M phosphate buffer at pH 6.8 and after drying covered with Pro-Texx (Scientific Products, McGaw Park, IL). The area of the sperm nuclei in the squashed preparations was determined using the formula for the area of an ellipse, after the lengths of the long and short axes had been measured.

In some cases, portions of the incubation mixture were fixed with Smith's solution, stained and sectioned for histological examination as previously described (18).

Autoradiographic Procedures: The squashed specimens were washed twice in cold 5% trichloroacetic acid, rinsed in running water for 1 h, air-dried, and coated with Kodak NTB2 emulsion. Slides were exposed for 14 to 17 d at 4°C and developed in Kodak D-19.

The relative amount of [^{3}H]dTTP incorporated by a sperm nucleus was expressed as the corrected grain count per nucleus. The corrected grain count per nucleus was obtained by counting the number of grains in a 100- or 400-μm^2 area in which the nucleus was located and subtracting from this value the number of grains found in an equivalent area surrounding the nucleus.

Preparation for Electron Microscopy: Freshly prepared sperm nuclei were incubated at 18°C in an undiluted ooplasmic preparation or in one of its cytoplasmic fractions. Aliquots of the incubation mixture were fixed overnight in ice-cold 2% glutaraldehyde buffered with 0.05 M phosphate buffer at pH 7.4. Nuclei were recovered by centrifugation at 1,500 *g* for 5 min, washed three times with phosphate buffer, postfixed for 2 h at room temperature in 1% osmium tetroxide, dehydrated through a graded concentration series of ethanol to propylene oxide, and embedded in Epon 812. Sections were made at 50 to 60 nm, stained with uranyl acetate, followed by lead citrate and examined at 60 kV on a Philips 201 electron microscope.

[1] *Abbreviations used in this paper:* [^{3}H]dTTP, [^{3}H]thymidine triphosphate.

RESULTS

Behavior of Sperm Chromatin in Ooplasmic Fractions

Crushing activated *R. pipiens* eggs by centrifugation at 9,000 *g* for 15 min resulted in a crude separation of ooplasmic components (Fig. 1*A*). Yolk, cortex and some pigment were found in the pellet. The supernatant obtained by this centrifugation consisted of two layers: an upper layer, which we call the "light ooplasmic fraction," and a more viscous and heavily pigmented lower layer, which we call the "heavy ooplasmic fraction." When 5 ml of eggs are crushed by centrifugation, ~1.4 to 1.6 ml of the light ooplasmic fraction and 0.6 to 0.8 ml of the heavy ooplasmic fraction is obtained. The centrifugation buffer is thought to contribute about one-third of the volume of the supernatants (17).

The lysolecithin-treated sperm nuclei were incubated either in the light or in the heavy ooplasmic fraction or in the buffer used in the preparation of these fractions (Table I). The sperm nuclei incubated in the heavy ooplasmic fractions underwent a series of changes in their morphology. In all cases, at the start of the incubation, the sperm chromatin, found in a long, thin form that is characteristic of the sperm nucleus, could be stained deeply with Giemsa (Fig. 2*A*). We refer to nuclei with this morphology as type A nuclei. Within 1 h, the nuclei changed to a round or oval shape, but their chromatin could still be stained deeply (type B nuclei; Fig. 2*B*). These nuclei then began to enlarge, as their peripheral chromatin became more decondensed, to form nuclei with an inner core deeply staining chromatin and a peripheral portion of lightly staining chromatin (type C nuclei; Fig. 2*C*). The process of nuclear enlargement gradually continued until the chromatin was completely decondensed, forming pronuclei whose chromatin was uniformly stained lightly by Giemsa (type D nuclei, Fig. 2*D*). The heavy ooplasmic fraction was able to transform both freshly prepared sperm nuclei and those that had been stored at −80°C in 30% glycerol into pronuclei (type D nuclei) during a 3-h incubation. This effect of the heavy fraction was observed not only with *Xenopus* sperm nuclei, but also with the nuclei isolated from *R. pipiens* sperm (Table I). In contrast, the morphology of the sperm nuclei did not change when they were incubated for 3 h in the buffer used for preparing the ooplasmic fractions or in the buffer that was diluted to one-third its original concentration. Similarly, the chromatin of sperm nuclei that were incubated in the light ooplasmic fraction remained condensed during a 3-h incubation, although in this case the nuclei often changed to a round or oval shape (Table I). These results clearly indicate that the heavy ooplasmic fraction contains the components that could induce the formation of the sperm pronucleus and that these components are lacking or greatly reduced in the light ooplasmic fraction.

We examined the effect of diluting the heavy ooplasmic fraction on its ability to decondense sperm nuclei (Table I).

TABLE I

Behavior of Sperm Nuclei during 3-h Incubation in Cell-free Preparations of Activated R. pipiens Eggs

Incubation conditions	No. of nuclei	Percentage of nuclei: Type A and B*	Type C*	Type D*	Other‡
Xenopus sperm					
Light fraction	905	68	4	1	27
Heavy fraction	464	1	16	82	1
Diluted heavy fraction (1/2)	703	17	32	47	4
Diluted heavy fraction (1/4)	655	91	8	0	1
Buffer	636	95	0	0	5
1/3 buffer	1,230	94	0	0	6
Rana sperm					
Heavy fraction	557	0	10	89	1
1/3 buffer	558	95	0	0	5

* Refer to text.
‡ Damaged or nonsperm nuclei.

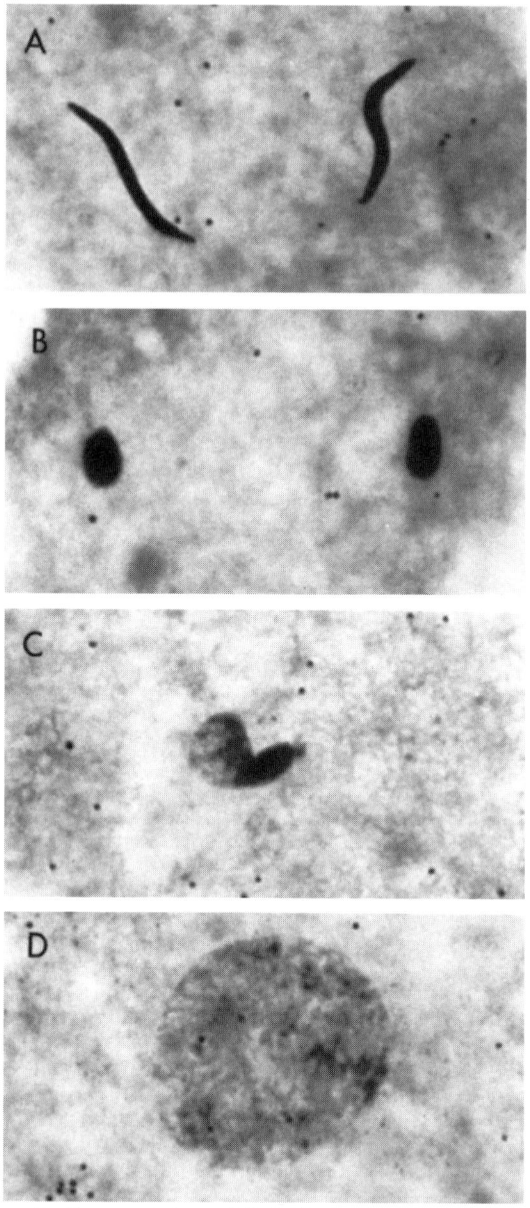

FIGURE 2 The morphology of *X. laevis* sperm nuclei at various times during incubation in ooplasmic preparations. (*A*) Condensed sperm nuclei (type A nuclei), 0 min. (*B*) Round sperm nuclei (type B nuclei), 60 min. (*C*) Partially decondensed sperm nucleus (type C nucleus), 90 min. (*D*) Completely decondensed sperm nucleus (type D nucleus), 180 min. × 1,000.

Sperm nuclei were incubated for 3 h in equal volumes of the undiluted heavy fraction or in the fraction after it was diluted to one-half or one-quarter of its original concentration with 1/3 buffer. The results show that the ability to induce pronuclear formation was dependent upon the concentration of the heavy fraction. In the undiluted heavy fraction, 82% of the sperm nuclei formed pronuclei within 3 h. On the other hand, only 47% of the sperm nuclei formed pronuclei in this fraction after it was diluted by one-half, and none of the sperm nuclei formed pronuclei in the fraction diluted to one-quarter (Table I).

As the sperm chromatin decondensed during incubation in the heavy ooplasmic fraction, the nuclei enlarged. This change could be quantitatively expressed as an increase in the area of the sperm nuclei in the squash preparations. The areas of pronuclei (type D nuclei) that formed during a 3-h incubation in undiluted or half-diluted heavy fractions was determined. The results showed that while the average area of the decondensed nuclei varied among experiments, in each experiment the mean area of the nuclei incubated in the undiluted heavy ooplasmic fraction was consistently about twice that of the nuclei incubated in the diluted fraction. For example, the average areas of 50 pronuclei formed during a 3-h incubation in undiluted preparations were 119 ± 50, 188 ± 97 and 352 ± 124 μm^2 in three different experiments, whereas the areas of those formed in half-diluted preparations in the same experiments were 61 ± 17, 86 ± 36, and 154 ± 53 μm^2, respectively. These results indicate that the ooplasmic components in the heavy fraction induce pronuclear formation and that their effect is concentration-dependent.

Assembly of the Nuclear Envelope

To further investigate the processes of pronuclear formation in vitro, we observed sperm nuclei incubated in the heavy ooplasmic fraction at an ultrastructural level. Previously, we found that treatment of *X. laevis* sperm with lysolecithin removed most of their plasma membrane and nuclear envelope, leaving the highly condensed chromatin intact (14). The morphology of these sperm nuclei remained unchanged for at least 3 h if they were incubated either in the buffer or in the buffer diluted to one-third of its initial concentration. In contrast, the sperm chromatin incubated in the heavy ooplasmic fraction was uniformly dispersed to become a less electron-dense, fibrous structure within 5 min. However, it should be noted that while the sperm chromatin was found to be dispersed within 5 min when it was examined ultrastructurally, no change in the chromatin could be discerned when the nuclei were examined under the light microscope at this time. Instead, the chromatin retained its elongate form and could still be stained deeply by Giemsa.

During incubation, a nuclear envelope began to assemble at the periphery of the sperm chromatin from vesicles that had previously been dispersed in the heavy ooplasmic fraction. From the morphology of vesicles observed near the periphery of the chromatin in samples incubated for 30 min, the sequence of changes that lead to the formation of the nuclear envelope was constructed (Fig. 3, *A–H*). Initially, the periphery of the dispersed chromatin was devoid of nuclear envelope and therefore directly exposed to the cytoplasmic components (Fig. 3*A*). Cytoplasmic vesicles having a diameter of 180 to 200 nm and containing electron-opaque material, were soon found at many sites along the periphery of the sperm chromatin (Fig. 3*B*). These vesicles fused to each other and flattened (Fig. 3*C*) to form the double membrane of the nuclear envelope. Electron-dense material (Fig. 3*E*) accumulated on the outer membrane of the flattened vesicles. Nuclear pore structures were formed in the flattened vesicles, possibly from the material that adhered to the outer membranes, at sites where the inner and outer membranes had coalesced (Fig. 3, *E–F*). Vesicles added to the periphery of the chromatin continued to fuse and flatten at the edges of the nascent nuclear envelope (Fig. 3*F*) until a continuous nuclear envelope, containing pores, enclosed the chromatin (Fig. 3*G*). The pores were also seen in sections tangential to the membrane (Fig. 3*H*). Although our observations clearly indicate that membrane vesicles that had a single membrane and contained electron-opaque material contributed membrane material to the formation of the nuclear envelope around the sperm chromatin, we cannot rule out the possibility that other membrane vesicles are also involved in nuclear envelope assembly.

The proportion of the perimeter of the sperm chromatin that was covered with nascent nuclear envelope increased with time during incubation in the cytoplasmic preparations. In nuclei examined at 30 min after incubation, only short fragments of nuclear envelope that covered less than one half of the chromatin contour, were found, although cytoplasmic vesicles had aligned at many sites on the chromatin periphery. By 60 min, however, 85% of the nuclei had at least half of the periphery of the chromatin covered with a nascent nuclear envelope, and by 90 min, practically all of the nuclei observed were enclosed by a continuous nuclear envelope.

Although nuclear envelope assembly around the sperm chromatin was usually complete by 90 min, no flattened membranes of any kind, resembling either the nuclear envelope or annulate lamellae, even in a fragmented form, were found in ooplasmic preparations that were examined after incubation for 180 min without sperm chromatin. Therefore, it is highly likely that the presence of sperm chromatin is required for the formation of the nuclear envelope from cytoplasmic membrane vesicles.

Ooplasmic Components Required for Nuclear Decondensation and Nuclear Envelope Assembly

The heavy ooplasmic fraction contains both soluble components and particulate material. Centrifugation of this fraction at 9,000 *g* for 30 min removed most of the pigment without a loss of the ability to induce pronuclear formation. However, as seen in Table II, the supernatant obtained from the heavy ooplasmic fraction after a centrifugation at 150,000 *g* for 120 min was unable to transform sperm nuclei into pronuclei during a 3-h incubation, although the nuclei often changed to a round or oval shape that could still be stained deeply by Giemsa. In contrast, pronuclei were formed when sperm nuclei were incubated for 3 h in the supernatant to which the particulate components, which had been sedimented in the fluffy layer during centrifugation (Fig. 1*B*), were returned and resuspended. These results, taken together with the ultrastructural observations described above, demonstrate that the presence of particulate components including cytoplasmic vesicles are prerequisite for the transformation of sperm nuclei into pronuclei.

To examine the role of soluble components in pronuclear formation, we incubated sperm nuclei with the fluffy part of the pellet after it had been resuspended in one of the following media: (*a*) the supernatant obtained by centrifugation of the

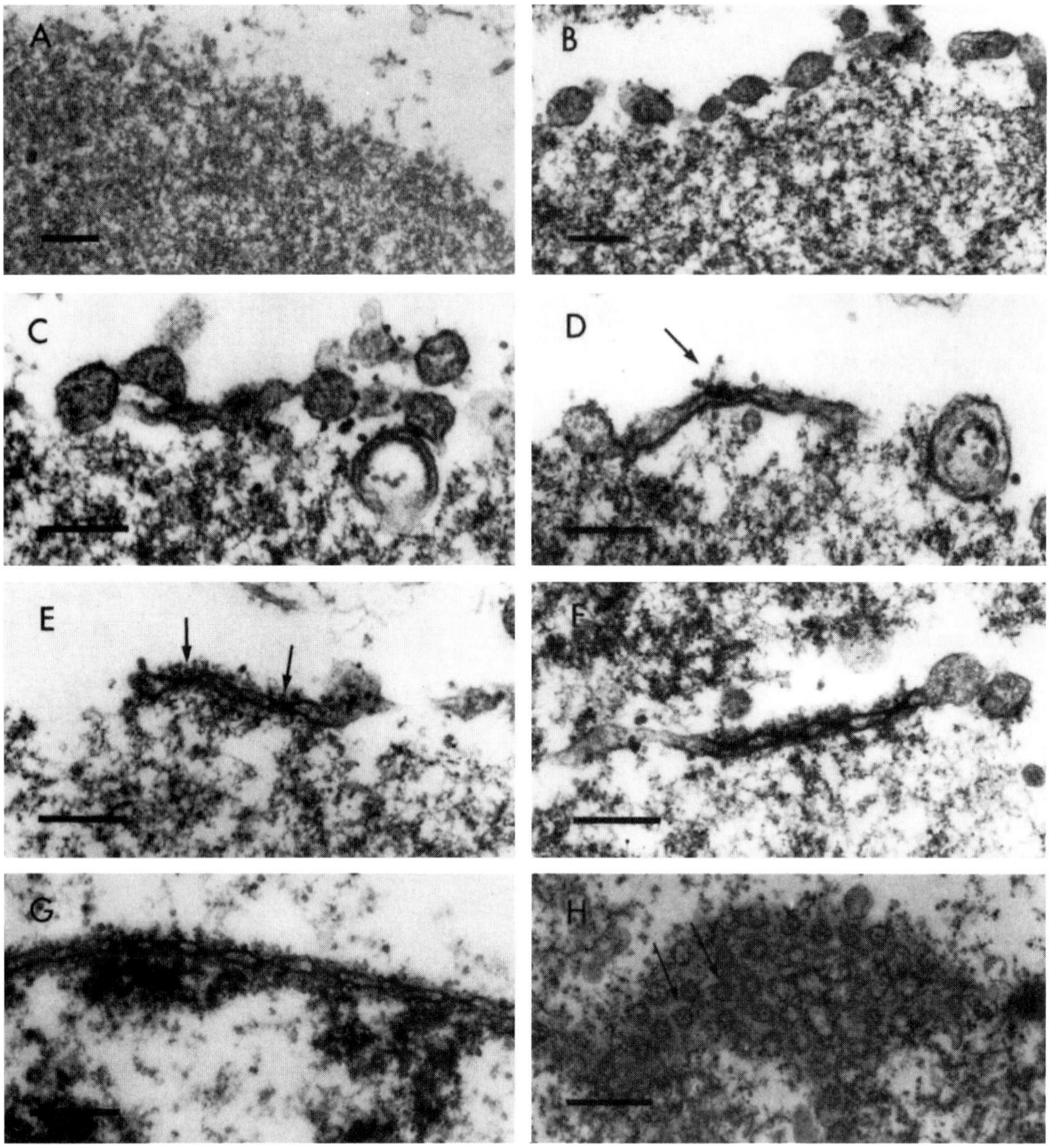

FIGURE 3 The sequence of changes resulting in the formation of a nuclear envelope. (*A*) The periphery of the sperm chromatin is devoid of nuclear envelope. × 35,500. (*B*) Cytoplasmic vesicles of ~200 nm in diameter accumulate at many sites along the periphery of the chromatin. × 35,500. (*C*) Vesicles in the process of fusing to each other and flattening to form the double membrane of the nuclear envelope. × 54,000. (*D*) Electron-dense material (arrow) accumulates on the outer membrane of the flattened vesicle. × 54,000. (*E*) Pores (arrows) form in the flattened vesicles, thereby forming short fragments of nuclear envelope. × 54,000. (*F*) Vesicles continue to fuse to the edges of the fragments of nascent nuclear envelope. × 54,000. (*G*) The nuclear envelope that entirely encloses the sperm chromatin. × 54,000. (*H*) Nuclear pores (arrows) found in a section tangential to a nuclear envelope. × 52,800. (*A*) 5-min incubation; (*B–F*) 30-min incubation; (*G–H*) 120-min incubation. Bars, 0.25 μm.

light ooplasmic fraction at 150,000 *g* for 2 h; (*b*) the heat-treated supernatant from the heavy ooplasmic fraction, obtained by centrifugation at 150,000 *g* for 2 h and then exposed to 60°C for 10 min; or (*c*) 1/3 buffer. In each of these cases, only a small percentage of the sperm nuclei were induced to form pronuclei, usually type C nuclei, except for the 5% that formed type D nuclei in the heat-treated supernatant (Table II). These results indicate that the particulate components are not sufficient to induce pronuclear formation. Rather, both soluble factors and particulate components contained in the heavy ooplasmic fraction are required. The soluble factors responsible for these effects appear to be large, heat-labile molecules, since neither the components in the heat-treated supernatant nor in the supernatant from the light ooplasmic fraction are able to complement the effect of the vesicular components that induces sperm chromatin decondensation.

We also determined the proportion of the sperm chromatin periphery that was covered by nascent nuclear envelope during a 2-h incubation in cytoplasmic fractions similar to those described above. As shown in Table III, continuous nuclear

TABLE II

Behavior of Sperm Nuclei during 3-h Incubation in Fractions of Cytoplasmic Preparations Obtained by Centrifugation at 150,000 g for 2 h

Cytoplasmic fraction	No. of nuclei	Percentage of nuclei: Type A and B*	Type C*	Type D*	Other‡
Heavy fraction (uncentrifuged)	2,705	1	5	81	12
Supernatant (heavy fraction)	2,023	83	3	0	14
Supernatant (heavy fraction) + pellet	2,337	0	6	87	7
Supernatant (light fraction) + pellet	1,185	85	5	0	10
Heat-treated supernatant§ (heavy fraction) + pellet	1,209	81	6	5	8
1/3 buffer + pellet	2,344	95	0	0	5

* Refer to text.
‡ Damaged or nonsperm nuclei.
§ 60°C for 10 min.

TABLE III

Nuclear Envelope Assembly on Sperm Chromatin Incubated for 2 h in Fractions of the Cytoplasmic Preparation Obtained by Centrifugation at 150,000 g for 2 h

Cytoplasmic fraction	No. of nuclei	Percentage of nuclei — Percentage of the chromatin perimeter lined by nuclear envelope: 0	<50	>50	100
Supernatant	50	100	–	–	–
Heat-treated* supernatant + pellet	50	98	2	–	–
1/3 buffer + pellet	60	92	8	–	–
Supernatant + pellet	39	–	5	20	75

* 100°C for 10 min.

envelopes were assembled around sperm chromatin only when it was incubated with both the untreated supernatant from the heavy fraction and the particulate cytoplasmic material. Sperm nuclei incubated in the supernatant, from which all particulate material had been removed by centrifugation, failed to form nuclear envelopes (Fig. 4*A*). Although these nuclei failed to form interphase nuclei, their chromatin became dispersed and less electron-dense, to the same extent as seen during the initial phase of sperm chromatin dispersion that was induced in the heavy cytoplasmic fraction containing vesicles. In contrast, when the sperm chromatin was incubated in the supernatants to which the vesicle-containing fraction was returned and resuspended, complete nuclear envelopes were found at the periphery of most nuclei (Fig. 4*B*). However, when the chromatin was incubated with the vesicle-containing fraction that had been resuspended either in 1/3 buffer or in the heat-treated supernatant exposed to 100°C for 10 min, the vesicles found at the chromatin periphery had neither fused to each other nor flattened to form a nuclear envelope (Fig. 4, *C–D*), except for a few cases in which only very small segments of nuclear envelope were found (Table III). Under these conditions, the sperm nuclei lacking a nuclear envelope failed to develop to a pronucleus, although the chromatin dispersed to about the same extent as the chromatin incubated in the soluble cytoplasmic components alone. Taken together, these results strongly suggest that interactions among the heat-labile, soluble cytoplasmic components, cytoplasmic vesicles, and chromatin are prerequisite for nuclear envelope assembly.

Other Activities of the Pronuclei

It was previously shown that the pronuclei that formed during a 3-h incubation in the heavy fraction containing [^{3}H]dTTP could synthesize DNA. The accumulation of radioactive label by the decondensed nuclei was examined by autoradiography. Both the completely decondensed sperm nuclei (type D nuclei) and nuclei whose chromatin had not yet decondensed completely, (type C nuclei), could incorporate the radioactive label. In contrast, sperm nuclei whose morphology remained unchanged during the incubation did not incorporate the label, nor did sperm nuclei that were incubated in 1/3 buffer containing [^{3}H]dTTP without the ooplasmic fraction.

Aphidicolin, a specific inhibitor of DNA polymerase-α, has been shown to inhibit the DNA synthesis required for chromosome replication (19). To examine whether or not the DNA synthesis observed here is related to chromosome replication, we tested the effect of aphidicolin on the incorporation of [^{3}H]dTTP by decondensed sperm nuclei. Sperm nuclei were incubated for 3 h in the heavy ooplasmic fraction after it was diluted by one-half with 1/3 buffer containing [^{3}H]-dTTP and either aphidicolin or DMSO, the vehicle solvent for aphidicolin. The ability of the heavy ooplasmic preparation to induce pronuclear formation was not affected by aphidicolin. However, aphidicolin inhibited the incorporation of [^{3}H]dTTP by the decondensed sperm nuclei. When aphidicolin was present at concentrations of 5 μg/ml or 25 μg/ml, <2% of the pronuclei, (type D nuclei), had grains counts that exceeded the background level. In contrast, when aphidicolin was not present, >98% of these pronuclei had grain counts that exceeded the background level. Therefore, we conclude that the DNA synthesis by the pronuclei results from the activity of DNA polymerase-α, the enzyme thought to be involved in chromosomal DNA replication in eucaryotes.

We have shown previously that the chromatin of pronuclei formed during a 3-h incubation in the undiluted heavy ooplasmic fraction often condensed again to form structures that resemble mitotic chromosomes when incubated for >3 h (14). Although these chromatin structures resemble chromosomes formed at metaphase, histological examination showed that the nuclear envelope was found to have remained intact, enclosing the chromosomes. The condensed chromatin was usually attached to the inside of the nuclear envelope. It should also be noted that the recondensation of the sperm chromatin to form mitotic chromosomes occurred only in the undiluted heavy ooplasmic fraction. Sperm nuclei incu-

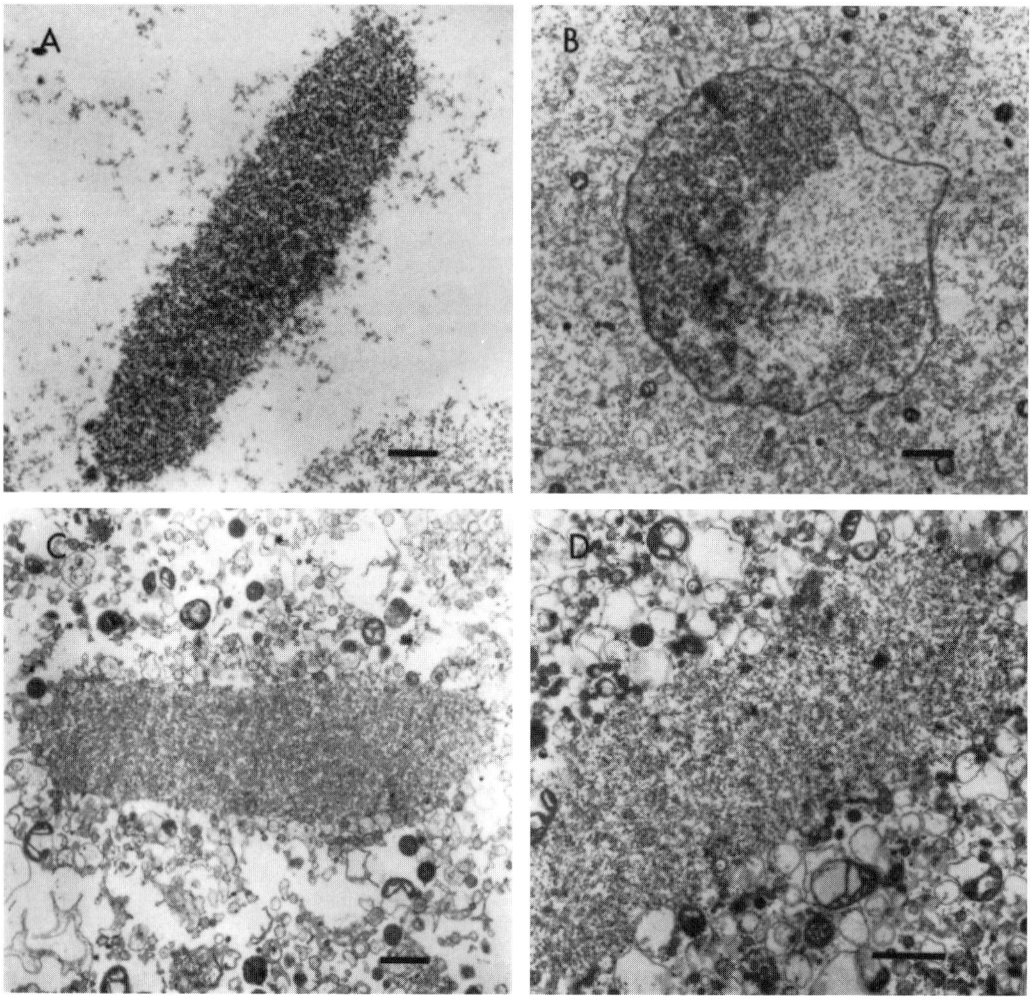

FIGURE 4 The extent of nuclear envelope assembly on lysolecithin-treated sperm nuclei during a 2-h incubation in the supernatant or vesicle-containing fraction obtained by centrifugation of the cytoplasmic preparation. (*A*) Supernatant alone, × 7,500. (*B*) Supernatant and pellet, × 7,500. (*C*) 1/3 buffer and pellet, × 7,500. (*D*) Heat-treated supernatant and pellet, × 10,700. Bars, 1.0 μm.

bated for 6 h in the heavy fraction diluted to one-half failed to condense chromosomes.

DISCUSSION

Numerous investigations have demonstrated that sperm or somatic cell nuclei transplanted into the cytoplasm of activated amphibian eggs are induced to enlarge and synthesize DNA (7–10). It would not be difficult to imagine that the cytoplasmic components acting upon the transplanted nuclei to induce their swelling are the same as those involved in the formation of the male pronucleus. In the present study, we have shown that a cytoplasmic preparation of *R. pipiens* eggs can induce, in vitro, a series of changes in sperm nuclear morphology that are similar to those occurring during pronuclear formation in the intact egg. These changes include (*a*) the initial rapid, but limited, dispersion of the highly condensed sperm chromatin, (*b*) the assembly of a nuclear envelope around the periphery of the sperm chromatin, (*c*) the enlargement of the sperm nuclei accompanied by an extensive decondensation of the sperm chromatin, (*d*) the initiation of DNA synthesis, and (*e*) the recondensation of the chromatin that results in the formation of structures resembling mitotic chromosomes within the nuclear envelope. These observations indicate that the egg cytoplasmic factors necessary for the formation of the sperm pronucleus and its following nuclear cycle can be kept active in this cell-free system.

Upon exposure to the heavy ooplasmic fraction, the sperm chromatin rapidly becomes dispersed and less electron-dense. This change in the sperm chromatin is clearly seen at the ultrastructural level, but it is difficult to observe in the squash preparations under a light microscope, where the sperm chromatin appeared to have been unchanged following a 5-min exposure to the ooplasmic preparations. Similar chromatin dispersion was also observed when sperm nuclei were incubated in the fresh soluble components, as well as in the soluble components that had been exposed to 100°C for 10 min. Furthermore, this type of chromatin dispersion also occurred in 1/3 buffer in which the vesicular components had been resuspended. Since no chromatin dispersion was observed when sperm nuclei were incubated in 1/3 buffer alone, it is likely that the factor responsible for the initial, rapid dispersion of sperm chromatin is a diffusible cytoplasmic substance contained in the vesicular components as well as in the soluble fraction. However, since the chromatin dispersion occurred even after the majority of proteins were denatured by the heat treatment, it is unlikely that specific, heat-labile cytoplasmic components are responsible for this rapid chromatin dispersion. Rather, a more likely explanation for the phenomenon may be that the initial dispersion of sperm chromatin results from the action of small, heat-stable molecules, including ions, in the cytoplasmic preparations.

A nuclear envelope was assembled in vitro around the periphery of the newly dispersed sperm chromatin. Our results indicate that the formation of the nuclear envelope requires the interaction of cytoplasmic vesicles, heat-labile soluble components and chromatin. The vesicles that we have shown to contribute membrane components to nuclear envelope assembly in this cell-free system, are morphologically similar to some of those that participate in nuclear envelope formation in living cells. Not only were similar vesicles found to be involved in the formation of the sperm pronuclear envelope in eggs of various species (4, 22, 23), but also in the reconstitution of the nuclear envelope at telophase in mitotically dividing cells (24–26). Therefore, the vesicles that participate in nuclear envelope assembly in the ooplasmic preparation may be involved in both the formation of the pronuclear envelope in the zygote, as well as, in the rapid reconstitution of the nuclear envelope in cleaving blastomeres. It may be hypothesized that these vesicles are stored in the frog egg cytoplasm for use as precursors of the nuclear envelope, just as the components required for chromosomal replication are stored in the egg cytoplasm for use during cleavage (10).

The origin of these vesicles is not entirely clear, although it has been supposed that in dividing cells they are derived from the endoplasmic reticulum as well as from remnants of the previous nuclear envelope broken down during mitosis (27). In the fertilized sea urchin egg the endoplasmic reticulum also appears to play a major role in providing components of the sperm pronuclear envelope (28). At present, the origin of the vesicles that contribute to nuclear envelope assembly in the frog egg is unknown. They may originate from the envelope of the germinal vesicle, which is fragmented when the germinal vesicle breaks down during meiotic maturation and later dispersed in the cytoplasm to become indistinguishable from the cisternae of the preexisting endoplasmic reticulum (29, 30). It has been shown that sperm nuclei fail to form pronuclei when they are exposed to the cytoplasm of mature oocytes from which the germinal vesicle has been removed, suggesting that the presence of germinal vesicle material is required for pronuclear formation (21). However, this does not necessarily mean that cytoplasmic membranes derived from sources other than the germinal vesicle are incompetent to assemble nuclear envelopes. In fact, oocytes whose germinal vesicles have been removed become capable of inducing pronuclear formation if they are reinjected with soluble components that were prepared from activated eggs by centrifuging the heavy ooplasmic fraction at 150,000 *g* for 2 h (31). Since the material that was injected in this experiment was completely devoid of particulate and membrane components, the nuclear envelopes of these pronuclei must have been formed from cytoplasmic membranes remaining within the enucleated oocyte.

The major proteinaceous, nonmembranous components of the nuclear envelope have been found to be localized in the nuclear pore complex, situated at the sites where the inner and outer nuclear membranes have joined, as well as in the nuclear lamina lying between the inner membrane of the nuclear envelope and the peripheral chromatin (for review, see references 27, 32). These structures are disassembled during mitosis and reassembled into the nuclear envelope when an interphase nucleus is reformed (33–36). During nuclear envelope assembly in our cell-free system, nuclear pore structures are formed after the membrane vesicles containing electron-opaque material have flattened, leaving the contents between the two membranes. Therefore, it is tempting to propose that at least some of the proteins constituting the nuclear pore complex and lamina originate from the contents of the vesicles, although the incorporation of soluble proteins into these structures is also likely.

Perhaps the most striking change in the sperm nuclei during incubation in the ooplasmic preparation is the dramatic increase in their size. Our results indicate that both soluble and vesicular cytoplasmic components are required to induce the sperm nuclei to form pronuclei and to enlarge. However, since both soluble and vesicular components are also required for nuclear envelope assembly, these results may indicate that nuclear envelope assembly is prerequisite for nuclear enlargement and further dispersion of the chromatin during pronuclear formation. The concept that the sperm nucleus can be induced to enlarge only after a nuclear envelope has formed is supported by the observations that during incubation in the ooplasmic preparation the assembly of the nuclear envelope around the sperm chromatin precedes the enlargement of the nuclei. In interphase nuclei, the pore complex of the nuclear envelope, the peripheral lamina, and the internal matrix form a nuclear skeleton (37). It may be hypothesized that during sperm pronuclear formation the nascent nuclear envelope serves as a support for the assembly of the nuclear skeleton and that assembly of the nuclear skeleton continues while the nuclei enlarge. A similar role could be proposed for the nuclear envelope during telophase when the nuclear envelope is reconstituted around the mitotic chromosomes before they decondense to form interphase nuclei.

When sperm nuclei are induced to enlarge after fertilization, sperm-specific nuclear proteins are replaced by histones and other proteins that were originally found in the egg cytoplasm (38–43). In amphibian eggs, cytoplasmic proteins also move into transplanted nuclei and are thought to induce them to enlarge (44–46). In our experiments, dilution of the ooplasmic preparation reduces both the percentage of nuclei that decondense and the extent to which the nuclei enlarge. This observation supports the idea that cytoplasmic proteins or other components in the ooplasmic preparations move into the sperm nuclei in a concentration-dependent manner to induce them to enlarge. Although at present, the identity and the mode of action of the proteins that move into the sperm nuclei is not known, we may assume that they include the germinal vesicle material found in the soluble ooplasmic fraction (31).

The initiation of DNA synthesis in dividing cells is regulated by cytoplasmic factors that are active in S-phase of the cell cycle. When the nuclei of nondividing cells are exposed to the cytoplasm of cells in S-phase, by cell fusion or nuclear transplantation, they are induced to synthesize DNA (6). Our results indicate that the sperm nuclei incubated in a cell-free preparation of egg cytoplasm are induced to synthesize DNA as well. The inhibition of DNA synthesis by aphidicolin indicates that it is dependent on DNA polymerase-α, the enzyme that is necessary for chromosomal DNA replication in eucaryotes (19). In intact eggs, the movement of cytoplasmic proteins into the transplanted nuclei may be necessary for DNA synthesis to be initiated (44–46). Similarly, DNA synthesis induced in the cell-free cytoplasmic preparations may depend on the migration into the sperm nuclei of the factors that initiate DNA synthesis in the zygote. However, at present, the possiblity cannot be ruled out that DNA polymerase-α and other enzymes that may play a role in the initiation of DNA synthesis are associated with the *Xenopus* sperm nuclei before their exposure to the ooplasmic preparations.

The cell fusion and nuclear transplantation experiments have also provided evidence that the condensation of chromosomes during mitosis also is controlled by cytoplasmic factors (6, 16, 46). These experiments clearly showed that the cytoplasm of cells in mitosis can induce the formation of mitotic chromosomes in interphase nuclei. When sperm nuclei are incubated in the ooplasmic preparation for 3 to 6 h, the chromatin of those nuclei that had once been induced to decondense to interphase could condense again to form structures that resemble mitotic chromosomes. This recondensation of chromatin does not appear to result from a degenerative change in the nuclei, since in some experiments the condensed chromosomes, if incubated further, could decondense again to return to interphase (14). Also, sperm nuclei were decondensed completely when incubated in preparations made from eggs immediately after activation, but, unlike those incubated in preparations made from eggs 1 h after activation, they did not recondense to form mitotic chromsomes during a 6-h incubation (unpublished results). Therefore, it may be a transient appearance of specific factors in the ooplasmic preparation made from eggs 1 h after activation that is responsible for the condensation of the decondensed chromatin to structures resembling metaphase chromosomes. In all probability, the putative ooplasmic factors responsible for the recondensation of chromatin in vitro are similar to those that regulate chromosome condensation during mitosis in intact cells. These factors may enter the decondensed sperm nuclei to induce chromatin condensation, since our observations indicate that, unlike the case in intact cells, chromosome condensation in the ooplasmic preparations is induced without breakdown of the nuclear envelope. Therefore, it may be that the factors responsible for chromosome condensation and for nuclear envelope breakdown during mitosis are different molecular entities that could act independently.

The results we present here suggest the possibility of analysing nuclear-cytoplasmic interactions during the cell cycle by using a cell-free preparation from amphibian eggs. Since this in vitro preparation is amenable to a wider range of manipulations than intact cells, it would be particularly useful for the biochemical investigation of the cell cycle.

We thank Mr. R. A. Villadiego of the Electron Microscope Laboratory of our department for sectioning the material for ultrastructural examination, Dr. J. Rossant for the gift of the aphidicolin, Dr. R. G. Korneluk for the *Xenopus*, and Mr. H. Clarke, Ms. M. Miller, and Ms. E. Shibuya for many helpful discussions during the course of this work. This research was supported by a grant to Y. Masui from National Science and Engineering Research Council of Canada and University of Toronto Open and Ontario Graduate Student Fellowships to M. Lohka.

Received for publication 26 September 1983, and in revised form 5 December 1983.

REFERENCES

1. Hinegardner, R. T., B. Rao, and D. E. Feldman. 1964. The DNA synthetic period during early cleavage development of the sea urchin egg. *Exp. Cell Res.* 36:53–61.
2. Graham, C. F., and R. W. Morgan. 1966. Changes in the cell cycle during early amphibian development. *Dev. Biol.* 14:439–460.
3. Anderson, K. V., and J. A. Lengyel. 1981. Changing rates of DNA and RNA synthesis in *Drosophila* embryos. *Dev. Biol.* 82:127–138.
4. Longo, F. J. 1973. Fertilization: a comparative ultrastructural review. *Biol. Reprod.* 9:149–215.
5. Longo, F. J., and M. Kunkle. 1978. Transformations of sperm nuclei upon insemination. *Curr. Topics Dev. Biol.* 12:149–184.
6. Johnson, R. T., and P. N. Rao. 1971. Nucleo-cytoplasmic interactions in the achievement of nuclear synchrony in DNA synthesis and mitosis in multinucleate cells. *Biol. Rev.* 46:97–155.
7. Graham, C. F., K. Arms, and J. B. Gurdon. 1966. Induction of DNA synthesis by frog egg cytoplasm. *Dev. Biol.* 14:439–460.
8. Graham, C. F. 1966. The regulation of DNA synthesis and mitosis in multinucleate frog eggs. *J. Cell Sci.* 1:363–374.
9. Leonard, R. A., N. J. Hoffner, and M. A. DiBerardino. 1982. Induction of DNA synthesis in amphibian erythroid nuclei in *Rana* eggs following conditioning in meiotic oocytes. *Dev. Biol.* 92:343–355.
10. Laskey, R. A., J. B. Gurdon, and M. Trendelenburg. 1979. Accumulation of materials involved in rapid chromosomal replication in early amphibian development. *In* Maternal Effects in Development. D. R. Newth and M. Ball, editors. Cambridge University Press, Cambridge. 65–80.
11. Barry, J. M., and R. W. Merriam. 1972. Swelling of hen erythrocyte nuclei in cytoplasm from *Xenopus* eggs. *Exp. Cell Res.* 71:90–96.
12. Kunkle, M., B. E. Magun, and F. J. Longo. 1978. Analysis of isolated sea urchin nuclei incubated in egg cytosol. *J. Exp. Zool.* 203:381–390.
13. Benbow, R. M., and C. C. Ford. 1975. Cytoplasmic control of nuclear DNA synthesis during early development of *Xenopus laevis*: a cell-free assay. *Proc. Natl. Acad. Sci. USA* 72:2437–2441.
14. Lohka, M. J., and Y. Masui. 1983. Formation *in vitro* of sperm pronuclei and mitotic chromosomes by amphibian ooplasmic components. *Science (Wash. DC).* 220:719–721.
15. Masui, Y. 1967. Relative roles of the pituitary, follicle cells and progesterone in the induction of oocyte maturation in *Rana pipiens. J. Exp. Zool.* 166:365–376.
16. Ziegler, D. H., and Y. Masui. 1973. Control of chromosome behavior in amphibian oocytes. I. The activity of maturing oocytes inducing chromosome condensation in transplanted brain nuclei. *Dev. Biol.* 35:283–292.
17. Masui, Y. 1983. Oscillatory activity of maturation promoting factor (MPF) in extracts of *Rana pipiens* eggs. *J. Exp. Zool.* 224:389–399.
18. Meyerhof, P. G., and Y. Masui. 1977. Ca and Mg control of cytostatic factors from *Rana pipiens* which causes metaphase and cleavage arrest. *Dev. Biol.* 61:214–229.
19. Ikegami, S., T. Tagushi, M. Ohashi, M. Oguro, H. Nagano, and Y. Mano. 1978. Aphidicolin prevents mitotic cell division by interfering with the activity of DNA polymerase-α. *Nature (Lond.)* 275:458–460.
20. Skoblina, M. N. 1976. Role of karyoplasm in the emergence of capacity of egg cytoplasm to induce DNA synthesis in transplanted sperm nuclei. *J. Embryol. Exp. Morphol.* 36:67–72.
21. Katagiri, C., and M. Moriya. 1976. Spermatozoan response to the toad egg matured after removal of germinal vesicle. *Dev. Biol.* 50:235–241.
22. Longo, F. J., and E. Anderson. 1968. The fine structure of pronuclear development and fusion in the sea urchin *Arbacia punctulata. J. Cell Biol.* 39:339–368.
23. Yanagimachi, R., and Y. D. Noda. 1970. Electron microscope studies on sperm incorporation in the golden hamster egg. *Am. J. Anat.* 128:429–462.
24. Chai, L. S., H. Weinfeld, and A. A. Sandberg. 1974. Ultrastructural changes in the nuclear envelope during mitosis of Chinese hamster cells: a proposed mechanism of nuclear envelope reformation. *J. Natl. Cancer Inst.* 53:1033–1048.
25. Gulyas, B. J. 1972. The rabbit zygote. III. Formation of the blastomere nucleus. *J. Cell Biol.* 55:533–541.
26. Longo, F. J. 1972. An ultrastructural analysis of mitosis and cytokinesis in the zygote of the sea urchin *Arbacia punctulata. J. Morphol.* 138:207–238.
27. Franke, W. W. 1974. Structure, biochemistry, and functions of the nuclear envelope. *Int. Rev. Cytol. Suppl.* 4:71–236.
28. Longo, F. J. 1976. Derivation of the membrane comprising the male pronuclear envelope in inseminated sea urchin eggs. *Dev. Biol.* 49:347–368.
29. Huchon, D., N. Crozet, N. Cantenot, and R. Ozon. 1981. Germinal vesicle breakdown in the *Xenopus laevis* oocyte: description of a transient microtubular structure. *Reprod. Nutr. Dev.* 21:135–148.
30. Szollosi, D., P. G. Calarco, and R. P. Donahue. 1972. The nuclear envelope: its breakdown and fate in mammalian oogonia and oocytes. *Anat. Rec.* 174:325–340.
31. Lohka, M. J., and Y. Masui. 1983. The germinal vesicle material required for sperm pronuclear formation is located in the soluble fraction of egg cytoplasm. *Exp. Cell Res.* 148:481–491.
32. Maul, G. G. 1977. The nuclear and the cytoplasmic pore complex: structure, dynamics, distribution, and evolution. *Int. Rev. Cytol. Suppl.* 6:75–186.
33. Maul, G. G. 1977. Nuclear pore complexes. Elimination and reconstruction during mitosis. *J.Cell Biol.* 74:492–500.
34. Gerace, L., A. Blum, and G. Blobel. 1978. Immunocytochemical localization of the major polypeptides of the nuclear pore complex-lamina fraction. *J. Cell Biol.* 79:546–566.
35. Gerace, L., and G. Blobel. 1980. The nuclear envelope lamina is reversibly depolymerized during mitosis. *Cell.* 9:277–287.
36. Jost, E., and R. T. Johnson. 1981. Nuclear lamina assembly, synthesis, and disaggregation during the cell cycle in synchronized HeLa cells. *J. Cell. Sci.* 47:25–52.
37. Berezney, R. 1979. Dynamic properties of the nuclear matrix. *in* The Cell Nucleus. H. Busch, editor Academic Press, Inc., New York. (Pt. D):413–456.
38. Carrol, A. G., and H. Ozaki. 1979. Changes in the histones of the sea urchin *Strongylocentrotus pupuratus* at fertilization. *Exp. Cell Res.* 119:397–315.
39. Das, N. K., J. Micou-Eastwood, and M. Alfert. 1975. Cytochemical studies of the protamine type in sperm nuclei after fertilization and the early embryonic histones of *Urechis caupo. Dev. Biol.* 43:333–339.
40. Ecklund, P. S., and L. Levine. 1975. Mouse sperm basic nuclear protein: electrophoretic characterization and fate after fertilization. *J. Cell Biol.* 66:25–262.
41. Rodman, T. C., F. H. Pruslin, H. P. Hoffman, and V. G. Allfrey. 1981. Turnover of basic chromosomal proteins in fertilized eggs: a cytoimmunochemical study of events in vivo. *J. Cell Biol.* 90:351–361.
42. Poccia, D., J. Salik, and G. Krystal. 1981. Transitions in histone variants of the male pronucleus following fertilization and evidence of a maternal store of cleavage-stage histones in the sea urchin egg. *Dev. Biol.* 82:287–296.
43. Kunkle, M., F. J. Longo, and B. E. Magun. 1978. Protein changes in maternally and paternally derived chromatin at fertilization. *J. Exp. Zool.* 203:371–380.
44. Arms, K. 1968. Cytonucleoproteins in cleaving eggs of *Xenopus laevis. J. Embryol. Exp. Morphol.* 20:367–374.
45. Merriam, R. W. 1969. Movement of cytoplasmic proteins into nuclei induced to enlarge and intiate DNA or RNA synthesis. *J. Cell Sci.* 5:333–349.
46. Hoffner, N. J., and M. A. DiBerardino. 1977. The acquisition of egg cytoplasmic non-histone proteins by nuclei during nuclear reprogramming. *Exp. Cell Res.* 108:421–427.
47. Gurdon, J. B. 1968. Changes in somatic cell nuclei inserted into growing and maturing amphibian oocytes. *J. Embryol. Exp. Morphol.* 20:401–414.

Görlich D., Prehn S., Laskey R.A., and **Hartmann E.** 1994. Isolation of a protein that is essential for the first step of nuclear protein import. *Cell* **79**: 767–778.

THE FLOW OF MACROMOLECULES IN BOTH DIRECTIONS between the nucleus and cytoplasm is required for many of the most basic processes of cell metabolism. Except for a short time during mitosis, all such flow must occur across the nuclear envelope and specifically through the nuclear pore complexes (1). The first hint that import of nuclear proteins involves a specific molecular pathway was the discovery of nuclear localization sequences (NLSs) in proteins that normally reside in the nucleus (2, 3). Most NLSs contain short runs of basic amino acids that are required for nuclear localization and, when transferred to a cytoplasmic protein, can cause its mislocalization to the nucleus. Just how the NLS works remained a mystery until the discovery of receptor proteins that bind the NLS and carry the nuclear protein to the nuclear pore complex. The first of these receptors to be isolated, importin, was described in the accompanying paper by Görlich et al. Subsequent studies identified numerous other receptors and clarified the complex pathways by which nuclear proteins are brought to the nuclear envelope and translocated through the nuclear pore complex into the nucleus (reviewed in 4).

---

1. Feldherr C.M., Kallenbach E., and Schultz N. 1984. Movement of a karyophilic protein through the nuclear pores of oocytes. *J. Cell Biol.* **99:** 2216–2222.
2. Dingwall C., Sharnick S.V., and Laskey R.A. 1982. A polypeptide domain that specifies migration of nucleoplasmin into the nucleus. *Cell* **30:** 449–458.
3. Kalderon D., Roberts B.L., Richardson W.D., and Smith A.E. 1984. A short amino acid sequence able to specify nuclear location. *Cell* **39:** 499–509.
4. Nakielny S. and Dreyfuss G. 1999. Transport of proteins and RNAs in and out of the nucleus. *Cell* **99:** 677–690.

Cell, Vol. 79, 767–778, December 2, 1994, Copyright © 1994 by Cell Press

Isolation of a Protein That Is Essential for the First Step of Nuclear Protein Import

Dirk Görlich,* Siegfried Prehn,† Ronald A. Laskey,* and Enno Hartmann‡
*Wellcome/CRC Institute
Tennis Court Road
Cambridge CB2 1QR
England
and Department of Zoology
University of Cambridge
Cambridge CB2 3EJ
England
†Institut für Biochemie
Humboldt Universität Berlin
10115 Berlin
Federal Republic of Germany
‡Max-Delbrück Centre for Molecular Medicine
Robert-Rössle-Strasse 10
13125 Berlin
Federal Republic of Germany

Summary

We have purified a cytosolic protein from Xenopus eggs that is essential for selective protein import into the cell nucleus. The purified protein, named importin, promotes signal-dependent binding of karyophilic proteins to the nuclear envelope. We have cloned, sequenced, and expressed a corresponding cDNA. Importin shows 44% sequence identity with SRP1p, a protein associated with the yeast nuclear pore complex. Complete, signal-dependent import into HeLa nuclei can be reconstituted by combining importin purified from Xenopus eggs or expressed in E. coli with Ran/TC4. Evidence for additional stimulatory factors is provided.

Introduction

Selective import of proteins into the cell nucleus occurs in two steps, both of which require the presence of a nuclear localization sequence (NLS; Newmeyer and Forbes, 1988; Richardson et al., 1988; Akey and Goldfarb, 1989). The first step is binding to the cytoplasmic surface of the nuclear pore complex. It does not require ATP or GTP. The second step is the energy-dependent translocation through the nuclear pore complex. Protein import into the nucleus has been reviewed extensively (Garcia-Bustos et al., 1991; Silver, 1991; Yamasaki and Lanford, 1992; Fabre and Hurt, 1994; Moore and Blobel, 1994).

Permeabilized cells (Adam et al., 1990) have been powerful tools for investigating nuclear import. The plasma membrane of cultured mammalian cells is selectively solubilized with a low concentration of digitonin, releasing the soluble contents of the cell, but leaving internal membranes such as the nuclear envelope intact. A fluorescent import probe can readily enter the leaky plasma membrane, and its uptake into the nucleus can be followed by fluorescence microscopy. The active import of karyophilic proteins was found to be absolutely dependent on the readdition of cytosol (Adam et al., 1990).

The biochemical approaches applied so far to identify factors involved in nuclear protein import can be classified into two main lines: first, the search for NLS-binding proteins by means of peptide cross-linking, affinity chromatography, or ligand-blotting techniques (for review see Yamasaki and Lanford, 1992), and second, fractionation of cytosol in respect to import activity (Newmeyer and Forbes, 1990; Moore and Blobel, 1992, 1993; Melchior et al., 1993; Adam and Adam, 1994).

The two steps of nuclear import, envelope binding and the subsequent energy-dependent translocation, were found to require different cytosolic fractions, called A and B, respectively (Moore and Blobel, 1992). Only one essential cytosolic component has been identified so far (Moore and Blobel, 1993; Melchior et al., 1993). It is the small GTP-binding protein Ran/TC4 from fraction B (Drivas et al., 1990; Bischoff and Pongstingl, 1991; Ren et al., 1993; Moore and Blobel, 1993; Melchior et al., 1993).

Here, we report the purification from Xenopus eggs, cDNA cloning, and sequencing of a cytosolic factor that is essential for the first step of nuclear protein import. It is a single polypeptide with an apparent molecular mass of 60 kDa on SDS gels. In its chromatographic properties and its effects on nuclear transport, it resembles the crude fraction A described by Moore and Blobel (1992). It mediates binding of karyophilic proteins to the nuclear envelope. Complete transport can be achieved with the purified protein and Ran/TC4 alone, although we present evidence for an additional stimulatory factor(s).

The recombinant 60 kDa protein expressed in bacteria is as active as the frog protein in restoring import activity of cytosol that has been depleted of the 60 kDa protein.

The 60 kDa protein is 64% identical with human Rch1 (Rag cohort 1; Cuomo et al., 1994) and 44% identical with SRP1p (suppressor of temperature-sensitive mutations of RNA polymerase I) in Saccharomyces cerevisiae (Yano et al., 1992). The SRP1 protein has previously been shown to be encoded by an essential gene (Yano et al., 1992, 1994) and to be a constituent of the yeast nuclear pore complex (Yano et al., 1992; Belanger et al., 1994).

Given the fact that the 60 kDa protein is essential for protein import into the cell nucleus, we suggest naming it importin.

Results

Purification of a 60 kDa Protein That Is Essential for Nuclear Protein Import

To identify further cytosolic factors required for nuclear import, we used a nuclear import system based on permeabilized cultured cells (Adam et al., 1990) and nucleoplasmin that had been labeled with fluorescein as a natural import probe.

On starting the fractionation of cytosol, the number of

Cell
768

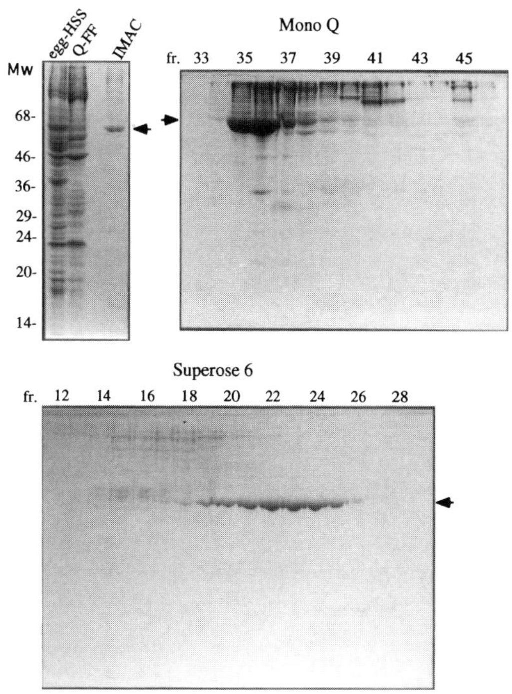

Figure 1. Purification of a 60 kDa Protein, Importin

Protein patterns of fractions at different stages of purification, visualized by SDS–polyacrylamide gel electrophoresis (SDS–PAGE) and Coomassie staining. A high speed supernatant of activated Xenopus eggs (egg-HSS) was applied to Q-Sepharose fast flow at 0.2 M NaCl. The protein fraction eluted with 1 M NaCl (Q-FF) was subjected to immobilized metal ion affinity chromatography (IMAC) on nitrilotriacetic acid–Sepharose charged with Ni^{2+}. IMAC represents the pool of active fractions eluting between 40–100 mM imidazole. Further purification was achieved on Mono Q (200–520 mM NaCl gradient). High activity was found in fractions 35 and 36, while less but still significant activity was found in fraction 37. Mono Q fractions 35–37 were applied to Superose 6. Significant activity was detected in fractions 18–26, and high activity was detected in fractions 21–24. Fraction 24 was used for the experiment shown in Figure 2, and fraction 23 was used for protein sequencing.

essential factors was unknown. To follow the purification of one component, we had to simplify the system. Therefore, bacterially expressed Ran/TC4 (100 μg/ml) was added to all assays, since it is known to be essential (Moore and Blobel, 1993; Melchior et al., 1993). Using Q-Sepharose fast flow (Pharmacia) as the initial fractionation step (see Figure 1), we first identified the most stringent binding conditions that still allowed depletion of an essential component, hoping that under these conditions only one, the most acidic, would bind (see legend to Figure 1). To ask which protein(s) has to be added back to restore activity, we used the Q-flowthrough (supplemented with Ran/TC4) as a depleted cytosol.

The eluate from the Q-Sepharose fast flow column was used for the further purification (see Figure 1). It was subjected to immobilized metal ion affinity chromatography (IMAC) on nickel(II)–nitrilotriacetic acid–Sepharose (Ni–NTA–Sepharose; Qiagen). This turned out to be the crucial purification step, allowing some 100-fold purification of the activity. After further ion exchange chromatography on Mono Q (Pharmacia) and gel filtration on Superose 6 (Pharmacia), a 60 kDa protein was purified to near homogeneity (Figure 1). Trace impurities were only evident in overloaded gels (as in Figure 1), and their pattern was variable in different active fractions and in different variants of the purification protocol. In contrast, we always observed a perfect correlation between the stimulation of import by the fractions and the presence of the 60 kDa band.

After the last column step, approximately 1 mg of purified 60 kDa protein was obtained from 1 g of protein in the starting material, corresponding to an approximately 15% yield as estimated by immunoblotting with antibodies raised against the 60 kDa protein (see below). The yield in terms of activity was in the same range (see Table 1). The abundance of the 60 kDa protein is approximately 0.7% of total protein in the postribosomal supernatant of Xenopus egg extract (this compares with some 1% Ran/TC4 for Xenopus oocyte extract; Moore and Blobel, 1993). The cytosolic concentration of the 60 kDa protein in Xenopus eggs is about 200 μg/ml (3 μM). Its maximum effect on import was reached within the same concentration range of 100–200 μg/ml (see below).

The determined N-terminal sequence of the 60 kDa protein did not match with any known protein in the data base. Given the fact that the 60 kDa protein is necessary and sufficient to restore the protein-import activity of Ni–NTA-Sepharose-depleted cytosol, we suggest naming it importin.

Importin elutes earlier from Superose 6 and Superdex 200 than one would expect for a 60 kDa protein. The apparent molecular mass of purified importin by means of gel filtration is close to 120 kDa, indicating either a very extended conformation or dimerization of this protein. In contrast, in crude cytosol, importin is exclusively found in higher molecular mass forms, ranging from 300 kDa to perhaps 1000 kDa (exclusion limit of Superdex 200; data not shown).

Importin, the 60 kDa Protein, Is Required for the First Step of Nuclear Protein Import

Having a purified protein in hand, we investigated its role in nuclear import (Figure 2). Unfractionated cytosol (at 4 mg/ml protein concentration) gave efficient transport of fluorescein-labeled bovine serum albumin (BSA)–NLS conjugate in the presence of an energy-regenerating system. In the absence of ATP (depleted by apyrase), transport was completely abolished. Instead, accumulation at the nuclear envelope was observed as reported previously (Newmeyer and Forbes, 1988; Richardson et al., 1988; Moore and Blobel, 1992).

Ran/TC4 alone had no effect whether or not an energy-regenerating system was present.

When both importin and Ran/TC4 were present at saturating concentrations (100 μg/ml each), the effects resembled those with unfractionated cytosol. In the presence of ATP, import occurred at a rate comparable with the cytosol

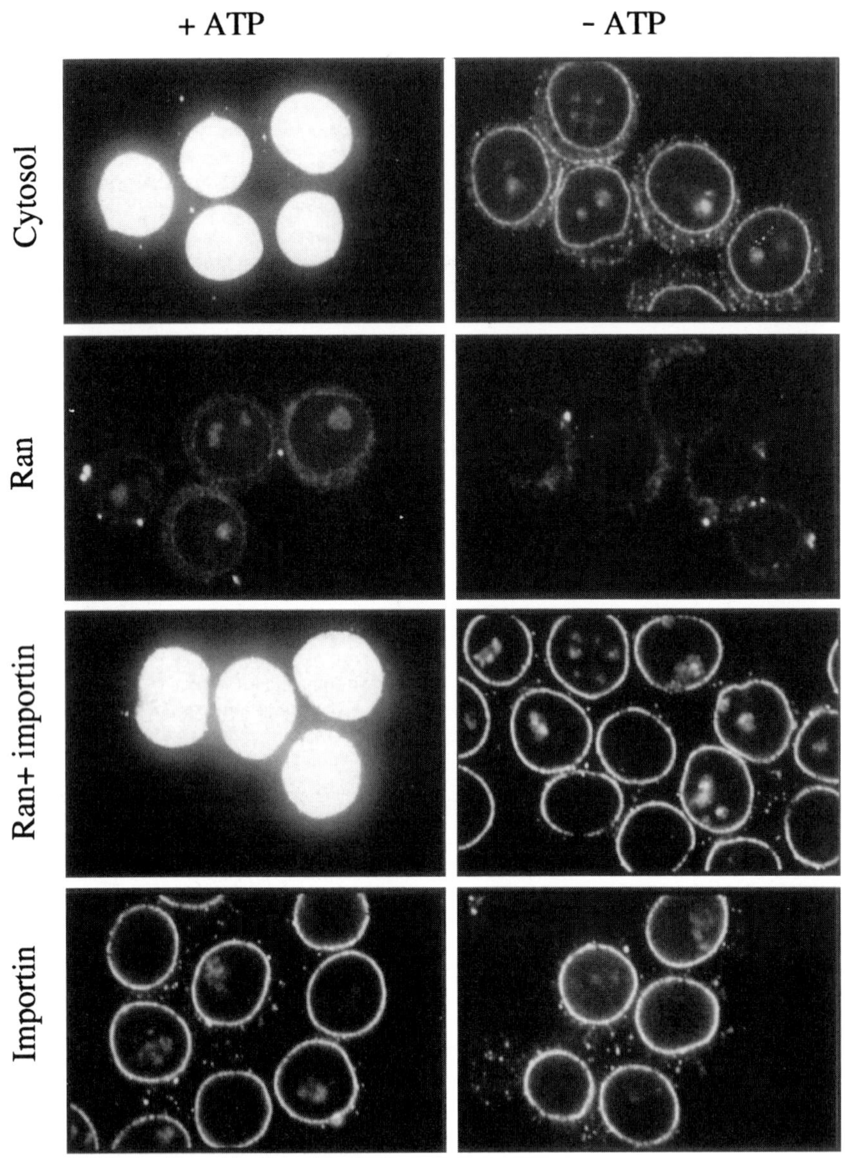

Figure 2. Effects of Importin on Nuclear Import

Confocal images after import reactions of fluorescein-labeled BSA–NLS conjugates into nuclei of permeabilized HeLa cells for 1 hr at 23°C.

Plus ATP indicates the presence of an energy-regenerating system (ATP plus GTP plus creatine phosphate and creatine kinase); minus ATP indicates the absence of the energy-regenerating system and inclusion of 50 U/ml apyrase; cytosol indicates the presence of a postribosomal supernatant of an activated Xenopus egg extract at a protein concentration of 4 mg/ml. Importin (Superose 6 fraction 24; see Figure 1) and recombinant Ran/TC4 were added at 100 μg/ml when indicated.

All panels were scanned and photographed under identical conditions. Therefore, the panels plus ATP/cytosol and plus ATP/Ran plus importin are overexposed.

control. In the absence of ATP, intense staining of the nuclear envelope was observed. When importin was the only cytosolic protein present, nuclear envelope binding was observed, but no import, whether or not the energy-regenerating system was present (Figure 2).

Envelope Binding Mediated by Importin Is Specific for NLSs

Next, we tested whether the envelope binding mediated by importin is specific. As seen in Figure 3, this phenomenon is crucially dependent on the addition of importin and

Cell
770

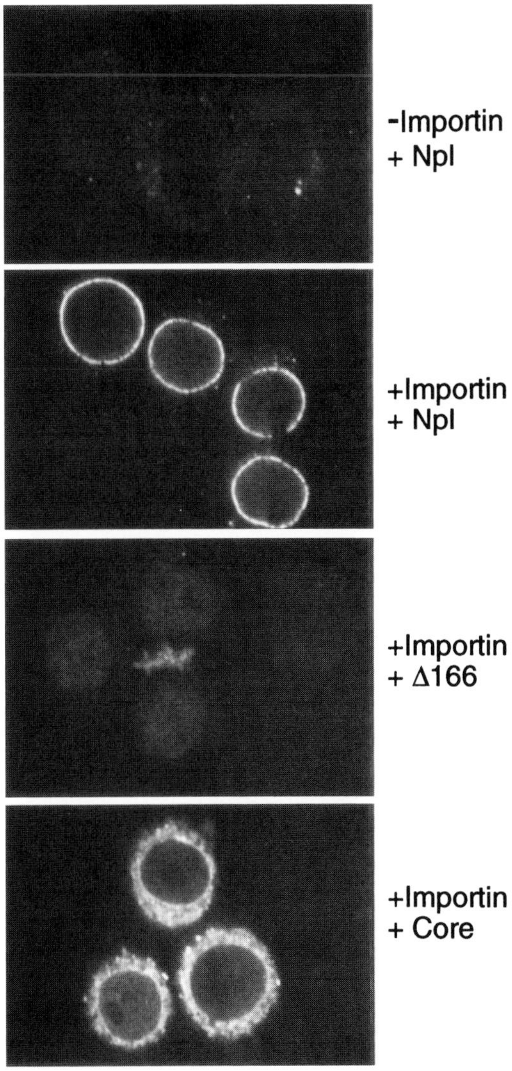

Figure 3. Envelope Binding of Fluorescein-Labeled Nucleoplasmin Depends Both on Importin and on an Intact NLS

Confocal scanning images after standard incubations with nuclei of permeabilized HeLa cells as indicated.

Minus or plus importin indicates the absence or presence of 100 μg/ml importin. Npl indicates 100 μg/ml fluorescent-labeled full-length nucleoplasmin. Δ166 is a fluorescent-labeled nucleoplasmin truncation consisting of the first 166 amino acid, thus lacking the downstream basic cluster of the NLS. Core indicates fluorescent-labeled nucleoplasmin core (first 149 amino acids), lacking the entire NLS. Note that the nucleoplasmin core stains the cytoplasmic remnants intensely.

All panels were scanned and photographed under identical conditions.

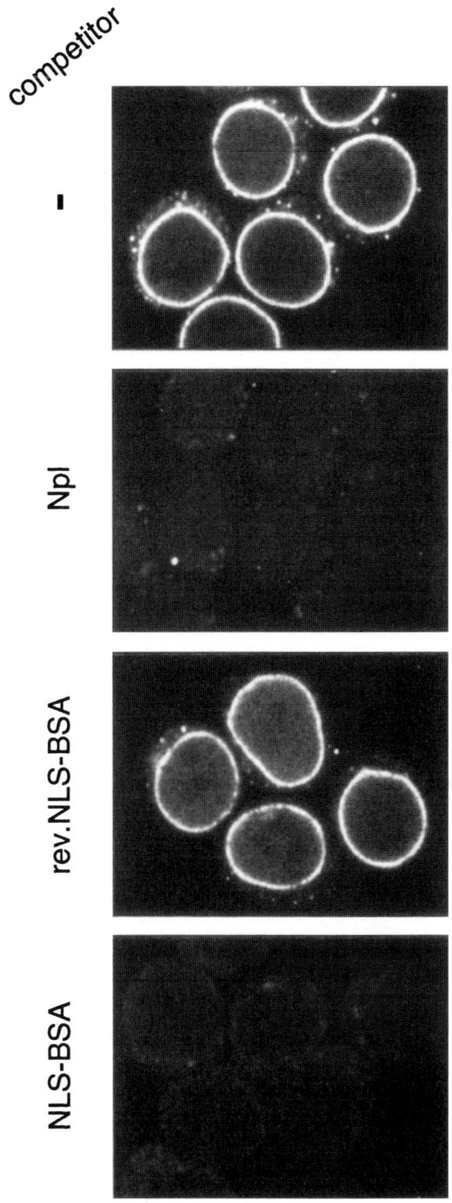

Figure 4. Functional NLSs Compete with Nucleoplasmin for Importin-Mediated Envelope Binding, Whereas Nonfunctional NLSs Do Not Compete

Confocal scanning images after import reactions using nuclei of permeabilized HeLa cells in the presence of 100 μg/ml importin, 50 μg/ml fluorescein-labeled full-length nucleoplasmin, and a 50-fold excess of unlabeled competitors as indicated. Npl indicates nucleoplasmin; rev.NLS–BSA indicates a (nonfunctional) reversed NLS peptide from SV40 large T antigen coupled to BSA; NLS–BSA indicates functional NLS peptides conjugated to BSA. All images were obtained under identical conditions.

the presence of an NLS within the import substrate. Envelope binding was not observed when truncated nucleoplasmin consisting of the first 166 amino acids (Δ166) was used. This lacks the downstream lysine cluster of the bipartite NLS (Robbins et al., 1991). Similarly, the nucleoplasmin core, lacking the entire NLS, failed to bind to the envelope. Instead, a bright background binding to the cytoplasmic remnants in the permeabilized cells was observed.

Competition between nucleoplasmin and BSA–T antigen–NLS conjugates for import into the nucleus has been

First Step of Nuclear Import
771

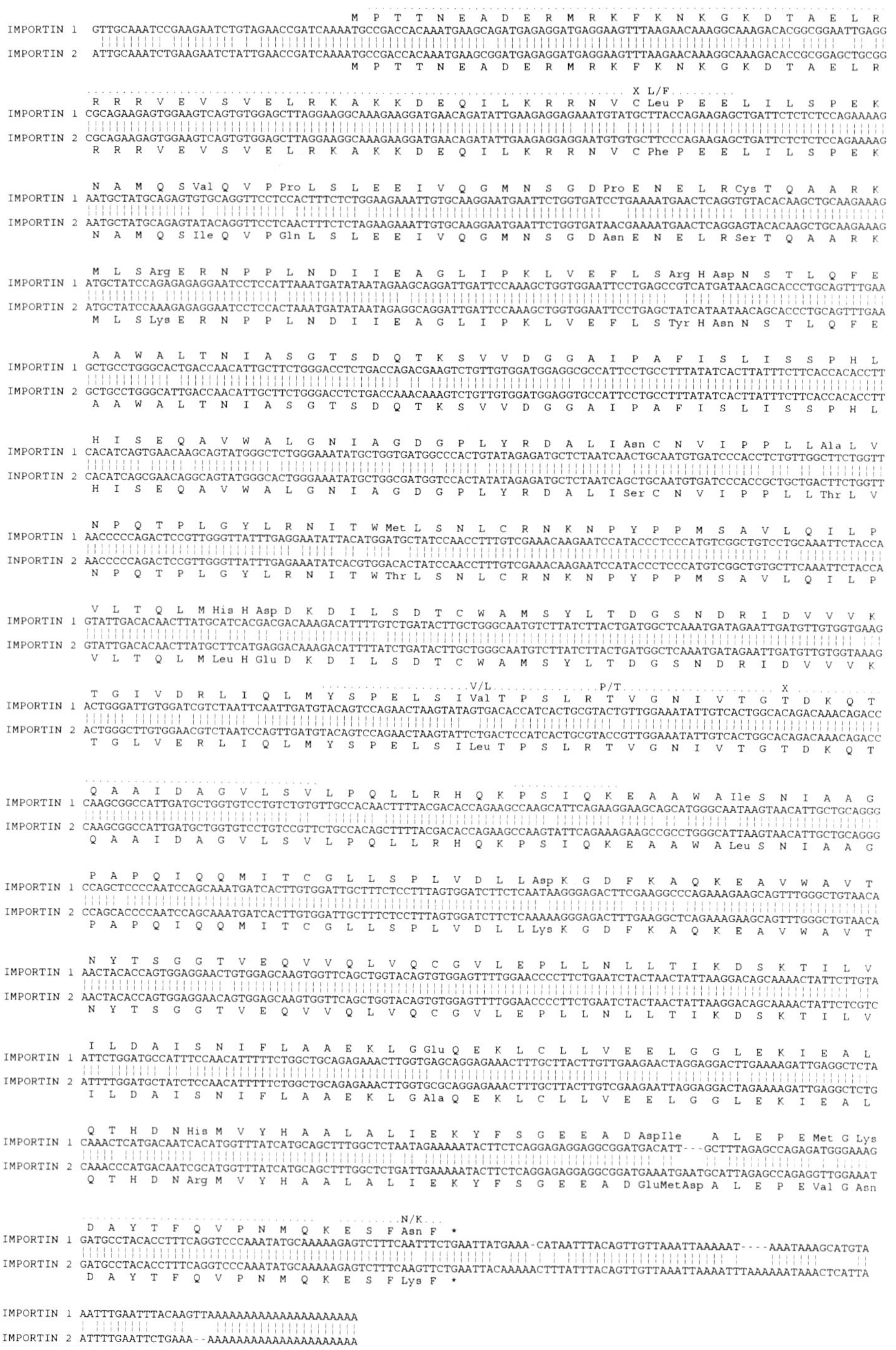

Figure 5. Sequence of Importin from Xenopus

A cDNA library from Xenopus ovary was screened with oligonucleotides corresponding to importin partial protein sequence. Several full-length clones were found, some of them varying in a few amino acid positions. The figure shows the alignment the cDNA sequences and the translated amino acid sequences of forms 1 and 2 of Xenopus importin. Amino acids are written in single letter code; three letter code was used in positions that are not identical in forms 1 and 2. Residues identified by direct peptide sequencing are indicated by a dotted line above; positions differing from the deduced amino acid sequence of either importin forms 1 or 2 are indicated by single letter code in this line.

Cell
772

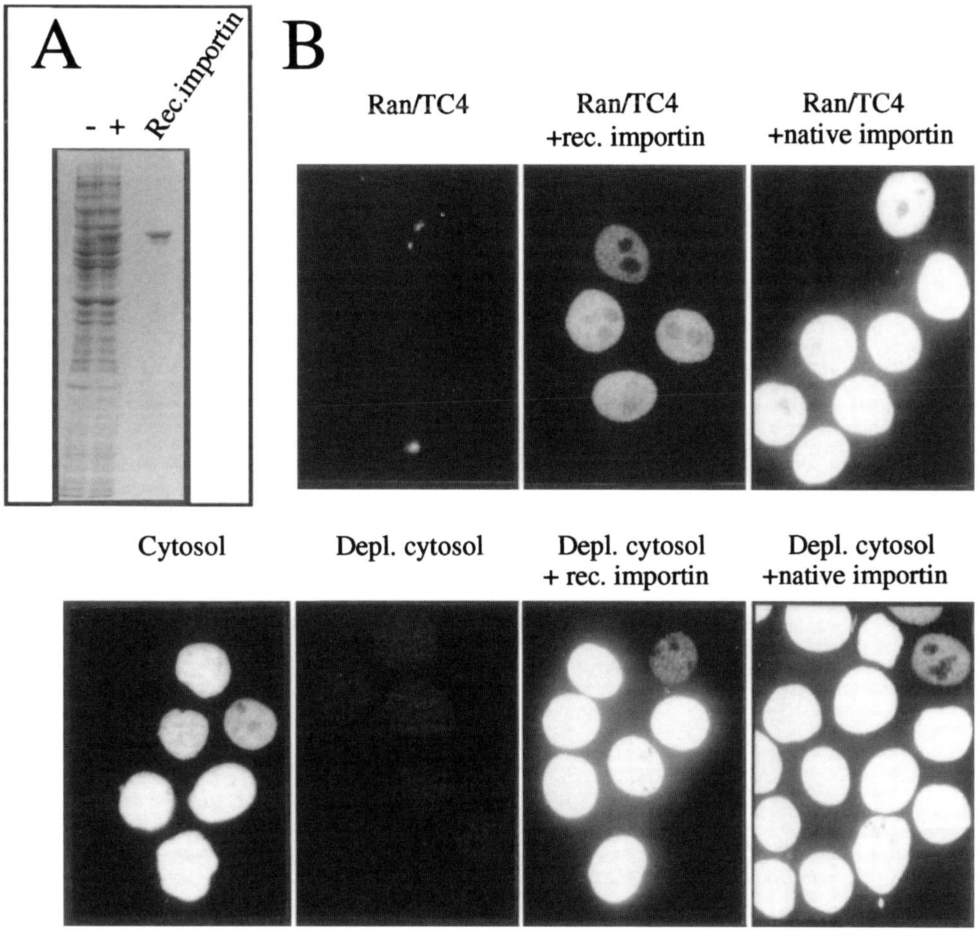

Figure 6. Expressed Recombinant Importin Is Active

(A) Recombinant expression of importin. Minus and plus indicate the total protein pattern after 3 hr induction with IPTG of E. coli strain BL21/Rep4 transformed either with pQE70 vector alone or with pQE70 that contained the cDNA insert coding for importin form 1. Rec.importin indicates the expressed recombinant importin after purification on Ni–NTA–Sepharose.

(B) Confocal sections through nuclei of permeabilized HeLa cells after import reactions with 50 μg/ml fluorescein-labeled nucleoplasmin and additions as indicated. Ran/TC4, 100 μg/ml; recombinant importin, 100 μg/ml; native importin, 100 μg/ml, importin purified from Xenopus eggs; depleted cytosol, Xenopus egg extract depleted of importin by passing it through Ni–NTA–Sepharose and used at a final protein concentration of 4 mg/ml; cytosol, Xenopus egg extract mock depleted by passing it through Sepharose CL-6B instead of Ni–NTA–Sepharose. Importin content of depleted cytosol was approximately 8% of the mock depleted.

All panels were scanned and photographed under identical conditions. Therefore, fluorescence intensity of nuclei in panels labeled depleted cytosol plus importin is above the linear range, whereas that of panel Ran/TC4 is below the detection limit.

reported previously (Michaud and Goldfarb, 1991). Therefore, we tested whether there is a similar competition between import substrates for importin-mediated binding to the nuclear envelope (Figure 4). In the absence of a competitor (actually in the presence of nucleoplasmin core; see Experimental Procedures), fluorescein-labeled nucleoplasmin bound to the nuclear envelope showing bright rings. In the presence of a 50-fold excess of unlabeled nucleoplasmin, envelope binding was completely abolished. Addition of the same excess of a BSA conjugate with reverse sequence NLS peptides from SV40 large T antigen did not diminish envelope binding of nucleoplasmin. As reported previously, the BSA–reverse NLS is not functional in nuclear import (Adam et al., 1989). In contrast, the functional NLS–BSA conjugate competed strongly and abolished envelope binding completely (Figure 4).

Thus, importin is essential for generating an NLS-binding site at the nuclear envelope. This binding site is specific for NLSs, and it is able to recognize both the bipartite signal of nucleoplasmin (Figures 3 and 4) and the targeting information in functional BSA–NLS conjugates (Figure 2; also Figure 4).

Molecular Cloning and Primary Structure of Xenopus Importin

We partially sequenced importin and designed degenerate oligonucleotide probes to screen a Xenopus ovary li-

Table 1. Dependence of Nuclear Import Efficiency on Importin Concentration

Additions	Transport
Ran	BG
Ran plus 30 μg/ml recombinant importin	20
Ran plus 100 μg/ml recombinant importin	53
Ran plus 250 μg/ml recombinant importin	69
Ran plus 30 μg/ml native importin	39
Ran plus 100 μg/ml native importin	96
Cytosol (30 μg/ml importin final)	100
Depleted cytosol (2.5 μg/ml importin final)	12
Depleted cytosol plus 30 μg/ml recominant importin	132
Depleted cytosol plus 100 μg/ml recombinant importin	273
Depleted cytosol plus 250 μg/ml recombinant importin	410
Depleted cytosol plus 30 μg/ml native importin	116
Depleted cytosol plus 100 μg/ml native importin	298

Standard import assays into nuclei of permeabilized cells were performed in the presence of 100 μg/ml flourescein-labeled nucleoplasmin, an energy-regenerating system, and additions as indicated. Ran indicates recombinant Ran/TC4 at 100 μg/ml; recombinant importin indicates importin form 1 expressed in E. coli; native importin indicates importin purified from activated Xenopus eggs; cytosol indicates mock-depleted cytosol in 4 mg/ml protein concentration; depleted cytosol indicates cytosol that had been depleted of importin by passing through Ni–NTA–Sepharose and used at 4 mg/ml protein concentration.

Import efficiency was evaluated from confocal images using the NIH Image software. For each sample, the mean fluorescence intensity of at leaset 50 nuclei was averaged. The value for Ran/TC4 alone was set to background (BG), the value for the cytosol control was set to 100.

brary for clones coding for it. Several full-length clones were found and sequenced. Importin mRNA seems to be abundant in oocytes, as the corresponding cDNA clones were found in the library at least at the frequency of actin-coding clones. To our surprise, the clones fell into different classes, deviating from each other in up to 22 amino acid positions out of 528, as seen for importin forms 1 and 2 in Figure 5. So far, four additional forms have been found. Three are closely related to form 1, deviating from it in at least three amino acid positions; one is closely related to form 2, differing from it in at least one position (data not shown). Whether this minimum of six cDNA forms represents more than two genes, or whether they reflect the existence of different alleles of two genes within the tetraploid species Xenopus laevis is unclear at present. Peptide sequencing confirmed that the purified importin is a mixture of various forms; as in positions that differ between forms 1 and 2, two different amino acids were indeed found (see Figure 5). It also seems that different forms of importin migrate at slightly different rates on high resolution SDS gels and elute differently from Ni–NTA–Sepharose probably owing to the difference in histidine content (data not shown; see Figure 5 and below).

Expressed Recombinant Importin Is Active in Import

To prove conclusively that importin is responsible for the activity, we expressed its cDNA (form 1) in Escherichia coli (see Figure 6A). We chose a prokaryotic host because it does not contain a nuclear import apparatus. Although the recombinant protein contains ten additional amino acids (a C-terminal histidine tag), it runs on SDS gels fractionally faster than the native protein isolated from Xenopus eggs (data not shown). A difference in posttranslational modifications is a possible explanation for this effect.

Figure 6B and Table 1 show that a single form of expressed recombinant importin can substitute for native importin isolated from Xenopus eggs in the import assay. As shown before, Ran/TC4 alone did not mediate any transport (see Figure 6B). Ran/TC4 plus saturating amounts of recombinant importin gave transport comparable to unfractionated cytosol at 4 mg/ml protein concentration, but this was only 30%–50% (in separate experiments) of the activity of native importin when tested under identical conditions (see Figure 6B and Table 1). A difference of activity between the recombinant and the native protein was also observed for envelope binding (data not shown). This quantitative difference in activity could be due to the lack of a modification. In that case, the recombinant protein should be able to rescue cytosol from which importin has been depleted. To test this, cytosol was depleted of importin by passing it through a Ni–NTA column (note that this depletion protocol is independent of the one used originally for assaying importin activity). Importin was depleted by about 92% as estimated from quantitative immunoblotting with peptide-specific anti-importin antibodies, S^{35} secondary antibodies, and a phosphoimager. On the autoradiogram, the cytosolic importin was resolved into two closely migrating bands (data not shown). The upper one (probably representing form 1) was more efficiently depleted than the lower one, thus reflecting differences in the histidine content (see Figure 5). Depletion of 92% of the cytosolic importin correlated well with the decrease in transport efficiency by 87% (see Table 1). Addition of a saturating amount of the recombinant protein resulted in a 40-fold restimulation of import efficiency. Equal amounts of recombinant and native protein yielded comparable rates of transport (Table 1 and Figure 6B). Thus, in crude depleted cytosol, the expressed recombinant importin and importin isolated from eggs have comparable activities.

The existence of additional stimulatory factors in the depleted cytosol is also evident (see Table 1 and Figure 6B) as suggested previously (Moore and Blobel, 1993, 1994). They provide a stimulation of transport for recombinant importin by a factor of 5–6 and for native importin by a factor of 2–3. These numbers are still underestimated, since the depleted cytosol contains a high concentration of nuclear proteins competing for import (e.g., nucleoplasmin, N1, histones). The difference in stimulation suggests that the recombinant protein needs still to be folded properly, posttranslationally modified, or both, by factors present in the egg extract in order to gain full activity. The fact that the rate of import mediated by the native importin and Ran/TC4 can be further stimulated indicates that at least one stimulatory activity plays a more direct role.

Structure and Evolutionary Conservation of Importin

Comparison of importin with sequences in the GenBank data base reveals strong homologies to the human Rch1

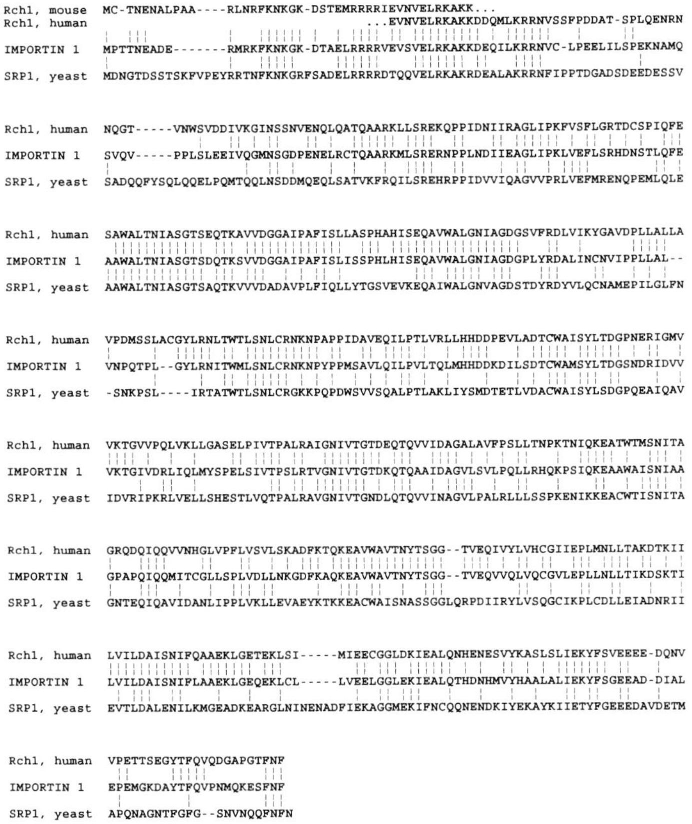

Figure 7. Comparison of Amino Acid Sequences of Xenopus Importin 1, Human and Mouse Rch1, and SRP1p from S. cerevisiae

Multiple alignment of the three proteins was performed with CLUSTALL within the PCGENE program package. The broken vertical lines indicate identical amino acid positions between two sequences. Note that the complete Rch1 sequence had been compiled from two incomplete sequences from mouse and human (Cuomo et al., 1994); however, in the overlapping region, both clones were reported to be identical.

protein (ACC U09559; Cuomo et al., 1994), to the SRP1 protein (ACC M75849) from S. cerevisiae (Yano et al., 1992), and to sequence fragments from rice (ACC D21306, ACC D23592), Arabidopsis thaliana (ACC T14159), maize (ACC T18817), Plasmodium falciparum (ACC T18047), Caenorhabditis elegans (ACC M76111, D28052), mouse (ACC X61876), and human (ACC T08580).

Xenopus importin is 64% identical with human Rch1, a protein found in a yeast two-hybrid system as an interaction partner of Rag1 (Cuomo et al., 1994). Since Rag1 contains an NLS, the interaction found might reflect the recognition of the NLS by Rch1.

SRP1p and importin share 44% identical amino acids. SRP1 was originally found as a suppressor of temperature-sensitive RNA polymerase I mutations (Yano et al., 1992). In contrast with the expected localization in the nucleolus, the protein was found to be a component of the yeast nuclear pore complex (Yano et al., 1992; Belanger et al., 1994). *SRP1* is an essential gene (Yano et al., 1992, 1994).

Xenopus importin, human Rch1, and yeast SRP1p have a characteristic structure in common: the N-terminal and C-terminal domains are very hydrophilic. The N-terminus contains conserved clusters of basic amino acids. These clusters resemble NLS, but preliminary immunofluorescence microscopy data suggest that most importin is found in the cytoplasm (A. D. Mills and D. G., unpublished data).

The middle part of importin/SRP1p consists of an 8-fold repetition of a hydrophobic motif. It was noticed previously that these motifs in SRP1p are similar to repeats (the so-called arm motifs) in other proteins namely in armadillo, β-catenin, plakoglobin, adenomatous polyposis coli (APC), p120, and smgGDS (Peifer et al., 1994; Yano et al., 1994). Interestingly, the GDP/GTP exchange factor smgGDS consists almost exclusively of the arm motif, and it was speculated that the arm domain might mediate G protein interactions in other proteins (Peifer et al., 1994). It remains to be clarified whether importin acts on the GTP cycle of Ran/TC4.

Importin is sensitive to alkylation with N-ethylmaleimide (NEM; data not shown). This is not surprising taking into account that importin form 1 contains eight cysteines; some of them also conserved in SRP1p and Rch1p.

Discussion

We have purified from Xenopus egg extract a 60 kDa protein, importin, that is essential for import of karyophilic proteins into the nucleus. We cloned, sequenced, and expressed this protein in a functional form in E. coli. The transport mediated by importin shows all the characteristic features reported for this process (for reviews see Garcia-Bustos et al., 1991; Silver, 1991; Yamasaki and Lanford,

1992; Fabre and Hurt, 1994; Moore and Blobel, 1994). It is dependent on an intact NLS, an energy-regenerating system, and the small GTP-binding protein Ran/TC4. It can be completely blocked by the lectin wheat germ agglutinin or by the nonhydrolyzable GTP analog GppNp (data not shown). Import mediated by importin occurs in two steps: signal-dependent binding to the nuclear envelope and subsequent translocation. Transport intermediates accumulate at the first stage in the presence of importin and in the absence of either nucleoside triphosphates or Ran/TC4.

Using importin purified from Xenopus eggs and Ran/TC4 as the sole cytosolic proteins, we have reconstituted efficient transport into the nuclei of permeabilized cells (see Figure 2). Furthermore, we have reconstituted transport using bacterially expressed recombinant importin, though it is less efficient than its native counterpart (Figure 6). We suspect that this difference is due to secondary modifications of the protein, but in both cases, transport can be further stimulated by at least one auxiliary factor. This may correspond to factor B2 of Moore and Blobel (1993).

Importin probably represents an active component in the crude fraction A described by Moore and Blobel (1992), as it was purified from a related source; it has similar chromatographic properties and identical effects in the import reaction. Both importin and fraction A are inactivated by ammonium sulfate precipitation or by alkylation with NEM.

In contrast, the relation of importin to nuclear import factor-1 (NIF-1; Newmeyer and Forbes, 1990) is unclear. NIF-1 was defined as the 30%–50% ammonium sulfate cut of Xenopus egg cytosol, a treatment resulting in poor recovery of importin activity. On the other hand, the facts that NIF-1 is sensitive to NEM, it was found in large complexes, and it was implicated in the first step of nuclear import suggest that importin might be an active component of NIF-1.

It is also possible, though not certain, that importin might be related to a 60 kDa protein that was found as the principal cross-linking partner to a wild-type but not to a mutant SV40 NLS peptide in rat liver nuclear envelopes (Adam et al., 1989). However, the authors needed to solubilize the nuclear envelopes with detergent to fully expose the peptide-binding site. Thus, the cross-linking partner might equally be a constituent of the intermembrane space (see Yamasaki and Lanford, 1992) rather than a receptor recognizing proteins coming from the cytosol. In any case, the 55 kDa protein subsequently purified by Adam and Gerace (1991) from bovine red blood cells seems to be different from importin. Thus, the ammonium sulfate precipitation or chromatography on phenyl–Sepharose used in the 55 kDa protein purification would have abolished the activity of importin. Importin is sensitive to alkylation with NEM, whereas p55 is not (Adam and Adam, 1994). Furthermore, the 55 kDa protein of Adam and Gerace (1991) behaves as a monomeric globular protein on sizing columns. In contrast, both purified native and recombinant importin have an apparent molecular mass close to 120 kDa on Superdex 200. Finally, as the nuclear import system of Adam and Gerace (1991) derived from the enucleated mammalian red blood cells is insensitive to GTP analogs (Adam and Adam, 1994), it appears to differ fundamentally from the nuclear import system described here, or by Moore and Blobel (1993) and by Melchior et al. (1993).

Xenopus importin shows 44% amino acid identity to the SRP1 protein from S. cerevisiae. SRP1 is essential for cell viability (Yano et al., 1992, 1994). When SRP1p was depleted from cells by switching off the *Gal7* promotor that controls the only *SRP1* gene in this strain, abolition of transcription, fragmentation of the nucleolus, and defects in both nuclear division and segregation were observed (Yano et al., 1994). These pleiotropic effects of SRP1p depletion as well as the pleiotropic phenotypes of *srp1* and *SRP1^{ts}* mutations (Yano et al., 1994) can be explained as consequences of transport defects.

In the light of 44% amino acid identity between importin and its yeast homolog SRP1p and the reported near identity between mouse and human Rch1 (Cuomo et al., 1994), the 63.7% sequence identity between frog importin and the human Rch1 appears surprisingly low. We found another, shorter human sequence tag with 55% sequence identity with Rch1 (corresponding to positions 234–347) in the data base (ACC T08580). Thus, a family of importin-like proteins might exist in one and the same organism, and the closest mammalian relative to the form of importin in Xenopus eggs might not have been identified yet. The same reservation might apply for yeast. It is tempting to speculate that the different members of the importin family fulfill different functions such as involvement in the import and export of different substrates such as proteins, RNPs, etc. On the other hand, there is the possibility of tissue specificity of different importin forms. The Xenopus egg importin might possibly be a special version to meet the extraordinary demands on nuclear transport during early development.

The actual pore passage of nucleoplasmin-coated gold seems to be preceded by binding to fiber elements extruding at the cytoplasmic face from the nuclear pore complex (Richardson et al., 1988). Importin is essential to generate the NLS-binding site at the nuclear envelope. That suggests that one site of the function of importins is the nuclear pore complex, the associated fibers, or both. In fact, beside its presence in a soluble pool (Yano et al., 1992; Belanger et al., 1994), the yeast homolog SRP1p is physically and functionally associated with the nuclear pore complex in the following ways: it colocalizes with Nup1p immunofluorescence microscopy (Yano et al., 1992), *srp1* genetically interacts with *nup1* (Belanger et al., 1994) and *nup2* (Yano et al., 1994), SRP1p interacts with Nup1p in a two-hybrid system, and SRP1p can be coimmunoprecipitated with Nup1p or Nup2p (Belanger et al., 1994).

Whether importin itself provides the entire NLS-binding site and whether it makes physical contact with the import substrate at all can only be tested by nearest neighbor analysis of import intermediates during nuclear envelope binding. Assays of NLS binding in simplified systems are notoriously error prone, perhaps because of the high density of positive charges in the signal (Yamasaki and Lanford, 1992). At present, it is not certain that the primary NLS recognition step occurs in the cytosol rather than at the nuclear pore. If NLS recognition does occur in the cytosol, importin is probably the best candidate for this role. However, one does have to consider two more possibilities. First, importin might have to bind first to the nuclear pore before it can bind an NLS. Second, importin might have no NLS-binding site itself, but might be a regulator of an NLS-binding site of the nuclear pore complex.

At least two functional states of the NLS-binding site at the nuclear pore complex must exist: a high affinity state for binding of the import substrate to the pore and a low affinity state to release it either into the nucleoplasm or to downstream components of the import machinery. It remains to be tested whether an import substrate remains bound to the primary NLS recognition site during its entire passage. The affinity of the NLS-binding site has to be regulated (e.g., by Ran/TC4) and might also involve changes in the spatial arrangement of the binding site, e.g., association and dissociation of some part of it. Importin would be an excellent candidate for this loosely bound nuclear pore protein.

Experimental Procedures

Recombinant Expression

The following proteins were expressed in E. coli strain Bl21/Rep4 from cDNA clones using the Qiaexpress system (Qiagen): full-length nucleoplasmin (Dingwall et al., 1987), two nucleoplasmin truncations consisting of the first 166 (Δ166) and the first 149 amino acids (core), Ran/TC4 (Dupree et al., 1992), and "importin" described here. The cDNAs coding for nucleoplasmin, its truncated derivatives, and importin were cloned into the SphI–BamHI sites of pQE70 and were expressed with C-terminal histidine tags. The Ran/TC4 gene was cloned into the SphI–HindIII sites of pQE32, and it was expressed with an N-terminal histidine tag.

Bacteria were grown at 37°C (nucleoplasmin, Δ166, core, Ran/TC4) or at 26°C (importin) in 400 ml 2TY medium to an optical density of 0.9 (600 nm) and were induced with 2 mM IPTG at the same temperature for 6 hr (nucleoplasmin, Δ166, core, Ran/TC4) or 3 hr (importin). PMSF was added to 2 mM, the culture was chilled on ice for 10 min, and cells were pelleted and resuspended in 15 ml of 0.2 M Tris–HCl (pH 8.0), 0.5 M NaCl, 20 μg/ml leupeptin, 10 μg/ml chymostatin, 4 μg/ml elastatinal, 5 mM β-mercaptoethanol. The suspension was subjected to two freeze–thaw cycles, and cells were finally disrupted by ultrasonic sound. The solution was clarified by a 20 min spin at 20,000 × g.

The recombinant proteins were purified from the corresponding homogenates either directly (importin) or from supernatants following ammonium sulfate precipitation (25% saturation for Ran/TC4 or 60% saturation for nucleoplasmin and its derivatives). The protein solutions were adjusted to 10 mM imidazole (pH 7.6) and were loaded onto a 5 ml Ni–NTA–Sepharose column (Qiagen) equilibrated in 500 mM NaCl, 10 mM imidazole (pH 7.0). The column was washed with 30 mM imidazole in the same buffer and finally eluted with a 40 ml gradient of 30–500 mM imidazole. Pooled fractions were adjusted to 20 mM DTT and were dialyzed against several changes of 20 mM HEPES–KOH, 250 mM sucrose, 2 mM DTT. Protein concentration was calculated by UV photometry at 280 and 288 nm using molar extinction coefficients calculated from the tyrosine and tryptophan content (Edelhoch, 1967). Final yield was 40–80 mg/l culture for nucleoplasmin and its derivatives, 25–35 mg/l for Ran/TC4, and 6 mg/l for importin. Since Ran/TC4 tends to aggregate at a high protein concentration, it was diluted to 1 mg/ml in 10% BSA before freezing.

Fluorescein-Labeling of Nucleoplasmin and Its Derivatives

Protein solutions were transferred to 100 mM potassium acetate by gel filtration on a PD10 column (Pharmacia). The concentration of free SH groups was kept in the range of 200–800 μM as determined with Ellmann's reagent (5,5'-dithio-bis(2-nitrobenzoic acid); Sigma). The solution was adjusted to 100 mM MES, 40 mM Tris (pH 7.0), 1.5 M potassium acetate; and fluorescein-5-maleimide (Calbiochem; 10 mg/ml in dimethylformamide) was added in a stochiometric amount to free SH groups. After 30 min, incubation at room temperature the reaction was quenched with 50 mM cysteine (pH 8.0). Free fluorochrome was removed by chromatography on a PD10 column equilibrated in 20 mM HEPES (pH 7.5). The average extent of labeling was usually one fluorescein per nucleoplasmin pentamer as determined from the absorptions at 280 and 490 nm.

BSA-Peptide Conjugates

SMCC-activated BSA was prepared by incubation of BSA at 20 mg/ml in 100 mM HEPES–KOH with a 100-fold molar excess of sulfosuccinimidyl 4-(N-maleimidomethyl) cyclohexane 1-carboxylate (sulfo-SMCC; Calbiochem) for 1 hr at room temperature. Excess cross-linker was removed by gel filtration on a PD10 column. A 50-fold molar excess of either the SV40 NLS peptide cgggPKKKRKVED or the reverse NLS peptide cgggDEVKRKKKP was added to the SMCC-activated BSA. The pH was readjusted to 7.5, and the reaction was allowed to proceed for 1 hr at 37°C. Noncoupled peptide was separated by gel filtration on a PD10 column equilibrated in 150 mM NaCl. The molar ratio of coupling was 20–30 peptides per BSA molecule as estimated from the electrophoretic mobility.

For fluorescein conjugation, the peptide conjugates were adjusted to 0.2 M $NaHCO_3$, and carboxyfluorescein-N-hydroxysuccinimide ester (Boehringer) dissolved in DMSO was added in a 2:1 molar ratio to BSA. After 30 min incubation at room temperature, the free fluorochrome was removed by gel filtration on a PD10 column and a subsequent precipitation with 60% v/v ethanol. The pellet was then dissolved in 150 mM NaCl.

Preparation of Digitonin Permeabilized HeLa Cells

The method used is based on the one described by Adam et al. (1990) and includes the modification by Leno et al. (1992) to solubilize cells in suspension rather than on coverslips and to freeze permeabilized cells after addition of 5% DMSO.

Digitonin (high purity, Calbiochem) was dissolved to 5% in boiling water. The solution was placed on ice overnight, and the precipitate was removed by centrifugation for 30 min in a microcentrifuge. The supernatant is defined as the 5% stock solution.

HeLa cells were grown in a 500 ml spinner culture. Cells were sedimented at 500 × g for 5 min, and the pellet was resuspended in 100 ml of cell dissociation solution (Sigma C5914) and incubated for 10 min at 37°C. Cells were sedimented and washed three times with cold permeabilization buffer: 50mM HEPES–KOH (pH 7.5), 5 mM magnesium acetate, 2 mM EGTA, 50 mM potassium acetate, 2 mM dithiothreitol, protease inhibitors (10 μg/ml leupeptin, 5 μg/ml chymostatin, 1 μg/ml elastatinal). Cells were finally resuspended in 50 ml of permeabilization buffer. Digitonin was added first to 20 μg/ml from a 100 μg/ml solution made up in permeabilization buffer. The digitonin concentration was then increased by 5 μg/ml increments, and after each addition, the permeabilization of the plasma membrane and the integrity of the nuclear membrane were checked with fluorescein-labeled nucleoplasmin core (inclusion into cells and exclusion from the nucleus). The optimum concentration was usually around 35 μg/ml digitonin. The permeabilization was stopped by addition of BSA to 1% final concentration. The cells were then pelleted, washed three times in permeabilization buffer, and finally resuspended in 10 ml of permeabilization buffer plus 1% BSA plus 5% DMSO and slowly frozen in small aliquots at −80°C.

Import Assay

The method used is similar to that described by Adam et al. (1990). The standard 20 μl assay contained: 20 mM HEPES–KOH, 100 mM potassium acetate, 2 mM magnesium acetate, 1 mM EGTA, 250 mM sucrose, 2 mM DTT, 10 mg/ml BSA, energy-regenerating system (unless otherwise indicated) consisting of 1 mM ATP, 0.2 mM GTP, 10 mM creatine phosphate, and 100 μg/ml creatine kinase, 100 μg/ml import probe (fluorescein-labeled nucleoplasmin or its derivatives or BSA–NLS conjugate), and either the fractions or unfractionated cytosol. The reaction was started by addition of some 2×10^4 nuclei (i.e., permeabilized cells) and was usually allowed to proceed for 60 min at room temperature.

As reported before (Moore and Blobel, 1992), there is high nonspecific binding of the fluorescent probe to the cytoplasmic remnants in the absence of cytosol or at low concentrations of protein fractions. BSA alone does not suppress this nonspecific binding, but the negative control substrates for nuclear import, namely nucleoplasmin core and reverse NLS–BSA conjugate, do suppress it. So routinely 5 mg/ml nucleoplasmin core was added to the assay in experiments with highly purified fractions.

The ionic conditions were found to have only a minor influence on import efficiency, as long as the ionic strength was kept between 70–150 mM and magnesium between 1–5 mM. There is no obvious difference in using HEPES or Tris as a buffer or sodium chloride or potassium acetate as the salt. Therefore, fractions with different ionic compositions were usually not dialyzed before assay, but salt concentration was adjusted to approximately 100 mM final with a compensating buffer. In contrast, imidazole is strongly inhibitory and was removed either by dialysis, gel filtration on G25, or ion exchange chromatography.

To follow the purification of importin, we measured the restimulation of import activity of cytosol that had been depleted of an essential activity by passing it through Q-Sepharose fast flow at 50 mM Tris–HCl (pH 7.4), 200 mM NaCl. To allow a high sample throughout, the import activity was evaluated by fluorescence microscopy of unfixed samples. Import was considered to be significant if the majority of nuclei had accumulated fluorescent nucleoplasmin clearly above the cytoplasmic concentration.

For photographs, import reactions were fixed on ice by dilution to 200 μl with 50 mM HEPES-KOH (pH 7.5), 100 mM potassium acetate, 250 mM sucrose, 1 mM magnesium acetate followed by addition of 200 μl of 8% paraformaldehyde, 100 mM HEPES-KOH (pH 7.5). After 5 min fixation on ice, the nuclei were spun through a 1 ml sucrose cushion (30% w/v in dilution buffer) onto polylysine-coated coverslips (3000 rpm, 6 min). The coverslips were washed briefly in water and mounted on a drop of Vectashield, and the edges of the coverslip were sealed with nail polish. The samples were then scanned with a Bio-Rad MRC600 confocal microsocope using the 60× objective lens under oil. The scanning conditions and exposure times in all subsequent photographic processes were identical for all samples in a given experiment.

To quantitate nuclear import, confocal image files were evaluated with the NIH Image program (version 1.52). For each sample, the mean fluorescence of at least 50 randomly chosen nuclei was averaged.

Purification of Importin

Xenopus eggs were collected, dejellied, and activated as described (Leno and Laskey, 1991). From this point, everything was done on ice or at 4°C. Finally, the eggs were washed in 50 mM HEPES-KOH, 100 mM sucrose, 5 mM DTT, 20 μg/ml leupeptin, 10 μg/ml chymostatin, 2 μg/ml elastatinal and homogenized in 3 vol of the same buffer. The homogenate was spun in a swinging bucket rotor to sediment particles larger than 60S (as calculated from the k-factors for the given rotors). The tubes were punctured to collect the middle layer representing the high speed supernatant (HSS).

Immediately before each column chromatographic step, the material was ultracentrifuged to sediment particles and aggregates larger than 30S.

HSS (260 ml; 1 g of protein) was adjusted to 20 mM Tris–HCl (pH 7.4), 150 mM NaCl and was applied to an 80 ml Q-Sepharose fast flow column (Pharmacia) equilibrated in 50 mM Tris–HCl (pH 7.4), 200 mM NaCl. The column was washed in 120 ml of equilibration buffer, and bound proteins were eluted with 50 mM Tris–HCl (pH 7.4), 1 M NaCl. The protein containing pool (80 ml, 300 mg protein) was adjusted to 5 mM imidazole and loaded onto a 5 ml Ni–NTA–Sepharose column (Qiagen) equilibrated in 50 mM Tris–HCl, 1 M NaCl, 5 mM imidazole. The column was subsequently washed with equilibration buffer and with 50 mM Tris–HCl, 1 M NaCl, 10 mM imidazole until the baseline was reached. Elution was performed with 10 ml steps of 15, 20, 40, 100, and 500 mM imidazole–HCl (pH 7.6) in 50 mM Tris–HCl, 100 mM NaCl, 5% glycerol. Protein containing fractions of each step were pooled and adjusted to 5 mM DTT. Aliquots (100 μl) were dialyzed against 50 mM HEPES-KOH (pH 7.5), 250 mM sucrose, 1 mM DTT and assayed for activity. Highest activity was found in the pools of the 40 and 100 mM imidazole steps.

The 40–100 mM pool (3.7 mg of protein) was diluted with an equal volume of 50 mM Tris–HCl (pH 7.5) and was applied to 1 ml Mono Q column (Pharmacia) equilibrated in 50 mM Tris–HCl (pH 7.4), 200 mM NaCl, 2mM DTT. Proteins were eluted with a 20 ml linear 200–520 mM NaCl gradient in 50 mM Tris–HCl (pH 7.4), 2 mM DTT. Highest activity was found in fractions 35 and 36 (see Figure 1), while less but still significant activity was found in fraction 37 (corresponding to some 300–350 mM NaCl elution position). Fractions 35–37 were pooled (1.5 mg of protein) and loaded onto a Superose 6 column (Pharmacia) equilibrated in 20 mM Tris–HCl (pH 7.4), 100 mM NaCl, 2 mM DTT. Significant activity was found in fractions 18–26 and high activity in fractions 21–24. Final yield was 1 mg of protein corresponding to an overall yield of 15% as estimated from quantitative immunoblotting with anti-importin antibodies. The antibodies were raised in rabbits against the peptide PTTNEADERMc (the N-terminus of importin) coupled to SMCC-activated keyhole limpet hemocyanin.

Size Determination of Importin

A 16/60 Hiload Superdex 200 column in 50 mM Tris–HCl (pH 7.4), 100 mM NaCl, 2% glycerol was calibrated (flow rate 1 ml/min) with the following markers: ATP (total volume), cytochrome c (12.5 kDa), ovalbumin (45 kDa), BSA (68 kDa), aldolase from rabbit muscle (158 kDa), catalase from beef liver (240 kDa), ferritin from horse spleen (450 kDa), and plasmid DNA (void volume). The running position of partially purified importin (after IMAC step, see Figure 1) and of expressed recombinant importin was determined by peak detection and was verified by PAGE and Coomassie staining. The running position of importin starting from a crude cytosol was detemined by immunoblotting. Apparent molecular weights were interpolated under the assumption of a semilogarithmic relationship between size and retention time.

Peptide Sequencing and Molecular Cloning

Purified importin was sequenced from its N-terminus and after cleavage with cyanogen bromide or Lys-C protease (Boehringer). Degenerate oligonucleotide probes were designed that correspond to the amino acid sequences MPTTNEA (atg ccn acn acn aa(t/c) ga(a/g) gc) and AKKDEQI (at(t/c) tg(t/c) tc(a/g) tc(t/c) tt(t/c) tt(t/c) ttn gc). Oligonucleotide probes were end-labeled with [γ-^{32}P]ATP and used to screen a Xenopus ovary cDNA library (Stratagene number 937652).

Several full-length clones were picked. The two shown in Figure 5 were sequenced on both stands. The sequences of importin forms 1 and 2 were deposited in the GenBank data base. Beside these clones, others were found that deviate in some positions from both forms 1 and 2. Sequences were analyzed with the following programs: sequence editing and alignments with PCGENE, identification of internal repeats with MCAFEE (Schuler et al., 1991) and tBLASTN algorithm data base screen for sequence homologies with NCBI mailserver (Altschul et al., 1990).

Depletion of Importin from Cytosol

The depletion protocol made use of the unusual property of importin to bind tightly to Ni–NTA–Sepharose. An egg extract was obtained by crushing activated Xenopus eggs in a swinging bucket rotor (as described by Leno and Laskey, 1991) and was respun for 20 min at 100,000 × g. Extract (5 ml) was transferred to depletion buffer (20 mM Tris–HCl [pH 7.4], 5 mM imidazole, 100 mM NaCl, 2 mM $MgCl_2$, 2 mM β-mercaptoethanol) by gel filtration on a 30 ml Sephadex G25 column. The protein containing pool was spun in a TLA100.4 rotor (Beckmann) at 80,000 rpm for 40 min. Half of the supernatant was passed through a 2.2 ml Ni–NTA–Sepharose column at 1 ml/hr (depleted cytosol), the other half was passed through the same volume of Sepharose CL-6B. Protein containing fractions were pooled and transferred into 20 mM HEPES-KOH (pH 7.4), 250 mM sucrose, 0.5 mM magnesium acetate, 2 mM DTT by gel filtration on Sephadex G25, adjusted to the same protein concentration (as estimated from UV absorbtion at 260 and 280 nm), and frozen in liquid nitrogen. Depletion efficiency (comparison between depleted and mock-depleted cytosol) was estimated by quantitative immunoblotting using peptide-specific antibody raised against the ten N-terminal amino acids of importin, ^{35}S-labeled secondary antibodies, a Molecular Dynamics phosphoimager, and recombinant importin of known concentration as a standard.

Acknowledgments

Correspondence should be addressed to D. G. We thank Dr. T. A. Rapoport, in whose lab part of the work has been performed, for generous support and helpful comments. We thank S. Kostka for protein sequencing, Dr. C. Dingwall for providing reagents, Drs. P. Dupree and V. Oikkonen for the kind gift of an MDCK Ran/TC4 clone, S. Knespel, E. Bürger, H. Wilkinson, and J. Robbins for technical help, T. Mills for advice in confocal microscopy and photography and for helpful discussions, Dr. P. Lemaire and N. Garret for providing a Xenopus oocyte library, and J. Makkerh and M. Madine for supplying cultured HeLa cells. We thank U. Kutay and Dr. T. Sommer for critical reading of the manuscript. This work was supported by the Cancer Research Campaign (SP1961), the Human Frontier Science Program Organization, the Deutsche Forschungsgemeinschaft (SFB 366), and a fellowship of the Deutscher Akademischer Austauschdienst, followed by a long-term fellowship of the Human Frontier Science Program Organization awarded to D. G.

Received July 13, 1994; revised September 19, 1994.

References

Adam, E. H., and Adam, S. A. (1994). Identification of cytosolic factors required for nuclear location sequence-mediated binding to the nuclear envelope. J. Cell Biol. *125*, 547–555.

Adam, S. A., and Gerace, L. (1991). Cytosolic proteins that specifically bind nuclear localization signals are receptors for nuclear import. Cell *66*, 837–847.

Adam, S. A., Lobl, T. J., Mitchell, M. A., and Gerace, L. (1989). Identification of specific binding proteins for a nuclear localization sequence. Nature *337*, 276–279.

Adam, S. A., Sterne-Marr, R., and Gerace, L. (1990). Nuclear import in permeabilized mammalian cells requires soluble cytoplasmic factors. J. Cell Biol. *111*, 807–816.

Akey, C. W., and Goldfarb, D. S. (1989). Protein import through the nuclear pore complex is a multistep process. J. Cell Biol. *109*, 971–1008.

Altschul, S. F., Gish, W., Miller, W., Myers, E. M., and Lipman, D. J. (1990). Basic local alignment search tool. J. Mol. Biol. *215*, 403–410.

Belanger, K. D., Kenna, M. A., Wei, S., and Davis, L. I. (1994). Genetic and physical interactions between SRP1p and nuclear pore complex proteins Nup1p and Nup2p. J. Cell Biol. *126*, 619–630.

Bischoff, F. R., and Pongstingl, H. (1991). Mitotic regulator protein RCC1 is complexed with a nuclear ras-related polypeptide. Proc. Natl. Acad. Sci. USA *88*, 10830–10834.

Cuomo, C. A., Kirch, S. A., Gyuris, J., Brent, R., and Oettinger, M. A. (1994). Rch1, a protein that specifically interacts with the RAG-1 recombination-activating protein. Proc. Natl. Acad. Sci. USA *91*, 6156–6160.

Dingwall, C., Dilworth, S. M., Black, S. J., Kearsey, S. E., Cox, L. S., and Laskey, R. A. (1987). Nucleoplasmin cDNA sequence reveals polyglutamic acid tracts and a cluster of sequences homologous to putative nuclear localization signals. EMBO J. *6*, 69–74.

Drivas, G. T., Shih, A., Coutavas, E., Rush, M. G., and D'Eustachio, P. (1990). Characterization of four novel RAS-related genes expressed in a human teratocarcinoma cell line. Mol. Cell. Biol. *10*, 1793–1798.

Dupree, P., Oikkonen, V. M., and Chavrier, P. (1992). Sequence of a canine cDNA clone encoding a Ran/TC4 GTP-binding protein. Gene *120*, 325–326.

Edelhoch, H. (1967). Spectroscopic determination of tryptophan and tyrosine in proteins. Biochemistry *6*, 1948–1954

Fabre, E., and Hurt, E. C. (1994). Nuclear transport. Curr. Opin. Cell Biol. *6*, 335–342

Garcia-Bustos, J., Heitman, J., and Hall, M. (1991). Nuclear protein localization. Biochim. Biophys. Acta *1071*, 83–101.

Leno, G. H., and Laskey, R. A. (1991). DNA replication in cell-free extracts from *Xenopus laevis*. Meth. Cell Biol. *36*, 561.

Leno, G. H., Downes, C. S., and Laskey, R. A. (1992). The nuclear membrane prevents replication of human G2 nuclei but not G1 nuclei in Xenopus egg extract. Cell *69*, 151–158.

Melchior, F., Paschal, B., Evans, E., and Gerace, L. (1993). Inhibition of nuclear protein import by nonhydrolyzable analogues of GTP and identification of the small GTPase Ran/TC4 as an essential transport factor. J. Cell Biol. *123*, 1649–1659.

Michaud, N., and Goldfarb, D. (1991). Muliple pathways in nuclear transport: the import of U2 snRNP occurs by a novel kinetic pathway. J. Cell Biol. *112*, 215–223.

Moore, M. S., and Blobel, G. (1992). The two steps of nuclear import, targeting to the nuclear envelope and translocation through the nuclear pore, require different cytosolic factors. Cell *69*, 939–950.

Moore, M. S., and Blobel, G. (1993). The GTP-binding protein Ran/TC4 is required for protein import into the nucleus. Nature *365*, 661–663.

Moore, M. S., and Blobel, G. (1994). A G protein involved in nucleocytoplasmic transport: the role of Ran. Trends Biochem. Sci. *19*, 211–216

Newmeyer, D. D., and Forbes, D. J. (1988). Nuclear import can be separated into distinct steps in vitro: nuclear pore binding and translocation. Cell *52*, 641–653.

Newmeyer, D. D., and Forbes, D. J. (1990). An N-ethylmaleimide-sensitive cytosolic factor necessary for nuclear protein import: requirement in signal-mediated binding to the nuclear pore. J. Cell Biol. *110*, 547–557.

Peifer, M., Berg, S., and Reynolds, A. B. (1994). A repeating amino acid motif shared by proteins with diverse cellular roles. Cell *76*, 789–791.

Ren, M., Drivas, G., D'Eustachio, P., and Rush, M. G. (1993). Ran/TC4: a small GTP-binding protein that regulates DNA synthesis. J. Cell Biol. *120*, 313–323.

Richardson, W. D., Mills, A. D., Dilworth, S. M., Laskey, R. A., and Dingwall, C. (1988). Nuclear protein migration involves two steps: rapid binding at the nuclear envelope followed by slower translocation through the nuclear pores. Cell *52*, 655–664.

Robbins, J., Dilworth, S. M., Laskey, R. A., and Dingwall, C. (1991). Two interdependent basic domains in nucleoplasmin targeting sequence: identification of a class of bipartite nuclear targeting sequence. Cell *64*, 615–623.

Schuler, G. D., Altschul, S. F., and Lipmann, D. J. (1991). A workbench for multiple alignment construction analysis. Protein Struct. Funct. Genet. *9*, 180–190.

Silver, P. A. (1991). How proteins enter the nucleus. Cell *64*, 489–497.

Yamasaki, L., and Lanford, R. E. (1992). Nuclear transport: a guide to import receptors. Trends Cell Biol. *2*, 123–127.

Yano, R., Oakes, M., Yamaghishi, M., Dodd, J. A., and Nomura, M. (1992). Cloning and characterization of SRP1, a suppressor of temperature-sensitive RNA polymerase I mutations in *Saccharomyces cerevisiae*. Mol. Cell. Biol. *12*, 5640–5651.

Yano, R., Oakes, M., Tabb, M. M., and Nomura, M. (1994). Yeast SRP1p has homology to armadillo/plakoglobin/β-catenin and participates in apparently multiple nuclear functions including the maintenance of the nucleolar structure. Proc. Natl. Acad. Sci. USA *91*, 6880–6884.

GenBank Accession Numbers

The accession numbers for the sequences reported in this paper, importin forms 1 and 2, are L36339 and L36340, respectively.

Note Added in Proof

During the further characterization of stimulatory factors, we found an activity that behaves as a second subunit of importin, but dissociates during the Ni-NTA-Sepharose step. N. Imamoto and Y. Yoneda recently informed us that they also have observed a similar activity that appears to correspond to these two subunits of importin.

Mitosis and Cell Cycle Control

Inoué S. and **Sato H.** 1967. Cell motility by labile association of molecules. The nature of mitotic spindle fibers and their role in chromosome movement. *J. Gen. Physiol.* **50:** 259–292.

This paper summarizes a great deal of work on live cells, visualized with a sensitive polarization microscope. It describes both the behavior of a living spindle in normal cells and the response of the spindle to a variety of experimental perturbations. Through these studies, Inoué and his collaborators demonstrated the lability of spindle fibers and the fact that some fibers shorten while others elongate as the chromosomes segregate, properties of the spindle that have subsequently been confirmed by more direct methods (1). This and related work demonstrated that spindle lability is essential for normal mitosis, but the authors went on to propose that microtubule disassembly actually provides the force for chromosome segregation. This idea was controversial for many years; some scientists favored Inoué's ideas while others thought that microtubule-dependent motor enzymes generated the forces for chromosome movement. It is now quite well established that both assembly-dependent processes and ATP-dependent motor enzymes contribute to mitotic movements (reviewed in 2). Indeed, there is evidence that a single microtubule-dependent motor enzyme can use either ATP hydrolysis or the free energy available from microtubule disassembly to generate forces for motility (3). Moreover, some motors shorten microtubules by using chemical energy to promote tubulin depolymerization (4). The mechanisms for mitosis are currently subjects for active research, and our understanding of the process is still at the stage of an increasing awareness of complexity. We know that multiple motor enzymes act in competing and complementary ways (5), but new factors are still being found that enhance (6) or antagonize (7) microtubule stability. A solid understanding of mitosis will require that we know not only the identity of all these players but also their roles in each mitotic event, a task that will require the interplay of information from many approaches to experimental biology.

---

1. Salmon E.D., Leslie R.J., Saxton W.M., Karow M.L., and McIntosh J.R. 1984. Spindle microtubule dynamics in sea urchin embryos: analysis using a fluorescein-labeled tubulin and measurements of fluorescence redistribution after laser photobleaching. *J. Cell Biol.* **99:** 2165–2174.
2. Inoué S. and Salmon E.D. 1995. Force generation by microtubule assembly/disassembly in mitosis and related movements. *Mol. Biol. Cell* **6:** 1619–1640.
3. Lombillo V.A., Stewart R.J., and McIntosh J.R. 1995. Minus-end-directed motion of kinesin-coated microspheres driven by microtubule depolymerization. *Nature* **373:** 161–164.
4. Desai A., Verma S., Mitchison T.J., and Walczak C.E. 1999. Kin I kinesins are microtubule-destabilizing enzymes. *Cell* **96:** 69–78.
5. Cottingham, F.R., and Hoyt M.A. 1997. Mitotic spindle positioning in *Saccharomyces cerevisiae* is accomplished by antagonistically acting microtubule motor proteins. *J. Cell Biol.* **138:** 1041–1053.
6. Cullen C.F., Deak P., Glover D.M., and Ohkura H. 1999. Mini spindles: A gene encoding a conserved microtubule-associated protein required for the integrity of the mitotic spindle in *Drosophila*. *J. Cell Biol.* **146:** 1005–1018.
7. Belmont L.D. and Mitchison T.J. 1996. Identification of a protein that interacts with tubulin dimers and increases the catastrophe rate of microtubules. *Cell* **84:** 623–631.

Cell Motility by Labile Association of Molecules

The nature of mitotic spindle fibers and their role in chromosome movement

SHINYA INOUÉ and HIDEMI SATO

From the Laboratories of Biophysical Cytology, Department of Biology, University of Pennsylvania, Philadelphia

ABSTRACT This article summarizes our current views on the dynamic structure of the mitotic spindle and its relation to mitotic chromosome movements. The following statements are based on measurements of birefringence of spindle fibers in living cells, normally developing or experimentally modified by various physical and chemical agents, including high and low temperatures, antimitotic drugs, heavy water, and ultraviolet microbeam irradiation. Data were also obtained concomitantly with electron microscopy employing a new fixative and through measurements of isolated spindle protein. Spindle fibers in living cells are labile dynamic structures whose constituent filaments (microtubules) undergo cyclic breakdown and reformation. The dynamic state is maintained by an equilibrium between a pool of protein molecules and their linearly aggregated polymers, which constitute the microtubules or filaments. In living cells under physiological conditions, the association of the molecules into polymers is very weak (absolute value of $\Delta F_{25°C} < 1$ kcal), and the equilibrium is readily shifted to dissociation by low temperature or by high hydrostatic pressure. The equilibrium is shifted toward formation of polymer by increase in temperature (with a large increase in entropy: $\Delta S_{25°C} \simeq 100$ eu) or by the addition of heavy water. The spindle proteins tend to polymerize with orienting centers as their geometrical foci. The centrioles, kinetochores, and cell plate act as orienting centers successively during mitosis. Filaments are more concentrated adjacent to an orienting center and yield higher birefringence. Astral rays, continuous fibers, chromosomal fibers, and phragmoplast fibers are thus formed by successive reorganization of the same protein molecules. During late prophase and metaphase, polymerization takes place predominantly at the kinetochores; in metaphase and anaphase, depolymerization is prevalent near the spindle poles. When the concentration of spindle protein is high, fusiform bundles of polymer are precipitated out even in the absence of obvious orienting centers. The shift of equilibrium from free protein molecules to polymer increases the length and number of the spindle microtubules or filaments. Slow depolymerization of the polymers, which can be brought about by low concentrations of colchicine or by gradual cooling, allows the filaments to shorten and perform work. The dynamic equilibrium controlled by orienting centers and other factors provides a plasusible mechanism by which chromosomes and other organelles, as well as the cell surface, are deformed or moved by temporarily organized arrays of microtubules or filaments.

In this paper, we shall attempt to relate the mitotic movement of chromosomes to the structure, physiology, and function of the mitotic spindle.

At each division of a eukaryotic cell a spindle is formed, and the chromosomes are oriented, aligned, and then separated regularly into two daughter cells. The spindle is disassembled when the task is done.

In a living cell, the fibrous elements of the mitotic spindle and its astral rays are weakly birefringent. The appearance and growth, and contraction and disappearance, of the fibers can therefore be observed in living cells with a sensitive polarizing microscope. The change in birefringence can also be measured with the polarizing microscope. Correlations can then be established between spindle fiber birefringence and alteration of fiber fine structure in cells altered physiologically or experimentally.

Many conditions and agents have been found in which spindle morphology and fiber birefringence are systematically and reversibly altered. From the various observations described or reviewed in this paper, we conclude that spindle fibers are composed of parallel arrays of thin filaments which are formed by a reversible association of globular protein molecules. The molecules in functioning spindle fibers are not stably aligned and cross-linked as in more stable fibers. Instead, spindle fibers are labile structures existing in a dynamic equilibrium with a large pool of unassociated molecules. The association of the molecules and formation of filaments are controlled by the activities of orienting centers and concentrations of active pool material.

We hypothesize that the contraction and elongation of the spindle fibers are responsible for regular mitotic movement of chromosomes. The spindle fibers, however, do not contract and elongate by folding and unfolding of polypeptide chains in the protein molecules. For example, the fiber birefringence remains unchanged during anaphase movement. Instead, the fibers are believed to elongate by increased addition and alignment of the molecules, which contribute to a pushing action. Conversely, they shorten and pull chromosomes as molecules are slowly removed from the filaments.

We view this hypothesis as providing a plausible mechanism, not only for mitotic movement of chromosomes, but also for movements of various other organelles and mechanical modulations of the cell surface.

In the following, we shall:

1. Review the visible changes of spindle fiber birefringence during normal mitosis
2. Discuss the mechanical integrity of the spindle fibers
3. Describe the dynamic nature of the spindle fibers
4. Discuss the dynamic equilibrium model
5. Describe the effect of heavy water on spindle fiber birefringence, protein content, and electron microscopic appearance
6. Provide evidence that the bulk of the spindle fiber protein comes from a preformed pool
7. Describe factors which control the formation and orientation of the fibers
8. Discuss the extension and contraction of spindle fibers
9. Discuss the nature of the spindle protein molecules
10. Relate spindle fiber birefringence to microtubules and filaments observed in fixed cells with the electron microscope
11. Discuss cell movements which appear to be brought about by labile association of molecules

1. VISIBLE CHANGES OF SPINDLE FIBER BIREFRINGENCE DURING MITOSIS

Spindle fibers and astral rays generally possess a positive birefringence whose magnitude is of the order of 1 mμ, or $\frac{2}{1000}$ of a wavelength of green light. With a sensitive polarizing microscope, they can be clearly visualized (Fig. 1) and their fine structural changes can be followed by interpreting the measured change in birefringence[1] (Bajer, 1961; Dietz, 1963; Forer, 1965, 1966; Inoué, 1953, 1964; Inoué and Bajer, 1961; Schmidt, 1937, 1939, 1941; Taylor, 1959; for a thorough discussion and general bibliography of mitosis, and for definition of terms, also see Schrader, 1953, and Mazia, 1961).

The change in spindle fiber birefringence during mitosis in a wide variety of animal and plant cells has been documented photographically (Inoué, 1964; also 1953). The same paper (Inoué, 1964) also describes a sequence of time-lapse motion pictures depicting the birefringence changes during mitosis and cytokinesis in several types of cells.

The major patterns of birefringence changes can be summarized as follows. In prophase before nuclear membrane breakdown, birefringent fibers are formed as astral rays in animal cells (Fig. 2 *A*) (Inoué and Dan, 1951; Forer, 1965; also see Cleveland, 1938, 1953, 1963), or as clear zone fibers in some plant cells (Fig. 3 *A*) (Inoué and Bajer, 1961; also see Bajer, 1957). In newt fibroblasts, linear growth of the birefringent central spindle has been measured

[1] In this paper, the term "birefringence" is used synonymously with retardation [i.e. $d(n_1 - n_2)$] and should not be interpreted to mean coefficient of birefringence ($n_1 - n_2$), or retardation per unit specimen thickness.

by Taylor (1959). In the central spindle formed between the separating asters, and in the clear zone material, the birefringence fiber axis (slow axis) corresponds to the direction in which more continuous fibers, sheath fibers, and chromosomal fibers appear. The central (continuous) spindle fibers and clear zone fibers thus establish a fine structural foundation for subsequent orientation of other fibers (also see Wada, 1965; however, see Inoué and Dan, 1951, for anomalous situations regarding axis of streak material vs. aster and spindle, and for changes to negative birefringence under presumed compression).

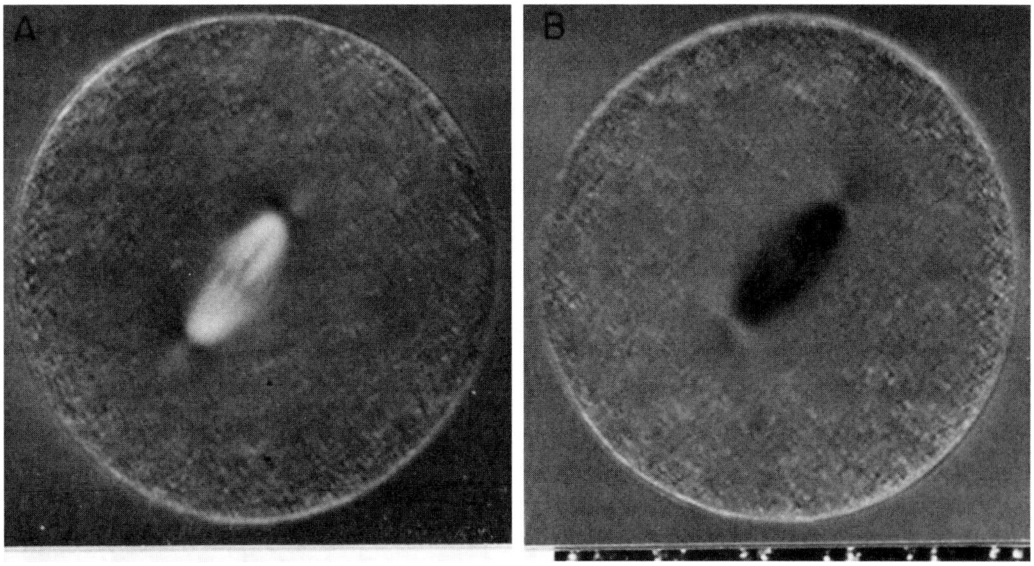

FIGURE 1. Mitotic spindle in arrested metaphase oocyte of *Pectinaria gouldi*. The over-all birefringence[1] of the spindle has been enhanced with 40% heavy water (D_2O)–sea water to approximately 3 mμ. Polarization microscopy. *A*, chromosomal fibers appear bright; *B*, compensator axis reversed, chromosomal fibers are dark. Scale interval, 10 μ.

In prometaphase following nuclear membrane breakdown (Fig. 3 *B*), more spindle material becomes oriented around the chromosomes to form sheath fibers in a crane fly, *Pales crocata* (Dietz, 1963; Went, 1966; also see Dietz, 1959, for special attributes of crane fly spindle poles), and in pollen mother cells of the Easter lily (Inoué, 1953, 1964). The birefringent sheath fibers represent a distinct accumulation of oriented material. In many cases, sheath fibers are not observed, and chromosomal fibers arise parallel to continuous fibers directly as in Fig. 3 *C*. In either case, the continuous or sheath fiber material appears to transform into chromosomal fibers. The birefringence of the former decreases as the birefringence of the latter increases (Fig. 3 *B* and *C*).

The birefringent chromosomal fibers are established parallel to the continuous fibers or sheath fibers of prometaphase. Their orientation in turn forecasts the direction of movement of the chromosomes (Figs. 2 *B–E* and 3 *B–E*). The chromosomal fibers are attached to the kinetochore of the chromosomes. In plant cells devoid of centrioles, the fiber birefringence is strongest adjacent to the kinetochore and weaker toward the poles (Fig. 3 *B–D*). Chromosomal spindle fibers with strong birefringence from kinetochore to poles are found in animal cells with active centrioles (Fig. 2 *B–D*).

In prometaphase, chromosomal spindle fibers also appear to participate in the orientation of the chromosomes and their alignment on the metaphase plate. (See Östergren, 1951, for a proposed mechanism of metaphase equilibrium and orientation.)

During prometaphase, fluctuation of birefringence is observed in the chromosomal spindle fibers of a grasshopper spermatocyte, *Dissosteira carolina*. In a time-lapse motion picture, the fluctuation of birefringence resembles that of northern lights (Inoué, 1964). It is interpreted to reflect the fluctuation in the amount of molecules oriented in the chromosomal spindle fibers at this stage. Such changes might contribute to the achievement of exact alignment of chromosomes on to the metaphase plate as their equilibrium position (Östergren, 1951).

In anaphase, the chromosomes are led by the strongly birefringent chromosomal spindle fibers to the poles of the spindle. The birefringence adjacent to the kinetochore and in much of the chromosomal spindle fiber remains unchanged for the major part of anaphase movement (Figs. 2 *C–D* and 3 *C–E*). The chromosomal spindle fiber may or may not shorten during most of the anaphase, depending on the exact mode of chromosome separation relative to spindle elongation.

Continuous fibers whose birefringence is extremely low in anaphase regain birefringence during late anaphase and telophase and establish a phragmoplast in the case of a plant cell (Fig. 3 *E*). The phragmoplast fibers guide the accumulation of vesicles onto its midplane (Inoué, 1953; Bajer, 1965). The vesicles fuse and form the primary cell wall (Fig. 3 *F*). (For electron microscopy of this process, see Frey-Wyssling et al., 1964, and Porter and Machado, 1960). In animal cells, continuous spindle fibers also become more strongly birefringent, and presumably more stable in telophase, forming the core of the stem (spindle rest, stem *Körper*) connecting the separating cells (Fig. 2 *F*). In this region electron micrographs show an especially large number of microtubules (e.g. Robinson and Gonatas, 1964).

Thus, what appears to be a transition of birefringent material from one component of the spindle to another is seen throughout division. Much of this is presumably brought about by successive activation and inactivation of orienting centers (Inoué, 1964). At the same time, the orientation of the birefringent fibers of an earlier stage appears to contribute to the orientation

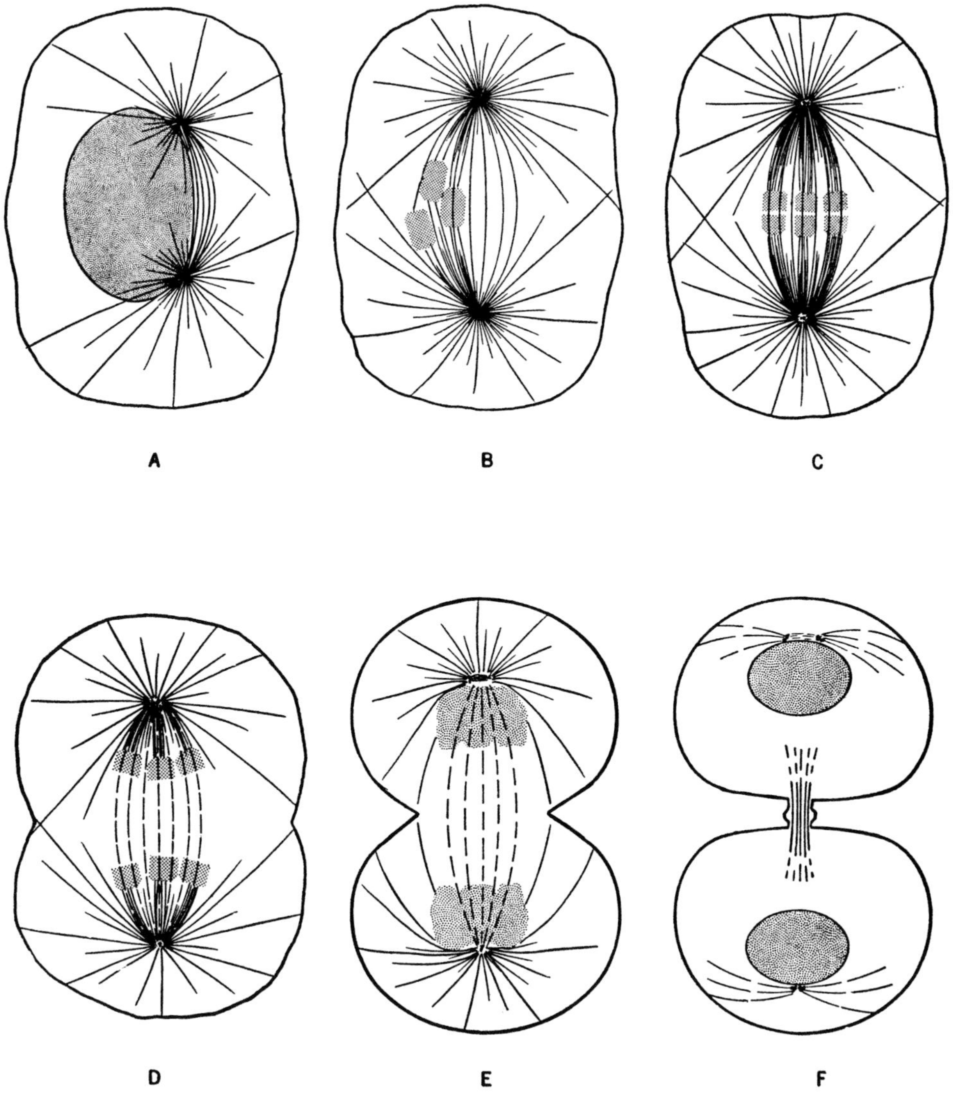

FIGURES 2 and 3. Schematic diagrams of mitosis. These figures were drawn specifically to illustrate change of birefringence distribution in various spindle regions. They are composites of observations on a variety of living cells undergoing normal mitosis. See references given in the text for photographic documentation.

Fig. 2 shows mitosis in spindles with centrioles, as commonly seen in animal cells.

of the subsequent fibers. (Also see Costello, 1961, for the role of centrioles in relating the axis of successive divisions.)

The successive fibers formed appear to co-orient and align chromosomes to the metaphase plate, pull them apart to the poles, and correlate cytokinesis with karyokinesis.

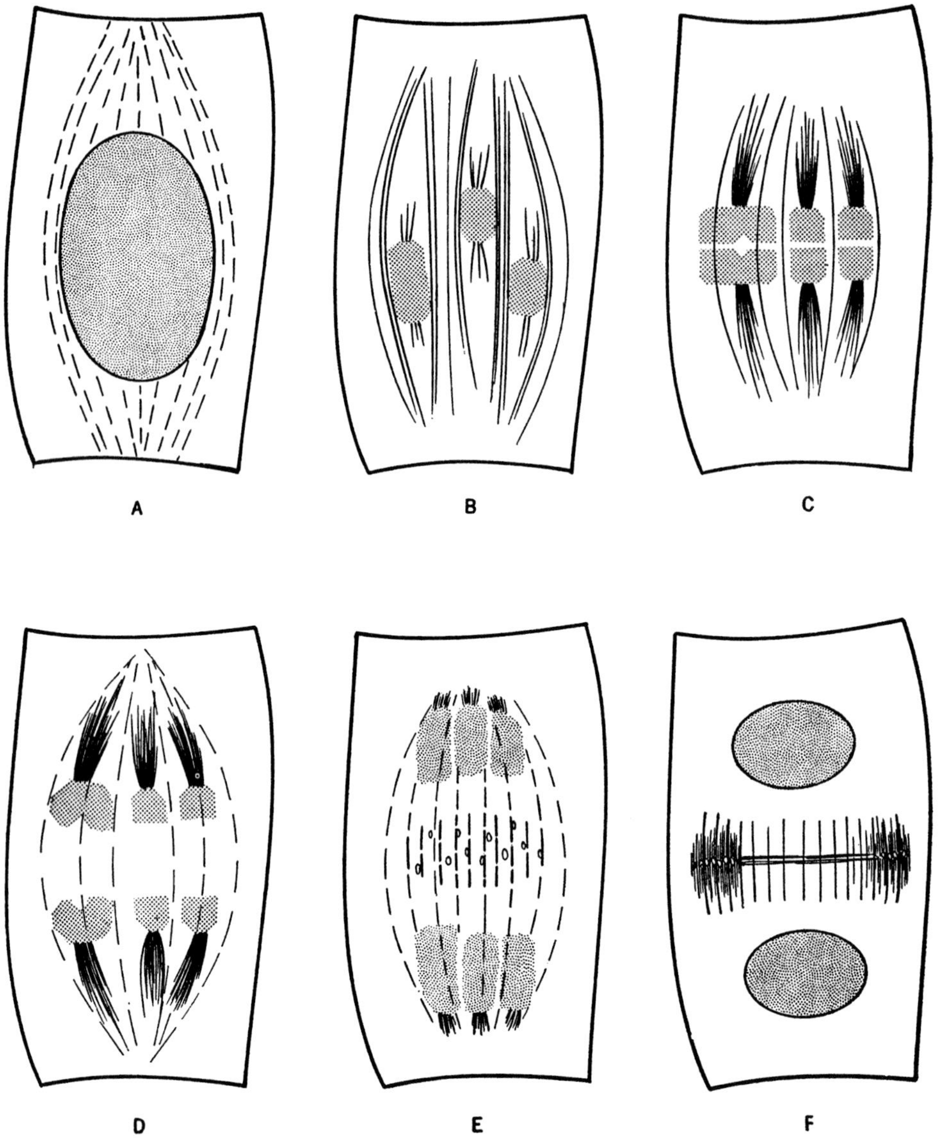

Fig. 3 shows mitosis in spindles without centrioles. Certain features shown in this illustration may appear in cells with the type of spindle shown in Fig. 2, as explained in the text.

2. MECHANICAL INTEGRITY OF SPINDLE FIBERS

Recently, the mechanical integrity of spindle fibers in prometaphase and anaphase movement was demonstrated directly by elegant micromanipulation experiments (Nicklas, 1965, 1967; also see earlier experiments of Carlson, 1952). Taking advantage of the Ellis micromanipulator (Ellis, 1962), Nicklas managed to pull and stretch chromosomes which were attached by their kinetochores to chromosomal spindle fibers. He was also able to sever the chromosomal spindle fiber, turn the chromosome, and point its kinetochore to the opposite spindle pole. The chromosome then moved toward the new pole, presumably acquiring a new spindle fiber attachment to that pole. The chromosomal spindle fibers are thus capable of resisting extension when the same force deforms the chromosome considerably. The fiber maintains a mechanical integrity within limits, but a new fiber can be reformed rapidly if the old one is broken.

The mechanical integrity of the chromosomal spindle fibers and the resistance of the central spindle to compression were also seen in centrifugal experiments performed by Shimamura (1940) and by Conklin (1917).

3. DYNAMIC NATURE OF SPINDLE FIBERS

The reformation of the spindle fibers following direct mechanical disruption by micromanipulation (Nicklas, 1965, 1967; also see Chambers, 1924) was described above. The spindle fibers have also been shown to reform rapidly after other means of disruption. Their birefringence is abolished in a matter of seconds by treatment with low temperature in *Chaetopterus* oocytes (Inoué, 1952 *a*) and in *Lilium* pollen mother cells (Inoué, 1964). Upon return to normal temperature, they recover in the course of a few minutes, after which chromosome movement can continue (Inoué, 1964). Low temperature disintegration of the spindle can be repeated for as many as 10 times in the same cell. In *Chaetopterus* oocytes, *Lilium* pollen mother cells, *Halistaura* developing eggs, and *Dissosteira* spermatocytes, recovery from low temperature treatment is not affected by the duration of chilling, even up to several hours or longer (Inoué, 1952 *b*, and unpublished data).

Colchicine, Colcemid, and many other drugs eliminate the spindle birefringence (Inoué, 1952 *a*; Inoué, Sato, and Ascher, 1965; also see Gaulden and Carlson, 1951, and Molé-Bajer, 1958). The effects are reversible, and *Pectinaria* oocytes treated with 10^{-5} M griseofulvin may recover their birefringent spindles in as little as 5.5–11 min when the cells are washed with normal sea water (Malawista and Sato, 1966).

High hydrostatic pressure is known to destroy the spindle organization reversibly (Pease, 1946; Zimmerman and Marsland, 1964; also see Marsland, 1966, on synergic action of colchicine and pressure). UV microbeam irradiation of portions of spindle fibers introduces temporary lesions which recover rapidly (Campbell and Inoué, 1965; Forer, 1965, 1966; Inoué and Sato, 1964; Izutsu, 1961 *a*, *b*).

The natural fluctuation of birefringence in prometaphase has already been described. The apparent assembly of the spindle material, first into the astral ray or clear zone material, then into the continuous fiber and sheath fiber material, from there to the chromosomal fiber, from chromosomal fiber to the continuous fiber, and, finally, to the phragmoplast fiber material, has also been described. We interpreted these changes to reflect the orderly assembly and disassembly of the same material, sequentially into different fibrous structures to perform different functions.

The various fibers of the spindle are thus extremely labile and their birefringence can be readily abolished. The spindle is also a dynamic structure, and, in most cases, the birefringence recovers rapidly when the external agent is removed. The birefringence can also be reduced or increased to another level by many agents and maintained at an equilibrium value.

4. THE DYNAMIC EQUILIBRIUM MODEL

In order to explain the dynamic nature of the spindle fiber and the response of the spindle birefringence and structure to various experimental alterations, Inoué (1959, 1960, 1964) earlier proposed a dynamic equilibrium model. In the model, spindle fibers are made up of oriented polymers which are in an equilibrium with dissociated molecules. The equilibrium is temperature-sensitive, the polymers dissociating at lower temperatures and the molecules associating to form oriented polymers at higher temperatures, up to a maximum.

Assuming that the birefringence measures the amount of oriented molecules, that the total amount of material which can be oriented is constant, and that the equilibrium between oriented and nonoriented material respresents a simple thermodynamic system, a van't Hoff plot was made of the equilibrium constant, log [(birefringence)/(maximum birefringence minus birefringence)] vs. $1/T$ (where T is absolute temperature). A straight line relationship was obtained which showed a very large *increase* in entropy (100 eu at 25°C) and a large heat of activation (29 kcal/mole) as the molecules polymerized. The standard free energy at 25°C was less than -1 kcal. Similar thermodynamic data have now been obtained, using *Pectinaria* oocytes with metaphase equilibrium spindles, in ordinary sea water and in sea water substituted with approximately 40% heavy water (Carolan et al., 1965, 1966).[2]

The high heat and the very high entropy increase associated with the formation of the polymers could be explained primarily by the loss in regularity of the water molecules, which is associated with the free protein molecules (see Kauzmann, 1957; Klotz, 1958, 1960; Robinson and Jencks, 1965; Scott and Berns, 1965; and Scheraga, 1967, for relevant discussions).

Higher temperatures are thought to dissociate bound water from the protein "subunit" molecules and allow them to interact closely (very likely by

[2] As of now, the thermodynamic data obtained comparing behavior of spindle birefringence in deuterated and normal sea water at different temperatures can be explained only by assuming an increased active pool size in D_2O.

hydrophobic interaction) and to associate. The low free energy reflects the weakness of the forces holding the "polymer" together.

The higher degree of alignment of material at higher temperature, as well as the thermodynamic parameters determined, resemble closely those obtained in the polymerization of tobacco mosaic virus (Lauffer et al., 1958) and in G- to F-actin transformation (Asakura et al., 1960; Grant, 1965; Oosawa et al., 1965).

Stevens and Lauffer (1965) have recently measured the buoyancy change of tobacco mosaic virus A-protein as it associated into virus-shaped rods. They concluded that in fact some 150 moles of water are dissociated from each 10^5 g of protein during their polymerization.

These simple model systems also respond to high hydrostatic pressure and to heavy water in a manner similar to the spindle in living cells (Ikkai and Ooi, 1966; Grant, 1965; Khalil et al., 1964).

5. EFFECT OF HEAVY WATER ON SPINDLE BIREFRINGENCE

In the dynamic equilibrium model, structured water is believed to dissociate from the protein molecules upon polymerization. Substitution of heavy water (D_2O) for normal water (H_2O) might then be expected to alter the equilibrium. Gross and Spindel (1960 *a*, *b*) and Marsland and Zimmerman (1965) have shown high concentrations of D_2O to "freeze" mitosis, or to overstabilize the gel structure of the spindle. Marsland and Hiramoto (1966) have shown the stiffness of the sea urchin egg to rise with D_2O.

According to Sidgwick (1950), heavy water molecules are held together more tightly than ordinary water molecules. Heavy water then may be expected to strip the protein molecules of ordered water and shift the equilibrium toward greater association of the molecules. This would enhance polymerization and birefringence of the spindle fibers (see Tomita et al., 1962, and Némethy and Scheraga, 1964, for other possible effects of heavy water on protein structure.)

This was, in fact, found to be so, as shown in Figs. 4–7, 9, and Table I, both in *Pectinaria* metaphase equilibrium oocyte and in dividing eggs of a sea urchin, *Lytechinus variegatus* (Inoué et al., 1963; Sato et al., 1966). When 45% D_2O was applied at the appropriate stage of mitosis, the spindle birefringence was found to increase at least 2-fold, and the volume occupied by the birefringent spindle, to increase as much as 10-fold at times. Fig. 7 illustrates the rapid change in the responsiveness of the spindle to D_2O (and temperature treatment) during mitosis in developing sea urchin eggs.

The action of heavy water is rapid, being 80% complete within 40 sec (see Tucker and Inoué, 1963, for determination of the rapid penetration rate of D_2O into sea urchin eggs). It is completely reversible, and the experiment can

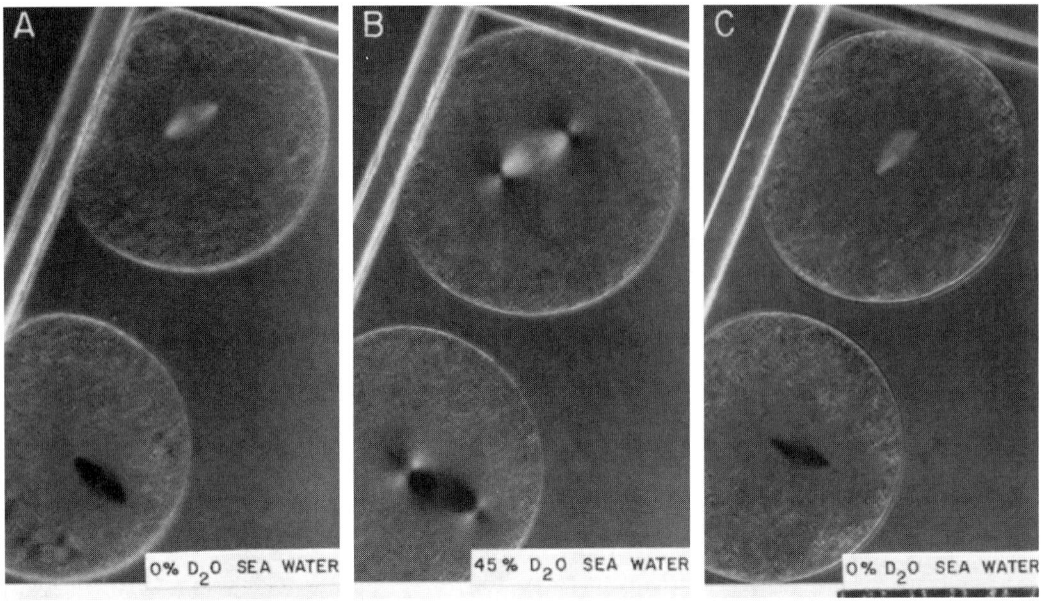

FIGURE 4. Reversible enhancement of spindle volume and birefringence by heavy water in arrested metaphase oocyte of *Pectinaria gouldi*. Cells supported by Butvar film and slightly compressed. Polarization microscopy. Scale interval, 10 μ. *A*, before D_2O–sea water perfusion; *B*, perfused for 2 min in 45% D_2O–sea water; *C*, approximately 3 min after return to normal sea water perfusion.

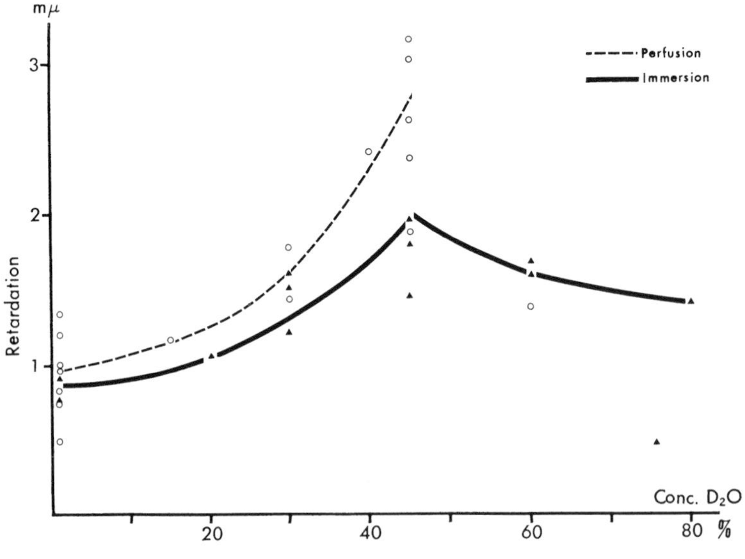

FIGURE 5. Effect of various concentrations of D_2O on the measured birefringence (retardation) of the mitotic spindle.

be repeated many times over on the same metaphase equilibrium cell. In many respects, it is similar to the effects of higher temperature and opposite to that of chilling.

Electron microscopy of spindles fixed with a new fixative, which prevents alteration of spindle birefringence during fixation (see section 10), showed the heavy water–treated spindle to have a larger bulk and more spindle filaments (Fig. 9 *B*) than the control (Fig. 9 *A*). The density of filaments was approximately equal to the control (Inoué, Kane, and Sato, unpublished data). Thus there is a parallel between birefringence observed in living cells under

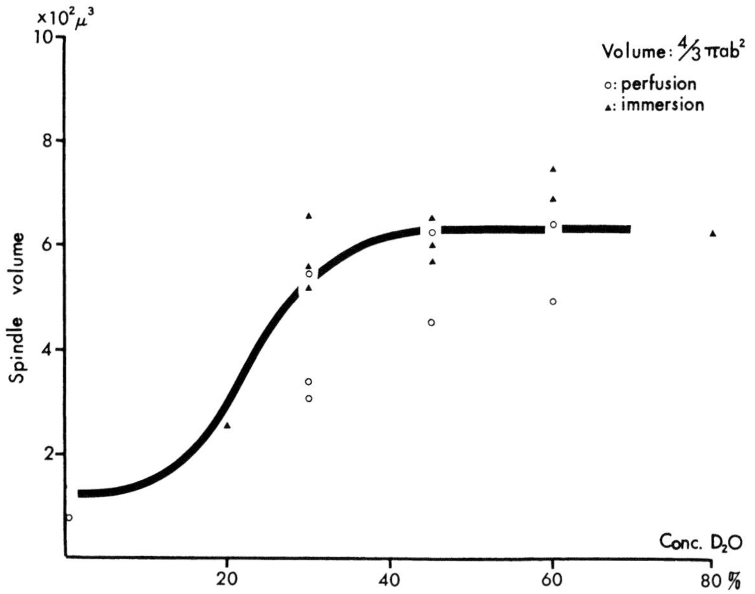

FIGURE 6. Spindle volume vs. D_2O concentration of *Pectinaria* oocyte. Temperature, 22° ± 1°C.

polarized light and the number of filaments (microtubules) observed with the electron microscope.

Similarly, a parallel increase is found in the quantity of the major (22S) protein extractable from an isolated spindle with and without heavy water treatment (Table I). In this connection, it should be pointed out that the total 22S protein extracted from each whole unfertilized egg treated with heavy water was no greater than that found in the untreated control eggs.

6. ASSEMBLY FROM POOL, NOT DE NOVO SYNTHESIS

Given the lability of the spindle fiber and its ability to recover rapidly from disruption and even to rapidly add more birefringent material, are the spindle fiber molecules synthesized rapidly each time they are required, or are they

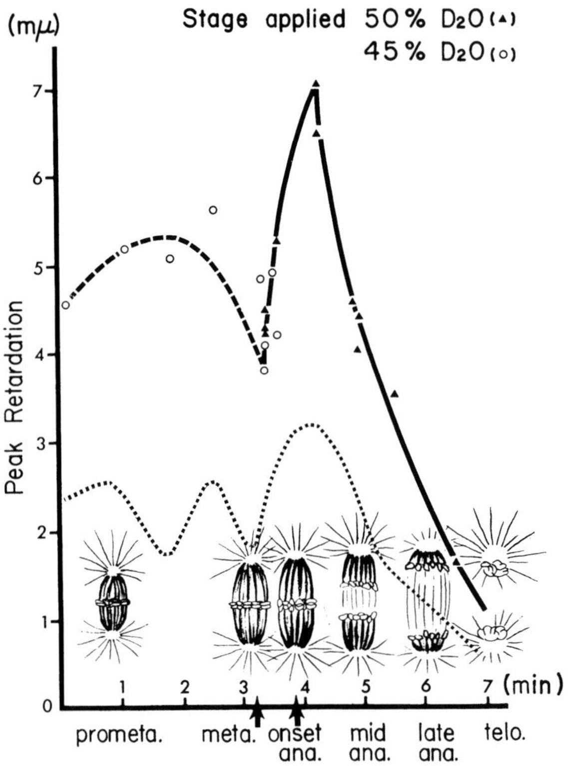

FIGURE 7. In developing eggs of sea urchin *Lytechinus variegatus* placed in D_2O–sea water, the spindle volume and birefringence increase, but with different sensitivities at different stages of mitosis. In 30–40% D_2O–sea water, mitosis proceeds. In 45–60% D_2O–sea water, mitosis is arrested unless the cell is exposed to D_2O during or after anaphase. At 50% D_2O, the retardation can be doubled and may reach a maximum of 7 mμ only if D_2O reaches the spindle at a particular 10 sec interval in early anaphase. (See Tucker and Inoué, 1963, for D_2O penetration rate.) The differential sensitivity of the spindle fiber birefringence to D_2O at different stages of mitosis closely parallels its sensitivity to chilling (Inoué, unpublished data). Heavy broken line and heavy black line in this figure indicate measured peak retardation achieved by spindle fiber with 45 and 50% D_2O–sea water at different stages of mitosis. Thin broken line indicates control.

TABLE I

ANALYTICAL ULTRACENTRIFUGATION DATA

Material as in Fig. 11 (Sato et al., 1966).

Concentration of D_2O in sea water	22S protein per spindle	Ratio*
%	*mg* × *10*$^{-7}$	
0	2.4–5.1‡	2.6–10.0
40	9.0–35.0	

* Ratios of 22S protein per spindle in D_2O vs. H_2O sea water eggs, calculated for each experiment.

‡ Average = 3.5.

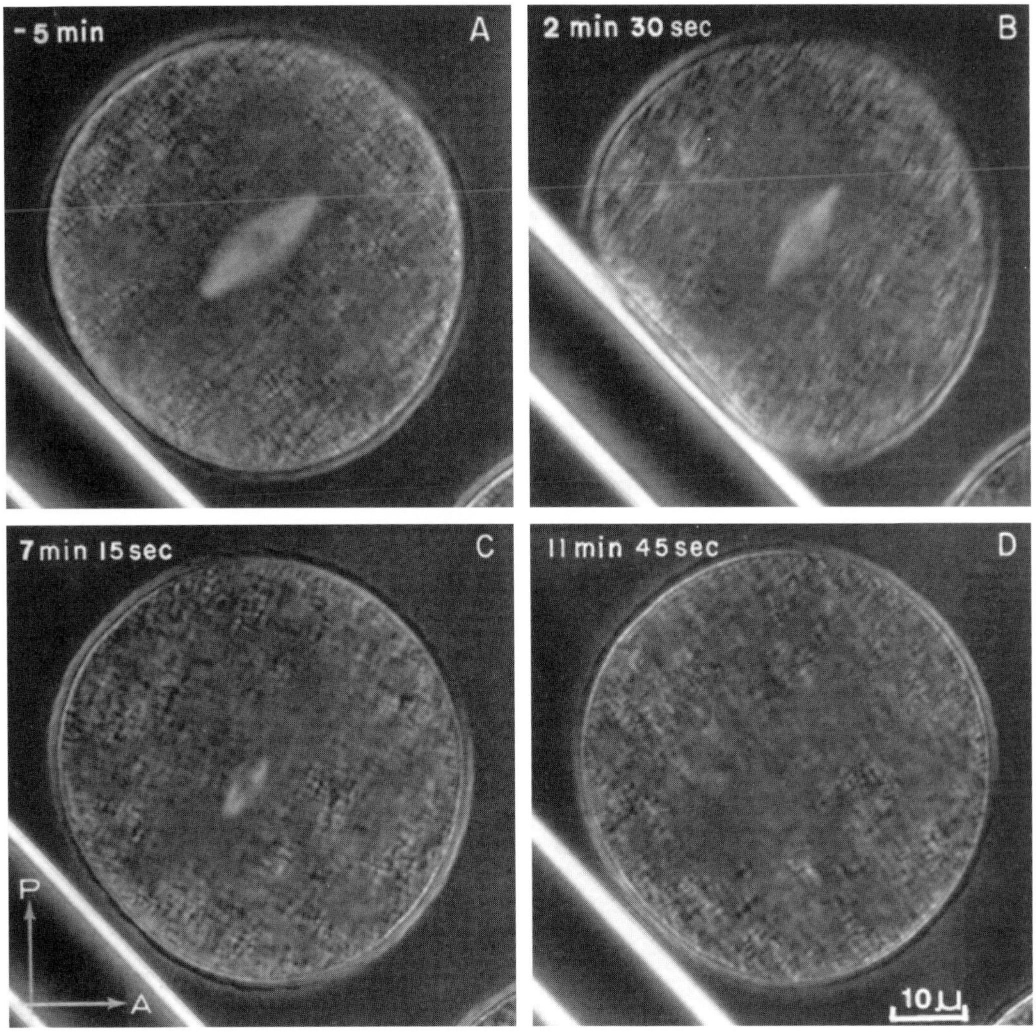

FIGURE 8. Ef.ect of 10^{-5} M Colcemid–sea water perfusion on the arrested metaphase spindle of *Pectinaria* oocyte. The eggs, without compression, were placed on a slide glass coated with a thin layer of Butvar film, to which the eggs adhere. They were surrounded by glass wool, which further prevents the eggs from flowing away during rapid perfusion. *A*, 5 min before the perfusion; *B*, 2 min 30 sec after the Colcemid perfusion was started; *C*, 7 min 15 sec in Colcemid, the miniature spindle is still visible; *D*, after 11 min 45 sec, the spindle has disappeared completely. *P*↑, vibrating direction of polarized light; *A*⟶, direction of analyzer; applies to all figures in polarized light.

available from a ready pool? Colchicine and analogues reversibly abolish the spindle birefringence as reported earlier (section 3). As shown in Fig. 8, the metaphase spindle of *Pectinaria* oocyte shrinks and loses its birefringence in the course of approximately 10 min with 10^{-5} M Colcemid in sea water.

As described in the case of *Chaetopterus* oocytes (Inoué, 1952 *a*), when low concentrations of colchicine or of Colcemid are applied, the chromosomal spindle fibers contract and pull the chromosomes to the cell surface to which they are closest. Fig. 9 *D* shows an electron micrograph of a spindle in the process of dissolution in Colcemid. The cell was fixed by the method described in section 10. Remnants of short fragments of spindle filaments could still be seen around the chromosome, and the chromosomes moved closer to the cell surface during contraction of the spindle (compare with control spindle, Fig. 9 *A*). Fig. 9 *C* shows a similar situation in a cell treated with cold. When cells treated with Colcemid are washed with normal sea water, their spindle recovers in the course of 45–60 min. The recovery is not shortened or prolonged by washing with heavy water.

If this recovery requires protein synthesis, recovery would be expected to be delayed or prevented by the addition of inhibitors of protein synthesis. In fact, addition of actinomycin D, puromycin, or chloramphenicol did not hinder recovery of the spindle to completion. Sato and Inoué (manuscript in preparation) describe the effects of these agents on regular mitosis and cleavage. For some unknown reason, puromycin and actinomycin D accelerated the recovery, as shown in Fig. 10 (Inoué et al., 1965). There is no suggestion, then, that protein synthesis is required during recovery of spindle filaments or microtubules.

As described earlier, heavy water increases the birefringence retardation and the volume of the birefringent spindle material (Figs. 4–7). It also increased both of these parameters in isolated spindles, provided the cells were treated with heavy water while still intact (Fig. 11). The amount of the major spindle (22S) protein in the isolated spindle increases approximately proportionately with the amount of birefringent material, as shown in Table I, while the total 22S protein per cell remains unchanged (Sato et al., 1966).

The rapid rise of birefringence and volume of the spindle in D_2O–sea water, and the parallel increase in the spindle protein content without increase in the same protein in the whole cell, suggests that a sizable pool of protein must preexist in the cell. Kane (1967) has, in fact, found a large quantity of the major spindle protein already present in the unfertilized sea urchin egg. The amount of 22S protein in the unfertilized whole egg is some 20 times greater than that extracted from the first mitotic metaphase spindle.

These observations fit our working hypothesis that there exists in the cell a large quantity of pool material which can be associated to form the filaments and fibers of the mitotic spindle.

7. CONTROL OF ORIENTATION

From observations of birefringence in the living spindle, and changes induced in spindle birefringence by ultraviolet microbeam irradiation (Campbell and Inoué, 1965; Forer, 1965, 1966; Inoué and Sato, quoted in Inoué, 1964),

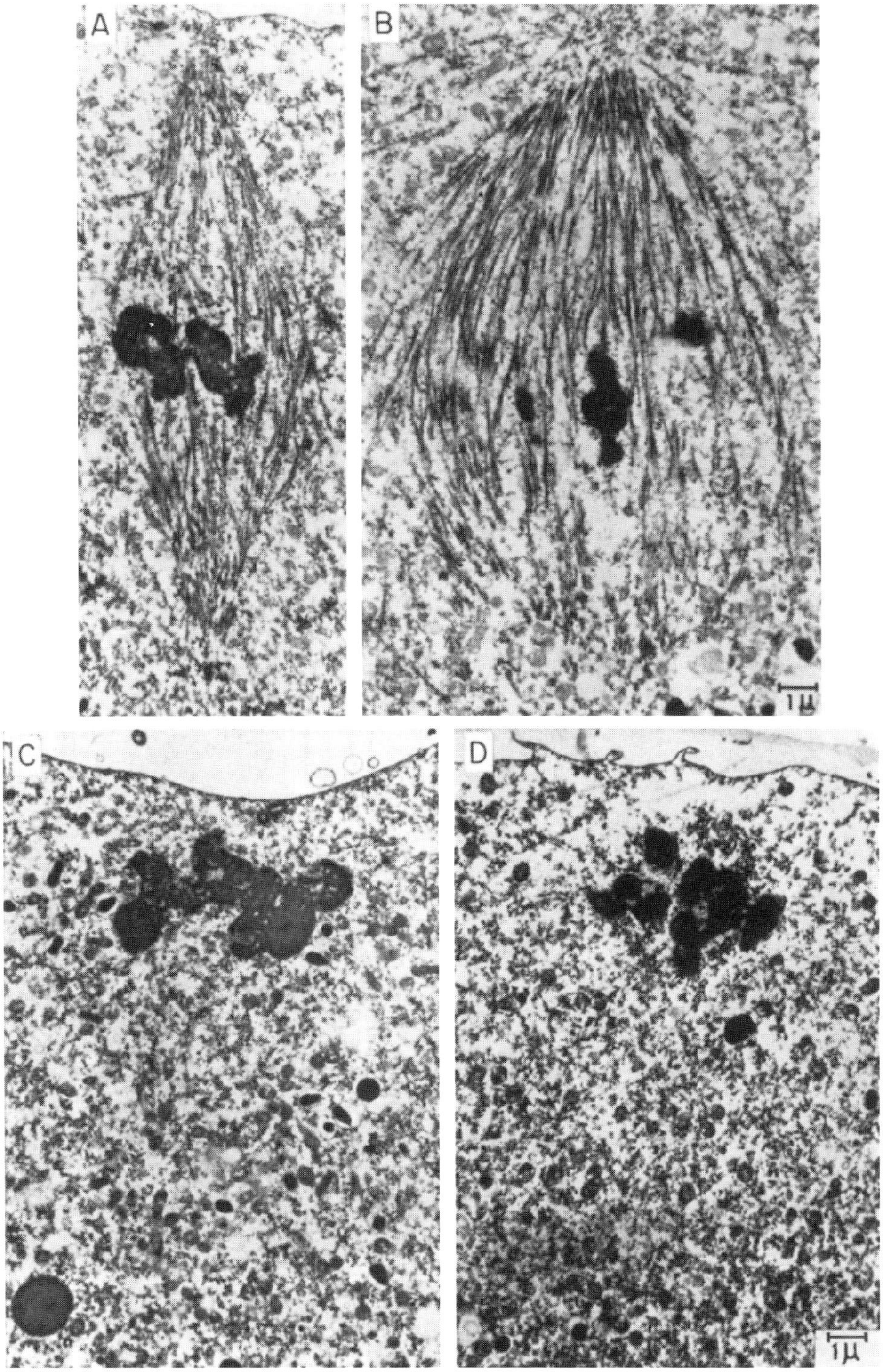

FIGURE 9

we believe that the spindle material can be assembled and oriented by three general mechanisms: (*a*) the activity of orienting centers, such as centrioles, kinetochores, and cell plate in the phragmoplast; (*b*) condensation of high concentration of monomer material into, for example, clear zone and sheath

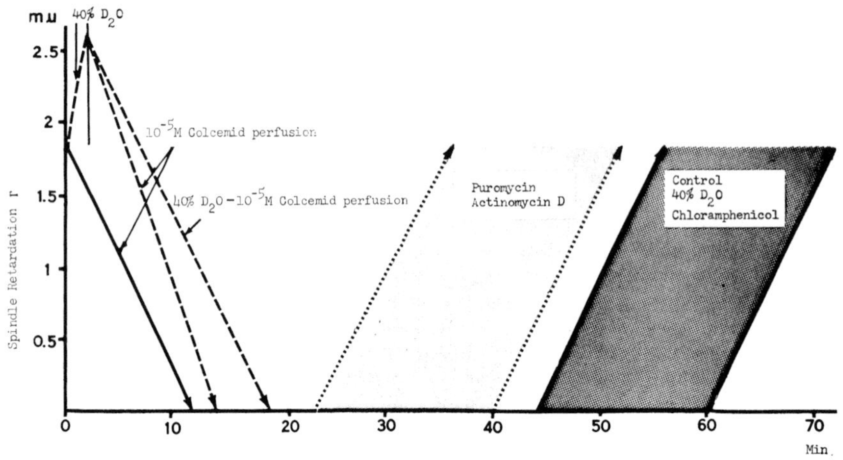

FIGURE 10. Summary of *Pectinaria* spindle behavior in 10^{-5} M Colcemid. Reduction of birefringence and spindle length are both delayed, but proceed at their same rates if 40% D_2O is mixed with Colcemid. Recovery upon washing, which was commenced 2 min after spindle birefringence became undetectable, is not accelerated by D_2O but is increased by the presence of puromycin or actinomycin D. No delay or inhibition of recovery is observed by the addition of these two antimetabolites or by chloramphenicol (Inoué et al., 1965).

fibers as seen in *Haemanthus*, *Lilium*, and *Pales* (also see Molé-Bajer, 1953 *a,b*); and (*c*) alignment parallel to other filaments or fibers already existing. Mechanism *a* has been discussed by Inoué (1964), and *b* and *c* have been discussed in section 1 of this paper.

FIGURE 9. Electron micrographs of arrested metaphase spindle of a *Pectinaria* oocyte under various experimental alterations. A new fixation procedure utilizing a mixture of glutaraldehyde and hexylene glycol was developed. This quantitatively maintains the birefringence of the spindle during fixation and, hence, preserves the ordered fine structure of the mitotic spindle (Inoué, Kane, and Sato, unpublished data). *A*, control. *B*, enlarged spindle in 45% D_2O. The total number of microtubles in the spindle is increased, but their density remains approximately the same as in the control. *C*, loss of spindle filaments. Metaphase chromosomes are pulled to the cell surface in a slowly chilled cell. *D*, process of spindle filament disintegration in 10^{-5} M Colcemid. Some fragments of the filaments are detectable around the chromosomes, which have been moved to the cell surface by the contracting spindle.

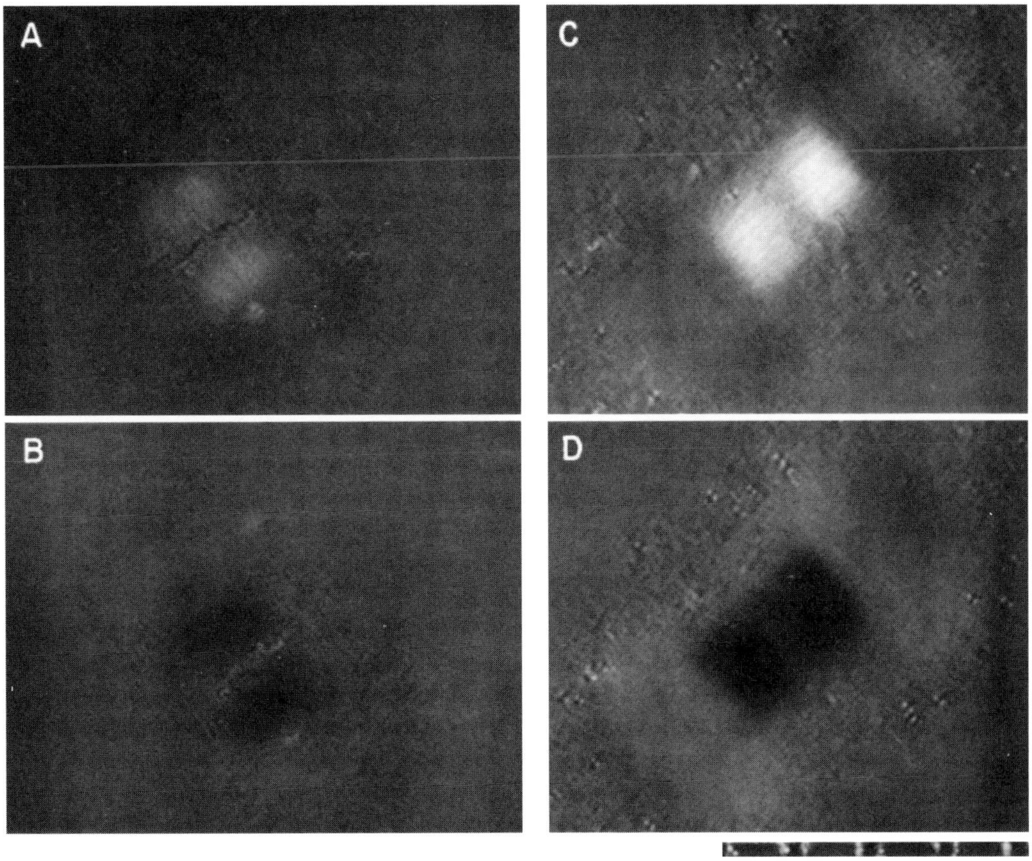

FIGURE 11. Mitotic spindles isolated with hexylene glycol (Kane, 1965) from eggs of *Arbacia punctulata* at first cleavage metaphase; polarization microscopy. *A*, control, chromosomal fibers appear bright. *B*, control, compensator axis reversed, chromosomal fibers dark. *C*, spindle isolated in 40% D_2O–sea water. Note increase of both spindle volume and birefringence. *D*, Same as *C*, compensator axis reversed. Scale interval, 10 μ.

In general, fibrogenesis may follow a pattern suggested by Rees (1951) and by Mercer (1952), or some slight modification of this mechanism: namely, by the linear or helical association of globular proteins to form filaments, which, in turn, condense into tactoids and then become cross-linked by chemical bonds to form coarser fibers. It is important, in the case of spindle fibers, not to have fibrogenesis proceed too far to establish stably cross-linked fibers, but to leave the oriented material in the labile intermediate state described before. With acids (Lewis, 1923; Kane, 1962), with dehydration (Molé-Bajer, 1953), and in the presence of cadmium salts or other heavy metals (Wada, 1965), one tends to push the equilibrium too far and acquire coarse aggregations, which then make the fibers unfunctional.

8. EXTENSION AND CONTRACTION OF SPINDLE FIBERS

We contend that the spindle fibers and filaments can elongate simply by condensation and polymerization of the spindle molecules. We believe that the elongation can provide forces to push chromosomes or to deform cell surfaces.

Many examples of elongation of spindle fibers with attendant chromosome movement occur naturally during formation of the spindle. It also occurs during recovery from cold, Colcemid, UV microbeam irradiation, and other treatments.

The polymerization that elongates the filaments could be brought about by a simple shift of equilibrium from free molecules to filaments by increased concentration of the free molecules, by the activation of orienting centers, or by mild dehydration. Increased polymerization by heavy water and elevated temperature have been described above. A slightly lower pH also tends to encourage further association, as suggested by Kane's data on the condition required for isolation of the spindle (Kane, 1962, 1965; Lewis, 1923; also see Anderson, 1956, and Klotz, 1958, for interesting relevant discussions).

Contraction, on the other hand, is believed to be brought about by slow removal of the molecules from the polymerized filaments. The fact that chromosomes are pulled poleward by the dissolving fibers in low concentrations of colchicine and by slow cooling has been described above and earlier (Inoué, 1952 *a*, 1964). Slow removal of the molecules from the filaments allows the filaments to reach a new equilibrium position rather than falling apart, thereby effectively shortening the fibers and resulting in a slow contraction. If the molecules were pulled out too fast, then the filaments would fall apart, the structure would simply collapse, and one would not achieve contraction. This was observed with rapid cooling or in the presence of higher concentrations of colchicine (Inoué, 1952 *b*).

It is proposed, in anaphase movement of chromosomes, that the slow removal of the material from the chromosomal spindle fibers, particularly toward the spindle pole region, is responsible for the shortening of the chromosomal spindle fibers and, thus, for the movement of the chromosomes (outside of the movement contributed by the elongation of the central spindle). The reason for the belief that depolymerization takes place primarily toward the pole arises from the observation of the distribution of birefringence during the normal course of division, and from the results of UV microbeam irradiation of the chromosomal spindle fibers (Forer, 1965, 1966; Inoué 1964; also see interpretation of Forer's work by Wada, 1965).

9. CHEMISTRY OF THE SPINDLE PROTEIN MOLECULES

Kane (1967) has prepared a pure 22S protein from isolated mitotic apparatuses and from whole cells. Stephens (1965, 1967) has characterized this protein. It is monodisperse, has a particle weight of 880,000, and can be broken down into units of 110,000 particle weight. The 22S material makes up more than 90% of the KCl-soluble proteins of the spindle after isolation with hexylene glycol. Its quantity in the spindle varies parallel with spindle birefringence and with the number of microtubules during D_2O treatment. It is, therefore, a likely candidate for the spindle fiber material. The 22S proteins extracted from the unfertilized whole egg and from the spindle show identical amino acid compositions (Stephens, 1965, 1967).

On the other hand, Sakai (1966) and Kiefer et al. (1966) isolated a 2.5S protein from the mitotic apparatus with a particle weight of approximately 34,000. This particle would appear to have a dimension better fitting the electron microscope periodicity seen in microtubules of spindle and other cellular structures. At this writing, it is not clear whether these are two distinctly different proteins or are the same material polymerized to different degrees (see Introduction by Mazia).

10. RELATION OF SPINDLE FIBER BIREFRINGENCE TO MICROTUBULES AND TO FILAMENTS OBSERVED BY ELECTRON MICROSCOPY

Earlier electron micrographs of dividing cells often failed to show any fibrous elements in the spindle region. The number of filaments or microtubules that were observed in the spindle fiber region was often very small. With the recognition that low temperature can rapidly destroy spindle birefringence and, hence, the oriented spindle material, and following introduction of improved fixatives, increasing numbers of microtubules and filaments have been observed in spindle regions which are birefringent in life. There now exists a good correlation with the distribution of spindle fiber birefringence and density of intact microtubules or filaments (compare Inoué, 1964, with de-Thé, 1964; Harris and Bajer, 1965; Porter, 1966; and Robbins and Gonatas, 1964). Also, when birefringence declines in isolated spindles, a parallel disruption of filaments in the electron micrographs is found (Kane and Forer, 1965).

Even with glutaraldehyde fixation, which generally shows large numbers of microtubules in the electron micrographs, we find that the *Pectinaria* oocyte spindle birefringence had declined by 50% in the fixative. Further reduction was found with neutral osmium fixation. However, mixture of these fixatives with the spindle-isolating medium, hexylene glycol, prevented the decline of birefringence. Furthermore, the elevated birefringence in cells treated with D_2O and the reduced birefringence of cells treated for a short period with Colcemid retained their values exactly when fixed in this new fixative (Inoué, Kane, and Sato, unpublished data). The number of spindle filaments thus observed with the electron microscope and the birefringence of the spindle fibers observed in the polarizing microscope showed a close

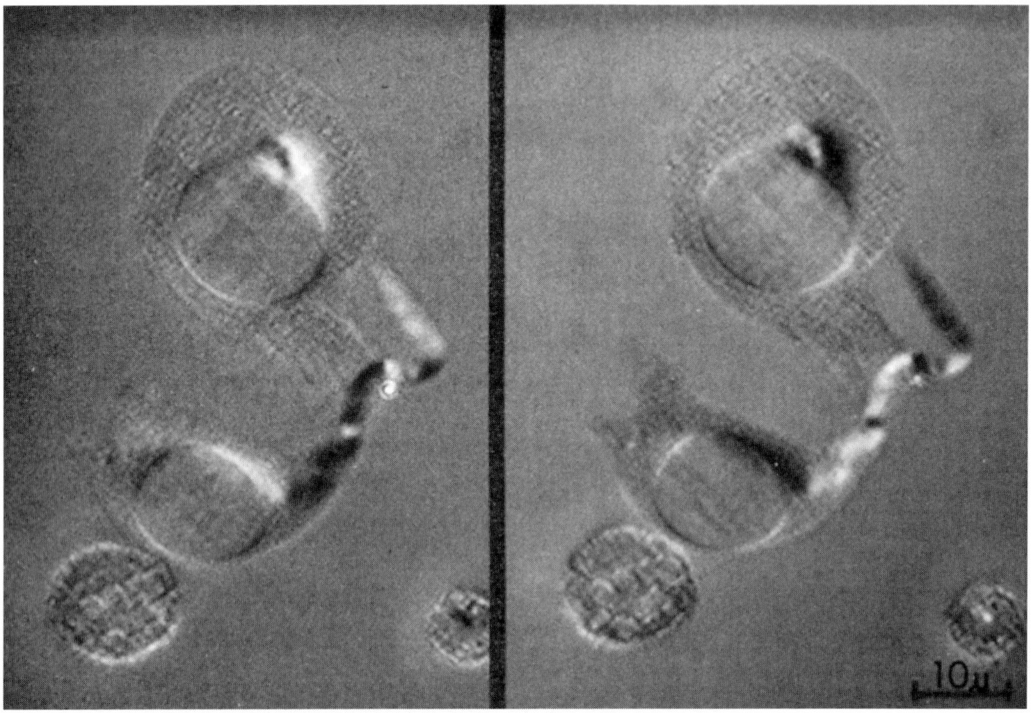

FIGURE 12. Polarization microscopy of Balb/C$^+$ renal tumor cell from the mesenteric vein of a mouse. Live cells were embedded in Kel-F 10 oil. The fibrous structures seen near the nucleus and cytoplasmic process have a positive sign of birefringence. They correspond to the aggregated bundle of microfilaments demonstrated in electron micrographs in Fig. 13.

parallel for D_2O-treated eggs (Fig. 9 *B*) and their untreated controls (Fig. 9 *A*).

Microtubules and filaments have been observed by electron microscopy following glutaraldehyde fixation, not only in the mitotic spindle, but also in various other regions of the cell (see review by Porter, 1966). Recently, we had the good fortune of observing renal tumor cells with Professor A. Claude. Examination of living cells under polarized light revealed positively birefringent bundles of material (Fig. 12) running parallel to the bundle of microfilaments, which Professor Claude had shown in his electron micrograph (Fig. 13) with a special fixative in 1960 (Claude, 1961 *a,b*). Wherever we have tested, we have found that the long axes of such microtubules and filaments lie parallel to the fiber axes of positively birefringent structures in living cells.

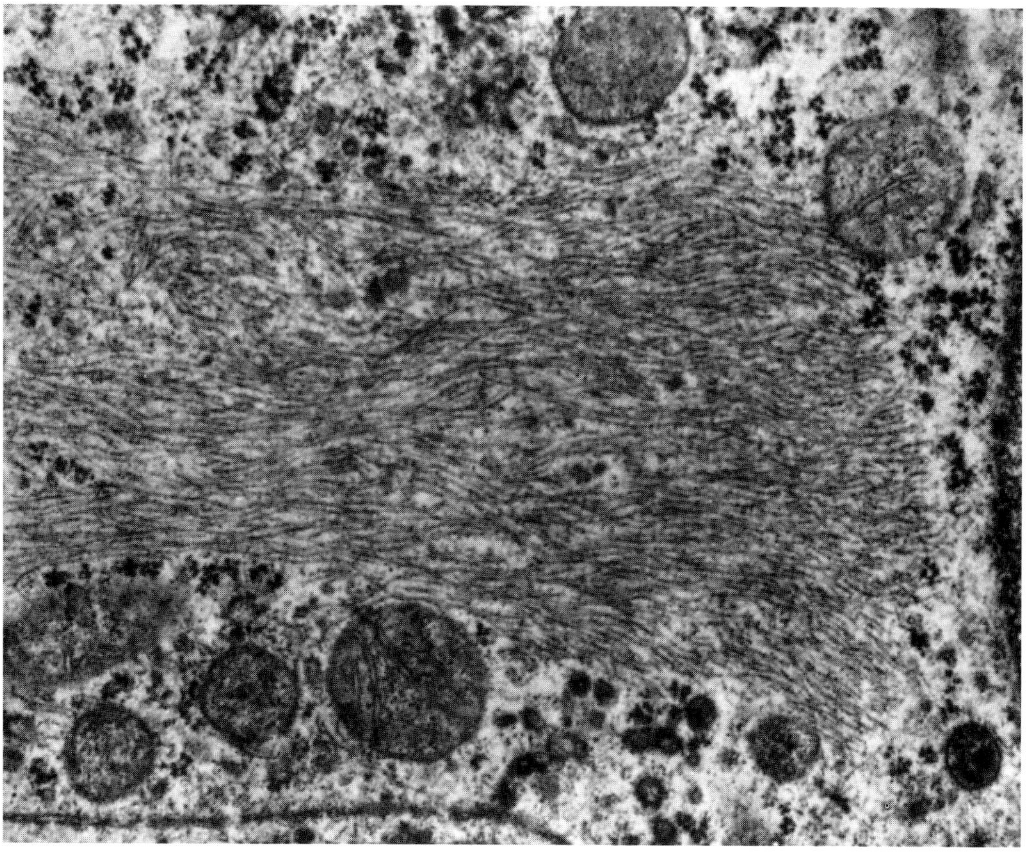

FIGURE 13. Electron micrograph of Balb/C^+ renal tumor cell of a mouse (Claude, 1961 *a,b*, 1965). × 33,400. *Figure reprinted by permission from La Biologie, Acquisitions Récentes, Centre International de Synthèse, Editions Aubier-Montaigne, Paris, 1965, 13.*

In general, conditions which give rise to decreased birefringence also give rise to the reduction of the microtubules and filaments. Conditions which give rise to increase in birefringence give rise to increased microtubules or filaments and, with D_2O, also to the 22S protein in the spindle. This correlation is found in spindle fibers and in axopods of *Actinosphaerium* and in other systems also (see sections 4 and 11).

It now seems reasonable, then, to assume that the measured birefringence of spindle fibers under a variety of conditions closely follows the number of microtubules or filaments making up the fibers (see Engelmann, 1875, 1906; Schmidt, 1937, 1941; and Picken, 1950, for general discussions relating birefringence or anisotropic fine structure to contraction).

11. CELL MOTILITY AND LABILE ASSOCIATION OF MOLECULES

In the foregoing discussions, the dynamic orientation equilibrium hypothesis was used to explain the organization and function of the spindle. A species of ubiquitous protein molecule available in a pool was reversibly assembled into filaments and fibers upon demand. The material could perform various mechanical functions in the mitotic apparatus during division of the cell, and, when not in division, the same material could be organized in different parts of the cell to perform other functions. It could maintain or alter cell shape. It could partake in pinocytosis, phagocytosis, ameboid movement, streaming, etc. In this connection it is interesting to recall that cells generally stop wandering, and round up, prior to mitosis as though localized mechanical modulations of the cell had disappeared. Cells also generally resume their various mechanical activities shortly following mitosis. The presence of microtubules and filaments in association with these activities has been described, among others: for streaming, by Wolfarth-Bottermann (1964), Nakajima (1964), and Nagai and Rebhun (1966); for melanophore pigment migration, by Bikle et al. (1966); for pinocytosis and ameboid movement, by Marshall and Nachmias (1964) and Nachmias (1964); and in differentiating cells associated with their shape changes, by Arnold (1966), Byers and Porter (1964), Taylor (1966), Tilney and Gibbins (1966), and Tilney et al. (1965, 1966) (also see review by Porter, 1966). In those cases tested, the labile microtubules responded to temperature, hydrostatic pressure, heavy water, and colchicine or their combinations in a manner virtually identical with the response of the spindle material, all being reversible as in the spindle (e.g. Tilney et al., 1965, 1966; Malawista, 1965; Marsland, 1965, 1966; Marsland and Hiramoto, 1966).

We are considering here primarily slow movements such as movements of chromosomes and changes in shape of developing cells. These generally involve velocities of 1 to a few micra per second. In general, the growth and dissolution of the filaments or microtubules are believed to impose an anisotropic modulation of the mechanical properties that could cause slow extension or contraction of local regions of cells. Faster movements may also arise secondarily. For example, cytoplasmic streaming in the slime mold *Physarum* could result from formation and dissolution of the birefringent fibers observed by Nakajima (1964) in their ectoplasm. Local changes in the intracellular pressure, i.e. an interplay between the growth and contraction of the filaments and the elastic properties of the slime mold surface, could indirectly drive the streaming.

We have outlined our current hypothesis regarding the relation of the birefringent structures to slow movement of cell parts by reversible association of molecules. An alternative interpretation would describe the observed changes in the birefringent structures as simply the formation and dissolution of the fibrous fine structure necessary to make the movement possible. The motor force itself could result from a separate mechanism, as in the sliding filament model of muscle contraction, or from a mechanism such as that proposed by Thornburg (1967), in which fine pulses travel along the length of the fine anisotropic filaments. The dynamic equilibrium hypothesis would then account for motile structure organization rather than for production of force per se. The distinction may, however, turn out to be more subtle than is apparent and awaits analysis of the various systems in further detail (for example, see Szent-Györgyi and Prior, 1966, for the behavior of actin filaments during muscle contraction.)

POSTSCRIPT

This article portrays the authors' best current picture of the molecular mechanism of reversible fibrogenesis and mitotic chromosome movements. Clearly, much information is still wanting, and the arguments are often less than tight. Nevertheless, we hope that the article may convey enough useful information to stimulate further experimentation and to bring to focus some of the critical questions about mitosis and spindle formation which now may be asked.

Supported in part by NIH Grant CA 10171 (formerly 04552) from the National Cancer Institute and by Grant GB 5120 (formerly 2060) from the National Science Foundation.
Much of our work reported here was done at the Department of Cytology, Dartmouth Medical School, and at the Marine Biological Laboratories, Woods Hole, Massachusetts. The authors gratefully acknowledge the efficient assistance provided by the devoted staff, as well as the encouragement and critical discussions offered by their colleagues and students.

REFERENCES

Anderson, N. G. 1956. Cell division. II. A theoretical approach to chromosomal movements and the division of the cell. *Quart. Rev. Biol.* **31**:243.

Arnold, J. M. 1966. Squid lens development in compounds that affect microtubules. *Biol. Bull.* **131**:383.

Asakura, S., M. Kasai, and F. Oosawa. 1960. The effect of temperature on the equilibrium state of actin solutions. *J. Polymer Sci.* **44**:35.

Bajer, A. 1957. Cine-micrographic studies on mitosis in endosperm. III. The origin of the mitotic spindle. *Exptl. Cell Res.* **13**:493.

Bajer, A. 1961. A note on the behavior of spindle fibres at mitosis. *Chromosoma.* **12**:64.

Bajer, A. 1965. Cine-micrographic analysis of cell plate formation in endosperm. *Exptl. Cell Res.* **37**:376.

BIKLE, D., L. G. TILNEY, and K. R. PORTER. 1966. Microtubules and pigment migration in the melanophores of *Fundulus heteroclitus* L. *Protoplasma.* **61**:322.

BYERS, B., and K. R. PORTER. 1964. Oriented microtubules in elongating cells of the developing lens rudiment after induction. *Proc. Natl. Acad. Sci. U.S.* **52**:1091.

CAMPBELL, R. D., and S. INOUÉ. 1965. Reorganization of spindle components following UV micro irradiation. *Biol. Bull.* **129**:401.

CARLSON, J. G. 1952. Microdissection studies of the dividing neuroblast of the grasshopper, *Chortophaga viridifasciata* (de Geer). *Chromosoma.* **5**:199.

CAROLAN, R. M., H. SATO, and S. INOUÉ. 1965. A thermodynamic analysis of the effect of D_2O and H_2O on the mitotic spindle. *Biol. Bull.* **129**:402.

CAROLAN, M., H. SATO, and S. INOUÉ. 1966. Further observations on the thermodynamics of the living mitotic spindle. *Biol. Bull.* **131**:385.

CHAMBERS, R. 1924. The physical structure of protoplasm as determined by microdissection and injection. *In* General Cytology. E. V. Cowdry, editor. University of Chicago Press, Chicago. 237.

CLAUDE, A. 1961 *a*. Mise en évidence par microscopie électronique d'un appareil fibrillaire dans le cytoplasme et le noyau de certaines cellules. *Compt. Rend.* **253**: 2251.

CLAUDE, A. 1961 *b*. Problems of fixation for electron microscopy. Results of fixation with osmium tetroxide in acid and alkaline media. *Ext. Pathol. Biol.* **9**:933.

CLAUDE, A. 1965. Naissance de la biologie moléculaire. Vingt années d'invention technique et de progrès dans l'exploration de la cellule. *In* La Biologie, Acquisitions Récentes. Centre International de Synthèse. Éditions Aubier-Montaigne, Paris. 13.

CLEVELAND, L. R. 1938. Origin and development of the achromatic figure. *Biol. Bull.* **74**:41.

CLEVELAND, L. R. 1953. Studies on chromosomes and nuclear division. *Trans. Am. Phil. Soc.* **43**:809.

CLEVELAND, L. R. 1963. Functions of flagellate and other centrioles in cell reproduction. *In* The Cell in Mitosis. L. Levine, editor. Academic Press, Inc., New York. 3.

CONKLIN, E. G. 1917. Effects of centrifugal force on the structure and development of the eggs of *Crepidula*. *J. Exptl. Zool.* **22**:311.

COSTELLO, D. P. 1961. On the orientation of centrioles in dividing cells, and its significance: a new contribution to spindle mechanics. *Biol. Bull.* **120**:285.

DE-THÉ, G. 1964. Cytoplasmic microtubules in different animal cells. *J. Cell Biol.* **23**:265.

DIETZ, R. 1959. Centrosomenfreie Spindelpole in Tipuliden-Spermatocyten. *Z. Naturforsch.* **146**:749.

DIETZ, R. 1963. Polarisationsmikroskipische Befunde zur chromosomeninduzierten Spindelbildung bei der Tipulide *Pales crocata* (Nematocera). *Zool. Anz.* **26** (Suppl.), 131.

ELLIS, G. W. 1962. Piezoelectric micromanipulators. Electrically operated micromanipulators add automatic high-speed movement to normal manual control. *Science.* **138**:84.

ENGELMANN, R. W. 1875. Contractilität und Doppelbrechung. *Arch. Ges. Physiol.* **11**:432.

Engelmann, R. W. 1906. Zur Theorie der Contractilität. *Sitzber. Kgl. Preuss. Akad. Wiss.* 694.

Forer, A. 1965. Local reduction of spindle fiber birefringence in living *Nephrotoma suturalis* (Loew) spermatocytes induced by ultraviolet microbeam irradiation. *J. Cell. Biol.* **25**:95.

Forer, A. 1966. Characterization of the mitotic traction system, and evidence that birefringent spindle fibers neither produce nor transmit force for chromosome movement. *Chromosoma.* **19**:44.

Frey-Wyssling, A., J. F. Lopez-Saez, and K. Mühlethaler. 1964. Formation and development of the cell plate. *J. Ultrastruct. Res.* **10**:422.

Gaulden, M. E., and J. G. Carlson. 1951. Cytological effects of colchicine on the grasshopper neuroblast in vitro with special reference to the origin of the spindle. *Exptl. Cell Res.* **2**:416.

Grant, R. J. 1965. Doctorate Thesis. Columbia University, New York.

Gross, P. R., and W. Spindel. 1960 *a*. Mitotic arrest by deuterium oxide. *Science.* **131**:37.

Gross, P. R., and W. Spindel. 1960 *b*. The inhibition of mitosis by deuterium. *Ann. N.Y. Acad. Sci.* **84**:745.

Harris, P. 1962. Some structural and functional aspects of the mitotic apparatus in sea urchin embryos. *J. Cell Biol.* **14**:475.

Harris, P., and A. Bajer. 1965. Fine structure studies on mitosis in endosperm metaphase of *Haemanthus katherinae* bak. *Chromosoma.* **16**:624.

Hiramoto, Y. 1964. Mechanical properties of the starfish oöcyte during maturation divisions. *Biol. Bull.* **127**:373.

Ikkai, T., and T. Ooi. 1966. The effects of pressure on F-G transformation of actin. *Biochemistry.* **5**:1551.

Inoué, S. 1952 *a*. The effect of colchicine on the microscopic and submicroscopic structure of the mitotic spindle. *Exptl. Cell Res.* Suppl. 2:305.

Inoué, S. 1952 *b*. Effect of temperature on the birefringence of the mitotic spindle. *Biol. Bull.* **103**:316.

Inoué, S. 1953. Polarization optical studies of the mitotic spindle. I. The demonstration of spindle fibers in living cells. *Chromosoma.* **5**:487.

Inoué, S. 1959. Motility of cilia and the mechanism of mitosis. *Rev. Mod. Phys.* **31**:402.

Inoué, S. 1960. On the physical properties of the mitotic spindle. *Ann. N.Y. Acad. Sci.* **90**:529.

Inoué, S. 1964. Organization and function of the mitotic spindle. *In* Primitive Motile Systems in Cell Biology. R. Allen and N. Kamiya, editors. Academic Press, Inc., New York. 549.

Inoué, S., and A. Bajer. 1961. Birefringence in endosperm mitosis. Chromosoma. **12**:48.

Inoué, S., and K. Dan. 1951. Birefringence of the dividing cell. *J. Morphol.* **89**:423.

Inoué, S., and H. Sato. 1964. See Inoué (1964). 580.

Inoué, S., H. Sato, and M. Ascher. 1965. Counteraction of Colcemid and heavy water on the organization of the mitotic spindle. *Biol. Bull.* **129**:409.

INOUÉ, S., H. SATO, and R. W. TUCKER. 1963. Heavy water enhancement of mitotic spindle birefringence. *Biol. Bull.* **125**:380.

IZUTSU, K. 1961 *a*. Effects of ultraviolet microbeam irradiation upon division in grasshopper spermatocytes. I. Results of irradiation during prophase and prometaphase I. *Mie Med. J.* **11**:199.

IZUTSU, K. 1961 *b*. Effects of ultraviolet microbeam irradiation upon division in grasshopper spermatocytes. II. Results of irradiation during metaphase and anaphase I. *Mie Med. J.* **11**:213.

KANE, R. E. 1962. The mitotic apparatus: isolation by controlled pH. *J. Cell Biol.* **12**:47.

KANE, R. E. 1965. The mitotic apparatus: physical-chemical factors controlling stability. *J. Cell Biol.* **25**:137.

KANE, R. E. 1967. The mitotic apparatus: identification of the major soluble component of the glycol-isolated mitotic apparatus. *J. Cell Biol.* **32**:243.

KANE, R. E., and A. FORER. 1965. The mitotic apparatus: structural changes after isolation. *J. Cell Biol.* **25**:31.

KAUZMANN, W. 1957. The physical chemistry of proteins. *Ann. Rev. Phys. Chem.* **8**:413.

KHALIL, M. T., R. A. SHALABY, and M. A. LAUFFER. 1964. Reversible polymerization of TMV protein in D_2O and versene solutions. Abstracts of the Biophysical Society 8th Annual Meeting. Chicago. TC3.

KIEFER, B., H. SAKAI, A. J. SOLARI, and D. MAZIA. 1966. The molecular unit of the microtubules of the mitotic apparatus. *J. Mol. Biol.* **20**:75.

KLOTZ, I. M. 1958. Protein hydration and behavior. *Science.* **128**:815.

KLOTZ, I. M. 1960. Non-covalent bonds in protein structure. *In Brookhaven Symp. Biol.* **13**:25.

LAUFFER, M. A., A. T. ANSEVIN, T. E. CARTWRIGHT, and C. C. BRINTON, JR. 1958. Polymerization-depolymerization of tobacco mosaic virus protein. *Nature.* **181**: 1338.

LEDBETTER, M. C., and K. R. PORTER. 1963. A "microtubule" in plant cell fine structure. *J. Cell Biol.* **19**:239.

LEWIS, M. R. 1923. Reversible gelation in living cells. *Bull. Johns Hopkins Hosp.* **34**:373.

MALAWISTA, S. E. 1965. On the action of colchicine: the melanocyte model. *J. Exptl. Med.* **122**:361.

MALAWISTA, S. E., and H. SATO. 1966. Vinblastine and griseofulvin reversibly disrupt the living mitotic spindle. *Biol. Bull.* **131**:397.

MARSHALL, J. M., and V. T. NACHMIAS. 1964. Cell surface and pinocytosis. *J. Cell Biol.* **13**:92.

MARSLAND, D. 1965. Partial reversal of the anti-mitotic effects of heavy water by high hydrostatic pressure: an analysis of the first cleavage division in the eggs of *Strongylocentrotus purpuratus*. *Exptl. Cell Res.* **38**:592.

MARSLAND, D. 1966. Anti-mitotic effects of colchicine and hydrostatic pressure; synergistic action on the cleaving eggs of *Lytechinus variegatus*. *J. Cell Physiol.* **67**:333.

MARSLAND, D., and Y. HIRAMOTO. 1966. Cell division: pressure-induced reversal of the antimeiotic effects of heavy water in the oocytes of the starfish, *Asterias forbesi* *J. Cell Physiol.* **67**:13.

MARSLAND, D., and A. M. ZIMMERMAN. 1965. Structural stabilization of the mitotic apparatus by heavy water, in the cleaving eggs of *Arbacia punctulata. Exptl. Cell Res.* **38**:306.

MAZIA, D. 1961. Mitosis and the physiology of cell division. *In* The Cell. J. Brachet and A. E. Mirsky, editors. Academic Press, Inc., New York. **3**:77.

MERCER, E. H. 1952. The biosynthesis of fibers. *Sci. Monthly.* **75**:280.

MOLÉ-BAJER, J. 1953 *a*. Influence of hydration and dehydration on mitosis. II. *Acta Soc. Botan. Polon.* **22**:33.

MOLÉ-BAJER, J. 1953 *b*. Experimental studies on protein spindles. *Acta Soc. Botan. Polon.* **22**:811.

MOLÉ-BAJER, J. 1958. Cine-micrographic analysis of C-mitosis in endosperm. *Chromosoma.* **9**:332.

NACHMIAS, V. T. 1964. Fibrillar structures in the cytoplasm of *Chaos chaos. J. Cell Biol.* **23**:183.

NAGAI, R., and L. I. REBHUN. 1966. Cytoplasmic microfilaments in streaming *Nitella* cells. *J. Ultrastruct. Res.* **14**:571.

NAKAJIMA, H. 1964. The mechanochemical system behind streaming in *Physarum. In* Primitive Motile Systems in Cell Biology. R. Allen and N. Kamiya, editors. Academic Press, Inc., New York. 111.

NÉMETHY, G., and H. A. SCHERAGA. 1964. Structure of water and hydrophobic bonding in proteins. IV. The thermodynamic properties of liquid deuterium oxide. *J. Chem. Phys.* **41**:680.

NICKLAS, R. B. 1965. Experimental control of chromosome segregation in meiosis. *J. Cell Biol.* **27**:117A.

NICKLAS, R. B. 1967. Chromosome micromanipulation. I. The mechanics of chromosome attachment to the spindle. II. Induced reorientation and the experimental control of segregation in meiosis. *Chromosoma.* **21**:1, 17.

OOSAWA, F., S. ASAKURA, S. HIGASHI, M. KASAI, S. KOBAYASHI, E. NAKANO, T. OHNISHI, and M. TANIGUCHI. 1965. Morphogenesis and motility of actin polymers. *In* Molecular Biology of Muscular Contraction. Igakushoin, Tokyo. 56.

ÖSTERGREN, G. 1951. The mechanism of co-orientation in bivalents and multivalents. *Hereditas.* **37**:85.

PEASE, D. C. 1946. High hydrostatic pressure effects upon the spindle figure and chromosome movement. II. Experiments on the meiotic divisions of *Tradescantia* pollen mother cells. *Biol. Bull.* **91**:145.

PICKEN, L. E. R. 1950. Discussion on morphology and fine structure. Fine structure and the shape of cells and cell-components. *Proc. Linnean Soc. London.* **162**:72.

PORTER, K. R. 1966. Cytoplasmic microtubules and their functions. *Ciba Found. Symp., Principles Biomol. Organ.* 308.

PORTER, K. R., and R. D. MACHADO. 1960. Studies on the endoplasmic reticulum. IV. Its form and distribution during mitosis in cells of onion root tip. *J. Biophys. Biochem. Cytol.* **7**:167.

REES, A. L. G. 1951. Directed aggregation in colloidal systems and the formation of protein fibers. *J. Phys. Chem.* **55**:1340.

ROBBINS, E., and N. K. GONATAS. 1964. The ultrastructure of a mammalian cell during the mitotic cycle. *J. Cell Biol.* **21**:429.

ROBINSON, D. R., and W. P. JENCKS. 1965. The effect of compounds of the urea-guanidinium class on the activity coefficient of acetyltetraglycine ethyl ester and related compounds. *J. Am. Chem. Soc.* **87**:2462, 2470, 2480.

SAKAI, H. 1966. Studies on sulfhydryl groups during cell division of sea-urchin eggs. VIII. Some properties of mitotic apparatus proteins. *Biochim. Biophys. Acta.* **112**:132.

SATO, H., S. INOUÉ, J. BRYAN, N. E. BARCLAY, and C. PLATT. 1966. The effect of D_2O on the mitotic spindle. *Biol. Bull.* **131**:405.

SCHERAGA, H. A. 1967. Contractility and conformation. *J. Gen. Physiol.* **50**(6, Pt. 2):5.

SCHMIDT, W. J. 1937. Die Doppelbrechung von Chromosomen und Kernspindel und ihre Bedeutung für das kausale Verständniss der Mitose. *Arch. Exptl. Zellforsch.* **19**:352.

SCHMIDT, W. J. 1939. Doppelbrechung der Kernspindel und Zugfasertheorie der Chromosomenbewegung. *Chromosoma.* **1**:253.

SCHMIDT, W. J. 1941. Die Doppelbrechung des Protoplasmas und ihre Bedeutung für die Erforschung seines submikroskopischen Baues. *Ergeb. Physiol. Biol. Chem. Exptl. Pharmakol.* **44**:27.

SCHRADER, F. 1953. Mitosis. The Movements of Chromosomes in Cell Division. Columbia University Press, New York. 2nd edition.

SCOTT, E., and D. S. BERNS. 1965. Protein-protein interaction. The phycocyanin system. *Biochemistry* **4**:2597.

SHIMAMURA T. 1940. Studies on the effect of the centrifugal force upon nuclear division. *Cytologia.* **11**:186.

SIDGWICK, N. V. 1950. Chemical Elements and Their Compounds. Clarendon Press, Oxford. **1**:44.

STEPHENS, R. E. 1965. Characterization of the major mitotic apparatus protein and its subunits. Doctorate Thesis. Dartmouth College, Hanover, N.H.

STEPHENS, R. E. 1967. The mitotic apparatus. Physical chemical characterization of the 22S protein component and its subunits. *J. Cell Biol.* **32**:255.

STEVENS, C. L., and M. A. LAUFFER. 1965. Polymerization-depolymerization of tobacco mosaic virus protein. IV. The role of water. *Biochemistry.* **4**:31.

SZENT-GYÖRGYI, A. G., and G. PRIOR. 1966. Exchange of adenosine diphosphate bound to actin in superprecipitated actomyosin and contracted myofibrils. *J. Mol. Biol.* **15**:515.

TAYLOR, A. C. 1966. Microtubules in the microspikes and cortical cytoplasm of isolated cells. *J. Cell Biol.* **28**:155.

TAYLOR, E. W. 1959. Dynamics of spindle formation and its inhibition by chemicals. *J. Biophys. Biochem. Cytol.* **6**:193.

THORNBURG, W. 1967. Mechanisms of biological motility. *In* Theoretical and Experimental Biophysics. A. Cole, editor. Marcel Dekker, New York **1**:77.

TILNEY, L. G., and J. R. GIBBINS. 1966. Microtubules and morphogenesis. The role of microtubules in the development of the primary mesenchyme in the sea urchin embryo. *Biol. Bull.* **131**:378.

Tilney, L. G., Y. Hiramoto, and D. Marsland. 1966. Studies on the microtubules in heliozoa. III. A pressure analysis of the role of these structures in the formation and maintenance of the axopodia of *Actinosphaerium nucleofilum* (Barrett). *J. Cell Biol.* **29**:77.

Tilney, L. G., and K. Porter. 1965. Studies on microtubules in heliozoa. I. The fine structure of *Actinosphaerium nucleofilum* (Barrett), with particular reference to the axial rod structure. *Protoplasma.* **60**:317.

Tomita, K., A. Rich, C. de Lozé, and E. R. Blout. 1962. The effect of deuteration on the geometry of the α-helix. *J. Mol. Biol.* **4**:83.

Tucker, R. W., and S. Inoué. 1963. Rapid exchange of D_2O and H_2O in sea urchin eggs. *Biol. Bull.* **125**:395.

Wada, B. 1965. Analysis of mitosis. *Cytologia.* **30** (Suppl.):1.

Went, M. A. 1966. The behavior of centrioles and the structure and formation of the achromatic figure. *Protoplasmatologia.* **6** (G 1):1.

Wohlfarth-Bottermann, K. E. 1964. Differentiations of the ground cytoplasm and their significance for the generation of the motive force of ameboid movement. *In* Primitive Motile Systems in Cell Biology. R. Allen and N. Kamiya, editors. Academic Press, Inc., New York. 79.

Zimmerman, A. M., and D. Marsland. 1964. Cell division: effects of pressure on the mitotic mechanisms of marine eggs (*Arbacia punctulata*). *Exptl. Cell. Res.* **35**:293.

Nicklas R.B. and **Koch C.A.** 1969. Chromosome micromanipulation III. Spindle fiber tension and the reorientation of mal-oriented chromosomes. *J. Cell Biol.* **43:** 40–50.

CHROMOSOMES ARE MOVED DURING MITOSIS by the spindle fibers that bind to them. Spindle fibers are readily detected by polarization optics in living cells (Inoué and Sato, this volume) or by electron microscopy in properly fixed ones (1), but the physiology of chromosome-spindle fiber attachment is not so easily studied. Special chromosomal regions called centromeres (Carbon and Clark, this volume) are required for spindle fiber attachment, but even a knowledge of the chromosomal sequences involved and the proteins that bind them to make the "kinetochore" (reviewed in 2) is insufficient to elucidate the character of spindle-chromosome attachment. In the paper presented here the authors used a conceptually straightforward but technically sophisticated method of in vivo micromanipulation to reorient chromosomes, break and remake their spindle attachments, and oppose their natural motions. Through these experiments they investigated the factors that influence the stability of a chromosome's attachment to the spindle. They demonstrated that a simple physical parameter, tension in the chromosomal spindle fiber, is essential for attachment stability. Further work from the Nicklas lab has shown that tension and the stability it induces are key for the fidelity of chromosome segregation. The biologically appropriate arrangement of a meiotic bivalent on the metaphase spindle of meiosis I induces tension and as a result is stable; all inappropriate arrangements are unstable and thus will break, allowing the chromosome to diffuse a while and try again. Correct spindle attachment, and therefore accurate chromosome segregation, is achieved by a selective mechanism, not a deterministic one. Later work from Nicklas' group also showed that tension at all kinetochores is essential for a meiotic cell to decide to enter anaphase (3). This feature appears to be an important part of the mitotic checkpoint process (Hartwell and Weinert, this volume). Through these experiments, Nicklas has elucidated the cellular mechanism by which chromosome segregation can be so accurate (4).

---

1. Roth L.E. and Daniels E.W. 1962. Electron microscopic studies of mitosis in amebae II. The giant ameba *Pelomyxa carolinensis J. Cell Biol.* **12:** 57–78.
2. Pluta A.F., Mackay A.M., Ainsztein A.M., Goldberg I.G., and Earnshaw W.C. 1995. The centromere: Hub of chromosomal activities. *Science* **270:** 1591–1594.
3. Li X. and Nicklas R.B. 1995. Mitotic forces control a cell-cycle checkpoint. *Nature* **373:** 630–632.
4. Nicklas R.B. 1997. How cells get the right chromosomes. *Science* **275:** 632–637.

CHROMOSOME MICROMANIPULATION

III. Spindle Fiber Tension and the Reorientation of Mal-Oriented Chromosomes

R. BRUCE NICKLAS and CAROL A. KOCH

From the Department of Zoology, Duke University, Durham, North Carolina 27706

ABSTRACT

Kinetochore reorientation is the critical process ensuring normal chromosome distribution. Reorientation has been studied in living grasshopper spermatocytes, in which bivalents with both chromosomes oriented to the same pole (unipolar orientation) occur but are unstable: sooner or later one chromosome reorients, the stable, bipolar orientation results, and normal anaphase segregation to opposite poles follows. One possible source of stability in bipolar orientations is the normal spindle forces toward opposite poles, which slightly stretch the bivalent. This tension is lacking in unipolar orientations because all the chromosomal spindle fibers and spindle forces are directed toward one pole. The possible role of tension has been tested directly by micromanipulation of bivalents in unipolar orientation to artificially create the missing tension. Without exception, such bivalents never reorient before the tension is released; a total time "under tension" of over 5 hr has been accumulated in experiments on eight bivalents in eight cells. In control experiments these same bivalents reoriented from a unipolar orientation within 16 min, on the average, in the absence of tension. Controlled reorientation and chromosome segregation can be explained from the results of these and related experiments.

INTRODUCTION

Controlled chromosome distribution to the daughter cells in mitosis depends upon kinetochore orientation and reorientation. Thus, kinetochore orientation (the association of a chromosome with a particular pole via chromosomal spindle fibers) determines the pole to which the chromosome will move in anaphase. But kinetochore *re*orientation is crucial to controlled distribution, because some mal-oriented chromosomes result from the initial orientation process in early prometaphase. In meiosis, for instance, bivalents occur with both half-bivalents oriented to the same pole. This "unipolar" orientation (see Fig. 1) would result in nondisjunction if it persisted; but instead, reorientation occurs, the normal, "bipolar" orientation (see Fig. 1) results, and orthodox segregation follows (e.g., refs. 10, 1).

The initial orientation process, and hence the flawless bipolar orientation of most chromosomes, is now understood (18). Equally certain, however, is our failure to understand the reorientation of mal-oriented chromosomes. Dietz (5) suggested that reorientation is effective not because a single reorientation certainly leads to bipolar orientation, but rather because only the bipolar orientation is stable, and any other orientation is unstable. That is, reorientation within a small fraction of the total prometaphase time is highly probable. Thus eventually the stable bipolar orientation is reached more or less by chance, and no

 Reprinted from THE JOURNAL OF CELL BIOLOGY, 1969, Vol. 43, No. 1, pp. 40–50 *Printed in U.S.A.*

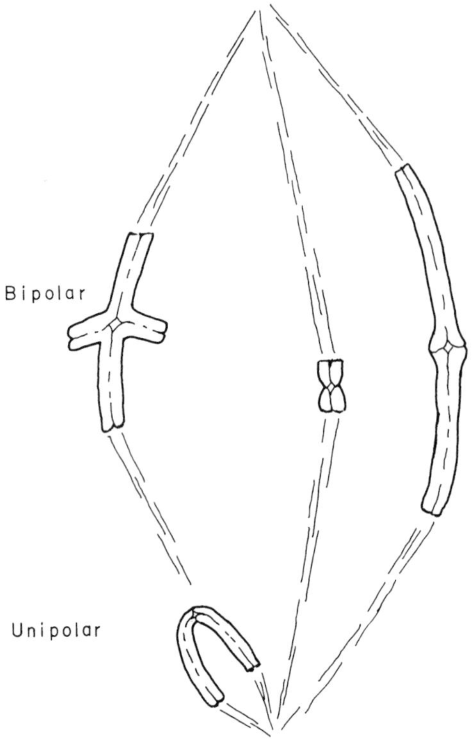

FIGURE 1 Semi-diagrammatic representation of orientation at prometaphase I in *Melanoplus*. Four bivalents, each composed of two chromosomes (half-bivalents), are shown. Chromosomal spindle fibers, represented by broken lines, run between the kinetochore of each half-bivalent and the pole to which that half-bivalent is oriented. Three bivalents (above) are shown in bipolar orientation, in each, the partner half-bivalents are oriented to opposite poles. One bivalent (below) is shown in unipolar orientation; both half-bivalents are oriented to the lower pole. The kinetochore appears as if it were terminal in all *Melanoplus* chromosomes.

further changes in orientation occur. This is exactly what is seen in living cells (1). Therefore the origin of differences in orientation stability is the key to understanding controlled chromosome distribution in eucaryotic cells.

Possible sources of differential orientation stability are readily suggested from the numerous other differences between mal-oriented and appropriately oriented chromosomes. Spindle fiber tension is one such difference: only in bipolar orientation is a chromosome or bivalent subjected to forces toward opposite poles. The presence of these opposed forces is readily recognized from the straightening or even stretching of the chromosomal material between the kinetochores oriented to opposite poles, and the absence of similar deformation in mal-oriented chromosomes is equally obvious. "Spindle fiber tension" signifies only that the tension forces are transmitted to chromosomes by chromosomal spindle fibers (e.g. ref. 19); spindle fiber *production* of the forces need not be assumed. Dietz (5, see p. 432) was the first to suggest that spindle tension might explain differences in orientation stability; he tentatively considered its attractiveness on theoretical grounds. We report here a direct test of the role of tension by micromanipulation experiments on living spermatocytes. A straightforward explanation of kinetochore reorientation is suggested by the results of these experiments.

MATERIAL AND METHODS

The grasshopper *Melanoplus differentialis* from a laboratory colony (see ref. 17) was used in these studies. Spermatocyte culture, micromanipulation, cinematographic recording, and analysis were carried out as described previously (19). Briefly, the cells were cultured (temperature range: 24° to 26°C) in a modified Ringer's solution. Bivalents in the first meiotic division were manipulated with the Ellis (6) piezoelectric micromanipulator equipped with a glass microneedle about 0.1 μ in diameter at the tip. The cells were observed and cinemicrography was carried out on a Zeiss inverted microscope with a 1.25 N.A. oil immersion, phase contrast objective.

All 20 cells considered in this report completed anaphase following micromanipulation; anaphase was normal in all cells, with the exception of experimentally induced nondisjunction of one bivalent in six cells. No exceptional results were found in the cells which failed to divide. The general influence of micromanipulations on spermatocytes has been considered earlier (18, 19). Here also, differential viability of micromanipulated and adjacent control cells has never been observed, nor have nonspecific effects on chromosome behavior.

RESULTS

The experiments to be described are possible because unipolar orientation of bivalents (see Fig. 1) in grasshopper spermatocytes is easily induced by micromanipulation (18). Reorientation to the normal bipolar orientation (see Fig. 1) follows within minutes and is identical with naturally occurring reorientation. In the present study, unipolar orientation to a given pole was always tested directly by placing a microneedle at the closed end of the bivalent and moving the needle toward the opposite pole. If the bivalent is unipolar, then the kinetochores will remain in position while the rest of the bivalent is slightly stretched (see Fig. 4 for an example, and Nicklas and Staehly, ref. 19, on the related test for bipolar orientation). Where no further operations intervene, such bivalents always move as expected—toward the pole to which both half-bivalents are oriented (e.g., Figs. 2 and 3 between 5.4 and 16.5 min).

Tension Experiments

If natural spindle tension toward opposite poles makes bipolar orientation stable, then artificial tension should stabilize unipolar orientations. This is the rationale for "tension experiments", which are feasible because a bivalent in unipolar orien-

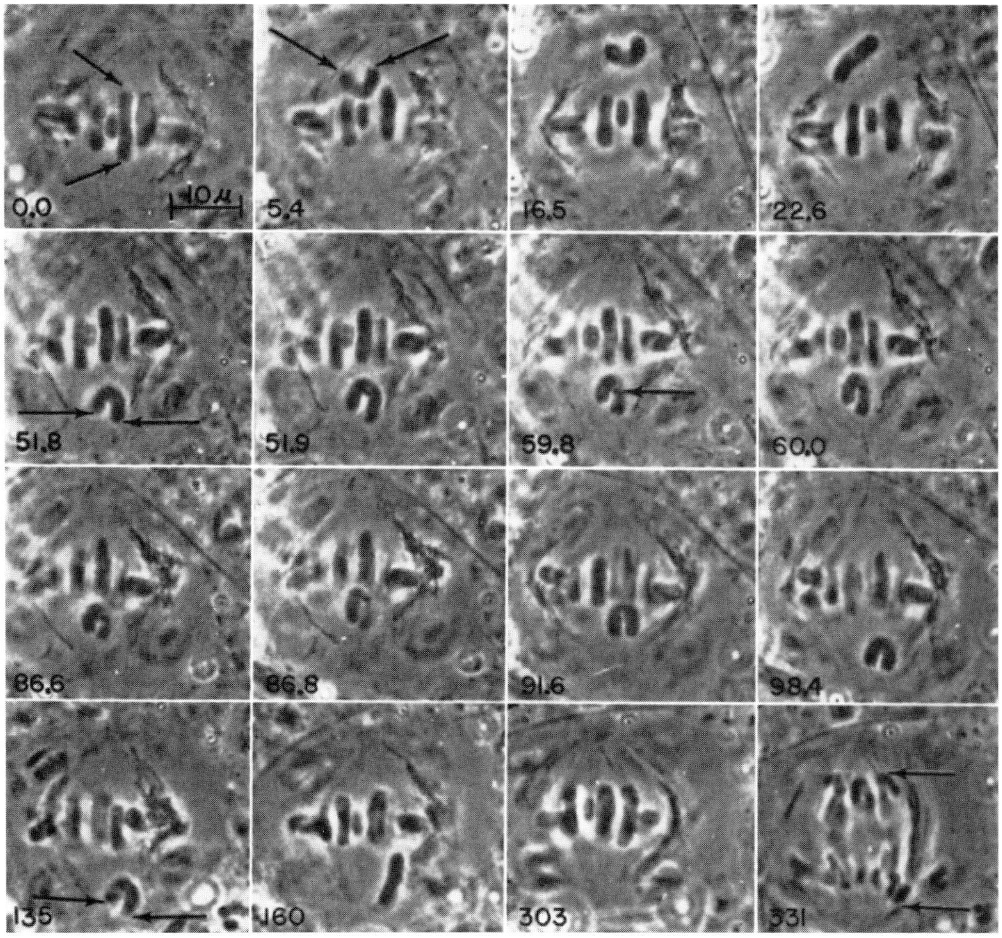

FIGURE 2 Prints from the cinématographic record of a tension experiment. The time in minutes is given on each print for comparison with the graph (Fig. 3). The pairs of arrows on several prints indicate the positions of the bivalent's kinetochores. The micromanipulation needle is rarely visible in still photographs, but its tip can be seen as a dark spot on the 51.9 through 86.8 min prints (see the arrow on the 59.8 min print). The bivalent, shown before manipulation on the 0.0 min print, was detached and unipolar orientation induced (5.4 min). A control experiment was then run with the needle close to, but not stretching, the bivalent; the bivalent promptly reoriented (16.5, 22.6 min). The bivalent was later detached again and unipolar orientation induced a second time (51.8 min). Tension toward the upper pole was then applied with the needle; note the slight stretching of the bivalent at 51.9 min. Tension was maintained for 39.7 min with periodic readjustments (compare 59.8 with 60.0 and 86.6 with 86.8 min). Reorientation did not occur during tension (59.8 to 91.6 min). A second control experiment (135 min) was followed by reorientation (160 min), congression (303 min), and normal anaphase (331 min). × 975.

tation is anchored to one pole by its chromosomal spindle fibers, and hence an artificial tension toward the opposite pole can be applied with a micromanipulation needle. The necessary controls, in which no tension is applied to the unipolar bivalent, can be done on the same bivalent before, and sometimes also after, a tension experiment.

A typical experiment is shown photographically in Fig. 2 and in graphical form in Fig. 3. The 0.0 min print (Fig. 2) shows the cell before the start of experimentation. A control experiment was then done as follows: the lower half-bivalent was detached and swung toward the upper pole; it soon oriented to that pole, producing unipolar orientation of the bivalent (5.4 min print). The needle was immediately placed within the U-shaped bivalent, nearly touching the bivalent at its closed end. The needle was moved as necessary to keep it in this position as the bivalent moved toward the pole (5.4 to 16.5) and reoriented (16.5 to 22.6 min), thus restoring the original bipolar orientation (Fig. 2, 22.6 min). The bivalent was then detached from the spindle again and unipolar orientation was induced a second time, this time toward the lower pole (51.8 min print). Then a tension experiment was begun by placing the needle at the closed end of the bivalent and moving the needle toward the upper pole (51.9 min). The tension thus created was evident from the increase in bivalent length (compare the 51.8 and 51.9 min prints). The position of the needle was adjusted repeatedly to maintain this slight tension, always gaged by the deformation of the bivalent when the tension is applied. "Before" and "after" pictures of such readjustments are exemplified by two pairs of prints: 59.8 and 60.0 min, and 86.6 and 86.8 min.

Tension was maintained for 39.7 min. The bivalent did not reorient in this interval. After removal of the needle (91.6 min print, Fig. 2) the bivalent moved toward the lower pole (98.4 min print, Fig. 2; 92 to 130 min, Fig. 3). The bivalent did not reorient for 38.2 min after tension was released; it was then detached, unipolarity was induced a third time (135 min, Fig. 2), and the needle was placed in the control position. Reorientation occurred after 15.7 min; a later stage is shown in

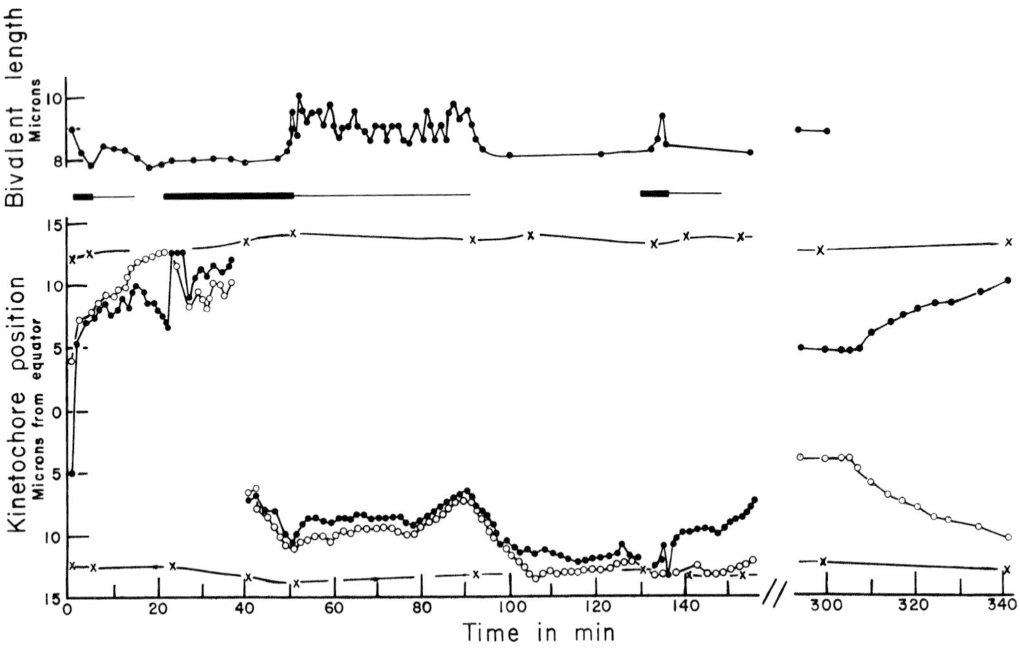

FIGURE 3 Graphical representation of the cell shown in Fig. 2. The heavy horizontal lines beneath the "Bivalent length" plot indicate when an operation to induce unipolarity was in progress; the thin lines indicate, from left to right, the durations of the first control, the tension experiment, and the second control. In the lower graph, the kinetochore positions of the two half-bivalents are indicated by open and closed circles; the positions of the poles are indicated by "X-es". The interval between 157 and 292 min is omitted to reduce the length of the plot.

Fig. 2 at 160 min. The bivalent then returned to the metaphase plate (303 min print), and a normal anaphase followed (331 min print).

The "bivalent length" plot in Fig. 3 provides a general guide to the forces acting on the bivalent. First, it reveals the slight tension normally exerted on bipolar bivalents at prometaphase: contrast the 9 μ length at 0.0 min when the bivalent was oriented normally, with the length of 8 to 8.5 μ when the bivalent had been detached or was unipolar (e.g., from 5 to 45 min, Fig. 3). After the conclusion of the experiments the original length of 9 μ was restored as the bivalent moved back toward the equator and was again under natural forces toward opposite poles (Fig. 3, 300 min). Second, measurements show that artificial tension (Fig. 3, 51 to 92 min) produced, on the average, the same 9μ bivalent length seen in normal bipolar orientation, although the length was far more variable, with a range of 8.5 to 10 μ.

The following general analysis of tension effects is based on experiments on eight bivalents in eight cells, including the cell shown in Fig. 2. Table I gives, for each cell, data in the order the experiments were performed. For the controls and the period after the tension experiment ("post-tension"), the time required for reorientation is given in minutes. Reorientation is recognized by the beginning of rapid motion toward the opposite pole of one or both half-bivalents.

Two new qualitative features emerge when the additional cells are considered. First, a second control experiment was not always performed (see

TABLE I

The Duration of Tension Experiments and the Time Required for Reorientation of Controls without Tension.

Time in minutes. A, anaphase intervened before reorientation; nondisjunction occurred; E, an operation intervened before reorientation. "Post-tension": see text.

Cell number	First control	Tension	Post-tension	Second control
1	11.6	39.7	38.2(E)	15.7
2	1.4	14.8	16.8	
3	19.5	40.0	20.0	22.6
4	13.3	41.5	37.5(A)	
5	11.8	50.3	30.2	
6	5.4	50.0		
7	13.6	46.2	62.0(A)	
8	22.1	29.1	11.3	20.0(A)

TABLE II

Tension Experiments and Controls: Average Time Required for Reorientation.

(Data from Table I)

	Reorientation		No reorientation		Average
	No.	Min per reorientation	No.	total min	min per reorientation
Controls	10	13.7	1	20.0	15.7
Tension	0	—	8	311.6	—
Post-tension	4	19.6	3	137.8	54.0

Table I) because the length of these experiments makes likely the intervention of anaphase or operator fatigue. Second, nondisjunction occurred in cells 4, 7, and 8 (Table I). Not included in the Table (because of the brevity of the tension experiment) is a cell in which anaphase began during the tension experiment itself. Thus, nondisjunction can be a direct consequence of the stable unipolar orientation induced by tension, and probably could be obtained routinely by maintaining tension until anaphase begins.

The general effects of tension become clear when average reorientation times are computed (Table II). Thus for the first and second controls combined, reorientation occurs within 15.7 min, on the average. Yet not a single reorientation occurred during a total of 311.6 min of tension applied to these same bivalents. Tension inhibition of reorientation is confirmed by t-test statistics; here each value for the duration of tension (Table I) is treated as a reorientation time, i.e., as if each experiment were terminated by reorientation (a similar assumption was applied to the second control experiment for cell 8; see Table I). This permits a highly conservative test of the null hypothesis of no increase in reorientation time in the presence of tension. The null hypothesis is rejected ($p > 0.0005$; $t = 5.75$; d.f. $= 17$).

Apparently, the inhibitory effect of tension sometimes persists after tension is released: the average "post-tension" reorientation time is 54.0 min (Table II). Again treating each value in Table I as if reorientation did occur, the null hypothesis of no increase in post-tension reorientation time versus the controls is rejected ($p < 0.01$; $t = 2.92$; d.f. $= 16$). Because in this instance the statistical treatment biases the test in an uncertain direction, certain demonstration of post-tension inhibition of reorientation is not claimed, but the result does justify considering these reorientations in a category separate from the controls.

The time required for reorientation in control experiments is so variable that for some purposes more values seemed desirable. Therefore an additional series of 11 control experiments was done on 6 bivalents in 6 cells. In four of these experiments the micromanipulation needle was not present after the test for unipolarity. No differences were observed from the usual control experiment in which reorientation occurs in the presence of the needle. Some of the variation in reorientation time may be due to micromanipulation, but most is probably of natural origin: in nine untreated crane fly spermatocytes each with a unipolar bivalent, the average time required for reorientation was 11 min with a range from 2 min to over 18 min (ref. 1, page 166).

The combined group of 22 controls was used for three analyses. First, the statistical comparison of controls versus tension duration was repeated, using the approach described above for the smaller group of controls. The mean time required for reorientation was 17.2 min for the combined group of controls. Again the null hypothesis of no increase in time required for reorientation was rejected ($p < 0.005$; $t = 3.55$; d.f. $= 28$). Second, the time required for reorientation in the controls was plotted as a function of the time at which reorientation occurred (minutes before anaphase). No effect was observed: in the interval from 5 hr before anaphase until its inception, there is no trend toward more or less rapid reorientation. This rules out a possible complication in interpreting results from separate experiments in one cell which may extend over 3 hr of prometaphase (e.g., the cell in Figs. 2 and 3). Third, we have considered the possible effect of position within the spindle on reorientation: is reorientation more likely near the poles or in the equatorial region? Bivalents in unipolar orientation move promptly to a pole, and since reorientation occurs only after 17 min on the average, reorientation is far more frequent near a pole than elsewhere on the spindle. But the critical datum is reorientation probability as a function of the time actually spent in each spindle region. This was determined by arbitrarily dividing each half of the spindle into a polar third and an equatorial two-thirds, and computing reorientation time for each region. The data are summarized in Table III for the 20 controls in which reorientation occurred. As expected, few, only 6 of 20, reorientations occurred in the equatorial region. However, the time required for reorientation is not different for the two spindle regions, at least for this rather small sample.

TABLE III
Time Required for Reorientation as a Function of Position on the Spindle

Spindle region	Total time in min	Number of reorientations	Time required (min per reorientation)
Equatorial	80.8	6	13.5
Polar	199.3	14	14.2

Direct Induction of Reorientation

The kinetochores of chromosomes under natural spindle tension are forced to point quite precisely toward a spindle pole. Thus it is possible that this constraint on kinetochore position, rather than the tension itself, produces stable orientation. This possibility has been examined experimentally by determining whether an enforced *change* in kinetochore position leads to a predictable change in orientation. Such an experiment is shown in Fig. 4. The upper half-bivalent (0.0 min print, arrow labeled "1") was first pushed toward the interpolar axis, thus tilting the bivalent. The needle was than placed near the middle of the bivalent and moved toward the upper pole, producing the inverted "J" configuration seen in the 10.5 and 10.7 min prints. Thus the kinetochoric end of half-bivalent number 1 was forced to face the lower pole, but neither half-bivalent was detached from the spindle. This configuration was established at 5.1 min and maintained for 9.0 min; the 14.1 min print shows the bivalent just after the needle was withdrawn. The altered orientation was documented by reinserting the needle and stretching the bivalent toward the upper pole (14.7 min print): the kinetochores of both half-bivalents remained in position and pointed toward the lower pole while the rest of the bivalent was deformed. The needle was then removed from the cell. Both half-bivalents then moved toward the lower pole (15 and 21 min prints), confirming the unipolar orientation. The bivalent then reoriented, returned to the metaphase plate, and divided normally in anaphase (30, 87, and 104 min prints).

Similar experiments have been performed on four additional cells. The results for cell 2 duplicated those in the cell in Fig. 4, but experiments on cell number 3 add final evidence that orienta-

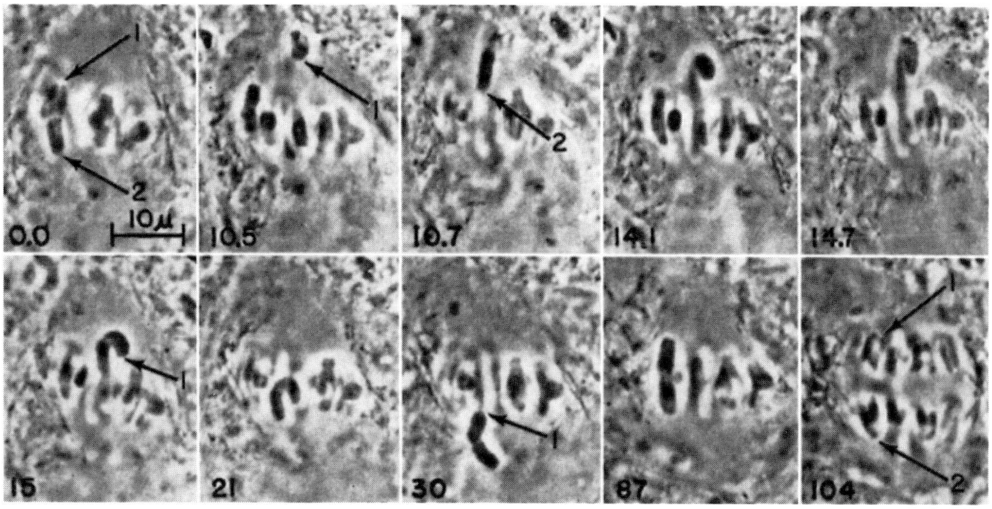

FIGURE 4 Prints from a cinématographic record illustrating direct induction of reorientation. The time in minutes is given on each print. Before the operation (0.0 min), the half-bivalent labeled "1" was oriented to the upper pole, half-bivalent "2" to the lower pole. Without detaching either half-bivalent, needle pressure was applied, and the kinetochoric end of half-bivalent 1 was forced to face the lower pole (10.5, 10.7 min) for 9 min. The needle was then withdrawn (14.1 min) and then reinserted, and the bivalent stretched to demonstrate unipolar orientation (14.7 min). The bivalent then moved toward the lower pole (15, 21 min), the number 1 half-bivalent reoriented (30 min), and normal anaphase followed (104 min). × 1000.

tion can be altered. Here the half-bivalent whose orientation was altered retained the new orientation while the other half-bivalent reoriented; thus, the half-bivalents segregated to different poles than they would have in the absence of experimental intervention. In the fourth and fifth cells of this experimental series, however, reorientation could not be induced in this fashion. No particular significance can be attached to these failures because the appropriate kinetochore position simply could not be produced and sustained. It is not known whether this represents important biological variation or insignificant technical failure. The positive result bears emphasis: whenever a half-bivalent's kinetochores can be forced to face the opposite pole for five minutes or longer, reorientation to that pole invariably follows. This orientation change is induced directly from one pole to the other without first detaching the bivalent from the spindle.

The experiment in a sixth cell merits separate treatment. The general experimental procedure duplicated that described for the cell in Fig. 4, but after tilting the bivalent on the spindle, much less force was applied toward the upper pole. This force was sufficient to swing the kinetochoric end of the upper half-bivalent nearly perpendicular to the spindle's axis (but not sufficient to make that end face the lower pole as in Fig. 4). The bivalent was held in this position for 40 min; after 20 min the upper half-bivalent had amphioriented (one chromatid oriented to one pole, the other to the opposite pole, as in mitosis). This orientation persisted, and in anaphase the lower half-bivalent moved normally to the lower pole, while the two chromatids of the upper half-bivalent separated and moved to opposite poles in late anaphase (this orientation and segregation pattern has been seen in earlier studies, e.g., Fig. 11 in ref. 18). This and related experiments with similar results were planned to test the prediction that decreasing the natural spindle tension on bivalents in bipolar orientation should lead to reorientation. In the experiment just described, the upper half-bivalent was relieved of natural spindle tension toward the lower pole and, consonant with the tension hypothesis of orientation stability, it was this half-bivalent which reoriented. But the outcome is ambiguous: the needle position which relieves tension also swings the kinetochoric end of the upper half-bivalent nearly perpendicular to the interpolar axis. Therefore, the observed amphiorientation might be due to direct induction, and the experiment is described in this section for that reason. Parenthetically, our numerous other attempts to induce reorientation by relieving tension should be noted. These have all failed, usually because natural spindle readjustments rapidly restore tension following needle-induced relaxation.

DISCUSSION

We conclude first that applied tension certainly inhibits the reorientation of bivalents in unipolar orientation. This is most dramatically evidenced by the total of over 5 hr with tension without a single reorientation, while in the absence of tension the same bivalents reoriented from unipolar orientation within 16 min on the average. Second, we conclude that this result is relevant to normal reorientation processes. Thus, the applied tension evidently approximates that produced by the spindle on bivalents in bipolar orientation as judged by the similar length increases; by this criterion artificial tension is more variable than, but does not often exceed, natural tension (Fig. 3). Also, unnatural factors other than tension, such as the presence of the needle near the bivalent, are duplicated in the controls, which reorient normally. Therefore, these experiments provide strong evidence that the differences in orientation stability which make possible controlled chromosome distribution arise from differences in spindle tension. Mal-oriented bivalents will reorient until bipolar orientation is achieved more or less by chance. They are then subjected to forces toward opposite poles normally associated only with bipolarity and which somehow confer stability.

Tension effectively stabilizes orientation even in cells where orientation is abnormally unstable. Dr. S. Alan Henderson has discovered that exposure to 10°C before examination at room temperature produces very frequent mal-orientation in *Melanoplus* spermatocytes, and the mal-oriented bivalents orient and reorient several times before achieving a stable, bipolar orientation. Unipolar orientations in cold treated cells are as clearly stabilized by artificial tension as the cells studied in this report, but the effect of tension is more striking because of the frequent reorientations immediately before and after the experiment (7).

We have identified spindle tension as the source of orientation stability, but is tension itself or some secondary effect of tension the immediate cause of stability? One secondary effect is altered position on the spindle: bivalents in unipolar orientation come to lie close to a pole, and either natural or artificial tension leads to a location nearer the equator. For instance, in Fig. 3 compare the distance from the pole in the presence of artificial tension (55 to 90 min) with that in the absence of tension (50 and 110 min). Position probably can be rejected as a significant influence, however. As argued earlier (18), polar position is at best an ancillary stimulus to reorientation. For the present (rather small) sample, reorientation is not more frequent near a pole than closer to the equator when computed on a minutes per reorientation basis (see Table III and associated text). The secondary effect of tension that cannot be eliminated as a possible cause of stability is its effect on the axial position of kinetochores. Thus tension forces the kinetochores into alignment with the pole-to-pole axis, with each kinetochore pointing directly at a pole, as long as tension is maintained and the spindle fibers are intact. The "direct induction of reorientation" experiments (see Fig. 4) suggest that this might indeed preclude reorientation, for changed orientation is readily induced by forcing a kinetochore to face a different pole. Thus, we are left with the alternatives that tension may act through influencing kinetochore position and/or directly by increasing spindle fiber stability.

The exact effect of the *absence* of tension in unstable orientations is equally obscure and this reveals our ignorance of the reorientation process. Reorientation involves both the loss of old spindle attachments and the formation of new ones (e.g., 18, 3). For instance, we do not know whether loss and reformation must occur sequentially or whether some connections to one pole persist while connections to the other pole are forming. If the latter is the case, then constraints on kinetochore position may be very important, but if loss must precede reformation, then attachment stability itself must be most critical. Moreover, if spindle attachments are lacking, a kinetochore orients preferentially to the pole it most nearly "faces" (18), and this makes reorientation indeed mysterious since the reorientation from unipolar to bipolar orientation involves at least a partial turning of a kinetochore to face a new pole. Merely a higher probability of spindle fiber loss will not suffice to explain such a reorientation: we would predict only repeated loss and reformation of spindle fiber connections to the same pole. We suggest that in addition to enhanced spindle fiber instability, the slight tilting of kinetochores with respect to the spindle axis is also involved. This might lead to some spindle connections to both poles, followed occasionally by motion toward the opposite pole, by increasing concentration of spindle connections toward that pole, and finally by stable bipolar orientation. The "direct induction" experiments are consistent with this suggestion, and the studies of Luykx (16) provide some ultrastructural evidence for bipolar microtubular arrays on single half-bivalents during early prometaphase. More decisive ultrastructural observations, on chromosomes in known stages of reorientation, currently are being obtained in collaboration with Dr. B. R. Brinkley.

Spindle tension may suffice as a general explanation of controlled chromosome reorientation. First, it is clearly applicable to ordinary chromosomes in mitosis. That reorientation occurs in mitosis is suggested by observations on "centrophilic" chromosomes of newt cells in tissue culture (2). Equal distribution of chromosomes in mitosis depends upon orientation of daughter kinetochores to opposite poles and the two daughter chromosomes are mechanically linked together until anaphase. Thus, all relevant mechanical properties are identical with those in meiotic bivalents, and, therefore, we postulate that stable orientation depends upon spindle tension produced following orientation to opposite poles. Reorientation will occur until this orientation is reached. Tension experiments have not been performed on mitotic chromosomes because the short distance between the kinetochores makes difficult both the operations and the observation of orientation patterns.

Second, meiotic units other than bivalents may be considered. Unpaired chromosomes (univalents) which orient both chromatid kinetochores to the same pole would be expected to reorient frequently, and this is observed with the unpaired x and y chromosomes of several tipulid files (4, 1), and with the x chromosome of *Melanoplus* (17). The former show significantly a transition to stable orientation after orientation of chromatids to opposite poles. Similar behavior when chromosomes, which normally form bivalents, are unpaired can be inferred from studies on fixed cells (reviewed by John and Lewis, ref. 12, pp. 52 ff).

Many alternative explanations of orientation in single chromosomes in mitosis and meiosis are possible but no satisfactory explanation of meiotic multivalent orientation has yet been proposed. The major orientation patterns in such linked associations of three or more chromosomes have conventionally been designated "alternate" or "adjacent"; some examples are diagrammed in Fig. 5. In adjacent orientations, the kinetochores of two or more adjacent chromosomes are oriented to the same pole (e.g. Fig. 5 *a* and *b*), while in alternate orientations (Fig. 5 *c*) there is a strict alternation of one chromosome to one pole, the next to the opposite pole, and so on, producing a characteristic "zig-zag" appearance of the multivalent at metaphase. Without exception, where multivalents persist as part of the normal cytogenetic system, they show alternate orientation in a very high proportion of the metaphase I cells. Our major task is to account for this preferential orientation. The general argument is that adjacent orientation is similar to unipolar orientation in

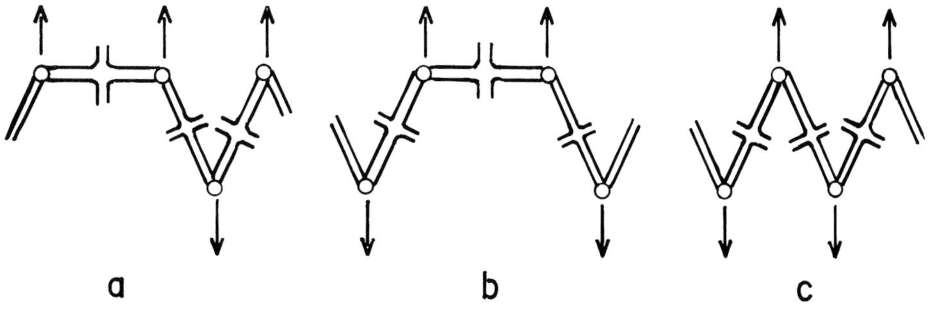

FIGURE 5 Three diagrams of orientation in multivalents, *a* and *b* showing adjacent, *c* showing alternate orientation. Kinetochores are represented by circles; the arrows indicate the direction of orientation. Chromosomes with median kinetochores and one chiasma per arm (as in *Oenothera* and *Rhoeo*) are represented in these examples. The diagrams may be regarded as representing either chains of four chromosomes or sections of still higher multivalents.

bivalents in reducing the tension toward opposite poles on chromosomes so oriented, while alternate orientation leads to uniform forces toward cpposite poles on each chromosone in the multivalent, and, therefcre, the alternate orientation is uniquely stable.

Trivalents of sex-chromosomes occur widely in animals, and in mantids it is known that their controlled segregation depends upon reorientation of L-shaped adjacent or linear orientations to V-shaped alternate orientations (10). Our tension experiments on bivalents mimic the V-configuration: the ends of the "V" are the kinetochores of the mal-oriented bivalent and the needle at the apex of the "V" assumes the role of the third chromosome's kinetochores (Dr. S. Alan Henderson first suggested this interpretation). The indefinite stability of this artificial trivalent is direct evidence for a role of spindle tension in natural trivalents. We have also attempted to imitate trivalents in adjacent crientation by applying the needle near the middle of a properly oriented bivalent and moving the needle toward a pole (cell 6, direct induction experiments, see p. 46). The prediction is specific: the half-bivalent oriented to that pole should reorient. This is observed, although reorientation to produce a stable V-shaped artificial trivalent has not been produced unless the needle is moved as far as shown in Fig. 4. Thus so far we have reproduced only one possible component of adjacent-trivalent instability which in natural trivalents would be an effect of variable forces at the middle chromosome's kinetochores on the *position* of the kinetochores of the chromosome oriented to the same pole.

Still higher multivalents are part of the normal cytogenetic system in such plant genera as *Oenothera* and *Rhoeo* (reviewed in ref. 12) where chains or rings of up to 14 chromosomes regularly show alternate orientation. Equally regular alternate orientation occurs in the recently discovered sex quadrivalent in the beetle *Cyrsylus volkameriae* (22). The stability of alternate orientation (Fig. *5 c*) and the instability of most adjacent orientations in higher multivalents are interpretable as effects of tension or its absence. For instance, if three interstitial chromosomes are oriented to the same pole, the middle chromosome is not subjected to bipolar forces and its reorientation is expected. Similarly, if two chromosomes at one end of a chain multivalent are oriented to the same pole (Fig. 5 *a*), the reorientation of the end chromosome is predicted. However, one class of adjacent orientations can not be interpreted at present: adjacent orientation of two interstitial chromosomes, as in Fig. 5 *b*, appears to produce the force distribution seen in alternate orientation (Fig. 5 *c*). Quadrivalents or higher multivalents have not yet been studied in living cells; only such studies will permit a final decision on the applicability of the tension hypothesis to multivalent orientation.

The discussion of multivalent orientation can be summarized as follows: we have experimental evidence suggesting a role cf tension in trivalents, and many, but not all, other configurations can be understood. An important feature of the interpretation is that no very special features of chromosomes or spindles are invoked to explain regular alternate orientation. Thus, ordinary bivalents in tension experiments behave as if they were part of a trivalent. In nature, newly arisen multivalents frequently show preferential alternate orientation (e.g. 15, 8, 20) as expected from our interpretation. Obviously, however, certain types of chromosome morphology, chiasma distribution, and general reorientation frequency are more or less conducive to regular attainment of alternate orientation. Therefore, it is not surprising that alternate orientation is not universal in multivalents of recent origin (for examples, see especially refs. 9, 1, and the review of John and Lewis, ref. 12), nor is it surprising that selection for alternate orientation is possible, as shown by studies of Thompson (21) and Lawrence (13, 14) on rye. However, it remains to be seen whether or not the tension hypothesis is adequate for the detailed explanation of these situations.

We close by stressing the experimentally verified role of spindle tension in the orientation of bivalents in meiosis and the easy extension of this explanation to mitotic chromosomes. However, the important details may be resolved, we already understand the general cytological basis of controlled reorientation and hence of controlled chromosome distribution. This cytological understanding is the necessary basis for progressing to an explanation of chromosome distribution in molecular terms. The molecular interpretation of tension and orientation stability may seem so mysterious that a possible route to understanding is worth outlining. Micromanipulation c early shows that each chromosome is individually associated with the spindle, probably by direct connections between its chromosomal spindle fibers and those of the rest of the spindle (19). These connections must produce, or at least transmit, the forces that

move chromosomes. Now in skeletal muscle, mechanical properties vary with functional status: active force production is associated with a resistance to stretch not present in the relaxed state. There is increasingly good reason to think that the resistance to stretch results from formation and/or altered properties of bridges between actin and myosin filaments (reviewed by Huxley and Hanson, ref. 11). No certain analogy between spindle and muscle is, or need be, suggested. Here, muscle merely provides a well-studied example of an association between force production and mechanical properties that could be related to the stability of the system. This suggests a possible molecular interpretation of one aspect of spindle tension: its direct effect on spindle stability may arise because tension is produced or transmitted by linkages which also determine the structural integrity, the stability, of spindle units.

This investigation was supported in part by research grant GM-13745 from the Division of General Medical Sciences, United States Public Health Service.
Received for publication 6 March 1969, and in revised form 19 May 1969.

REFERENCES

1. BAUER, H., R. DIETZ, and C. RÖBBELEN. 1961. Die Spermatocytenteilungen der Tipuliden. III. Das Bewegungsverhalten der Chromosomen in Translokationsheterozygoten von *Tipula oleracea*. *Chromosoma*. **12**:116.
2. BLOOM, W., R. E. ZIRKLE, and R. B. URETZ. 1955. Irradiation of parts of individual cells. III. Effects of chromosomal and extrachromosomal irradiation on chromosome movements. *Ann. N.Y. Acad. Sci.* **59**:503.
3. BRINKLEY, B. R., and R. B. NICKLAS. 1968. Ultrastructure of the meiotic spindle of grasshopper spermatocytes after chromosome micromanipulation. *J. Cell Biol.* **39**:16a.
4. DIETZ, R. 1956. Die Spermatocytenteilungen der Tipuliden. II. Graphische Analyse der Chromosomenbewegung währen der Prometaphase I in Leben. *Chromosoma*. **8**:183.
5. DIETZ,R. 1958. Multiple Geschlechchromosomen bei den cypriden Ostracoden, ihre Evolution und ihr Teilungsverhalten. *Chromosoma*. **9**:359.
6. ELLIS, G. 1962. Piezoelectric micromanipulators. *Science (Washington)*. **138**:84.
7. HENDERSON, S. A., R. B. NICKLAS, and C. A. KOCH. 1969. Temperature-induced orientation instability during meiosis: an experimental analysis *J. Cell Sci.* In press.
8. HEWITT, G., and G. SCHROETER. 1968. Population cytology of *Oedaleonotus*. I. The karyotypic facies of *Oedaleonotus enigma* (Scudder). *Chromosoma*. **25**:121.
9. HUGHES-SCHRADER S. 1943. Meiosis without chiasmata—in diploid and tetraploid spermatocytes of the mantid *Callimantis antillarum* Saussure. *J. Morphol.* **73**:111.
10. HUGHES-SCHRADER, S. 1943. Polarization, kinetochore movements, and bivalent structure in the meiosis of male mantids. *Biol. Bull. (Woods Hole)*. **85**:265.
11. HUXLEY, H. E., and J. HANSON. 1960. The molecular basis of contraction in cross-striated muscles. *In* The structure and function of muscle. G. H. Bourne, editor. Academic Press Inc., New York. **1**:183.
12. JOHN, B., and K. R. LEWIS. 1965. The meiotic system. *Protoplasmatol. Handb. Protoplasmaforsch.* VI/F/1. Springer-Verlag. Wein. 1.
13. LAWRENCE, C. W. 1958. Genotypic control of chromosome behavior in rye. VI. Selection for disjunction frequency. *Heredity*. **12**:127.
14. LAWRENCE, C. W. 1963. The orientation of multiple associations resulting from interchange heterozygosity. *Genetics*. **48**:347.
15. LEWIS, K. R., and B. JOHN. 1957. Studies on *Periplaneta americana*. II. Interchange heterozygosity in isolated populations. *Heredity*. **11**:11.
16. LUYKX, P. 1965. Kinetochore-to-pole connections during prometaphase of the meiotic divisions in *Urechis* eggs. *Exp. Cell Res.* **39**:658.
17. NICKLAS, R. B. 1961. Recurrent pole-to-pole movements of the sex chromosome during prometaphase I in *Melanoplus differentialis* spermatocytes. *Chromosoma*. **12**:97.
18. NICKLAS, R. B. 1967. Chromosome micromanipulation II. Induced reorientation and the experimental control of segregation in meiosis. Chromosoma. **21**:17.
19. NICKLAS, R. B., and C. A. STAEHLY. 1967. Chromosome micromanipulation I. The mechanics of chromosome attachment to the spindle. *Chromosoma*. **21**:1.
20. SYBENGA, J. 1968. Orientation of interchange multiples in *Secale cereale*. *Heredity*. **23**:73.
21. THOMPSON, J. B. 1956. Genotypic control of chromosome behavior in rye. II. Disjunction at meiosis in interchange heterozygotes. *Heredity*. **10**:99.
22. VIRKKI, N. 1968. A chiasmate sex quadrivalent in the male of an alticid beetle, *Cyrsylus volkameriae*. *Can. J. Genet. Cytol.* **10**:898.

Hartwell L.H., Culotti J., Pringle J.R., and **Reid B.J.** 1974. Genetic control of the cell division cycle in yeast. *Science* **183:** 46–51.

It had long been recognized that cell proliferation was based on a cycle of alternating phases of growth and division. The important descriptive literature that characterized many of the underlying processes was reviewed by JM Mitchison in 1972 (1). An understanding of the control pathways for cell cycle regulation came comparatively slowly, however, since it depended on both the descriptive work and on some sense of the genes involved. The latter advanced a major step with the publication of the paper reproduced here. Hartwell brought a geneticist's point of view to cell cycle study. He used the physiology of budding yeast to screen for temperature sensitive mutants in which some essential aspect of cell cycle progression was not properly executed, e.g., the onset of DNA synthesis, the separation of the chromosomes, or the completion of cytokinesis. To eliminate mutations that simply shut down metabolism, he required that cell growth would continue at the restrictive temperature, as shown by continued enlargement of the cell's bud, even though some aspect of cell cycle progression was arrested. The phenotypes of such mutants in the "cell division cycle" (cdc) provided a framework for thinking about control of the cell cycle. The actions of some genes were found to be dependent on the prior function of others, in the sense that one was required for the cell to progress to the point that the function of the second was important or even possible. Other genes appeared to fall into independent pathways, patterns familiar from the study of phage morphogenesis. For example, some strains that were temperature sensitive for cytokinesis would continue at the restrictive temperature to form buds and initiate additional rounds of DNA synthesis and mitosis. From these and other studies on the interactions among the cdc mutants (2), considerable progress was made in figuring out the logic of cell cycle control. It did, however, require the emergence of molecular methods by which these genes could be cloned, sequenced, expressed, and their products studied biochemically to reach a deep understanding of cell cycle control (see, e.g., Gautier et al., this volume).

---

1. Mitchison J.M. 1972. *The biology of the cell cycle*, Cambridge University Press, New York.
2. Nurse, P., Thuriaux P., and Nasmyth K. 1976. Genetic control of the cell division cycle in the fission yeast *Schizosaccharomyces pombe*. *Mol. Gen. Genet.* **146:** 167–178

Reprinted from
11 January 1974, Volume 183, pp. 46-51

SCIENCE

Genetic Control of the Cell Division Cycle in Yeast

Leland H. Hartwell, Joseph Culotti,
John R. Pringle, Brian J. Reid

Genetic Control of the Cell Division Cycle in Yeast

A model to account for the order of cell cycle events is deduced from the phenotypes of yeast mutants.

Leland H. Hartwell, Joseph Culotti, John R. Pringle, Brian J. Reid

Mitotic cell division in eukaryotes is accomplished through a highly reproducible temporal sequence of events that is common to almost all higher organisms. An interval of time, *G1*, separates the previous cell division from the initiation of DNA synthesis. Chromosome replication is accomplished during the DNA synthetic period, *S*, which typically occupies about a third of the cell cycle. Another interval of time, *G2*, separates the completion of DNA synthesis from prophase, the beginning of mitosis, *M*. A dramatic sequence of changes in chromosome structure and of chromosome movement characterizes the brief mitotic period that results in the precise separation of sister chromatids to daughter nuclei. Mitosis is followed by cytokinesis, the partitioning of the cytoplasm into two daughter cells with separate plasma membranes. In some organisms the cycle is completed by cell wall separation.

Each of these events occurs during the cell division cycle of the yeast, *Saccharomyces cerevisiae (1)* (Fig. 1). However, two features which distinguish the cell cycle of *S. cerevisiae* from most other eukaryotes are particularly useful for an analysis of the gene functions that control the cell division cycle. First, the fact that both haploid and diploid cells undergo mitosis permits the isolation of recessive mutations in haploids and their analysis by complementation in diploids. Second, the daughter cell is recognizable at an early stage of the cell cycle as a bud on the surface of the parent cell. Since the ratio of bud size to parent cell size increases progressively during the cycle, this ratio pro-

Dr. Hartwell is a professor, Mr. Culotti and Mr. Reid are graduate students, and Dr. Pringle is a postdoctoral fellow in the department of genetics, at the University of Washington, Seattle 98195.

vides a visual marker of the position of the cell in the cycle.

We have taken advantage of these features of the *S. cerevisiae* cell cycle in the isolation and characterization of 150 temperature-sensitive mutants of the cell division cycle (*cdc* mutants). These mutants are temperature-sensitive in the sense that they are unable to reproduce at 36°C (the restrictive temperature) but do grow normally at 23°C (the permissive temperature); the parent strain from which they were derived reproduces at both temperatures. These mutations define 32 genes, each of whose products plays an essential role in the successful completion of one event in the mitotic cycle (*2*). Although our genetic dissection of the cell cycle is in its early stages, the phenotypes of the mutants already examined provide information on the interdependence of events in the cycle. We shall discuss the conclusions that can be derived from the mutant phenotypes in the context of the following question: How are the events bud emergence, initiation of DNA synthesis, DNA synthesis, nuclear migration, nuclear division, cytokinesis, and cell separation coordinated in the yeast cell cycle so that their sequence is fixed? While it is not necessarily the case that all events in the cell cycle are ordered relative to one another in a fixed sequence, it is reasonable to assume that these events are, since their proper order is essential for the production of two viable daughter cells.

It has been pointed out by Mitchison that two possible mechanisms exist for ordering a fixed sequence of cell cycle events relative to one another (*3*). First, there may be a direct causal connection between one event and the next. In this case, it would be necessary for the earlier event in the cycle to be completed before the later event could occur. For example, the "product" of the earlier event might provide the "substrate" for the later event, as in a biochemical pathway, or the completion of the earlier event might activate the occurrence of the later event. We shall refer to this model as the "dependent pathway model" (Fig. 2).

A second possibility is that there is not a direct causal connection between two events, but that they are ordered by signals from some master timing mechanism. In this model it would not be necessary for the earlier event to be completed before the later event could occur, although the two events

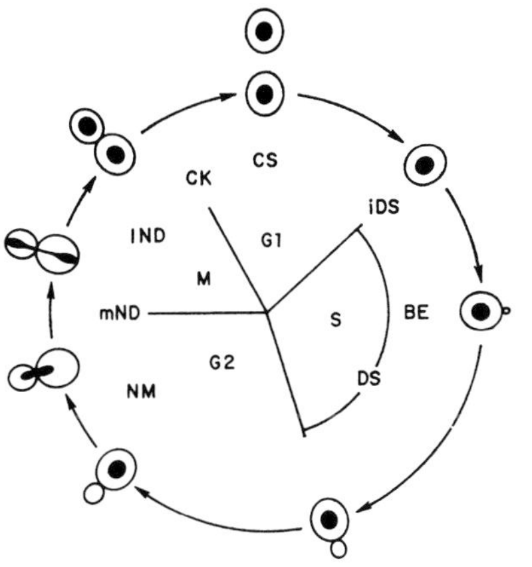

Fig. 1. The sequence of events in the cell division cycle of yeast: *iDS*, initiation of DNA synthesis; *BE*, bud emergence; *DS*, DNA synthesis; *NM*, nuclear migration; *mND*, medical nuclear division; *lND*, late nuclear division; *CK*, cytokinesis; *CS*, cell separation. Other abbreviations: *G1*, time interval between previous cytokinesis and initiation of DNA synthesis; *S*, period of DNA synthesis; *G2*, time between DNA synthesis and onset of mitosis; and *M*, the period of mitosis.

would normally occur in the proper order because of the activity of the timer. This model has appeared frequently in the literature in one guise or another, and two specific ideas have been presented concerning a possible timing mechanism. One invokes the accumulation of a specific division protein (*4*) and another a temporal sequence of genetic transcriptions (*5*). We shall refer to this model as the "independent pathways model" (Fig. 2).

It is important to note that these two possible models relate, strictly speaking, to the dependence or independence of events in the cell cycle taken two at a time. It is quite possible that the cell cycle is controlled by a combination of the two models, with some events related to one another in a dependent

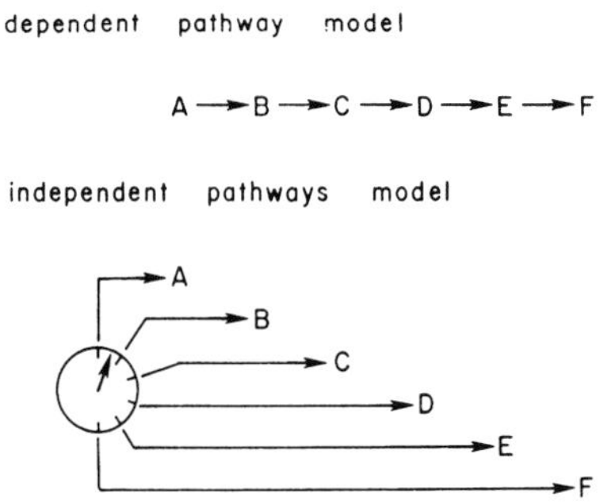

Fig. 2. Two models to account for the ordering of cell cycle events.

pathway and others in independent pathways.

It should be possible to distinguish between these fundamentally different models by specifically inhibiting one and only one event of the cell cycle. If an event is dependent upon the prior occurrence of an earlier event, a specific block of the earlier event should prevent the occurrence of the later event. If, on the other hand, the two events are independent of one another, then a specific block of the earlier event should not prevent the occurrence of the subsequent event. Indeed, studies employing inhibitors that act specifically on one event of the cycle, such as DNA synthesis or mitosis, have already provided some information on the interdependence of cell cycle events. However, the temperature-sensitive *cdc* mutants of *S. cerevisiae* permit more detailed conclusions, both because of the greater number of specific cell cycle blocks in a single organism and because of the greater assurance that a single gene defect directly affects one and only one event in the cell cycle.

Mutations Affecting the Cell Cycle

Cell division cycle mutants of *S. cerevisiae* were detected among a collection of temperature-sensitive mutants by looking for mutants in which development was arrested at the restrictive temperature at a specific stage in the cell cycle, as evidenced by the cellular and nuclear morphology (*6*). The phenotype of each mutant class is described in Table 1 by the sequence of events that occurs in a cell when it is shifted from the permissive temperature to the restrictive temperature at the beginning of the cell division cycle (that is, at cell separation, see *CS* in Fig. 1). The initial defect in a mutant is defined as the first cell cycle event (among those which can presently be monitored) that fails to take place at the restrictive temperature. The events for which initial defects have been found in mutants include the initiation of DNA synthesis, bud emergence, DNA synthesis, medial nuclear division, late nuclear division, cytokinesis, and cell separation. Information on the interdependence of steps in the cell division cycle is obtained by observing which events in the first cell cycle at the restrictive temperature occur or do not occur after arrest at the initial defect.

Two dependent pathways in the cycle. The model of the cell division cycle presented in Fig. 3 can be derived from the phenotypes of the mutants (Table 1) by the following reasoning. First, let us compare the phenotypes of these mutants with the predictions of the dependent pathway model. Working backwards through the cell cycle we see that this model is adequate for the sequence: cell separation, cytokinesis, late nuclear division, medial nuclear division, DNA synthesis, and the initiation of DNA synthesis. A mutant with an initial defect in any one of these six processes fails to complete any of the other events in this group which normally occurs later in the cycle. The simplest explanation of these observations is that these six events comprise a dependent pathway in which the completion of each event is a necessary prerequisite for the occurrence of the immediate succeeding event (Fig. 3).

In contrast, although bud emergence and DNA synthesis normally occur at about the same time in the cell cycle, they must be on separate pathways (Fig. 3) since they are independent of one another. Mutants defective in the initiation of DNA synthesis (*cdc* 4 and *cdc* 7) or in DNA synthesis (*cdc* 8 and *cdc* 21) undergo bud emergence, and mutants defective in bud emergence (*cdc* 24) undergo DNA synthesis. Furthermore, inhibitors that block DNA synthesis (hydroxyurea and trenimon) do not inhibit bud emergence (*7*).

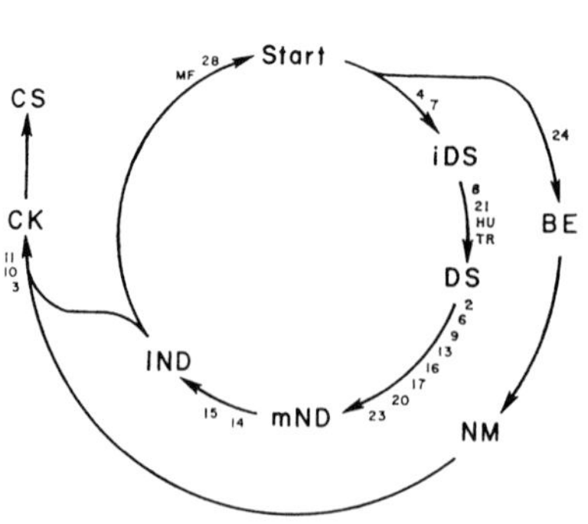

Fig. 3. The circuitry of the yeast cell cycle. Events connected by an arrow are proposed to be related such that the distal event is dependent for its occurrence upon the prior completion of the proximal event. The abbreviations are the same as in Fig. 1. Numbers refer to *cdc* genes that are required for progress from one event to the next; *HU* and *TR* refer to the DNA synthesis inhibitors hydroxyurea and trenimon, respectively; *MF* refers to the mating factor, α factor.

Although we do not have mutants with initial defects in nuclear migration, it is apparent that this event, like the bud emergence event, occurs in all mutants defective in the initiation of DNA synthesis and in DNA synthesis (Table 1). Nuclear migration also occurs when DNA synthesis is inhibited with hydroxyurea or trenimon (*7, 8*). Nuclear migration is therefore independent of initiation of DNA synthesis and DNA synthesis. Furthermore, since the nucleus normally migrates into the neck between the bud and parent cell, it seems reasonable to suppose that nuclear migration is dependent upon bud emergence. We propose, therefore, that nuclear migration is an event on the same pathway as bud emergence and subsequent to it on this pathway (Fig. 3).

Finally, medial and late nuclear division are completed in the mutant defective in bud emergence (*cdc* 24) but neither cytokinesis nor cell separation occurs in this mutant. These observations suggest that the separate pathway that leads to bud emergence and nuclear migration joins the first pathway at the event of cytokinesis (Fig. 3). Thus, cytokinesis and cell separation are dependent upon bud emergence as well as upon nuclear division.

A common step controls both pathways. Although bud emergence is not necessary for the initiation of DNA synthesis, and vice versa, the product of gene *cdc* 28 is required for both processes (Table 1). Furthermore, the mating factor produced by cells of mating type α (α factor) blocks both bud emergence and the initiation of DNA synthesis in cells of mating type *a* (*9, 10*). One hypothesis to explain these observations is that the two pathways leading, respectively, to bud emergence and to initiation of DNA synthesis diverge from a common pathway, and that both the *cdc* 28 gene product and the α factor sensitive step are elements of this common pathway.

A prediction of this hypothesis is that the α factor sensitive step and the step mediated by the *cdc* 28 gene product should precede and be required for the *cdc* 4 and *cdc* 7 mediated steps that lead to the initiation of DNA synthesis and for the *cdc* 24 mediated step that leads to bud emergence. Both of these predictions have been confirmed (*11*). We assume, therefore, that the *cdc* 28 gene product and the α factor sensitive step mediate some early event or events in the cell cycle that are necessary prerequisites for both of the dependent pathways described above.

Table 1. Summary of mutant phenotypes. Cells were shifted from 23° to 36°C at the time of cell separation. Abbreviations are as in Fig. 1. A minus sign indicates that an event does not occur, a plus indicates that the event occurs once, and a double plus indicates that the event occurs more than once.

*cdc**	Initial defect	Events completed at restrictive temperature								
		BE	iDS	DS	NM	mND	lND	CK	CS	Reference
28	Start	–	–	–	?	–	–	–	–	(*18*)
24	BE	–	++	++	?	++	++	–	–	(*2*)
4	iDS	++	–	–	+	–	–	–	–	(*16, 18*)
7	iDS	+	–	–	+	–	–	–	–	(*18*)
8	DS	+	+	–	+	–	–	–	–	(*16, 18*)
21	DS	+	+	–	+	–	–	–	–	(*18*)
2	mND	+	+	+	+	–	–	–	–	(*17*)
6	mND	+	+	+	+	–	–	–	–	(*17*)
9	mND	+	+	+	+	–	–	–	–	(*17*)
13	mND	+	+	+	+	–	–	–	–	(*17*)
16	mND	+	+	+	+	–	–	–	–	(*2*)
17	mND	+	+	+	+	–	–	–	–	(*2*)
20	mND	+	+	+	+	–	–	–	–	(*2*)
23	mND	+	+	+	+	–	–	–	–	(*2*)
14	lND	+	+	+	+	+	–	–	–	(*17*)
15	lND	+	+	+	+	+	–	–	–	(*17*)
3	CK	++	++	++	++	++	++	–	–	(*19*)
10	CK	++	++	++	++	++	++	–	–	(*19*)
11	CK	++	++	++	++	++	++	–	–	(*19*)
	CS	++	++	++	++	++	++	++	–	

* Although mutations in 32 *cdc* genes have been discovered, only 19 of these genes are included here for consideration in developing a model of the cell cycle. Most of those not included were left out because they progress through several cycles at the restrictive temperature before development is arrested and this prevents an analysis of DNA synthesis during their terminal cycle. The mutant *cdc* 1 (*19*) was excluded because macromolecule synthesis, as well as bud emergence, is rapidly arrested in this mutant at the restrictive temperature, and we suspect that this inhibition of growth prevents the occurrence of some events which are not normally dependent upon bud emergence, but which are dependent on growth.

We shall term this event "start" (Fig. 3). In principle, completion of the "start" event can be monitored by the acquisition of insensitivity to α factor in haploids of mating type *a* or by the acquisition of insensitivity to temperature in a *cdc* 28 mutant, although this is not possible in all experimental situations.

Several observations suggest that "start" is in fact the beginning of the yeast cell cycle. First, stationary phase populations obtained by limiting any one of several nutrients (glucose, ammonia, sulfate, phosphate) consist almost exclusively of cells which are arrested at a point in the cell cycle after cell separation, but prior to bud emergence and the initiation of DNA synthesis (*12*). Stationary phase cells of mating type *a* do not undergo bud emergence after inoculation into fresh medium in the presence of α factor. It appears, therefore, that as yeast cells exhaust their nutrients they finish cell cycles and become arrested prior to "start" in the cell cycle.

Similarly, when cultures are grown with limited glucose in a chemostat there is a striking correlation between the generation time and the proportion of unbudded cells in the population (*13*). The increase in the proportion of unbudded cells as the generation time increases suggests that the unbudded cells delay the "start" of new cycles until some requirement for growth or for the accumulation of energy reserves (*14*) has been met, and that, as expected, the time necessary to meet this requirement is a function of the rate of supply of glucose.

Finally, passing "start" in the cell cycle appears to represent a point of commitment to division, as opposed to mating, for haploid cells. If a cell of mating type *a* is beyond "start" in the cell cycle at the time of exposure to α factor, it proceeds through the cell cycle to cell separation at a normal rate, and then both daughter cells become arrested at "start" (*10*). Furthermore, *cdc* mutants arrested at various positions in the cycle are unable to mate with cells of opposite mating type, with two exceptions: mutants that are arrested at "start" (*cdc* 28), and mutants that repeatedly pass through "start" at the restrictive temperature (as evidenced by the attainment of a multinucleate state, Table 1) appear to mate relatively well (*15*).

Events Necessary for "Start"

Let us consider what events of one cell cycle must be completed in order to permit the "start" event of the next cell cycle. Since in these experiments we could not monitor "start" directly, our conclusions regarding this event must be considered tentative. However, it appears from the following observations that an initial defect in cell separation, cytokinesis, or bud emergence does not prevent a cell from undergo-

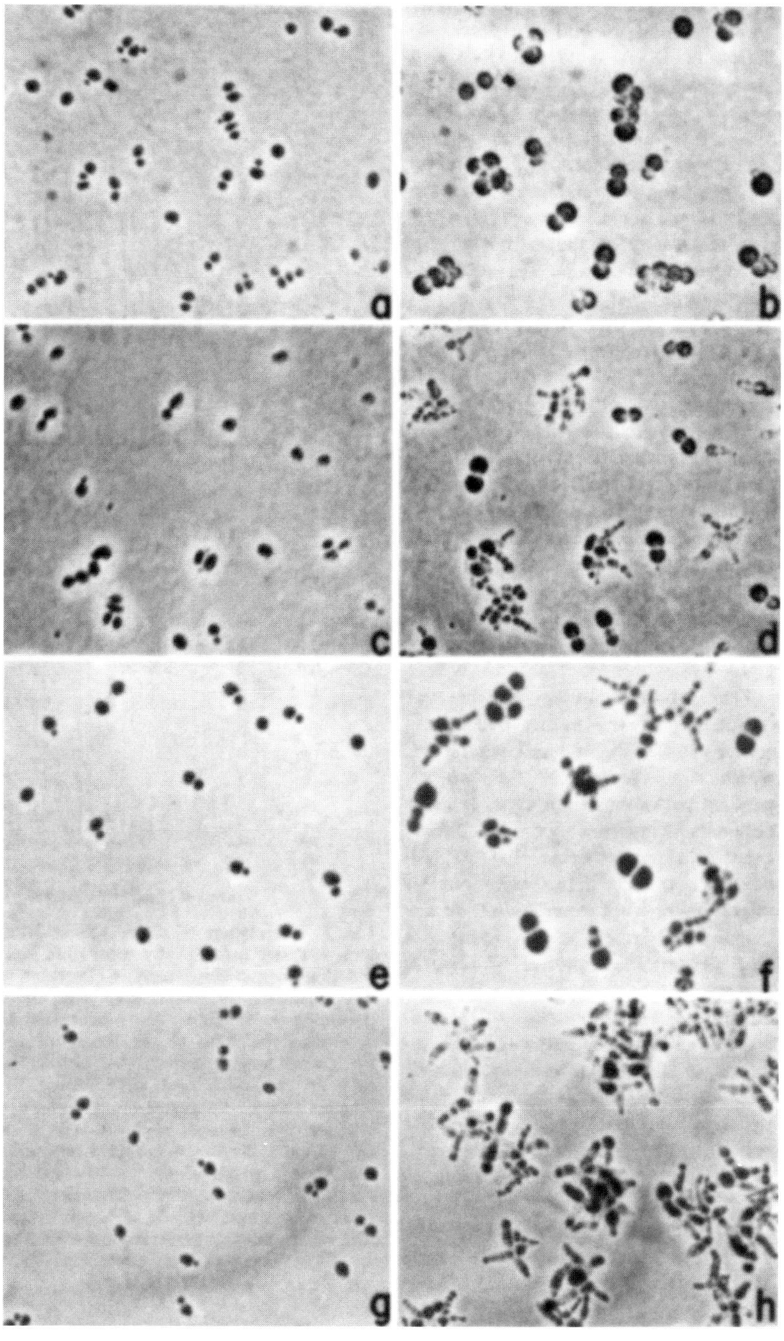

Fig. 4. Time-lapse photographs of diploid strains homozygous for two *cdc* mutations. Cells were grown at the permissive temperature (23°C) and shifted onto agar plates at the restrictive temperature (36°C). The cells were photographed at the time of the temperature shift, and at successive intervals while the plate was maintained at 36°C. (a) *cdc* 4 *cdc* 24 (strain RD314, 182-1-1, 2) at time 0; (b) *cdc* 4 *cdc* 24 after 6 hours at 36°C; (c) *cdc* 4 *cdc* 8 (strain RD314, 198, 3) at time 0; (d) *cdc* 4 *cdc* 8 after 6 hours at 36°C; (e) *cdc* 4 *cdc* 13 (strain RD314, 428, 3) at time 0; (f) *cdc* 4 *cdc* 13 after 7 hours at 36°C; (g) *cdc* 4 (strain 314D5) at time 0; (h) *cdc* 4 after 11 hours at 36°C.

ing the "start" event of subsequent cell cycles.

First, although mutants defective in cell separation have not been extensively studied, they have been isolated from mutagenized cultures after selecting for large cell aggregates by filtration through nylon mesh (*8*). A defect in cell separation does not appear to be lethal, and cells can go through an indefinite number of cell cycles despite a failure to complete this event.

Second, mutants defective in cytokinesis (*cdc* 3, *cdc* 10, *cdc* 11) undergo multiple rounds of bud emergence, initiation of DNA synthesis, DNA synthesis, and nuclear division, frequently attaining an octanucleate stage. These mutants do not continue to go through cell cycles indefinitely, and the reason for their eventual cessation of development is unknown, although it is the case that many of the cells lyse after extended incubation at the restrictive temperature.

Finally, mutants defective in bud emergence frequently undergo additional nuclear cycles in that about 50 percent of the cells in a diploid strain homozygous for the *cdc* 24 lesion become tetranucleate at the restrictive temperature. Haploid *cdc* 24 mutants usually stop development at the binucleate stage, and rarely become tetranucleate, but an analysis of DNA synthesis in the haploid cells at the restrictive temperature suggests that many of the cells synthesize a second round of DNA (*8*).

These observations suggest that the completion of the events bud emergence, cytokinesis, and cell separation, which comprise one of the two dependent pathways in the cycle, is not a necessary condition for the "start" event in a second cycle (Fig. 3). Although a mutant defective in nuclear migration has not been found we anticipate that arrest at this event will also permit the start of subsequent cycles.

With one exception, none of the mutants blocked in the pathway from initiation of DNA synthesis to late nuclear division show evidence of going through additional cell cycles after being arrested at the sites of their initial defects. We interpret this result to mean that it is necessary to complete these events in order to undergo the "start" event in a subsequent cell cycle (Fig. 3).

A Timer Controls Bud Emergence

Mutants defective in the *cdc* 4 gene (required for initiation of DNA synthesis) are exceptional in that they continue bud emergence for multiple cycles at the restrictive temperature, attaining as many as five buds on a single mononucleate cell (*16*). These successive cycles of budding continue despite the fact that the initiation of DNA synthesis, DNA synthesis, nuclear division, cytokinesis, and cell separation are not occurring. Furthermore,

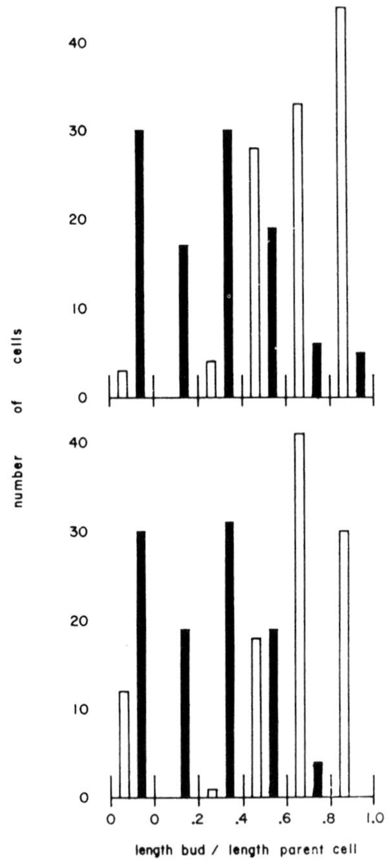

Fig. 5. Correlation of phenotype at 36°C with bud size at time of temperature shift for double mutant strains. Cells from a large number of photographs like those presented in Fig. 4 were measured to determine the ratio of the length of the bud to the length of the parent cell at the time of the shift, and were scored for whether they developed a morphology characteristic of cells with the *cdc* 8 or *cdc* 13 mutation (a single round bud, solid bars), or a morphology characteristic of the *cdc* 4 mutation (one to five elongated buds on a single cell, open bars). (Top) The results for *cdc* 4 *cdc* 8 (strain RD314, 198, 3); (bottom) the results for *cdc* 4 *cdc* 13 (strain RD314, 428, 3).

the time interval between successive budding events in *cdc* 4 mutants at the restrictive temperature maintains a periodicity of about one cell cycle time. This observation suggests that some type of intracellular timer initiates the successive cycles of budding, and that this timer can run independently of many of the cell cycle events.

The unusual behavior of mutants defective in *cdc* 4, and the surprising conclusion that their phenotype suggests, prompted us to consider the possibility that this phenotype might not be reflecting normal control mechanisms, but might be a result of an artifact. For example, the putative buds on *cdc* 4 mutants might not be the result of normal bud emergence events, but might be caused by some unrelated morphologic alteration. Alternatively, they might be the result of normal bud emergence events, but these events might be activated in an anomolous way by the abnormal *cdc* 4 gene product. Although we cannot completely rule out the hypothesis of artifact in the behavior of *cdc* 4 mutants, the properties of a few double *cdc* mutants do eliminate some possible sources of error. Double mutant strains containing a defect in the initiation of DNA synthesis (*cdc* 4) as one mutation, and a defect in bud emergence (*cdc* 24), DNA synthesis (*cdc* 8), or medial nuclear division (*cdc* 13) as the second mutation, were constructed and examined by time-lapse photomicroscopy (Fig. 4). A diploid strain carrying only the homozygous *cdc* 4 mutation is shown in Fig. 4, g and h, for comparison.

The double mutant strain defective in *cdc* 4 and *cdc* 24 does not undergo multiple rounds of bud emergence at the restrictive temperature (Fig. 4, a and b). This result indicates that mutants defective in *cdc* 4 require a functional *cdc* 24 gene product in order to display the phenotype of repeated bud emergence and this phenotype is not, therefore, unrelated to the normal budding process.

The double mutant strains harboring lesions in *cdc* 4 and *cdc* 8, or in *cdc* 4 and *cdc* 13, exhibit an unusual pattern of development at the restrictive temperature (Fig. 4, c and d and e and f) (*16*). The result is striking in that the populations of cells from both double mutant strains behave heterogeneously. Some cells continue periodic bud emergence (characteristic of a defect in *cdc*

4 alone), and other cells terminate development with a single large bud on each parent cell (characteristic of a defect in *cdc* 8 or *cdc* 13 alone). Furthermore, the phenotype that a particular cell exhibits is correlated with the position of that cell in the cell division cycle at the time of the shift to the restrictive temperature. This correlation is evident in Fig. 4, c and d and e and f, and is recorded for a larger number of cells in Fig. 5. In both double mutant strains most of the cells that do not continue bud emergence were either unbudded or had small buds, while most of those that do continue bud emergence were unbudded or had large buds. These observations are interpreted to mean that the former class of cells are those that block at the *cdc* 8 or *cdc* 13 mediated processes, DNA synthesis, and medial nuclear division, respectively, while the latter class of cells are those that block at the *cdc* 4 mediated process, initiation of DNA synthesis. This interpretation is consistent with our previous determinations of the time of function of these gene products (*16, 17*). We may conclude, therefore, that continued bud emergence in a mutant strain is not due to the lesion in gene *cdc* 4 per se, but is merely a result of the cell's position in the cell division cycle at the time it is arrested.

These results seem to us to be best interpreted by the hypothesis of a timer that controls bud emergence and that can express itself at only one discrete stage in the cell cycle, the stage of arrest in the *cdc* 4 mutant. The role of this timer in the normal cell cycle, and, in particular, its relation to the "start" event, are at present unclear. The action of the timer might be a prerequisite for, be dependent upon, or be part of the "start" event.

Implications of the Model

Let us return now to the question we posed at the outset: How are the events of the cell cycle coordinated so that their sequence remains invariant? The phenotypes of the *cdc* mutants suggest that the following events are ordered in a single dependent pathway: "start," initiation of DNA synthesis, DNA synthesis, medial nuclear division, late nuclear division, cytokinesis, and cell separation. Hence, the temporal sequence of these events is easily accounted for by the fact that no event in this pathway can occur without the prior occurrence of all preceding events. A second dependent pathway is comprised of the events "start," bud emergence, nuclear migration, cytokinesis, and cell separation. Thus, the temporal sequence of these five events is also assured. Furthermore, the integration of the two pathways is accomplished by the facts that both diverge from a common event, "start," and that both converge on a common event, cytokinesis.

Although evidence was found for the existence of a timer that controls bud emergence, there is no indication that this timer plays any role in *coordinating* different events of the cell cycle. It is conceivable that the timer serves to phase bud emergence with respect to the events of the DNA synthesis and nuclear division pathway, but it seems to us that the joint dependence of bud emergence and initiation of DNA synthesis on "start" is sufficient to explain the coordination between the two pathways. Although the function of the timer in the cell cycle is unknown, we favor the idea that the timer is either phasing successive "start" events, perhaps by monitoring cell growth, or is phasing successive bud emergence events in order to limit the cell to one such event per cycle. A variation of the dependent pathway model appears to be sufficient, therefore, to account for the coordination of cell cycle events, and it does not appear to be necessary to invoke the model of independent pathways with a central timing mechanism.

Applicability to other organisms. The events that comprise the cell division cycle have their origin in a distant evolutionary past common to all eukaryotic organisms. The complexity of this process suggests that a high degree of conservation of its basic elements might be expected. In this context, it is interesting to note that the only events of the *S. cerevisiae* cell cycle that are not common to most eukaryotes, bud emergence and nuclear migration, are on a separate pathway from the other events, as if they were appendages added to the basic plan. We would not be surprised, therefore, if in most eukaryotes an event, "start," activates and acts as a point of commitment for the dependent pathway of events leading from the initiation of DNA synthesis, to DNA synthesis, to successive stages of nuclear division, and finally culminating in cytokinesis and, where applicable, cell wall separation. Furthermore, the completion of some stages of nuclear division, but not cytokinesis or cell separation, may in general be necessary in one cell cycle for the "start" of the next cell cycle.

References and Notes

1. D. H. Williamson, in *Cell Synchrony*, I. L. Cameron and G. M. Padilla, Eds. (Academic Press, New York, 1966), p. 81; L. H. Hartwell, *Annu. Rev. Genet.* **4**, 373 (1970).
2. L. H. Hartwell, R. K. Mortimer, J. Culotti, M. Culotti, *Genetics* **74**, 267 (1973).
3. J. M. Mitchison, *The Biology of the Cell Cycle* (Cambridge Univ. Press, New York, 1972), p. 244
4. E. Zeuthen and N. E. Williams, in *Nucleic Acid Metabolism, Cell Differentiation, and Cancer Growth*, E. V. Cowdry and S. Seno, Eds. (Pergamon, Oxford, 1969), p. 203.
5. D. Prescott, *Recent Results Cancer Res.* **17**, 79 (1969); H. Halvorson, J. Gorman, P. Tauro, R. Epstein, M. LaBerge, *Fed. Proc.* **23**, 1002 (1964).
6. L. H. Hartwell, J. Culotti, B. Reid, *Proc. Natl. Acad. Sci. U.S.A.* **66**, 352 (1970).
7. M. L. Slater, *J. Bacteriol.* **113**, 263 (1973); L. Jaenicke, K. Scholz, M. Donike, *Eur. J. Biochem.* **13**, 137 (1970).
8. L. H. Hartwell, unpublished results.
9. E. Throm and W. Duntze, *J. Bacteriol.* **104**, 1388 (1970).
10. E. Bücking-Throm, W. Duntze, L. H. Hartwell, T. Manney, *Exp. Cell Res.* **76**, 99 (1973); L. H. Hartwell, *ibid.*, p. 111.
11. L. Hereford and L. H. Hartwell, in preparation.
12. D. H. Williamson and A. W. Scopes, *Exp. Cell Res.* **20**, 338 (1960); J. R. Pringle, R. J. Maddox, L. H. Hartwell, in preparation.
13. H. K. von Meyenberg, *Pathol. Microbiol.* **31**, 117 (1968); C. Beck and H. K. von Meyenberg, *J. Bacteriol.* **96**, 479 (1968).
14. M. T. Küenzi and A. Fiechter, *Arch. Mikrobiol.* **64**, 396 (1969); *ibid.* **84**, 254 (1972).
15. B. J. Reid, in preparation.
16. L. H. Hartwell, *J. Mol. Biol.* **59**, 183 (1971).
17. J. Culotti and L. H. Hartwell, *Exp. Cell Res.* **67**, 389 (1971).
18. L. H. Hartwell, *J. Bacteriol.* **115**, 966 (1973).
19. ———, *Exp. Cell Res.* **69**, 265 (1971).
20. Supported by research grant 6M17709 from the Institute of General Medical Sciences to J.C., J.R.P. and B.J.R.

Gautier J., Norbury C., Lohka M., Nurse P., and **Maller J.** 1988. Purified maturation-promoting factor contains the product of a *Xenopus* homolog of the fission yeast cell cycle control gene *cdc2*$^+$ *Cell* **54:** 433–439.

Masui and Markert identified a "Maturation Promoting Factor" (MPF) in amphibian eggs (1). The ability of this factor to induce complex cellular behavior, like egg maturation or the onset of meiosis, drew the attention of several labs interested in the control of cell cycle progression. Meanwhile, genetic analyses of cell cycle control in both budding and fission yeasts were identifying the relevant genes and the interactions among them (Hartwell et al., this volume, and reviewed in 2). The work reproduced here (see also 3) showed that a putative kinase of 34 kD, encoded by the *cdc2*$^+$ gene in fission yeast, was almost certainly the same protein as the 34 kD subunit of purified MPF. Contemporaneous work on the regulation of this kinase through its phosphorylation and dephosphorylation suggested that the enzyme lay at the heart of a cell's ability to begin mitotic chromosome segregation at a suitable time (reviewed in 2). This advance established an invaluable connection between the genetic work on cell cycle control and the biochemical and cell biological work on MPF regulation of eggs and blastomeres. Shortly thereafter, a "cyclin" was identified as an essential regulatory cofactor for the p34 kinase subunit (4). Cyclin had previously been identified as a protein that accumulates during interphase and is destroyed at each embryonic cleavage (5). The demonstration that the same p34 kinase could form complexes with either of two cyclins, and that the mitotic degradation of the cyclin subunit by proteolysis inactivated the kinase (6), established our current view of the basics for cell cycle regulation.

---

1. Masui Y. and Markert C.L. 1971. Cytoplasmic control of nuclear behavior during meiotic maturation of frog oocytes. *J. Exp. Zool.* **177:** 129–145.
2. Nurse P. 1990. Universal control mechanism regulating onset of M-phase. *Nature* **344:** 503–508.
3. Dunphy W.G., Brizuela L., Beach D., and Newport J. 1988. The *Xenopus* cdc2 protein is a component of MPF, a cytoplasmic regulator of mitosis. *Cell* **54:** 423–431.
4. Labbe J.C., Capony J.P., Caput D., Cavadore J.C., Derancourt J., Kaghad M., Lelias J.M., Picard A., and Doree M. 1989. MPF from starfish oocytes at first meiotic metaphase is a heterodimer containing one molecule of cdc2 and one molecule of cyclin B. *EMBO J.* **8:** 3053–3058.
5. Evans T., Rosenthal E.T., Youngblom J., Distel D., and Hunt T. 1983. Cyclin: A protein specified by maternal mRNA in sea urchin eggs that is destroyed at each cleavage division. *Cell* **33:** 389–396.
6. Draetta G., Luca F., Westendorf J., Brizuela L., Ruderman J., and Beach D. 1989. Cdc2 protein kinase is complexed with both cyclin A and B: Evidence for proteolytic inactivation of MPF. *Cell* **56:** 829–838.

Cell, Vol. 54, 433–439, July 29, 1988, Copyright © 1988 by Cell Press

Purified Maturation-Promoting Factor Contains the Product of a Xenopus Homolog of the Fission Yeast Cell Cycle Control Gene *cdc2*+

Jean Gautier,* Chris Norbury,† Manfred Lohka,* ‡ Paul Nurse,† and James Maller*
*Department of Pharmacology
University of Colorado School of Medicine
Denver, Colorado 80262
†ICRF Cell Cycle Control Laboratory
Microbiology Unit
Department of Biochemistry
University of Oxford
Oxford OX13QU, England

Summary

In the fission yeast S. pombe, the M_r = 34 kd product of the *cdc2*+ gene ($p34^{cdc2}$) is a protein kinase that controls entry into mitosis. In Xenopus oocytes and other cells, maturation-promoting factor (MPF) appears in late G2 phase and is able to cause entry into mitosis. Purified MPF consists of two major proteins of $M_r \approx 32$ kd and 45 kd and expresses protein kinase activity. We report here that antibodies to S. pombe $p34^{cdc2}$ are able to immunoblot and immunoprecipitate the ≈32 kd component of MPF from Xenopus eggs. The $M_r \approx 32$ kd and 45 kd proteins exist as a complex that expresses protein kinase activity. These findings indicate that a Xenopus $p34^{cdc2}$ homolog is present in purified MPF and suggest that $p34^{cdc2}$ is a component of the control mechanism initiating mitosis generally in eukaryotic cells.

Introduction

One of the most fundamental problems in cell biology concerns the control of the eukaryotic cell cycle. Changes in the regulation of the cycle are clearly evident in neoplastic cells and during early development, implicating cell cycle control as a major target for understanding the basis of both cancer and morphogenesis. In most eukaryotes, the cell cycle consists of several phases, in the simplest case a mitotic (M) phase in which chromosomes are distributed equally to daughter cells and an S phase in which DNA is replicated. While cells in early embryos such as Xenopus exhibit only M and S (Graham and Morgan, 1966), most cells exhibit two additional phases involving a G1 phase between mitosis and S phase and a G2 phase between S phase and mitosis. Two general control points have been identified in the cell cycle, one in G1 acting over entry into S phase and one in late G2 acting over entry into mitosis. These controls have been described in a wide range of cells, from yeast to humans, suggesting that they are fundamental features of the cell cycle.

Analysis of the G2/M control point has been primarily carried out with cells of the fission yeast Schizosaccharomyces pombe (Nurse, 1985 for review) and of the frog Xenopus laevis, although mammalian cells can also be arrested in G2 under certain experimental conditions (Gelfant, 1962; Pedersen and Gelfant, 1970; Nose and Katsuta, 1975; Stambrook and Velez, 1976; Melchers and Lernhardt, 1985). In fission yeast, a regulatory network of genes has been identified as controlling entry into mitosis. A protein kinase designated $p34^{cdc2}$ encoded by the *cdc2*+ gene is controlled by a number of positive (*cdc25*+, *nim1*+), and negative (*wee1*+) regulatory factors (Nurse, 1975, 1985; Nurse and Thuriaux, 1980; Russell and Nurse, 1986; Simanis and Nurse, 1986). These factors act together to effect an orderly transition from G2 into mitosis and are thought to respond to various signals important for mitotic initiation, such as those generated by changes in cell size and growth rate (Fantes and Nurse, 1977) and the completion of S phase. The *nim1*+ and *wee1*+ genes encode proteins with consensus sequences for protein kinases (Russell and Nurse, 1987a, 1987b), implicating protein phosphorylation in the activation of the $p34^{cdc2}$ protein kinase for the initiation of mitosis. The *cdc2*+ gene is structurally similar and functionally interchangeable with the cell cycle gene *CDC28* in the distantly related budding yeast Saccharomyces cerevisiae (Beach et al., 1982). In addition, a human homolog of *cdc2*+ has been isolated by complementation, which is able to substitute for the functions encoded by *cdc2*+ in yeast, suggesting that elements of the mitotic control are conserved in all eukaryotic cells (Lee and Nurse, 1987). This idea is further supported by the ability of antibody to $p34^{cdc2}$ to recognize homologs in S. cerevisiae and in HeLa cells (Draetta et al., 1987).

In frogs, information about events at the G2/M transition point has come largely from studies with oocytes (Maller, 1985, for review). Xenopus oocytes are arrested physiologically in late G2, and reentry into the cell cycle occurs in response to a variety of mitogenic agents, notably progesterone, insulin, or IGF_1. Several hours later, shortly before germinal vesicle (nuclear) breakdown, an activity appears in the cytoplasm that is believed to be directly responsible for triggering the initiation of mitotic events. This activity, known as maturation-promoting factor (MPF), was described as a factor present in mature oocyte cytoplasm that upon microinjection could cause G2-arrested oocytes to enter meiosis even in the absence of protein synthesis (Masui and Markert, 1971; Smith and Ecker, 1971). Subsequent work has revealed that an activity with identical properties appeared during M phase in a variety of oocytes, in somatic cells entering mitosis (Kishimoto et al., 1982, 1984; Sunkara et al., 1979), and in budding yeast (Tachibana et al., 1987). These results support the concept that MPF is a fundamental and universal regulator of entry into M phase.

Other evidence strongly suggested that the biochemical basis of MPF action, like $p34^{cdc2}$, was protein phosphorylation. When MPF appears in cells or is injected into

‡ Present address: Department of Cellular and Structural Biology, University of Colorado School of Medicine, Denver, Colorado 80262.

oocytes of various species, there is an immediate increase in total protein phosphorylation (Maller et al., 1977; Doree et al., 1983), which includes the appearance of new phosphoproteins (Maller and Smith, 1985). Moreover, MPF activity cannot be extracted from cells in the absence of several putative phosphatase inhibitors, notably β-glycerophosphate, indicating that MPF might contain a protein kinase or an activator of a protein kinase. Purification of MPF has provided support for this hypothesis.

Initially, little progress was made in purifying MPF using the Xenopus oocyte microinjection assay, partly because activity could only be detected in concentrated fractions (Wu and Gerhart, 1980). Extracts from unfertilized Rana eggs with MPF activity were shown to cause sperm nuclei to form chromosomes in vitro (Lohka and Masui, 1984), and this system was extended to Xenopus eggs by Lohka and Maller (1985), who showed that the addition of partially purified MPF to nuclei resulted in nuclear envelope breakdown, chromosome condensation, and spindle formation. In addition, the phosphoproteins observed to accompany MPF action in vivo were also found to become phosphorylated in the cell-free system after MPF addition (Lohka et al., 1987). The development of an MPF-dependent cell-free system for early mitotic events formed the basis for a new assay to assist in MPF purification. By monitoring breakdown of pronuclei assembled in vitro, we recently purified MPF 3500-fold from unfertilized Xenopus eggs (Lohka et al., 1988). Purified MPF was active not only in causing nuclear breakdown in the cell-free system but also in causing germinal vesicle breakdown when microinjected into cycloheximide-treated Xenopus oocytes. The purified preparation consisted largely of two polypeptides of apparent molecular weight 32,000 and 45,000. The purified MPF expressed a protein kinase activity that phosphorylated the $M_r \approx 45{,}000$ protein as well as histone H1, casein, and phosphatase inhibitor 1 (Lohka et al., 1988).

Because p34^{cdc2} and MPF both appear to act as fundamental regulators of entry into mitosis and are found in a wide range of eukaryotic cells, it is important to establish whether these two activities are related. This is of particular interest given that purified MPF contains a protein of similar molecular weight to p34^{cdc2}. We report here that antibodies against p34^{cdc2} both immunoblot and immunoprecipitate the ≈32 kd component of MPF, indicating that these two proteins are very closely related and probably similar in function.

Results

cdc2 Antibody Immunoblots an ≈34 kd Protein in Purified MPF

Previous studies have shown that the predicted amino acid sequence of the human *CDC2* homolog exhibits 63% overall identity with that of S. pombe *cdc2*$^+$ (Lee and Nurse, 1987) and that certain regions in p34^{cdc2} are perfectly conserved between yeast and humans. One such region, of 16 amino acid residues (EGVPSTAIREISLLKE), was used to generate a rabbit polyclonal antibody (Lee and Nurse, 1987). Immunoblotting of extracts from a range of eukaryotic cells has shown that this antibody (called PSTAIR) specifically recognizes a single protein of M_r of approximately 34,000. This antibody also detects a single protein in Xenopus egg extracts. Immunoblotting of crude ammonium sulfate fractions of egg extract revealed a single reactive component migrating with an $M_r \approx 34{,}000$ (Figure 1a). In the 0%–34% ammonium sulfate fraction, which specifically precipitates active MPF (Wu and Gerhart, 1980; Lohka et al., 1988), p34^{cdc2} migrated slightly more slowly (Figure 1a, lane 1) than in the 34%–100% fraction (Figure 1a, lane 2). No reaction occurred with normal rabbit serum, and the reaction was fully blocked by preincubation of the antibody with the synthetic peptide (Figure 1a, lane 3). More importantly, the antibody also specifically recognized the $M_r \approx 34{,}000$ protein in a highly purified MPF fraction eluted from a Mono S column (Figure 1b, lane 2). This protein was found to comigrate with p34^{cdc2} encoded by the human homolog of *cdc2*$^+$ (Figure 1b, lane 3). The fractions that retained full MPF activity across the final Mono S column also contained the $M_r \approx 34{,}000$ protein (Figure 1c, lanes 11 and 12). The same fractions also contained the protein found in purified MPF and previously described to be of $M_r \approx 32{,}000$ (Lohka et al., 1988). Fraction 13 contained partial MPF activity and also contained the $M_r \approx 34{,}000$ protein.

cdc2 Antibody Immunoprecipitates a 34 kd Protein in Purified MPF

To obtain further evidence that p34^{cdc2} is related to the $M_r \approx 32{,}000$ protein in MPF, immunoprecipitation of purified MPF was investigated. Previous studies have shown that incubation of purified MPF with [γ-^{32}P]ATP results in phosphorylation of the $M_r \approx 45{,}000$ protein in the final purified Mono S preparation (Lohka et al., 1988). Upon long autoradiographic exposures, labeling of the $M_r \approx 32{,}000$ component is also evident, and this labeling is more efficient at earlier stages of the purification in the peak fraction from the TSK 3000 SW column. Peak fractions of MPF from both the TSK 3000 SW and Mono S columns were incubated with [γ-^{32}P]ATP and then subjected to immunoprecipitation with the PSTAIR antibody. As shown in Figures 2a and 2b, a labeled protein of $M_r \approx 34{,}000$ was precipitated from the MPF peak fractions of both columns. Hydrolysis of the labeled protein revealed that the majority of phosphorylation was on threonine residues (data not shown). No immunoprecipitation was seen with normal rabbit serum nor if peptide was preincubated with the antibody prior to additon of the peak fractions (Figures 2a and 2b). Because the PSTAIR antibody efficiently immunoprecipitates material from the Mono S column, which is only labeled inefficiently, it is possible that the antibody preferentially precipitates the phosphorylated form of the Xenopus protein. A second antibody raised against the carboxy-terminal region of human p34^{cdc2} also immunoprecipitated the $M_r \approx 34{,}000$ protein (data not shown). These results, together with the immunoblotting data, indicate that purified MPF contains a protein closely related to p34^{cdc2}. The immunoprecipitates were also found to contain the $M_r \approx 45{,}000$ protein component of purified MPF (Figures 2a and 2b). This protein was not immunoprecipitated with

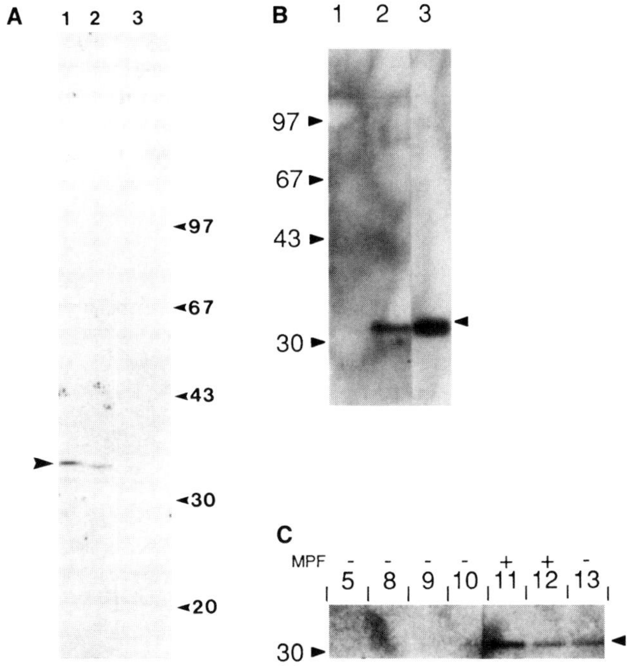

Figure 1. Immunoblotting of MPF with Antibody to p34^{cdc2}

(A) Ammonium sulfate fractions of high speed supernatants from unfertilized eggs. High speed supernatants from unfertilized Xenopus eggs were prepared as described previously (Lohka et al., 1988) and subjected to 0%–34% ammonium sulfate precipitation. Aliquots of the precipitated and soluble fractions were resolved on SDS polyacrylamide gels and immunoblotted with PSTAIR antibody as described in Experimental Procedures. An autoradiograph of the blot is shown. The numbers to the right indicate the positions to which molecular weight standards migrated. Lane 1, 0%–34% ammonium sulfate precipitated fraction; lane 2, 0%–34% ammonium sulfate soluble fraction; lane 3, 0%–34% ammonium sulfate precipitated fraction with antibody previously incubated with the 16mer synthetic peptide corresponding to the conserved sequence of p34^{cdc2}.
(B) MPF purified through Mono S chromatography. Immunoblotting was performed as described above. Lane 1, total E. coli protein (negative control); lane 2, a peak MPF fraction from a Mono S column; lane 3, total protein from ICRF23 human fibroblasts.
(C) MPF-positive and negative Mono S fractions. Immunoblotting was performed as described in (A). Each lane contains an aliquot of protein from the Mono S fraction series described by Lohka and Maller (1988); the number of each fraction is indicated, together with the presence (+) or absence (−) of full MPF activity, as judged by the ability to induce nuclear envelope breakdown (NEBD) in vitro. "+" indicates induction of 100% NEBD within 60 min; lower levels of activity were detected in fractions to either side of fractions 11 and 12.

normal rabbit serum if peptide was preincubated with the antibody prior to addition of the peak fractions. This suggests that the two proteins exist as a complex, resulting in their coimmunoprecipitation. The immunoprecipitate also had protein kinase activity, shown when the peak TSK 3000 SW fraction was immunoprecipitated with the PSTAIR antibody, and the immune precipitate incubated with histone H1 and [γ-^{32}P]ATP. Phosphorylation of histone H1 was evident, which could be completely blocked by preincubation of the antibody with peptide (Figure 2c). This result is consistent with the observations that both MPF and p34^{cdc2} have protein kinase activity (Simanis and Nurse, 1986; Lohka et al., 1988).

p34^{cdc2} Peptide Accelerates MPF Activity

These results suggest that MPF contains a Xenopus homolog of p34^{cdc2}. To examine further the relationship between p34^{cdc2} and MPF activity, we attempted to immunodeplete p34^{cdc2} from purified MPF in order to test the ability of the depleted activity to induce nuclear envelope breakdown and chromosome condensation in vitro. Unfortunately, the cross reactive antibodies were unable to completely immunodeplete p34^{cdc2} from purified MPF (data not shown), which prevented us from further pursuing this experimental approach. We next tested the effects on MPF activity of the peptide corresponding to the 16 amino acid region of p34^{cdc2} perfectly conserved from yeasts to humans. Because this region of the protein is so conserved, it seems likely that it interacts with another component in the mitotic control regulatory network that might influence p34^{cdc2} function. Conceivably, addition of the peptide could compete with p34^{cdc2} for interaction with another component and thus influence entry into mitosis. Addition of the peptide alone did not cause any change in nuclear morphology. However, when the peptide was added together with MPF to pronuclei assembled in vitro in Xenopus egg extracts, there was a marked acceleration of the rate at which MPF caused nuclear breakdown and chromosome condensation (Figure 3). In the presence of peptide, MPF-induced nuclear envelope breakdown and chromosome condensation occurred 30 min earlier relative to controls utilizing buffer or another peptide with no homology to p34^{cdc2}.

Discussion

The results in this paper provide evidence that a Xenopus protein homologous to S. pombe p34^{cdc2} is a component

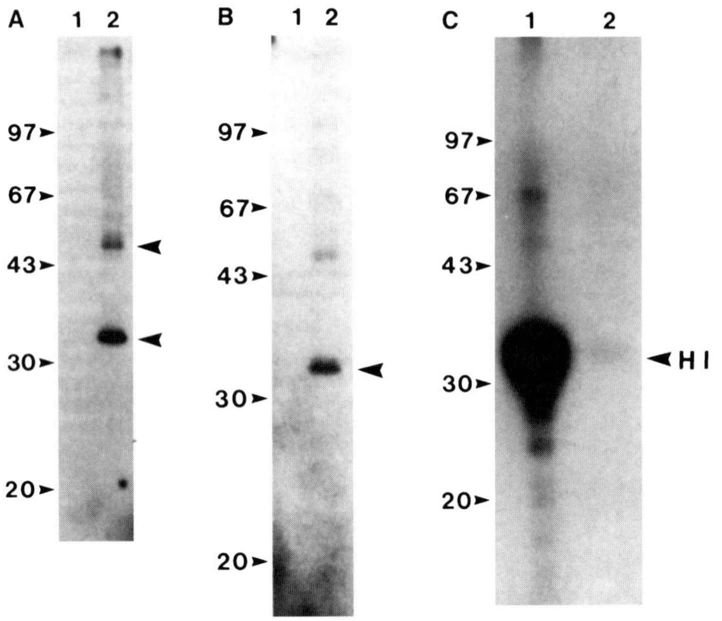

Figure 2. Immunoprecipitation of MPF by Antipeptide Antibody

(A) MPF purified through TSK chromatography. A peak MPF fraction from a TSK 3000 SW column was incubated with [γ-^{32}P]ATP, immunoprecipitated with antibody that had (lane 1) or had not (lane 2) been preincubated with the 16mer peptide corresponding to the conserved region of p34^{cdc2} as described under Experimental Procedures, and subjected to SDS gel electrophoresis. An autoradiograph of the gel is shown.

(B) MPF purified through Mono S chromatography. A peak MPF fraction from a Mono S column was incubated with [γ-^{32}P]ATP immunoprecipitated as described in (A) with antibody that had (lane 1) or had not (lane 2) been preincubated with the 16mer peptide.

(C) H1 kinase activity of immunoprecipitated MPF. MPF from a peak fraction of TSK 3000 SW columns was immunoprecipitated, and the immunoprecipitate was used to phosphorylate histone H1 as described in Experimental Procedures. An autoradiograph of the reaction products is shown after SDS gel electrophoresis. The antibody either had (lane 2) or had not (lane 1) been preincubated with the 16mer peptide.

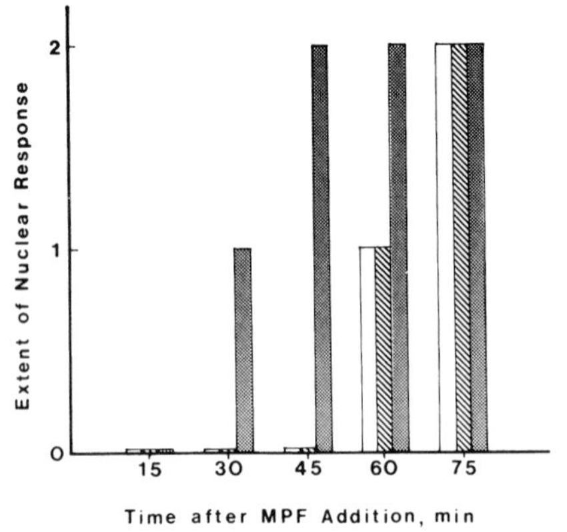

Figure 3. Acceleration of Nuclear Envelope Breakdown and Chromosome Condensation by a p34^{cdc2} Peptide

Pronuclei were assembled in vitro in Xenopus egg extracts and incubated for 20 min with 40 ng of the synthetic peptide corresponding to a conserved domain in p34^{cdc2}. A second addition of 40 ng of peptide was then made along with 2 U of MPF activity and the time course of nuclear events monitored as described in Experimental Procedures. Open bars, buffer addition; cross-hatched bars, control peptide addition; filled bars, p34^{cdc2} peptide. Data are expressed as the extent of nuclear response in arbitrary units, where 0 represents less than 20% pronuclear breakdown (Lohka et al., 1988); 1 represents approximately 50% nuclear breakdown and chromosome condensation, and 2 represents induction of these events in 90%–100% of the pronuclei.

of MPF. An antibody directed against p34^{cdc2} detects the 32–34 kd protein component of highly purified MPF (Figure 1). This antibody is very specific, having been raised against a domain perfectly conserved from the yeasts to humans, and detects only a single protein of approximately 34 kd in crude extracts of a range of eukaryotic cells. Thus, two independent lines of work have led to the identification of a single protein species as a key component in the initiation of mitosis. The biochemical approach in Xenopus has led to the purification of a homolog of p34^{cdc2} as a component of MPF, and the genetic approach in fission yeast has led to the cloning of the *cdc2*$^{+}$ gene encoding p34^{cdc2} as an element in the regulatory gene network controlling mitotic initiation.

Purified fractions of MPF contain not only a $M_r \approx 32$ kd protein but also a $M_r \approx 45$ kd protein, suggesting that both may be needed for MPF activity. Our immunoprecipitation results demonstrate that these two proteins can exist as a complex. The immunoprecipitated p34^{cdc2} can also be labeled with [γ-^{32}P]ATP in vitro. Incubation of the TSK 3000 and Mono S fractions with [γ-^{32}P]ATP leads to the formation of labeled p34^{cdc2} and labeled 45 kd protein in immunoprecipitates (Figure 2). It is likely that this labeling of p34^{cdc2} is inefficient, as the majority of the p34^{cdc2} present before immunoprecipitation is not highly labeled compared with that observed for the 45 kd protein (Lohka et al., 1988). The ability of MPF immunoprecipitated with cdc2 antibody to express H1 histone kinase activity provides evidence this activity is due to the $p34^{cdc2}$ component since $p34^{cdc2}$ from S. pombe has also been found to phosphorylate H1 histones (Moreno and Nurse, unpublished data). The phosphorylation of H1 histone may be physiologically significant since phosphorylation of this protein is known to increase in mitotic cells (Ajiro et al., 1983; Mueller et al., 1985).

The 45 kd protein may be of interest for two reasons. Firstly, *nim1*$^+$, a putative protein kinase activator of $p34^{cdc2}$ functions, is predicted to encode a protein of M_r $\approx$ 45 kd (Russell and Nurse, 1987b). A Xenopus homolog of this protein could be necessary to activate $p34^{cdc2}$ in the purified MPF complex. Secondly, a potential substrate has been identified for $p34^{CDC28}$ (the budding yeast homolog of $p34^{cdc2}$) of approximate $M_r \approx 40$ kd (Mendenhall et al., 1987). In purified Mono S preparations of MPF, the 45 kd protein only becomes phosphorylated in fractions that also contain $p34^{cdc2}$ (Lohka et al., 1988), and it is therefore possible that the 45 kd protein is a substrate. Given that the amount of $p34^{cdc2}$ in yeast does not change during the cell cycle, it is possible that the association of Xenopus $p34^{cdc2}$ with the $M_r \approx 45$ kd protein regulates MPF activity.

Immunoblotting data (Figure 1) show that the cdc2 antibody recognized $p34^{cdc2}$ in ammonium sulphate fractions with MPF activity as well as those without detectable activity. It was notable that the $p34^{cdc2}$ in active MPF fractions migrated on a 1-dimensional gel more slowly than in nonactive fractions, suggesting that a modified, possibly phosphorylated, form of $p34^{cdc2}$ might be enriched in active MPF fractions. Consistent with this idea, MPF is known to be stored in an inactive form in Xenopus oocytes (Reynhout and Smith, 1974; Cyert and Kirchner, 1988). In addition, it should be remembered that the *cdc2*$^+$ gene in fission yeast is required at the G1/S boundary of the cell cycle as well as at G2/M (Nurse, 1985). It is possible that different modified forms of $p34^{cdc2}$ are required at both points in the cell cycle and that only the form arising in G2 complexed with the 45 kd component has MPF activity.

The importance of $p34^{cdc2}$ in MPF activity is supported by the finding that a peptide corresponding to a highly conserved domain in $p34^{cdc2}$ can accelerate MPF action, strengthening our view that $p34^{cdc2}$ and a component of MPF are related functionally. It is conceivable that the peptide could have a direct activating effect on an M-phase regulatory component. However, MPF inactivating systems are known to occur in vivo (Gerhart et al., 1984) and in vitro (Cyert and Kirschner, 1988), and it is also possible that the peptide could compete for an inactivating system, allowing MPF to work at an effective higher dose. In the cell-free system, higher doses of MPF are able to cause more rapid nuclear envelope breakdown and chromosome condensation. The fact that the effect is largely on the time course of nuclear breakdown could also reflect the rapid destruction of the peptide in egg extracts, as we have observed previously with other synthetic peptides. In any case, the ability of the peptide to affect MPF action supports the hypothesis that cdc2 activity is involved in MPF action.

The linking together of Xenopus and fission yeast mitotic controls will enable both biochemical and genetic approaches to be combined in the study of initiation of mitosis. Proteins associated with MPF and identified as potential substrates in Xenopus can be compared with gene products thought to interact with $p34^{cdc2}$ function in yeast. These studies should be useful in establishing how the G2 to mitosis transition is controlled and how the early events of mitosis are implemented. We are encouraged in our belief that the basic mechanism of mitotic initiation is conserved in all eukaryotic cells and that $p34^{cdc2}$ is a key component of the control exerted by MPF.

Experimental Procedures

MPF Purification

MPF was purified from unfertilized Xenopus eggs as previously described (Lohka et al., 1988).

Antibody Preparation and Immunochemical Assays

Antisera and affinity-purified antibody were prepared as previously described (Lee and Nurse, 1987). For immunoblotting, proteins were transferred from 10% polyacrylamide gels (Laemmli, 1970) containing 10% glycerol to nitrocellulose filters using a semidry blotting system (Nova-Blot, LKB) at 0.8 mA/cm^2 for 55 min in 50 mM Tris, 39 mM glycine, 0.0375% SDS, and 20% methanol. The nitrocellulose filter was soaked in blocking solution (10% dry milk in phosphate buffered saline [PBS] for 1 hr at room temperature under constant agitation, and then the filter was incubated overnight at 4°C with the affinity-purified antibody in PBS containing 10% dry milk and 0.3% Tween 20. The filter was washed several times at room temperature with PBS containing 0.3% Tween 20 and then incubated with the same solution containing [^{125}I]-protein A (10^6 dpm/ml) for 1 hr at room temperature. The filter was washed several times in PBS/Tween 20 with the last rinse being in PBS containing 0.3% Tween 20, 1% Triton X100, and 0.05% SDS for 10 min. Control experiments were performed using affinity-purified antibody previously incubated for 1 hr at 4°C with the 16mer peptide (1 ng/µl of affinity-purified serum).

Immunoprecipitations using anti-cdc2 antibody were performed using either [γ-^{32}P]ATP labeled samples or unlabeled samples for subsequent H1 kinase assays. In a typical experiment, 50 µl of MPF from either the TSK or the Mono S chromatography step was incubated in 25 µl of protein A–Sepharose (Pharmacia) for 1 hr at 4°C under constant agitation. The protein A–resin had previously been equilibrated in RIPA buffer (20 mM Tris [pH 7.4], 5 mM EDTA, 100 mM NaCl, 1% Triton X-100) containing 300 mM polymethylsulfonyl fluoride (PMSF), 5 µg/ml of leupeptin, 1 mM Na pyrophosphate, 1 mM Na vanadate, 10 mM Na fluoride, 1 mM EGTA, and 1 mM β-glycerophosphate. The mixture was centrifuged at 15,000 × g for 1 min, and the supernatant was used for immunoprecipitation. In case of ^{32}P-labeling, prior to immunoprecipitation, the sample was incubated for 15 min at 30°C with 1 × 10^8 cpm of [γ-^{32}P]ATP, at a final ATP concentration of 0.1 mM. The reaction was stopped by the addition of EDTA to a final concentration of 10 mM. In all cases, the sample was then incubated overnight at 4°C under moderate agitation with 3 µl of affinity-purified serum, after which 25 µl of protein A–Sepharose was added and the sample incubated for 1 hr. Following centrifugation and removal of the supernatant, the protein A–resin was washed quickly two times with RIPA buffer and a third time for 1 hr, then washed twice in RIPA containing 1 M NaCl, and twice in RIPA containing 100 mM NaCl. After removing the supernatant, the resin was processed either for electrophoresis (in the case of ^{32}P-labeled extract) or for H1 kinase assay.

Protein Kinase Assays

The histone H1 assay was performed in a final volume of 40 µl containing 20 mM HEPES, 30 mM β-mercaptoethanol, 0.1 mg/ml of BSA, 10 mM MgCl, 0.5 mg/ml of rat thymus histone H1 (gift of Dr. T. A. Langan, Department of Pharmacology), and 0.1 mM ATP containing 1 × 10^8 cpm of [γ-^{32}P]ATP. Reactions were incubated at 30°C for 15 min and stopped by the addition of ¼ vol of 5× electrophoresis sample buffer.

Control experiments were done using affinity purified antibody that had been previously incubated with 1 ng of the 16mer peptide per μl of antibody for 1 hr at 4°C.

Assay of Peptide Effect on MPF Activity

Extracts able to cause pronuclear formation in vitro were prepared as previously described (Lohka and Maller, 1985) and incubated with demembranated sperm nuclei for 1 hr at 18°C. At this stage, 1 μl of the peptide was added to 12.5 μl of the pronuclear suspension, and the mixture was incubated for 20 min at 18°C. Then another 1 μl aliquot of the peptide was added with 25 μl of MPF containing 2 U of activity, and the progression of nuclear envelope breakdown was monitored by taking 5 μl aliquots of the mixture every 15 min, mixing with an equal volume of a 5 μg/ml of 4,6-Diamidino-2-phenylindole (DAPI) solution, and observing nuclear morphology by phase contrast and fluorescence microscopy. Peptide concentrations ranging from 4–140 ng/μl were assayed. The most effective peptide concentration was found to be 30–40 ng/μl. The control for MPF activity was performed by adding MPF to the extract with either buffer or a control peptide (AETAAAAKFLRAAA) whose sequence was unrelated to $p34^{cdc2}$. Data are expressed as the extent of nuclear response in arbitrary units, where 0 represents less than 20% pronuclear breakdown (Lohka et al., 1988); 1 represents approximately 50% nuclear breakdown and chromosome condensation, and 2 represents induction of these events in 90%–100% of the pronuclei.

Acknowledgments

We thank Karen Eckart for secretarial assistance. This work was supported by grants to J. L. M. from the National Institutes of Health and the American Cancer Society and by ICRF funding to P. N.

The costs of publication of this article were defrayed in part by the payment of page charges. This article must therefore be hereby marked "*advertisement*" in accordance with 18 U.S.C. Section 1734 solely to indicate this fact.

Received June 10, 1988.

References

Ajiro, K., Nishimoto, T., and Takahashi, T. (1983). Histone H1 and H3 phosphorylation during premature chromosome condensation in a temperature sensitive mutant ts BN2 of baby hamster kidney cells. J. Biol. Chem. *258*, 4534–4538.

Beach, D., Durkacz, B., and Nurse, P. (1982). Functionally homologous cell cycle control genes in fission yeast and budding yeast. Nature *300*, 706–709.

Cyert, M. S., and Kirschner, M. W. (1988). Regulation of MPF activity in vitro. Cell *53*, 185–195.

Doree, M., Peaucellier, G., and Picard, A. (1983). Activity of the maturation-promoting factor and the extent of protein phosphorylation oscillate simultaneously during meiotic maturation of starfish oocytes. Dev. Biol. *99*, 489–501.

Draetta, G., Brizuela, L., Potashkin, J., and Beach, D. (1987). Identification of p34 and p13, human homologs of the cell cycle regulators of fission yeast encoded by *cdc2*$^+$ and *suc1*$^+$. Cell *50*, 319–325.

Fantes, P., and Nurse, P. (1977). Control of cell size in fission yeast by a growth modulated size control over nuclear division. Exp. Cell Res. *107*, 377–386.

Gelfant, S. (1962). Initiation of mitosis in relation to the cell division cycle. Exp. Cell Res. *26*, 395–403.

Gerhart, J., Wu, M., and Kirschner, M. W. (1984). Cell cycle dynamics of an M-phase cytoplasmic factor in Xenopus laevis oocytes and eggs. J. Cell Biol. *98*, 1247–1255.

Graham, C. F., and Morgan, R. W. (1966). Changes in the cell cycle during early amphibian development. Dev. Biol. *14*, 349–381.

Kishimoto, T., Kuriyama, R., Kondo, H., and Kanatani, H. (1982). Generality of the action of various maturation-promoting factors. Exp. Cell Res. *137*, 121–126.

Kishimoto, T., Yamazaki, K., Kato, Y., Koide, S. S., and Kanatani, H. (1984). Induction of starfish oocyte maturation by maturation-promoting factor of mouse and surf clam oocytes. J. Exp. Zool. *231*, 293–295.

Laemmli, U. K. (1970). Cleavage of structural proteins during the assembly of the head of bacteriophage T4. Nature *227*, 680–685.

Lee, M. G., and Nurse, P. (1987). Complementation used to clone a human homolog of the fission yeast cell cycle control gene *cdc2*$^+$. Nature *327*, 31–35.

Lohka, M. J., and Maller, J. L. (1985). Induction of nuclear envelope breakdown, chromosome condensation and spindle formation in cell-free extracts. J. Cell Biol. *101*, 518–523.

Lohka, M. J., and Masui, Y. (1984). Effects of Ca^{2+} ions on the formation of metaphase chromosomes and sperm pronuclei in cell-free preparations from unactivated Rana pipiens eggs. Dev. Biol *103*, 434–442.

Lohka, M. J., Kyes, J. L., and Maller, J. L. (1987). Metaphase protein phosphorylation in Xenopus laevis eggs. Mol. Cell. Biol. *7*, 760–768.

Lohka, M. J., Hayes, M. K., and Maller, J. L. (1988). Purification of maturation-promoting factor, an intracellular regulator of early mitotic events. Proc. Natl. Acad. Sci. USA *85*, 3009–3013.

Maller, J. (1985). Regulation of amphibian oocyte maturation. Cell. Diff. *16*, 211–221.

Maller, J. L., and Smith, D. S. (1985). Two-dimensional polyacrylamide gel analysis of changes in protein phosphorylation during maturation of Xenopus oocytes. Dev. Biol. *109*, 150–156.

Maller, J. L., Wu, M., and Gerhart, J. C. (1977). Changes in protein phosphorylation accompanying maturation of Xenopus laevis oocytes. Dev. Biol. *58*, 295–312.

Masui, Y., and Markert, C. L. (1971). Cytoplasmic control of nuclear behavior during meiotic maturation of frog oocytes. J. Exp. Zool. *177*, 129–146.

Melchers, F., and Lernhardt, W. (1985). Three restriction points in the cell cycle of activated murine B lymphocytes. Proc. Natl. Acad. Sci. USA *82*, 7681–7685.

Mendenhall, M. D., Jones, C. A., and Reed, S. I. (1987). Dual regulation of the yeast CDC28–P40 protein complex: cell cycle, pheromone, and nutrient limitation effects. Cell *50*, 927–935.

Mueller, R. D., Yasuda, H., and Bradbury, E. M. (1985). Phosphorylation of histone H1 through the cell cycle of Physarum polycephalum: 24 sites of phosphorylation at metaphase. J. Biol. Chem. *260*, 5081–5086.

Nose, K., and Katsuta, H. (1975). Arrest of cultured rat liver cells in G2 phase by the treatment with dibutyryl cAMP. Biochem. Biophys. Res. Commun. *64*, 983–988.

Nurse, P. (1975). Genetic control of cell size at cell division in yeast. Nature *256*, 547–551.

Nurse, P. (1985). Cell cycle control genes in yeast. Trends Genet. *1*, 51–55.

Nurse, P., and Thuriaux, P. (1980). Regulatory genes controlling mitosis in the fission yeast Schizosaccharomyces pombe. Genetics *96*, 627–637.

Pedersen, T., and Gelfant, S. (1970). G2-population cells in mouse kidney and duodenum and their behavior during the cell division cycle. Exp. Cell Res. *59*, 32–36.

Pringle, J. R., and Hartwell, L. H. (1981). The Saccharomyces cerevisiae cell cycle. In Molecular Biology of the Yeast Saccharomyces: Life Cycle and Inheritance, J. N. Strathern, E. W. Jones, and J. R. Broach, eds. (Cold Spring Harbor, New York: Cold Spring Harbor Laboratory), pp. 97–142.

Reynhout, J. K., and Smith, L. D. (1974). Studies of the appearance and nature of a maturation-inducing factor in the cytoplasm of amphibian oocytes exposed to progesterone. Dev. Biol. *25*, 232–247.

Russell, P., and Nurse, P. (1986). *cdc25*$^+$ functions as an inducer in the mitotic control of fission yeast. Cell *45*, 145–153.

Russell, P., and Nurse, P. (1987a). Negative regulation of mitosis by *wee1*$^+$, a gene encoding a protein kinase homolog. Cell *49*, 559–567.

Russell, P., and Nurse, P. (1987b). The mitotic inducer *nim1*$^+$ functions in a regulatory network of protein kinase homologs controlling initiation of mitosis. Cell *49*, 569–576.

$p34^{cdc2}$ Is Present in MPF
439

Simanis, V., and Nurse, P. (1986). The cell cycle control gene *cdc2*⁺ of fission yeast encodes a protein kinase potentially regulated by phosphorylation. Cell *45*, 261–268.

Smith, L. D., and Ecker, R. E. (1971). The interaction of steroids with Rana pipiens oocytes in the induction of maturation. Dev. Biol. *25*, 233–247.

Stambrook, P., and Velez, C. (1976). Reversible arrest of Chinese hamster V79 cells in G2 by dibutyryl cyclic AMP. Exp. Cell Res. *99*, 57–62.

Sunkara, P. S., Wright, D. A., and Rao, P. N. (1979). Mitotic factors from mammalian cells induce germinal vesicle breakdown and chromosome condensation in amphibian oocytes. Proc. Natl. Acad. Sci. USA *76*, 2799–2802.

Tachibana, K., Yanagashima, N., and Kishimoto, T. (1987). Preliminary characterization of maturation-promoting factor from yeast Saccharomyces cerevisiae. J. Cell Sci. *88*, 273–281.

Wu, M., and Gerhart, J. C. (1980). Partial purification and characterization of the maturation-promoting factor from eggs of Xenopus laevis. Dev. Biol. *79*, 465–477.

Hartwell L.H. and **Weinert T.A.** 1989. Checkpoints: Controls that ensure the order of cell cycle events. *Science* **246:** 629–634.

Two views of cell cycle control emerged from experimentally distinct studies on different kinds of organisms. The cleavage cycles of animal zygotes drew attention to the clock-like behavior of blastomeres and even of cytoplasm extracted from them (reviewed in 1). Such cells contain large amounts of presynthesized material, thanks to the maternal dowry, so only DNA and a few key proteins must be synthesized as they move from one cycle to the next. Other cells, like micro-organisms and the somatic cells of many plants and animals, must carry out both RNA and protein synthesis before they are ready to initiate DNA synthesis; most such cells also make new macromolecules before they enter mitosis. Students of these organisms tended to think about pathways of dependency, such as the ones identified in budding and fission yeasts (Hartwell et al, this volume, and reviewed in 2). Whereas students of both kinds of cells recognized the fruitful contrasts exposed by these differences, it was the paper reproduced here that provided insight into the ways that cells hold back from cell cycle progression until an essential process has been satisfactorily completed. By analyzing the genes required to delay mitosis until DNA damage had been repaired, or until DNA synthesis had been completed, these workers identified a kind of quality control that they called a "checkpoint". This paper defines a checkpoint as a process that is not essential for cell cycle progression but that monitors a job being done. When errors occur, they are identified by the checkpoint, leading to an arrest in cell cycle progression until the error is corrected, thus keeping the cell out of trouble. This paper also drew attention to a literature that implied the existence of another checkpoint at mitosis, a quality control by which cells could recognize an improperly formed spindle and take steps to correct it, or at least give itself more time to do the job right before starting anaphase. These ideas have been developed into new fields of cell cycle research, both for mitosis (e.g. 3,4) and for DNA synthesis (reviewed in 5). Checkpoints have taken on additional importance with the discovery that tumor suppressor genes may encode checkpoint proteins, suggesting that further study of cell cycle control will contribute significantly to the understanding of human cancers (6,7).

---

1. Murray A.M. and Kirschner M.W. 1989. Dominos and Clocks: The union of two views of the cell cycle. *Science* **246:** 614–621.
2. Nurse P. 1990. Universal control mechanism regulating onset of M-phase. *Nature* **344:** 503–508.
3. McIntosh J.R. 1991. Structural and mechanical control of mitotic progression. *Cold Spring Harbor Symp. Quant. Biol.* **56:** 613-619.
4. Chen R.H., Waters J.C., Salmon E.D., and Murray A.W. 1996. Association of spindle assembly checkpoint component XMAD2 with unattached kinetochores. *Science* **274:** 242–246.
5. Weinert, T. 1998. DNA damage and checkpoint pathways: Molecular anatomy and interactions with repair. *Cell* **94:** 555–558.
6. Hartwell L.H. and Kastan M.B. 1994. Cell cycle control and cancer. *Science* **266:** 1821–1828.
7. Cahill D.P., Lengauer C., Yu J., Riggins G.J., Willson J.K., Markowitz S.D., Kinzler K.W., and Vogelstein B. 1998. Mutations of mitotic checkpoint genes in human cancers. *Nature* **392:** 300–303.

Checkpoints: Controls That Ensure the Order of Cell Cycle Events

LELAND H. HARTWELL AND TED A. WEINERT*

The events of the cell cycle of most organisms are ordered into dependent pathways in which the initiation of late events is dependent on the completion of early events. In eukaryotes, for example, mitosis is dependent on the completion of DNA synthesis. Some dependencies can be relieved by mutation (mitosis may then occur before completion of DNA synthesis), suggesting that the dependency is due to a control mechanism and not an intrinsic feature of the events themselves. Control mechanisms enforcing dependency in the cell cycle are here called checkpoints. Elimination of checkpoints may result in cell death, infidelity in the distribution of chromosomes or other organelles, or increased susceptibility to environmental perturbations such as DNA damaging agents. It appears that some checkpoints are eliminated during the early embryonic development of some organisms; this fact may pose special problems for the fidelity of embryonic cell division.

THE CELL CYCLE IS OFTEN CONSIDERED TO BE COMPOSED of four phases, the gap before DNA replication (G_1), the DNA synthetic phase (S), the gap after DNA replication (G_2), and the mitotic phase, which culminates in cell division (M). Although this formulation is useful and can serve as an organizing principle, the cell cycle is seen on closer examination to be more complex. A large number of macromolecular components are assembled, activated, or moved; a sequence of events involving one organelle (such as the centrosome) may occur throughout all four stages of the cell cycle, and independent sequences are coordinated with one another.

Biochemical, genetic, and cytological research in the last decade has greatly increased our appreciation of the structural and functional complexity involved in numerous cell cycle processes such as DNA replication (*1, 2*), chromosome organization (*1, 3*), centrosome duplication and movement (*4*), the dynamic organization of the mitotic spindle (*5*), chromosome movement on the spindle (*6*), nuclear envelope breakdown and reformation (*7*), organelle duplication and distribution (*8*), and the establishment of the site of cell division (*9*). Each of these processes involves cellular organelles that are present in small numbers and whose accurate reproduction and distribution is important for cell viability. While executing these events, the cell must avoid or correct errors that lead to the production of nonfunctional organelles and those that lead to the production of or distribution of the wrong number of organelles.

Recent results comparing somatic and embryonic cell cycles (*10*) have revealed basic similarities as well as striking differences in how events are controlled. In the early embryonic cell divisions of *Xenopus*, the initiation and order of events of the cell cycle is determined by cyclic activation of maturation promoting factor (MPF), and the events occur independently of one another. This subject is reviewed by Murray and Kirshner in this issue of *Science* (*11*). In the cell cycle of yeast and the somatic divisions of many metazoan organisms, an additional principle appears to operate. Although MPF plays an initiating role in the somatic cell cycle, the order of events is ensured by dependent relationships; the initiation of late events is dependent on the completion of early events. We think dependent relationships seen in somatic cells are a key element in understanding the high fidelity of organelle reproduction and distribution during cell division. The purpose of this article is to consider how these dependent relationships are achieved and the consequences incurred by the cell upon their elimination.

Dependent Relationships in the Cell Cycle

The existence of dependent relationships is not usually apparent simply by observing the normal cell cycle. Dependencies are revealed by perturbations of specific events, that is, by the application of chemicals, by the study of mutants that specifically inhibit one event in the cell, or by surgical and cell fusion techniques (*12*). For example, specific chemical inhibition of DNA replication prevents nuclear division and cell division in most cell types including bacteria, fungi, and vertebrate somatic cells. Mutants that block specific events in the cell cycle provide additional resolution of dependent relationships. In the yeasts *Saccharomyces cerevisiae* (*13*) and *Schizosaccharomyces pombe* (*14*), many mutants have been isolated and characterized that appear to have defects at specific stages of the cell cycle. In *S. cerevisiae*, temperature-sensitive mutations exist that block bud formation, spindle pole body enlargement and division (*15*), spindle pole body separation and migration (*16*), tubulin assembly (*16*), spindle elongation, initiation of DNA replication, DNA elongation (*17*), DNA ligation (*18*), chromatin assembly (*19*), bipolar association of chromosomes (*20*), sister chromatid decatenation (*21*), sister chromatid separation (*22*), nuclear division, and cytokinesis (*13*). The phenotypes of the mutants suggest that most of these events are ordered into a few dependent pathways. For example, the sequence of events that encompass spindle pole body duplication and segregation as well as those comprising chromo-

The authors are in the Department of Genetics, University of Washington, Seattle, WA 98195.

*Present address: Department of Molecular and Cellular Biology, University of Arizona, Tucson, AZ 85721.

some replication and segregation constitute one dependent pathway.

What principles does the cell use to establish an ordered pathway of events? Does the existence of such order imply the existence of control mechanisms that enforce order? Since many of the events of interest in the cell cycle are those involving the assembly of macromolecular complexes, it may be informative to consider the principles that have been gleaned from the extensive investigations of another case of macromolecular assembly—the formation of a bacteriophage particle (*23*). Bacteriophage T4 is constructed from three components—the head, tail, and tail fibers—each of which is assembled by an invariant pathway. All structural proteins are synthesized at the same time, and unassembled proteins remain unassociated until the partially assembled structure becomes ready for their addition. At this stage, reactive sites that accommodate the assembly of the next component are created by the addition of the previous protein in the assembly pathway. For example, tail tube subunits do not associate with themselves until the baseplate is assembled; at this stage some tail tube subunits associate with the baseplate, and these provide a seed for the polymerization of additional identical subunits to form a long helical polymer. Similarly, assembly of the bacteriophage T4 baseplate itself occurs by an invariant pathway where each step utilizes the previous step as substrate and in turn provides the substrate for the next step (*24*). The important lesson from bacteriophage assembly is that a very complicated series of morphogenetic events can be ordered by a principle intrinsic to the components themselves without requirement for extrinsic control mechanisms. We shall refer to this type of ordered pathway as substrate-product ordered.

The maturation of bacteriophage λ DNA during packaging into the phage head provides an example of dependence due to a control mechanism. Concatameric DNA is not cut by the enzyme terminase, unless phage proheads are present to package the DNA. This dependence of DNA cutting on the presence of proheads could have been enforced by substrate-product order if terminase was activated by binding to proheads; however, this is not the mechanism, rather a trans-acting inhibitor prevents terminase from cutting in the absence of proheads, and this dependence on proheads can be relieved by eliminating the inhibitor (*25*).

Likewise, the dependency of late events in the cell cycle on the completion of early events may be due to either substrate-product order or to a control mechanism (Fig. 1). If, for example, replicated chromosomes are essential substrates for mitosis, then the dependence is due to substrate-product order; alternatively, if the dependence is due to an inhibitor of mitosis produced in response to unreplicated chromosomes, then we would say the dependence is due to control. Control might also be exerted by an activator; for example, completion of DNA synthesis might produce an activator of mitosis. Since it is difficult to distinguish control by activation from substrate-product order by an empirical test, we will concentrate our discussion of control mechanisms on those that act by inhibition. We use the term control to include regulation at any level: transcription, translation, or posttranslation.

How can one distinguish extrinsic control by inhibition from substrate-product order? The existence of a control mechanism is suggested when one finds chemicals, mutants, or other conditions that relieve a dependent relationship; that is, conditions that permit a late event to occur even when an early, normally prerequisite event, is prevented. We term such an observation "relief of dependence." This argument rests on the assumption that if dependence is intrinsic to the mechanism of assembly as it is for the bacteriophage T4 tail, then one would not be likely to relieve this dependence by mutational inactivation of a gene product. For example, one would not likely find a mutation that would relieve the need for an early protein in the assembly sequence [however, see (*26*)]. Similarly, if the dependence of nuclear division on DNA replication in the cell cycle is due to an intrinsic requirement of replicated chromosomes in the nuclear division machinery, we would not expect to relieve this dependence by mutation. We are aware of the fact that this empirical criterion cannot be taken as rigorous evidence for the existence of a control mechanism. One can imagine rare mutations that will alter a protein in such a way as to permit it to assemble without its normal substrate, but most mutations will eliminate a function. We suggest that a finding of "relief of dependence" provides prima facie evidence for control, especially when the mutation is shown to eliminate the function of a protein. By the relief of dependence criterion, a number of control mechanisms have recently been identified in the cell cycle. We call these control mechanisms checkpoints (*27*), because they appear to have the role of checking to see that prerequisites (such as DNA replication in the case of mitosis) have been properly satisfied.

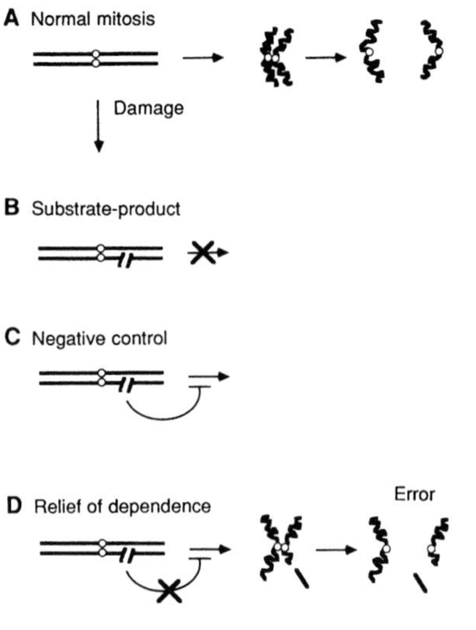

Fig. 1. Models illustrating the dependence of mitosis on replicated and undamaged DNA. (**A**) A replicated, undamaged chromosome passes through chromosome condensation and segregation. (B, C, and D) A chromosome sustaining a double-strand break is illustrated; however, a gap or a single-strand lesion remaining after DNA replication would be expected to have the same consequences. (**B**) The broken chromosome is an inadequate substrate for chromosome condensation (or for some later step) and mitosis is blocked. (**C**) The DNA break creates a signal that inhibits chromosome condensation. (**D**) A mutation (like *rad9*) eliminates the negative inhibition so the damaged chromosome passes through mitosis and an error occurs.

Checkpoints in the Cell Cycle

We discuss in this section a few cases where sufficient evidence exists to suggest that the dependence of a late event in the cell cycle on an early event is due to a checkpoint. We will describe first in some detail a control mechanism that we have worked on, the *RAD9* system in yeast, which is responsible for making mitosis dependent on the completion of DNA replication, and then discuss more briefly a few other control systems.

Dependence of mitosis on DNA synthesis. In yeast, mammalian tissue culture cells, *Aspergillus*, and many other eukaryotic organisms, arrest of DNA synthesis by specific inhibitors or by mutational inactivation of replication enzymes prevents mitosis. In yeast, the *RAD9* gene specifies a component of this control system (*27*).

Temperature-sensitive mutants defective in some DNA replication functions (DNA polymerase I, *cdc17*; DNA polymerase III, *cdc2*; and DNA ligase, *cdc9*) do not normally undergo mitosis at the restrictive temperature. Dependency of mitosis on the completion of DNA synthesis is relieved, however, by a complete deficiency of *RAD9*; if these same *cdc* mutants have a *rad9* gene defect, then the cells continue through mitosis into the next cell cycle at the restrictive temperature (*28*). A mutant temperature-sensitive for

DNA ligase illustrates the effect of the *RAD9* checkpoint (Fig. 2). The analysis exploits the fact that in *S. cerevisiae* different phases of the cell cycle are accompanied by distinctive changes in daughter bud morphology. During incubation at the restrictive temperature for 3 hours, G_1 (unbudded) cells containing the *cdc9* mutation became blocked before mitosis (large budded cells) (Fig. 2, A and B). Arrested cells had completed the bulk of DNA synthesis [confirmed by analysis of DNA content by flow cytometry (*27*)], but become arrested presumably because many unligated single-strand lesions remain in the DNA (*28*). In contrast, *cdc9* G_1 cells that also have the *rad9* defect typically proceed past mitosis and enter the next cell division at the restrictive temperature (Fig. 2, C and D). Analysis of the nuclear morphology of growing cells shifted to the restrictive temperature confirms that *cdc9* cells are blocked before chromosome separation in mitosis (Fig. 2E), whereas *cdc9-rad9* cells are not arrested, but display a distribution of cells in different phases of the cell cycle.

The *cdc9-rad9* double mutant illustrates a principle that we think may apply to many checkpoints, that elimination of the checkpoint may have catastrophic or subtle consequences depending on prevailing conditions. Temporary inactivation of DNA ligase activity is not lethal for most cells, as shown by the ability of *cdc9* mutant cells to retain viability after a brief incubation at the restrictive temperature. In the absence of *RAD9*, however, DNA ligase–deficient cells die much more rapidly at the restrictive temperature (Fig. 3) (*29*). Therefore, relief of the dependence of mitosis on DNA synthesis is lethal when completion of DNA synthesis is blocked. Alternatively, if cells are not perturbed by an interruption of DNA replication (or by extrinsic DNA damage, see below), then *RAD9* is not an essential function; cells have indistinguishable growth properties whether the *RAD9* gene is intact or defective (*27*). The only effect of complete deficiency of the *RAD9* checkpoint is that *rad9* cells lose chromosomes spontaneously at a rate 21 times higher than that of wild-type cells [rate of loss of one chromosome from a disome strain; 6.0×10^{-4} in a *rad9* deletion and 2.9×10^{-5} in Rad^+ (*27*)]. We imagine that these two phenotypes, cell death or chromosome loss, are extremes of the same primary role of this checkpoint, to ensure that chromosomes have been fully replicated and are intact before mitosis is initiated. Thus the order of DNA synthesis and mitosis is apparently established independently of the *RAD9* checkpoint, but the *RAD9* product ensures the order if DNA synthesis is interrupted. An attractive hypothesis to explain how *RAD9* inhibits mitosis is to suggest that it negatively regulates a function essential for mitosis. It will be of interest to determine whether the *RAD9* gene product interacts with MPF or other components known to play essential roles in the initiation of mitosis.

The *RAD9* control system was initially identified in a search for mutations that permit cell division of cells with defective genomes due to damage induced by x-irradiation (*27*). Mutations in the *RAD9* gene allow cells with DNA damage to proceed through cell division, whereas irradiated wild-type cells arrest in G_2 until the damage is repaired. Mitosis in most other eukaryotic cells (*30*) is also dependent on undamaged DNA, since x-irradiation and other DNA damaging agents arrest cells before mitosis. The dependence of mitosis on completion of DNA replication is relieved in mammalian somatic cells by caffeine (*31*). Like *rad9* defects in *S. cerevisiae*, caffeine treatment of irradiated mammalian cells permits their entry into mitosis (and decreases cell viability) and yet has no observable effect on this transition in unirradiated cells (*32*).

Temperature-sensitive recessive mutations in mouse BHK cells (*tsBN2*) and *Aspergillus* (*bimE7*) qualify as checkpoints since both relieve dependency of mitosis on DNA synthesis (*33*). In both mutants, cells blocked in DNA synthesis enter mitosis without completing DNA synthesis when mutant cells are shifted to the restrictive temperature. Death of cells carrying these mutations is probably not due solely to relief of dependence, as *bimE7* mutants shifted to the restrictive temperature are arrested in mitosis. The cellular functions are unknown, but deficiency for either must render inactive a checkpoint that prevents mitosis should DNA synthesis be blocked. Both are suggested to be negative regulators of additional functions essential for mitosis.

Dependence of anaphase on metaphase and delay as evidence of a checkpoint. In the examples of dependence cited above, inhibition of

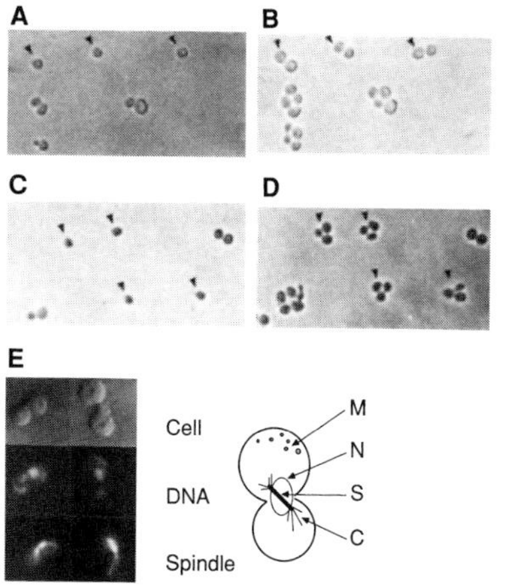

Fig. 2. DNA ligase-defective cells are arrested before mitosis only if the *RAD9* gene is present. Progression in the cell cycle of individual cells was determined by photomicroscopy. (**A** and **B**) *cdc9* and (**C** and **D**) *cdc9-rad9* cells were grown at the permissive temperature then plated on thin agar slabs. Photographs of the same fields were made at the time cells were initially shifted to the restrictive temperature for DNA ligase (A and C) and 3 hours later (B and D). Unbudded cells have not replicated DNA and require ligase activity in the first cell cycle. The fate of G_1 (unbudded) cells was quantitated by analysis of several fields; 78% (49 out of 63) of G_1 *cdc9* cells are arrested with one large bud, whereas only 21% (12 out of 57) of G_1 *cdc9-rad9* cells are arrested with one large bud. Most *cdc9-rad9* G_1 cells (79%, 45 out of 57) generated at least a third bud indicating the original G_1 cell had proceeded through mitosis and one progeny cell had initiated bud formation (and DNA synthesis) in the next cell cycle. The fates of cells that were budded at the time of ligase inactivation were also analyzed [note three in (A) and two in (C)]. Most of the budded cells have completed replication and do not require ligase activity in the first cell cycle but do require ligase for the second cell cycle. Eighty percent of budded *cdc9* cells generate two cells each arresting with one large bud, whereas 45% (45 out of 99) of budded *cdc9-rad9* cells generate two cells each with one large bud and 55% (54 out of 99) proceed to a subsequent cell cycle. Combining data from unbudded and budded cells 79% of *cdc9* cells and only 36% of *cdc9-rad9* cells are blocked as large budded cells after 3 hours. Cytological examination (described below) shows that even the few *cdc9-rad9* cells that are arrested with one large bud have proceeded past mitosis, whereas most of the *cdc9* cells have arrested at mitosis. About 40% of the *cdc9-rad9* cells at the time of plating were inviable and did not change their shape after 3 hours [note unchanged large budded cell in the corner of (C) and (D)]; these cells were excluded from quantitative analysis. (**E**) DNA ligase inactivation results in the arrest of cells before mitosis. *cdc9* cells grown at the permissive temperature (23°C) were shifted to restrictive temperature (36°C) for 3 hours, fixed, and analyzed for cell (top), nuclear (middle), and microtubule (bottom) morphology. Two cells are shown. M, mitochondria; N, nucleus; S, spindle; C, cytoplasmic microtubules. Cell morphology was determined by differential-interference-contrast microscopy and nuclear and microtubule morphologies, by epifluorescence from the DNA-binding dye DAPI (4′,6-diamidino-2-phenylindole dihydrochloride) and from FITC-conjugated antibodies to antibodies to tubulin, as described (*38*). The position of the nuclei and the presence of a short spindle at the neck of each large budded cell has been described (*36*). When cells were shifted from growth at 23°C to the restrictive temperature for *cdc9*, the percentage of cells with DNA at the neck of a large budded cell increased from 11 to 84% for *cdc9* and only from 11 to 32% for *cdc9-rad9*. As expected, the distribution of cells in other parts of the cell cycle showed that most *cdc9* cells were arrested before mitosis, whereas most *cdc9-rad9* cells were not arrested before mitosis (*27*). Strains: 598-3 *MATα*, *cdc9-8 ura3 ade2 ade3 leu1 can1 cyh2 sap3 trp1 SCE::URA3* (provided by L. Kadyk), and 7851-3-2 *MATa cdc9-8 rad9::TRP1 trp1 ura3 leu2 ade3 ade2 his3 leu1*. Both are congenic with A364a. The *RAD9* deletion we constructed by deleting 93% of the internal coding sequence and inserting a DNA fragment containing the *TRP1* gene, and will be described in detail elsewhere.

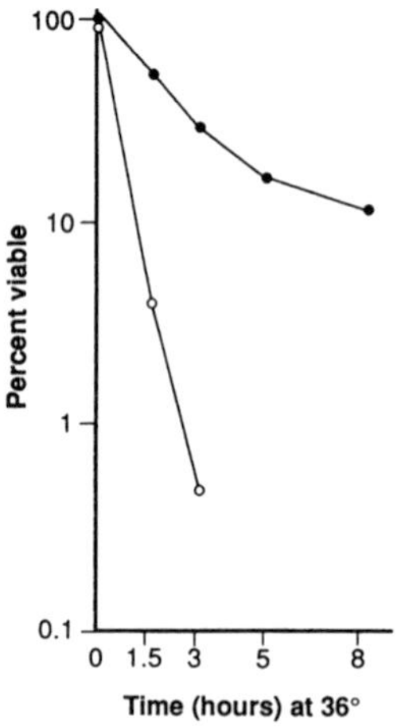

Fig. 3. Rapid loss of viability in cells defective both for DNA ligase and for the *RAD9* gene. *cdc9* (●) and *cdc9-rad9* (○) cells growing at 23°C were shifted to the restrictive temperature and viability was determined by plating for viable colonies at the permissive temperature (23°C). The cell viability reported is relative to viability at the time of temperature shift. Results were reproducible in separate experiments and with other congenic strains.

an early event resulted in complete arrest of a late event. It is clear in other cases that inhibition of an early event merely delays a late event. Unless careful studies are carried out the delay might be overlooked. In cases where delay occurs, a checkpoint may exist. The delay in the late event indicates its dependence on the early event; moreover, the eventual occurrence of the late event in the absence of the early event indicates that relief of dependence eventually occurs spontaneously making it unlikely that dependence is due to a substrate-product relationship.

A control mechanism is suggested by the observation that when a chromosome lags in finding its way to the metaphase plate, anaphase is delayed, often until the lagging chromosome arrives at the metaphase plate (*34*). Perhaps a similar checkpoint is present in yeast cells, since cell division is greatly delayed in those divisions at which loss of centromere-containing plasmids occurs (*35*); the delay of division occurs at a specific stage in the cell cycle probably corresponding to mitosis.

Dependence of centrosome duplication on DNA synthesis. In yeast (*36*) and mammalian cells (*37*), inhibition of DNA synthesis by temperature-sensitive mutations in replication enzymes or by aphidicolin, respectively, arrest spindle pole body or centrosome duplication at a stage characteristic of metaphase. A mutation of yeast, *esp1*, blocks DNA replication and nuclear division but not spindle pole body duplication, so that nuclei accumulate with as many as eight spindle pole bodies (*38*); thus, this essential gene plays a role in preventing spindle pole body duplication in the absence of DNA synthesis.

Dependence of replicon initiation on other replicons. In mammalian cells large contiguous regions of DNA, corresponding roughly to the cytogenetic bands of chromosomes, replicate coordinately as a result of the fact that large arrays containing as many as a hundred replicons initiate replication synchronously (*39*). The initiation of DNA synthesis has been found to be exquisitely sensitive to single-strand lesions (*40*); one single-strand break may inactivate initiation in as many as a hundred replicons, presumably those adjacent to one another. This result is surprising, since the chromosome is thought to be organized into topological domains corresponding roughly to single replicons. The coordinate inhibition of initiation in clusters of replicons is likely due to a control mechanism, since cells from individuals with the genetic disease, ataxia telangiectasia, a syndrome predisposing individuals to cancer, are resistant to this inhibition (*41*). In these cells dependency of DNA synthesis on intact template is relieved and might be the cause of radiation sensitivity. In unirradiated wild-type cells, this control may prevent reinitiation of any unligated strands remaining from the previous S phase; replication of a chromatid containing a single-strand lesion would generate a double-strand break leading to the production of an acentric chromosome fragment.

Dependence of DNA reinitiation on mitosis. Mammalian cells inhibited in mitosis by colchicine (*42*) or mutations (*43*) delay for many hours but eventually reconstitute an interphase nucleus and replicate their DNA without completing chromosome segregation. These observations may identify a checkpoint that makes reinitiation of DNA replication dependent on mitosis. This dependence has been reproduced in vitro in extracts of *Xenopus* eggs where added DNA is assembled into a nucleus and replicates once (*44*); nuclei can rereplicate without completing mitosis if they are treated with agents that make them permeable, an observation that has suggested a specific model for this dependence (*45*).

Dependence of mitosis on growth. The product of the *WEE1* gene of *S. pombe* is not essential, but its presence delays mitosis; deletion of *WEE1* leads to cells that are smaller than normal at mitosis, and increased dosage leads to proportionately larger cells at mitosis (*46*). This observation may identify a checkpoint that integrates growth and division (*11*).

Checkpoints in bacteria. In *Escherichia coli*, cell division is dependent on completion of DNA replication (*47*) and on the presence of undamaged DNA (*48*) as well. The SOS regulatory system of *E. coli*, like the *RAD9* system of yeast, is responsible for the arrest (dependency) of cell division in response to an inhibition of DNA replication and in response to DNA damage (*49*). DNA damage is recognized (directly or indirectly) by the RecA protein, which in its activated form stimulates proteolytic cleavage of the LexA protein, an inhibitor of *SulA* gene transcription. SulA protein inhibits cell division possibly by inhibiting the FtsZ protein. Null mutations of *SulA* are insensitive to division arrest by DNA damage, thus meeting the criterion of a checkpoint. Deficiency for *SulA* generates a higher frequency of anucleate cells when DNA synthesis is inhibited than is the case for wild-type cells (*50*).

A checkpoint that couples F plasmid replication and segregation to *E. coli* cell division has been identified (*51*). A gene, *CcdB*, carried by stable mini-F plasmids inhibits *E. coli* cell division when replication of the plasmid is inhibited. Cells containing plasmids lacking this function produce plasmid-free cells at much higher frequencies than those containing this function.

Some Embryonic Cell Cycles Lack Some Checkpoints

Many experimental results suggest that some early embryonic cell cycles are controlled differently than somatic cell cycles (*11*); the differences may be attributable to the existence of fewer checkpoints in some embryonic cell cycles.

Perhaps the most definitive evidence for a difference in the control of the cell cycle between somatic cells and some embryonic systems arises from the consequences of inhibiting DNA synthesis. As discussed above, chromosome condensation, elaboration of the mitotic spindle, and cytokinesis are all prevented in yeast and other somatic cells when DNA synthesis is inhibited. However, in *Drosophila* one or more aberrant nuclear divisions may occur after inhibition of DNA synthesis (*52*); centrosomes continue dividing for many divisions up to the time of cellular blastoderm (*53*), and, although cytokinesis is not completed, the egg surface undergoes periodic budding. Thus, some of the events of cell division are being activated. In early *Xenopus* embryos (*54*), cell division continues at a normal rate after inhibition of DNA synthesis until the mid-blastula transition; anucleate cells are the products of these divisions.

Some embryonic cells differ from somatic cells in their response to broken DNA. Although somatic cells arrest in G_2 in response to x-irradiation, some embryos seem to be insensitive to broken DNA since *Rana pipiens* oocytes fertilized with heavily irradiated sperm divide to produce haploid organisms (*55*) and early cleavage stage *Drosophila* embryos continue nuclear division, DNA replication, and centrosome duplication for several cycles with little or no mitotic delay after x-irradiation (*56*). We speculate that the *Rana pipiens* and *Drosophila* analogs of the *RAD9* checkpoint are inactive in these embryos.

The *gnu* mutation of *Drosophila* has a phenotype that supports the idea that many cell cycle events are independent of one another in the embryo (*57*); *gnu* embryos replicate DNA and centrosomes without undergoing nuclear division. In contrast, yeast mutations that block nuclear division also prevent the next round of DNA replication (*58*) and spindle pole body duplication (*36*).

Quite dramatic results have been obtained in the study of enucleated embryos where centrosome duplication and cortical contractions characteristic of cytokinesis continue for many cycles, often with normal kinetics (*55, 59, 60*). However, these results are not necessarily informative with respect to our search for the presence of checkpoints, since these control mechanisms probably require signals generated by the failure of one event that are received by and serve to inhibit some other event; with the nucleus or another organelle removed it is unavailable to send signals. For example, sea urchin embryos have been reported to continue to prematurely condense the chromosomes of fertilizing sperm after enucleation but not after inhibition of DNA synthesis with aphidicolin (*62*).

Fidelity in the Embryonic Cell Cycle

The function of checkpoints in the cell cycle is to ensure the completion of early events before late events begin. When checkpoints are eliminated by mutation or other means, cell death, infidelity of chromosome transmission, or increased susceptibility to environmental perturbations (like DNA damaging agents) result.

Since some early embryos (*Drosophila, Xenopus*) lack the checkpoint that makes mitosis dependent on DNA replication, the considerations discussed above would predict that the embryonic cell divisions in these organisms might occur with less fidelity than the somatic cell divisions of the same organism. To our knowledge, there is no data available at present on the fidelity of chromosome or other organelle distribution during embryonic cell division in comparison to somatic cell divisions of the same organism; we will, nevertheless, pursue the implications of this thought.

If this prediction of a lower fidelity in some embryonic divisions is verified, we will need to consider why embryonic development has sacrificed fidelity. We suggest that checkpoints would delay cell division in those cells where stochastic problems require correction and would lead to asynchrony in a cell population. It appears that those embryonic systems that have eliminated the checkpoint, making mitosis dependent on DNA replication, are the systems where rapid and synchronous division is evident. Perhaps checkpoints have been eliminated because synchrony and speed were important for embryonic development.

What is the value of fidelity to metazoans and what price would the embryo pay in sacrificing fidelity? Developmental abnormalities (*63*) and cancer (*64*) seem to be much greater risks in organisms with chromosomal aneuploidy. It seems unlikely that embryonic systems would incur these risks. We suspect therefore that these embryonic systems have evolved some compensating system to allay increased risk of cancer and developmental aberrations. We suggest that embryonic systems utilizing rapid synchronous divisions will detect and eliminate aneuploid cells and possibly cells that have incurred errors in the segregation of other organelles. A likely period to look for the deliberate elimination of defective cells is at the time when cell divisions slow down, transcription begins, groups of cells become asynchronous, and gastrulation commences (the mid-blastula transition in *Xenopus* and division fourteen of *Drosophila*). It may be pertinent that nuclei of *Drosophila* embryos that fail to divide or that merge with other nuclei fall into the interior of the egg at this time in embryogenesis and do not contribute to the larval cells (*65*). The *Drosophila* embryo has the capacity to replace lost nuclei even when the number lost is quite large; when nuclei are inactivated by ultraviolet irradiation before the fourteenth division, compensatory divisions occur that approximately restore the appropriate number of nuclei before cellularization (*66*). The existence of a monitoring system for aneuploidy does not seem too difficult to imagine in view of the phenomena of dosage compensation (*67*) and X chromosome inactivation (*68*), situations in which differences in chromosome ratios are detected.

A variety of abnormal cells arising from infidelity of the mitotic process have been detected in humans including aneuploidy, gene amplification, and multipolar mitoses. Sporadic cases of mitotic infidelity may not merit special attention since many causes are possible. However, when mitotic infidelity is rampant and reproducible as it is in many types of human tumors it may be fruitful to consider perturbations of the checkpoints that normally ensure mitotic fidelity as potential causes.

REFERENCES AND NOTES

1. C. S. Newlon, *Microbiol. Rev.* **52**, 568 (1988).
2. A. Kornberg, *DNA Replication* (Freeman, San Francisco, 1988).
3. S. M. Gasser and U. K. Laemmli, *Trends Genet.* **3**, 16 (1987).
4. D. Mazia, *Int. Rev. Cytol.* **100**, 49 (1987); J. R. McIntosh, *Mod. Cell Biol.* **2**, 115 (1983).
5. J. R. McIntosh, in *Aneuploidy: Etiology and Mechanisms*, V. L. Dellarco, P. E. Voytek, A. Hollaender, Eds. (Plenum, New York, 1985), pp. 197–229.
6. T. J. Mitchison, *Annu. Rev. Cell Biol.* **4**, 527 (1988).
7. J. W. Newport and D. J. Forbes, *Annu. Rev. Biochem.* **56**, 535 (1987).
8. C. W. Birky, *Int. Rev. Cytol.* **15**, 49 (1983).
9. J. M. Mitchison and P. Nurse, *J. Cell Sci.* **75**, 357 (1985).
10. We use the term early embryonic cell cycle to refer to the early divisions of organisms like *Xenopus* and *Drosophila* in which the divisions are rapid, synchronous, occur without growth, and are under the control of the maternal genome—that is, those before cycle fourteen of *Drosophila* and those before the mid-blastula transition of *Xenopus*. We use the term somatic cell cycle to refer to the divisions of eukaryotic microorganisms as well as most metazoan cells, excluding only those defined as early embryonic.
11. A. W. Murray and M. W. Kirshner, *Science* **246**, 614 (1989).
12. P. N. Rao, in *Premature Chromosome Condensation: Application in Basic, Clinical and Mutation Research*, P. N. Rao, R. T. Johnson, K. Sperling, Eds. (Academic Press, New York, 1982), pp. 1–42.
13. J. R. Pringle and L. H. Hartwell, in *Molecular Biology of the Yeast Saccharomyces: Life Cycle and Inheritance*, J. N. Strathern, E. Jones, J. Broach, Eds. (Cold Spring Harbor Laboratory, Cold Spring Harbor, NY, 1981), vol. 1, pp. 97–142.
14. T. Hirano and M. Yanagida, in *Molecular and Cell Biology of Yeasts*, E. F. Walton, Ed. (Blackie, Glasglow, Scotland, 1988), pp. 223–245.
15. P. Baum, C. Furlong, B. Byers, *Proc. Natl. Acad. Sci. U.S.A.* **83**, 5512 (1986); M. Rose and G. Fink, *Cell* **48**, 1047 (1987).
16. T. C. Huffaker, J. H. Thomas, D. Botstein, *J. Cell Biol.* **106**, 1997 (1988).
17. M. Budd and J. L. Campbell, *Proc. Natl. Acad. Sci. U.S.A* **84**, 2838 (1987); K. C. Sitney, M. E. Budd, J. L. Campbell, *Cell* **56**, 599 (1989).
18. D. G. Barker, A. L. Johnson, L. H. Johnston, *MGG* (*Mol. Gen. Genet.*) **200**, 458 (1985).
19. M. Han, M. Chang, U. J. Kim, M. Grunstein, *Cell* **48**, 589 (1987).
20. J. H. Thomas and D. Botstein, *ibid.* **44**, 65 (1986).
21. C. Holm, T. Goto, J. C. Wang, D. Botstein, *ibid.* **41**, 553 (1985).
22. D. Koshland and L. H. Hartwell, *Science* **238**, 1713 (1987).
23. P. B. Berget, in *Virus Structure and Assembly*, S. Casjens, Ed. (Jones and Bartlett, Boston, 1985), pp. 149–168.
24. P. B. Berget and J. King, in *Bacteriophage T4*, C. K. Mathews, E. M. Kutter, G. Mosig, P. B. Berget, Eds. (American Society for Microbiology, Washington, DC, 1983), pp. 246–258.
25. H. Murialdo and W. L. Fife, *Genetics* **115**, 3 (1987).
26. In some mutants defective in genes needed for T4 head morphogenesis, the major head subunit assembles into aberrant structures; most of these structures are easily distinguished from normal heads. [See U. K. Laemmli, E. Molbert, M. Show, E. Kellenberger, *J. Mol. Biol.* **49**, 99 (1970).
27. T. A. Weinert and L. H. Hartwell, *Science* **241**, 317 (1988); unpublished observations.

28. L. H. Johnston and K. A. Nasmyth, *Nature* **274**, 891 (1978).
29. R. H. Schiestl *et al.*, *Mol. Cell. Biol.* **9**, 1882 (1989).
30. G. Olivieri and A. Micheli, *Mutat. Res.* **122**, 65 (1983); V. W. Burns, *Radiat. Res.* **4**, 411 (1956); C. Lucke-Huhle, E. A. Blakely, P. Y. Chang, C. A. Tobias, *ibid.* **79**, 97 (1979); B. S. Baker, A. T. C. Carpenter, M. Gatti, *Prog. Top. Cytogenet.* 7, 273 (1987); J. G. Carlson, *Genetics* **23**, 596 (1938).
31. C. Lau and A. Pardee, *Proc. Natl. Acad. Sci. U.S.A.* **79**, 2942 (1982); P. M. Busse, S. K. Bose, R. W. Jones, L. J. Tolmach, *Radiat. Res.* **76**, 292 (1978).
32. However, a significant distinction between the effects of caffeine and those of *RAD9* deficiency is that cells blocked in the middle of DNA synthesis can enter mitosis after caffeine treatment, whereas *rad9*-deficient cells blocked in early S phase remain arrested; only cells blocked late in S enter mitosis [T. A. Weinert and L. H. Hartwell, unpublished observations; R. Schlegel and A. B. Pardee, *Science* **232**, 1264 (1986)].
33. S. A. Osmani, D. B. Engle, J. H. Doonan; N. R. Morris, *Cell* **52**, 241 (1988); T. Nishimoto, R. Ishida, K. Ajiro, S. Yamamoto, T. Takahashi, *J. Cell. Physiol.* **109**, 299 (1981).
34. A. Bajer and J. Mole-Bajer, *Chromosoma* **7**, 558 (1956); C. L. Rieder and S. P. Alexander, in *Aneuploidy: Mechanisms of Origin*, M. A. Resnick Ed. (Liss, New York, 1989); R. E. Zirkle, *Radiat. Res.* **41**, 516 (1970).
35. A. W. Murray and J. W. Szostak, personal communication.
36. B. Byers and L. Goetsch, *Cold Spring Harbor Symp. Quant. Biol.* **38**, 123 (1974).
37. J. B. Rattner and S. G. Phillips, *J. Cell Biol.* **57**, 359 (1973); R. Kuriyama and G. G. Borisy, *ibid.* **91**, 841 (1981).
38. P. Baum, C. Yip, L. Goetsch, B. Byers, *Mol. Cell. Biol.* **8**, 5386 (1988).
39. R. Hand, *Cell Biol.* **2**, 389 (1979); Y. F. Lau and F. E. Arrighi, *Chromosoma* **83**, 721 (1981).
40. I. Watanabe, *Radiat. Res.* **58**, 541 (1974); B. Painter and B. R. Young, *Biochim. Biophys. Acta* **418**, 146 (1976); L. F. Povirk, *J. Mol. Biol.* **114**, 141 (1977).
41. R. B. Painter and B. R. Young, *Proc. Natl. Acad. Sci. U.S.A.* 77, 7315 (1980); Y. Shiloh, E. Tabor, Y. Becker, *KROC Found. Ser.* **19**, 111 (1985); R. B. Painter, *ibid.*, p. 89.
42. H. N. Barber and H. G. Callan, *Proc. R. Soc. London Ser. B* **131**, 258 (1942).
43. R. J. Wang and L. Yin, *Exp. Cell Res.* **101**, 331 (1976).
44. J. J. Blow and R. A. Laskey, *Nature* **332**, 546 (1988).
45. R. A. Laskey, M. P. Fairman, J. J. Blow, *Science* **246**, 609 (1989).
46. P. Russell and P. Nurse, *Cell* **49**, 559 (1987).
47. W. E. Howe and D. W. Mount, *J. Bacteriol.* **124**, 1113 (1975).
48. J. W. Little and D. W. Mount, *Cell* **29**, 11 (1982).
49. G. C. Walker, *Microbiol. Rev.* **48**, 60 (1984).
50. A. Jaffe, R. D'Ari, V. Norris, *J. Bacteriol.* **165**, 66 (1986).
51. T. Ogura and S. Hiraga, *Proc. Natl. Acad. Sci. U.S.A.* **80**, 4784 (1983).
52. B. A. Edgar and G. Schubiger, *Cell* **44**, 871 (1987).
53. J. W. Raff and D. M. Glover, *J. Cell Biol.* **107**, 2009 (1988).
54. D. Kimelman, M. Kirschner, T. Scherson, *Cell* **48**, 399 (1987); M. Dasso and J. Newport, personal communication.
55. R. Briggs, E. U. Green, T. J. King, *J. Exp. Zool.* **116**, 455 (1951).
56. A. Schneider-Minder, *Int. J. Radiat. Biol.* **11**, 1 (1966).
57. M. Freeman, C. Nusslein-Volhard, D. M. Glover, *Cell* **46**, 457 (1986).
58. J. Culotti and L. H. Hartwell, *Exp. Cell Res.* **67**, 389 (1971).
59. K. Hara, P. Tydeman, M. Kirschner, *Proc. Natl. Acad. Sci. U.S.A.* 77, 462 (1980).
60. E. Schierenberg and W. B. Wood, *Dev. Biol.* **107**, 337 (1985).
61. C. Killian, C. Bland, J. Kuzava, D. Nishioka, *Exp. Cell Res.* **158**, 519 (1985).
62. G. Sluder, F. J. Miller, C. L. Rider, *J. Cell Biol.* **103**, 1873 (1986).
63. C. J. Epstein, *The Consequences of Chromosome Imbalance: Principles, Mechanisms, and Models*, vol. 18 of Developmental and Cell Biology Series (Cambridge Univ. Press, New York, 1986).
64. J. J. Yunis, *Science* **221**, 227 (1983); J. M. Bishop, *ibid.* **235**, 305 (1987); F. Mitelman, *Nature* **310**, 325 (1984); E. M. Kuhn and E. Therman, *Cancer Genet. Cytogenet.* **22**, 1 (1986); R. Holliday, *Trends Genet.* **5**, 42 (1989).
65. W. Sullivan, J. S. Minden, B. M. Alberts, *J. Cell Biol.*, in press; J. S. Minden, D. A. Agard, J. W. Sedat, B. M. Alberts, *ibid.* **109**, 505 (1989).
66. G. Yasuda and G. Schubiger, personal communication.
67. E. Jaffe and C. Laird, *Trends Genet.* **2**, 316 (1986).
68. S. G. Grant and V. M. Chapman, *Annu. Rev. Genet.* **22**, 199 (1988).
69. We thank B. Brewer, B. Byers, F. Cross, W. Fangman, N. Hollingsworth, C. Mann, C. Manoil, and G. Schubiger for criticism of the manuscript; M. Dasso, L. Edgar, A. Murray, J. Newport, C. Rieder, G. Schubiger, J. Szostak, B. Sullivan, and G. Yasuda for unpublished information. Our work was supported by grants from the NIH (GM17709) Institute of General Medical Sciences and the American Business Foundation for Cancer Research. T.A.W. was supported by a fellowship from the Jane Coffin Childs Foundation.

Cell Membrane and Extracellular Martrix

Farquhar M.G. and **Palade G.E.** 1963 Junctional complexes in various epithelia. *J. Cell Biol.* **17:** 375–412.

DURING THE 1950S AND 1960S, methods were discovered that allowed electron microscopists to view many aspects of a cell's internal structure. With these methods, the complexity of the cell's surface organization also became obvious. The paper reproduced here used the best techniques of the day to prepare samples of animal tissue for examination in the EM, revealing both the diversity of surface organization on epithelial cells and the pleasing fact that certain motifs were common to several tissue types. Farquhar and Palade described a constellation of structures that occurred together at the interface between adjacent cells in an epithelial sheet. From the context of these structures they were able to make strong inferences about the likely functions of elements they called the zonula occludens (tight junction), the zonula adherens (adherence junction), and the desmosome. The work both defined some of the important specializations of the cell surface and showed that there was considerable conservation of these specializations across a range of distinct tissues. Shortly after this paper, other workers used related methods to define and characterize additional specializations on and between cells (1). These morphological studies identified cell surface structures, but it was some time before biochemical and molecular methods identified the protein subunits of these membrane specializations and the genes that encode them (2–4).

---

1. Revel J.P. and Karnovsky M.J. 1967. Hexagonal array of subunits in intercellular junctions of the mouse heart and liver. *J. Cell Biol.* **33:** 7–12.
2. Goodenough,D.A. and Stoeckenius W. 1972. The isolation of the mouse hepatocyte gap junctions: Preliminary chemical characterization and X-ray diffraction. *J. Cell Biol.* **54:** 646–656.
3. Stevenson B.R. and Goodenough D.A. 1984. Zonulae occludentes in junctional complex-enriched fractions from mouse liver. Preliminary morphological and biochemical characterization. *J. Cell Biol.* **98:** 1209–1221.
4. Furuse M., Hirase T., Itoh M., Nagafuchi A., Yonemura S., and Tsukita S. 1993. Occludin: A novel integral membrane protein localizing at tight junctions. *J. Cell Biol.* **123:** 1777–1788.

JUNCTIONAL COMPLEXES IN VARIOUS EPITHELIA

MARILYN G. FARQUHAR, Ph.D., and GEORGE E. PALADE, M.D.

From The Rockefeller Institute. Dr. Farquhar's present address is Department of Pathology, University of California School of Medicine, San Francisco

ABSTRACT

The epithelia of a number of glands and cavitary organs of the rat and guinea pig have been surveyed, and in all cases investigated, a characteristic tripartite junctional complex has been found between adjacent cells. Although the complex differs in precise arrangement from one organ to another, it has been regularly encountered in the mucosal epithelia of the stomach, intestine, gall bladder, uterus, and oviduct; in the glandular epithelia of the liver, pancreas, parotid, stomach, and thyroid; in the epithelia of pancreatic, hepatic, and salivary ducts; and finally, between the epithelial cells of the nephron (proximal and distal convolution, collecting ducts). The elements of the complex, identified as *zonula occludens* (tight junction), *zonula adhaerens* (intermediary junction), and *macula adhaerens* (desmosome), occupy a juxtaluminal position and succeed each other in the order given in an apical-basal direction.

The *zonula occludens* (tight junction) is characterized by fusion of the adjacent cell membranes resulting in obliteration of the intercellular space over variable distances. Within the obliterated zone, the dense outer leaflets of the adjoining cell membranes converge to form a single intermediate line. A diffuse band of dense cytoplasmic material is often associated with this junction, but its development varies from one epithelium to another.

The *zonula adhaerens* (intermediate junction) is characterized by the presence of an intercellular space ($\sim$200 A) occupied by homogeneous, apparently amorphous material of low density; by strict parallelism of the adjoining cell membranes over distances of 0.2 to 0.5 μ; and by conspicuous bands of dense material located in the subjacent cytoplasmic matrix.

The desmosome or *macula adhaerens* is also characterized by the presence of an intercellular space ($\sim$240 A) which, in this case, contains a central disc of dense material; by discrete cytoplasmic plaques disposed parallel to the inner leaflet of each cell membrane; and by the presence of bundles of cytoplasmic fibrils converging on the plaques.

The *zonula occludens* appears to form a continuous belt-like attachment, whereas the desmosome is a discontinuous, button-like structure. The *zonula adhaerens* is continuous in most epithelia but discontinuous in some. Observations made during experimental hemoglobinuria in rats showed that the hemoglobin, which undergoes enough concentration in the nephron lumina to act as an electron-opaque mass tracer, does not penetrate the intercellular spaces beyond the *zonula occludens*. Similar observations were made in pancreatic acini and ducts where discharged zymogen served as a mass tracer. Hence the tight junction is impervious to concentrated protein solutions and appears to function as a diffusion barrier or "seal." The desmosome and probably also the *zonula adhaerens* may represent intercellular attachment devices.

Reprinted from The Journal of Cell Biology, 1962, Vol. 17, No. 2, pp. 375–412
Printed in U.S.A.

Specialized intercellular junctions, known as desmosomes and terminal bars, have been studied extensively during the last few years (*cf.* 1–4) by electron microscopy. As originally suggested by the light microscopical observations of Bizzozero (5) and Schaffer (6), these junctions are now recognized as local modifications of the surface of adjacent yet separate cells, rather than as intercellular bridges (*cf.* 4). As far as their functional significance is concerned, two distinct aspects should be considered: first, their function in cell-to-cell adhesion, and, secondly, their effect upon epithelial permeability. Earlier studies (2–4, 7, 8) have primarily or exclusively stressed the first aspect, as evidenced by the frequent use of the terms "adhesion plate" (9) and "attachment belt" (10) as synonyms for desmosome and terminal bar. The second aspect has only recently received some attention (11–15) in several laboratories, including ours (16, 17).

For this study we have surveyed a number of epithelia, particularly those of the mucosae of cavitary organs, and in all cases investigated we have found a characteristic junctional complex between adjacent epithelial cells. The arrangement varies in detail from one epithelium to another but typically consists of three successive components to be described as tight junction (*zonula occludens*), intermediate junction (*zonula* or *fascia adhaerens*), and desmosome (*macula adhaerens*). The tight junction is located the closest to the lumen and the desmosome the farthest away from it.

Experiments in which concentrated solutions of hemoglobin were used as a mass tracer indicate that it is the tight junction, rather than the desmosome, which acts as a barrier to the diffusion of the tracer, and suggest that this element of the junctional complex seals off the lumen from the intercellular spaces.

MATERIALS AND METHODS

Materials

Our observations were carried out on the following species and epithelia:

RAT: The absorptive epithelia of the stomach, jejunum, and colon; the glandular epithelia of the stomach, pancreas, liver, parotid, and thyroid; the duct epithelia of the liver, pancreas, and parotid; and the nephron epithelium (proximal and distal convolution, and collecting ducts).

GUINEA PIG: The absorptive epithelia of the segments of the digestive tract mentioned above for the rat, plus that of the gall bladder; the glandular epithelia of the stomach, liver, and pancreas; the duct epithelia of the pancreas and liver; the epithelia of the mucosae of the uterus and oviduct.

Tracer Experiments

The experiments in which hemoglobin was used as a mass tracer were carried out on 3 normal and 2 nephrotic (*cf.* 16) rats. Two injections, each of 1 gm of 2 × crystallized bovine hemoglobin (Pentex Inc., Kankakee, Illinois) in ~5 ml saline, were administered intraperitoneally, the first at 24 and the last at 16 hours before collecting kidney specimens. Two additional rats (one normal, the other nephrotic) received hemoglobin infusions of approximately 4 gm in 15 ml saline over 15 minutes, kidney specimens being fixed 7 to 8 minutes after the completion of the infusion. Similar results were obtained in all cases (see page 396).

Fixation and Processing of Tissue

For intestine, stomach, gall bladder, uterus, and oviduct, fixation was started by injecting the fixative into the corresponding central cavity as recommended by Palay and Karlin (18). In the case of the pancreas, liver, parotid, and thyroid, pieces of tissue were removed directly from the anesthetized animals, and immersed in a drop of fixative in which they were subsequently trimmed. The fixation of the kidney was initiated by injecting the fixative *in situ* as previously described (19). All tissues were fixed for 1 to 1½ hours in 1 or 2 per cent osmium tetroxide (OsO_4) buffered at pH 7.6 with either acetate-Veronal (20) or phosphate buffer (21). Several tissues (intestine and kidney) were also fixed in potassium permanganate (22). Most specimens were subsequently dehydrated in graded ethanols and finally embedded either in methacrylate (1/5, methyl/butyl), Epon (23), or Araldite (23).

Some blocks of each tissue were stained by placing them for 30 minutes in 1 per cent phosphotungstic acid in absolute ethanol prior to infiltration (*cf.* 23). Other specimens were dehydrated in acetone and stained in potassium permanganate before embedding (24). Thin sections prepared from all these blocks were stained with uranyl acetate (25) or lead hydroxide according to Karnovsky (26). Finally some preparations were "doubly stained," first in uranyl acetate then in lead hydroxide.

Sections cut from Epon- or Araldite-embedded specimens were examined directly after staining, whereas those cut from methacrylate-embedded tissues were "sandwiched" with carbon prior to examination in the electron microscope (27).

Microscopy

Micrographs were taken at original magnifications of 12,000 to 40,000 with a Siemens Elmiskop I, operating at 60 or 80 kv, with a double condenser, and a 50 μ objective aperture.

OBSERVATIONS

Structure of the Junctional Complex

In all epithelia studied we have found a tripartite junctional complex located along the lateral surfaces of adjacent cells, in close proximity to the lumen. Although there are characteristic variations in detailed arrangement from one situation to another, similar complexes have been regularly encountered in all the epithelia listed under Materials and Methods. The complex occurs in a typical form in the epithelium of the intestinal mucosa (jejunum) where its detailed arrangement will be described. Since many of the structural details are similar to those encountered in other epithelia, they will be illustrated with micrographs taken not only from the intestine but from other organs as well.

Junctional Complex of the Intestinal Epithelium

Three morphologically distinct types of surface modifications can be identified along the sides of adjoining intestinal epithelial cells. Beginning nearest the lumen and proceeding away from it along the intercellular spaces, there is first a tight junction, followed by an intermediate junction, which in turn is frequently followed by a typical desmosome (Fig. 1).

TIGHT JUNCTION (ZONULA OCCLUDENS): The first element of the complex is located immediately behind the line of reflexion of the plasma membrane from the apical to the lateral surface of the cell body. Hence it is situated close to the lumen at the base of the striated border (Figs. 1, 2, and 4). At relatively low magnification, this junction is seen as an area of what appears to be an extreme narrowing of the intercellular "gap" which at this level measures only 70 to 90 A and sometimes is bisected by a faint intermediate line (Figs. 1 and 2). At higher magnifications it is apparent that the tight junction is actually a region in which the membranes of adjoining cells come together and fuse,[1] with resultant obliteration of the intercellular space, the intermediate line representing the merged outer leaflets of the adjoining cell membranes (Figs. 3 to 5). As originally described by Zetterqvist (29), and Sjöstrand and Zetterqvist (30), the plasma membrane on the luminal surface of the intestinal epithelial cell has a triple-layered structure readily visible in OsO_4-fixed tissues whenever the cell membrane is sectioned perpendicularly to its surface (Figs. 1, 2, and 4). More precisely, the membrane is composed of two dense layers ~40 A each, separated by a lighter ~30 A layer. The outer layer is slightly less dense and in places thinner than the inner one (Fig. 1). This slight asymmetry of the membrane is almost completely removed by $KMnO_4$ fixation (31, 32) or by $KMnO_4$ staining (see Fig. 2). The total thickness of the structure is ~110 A, or slightly greater than that of the usual "unit membrane" of Robertson (31, 32). In thin sections cut normally to the plane of the junction, the three layers of the cell membrane on the luminal surface of each cell can be followed to their point of mergence, usually located where the cell membranes turn from the apical to the lateral surfaces of the adjacent cells (Figs. 3 to 5). At this point the two outer leaflets of the converging membranes join to become the intermediate line of the junction. Thus each tight junction has a structure comparable to that of a single myelin lamella and corresponds to an "external compound membrane" of Robertson (33). Moreover, as in the case of myelin lamellae, in OsO_4-fixed tissues the intermediate line of the junction, which represents the fused outer leaflets of the two membranes, appears thinner and less dense than the inner leaflets on either side (Fig. 4). It becomes, however, as thick and as dense as the latter after $KMnO_4$ staining. Rather frequently this fusion line is discontinuous: it appears more as a series of dots than as a continuous structure, a zone of discontinuity frequently occurring immediately behind the point of mergence.

In normal sections, the tight junction follows a straight or slightly wavy course and extends in depth from 0.2 to 0.5 μ. At the basal end of the junction, the fusion line splits again at a low angle

[1] By membrane fusion we mean the mergence of the outer leaflets of the apposed membranes into a single leaflet of about the same thickness (20 to 30 A) as each of the contributory layers. The term is therefore used in a more general sense than by Robertson (28) for whom membrane "fusion" implies extensive elimination of the material of the outer leaflets (supposedly protein or mucoprotein).

into two lines: the emerging outer leaflets of the lateral cell membranes (Fig. 5)

Variations in this general arrangement are sometimes encountered and consist of one or several focal splittings of the fusion line within the junction followed by their refusion (Fig. 5). In some cases, the space separating the inner leaflets from one another is narrowed in places to <90 A with no intermediate line visible (Fig. 7). In such instances a reorganization of the fused membranes may be involved (see footnote 1).

In general there are no visible fibrillar differentiations in the subjacent cytoplasmic matrix along this element of the junctional complex although sometimes a thin accompanying zone of diffuse densification is encountered (Figs. 3 and 4).

Abbreviations for Figures

D, desmosome, *macula adhaerens*
L, lumen
Za, *Zonula adhaerens*
Zo, *Zonula occludens*
bb, brush border
cm, cell membrane
fl, fusion line in *zonula occludens*
il, inner leaflet of the cell membrane
ol, outer leaflet of the cell membrane
mv, microvilli
v, vacuole
z, zymogen granule

Figure 1

Junctional complex between two cells in the epithelium of the intestinal mucosa (rat). The tight junction (*zonula occludens*), located nearest the lumen, extends from arrow *1* to arrow *2*. The narrowing of the apparent intercellular "gap" (~90 A) is clearly visible, but the fusion line of the two apposed membranes cannot be clearly distinguished at this magnification. Note that there is relatively little accumulation of dense cytoplasmic material along this part of the complex.

The intermediate junction (*zonula adhaerens*) extends from arrow *2* to arrow *3*. A relatively wide intercellular space (~200 A) is maintained throughout the junction. Extensive condensation of cytoplasmic fibrils occurs as a fine feltwork along either side of the junction. This condensation is continuous with the terminal web (*tw*) into which the filamentous rootlets (*r*) of the microvilli penetrate. Plate-like densifications within the cytoplasmic feltwork can be seen along part of the junction, especially along the right side (*pi*).

The limits of a desmosome are marked by arrows *4* and *5*. This element is characterized by a wide intercellular space (~240 A) bisected by an intermediate line (*id*). Bundles of cytoplasmic fibrils (*fd*), coarser (diameter ≃80 A) and more distinct than those of the terminal web, converge into dense plates (*pd*) on each side of the desmosome. These plates are separated from the inner leaflets of the cell membrane by a zone of low density. Similar fibrils (*ff*) appear throughout the remainder of the field below the terminal web.

Between the intermediate junction and the desmosome, the two apposed cell membranes are separated by an irregular space of varying width and show membrane invaginations and associated vesicles (*v*). The trilaminar structure of the cell membrane (*cm*) shows clearly along the microvilli (*mv*), (wherever the membrane is sectioned normally), and within the desmosome. It can also be made out, though less regularly, along the lateral cell margins (*e.g.*, at unnumbered arrow). Note that the luminal membrane is nearly symmetrical, the outer leaflet being only slightly thinner and less dense than the inner leaflet, whereas the lateral membrane is definitely asymmetrical. The total thickness of all three layers is about 110 A along the apical surface of the cell but measures only about 70 to 80 A along the lateral intercellular spaces. Note also the fluffy dense material (*fm*) (probably mucus) associated with the tips and sides of the microvilli.

Specimen fixed in 2 per cent OsO_4 in acetate-Veronal buffer (pH 7.6); and embedded in Epon. $Pb(OH)_2$-stained section. × 96,000.

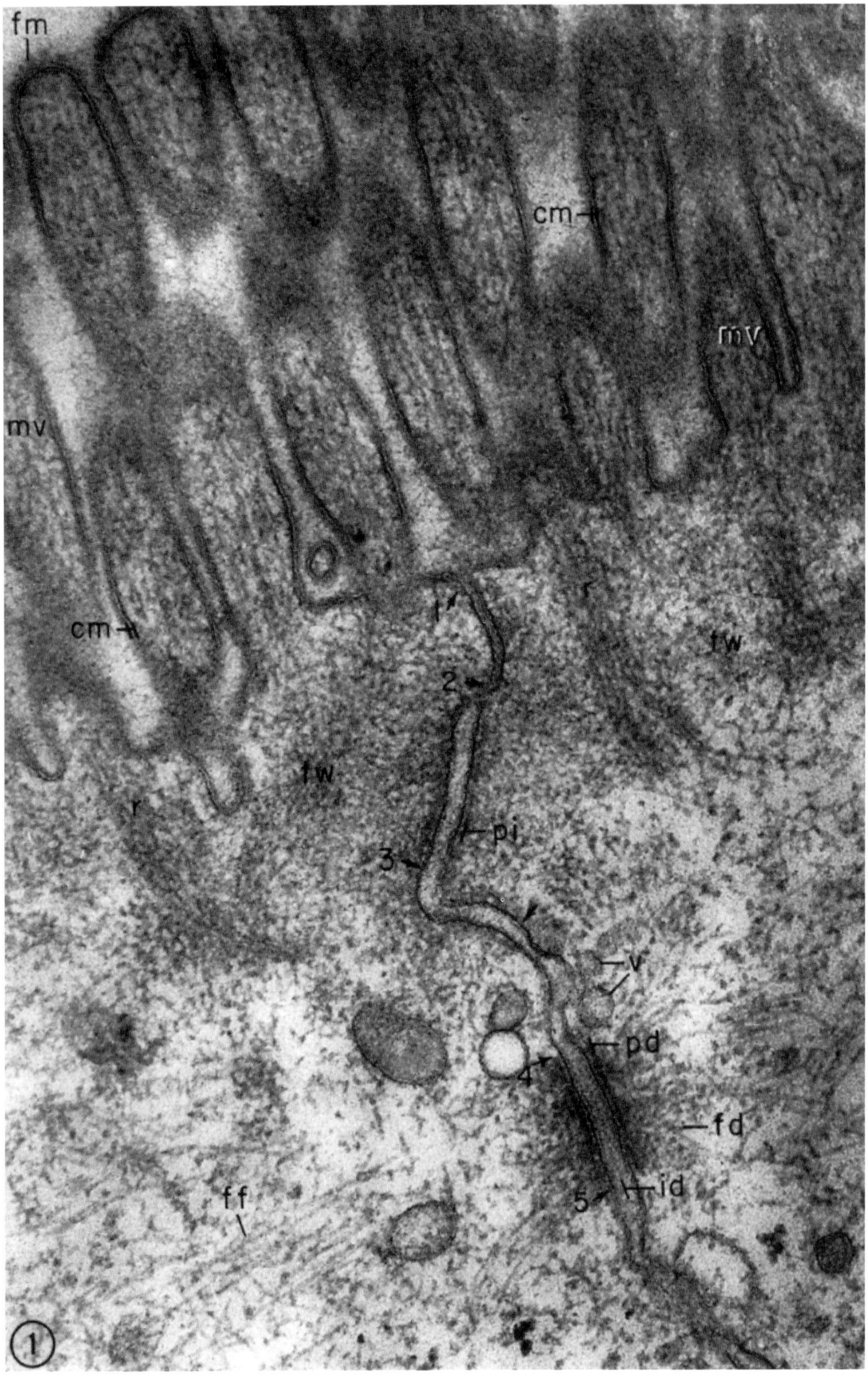
fm
cm
mv
mv
cm
1
2
fw
fw
r
r
pi
3
v
pd
4
fd
ff
5
id
1

Tight junctions are seen in the described location whenever the plane of section approaches normal to the plane of the intercellular boundary; moreover, oblique or grazing sections, cutting close to the luminal surface of the epithelium (Figs. 2 and 6), reveal that such junctions run without interruption for considerable distances along the apical perimeter of the cells. Accordingly it can be assumed that these junctions are continuous structures which surround each cell and bind together the plasma membranes of the entire epithelium. Admittedly neither a complete "belt" around a given cell, nor belt continuity throughout the epithelium has yet been demonstrated; however, it should be stressed that a "loose" type of junction has never been seen so far in the location described, and that points of continuity from one belt to another are frequently seen in favorably oriented sections (Fig. 10).

In view of these findings we propose that the first element of the junction complex be called *zonula occludens*, which means closing belt in Latin, the language used in the international morphological nomenclature. The term describes better than "tight junction" the main structural features of this element.

INTERMEDIATE JUNCTION (ZONULA ADHAERENS): The second element of the junctional complex is located immediately behind the tight junction, between the latter and the desmosome—hence the name here used (Fig. 1). It extends in depth for ~0.3 to 0.5 μ, and is characterized by the presence of a true intercellular space, which measures ~200 A across, and by a conspicuous densification of the subjacent cytoplasmic matrices. In normal sections the plasma membranes of the adjoining cells clearly show a triple-layered structure. However, here as well as along the rest of the lateral surface of the cells (Fig. 1), the outer leaflet is less dense and less clearly outlined than the inner one, to the extent that it is sometimes difficult to demonstrate. In addition, at this level the cell membrane measures ~80 A, the inner, middle, and outer layers accounting for ~30 A, ~25 A, and ~25 A, respectively; these dimensions remain essentially the same along the entire lateral cell surface (*cf.* Fig. 1). The cell membrane is, therefore, thinner and more asymmetrical on the lateral than on the apical aspect of the cell (*cf.* 34). Throughout the intermediate junction, the apposed membranes run strictly parallel to one another over a straight or only slightly wavy course. This arrangement contrasts markedly with the meandering and

FIGURE 2

Section cutting normally through a tight junction (*zonula occludens*) close to the luminal surface of the intestinal epithelium (rat). The junction extends diagonally across the micrograph from the lower left to the upper right corner (arrows) over a distance of 6 μ. The narrow apparent gap, (*i.e.* the space separating the inner leaflets of the apposed cell membranes), is visible all along the junction, but the fusion line cannot be clearly distinguished at this magnification. The marker to the right indicates the dimensions of a usual intercellular space at the same magnification.

The triple-layered structure of the epithelial cell membrane (total thickness 110 A) is visible along the free cell margin and around the microvilli (*mv*) which are cut mostly in cross-section. Note that the usual slight asymmetry of the apical cell membrane has been virtually removed by $KMnO_4$-staining (compare with Fig. 1).

Specimen fixed in 1 per cent OsO_4 in phosphate buffer (pH 7.6), dehydrated in acetone, stained in block with $KMnO_4$, and embedded in Epon. $Pb(OH)_2$-stained section. × 60,000.

FIGURE 3

Tight junction between two epithelial cells (mucous surface cells) of the gastric mucosa. The fusion line (*fl*) which represents the fused outer leaflets (*ol*) of the adjoining cell membranes is clearly visible all along the junction. In the case of the gastric epithelium, the outer leaflet of the cell membrane is slightly denser and thicker than in other epithelia, and accordingly the fusion line is more readily visible. Note the slight condensation of cytoplasmic material along the junction.

Specimen fixed in 1 per cent OsO_4 in phosphate buffer (pH 7.6) and embedded in Epon. $Pb(OH)_2$-stained section. × 230,000.

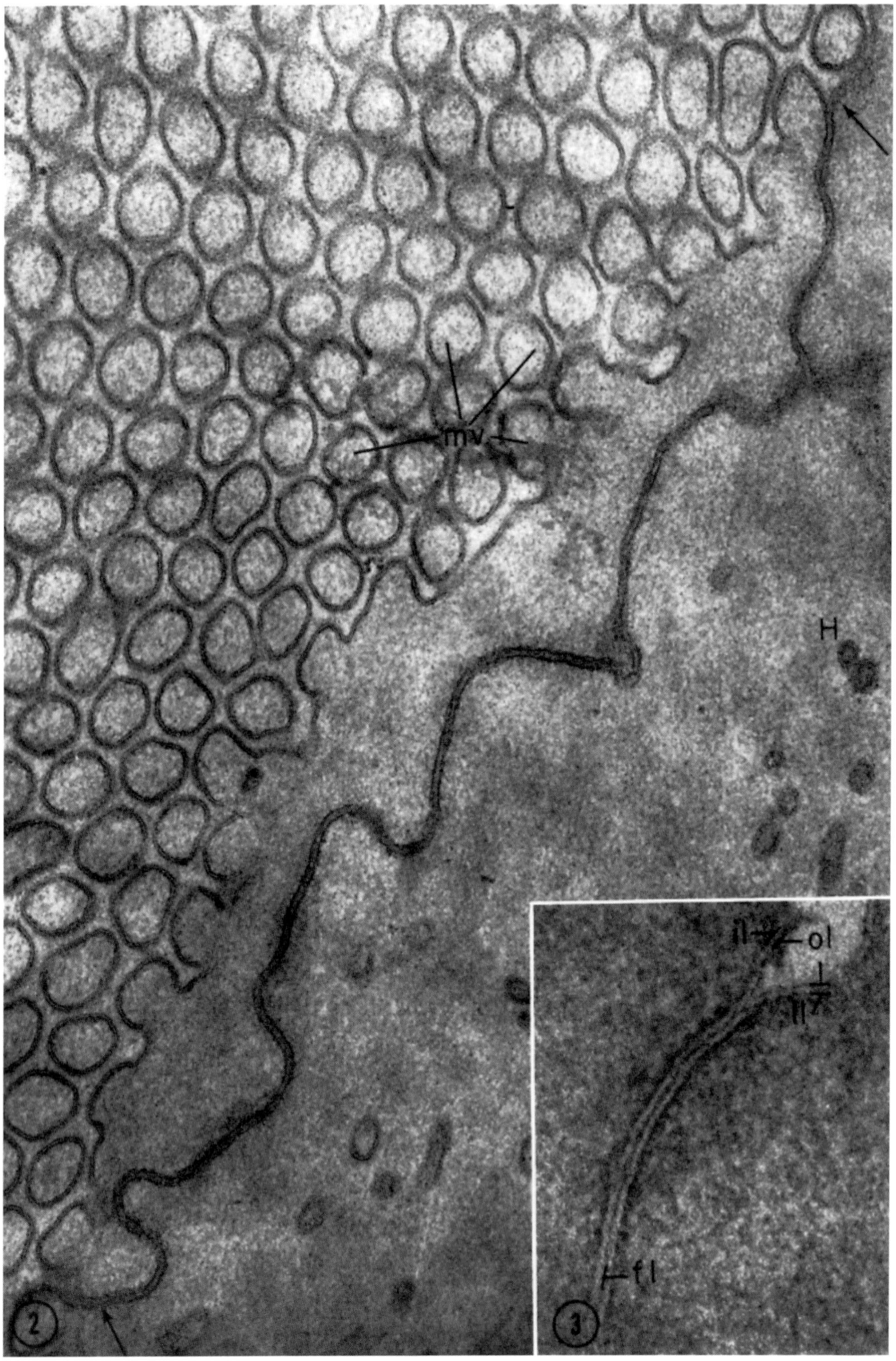
mv
H
il
ol
il
fl
2
3

sometimes divergent course followed by the same membranes in the basal direction beyond the junction.

The intercellular space of the intermediate junction is occupied by a homogeneous, apparently amorphous material of moderate density occasionally bisected by a faint band of higher density.

The zones of cytoplasmic densification along this junctional element have a tightly matted fine fibrillar texture (Fig. 1) with most of the fibrils seemingly running parallel to the cell surface. Occasionally the mat is further condensed into a plate running parallel to the cell membrane from which it is separated by a narrow light zone (Figs. 1 and 20). The fibrillar material in these dense zones appears to be a local condensation of the terminal web described Puchtler and Leblond (35).

Like the tight junction, this second element of the junctional complex can be readily identified in its characteristic position whenever the section approaches a plane normal to the boundary of the cell. It can also be followed without interruption over relatively long distances in oblique or grazing sections (*cf.* Figs 17 and 20). Hence we assume that it also forms a continuous belt around each cell and a continuous two-dimensional network throughout the epithelium, at least in the intestine.

We propose to call this element of the junctional complex *zonula adhaerens*, *i.e.* adhering-belt, or *fascia adhaerens*, *i.e.* adhering band. The first form applies to situations similar to one described in the intestinal epithelium; the second to those encountered in other epithelia (see below) in which the intermediary junction is not continuous around the perimeter of each cell. The new terms refer to structural and functional rather than topographical details, hence their application is more general. The intermediate position of the junction, stressed by the term originally used, is not apparent in many sections, since the desmosomes (see below) are discontinuous and frequently sparse structures. Moreover sometimes they seem to be altogether absent.

THE DESMOSOME (MACULA ADHAERENS): The third element of the junctional complex is usually located at a distance of 0.2 μ or more from the basal end of the intermediate junction. It is characterized by the presence of laminar densities in the intercellular space, and by high local concentrations of dense amorphous and fibrillar material in the subjacent cytoplasmic matrix (Fig. 1).

Normal sections show that the desmosome con-

FIGURES 4 AND 5

Detailed structure of tight junctions (*zonulae occludentes*) in the epithelium of the intestinal (Fig. 4) and gastric (Fig. 5) mucosa (rat) shown at high magnification. Each tight junction (between arrows *1* and *2*) can be recognized as an area in which the lateral cell membranes of two adjoining cells come together and fuse with resultant obliteration of the intercellular space. Because of the favorable orientation of the sections, the three-layered membranes which cover the luminal surface of the two adjoining cells can be followed to their point of mergence (arrow *1*), located near the point where the membranes turn from the apical to the lateral surfaces of the cells. At this point the two outer leaflets (*ol*) of the converging membranes fuse to form the intermediate or fusion line (*fl*) of the junction. In Fig. 4 the fusion line appears thinner and less dense than the inner leaflets on either side. In Fig. 5, the fusion line is denser and thicker presumably because the outer leaflet of the cell membrane of gastric epithelial cells is noticeably thicker and denser than the corresponding leaflet in the intestinal epithelium. In Fig. 4 the membranes remain fused throughout the junction while in Fig. 5 there are several focal splittings of the fusion line within the junction followed by their refusion. The apparent discontinuity in the outline of the leaflets ~30 mμ before the merging point in Fig. 5 is due to the twisting of the membranes within the thickness of the section.

An intermediate junction (*zonula adhaerens*) (arrows *2* to *3*) is included in Fig. 5. At its level the intercellular space is widened to ~200 A, and is occupied by a moderately dense, amorphous material.

Specimens fixed in 1 per cent OsO_4 in phosphate buffer (pH 7.6) and embedded in Epon. $Pb(OH)_2$-stained sections. Fig. 4, × 200,000; Fig. 5, × 220,000.

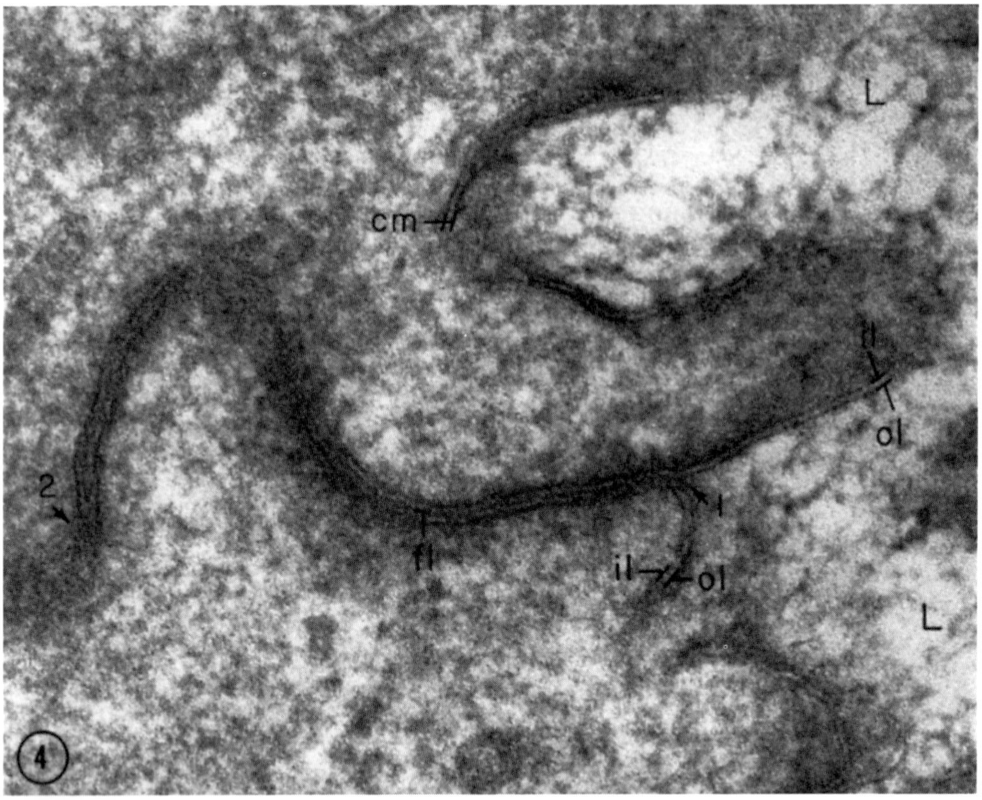
L
cm
il
ol
2
fl
l
il
ol
L
4

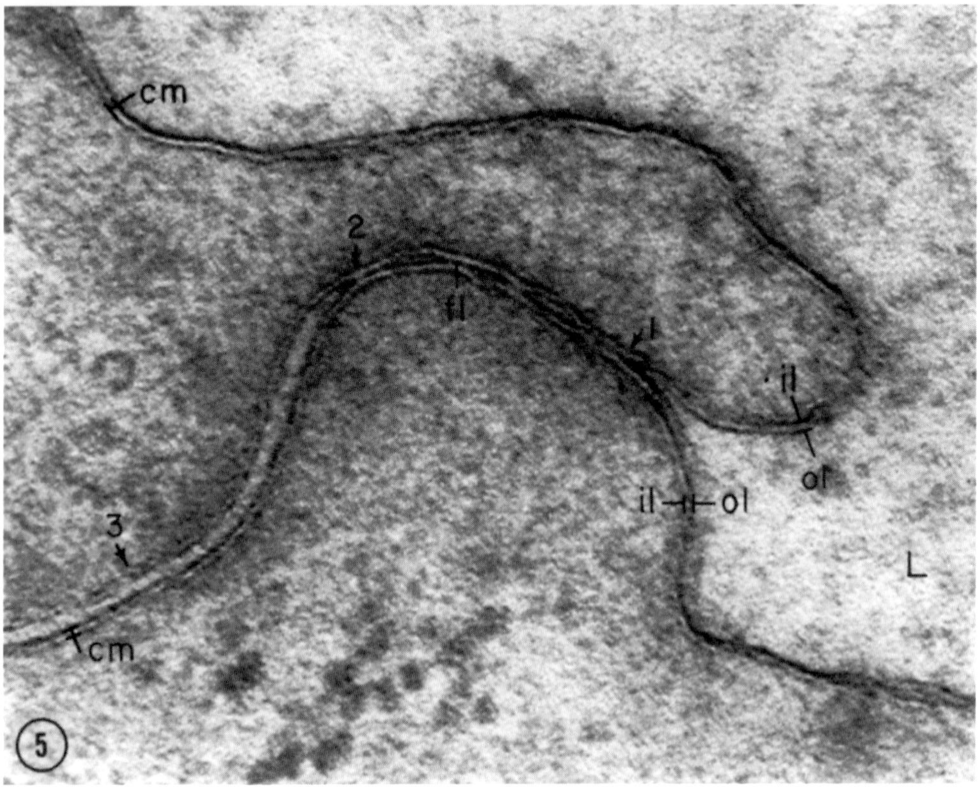
cm
2
fl
l
il
ol
il
ol
3
L
cm
5

sists of 2 straight plaques of dense material each disposed parallel to the inner leaflet of the corresponding plasma membrane (*cf.* 1, 3, 7, 36–38) and separated therefrom by a light space of ~40 A. The two apposed plaques are aligned in perfect phase, measure 0.2 to 0.3 μ in length, and appear to be of circular or oval shape. They are arranged in strict parallelism with respect to one another and seem to be rather rigid; they are rarely curved, but the cell membrane may be sharply bent at their margins. The unit membrane can be clearly followed through the desmosome, the only visible modification being a densification of the inner leaflet in phase with the plaques (Figs. 1, 8, and 9).

At this level the intercellular space measures ~240 A across and is usually occupied by a disc of moderately dense material bisected by a denser central layer referred to in the literature as "intercellular contact layer" (3)[2] or "median stratum" (38).

Bundles of cytoplasmic fibrils, coarser and more distinct than those of the terminal web, converge on the inner aspect of each desmosomal plaque (Fig. 1), sometimes obscuring its outlines (*cf.* Fig. 12). Most of these fibers approach the plaque at a high angle, but occasionally the bundles run parallel to it and fray out into the cytoplasmic matrix at its margins. Within each convergent bundle of fibrils one or two plate-like zones of further densification are frequently seen running parallel to the main plaque.

In contradistinction to the first and second element of the complex, the desmosome is not present in all normal sections, is not continuous over long distances, and therefore appears to be a discontinuous, button-like—rather than a continuous, belt-like—type of attachment (Fig. 12). In fact more than one row of desmosomes probably occur along the sides of the cells, since several such structures can frequently be detected along a given intercellular space.

In keeping with the rest of our nomenclature, we propose the term *macula adhaerens*, *i.e.* adhering spot, as an alternate for desmosome. It has the advantage of describing one of the main features of the structure, *i.e.* its discontinuous button-like character, and of distinguishing it from other "bodies" or structural entities involved in cell-to-cell attachment.

GENERAL REMARKS: The junctional complex described is encountered among all cells of the intestinal epithelium irrespective of their differentiation: for instance, it binds absorptive epithelial cells to goblet cells (Fig. 18) or to other glandular cells in the crypts. So far, no evidence has been obtained about the fate of the junctional complex during sloughing off of the epithelium or leucocyte migration. As a rule, the cell membrane directly involved in junctional elements is free of invaginations or associated pinocytic vesicles; however, such structures are encountered along

[2] Originally Odland (3) assumed that the "intercellular contact layer" was flanked by two additional extracellular dense bands referred to as "intermediate dense layers." However, as shown by Karrer (38) and confirmed by us, the latter actually correspond to the outer leaflet of the apposed cell membranes which may appear slightly denser at this level than along the rest of the lateral aspect of the cell.

FIGURE 6

Oblique section through a gastric gland (rat) near the luminal surface of two parietal cells. A tight junction is cut over a long distance and extends diagonally across the field from the lower left to upper right (arrows). The fusion line (*fl*) can be followed in most places throughout the junction. The upper left corner is occupied by a zymogenic cell. Note that in this case, as in that of the covering epithelia of the gastric mucosa (Fig. 5), the usual asymmetry of the apical cell membrane is reversed; *i.e.*, the outer leaflet appears denser and thicker than the inner one.

FIGURE 7

Tight junction between two epithelial cells of the gastric mucosa (rat) showing an area (arrow) at the initial point of membrane fusion where the space separating the inner leaflets of the fused membranes is narrowed to 100 A and no intermediate line is visible.

Specimens fixed in 1 per cent OsO_4 phosphate buffer (pH 7.6) and embedded in Epon. $Pb(OH)_2$-stained sections. Fig. 6, × 145,000; Fig. 7, × 185,000.

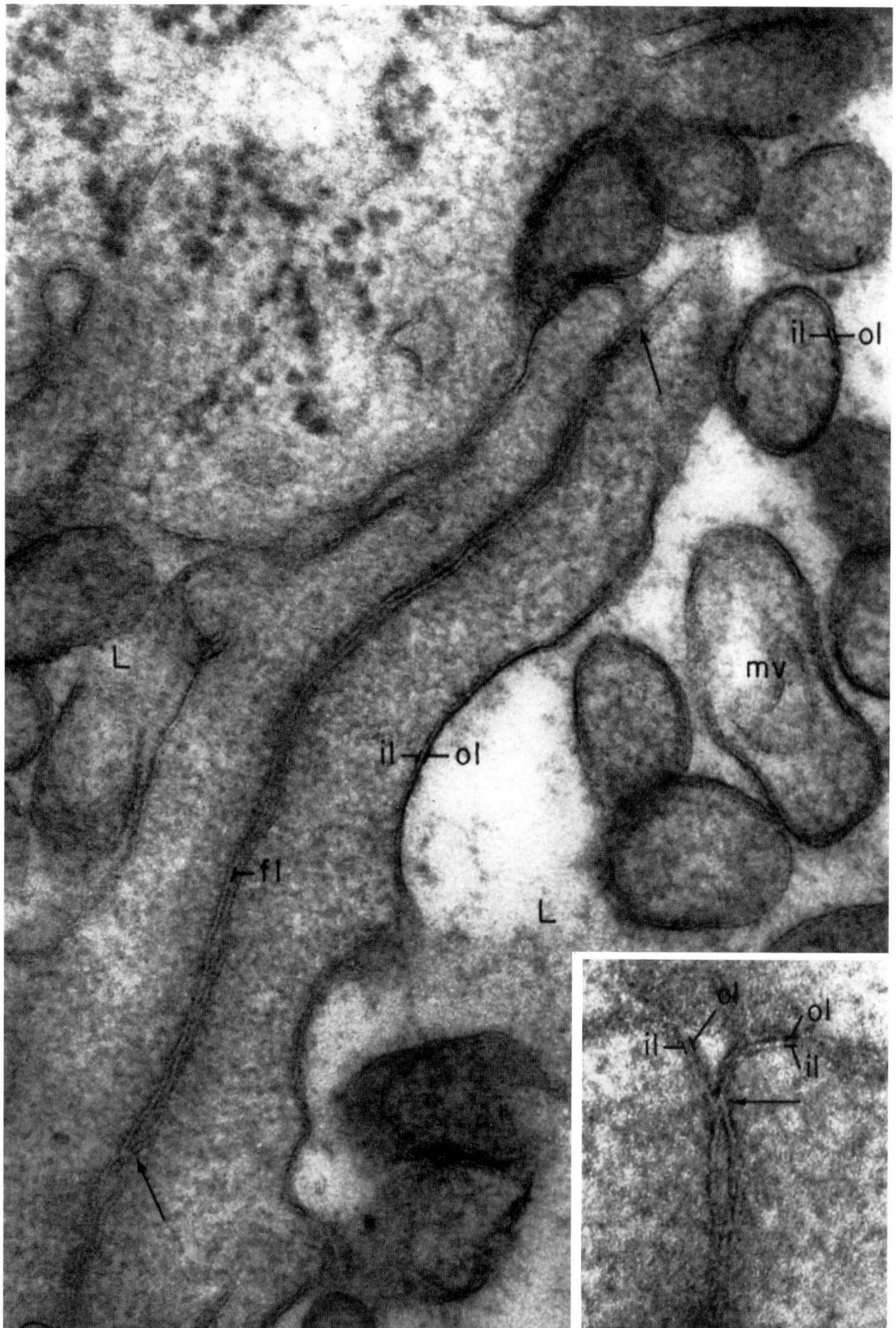
il
ol
L
mv
il
ol
fl
L
6
ol
il
ol
il
7

the membrane which connects one element of the complex to another (Fig. 1).

Hypotonic treatment, carried out by injecting distilled water into the lumen of an isolated intestinal loop for 15 to 30 minutes prior to OsO_4 fixation, does not cause opening of the tight junctions. It is noteworthy that in such specimens the outer leaflet of the apical cell membrane appears thicker and denser than in untreated specimens; consequently the fusion line of the tight junctions is more readily visible.[3]

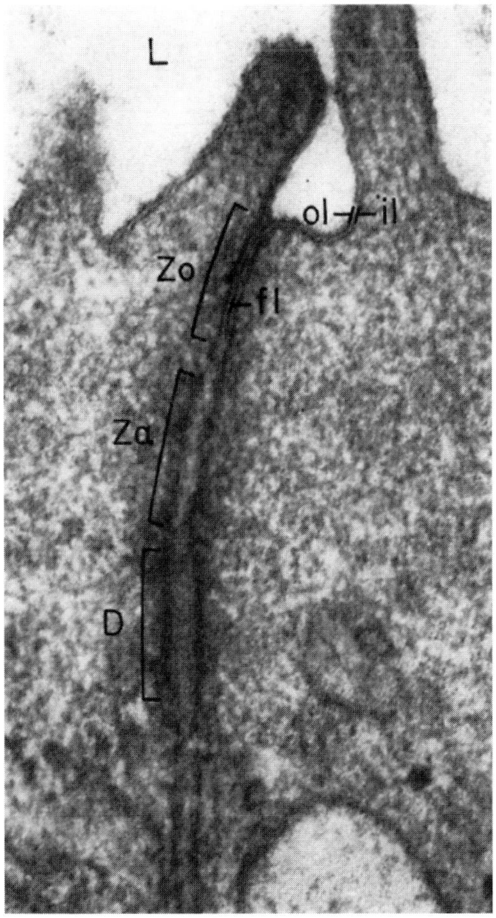

Junctional Complexes in the Epithelia of Other Segments of the Digestive Tract

The complexes joining the epithelial cells of the gastric or colonic mucosae are entirely similar to those described in the epithelium of the jejunal mucosa, only the fusion line of the *zonula occludens* is more easily demonstrated (Figs. 3 and 5). This

[3] In a systematic study of the effect of hypotonic treatment upon the membrane of the microvilli, Millington and Finean (39) noted an increase in the thickness of the membrane to 140 A after longer intervals than used by us. They do not mention the increase in thickness of the outer leaflets which in their case may have been obscured by PTA staining. Their observations indicate that after PTA staining the asymmetry of the membrane is actually reversed, the outer leaflet being slightly denser and thicker (by ~10 A) than the inner one.

Figure 8

Junctional complex between two epithelial cells of the mucosa of the gall bladder (guinea pig), showing the occluding zonule (*Zo*) followed by adhering zonule (*Za*), followed by the desmosome (*D*). The trilaminar structure of the apical cell membrane can be clearly distinguished and traced into the occluding zonule. The fusion line (*fl*) is also visible within the junction.

Specimen fixed in 1 per cent OsO_4 in phosphate buffer (pH 7.6) and embedded in Araldite. $Pb(OH)_2$-stained section. × 155,000.

Figure 9

Junctional complex between two epithelial cells of a thyroid follicle (rat).

The occluding zonule stretches from arrow *1* to arrow *4* with an interruption between arrows *2* and *3*, probably due to the fact that at this level the zonule is bent and the section cuts "below" its basal margin. The fusion line (*fl*) is visible within both segments of the zonule but the merging of the outer leaflets (*ol*) of the cell membranes into this line is not clearly shown (see Fig. 11). An adhering zonule can be recognized between arrows *5* and *6*, and a desmosome between arrows *7* and *8*. A few mμ to the left of arrow *7* the section probably touches the margin of another desmosome.

Specimen fixed in 2 per cent OsO_4 in acetate-Veronal buffer (pH 7.6), dehydrated in acetone, stained in block with $KMnO_4$, and embedded in Epon. Section stained in uranyl acetate and $Pb(OH)_2$. × 145,000.

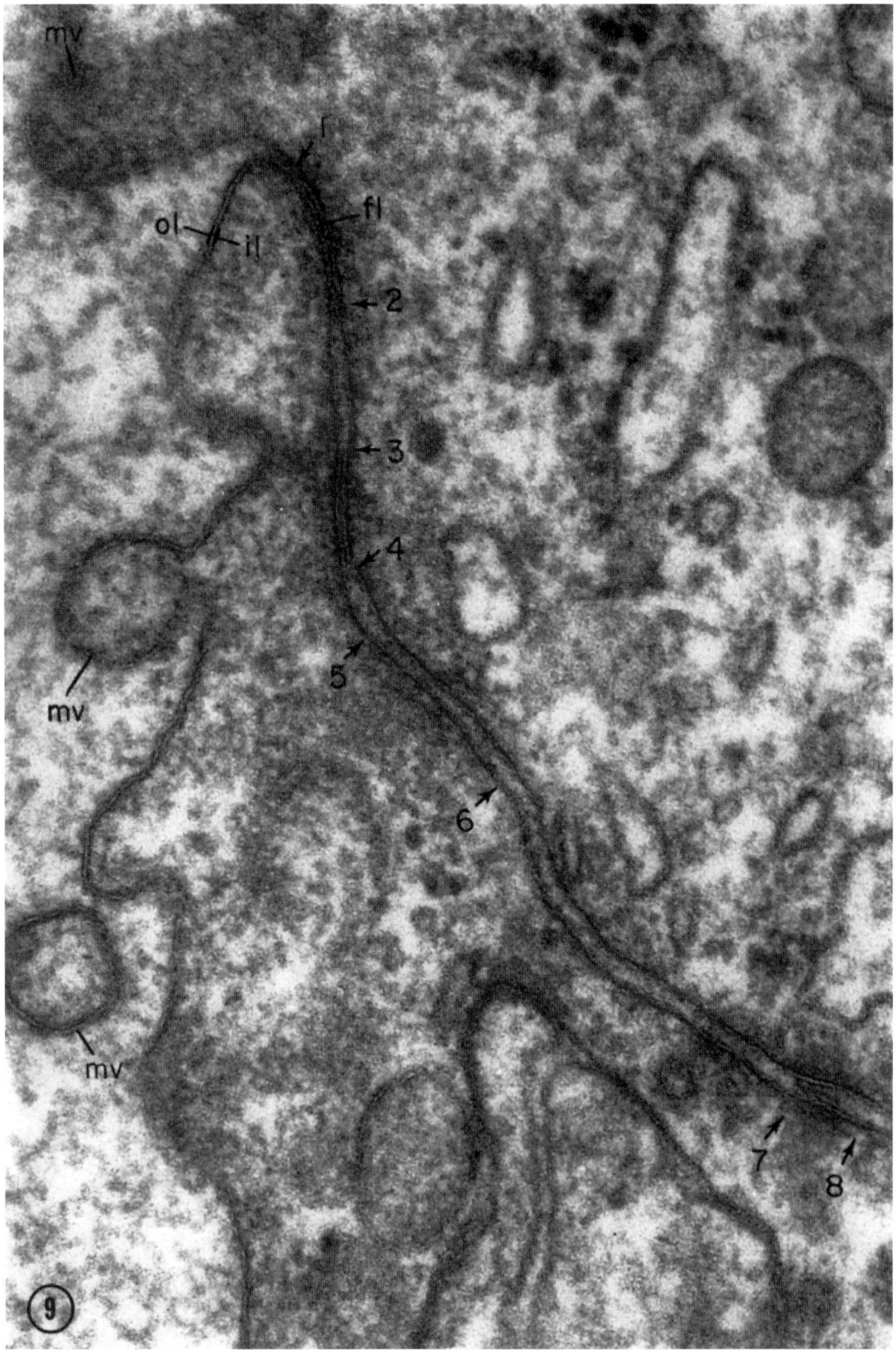
mv
l
fl
ol
il
2
3
4
mv
5
6
mv
7
8
9

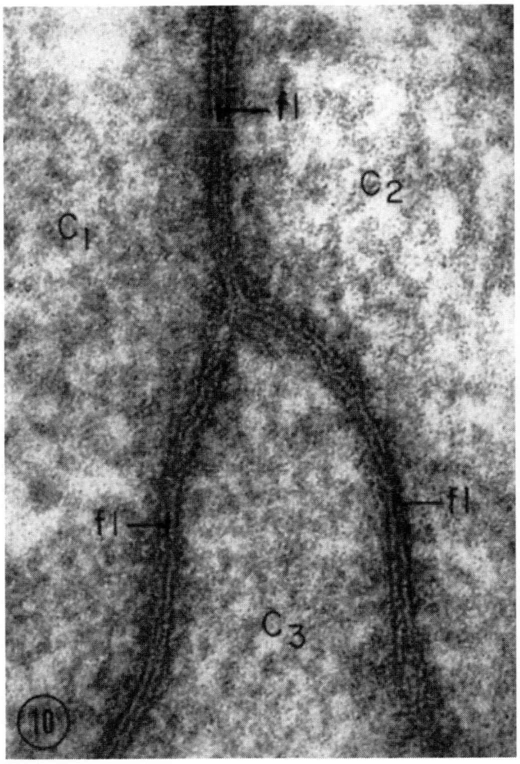

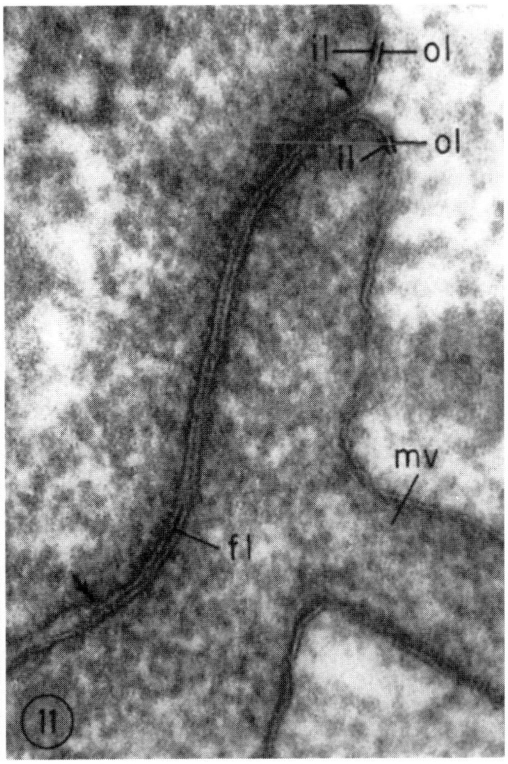

FIGURE 10

Thyroid follicle (rat). The section cuts through the epithelium parallel and immediately basal to the luminal surface.

The occluding zonules of three adjacent cells (C_1, C_2, C_3) merge into an inverted Y figure. Fusion lines (*fl*) are seen in each branch of the Y and appear to be continuous from one branch to another. × 190,000.

FIGURE 11

Zonula occludens in a thyroid follicle (rat).

The tight element of the junctional complex extends between the arrows. The mergence of the outer leaflets (*ol*) of the cell membranes into a fusion line (*fl*) is clearly shown. Note the bands of condensed cytoplasmic matrix associated with the occluding zonule and continuing beyond it. Specimens and sections prepared as for Fig. 9. × 130,000.

is due to the fact that, especially in the epithelium of the gastric mucosa, the outer leaflet of the apical cell membrane is slightly denser and thicker than the inner leaflet. Hence the asymmetry of the membrane is reversed in comparison to that already described for the apical membrane of intestinal epithelia. The lateral cell membrane in the gastric epithelium shows the same type of "reversed" asymmetry (Fig. 5).

JUNCTIONAL COMPLEXES IN GLAND AND DUCT EPITHELIA

Entirely similar complexes are encountered in the gastric glands (Fig. 6) and the gall bladder epithelium (Fig. 8), except that in the latter the depth of the *zonula adhaerens* is more variable.

A slight variant occurs in the thyroid (Figs. 9 to 12), the liver (Figs. 13 and 14), the pancreas (Figs. 15 to 17), and the parotid. In all these cases,

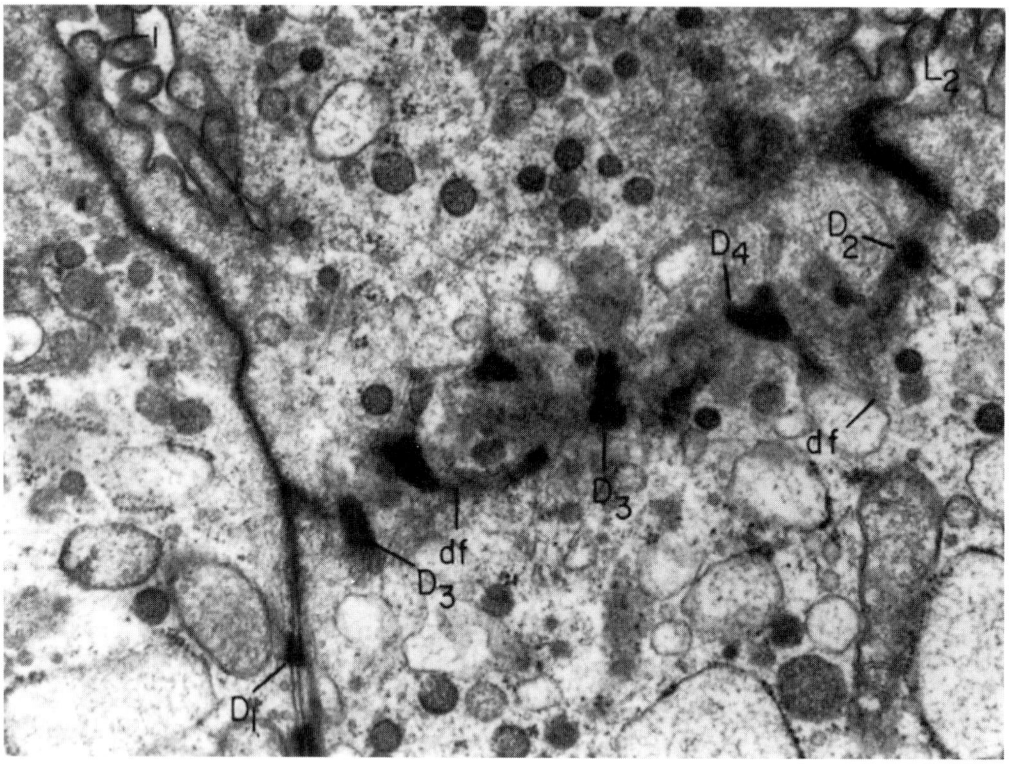

FIGURE 12

Thyroid follicle (rat).

Oblique section which touches the follicular lumen at L_1 and L_2; cuts normally through a junctional complex from L_1 to the desmosome marked D_1; and grazes the surface of one of the cells joined in the complex (upper part of the field).

The dense more or less circular (D_2), oval (D_3) and "lobated" (D_4) dense regions are desmosomal plates partially or entirely included in the thickness of the section. The micrograph shows the discontinuous, "macular" character of the desmosomes, and the bundles of fibrils (*df*) that converge on their plates.

Specimen and section prepared as for Fig. 9. × 26,000.

the outer leaflet of the plasma membrane appears as a faint discontinuous band after OsO_4 fixation, and correspondingly the fusion line of the tight junction is either poorly outlined or entirely invisible (Figs. 15 and 16). Staining the tissue in block with phosphotungstic acid or in sections with uranyl or lead does not substantially improve the visualization of the unit membrane; however, staining in block with $KMnO_4$ (24) or double staining the sections (first with uranyl and then with lead) increases the density of the outer leaflet and fusion line (Figs. 9 to 11, 14, 29, and 30). From such preparations it can be clearly appreciated that the luminal membrane of these cells differs from the apical membrane of the intestinal epithelia: it is more asymmetrical, for the two dense leaflets are more unequal in thickness and density; and it is finer, since the over-all thickness of the unit does not exceed 70 A.

In the case of most glandular epithelia (Figs. 9 to 11 and 13 to 17) as well as the epithelium of the gall bladder (Fig. 8), a high concentration of dense material is visible in the cytoplasmic matrix along the *zonula occludens* and continues without noticeable change along the *zonula adhaerens*.

In glandular epithelia, adhering zonules although generally present show considerable variation in the extent of their development. It is probable that in some cases they are discontinuous, therefore *fasciae* rather than *zonulae*. Occasionally they are missing and in such cases the occluding zonule may be immediately followed by a typical desmosome. Desmosomes are present in all these epithelia but their size and frequency vary noticeably from one epithelium to another: they are relatively small (Fig. 13) and infrequent in the liver, and unusually large and frequent in the thyroid (Figs. 9 and 12) and the parotid.

In all these cases, the same complex is used to join together all types of cells in a given epithelium irrespective of their differentiation. In the liver, for example, the same complex appears between parenchymatous liver cells on each side of the bile capillaries; between parenchymatous cells and duct cells at the confluence of bile capillaries and bile ducts; and between duct cells along the duct epithelium. In the gastric glands, typical junctional complexes bind parietal cells to either zymogenic cells, or mucous neck cells; and in the pancreas, they join—in succession—acinar cells to centroacinar (Fig. 16) and centroacinar to duct cells.

Junctional Complexes in the Epithelium of the Uterine and Oviductal Mucosae

The junctional complexes in the columnar epithelia of the mucosae of the uterus and oviduct are similar to those found in the gland and duct epithelia described above, except for a more pronounced variability in the development of the adhering zonules. In many cases the latter are rudimentary or altogether missing.

Junctional Complexes in the Nephron

Considerably more variation is found in the nephron epithelium in which the organization of the junctional complexes varies typically from one segment to another.

PROXIMAL TUBULE: In the proximal convolution, the occluding zonule is extremely shallow, being as a rule reduced to 200 to 400 A in depth (Figs. 19, 20, 24 to 26). Even so, it is clearly

FIGURE 13

Bile capillary between two parenchymatous liver cells (guinea pig). On each side, an occluding zonule can be easily recognized by the apparent narrowing of the intercellular "gap" between arrows *1* and *2*. Each tight junction is followed by an adhering zonule extending between arrows *2* and *3* and, farther away, by a desmosome marked *D*. A condensation of finely fibrillar material occurs mainly along the adhering zonules on each side of the bile capillary.

Specimen fixed in 1 per cent OsO_4 in phosphate buffer (pH 7.6) and embedded in Epon. Section stained in uranyl acetate and $Pb(OH)_2$. × 65,000.

FIGURE 14

Guinea pig liver, bile capillary.

An occluding zonule, with a distinct fusion line (*fl*) and associated bands of dense cytoplasmic material (*db*), closes the luminal end of an intercellular space. Microvilli protrude into the lumen, and small vesicles (*v*) limited by a distinct unit membrane, occur in the cytoplasm.

Specimen fixed in 1 per cent OsO_4 in phosphate buffer (pH 7.6), dehydrated in acetone, stained in block with $KMnO_4$, and embedded in Epon. Section stained with uranyl acetate and $Pb(OH)_2$. × 145,000.

FIGURE 15

An occluding zonule extends from arrow *1* to arrow *2* and closes the luminal end of the intercellular space in a pancreatic acinus (guinea pig).

The convergence of the outer leaflets (*ol*) of the two merging cell membranes can be seen quite clearly, but the fusion line within the zonule is barely visible in this preparation stained only with $Pb(OH)_2$ (compare with Fig. 14). The two wider spots within the zonule probably represent focal splittings.

Specimen fixed in 1 per cent OsO_4 in phosphate buffer (pH 7.6) and embedded in Epon. × 135,000.

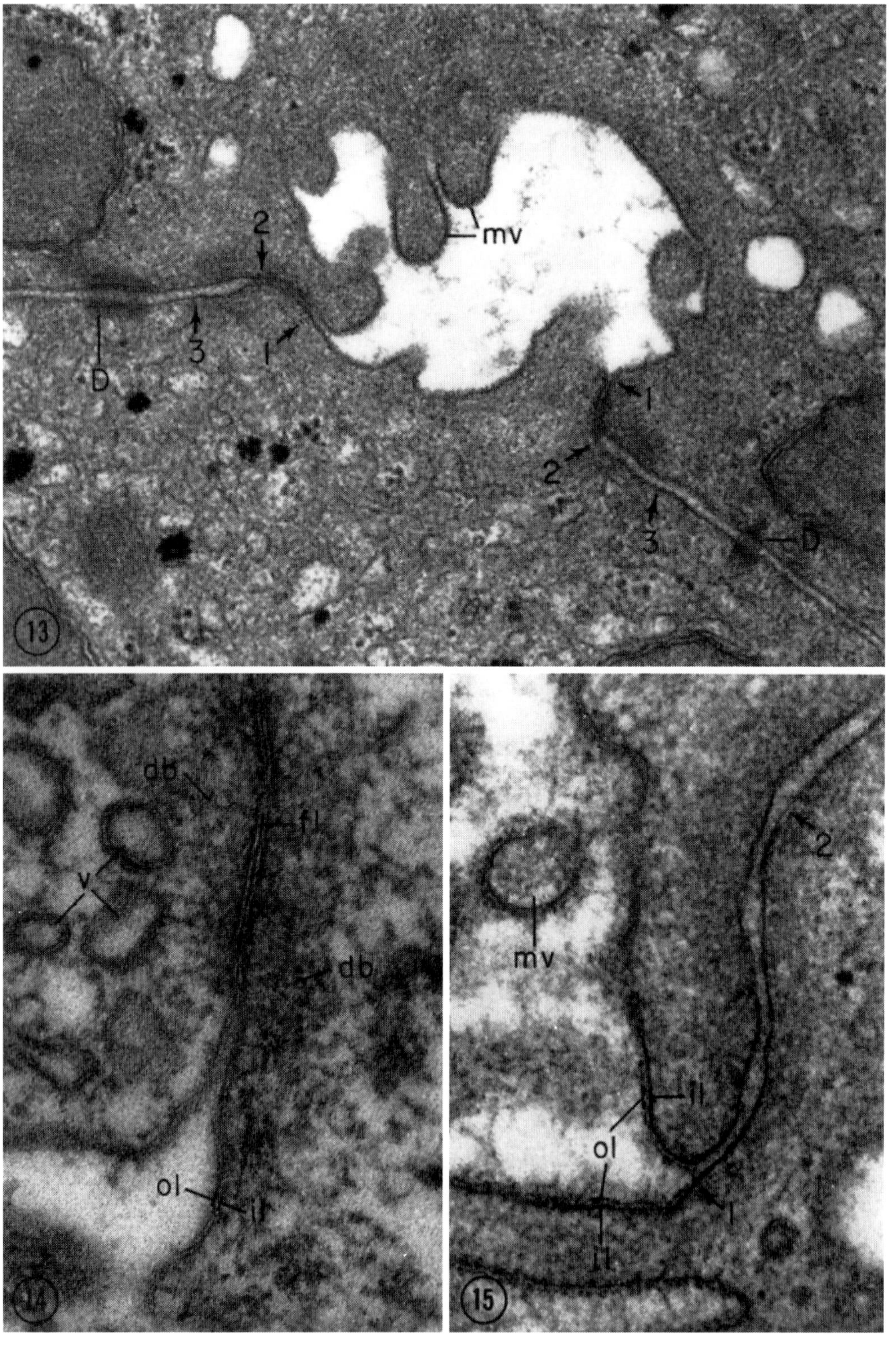
mv
2
3
1
D
2
1
2
3
D
13
db
fl
v
db
ol
il
14
2
mv
il
ol
1
il
15

recognizable, especially in sections normal to the luminal surface, as a localized stricture (~90 A) occurring near the lumen along the otherwise relatively open intercellular spaces. This stricture represents the space between the inner leaflets of the apposed cell membranes. The fusion line in this miniature tight junction is difficult to demonstrate, for as already noted by Miller (11), the outer leaflet of the apical plasma membrane is only rarely visible in OsO_4-fixed proximal convolutions. It can, however, be demonstrated in $KMnO_4$-fixed or stained specimens (Fig. 22). In such preparations the total thickness of the membrane, like that of gland and duct epithelia, does not exceed ~70 A. The adhering zonules, which follow immediately behind the tight strictures, are deep and usually associated with a continuous, relatively broad band of dense cytoplasmic material (Figs. 19, 20, and 24 to 26). At their level the intercellular space is occupied by an amorphous or finely fibrillar material of moderate density, whereas beyond the junctions the space appears "empty" (Figs. 24 to 26).

DISTAL TUBULE AND COLLECTING DUCT: The situation is almost reversed in the distal convolution and in the collecting tubules: the occluding zonule is generally well defined and extensively developed. It measures up to 0.3 μ in depth (Figs. 21, 23, and 27), and its fusion line is readily demonstrated (Figs. 23 and 27), for the outer leaflet of the apical cell membrane is well outlined in OsO_4-fixed specimens. Moreover the apical membrane is relatively thick: it measures ~110 A like the corresponding membrane in intestinal epithelia. Bands of dense cytoplasmic material accompany the occluding zonule along its entire course (Figs. 21 and 23) and continue along the second element of the complex when present (Figs. 23 and 27) (see below). This second element, however, shows marked variation in its development from one intercellular space to another: in some instances it is well defined, measures up to 0.2 μ in depth and contains, as in the proximal convolution, a moderately dense material in its gap (Figs. 23 and 27); in others, the junction is shallow and poorly developed; and finally in many cases no distinct differentiation of the cell surface can be recognized in the expected location (Fig. 21). It follows that in the distal convolution the second element of the junctional complex is not a continuous structure around each cell and throughout the epithelium; *i.e.*, a zonule. Accordingly it deserves the designation *fascia adhaerens*. As already noted, a similar situation seems to prevail in the epithelium of the mucosae of the uterus and oviduct and in that of the thyroid.

All along the epithelium of the nephron tubule, and especially in the distal convolutions, desmosomes are small and definitely less frequent than in other epithelia examined; the interval between

FIGURE 16

Section through the lumen of a pancreatic acinus (guinea pig) showing three junctional complexes between contiguous cells. Each occluding zonule runs from the lumen to the first arrow along the intercellular space and is immediately followed by an adhering zonule whose end is marked by the second arrow. Small desmosomes (D_1, D_2, D_3) complete each junctional complex.

The clearest image is given by the complex that includes D_1; the adhering zonule that precedes D_3 shows a "gap" narrowing in its middle portion, where the section presumably cuts again through the occluding zonule. The obliquity of the section blurs the image of the adhering zonule and of most of the occluding zonule that precede D_2; hence the position of the first arrow on this complex is uncertain.

Associated bands of dense cytoplasmic material begin at the occluding zonules and continue uninterrupted along the adhering zonules, becoming thicker at the level of the latter.

Note that although the upper cell is centroacinar, and the lower cells acinar (as indicated by the zymogen granules (Z) they contain), they are attached to one another by similar complexes.

Specimen fixed in 1 per cent OsO_4 in phosphate buffer (pH 7.6) and embedded in Epon. Section stained in uranyl acetate and $Pb(OH)_2$. × 75,000.

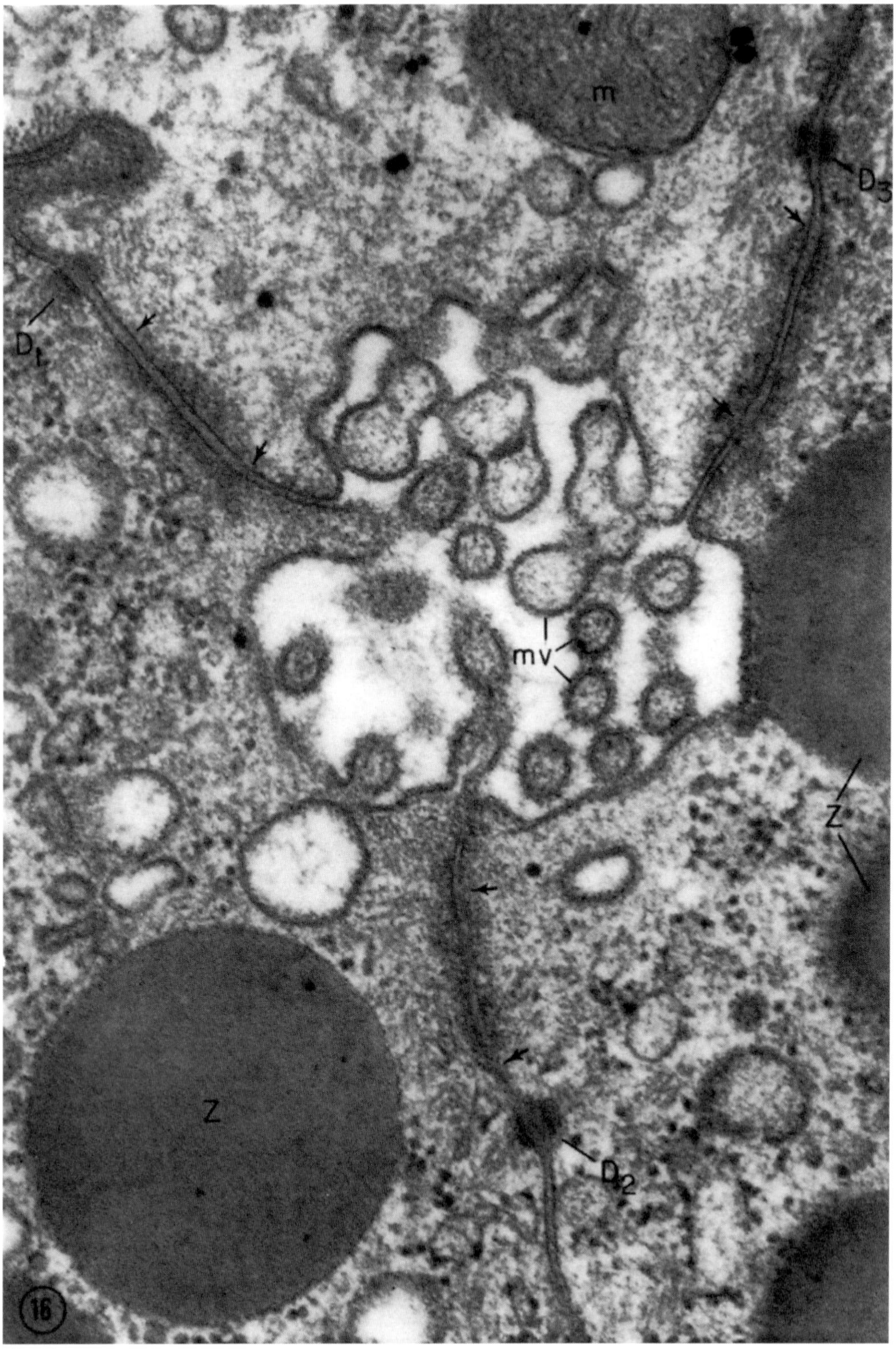
m
D3
D1
mv
Z
Z
D2
16

them and the adhering zonules or fasciae is more variable and generally greater (Fig. 19).

GLOMERULAR EPITHELIUM: A far reaching exception to the junctional pattern we have described is encountered at the beginning of the nephron where the cells of the visceral glomerular epithelium appear to be joined together in an entirely different manner: the junction zone is

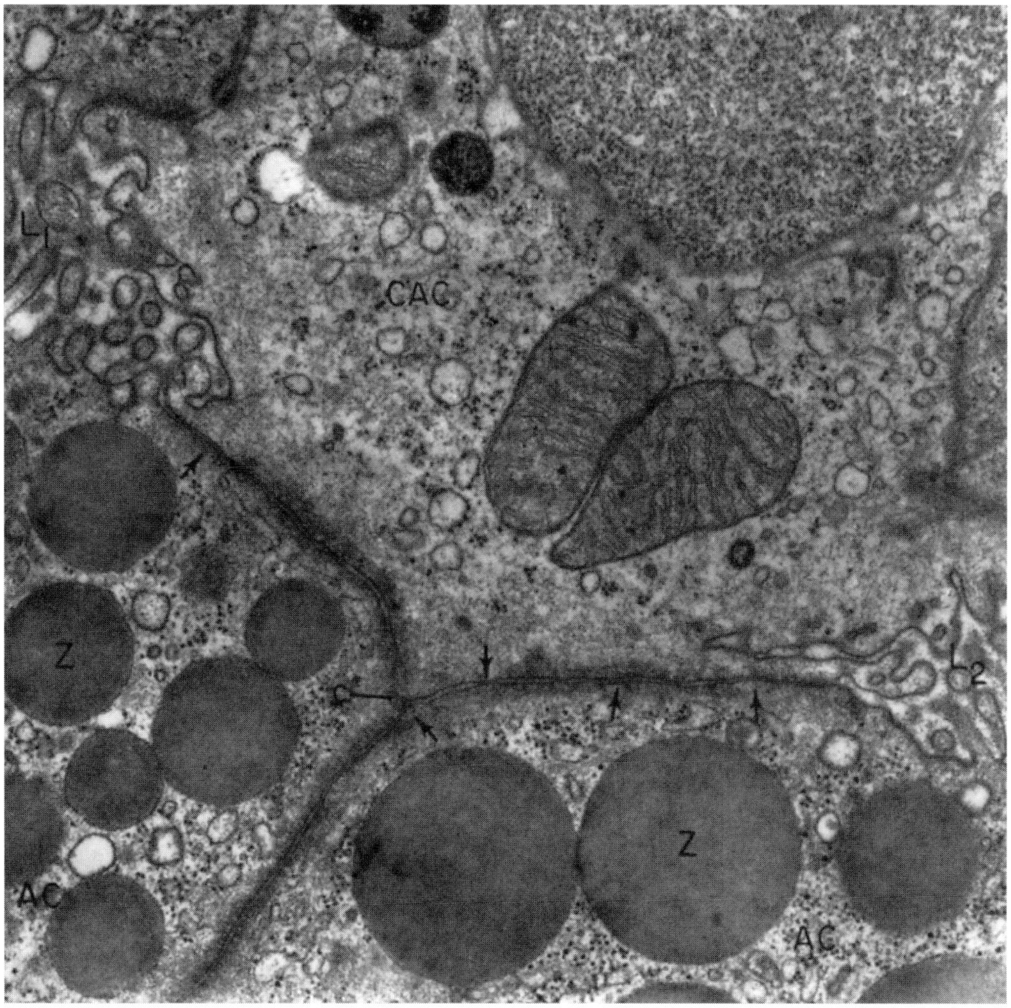

FIGURE 17

Pancreatic acinus (guinea pig).

Between the acinar lumina L_1 and L_2, the section cuts broadly through junctional complexes at the level of occluding and adhering zonules. The latter can be recognized by the apparent narrowing of the intercellular gap (arrows). The micrograph shows clearly the broad band of dense cytoplasmic material associated with both the occluding and adhering zonules. Note the continuity of these elements of the complex in the plane of the epithelium and the continuity of the associated bands of condensed cytoplasmic material over the point of confluence of three intercellular spaces (*c*). *AC*, acinar cell; *CAC*, centroacinar cell.

Specimen fixed in 1 per cent OsO_4 in phosphate buffer (pH 7.6) and embedded in Epon. Section stained in uranyl acetate and $Pb(OH)_2$. × 26,000.

reduced in depth (19),[4] and the junction line is extremely tortuous, because of the well known extensive interdigitation of the foot processes of the epithelial cells. No junctional elements directly comparable to an occluding or adhering zonule are encountered in mature glomeruli. The zone of closest apposition, *i.e.* the urinary slit, has some structural features reminiscent of desmosomes: in grazing sections it appears as a relatively large gap (200 to 250 A) bisected by a distinct intermediate line as previously described by us in the rat (19) and more recently confirmed by Rhodin (40) in the mouse and by Trump and Benditt (41) in the human glomerulus. However, this intermediate line is a filamentous two-dimensional structure, not a plate, as in the case of the desmosome. Moreover, the type of associated cytoplasmic fibrillar differentiation and the continuous character of the urinary slit are comparable to those of the adhering zonules.

[4] We have previously considered this junction to have a depth of about 40 A—that is, the thickness of the slit membrane. Another possible interpretation is that the junction extends from the level of the basement membrane to the slit membrane (~500 A).

Results of Experiments Utilizing Mass Tracers

To obtain some information on the permeability characteristics of the various elements of the junctional complex, we took advantage of the progressive concentration undergone by hemo-

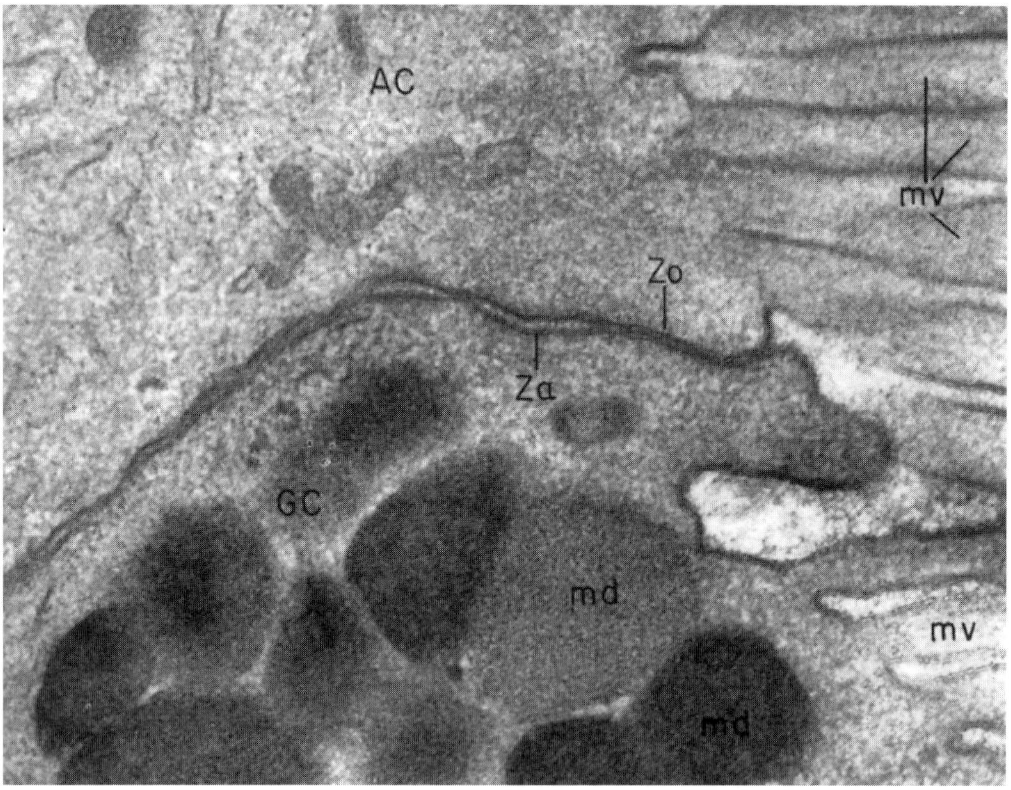

FIGURE 18

This figure illustrates a junctional complex with an occluding zonule (*Zo*) and an adhering zonule (*Za*) between an absorptive epithelial cell (*AC*) and a goblet cell (*GC*) in the epithelium of the intestinal mucosa (rat). The dense mucus droplets of the goblet cell are marked *md*.

Specimen fixed in 1 per cent OsO_4 in phosphate buffer (pH 7.6) and embedded in Epon. $Pb(OH)_2$-stained section. × 90,000.

globin in the renal tubules during experimental hemoglobinuria. Hemoglobin injected intravenously or intraperitoneally filters through the glomeruli, is concentrated by reabsorption of water in the tubules, and eventually forms a dense homogeneous mass that fills the tubular lumina (*cf.* Miller, 11). The concentration of the intraluminal hemoglobin—and therefore its electron opacity—varies considerably not only along a given tubule, but also among equivalent segments of different nephrons. In extreme instances, when the hemoglobin-containing filtrate is concentrated in the lumen of the proximal segments, distal segments, and collecting ducts to a high density, it can be used as a mass tracer to explore the outline of the tubule lumen and its connections (11). In all segments of the tubule examined the dense mass can be followed along the cellular margins down to the level of the occluding zonules (Figs. 24 to 27). Beyond this level, the intercellular spaces appear free of the tracer. Since the density of the material in the junctional gaps is considerably lower than that of the luminal hemoglobin, and since there is no evidence of a concentration gradient down the junctional complex, it can be assumed that the penetration of hemoglobin molecules is effectively stopped along the line of fusion of the adjoining cell membranes in the tight junction (Fig. 27). In any case it is clear that the concentrated tracer does not reach the adhering zonules or fasciae, the desmosomes, and the intercellular spaces beyond the level of the occluding zonules.

Seeking another situation in which the permeability characteristics of the junctional complex could be investigated, we studied pancreatic acini and ducts during zymogen discharge, occurring either spontaneously during prolonged fasting (hunger secretion); or induced physiologically, upon chyme entry into the duodenum; or experimentally, upon the administration of carbamylcholine. The discharged zymogen acts as a natural mass tracer which frequently fills the acinar and duct lumina, clearly outlining every surface detail with its continuous, homogeneous, electron-opaque substance (Figs. 28 to 30). In our material, this mass tracer was found to stop regularly at the point of mergence of the cell membranes that marks the beginning of the tight junctions (Figs. 29 and 30). No dense material has been seen so far in the intercellular spaces beyond the occluding zonules of either acini or ducts (Figs. 28 to 30).

Thus in the two examples provided it is demonstrated that the *zonulae occludentes* are impervious to concentrated protein solutions.

DISCUSSION

Significance of Results

From a morphological standpoint, our findings demonstrate the regular occurrence in various epithelia, especially in those of the mucosae lining cavitary organs, of a characteristic junctional complex whose components bear a constant relationship to each other and to the lumen of the

FIGURES 19 AND 20

Junctional complexes in the epithelium of the proximal convolution (rat kidney). In this part of the nephron, the tight junction is very shallow and appears only as a short stricture along the otherwise "open" intercellular space. (See also Figs. 24 to 26.) In Fig. 19 an occluding zonule (*Zo*) is followed by an adhering zonule (*Za*) which in turn is followed at some distance by a small desmosome (*D*). Luminal to the desmosome, the cell membranes are relatively closely apposed and follow a more or less parallel course, whereas basal to it the intercellular space expands greatly to form a large intercellular lake (*Il*).

In Fig. 20, where the plane of the section is oblique and cuts close to the luminal surface of the cells, two adhering zonules (*Za*) are cut broadly and the prominent condensation of cytoplasmic material found along them is clearly demonstrated. A large absorption droplet is marked *ab* and the microvilli of the brush border *bb*.

Specimens fixed in 1 per cent OsO_4 in acetate-Veronal buffer (pH 7.6) and embedded in Epon. $Pb(OH)_2$-stained sections. Fig. 19, × 42,000; Fig. 20, × 46,000.

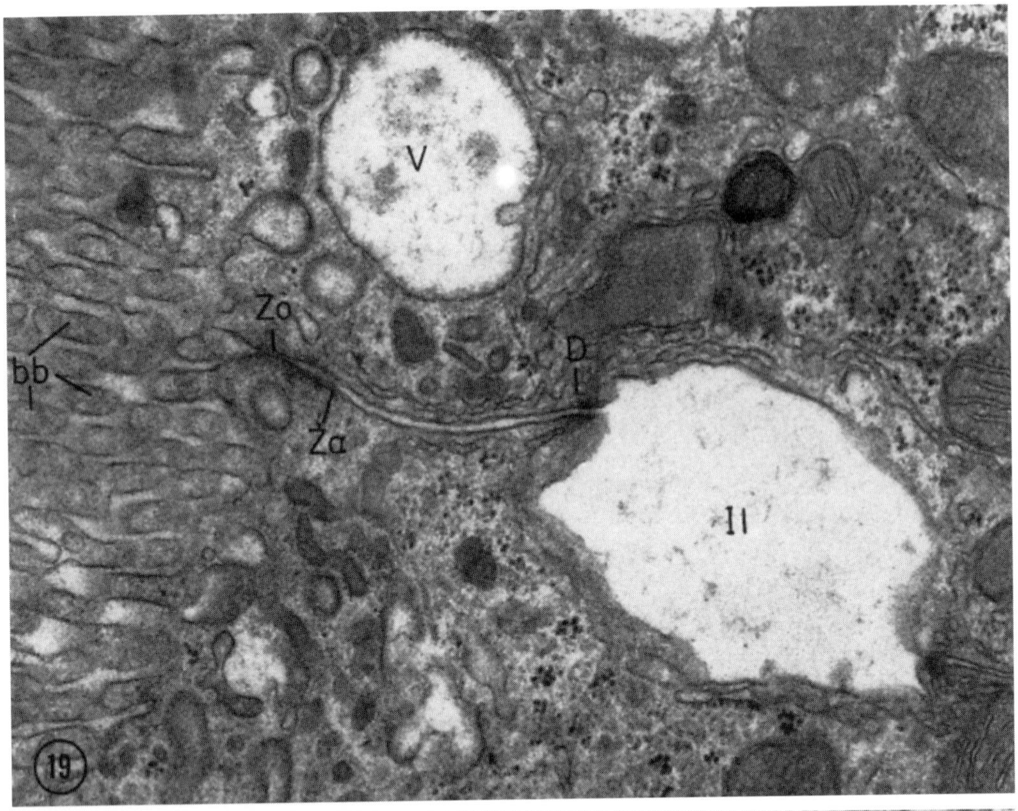
V
Zo
D
bb
Za
Il
19

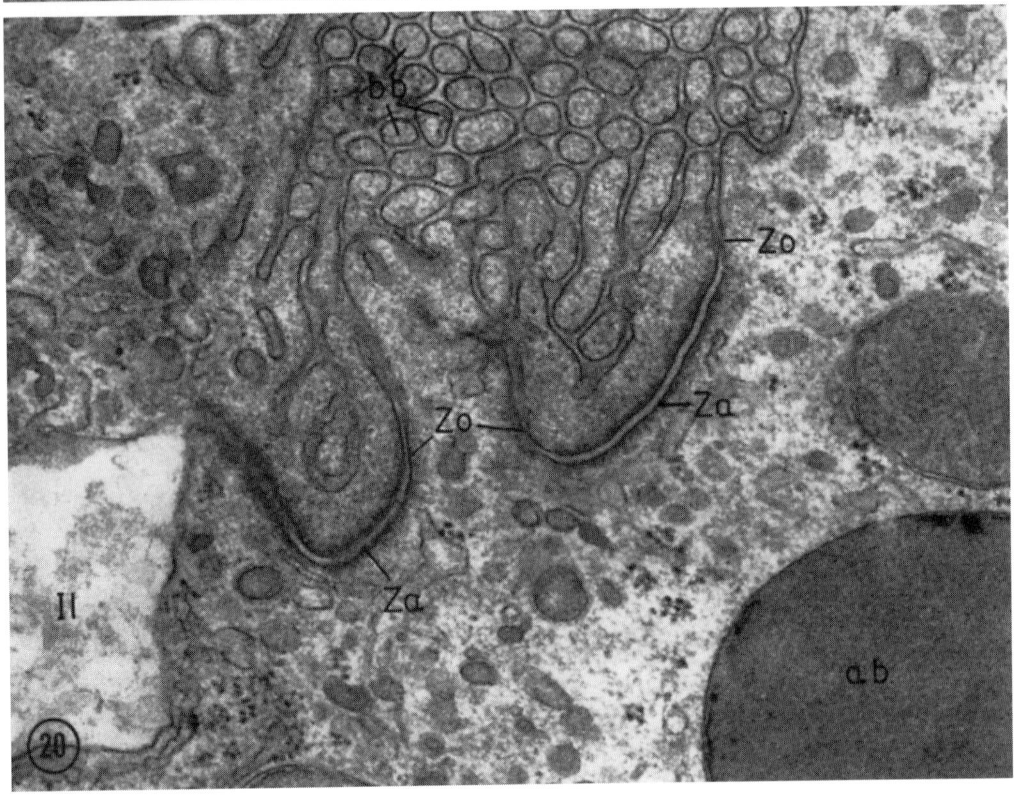
bb
Zo
Za
Zo
Za
Il
a.b
20

organ. Indeed, with some variations already discussed, we have found this complex in all organs so far examined. It can therefore be considered to be of widespread, probably universal occurrence in such epithelia. From a functional standpoint, we have demonstrated that it is the occluding zonule of this complex which acts as a barrier to the free passage of concentrated protein solutions from the lumina to the intercellular spaces.

Morphology

ZONULA OCCLUDENS (TIGHT JUNCTION): We have already pointed out that the "tight junction" corresponds to Robertson's "external compound membrane" and is basically similar in structure to a single lamella of either compact (31, 32) or loose (42) myelin. However, the occurrence of such areas of membrane fusion in other locations has not been generally appreciated, probably because the existing descriptions have appeared mostly as incidental observations. For example, in the epithelium of Brunner's glands, Moe (43) noted that in some places, particularly near the apical part of the cells, "the lateral cell membranes seem to fuse, the outer dense layers of the membranes merging into each other." Choi (44) also recognized "areas of fusion of the lateral cell membranes" immediately luminal to "terminal bars" in the epithelium of the toad bladder. Junctions characterized by fusion of the lateral cell membranes have also been found in other locations, not necessarily connected with visceral lumina, and given a variety of names. They were described by Karrer, under the name of "quintuple-layered cell interconnections," between cells of the human cervical epithelium (38), and between striated muscle fibers in the wall of thoracic veins (45); by Robertson as "external compound membranes" near the luminal boundary of intestinal epithelial cells (31), and between endothelial cells in the blood capillaries of developing brain (32); by Gray as areas of "closed contact" between glial processes surrounding cerebral capillaries (46); by Muir and Peters as "quintuple-layered membrane junctions" between endothelial cells of blood capillaries in a number of tissues (13); by Peters as "quintuple-layered units" between glial cells and between myelin sheaths and glia in the optic nerve (47); by Devis and James as "quintilinear regions" between fibroblasts in tissue culture (48); and, most recently, by Dewey and Barr as "nexuses" between smooth muscle cells in the jejunum (49). Finally, we have occasionally encountered "close contacts" of the same type between erythrocytes and endothelial cells in the blood capillaries of the rat and guinea pig. In all these locations the structure described is characterized, like our *zonula occludens*, by fusion of the cell membranes and obliteration of the intercellular gap with the formation of an intermediate line representing the fused outer leaflets of the adjoining membranes; hence it is clear that, regardless of the name applied, all these appearances represent basically the same structure,

FIGURE 21

Section through a distal convolution (rat kidney) showing two junctional complexes between contiguous cells. In both cases the only element of the junctional complex encountered along the intercellular spaces is the occluding zonule (Zo_1, Zo_2) which in this epithelium usually extends, as it does here, for distances of 0.1 to 0.2 μ. Adhering zonules appear to be absent, and no desmosomes are visible throughout this entire field. Note that the intercellular space closed by Zo_2 can be followed throughout the section all the way from the luminal to the basal surface of the cell. The latter is covered by a basement membrane (*B*). The apposing cell membranes are closely approximated ($\sim$200 to 250 A) and run a virtually parallel course from the end of the occluding zonule nearly to the cell base where they expand to limit a broader, more irregular space (*Il*). The deep infoldings of the cell membrane, the basal compartments they form, the high frequency of mitochondria, and their relationship to the compartments are well demonstrated.

Specimen fixed in 1 per cent OsO_4 in an acetate-Veronal buffer and embedded in Epon. $Pb(OH)_2$-stained section. $\times$ 34,000.

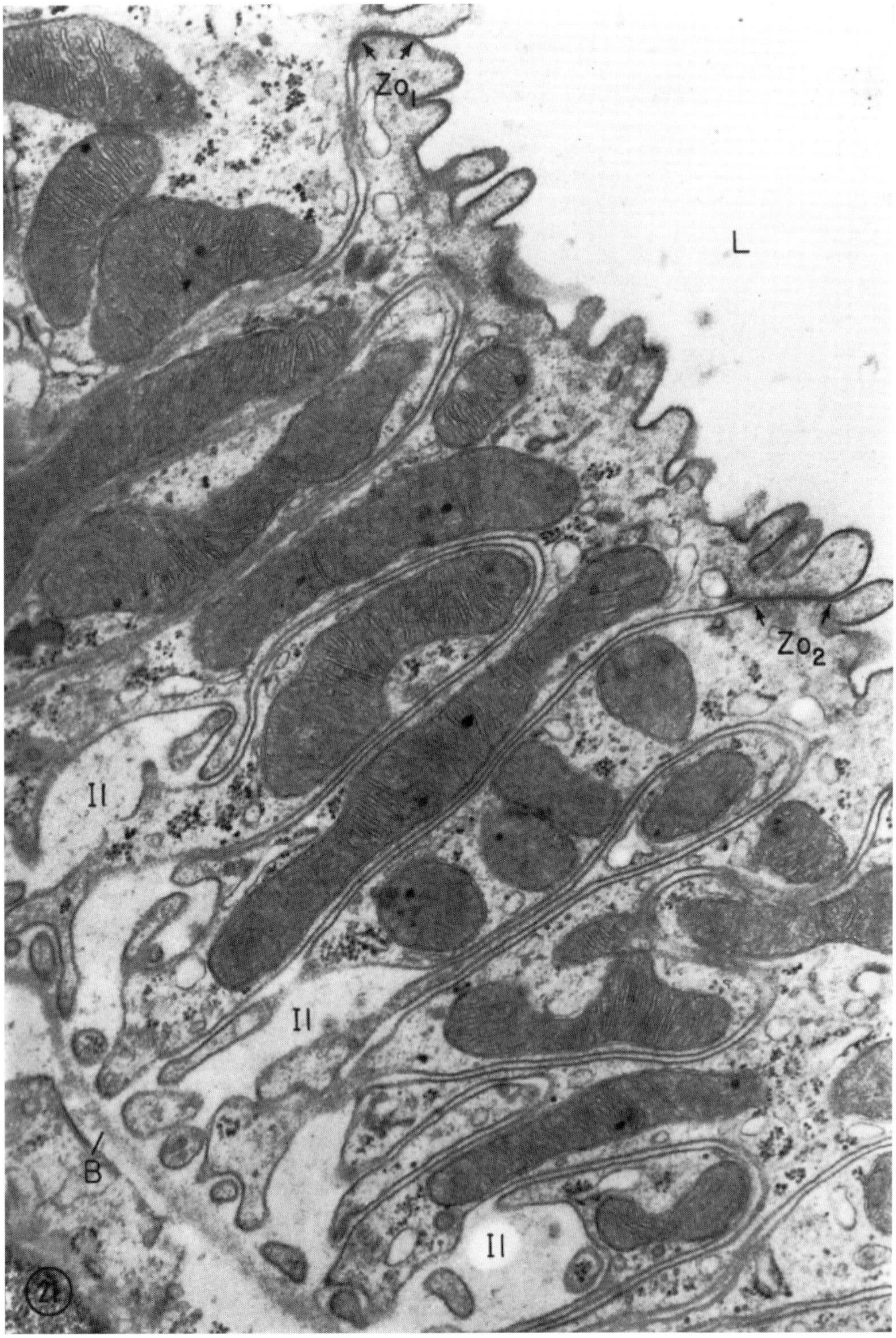

Zo1
L
Zo2
Il
Il
B
Il
21

although in three dimensions many of them are probably discontinuous and hence maculae or fasciae, rather than zonulae.

ZONULA ADHAERENS (INTERMEDIATE JUNCTION): We can find in the literature no clear description of the adhering zonule as a distinct entity. However, junctions of similar structure were clearly illustrated by Fawcett (2) in a renal carcinoma of the frog under the name of "terminal bars." In fact, from a perusal of the literature it is clear that, although this type of junction has been illustrated and sometimes designated as "terminal bar" by many microscopists (*e.g.* 18), it has not been clearly distinguished from the usual desmosomes.

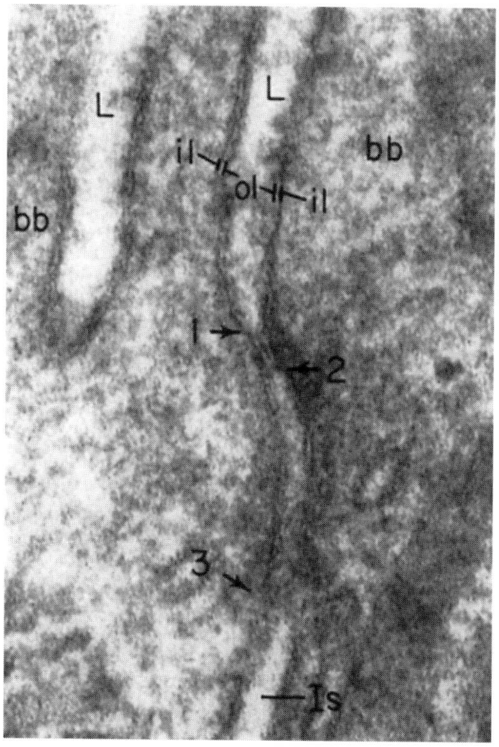

FIGURE 22

Junctional complex in the proximal convolution (rat kidney). The occluding zonule extends from arrow *1* to arrow *2*, is very shallow, but shows quite clearly a fusion line along its entire length. The adhering zonule, seen between arrows *2* and *3*, is partially blurred by the obliquity of the section. The intercellular space beyond the adhering zonule is marked *Is*.

Specimen fixed in 2 per cent OsO_4 in acetate-Veronal buffer (pH 7.6), dehydrated in acetone, stained in block with $KMnO_4$, and embedded in Araldite. Section doubly stained with uranyl acetate and $Pb(OH)_2$. × 150,000.

MACULA ADHAERENS (DESMOSOME): Desmosomes have been described in nearly all epithelia examined with the electron microscope (*cf.* Fawcett, 4). Their detailed structure has been studied in a number of tissues, particularly the epidermis (3, 36), cervical epithelium (38), mesothelium (7), and vascular endothelium in the swim bladder of the toad fish (4). An examination of the micrographs published by others shows, however, that this term has not been restricted to junctions having the structure of classical desmosomes. At various times it has been used to designate elements having the morphology of each of the types of junctions we have described in this paper. Further confusion stems from the fact that during the past few years the terms "desmosome" and "terminal bar" have been used more or less interchangeably. This leads directly to the next question in our discussion, as to which of the three elements of our junctional complex corresponds to the "terminal bar" of classical histology.

TERMINAL BARS: In the histological literature, it is generally assumed that terminal bars are bands of condensed intercellular cement which hold together the cells of an epithelium and seal up the intercellular spaces, thereby preventing or hindering matter from reaching or leaving the corresponding lumen along these spaces (50, 51). The concept has an exclusively morphological basis: the demonstration—by iron hematoxylin staining—of a chromophil material located around the cells, presumably in the intercellular spaces. In grazing sections this material is seen as a nearly hexagonal network outlining the cells, whereas in normal sections it appears as a series of dots located between adjacent cells, immediately below the free surface of the epithelium (*cf.* 6, 2). Although not experimentally proved, the functional implications of the concept turned out to be strong enough to affect the nomenclature. The bars were originally named *Schlussleisten* (50), *bandelettes de fermeture* or *bandelettes obturantes* (*cf.* 52), *i.e.* closing or locking bands, a meaning still implied, though less clearly expressed, in the English equivalent "terminal bar."

From the beginning, some histologists have regarded the bars as a special type of desmosome (6, 53) and recently this view has been restated and further developed by Fawcett (3, 4) and Fawcett and Selby (54) who showed that at the sites where terminal bars are observed in the light microscope, a structure similar to that of desmosomes could be resolved by electron microscopy.[5] For this reason they concluded that terminal bars and desmosomes are indistinguishable except for their shape: the former are strips or bands that "may extend the whole width of the cell," whereas the latter take the form of round plaques or discs (54). During the last few years this view has been widely accepted and, as a result, the terms desmosome and terminal bar have been used more or less interchangeably in describing attachment structures in columnar epithelia (*cf.* 12, 18, 29, 43, 44). We believe that this trend should be reexamined in view of a number of recent findings, including the ones here reported. It is clear that the occluding zonule is the element of the complex best suited to close up the intercellular spaces, a function originally ascribed to the terminal bar. It is also reasonably certain that the hematoxylin staining and silver impregnation of these bars are primarily due to the associated dense band of cytoplasmic material, rather than to any special property of the cell membrane or intercellular material involved (2, 55). Our survey shows that this dense band is characteristically associated with either the adhering zonule or the occluding zonule or both, depending on the epithelium. Finally electron microscope evidence, contributed by many (*cf.* 2, 4), including us, clearly indicates that the desmosomes are discontinuous structures located away from the lumina, and varying greatly in relative frequency. Hence it is unlikely that they contribute regularly or frequently to the terminal bar image seen in the light microscope. It follows that in the large majority of cases the electron microscopical equivalent of the terminal bar must be limited to the first two elements of the junctional complex, and that the respective contribution of each of these elements must vary from one epithelium to another. For instance, in the proximal convolution of the nephron the adhering zonule undoubtedly represents the terminal bar, for the occluding zonule is extremely shallow: its depth is

[5] Fawcett showed that adjoining cells differentiate peripheral bands (in three dimensions, plates) of high density along desmosomes as well as along terminal bars. He originally ascribed this appearance to a thickening of the apposed cell membranes, but on improved preparations recently established that the bands actually represent an accumulation of dense material in the subjacent cytoplasmic matrix (4).

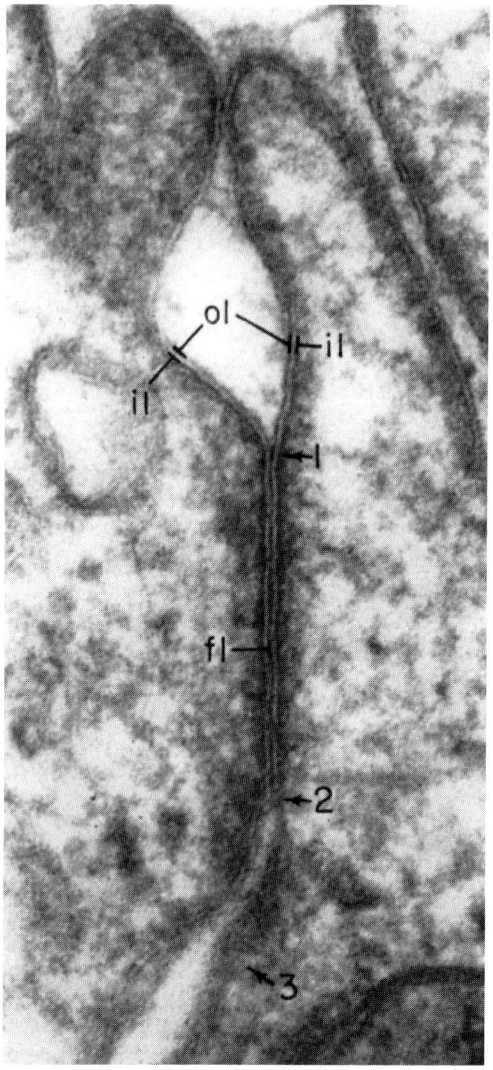

FIGURE 23

High magnification showing details of a junctional complex in the distal convolution (rat kidney). The tight junction is quite deep, extending between arrows *1* and *2*. The fusion line (*fl*) is visible along most of its length. The triple-layered structure of the apical cell membrane can be seen along most of the luminal cell surface. Part of a shallow adhering zonule appears between arrows *2* and *3*. Dense cytoplasmic material extends along the occluding as well as the adhering zonule.

Specimen fixed in 2 per cent OsO_4 in acetate-Veronal buffer (pH 7.6), dehydrated in acetone, stained in block with $KMnO_4$, and embedded in Araldite. Section doubly stained with uranyl acetate and $Pb(OH)_2$. × 150,000.

actually below the limit of resolution of the light microscope. In the intestinal epithelium the main component of the terminal bar is probably also the adhering zonule, although some contribution by the tight junction is not excluded: the tight component has enough depth to be seen in the light microscope, but only a limited amount of associated dense cytoplasmic material. In many other epithelia, especially in those of the glands and ducts examined and in those of the distal segments of the nephron, the contribution of the occluding zonule to the terminal bar must be substantial if not predominant, for the band of associated cytoplasmic material has considerable depth as well as density. Moreover, in certain cases the second element of the complex is poorly developed (*e.g.* gall bladder, thyroid, liver, uterus, and oviduct) and in some instances it can be entirely missing (distal convoluted tubule, thyroid, and uterus). Our survey suggests that a poor development of the adhering zonule occurs concomitantly with an extensive development of the occluding element, but further work is needed to establish this possible correlation.

Our observations indicate, therefore, that the term "terminal bar" does not designate a single, well defined structure at current levels of microscopical resolution. Moreover, they show that the function implied by this term is carried out by a structure (belt of membrane fusion) other than that responsible for chromophilia (belt or band of cytoplasmic condensation) and that these two structures are not necessarily coincidental. Since this situation can only end in confusion, we suggest that the use of the term "terminal bar" be discontinued in electron microscopical cytology and histology, and propose that the old and the newly recognized elements of the junctional complex be designated by the English names or their Latin equivalents we have already introduced and discussed under Results.

Junctional Complex and Epithelial Permeability

The assumption that intercellular junctions play a role in epithelial permeability can be traced back to Bonnet (50) and Zimmermann (51) who suggested that terminal bars may act as a diffusion barrier between the corresponding lumen and intercellular spaces. Direct support for this assumption was recently provided by Miller (11) who showed that in experimental hemoglobinuria the hemoglobin concentrated in the nephron lumina does not penetrate beyond the "terminal bars." Comparable evidence was obtained by Kaye *et al.* (14, 15) on a simple squamous epithelium (the corneal endothelium) using particulate (and, as such, less favorable) tracers which also did not penetrate the "terminal bars." In neither instance was the structure of this barrier further defined. Our studies clarify Miller's results by demonstrating that the *zonula occludens* is the barrier that restricts the flow of the mass tracer along the intercellular spaces, and by providing a reasonable structural basis, *i.e.* the fusion of the apposed cell membranes, for his and our observations. Moreover, we extended this type of evidence to other epithelia, namely those of the pancreatic acini and ducts, using discharged zymogen as a natural mass tracer.

FIGURE 24

Portion of a proximal convoluted tubule from a rat with experimental hemoglobinuria. A dense continuous mass of concentrated hemoglobin completely fills the tubular lumen and outlines the microvilli of the brush border (*bb*) and the various invaginations of the cell surface. Some of these invaginations, and the vesicles they form, appear surrounded by a distinct layer (*x*) of condensed cytoplasmic material. Where two cells meet, the dense mass can be followed along the cellular margins only down to the level of the occluding zonules (*Zo*); it does not appear in the adhering zonules (*Za*) or in the distended intercellular spaces (*Il*) below this level.

A number of membrane limited bodies (*ab*) of varying sizes and with a density similar to, or somewhat greater than that of the hemoglobin in the lumen are present in the cytoplasm. They undoubtedly represent protein absorption droplets filled with ingested hemoglobin.

Specimen fixed in 1 per cent OsO_4 in acetate-Veronal buffer (pH 7.6), dehydrated in alcohol, stained in block with PTA, embedded in Epon. $Pb(OH)_2$-stained section. $\times$ 50,000.

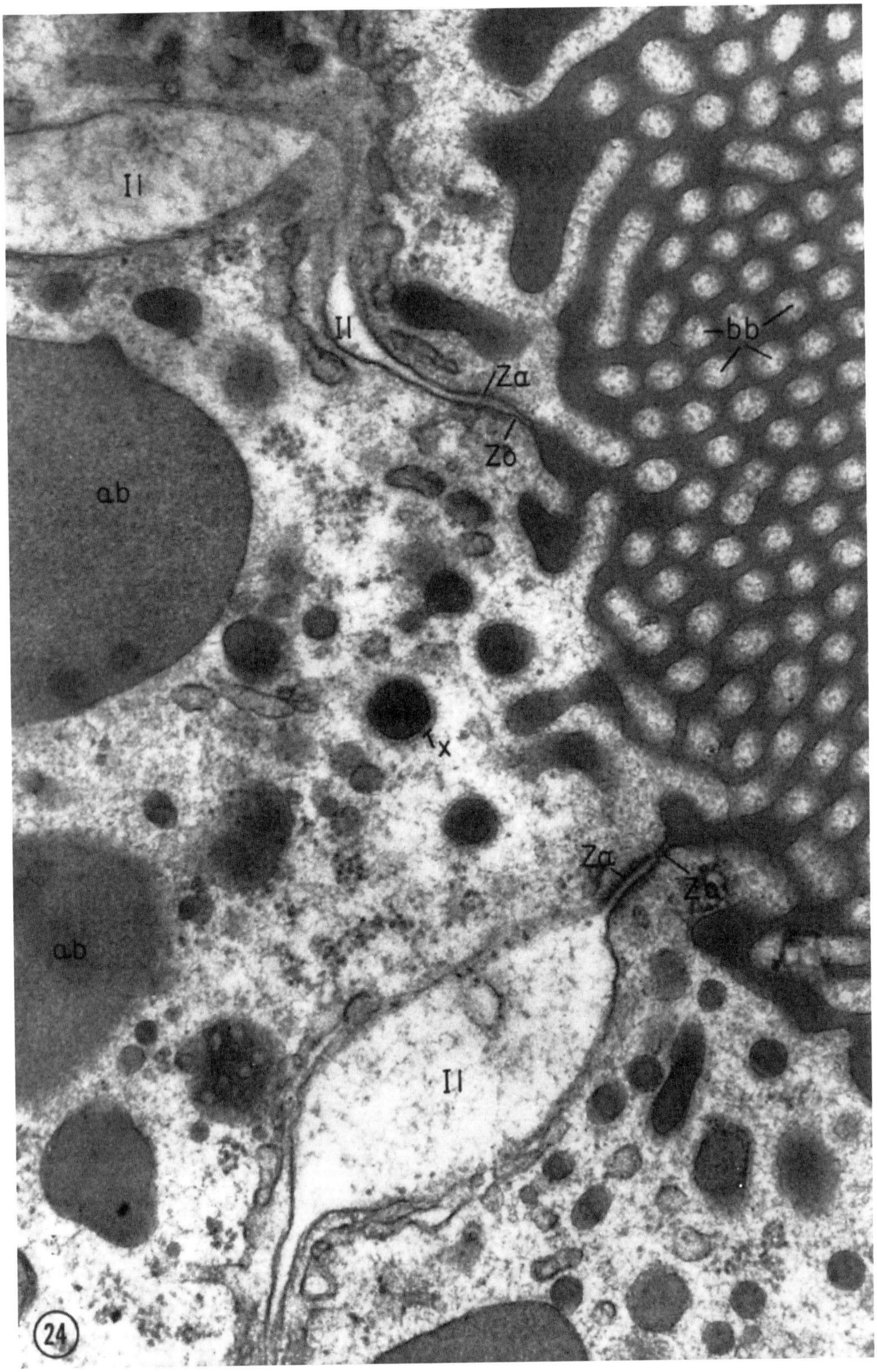
Il
Il
bb
Za
Zo
ab
x
ab
Za
Za
Il
24

Indirect evidence suggests that occluding zonules may also be impermeable to small molecules and possibly water. They seal up the intercellular spaces of the epithelium in organs which, as part of their function, maintain marked chemical and electrical potential gradients between the lumen and the subepithelial spaces. Recent work on amphibian kidney (56, 57) and urinary bladder (58), for instance, suggests that such gradients are created by the activity of the corresponding epithelia.[6] It should be added that they are successfully maintained, notwithstanding the multitude of intercellular spaces that break the continuity of the epithelium. It is, therefore, reasonable to assume that back diffusion along the intercellular spaces is minimized or prevented. In the interpretation of physiological data, the spaces are either ignored or tacitly assumed to be closed at their luminal end by a cement substance, an assumption which follows old morphological ideas about the nature and topography of the terminal bars. Our observations indicate that the *zonula occludens* must be the site of maximal constraint to diffusion along the intercellular spaces, for the latter appear to be obliterated at the level of this junctional element, at least according to morphological evidence. It should be stressed, however, that the intercellular spaces are closed off near the lumen by the fusion of the adjoining cell membranes, and not by the interposition of a resolvable layer of cement, as previously assumed in the light microscopical literature. This finding is timely for, according to recent work (2, 55), the basis for the whole intercellular cement hypothesis seems to be topographically in error; the hematoxylin-stained or silver-impregnated material of the terminal bar appears to be intra-, not extracellular in location.

Within the past few years a number of researchers have speculated on the possible influence of intercellular junctions, characterized by membrane fusion and obliteration of intercellular spaces, upon epithelial permeability. Robertson suggested that such junctions may serve to regulate diffusion along the intercellular spaces (32). Peachey and Rasmussen (12) postulated that the zones of extreme narrowing found along the intercellular spaces in the epithelium of the toad bladder may assure an "almost leakproof bladder." Gray (46) suggested that similar junctions ("close contacts") among the glial cells surrounding cerebral capillaries serve to seal off the extracellular spaces and route water and metabolites across

[6] A similar situation applies in the frog skin (59). In this situation, however, the geometry of the epithelium is complicated by stratification, and information about intercellular junctions is still incomplete (*cf.* 1, 60).

FIGURES 25 TO 27

Higher magnifications of intercellular junctions between kidney tubule cells from rats with experimental hemoglobinuria. Figs. 25 and 26 are from the proximal convolution, and Fig. 27 is from a distal tubule. In all the figures it can be clearly seen that the hemoglobin concentrated in the tubular lumen penetrates only down to the level of the occluding zonule and does not reach the adhering zonule or intercellular space (*Is*) beyond. The penetration of the tracer appears to be effectively stopped at the initial fusion point of the adjoining cell membranes that marks the beginning of the occluding zonule (Fig. 27). Note that in the case of the proximal tubule (Figs. 25 and 26) the occluding zonule is very shallow and appears only as a focal (~300 A) stricture (between arrows *1* and *2*) along the otherwise open intercellular spaces, whereas the adhering zonule (between arrows *2* and *3*) is quite deep (~2500 A) and is associated with a dense condensation of cytoplasmic material. In the distal convolution (Fig. 27) the situation is reversed, the occluding zonule (arrows *1* to *2*) is deep (~3000 A) and is followed by a relatively shallow (~800 A) adhering zonule (arrows *2* to *3*). Dense cytoplasmic material is associated with both the occluding and adhering zonules. In the case of the proximal convolution the outer leaflet of the cell membrane and the fusion line are rarely visible in this type of preparation, whereas in the distal segment both leaflets can be clearly seen and followed into the occluding zonule.

Specimens in Figs. 25 and 26 fixed in 2 per cent OsO_4 in acetate-Veronal buffer (pH 7.6) with sucrose and embedded in Araldite. $Pb(OH)_2$-stained sections. The specimen preparation for Fig. 26 is the same as for Fig. 24. Fig. 25, × 135,000; Fig. 26, × 96,000; Fig. 27, × 150,000.

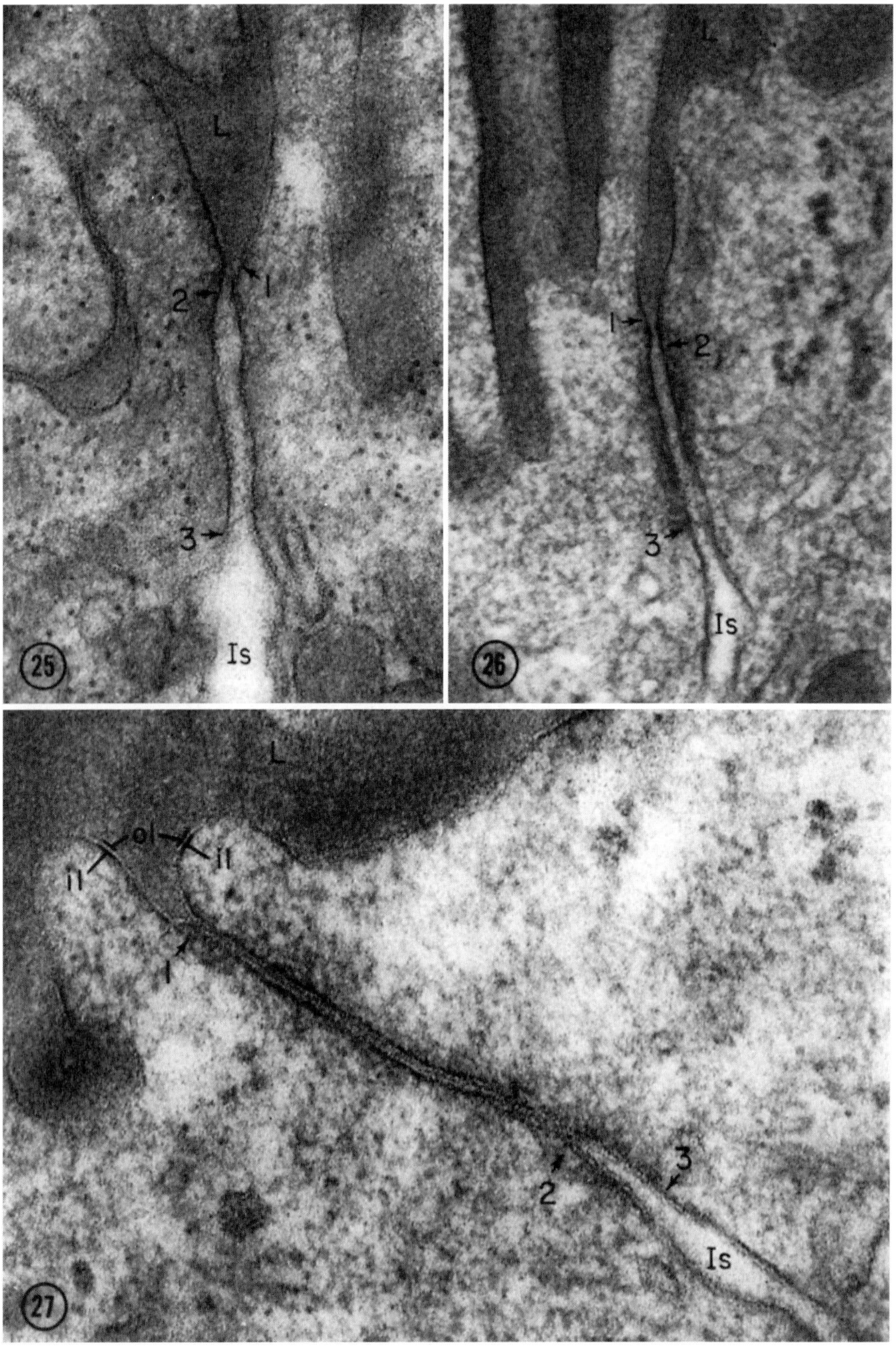
L
2
1
3
Is
25
L
1
2
3
Is
26
L
il
ol
il
1
2
3
Is
27

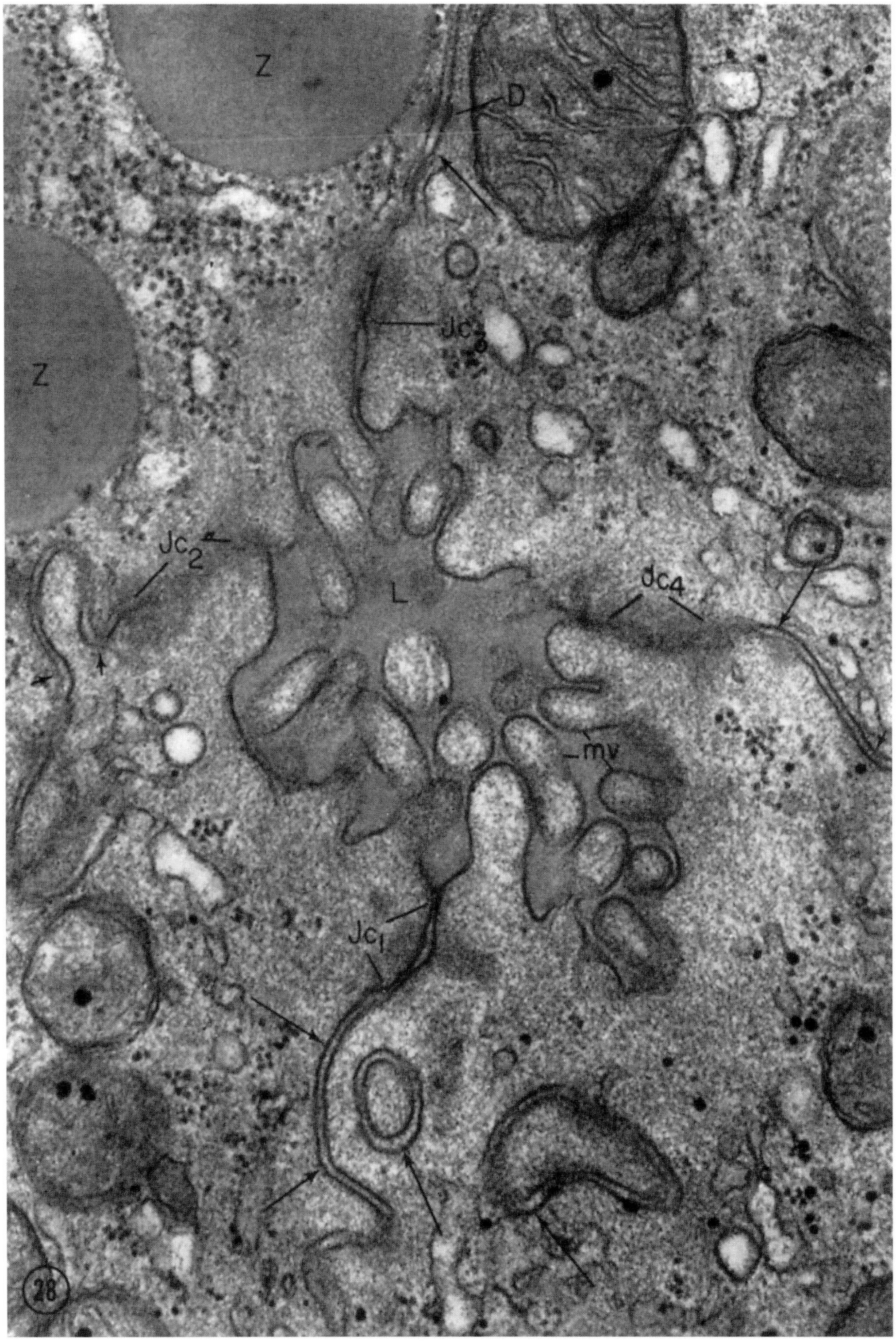
Z
D
Z
Jc3
Jc2
L
Jc4
mv
Jc1
28

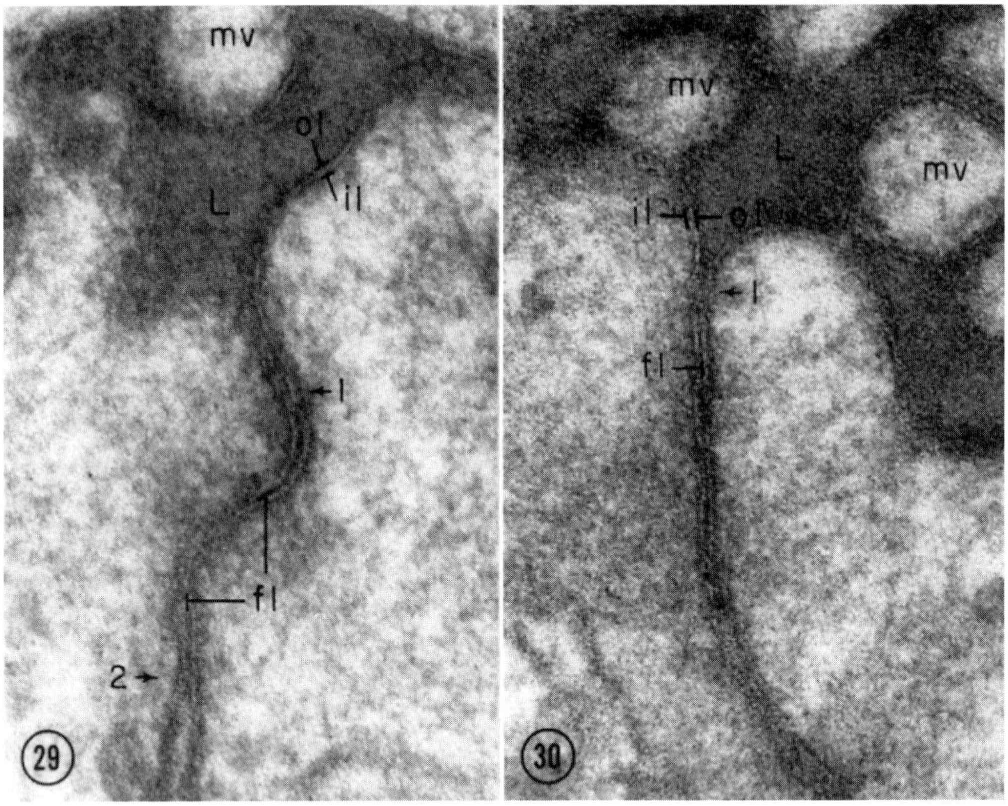

FIGURES 29 AND 30

In these figures the dense mass of discharged zymogen is stopped at the point of mergence of the membranes of two apposed cells into an occluding zonule (arrow *1*). The fusion line (*fl*) is visible in both cases. In Fig. 29 the intercellular space beyond the occluding zonule (arrow *2*) appears free of discharged zymogen.

Specimens fixed in 1 per cent OsO_4 in phosphate buffer (pH 7.6), dehydrated in acetone, stained in block with $KMnO_4$, and embedded in Epon. $Pb(OH)_2$-stained sections. × 200,000.

FIGURE 28

This figure together with the following ones (Figs. 29 and 30) illustrate the behavior of discharged zymogen in the acinar and duct lumina of guinea pig pancreas.

In Fig. 28 discharged zymogen, which matches in density the content of zymogen granules (*Z*) completely fills a glandular lumen (*L*) bounded by an acinar and three centroacinar or duct cells. The material that occupies the intercellular spaces beyond the junctional complexes marked Jc_1 to Jc_4 is of noticeably lower density (arrows); the difference shows to better advantage in normally or nearly normally sectioned spaces (long arrows). A single junctional complex (Jc_1) is normally cut. At its level it is clear that the mass of discharged zymogen stops at the occluding zonule. The situation at the other junctional complex is obscured by the obliquity of their section.

Specimen fixed in 1 per cent OsO_4 in phosphate buffer (pH 7.6) and embedded in Epon. Section stained in uranyl acetate and $Pb(OH)_2$. × 58,000.

glial cells. Finally, Muir and Peters (13) recently found similar junctions in the endothelium of blood capillaries and predicted that comparable arrangements will be detected "wherever a sheet of cells separates two zones of different constitution in order to prevent intercellular diffusion of ions and small molecules."

Our studies, initiated and carried out independently of those of Muir and Peters, confirm their hypothesis as far as morphological aspects are concerned, and provide preliminary evidence for the physiological speculations of the other authors (12, 32, 46, 47) previously mentioned. Moreover, our findings suggest that the elements described in the junctional complexes of epithelia are functionally specialized: the occluding zonule as a diffusion barrier or "seal," and the desmosomes (adhering maculae) as intercellular attachment devices, with the adhering and probably the occluding zonules participating also in the latter function.[7] Further work on the junctions found in simple and stratified squamous epithelia, especially in vascular endothelia[8] may further clarify this assumption.

Finally, it should be mentioned that at least in one case there is evidence that one type of junction can be replaced by another in relation to certain changes in function. Our previous studies (16, 19) have shown that a special type of junction, *i.e.* the urinary slits of the renal glomerular epithelium, are progressively replaced by "tight junctions" during the development of an experimental nephrosis in rats. In a reversed situation, Vernier and Birch-Andersen (61) have recently shown that the "tight junctions" present in the visceral epithelium of immature human glomeruli, are replaced during maturation by normal urinary slits. These findings are consistent with the assumption that these slits, whose partial resemblance to desmosomes and adhering zonules has already been discussed, are permeable to the glomerular filtrate, whereas "tight junctions" are not.

Comments on Structural Variations Encountered in Cell Membranes

The current prevailing tendency is to regard all cell membranes (as well as those of most intracellular organelles) as similar if not identical in structure. This trend is due mainly to the extensive studies of Robertson (28, 31–33, 62) who has shown that all cellular membranes have a similar trilaminar appearance in $KMnO_4$-fixed tissues, and hence postulated that they are all composed of a single bimolecular leaflet of polar lipids, sandwiched between two dissimilar layers of non-lipid material. His concept has derived additional strength from its agreement with older, indirect evidence concerning the molecular architecture of the cell membrane (*cf.* 63). So strong has been the concentration on Robertson's hypothesis of the "unit membrane" that relatively little attention has been paid to variations in membrane structure visible after OsO_4 fixation.

Our survey, carried out primarily on OsO_4-fixed tissues, confirms that the cell membrane has the same trilaminar structure in all epithelial cells investigated, but in addition reveals that this basic pattern shows significant variations primarily in respect to the total thickness of the unit and the thickness and density of its outer leaflet. These differences seem to us of sufficient importance to warrant separate discussion.

Based on their appearance after OsO_4 fixation, the cell membranes we have studied can be separated into two distinct types. The first is limited in distribution to the luminal (apical) aspects of some absorptive epithelia (intestine, stomach, colon, and gall bladder) and certain segments of the nephron (see below); it is relatively thick ($\sim$110 A), and its outer layer shows up clearly, especially after lead staining, wherever the membrane is normally sectioned. As a result, the membranes in this category appear nearly symmetrical, the outer leaflet being only slightly less dense than the inner one. The other type of membrane occurs more widely: it is found along the lateral and basal surfaces of the absorptive epithelia just

[7] Preliminary observations indicate that epithelial cells remain attached to one another at the level of the desmosomes, as well as at that of the adhering and occluding zonules, when the cell bodies retract and the intercellular spaces greatly enlarge, under the influence of hypertonic treatment prior to or during fixation.

[8] Our findings indicate the presence of tight junctions in the peritoneal mesothelium and, in confirmation and extension of references 32, 13, and 47, in the endothelium of all types of blood capillaries. In the latter case the occluding zonule frequently takes the form of a focal stricture or "pinch" similar to the situation described above in the proximal convolution, the adhering zonule is missing or poorly developed, and desmosomes are generally absent, at least in mammalian material.

mentioned, and along the entire perimeter of all other epithelial cells studied, *i.e.* gland and duct epithelia of the liver, pancreas, and thyroid, epithelia of the uterus and oviduct, mesothelia and capillary endothelia. In these locations the total thickness of the plasmalemma, like that of Schwann cells (28, 31, 32), is significantly less (70 to 80 A); furthermore, the membrane is highly asymmetrical: even after lead or uranyl staining, the outer leaflet is generally much finer and less dense than the inner one. In certain cases it is indeed hardly visible. The same layer is readily visualized, however, in tissues that have been fixed in OsO_4 and stained in block with $KMnO_4$ prior to embedding (24) or in sections which have been doubly stained, first with uranyl and then with lead. It should further be noted that within each of these groups there is a certain amount of further variation according to cell type: in the first group, for instance, the luminal membrane of gastric and, in some cases, colonic epithelia has an outer leaflet which is thicker and denser than the inner one; hence in this situation the usual asymmetry of the cell membrane is reversed. In the second group, the cell membrane of gland and duct cells is more asymmetrical than the membrane covering the lateral and basal aspects of absorptive epithelia; furthermore, the luminal membrane of many gland and duct epithelia (*e.g.* thyroid, liver, and parotid) is less asymmetrical than the lateral and basal membranes of the same cells (*cf.* Fig. 15.)

It will be noted that while membranes of the first type were found to occur, so far, mainly along the luminal surfaces of absorptive epithelia, not *all* absorptive epithelia possess this type of luminal membrane. In this respect, the most remarkable variation is encountered along the various segments of the nephron where the luminal cell membrane of glomerular epithelia (*cf.* 19) and the luminal membrane of the cells of distal tubules and collecting ducts are of the first (thick, nearly symmetrical type) whereas the entire membrane of the cells lining the proximal tubule and the lateral and basal membrane of the cells lining the distal tubule and collecting duct are of the second, (thin, symmetrical) type.

There are in the literature several comments on structural differences encountered in OsO_4-fixed cell membranes: Sjöstrand (34) has stressed the fact that the plasma membrane can appear rather different along various parts of the surface of the same cell. Wissig (64) has also called attention to differences between the apical and basal or lateral plasmalemmata in various types of nephron epithelia; moreover he has utilized these differences to classify pinocytic vesicles into different groups, based on their type of limiting membrane. Mercer (65) has noted that a typical stratification is regularly seen in the surface membrane of amebae, but is less apparent in other membranes of the same cell. We have called attention to the fact that the cell membrane of one cell type, the glomerular epithelial cell, differs from that of other cell types in the same section (19). Several investigators (19, 66–69) have shown that the plasmalemmata of certain cell types (or of certain aspects of the same cell, 67, 69) respond selectively to staining with PTA.

Although Robertson has attached little importance to such differences and attributed them to deficiencies in OsO_4 fixation,[9] it seems to us that they deserve closer scrutiny. The different patterns described are consistent and, as such, probably valid. Such findings do not challenge the unit membrane hypothesis; they suggest only that the composition of the membrane layers varies with the cell type and the environment the cell faces. It is hoped that the morphological differences described will eventually be correlated with distinctive functional characteristics of the plasmalemma in different locations (and of the limiting membranes of various cellular organelles), but for the moment such a correlation is not yet possible.

Part of this work was presented at the First Annual Meeting of the American Society for Cell Biology, Chicago, November 2 to 4, 1961 (17).

This investigation was supported by a research grant (H-5648) from the National Institutes of Health, United States Public Health Service.

Received for publication, August 20, 1962.

[9] In the case of OsO_4-fixed intestinal epithelium, Robertson agreed (*cf.* 32) that the differences in the appearance of the membranes lining the microvilli, on the one hand, and the remainder of the cell surface, on the other, "no doubt reflect underlying chemical differences in the components of the membranes."

REFERENCES

1. PORTER, K. R., Observations on the fine structure of animal epidermis, *Proc. Internat. Conf. Electron Microscopy*, London, 1956, 539.
2. FAWCETT, D. W., Structural specializations of the cell surface, *in* Frontiers in Cytology, (S. L. Palay, editor), New Haven, Yale University Press, 1958, 19.
3. ODLAND, G. F., The fine structure of the interrelationship of cells in the human epidermis, *J. Biophysic. and Biochem. Cytol.*, 1958, **4**, 529.
4. FAWCETT, D. W., Intercellular bridges, *Exp. Cell Research*, 1961, suppl. **8**, 174.
5. BIZZOZERO, G., Osservazioni sulla struttura degli epiteli pavimentosi stratificati, *Rend. Ist. Lombardo Sc. e Lettere*, 1870, series II, **3**, 675.
6. SCHAFFER, J., Das Epitheligewebe, *in* Handbuch der mikeroskopische Anatomie des Menschen, (W. von Mollendorff, editor), Berlin, Julius Springer, 1927, **2**, pt. 1, 35.
7. HAMA, K., The fine structure of the desmosomes in frog mesothelium, *J. Biophysic. and Biochem. Cytol.*, 1960, **7**, 575.
8. WOOD, R. L., Intercellular attachment in the epithelium in *Hydra* as revealed by electron microscopy, *J. Biophysic. and Biochem. Cytol.*, 1959, **6**, 343.
9. PALADE, G. E., and PORTER, K. R., Studies on the endoplasmic reticulum. I. Its identification in cells *in situ*, *J. Exp. Med.*, 1954, **100**, 641.
10. BENNETT, H. S., LUFT, J. H., and HAMPTON, J. C., Morphological classifications of vertebrate blood capillaries, *Am. J. Physiol.*, 1959, **196**, 381.
11. MILLER, F., Hemoglobin absorption by the cells of the proximal convoluted tubule in mouse kidney, *J. Biophysic. and Biochem. Cytol.*, 1960, **8**, 689.
12. PEACHEY, L. D., and RASMUSSEN, H., Structure of the toad's urinary bladder as related to its physiology. *J. Biophysic. and Biochem. Cytol.*, 1961, **10**, 529.
13. MUIR, A. R., and PETERS, A., Quintuple-layered membrane junctions at terminal bars between endothelial cells, *J. Cell Biol.*, 1962, **12**, 443.
14. KAYE, G. I., and PAPPAS, G. D., Studies on the cornea. I. The fine structure of the rabbit cornea and the uptake and transport of colloidal particles by the cornea *in vivo*, *J. Cell Biol.*, 1962, **12**, 457.
15. KAYE, G. I., PAPPAS, G. E., DONN, A., and MALLETT, N., Studies on the cornea. II. The uptake and transport of colloidal particles by the living rabbit cornea *in vitro*, *J. Cell Biol.*, 1962, **12**, 481.
16. FARQUHAR, M. G., and PALADE, G. E., Glomerular permeability. II. Ferritin transfer across the glomerular capillary wall in nephrotic rats, *J. Exp. Med.*, 1961, **114**, 699.
17. FARQUHAR, M. G., and PALADE, G. E., Tight intercellular junctions, Abstracts of papers presented at the First Annual Meeting of the American Society for Cell Biology, Chicago, November 2 to 4, 1961, 57.
18. PALAY, S. L., and KARLIN, L. J., An electron microscopic study of the intestinal villus. I. The fasting animal, *J. Biophysic. and Biochem. Cytol.*, 1959, **5**, 363.
19. FARQUHAR, M. G., WISSIG, S. L., and PALADE, G. E., Glomerular permeability. I. Ferritin transfer across the normal glomerular capillary wall, *J. Exp. Med.*, 1961, **113**, 47.
20. CAULFIELD, J. B., Effect of varying the vehicle for OsO_4 fixation, *J. Biophysic. and Biochem. Cytol.*, 1957, **3**, 827.
21. MILLONIG, G., Advantages of a phosphate buffer for OsO_4 solutions in fixation, *J. Appl. Physics*, 1961, **32**, 1637.
22. LUFT, J. H., Permanganate—a new fixative for electron microscopy, *J. Biophysic. and Biochem. Cytol.*, 1956, **2**, 799.
23. LUFT, J. H., Improvements in epoxy embedding methods, *J. Biophysic. and Biochem. Cytol.*, 1961, **9**, 409.
24. PARSONS, D. F., A simple method for obtaining increased contrast in Araldite sections by using postfixation staining of tissues with potassium permanganate, *J. Biophysic. and Biochem. Cytol.*, 1961, **11**, 492.
25. WATSON, M. L., Staining of tissue sections for electron microscopy with heavy metals, *J. Biophysic. and Biochem. Cytol.*, 1958, **4**, 475.
26. KARNOVSKY, M. J., Simple methods for "staining with lead" at high pH in electron microscopy, *J. Cell Biol.*, 1961, **11**, 729.
27. WATSON, M. L., Reduction of heating artifacts in thin sections examined in the electron microscope, *J. Biophysic. and Biochem. Cytol.*, 1957, **3**, 1017.
28. ROBERTSON, J. D., Ultrastructure of excitable membranes and the crayfish median-giant synapse, *Ann. New York Acad. Sc.*, 1961, **94**, 339.
29. ZETTERQVIST, H., The ultrastructural organization of the columnar absorbing cells of the mouse intestine, Stockholm, Aktiebolaget Godvil, 1956.
30. SJÖSTRAND, F. S., and ZETTERQVIST, H., Functional changes of the free cell surface membrane of the intestinal absorbing cells, Proceedings of the Stockholm Congress on Electron Micros-

copy, Stockholm, Almqvist and Wiksell, 1956, 150.

31. Robertson, J. D., The molecular structure and contact relationships of cell membranes, *Prog. Biophysics and Biophysic. Chem.*, 1960, **10**, 343.
32. Robertson, J. D., The unit membrane, *in* Electron Microscopy in Anatomy, (J. D. Boyd, F. R. Johnson, and J. D. Lever, editors), London, Edward Arnold and Co., 1961, 55.
33. Robertson, J. D., Structural alterations in nerve fibers produced by hypotonic and hypertonic solutions, *J. Biophysic. and Biochem. Cytol.*, 1958, **4**, 349.
34. Sjöstrand, F. S., Fine structure of cytoplasm: The organization of membranous layers, *Rev. Mod. Physics*, 1959, **31**, 301.
35. Puchtler, H., and Leblond, C. P., Histochemical analysis of cell membranes and associated structures as seen in the intestinal epithelium, *Am. J. Anat.*, 1958, **102**, 1.
36. Hortsmann, E., and Knoop, A., Elektronenmikroskopische studie an der Epidermis. I. Rattenfote, *Z. Zellforsch u. Mikroskop. Anat.*, 1958, **47**, 348.
37. Pillai, E. A., Guenin, H. A., and Gautier, A. Les liaisons cellulaires dans l'épiderme du Triton normal au microscope électronique. *Bull. Soc. Vaudoise Sc. Nat.*, 1960, **67**, 215.
38. Karrer, H. E., Cell interconnections in normal human cervical epithelium, *J. Biophysic. and Biochem. Cytol.*, 1960, **7**, 181.
39. Millington, P. F., and Finean, J. B., Electron microscope studies of the structure of the microvilli on principal epithelial cells of rat jejunum after treatment in hypo- and hypertonic saline, *J. Cell Biol.*, 1962, **14**, 125.
40. Rhodin, J. A. G., The diagram of capillary endothelial fenestrations, *J. Ultrastruct. Research*, 1962, **6**, 171.
41. Trump, B. F., and Benditt, E. P., Electron microscopic studies of human renal disease; observations of normal visceral glomerular epithelium and its modification in disease, *Lab. Inv.*, 1962, **11**, 753.
42. Rosenbluth, J., and Palay, S. L., The fine structure of nerve cell bodies and their myelin sheaths in the eighth nerve ganglion of the goldfish, *J. Cell Biol.*, 1961, **9**, 853.
43. Moe, H., The ultrastructure of Brunner's glands of the cat, *J. Ultrastruct. Research*, 1960, **4**, 58.
44. Choi, J. K., Light and electron microscopy of the toad urinary bladder, *Anat. Rec.*, 1961, **139**, 214, (abstract).
45. Karrer, H. E., The striated musculature of blood vessels. II. Cell interconnections and cell surface, *J. Biophysic. and Biochem. Cytol.*, 1960, **8**, 135.
46. Gray, E. G., Ultra-structure of synapses of cerebral cortex and of certain specializations of neuroglial membranes, *in* Electron Microscopy in Anatomy, (J. D. Boyd, F. R. Johnson, and J. D. Lever, editors), Edward Arnold and Co., London, 1961, 54.
47. Peters, A., Plasma membrane contacts in the central nervous system, *J. Anat.*, 1962, **96**, pt. 2, 237.
48. Devis, R., and James, D. W., Electron microscopic appearance of close relationships between adult guinea pig fibroblasts in tissue culture, *Nature*, 1962, **194**, 695.
49. Dewey, M. M., and Barr, L., Intercellular connection between smooth muscle cells: the Nexus, *Science*, 1962, **137**, 670.
50. Bonnet, R., Uber die "Schlussleisten" der Epithelium, *Deut. Med. Wochenschr.*, 1895, **21**, 58.
51. Zimmerman, K. W., Zur Morphologie der Epithelzellen der Saugetierniere, *Arch. mikr. Anat.*, 1911, **78**, 199.
52. Chèvremont, M., Notions de Cytologie et Histologie, Liège, Desoer, 1956.
53. Dahlgren, U., and Kepner, W., Principles of Animal Histology, New York, Macmillan Co., 1925.
54. Fawcett, D. W., and Selby, C. C., Observations on the fine structure of the turtle atrium, *J. Biophysic. and Biochem. Cytol.*, 1958, **4**, 63.
55. Buck, R. C., The fine structure of endothelium of large arteries, *J. Biophysic. and Biochem. Cytol.*, 1958, **4**, 187.
56. Giebisch, G., Electrical potential measurements on single nephrons of Necturus, *J. Cellular and Comp. Physiol.*, 1958, **51**, 221.
57. Windhager, E. E., Whittemburg, G., Oken, D. E., Schatzmann, H. J., and Solomon, A. K., Single proximal tubules of the Necturus kidney. III. Dependence of H_2O movement on NaCl concentration, *Am. J. Physiol.*, 1959, **197**, 313.
58. Leaf, A., Some actions of neurohypophyseal hormones on a living membrane, *J. Gen. Physiol.*, 1960, **43**, No. 5, pt. 2, 175.
59. Ussing, H. H., The frog skin potential, *J. Gen. Physiol.*, 1960, **43**, No. 5, pt. 2, 135.
60. Ottoson, D., Sjöstrand, F., Stenstrom, S., and Svaetichin, G., Microelectrode studies on the EMF of the frog skin related to electron microscopy of the dermo-epidermal junction, *Acta Physiol. Scand.*, 1953, **26**, 611.
61. Vernier, R. L., and Birch-Andersen, A., personal communication from R. L. Vernier.
62. Robertson, J. D., New unit membrane organelle of Schwann cells, *in* Biophysics of Physiological and Pharmacological Actions, Washington,

D. C., American Association for the Advancement of Science, 1961.

63. Davson, H., and Danielli, J. F., The Permeability of Natural Membranes, London, Cambridge University Press, 1952.
64. Wissig, S. L., Structural differentiations in the plasmalemma and cytoplasmic vesicles of selected epithelial cells, *Anat. Rec.*, 1962, **142,** 292.
65. Mercer, E. H., An electron microscope study of Amoeba proteus, *Proc. Roy. Soc. Lond., Series B*, 1959, **150,** 216.
66. Marinozzi, V., and Gautier, A., Essais de cytochimie ultrastructurales. Du rôle de l'osmium réduit dans les "colorations" électronique, *Compt. rend. Acad. sc.*, 1961, **253,** 1180.
67. Steiner, J. W., and Carruthers, J. S., Studies on the fine structure of the terminal branches of the biliary tree I. The morphology of normal bile canaliculi, bile pre-ductules (Ducts of Hering) and bile ductules, *Am. J. Path.*, 1961, **38,** 639.
68. Movat, H. Z., and Steiner, J. W., Studies of nephrotoxic nephritis. I. The fine structure of the glomerulus of the dog, *Am. J. Clin. Path.*, 1961, **36,** 289.
69. Latta, H., The plasma membrane of glomerular epithelium, *J. Ultrastruct. Research*, 1962, **6,** 407.

Branton D. 1966. Fracture faces of frozen membranes. ***Proc. Natl. Acad. Sci.*** **55: 1048–1056.**

THE MOLECULAR ORGANIZATION OF CELLULAR MEMBRANES is of key importance for cells, both because such membranes define the cell's margin and divide the cell into compartments, and because many reactions essential for cell metabolism occur in or on membranes. The idea of a lipid bilayer as the framework for biological membranes was proposed early in the 20th century (1). Much physiological, biochemical, and structural work supported this idea (for an historical review, see 2), but other arrangements of lipids were still seriously considered as late as the 1960s. Branton's work with electron microscopy of cells that had been prepared for study by rapid freezing and freeze fracturing, followed by shadowing with heavy metals, clearly showed that each membrane contained a plane of weakness along which the fracture was likely to travel. Others pioneered the use of rapid freezing and freeze-fracture (3), but it was Branton's work that first led to a correct interpretation of the resulting images of membrane structure. He identified the inner and outer faces of the membrane, as well as the planes that represented the interior of the membrane's lipid layer. Particles were often seen to penetrate the latter layer from both sides of the membrane, providing morphological evidence for trans-membrane proteins and the idea that some membrane proteins might be folded with hydrophobic amino acids exposed to the hydrocarbon chains of the membrane's interior. Branton's subsequent work (4) demonstrated experimentally that the plane of fracture generally followed the lipid bilayer, cleaving it into two separate leaflets. Thus the architecture of biological membranes was firmly established, and a framework was built within which one could understand the complexity of membrane structure and function (Singer and Nicolson, this volume).

---

1. Gorter E. and Grendel F. 1925. On biomolecular layers of lipoids on the chromocytes of the blood. *J. Exp. Med.* **41:** 439.
2. Robertson J.D. 1981. Membrane structure. *J. Cell Biol.* **91:** 189–204.
3. Moor H., Muhlenthaler K., Waldner H., and Frey-Wyssling A. 1961. A new freezing ultramicrotome *J. Biophys. Biochem. Cytol.* **10:** 1–14.
4. Dreamer D. and Branton D. 1967. Fracture planes in an ice-bilayer model membrane system. *Science* **158:** 655–657.

Reprinted from the PROCEEDINGS OF THE NATIONAL ACADEMY OF SCIENCES
Vol. 55, No. 5, pp. 1048–1056. May, 1966.

*FRACTURE FACES OF FROZEN MEMBRANES**

By DANIEL BRANTON

DEPARTMENT OF BOTANY, UNIVERSITY OF CALIFORNIA, BERKELEY

Communicated by Melvin Calvin, March 11, 1966

The biological membrane, according to one widely accepted concept, has as its framework a bimolecular leaflet which under appropriate conditions can be seen in the electron microscope as two dark 20-Å-thick layers separated by a lighter 35-Å-thick layer.[1] Well-known theories and evidence[2–9] suggest that this structure is composed of a bimolecular leaflet of oriented lipid molecules sandwiched between two layers of protein. Though Robertson[10] has formalized these ideas as the basis of his generalized unit membrane concept, new chemical[11–13] and structural[10, 12, 14–16] evidence requires that other molecular arrangements also be considered. This has been the case in several recently proposed membrane models. Though some of these models take as their starting point the general notion of a bimolecular leaflet[17, 18] and others take as a starting point a repeating particulate subunit,[11, 14, 16] they all emphasize the possibility of dynamic interrelations between the several membrane components and explicitly deny the notion of a biological membrane which is spatially and temporally uniform.

The structural implications of these recent models are difficult to study, as there are few high-resolution techniques which can be used to examine rapidly changing forms. However, the recently improved freeze-etching technique[19, 20] should provide a direct view of membrane structure. Since this method does not involve the use of chemical fixatives or stains and the freeze-fixation employed need not kill the cells,[21] one can study membranes as they respond to a given physiological environment, rather than as they respond to a fixative environment. Furthermore, surfaces exposed in freeze-etching are three-dimensional fractures in which spatially extended areas of membranes can be examined.

Initial experiments with freeze-etching have demonstrated its applicability to a wide variety of biological specimens.[21–25] This report presents a more detailed interpretation of the fractured membrane faces exposed by this technique. Preliminary observations of freeze-etched root tip cells[24] revealed membrane faces whose morphological features could not be equated with the known features of membrane surfaces. The investigations reported here have explained this by demonstrating that what is usually considered as the true membrane surface (the interface between a membrane and any contiguous protoplasm, cell wall, or vacuolar material) is rarely seen in freeze-etched preparations. Instead, the fracture process splits the membrane and exposes an internal membrane face.

Materials and Methods.—Adventitious roots of onion sets (*Allium cepa* L., var. White Globe) were used for most of the experiments. They were grown in 20% glycerol.[24] Other experiments involved *Porphyridium cruentum* (Ag.) Naeg. or *Saccharomyces* sp. The unicellular red alga *P. cruentum* (Indiana University culture collection) was grown as described elsewhere.[26] *Saccharomyces* sp. cells (Fleischmann's yeast cakes) were separated by centrifugation from the starch used as a binder in the cakes and suspended in fresh tap water for 2 hr.

In preparation for freeze-etching, onion roots were cut in half lengthwise. A 1–2-mm piece of the half tip was placed in a thin syrup of gum arabic dissolved in 20% glycerol, transferred to a 3-mm copper disk, and then rapidly frozen in liquid Freon 22 (chlorodifluoromethane). Gum arabic helped to cement the frozen root tip to the copper disk but had no observable effect on cell ultrastructure. The *Porphyridium* and *Saccharomyces* cells were collected by centrifugation. Small droplets of the cell pellet were placed on copper disks and frozen in Freon 22.

The frozen specimens were freeze-etched (Fig. 1) as described by Moore *et al.*[20, 22] In some experiments the amount of etching applied to the fractured surface was varied by manipulating either the etching time or the etching temperature, or both. "Normal etching" was accomplished by leaving the freshly fractured specimens *in vacuo* (less than 2×10^{-6} torr) for 1–5 min at −100°C. "No etching" was accomplished by keeping the temperature of freshly fractured surfaces below −165°C and by replicating the surfaces as rapidly as possible (less than 10 sec) after they were fractured. "Deep etching" was accomplished by leaving the freshly fractured surfaces *in vacuo* for 10 min at −95°C.

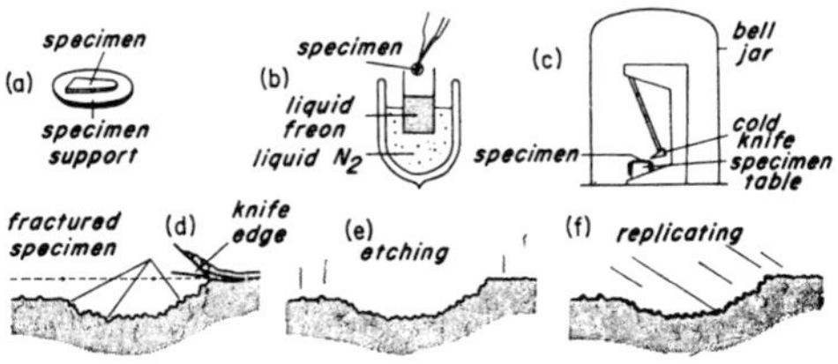

FIG. 1.—Freeze-etching. (*a*) The fresh specimen was placed on a Cu disk, (*b*) rapidly frozen in liquid Freon 22, and (*c*) placed in the precooled freeze-etching vacuum chamber. (*d*) The frozen specimen was fractured with a microtome knife at −185°C, and in some cases (*e*) the freshly fractured surface was etched. (*f*) The surface was shadowed and replicated with Pt and C.

All micrographs have been printed so that shadows appear as light areas and have been mounted for publication with shadows extending from bottom to top.

Results.—Basic observations: A consistent feature in all freeze-etched preparations is the presence of a small ridge (Figs. 2–4) at the base of most exposed mem-

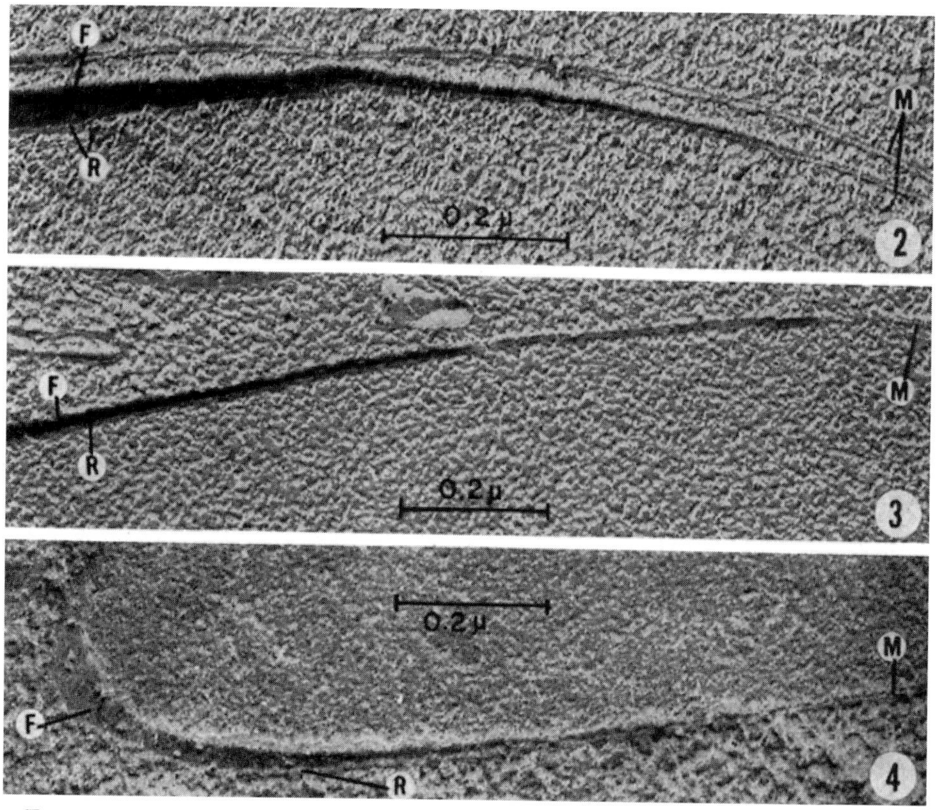

FIGS. 2–4.—FIG. 2.: Endoplasmic reticulum in onion root tip. FIG. 3: Vacuolar membrane face in onion root tip; view from inside the vacuole. FIG. 4: Vacuolar membrane face in onion root tip; view from outside the vacuole. In Figs. 2, 3, and 4, the fractures are tangent to the membrane surfaces on the left and almost perpendicular to the membrane surfaces on the right. The small ridge (*R*) at the base of an exposed membrane face (*F*) on the left is continuous with one of two ridges which forms the typical freeze-etch image of a single membrane (*M*) on the right.

brane faces. After careful scrutiny of a large number of photomicrographs, it became apparent that the small ridge was in fact continuous with and identical to one of the ridges that had previously been assumed to represent part of a unit membrane.[22, 24] This same type of fracture was observed in freeze-etched preparations of the plasma, nuclear, vacuolar, and dictyosomal membranes. Figure 5 is an interpretative diagram of what can be seen in the micrographs of Figures 2–4. It implies that during fracturing, membranes are split to expose either one or the other

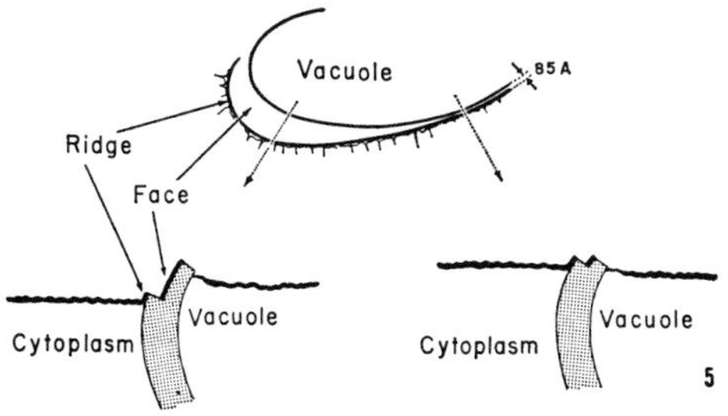

FIG. 5.—*Top,* a representation of Fig. 4. *Bottom,* diagrams of imaginary sections perpendicular to the plane of the page through the fractured tissue along the *dashed arrows.* These diagrams assume arrival of shadow-replica material from upper left, and show why fracture of an inclined *ca.* 75-Å-wide single membrane frequently produces the freeze-etch image seen in Figs. 2–4.

of two nonetchable inner faces. Since the diagram is intended as a generalized scheme, it also implies that freeze-etching will show neither the true membrane surfaces nor the surfaces of any materials contiguous to membranes. These implications have been verified by the experiments and observations described below.

Etching variations: Two sets of onion root tips were freeze-etched. In order to control the amount of water sublimed from the fractured surface, one set received no etching, whereas the other set received normal etching. Figure 6 shows cells which received no etching, and Figure 7 shows cells which received normal etching. Comparison reveals a striking similarity in the over-all appearance of the membrane faces in spite of clear differences in the textural appearance of the rest of the protoplasm. Whereas etching had little apparent effect on the membrane faces, it gave the protoplasmic matrix a distinct pebbled appearance. In other words, water was sublimed from the fractured cytoplasmic matrix, karyoplasm, and vacuole, but not from any of the faces along which the fractures followed membrane contours. This suggested the prediction, diagrammed in Figure 8, that deep etching would reveal more of the true membrane surface than had normal etching. Since etching removes primarily water and not other cell constituents, the effects of deep etching could best be seen in yeast cells which had been suspended in plain tap water for 2 hr. Figure 9 is the result of such an experiment and shows that with deep etching a greater portion of the true membrane surface is exposed. As anticipated, this is particularly noticeable at the edge of vacuoles, because of the high water content, and consequent etchability, of vacuolar fluid.

Surface features: If freeze-etching splits membranes so as to reveal inner faces rather than the true surface, structural features known to exist on membrane surfaces should not be visible on membrane faces exposed in freeze-etching. Ribosomal particles, frequently associated with the surface of the endoplasmic reticulum in chemically fixed preparations, cannot be seen on the endoplasmic reticulum faces revealed in freeze-etching. Small particles, averaging *ca.* 85 Å in diameter, are

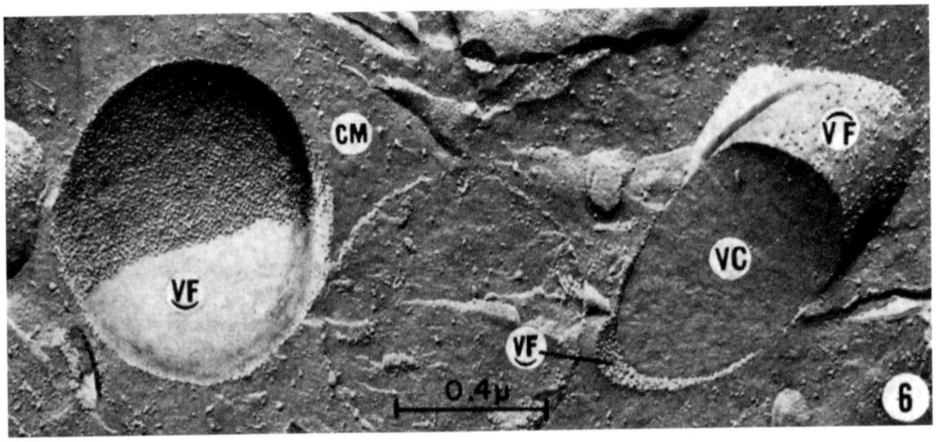

Fig. 6.—Part of an onion root tip cell with no etching. With the exception of a few small protuberances, the surface of the cytoplasmic matrix (*CM*) and vacuolar contents (*VC*) is relatively smooth. Both concave faces ($\underset{\smile}{VF}$) and convex faces ($\widehat{VF}$) of vacuolar membranes have been exposed. The vacuole in the upper right has been partially fractured and shows that the concave and convex vacuolar membrane faces had been apposed before fracturing. See Fig. 7.

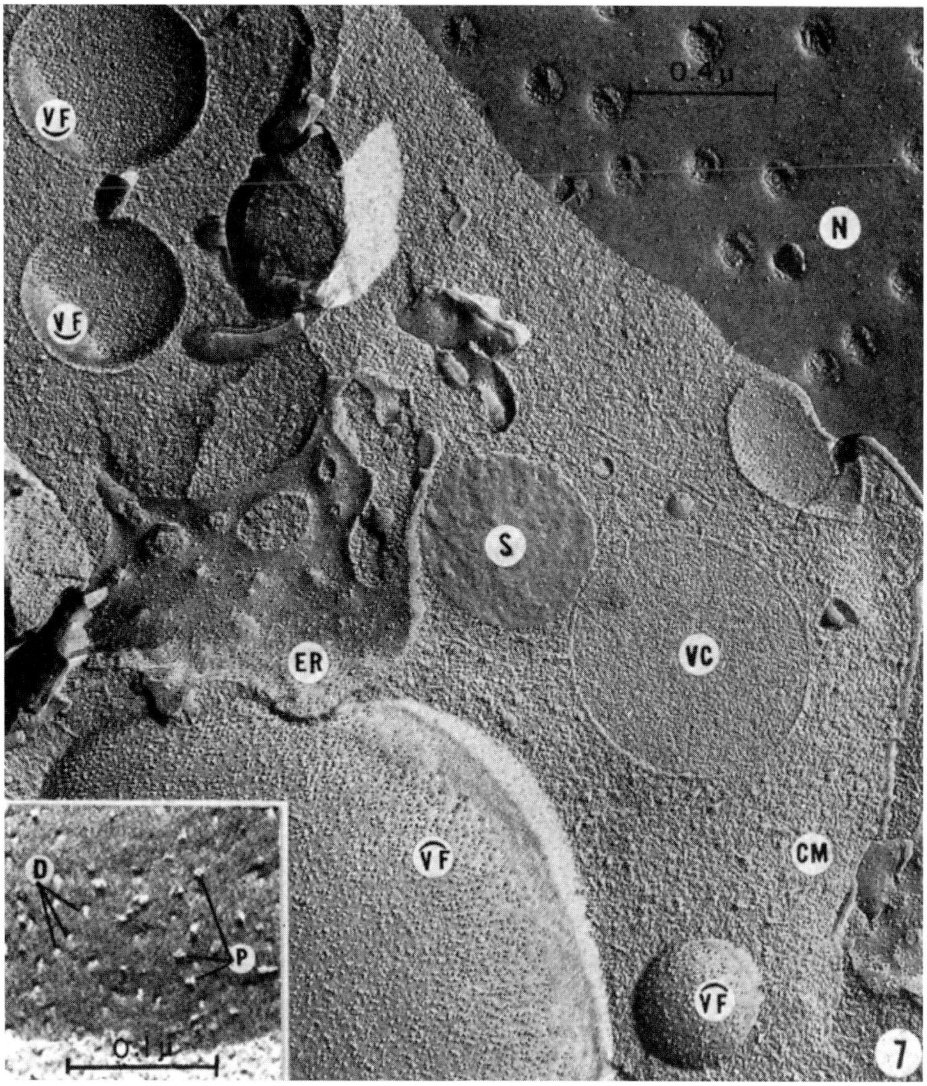

FIG. 7.—Part of an onion root tip cell with normal etching. Surfaces of the cytoplasmic matrix (*CM*) and vacuolar contents (*VC*) are pebbled as the result of etching, but vacuolar membrane faces (*VF*) are similar to those in Fig. 6, indicating that they are nonetchable. Note the smooth, non-etchable appearance of lipoidal material in the spherosomes (*S*) and compare with the smooth portions of membrane faces, including nuclear membrane (*N*) and endoplasmic reticulum (*ER*). Compare with Fig. 6. *Inset* shows a typical vacuolar membrane face and associated small particles (*P*), as well as depressions (*D*) where some of these particles were pulled off during the fracture process.

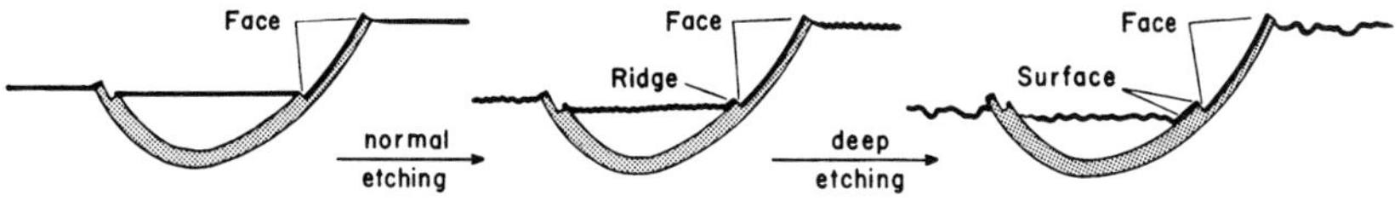

FIG. 8.—Interpretation of the etching procedure.

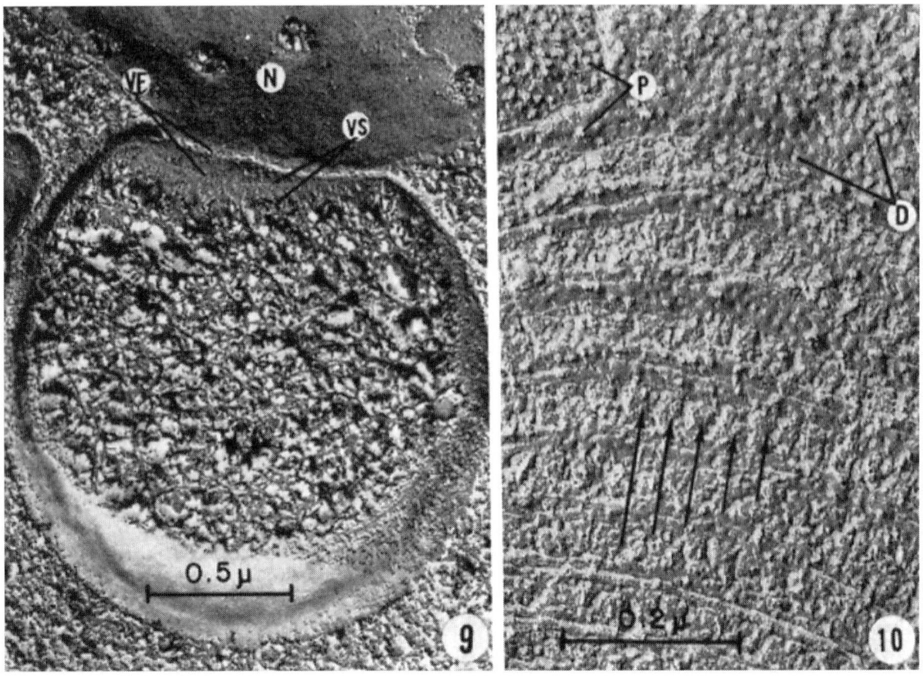

FIG. 9.—Deep etching in a yeast cell. Nuclear (*N*) and vacuolar membrane faces (*VF*) are similar to those seen in normally etched preparations (Fig. 7), but, as predicted in Fig. 8, a portion of the true vacuolar membrane surface (*VS*) has been exposed by the deep etching. Cf. with Fig. 3.

FIG. 10.—Freeze-etched *P. cruentum*, normal etching. Note the regular array of large particles (*arrows*) between the chloroplast membranes, but the absence of these particles on exposed membrane faces. Only smaller particles (*P*) and depressions (*D*) are on the exposed faces.

seen in varying numbers on all of the membrane faces in onion root tip cells (Fig. 7). Similar particles have been reported in other freeze-etch studies.[21–24] As previously suggested,[24] these small particles cannot be equated with ribosomes because they are too small and are found on membranes (including Golgi membranes) which are known from fixed and sectioned preparations to be devoid of ribosomes.[27]

Photomicrographs of *Porphyridium cruentum*[26] show that in these cells 320-Å particles are attached in extremely orderly arrays on the outer surfaces of the chloroplast membranes. Our own sections of chemically fixed *P. cruentum* confirmed these findings. The extraordinary regularity and large size of these particles suggested their use as markers of the outer membrane surface. Figure 10 illustrates the appearance of the plastid membranes in freeze-etched *P. cruentum*. Although the 320-Å particles are seen between the chloroplast membranes, neither these particles nor any depressions out of which they might have been fractured are evident on the tangentially fractured membranes; only smaller, randomly distributed particles and depressions similar to those found in other plant chloroplasts[28] are visible on these fractured faces. These observations confirm the proposition that freeze-etching does not normally expose the true outer membrane surface and indicate that freeze-etching does expose a hitherto unseen inner membrane face.

Discussion.—Three lines of evidence suggest that during freeze-etching, membranes are split in half, revealing either of two internal membrane faces. The first evidence for this type of fracture was encountered when it became possible to follow the contours of a single membrane which had been fractured almost normally in one part and tangentially in another part. In such preparations (Figs. 2–4) it became clear that the *ca.* 85-Å-thick rim representing a normally fractured membrane was an image formed by the confluence of two ridges, one bordering the base, the other forming the top, of the tangentially fractured portion. It appears unlikely that either of these ridges alone can be considered the entire membrane, as this would reduce the thickness of the biological membrane to less than 40 Å. For similar reasons these confluent ridges cannot be mere eutectic mixtures or organized, but nonmembrane, cytoplasmic components, as this would reduce the dimension of the biological membrane to that of a Euclidean plane.

A second line of evidence arises from the fact that in vitrified cells all of the fracture planes which follow membrane surfaces are nonetchable. Figure 6 shows convex faces over the top of vacuoles and concave faces out of which vacuoles have been removed. Comparison of these faces with analogous faces in Figure 7 shows that all exposed membrane faces appear to be identical whether or not they are etched. Such a result indicates that the fracture process in freeze-etching exposes two nonetchable membrane faces. If the fracture did not split the membrane but separated it from contiguous vacuolar or cytoplasmic material, only one of the exposed faces would be that of the membrane while the other would be that of some cell material such as protoplasmic matrix or vacuolar fluid. If this were the situation, at most one of the two exposed faces—that of the membrane—would be nonetchable. The other face would be that of protoplasmic matrix, vacuolar fluid, etc., and therefore would be etchable (cf. Fig. 6 with Fig. 7).

The third line of evidence rests on the observation that freeze-etched membrane faces do not show the structural features associated with the true membrane surfaces. The absence of the 320-Å particles on freeze-etched *P. cruentum* chloroplast membrane faces, in spite of the demonstrable presence of these particles in chemically fixed and in freeze-etched material, is graphic evidence that the membrane faces exposed after normal freeze-etching are not true membrane surfaces. A completely analogous situation has been observed in attempts to view the outer cell wall surface of yeast and bacterial cells. The outer wall surface of these unicells is rarely exposed by the fracture process but it can be exposed by deep etching.

Though at first it may appear surprising that membranes should split in two, consideration of the same types of evidence which led Danielli and Harvey[4] to postulate the presence of proteins on membrane surfaces leads to the prediction that fractures in vitrified cells might not occur along the true membrane surface. Danielli and Harvey postulated the existence of protein layers on membrane surfaces to account for the lower interfacial tension observed at surfaces of living cells. Though recent evidence[11, 12] suggests that in some membrane systems lipids with highly polar groups may assume the emulsifying roles which Danielli attributed to proteins, any adsorbed emulsifier would not only reduce interfacial tension, but would also provide the membrane interface with mechanical stability.[29] A macromolecular interface such as the surface of a membrane can interact with the surrounding aqueous phase by various types of polar bonds, and must be encased in a thin layer of bound water molecules whose properties merge, gradually, with those of the bulk phase.[18] As a result, it appears unlikely that the surface of a membrane with low interfacial tension vis-à-vis any contiguous protoplasmic material would present a sharp discontinuity of the sort which would make it uniquely liable to mechanical rupture while in the vitrified state.

Small particles, averaging *ca.* 85 Å in diameter, are seen on many freeze-etch membrane faces (see figures here and in refs. 21–25). One important difficulty presently under study, and one which should serve to emphasize the tentative nature of all freeze-etch interpretations so far proposed, is the absence of an adequate number of depressions (Figs. 7 and 11) into which these particles can be fitted. Although the nature of these small particles is currently being investigated, the tentative hypothesis adopted is that these substructures represent units within which membrane components have assumed globular or micellar configurations (Fig. 11). Such particles may represent lipoprotein associations analogous, perhaps, to those which have been reported in mitochondria,[11, 15] plastids,[12, 16] and other membranes.[14] According to this interpretation, the smooth regions between the 85-Å particles would represent regions in which the membrane components exist as an extended bilayer. The appearance and nonetchability of these smooth faces is similar to that of lipid material seen in spherosomes (Fig. 7). Thus, these smooth regions appear to be free of water and may be lipid faces.

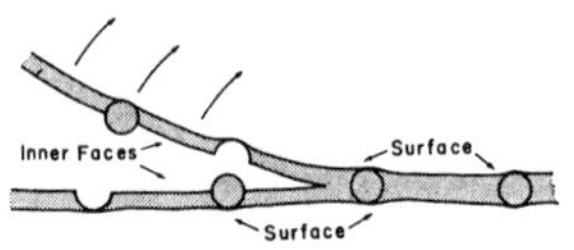

FIG. 11.—Interpretation of the fracture process. The inner membrane faces are seen in normal freeze-etching. Either particles, or depressions out of which the particles have been fractured, are seen (see *inset*, Fig. 7).

Comparative studies show that the number and manner in which the small particles are associated within a given membrane is a function of the type of cell organelle examined.[22, 24] For example, the freeze-etch results shown here indicate that the nuclear membrane in onion root tips (Fig. 7) appears to exist primarily as an extended bilayer, whereas similar freeze-etching studies of chloroplast membranes[28] show that they are composed almost entirely of the globular substructures. The function of these particles in different membranes as well as the environmental factors which modify their configuration is under study.

Summary.—Fracture planes within frozen cell membranes have been examined by freeze-etching. The frozen membrane is fractured so as to expose inner membrane faces. Examination of these faces suggests that the biological membrane is organized in part as an extended bilayer and in part as globular subunits. The relative proportion of the membrane which exists in either of these organizational modes varies among the different cell organelles.

The author gratefully acknowledges the expert assistance of Miss Susan Whytock.

* Research supported by National Science Foundation grant GB 2365.

1 Elbers, P. F., *Recent Progr. Surface Sci.*, **2**, 443 (1964).
2 Overton, E., *Jahrb. Wiss. Bot.*, **34**, 669 (1900).
3 Gorter, E., and F. Grendel, *J. Exptl. Med.*, **41**, 439 (1925).
4 Danielli, J. F., and E. N. Harvey, *J. Cellular Comp. Physiol.*, **5**, 483 (1935).
5 Danielli, J. F., and H. Davson, *J. Cellular Comp. Physiol.*, **5**, 495 (1935).
6 Schmitt, F. O., R. S. Bear, and G. L. Clark, *Radiology*, **25**, 131 (1935).
7 Schmidt, W. J., *Z. Zellforsch. Mikroskop. Anat.*, **23**, 657 (1936).
8 Fernández-Morán, H., and J. B. Finean, *J. Biophys. Biochem. Cytol.*, **3**, 725 (1957).
9 Robertson, J. D., *Biochem. Soc. Symp.*, **16**, 3 (1959).
10 Robertson, J. D., *Symp. Soc. Study of Development and Growth*, **22**, 1 (1964).

[11] Green, D. E., and S. Fleischer, *Biochim. Biophys. Acta*, **70**, 554 (1963).
[12] Park, R. B., *J. Cell Biol.*, **27**, 151 (1965).
[13] Maddy, A. H., and B. R. Malcolm, *Science*, **150**, 1616 (1965).
[14] Sjøstrand, F. S., *J. Ultrastruct. Res.*, **9**, 340 (1963).
[15] Fernández-Morán, H., T. Oda, P. V. Blair, and D. E. Green, *J. Cell Biol.*, **22**, 63 (1964).
[16] Weir, T. E., A. H. P. Engelbrecht, A. Harrison, E. B. Risley, *J. Ultrastruct. Res.*, **13**, 92 (1965).
[17] Hechter, O., *Bull. Neurosci. Res. Progr.*, **2**, 36 (1964).
[18] Kavanau, J. L., *Structure and Function in Biological Membranes* (San Francisco: Holden-Day, Inc., 1965).
[19] Steere, R. L., *J. Biophys. Biochem. Cytol.*, **7**, 167 (1957).
[20] Moor, H., K. Muhlethaler, H. Waldner, and A. Frey-Wyssling, *J. Biophys. Biochem. Cytol.*, **10**, 1 (1961).
[21] Moor, H., *Z. Zellforsch. Mikroskop. Anat.*, **62**, 546 (1964).
[22] Moor, H., and K. Muhlethaler, *J. Cell Biol.*, **17**, 609 (1963).
[23] Moor, H., C. Ruska, and H. Ruska, *Z. Zellforsch. Mikroskop. Anat.*, **62**, 581 (1964).
[24] Branton, D., and H. Moor, *J. Ultrastruct. Res.*, **11**, 401 (1965).
[25] Jost, M., *Arch. Mikrobiol.*, **50**, 211 (1965).
[26] Gantt, E., and S. F. Conti, *J. Cell Biol.*, **26**, 365 (1965).
[27] Whaley, W. G., H. H. Mollenhauer, and J. H. Leech, *Am. J. Bot.*, **47**, 401 (1960).
[28] Branton, D., and R. B. Park, in preparation.
[29] Adam, N. K., *Physics and Chemistry of Surfaces* (London: Oxford University Press, 1941).

Singer S.J. and Nicolson G.L. 1972. The fluid mosaic model of the structure of cell membranes. *Science* **175:** 720–731.

STRUCTURAL WORK ON BIOLOGICAL MEMBRANES had revealed their organization as lipid bilayers (Branton, this volume). The ability of the plasma membrane to support robust junctions between cells (Farquhar and Palade, this volume) argued for stability in the behavior of membrane specializations. However, experiments based on the fusion of two cell types, each of whose surface antigens had been labeled with a different fluorophore, indicated that membranes behaved as miscible fluids (1). It was therefore difficult to rationalize the complexity of membrane behavior and function with a single concept of membrane architecture. The "fluid mosaic" model for membranes, developed in the paper reproduced here, gave cell biologists a simple and fundamentally sound way of thinking about membrane organization. It unified ideas about the structure and function of membranes with their underlying chemistry, and it identified ways in which membranes could serve so many cellular functions. It also offered a rational way in which to conceive how trans-membrane proteins might be sorted during vesicular transport. Although the ideas presented were heavily grounded in previously published work, this paper clarified thought on the organization of molecules in membranes and greatly advanced progress on membrane cell biology. Its predictions about membrane structure and behavior were largely borne out in subsequent experimental work (2–4).

---

1. Frye L.D. and Edidin M. 1970. The rapid intermixing of cell surface antigens after formation of mouse-human heterokaryons. *J. Cell Sci.* **7:** 319–335.
2. Jacobson K., Derzko Z., Wu E.S., Hou Y., and Poste G. 1976. Measurement of the lateral mobility of cell surface components in single, living cells by fluorescence recovery after photobleaching. *J. Supramol. Struct.* **5:** 565–576.
3. Axelrod D., Koppel D.E., Schlessinger J., Elson E., and Webb W.W. 1976. Mobility measurement by analysis of fluorescence photobleaching recovery kinetics. *Biophys. J.* **16:** 1055–1069.
4. Kirschner D.A., Hollingshead C.J., Thaxton C., Caspar D.L.D., and Goodenough D.A. 1979. Structural states of myelin observed by X-ray diffraction and freeze-fracture electron microscopy. *J. Cell Biol.* **82:** 140–149.

Reprinted from
18 February 1972, Volume 175, pp. 720-731

SCIENCE

The Fluid Mosaic Model of the Structure of Cell Membranes

S. J. Singer and Garth L. Nicolson

The Fluid Mosaic Model of the Structure of Cell Membranes

Cell membranes are viewed as two-dimensional solutions of oriented globular proteins and lipids.

S. J. Singer and Garth L. Nicolson

Biological membranes play a crucial role in almost all cellular phenomena, yet our understanding of the molecular organization of membranes is still rudimentary. Experience has taught us, however, that in order to achieve a satisfactory understanding of how any biological system functions, the detailed molecular composition and structure of that system must be known. While we are still a long way from such knowledge about membranes in general, progress at both the theoretical and experimental levels in recent years has brought us to a stage where at least the gross aspects of the organization of the proteins and lipids of membranes can be discerned. There are some investigators, however, who, impressed with the great diversity of membrane compositions and functions, do not think there are any useful generalizations to be made even about the gross structure of cell membranes. We do not share that view. We suggest that an analogy exists between the problems of the structure of membranes and the structure of proteins. The latter are tremendously diverse in composition, function, and *detailed* structure. Each kind of protein molecule is structurally unique. Nevertheless, generalizations about protein structure have been very useful in understanding the properties and functions of protein molecules. Similarly, valid generalizations may exist about the ways in which the proteins and lipids are organized in an intact membrane. The ultimate test of such generalizations, or models, is whether they are useful to explain old experiments and suggest new ones.

Singer (*1*) has recently examined in considerable detail several models of the gross structural organization of membranes, in terms of the thermodynamics of macromolecular systems and in the light of the then available experimental evidence. From this analysis, it was concluded that a mosaic structure of alternating globular proteins and phospholipid bilayer was the only membrane model among those analyzed that was simultaneously consistent with thermodynamic restrictions and with all the experimental data available. Since that article was written, much new evidence has been published that strongly supports and extends this mosaic model. In particular, the mosaic appears to be a fluid or dynamic one and, for many purposes, is best thought of as a two-dimensional oriented viscous solution. In this article, we therefore present and discuss a fluid mosaic model of membrane structure, and propose that it is applicable to most biological membranes, such as plasmalemmal and intracellular membranes, including the membranes of different cell organelles, such as mitochondria and chloroplasts. These membranes are henceforth referred to as functional membranes. There may be some other membrane-like systems, such as myelin, or the lipoprotein membranes of small animal viruses, which we suggest may be rigid, rather than fluid, mosaic structures, but such membrane systems are not a primary concern of this article.

Our objectives are (i) to review briefly some of the thermodynamics of macromolecular, and particularly membrane, systems in an aqueous environment; (ii) to discuss some of the properties of the proteins and lipids of functional membranes; (iii) to describe the fluid mosaic model in detail; (iv) to analyze some of the recent and more direct experimental evidence in terms of the model; and (v) to show that the fluid mosaic model suggests new ways of thinking about membrane functions and membrane phenomena.

Dr. Singer is a professor of biology at the University of California at San Diego, La Jolla. Dr. Nicolson is a research associate at the Armand Hammer Cancer Center of the Salk Institute for Biological Studies, La Jolla, California.

Thermodynamics and Membrane Structure

The fluid mosaic model has evolved by a series of stages from earlier versions (*1–4*). Thermodynamic considerations about membranes and membrane components initiated, and are still central to, these developments. These considerations derived from two decades of intensive studies of protein and nucleic acid structures; the thermodynamic principles involved, however, are perfectly general and apply to any macromolecular system in an aqueous environment. These principles and their application to membrane systems have been examined in detail elsewhere (*1*) and are only summarized here. For our present purposes, two kinds of noncovalent interactions are most important, *hydrophobic* (*5*) and *hydrophilic* (*1*). By hydrophobic interactions is meant a set of thermodynamic factors that are responsible for the sequestering of hydrophobic or nonpolar groups away from water, as, for example, the immiscibility of hydrocarbons and water. To be specific, it requires the expenditure of 2.6 kilocalories of free energy to transfer a mole of methane from a nonpolar medium to water at 25°C (*5*). Free energy contributions of this magnitude, summed over the many nonpolar amino acid residues of soluble proteins, are no doubt of primary importance in determining the conformations that protein molecules adopt in aqueous solution (*6*), in which the nonpolar residues are predominantly sequestered in the interior of the molecules away from contact with water. By hydrophilic interactions is meant a set of thermodynamic factors that are responsible for the preference of ionic and polar groups for an aqueous rather than a nonpolar environment. For example, the free energy required to transfer a mole of zwitterionic glycine from water to acetone is about 6.0 kcal at 25°C, showing that ion pairs strongly prefer to be in water than in a nonpolar medium (*1*). These and related free energy terms no doubt provide the reasons why essentially all the ionic residues of protein molecules are observed to be in contact with water, usually on the outer surface of the molecule, according to x-ray crystallographic studies. Similar thermodynamic arguments apply to saccharide residues (*1*). It requires the expenditure of substantial free energy to transfer a simple saccharide from water to a nonpolar solvent, and such residues will therefore be in a lower free energy state in contact with water than in a less polar environment.

There are other noncovalent interactions, such as hydrogen bonding and electrostatic interactions, which also contribute to determine macromolecular structure. However, with respect to gross structure, with which we are now concerned, these are very likely of secondary magnitude compared to hydrophobic and hydrophilic interactions.

The familiar phospholipid bilayer structure illustrates the combined effects of hydrophobic and hydrophilic interactions. In this structure (Fig. 1) the nonpolar fatty acid chains of the phospholipids are sequestered together away from contact with water, thereby maximizing hydrophobic interactions. Furthermore, the ionic and zwitterionic groups are in direct contact with the aqueous phase at the exterior surfaces of the bilayer, thereby maximizing hydrophilic interactions. In the case of zwitterionic phospholipids such as phosphatidylcholine, dipole-dipole interactions between ion pairs at the surface of the bilayer may also contribute to the stabilization of the bilayer structure.

In applying these thermodynamic principles to membranes, we recognize first that of the three major classes of membrane components—proteins, lipids, and oligosaccharides—the proteins are predominant. The ratio by weight of proteins to lipids ranges from about 1.5 to 4 for those functional membranes which have been well characterized [compare (*7*)]. A substantial fraction of this protein most probably plays an important role in determining the structure of membranes, and the structural properties of these proteins are therefore of first-order importance. Membrane proteins are considered in some detail in the following section. At this juncture, the significant point is that if hydrophobic and hydrophilic interactions are to be maximized and the lowest free energy state is to be attained for the intact membrane in an aqueous environment, the nonpolar amino acid residues of the proteins—along with the fatty acid chains of the phospholipids—should be sequestered

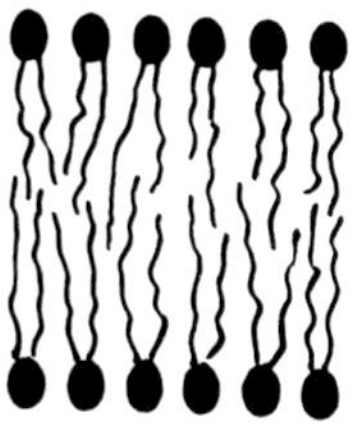

Fig. 1. A phospholipid bilayer: schematic cross-sectional view. The filled circles represent the ionic and polar head groups of the phospholipid molecules, which make contact with water; the wavy lines represent the fatty acid chains.

(to the maximum extent feasible) from contact with water, while the ionic and polar groups of the proteins—along with those of the lipids and the oligosaccharides—should be in contact with the aqueous solvent. These requirements place restrictions on models of membrane structure; in particular, they render highly unlikely the classical model of a trilaminar arrangement of a continuous lipid bilayer sandwiched between two monolayers of protein. The latter model is thermodynamically unstable because not only are the nonpolar amino acid residues of the membrane proteins in this model perforce largely exposed to water but the ionic and polar groups of the lipid are sequestered by a layer of protein from contact with water. Therefore, neither hydrophobic nor hydrophilic interactions are maximized in the classical model.

Some Properties of Membrane Components

Peripheral and integral proteins. It seems both reasonable and important to discriminate between two categories of proteins bound to membranes, which we have termed *peripheral* and *integral* proteins (*1*). Peripheral proteins may be characterized by the following criteria. (i) They require only mild treatments, such as an increase in the ionic strength of the medium or the addition of a chelating agent, to dissociate them molecularly intact from the membrane; (ii) they dissociate free of lipids; and (iii) in the dissociated state they are relatively soluble in neutral aqueous buffers. These criteria suggest that a peripheral protein is held to the membrane only by rather weak noncovalent (perhaps mainly electrostatic) interactions and is not strongly associated with membrane lipid. The cytochrome c of mitochondrial membranes, which can be dissociated free of lipids by high salt concentrations, and the protein spectrin (*8*) of erythrocyte membranes, which can be removed by chelating agents under mild conditions, are examples of membrane proteins that satisfy the criteria for peripheral proteins. On the other hand, the major portion (> 70 percent) of the proteins of most membranes have different characteristics, which may be assigned to integral proteins: (i) they require much more drastic treatments, with reagents such as detergents, bile acids, protein denaturants, or organic solvents, to dissociate them from membranes; (ii) in many instances, they remain associated with lipids when isolated; (iii) if completely freed of lipids, they are usually highly insoluble or aggregated in neutral aqueous buffers (*9*).

The distinction between peripheral and integral proteins may be useful in several regards. It is assumed that only the integral proteins are critical to the structural integrity of membranes. Therefore, the properties and interactions of peripheral proteins, while interesting in their own right, may not be directly relevant to the central problems of membrane structure. The properties of cytochrome c, for example, may not be typical of mitochondrial membrane proteins. Furthermore, the biosynthesis of peripheral and integral proteins and their attachment to the membrane may be very different processes. This is not the appropriate occasion to discuss membrane biogenesis in any detail, but it may be significant that, although cytochrome c is a mitochondrial protein, it is synthesized on cytoplasmic rather than mitochondrial ribosomes; in fact only a small fraction of the total mitochondrial protein (perhaps only the integral proteins of the inner mitochondrial membrane?) appears to be synthesized on mitochondrial ribosomes (*10*). In any event, because of the relatively unimportant membrane structural role assigned to the peripheral proteins, they are not a primary concern of this article.

Properties of integral proteins. Since the proteins we have classified as integral, according to the criteria specified, constitute the major fraction of membrane proteins, we assume that the properties to be discussed apply to the integral proteins.

1) For several well-characterized membrane systems, including erythrocyte and other plasma membranes, and mitochondrial membranes, the proteins have been shown to be grossly heterogeneous with respect to molecular

weights (*11*). There is no convincing evidence that there exists one predominant type of membrane protein that is specifically a structural protein; recent reports to the contrary have been withdrawn. We consider this heterogeneity to be more significant for a general model of membrane structure than the fact that in a few specialized instances, as in the case of disk membranes of retinal rod outer segments (*12, 13*), a single protein species predominates. A satisfactory membrane model must be capable of explaining the heterogeneity of the integral membrane proteins.

2) The proteins of a variety of intact membranes, on the average, show appreciable amounts of the α-helical conformation, as was first shown by Ke (*14*), Wallach and Zahler (*4*), and Lenard and Singer (*3*). For example, circular dichroism measurements of aqueous suspensions of intact and mechanically fragmented human erythrocyte membranes (provided that we take into account certain optical anomalies of these measurements) reveal that about 40 percent of the protein is in the right-handed α-helical conformation (*15*). Most soluble globular proteins whose circular dichroism spectra have been obtained exhibit a smaller fraction of α-helix in their native structures. This suggests that the integral proteins in intact membranes are largely globular in shape rather than spread out as monolayers. On the other hand, a membrane model in which such globular proteins are attached to the *outer* surfaces of a lipid bilayer (*16*) would not be satisfactory because, among other reasons, it would require membrane thicknesses much larger than the 75 to 90 angstroms generally observed. A model in which globular protein molecules are intercalated within the membrane would, however, meet these restrictions.

The phospholipids of membranes. There is now substantial evidence that the major portion of the phospholipids is in bilayer form in a variety of intact membranes. For example, differential calorimetry of intact mycoplasma membranes shows that they undergo a phase transition in a temperature range very similar to that of aqueous dispersions of the phospholipids extracted from the membranes (*16, 17*). Thus the structures of the lipid in the membrane and of the lipid in isolated aqueous dispersion are closely similar; presumably the latter is the bilayer form. This conclusion is supported by x-ray diffraction (*18*) and spir -label studies (*19*) on similar membrane preparations.

The bilayer character of membrane lipids rules out models such as that of Benson (*20*) in which the proteins and lipids form a single-phase lipoprotein subunit that is repeated indefinitely in two dimensions to constitute the membrane. In such a model, most of the lipids would be expected to have distinctly different properties from those of a bilayer.

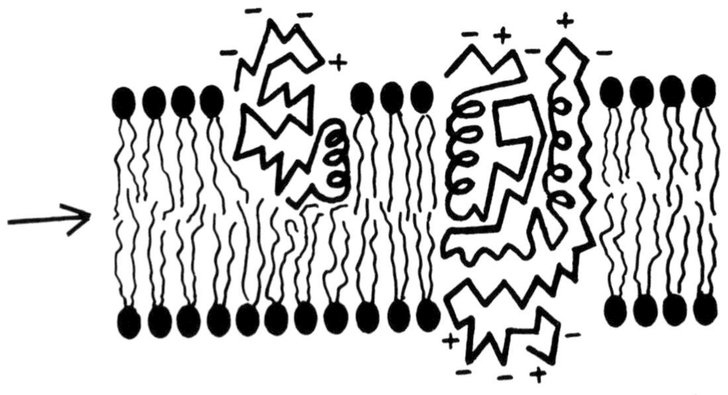

Fig. 2. The lipid-globular protein mosaic model of membrane structure: schematic cross-sectional view. The phospholipids are depicted as in Fig. 1, and are arranged as a discontinuous bilayer with their ionic and polar heads in contact with water. Some lipid may be structurally differentiated from the bulk of the lipid (see text), but this is not explicitly shown in the figure. The integral proteins, with the heavy lines representing the folded polypeptide chains, are shown as globular molecules partially embedded in, and partially protruding from, the membrane. The protruding parts have on their surfaces the ionic residues (— and +) of the protein, while the nonpolar residues are largely in the embedded parts; accordingly, the protein molecules are amphipathic. The degree to which the integral proteins are embedded and, in particular, whether they span the entire membrane thickness depend on the size and structure of the molecules. The arrow marks the plane of cleavage to be expected in freeze-etching experiments (see text). [From Lenard and Singer (*3*) and Singer (*1*)]

Two qualifications should be stressed, however, concerning the bilayer form of membrane lipids. (i) None of the evidence so far obtained for the bilayer form permits us to say whether the bilayer is *continuous* or *interrupted* (*1*). The calorimetrically observed phase transitions, for example, occur over a broad temperature interval, allowing the possibility that the cooperative unit involved in the phase transition is quite small, consisting perhaps of only 100 lipid molecules on the average. (ii) None of the experiments mentioned above is sufficiently sensitive and quantitative to prove whether 100 percent of the phospholipid is in the bilayer form. It is therefore not excluded that some significant fraction of the phospholipid (perhaps as much as 30 percent) is physically in a different state from the rest of the lipid.

Protein-lipid interactions in membranes. Several kinds of experiments indicate that protein-lipid interactions play a direct role in a variety of membrane functions. Many membrane-bound enzymes and antigens require lipids, often specific phospholipids, for the expression of their activities [see table 2 in (*21*)]. Furthermore, the nature of the fatty acids incorporated into phospholipids affects the function of certain membrane-bound proteins in bacterial membranes (*22*).

On the other hand, the calorimetric data discussed above give no significant indication that the association of proteins with the phospholipids of intact membranes affects the phase transitions of the phospholipids themselves. Experiments with phospholipase C and membranes have shown that the enzymic release of 70 percent of the phosphorylated amines from intact erythrocyte membranes profoundly perturbs the physical state of the residual fatty acid chains, but has no detectable effect (as measured by circular dichroism spectra) on the average conformation of the membrane proteins (*2*). Such results therefore suggest that the phospholipids and proteins of membranes do not interact strongly; in fact, they appear to be largely independent.

This paradox, that different types of experiments suggest strong protein-lipid interactions on the one hand, and weak or no interactions on the other, can be resolved in a manner consistent with all the data if it is proposed that, while the largest portion of the phospholipid is in bilayer form and not strongly coupled to proteins in the membrane, a small fraction of the lipid is more tightly coupled to protein. With any one membrane protein, the tightly coupled lipid might be specific; that is, the interaction might require that the phospholipid contain specific fatty acid chains or particular polar head groups. There is at present, however, no satisfactory direct evidence for such a distinctive lipid fraction. This problem is considered again in connection with a discussion of the experiments of Wilson and Fox (*23*).

Fluid Mosaic Model

Mosaic structure of the proteins and lipids of membranes. The thermodynamic considerations and experimental results so far discussed fit in with the idea of a mosaic structure for membranes (*1–3, 24*) in which globular molecules of the integral proteins (perhaps in particular instances attached to oligosaccharides to form glycoproteins, or interacting strongly with specific lipids to form lipoproteins) alternate with sections of phospholipid bilayer in the cross section of the membrane (Fig. 2). The globular protein molecules are postulated to be amphipathic (*3, 4*) as are the phospholipids. That is, they are structurally asymmetric, with one highly polar end and one nonpolar end. The highly polar region is one in which the ionic amino acid residues and any covalently bound saccharide residues are clustered, and which is in contact with the aqueous phase in the intact membrane; the nonpolar region is devoid of ionic and saccharide residues, contains many of the nonpolar residues, and is embedded in the hydrophobic interior of the membrane. The amphipathic structure adopted by a particular integral protein (or lipoprotein) molecule, and therefore the extent to which it is embedded in the membrane, are under thermodynamic control; that is, they are determined by the amino acid sequence and covalent structure of the protein, and by its interactions with its molecular environment, so that the free energy of the system as a whole is at a minimum. An integral protein molecule with the appropriate size and structure, or a suitable aggregate of integral proteins (below) may transverse the entire membrane (*3*); that is, they have regions in contact with the aqueous solvent on both sides of the membrane.

It is clear from these considerations that different proteins, if they have the appropriate amino acid sequence to adopt an amphipathic structure, can be integral proteins of membranes; in this manner, the heterogeneity of the proteins of most functional membranes can be rationalized.

The same considerations may also explain why some proteins are membrane-bound and others are freely soluble in the cytoplasm. The difference may be that either the amino acid sequence of the particular protein allows it to adopt an amphipathic structure or, on the contrary, to adopt a structure in which the distribution of ionic groups is nearly spherically symmetrical, in the lowest free energy state of the system. If the ionic distribution on the protein surface were symmetrical, the protein would be capable of interacting strongly with water all over its exterior surface, that is, it would be a monodisperse soluble protein.

The mosaic structure can be readily diversified in several ways. Although the nature of this diversification is a matter of speculation, it is important to recognize that the mosaic structure need not be restricted by the schematic representation in Fig. 2. Protein-protein interactions that are not explicitly considered in Fig. 2 may be important in determining the properties of the membrane. Such interactions may result either in the specific binding of a peripheral protein to the exterior exposed surface of a particular integral protein or in the association of two or more integral protein subunits to form a specific aggregate within the membrane. These features can be accommodated in Fig. 2 without any changes in the basic structure.

The phospholipids of the mosaic structure are predominantly arranged as an interrupted bilayer, with their ionic and polar head groups in contact with the aqueous phase. As has been discussed, however, a small portion of the lipid may be more intimately associated with the integral proteins. This feature is not explicitly indicated in Fig. 2. The thickness of a mosaic membrane would vary along the surface from that across a phospholipid bilayer region to that across a protein region, with an *average* value that could be expected to correspond reasonably well to experimentally measured membrane thicknesses.

Matrix of the mosaic: lipid or protein? In the cross section of the mosaic structure represented in Fig. 2, it is not indicated whether it is the protein or the phospholipid that provides the matrix of the mosaic. In other words, which component is the mortar, which the bricks? This question must be answered when the third dimension of the mosaic structure is specified. These two types of mosaic structure may be expected to have very different structural and functional properties, and the question is therefore a critical one. It is our hy-

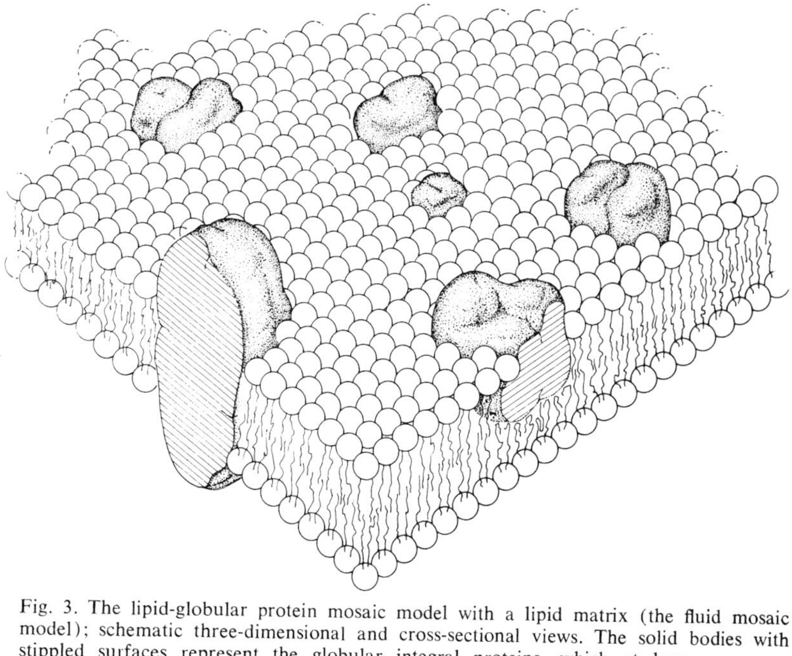

Fig. 3. The lipid-globular protein mosaic model with a lipid matrix (the fluid mosaic model); schematic three-dimensional and cross-sectional views. The solid bodies with stippled surfaces represent the globular integral proteins, which at long range are randomly distributed in the plane of the membrane. At short range, some may form specific aggregates, as shown. In cross section and in other details, the legend of Fig. 2 applies.

pothesis that functional cell membranes have a long-range mosaic structure with the *lipids* constituting the matrix, as is shown in Fig. 3. Supporting evidence is discussed later. At this point, let us consider some of the consequences of this hypothesis.

1) There should generally be no long-range order in a mosaic membrane with a lipid matrix. By long range, we mean over distances of the order of a few tenths of a micrometer and greater. Suppose we have a membrane preparation containing many different protein species, and suppose further that 10,000 molecules of protein *A* are present in the membrane of a single cell or organelle. How is protein *A* distributed over the membrane surface? If the membrane proteins formed the matrix of the mosaic, defined by specific contacts between the molecules of different integral proteins, protein *A* might be distributed in a highly ordered, two-dimensional array on the surface. On the other hand, if lipid formed the matrix of the mosaic, there would be no long-range interactions *intrinsic* to the membrane influencing the distribution of *A* molecules, and they should therefore be distributed in an aperiodic random arrangement on the membrane surface.

The absence of long-range order should not be taken to imply an absence of short-range order in the membrane. It is very likely that such short-range order does exist, as, for example, among at least some components of the electron transport chain in the mitochondrial inner membrane. Such short-range order is probably mediated by specific protein (and perhaps protein-lipid) interactions leading to the formation of stoichiometrically defined aggregates within the membrane. However, in a mosaic membrane with a lipid matrix, the long-range distribution of such aggregates would be expected to be random over the entire surface of the membrane.

The objection may immediately be raised that long-range order clearly exists in certain cases where differentiated structures (for example, synapses) are found within a membrane. We suggest, in such special cases, either that short-range specific interactions among integral proteins result in the formation of an unusually large two-dimensional aggregate or that some agent extrinsic to the membrane (either inside or outside the cell) interacts multiply with specific integral proteins to produce a clustering of those proteins in a limited area of the membrane surface. In other words, we suggest that long-range random arrangements in membranes are the norm; wherever nonrandom distributions are found, mechanisms must exist which are responsible for them.

2) It has been shown that, under physiological conditions, the lipids of functional cell membranes are in a fluid rather than a crystalline state. (This is not true of myelin, however.) This evidence comes from a variety of sources, such as spin-labeling experiments (*25*), x-ray diffraction studies (*18*), and differential calorimetry (*16, 17*). If a membrane consisted of integral proteins dispersed in a fluid lipid matrix, the membrane would in effect be a two-dimensional liquid-like solution of monomeric or aggregated integral proteins (or lipoproteins) dissolved in the lipid bilayer. The mosaic structure would be a dynamic rather than a static one. The integral proteins would be expected to undergo translational diffusion within the membrane, at rates determined in part by the effective viscosity of the lipid, unless they were tied down by some specific interactions intrinsic or extrinsic to the membrane. However, because of their amphipathic structures, the integral proteins would maintain their molecular orientation and their degree of intercalation in the membrane while undergoing translational diffusion in the plane of the membrane (as discussed below).

In contrast, if the matrix of the mosaic were constituted of integral proteins, the long-range structure of the membrane would be essentially static. Large energies of activation would be required for a protein component to diffuse in the plane of the membrane from one region to a distant one because of the many noncovalent bonds between the proteins that would have to be simultaneously broken for exchange to take place. Therefore, a mosaic membrane with a protein matrix should make for a relatively rigid structure with essentially no translational diffusion of its protein components within the membrane.

From the discussion in this and the previous section, it is clear that the fluid mosaic model suggests a set of structural properties for functional membranes at least some of which can be tested experimentally. In an earlier article (*1*), a large body of experimental evidence was examined for its relevance to models of membrane structure. It was concluded that a mosaic structure was most consistent with the available evidence. Some more recent results, however, bear even more directly on the problem, and only this evidence is discussed below.

Some Recent Experimental Evidence

Evidence for proteins embedded in membranes. One proposal of the fluid mosaic model is that an integral protein is a globular molecule having a significant fraction of its volume embedded in the membrane. The results of recent freeze-etching experiments with membranes strongly suggest that a substantial amount of protein is deeply embedded in many functional membranes. In this technique (*26*) a frozen specimen is fractured with a microtome knife; some of the frozen water is sublimed (etched) from the fractured surface if desired; the surface is then shadow cast with metal, and the surface replica is examined in the electron microscope. By this method the topography of the cleaved surface is revealed. A characteristic feature of the exposed surface of most functional membranes examined by this technique, including plasmalemmal, vacuolar, nuclear, chloroplast, mitochondrial, and bacterial membranes (*27, 28*), is a mosaic-like structure consisting of a smooth matrix interrupted by a large number of particles. These particles have a fairly characteristic uniform size for a particular membrane, for example, about 85-Å diameter for erythrocyte membranes. Such surfaces result from the cleavage of a membrane along its *interior* hydrophobic face (*29*). This interior face (Fig. 2) corresponds to the plane indicated by the arrow. If cleavage were to occur smoothly between the two layers of phospholipid in the bilayer regions, but were to circumvent the protein molecules penetrating the mid-plane of the membrane, then the alternating smooth and particulate regions observed on the freeze-etch surfaces can be readily explained by a mosaic structure for the membrane (Fig. 2), provided that the particles can be shown to be protein in nature. That the particles are indeed protein has been suggested by recent experiments (*30*).

Another consequence of the mosaic model, suggested from its inception (*3*), is that certain integral proteins possessing the appropriate size and structure may span the entire thickness of the membrane and be exposed at both membrane surfaces. Chemical evidence that a trans-membrane protein, whose molecular weight is about 100,000, is present in large amounts in the human erythrocyte membrane has been obtained by two independent methods—one involving proteolysis of normal compared to everted membranes (*31*), and the other specific chemical labeling of the membrane proteins (*32*).

Distribution of components in the plane of the membrane. A prediction of the fluid mosaic model is that the two-dimensional long-range distribution of any integral protein in the plane of the membrane is essentially random. To test this prediction, we have developed and applied electron microscopic techniques to visualize the distribution of specific membrane antigens over large areas of their membrane surfaces (*33*) and have so far studied the distribution of the $Rh_0(D)$ antigen on human erythrocyte membranes (*34*), and of H-2 histocompatibility alloantigens on mouse erythrocyte membranes (*35*).

In the case of the $Rh_0(D)$ antigen, for example, cells of O, Rh-positive type were reacted with a saturating amount of ^{125}I-labeled purified human antibody to $Rh_0(D)$ [anti-$Rh_0(D)$], and the treated (sensitized) cells were lysed at an air-water interface, so that the cell membranes were spread out flat. The flattened membranes, after being picked up on an electron microscope grid, were treated with the specific "indirect stain," ferritin-conjugated goat antibodies specific for human γ-globulin. Thus, wherever the human anti-Rh_0 (D) molecules were bound to the Rh_0 (D) antigen on the membrane surface, the ferritin-labeled goat antibodies became specifically attached. In other words, the human γ-globulin antibody now functioned as an antigen for the goat antibodies (Fig. 4). The ferritin was distributed in discrete clusters, each containing two to eight ferritin molecules within a circle of radius about 300 Å. The numbers of such clusters per unit area of the membrane surface corresponded to the number of ^{125}I-labeled human anti-$Rh_0(D)$ molecules bound per unit area. This indicates that each ferritin cluster was bound to a single anti-$Rh_0(D)$ molecule, and a cluster represents the number of goat antibody molecules bound to a single human γ-globulin molecule. Each cluster therefore corresponds to a single $Rh_0(D)$ antigen site (*36*) on the membrane. Since the clusters were distributed in a random array, we conclude that the $Rh_0(D)$ antigen, which exhibits properties of an integral protein (*37*), is molecularly dispersed and is distributed in a random two-dimensional array on the human erythrocyte membrane.

Similar experiments were carried out with the H-2 alloantigenic sites on mouse erythrocyte membranes. In this case (Fig. 5) the clusters of ferritin molecules of the indirect stain were not isolated, as in the case of the $Rh_0(D)$ antigen, but instead occurred in patches. The patchy distribution of the H-2 histocompatibility alloantigenic sites had earlier been observed by different techniques (*38*), but the two-dimensional distribution of the patches could not be ascertained. In our experiments, the patches contained variable numbers of clusters, and were arranged in an irregular two-dimensional array on the membrane surface. The histocompatibility antigen appears to be glycoprotein in nature (*39*). The long-range distribution of both the $Rh_0(D)$ and H-2 histocompatibility antigens on their respective membrane surfaces, therefore,

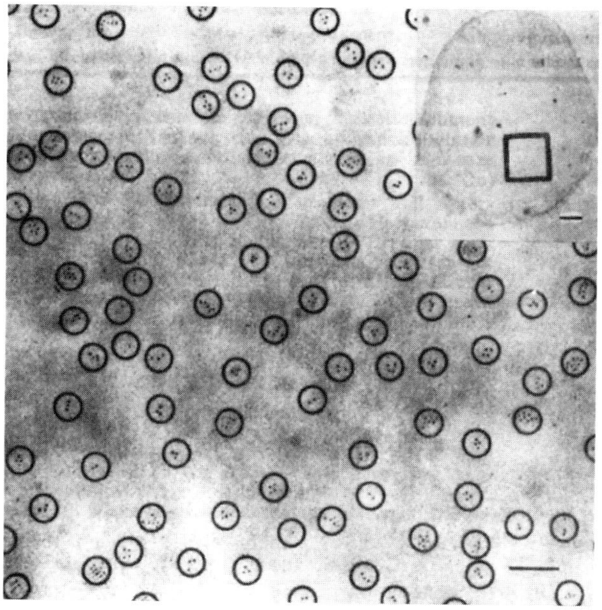

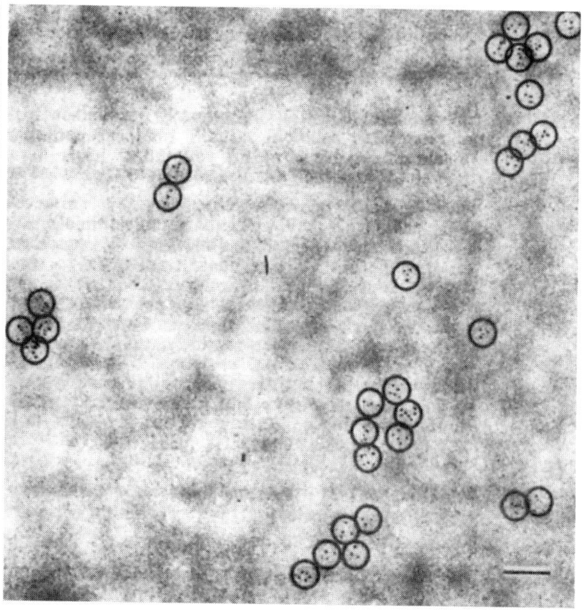

Fig. 4 (left). The outer membrane surface of an Rh-positive human erythrocyte sensitized with human anti-$Rh_0(D)$ and stained with ferritin-conjugated goat antibody to human γ-globulin. The cells were first labeled to saturation with purified ^{125}I-labeled human antibody to $Rh_0(D)$ and then lysed at an air-water interface. The erythrocyte membrane ghosts, flattened by surface forces (inset, low magnification) were picked up on a coated, electron microscope grid and indirectly stained with ferritin-conjugated goat antibodies to human γ-globulin. The ferritin appears bound to the membrane in discrete clusters of two to eight ferritin-conjugates; each cluster is circumscribed by a circle of radius 300 Å. The number of such clusters per cell (9300) is equal within experimental error to the number of ^{125}I-labeled human antibody to $Rh_0(D)$ molecules bound per cell (10,200). Each cluster therefore corresponds to an individual $Rh_0(D)$ antigenic site. Scale is 0.1 μm; inset scale is 1 μm. [From Nicolson, Masouredis, and Singer (*34*)] Fig. 5 (right). The outer membrane surface on a mouse erythrocyte (H-2^b) sensitized with alloantibodies against H-2^b histocompatibility antigens and stained with ferritin-conjugated antibodies against 7*S* mouse γ-globulin. The procedures are the same as listed in the legend to Fig. 4. The ferritin-antibody clusters are present in randomly spaced "patches" of variable size on the membrane surface. Scale is 0.1 μm. [From Nicolson, Hyman, and Singer (*35*)]

are in accord with the prediction of the fluid mosaic model that the integral proteins of membranes are randomly arranged in two dimensions.

The particles on the inner membrane faces revealed by freeze-etching experiments, which (as discussed above) are probably protein in nature, are generally also relatively randomly distributed in two dimensions.

Evidence that proteins are in a fluid state in intact membranes. An important series of experiments has been carried out (*12, 40–44*) with receptor disk membranes from the retina of the frog. This membrane system is unusual in that it contains as its predominant, if not only, protein component the pigment rhodopsin. In electron microscopy of the negatively stained surfaces of the dried membranes, a somewhat tightly packed and ordered array of particles (about 40 Å) was observed. These particles are the individual rhodopsin molecules. Although the earlier studies suggested that there was a long-range order in the distribution of the particles (*40*), more recent x-ray diffraction data (*42*) on pellets of wet, receptor disk membranes showed that only a few orders of reflection were observed corresponding to the spacings of the rhodopsin molecules in the plane of the membrane. This indicated that a noncrystalline, aperiodic arrangement of the particles existed in the plane of the membrane. Furthermore, the temperature dependence of the characteristics of the x-ray diffraction maxima were consistent with the suggestion that the particles were in a planar liquid-like state in the intact membrane. Additional support for the existence of this liquid-like state was the observation that the absorption of a foreign protein (bovine serum albumin) to the membrane could definitely alter the x-ray spacings due to the rhodopsin particles; that is, the distribution of the rhodopsin molecules in the plane of the membrane was radically altered by the weak binding of the albumin. This alteration would not be expected if a rigid lattice structure of the rhodopsin molecules, or aggregates, were present in the plane of the membrane.

These studies are particularly noteworthy because they involved a membrane which, by conventional electron microscopic techniques, appears to show long-range periodicity over its surface. Other specialized membranes have also exhibited ordered electron micrographic images of their surfaces [compare (*43*)]. However, it is likely that a very concentrated two-dimensional fluid solution of identical protein molecules will appear, when dried, to be arranged in an ordered array, particularly when optical tricks are used to enhance the apparent order (*43*). What is really a fluid phase may therefore artifactually be made to appear as a crystalline solid. This appears to be the situation with the retinal receptor disk membranes.

A major contribution to membrane studies has been made by Frye and Edidin (*44*), who investigated the membrane properties of some cell fusion heterokaryons. Human and mouse cells in culture were induced to fuse with one another, with Sendai virus as the fusing agent. The distribution of human and mouse antigenic components of the fused cell membranes was then determined by immunofluorescence, with the use of rabbit antibodies directed to the whole human cells, mouse antibodies directed against the H-2 alloantigen on the mouse cell membranes, and, as indirect stains, goat antiserum to rabbit γ-globulin and goat antiserum to mouse γ-globulin labeled with two different fluorescent dyes. Shortly after cell fusion, the mouse and human antigenic components were largely segregated in different halves of the fused cell membranes; but after about 40 minutes at 37°C the components were essentially completely intermixed. Inhibitors of protein synthesis, of adenosine triphosphate (ATP) formation, and of glutamine-dependent synthetic pathways, applied before or after cell fusion, had no effect on the rate of this intermixing process, but lowering the temperature below 15°C sharply decreased it.

Frye and Edidin (*44*) suggest that the intermixing of membrane components is due to diffusion of these components within the membrane, rather than to their removal and reinsertion, or to the synthesis and insertion of new copies of these components, into the heterokaryon membrane. An unexplained finding of these experiments was the fairly frequent occurrence, at early and intermediate times after cell fusion, of heterokaryon membranes in which the human antigenic components were uniformly distributed over the membrane surface but the mouse components were still largely segregated to about half the membrane ($M_{1/2}$-H_1 cells). On the other hand, the reverse situation, with the mouse antigenic components uniformly spread out over the membrane and the human components segregated (M_1-$H_{1/2}$), was only rarely observed. This result can now be explained by a diffusion mechanism for the intermixing process, as follows. The antibodies to the human cell membrane were no doubt directed to a heterogeneous set of antigens, whereas the antibodies to the mouse cell were directed specifically to the histocompatibility alloantigen. However, the histocompatibility antigens occur as large aggregates in the membrane (Fig. 5), and might therefore be expected to diffuse more slowly than a complex mixture of largely unaggregated human antigens in the membrane. Thus, at appropriate intermediate times after cell fusion, significant numbers of ($M_{1/2}$-H_1) but not of (M_1-$H_{1/2}$) fused cells might appear, to be converted at longer times to cells with completely intermixed components.

A rough estimate may be made of the average effective diffusion constant required of the membrane components to account for the kinetics of intermixing in the Frye-Edidin experiments. Taking the average distance of migration, x, to have been about 5 micrometers in a time, t, of 40 minutes gives an apparent diffusion constant, $D = x^2/2t$, of 5×10^{-11} cm²/sec. For comparison, the diffusion constant of hemoglobin in aqueous solutions is about 7×10^{-7} cm²/sec. The apparent effective viscosity of the membrane fluid phase is therefore about 10^3 to 10^4 times that of water.

The Frye-Edidin experiments can be rationalized by the fluid mosaic model of membrane structure as being the result of the free diffusion and intermixing of the lipids and the proteins (or lipoproteins) within the fluid lipid matrix.

Some experiments, however, appear to suggest that the lipids of membranes are not readily interchangeable within the membrane and are therefore not free to diffuse independently. For example, Wilson and Fox (*23*) have studied the induction of β-galactoside and β-glucoside transport systems in mutants of *Escherichia coli* that cannot synthesize unsaturated fatty acids. Such fatty acids can be incorporated into phospholipids, however, if they are supplied in the growth medium. When cells were grown in particular fatty acid supplements and induced for the synthesis of the transport systems, the effect of temperature on the transport rate was characteristic of that fatty acid. If, then, the cells were first grown in medium containing oleic acid and then shifted to growth in a medium supplemented with linoleic acid during a brief period of induction of either of the transport systems, the effect of temperature on transport was said to be characteristic of cells grown continually in the linoleic acid medium. In other words, although most of the phopholipids of the membrane contained oleic acid chains, these did not appear to exchange with the newly synthesized small amounts of phospholipids containing linoleic acid chains. These experiments, however, do not necessarily contradict the thesis that most of the phospholipids of membranes are freely diffusible and, hence, exchangeable. For example, each of the two transport systems might be organized in the membrane as a specific protein aggregate containing intercalated and strongly bound phospholipid components. If such lipoprotein aggregates had first to be assembled in order to be incorporated into the bulk lipid matrix of the membrane, the results of Wilson and Fox would be anticipated. In particular, the small fraction of the membrane phospholipid that was strongly bound, and perhaps segregated in such aggregates from the bulk of the membrane lipid, might not exchange rapidly with the bulk lipid. The Wilson-Fox experiments therefore do not require that the major part of the membrane phospholipid be static, but only that a small fraction of the lipids be structurally differentiated from the rest. The structural differentiation of some of the membrane lipid by strong binding to integral proteins is a possibility that was discussed above.

The observations of Wilson and Fox, that there is a significant coupling of lipid and protein incorporation into membranes, appear to be a special case. The experiments of Mindich (*45*) demonstrate that more generally lipid and protein incorporation into bacterial membranes can occur independently, and that quite wide variations in the ratio of lipids and proteins in the membrane can be produced in vivo, as might be expected from the fluid mosaic model of membrane structure.

The asymmetry of membranes. A substantial amount of evidence has accumulated showing that the two surfaces of membranes are not identical in composition or structure. One aspect of this asymmetry is the distribution of oligosaccharides on the two surfaces of membranes. There exist plant proteins, called lectins or plant agglutinins, which bind to specific sugar residues, and, as a result, can cause the agglutination of cells bearing the sugar residues on their surfaces. By conjugating several such agglutinins to ferritin, we have been able to visualize the distribution of oligosaccharides on membranes in the electron microscope (*33*). For example, the ferritin conjugate of concanavalin A, a protein agglutinin that binds specifically to terminal α-D-glucopyranosyl or α-D-mannopyranosyl residues (*46*), attaches specifically to the outer surface of erythrocyte membranes and not at all to the inner cytoplasmic surface (*33*). A similar, completely asymmetric distribution of ferritin conjugates of ricin (a protein agglutinin) on the membranes of rabbit erythrocytes is shown in Fig. 6. Ricin binds specifically to terminal β-D-galactopyranosyl and sterically related sugar residues (*47*). Such asymmetry has now been observed with several ferritin-conjugated agglutinins and a number of different mammalian cell plasma membranes (*48*). These findings extend earlier results obtained by different methods (*49*).

The foregoing observations bear on many problems, including cell-cell interactions and membrane biogenesis (*50*). In the context of this article, however, the absence of oligosaccharides on inner membrane surfaces indicates that rotational transitions of the glycoproteins of erythrocyte and other plasma membranes from the outer to the inner surfaces must occur at only negligibly slow rates. This conclusion probably applies to membrane proteins other than glycoproteins; for example, the Na,K-dependent and Mg-dependent adenosine triphosphatase activities of erythrocyte membranes are exclusively localized to the inner cytoplasmic surfaces (*51*). Individual molecules of spin-labeled zwitterionic and anionic phospholipids also exhibit very slow inside-outside transitions in synthetic vesicles of phospholipid bilayers (*52*). The very slow or negligible rates of such transitions can be explained by the mosaic model and the thermodynamic arguments already discussed. If the integral proteins (including the glycoproteins) in intact membranes have, like the phospholipids, an amphipathic structure, a large free energy of activation would be required to rotate the ionic and polar regions of the proteins through the hydrophobic interior of the membrane to the other side.

To accommodate the fluid mosaic model to these conclusions concerning asymmetry, we specify that, while the two-dimensional translational diffusion of the integral proteins and the phospholipids of membranes occurs freely,

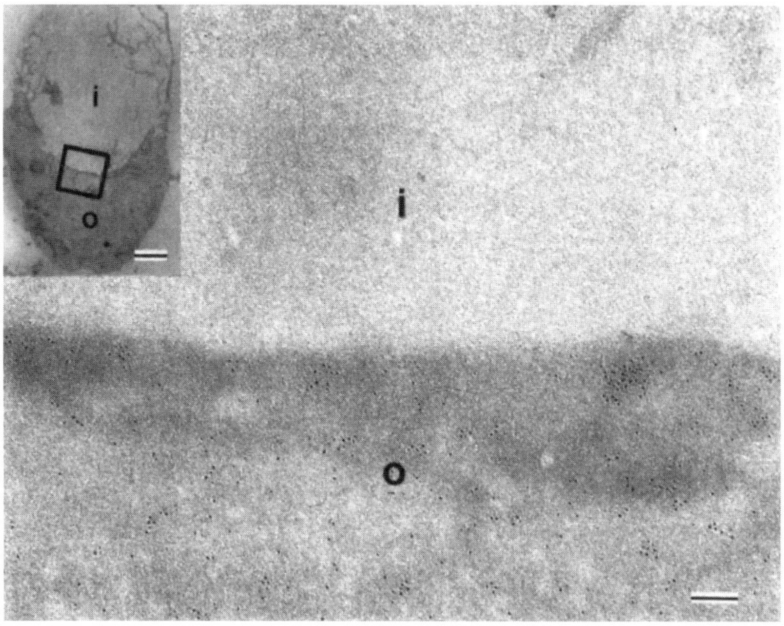

Fig. 6. The inner (*i*) and outer (*o*) membrane surfaces of a rabbit erythrocyte membrane that has been stained with ferritin-conjugated ricin. In preparing membrane specimens such as are shown in Figs. 4 and 5, occasionally a cell lyses with membrane rupture such that both inner and outer surfaces of the membrane are exposed. In this case the mounted membrane was stained with ferritin conjugated to ricin, a plant agglutinin that specifically binds to terminal β-D-galactopyranosyl and sterically related terminal sugar residues in oligosaccharides. The ferritin-agglutinin is found on the outer membrane surface only. The scale is equivalent to 0.1 μm; the insert scale is equivalent to 1 μm.

the rotational diffusion of these components is generally restricted to axes perpendicular to the plane of the membrane; that is, in general, molecular tumbling does not occur at significant rates within the membrane. The asymmetry of the membrane introduces another factor into the problem of translational diffusion of membrane components. In the experiments of Frye and Edidin (*44*) only those membrane antigens exposed at the outer surface of the membrane were labeled by fluorescent antibodies, and the conclusion that these particular antigens were mobile in the plane of the membranes therefore, strictly speaking, applies only to those components accessible at the outer surface. Whether components confined to the inner surfaces also intermix and diffuse should be separately established.

Thus, recent evidence obtained with many experimental methods and different kinds of functional membrane systems is entirely consistent with the predictions of the fluid mosaic model of membrane structure and provides strong support for the model. It seems amply justified, therefore, to speculate about how a fluid mosaic structure might carry out various membrane functions, and to suggest specific mechanisms for various functions that can be subjected to experimental tests.

The Fluid Mosaic Model and Membrane Functions

The hypothesis that a membrane is an oriented, two-dimensional, viscous solution of amphipathic proteins (or lipoproteins) and lipids in instantaneous thermodynamic equilibrum, leads to many specific predictions about the mechanisms of membrane functions. Rather than catalog a large number of these, we suggest some directions that such speculations may usefully take. Among these problems are nerve impulse transmission, transport through membranes, and the effects of specific drugs and hormones on membranes (*1*). The fluidity of the mosaic structure, which introduces a new factor into such speculations, is emphasized here. This new factor may be stated in general form as follows. The physical or chemical perturbation of a membrane may affect or alter a particular membrane component or set of components; a redistribution of membrane components can then occur by translational diffusion through the viscous two-dimensional solution, thereby allowing new thermodynamic interactions among the altered components to take effect. This general mechanism may play an important role in various membrane-mediated cellular phenomena that occur on a time scale of minutes or longer. Much more rapidly occurring phenomena, such as nerve impulse transmission, would find the mosaic structure to be a static one, insofar as translational diffusion of the membrane components is concerned. In order to illustrate the concepts involved, we discuss two specific membrane phenomena.

Malignant transformation of cells and the "exposure of cryptic sites." Normal mammalian cells grown in monolayer culture generally exhibit "contact inhibition"; that is, they divide until they form a confluent monolayer and they then stop dividing. Cells that have become transformed to malignancy by oncogenic viruses or by chemical carcinogens lose the property of contact inhibition; that is, they overgrow the monolayer. For some time, this experimental finding has been thought to reflect the difference between the normal and the malignant states in vivo, and to be due to differences in the surface properties of normal and malignant cells. Much excitement and investigative activity therefore attended the demonstration (*53, 54*) that malignant transformation is closely correlated with a greatly increased capacity for the transformed cells to be agglutinated by several saccharide-binding plant agglutinins. Furthermore, mild treatment of normal cells with proteolytic enzymes can render them also more readily agglutinable by these protein agglutinins. Burger (*54*) has suggested, therefore, that the agglutinin-binding sites are present on the membrane surfaces of normal cells but are "cryptic" (Fig. 7A) (that is, they are shielded by some other membrane components from effectively participating in the agglutination process), and that proteolytic digestion of normal cells or the processes of malignant transformation "exposes" these cryptic sites on the membrane surface. In some cases, quantitative binding studies have indeed indicated that no significant change in the numbers of agglutinin-binding sites on the membrane accompanies either mild proteolysis of normal cells or malignant transformation (*55*).

An alternative explanation of these phenomena (Fig. 7B), based on the fluid mosaic model of membrane structure, may be proposed. Consider first the proteolysis experiments with normal cells. Suppose that the integral glycoproteins in the normal cell membrane are molecularly dispersed in the fluid mosaic structure. It is likely that mild proteolysis would preferentially release a small amount of glycopeptides and other polar peptides from these proteins because these are the most exposed portions of the integral proteins at the outer surface of the membrane (Figs. 2 and 3). The remaining portions of these proteins may still contain a large fraction of their original oligosaccharide chains after the limited proteolysis, but the release of some of the more polar structures would make the remaining portions more hydrophobic. As these more hydrophobic glycoproteins diffused in the membrane, they might then aggregate in the plane of the membrane. The result would be a *clustering* of the agglutinin-binding sites on the enzyme-treated cell surface, as compared to the normal untreated surface. Such clustering (with no increase, or perhaps even a decrease in the total numbers of sites because of digestion) could enhance the agglutination of the treated cells, as compared to that of normal cells, because it would increase the probability of agglutinin bridges forming between the surfaces of two cells.

In malignant transformation, distinct chemical changes in the glycolipids and the glycoproteins of the cell membrane are known to occur (*56*), and the enhanced agglutinability of the transformed cells may be much more complicated than is the case in the proteolysis of normal cells. If, however, the two phenomena do have a basic feature in common, it could be a similar clustering of saccharide-binding sites on the transformed and the enzyme-treated normal cells. In malignant transformation, such clustering could be the result of the chemical changes in the membrane mentioned above; or some virus-induced gene product (*57*) may be incorporated into the cell membrane and serve as a nucleus for the aggregation of the agglutinin-binding glycoproteins within the membrane.

These suggestions can be tested experimentally by the use of ferritin-conjugated agglutinins (*33*) as already discussed (Fig. 6). The prediction is that with normal cells subjected to mild proteolysis, and also with malignant transformed cells, the total number of ferritin-agglutinin particles specifically bound to the outer surfaces of the cells might not be greatly different from those of normal cells, but larger clusters of ferritin particles would be found.

Cooperative phenomena in membranes. By a cooperative phenomenon we mean an effect which is initiated at one site on a complex structure and transmitted to another remote site by some structural coupling between the two sites. A number of important membrane phenomena may fall into this category. However, before enumerating them, we should first discriminate between two types of cooperative effects that may occur. These can be termed *trans* and *cis*. *Trans* effects refer to cooperative (allosteric) changes that have been postulated to operate at some localized region on the membrane surface, to transmit an effect from one side of the membrane to the other (*58*). For example, an integral protein may exist in the membrane as an aggregate of two (or more) subunits, one of which is exposed to the aqueous solution at the outer surface of the membrane, and the other is exposed to the cytoplasm at the inner surface. The specific binding of a drug or hormone molecule to the active site of the outward-oriented subunit may induce a conformational rearrangement within the aggregate, and thereby change some functional property of the aggregate or of its inward-oriented subunit. By *cis* effects, on the other hand, we refer to cooperative changes that may be produced over the *entire* membrane, or at least large areas of it, as a consequence of some event or events occurring at only one or a few localized points on the membrane surface. For example, the killing effects of certain bacteriocins on bacteria (*59*), the lysis of the cortical granules of egg cells upon fertilization of eggs by sperm (*60*), and the interaction of growth hormone with erythrocyte membranes (*61*) are cases which may involve transmission and amplification of localized events over the entire surface of a membrane. These phenomena may not all occur by the same or related mechanisms, but in at least two experimental studies, that involving the interaction of colicin E_1 with intact *Escherichia coli* cells (*62*), and that of human growth hormone and isolated human erythrocyte membranes (*61*), there is substantial evidence that long-range *cis*-type cooperative effects intrinsic to the membranes are involved.

The question we now address is, How might such *cis* effects work? Changeux and his co-workers (*63*) have proposed an extension to membranes of the Monod-Wyman-Changeux allosteric model of protein cooperative phenomena, using as a model of membrane structure an infinite two-dimensional aggregate of identical lipoprotein subunits [as, for example, the model described by Benson (*20*)]. In this theoretical treatment, the individual subunits are capable of existing in either of two conformational states, one of which has a much larger binding affinity for a specific ligand than does the other. The binding of a single ligand molecule to any one subunit then triggers the cooperative conversion of many of the subunits to the ligand-bound conformation, in order to maximize the interactions among the subunits.

This theory as presented relies on the membrane model used. If, however, the membrane is not a two-dimensional aggregate of lipoprotein subunits, but is instead a fluid mosaic of proteins and lipids, the physical situation would be quite different. The basic theory of Changeuex *et al.* (*63*) might still be formally applicable, but with important changes in physical significance. It is possible, for example, that a particular integral protein can exist in either of two conformational states, one of which is favored by ligand binding; in its normal unbound conformation the integral protein is monomolecularly dispersed within the membrane, but in the conformation promoted by ligand binding, its aggregation is thermodynamically favored. The binding of a ligand molecule at one integral protein site, followed by diffusion of the non-liganded protein molecules to it, might then lead to an aggregation and simultaneous change in conformation of the aggregated protein within the membrane. This mechanism could result in a long-range *cis*-type cooperative phenomenon, if the eventual aggregate size was very large and if its presence produced local perturbations in the properties of the membrane. However,

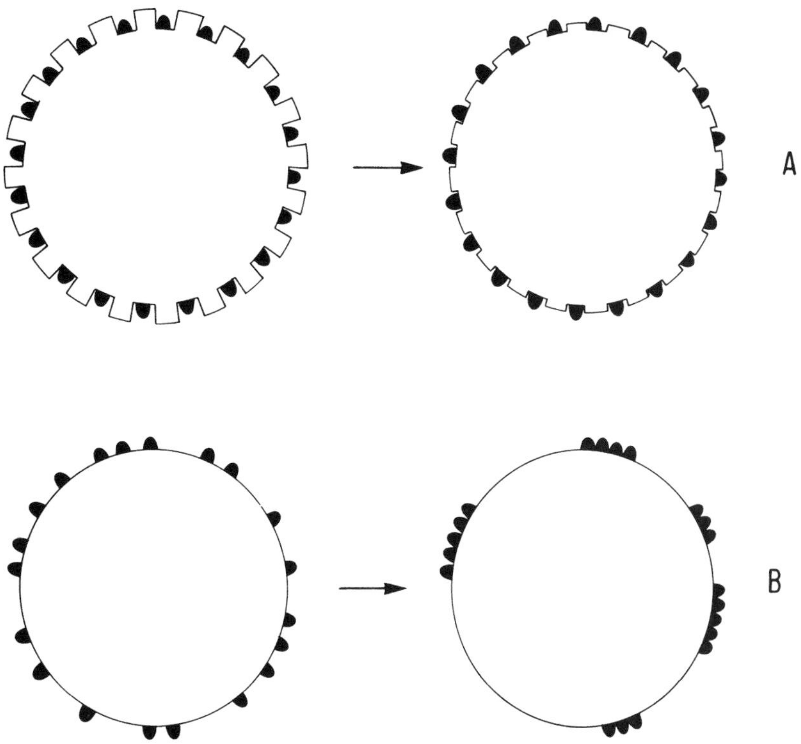

Fig. 7. Two different mechanisms to explain the findings that either malignantly transformed cells or normal cells that are subjected to mild proteolysis become much more readily agglutinable by several plant agglutinins. (A) The mechanism of Burger (*54*): agglutinin-binding sites that are present on the surfaces of normal cells, but are obstructed ("cryptic sites"), are exposed by proteolysis or the processes of malignant transformation. (B) The redistribution mechanism (see text): the agglutinin sites on normal cell surfaces are largely monomolecularly dispersed in the fluid mosaic structure, but on proteolysis or malignant transformation, they diffuse and aggregate in clusters. The probability of agglutination of two such modified cells is enhanced by the clustering of binding sites.

the transition would occur at a rate and over a time period determined by the rate of diffusion of the molecules of the integral protein in the fluid mosaic membrane. This time period is likely to be relatively long, of the order of minutes (*44*), as already mentioned. On the other hand, if *cis*-type cooperative effects occurred in a lipoprotein subunit model according to the mechanism postulated by Changeux *et al.* (*63*), one would expect the cooperative change to be much faster. Conformation changes in the soluble allosteric protein aspartyltranscarbamylase, for example, have half-times of the order of 10 milliseconds (*64*). It is therefore of some interest that in the studies of the interaction of colicin E_1 and *E. coli* the fluorescence changes that marked the apparent *cis*-type cooperative transitions in the cell membrane occurred over intervals of one to several minutes (*62*). If this suggested mechanism for the colicin effect is valid, one would predict that (i) freeze-etching experiments on the colicin-treated bacteria (*28*) might reveal an aggregation of normally dispersed particles at the inner membrane face, or (ii) changes in membrane fluidity, such as would be produced by suitable changes in temperature or by different compositions of membrane phospholipids (*65*), might markedly affect the kinetics of the fluorescence changes that are observed on addition of the colicin to the bacteria.

In this discussion of membrane functions, some detailed mechanisms to account for two membrane phenomena have been presented. It may well turn out that these mechanisms are incorrect. Our object has been not so much to argue for these specific mechanisms, as to illustrate that the fluid mosaic model of membrane structure can suggest novel ways of thinking about membrane functions—ways that are amenable to experimental tests. Other membrane phenomena may be influenced by similar diffusional mechanisms: for example, cell-cell and cell-substrate interactions, where the apposition of intense local electric fields to a cell membrane may affect the distribution of electrically charged integral proteins within the membranes; or the specific binding of multivalent antibody to cell surface antigens, where the simultaneous binding of one antibody molecule to several molecules of the antigen may induce rearrangements of the distribution of the antigen in the plane of the membrane, an effect that may be involved in the phenomenon of antigenic modulation (*66*). There are other specific examples as well.

It may well be that a number of critical metabolic functions performed by cell membranes may require the translational mobility of some important integral proteins. This could be the ultimate significance of the long-standing observation (*67*) that the membrane lipids of poikilothermic organisms contain a larger fraction of unsaturated fatty acids the lower their temperature of growth. Appropriate enzymes apparently carry out the necessary biochemical adjustment (*68*) that keeps the membrane lipids in a fluid state at the particular temperature of growth; if these enzymes are not functional, for example, because of mutations, the organism—to grow at the lower temperature (*65*)—must be supplied with the unsaturated fatty acid exogenously. While it has been suggested before that the maintenance of lipid fluidity may be important to carrier mechanisms operating across a functional membrane, it is also possible that the real purpose of fluidity is to permit some critical integral proteins to retain their translational mobility in the plane of the membrane, as an obligatory step in their function.

Summary

A fluid mosaic model is presented for the gross organization and structure of the proteins and lipids of biological membranes. The model is consistent with the restrictions imposed by thermodynamics. In this model, the proteins that are integral to the membrane are a heterogeneous set of globular molecules, each arranged in an *amphipathic* structure, that is, with the ionic and highly polar groups protruding from the membrane into the aqueous phase, and the nonpolar groups largely buried in the hydrophobic interior of the membrane. These globular molecules are partially embedded in a matrix of phospholipid. The bulk of the phospholipid is organized as a discontinuous, fluid bilayer, although a small fraction of the lipid may interact specifically with the membrane proteins. The fluid mosaic structure is therefore formally analogous to a two-dimensional oriented solution of integral proteins (or lipoproteins) in the viscous phospholipid bilayer solvent. Recent experiments with a wide variety of techniques and several different membrane systems are described, all of which are consistent with, and add much detail to, the fluid mosaic model. It therefore seems appropriate to suggest possible mechanisms for various membrane functions and membrane-mediated phenomena in the light of the model. As examples, experimentally testable mechanisms are suggested for cell surface changes in malignant transformation, and for cooperative effects exhibited in the interactions of membranes with some specific ligands.

Note added in proof: Since this article was written, we have obtained electron microscopic evidence (*69*) that the concanavalin A binding sites on the membranes of SV40 virus–transformed mouse fibroblasts (3T3 cells) are more clustered than the sites on the membranes of normal cells, as predicted by the hypothesis represented in Fig. 7B. There has also appeared a study by Taylor *et al.* (*70*) showing the remarkable effects produced on lymphocytes by the addition of antibodies directed to their surface immunoglobulin molecules. The antibodies induce a redistribution and pinocytosis of these surface immunoglobulins, so that within about 30 minutes at 37°C the surface immunoglobulins are completely swept out of the membrane. These effects do not occur, however, if the bivalent antibodies are replaced by their univalent Fab fragments or if the antibody experiments are carried out at 0°C instead of 37°C. These and related results strongly indicate that the bivalent antibodies produce an aggregation of the surface immunoglobulin molecules in the plane of the membrane, which can occur only if the immunoglobulin molecules are free to diffuse in the membrane. This aggregation then appears to trigger off the pinocytosis of the membrane components by some unknown mechanism. Such membrane transformations may be of crucial importance in the induction of an antibody response to an antigen, as well as in other processes of cell differentiation.

References and Notes

1. S. J. Singer, in *Structure and Function of Biological Membranes*, L. I. Rothfield, Ed. (Academic Press, New York, 1971), p. 145.
2. M. Glaser, H. Simpkins, S. J. Singer, M. Sheetz, S. I. Chan, *Proc. Nat. Acad. Sci. U.S.* **65**, 721 (1970).
3. J. Lenard and S. J. Singer, *ibid.* **56**, 1828 (1966).
4. D. F. H. Wallach and P. H. Zahler, *ibid.*, p. 1552.
5. W. Kauzmann, *Advan. Protein Chem.* **14**, 1 (1959).

6. C. Tanford, *J. Amer. Chem. Soc.* **84**, 4240 (1962).
7. E. D. Korn, *Annu. Rev. Biochem.* **38**, 263 (1969).
8. V. T. Marchesi and E. Steers, Jr., *Science* **159**, 203 (1968).
9. S. H. Richardson, H. O. Hultin, D. E. Green, *Proc. Nat. Acad. Sci. U.S.* **50**, 821 (1963).
10. M. Ashwell and T. S. Work, *Annu. Rev. Biochem.* **39**, 251 (1970).
11. D. Haldar, K. Freeman, T. S. Work, *Nature* **211**, 9 (1966); E. D. Kiehn and J. J. Holland, *Proc. Nat. Acad. Sci. U.S.* **61**, 1370 (1968); D. E. Green, N. F. Haard, G. Lenaz, H. I. Silman, *ibid.* **60**, 277 (1968); S. A. Rosenberg and G. Guidotti, in *Red Cell Membrane*, G. A. Jamieson and T. J. Greenwalt, Eds. (Lippincott, Philadelphia, 1969), p. 93; J. Lenard, *Biochemistry* **9**, 1129 (1970).
12. J. K. Blasie, C. R. Worthington, M. M. Dewey, *J. Mol. Biol.* **39**, 407 (1969).
13. D. Bownds and A. C. Gaide-Huguenin, *Nature* **225**, 870 (1970).
14. B. Ke, *Arch. Biochem. Biophys.* **112**, 554 (1965).
15. M. Glaser and S. J. Singer, *Biochemistry* **10**, 1780 (1971).
16. J. M. Steim, M. E. Tourtellotte, J. C. Reinert, R. N. McElhaney, R. L. Rader, *Proc. Nat. Acad. Sci. U.S.* **63**, 104 (1969).
17. D. L. Melchoir, H. J. Morowitz, J. M. Sturtevant, T. Y. Tsong, *Biochim. Biophys. Acta* **219**, 114 (1970).
18. D. M. Engelman, *J. Mol. Biol.* **47**, 115 (1970); M. H. F. Wilkins, A. E. Blaurock, D. M. Engelman, *Nature* **230**, 72 (1971).
19. M. E. Tourtellotte, D. Branton, A. Keith, *Proc. Nat. Acad. Sci. U.S.* **66**, 909 (1970).
20. A. A. Benson, *J. Amer. Oil Chem. Soc.* **43**, 265 (1966).
21. D. Triggle, *Recent Progr. Surface Sci.* **3**, 273 (1970).
22. H. U. Schairer and P. Overath, *J. Mol. Biol.* **44**, 209 (1969).
23. G. Wilson and C. F. Fox, *ibid.* **55**, 49 (1971).
24. J. Lenard and S. J. Singer, *Science* **159**, 738 (1968); G. Vanderkooi and D. E. Green, *Proc. Nat. Acad. Sci. U.S.* **66**, 615 (1970).
25. W. L. Hubbell and H M. McConnell, *Proc. Nat. Acad. Sci. U.S.* **61**, 12 (1968); A. D. Keith, A. S. Waggoner, O. H. Griffith, *ibid.*, p. 819.
26. R. L. Steere, *J. Biophys. Biochem. Cytol.* **3**, 45 (1957); H. Moor, K. Mühlethaler, H. Waldner, A. Frey-Wyssling, *ibid.* **10**, 1 (1961).
27. D. Branton, *Annu. Rev. Plant Physiol.* **20**, 209 (1969). J. M. Wrigglesworth, L. Packer, D. Branton [*Biochim. Biophys. Acta* **205**, 125 (1970)] find no evidence by freeze-etching for the existence of the stalked-knobs on mitochondrial inner membranes that are observed with negatively stained preparations.
28. M. E. Bayer and C. C. Remsen, *J. Bacteriol.* **101**, 304 (1970).
29. P. Pinto da Silva and D. Branton, *J. Cell Biol.* **45**, 598 (1970); T. W. Tillack and V. T. Marchesi, *ibid.*, p. 649.
30. P. Pinto da Silva, S. D. Douglas, D. Branton, *Abstracts of the 10th Meeting of the American Society for Cell Biology, San Diego, Calif., November 1970* (1970), p. 159; T. W. Tillack, R. E. Scott, V. T. Marchesi, *ibid.*, p. 213.
31. T. L. Steck, G. Fairbanks, D. F. H. Wallach, *Biochemistry* **10**, 2617 (1971).
32. M. S. Bretscher, *Nature* **231**, 225 (1971); *J. Mol. Biol.* **59**, 351 (1971).
33. G. L. Nicolson and S. J. Singer, *Proc. Nat. Acad. Sci. U.S.* **68**, 942 (1971).
34. G. L. Nicolson, S. P. Masouredis, S. J. Singer, *ibid.*, p. 1416.
35. G. L. Nicolson, R. Hyman, S. J. Singer, *J. Cell Biol.* **50**, 905 (1971).
36. R. E. Lee and J. D. Feldman, *ibid.* **23**, 396 (1964).
37. F. A. Green, *Immunochemistry* **4**, 247 (1967); *J. Biol. Chem.* **243**, 5519 (1968).
38. W. C. Davis and L. Silverman, *Transplantation* **6**, 536 (1968); T. Aoki, U. Hämmerling, E. de Harven, E. A. Boyse, L. J. Old, *J. Exp. Med.* **130**, 979 (1969).
39. A. Shimada and S. G. Nathenson, *Biochemistry* **8**, 4048 (1969); T. Muramatsu and S. G. Nathenson, *ibid.* **9**, 4875 (1970).
40. J. K. Blasie, M. M. Dewey, A. E. Blaurock, C. R. Worthington, *J. Mol. Biol.* **14**, 143 (1965).
41. M. M. Dewey, P. K. Davis, J. K. Blasie, L. Barr, *ibid.* **39**, 395 (1969).
42. J. K. Blasie and C. R. Worthington, *ibid.*, p. 417.
43. R. C. Warren and R. M. Hicks, *Nature* **227**, 280 (1970).
44. C. D. Frye and M. Edidin, *J. Cell Sci.* **7**, 313 (1970).
45. L. Mindich, *J. Mol. Biol.* **49**, 415, 433 (1971).
46. R. D. Poretz and I. J. Goldstein, *Biochemistry* **9**, 2870 (1970).
47. R. G. Drysdale, P. R. Herrick, D. Franks, *Vox Sang.* **15**, 194 (1968).
48. G. L. Nicolson and S. J. Singer, in preparation.
49. E. H. Eylar, M. A. Madoff, O. V. Brody, J. L. Oncley, *J. Biol. Chem.* **237**, 1962 (1962); E. L. Benedetti and P. Emmelot, *J. Cell Sci.* **2**, 499 (1967).
50. G. L. Nicolson and S. J. Singer, in preparation.
51. V. T. Marchesi and G. E. Palade, *J. Cell Biol.* **35**, 385 (1967).
52. R. K. Kornberg and H. M. McConnell, *Biochemistry* **10**, 1111 (1971).
53. M. M. Burger and A. R. Goldberg, *Proc. Nat. Acad. Sci. U.S.* **57**, 359 (1967); M. Inbar and L. Sachs, *ibid.* **63**, 1418 (1969).
54. M. M. Burger, *ibid.* **62**, 994 (1969).
55. B. Sela, H. Lis, N. Sharon, L. Sachs, *Biochim. Biophys. Acta*, in press; B. Ozanne and J. Sambrook, *Nature* **232**, 156 (1971).
56. S. Hakamori and W. T. Murakami, *Proc. Nat. Acad. Sci. U.S.* **59**, 254 (1968); P. T. Mora, R. O. Brady, R. M. Bradley, V. M. McFarland, *ibid.* **63**, 1290 (1969); C. A. Buck, M. C. Glick, L. Warren, *Biochemistry* **9**, 4567 (1970); H. C. Wu, E. Meezan, P. H. Black, P. W. Robbins, *ibid.* **8**, 2509 (1969).
57. T. L. Benjamin and M. M. Burger, *Proc. Nat. Acad. Sci. U.S.* **67**, 929 (1970).
58. T. R. Podleski and J.-P. Changeux, in *Fundamental Concepts in Drug-Receptor Interactions*, D. J. Triggle, J. F. Danielli, J. F. Moran, Eds. (Academic Press, New York, 1969), p. 93.
59. M. Nomura, *Proc. Nat. Acad. Sci. U.S.* **52**, 1514 (1964).
60. D. Epel, B. C. Pressman, S. Elsaesser, A. M. Weaver, in *The Cell Cycle: Gene-Enzyme Interactions*, G. N. Padilla, G. L. Whitson, I. L. Camerson, Eds. (Academic Press, New York, 1969), p. 279.
61. M. Sonenberg, *Biochem. Biophys. Res. Commun.* **36**, 450 (1969); *Proc. Nat. Acad. Sci. U.S.* **68**, 1051 (1971).
62. W. A. Cramer and S. K. Phillips, *J. Bacteriol.* **104**, 819 (1970).
63. J. P. Changeux, J. Thiéry, Y. Tung, C. Kittel, *Proc. Nat. Acad. Sci. U.S.* **57**, 335 (1967).
64. J. Eckfeldt, G. G. Hammes, S. C. Mohr, C. W. Wu, *Biochemistry* **9**, 3353 (1970).
65. D. F. Silbert and P. R. Vagelos, *Proc. Nat. Acad. Sci. U.S.* **58**, 1579 (1967).
66. E. A. Boyse and L. J. Old, *Annu. Rev. Genet.* **3**, 269 (1969).
67. E. F. Terroine, C. Hofferer, P. Roehrig, *Bull. Soc. Chim. Biol.* **12**, 657 (1930); G. Frankel and A. S. Hopf, *Biochem. J.* **34**, 1085 (1940).
68. M. Sinensky, *J. Bacteriol.* **106**, 449 (1971).
69. G. L. Nicolson, *Nature* **233**, 244 (1971).
70. R. B. Taylor, W. P. H. Duffus, M. C. Raff, S. dePetris, *ibid.*, p. 225.
71. The original studies reported in this article were supported by grant GM 15971 from the National Institutes of Health (to S.J.S.).

Meier S. and **Hay E.D.** 1975. Stimulation of corneal differentiation by interaction between cell surface and extracellular matrix. I. Morphometric analysis of transfilter "induction." *J. Cell Biol.* **66**: 275–291.

THE ABILITY OF ONE EMBYONIC TISSUE to induce changes in another was well established by early experimental work on a variety of embryos. Grobstein suggested that molecules of extracellular matrix (ECM) formed by the inducing tissue might either be or contain the relevant signals, an idea supported by later experimental work (1, 2). The paper presented here took a rigorous experimental approach to the question of whether a cell's surface had to touch the polymers of the ECM, or whether a diffusing signal molecule was sufficient. The work combined then modern methods for cell culture with high quality electron microscopy and morphometry to show that collagen synthesis is significantly stimulated in corneal cells by direct contact with extracellular fibrous material. These results anticipated the discovery of plasma membrane proteins that can bind the fibers of ECM and form links between the cell's environment and its cytoskeleton (3,4). The recognition that the same proteins, now called integrins, can also participate in the transmission of signals to help regulate cell behavior (reviewed in 5) has led in turn to the more recent understanding that such signals can help to control gene expression and cell differentiation (Streuli et al. this volume).

---

1. Hauschka S.D. and Konigsberg I.R. 1966. The influence of collagen on the development of muscle clones. *Proc. Natl. Acad. Sci.* **55:** 119–123
2. Kosher R.A., Lash J.A., and Minor R.R. 1975. Environmental enhancement of in vitro chondrogenesis. IV. Stimulation of somite chondrogenesis by exogenous chondromucoprotein. *Dev. Biol.* **35:** 210–220.
3. Tamkun J.W., DeSimone D.W., Fonda D., Patel R.S., Buck C., Horwitz A.F., and Hynes R.O. 1986. Structure of integrin, a glycoprotein involved in the transmembrane linkage between fibronectin and actin. *Cell* **46:** 271–282.
4. Ruoslahti E. and Pierschbacher M.D. 1987. New perspectives in cell adhesion: RGD and integrins. *Science* **238:** 491–497.
5. Hynes R.O. 1992. Integrins: Versatility, modulation, and signaling in cell adhesion. *Cell* **69:** 11–25.

STIMULATION OF CORNEAL DIFFERENTIATION BY INTERACTION BETWEEN CELL SURFACE AND EXTRACELLULAR MATRIX

I. Morphometric Analysis of Transfilter "Induction"

STEPHEN MEIER and ELIZABETH D. HAY

From the Department of Anatomy, Harvard Medical School, Boston, Massachusetts 02115. Dr. Meier's present address is the Department of Anatomy, University of Southern California School of Medicine, Los Angeles, California 90033.

ABSTRACT

The present study was undertaken to determine whether or not physical contact with the substratum is essential for the stimulatory effect of extracellular matrix (ECM) on corneal epithelial collagen synthesis. Previous studies showed that collagenous substrata stimulate isolated epithelia to produce three times as much collagen as they produce on noncollagenous substrate; killed collagenous substrata (e.g., lens capsule) are just as effective as living substrata (e.g., living lens) in promoting the production of new corneal stroma in vitro. In the experiments to be reported here, corneal epithelia were placed on one side of Nucleopore filters of different pore sizes and killed lens capsule on the other, with the expectation that contact of the reacting cells with the lens ECM should be limited by the number and size of the cell processes that can traverse the pores. Transfilter cultures were grown for 24 h in [^{3}H]proline-containing media and incorporation of isotope into hot trichloroacetic acid-soluble protein was used to measure corneal epithelial collagen production. Epithelial collagen synthesis increases directly as the size of the pores in the interposed filter increases and decreases as the thickness of the filter layer increases. Cell processes within Nucleopore filters were identified with the transmission electron microscope with difficulty; with the scanning electron microscope, however, the processes could easily be seen emerging from the undersurface of even 0.1-μm pore size filters. Morphometric techniques were used to show that cell surface area thus exposed to the underlying ECM is linearly correlated with enhancement of collagen synthesis. Epithelial cell processes did not pass through ultrathin (25-μm thick) 0.45-μm pore size Millipore filters nor did "induction" occur across them. The results are discussed in relation to current theories of embryonic tissue interaction.

Recent investigations of embryonic tissue interaction in vitro have called renewed attention to Grobstein's longstanding hypothesis that extracellular matrix (ECM) produced by the "inducer" either is or at least contains the material which stimulates or stabilizes the differentiation of the reacting tissue. In investigations of notochord-somite tissue recombinations, Kosher et al (20) have found that chondroitin sulfate proteoglycan added to the culture medium effectively substitutes for notochord in promoting somite chondrogenesis, recalling the demonstration by Nevo and Dorfman (28) that chondromucoprotein added to the culture medium of suspended chondrocytes has a positive feedback on its own synthesis. Matrix molecules produced by embryonic chick notochord and neural tube at the time of notochord-somite interaction include chondroitin sulfate as well as collagen (6, 34, 15, 20). During embryonic chick development in vivo, cephalic neural crest mesenchyme which migrates to a position under the pigmented epithelium differentiates into scleral cartilage. Newsome (29) has shown that clones of unexpressed neural crest cells undergo chondrogenesis when cultured on Millipore filter containing ECM secreted by pigment epithelium.

In investigations of the enhancement of corneal differentiation by lens in vitro, we found that the killed collagenous lens capsule is as effective as living lens in stimulating corneal epithelial synthesis of corneal matrix molecules and overt differentiation of corneal stroma (7, 8, 14, 24, 25). Enzymatically isolated epithelia, when placed directly on a dead lens capsule, produce a morphologically identifiable stroma at the lens-epithelium interface and within the epithelium itself, in intercellular clefts between adjacent eipthelial cells (25). Lens capsule extracted in NaOH to remove noncollagenous protein, collagen-rich vitreous humor, frozen-killed corneal stroma, killed cartilage matrix, rat tail tendon collagen gels, and pure cartilage (chondrosarcoma) collagen are almost equally effective in stimulating the corneal epithelium to produce a corneal stroma similar in biochemical composition and in ulstructure to the primary stroma synthesized by corneal epithelium in vivo. Corneal epithelia cultured on noncollagenous substrata such as glass, plastic, albumin, keratin, or Millipore filter, synthesize a base-line level of collagen and glycosaminoglycan (GAG) which is only one-third of the level achieved by cultures grown on collagenous substrata and produce no overt corneal stroma.

Since highly purified chondrosarcoma collagen was as effective as the collagenous lens capsule in promoting stroma production and since none of the noncollagenous substrata stimulated corneal differentiation, we concluded that the collagen in the adjacent ECM is one of the active factors controlling differentiation of the corneal epithelium in vitro and perhaps in vivo as well (25). This work supports the idea that collagen can have a direct effect on cell differentiation, as proposed originally by Konigsberg and Hauschka (18) who demonstrated that collagen is required in vitro as a substratum for the development of differentiated muscle from isolated embryonic chick myoblasts. That GAG also plays a role in corneal development is suggested by the fact that heparan and chondroitin sulfates, when added to the culture medium, double GAG synthesis by epithelia; however, the addition of polysaccharides does not influence collagen synthesis nor promote visible stroma differentiation (26).

The availability of thin Nucleopore filters (32) that are readily penetrated by cell processes (35) allowed us to modify our in vitro system to study the role of cell-ECM contact in the stimulation of collagen synthesis by lens extracellular matrix. Nucleopore filters of various pore sizes were interposed between corneal epithelia and killed lens capsules (Fig. 1) and both the cell surface area exposed by cytoplasmic processes traversing the

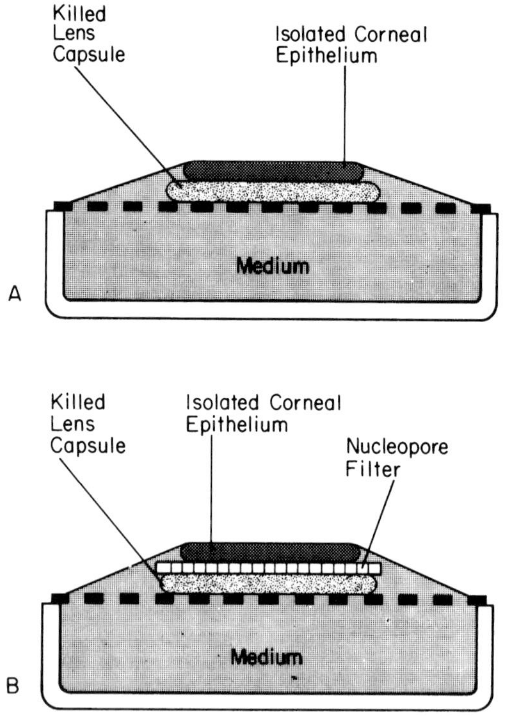

FIGURE 1 Diagrams of the cultures. (*a*) Epithelium cultured directly on killed lens capsule. (*b*) Epithelium separated from the killed lens capsule by a Nucleopore filter barrier to evaluate the role of cell contact with ECM in stimulating corneal differentiation.

filter and epithelial collagen production were measured. The resulting data show that direct cell contact with ECM is responsible for the stimulatory effect of collagen on stroma production by corneal epithelium.

MATERIALS AND METHODS

Organ Culture

Corneas were dissected from 5-day old White Leghorn chick embryos (Spafas, Norwich, Conn.) and corneal epithelia were isolated by trypsin-collagenase digestion as previously described (25). Individual epithelia were cultured basal-side down on one of the following substrata:(*a*) 0.1-μm, 0.2-μm, 0.4-μm, or 0.8-μm pore size (8-μm thick) Nucleopore filters which were washed and autoclaved (Arthur H. Thomas Co., Philadelphia, Pa.); (*b*) ultrathin (25-μm thick) Millipore filters (type HA, pore size 0.45 μm, Millipore Corp., Bedford, Mass.) which were washed and either autoclaved or sterilized by UV light: *c*) 12-day old embryonic chick lens, killed by autoclaving or by repeated (10 times) freeze-thawing in distilled water. In the latter case, most of the lens cell debris was removed because the capsules broke; these substrata will be referred to as lens capsule preparations. Epithelia were also cultured transfilter from the lens substratum. In this case, an epithelium was placed on a Nucleopore or Millipore filter (either side), which in turn was placed on a frozen-killed lens; the epithelium and lens thus were separated by a filter barrier (Fig. 1). Epithelia, covered by a meniscus of medium, were grown on the various substrata at the air-liquid interface on metal grids in Falcon dishes (No.3010, BioQuest Div, Becton, Dickinson & Co., Cockeysville, Md.). The medium was Ham's F-12 stock supplemented with 10% fetal calf serum (Grand Island Biological Corp., Grand Island, N.Y.), 0.25% whole embryo extract (5), and antibiotics 100 U'ml pencillin, 100 μg/ml streptomycin, and 0.25 μg/ml Fungizone (E. R. Squibb & Sons, Princeton, N.J.). Cultures were incubated for 24 h (unless otherwise specified) at 38°C in a humidified gas mixture (95% air, 5% CO_2).

Isotope Incorporation

Collagen synthesis was estimated by measuring the amount of [^{3}H]proline (G, 6 Ci/mmol, New England Nuclear, Boston, Ma., 5μCi/ml medium) that was incorporated during the 24-h culture period into hot trichloroacetic acid (TCA)-soluble material. Eight epithelia were set up in one organ culture dish and each biochemical determination of collagen synthesis is the result of their pooled effort. Isotopically labeled tissues plus substrata were rinsed in regular Hanks' solution, sonicated in ice-cold 5% TCA, and the insoluble material was collected by centrifugation at 4°C. The precipitate was resuspended in 5% TCA and heated at 90°C for 45 min. After cooling to room temperature, the samples were centifuged and the radioactivity in the neutralized supernate was measured in 10 ml of Aquasol (USV Pharmaceutical Corp.) using a Beckman LS-150 liquid scintillation counter (Beckman Instruments Inc., Fullerton, Calif.). It has been previously shown (25) that at least 85% of the radioactivity in the hot TCA-soluble material is sensitive to a highly purified, protease-free collagenase (30). Since there is little degradation of newly produced collagen during the 24-h culture period (25), the measurement of newly produced collagen reflects fairly accurately the amount synthesized in this period.

In order to determine what proportion of the [^{3}H]-proline in the hot TCA extract was converted to [^{3}H]hydroxyproline, cultures were labeled with 10 μCi/ml [2, 3-^{3}H]proline (N, 25 Ci/mmol, New England Nuclear) for 24 h. The cultures were pooled (16–24 epithelia plus substrata) and collagen was extracted with hot TCA as described above. The TCA was removed with ether and the dried precipitate was hydrolyzed overnight in 6 N HCl. The dried hydrolyzate was dissolved in distilled water containing 0.5 mg/ml proline and hydroxyproline as carrier. Tritiated proline was converted to [^{3}H]pyrrole according to the method of Switzer and Summer (33), extracted in toluene, and counted in 10 ml Aquasol. The remaining aqueous portion of the sample was boiled, and [^{3}H]hydroxyproline extracted in toluene and counted. For epithelia cultured on filters (without lens), the hypro-to-pro ratio was 0.63 ± 0.02 (*n*, 4), while the hypro-to-pro ratio of epithelia cultured on lens was 0.74 ± 0.04 (*n*, 3).

Although each experiment consists of eight epithelia, the results are expressed as collagen synthesis per epithelium in Table II, because the exposed surface area was calculated on a per epithelium basis. 5-day old corneal epithelium behaves consistently in culture; DNA accumulates at a level only slightly less than that which would have occurred in the same 24-h period in vivo (25). We conclude that there is relatively little cell death in this in vitro system and that cell numbers are comparable from epithelium to epithelium in the time period studied (25). Moreover, 5-day corneal epithelia cultured for 24 h on noncollagenous substrata, such as filters, accumulate DNA, RNA, and noncollagenous protein at approximately the same level (25).

Electron Microscopy

For transmission electron microscopy, cultures were fixed in gluteraldehyde-formaldehyde followed by osmium tetroxide, stained en bloc with uranyl acetate, and embedded in Araldite as described previously (16). For scanning electron microscopy, epithelium plus filter were first removed from the underlying lens and then processed as above, except that after fixation in osmium tetroxide and dehydration in ethanol, they were critical-point dried in a Samdri apparatus by use of liquid CO_2. The undersurface of filters not in contact with lens was also studied. Dried specimens were glued epithelium-side down on aluminum chucks and coated with a palladium-gold (60:40) alloy (Ladd Research Industries, Inc., Burlington, Vt.). Specimens were scanned at 25 kV in a JEOL SMU-3 scanning electron microscope.

Morphometric Methods

Examination of the filter undersurface revealed cell processes emerging from pores in the filter in an area restricted roughly by the size of the epithelium found on the opposite side of the filter. After a reconnaissance scan, the specimen was removed from the scope, and a 400-mesh copper grid (Ted Pella Co., Tustin, Calif.) was placed on the filter over the cell processes. An area three grid squares across and three grid squares down (nine grid squares total, total area 25,186 μm^2, open area 10,712 μm^2) exceeded the area of the filter covered by an individual epithelium, as outlined by underlying cell processes. A systematic random sample area of the lower left-hand corner of each of the nine grid squares was photographed at a magnification of 10,000. Since the smallest cell process was about 0.2 mm wide at this magnification, a graph grid was constructed with vertical and horizontal lines 0.2-mm apart. Therefore, the smallest cell process could not span two consecutive intersections of lines in the graph in any direction. The graph grid was placed over each photograph and the number of times that a line intersection fell on a cell process was recorded as a "hit." A hit was assigned the area of one graph square (0.04 mm^2). Thus, using a sterological point-counting technique (3), we analyzed equivalent test areas of the undersurface of filters. The thickness of Nucleopore filters was determined from thick sections by direct measurement employing a calibrated stage micrometer.

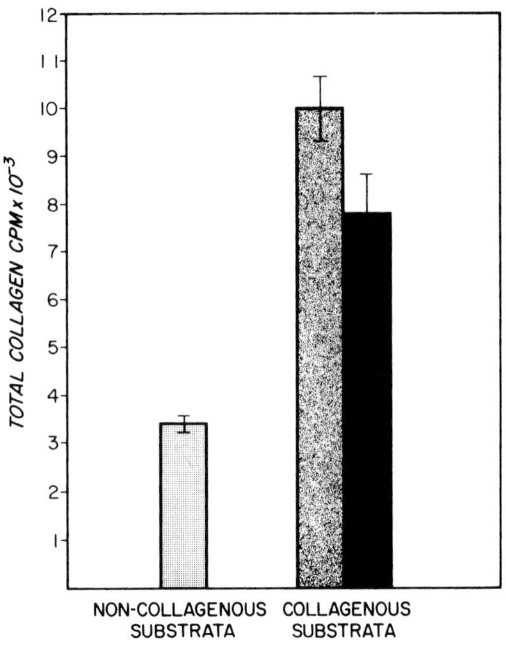

FIGURE 2 Epithelium separated from the killed lens capsule by a Nucleopore filter ■ produces about 75% of the amount of collagen produced by epithelium in direct contact with lens ECM ▒. Epithelia cultured on collagenous substrata such as lens capsule, concentrated chondrosarcoma collagen gels, and rat tail tendon gels synthesize nearly three times as much collagen in the 24-h culture period as do epithelia cultured on noncollagenous substrata such as Millipore filter, glass, and plastic □ (25). Since epithelial collagen production on a variety of noncollagenous substrata is consistently lower, we refer to this as the base-line level. Epithelia cultured directly on Nucleopore filter without an underlying lens produce collagen at this base-line level.

RESULTS

Collagen Synthesis in Nucleopore Transfilter Cultures

In the initial experiment, enzymatically isolated corneal epithelia were grown on killed lens capsule either in direct contact with or separated from the capsule by Nucleopore filters with 0.8-μm size holes. Collagen synthesis during the 24-h culture period was determined by measuring the incorporation of [^{3}H]proline into hot TCA-soluble protein. It was found that an epithelium separated from the lens by the large pore-size Nucleopore filter produced almost 75% as much collagen as when the epithelium was in direct contact with the lens. Epithelium cultured directly on Nucleopore filter alone synthesized collagen at the base-line level (Fig. 2) common to all other noncollagenous substrata previously examined (25).

We reasoned that if cell processes were penetrating the Nucleopore filter (35) to contact the underlying lens, then increasing the distance between the lens and the epithelium should interfere with the ability of the lens to stimulate collagen synthesis. Since Nucleopore filters are made in only one thickness (9 μm ± 1.2 μm), we increased the distance between the lens and the epithelium by stacking 0.8-μm pore size filters. Again, collagen production was measured over a 24-h period by isotope incorporation into hot TCA-soluble protein. As can be seen in Fig. 3, total collagen production during the 24-h culture period decreases as the distance between the lens and the epithelium increases, falling to 75% of the direct

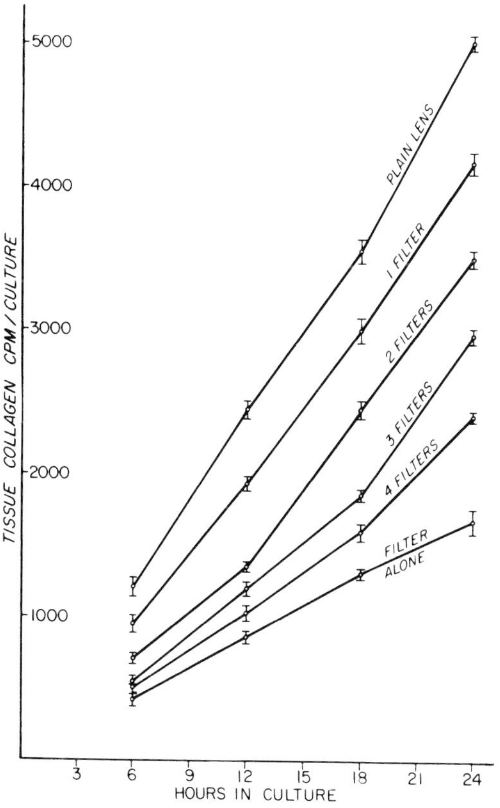

FIGURE 3 Corneal epithelium was grown on one plain Nucleopore filter, 0.8-μm pore size (bottom curve), on stacks of 0.8-μm pore size filters containing lens capsule on the side opposite the epithelium, and directly on lens capsule (top curve). The bottom curve corresponds to the base-line level synthetic activity and the top curve to the maximal "induced state." Since the stimulation of synthetic activity decreases with increasing filter thickness, the results are consistent with the idea that cell processes traverse the filter to contact the lens capsule on the other side. Cultures were grown continuously in the presence of 5μCi/ml [^{3}H]proline and harvested at various times up to 24 h in vitro. Vertical bars indicate the standard deviation about the mean for four determinations.

contact level when one filter (9-μm thick) is interposed, and to only 25% more than the baseline level when four filters (together 36 μm thick) are interposed. The size of direct channels available to cellular processes will be a function of the degree of overlap of holes in each filter. By cutting out the holes in photographs of single filters and arranging the pictures in random stacks, we were able to determine the size of direct channel space through a variety of combinations of stacked filters. We found that many channel sizes were created, some as large as 0.4 μm across.

In order to investigate the relation of channel size to transfilter stimulation of collagen synthesis by ECM, we cultured corneal epithelium on individual 0.8-μm, 0.4-μm, 0.2-μm, and 0.1-μm pore size Nucleopore filters. We expected the decreasing size of the pores would so decrease penetration of cell processes through the filters that epithelial cell contact with the extracellular matrix of the lens on the other side of the filter would be limited. If the level of collagen synthesis depends on epithelial cell contact with the lens capsule, then total synthesis should decrease as the size of the pores in the interposed filter decreases. This proved to be the case (Fig. 4). When a 0.1μm pore size filter is placed between an epithelium and the lens, collagen synthesis is greatly reduced, being only 20% greater than the nonstimulated (baseline) level. The level achieved is much greater with larger pores. In all cases where the level of collagen synthesis was greater than the nonstimulated (base-line) level, we were able to identify with the electron microscope a visible stroma of collagen fibrils deposited in clefts within the epithelium (Fig. 5) and on the epithelial undersurface next to the filter.

Electron Microscopy of Nucleopore Cultures

In order to establish that epithelial cell processes do traverse the filter to contact the lens "inductor," transfilter cultures were fixed after 24 h and thin sections were examined with the transmission electron microscope (TEM). Fig. 6 shows a cell process penetrating a channel in a 0.8-μm pore size filter. The process enters on the upper surface of the filter and, filling the cylindrical hole from edge to edge with cytoplasm, seems to probe along the channel walls in the direction of least resistance. The channels themselves pass straight through the filters, but they are not always perpendicular to the filter surface. Occasionally, the channels cross each other if the penetration angle is sufficiently oblique. The cell process illustrated in Fig. 6 seems to have coursed upward into a chamber made available by such a crisscrossing of channels in the filter. Those cell processes which emerge onto the other side of the filter may spread on the filter undersurface (Fig. 7) or along the lens capsule if they contact it. Seen in thin section, they are truly cell processes, since they are bound by a unit

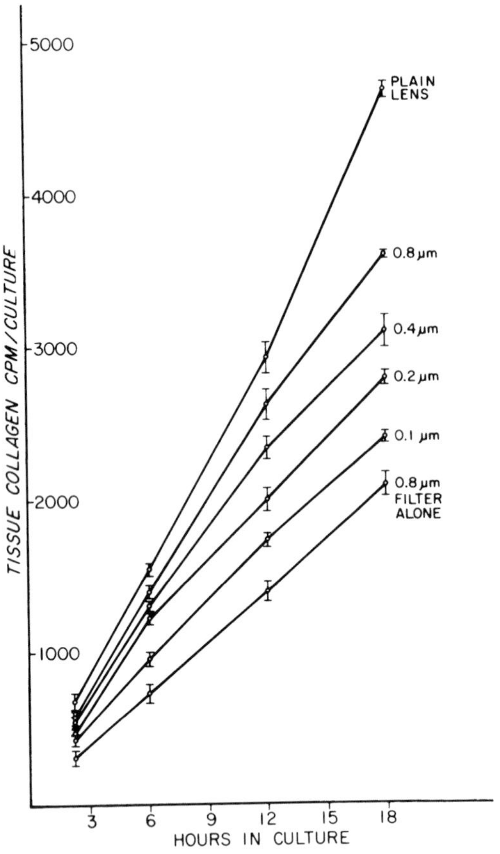

FIGURE 4 Experiment similar to that illustrated in Fig. 3, but in this case single filters of progressively smaller pore size were interposed between the lens capsule and corneal epithelium. Again, vertical bars indicate the standard deviation about the mean for four determinations. Since the stimulation of epithelial collagen synthesis decreases as the size of the pores in the interposed filter decreases, it is likely that smaller pore size filters limit epithelial-ECM contact.

membrane and contain cytoplasm filled with microfilaments (Fig. 7).

It proved difficult to quantitate the amount of transfilter cell surface from the small sample of thin sections that was practical to examine with the TEM. Therefore, we decided to examine the whole undersurface of the filter with the scanning electron microscope (SEM). As seen with the SEM, cytoplasmic processes emerging from the pores onto the undersurface of a 0.8-μm pore size filter appear to be large bulbous protrusions (Fig. 8). These processes can even be seen penetrating the small 0.1-μm pore size filter (fig. 9). If a lens capsule is present under the filter, the cell processes tend to be long and slender rather than bulbous.

Examination by SEM of the undersurface of 0.1-μm, 0.2-μm, 0.4-μm, and 0.8-μm pore size filters cultured for 24 h with corneal epithelium indicates there are obvious differences in the amount of cell surface exposed by cytoplasmic process penetration through the filters (Fig. 10). More and larger cell processes seem to have passed through the large pore size filters than the small pore size filters. To support this conclusion, it was necessary to calculate objectively the relative amount of cell surface which was exposed on the undersurface of the filter for each pore size filter used in the transfilter experiments.

Morphometric Analysis of Transfilter Cell Processes

As described in Materials and Methods, we employed a stereological point counting technique to analyze equivalent test areas of the underside of various pore size filters on which corneal epithelia were grown. An initial comparison was made between cultures of epithelia grown for 24 h on 0.4-μm pore size filters alone and cultures grown on the same size filter opposite lens for 24 h. Photographs taken at the same magnification in the SEM were overlaid with the analyzing graph grid and the number of times grid intersections fell on cell processes was translated into surface area for five different epithelial cultures in each experiment. The results in Table I indicate that the same amount of cell surface area (as estimated after critical point drying) appears on the undersurface of a filter cultured with an epithelium alone as appears on the undersurface of a filter with epithelium on the upper surface and lens on the undersurface.

Since the dead lens did not apparently promote the penetration of epithelial cell processes through the filter in transfilter cultures, we grew epithelia on various pore size filters without lens for 24 h so as to be able to analyze the amount of transfilter cell surface without the necessity of removing the lens for viewing. In addition, we measured the porosity for each pore size filter, that is the actual amount of open space the filter offers for epithelial process penetration, by counting the number of grid intersections which fell on the open pores of the filters. The results in Table II indicate that the mean cytoplasmic area per epithelium on the

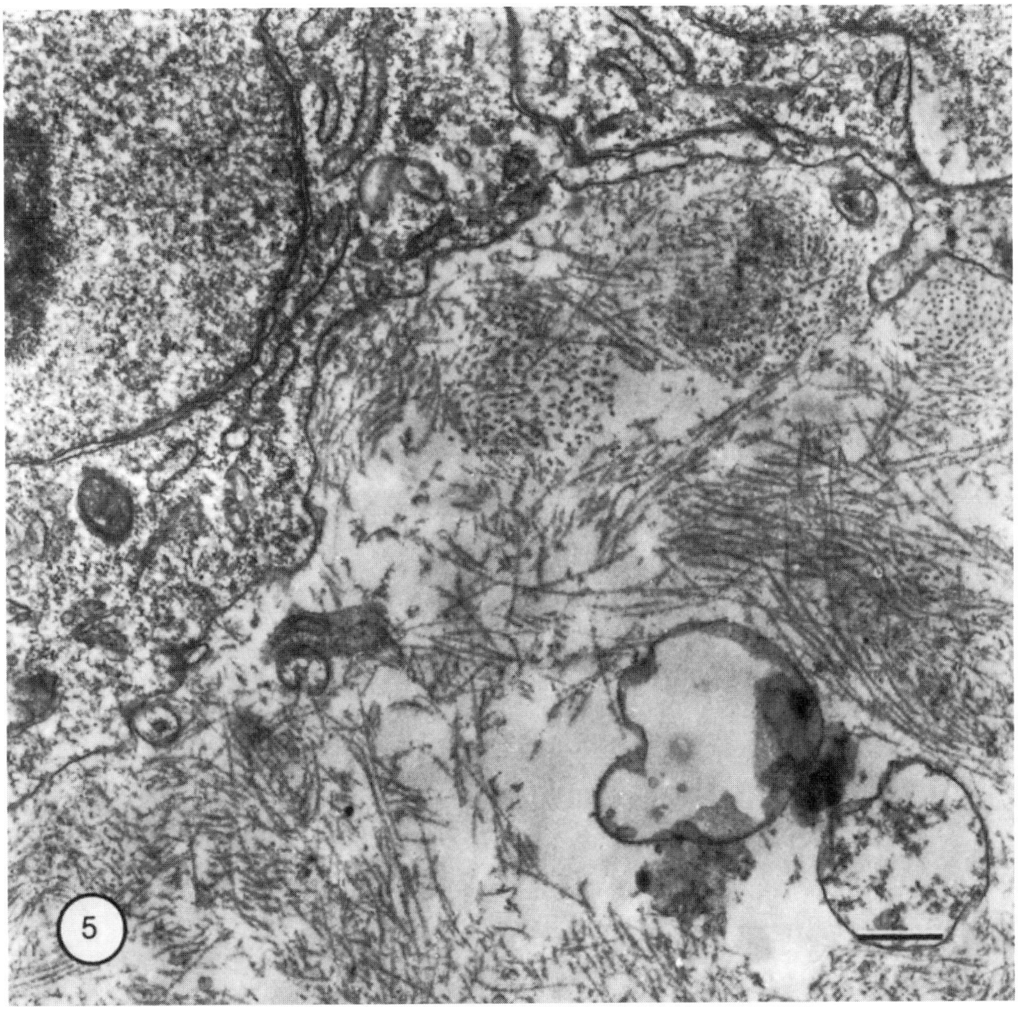

FIGURE 5 Electron micrograph showing deposition of collagen fibrils (corneal stroma) within an intercellular cleft in an epithelium grown on a 0.4-μm pore size filter for 24 h. Interestingly, the new stroma polymerizes mainly on the epithelial side of the filter, rather than next to the lens capsule on the other side. A few poorly preserved cell processes are entrapped in the new stroma. The bar represents 0.5 μm. $\times$ 23,000.

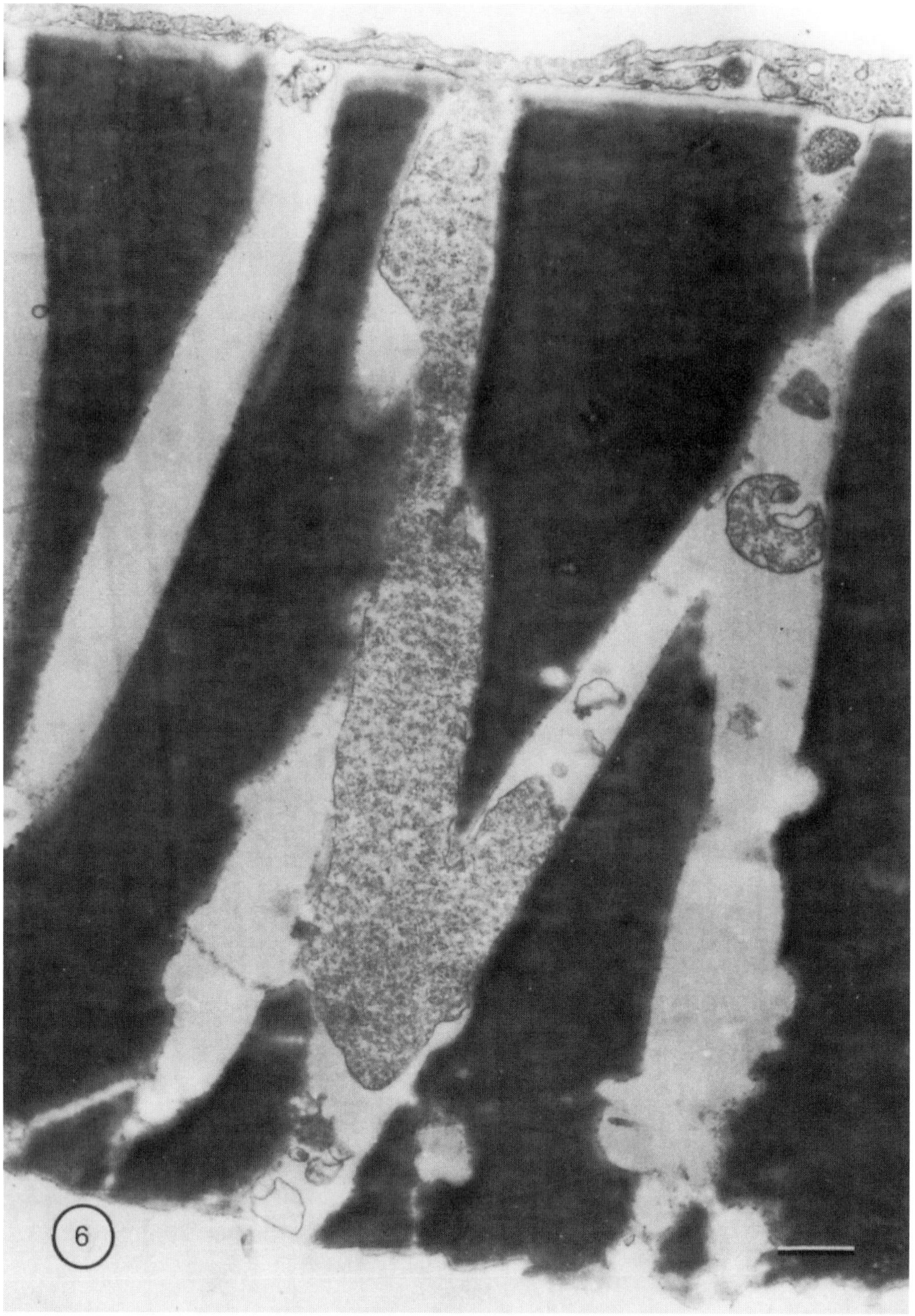

FIGURE 6 Transmission electron micrograph showin a corneal epithelial cell process extending into a channel in a 0.8-μm pore size Nucleopore filter. The bar represents 0.5 μm. × 23,000.

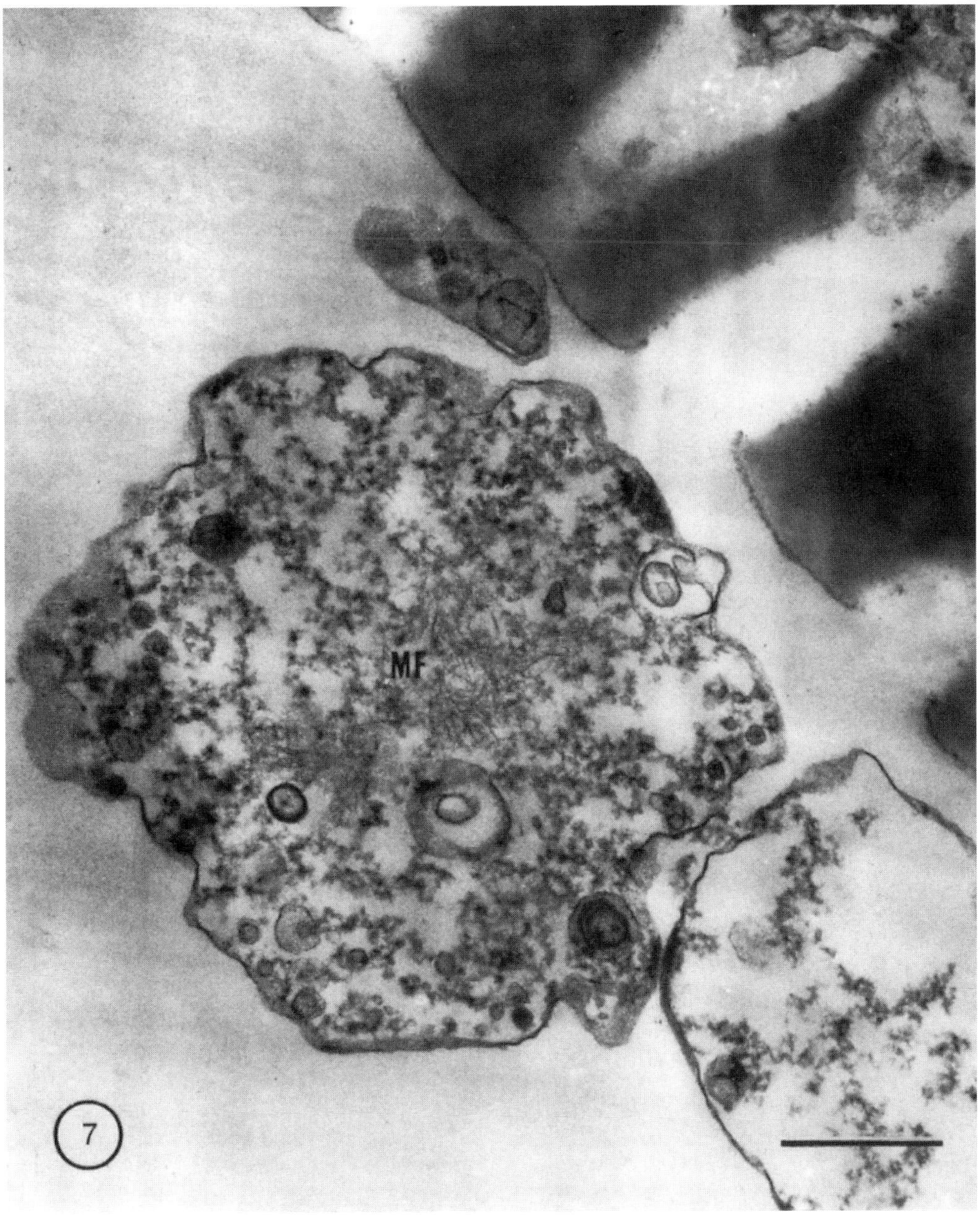

FIGURE 7 Transmission electron micrograph showing an epithelial cell process which has emerged onto the undersurface of 0.8-μm pore size Nucleopore filter. These processes typically are bound by a unit membrane and they contain microfilaments (MF). The bar represents 0.5 μm. $\times$ 47,000.

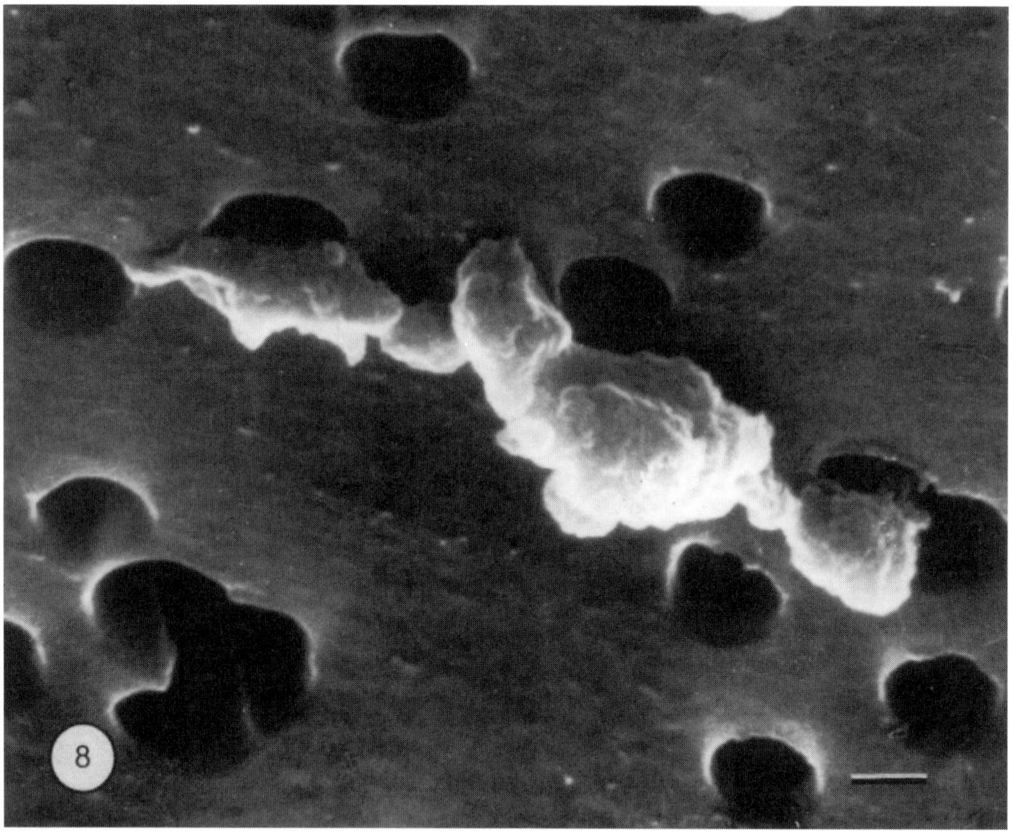

FIGURE 8 Scanning electron micrograph of the undersurface of a 0.8-μm pore size Nucleopore filter showing the emergence of cell processes derived from the corneal epithelium cultured in the other side of the filter. The bar represents 0.5 μm. × 20,000.

undersurface of the filter decreases directly as the size of the pores in the filter decreases. The porosity (open space) of the filters, on the other hand, does not differ as greatly among the various pore size filters. For example, the porosity of a 0.4-μm pore size filter is similar (due to a larger number of pores) to that of a 0.8-μm pore size filter (Table II). While little correlation exists between collagen synthesis and filter porosity, a strong correlation can be made between the degree of collagen production and the actual exposed surface area created by the transfilter cell processes. A plot of the log of the data in columns 3 and 4 of Table II indicates a linear relation between the exposed surface area and the level of increased collagen synthesis for each pore size filter examined (Fig. 11).

Examination of Millipore Transfilter Cultures

In order to compare our transfilter system to those used by Grobstein (10) and other earlier workers studying transfilter induction, we cultured corneal epithelia on ultrathin (25-μm thick) 0.45-μm pore size Millipore filters placed on dead lens capsules. As can be seen from Table III, collagen production by corneal epithelium on Millipore filter cultured transfilter to lens occurs at the base-line (nonstimulated) level. Even after 48 h on Millipore filter, corneal epithelium cultured transfilter to lens synthesizes collagen only at the base-line level, suggesting that cell processes have not crossed the filter to contact the lens.

Careful examination of the undersurface of

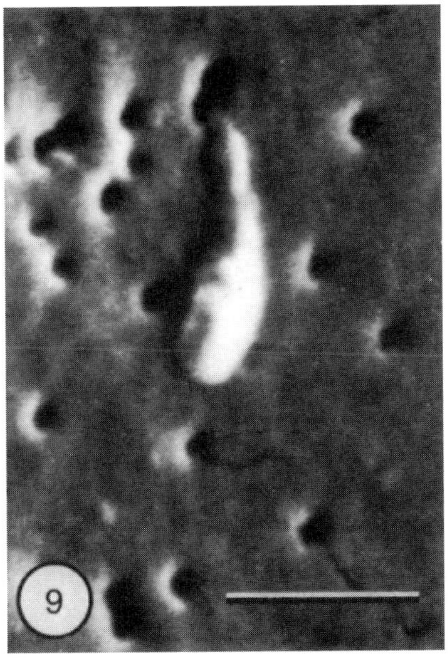

FIGURE 9 Scanning electron micrograph of the undersurface of a 0.1-μm pore size Nucleopore filter. A small cell process from the epithelium on the other side has emerged. The bar represents 0.5 μm. $\times$ 50,000.

ultrathin Millipore filters bearing epithelia on the opposite side failed to reveal penetration of cellular processes through the filters (0/40 cases). The filter undersurface appears irregularly contoured, with interconnecting strands of cellulose meshing to form holes and channels of a variety of sizes and shapes (Fig. 12). Filters which were sterilized by UV treatment or short-term autoclaving (5 min) were indistinguishable from those which were not sterilized at all. However, longer periods of autoclaving (15 min) caused the filters to ripple and shrink and the pores and channels in these filters seemed greatly reduced (Fig. 12).

DISCUSSION

The response of corneal epithelium in vitro to direct contact with extracellular matrix (ECM) is heightened collagen and GAG synthesis and the production of a biochemically and morphologically defined stroma. Isolated epithelia cultured on noncollagenous substrata synthesize collagen at a low (base-line) level and fail to produce a stroma (25). In the present work, study of the nature of the interaction of the corneal epithelium with underlying extracellular matrix was facilitated by the ready invasion across Nucleopore filters of naked cell processes extruded by isolated epithelium. By interposing filters which allowed varying degrees of contact between the lens extracellular matrix and the epithelial cells, we were able to demonstrate intermediate levels of epithelial collagen synthesis which were above the base-line level but below the level achieved on unimpeded contact with lens capsule. Variable amounts of orthogonally arranged striated collagen fibrils were produced by the epithelia that synthesized collagen at intermediate levels.

Since epithelia cultured on Nucleopore filters alone fail to produce a stroma, it seemed likely that transfilter stimulation of collagen synthesis was due to the contact of the epithelial cell processes with the lens. Therefore, we measured by morphometric techniques the amount of epithelial surface exposed by cell processes at intermediate levels of corneal stroma production. By this means, a linear relationship was demonstrated between the level of epithelial collagen synthesis and the total area of cell surface contacting lens extracellular matrix. Thus, we conclude that a major anabolic activity of corneal cells, collagen synthesis, is regulated by cell surface-ECM interaction.

It is unlikely that transfilter stimulation of epithelial collagen production by ECM could be due to the diffusion of molecules originating from the dead lens. For such a situation, a diffusion gradient would have to be maintained over a short distance (10–15 μm) through the filter for at least 24 h. This is improbable because the meniscus covering the tissue connects the epithelium with the medium below and permits the diffusion of substances in all directions and to all parts of the culture system. More importantly, there was no correlation between collagen production and filter porosity, that is the total open space in the filter. For example, since there are more pores in the 0.4-μm pore size filter than in the 0.8-μm pore size filter, the porosity was nearly the same (19.0% $\pm$ 3.2%) for both filters, yet transfilter epithelial collagens production on 0.8-μm pore size filters was 33% greater than that on 0.4-μm pore size filters. Moreover, there was no enhanced epithelial stroma synthesis by lens in transfilter experiments utilizing 0.45-μm pore size Millipore filters; here the porosity of the filters is greater than 75%. The amount of cell surface exposed by cell processes

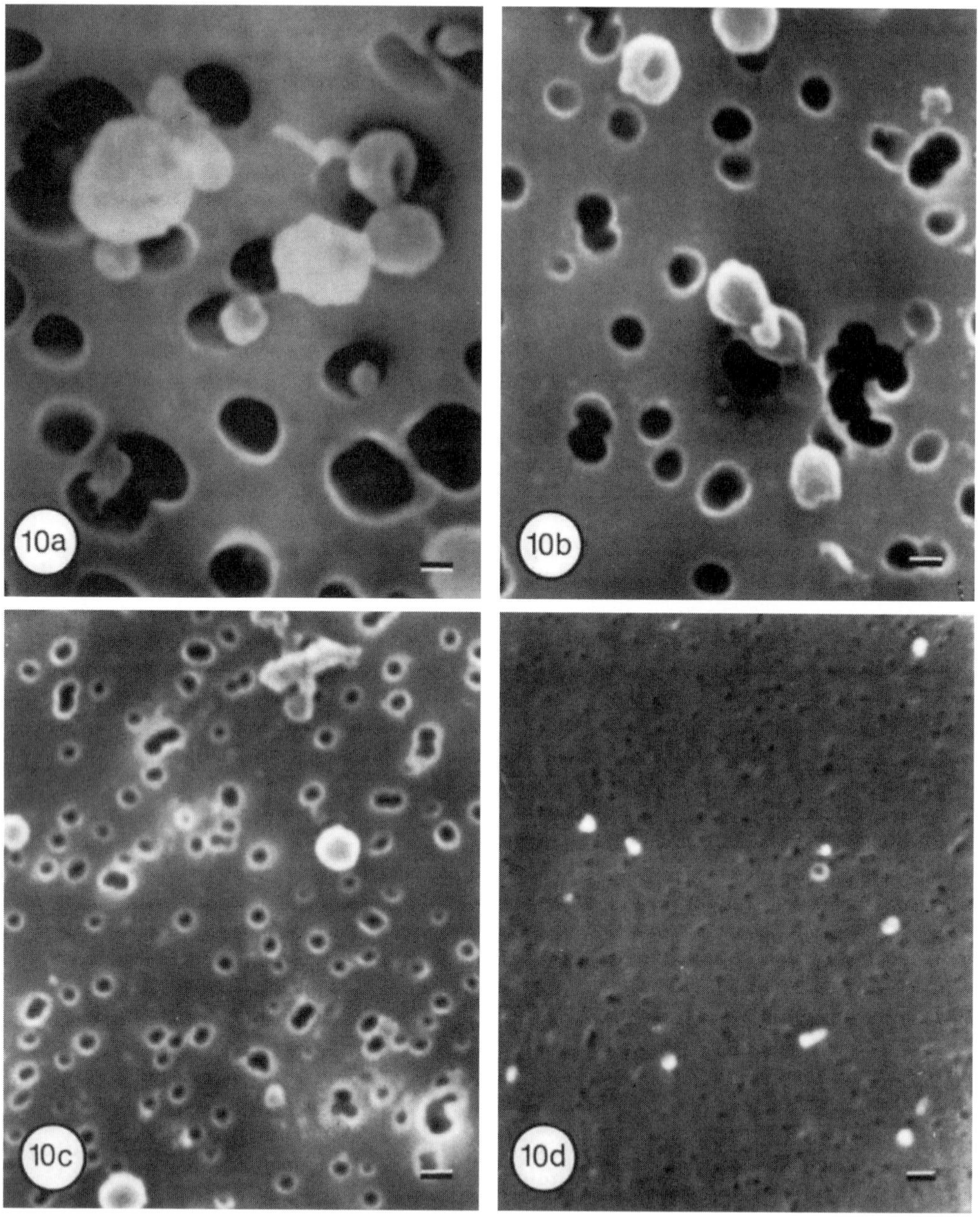

FIGURE 10 Scanning electron micrographs of cell processes emerging onto the under surface of 0.8-μm pore size (*a*) 0.4-μm pore size (*b*), 0.2-μm pore size (*c*), and 0.1μm pore size (*d*) Nucleopore filters. Photographs of this type were analyzed with the graph grid overlay to determine the amount of cell surface area created by cell processes which penetrate the filters. The bar is equivalent to 0.5 μm. × 8,500.

TABLE I

*A Comparison of the Exposed Cell Surface of Epithelia Cultured on 0.4-μm Pore Size Nucleopore Filters Alone and Transfilter to Lens**

Substratum	Exposed surface area
	μm²
0.4-μm pore size Nucleopore filter	4,447 + 1,122
0.4-μm pore size Nucleopore filter on lens capsule	4,406 ± 647

* Results are expressed per epithelium as the mean of five determinations ± the standard deviation.

TABLE II

*The Relation of Pore Size to Exposed Cell Surface in Transfilter Stimulation of Epithelial Collagen Synthesis by ECM**

Pore size	Porosity (open space)	Exposed surface area	Collagen synthesis
μm	%	*μm²*	*cpm*
0.8	21.0 ± 1.4	8,922 ± 1,842	566 ± 42
0.4	17.1 ± 1,6	4,447 ± 1,122	452 ± 19
0.2	12.3 ± 1.4	1,436 ± 406	368 ± 24
0.1	8.5 ± 1.2	350 ± 108	312 ± 11

* Results are expressed per epithelium as the mean of five determinations ± the standard deviation. Cultures were labeled for 24 h with 5 μCi/ml [³]proline.

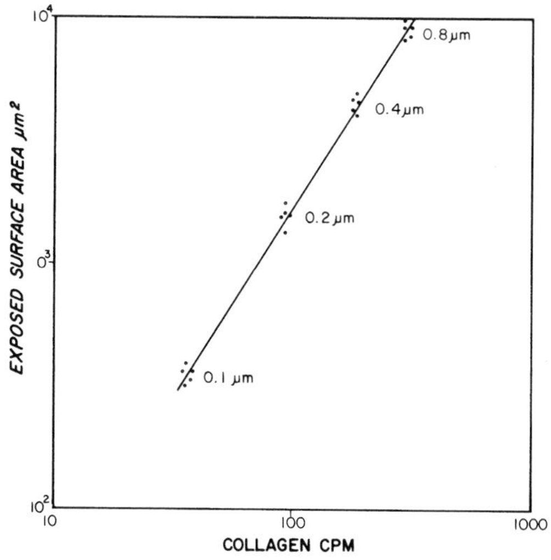

FIGURE 11 A plot on a log-log scale of the numerical data in columns 3 and 4 of Table II. The pore size of the filters is indicated. A linear relationship is shown between the exposed surface area of the emerging epithelial cell processes and the level of increased collagen synthesis. The level of increased synthesis is expressed as counts per minute above the base-line (265 cpm) level per epithelium; exposed surface area as square micrometers (Table II).

TABLE III

*The Effect of Lens Capsule on Transfilter Collagen Production by Corneal Epithelium Cultured on Ultrathin Millipore Filter**

Time in culture	Substratum	Collagen synthesis
h		*cpm*
24	Millipore filter	2406 ± 264
	Millipore filter on lens capsule	2475 ± 301
	Lens capsule	6198 ± 102
48	Millipore filter	2961 ± 299
	Millipore filter on lens capsule	3125 ± 314
	Lens capsule	12,842 ± 864

* Results are expressed as the mean of four determinations ± the standard deviation (eight epithelia per determination). Cultures were labeled for the last 24 h in vitro with 5 μCi/ml [³H]proline.

which penetrate the various pore size Nucleopore filters is linearly correlated with the level of enhanced collagen synthesis.

The quantitative relation demonstrated between pore size and exposed surface area, on the one hand, and the level of transfilter stimulation of collagen synthesis, on the other, invites a consideration of the nature of the response of the epithelium to contact with ECM. A plot of the log of the cell surface area exposed by cell processes traversing the filter versus the log of increased collagen synthesis is a straight line. This result implies that only a small amount of increased epithelial contact with lens ECM is necessary to elicit a significant increase in collagen production above the base-line level. For instance, the exposed cell surface area of epithelia cultured on 0.8-μm pore size filters is only about 30% of the total epithelial basal surface; however, collagen synthesis by 0.8-μm pore size transfilter cultures is stimulated to a level nearly 75% of the direct contact control. It is tempting to think that epithelial cell processes might contain

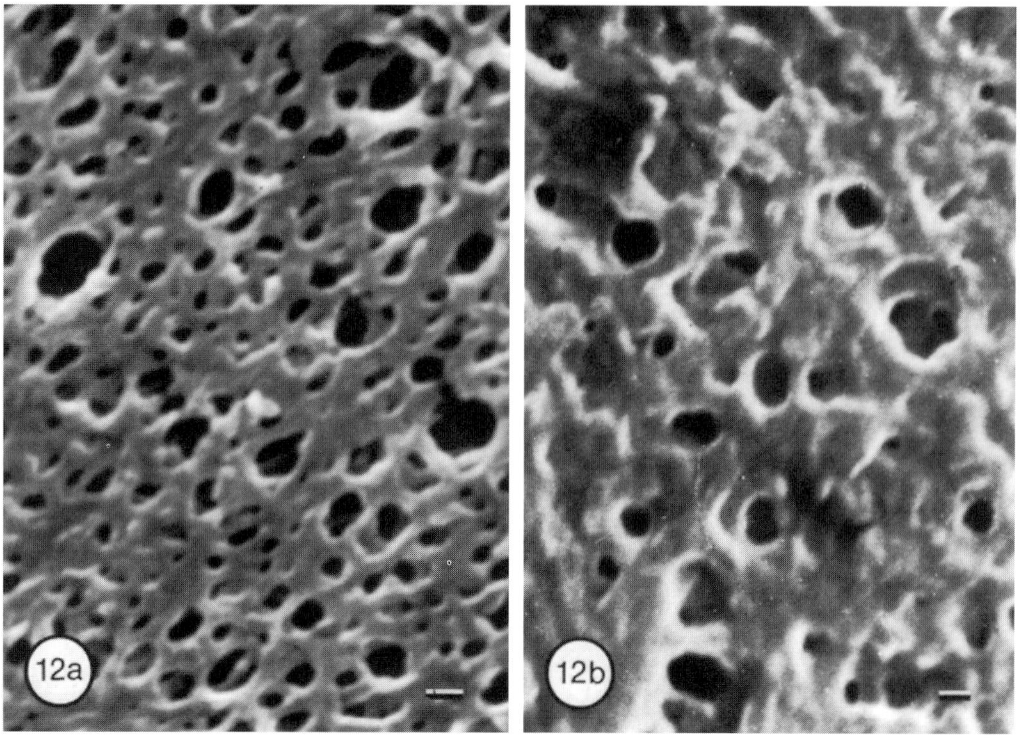

FIGURE 12 Scanning electron micrograph of the undersurface of ultrathin 0.45-μm pore size Millipore filters which bear corneal epithelia on the other side. (*a*) The undersurface of a briefly (5 min) autoclaved filter from a 48-h culture. The filter undersurface is free of epithelial cell processes. (*b*) The undersurface of a Millipore filter which was autoclaved for 15 min before being used as a substratum for transfilter corneal culture. The open spaces between interconnecting strands of cellulose are smaller. Again, there is no indication of epithelial process penetration. The bar represents 0.5 μm. (*a*) × 10,000; (*b*) × 7,000.

specific sites for collagen-ECM interaction and that the stimulatory mechanism is easily saturated. Whether enhanced collagen production can be attributed to the synthetic activity of only those epithelial cells which make contact with lens ECM or instead is the result of the pooled effort of every epithelial cell remains an open question which is difficult to answer because of the base-line synthetic activity of all of the cells.

Ancillary evidence to support the idea that it is cell surface contact with lens ECM which mediates enhanced corneal stroma production is derived from the filter-stacking experiments. We observed that the ability of the lens to promote epithelial stroma production is inversely related to the distance between the lens and the epithelium; the greater the transfilter distance, the smaller the stimulation of epithelial collagen synthesis by lens at any given time. We believe that lens-promoted epithelial collagen synthesis across stacked filters is dependent on the time it takes for a cell process to traverse the filter and thus is also related to the final amount of cell surface which emerges to contact the lens.

When corneal epithelia are isolated by treatment with trypsin-collagenase or EDTA, they are stripped of their underlying basement lamina (8). The epithelium responds to such treatment by extending blebs of cytoplasm from its basal surface. It is not surprising, then, that freshly isolated epithelium, cultured basal-side down on Nucleopore filters, can send cytoplasmic processes down into the pores and channels of the substratum. The probing of the filter by epithelial cell processes involves motile activity on the part of the living tissue. Seen in thin sections, the cell processes often contain numerous microfilaments, elements which have been implicated in cytoplasmic motility in many tissues (for review, see reference 31). The same number of cell processes penetrate a 0.4-μm pore size filter alone as when the epithelium is cultured transfilter to lens. The dead lens, then, does not promote the cytoplasmic invasion across the filters; the invasion is inherent to the basally disrupted epithelium.

By taking advantage of the ability of the scanning electron microscope to examine total undersurface of the transfilter cultures, we were able to establish conclusively that cell processes as small as 0.1-μm in diameter are able to travel across a 10-μm thick filter. This observation would have been nearly impossible to make by analysis of thin sections with the transmission electron microscope, because the chance of finding a cell process completely traversing a 0.1-μm pore size filter in thin sections is remote. This formidable task is further complicated by poor preservation of cell structure deep within Nucleopore filters. The suggestion that 0.1-μm pore size Nucleopore filter excluded cell processes from spinal cord-metanephros transfilter cultures (35) merits re-evaluation, and the claim of Bray (4) and others that cells cannot extend long processes 0.1-μm in diameter is clearly contradicted by our study.

The demonstration here and by Wartiovaara et al. (34) that cell processes of small dimension traverse thin Nucleopore filters in no way contradicts the conclusions of Grobstein and others that cell processes do not traverse Millipore filters (12,11,17). Nucleopore filters are solid polycarbonate disks which are perforated by chemically etched, uniform-size holes (32). Although a cell process could migrate in any direction, as evidenced in cases where two channels intersect within the filter, the majority of the processes which enter the pores on the epithelial surface are forced to traverse the filter and emerge on the undersurface because most of the pores pass straight across the filter. Millipore filters, on the other hand, are composed of intersecting strands of cellulose which mesh to create a thin sponge. Cell processes readily enter on the epithelial side and may migrate into any of the numerous interconnecting spaces of Millipore filters, but usually no deeper than the first 5-μm of the filter, as determined from examination of thin sections (8). Here, we found that a 0.45-μm pore size Millipore filter (25-μm thick) interposed between the lens and corneal epithelium blocked enhanced corneal collagen synthesis by lens, and we observed no cell processes emerging from the Millipore filter undersurface, even after 48 h of culture. We conclude that Millipore filters are not conducive to the direct passage of cell processes across them.

This study provides the best evidence to date supporting Grobstein's original theory that embryonic induction, as measured in the in vitro system he introduced, may involve "interaction between intercellular materials or matrices of cells, rather than direct interaction between the cells themselves in the sense of exchange between cytoplasms or contact of their cytoplasmic boundaries" (reference 10, p. 251). In closing, however, it seems appropriate to question the continued use of the term "induction" in this context. The tissue isolation approach (27) adapted by Grobstein (9) which has been used so extensively to study so-called embryonic induction in recent years (13, 22, 2, 7, 8, 23, 25, 26) is an artificial recreation of the in vivo situation because the reacting tissues are separated by treatment with enzymes or EDTA from their own extracellular microenvironment which contains matrix molecules now known to stabilize the expression of the differentiated state. These molecules are produced by both the "inducer" and the reacting tissue in the systems which have received the most attention in recent years. The active molecules (collagen, GAG) can be products of either the inducer or the induced in the case of the cornea (8, 25, 26); in cartilage (21, 20), they are the same products as produced by chondrocytes (28). Thus, it is likely that what is being measured in these and in many other in vitro systems is the ability of the so-called reacting tissue to recover enough collagen, GAG, and other critical molecules in its glycocalyx (1) to proceed with its differentiation after the tissue isolation procedure.

In the end, the term "embryonic induction" will surely be abandoned as mechanisms of cytodifferentiation become more clearly understood. In the present study, we have shown that transfilter "embryonic induction" as studied in vitro with Nucleopore filters can be accomplished via cell processes in contact not with other cells but with the "inductor's" ECM. The real significance of the work, however, is the demonstration of a role of cell surface-ECM interaction in the stimulation of ECM synthesis by the reacting tissue. In the next paper of this series, we will examine this cell-extracellular matrix interaction in more detail.

This paper is dedicated to Professor Etienne Wolff on the occasion of his retirement.

We are grateful to Robert P. Bolender for his invaluable help in planning the morphometric analysis and to Kathleen Kiehnau for her superb technical assistance. We thank the JEOL Ltd. for their cooperation, and especially Mr. Joseph Geller for his assistance.

This research was supported by United States Public Health Service grant number HD-00143. Stephen Meier is United States Public Health Service Fellow number DE-00383.

Received for publication 27 January 1975, and in revised form 20 March 1975.

REFERENCES

1. BENNETT, H. S. 1963. Morphological aspects of extracellular polysaccharides. *J. Histochem. Cytochem.* **11**:14–23.
2. BERNFIELD, M. R., and N. K. WESSELLS. 1970. Intra- and extracellular control of epithelial morphogenesis. *In* Changing Synthesis in Development. M. Runner, editor. Academic Press, Inc., New York. 195–249.
3. BOLENDER, R. P., and E. R. WEIBEL. 1973. A morphometric study of the removal of phenobarbital-induced membranes from hepatocytes after cessation of treatment. *J. Cell Biol.* **56**:746–959.
4. BRAY, D. 1973. Model for membrane movements in the neural growth cone. *Nature.* (*Lond.*). **244**:93–95.
5. CAHN, R. D., H. G. COON, and M. B. CAHN. 1968. Growth of differentiated cells: cell culture and cloning techniques. *In* Methods in Developmental Biology. F. Wilt and N. Wessells, editors. Thomas Y. Crowell Company, New York. 493–530.
6. COHEN, A. M., and E. D. HAY. 1971. Secretion of collagen by embryonic neuroepithelium at the time of spinal cord-somite interaction. *Dev. Biol.* **26**:578–605.
7. DODSON, J. W., and E. D. HAY. 1971. Secretion of collagenous stroma by isolated epithelium grown *in vitro*. *Exp. Cell Res.* **65**:215–220.
8. DODSON, J. W., and E. D. HAY. 1974. Secretion of collagen in corneal epithelium. II. Effect of the underlying substratum on secretion and polymerization of epithelial products. *J. Exp. Zool.* **189**:51–72.
9. GROBSTEIN, C. 1953. Epithelio-mesenchymal specificity in the morphogenesis of mouse sub-mandibular rudiments *in vitro*. *J. Exp. Zool.* **124**:383–414.
10. GROBSTEIN, C. 1955. Tissue interaction in the morphogenesis of mouse embryonic rudiments *in vitro*. *In* Aspects of Synthesis and Order in Growth. D. Rudnick, editor. Princeton University Press, Princeton, N.J. 233–256.
11. GROBSTEIN, C. 1961. Cell contact in relation to embryonic induction. *Exp. Cell Res. Suppl.* **8**:234–245.
12. GROBSTEIN, C., and A. J. DALTON. 1957. Kidney tubule induction in mouse metanephrogenic mesenchyme without cytoplasmic contact. *J. Exp. Zool.* **135**:57–73.
13. GROBSTEIN, C., and H. HOLTZER. 1955. *In vitro* studies of cartilage induction in mouse somite mesoderm. *J. Exp. Zool.* **128**:333–357.
14. HAY, E. D., and J. W. DODSON. 1973. Secretion of collagen by corneal epithelium. I. Morphology of the collagenous products produced by isolated epithelia grown on frozen-killed lens. *J. Cell Biol.* **71**:152–168.
15. HAY, E. D., and S. MEIER. 1974. Glycosaminoglycan synthesis by embryonic inductors: neural tube, notochord, and lens. *J. Cell Biol.* **62**:889–898.
16. HAY, E. D., and REVEL, J. P. 1969. Fine structure of the developing avian cornea. *In* Monograph in Developmental Biology. A. Wolski and P. S. Chen, editors. S. Karger AG., Basel, Vol. 1. 1–144.
17. HILFER, S. R. 1968. Cellular interactions in the genesis and maintenance of thyroid characteristics. *In* Epithelial-mesenchymal Interactions. R. Fleischmajer and R. E. Billingham, editors. Williams & Wilkins Company, Baltimore, Md. 177–199.
18. KONIGSBERG, I. R., and S. D. HAUSCHKA. 1965. Cell and tissue interactions in the reproduction of cell type. *In* Reproduction: Molecular, Subcellular, and Cellular. M. Locke, editor. Academic Press, Inc., New York. 243–290.
19. KOSHER, R. A., and J. W. LASH. 1975. Notochordal stimulation of *in vitro* somite chondrogenesis before and after enzymatic removal of perinotochordal materials. *Dev. Biol.* **42**:362–378.
20. KOSHER, R. A., J. W. LASH, and R. R. MINOR. 1973. Environmental enhancement of *in vitro* chondrogenesis. IV. Stimulation of somite chondrogenesis by exogenous chondromucoprotein. *Dev. Biol.* **35**:210–220.
21. LASH, J. W. 1968. Somitic mesenchyme and its response to cartilage induction. *In* Epithelial-Mesenchymal Interactions. M. Locke, editor. Williams & Wilkins Company, Baltimore, Md. 165–172.
22. LASH, J., S. HOLTZER, and H. HOLTZER. 1957. Experimental analysis of development of spinal column. VI. Aspects of cartilage induction. *Exp. Cell Res.* **13**:292–303.
23. LEVINE, S., R. PICTET, and W. J. RUTTER. 1973. Control of cell proliferation and cytodifferentiation by a factor reacting with the cell surface. *Nat. New Biol.* **246**:49–52.
24. MEIER, S., and E. D. HAY. 1973. Synthesis of sulfated glycosaminoglycans by embryonic epithelium. *Dev. Biol.* **35**:318–331.
25. MEIER, S., and E. D. HAY. 1974. Control of corneal differentiation by extracellular materials. Collagen as a promoter and stabilizer of epithelial stroma production. *Dev. Biol.* **38**:249–270.
26. MEIER, S., and E. D. HAY. 1974. Stimulation of extracellular matrix synthesis in the developing

cornea by glycosaminoglycans. *Proc. Natl. Acad. Sci. U. S. A.* **71**:2310–2313.

27. MOSCONA, A. 1952. Cell suspensions from organ rudiments from chick embryos. *Exp. Cell Res.* **3**:535–539.
28. NEVO, A., and A. DORFMAN. 1972. Stimulation of chondromucoprotein synthesis in chondrocytes by extracellular chondromucoprotein. *Proc. Natl. Acad. Sci. U. S. A.* **69**:2069–2072.
29. NEWSOME, D. 1975. *In vitro* induction of cartilage in embryonic chick neural crest cells by products of retinal pigmented epithelium. *Dev. Biol.* In Press.
30. PETERKOFSKY, B., and R. DIEGELMANN. 1971. Use of a mixture of proteinase-free collagenases for the specific assay of radioactive collagen in the presence of other proteins. *Biochemistry.* **10**:988–994.
31. POLLARD, T. D., and R. R. WEIHING. 1973. Cytoplasmic action and myosin and cell movement. CRC Critical Reviews of Biochemistry. **2**:1–65.
32. PORTER, M. C. 1974. A novel membrane filter for the laboratory. *Am. Lab. (Greens Farms, Conn.).* **6**:63–76.
33. SWITZER, B. R., and G. K. SUMMER. 1971. Improved method for hydroxyproline analysis in tissue hydrolyzates. *Anal. Biochem.* **39**:487–491.
34. TRELSTAD, R. L., A. H. KANG, A. M. COHEN, and E. D. HAY. 1973. Collagen synthesis *in vitro* by embryonic spinal cord epithelium. *Science (Wash. D.C.).* **179**:295–297.
35. WARTIOVAARA, J., S. NORDLING, E. LEHTONEN, and L. SAXEN. 1974. Transfilter induction of kidney tubules: correlation with cytoplasmic penetration into Nucleopore filters. *J. Embryol. Exp. Morphol.* **31**:667–682.

Yoshida C. and **Takeichi M.** 1982. Teratocarcinoma cell adhesion: Identification of a cell-surface protein involved in calcium-dependent cell aggregation. *Cell* **28:** 217–224.

CLASSIC STUDIES IN THE LABORATORIES of Holtfreter and of Moscona demonstrated the importance of graded affinities among cells for the ability of embryonic cells to associate meaningfully during development. Steinberg (1) was able to reconsitute tissues from dissociated embryos, showing a spectrum of intercellular affinities that included both homotypic and heterotypic binding. Early work distinguished between cell adhesions that depended upon Ca^{++} and those that did not, but it was difficult to identify the molecules responsible for the adhesion processes. The mutual association of embryonic retinal cells served as the experimental system from which to purify a membrane-associated glycoprotein important for neural cell adhesion (an N-CAM) (2). A similar approach, based on antibodies that could interfere with specific cell adhesion, served in the paper reprinted here to identify the first Ca^{++}-dependent cell adhesion molecule and facilitate its purification. Analogous approaches in many labs (e.g., 3) have led to the identification of a large family of Ca^{++}-dependent cell adhesion molecules, now called the cadherins, which are of wide-spread importance as regulators of cell and tissue morphogenesis (reviewed in 4). In recent years it has been shown that the functional disruption of such molecules (e.g., E-cadherin), can promote a metastatic phenotype for some cell types (5), suggesting that cell adhesion molecules play an important role in maintaining normal tissue order and behavior.

---

1. Steinberg M.S. 1963. Reconstruction of tissues by dissociated cells. *Science* **141:** 401–408.
2. Thiery J.P., Brackenbury R., Rutishauser U., and Edelman G.M. 1977. Adhesion among neural cells of the chick embryo. II. Purification and characterization of a cell adhesion molecule from neural retina. *J. Biol. Chem.* **252:** 6841–6845.
3. Damsky C.H., Knudsen K.A., Dorio R.J., and Buck C.A. 1981. Manipulation of cell-cell and cell-substratum interactions in mouse mammary tumor epithelial cells using broad spectrum antisera. *J. Cell. Biol.* **89:** 173–184.
4. Takeichi M. 1991. Cadherin cell adhesion receptors as a morphogenetic regulator. *Science* **251:** 1451–1455.
5. Birchmeier W., Weidner K.M., and Behrens J. 1996. Molecular mechanisms leading to loss of differentiation and gain of invasiveness in epithelial cells. *J. Cell Sci.* **17:** 159–164.

Cell, Vol. 28, 217–224, February 1982, Copyright © 1982 by MIT

Teratocarcinoma Cell Adhesion: Identification of a Cell-Surface Protein Involved in Calcium-Dependent Cell Aggregation

Chikako Yoshida and Masatoshi Takeichi
Department of Biophysics
Faculty of Science
Kyoto University
Kitashirakawa, Sakyo-ku
Kyoto 606, Japan

Summary

Teratocarcinoma cells have a Ca^{2+}-dependent cell–cell adhesion site (t-CDS) that is unique in being inactivated with trypsin in the absence of Ca^{2+} but not in the presence of Ca^{2+}. Fab fragments of antibodies raised against teratocarcinoma F9 cells dissociated by treatment with trypsin and calcium (anti-TC-F9) inhibit the aggregation of teratocarcinoma cells mediated by t-CDS. This inhibitory effect of Fab is removed when anti-TC-F9 is absorbed with F9 cells treated with trypsin and calcium (TC-F9), but not when it is absorbed with F9 cells treated with trypsin and EGTA (TE-F9). Comparisons of cell-surface antigens reactive to anti-TC-F9 in TC-F9 cells with those in TE-F9 cells reveal that only one component, with an approximate molecular weight of 140,000 (p140), is detected specifically on the surface of TC-F9 cells. When TC-F9 cells are retrypsinized in the absence of Ca^{2+}, a substance with an approximate molecular weight of 34,000 (p34) is released that can neutralize the aggregation-inhibitory effect of the Fab. This p34 interferes with the immunoprecipitation of p140 with anti-TC-F9, suggesting that p34 is a tryptic fragment of p140. Anti-TC-F9 Fab causes the dissociation of the monolayers of teratocarcinoma cells. This effect is removed by absorption of the Fab with p34 as well as with TC-F9 cells, but not with TE-F9 cells. These results suggest that p140 is essential for the function of t-CDS, and that this is an actual cell-adhesion molecule active in the establishment of monolayers of teratocarcinoma cells.

Introduction

Identification of molecules participating in cell–cell adhesion in animal tissues is essential for the elucidation of the molecular mechanisms in cell adhesion and cell recognition. It is of particular importance to characterize the cell-adhesion molecules present in embryonic cells, the intercellular recognition of which is the basic premise of animal morphogenesis in development. In biochemical studies on cell adhesion in mammalian embryos, teratocarcinoma (embryonal carcinoma) cells certainly provide an ideal model system, since they resemble early embryonic cells in several respects (Martin, 1980).

We have previously set forth a hypothesis of dual mechanisms to explain the initial step of cell–cell adhesion in animal cells (Takeichi, 1977, 1981; Urushihara et al., 1977, 1979; Takeichi et al., 1979; Aoyama et al., 1980; Uéda et al., 1980). This hypothesis, which explains cell–cell adhesion as a function of two distinct adhesion sites, a Ca^{2+}-dependent site (CDS) and a Ca^{2+}-independent site (CIDS) (Urushihara and Takeichi, 1980), is well supported by work carried out in other laboratories (Grunwald et al., 1980; Brackenbury et al., 1981; Magnani et al., 1981; Thomas and Steinberg, 1981; Thomas et al., 1981a, 1981b). It has been shown that both teratocarcinoma cells and early embryonic mouse cells also possess these two kinds of adhesion sites, having similar properties to those in several types of differentiated cells (Atsumi and Uno, 1979; Takeichi et al., 1981; S. Ogou, T. S. Okada and M. Takeichi, manuscript submitted). Teratocarcinoma and early embryonic cells, however, do not crossadhere via the CDS to some differentiated cells such as fibroblasts. Furthermore, the CDS of teratocarcinoma cells (t-CDS) can be distinguished from that of fibroblasts (f-CDS) in immunologic specificity. These observations suggested that CDS undergoes some developmental changes during the process of cell differentiation. To facilitate future studies of the role of CDS in cell recognition in animal morphogenesis, we have attempted to identify the adhesion molecules functionally expressed as t-CDS in teratocarcinoma cells on the basis of the following rationale.

CDS shows a unique protease sensitivity, in that it is readily inactivated by the incubation of cells in a relatively low concentration of proteases such as trypsin in the absence of Ca^{2+}, but is not inactivated with proteases in the presence of Ca^{2+} (Takeichi, 1977). For example, F9 teratocarcinoma cells (Bernstine et al., 1973) dissociated by treatment with 0.01% trypsin containing 1 mM Ca^{2+} (TC) are capable of aggregation in the presence of Ca^{2+} without any lag period, while those dissociated with 0.01% trypsin containing 1 mM EGTA (TE) can not aggregate until certain cell-surface components are restored by protein synthesis (Atsumi and Takeichi, 1980; Takeichi et al., 1981). Fab fragments of antibodies raised against F9 cells dissociated with TC (anti-TC-F9) inhibit the aggregation of teratocarcinoma cells mediated by t-CDS (Takeichi et al., 1981). This aggregation-inhibitory effect of Fab is removed when it is absorbed with F9 cells dissociated with TC (TC-F9), but not with F9 cells dissociated with TE (TE-F9). Thus it is expected that we can detect the molecules responsible for the function of t-CDS by looking for the particular components bound to anti-TC-F9 present in TC-F9 but not in TE-F9 cells. The experiments described show that there is a TC-F9-specific cell-surface protein, which neutralizes the aggregation-inhibitory effect of anti-TC-F9 Fab in a specific manner.

Results

Immunologic Detection of TC-F9-Specific Antigens

F9 cells harvested and dissociated from monolayer

cultures by treatment with TC and with TE were used as cells with t-CDS (TC-F9 cells) and without t-CDS (TE-F9 cells), respectively, throughout the experiments.

Proteins in the whole-cell lysate of TC-F9 and TE-F9 cells were subjected to SDS-polyacrylamide gel electrophoresis and transferred to nitrocellulose. The cellular components that reacted with anti-TC-F9 were detected with ^{125}I-labeled protein A. Figure 1A shows that one component, with an approximate molecular weight of 140,000 (p140), was present only in TC-F9 cells, representing a TC-F9-specific antigen. A component with a molecular weight of 52,000 (p52) was found predominantly in TC-F9 cells, though a small amount of this component seemed present in TE-F9 cells. All other antigens bound to anti-TC-F9 were present both in TC-F9 and TE-F9 cells in nearly equal amounts.

Immunoprecipitation of TC-F9-Specific Surface Antigens

We next examined whether TC-F9-specific antigens were actually present on the surface of cells as target molecules whose recognition by anti-TC-F9 blocks the function of t-CDS. The following experiments were designed to detect cell-surface antigens selectively by immunoprecipitation techniques.

The first step in these experiments was to absorb anti-TC-F9 exhaustively with TE-F9 cells so as to exclude antibodies not related to t-CDS. As shown in Table 1, anti-TC-F9 absorbed with a large number of TE-F9 cells still reacted strongly with TC-F9 cells but hardly at all with TE-F9 cells. Hereafter, this absorbed anti-TC-F9 will be referred to as anti-TC-F9^{te}.

The exhaustive absorption of anti-TC-F9 Fab also removed the reactivity to the surface of TE-F9 cells, leaving the binding capacity to TC-F9 cells, as judged by indirect immunofluorescent antibody staining with fluorescein-labeled antirabbit IgG. This absorbed Fab still inhibited cell–cell adhesion by the t-CDS of teratocarcinoma cells as strongly as the nonabsorbed Fab (see Figure 5), as expected from our previous results (Takeichi et al., 1981). Thus the anti-TC-F9 was successfully absorbed with TE-F9 cells, leaving the antibodies with the capacity to block the function of t-CDS.

The second step was to establish an ideal system for the immunoprecipitation of membrane proteins. To detect by immunoprecipitation any antigens present in cells and reactive to given antibodies, we must

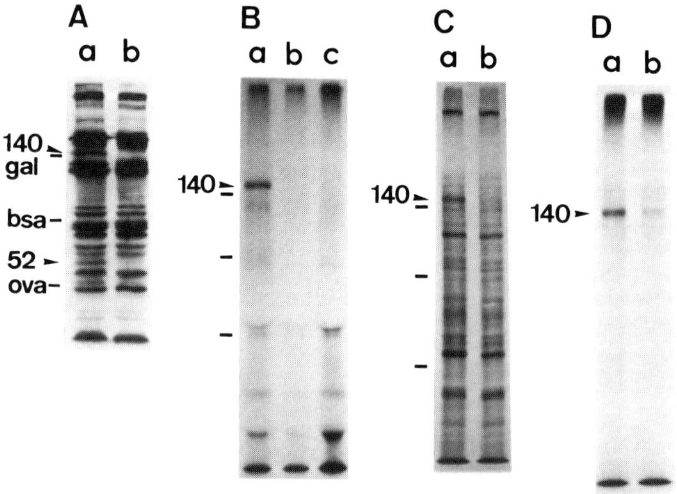

Figure 1. Immunologic Detection of TC-F9-Specific Antigens

(A) Total cellular antigens reactive with anti-TC-F9 in TC-F9 cells (lane a) and TE-F9 cells (lane b). The lysates of 2 × 10^5 TC-F9 and TE-F9 cells solubilized in 20 μl of SDS sample buffer were subjected to SDS-polyacrylamide gel electrophoresis and transferred to nitrocellulose. Electrophoretic blots were incubated with anti-TC-F9, and subsequently with ^{125}I-labeled protein A. gal: β-galactosidase from E. coli. bsa: bovine serum albumin. ova: ovalbumin. The positions of these markers are shown by bars in (B) and (C).

(B) Immunoprecipitate with anti-TC-F9^{te} from the lysate of radioiodinated TC-F9 cells (lane a) and TE-F9 cells (lane b). A control sample (lane c) was prepared with preimmune serum IgG from the lysate of TC-F9 cells.

(C) Immunoprecipitate with anti-TC-F9^{te} from the lysate of ^{35}S-methionine-labeled F9 cells (lane a) and a control sample prepared with preimmune serum IgG (lane b).

(D) Immunoprecipitation of p140 with anti-TC-F9^{te} in the presence of the p34 fraction of the TC/TE extract. The p34 fraction of the TC/TE extract derived from 3 × 10^8 cells was lyophilized and mixed with 100 μg of anti-TC-F9^{te} IgG. This absorbed IgG preparation was used for the immunoprecipitation from the radioiodinated TC-F9 lysate (lane b). A control sample (lane a) was prepared with anti-TC-F9^{te} mixed with the Sephacryl fraction of the TE/TE extract corresponding in position to the p34 fraction of the TC/TE extract. The ratio of radioactivity of p140 detected in lanes a and b was about 2:1, as estimated by counting, with a gamma counter, the radioactivity of each band cut out of the gel. Antigen–antibody complexes were collected by bacterial immunoadsorbents in this experiment.

Table 1. Absorption of Anti-TC-F9 with TE-F9 Cells

Number of TE-F9 Cells Used for Absorption	Binding of Antibodies (cpm) to	
	TC-F9 cells	TE-F9 cells
5 × 10^7	2798	1021
1 × 10^8	2284	625
Control	425	315

For absorption, 50 μg of anti-TC-F9 IgG in 0.25 ml HMF (see Experimental Procedures) were incubated with the indicated numbers of TE-F9 cells for 60 min at 4°C. The supernatant collected by centrifugation was used as the absorbed anti-TC-F9. For assaying the binding of the absorbed anti-TC-F9 to the surface of cells, we added 1 × 10^6 TC-F9 or TE-F9 cells to the absorbed antibody preparations and incubated them for 60 min at 4°C. After washing, the cells were incubated with 0.01 μCi ^{125}I-labeled protein A in 100 μl HMF for 30 min at 4°C. The radioactivity bound to the cells was counted by a gamma counter. Control samples were prepared with preimmune serum IgG. The data represent averages of triplicate samples.

solubilize all cellular components. While SDS is the best detergent for the solubilization of cellular components, it is doubtful if all immunologic reactions take place in this detergent. However, we found that in a solution containing both SDS and the nonionic detergent Nonidet-P40 (NP40), both immunoprecipitation and IgG–protein A binding take place. We applied this finding in preparing the cell lysates used for immunoprecipitation in the following steps.

TC-F9 and TE-F9 cells, whose surface proteins had been selectively labeled by lactoperoxidase-catalyzed radioiodination, were solubilized in SDS, and this was followed by the addition of NP40. To these cell lysates was added anti-TC-F9^{te}, and immunoprecipitates, formed by the subsequent addition of antirabbit IgG, were collected and analyzed by SDS-polyacrylamide gel electrophoresis. Figure 1B shows that only p140 was precipitated with anti-TC-F9^{te} from the TC-F9 cell lysate, whereas no radioiodinated material was immunoprecipitated specifically with anti-TC-F9^{te} from the TE-F9 cell lysate. All other radioactive materials detected in the gels are considered nonspecifically coprecipitated with immunoprecipitates, since a control sample prepared with a preimmune serum IgG contained the same components. We avoided this contamination by using bacterial immunoadsorbents as shown in Figure 1D.

The next experiment was designed to pick up only antigens exposed at the surface of cells from all cellular components metabolically labeled with ^{35}S-methionine. The radiolabeled F9 cells in a monolayer culture were first incubated with anti-TC-F9^{te} for 60 min at 4°C, washed to remove the unbound antibodies and then solubilized in SDS and NP40 as described above. To this cell lysate, we added the bacterial immunoadsorbents to collect the antigen–antibody complexes. Figure 1C shows that only p140 was detected as an antigen reactive with anti-TC-F9^{te}. All other components in the gels were probably introduced into the samples by nonspecific adsorption on the bacteria, since the sample prepared with the preimmune serum IgG contained exactly the same components.

We obtained the same results by using TC-F9 cells labeled with ^{35}S-methionine in the above test. p52 was not immunoprecipitated in the above experiments. The antibodies to p52 seemed to be absorbed with TE-F9 cells, since p52 was not clearly detected with anti-TC-F9^{te} on nitrocellulose to which whole-TC-F9-cell proteins had been transferred as described above (data not shown).

The above results also show that p140 is not a tryptic fragment of a larger molecule, since this component was detectable in nontrypsinized cells.

Tryptic Fragments of Target Molecules of Anti-TC-F9

TC-F9 cells were again treated with trypsin in the presence of EGTA. The supernatant of this trypsin extract (TC/TE extract) neutralized the aggregation-inhibitory effect of anti-TC-F9 Fab (Figure 2). The trypsin extract obtained by retreatment of TE-F9 cells with trypsin and EGTA (TE/TE extract) contained no such activity.

When the TC/TE extract was fractionated through a Sephacryl S-200 column, the Fab-neutralizing activity migrated as a single peak with an approximate molecular weight of 34,000 (Figure 3). This activity, however, was several times lower than that in the original unfractionated sample. Since we could not detect the same activity in the other fractions, it is probable that the antibody-binding capacity of the antigens decreases with fractionation. This capacity seemed to undergo further reduction following lyophilization of the fractionated materials, although some activity was still detectable.

To identify the molecules in the trypsin extracts reactive with anti-TC-F9^{te}, we subjected the unfractionated TC/TE and TE/TE extracts to SDS-polyacrylamide gel electrophoresis, and blotted them onto nitrocellulose for immunologic detection of antigens. Figure 4 shows that a component with a molecular weight of 34,000 (p34) was detected in the TC/TE but not in the TE/TE extract. When the same test was carried out for the Sephacryl-active fraction of the TC/TE extract, which neutralizes the aggregation-inhibitory effect of anti-TC-F9 Fab, p34 was the only detectable component (Figure 4).

To determine the relation between p140 and p34, we added the concentrated sample of the Sephacryl p34 fraction of the TC/TE extract to anti-TC-F9^{te}, and used this mixture for the immunoprecipitation to detect p140 from the lysate of radioiodinated TC-F9 cells. Figure 1D shows that the amount of immunoprecipitated p140 was greatly reduced by the previous addition of the p34 fraction of the TC/TE extract to anti-TC-F9^{te}. The failure of added TC/TE extract to inhibit

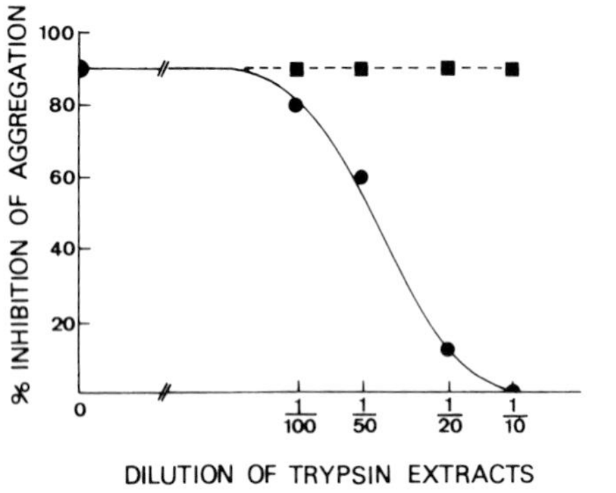

Figure 2. Absorption of Anti-TC-F9 Fab with Trypsin Extracts

Trypsin extract (TC/TE or TE/TE) dialyzed against HMF (see Experimental Procedures) was serially diluted and mixed with 0.17 mg Fab in 0.45 ml HMF (final volume). After 10 min, 1 × 10^5 F9 cells dissociated with pronase and Ca^{2+}, suspended in 50 μl HMF, were added, and the mixture was incubated for 60 min to bring about aggregation. The percentage of inhibition of aggregation by Fab was calculated as described previously (Urushihara and Takeichi, 1980). One milliliter of the undiluted trypsin extract was collected from 1.5 × 10^8 cells. (●——●) TC/TE extract. (■ - - ■) TE/TE extract.

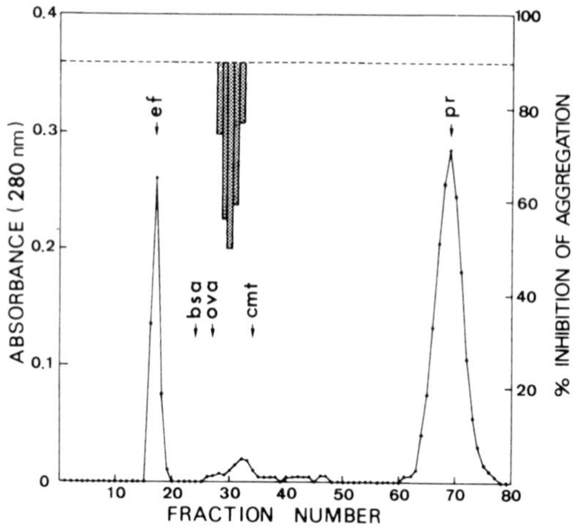

Figure 3. Fractionation of TC/TE Extract by Sephacryl S-200

To assay the activity of each fraction (5 ml) in neutralizing the aggregation-inhibitory effect of anti-TC-F9 Fab, we incubated 1 × 10^5 F9 cells dissociated with pronase and Ca^{2+} in 0.5 ml of each fraction supplemented with 0.17 mg Fab and 1 mM $CaCl_2$, and allowed them to aggregate for 60 min. Bars: the percentage of inhibition of aggregation by Fab. Arrows: positions of the excluded fraction (ef) and the molecular weight markers bovine serum albumin (bsa), ovalbumin (ova), α-chymotrypsinogen A (cmt) and phenol red (pr).

immunoprecipitation of p140 completely may be due to the lower antibody-binding capacity of the fractionated material.

The Fab-neutralizing activity of TC/TE extract was not removed by lectins conjugated to Sepharose beads, such as concanavalin A, peanut agglutinin, Dolichos biflorus agglutinin, Ulex europaeus agglutinin and soybean agglutinin.

Dissociation of Teratocarcinoma Cell Monolayer with Anti-TC-F9 Fab

The addition of anti-TC-F9 Fab to monolayer cultures of AT805 and F9 teratocarcinoma cells caused disruption of the contact between neigboring cells and rounding-up of the cells, but it did not cause their detachment from the culture substratum (Figure 5). AT805 cells responded more sensitively to Fab than F9 cells. Absorption of Fab with TC-F9 cells or with the Sephacryl p34 fraction of the TC/TE extract removed the cell-dissociation effect of the Fab, but absorption with TE-F9 cells or with the TE/TE extract failed to do this (Figure 5).

Discussion

Target molecules recognized by anti-TC-F9 that inhibit the aggregation of teratocarcinoma cells via t-CDS should be present in TC-treated but not in TE-treated cells, as suggested by our previous absorption experiments (Takeichi et al., 1981). Among a number of radioiodinatable cell-surface proteins of teratocarcinoma cells, p140 was the only component present in TC-treated but not TE-treated cells. We have now demonstrated that p140 is actually immunoreactive with anti-TC-F9; and no other protein exposed on the surface of cells reacts with this antibody preabsorbed with TE-F9 cells. It is therefore likely that p140 is the target molecule of anti-TC-F9 in blocking the function of t-CDS.

This conclusion is supported by another approach in the identification of the target of anti-TC-F9. A substance, p34, released from TC-F9 cells when they were retrypsinized in the absence of Ca^{2+}, neutralized the aggregation-inhibitory effect of anti-TC-F9 Fab. This substance interfered with the immunoprecipitation of p140 with anti-TC-F9, indicating that p34 and p140 share common antibody-binding sites. Since the disappearance of p140 from the cell surface coincided with the release of p34 from the cells, it is most likely that p34 is a tryptic fragment of p140.

All these results indicate that p140 is a component essential for the function of t-CDS. There is the possibility, however, that p140 is not the molecule actually representing t-CDS, but that the binding of antibodies to this molecule results in masking t-CDS in a nonspecific manner. We observed that the aggregation-inhibitory effect of anti-TC-F9 Fab is abolished by absorption with the p34 fraction of the trypsin extract from TC-F9 cells. This absorption should leave abundant antibodies to react with the surface of F9 cells in anti-TC-F9, since the anti-TC-F9 contains antibodies commonly reactive with both TC-F9 and TE-F9 cells. This indicates that the binding of antibodies to the cell surface does not always block the function of t-CDS.

It therefore seems more natural to suppose that p140 is a molecule directly involved in cell–cell adhesion mediated by t-CDS. The possibility still remains, however, that p140 is a molecule structurally associated with t-CDS in some specific manner, and not the cell-adhesion molecule per se.

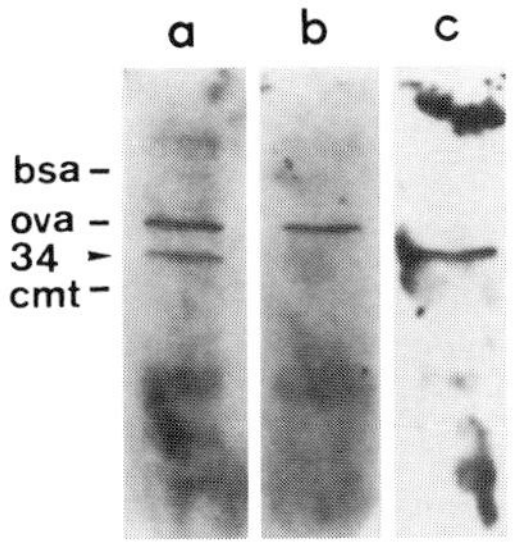

Figure 4. Immunologic Detection of Antigens to Anti-TC-F9[te] in Trypsin Extracts

TC/TE extract (lane a) and TE/TE extract (lane b), collected from 1.5×10^8 cells, and the Sephacryl fraction of TC/TE extract with Fab-neutralizing activity originally derived from 3×10^8 cells (lane c), were lyophilized and subjected to electrophoresis. The blots of these samples on nitrocellulose were incubated with anti-TC-F9[te], and subsequently with ^{125}I-labeled protein A. Molecular weight markers were the same as used in Figure 1, except for the addition of α-chymotrypsinogen A (cmt).

We found another component, p52, that is predominantly present in TC-F9 cells; it is therefore possible that this molecule is also involved in t-CDS. However, the correlation between the presence of p52 and that of t-CDS was not so clear as in the case of p140, since TE-F9 cells seemed to be capable of absorbing the antibodies against p52. We need further studies to determine the relation of p52 with t-CDS.

Figure 6 shows a tentative model explaining the mode of action of p140 in binding cells together and also the process of trypsin cleavage of t-CDS. We have not yet determined whether p140 reacts with a receptor molecule in a lock-and-key manner, or whether it reacts with an identical counterpart, as shown in Figure 6. The role of Ca^{2+} in rendering t-CDS active and resistant to protease also remains to be determined. Possibly, Ca^{2+} induces conformational changes in p140, or aggregation of p140 molecules in the cell membrane, as shown in Figure 6.

The effect of antibodies raised against teratocarcinoma cells in inducing the rounding-up of teratocarcinoma cells in monolayer cultures and the blastomeres of compacted mouse embryos was originally reported by Kemler et al. (1977), Johnson et al. (1979), Ducibella (1980), Hyafil et al. (1980) and Nicolas et al. (1981). In the present study, we neu-

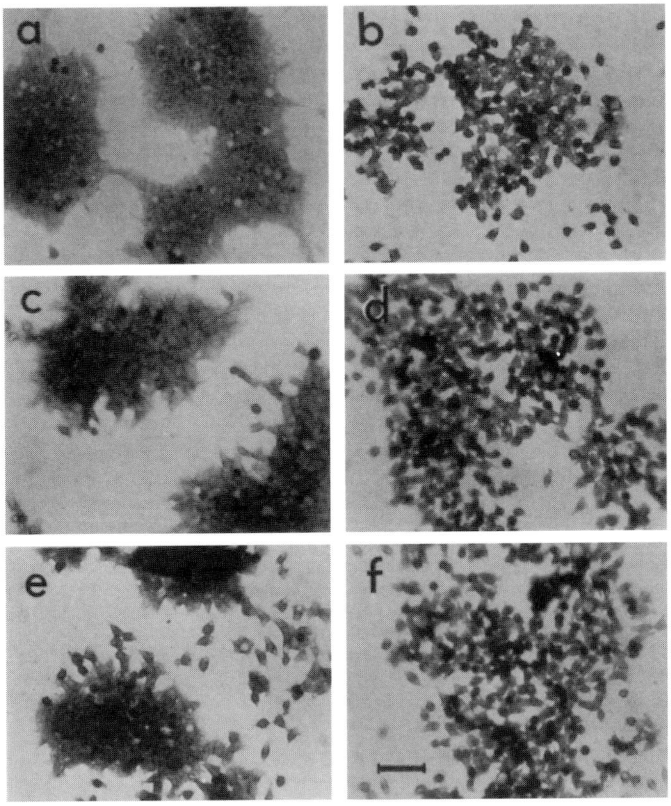

Figure 5. Effect of Anti-TC-F9 Fab on the Morphology of Teratocarcinoma Cells

AT805 cells were cultured in wells of a Nunclon 96-well Micro Test plate for 2 days. To each well were added 33 μg anti-TC-F9 Fab in 100 μl of culture medium supplemented with additional components whose effect was to be tested. After 4 hr of culture, the cells were fixed and stained with Giemsa. (a) Control. (b) Unabsorbed anti-TC-F9 Fab. (c) Anti-TC-F9 Fab absorbed with TC-F9 cells (33 μg Fab absorbed with 2×10^6 cells). (d) Anti-TC-F9 Fab absorbed with TE-F9 cells (33 μg Fab absorbed with 1×10^8 cells). (e) Anti-TC-F9 Fab mixed with the lyophilized Sephacryl p34 fraction of TC/TE extract derived from 2×10^8 cells. (f) Anti-TC-F9 Fab mixed with unfractionated TE/TE extract derived from 2×10^8 cells. Bar = 50 μm.

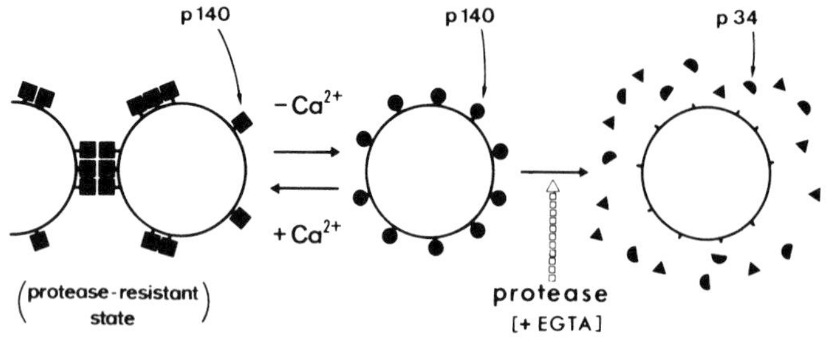

Figure 6. A Tentative Model Illustrating the Mode of Action of t-CDS in Binding Cells and the Process of Disruption of t-CDS by Treatment with Protease

tralized this effect of anti-TC-F9 Fab by absorbing it not only with TC-F9 cells but also with the p34 fraction of the TC/TE extract. However, this effect could not be neutralized with TE-F9 cells. These results strongly suggest that the t-CDS–p140 system for cell adhesion is actually responsible for the maintenance of the mutual adhesion of teratocarcinoma cells in monolayer culture. However, Hyafil et al. (1980) identified a glycoprotein with a molecular weight of 84,000 (gp84) that neutralizes the capacity of Fab fragments of their anti-F9 antibodies to induce the decompaction of teratocarcinoma cells and compacted embryos. This neutralizing effect of gp84 is apparently the same as that of t-CDS observed here. There are three possible explanations for reconciling this result with ours.

First, gp84 is possibly one of the tryptic fragments of p140. According to Hyafil et al., they extracted gp84 by trypsinizing the membrane preparation of teratocarcinoma cells in Eagle's minimal essential medium (supposedly containing Ca^{2+}). Since this procedure for the trypsin extraction of Fab targets is completely different from ours, there may have resulted the production of tryptic fragments of p140 with a molecular size different from that of p34. The second possibility is that p140 and gp84 represent different domains of a t-CDS molecule, and that both are essential for the function of t-CDS. Masking of either molecule with antibodies may cause the suppression of t-CDS action. Third, the adhesion between teratocarcinoma cells may require multiple steps or factors. This is actually the case, since the presence of two distinct cell–cell adhesion sites, CDS and CIDS, in teratocarcinoma cells was demonstrated (Atsumi and Uno, 1979). It is therefore possible that p140 and gp84 represent independent systems for cell adhesion; the inhibition of either system may cause the breakdown of cell adhesion. Further biochemical and immunologic studies on each molecule will be necessary to answer these questions.

As already reported, certain types of differentiated cells, such as fibroblasts, prepared so as to leave only CDS (f-CDS) intact do not crossadhere to teratocarcinoma cells by CDS; and anti-TC-F9 Fab does not inhibit the aggregation of fibroblasts (Takeichi et al., 1981). The results of another study we carried out suggested that a cell-surface protein with a molecular weight of 150,000 (p150) may be involved in the aggregation of fibroblasts by f-CDS (Takeichi, 1977; Takeichi et al., 1981). Thomas et al. (1981a) reported that neural retina and heart cells of chicken embryos crossadhere to each other by CDS, but the CDS of these two cell types shows immunologic distinctions. These findings suggest the possibility that different cell types express different CDS molecules. Future studies of the possible molecular heterogeneity of CDS will perhaps elucidate the molecular mechanisms of specific cell adhesions as observed in a variety of developmental systems. Some other molecules implicated in cell adhesion are, for example, the 125,000 dalton component involved in CIDS of fibroblasts (Urushihara and Takeichi, 1980); CAM (Thiery et al., 1977) and cognin (Hausman and Moscona, 1976) for neural retina cell aggregation; the 68,000 dalton component for liver cell aggregation (Bertolotti et al., 1980); a carbohydrate-binding component for teratocarcinoma adhesion (Grabel et al., 1979); the 140,000 dalton group of glycoproteins for fibroblast adhesion to the substratum (Wylie et al., 1979); contact sites (Müller and Gerisch, 1978; Bozzaro et al., 1981) and gp150 (Geltosky et al., 1979) for slime mold adhesion; and fibronectin (Akiyama et al., 1981). It will be interesting to know what molecular similarities, if any, may exist among these various cell-adhesion molecules, in elucidating the general mechanisms of cell adhesion.

Experimental Procedures

Cell Preparation and Cell Aggregation

Nullipotent teratocarcinoma F9 and multipotent teratocarcinoma AT805 (Takeichi et al., 1981) cells were cultured in Dulbecco's modified Eagle's minimal essential medium supplemented with 6%–8% fetal calf serum. We prepared F9 cells with t-CDS (TC-F9 cells) by treating the cell monolayers with 0.01% crystallized trypsin (Type I; Sigma) in HEPES-buffered Ca^{2+}- and Mg^{2+}-free saline (pH 7.4) (HCMF; see Takeichi [1977] for the ionic composition) supplemented with 1 mM $CaCl_2$ (HMF) for 30 min at 37°C on a gyratory shaker, as described previously (Takeichi et al., 1981). F9 cells without t-CDS (TE-F9 cells) were obtained by the same treatment as TC-F9 cells, except that 1 mM EGTA was added to the trypsin solution instead of 1 mM $CaCl_2$. The trypsinized cells were washed once with HCMF containing 0.05% soybean trypsin inhibitor and twice with HCMF at 4°C, and used in the experiments. On certain occasions, HCMF supplemented with 1 mM EGTA and 50 Kallikrein inhibitor units Aprotinin (Sigma) per milliliter was used for the second washing of cells to facilitate removal of Ca^{2+}, particularly before lactoperoxidase-catalyzed radioiodination of the cells.

In preparing F9 cells with t-CDS for the aggregation assay in particular, we dissociated cells with 0.01% pronase in HMF, since the complete dissociation of F9 cells leaving t-CDS intact was possible only with this enzyme, as described by Takeichi et al. (1981). We assayed the Ca^{2+}-dependent aggregation of these F9 cells by incubating 1×10^5 cells in 0.5 ml HMF supplemented with additional substances whose effect was tested in a well of a 24-well plate (Linbro) for 60 min at 37°C on a gyratory shaker (80 rpm), as described by Urushihara et al. (1979). To measure the degree of cell aggregation, we counted the particle number in cell suspensions in each well by a Coulter counter, and determined the aggregation index N_t/N_0, where N_0 is the total particle number before incubation and N_t is the total particle number after incubation for t minutes. (See Takeichi [1977] for the technical details concerning the methods of measurement of cell aggregation.)

Radiolabeling of Cells

The surface proteins of dissociated cells were radioiodinated by the lactoperoxidase method (Hubbard and Cohn, 1972) under Ca^{2+}-free conditions (Takeichi, 1977). Total cellular proteins were metabolically labeled by incubation of monolayer cultures of cells for 16 hr in 100 μCi/ml ^{35}S-methionine (1300 Ci/mmole; Amersham) in a methionine-free Eagle's minimal essential medium supplemented with 10% fetal calf serum.

Preparation of Trypsin Extracts

TC-F9 or TE-F9 cells (3×10^8) were retreated with 2 ml of 0.01% trypsin with 1 mM EGTA for 30 min at 37°C. After adding 200 μl of 0.5% soybean trypsin inhibitor, we collected the supernatant by centrifugation and dialyzed it against HMF. For fractionation of trypsin extracts, 2.2 ml of the sample prior to dialysis were applied to a Sephacryl S-200 column (Pharmacia) (3.6 cm in diameter and 90 cm in length) and eluted with glucose-free HCMF supplemented with 0.01% NaN_3. The fractions (5 ml each) with the capacity to neutralize the aggregation-inhibitory effect of anti-TC-F9 Fab were pooled, dialyzed against 0.01 M ammonium acetate and lyophilized.

Immunologic Detection of Antigens on Nitrocellulose

Cells and trypsin extracts were first subjected to SDS-polyacrylamide gel electrophoresis. The proteins were then transferred to nitrocellulose sheets (membrane filter BA85; 0.45 μm; Schleicher and Schüll) according to the method of Towbin et al. (1979). The electrophoretic blots were soaked in phosphate-buffered saline containing 1.5% bovine serum albumin and 1.5% ovalbumin (albumin–PBS) for 1 hr to saturate additional protein-binding sites. They were then incubated overnight with antibodies prepared in albumin–PBS (about 1 mg IgG in 1 ml for each electrophoretic lane). The sheet was washed in phosphate-buffered saline for 1 hr and incubated with 0.125 μCi ^{125}I-labeled protein A (30 mCi/mg; Amersham) in 1 ml albumin–PBS for 2 hr, and washed with phosphate-buffered saline. This sheet was subjected to autoradiography.

Immunoprecipitation

Two different methods were used. In the first method, radiolabeled cells (6×10^6) were dissolved in 0.5 ml of 0.5% SDS in 10 mM Tris–HCl (pH 7.5) (SDS lysis buffer). After 10 min, 0.5 ml of 2% NP40 in 10 mM Tris–HCl (pH 7.5) (NP40 lysis buffer) supplemented with 1 μg/ml DNAase I (DN-CL; Sigma) was added. Both lysis buffers contained 1 mM phenylmethylsulfonyl fluoride and 1 mM *p*-tosyl-L-arginine methyl ester. When the DNA gel was digested (within a few minutes), a trace amount of insoluble material was removed by centrifugation. To this sample was added about a 100 μg IgG fraction of anti-TC-F9 or preimmune rabbit serum in 0.5 ml HMF, and after 30 min 100 μl of goat antirabbit IgG (14.8 mg antibody per milliliter; Miles) were added. After 60 min, the precipitates were sedimented and washed several times with a 1:1 mixture of SDS and NP40 lysis buffers, and finally washed with 0.125 mM Tris–HCl (pH 6.8). The precipitates were solubilized in 2% SDS, 5% mercaptoethanol and 20% glycerol in 0.125 mM Tris–HCl (pH 6.8) (SDS sample buffer) by incubation at 37°C for several hours, and then subjected to electrophoresis. In some cases, the antigen–antibody complexes were collected with bacterial immunoadsorbent 30 min following the addition of the first antibody, as described previously (Urushihara and Takeichi, 1980) and also below.

In the second method, radiolabeled cells (2 to 5×10^6) were first incubated in 100 μg IgG of anti-TC-F9 or preimmune serum in 0.7 ml Hanks' saline buffered with 10 mM HEPES (pH 7.4) at 4°C for 60 min. The cells were then washed several times with HMF and solubilized in 0.5 ml SDS lysis buffer, followed by the addition of 0.5 ml NP40 lysis buffer, as described above. To this were added 10 μl (in wet volume) of formalin-fixed and heat-killed Staphylococcus aureus prepared according to the method of Kessler (1975). After incubation for 30 min under constant agitation, the bacteria were washed several times with a 1:1 mixture of SDS and NP40 lysis buffers. Antigens were released from the adsorbents by boiling in SDS sample buffer for 3 min, and were then subjected to electrophoresis.

Electrophoresis and Autoradiography

SDS-polyacrylamide gel electrophoresis was carried out according to the method of Laemmli (1970), with slight modifications (Takeichi, 1977). All samples were analyzed both with 7.5% and 15% acrylamide gels, although the data in the text are shown with gels of one or the other acrylamide concentration. Autoradiography and fluorography of the dried gels were performed as described by Urushihara and Takeichi (1980).

Acknowledgments

We are indebted to Prof. T. S. Okada for his encouragement and support during this project. We should like to thank Drs. M. S. Steinberg and J. Wartiovaara for their critical comments on the manuscript. We also thank Dr. M. Kawano for his gift of lectin-conjugated Sepharose beads, and Ms. A. Yabe for her assistance in the preparation of this manuscript. The radioiodination experiments were carried out in the Radioisotope Research Center of Kyoto University. This work was supported by research grants from the Ministry of Education, Science and Culture of Japan.

The costs of publication of this article were defrayed in part by the payment of page charges. This article must therefore be hereby marked "*advertisement*" in accordance with 18 U.S.C. Section 1734 solely to indicate this fact.

Received October 14, 1981; revised November 30, 1981

References

Akiyama, S. K., Yamada, K. M. and Hayashi, M. (1981). The structure of fibronectin and its role in cellular adhesion. J. Supramol. Struct. *16*, 345–358.

Aoyama, H., Okada, T. S. and Takeichi, M. (1980). Analysis of the cell adhesion mechanisms using somatic cell hybrids: I. Aggregation of hybrid cells between adhesive V79 and nonadhesive Ehrlich's ascites tumor cells. J. Cell Sci. *43*, 391–406.

Atsumi, T. and Takeichi, M. (1980). Cell association pattern in aggregates controlled by multiple cell–cell adhesion mechanisms. Dev. Growth Diff. *22*, 133–142.

Atsumi, T. and Uno, K. (1979). Clonal teratocarcinoma stem cells have similar adhesion mechanisms to cells from differentiated tissues. Cell Struct. Funct. *4*, 388.

Bernstine, E. G., Hooper, M. L., Grandchamp, S. and Ephrussi, B. (1973). Alkaline phosphatase activity in mouse teratoma. Proc. Nat. Acad. Sci. USA *70*, 3899–3903.

Bertolotti, R., Rutishauser, U. and Edelman, G. M. (1980). A cell surface molecule involved in aggregation of embryonic liver cells. Proc. Nat. Acad. Sci. USA *77*, 4831–4835.

Bozzaro, S., Tsugita, A., Janku, M., Monok, G., Opatz, K. and Gerisch, G. (1981). Characterization of a purified cell surface glycoprotein as a contact site in *Polysphondylium pallidum*. Exp. Cell Res. *134*, 181–191.

Brackenbury, R., Rutishauser, U. and Edelman, G. M. (1981). Distinct calcium-independent and calcium-dependent adhesion systems of chicken embryo cells. Proc. Nat. Acad. Sci. USA *78*, 387–391.

Ducibella, T. (1980). Divalent antibodies to mouse embryonal carcinoma cells inhibit compaction in the mouse embryo. Dev. Biol. *79*, 356–366.

Geltosky, J. E., Weseman, J., Bakke, A. and Lerner, R. A. (1979). Identification of a cell surface glycoprotein involved in cell aggregation in D. discoideum. Cell *18*, 391–398.

Grabel, L. B., Rosen, S. D. and Martin, G. R. (1979). Teratocarcinoma stem cells have a cell surface carbohydrate-binding component implicated in cell-cell adhesion. Cell *17*, 477–484.

Grunwald, G. B., Geller, R. L. and Lilien, J. (1980). Enzymatic dissection of embryonic cell adhesive mechanisms. J. Cell Biol. *85*, 766–776.

Hausman, R. E. and Moscona, A. A. (1976). Isolation of retina-specific cell-aggregating factor from membranes of embryonic neural retina tissue. Proc. Nat. Acad. Sci. USA *73*, 3594–3598.

Hubbard, A. L. and Cohn, Z. A. (1972). The enzymatic iodination of the cell membrane. J. Cell Biol. *55*, 390–405.

Hyafil, F., Morello, D., Babinet, C. and Jacob, F. (1980). A cell surface glycoprotein involved in the compaction of embryonal carcinoma cells and cleavage stage embryos. Cell *21*, 927–934.

Johnson, M. H., Chakraborty, J., Handyside, A. H., Willison, K. and Stern, P. (1979). The effect of prolonged decompaction on the development of the preimplantation mouse embryo. J. Embryol. Exp. Morphol. *54*, 241–261.

Kemler, R., Babinet, C., Eisen, H. and Jacob, F. (1977). Surface antigen in early differentiation. Proc. Nat. Acad. Sci. USA *74*, 4449–4452.

Kessler, S. W. (1975). Rapid isolation of antigens from cells with a Staphylococcal protein A-antibody adsorbent: parameters of the interaction of antibody-antigen complexes with protein A. J. Immunol. *115*, 1617–1624.

Laemmli, U. K. (1970). Cleavage of structural proteins during the assembly of the head of bacteriophage T4. Nature *227*, 680–685.

Magnani, J. L., Thomas, W. A. and Steinberg, M. S. (1981). Two distinct adhesion mechanisms in embryonic neural retina cells. I. A kinetic analysis. Dev. Biol. *81*, 96–105.

Martin, G. R. (1980). Teratocarcinomas and mammalian embryogenesis. Science *209*, 768–776.

Müller, K. and Gerisch, G. (1978). A specific glycoprotein as the target site of adhesion blocking Fab in aggregation of *Dictyostelium* cells. Nature *274*, 445–449.

Nicolas, J., Kemler, R. and Jacob, F. (1981). Effects of anti-embryonal carcinoma serum on aggregation and metabolic cooperation between teratocarcinoma cells. Dev. Biol. *81*, 127–132.

Takeichi, M. (1977). Functional correlation between cell adhesive properties and some cell surface proteins. J. Cell Biol. *75*, 464–474.

Takeichi, M. (1981). Identification of cell-to-cell adhesion molecules of chinese hamster fibroblasts. In Cancer Cell Biology, Gann Monograph on Cancer Research *25*, T. Nagayo and W. Mori, eds. (Tokyo: Japan Scientific Societies Press), pp. 3–8.

Takeichi, M., Ozaki, H. S., Tokunaga, K. and Okada, T. S. (1979). Experimental manipulation of cell surface to affect cellular recognition mechanisms. Dev. Biol. *70*, 195–205.

Takeichi, M., Atsumi, T., Yoshida, C., Uno, K. and Okada, T. S. (1981). Selective adhesion of embryonal carcinoma cells and differentiated cells by Ca^{2+}-dependent sites. Dev. Biol. *87*, 340–350.

Thiery, J.-P., Brackenbury, R., Rutishauser, U. and Edelman, G. M. (1977). Adhesion among neural cells of the chick embryo. II. Purification and characterization of a cell adhesion molecule from neural retina. J. Biol. Chem. *252*, 6841–6845.

Thomas, W. A. and Steinberg, M. S. (1981). Two distinct adhesion mechanisms in embryonic neural retina cells. II. An immunological analysis. Dev. Biol. *81*, 106–114.

Thomas, W. A., Edelman, B. A., Lobel, S. M., Breitbart, A. and Steinberg, M. S. (1981a). Two chick embryonic adhesion systems: molecular vs. tissue specificity. J. Supramol. Struct. *16*, 15–27.

Thomas, W. A., Thomson, J., Magnani, J. L. and Steinberg, M. S. (1981b). Two distinct adhesion mechanisms in embryonic neural retina cells. III. Functional specificity. Dev. Biol. *81*, 379–385.

Towbin, H., Staehelin, T. and Gordon, J. (1979). Electrophoretic transfer of proteins from polyacrylamide gels to nitrocellulose sheets: procedure and some applications. Proc. Nat. Acad. Sci. USA *76*, 4350–4354.

Uéda, K., Takeichi, M. and Okada, T. S. (1980). Differences in the mechanisms of cell-cell and cell-substrate adhesion revealed in a human retinoblastoma cell line. Cell Struct. Funct. *5*, 183–190.

Urushihara, H., and Takeichi, M. (1980). Cell-cell adhesion molecule: identification of a glycoprotein relevant to the Ca^{2+}-independent aggregation of Chinese hamster fibroblasts. Cell *20*, 363–371.

Urushihara, H., Ueda, M. J., Okada, T. S. and Takeichi, M. (1977). Calcium-dependent and -independent adhesion of normal and transformed BHK cells. Cell Struct. Funct. *2*, 289–296.

Urushihara, H., Ozaki, H. S. and Takeichi, M. (1979). Immunological detection of cell surface components related with aggregation of Chinese hamster and chick embryonic cells. Dev. Biol. *70*, 206–216.

Wylie, D. E., Damsky, C. H. and Buck, C. A. (1979). Studies on the function of cell surface glycoproteins. I. Use of antisera to surface membranes in the identification of membrane components relevant to cell-substrate adhesion. J. Cell Biol. *80*, 385–402.

Streuli C.H., Bailey N., and **Bissell M.J.** 1991. Control of mammary epithelial differentiation: Basement membrane induces tissue-specific gene expression in the absence of cell-cell interaction and morphological polarity. *J. Cell Biol.* **115:** 1383–1395.

MANY OBSERVATIONS HAVE SUPPORTED the idea that an animal cell's environment has impact on its ability to differentiate and function. In addition to the inductive effects of adjacent embryonic cells or tissues, described in the classic literature on animal development (Meier and Hay, this volume), cell biologists have shown the importance of extracellular matrix (ECM) for the proper behavior of adult animal cells (1-3) and have discovered enzymes that modify or degrade existing ECM (4). More recent studies on synovial fibroblasts have demonstrated that interactions of extracellular ligands with membrane proteins of the "integrin" family can induce the expression of genes that encode enzymes for degrading ECM (5). The paper reprinted here showed that cultured cells from mammary epithelium could be induced to differentiate, turning on the expression of an uncounted number of genes, by the combined action of steroid hormones and the binding of components from basement membrane. In the assay used, single cells growing in a laminin-rich matrix turned on the synthesis of β-casein, while those in a matrix of collagen-I did not. This evidence suggested that the inductive signal was mediated though a receptor specific for laminin. Subsequent work has shown that several receptors are important in this system, suggesting the existence of an information processing network that assesses the composition and structure of the ECM and helps to define patterns of cell growth and differentiation.

---

1. Kleinman H.K., Klebe R.J., and Martin G.R. 1981. Role of collagenous matrices in the adhesion and growth of cells. *J. Cell Biol.* **88:** 473–485.
2. Humphries M.J., Olden K., and Yamada K.M. 1986. A synthetic peptide from fibronectin inhibits experimental metastasis of murine melanoma cells. *Science* **233:** 467–470.
3. Liotta L.A., Rao C.N., and Wewer U.M. 1986. Biochemical interactions of tumor cells with the basement membrane. *Annu. Rev. Biochem.* **55:** 1037–1057.
4. Gross J. and Nagai Y. 1965. Specific degradation of the collagen molecule by tadpole collagenolytic enzyme. *Biochemistry* **54:** 1197–1204.
5. Werb Z., Tremble P.M., Behrendtsen O., Crowley E., and Damsky C.H. 1989. Signal transduction through the fibronectin receptor induces collagenase and stromelysin gene expression. *J. Cell Biol.* **109:** 877–889.

Control of Mammary Epithelial Differentiation: Basement Membrane Induces Tissue-specific Gene Expression in the Absence of Cell–Cell Interaction and Morphological Polarity

Charles H. Streuli, Nina Bailey, and Mina J. Bissell

Cell and Molecular Biology Division, Lawrence Berkeley Laboratory, Berkeley, California 94720

Abstract. Functional differentiation in mammary epithelia requires specific hormones and local environmental signals. The latter are provided both by extracellular matrix and by communication with adjacent cells, their action being intricately connected in what appears to be a cascade of events leading to milk production. To distinguish between the influence of basement membrane and that of cell–cell contact in this process, we developed a novel suspension culture assay in which mammary epithelial cells were embedded inside physiological substrata. Single cells, separated from each other, were able to assimilate information from a laminin-rich basement membrane substratum and were induced to express β-casein. In contrast, a stromal environment of collagen I was not sufficient to induce milk synthesis unless accompanied by cell–cell contact. The expression of milk proteins did not depend on morphological polarity since E-cadherin and α_6 integrin were distributed evenly around the surface of single cells. In medium containing 5 μM Ca^{2+}, cell–cell interactions were impaired in small clusters and E-cadherin was not detected at the cell surface, yet many cells were still able to produce β-casein. Within the basement membrane substratum, signal transfer appeared to be mediated through integrins since a function-blocking anti–integrin antibody severely diminished the ability of suspension-cultured cells to synthesize β-casein. These results provide evidence for a central role of basement membrane in the induction of tissue-specific gene expression.

REGULATION of differentiation in complex tissues is determined not only by growth factors and hormones, but also by intercellular communication and by interactions between cells and their extracellular matrix (Stoker et al., 1990). In epithelial tissue, part of the extracellular matrix (ECM)[1] occurs in the form of a basement membrane, which provides positional information for cells and cues for organizing intracellular structure, as well as signals that regulate cellular behaviour. An ideal model for studying how ECM signals are transduced to control tissue-specific function is the simple epithelium of mammary gland. In adult animals, the cells undergo developmental changes during pregnancy and become highly secretory at lactation. The great advantage of this system is that in culture, mammary epithelial cells regain their differentiated phenotype only under suitable hormonal and substratum conditions. The model can therefore be used to understand the mechanism by which tissue-specific genes are expressed (Bissell and Hall, 1987; Streuli and Bissell, 1991).

Considerable evidence now indicates that basement membrane plays a significant role in regulating mammary phenotype. In vivo, the alveolar epithelium of rodent mammary gland is surrounded by a basement membrane; its dissolution during involution correlates with functional regression of the gland (Talhouk et al., 1991). In culture, on stromal collagen I substrata, mammary cells synthesize and secrete milk proteins only if an endogenous basement membrane is deposited basally to the cell layer; this occurs if collagen I gels are released into the culture medium (Streuli and Bissell, 1990). More complete differentiation can be achieved on the basement membrane matrix derived from Engelbreth-Holm-Swarm tumour (EHS matrix, or 'matrigel') where, in addition to high level milk production, the cells undergo morphogenetic changes and form spherical structures that resemble the alveoli of lactating mammary gland with striking fidelity (Li et al., 1987; Barcellos-Hoff et al., 1989; Chen and Bissell, 1989; Aggeler et al., 1991). However, under conditions where basement membranes are not present, such as on nonreleased collagen gels or on plastic dishes, mammary epithelial cells express little or no milk proteins even when cultured with lactogenic hormones.

The inability of mammary epithelium to function correctly in the absence of basement membrane strengthens the argument for its involvement in the induction and maintenance of differentiation (Bissell and Barcellos-Hoff, 1987; Streuli and Bissell, 1991). However, biochemical function in cells cultured on either floating collagen gels or basement

1. *Abbreviations used in this paper*: DAPI, 4,6-diamidino-2-phenylindole; ECM, extracellular matrix; EHS, Engelbreth-Holm-Swarm.

The Journal of Cell Biology, Volume 115, Number 5, December 1991 1383–1395 membrane substrata is accompanied by cytostructural alterations and by increased interactions between adjacent cells. Because such interactions might influence tissue-specific gene expression, it has now become crucial to address the issue of whether these local environmental cues are indeed required for mammary cell differentiation.

We have therefore developed a novel suspension culture assay, firstly to distinguish between the influence of basement membrane and that of cell–cell contact in the control of mammary phenotype, and secondly to identify whether or not basement membrane can signal changes in gene expression via integrins. Here we present evidence that basement membrane regulates the synthesis of milk proteins in mammary epithelium directly, and that integrins are involved in this process. We also discuss the contribution of cell–cell interactions, and suggest that these signals have additional roles in the differentiation of a functional epithelium.

Materials and Methods

Substrata and Antibodies

EHS matrix was prepared from EHS tumors passaged in C57BL mice by urea extraction at 4°C (Kleinman et al., 1986). This material was stored for up to four weeks at 0°C. Each preparation contained 5–10 mg per ml protein, and was routinely assayed for purity and lack of protein degradation on 5% reducing SDS-polyacrylamide gels followed by silver staining. Growth factor–depleted EHS matrix was prepared from freshly made EHS matrix by precipitation twice with 20% $(NH_4)_2SO_4$ on ice followed by dialysis, using the protocol of Taub et al. (1990), and was used immediately. Collagen I was prepared from rat tails (Lee et al., 1984) and was stored at a concentration of 2–3 mg per ml in 0.1% acetic acid at 4°C. For use in the suspension assay, it was dialyzed against three changes of water (total time 48 h, 4°C) and adjusted to 1 × F12 medium immediately before use. Under these conditions, toxicity due to salt and pH imbalance was minimized, and the collagen solution remained workable for two to three hours. To form base layers for either the primary epithelial cultures or for the cell-embedded gels, the substrata were spread at 15.6 μl per cm^2, and gelled at 37°C.

Polyclonal rabbit antiserum to mouse milk was obtained by immunization with skimmed milk proteins as previously described (Lee et al., 1985). A mouse mAb to rat β-casein was a gift from Dr. C. Kaetzel (Institute of Pathology, Case Western Reserve University, Cleveland, OH) (Kaetzel and Ray, 1984). The rat mAb to α_6 integrin (GoH3; ascites) was a gift from Dr. A. Sonnenberg (Lab. for Experimental and Clinical Immunology, University of Amsterdam, The Netherlands) (Sonnenberg et al., 1987). The rat mAbs to E-cadherin (ECCD-1, ECCD-2; culture supernatant) were generously supplied by Dr. M. Takeichi, Dept. Biophysics, Kyoto University, Japan (Yoshida-Noro et al., 1984; Shirayoshi et al., 1986). The rat mAb to β_4 integrin (346-11A) was a gift from Dr. S. Kennel (Biology Division, Oak Ridge National Laboratory, Oak Ridge, TN) (Kennel et al., 1989). Polyclonal goat antiserum to baby hamster kidney fibroblast β_1 integrins (anti ECM-R) was the kind gift of Dr. C. Damsky (Dept. Anatomy and Stromatology, University of California, San Francisco, CA) (Knudsen et al., 1981; Damsky et al., 1982). Polyclonal rabbit antiserum to mouse laminin was obtained from E. Y. Labs (San Mateo, CA).

Cell Culture

Primary mammary epithelia were prepared from 14.5-d pregnant CD-1 mice and cultured on physiological substrata as described previously (Lee et al., 1985; Barcellos-Hoff et al., 1989) in the presence of lactogenic hormones and, for the first 36 h, 1 mg per ml fetuin (Sigma Chemical Co., St. Louis, MO) and 10% FCS. The initial plating density was 4.7 × 10^5 cells per cm^2. Insulin (5 μg/ml) and lactogenic hormones (1 μg/ml hydrocortisone, 3 μg/ml prolactin) were included in all experiments, and the medium was changed every day. The secondary cultures were derived from primaries cultured for five or six days on plastic dishes. These essentially nonfunctional cells were also cultured with lactogenic hormones. Monolayers were washed briefly with STV (0.05% trypsin, 0.02% EDTA in saline A) and trypsinized in STV for 10–15 min, at which time serum was added to inactivate the enzyme. In the serum-free experiments, trypsinization was stopped with 5 mg per ml soybean trypsin inhibitor (Gibco Laboratories, Grand Island, NY). The cells were washed with F12 medium and counted. Mild trituration broke up large cell aggregates and resulted in populations containing 55–65% single cells.

Secondary cultures were plated at 4.7 × 10^5 cells per cm^2 onto various substrata using identical conditions to those for the primaries. For the suspension assays, cells were embedded inside physiological substrata, such as the EHS matrix or a collagen I gel. Freshly harvested cells (10^6) were pelleted for 2 min at 1,300 *g* in sterile microfuge tubes, chilled on ice, immediately resuspended with mild trituration to a density of 2.2–4.5 × 10^6 cells per cm^3 in EHS matrix or collagen I at 0°C, and 100-μl aliquots were plated onto preformed base layers of the corresponding substratum. At the latter density the cells were well separated from each other, being on average at least five cell diameters apart. Some cells reaggregated immediately after trypsinization and remained as clusters during the course of the experiment. The cell-containing substrata were gelled (30–60 min, 37°C) and then covered with hormone-containing medium. After six days in culture the majority of cells were still isolated from each other; at six days, 44–60% of all the single cells plus clusters were single, 16–31% were doublets, and the remainder existed in larger clusters. Cells induced to differentiate were detected by immunohistochemistry, either directly in the whole mounts, or in cryosectioned cultures. Cell viability was confirmed by autoradiography performed on sections of suspension cultures that had been pulse chased with ^{35}[S]methionine before fixation.

In some experiments, heat-inactivated (1 h, 56°C) anti–ECM-R antibody was included in the substratum (1%) and in the culture medium (0.5%). At a similar concentration (0.6–1%), this antibody blocked B16-BL6 melanoma cell adhesion to basement membrane matrices (Kramer et al., 1990), and blastocyst outgrowth on laminin and fibronectin (Sutherland et al., 1988). It also blocked adhesion of PC12 cells to both laminin and collagen IV substrata, and prevented outgrowth of their neuronal processes on laminin (Tomaselli et al., 1987). In addition, the binding of mouse mammary cells to laminin-coated dishes was inhibited efficiently by the anti ECM-R antibody (not shown).

Immunohistochemistry

Immunostaining was carried out according to general principles in Harlow and Lane (1988). Non-specific antibody-binding sites were blocked and all washes were done in PBS containing 0.1% BSA, 0.2% Triton X-100, 0.05% Tween 20, 0.05% NaN_3 (at least 1 h, room temperature). Sections or whole mounts were incubated with primary antibodies for 60 min and subsequent antibodies for 30 min. Primary rat antibodies were detected with Texas red anti–rat IgG (Caltag Labs, South San Francisco, CA), mouse antibodies with either FITC anti–mouse IgG (Amersham Corp.), or biotinylated anti–mouse IgG (Amersham Corp.) followed by Texas red streptavidin (Amersham Corp.), and rabbit antibodies were detected with biotinylated anti–rabbit IgG (Amersham Corp.) followed by Texas red streptavidin. In all experiments nuclei were counter stained (3 min) during the final wash with 0.5 μg per ml DAPI (4,6-diamidino-2-phenylindole; Sigma Chemical Co.). The fluorescence optics were Zeiss (Oberkochen, Germany), and photography was with EL400 or TX400 film (Eastman Kodak Co., Rochester, NY).

For the whole mounts, cells were cultured inside the relevant substratum on coverslips, fixed with 2% paraformaldehyde in PBS (20 min, room temperature), quenched with 0.1 M glycine (3 × 20 min), the cell membranes permeabilized in 1% Triton X-100 (90 s), and the cells stained with the β-casein mAb. All other suspension cultures were fixed with 2% paraformaldehyde, quenched with 0.1 M glycine, and equilibrated first with sucrose and subsequently with Tissue-Tek OCT compound (Miles Scientific Div., Elkhart, IN), before freezing on a dry ice/ethanol bath. Sections were cut with a cryotome (E. Leitz, Inc., Rockleigh, NJ), collected on gelatin-coated coverslips and air dried before immunostaining with the relevant antibody. To reduce nonspecific background, the detection antibodies were adsorbed onto a thin layer of gelled EHS matrix before use (60 min, 37°C). The counting procedure for β-casein–staining cells was as follows. Single cells or cell clusters were located with Nomarski optics in conjunction with DAPI fluorescence, nuclei then were viewed with fluorescence alone to count the number of cells in the cluster, and the single cells or clusters were finally scored for casein expression using the FITC or Texas red fluorescence. Optical sectioning through the thick sections or whole mounts ensured that the scored single cells were actually single. Although some clusters expressing β-casein contained a few nonstaining cells, such clusters were scored as being positive. In most experiments, enough fields were viewed to score more than 200 single cells plus clusters.

Alveolar cultures of mammary cells on EHS matrix were fixed with paraformaldehyde, quenched, and embedded in OCT before freezing and cryosectioning. For this purpose, 4-μm sections were cut at an angle oblique to the plane of the culture dish. Monolayer cultures, grown on acid-washed glass coverslips, were stained after a 5 min methanol/acetone (1:1) fixation at −20°C.

Protein Analyses

For immunoblotting, cells were harvested 24 h after the previous media change. Media fractions were collected and cleared (1 min, 16,000 *g*). Luminal fractions, containing material secreted inside the cultured alveoli, were extracted with 2.5 mM EGTA as described (Barcellos-Hoff et al., 1989) and cleared. These fractions were then diluted in sample buffer (10% glycerol, 50 mM Tris-HCl, 2% SDS, 0.02% bromophenol blue, 5% mercaptoethanol, pH 6.8). The remaining cell fractions were scraped directly into sample buffer. The samples from different culture conditions were normalized by counting cells (in triplicate) on identical dishes, and adjusting each of the media, luminal, and cell fractions to an equivalent volume. Material from only ~1,000 cells was resolved on 0.75-mm-thick 13% polyacrylamide gels under reducing conditions, and transferred to Immobilon-P membranes (Millipore Continental Water Systems, Bedford, MA) using a dry blot apparatus. The membranes were incubated overnight in a wash buffer containing 100 mM Tris-HCl, 150 mM NaCl, 0.3% Tween 20, the nonspecific sites blocked in the same buffer containing 2% BSA, and the milk proteins detected with rabbit anti-mouse milk antiserum, followed by alkaline phosphatase-conjugated anti-rabbit IgG (Caltag Labs), and subsequently BCIP/NBT (bromo chloro indolyl phosphate/nitro blue tetrazolium; Sigma Chemical Co.). Mouse milk (50 ng) was routinely included in the gels as a control. Biotinylated size markers were detected with alkaline phosphatase conjugated streptavidin (Zymed).

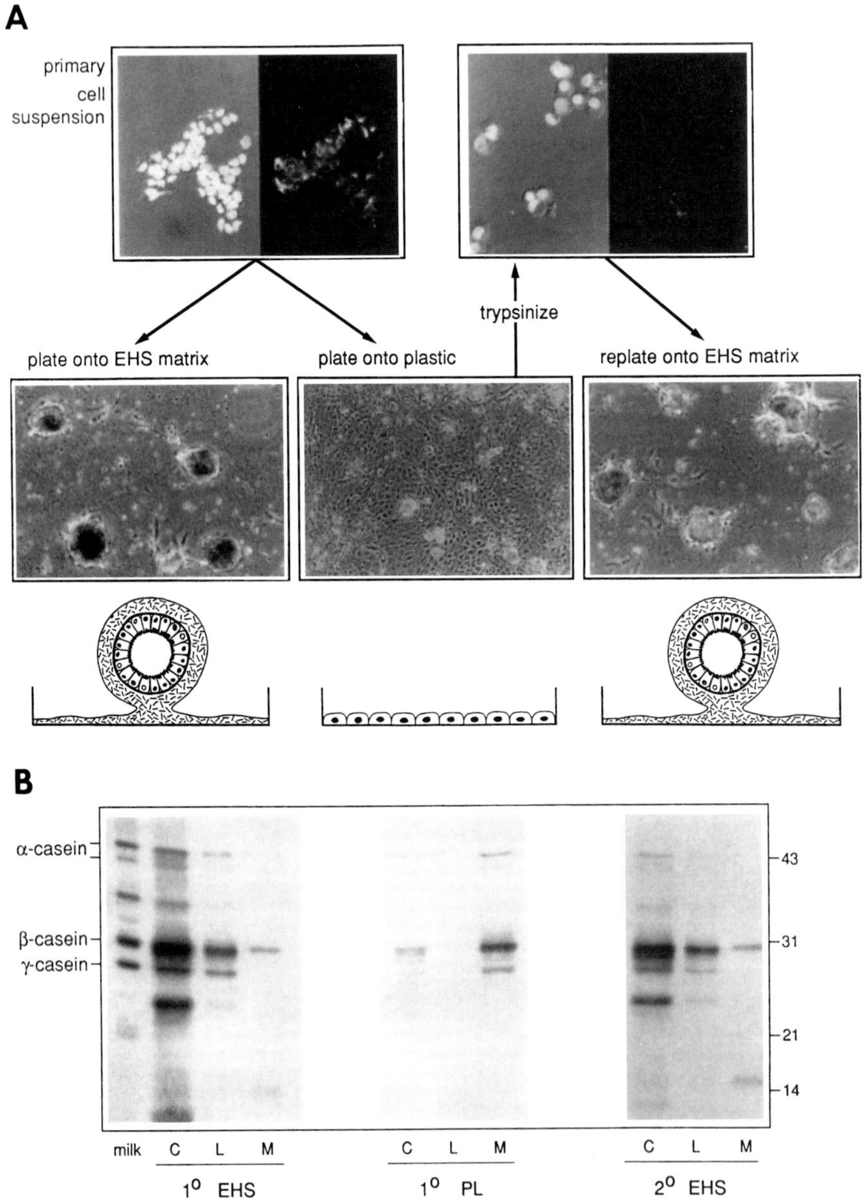

Figure 1. Single mammary cells can redifferentiate on a basement membrane matrix. (*A*) Cells isolated from pregnant mammary gland existed as aggregates, and many cells (~70%) expressed milk proteins, here assayed by immunofluorescence of cytospins using mAb specific for β-casein (*top left*, Nomarski images of the cells are also shown, their nuclei stained with DAPI). After six days in culture on EHS matrix, they formed spherical structures (shown by phase-contrast microscopy and as a schematic cross section). On a plastic substratum, a squamous monolayer resulted. These cells could be trypsinized to single cell suspensions (*top right*), of which only 6–17% express β-casein in cytospin analysis; in this field a small cluster of three cells was weakly positive. The cells aggregated and formed alveolar, casein-expressing spherical structures when replated on EHS matrix and cultured for six days (*right panels*). (*B*) Secondary cultures (2°) maintained for six days on EHS matrix (*EHS*) accumulated quantities of milk comparable to primary cultures (*1°*). This immunoblot shows caseins in the cell fractions (*C*) and those secreted into lumina (*L*) of spheres. The high levels in the former were largely due to precipitation of secreted milk within the lumina and its consequent inefficient extraction. Very small amounts of milk were secreted basally into the medium (*M*) of these cultures. Cells cultured on plastic dishes (*PL*) produced low overall levels of milk; what is seen was due to synthesis by cells remaining in aggregates (unpublished data). In each case, material from ~1,000 cells was resolved by SDS-PAGE; note that the level of milk production by only this number of cells cultured on the EHS matrix exceeded 50 ng of mouse milk. Size markers (in kD) are indicated.

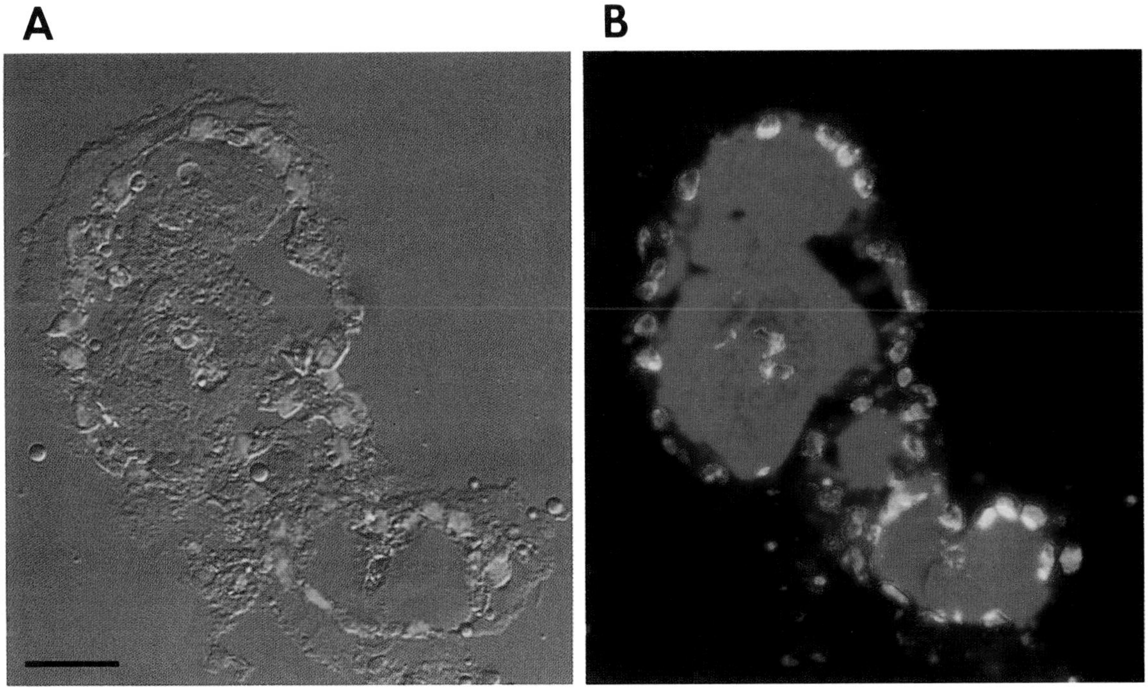

Figure 2. β-casein accumulates in the lumina of alveolar cultures. 4-μm sections of six day EHS matrix cultures were stained by immunofluorescence using a mAb specific for β-casein. Views of DAPI-staining nuclei are superimposed either on (*A*) a Nomarski image of a cryosection, or on (*B*) an image of the β-casein localized by immunofluorescence. Bar, 30 μm.

Immunoprecipitation was carried out as described previously (Streuli and Bissell, 1990), except that cells were labeled for 4 h with 0.25 mCi per ml Tran^{35} S-label (ICN Biomedicals, Irvine, CA), and the proteins were precipitated with rabbit anti-mouse milk antiserum followed by protein A sepharose (Sigma Chemical Co.), before separating on 13% polyacrylamide gels under reducing conditions.

Results

Differentiation of Mammary Epithelial Cells in Culture

The initial aim of these studies was to see whether single mammary epithelial cells, separated from each other, could be induced to synthesize milk proteins in suspension culture. First, however, it was necessary to show that single mammary cells would redifferentiate in culture if they were brought together under optimal conditions. Primary preparations of epithelial cells recovered from collagenase digests of midpregnant mouse mammary gland existed almost exclusively as aggregates (Fig. 1 *A, top left*). More vigorous cell-separation techniques resulted in unacceptable mortality. When plated on plastic dishes, the cell aggregates spread to form a squamous cobblestone monolayer (Fig. 1 *A*) and did not deposit their own basement membrane (Streuli and Bis-

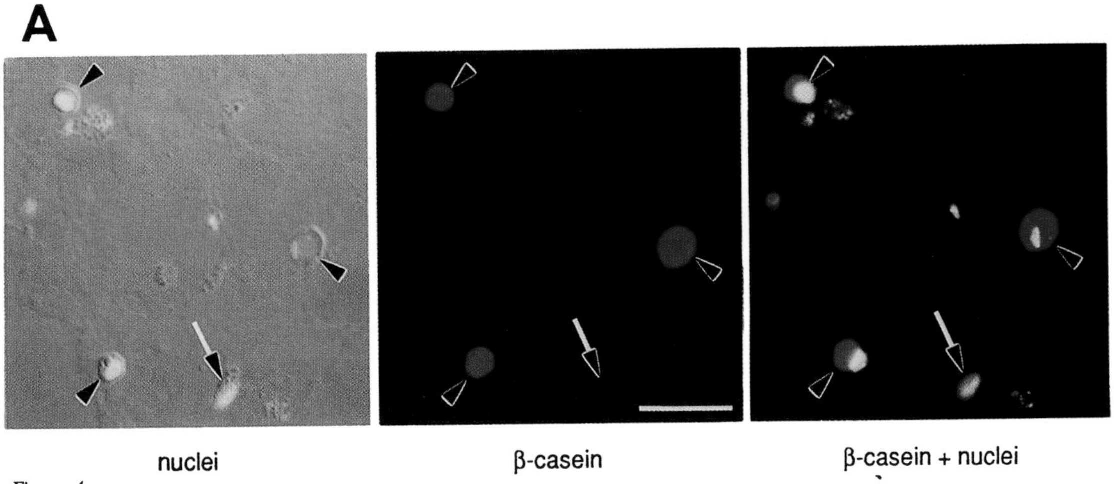

Figure 4.

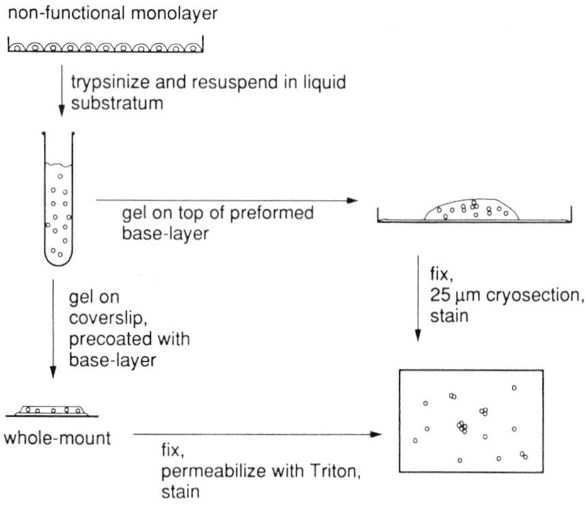

Figure 3.

sell, 1990). They were unable to synthesize milk proteins even in the presence of lactogenic hormones (Fig. 1 *A, top right*; and Fig. 1 *B*) (Lee et al., 1984; Li et al., 1987), and therefore represented a potentially useful starting cell population on which to test the effect of basement membrane. These cultures were easily trypsinizable, producing suspensions that contained mostly single cells (55–65%) as well as small clusters of two or more cells. When replated onto EHS matrix, the cells eventually aggregated, underwent alveolar morphogenesis (Fig. 1 *A*), and secreted milk proteins vectorially (Figs. 1 *B* and 2). The extent to which differentiation could be reestablished compared well with that of primary cultures on the EHS matrix (Fig. 1 *B*).

Differentiation Can Be Induced in Single Mammary Cells Cultured within Physiological Substrata

To ask whether mammary cells could differentiate in the absence of both cell–cell interactions and organized intracellular polarity, a single cell suspension assay was developed. For this purpose, cells were embedded inside either EHS matrix or a collagen I gel. The chemical properties of these substrata are ideal for maintaining cells in suspension culture, since they are viscous at 0°C and gel rapidly at 37°C to form malleable lattices with high-order supramolecular structures. Mammary epithelial cells, isolated from primary monolayers with low levels of function cultured on plastic dishes, were embedded so that they were well separated from one another at a cell density of 2.2–4.5 × 10^6 cells per cm^3. Cell-containing droplets were allowed to gel on preformed base layers of either the corresponding substratum or collagen I, and were covered with growth medium (Fig. 3). Most cells cultured inside EHS matrix remained viable during the culture period. The cells were rounded, had smooth refractile borders, and their nuclei retained distinct nucleoli. Cells that became granular and had dispersed or absent nuclei were disregarded in subsequent analyses.

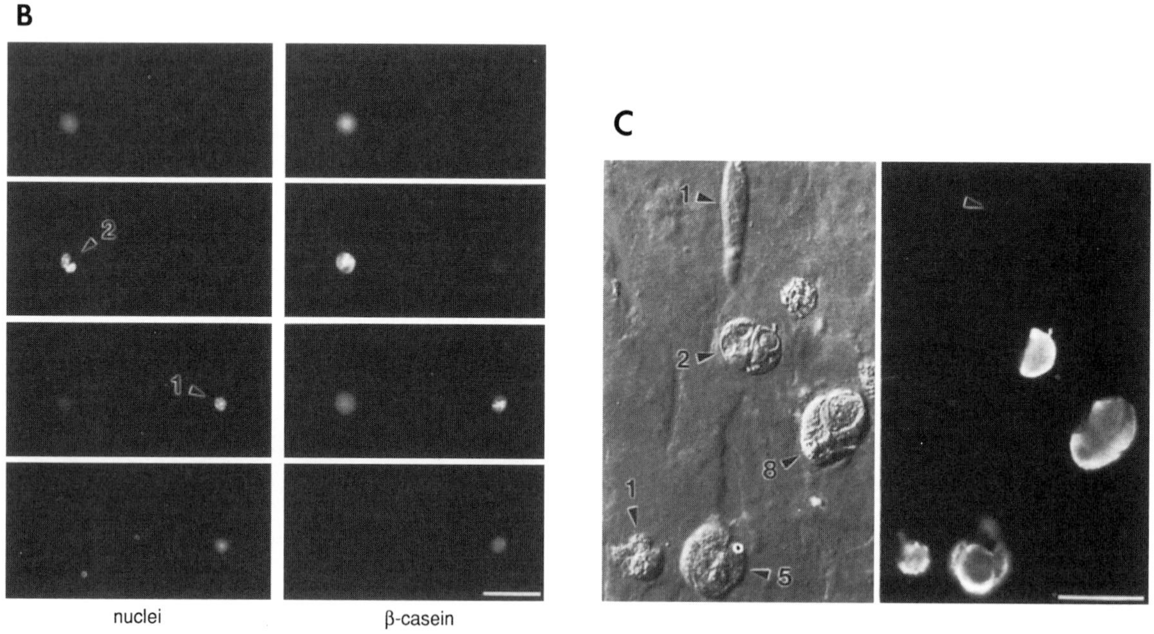

Figure 4. Mammary cells in EHS matrix suspension culture express β-casein. (*A*) Six day EHS matrix suspension cultures were stained by immunofluorescence using a β-casein–specific mAb. A cyrosection shows three cells that strongly expressed β-casein (*arrowheads*), and one negative cell (*arrow*). The nuclei were counterstained with DAPI. (*B*) In a whole mount of cells embedded inside EHS matrix, one field of view was optically sectioned by photographing at different levels of focus. This type of analysis proves that single cells do indeed express β-casein. Here, one single cell and one cluster of two cells were both positive. (*C*) In this view of a thick 25-μm cryosection, several clusters are visible. These clusters existed at the time of suspending the cells. By six days, the cells have coalesced and they produced high levels of casein. The number of nuclei in each cluster are indicated, and one nonstaining cell is shown (*arrowhead*). Bars: (*A*) 50 μm; (*B*) 50 μm; (*C*) 30 μm.

Table I. β-casein Expression in EHS Matrix Suspension Culture

	Percentage of β-casein staining			
Experiment	Single cells	2 cell clusters	>4 cell clusters	Total¶
1	54 (307)§	86 (102)§	96 (94)§	(590)
2	64 (133)	92 (39)	100 (14)	(223)
3	58 (342)	78 (127)	95 (86)	(679)
4	71 (98)	94 (34)	95 (22)	(180)
5‡	55 (114)	72 (54)	98 (45)	(260)

Each experiment is from an independent culture derived from an independent preparation of primary mammary epithelial cells.
‡ Cells cultured in whole mounts were analyzed in this experiment.
§ Numbers of single cells, 2 cell clusters, or >4 cell clusters counted.
¶ Total numbers of single cells + cell clusters counted; this includes the numbers of 3 and 4 cell clusters, which for the sake of clarity, are not shown in the table.

Assays for mammary function were carried out on individual cells in situ, using a mAb for β-casein as a suitable marker for biochemical differentiation (Kaetzel and Ray, 1984). In most experiments gels were sectioned, stained by indirect immunofluorescence, and counterstained with DAPI to localize nuclei.

Cryosections of embedded cultures revealed that single mammary cells within the EHS matrix could express β-casein (Fig. 4). In five independent experiments, 54–71% of the single cells were induced (Table I). Single cells were well represented within 25-μm sections, which were approximately three times thicker than the average cell diameter of 10 μm; occasionally, whole mounts were used to confirm that single cells were being studied (Fig. 4 *B*). The appearance of β-casein within cells was slightly punctate, but homogeneous. A sharp boundary of fluorescence existed at the cell periphery, suggesting that the casein was not secreted.

Small clusters of two or more cells also expressed β-casein; in some cases only one of a pair of cells was functional (Fig. 4 *C*). Large clusters stained strongly, and occasionally formed alveolar structures. In some of these it appeared that milk products were secreted vectorially into a luminal space (not shown), which may explain the greater intensity of fluorescence. The percentage of positive clusters was higher than that of single cells (Table I). However, since not every cell within each cluster stained for casein, this apparent enhancement of staining may have been due in part to random assortment of positive cells within the clusters.

Time course analysis was performed to confirm the inducibility of casein expression. The percentage of casein-positive cells rose dramatically during the first few days of culture (Table II), with the kinetics of induction resembling

Table II. Induction of β-casein Expression in EHS Suspension Culture

	Percentage of β-casein staining			
Day	Single cells	2 cell clusters	>4 cell clusters	Total
1	4 (191)	11 (46)	48 (29)	(316)
2	15 (115)	34 (53)	84 (25)	(221)
4	34 (127)	86 (37)	94 (18)	(212)
6	63 (134)	89 (36)	100 (22)	(231)
9	60 (136)	81 (46)	95 (38)	(245)

the onset of function in alveolar cultures (Barcellos-Hoff et al., 1989; Aggeler et al., 1991). For single cells, biochemical differentiation required at least 1–2-d exposure to a basement membrane environment.

The requirement for a basement membrane was assessed by studying the behavior of mammary cells embedded within a collagen I gel. In this stromal environment, the cells responded very differently. Most single cells were unable to synthesize β-casein (Table III; Fig. 5 *A*). In contrast, many of the clusters were positive, showing that cell–cell interaction was necessary for functional differentiation when cells were embedded inside collagen I. However, since the positive cells, especially those in aggregates, were frequently associated with some laminin visible by immunofluorescence at the cell periphery (Fig. 5 *B*), the enhancement of β-casein expression was most probably because of an interaction with newly synthesized basement membrane components, as could be inferred from previous studies (Streuli and Bissell, 1990).

The ability of single cells to express β-casein when cultured inside EHS matrix implies that, as long as basement membrane components are present, cell–cell contact is not essential for biochemical differentiation. We then investigated whether or not morphological polarity was necessary for β-casein expression.

The Expression of Milk Proteins Is Not Coupled to a Polarized Phenotype

E-cadherin (uvomorulin) forms part of the epithelial intercellular adhesion system (Nagafuchi et al., 1987; Gumbiner et al., 1988; McNeill et al., 1990; Takeichi, 1991). This protein is located along lateral surfaces of cells in the highly polarized epithelium of mammary gland (not shown), a distribution that was also apparent between adjacent cells of "alveoli" produced in culture (Fig. 6 *A*). In suspension culture it was concentrated at sites of contact, indicating that in clusters the cells interacted closely with one another and established certain aspects of a polarized phenotype (Fig. 6 *B*). However in single cells, many of which were able to express β-casein, E-cadherin appeared in patches around the cell (Fig. 6, *C* and *D*), indicating a lack of morphological polarity.

To further rule out cell–cell interaction and polarity as mediators of milk protein expression, embedded cells were cultured in medium containing only 5 μM Ca^{2+}. Such cells remained rounded and the clusters failed to coalesce. Neither the single cells nor the clusters showed any evidence of E-cadherin staining (Fig. 7 *A*). Yet, many of the same cells were still able to synthesize substantial levels of β-casein (Fig. 7 *B*). This applied both to secondary cultures and to CID-9 cells, a recently isolated mouse mammary cell strain that exhibits ECM-dependent differentiation (Schmidhauser et al., 1990). Thus, it can be concluded that neither cell–cell interactions, nor morphological polarization, are necessary for basement membrane-directed induction of β-casein expression.

Basement Membrane Can Direct Differentiation in the Absence of Serum or Other Growth Factors

There is an increasing awareness that growth and differentiation factors can be bound by ECM and be presented to cells

Table III. Effect of Substratum on Differentiation of Single Cells or Cell Clusters

Experiment	Substratum	Percentage of β-casein staining			
		Single cells	2 cell clusters	>4 cell clusters	Total
1	EHS matrix	54 (307)	86 (102)	96 (94)	(590)
	collagen I	4 (369)	34 (70)	74 (50)	(557)
2	EHS matrix	71 (98)	94 (34)	95 (22)	(180)
	collagen I	7 (175)	51 (69)	74 (42)	(320)
3*	EHS matrix	55 (114)	72 (54)	98 (45)	(260)
	collagen I	7 (81)	23 (31)	75 (25)	(155)

* Cells cultured in whole mounts were analyzed in this experiment.

in this form (Bradley and Brown, 1990; Klagsbrun, 1990; Rathjen et al., 1990; Ruoslahti and Yamaguchi, 1991). It was therefore necessary to test whether factors potentially present in EHS matrix (Vigny et al., 1988) or in serum contributed to milk production. If already active, or if activated in culture, such factors could participate in the control of mammary differentiation. SDS-PAGE detectable low molecular weight proteins were removed from EHS matrix using an ammonium sulfate precipitation protocol shown previously to deplete the matrix of factors such as TGFα, TGFβ, and EGF (Taub et al., 1990). Primary mammary cells formed bona fide spherical structures on this substratum (not shown) and synthesized high levels of milk proteins (Fig. 8) in the absence of serum. Cryosections with similar immunofluorescence profiles to those in Fig. 2 revealed that β-casein was secreted vectorially into alveolar lumina. Fur-

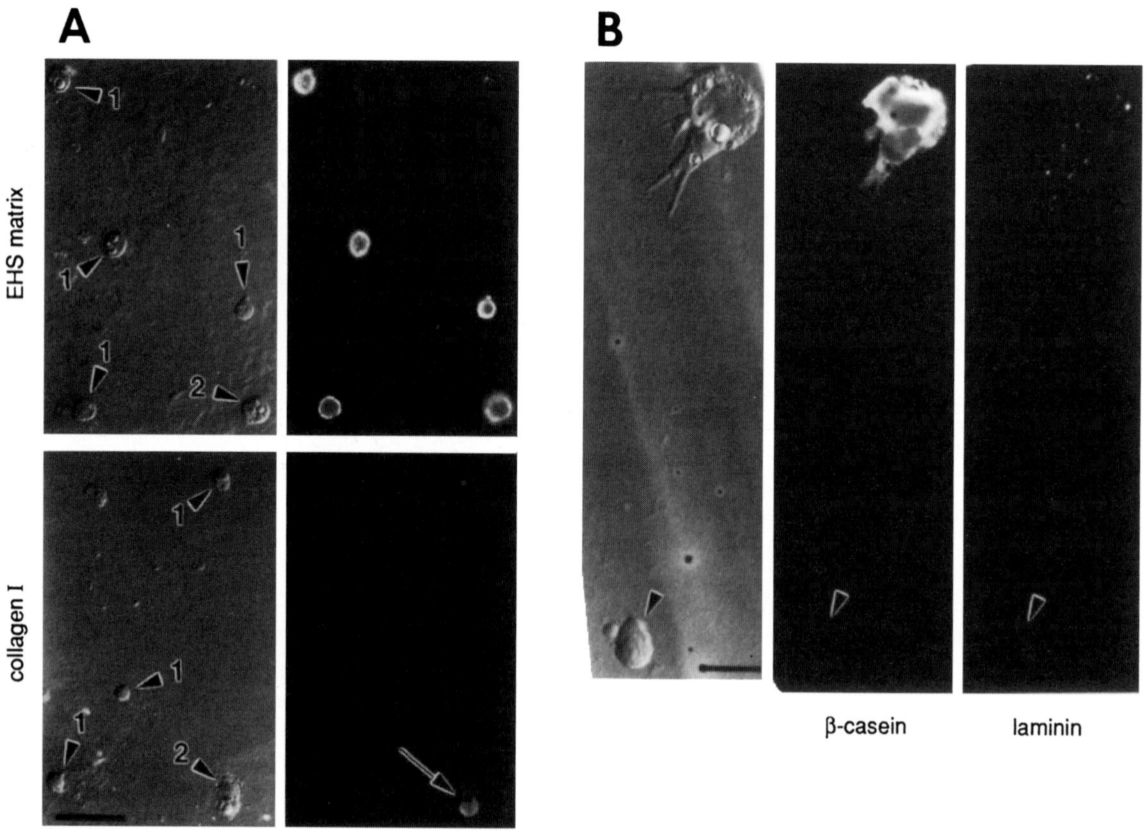

Figure 5. Single cells cultured inside collagen I do not express β-casein. (*A*) 25-μm cryosections of cells embedded inside either EHS matrix or collagen I were stained for their ability to express β-casein. Although the majority of cells inside the basement membrane gel were positive, single cells inside collagen I were not induced to express this gene; positives only existed in cell aggregates, and even then the staining was much weaker than that in EHS matrix embedded cells (*arrow*). These fluorescence micrographs were exposed for equivalent amounts of time. This number of nuclei in each cluster are indicated. (*B*) A section of cells cultured inside collagen I was stained with antibodies for β-casein and laminin. At the top of the field, laminin stipples are visible in close association with a casein-expressing two cell cluster. At the bottom is a single cell (*arrowhead*) that expressed neither laminin nor β-casein. Bars: (*A*) 50 μm; (*B*) 20 μm.

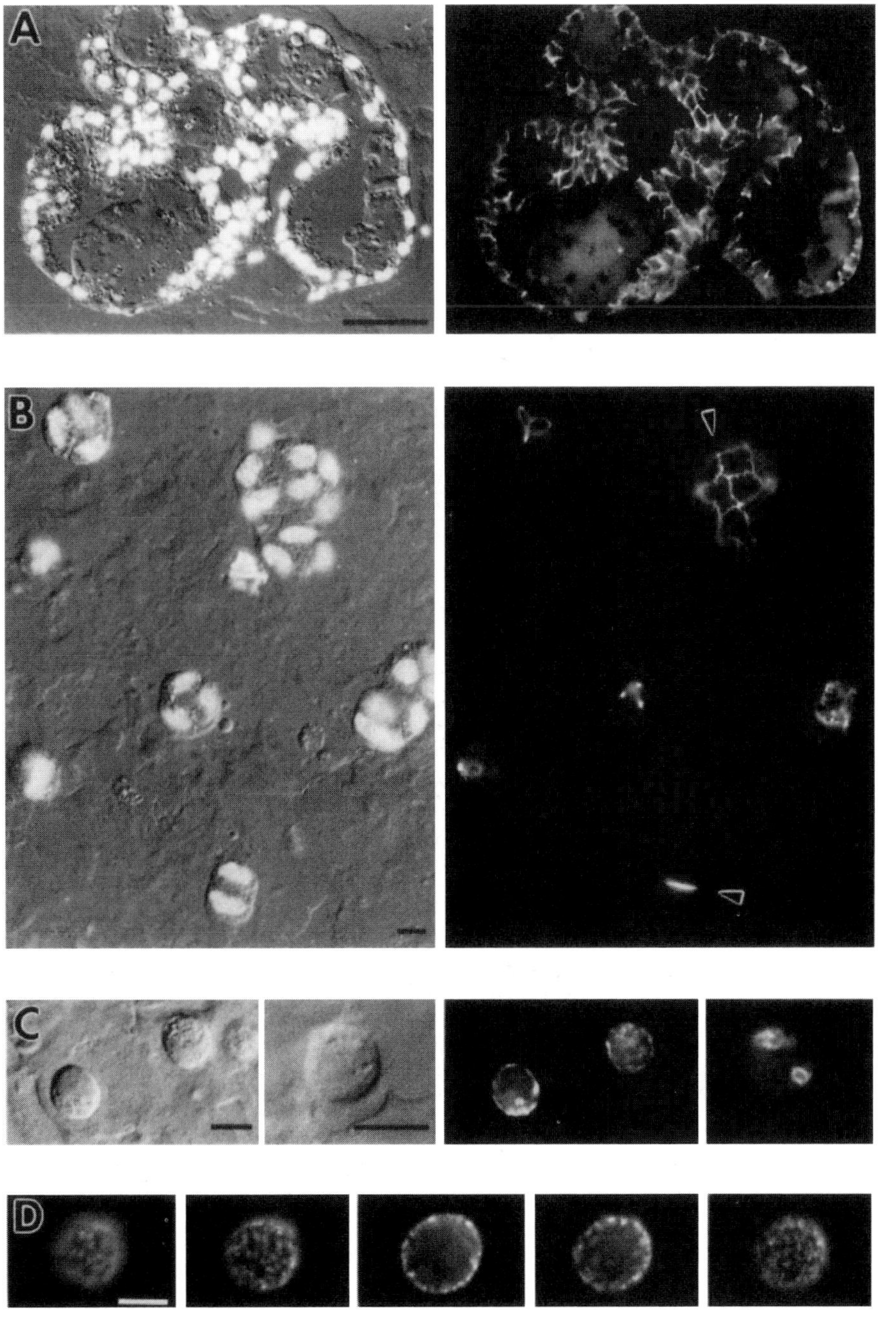

Figure 6. Distribution of E-cadherin on mammary cells in EHS matrix culture. (*A*) The highly polarized epithelium of cells cultured as 'alveoli' on a reconstituted basement membrane segregated E-cadherin mostly to lateral cell surfaces. Here, a 4-μm cryosection of primary mammary cells cultured on EHS matrix for six days was stained by indirect immunofluorescence with ECCD-2, a mAb specific for E-cadherin. In the Nomarski image, DAPI-stained nuclei are visible. Compare the morphology of this structure with that shown in Fig. 2. (*B–D*) 25-μm cryosections of embedded mammary cells were stained by immunofluorescence with ECCD-2. (*B*) Note that E-cadherin accumulated at regions of cell–cell interaction (*arrowheads*). (*C*) Single cells show patching of E-cadherin to several sites around each cell. (*D*) Such patching is seen more clearly in another view of a single cell, photographed at different levels of focus. Bars: (*A*) 50 μm; (*B–D*) 10 μm.

thermore, cells on plastic dishes that had not been exposed to growth factors for three days, and then suspension cultured in factor-depleted EHS matrix in the absence of serum for a further six days, were still induced to express β-casein; indeed, in one experiment 93% of single cells and 100% of cell clusters were casein-positive under these conditions (compare with Table I).

These results suggest that apart from a requirement for lactogenic hormones, biochemical differentiation of mammary cells is independent of signals from growth factors, although it is not yet possible to design experiments that rule out involvement of autocrine growth factors which may be ECM bound. However, if factors are produced endogenously they are induced to do so by the basement membrane matrix, since collagen I is not effective at causing single cells to function. The results presented below demonstrate further that basement membrane itself provides direct signals for tissue-specific gene expression, since ECM receptors are involved in this process.

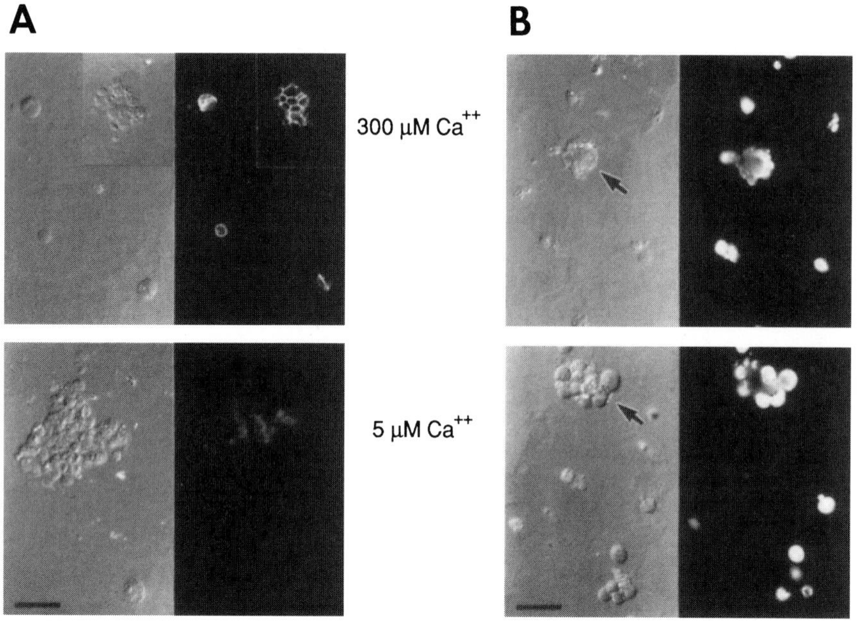

Figure 7. Single cells can be induced to express β-casein even in 5 μM Ca^{2+}. (*A*) E-cadherin was localized to regions of contact in mammary cells suspension-cultured in EHS matrix, but only in normal medium with 300 μM Ca^{2+}. In medium containing only 5 μM Ca^{2+} it was almost completely absent, even in large clusters of cells. The fluorescence micrographs were exposed for equivalent amounts of time. (*B*) In contrast β-casein is readily visible, even at low concentrations of Ca^{2+}. In two experiments, the percentage of positive single cells was similar under both conditions. Note that while the cell clusters inside normal EHS matrix coalesced, those cultured in 5 μM Ca^{2+} often failed to do so (*arrows*; these clusters each contained more than 10 cells). Bars, 50 μm.

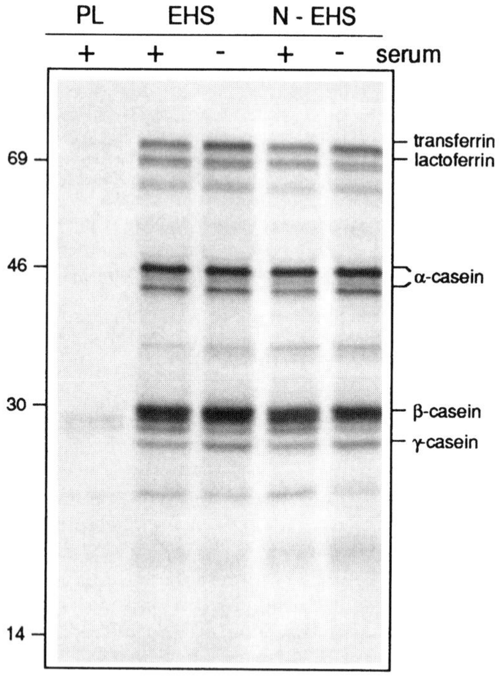

Figure 8. Milk protein expression is induced by basement membrane in the absence of matrix-associated soluble factors. In the presence of lactogenic hormones, growth factor–depleted EHS matrix (*N-EHS*) was permissive for the expression of a full range of milk proteins by primary mammary epithelial cells. Differentiation was not adversely affected by the complete absence of serum. Here, cell fractions, including material vectorially secreted into the alveolar lumina, were isolated from pulse-labelled cultures, immunoprecipitated with a milk-specific polyclonal antiserum, and the milk proteins were separated by SDS-PAGE. Size markers (in kD) are indicated.

Integrins Are Involved in the Control of Mammary Differentiation

Little is known about the types of integrins in mammary cells. One receptor for the major basement membrane component laminin is the α_6 integrin heterodimer (Sonnenberg et al., 1990), which in mammary gland (not shown) and in cultured alveoli (Fig. 9 *A*) is predominantly basal. In EHS matrix suspension culture, this ECM receptor was localized uniformly around the periphery of single cells and clusters, indicative of an interaction with the surrounding substratum (Fig. 9 *B*). In monolayer cultures on plastic dishes there was faint intercellular staining although none was seen basally (not shown), reflecting the absence of a basement membrane (Streuli and Bissell, 1990).

An α_6-specific mAb (GoH3), shown previously to block binding of primary mammary epithelial cells (not shown) and other cell types to laminin (Sonnenberg et al., 1988, 1990; Hall et al., 1990; Lotz et al., 1990), had no effect on the ability of single cells to produce β-casein (Table IV, experiment *1*). Since we did not know which specific integrins are involved in the interaction between mammary cells and ECM, we took advantage of an antibody that could broadly block such interactions via β_1 integrins (Tomaselli et al., 1987, 1988).

When single mammary cells were cultured in the presence of such an anti-integrin antibody (anti ECM-R; Knudsen et al., 1981; Damsky et al., 1982), β-casein synthesis was reduced dramatically (Table IV, experiments *2–4*). A few single cells and individual cells within clusters escaped this block, possibly due to antibody accessibility; however, even those cells that were positive, stained weakly in comparison to those in the control experiments. In single, embedded CID-9 cells, β-casein synthesis was blocked completely (Table IV, experiment *5*), showing that integrin-mediated signalling was not restricted to secondary cultures of mammary

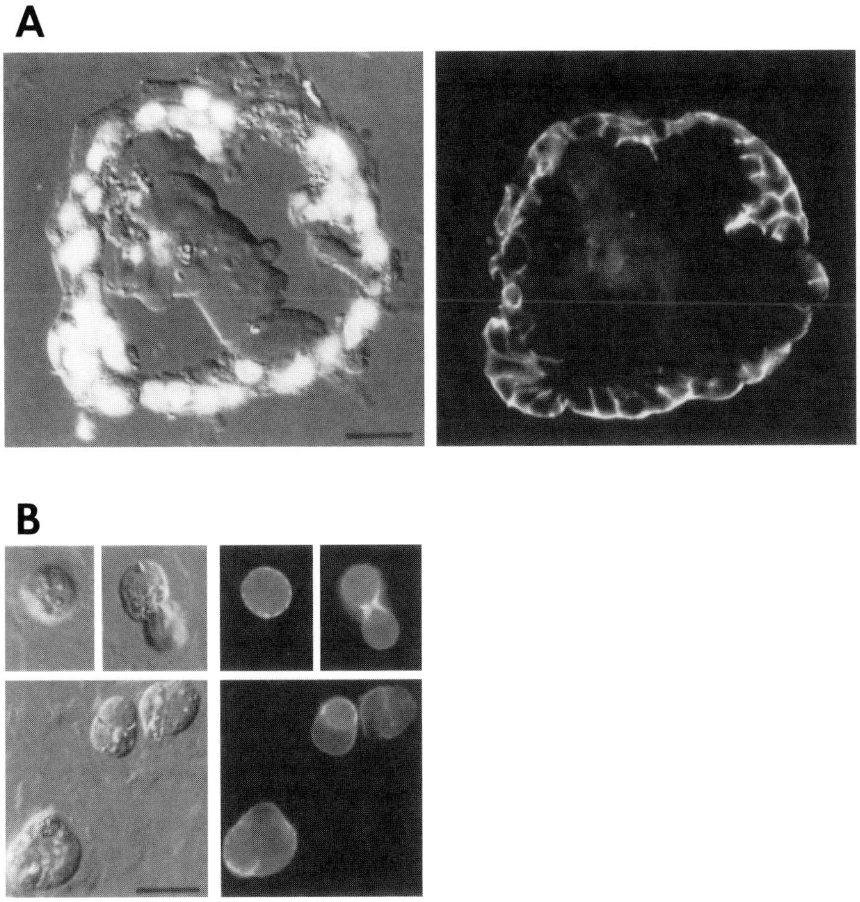

Figure 9. α_6 Integrin in mammary epithelial cells. (*A*) The basal distribution of α_6 integrin is evident in this 4-μm cryosection of primary mammary cells cultured for six days on EHS matrix. The section was stained by indirect immunofluorescence with GoH3, a mAb specific for α_6 integrin. This integrin was also localized to some extent at regions of cell–cell contact. Compare the morphology of this alveolar structure with that shown in Fig. 2. (*B*) α_6 integrin was located at peripheral regions of mammary cells embedded inside EHS matrix. Here, 25-μm cryosections were stained by immunofluorescence with GoH3. Bars, 20 μm.

epithelia, but applied also to functional cell strains. The anti ECM-R antibody also inhibited β-casein expression in the few positive single cells cultured in the collagen I gel system. In contrast, antibodies that bound to and interfered with the normal activity of other surface molecules such as E-cadherin (ECCD-1; Yoshida-Noro et al., 1984) had no effect on the expression of β-casein in suspension cultured cells (Table IV, experiments *4* and *5*). Since both this and the GoH3 antibodies failed to interfere with differentiation, their presence within the EHS matrix (and indeed, on the surface of embedded cells) was confirmed by immunostaining (not shown). Finally, to show that the anti ECM-R antibody was specific and did not disrupt the normal pattern of cell–cell contact in clusters, the distribution of E-cadherin was examined and was found to be similar to that in controls.

Discussion

Two major conclusions can be derived from this work. First, the basement membrane has a central role in controlling tissue-specific gene expression in mammary epithelial cells, and integrins are involved in this signal transduction. Second, cell–cell interaction is not necessary for the synthesis of β-casein in the presence of a basement membrane, but such interaction enhances β-casein production in a stromal matrix.

Basement Membrane Directs β-Casein Expression

Evidence that basement membrane itself is sufficient to direct expression of β-casein in the presence of lactogenic hormones comes from two observations. (*a*) Single and separated mammary cells, which have lost the capacity to synthesize milk proteins after being cultured on plastic dishes, could be induced to reexpress this gene in the presence of a laminin-rich, basement membrane matrix, but not in the presence of other physiological substrata such as collagen I. ECM-bound growth factors do not appear to be the determining element in this process; support for this comes from our work with collagen I cultures where β-casein is expressed only if an endogenously produced basement membrane is deposited after floatation of the collagen I gel (Streuli and Bissell, 1990). However, our studies do not preclude a role for autocrine, ECM-bound growth factors; (*b*) an anti-integrin antiserum that interferes with the cell–ECM interaction blocks biochemical differentiation in mammary epithelial cells. The antibody used in this study is known to block attachment of cells to laminin, collagen IV, and fibronectin substrata by interfering with β_1 integrin function (Tomaselli et al., 1987, 1988), indicating that this class of ECM receptors transduces the signals controlling β-casein expression.

It is now clear that a major control point in the expression of β-casein occurs at the transcriptional level. Convincing

Table IV. Effect of Antibodies on Differentiation of Single Cells or Cell Clusters Cultured for Six Days inside EHS Matrix

Experiment	Antibody	Percentage of β-casein staining			
		Single cells	2 cell clusters	>4 cell clusters	Total
1	None	86 (106)	94 (51)	100 (14)	(200)
	GoH3	83 (83)	85 (40)	92 (13)	(162)
2	None	51 (169)	84 (45)	91 (33)	(281)
	Anti ECM-R‡	11¶(187)	17¶(18)	65¶(23)	(240)
3	None	71 (98)	94 (34)	95 (22)	(180)
	Anti ECM-R	9¶(96)	15¶(20)	75¶(8)	(133)
4	Goat serum	77 (62)	83 (29)	86 (7)	(119)
	ECCD-1§	74 (62)	88 (25)	100 (5)	(118)
	Anti ECM-R	16¶(122)	23¶(64)	43¶(7)	(205)
5*	None	27 (205)	39 (105)	89 (104)	(508)
	ECCD-1	24 (108)	32 (53)	79 (56)	(268)
	Anti ECM-R	0 (210)	0 (62)	30¶(113)	(470)

* CID-9 cells were used in this experiment.
‡ Anti ECM-R is a pan-specific anti β_1 integrin antibody.
§ ECCD-1 is a function-blocking anti E-cadherin antibody; in our hands, ECCD-1 caused cells cultured as monolayers on plastic dishes to separate from each other within a few hours.
¶ Clusters were scored positive even if only one of the group stained; the level of staining in single cells was often lower than in the controls.

evidence exists for the presence of an ECM-dependent regulatory element in the upstream region of the β-casein gene (Schmidhauser et al., 1990; and unpublished results). Thus, the consequence of an ECM-integrin interaction in these cells is an intracellular cascade of events leading to a dramatic increase in the expression of tissue-specific genes.

Our results widen the spectrum of biological processes that are integrin mediated (Albeda and Buck, 1990; Ruoslahti and Giancotti, 1990). Two other examples of altered gene expression resulting from signal transduction through integrins have been documented (Adams and Watt, 1989; Werb et al., 1989). In addition, many developmental events in invertebrates (Leptin et al., 1989; Zusman et al., 1990; Volk et al., 1990) and in vertebrates involve integrins. These include the migration of neural crest cells (Bronner-Fraser, 1986) and myoblasts (Jaffredo et al., 1988), myogenesis (Menko and Boettiger, 1987; Volk et al., 1990), trophoblast outgrowth (Sutherland et al., 1988), kidney development (Sorokin et al., 1990), and the outgrowth of neurites in developing neural tissue (Reichardt et al., 1989).

A number of integrins are candidates for the signal transduction process; several different integrin heterodimers for example, are receptors for epitopes contained within basement membranes (Mercurio, 1990). Despite the basal distribution of α_6 integrin in the mammary cultures, our results suggest that this integrin may not be responsible for basement membrane-mediated signalling; but given the fact that we have tested only one α_6 integrin antibody, we cannot rule out its involvement at this stage. $\alpha_3\beta_1$ integrin is an alternative possibility since it is recognized by the anti ECM-R antiserum in PC 12 cells (Tomaselli et al., 1987; Reichardt et al., 1989), and is expressed in our mammary epithelial cultures (unpublished data). However, the precise identity of the signalling molecule will have to await the production of additional rodent-specific anti-integrin reagents.

β-Casein Expression Is Independent of Cell-Cell Contact

The ability to express β-casein in cells that were separated from each other, and suspended within a basement membrane matrix, implies that some features of differentiation are independent of cell-cell contact. This conclusion is supported by our experiments in which casein production: (*a*) could still be induced in the presence of only 5 μM Ca^{2+}; but (*b*) was largely inhibited by anti integrin antibodies even in clusters of cells, although contact molecules such as E-cadherin were located normally. In all these studies cells were plated on average at least five cell diameters apart, and at both this and at lower cell densities β-casein was synthesized.

In one other system, that of keratinocytes, terminal differentiation of individual cells can occur in the absence of cell-cell contact (Green, 1977; Watt et al., 1988). In this case, however, it is the detachment from the substratum and a reduction in the ability of integrins to bind certain ECM components that triggers a phenotypic response (Adams and Watt, 1990). Thus, the situation in stratified epithelium contrasts with that in a glandular tissue such as mammary gland.

Cell-cell contact is required, however, for full differentiation in secretory epithelial monolayers. Indeed, contact is essential for the onset of morphological and functional polarity (Rodriguez-Boulan and Nelson, 1989; Gumbiner, 1990), and has been implicated previously for mammary differentiation. Cells cultured on collagen I gels for example, only synthesize milk proteins when the gels are floated into the medium, an event that triggers profound changes in cytostructure and induces a high degree of cell-cell interaction (Emerman and Pitelka, 1977). Although this results in the facility for secretion, it also results in the deposition of an intact basement membrane (Streuli and Bissell, 1990). As is the case with MDCK cells (Wang et al., 1990), this would contribute to full polarization of the monolayer. From the data presented in this paper, we now believe that it is this basement membrane which provides the critical signals for biochemical differentiation in mammary cells. Accordingly, in collagen I suspension culture where there is no exogenous basement membrane, essentially only those cells in clusters are able to express β-casein. Thus, in a stromal environment cell–cell contact does appear to be necessary for function, and indeed, the cell-associated laminin (and other ECM components) in such clusters may be relevant to the observed tissue-specific gene expression.

β-Casein Expression Is Independent of Morphological Polarity

Single cells embedded inside the basement membrane matrix remain rounded and do not appear to be polarized at the morphological level. β-casein is distributed uniformly and does not show any apical–basal segregation. Neither E-cadherin nor α_6 integrin appear to be directed to polar (lateral or basal) regions; indeed the latter is distributed evenly around the cell surface. Since the formation of a polarized epithelium depends upon the segregation of plasma membrane into distinct regions that define apical and basal surfaces (Rodriguez-Boulan and Nelson, 1989; Gumbiner, 1990), our results indicate that the expression of tissue-specific genes such as β-casein can be dissociated functionally from the establishment of morphological polarity.

Two further lines of evidence support this conclusion. Firstly, mammary cells cultured as flat monolayers on plastic dishes establish certain elements of polarity, yet such cultures largely lose their capacity to express milk proteins: E-cadherin, required for polarization of basal-lateral membranes (Nelson et al., 1990), and the Na^+, K^+-ATPase (McNeill et al., 1990) were both present on the lateral cell surfaces of our primary monolayer cultures; ZO-1, essential for the formation of tight junctions (Siliciano and Goodenough, 1988), appeared at the apical region between adjacent cells (our unpublished data); further, an actin cytoskeleton with associated proteins such as PAS-O could be organized apically in a mammary cell line cultured on plastic dishes (Parry et al., 1990). Secondly, in 5 μM Ca^{2+} medium, E-cadherin was not evident in cultures of cells embedded within EHS matrix. These cells therefore lacked polarity even within the cell clusters, yet β-casein was still synthesized.

Formation of a Fully Functional Epithelium

Our results allow us to propose a hierarchical model for the establishment and maintenance of a fully differentiated mammary epithelium. Interactions between epithelial cells provide signals for both the correct organization of intracellular cytostructure and the deposition of a basement membrane. In turn, the basement membrane contributes to the generation of a fully polarized, and tight, epithelium, enclosing a lumen into which milk can be secreted. Ultimately, the basement membrane itself provides molecular signals that regulate tissue-specific gene expression and, in mammary gland, allows the synthesis and secretion of β-casein.

The authors are grateful to Drs. Caroline Damsky, Charlotte Kaetzel, Stephen Kennel, Karen Knudsen, James Nelson, Arnoud Sonnenberg, and Masatoshi Takeichi for the kind gifts of antibodies, and to Drs. Caroline Damsky, James Nelson, Gordon Parry, Zena Werb, and our colleagues in the Laboratory of Cell Biology at Lawrence Berkeley Laboratory for critical review of the manuscript. C. H. Streuli is a Fogarty International Research Fellow.

This work was supported by the Health Effects Research Division, Office of Health and Environmental Research, US Department of Energy, Contract DE-AC-03-76SF00098, and a gift for research from Monsanto Company, Inc.

Received for publication 20 June 1991 and in revised form 5 August 1991.

References

Adams, J. C., and F. M. Watt. 1989. Fibronectin inhibits the terminal differentiation of human keratinocytes. *Nature (Lond.).* 340:307–309.

Adams, J. C., and F. M. Watt. 1990. Changes in keratinocyte adhesion during terminal differentiation: reduction in fibronectin binding precedes $\alpha_5\beta_1$ integrin loss from the cell surface. *Cell.* 63:425–435.

Aggeler, J., J. Ward, L. M. Blackie, M. H. Barcellos-Hoff, C. H. Streuli, and M. J. Bissell. 1991. Cytodifferentiation of mouse mammary epithelial cells cultured on a reconstituted basement membrane reveals striking similarities to development *in vivo*. *J. Cell Sci.* 99:407–417.

Albelda, S. M., and C. A. Buck, 1990. Integrins and other cell adhesion molecules. *FASEB (Fed. Am. Soc. Exp. Biol.) J.* 4:2868–2880.

Barcellos-Hoff, M. H., J. Aggeler, T. G. Ram, and M. J. Bissell. 1989. Functional differentiation and alveolar morphogenesis of primary mammary cultures on reconstituted basement membrane. *Development.* 105:223–235.

Bissell, M. J., and M. H. Barcellos-Hoff. 1987. The influence of extracellular matrix on gene expression: is structure the message? *J. Cell Sci. Suppl.* 8:327–343.

Bissell, M. J., and G. Hall. 1987. Form and function in the mammary gland: the role of extracellular matrix. *In* "The Mammary Gland". M. Neville and C. Daniel, editors. Plenum Press Publishing Corp. New York. 97–146.

Bradley, R. S., and A. M. Brown. 1990. The proto-oncogene int-1 encodes a secreted protein associated with the extracellular matrix. *EMBO (Eur. Mol. Biol. Organ.) J.* 9:1569–1575.

Bronner-Fraser, M. 1986. An antibody to a receptor for fibronectin and laminin perturbs cranial neural crest development in vivo. *Dev. Biol.* 117:526–536.

Chen, L. H., and M. J. Bissell. 1989. A novel regulatory mechanism for whey acidic protein gene expression. *Cell Reg.* 1:45–54.

Damsky, C. H., K. A. Knudsen, and C. A. Buck. 1982. Integral membrane glycoproteins related to cell substratum adhesion in mammalian cells. *J. Cell Biochem.* 18:1–13.

Emerman, J. T., and D. R. Pitelka. 1977. Maintenance and induction of morphological differentiation in dissociated mammary epithelium on floating collagen membranes. *In Vitro.* 13:316–328.

Green, H. 1977. Terminal differentiation of cultured human epidermal cells. *Cell.* 11:405–416.

Gumbiner, B. 1990. Generation and maintenance of epithelial cell polarity. *Curr. Opin. Cell Biol.* 2:881–887.

Gumbiner, B., B. Stevenson, and A. Grimaldi. 1988. The role of the cell adhesion molecule uvomorulin in the formation and maintenance of the epithelial junctional complex. *J. Cell Biol.* 107:1575–1587.

Hall, D. E., L. F. Reichardt, E. Crowley, B. Holley, H. Moezzi, A. Sonnenberg, and C. H. Damsky. 1990. The alpha 1/beta 1 and alpha 6/beta 1 integrin heterodimers mediate cell attachment to distinct sites on laminin. *J. Cell Biol.* 110:2175–2184.

Harlow, E., and D. Lane. 1988. Antibodies: a laboratory manual. Cold Spring Harbor Laboratory, Cold Spring Harbor, NY. 726 pp.

Jaffredo, T., A. F. Horwitz, C. A. Buck, P. M. Rong, and F. Dieterlen-Lievre. 1988. Myoblast migration is specifically inhibited in the chick embryo by grafted CSAT hybridoma cells secreting an anti-integrin antibody. *Development.* 103:431–446.

Kaetzel, C. S., and D. B. Ray. 1984. Immunochemical characterization with monoclonal antibodies of three major caseins and α-lactalbumin from rat milk. *J. Dairy Sci.* 67:64–75.

Kennel, S. J., L. J. Foote, R. Falcioni, A. Sonnenberg, C. D. Stringer, C. Crouse, and M. E. Hemler. 1989. Analysis of the tumor-associated antigen TSP-180. *J. Biol. Chem.* 264:15515–15521.

Klagsbrun, M. 1990. The affinity of fibroblast growth factors for heparin; FGF-heparan sulfate interactions in cells and extracellular matrix. *Curr. Opin. Cell Biol.* 2:857–863.

Kleinman, H. K., M. L. McGarvey, J. R. Hassell, V. L. Star, F. B. Cannon, G. W. Laurie, and G. R. Martin. 1986. Basement membrane complexes with biological activity. *Biochemistry.* 25:312–318.

Knudsen, K. A., P. E. Rao, C. H. Damsky, and C. A. Buck. 1981. Membrane glycoproteins involved in cell-substratum adhesion. *Proc. Natl. Acad. Sci. USA.* 78:6071–6075.

Kramer, R. H., K. A. McDonald, E. Crowley, D. M. Ramos, and C. H. Damsky. 1990. Melanoma cell adhesion to basement membrane mediated by integrin-related complexes. *Cancer Res.* 49:393–402.

Lee, E. Y.-H., G. Parry, and M. J. Bissell. 1984. Modulation of secreted proteins of mouse mammary epithelial cells by the collagenous substrata. *J. Cell Biol.* 98:146–155.
Lee, E. Y.-H., W.-H. Lee, C. S. Kaetzel, G. Parry, and M. J. Bissell. 1985. Interaction of mouse mammary epithelial cells with collagen substrata: Regulation of casein gene expression and secretion. *Proc. Natl. Acad. Sci. USA.* 82:1419–1493.
Leptin, M., T. Bogaert, R. Lehmann, and M. Wilcox. 1989. The function of PS integrins during drosophila embryogenesis. *Cell.* 56:401–408.
Li, M. L., J. Aggeler, D. A. Farson, C. Hatier, J. Hassell, and M. J. Bissell. 1987. Influence of a reconstituted basement membrane and its components on casein gene expression and secretion in mouse mammary epithelial cells. *Proc. Natl. Acad. Sci. USA.* 84:136–140.
Lotz, M. M., C. A. Korzelius, and A. M. Mercurio. 1990. Human colon carcinoma cells use multiple receptors to adhere to laminin: involvement of $\alpha_6\beta_4$ and $\alpha_2\beta_1$ integrins. *Cell Reg.* 1:249–257.
McNeill, H., M. Ozawa, R. Kemler, and W. J. Nelson. 1990. Novel function of the cell adhesion molecule uvomorulin as an inducer of cell surface polarity. *Cell.* 62:309–316.
Menko, S. A., and D. Boettiger. 1987. Occupation of the extracellular matrix receptor, integrin, is a control point for myogenic differentiation. *Cell.* 51:51–57.
Mercurio, A. M. 1990. Laminin: multiple forms, multiple receptors. *Curr. Opin. Cell Biol.* 2:845–849.
Nagafuchi, A., Y. Shirayoshi, K. Okazaki, K. Yasuda, and M. Takeichi. 1987. Transformation of cell adhesion properties by exogenously introduced E-cadherin cDNA. *Nature (Lond.).* 329:341–343.
Nelson, W. J., E. M. Shore, A. Z. Wang, and R. W. Hammerton. 1990. Identification of a membrane-cytoskeletal complex containing the cell adhesion molecule uvomorulin (E-cadherin), ankyrin, and fodrin in Madin-Darby canine kidney epithelial cells. *J. Cell Biol.* 110:349–357.
Parry, G., J. C. Beck, L. Moss, J. Bartley, and G. K. Ojakian. 1990. Determination of apical membrane polarity in mammary epithelial cell cultures: the role of cell-cell, cell-substratum, and membrane-cytoskeleton interactions. *Exp. Cell Res.* 188:302–311.
Rathjen, P. D., S. Toth, A. Willis, J. K. Heath, and A. G. Smith. 1990. Differentiation inhibiting activity is produced in matrix-associated and diffusible forms that are generated by alternate promoter usage. *Cell.* 62:1105–1114.
Reichardt, L. F., J. L. Bixby, D. E. Hall, M. J. Ignatius, K. M. Neugebauer, and K. J. Tomaselli. 1989. Integrins and cell adhesion molecules: neuronal receptors that regulate axon growth on extracellular matrices and cell surfaces. *Dev. Neuroscience.* 11:332–347.
Rodriguez-Boulan, E., and W. J. Nelson. 1989. Morphogenesis of the polarized epithelial cell phenotype. *Science (Wash. DC).* 245:718–725.
Ruoslahti, E., and F. G. Giancotti. 1990. Integrins and tumor cell dissemination. *Cancer Cells.* 1:119–126.
Ruoslahti, E., and Y. Yamaguchi. 1991. Proteoglycans as modulators of growth factor activities. *Cell.* 64:867–869.
Schmidhauser, C., M. J. Bissell, C. A. Myers, and G. F. Casperson. 1990. Extracellular matrix and hormones transcriptionally regulate bovine β-casein 5′ sequences in stably transfected mouse mammary cells. *Proc. Natl. Acad. Sci. USA.* 87:9118–9122.
Shirayoshi, Y., A. Nose, K. Iwasaki, and M. Takeichi. 1986. N-linked oligosaccharides are not involved in the function of a cell-cell binding glycoprotein E-cadherın. *Cell Struct. Funct.* 11:245–252.
Siliciano, J. D., and D. A. Goodenough. 1988. Localization of the tight junction protein, ZO-1, is modulated by extracellular calcium and cell-cell contact in Madin-Darby canine kidney epithelial cells. *J. Cell Biol.* 107:2389–2399.
Sonnenberg, A., H. Janssen, F. Hogervorst, J. Calafat, and J. Hilgers. 1987. A complex of platelet glycoproteins Ic and IIa identified by a rat monoclonal antibody. *J. Biol. Chem.* 262:10376–10383.
Sonnenberg, A., P. W. Modderman, and F. Hogervorst. 1988. Laminin receptor on platelets is the integrin VLA-6. *Nature (Lond.).* 336:487–489.
Sonnenberg, A., C. J. Linders, P. W. Modderman, C. H. Damsky, M. Aumailley, and R. Timpl. 1990. Integrin recognition of different cell-binding fragments of laminin (P1, E3, E8) and evidence that alpha 6 beta 1 but not alpha 6 beta 4 functions as a major receptor for fragment E8. *J. Cell Biol.* 110:2145–2155.
Sorokin, L., A. Sonnenberg, M. Aumailley, R. Timpl, and P. Ekblom. 1990. Recognition of the laminin E8 cell-binding site by an integrin possessing the alpha 6 subunit is essential for epithelial polarization in developing kidney tubules. *J. Cell Biol.* 111:1265–1273.
Sutherland, A. E., P. G. Calarco, and C. H. Damsky. 1988. Expression and function of cell surface extracellular matrix receptors in mouse blastocyst attachment and outgrowth. *J. Cell Biol.* 106:1331–1348.
Stoker, A., C. H. Streuli, M. Martins-Green, and M. J. Bissell. 1990. Designer microenvironments for the analysis of cell and tissue function. *Curr. Opin. Cell Biol.* 2:864–874.
Streuli, C. H., and M. J. Bissell. 1990. Expression of extracellular matrix components is regulated by substratum. *J. Cell Biol.* 110:1405–1415.
Streuli, C. H., and M. J. Bissell. 1991. Mammary epithelial cells, extracellular matrix and gene expression. *In* Regulatory Mechanisms in Breast Cancer. M. E. Lippman and R. Dickson, editors. Kluwer Academic Publishers, Norwell, MA. 365–381.
Takeichi, M. 1991. Cadherin cell adhesion receptors as a morphogenetic regulator. *Science (Wash. DC).* 251:1451–1455.
Talhouk, R. S., J. R. Chin, E. N. Unemori, Z. Werb, and M. J. Bissell. 1991. Proteinases of the mammary gland: developmental regulation in vivo and vectorial secretion in culture. *Development.* 112:439–449.
Taub, M., Y. Wang, T. M. Szczesny, and H. K. Kleinman. 1990. Epidermal growth factor or transforming growth factor α is required for kidney tubulogenesis in matrigel cultures in serum-free medium. *Proc. Natl. Acad. Sci. USA.* 87:4002–4006.
Tomaselli, K. J., C. H. Damsky, and L. F. Reichardt. 1987. Interactions of a neuronal cell line (PC12) with laminin, collagen IV, and fibronectin: identification of integrin-related glycoproteins involved in attachment and process outgrowth. *J. Cell Biol.* 105:2347–2358.
Tomaselli, K. J., C. H. Damsky, and L. F. Reichardt. 1988. Purification and characterization of mammalian integrins expressed by a rat neuronal cell line (PC12): evidence that they function as alpha/beta heterodimeric receptors for laminin and type IV collagen. *J. Cell Biol.* 107:1241–1252.
Vigny, M., M. P. Ollier-Hartmann, M. Lavigne, N. Fayein, J. C. Jeanny, M. Laurent, and Y. Courtois. 1988. Specific binding of basic fibroblast growth factor to basement membrane-like structures and to purified heparan sulfate proteoglycan of the EHS tumor. *J. Cell. Physiol.* 137:321–328.
Volk, T., L. I. Fessler, and J. H. Fessler. 1990. A role for integrin in the formation of sarcomeric cytoarchitecture. *Cell.* 63:525–536.
Wang, A. Z., G. K. Ojakian, and W. J. Nelson. 1990. Steps in the morphogenesis of a polarized epithelium. *J. Cell Sci.* 95:137–151.
Watt, F. M., P. W. Jordan, and C. H. O'Neill. 1988. Cell shape controls terminal differentiation of human epidermal keratinocytes. *Proc. Natl. Acad. Sci. USA.* 85:5576–5580.
Werb, Z., P. M. Tremble, O. Behrendtsen, E. Crowley, and C. H. Damsky. 1989. Signal transduction through the fibronectin receptor induces collagenase and stromelysin gene expression. *J. Cell Biol.* 109:877–889.
Yoshida-Noro, C., N. Suzuki, and M. Takeichi. 1984. Molecular nature of the calcium-dependent cell-cell adhesion system in mouse teratocarcinoma and embryonic cells studied with mouse monoclonal antibody. *Dev. Biol.* 101:19–27.
Zusman, S., R. S. Patel-King, C. ffrench-Constant, and R. Hynes. 1990. Requirements for integrins during drosophila development. *Development.* 108:391–402.

Protein Synthesis and Membrane Traffic

Jamieson J.D. and **Palade G.E.** 1967. Intracellular transport of secretory proteins in the pancreatic exocrine cell. II. Transport to condensing vacuoles and zymogen granules. *J. Cell Biol.* **34:** 597–615.

One of the most fruitful approaches to the study of cell biology is the identification of a cell type or organism in which a common or even universal process is carried out with particular speed, extent, or strength. A detailed study of that cell type can then provide insight into the more subtle manifestations of the homologous processes in other cells. The study of zymogen secretion by epithelial cells of the exocrine pancreas is a nice example of this principle. Work by several pioneers (e.g., 1, 2) showed the value of these "acinar" cells for experimental work. Their work allowed zymogen secretion to be dissected effectively, even with the relatively insensitive experimental techniques available in the 1960s. The paper reproduced here took this study to fruition by employing cell fractionation and autoradiography at both the light and electron microscopic levels. It followed the path of labeled amino acids from their incorporation into protein in the endoplasmic reticulum, through the Golgi complex and condensing vacuoles, and on into the zymogen granules. The paper was a technical triumph in its day and remains a model of thoroughness and care in the experimental study of a complex cellular process. More recently, similar results have been obtained with less specialized cells, studied with more sensitive methods (3). Such continuity encourages the belief that we now understand the pathways by which protein secretion is accomplished.

---

1. Siekevitz P. and Palade G.E. 1960. A cytochemical study on the pancreas of the guinea pig. V. In vivo incorporation of leucine-C14 into the chymotrypsinogen of various cell fractions. *J. Biophys. Biochem. Cytol.* **7:** 619–630.
2. Greene L.J., Hirs C.H.W., and Palade G.E. 1963. On the protein composition of bovine pancreatic zymogen granules. *J. Biol. Chem.* **238:** 2054–2070.
3. Quinn P., Griffiths G., and Warren G. 1984. Density of newly synthesized plasma membrane proteins in intracellular membranes. II. Biochemical studies. *J. Cell Biol.* **98:** 2142–2147.

INTRACELLULAR TRANSPORT OF SECRETORY PROTEINS IN THE PANCREATIC EXOCRINE CELL

II. Transport to Condensing Vacuoles and Zymogen Granules

JAMES D. JAMIESON and GEORGE E. PALADE

From The Rockefeller University, New York 10021

ABSTRACT

In the previous paper we described an in vitro system of guinea pig pancreatic slices whose secretory proteins can be pulse-labeled with radioactive amino acids. From kinetic experiments performed on smooth and rough microsomes isolated by gradient centrifugation from such slices, we obtained direct evidence that secretory proteins are transported from the cisternae of the rough endoplasmic reticulum to condensing vacuoles of the Golgi complex via small vesicles located in the periphery of the complex. Since condensing vacuoles ultimately become zymogen granules, it was of interest to study this phase of the secretory cycle in pulse-labeled slices. To this intent, a zymogen granule fraction was isolated by differential centrifugation from slices at the end of a 3-min pulse with leucine-^{14}C and after varying times of incubation in chase medium. At the end of the pulse, few radioactive proteins were found in this fraction; after +17 min in chaser, its proteins were half maximally labeled; they became maximally labeled between +37 and +57 min. Parallel electron microscopic radioautography of intact cells in slices pulse labeled with leucine-^{3}H showed, however, that zymogen granules become labeled, at the earliest, +57 min post-pulse. We assumed that the discrepancy between the two sets of results was due to the presence of rapidly labeled condensing vacuoles in the zymogen granule fraction. To test this assumption, electron microscopic radioautography was performed on sections of zymogen granule pellets isolated from slices pulse labeled with leucine-^{3}H and subsequently incubated in chaser. The results showed that the early labeling of the zymogen granule fractions was, indeed, due to the presence of highly labeled condensing vacuoles among the components of these fractions.

INTRODUCTION

In the preceding paper (Jamieson and Palade, 1967), we have described the functional and morphological characteristics of an in vitro[1] system of guinea pig pancreatic slices in which the secretory proteins produced by the acinar cell can be conveniently labeled with radioactive amino acids in short, well defined pulses. Because of such labeling, the kinetics of intracellular transport of newly synthesized secretory proteins can be profitably investigated. Rough and smooth microsomal fractions were isolated from the incubated slices by isopycnic centrifugation in a continuous sucrose density gradient. The first fraction consists of healed fragments of the rough endoplasmic reticulum (ER), while the second represents primarily the peripheral elements of the Golgi complex. Kinetic experiments performed on these microsomal subfractions after 3 min of pulse labeling and after various times of incubation in a chase medium indicated that newly synthesized secretory proteins were transported from their site of synthesis in the rough ER to cumuli of smooth-surfaced vesicles at the periphery of the Golgi complex.

[1] In this paper, in vitro will refer to incubated pancreatic slices while in vivo will refer to the gland in situ, i.e., in the animal.

Reprinted from THE JOURNAL OF CELL BIOLOGY, 1967, Vol. 34, No. 2, pp. 597–615 *Printed in U.S.A.*

Since in vivo, secretory proteins are transported to, and finally stored in zymogen granules (Keller and Cohen, 1961; Greene, Hirs, and Palade, 1963), experiments were performed on pulse-labeled slices to determine if acinar cells could carry out in vitro the remaining steps of intracellular transport, viz., the transfer of secretory proteins to the condensing vacuoles of the Golgi complex and their storage in zymogen granules prior to discharge. The results of combined cell fractionation and radioautographic studies presented in this paper show that the acinar cells of pancreatic slices can carry out all the operations involved in the secretory process up to and including the discharge of the zymogen granule content into the duct system.

METHODS

Techniques for preparing and pulse-labeling guinea pig pancreatic slices, for isolating cell fractions from homogenized slices, and for assaying these fractions chemically and radiochemically are described in the previous paper (Jamieson and Palade, 1967).

Slices to be used for light and electron microscopic radioautography were pulse labeled in incubation medium containing 200 μc/ml L-leucine-4,5-^{3}H (40 μM) and incubated post-pulse in a chase medium containing L-leucine-^{1}H at a concentration of 2.0 mM (50× excess). Light and electron microscopic radioautography were performed as described by Caro and van Tubergen (1962) on tissue fixed and embedded in Epon as outlined in the preceding paper.

LIGHT MICROSCOPIC RADIOAUTOGRAPHY: 0.5-μ thick Epon sections were mounted on glass slides and coated with Ilford L-4 emulsion. The preparations were exposed for 4–7 days at 4°. After photographic processing, the sections were stained through the emulsion with 1% methylene blue in 1% sodium borate (Richardson, Jarett, and Finke, 1960), a procedure which does not remove or displace the silver grains. The preparations were coated with a clear plastic aerosol spray (Jennings, Farquhar, and Moon, 1959) and were examined and photographed under the oil immersion lens using bright-field illumination. With this technique, a satisfactory image of both tissue structure and radioautographic grains can be obtained.

ELECTRON MICROSCOPIC RADIOAUTOGRAPHY: Thin sections of cells or cell fractions were mounted on grids covered with a carbon-coated formvar film. Ilford L-4 emulsion was applied by the loop method (Caro and van Tubergen, 1962). Exposure times were from 2–4 wk. After photographic processing, the sections were doubly stained with uranyl acetate and lead citrate (Venable and Coggeshall, 1965). This staining schedule does not produce any detectable loss or displacement of silver grains.

MATERIALS: L-leucine-4,5-^{3}H (specific activity 5 c/millimole) was obtained from New England Nuclear Corporation, Boston. Ilford Nuclear Research emulsion L-4 was obtained from Ilford Limited, Ilford, Essex, England.

TABLE I

Chemical Composition of Zymogen Granule Fractions

Source	mg RNA/mg protein	μg phospholipid-P / mg protein
Guinea pig* pancreatic slice	0.01–0.03	1.3–3.3
Guinea pig‡	0.16	—
Dog§	0.005	1.1
Ox‖	0.006	2.2
Mouse¶	0.04	—

* These values are the range from three experiments.
‡ Siekevitz and Palade, 1958 *a* (isolated by discontinuous gradient centrifugation).
§ Hokin, 1955.
‖ Greene et al., 1963.
¶ Van Lancker and Holtzer, 1959.

RESULTS

Intracellular transport in the slice system was investigated in parallel by cell fractionation and radioautography.

Cell Fractionation Studies

ISOLATION AND CHARACTERIZATION OF THE ZYMOGEN GRANULE FRACTION: So that we could follow the transport of secretory proteins into zymogen granules, it was necessary to isolate these granules in a relatively clean cell fraction. The method used is described in the preceding paper (Jamieson and Palade, 1967) and is based, with minor modifications, on that proposed by Hokin (1955) for the isolation of a zymogen granule fraction from the dog pancreas.

CHEMICAL COMPOSITION OF THE ZYMOGEN GRANULE FRACTION: Table I gives the RNA/protein ratio of the zymogen granule fraction isolated by us from guinea pig pancreatic slices and compares it with that of zymogen granule fractions isolated from guinea pig pancreas by discontinuous gradient centrifugation, and from dog, ox, and mouse pancreas by differential centrifugation. RNA was determined in all cases

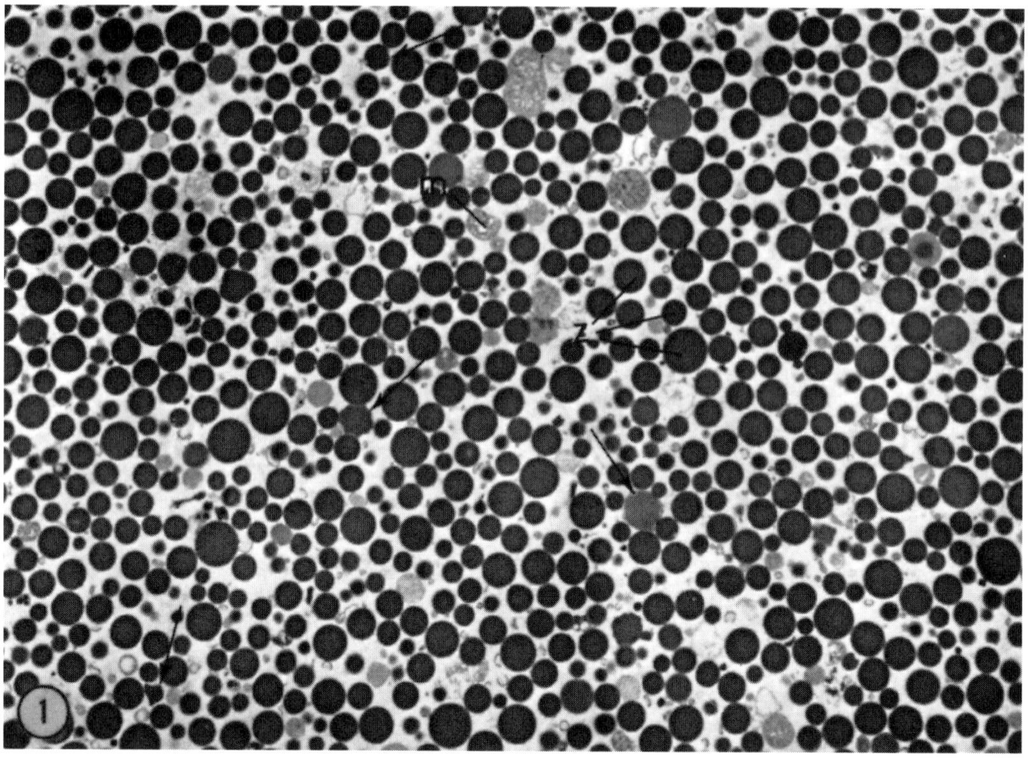

FIGURE 1 Low magnification electron micrograph of the zymogen granule fraction. This field is representative of all levels in the pellet. Zymogen granules (*z*) comprise the majority of structures. A smaller population (~ 5%) of condensing vacuoles (arrows), identified by their scalloped profiles, content of lesser electron opacity, or both features, can be recognized in the fraction. A few mitochondria are seen (*m*). × 3,000.

by the orcinol reaction (Mejbaum, 1939). Insofar as the RNA content reflects residual microsomal contamination, the zymogen granule fraction isolated from pancreatic slices compares favorably to fractions isolated from dog and ox pancreas and is noticeably less contaminated than that isolated by density gradient centrifugation. The phospholipid-phosphorus content of the fractions is given for comparison.

MORPHOLOGY OF THE ZYMOGEN GRANULE FRACTION: Thin sections of fixed and embedded pellets of this fraction, cut to include the entire depth of the pellet in the axis of centrifugation, show that it consists mainly of zymogen granules with a few contaminating mitochondria (Fig. 1) and rough microsomes (Fig. 2). The contaminants were evenly scattered from top to bottom throughout the pellet, which did not show any detectable stratification. Zymogen granules isolated from incubated slices were identical to those prepared from fresh pancreatic tissue (Siekevitz and Palade, 1958 *a*; Greene, Hirs, and Palade, 1963) and were characterized by their spherical shape, highly dense, homogeneous, and apparently amorphous content, and a limiting unit membrane which measures ~100 A in thickness, and which can be seen only when sectioned normally.

In addition to zymogen granules, the pellet contains bodies of comparable size but of irregular shape which gives them angular or scalloped profiles (Figs. 1, 2). Like zymogen granules, they are limited by a unit membrane and have an amorphous, homogeneous content of equal or lower density. On account of their morphology, these bodies are identified as condensing vacuoles of the Golgi complex. They represent ~5% of the total particle population of the zymogen granule fraction.

KINETICS OF LABELING OF THE ZYMOGEN GRANULE FRACTION: To determine the ki-

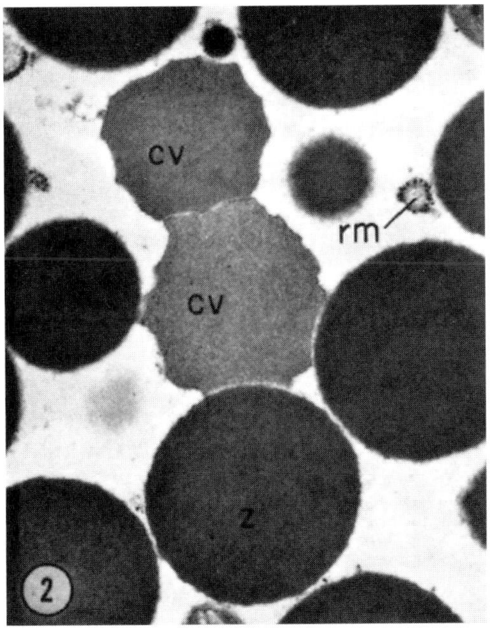

FIGURE 2 Higher magnification field from the zymogen granule fraction. A few contaminating rough microsomes (*rm*) are seen. *cv*, condensing vacuoles; *z*, zymogen granule. × 23,000.

netics of labeling of the zymogen granule fraction, we pulse labeled sets of slices as before (Jamieson and Palade, 1967) with leucine-^{14}C and fractionated them at the end of the pulse and after various times of incubation in chase medium. For these experiments, the fractionation scheme was simplified to the extent that only a zymogen granule fraction and a total microsomal fraction were isolated by differential centrifugation.

1. LABELING OF CELL FRACTIONS DURING 1-HR INCUBATION: The results of a typical short-term experiment are presented in Fig. 3. The specific radioactivity of the proteins of the zymogen granule fraction is negligible initially (i.e., compared to that of microsomal proteins); after +17 min of incubation in chase medium,[2] its proteins are half-maximally labeled and become maximally labeled between +37 and +57 min of incubation. In other cell fractionation experiments not shown, a small but significant amount of labeled protein appeared in the zymogen granule fraction as early as +7 min following the pulse. The specific radioactivity curve of microsomal proteins declines rapidly after the pulse. The unchanging specific activity of total homogenate proteins indicates an effective chase. The specific activity of the proteins of the post-microsomal supernate remains constant throughout incubation, signifying negligible transfer of labeled proteins from cell particulates to the cytoplasmic matrix, i.e., the cell compartment which is the main contributor to the post-microsomal supernate. After +17 min of incubation, labeled proteins begin to appear in the incubation medium.

2. LABELING OF CELL FRACTIONS DURING 3 HR OF INCUBATION: Fig. 4. shows the kinetics of labeling of proteins in cell fractions isolated from sets of pulse-labeled slices incubated in chase medium for +1, +2, and +3 hr. As before, the specific radioactivity curve of microsomal proteins falls rapidly during the chase. Labeling of proteins of the zymogen granule fraction is essentially complete after +1 hr of incubation, rises slightly during the next hr, and then begins to fall. The specific radioactivity of proteins in the incubation medium shows a slow progressive rise during 3 hr. The specific radio-

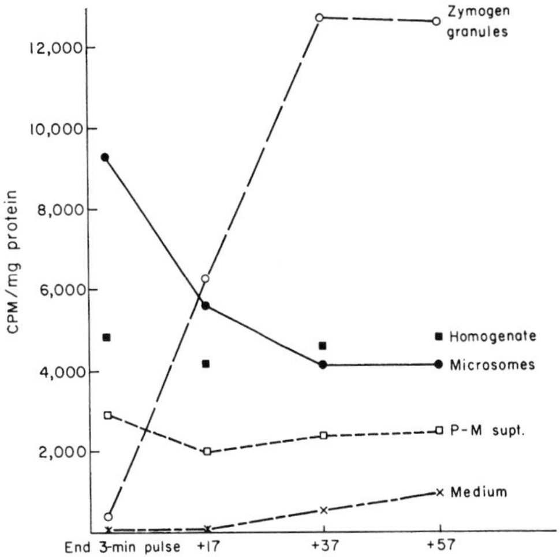

FIGURE 3 Specific radioactivity of proteins in cell fractions isolated from pancreatic slices pulse labeled for 3 min and incubated up to +57 min in chase. Pulse medium: Krebs-Ringer bicarbonate, supplemented with amino acids (minus leucine-^{12}C) and containing 1 μc/ml (4 μM) L-leucine-^{14}C. Chase medium: as above with L-leucine-^{12}C, 0.4 mM. *P-M supt.*, post-microsomal supernate.

[2] The times of incubation in chase medium following 3-min pulse labeling will be referred to hereafter as +17 min of incubation, +37 min of incubation, etc.

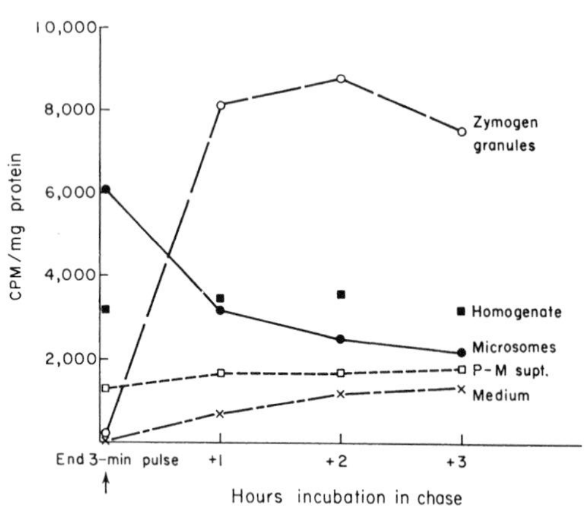

FIGURE 4 Labeling of cell fractions from pulse-labeled slices during 3 hr of incubation. For composition of media, see Fig. 3.

activity of proteins in both the total homogenate and the post-microsomal supernate remains constant throughout incubation.

These cell fractionation data indicate that the zymogen granule fraction begins to accumulate radioactive proteins as early as +7 min after pulse labeling and that it becomes maximally labeled after +37–+57 min of incubation. As was demonstrated in the previous paper (Jamieson and Palade, 1967), secretory proteins are transferred in time from the cisternae of the rough ER to the small vesicles at the periphery of the Golgi complex: the labeled proteins which accumulate in the zymogen granule fraction have come, therefore, from these small vesicles. The kinetics of labeling of zymogen granule fractions in vitro are similar to those of labeling of the same fractions in vivo, as shown by data obtained in the guinea pig (Siekevitz and Palade, 1958 *b*) and the mouse pancreas (Morris and Dickman, 1960). Hansson (1959) found, however, a somewhat slower in vivo labeling of zymogen granule fractions in guinea pig pancreas.

After ~+20 min of incubation, labeled proteins begin to appear in small amounts in the incubation medium. Although some of them probably represent leakage from damaged cells, especially at the early times, the radioautographic studies reported later in this paper indicate that exocrine cells in vitro are able to discharge labeled proteins in the acinar lumina beginning about 1 hr after pulse-labeling. Junqueira, Hirsch, and Rothschild (1955) and Hansson (1959) have reported that in vivo longer times elapse (~150 min) before labeled proteins appear in the pancreatic juice.

Radioautographic Studies

Intracellular transport was also studied by radioautography taking advantage of the fact that the slice system allows heavy labeling of secretory proteins during a short (3-min) pulse with L-leucine-4,5-^{3}H. The conditions are distinctly better than in the whole animal in which neither an equally high dose of labeled amino acids can be administered nor an equally effective chase can be achieved.

The results obtained on slices can be used to complete and correct—if necessary—previous findings on the intact animal. In addition, they represent an independent line of evidence needed for a reliable interpretation of the kinetic data obtained from cell fractionation studies.

INCUBATION CONDITIONS: Slices were pulse labeled for 3 min with L-leucine-^{3}H (200 μc/ml; 40 μM) and incubated post-pulse for further +7, +17, +37, +57, and +117 min in chase medium containing 2.0 mM unlabeled L-leucine (50x excess). Determinations of TCA-soluble and insoluble radioactivity on homogenates of these slices at the end of the pulse and after chase incubation indicated that even with this relatively massive dose of tracer, a true pulse label was obtained.

To ascertain that the incorporated label is retained in the tissue during processing for electron microscopy, we treated specimen blocks from slices at the end of the pulse and at various times during the chase as follows. After fixation, dehydration, and infiltration with Epon, the blocks were hydrolyzed in 1 N NaOH (with heating) and aliquots assayed for protein and radioactivity. It was found that the specific radioactivity of proteins from these blocks was identical to that of proteins from slices labeled under the same conditions but homogenized, precipitated with TCA, and assayed for incorporated radioactivity in the usual way. These results indicate that the tissue-processing schedule extracts negligible protein-associated radioactivity although it effectively washes out soluble, nonincorporated label (cf. Caro and Palade, 1964). We assume, then, that the radioautographic grains mark only label incorporated into protein.

LIGHT MICROSCOPIC RADIOAUTOGRAPHY:

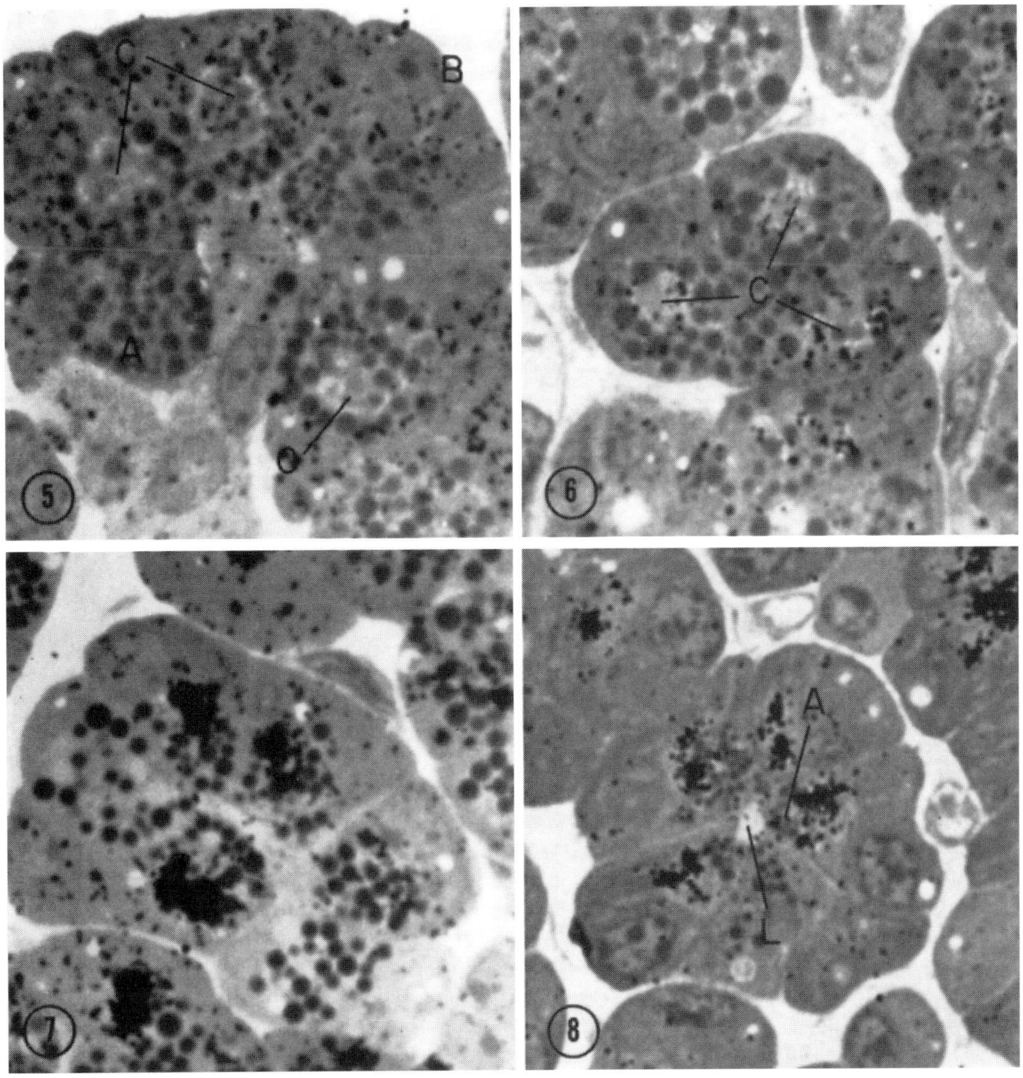

FIGURES 5–9 Light microscopic radioautograph of slices pulse labeled for 3 min with L-leucine-^{3}H (Fig. 5) and incubated in chase for further +17 min (Fig. 6), +37 min (Fig. 7), +57 min (Fig. 8), and +117 min (Fig. 9). *B*, base of cell; *C*, centrosphere region (Golgi complex); *L*, acinar lumen; *D*, collecting duct; *A*, cell apex with zymogen granules. Exposure time 7 days. × 1,750.

Epon sections, 0.5 μ thick, from tissue at the end of the pulse (3-min) and after +7, +17, +37, +57, and +117 min of incubation in chase medium were prepared for radioautography as described under Methods.

At the end of the pulse (Fig. 5), radioautographic grains were generally localized over the basal cytoplasm of the exocrine cell. No grains were seen to overlie either the centrosphere region (Golgi complex) or zymogen granules. The few grains among zymogen granules probably marked rough ER elements scattered in the apical regions of the cell. After +7 min, and especially after +17 min (Fig. 6), however, few grains were seen over the basal cytoplasm; the majority were located centrally in the cell and were often seen to outline the pale-staining centrosphere region. After +37 min, the grains were even more concentrated over the centrosphere region of the cells (Fig. 7) with few grains re-

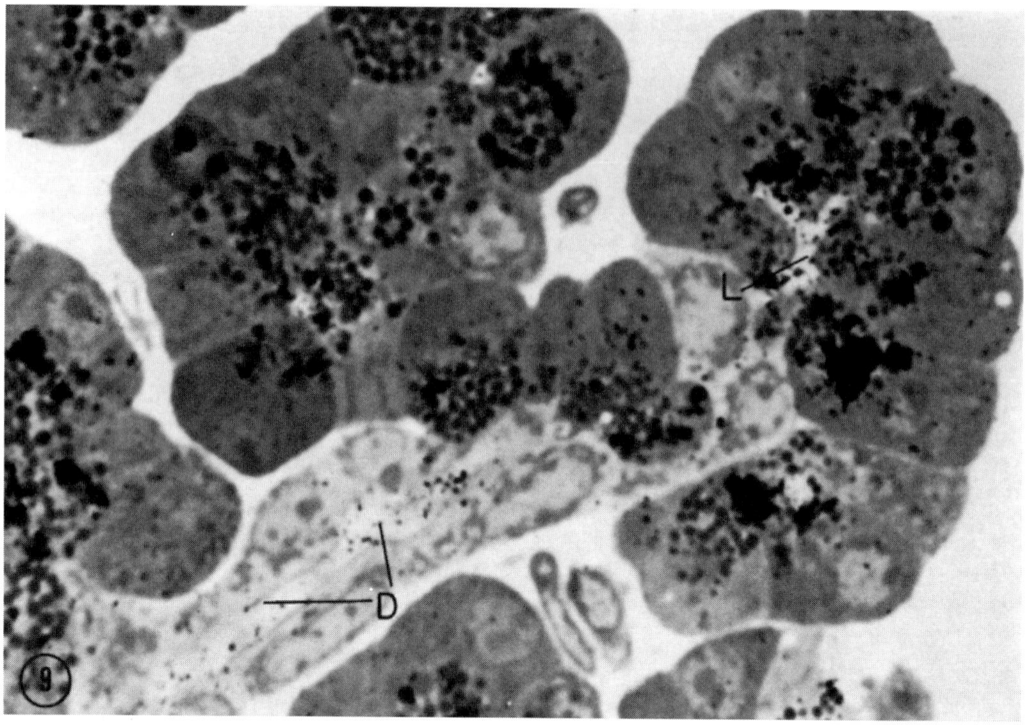

maining over the basal region. At this time, zymogen granules in the apical region of the cells were generally unlabeled. At +57 min (Fig. 8), the zymogen granules in the cell apex began to accumulate label and after +117 min were highly labeled. At this time, numerous grains were also found over acinar lumina and larger collecting ducts of the gland (Fig. 9).

From these results, it is clear that the initial site of incorporation of label in the pancreatic acinar cell is the basal ergastoplasm. During post-pulse incubation in chase medium, the labeled proteins move as a well defined wave through the cells and eventually appear in large amounts in the duct system after 2 hr of incubation.

It should be noted that acini were not uniformly labeled throughout the thickness of the slice. A gradient of labeling from the periphery to the center of the slice was found, though only a few acini, farthest from the surface of the slice, were free of label. Apparently, the thickness of the slice does, to some extent, present a barrier to the penetration of label. Nevertheless, at any time the location of grains over cell structures was the same throughout the depth of the slice, indicating that the penetration barrier merely affected the intensity of the radioautographic response without introducing asynchronous labeling. Thus, the gradient does not alter the intracellular transport schedule.

ELECTRON MICROSCOPIC RADIOAUTOGRAPHY: Since the resolution achieved by light microscopic radioautography does not allow us to determine the relationship of silver grains to the individual elements of the centrosphere and apical regions, electron microscopic radioautography was performed on thin sections cut from the same material used for light microscopic radioautography.

At the end of the pulse (Fig. 10), the radioautographic grains were localized exclusively over elements of the rough ER. No grains were seen over elements of the Golgi complex or zymogen granules.

After +7 min (Fig. 11), although the majority of the label still marked the rough ER, radioautographic grains began to appear over the clusters of small, smooth-surfaced vesicles located at the periphery of the Golgi complex. Little of the label progressed as far as condensing vacuoles at this time.

At +17 min, the situation was intermediate

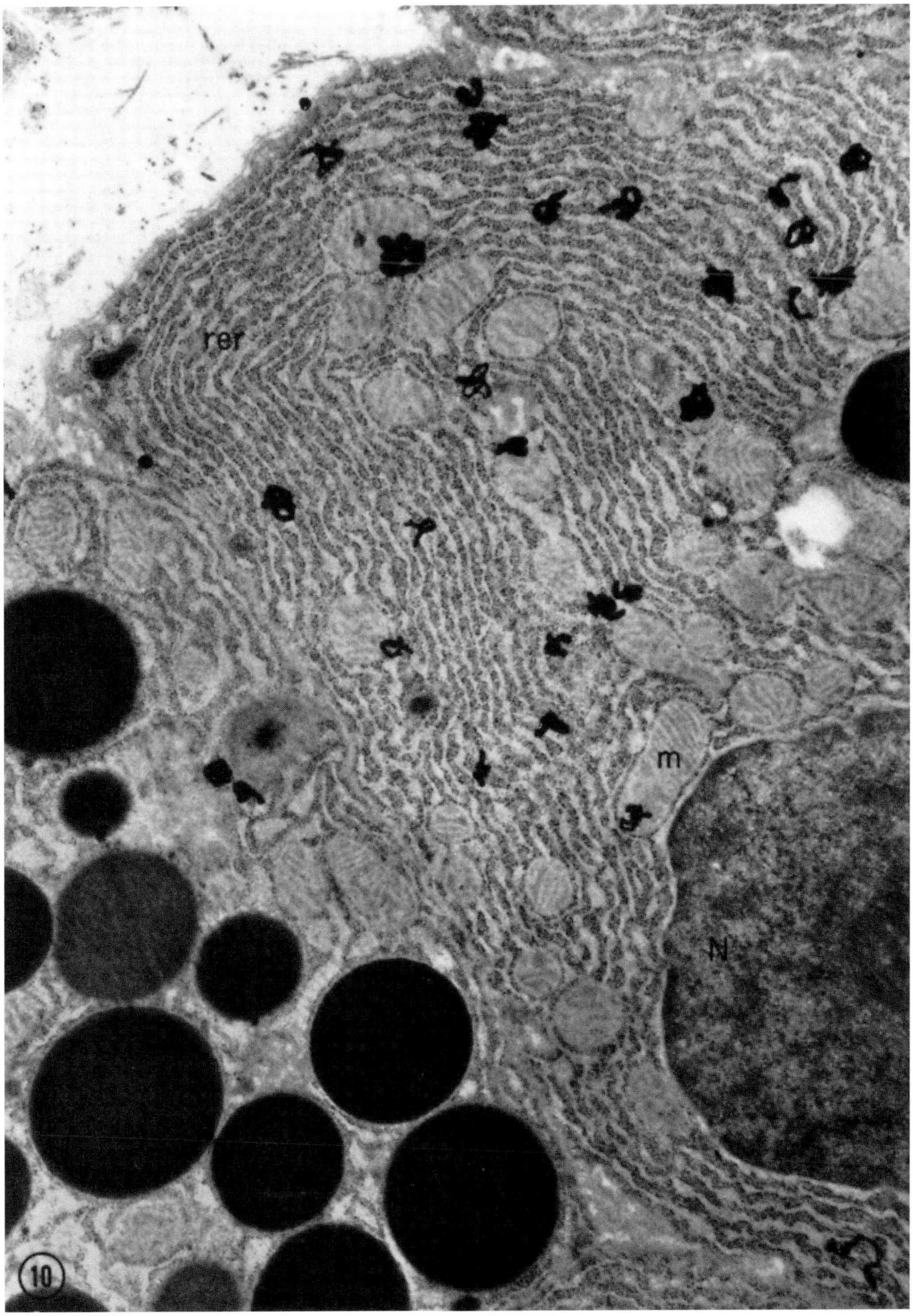

FIGURE 10 Electron microscopic radioautograph of an acinar cell at the end of pulse labeling for 3 min with L-leucine-^3H. The radioautographic grains are located almost exclusively over elements of the rough ER (*rer*). A few grains partly overlie mitochondria but, for reasons discussed in the text, most likely label the adjacent rough ER. *m*, mitochondrion; *N*, nucleus. × 17,000.

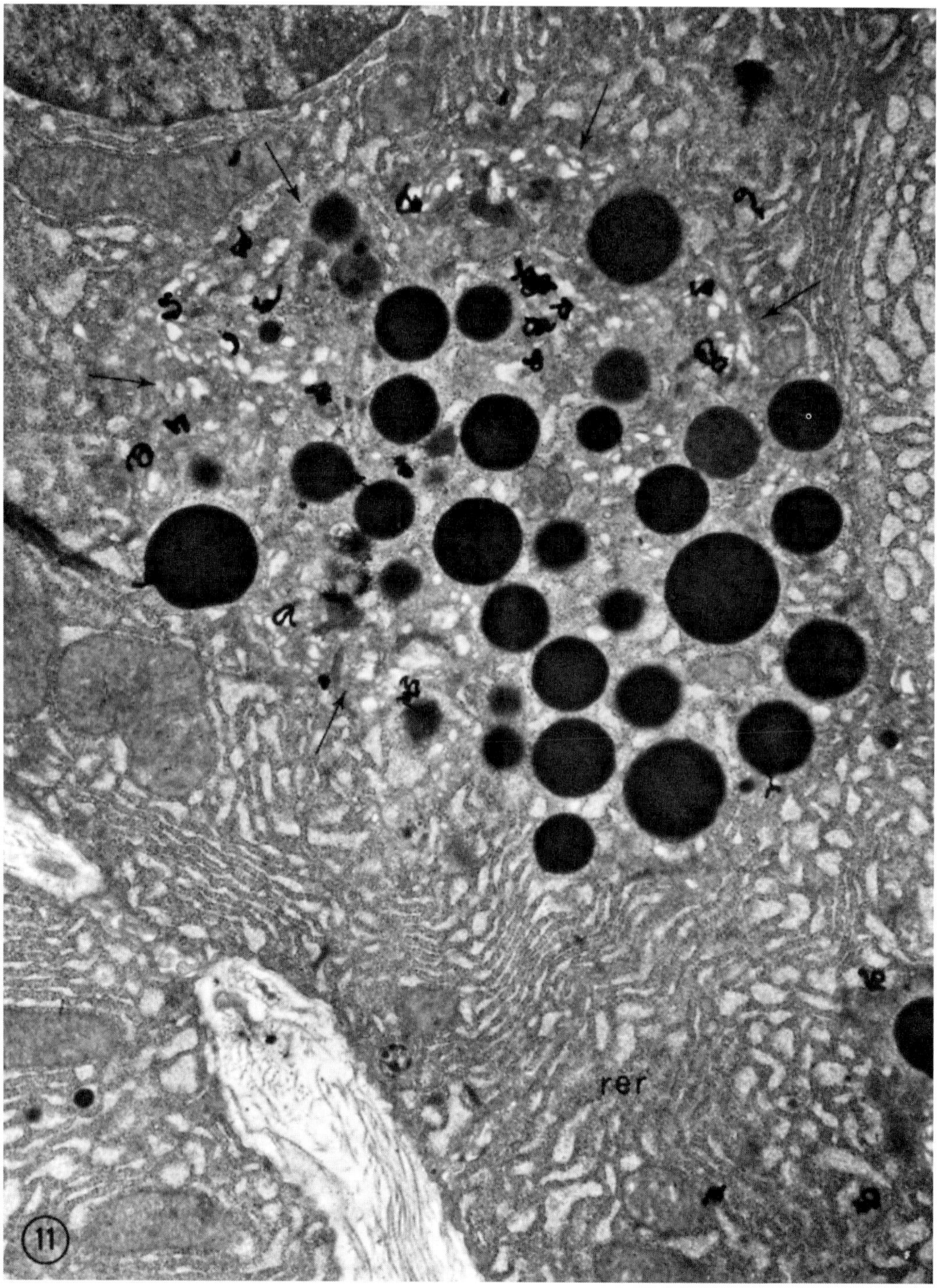

FIGURE 11 Electron microscopic radioautograph of an acinar cell after +7 min of incubation post-pulse. The majority of the label marks the periphery of the Golgi complex (arrows) with little remaining in the rough ER (*rer*). × 17,000.

between that at +7 and +37 min: most of the label had migrated from the basal rough ER and was localized over the periphery of the Golgi complex, while part of it had already reached condensing vacuoles.

After +37 min (Fig. 12), few grains were still found at the periphery of the Golgi complex; most of them now were preferentially and highly concentrated over the condensing vacuoles of the complex. At this time, a small number of grains was already located over zymogen granules.

After +57 min, a larger proportion of the label was found over zymogen granules, especially those adjacent to the Golgi region. Some label remained over the condensing vacuoles though few grains still labeled the periphery of the Golgi complex. By +117 min (which was the longest time point examined by radioautography), the grains heavily labeled zymogen granules grouped near the cell apex and numbers of them were present over fibrous material (presumably discharged secretory protein) in the acinar lumina and collecting ducts of the gland (Fig. 13). The condensing vacuoles of the Golgi complex were now practically devoid of label.

At all times longer than +17 min, a small number of grains persisted over the regions of the cell occupied by rough ER. No significant labeling of centroacinar cells, islet cells, or duct cells was observed at any time. Background label was negligible in all experiments.

Quantitation of radioautographic grain distribution over various components of acinar cells was made at each time point by counting grains on a series of low magnification electron micrographs taken at random (Table II). Analysis ot these data gives information about the quantity of labeled material in the compartment represented by the corresponding cell component rather than about the specific radioactivity of its content.[3] Thus, grain counts cannot be compared directly to specific radioactivity data derived from cell fractionation experiments, although similar patterns of labeling occur.

At the end of the pulse, radioautographic grains were associated almost exclusively with the rough ER of the acinar cell; during chase incubation, the numbers of grains over the rough ER progressively decreased. This is consistent with the kinetics of labeling of the rough microsomes which are derived from the rough ER (cf. Jamieson and Palade, 1967). At the end of +117 min, ~20% of the grains were still located over the rough ER. This label probably identifies non-exportable proteins with a long turnover time (see Siekevitz and Palade, 1960; Warshawsky, Leblond, and Droz, 1963).

Concurrently with the decrease in grains over the ER, there was a progressive rise in grains over elements of the Golgi complex. Thus, at +7 min, ~47% of the grains were over this region, with 43% located over the small vesicles at the periphery of the complex. By +17 min, ~58% of the grains were located over the Golgi complex, with the majority still clustered over peripheral elements of the complex. Following +37 min of incubation, the majority of the label (~49%) had migrated into condensing vacuoles. At this time, a small but significant number of zymogen granules was labeled. Labeling of the zymogen granules progressively increased during the next two time points examined (+57 and +117 min) with an over-all decrease over condensing vacuoles. During the same times, a significant percentage of the grains was located in the acinar lumina.

At all time points, the labeling of mitochondria and nuclei was low and generally decreased during the chase.

From radioautography we can conclude that pancreatic acinar cells incubated in vitro are able to perform the entire sequence of operations involved in the secretory cycle and that the timetable of these events is similar to that established in the whole animal by the radioautographic studies of Caro and Palade (1964).

According to our radioautographic data, at +37 min only a small proportion of the label had progressed as far as zymogen granules, the majority being located over condensing vacuoles. From the labeling kinetics of the zymogen granule fraction (cf. Fig. 3), however, protein radioactivity appears in this fraction as early as +17 min and the fraction is maximally labeled between +37 and +57 min. Because zymogen granule fractions are known (this work and Greene et al., 1963) to contain a small population of condensing vacuoles, we assumed that this discrepancy was only apparent and reflected the presence of con-

[3] It should be noted that the method of analyzing radioautographic data reported here (Table II) is valid since the total amount of label remains constant in the tissue during the times of observation (efficient chase incubation).

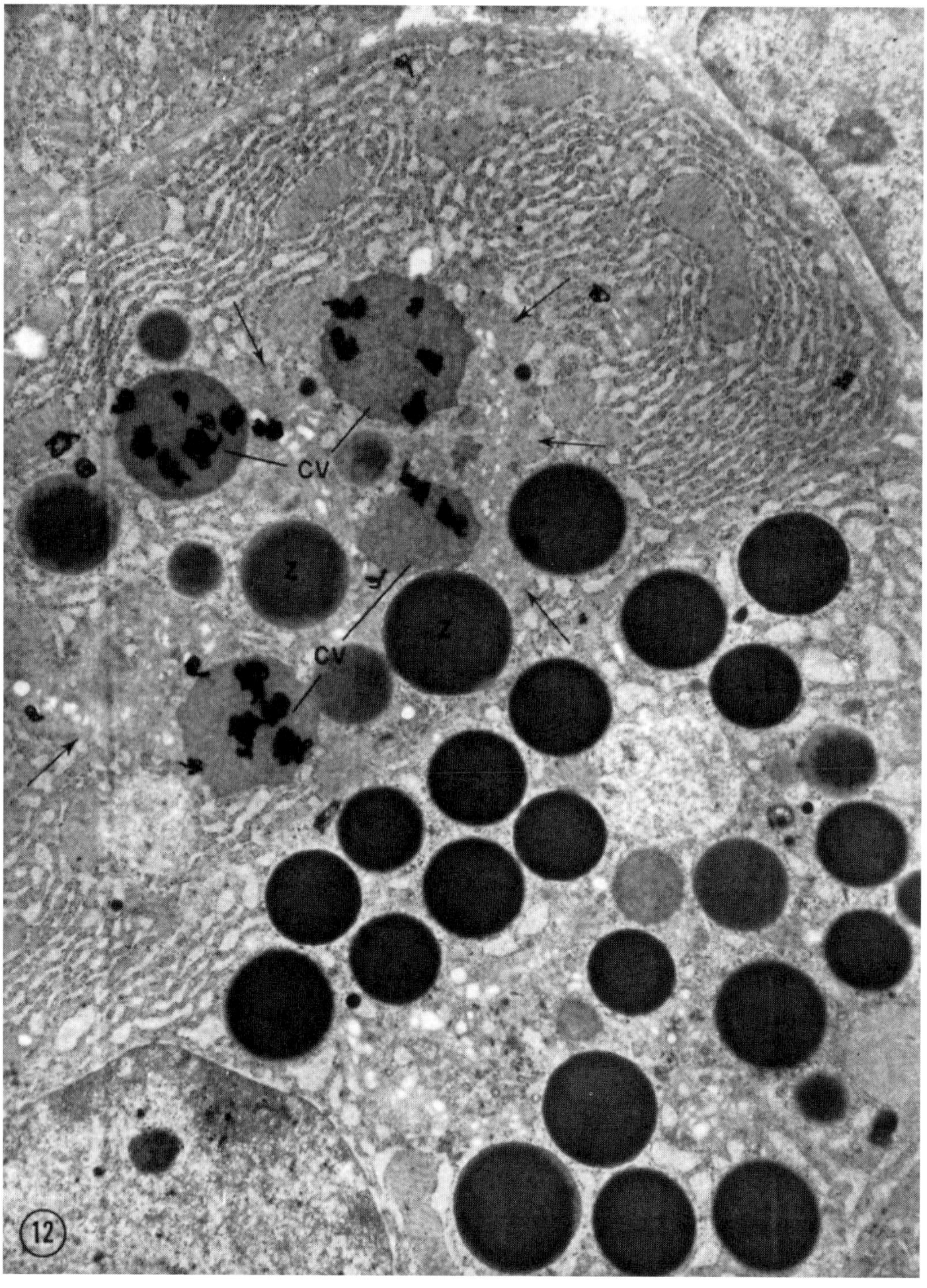

FIGURE 12 Portion of an acinar cell pulse labeled as before and incubated post-pulse for +37 min. Arrows indicate the periphery of the Golgi complex. Radioautographic grains are highly concentrated over condensing vacuoles (*cv*) of the Golgi complex. Zymogen granules (*z*) are unlabeled. × 13,000.

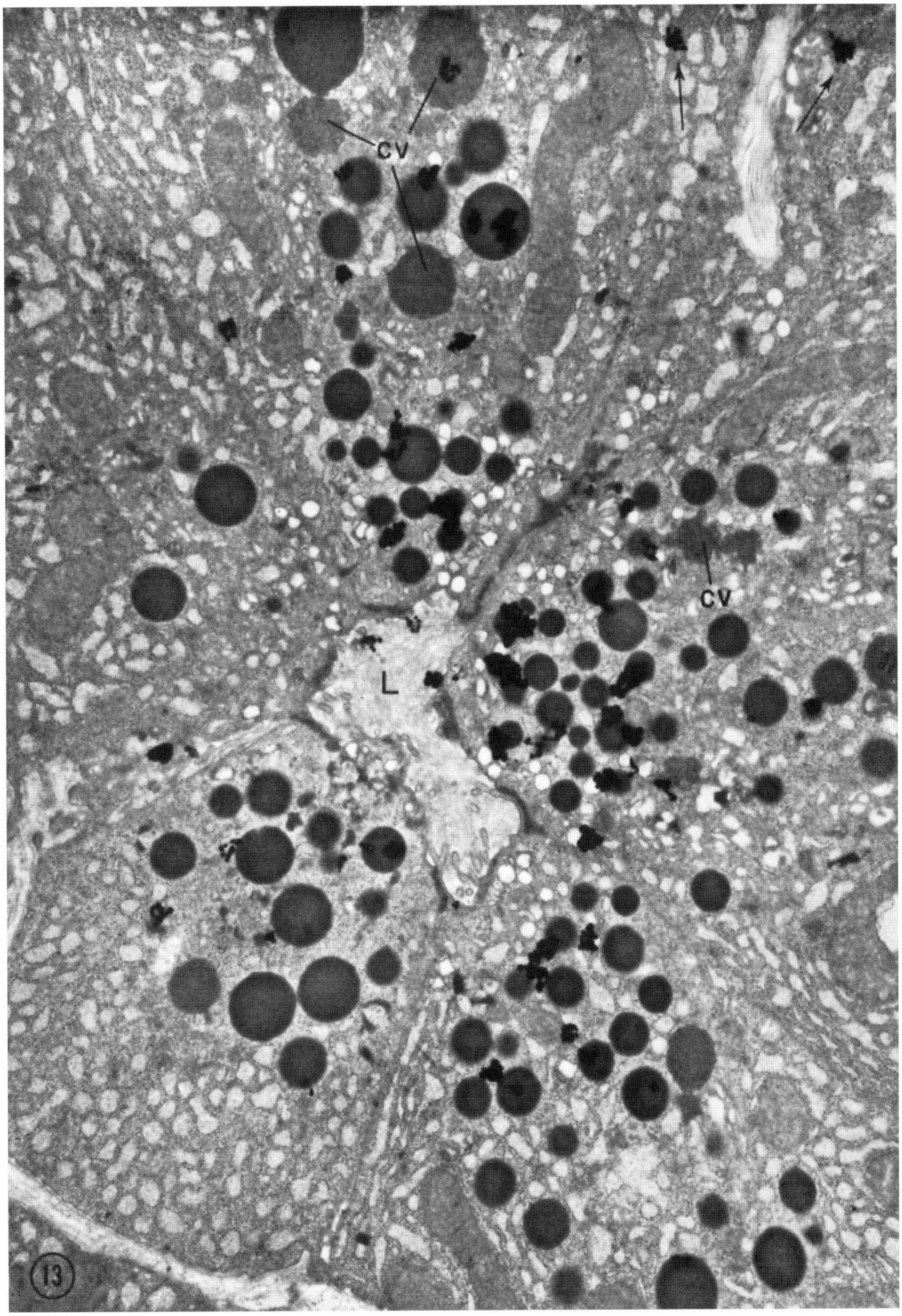

FIGURE 13 Electron microscopic radioautograph of several acinar cells incubated post-pulse for +117 min. Radioautographic grains are located primarily over zymogen granules near the cell apex. Condensing vacuoles (*cv*) are practically devoid of label. Note residual labeling of the rough ER (arrows). Some label marks secretory material in the acinar lumen (*L*). × 13,000.

TABLE II

Distribution of Radioautographic Grains over Cell Components

	% of radioautographic grains					
	3-min (pulse)	Chase incubation				
		+7 min	+17 min	+37 min	+57 min	+117 min
Rough endoplasmic reticulum	**86.3**	43.7	37.6	24.3	16.0	20.0
Golgi complex*						
Peripheral vesicles	2.7	**43.0**	37.5	14.9	11.0	3.6
Condensing vacuoles	1.0	3.8	19.5	**48.5**	35.8	7.5
Zymogen granules	3.0	4.6	3.1	11.3	32.9	**53.6**
Acinar lumen	0	0	0	0	2.9	7.1
Mitochondria	4.0	3.1	1.0	0.9	1.2	1.8
Nuclei	3.0	1.7	1.2	0.2	0	1.4
No. of grains counted	300	1146	587	577	960	1140

The boldfaced numbers indicate maximum accumulation of grains over the corresponding cell component.

* At no time were significant numbers of grains found in association with the flattened, piled cisternae of the complex.

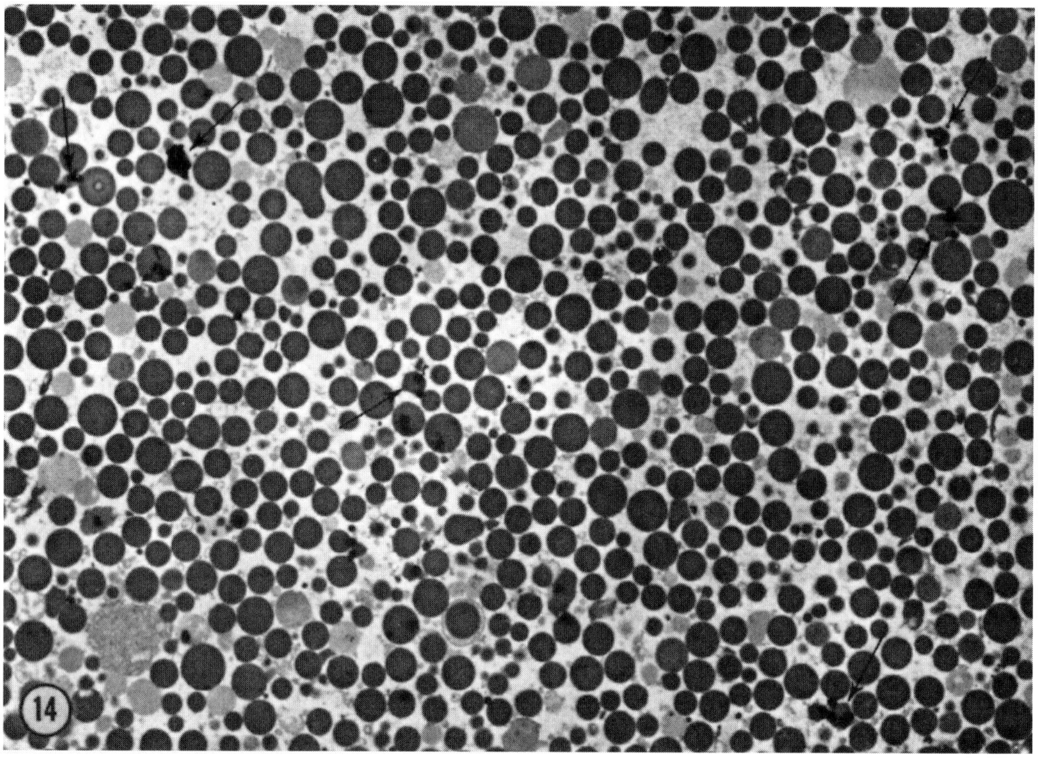

FIGURE 14 Low magnification electron microscopic radioautograph of the zymogen granule pellet derived from slices pulse labeled for 3 min with leucine-^{3}H and incubated +37 min in chase. The arrows indicate highly labeled condensing vacuoles. The majority of the zymogen granules is unlabeled. × 3,000.

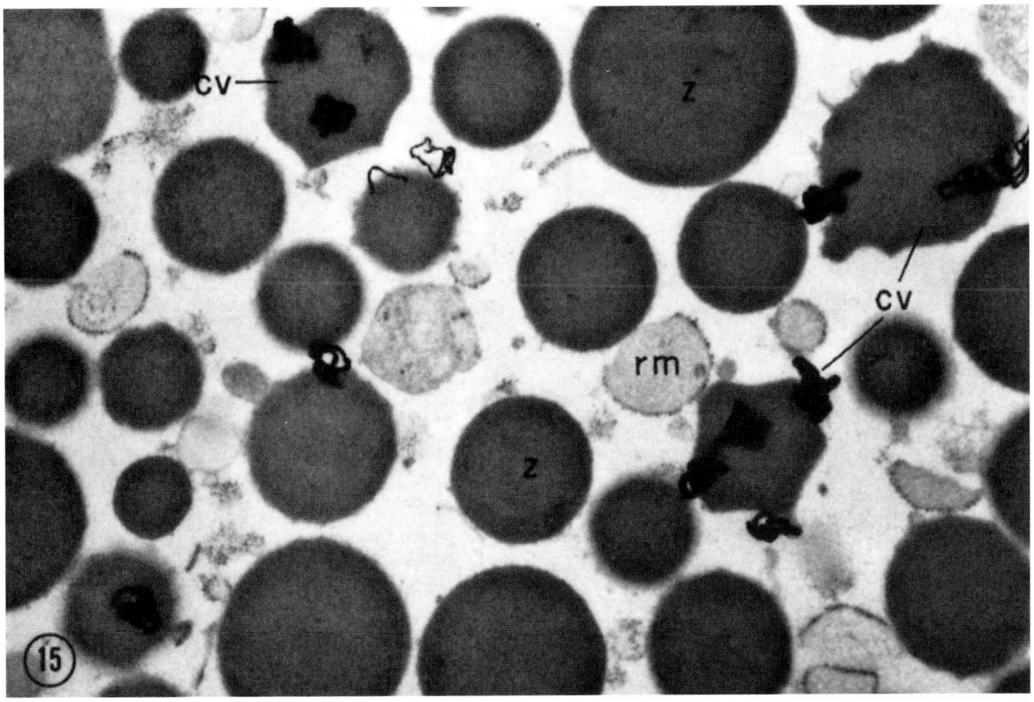

FIGURE 15 Higher magnification view of the same pellet seen in Fig. 14, showing the localization of radioautographic grains over condensing vacuoles (*cv*). *z*, zymogen granules; *rm*, rough microsomes. × 13,000.

densing vacuoles highly labeled at early times after the pulse. Since at present no satisfactory method exists for the separation of condensing vacuoles from zymogen granules, this assumption was tested by performing electron microscopic radioautography on thin sections of zymogen granule pellets derived from slices pulse labeled with leucine-^{3}H for 3 min and incubated for +17 and +37 min in chase medium.

For these studies, the conditions of pulse labeling and incubation in chase medium were identical to those used for tissue radioautography. Thin sections of zymogen granule pellets from each of the three time points were cut through the entire depth of the pellet in the direction of the centrifugal field and mounted on bar-grids which were subsequently processed for electron microscopic radioautography as described under Methods. Grids from the three time points were coated with emulsion at the same time (in order to ensure equal distribution of photographic grains over the specimens), exposed for 3–4 wk, and finally processed photographically. Sequential micrographs were taken in the electron microscope at low magnification (× 750) so as to include the entire depth of the pellet.

At the end of the pulse, no radioautographic grains were found over any component of the zymogen granule pellet. After +17 min, and especially after +37 min of incubation, numerous grains were localized over recognizable condensing vacuoles in the zymogen granule pellets. A typical low magnification field, taken for illustration from the +37-min time point, is shown in Fig. 14. Even at this magnification, radioautographic grains are seen to be preferentially located over condensing vacuoles which are identified in the pellet, as in the cell, by their angular profiles. This is seen to advantage in Fig. 15, which is a higher magnification view of the same pellet.

To quantitate these results, we enlarged photographically the low magnification micrographs to 16 x 20 in. A grid of this size, subdivided into 10- x 10-cm squares, was drawn on a thin plastic sheet and overlaid on the photographic prints. The numbers of labeled condensing vacuoles and

TABLE III

Labeled Condensing Vacuoles and Zymogen Granules in Zymogen Granule Fraction

	3-min (pulse)	Pulse + chase incubation 3 + 17 min	3 + 37 min
Condensing vacuoles labeled (%)	0	7.3	14.0
Zymogen granules labeled (%)	0	0.15	0.53
Labeled condensing vacuoles as % total labeled condensing vacuoles + zymogen granules	0	73	65
Condensing vacuoles as % total condensing vacuoles + zymogen granules	4.95	4.06	6.60

zymogen granules were recorded within each 10-cm square (Table III).

No labeled condensing vacuoles or zymogen granules were found in the micrographs from the 3-min zymogen granule pellet. By +17 min, 7.3% of all condensing vacuoles were labeled and at +37 min the number had increased to 14.0%. At this time, the number of radioautographic grains per vacuole was also considerably greater than at +17 min. At +17 min, and especially at +37 min, a small but significant number of zymogen granules was labeled, though at these time points the majority of labeled particles was condensing vacuoles. Contaminating rough microsomes and mitochondria were not labeled at any time.

These data thus support the original assumption that early labeling of the zymogen granule fraction is accounted for by highly labeled condensing vacuoles and satisfactorily explain the apparent discrepancy between cell fractionation data and electron microscopic radioautography.

DISCUSSION

General Premises Used to Interpret the Experimental Data

In interpreting the results of the experiments reported in this and the previous paper, we have assumed that the label identifies newly synthesized secretory proteins in all cell compartments examined, including smooth microsomes. This assumption is justified by the data of Siekevitz and Palade (1960) who showed that the rate of labeling of secretory proteins in the guinea pig pancreas is five to seven times faster than that of proteins retained in the exocrine cell; it is compatible with our radioautographic data which indicate that the majority of the label moves as a single wave from one cell compartment to another; and it is supported by the results of experiments which show that a large and comparable amount of labeled proteins can be extracted by saline-bicarbonate from all fractions examined (rough and smooth microsomes, zymogen granule fraction). This treatment is known to solubilize digestive enzymes and enzyme precursors from zymogen granules (Greene et al., 1963). Incorporation of label into nonexportable protein undoubtedly takes place, but can be considered negligible within the time limits involved in these experiments.

Results of Present Study

SYNTHESIS AND SEGREGATION OF SECRETORY PROTEINS: In the present study, short (2 – 3-min) pulse labeling of secretory proteins coupled with efficient chase incubation has allowed us to demonstrate unequivocally by electron microscopic radioautography and cell fractionation that the single site of incorporation of label into proteins is in the rough ER. The general location of this site has been identified by radioautography, and the structure involved—rough microsomes derived from the rough ER—has been pinpointed by cell fractionation. The existence of a second site of synthesis in the Golgi complex (Sjöstrand, 1962) is excluded since, by the time labeled proteins appear in the complex, the intracellular pool of soluble precursors has effectively been diluted. Further, the label which appears in Golgi elements can be accounted for by label lost from the rough ER (Jamieson and Palade, 1967). The existence of a second site could not be ruled out by previous radioautographic work on intact animals (Caro and Palade, 1964). Thus, the present results put on a factual basis the hypothesis of Siekevitz and Palade (1958 *b*) according to which secretory proteins are synthesized only in the rough ER and subsequently transported to other cell compartments.

Our radioautographic studies indicate that at the end of the pulse only ~3% of the label was located over mitochondria and ~3% over nuclei; during incubation in the chase medium, the percentage of grains located over these structures progressively decreased in parallel with the loss of label from the rough ER. It is, therefore, likely that the labeling of mitochondria and nuclei reflects labeling of the adjacent rough ER. Accordingly, the data do not support the assumption of Straub (1958) that mitochondria participate in the synthesis of pancreatic secretory proteins. Similarly, the limited amount of label located over elements of the Golgi complex and zymogen granules at the end of the pulse should be taken as a maximum figure which reflects to a large extent labeling of adjacent rough ER elements.

Our results do not provide further insight into either the actual site of protein synthesis in the rough ER, or the fate of the protein immediately after synthesis. Earlier evidence (Siekevitz and Palade, 1960) had indicated that α-chymotrypsinogen is synthesized on attached ribosomes, and recently Redman, Siekevitz, and Palade (1966) have shown that another secretory protein, α-amylase, is synthesized in the attached ribosomes of pigeon pancreatic microsomes incubated in vitro. In addition, these investigators have demonstrated that newly synthesized amylase is preferentially transported to the content of the microsomal vesicles, which is the in vitro equivalent of the content of the cisternal space. Using liver microsomes, Redman and Sabatini (1966) have also shown that peptides labeled in vitro and discharged by puromycin from attached ribosomes are preferentially transported into the cisternal space, indicating that from the inception of its synthesis, the peptide chain is destined to reach this space. It can be tentatively assumed that these findings (Redman et al., 1966; Redman and Sabatini, 1966) apply to all secretory proteins produced by the exocrine cell.

TRANSPORT OF SECRETORY PROTEINS THROUGH THE PERIPHERY OF THE GOLGI COMPLEX: The results of kinetic experiments performed on smooth microsomal fractions (Jamieson and Palade, 1966, 1967), isolated from pancreas slices at the end of the pulse and after various times of incubation in the chase medium, provide direct evidence that secretory proteins pass through the small peripheral vesicles of the Golgi complex in transit from their site of synthesis and segregation in the rough ER to their site of concentration in condensing vacuoles. The cell fractionation data also indicate that the specific radioactivity of proteins in the post-microsomal supernate (i.e., of proteins located mainly in the cytoplasmic matrix or cell sap) remains constant at all times examined, indicating negligible equilibration between labeled proteins of the particulate components of the cell and the cytoplasmic matrix.[4] These results do not support the assumption (Laird and Barton, 1958; Redman and Hokin, 1959; Morris and Dickman, 1960) that secretory proteins leave the microsomes and pass, in soluble form, through the cytoplasmic matrix to zymogen granules or the acinar lumen.

Cell fractionation findings are in good agreement with radioautographic data obtained on intact cells labeled either in slices or in the whole animal (Caro and Palade, 1964). The data show that, at the time labeled protein becomes concentrated in the smooth microsomes (between +7 and +17 min), radioautographic grains are concentrated over a region of the Golgi complex containing numerous small, smooth-surfaced vesicles; at this time, no other smooth-surfaced vesicles in the cell (e.g., small clusters of vesicles located among the rough ER cisternae and under the plasmalemma, and small apical vacuoles) or potential sources of smooth vesicles (cell membrane or mitochondria) are labeled. The results of cell fractionation and radioautography thus support and complement each other: radioautography indicates that the smooth microsomal fraction must contain peripheral elements of the Golgi complex while cell fractionation shows that the label is localized to the small vesicles of this region rather than to the surrounding cytoplasmic matrix. The majority of the radioactive protein in the smooth microsomes can be extracted by mild alkaline treatment, indicating that the label is located within the cavity of the small Golgi vesicles and not in their limiting membrane.

From these combined data, it is clear that the small vesicles of the Golgi complex mediate the transport of secretory proteins from the rough ER to condensing vacuoles. This implies transport in bulk or in mass since each vesicle must contain and carry a large number of protein molecules.

[4] From recovery experiments, not shown in this paper, it could be calculated that if the total labeled proteins of the microsomal fraction were transferred completely to the post-microsomal supernate during chase incubation, then the specific radioactivity of proteins in the supernate should increase 2.5- to 3-fold. This did not occur (see Fig. 3).

The transport must also be discontinuous; otherwise, concentration of the solution of secretory proteins at the next step (condensing vacuoles) would not be possible. With these restrictions in mind, we can inquire into cellular mechanisms possibly involved in this operation. It should be recalled that the transitional elements of the rough ER adjacent to the periphery of the Golgi complex are limited by part rough- and part smooth-surfaced membranes. The latter are frequently in contact with smooth-surfaced vesicles or protrude as smooth-surfaced blebs toward the Golgi complex. These blebs are comparable in size to the small, smooth-surfaced vesicles of the region. Likewise, the smooth-surfaced membrane bounding the condensing vacuoles is often in contact with small vesicles or is thrown into small surface blebs. These images suggest repeated fission of, and fusion with, vesicles at both terminals. It can be assumed, then, that secretory proteins are transported from the transitional elements of the rough ER to condensing vacuoles through intermittent tubular connections, or, alternatively, that transport is effected by discrete vesicles that shuttle between the two compartments. Both possibilities are compatible with the data, but the second is more consistent with the fine structural details of the Golgi complex.

The model proposed for the transport of secretory proteins within the periphery of the Golgi complex is reminiscent of the transport of macromolecules in bulk across the vascular endothelium by pinocytic vesicles (Palade and Bruns, 1964; Karnovsky, 1965), though in the case of the pancreatic exocrine cell, transport in bulk operates between two intracellular compartments, the rough ER and the condensing vacuoles.

TRANSPORT OF LABELED PROTEINS TO CONDENSING VACUOLES, ZYMOGEN GRANULES, AND THE ACINAR LUMEN: The radioautographic studies of Caro and Palade (1964) have firmly established that proteins are transported to condensing vacuoles of the Golgi complex where they are intensively concentrated, the condensing vacuoles being converted in the process into zymogen granules. We have confirmed the position of the condensing vacuoles in the secretory cycle by electron microscopic radioautography of pancreatic slices and have demonstrated that the time taken to accumulate labeled proteins within condensing vacuoles is approximately the same in vitro as in vivo. Further, because of the short, well defined label which passes through the cell as a single wave, the arrival and concentration of radioactive proteins in condensing vacuoles are more strikingly demonstrated than was possible in the whole animal.

Earlier cell fractionation (Siekevitz and Palade, 1958 *b*) and radioautographic data (Caro and Palade, 1964) were not in agreement concerning the rate of transport of labeled secretory proteins into zymogen granules. According to the former, the zymogen granule fraction accumulates radioactive protein as early as 10 min after the injection of tracer, whereas radioautography indicated that zymogen granules were not labeled until ~1 hr after injection of tracer. In the present experiments, a similar discrepancy was observed. Since the zymogen granule pellet contains a small population of condensing vacuoles, we assumed that the discrepancy was due to the presence of early and intensive labeling of condensing vacuoles. We tested this possibility by electron microscopic radioautography of sections of the zymogen granule pellet at various times after pulse labeling. The results substantiated the assumption and pointed out the important contribution made by condensing vacuoles to the early labeling of the zymogen granule fraction. Evidently, a correction factor must be applied to previous cell fractionation data which indicated that zymogen granules become labeled shortly after the administration of the tracer (Siekevitz and Palade, 1958 *b*; Hansson, 1959; Morris and Dickman, 1960).

It should be noted here that Schramm and Bdolah (1964) have concluded, on the basis of cell fractionation studies carried out on pulse-labeled rat parotid slices incubated in vitro, that newly synthesized amylase moves from the microsomal fraction sequentially into fractions containing granules of increasing size or density. Since the pancreatic and salivary exocrine cells are similar in organization (cf. Parks, 1961, 1962) it is probable that here, too, early labeling of the granule fractions is due to the presence of labeled condensing vacuoles.

Radioautographic observations on intact cells of slices after 1 hr and especially after 2 hr postpulse incubation conclusively demonstrate the transformation of condensing vacuoles into zymogen granules and, finally, the discharge of protein into the acinar lumen. The terminal steps in the secretory cycle are events well documented by previous work (Palade, 1959; Caro and Palade, 1964). Hence, we now have a reasonably complete elucidation of the entire intracellular pathway of secretory proteins from their site of synthesis on attached ribosomes to their ultimate discharge into the duct system of the gland. According to the radioautographic data, our interpretations probably apply to the pathway and timetable of transport of all digestive enzymes in the guinea pig pancreas. Exceptions for exportable proteins which represent a small fraction of the total output in the guinea pig or for any protein in other species are improbable but not excluded (cf. Redman and Hokin, 1959; Morris and Dickman, 1960).

We should mention that our radioautographic studies indicate that the piled Golgi cisternae located at the periphery of the complex do not seem to play a direct role in the transport and concentration of labeled secretory proteins. Nevertheless, in the exocrine pancreas of other species (rat, mouse), condensing vacuoles appear to form at the expense of the innermost cisternae of the piles. A comparable situation exists in other secretory cells where the function of the condensing vacuoles is taken over by the piled cisternae of the Golgi complex (Kurosumi, 1963, Bainton and Farquhar, 1966; Smith and Farquhar, 1966; Essner and Novikoff, 1962).

To what extent does the well established sequence of operations involved in the pancreatic secretory cycle apply to secretory processes in other cells? The same sequence seems to operate in glandular cells specialized in the synthesis of proteins for export, provided the secretory product is quantized and stored temporarily in the cell in concentrated form. For example, radioautographic evidence suggests that the model applies to the mammary gland (Wellings and Philp, 1964), and to myelocytes (Fedorko and Hirsch, 1966). In cells in which there is no intracellular accumulation of secretory product or in which the latter accumulates within the ER cisternae (Ross and Benditt, 1965; Revel and Hay, 1963; Rifkind et al., 1962), the extent to which the Golgi complex is involved is not clear.

In conclusion, we now have a satisfactory general understanding of the secretory cycle of the pancreatic exocrine cell based on experimentally established facts from a number of studies. In addition, these studies provide direct or indirect evidence that several types of intracellular transport, some of them new, participate in the cycle. These include: vectorial transport of newly synthesized proteins across the membranes of the rough ER (Redman et al., 1966; Redman and Sabatini, 1966); transport in bulk of materials between cell compartments; concentration of cell products within membrane-bounded structures possibly by intracellular ion pumps; and finally, transport of secretory proteins from zymogen granules to the acinar lumina (Palade, 1959). The forces, molecular events, and control mechanisms operating at each step remain to be elucidated by future work.

Part of this work was supported by a United States Public Health Service Research Grant AM 10928–01.

Received for publication 20 February 1967.

REFERENCES

Bainton, D. F., and M. G. Farquhar. 1966. *J. Cell Biol.* **28:** 277.

Caro, L. G., and G. E. Palade. 1964. *J. Cell Biol.* **20:** 473.

Caro, L. G., and R. P. van Tubergen. 1962. *J. Cell Biol.* **15:** 173.

Essner, E., and A. B. Novikoff. 1962. *J. Cell Biol.* **15:** 289.

Fedorko, M. E., and J. G. Hirsch. 1966. *J. Cell Biol.* **29:** 307.

Greene, L. J., C. H. W. Hirs, and G. E. Palade. 1963. *J. Biol. Chem.* **238:** 2054.

Hansson, E. 1959. *Acta Physiol. Scand.* **46** *(Suppl.)*: 161.

Hokin, L. E. 1955. *Biochim. Biophys. Acta.* **18:** 379.

Jamieson, J. D., and G. E. Palade. 1966. *Proc. Natl. Acad. Sci.* **55:** 424.

Jamieson, J. D., and G. E. Palade. 1967. *J. Cell Biol.* **34:** 577.

Jennings, B. M., M. G. Farquhar, and H. D. Moon. 1959. *Am. J. Pathol.* **35:** 991.

Junqueira, L. C. U., G. C. Hirsch, and H. A. Rothschild. 1955. *Biochem. J.* **61:** 275.

Karnovsky, M. J. 1965. *J. Cell Biol.* **27:** 49A. (Abstr.)

Keller, P. J., and E. Cohen. 1961. *J. Biol. Chem.* **236:** 1407.

Kurosumi, K. 1963. *In* International Symposium for Cellular Chemistry. S. Seno and E. V. Cowdry, editors. Japan Society for Cell Biology, Okayama, Japan. p. 259.

Laird, A. K., and A. D. Barton. 1958. *Biochim. Biophys. Acta.* **27:** 12.

Mejbaum, W. 1939. *Z. Physiol. Chem.* **258:** 117.

Morris, A. J., and S. R. Dickman. 1960. *J. Biol. Chem.* **235:** 1404.

Palade, G. E. 1959. *In* Subcellular Particles. T.

Hayashi, editor. Ronald Press Co., New York. p. 64.
PALADE, G. E., and R. R. BRUNS. 1964. *In* Small Blood Vessel Involvement in Diabetes Mellitus. M. D. Siperstein, A. R. Colwell, Sr., and K. Meyer, editors. American Institute of Biological Sciences, Washington, D. C. p. 39.
PARKS, H. F. 1961. *Am. J. Anat.* **108:** 303.
PARKS, H. F. 1962. *J. Ultrastruct. Res.* **6:** 449.
REDMAN, C. M., and L. E. HOKIN. 1959. *J. Biophys. Biochem. Cytol.* **6:** 207.
REDMAN, C. M., and D. D. SABATINI. 1966. *Proc. Natl. Acad. Sci.* **56:** 608.
REDMAN, C. M., P. SIEKEVITZ, and G. E. PALADE. 1966. *J. Biol. Chem.* **241:** 1150.
REVEL, J.-P., and E. D. HAY. 1963. *Z. Zellforsch.* **61:** 110.
RICHARDSON, K. C., L. JARETT, and E. H. FINKE. 1960. *Stain Technol.* **35:** 313.
RIFKIND, R. A., E. F. OSSERMAN, K. C. HSU, and C. MORGAN. 1962. *J. Exptl. Med.* **116:** 423.
ROSS, R., and E. P. BENDITT. 1965. *J. Cell Biol.* **27:** 83.
SCHRAMM, M., and A. BDOLAH. 1964. *Arch. Biochem. Biophys.* **104:** 67.
SIEKEVITZ, P., and G. E. PALADE. 1958 *a*. *J. Biophys. Biochem. Cytol.* **4:** 203.
SIEKEVITZ, P., and G. E. PALADE. 1958 *b*. *J. Biophys. Biochem. Cytol.* **4:** 557.
SIEKEVITZ, P., and G. E. PALADE. 1960. *J. Biophys. Biochem. Cytol.* **7:** 619.
SJÖSTRAND, F. S. 1962. *In* Ciba Foundation Symposium on the Exocrine Pancreas. A. V. S. de Reuck and M. P. Cameron, editors. J. & A. Churchill Ltd., London. p. 1.
SMITH, R. E., and M. G. FARQUHAR. 1966. *J. Cell Biol.* **31:** 319.
STRAUB, F. B. 1958. *Symp. Soc. Exptl. Biol.* **12:** 176.
VAN LANCKER, J. L., and R. L. HOLTZER. 1959. *J. Biol. Chem.* **234:** 2359.
VENABLE, J. H., and R. COGGESHALL. 1965. *J. Cell Biol.* **25:** 407.
WARSHAWSKY, H., C. P. LEBLOND, and B. DROZ. 1963. *J. Cell Biol.* **16:** 1.
WELLINGS, S. R., and J. R. PHILP. 1964. *Z. Zellforsch.* **61:** 871.

Blobel G. and **Dobberstein B.** 1975. Transfer of proteins across membranes. I. Presence of proteolytically processed and unprocessed nascent immunoglobulin light chains on membrane-bound ribosomes of murine myeloma. *J. Cell Biol.* **67**: 835–851.

DURING THE 1950S AND 1960S, Palade's group showed that proteins destined either for secretion or for incorporation into membranes were translated on ribosomes that were bound to membranes of the endoplasmic reticulum (ER). By contrast, proteins destined for the cytosol were translated on free polysomes. The mechanisms by which cells could distinguish the messenger RNAs of these classes remained to be defined. The paper reprinted here tested the hypothesis, proposed a few years earlier (1), that a nascent protein to be secreted or inserted into the membrane contained an address, a "signal sequence" of amino acids that would distinguish it from proteins that should be translated on free polysomes. An in vitro assay was essential to determine whether proteins with signals still attached were detectably bigger than their mature counterparts and to see if the proteolytic removal of the signal occurred co-translationally or after synthesis was complete. This paper established the necessary assay. It has subsequently been used to show that many membrane proteins contain the hypothesized signal and to characterize the essential components and regulatory molecules of co-translational protein insertion into membranes. Such work has identified a ribonucleoprotein, the signal recognition particle, which discriminates between mRNAs with and without signals, halting polypeptide elongation for those with signals until their ribosomes have docked on the ER membrane (2). It has also led to recognition of a protein conducting channel in the ER membrane (3) and the identification of its protein components (4,5).

---

1. Blobel G. and Sabatini D.D. 1971 Ribosome-membrane interaction in eukarayotic cells. In Biomembranes (ed. L.A. Manson), vol. 2, pp. 193–195. Plenum Press, New York.
2. Walter P., Ibrahimi I., and Blobel G. 1981. Translocation of proteins across the endoplasmic reticulum. I. Signal recognition protein (SRP) binds to in-vitro-assembled polysomes synthesizing secretory protein. *J. Cell. Biol.* **91:** 545–550.
3. Simon S.M. and Blobel G. 1991. A protein-conducting channel in the endoplasmic reticulum. *Cell* **65:** 371–380.
4. Deshaies R.J. and Schekman R. 1987. A yeast mutant defective at an early stage of import of secretory protein precursors into the endoplasmic reticulum. *J. Cell Biol.* **105:** 633–645.
5. Görlich D. and Rapoport T.A. 1993. Protein translocation into proteoliposomes reconstituted from purified components of the endoplasmic reticulum membrane. *Cell* **75:** 615–630.

TRANSFER OF PROTEINS ACROSS MEMBRANES

I. Presence of Proteolytically Processed and Unprocessed Nascent Immunoglobulin Light Chains On Membrane-Bound Ribosomes of Murine Myeloma

GÜNTER BLOBEL and BERNHARD DOBBERSTEIN

From The Rockefeller University, New York 10021

ABSTRACT

Fractionation of MOPC 41 DL-1 tumors revealed that the mRNA for the light chain of immunoglobulin is localized exclusively in membrane-bound ribosomes. It was shown that the translation product of isolated light chain mRNA in a heterologous protein-synthesizing system in vitro is larger than the authentic secreted light chain; this confirms similar results from several laboratories. The synthesis in vitro of a precursor protein of the light chain is not an artifact of translation in a heterologous system, because it was shown that detached polysomes, isolated from detergent-treated rough microsomes, not only contain nascent light chains which have already been proteolytically processed in vivo but also contain unprocessed nascent light chains. In vitro completion of these nascent light chains thus resulted in the synthesis of some chains having the same mol wt as the authentic secreted light chains, because of completion of in vivo proteolytically processed chains and of other chains which, due to the completion of unprocessed chains, have the same mol wt as the precursor of the light chain.

In contrast, completion of the nascent light chains contained in rough microsomes resulted in the synthesis of only processed light chains. Taken together, these results indicate that the processing activity is present in isolated rough microsomes, that it is localized in the membrane moiety of rough microsomes, and, therefore, that it was most likely solubilized during detergent treatment used for the isolation of detached polysomes. Furthermore, these results established that processing in vivo takes place before completion of the nascent chain.

The data also indicate that in vitro processing of nascent chains by rough microsomes is dependent on ribosome binding to the membrane. If the latter process is interfered with by aurintricarboxylic acid, rough microsomes also synthesize some unprocessed chains.

The data presented in this paper have been interpreted in the light of a recently proposed hypothesis. This hypothesis, referred to as the signal hypothesis, is described in greater detail in the Discussion section.

Biological membranes present a diffusion barrier for macromolecules such as proteins, but transfer of a large number of specific proteins across membranes is an important physiological activity of virtually all cells. Segregation by a membrane is required not only for secretory proteins but also for lysosomal and peroxysomal proteins and for certain mitochondrial or chloroplast proteins synthesized in the cytoplasm. Transfer of proteins across membranes may even be required for some intramembrane proteins, e.g., if the site of insertion into the membrane were separated from the site of synthesis by the lipid bilayer. The discovery of an abundance of ribosome membrane junctions, particularly in secretory cells in the mid 50's (24, 25) and the demonstration in the mid 60's (1, 26, 27, 29) that nascent chains synthesized on membrane-bound ribosomes are vectorially discharged across the membrane, suggested that the ribosome membrane junction may function in the transfer of proteins across the membrane; by topologically linking the site of synthesis with the site of transfer, the protein would transverse the membrane only in status nascendi in an extended form before assuming its native structure, thus maintaining the membrane's role as a diffusion barrier to proteins.

However, the function of the ribosome membrane junction in the transfer of proteins across the membrane did not explain the cell's ability to determine which proteins can traverse via the ribosome membrane junction. Data accumulated in the late 60's (reviewed in reference 28) indicated that mRNA's for some secretory proteins are translated almost exclusively on membrane-bound ribosomes while mRNA's for some cytosol proteins are translated on free ribosomes. In an attempt to explain this dichotomy, a hypothesis was suggested in 1971 (5). It was postulated that all mRNA's to be translated on bound ribosomes contain a unique sequence of codons to the right of the initiation codon (henceforth referred to as the signal codons); translation of the signal codons results in a unique sequence of amino acid residues on the amino terminal end of the nascent chain (henceforth referred to as the signal sequence); the latter triggers attachment of the ribosome to the membrane. A somewhat more detailed version of this hypothesis, henceforth referred to as the signal hypothesis, is presented in this paper.

Thus far, the most compelling support for the signal hypothesis derives from data reported by several laboratories on the in vitro translation of IgG light chain mRNA isolated from murine myelomas. The translation product of this mRNA is larger than the authentic light chain (15, 19, 23, 32, 34–36). Furthermore, it was established by peptide mapping and partial NH_2-terminal sequence analysis (33) that the in vitro translation product contains an extra sequence of ~20 amino acid residues at the NH_2 terminus. It has been suggested independently that this extra sequence, which is subsequently removed, may function in the binding of the ribosome to the membrane (23). We therefore have chosen the murine myeloma as a model system for these studies. In this paper we present fractionation and in vitro protein synthesis experiments designed to examine some aspects of the signal hypothesis.

METHODS AND MATERIALS

All experiments were carried out with a MOPC 41 murine myeloma obtained from Litton Bionetics, Kensington, Md. The light chain produced by this tumor proved to be ~2,000 daltons smaller than that produced by the original MOPC 41 obtained from Dr. M. Potter of the National Institutes of Health, Bethesda, Md. We assume that this change results from the selection of a deletion mutant clone during our initial transfers, and hereafter we will refer to this myeloma as MOPC 41 DL-1 (see also companion paper).

Fractionation of MOPC 41 DL-1 Tumor

All sucrose solutions used in cell fractionation contained 50 mM triethanolamine·HCl pH 7.4 at 20°C, 50 mM KCl, and 5 mM $MgCl_2$ (TeaKM).[1]

Excised tumors freed of necrotic portions were passed through an ice-cold tissue press and were homogenized in ~3 vol of 0.25 M sucrose with a few strokes in a Potter-Elvehjem homogenizer. All subsequent operations were performed between 1–4°C. The homogenate was centrifuged for 10 min at 10,000 g_{av} in an angle rotor to yield a postmitochondrial supernate. The latter was layered over 2.0 ml of 1.3 M sucrose and centrifuged for 15 min at 105,000 g_{av} in a Spinco no. 40 rotor (Beckman Instruments, Inc., Spinco Div., Palo Alto, Calif.). The supernate was aspirated with a syringe and subsequently used for the preparation of free ribosomes. The pellet, resuspended by homogenization in 2.3 M sucrose, was used to isolate rough microsomes by the following procedure: 3.5 ml of the suspension was loaded at the

[1] *Abbreviations used in this paper:* AR, autoradiography; ATA, aurintricarboxylic acid; DTT, dithiothreitol; ER, endoplasmic reticulum; L^o, derived large ribosomal subunits; PAGE, polyacrylamide gel electrophoresis; S^N, native small ribosomal subunits; SDS, sodium dodecyl sulfate; TeaKM, 50 mM triethanolamine·HCl pH 7.4 at 20°C, 50 mM KCl, and 5 mM $MgCl_2$.

bottom of a tube fitting the SB 283 rotor of the IEC centrifuge (Damon/IEC Div., Damon Corp., Needham Heights, Mass.) and overlayed with 3-ml aliquots each of 1.75 M, 1.5 M, and 0.25 M sucrose. The discontinuous gradient was centrifuged for 12 h at 190,000 g_{av}. The material banding in the 1.75 M sucrose layer was removed with a syringe, diluted with 1 vol of TeaKM, layered over a 2-ml cushion of 1.3 M sucrose, and centrifuged for 30 min at 105,000 g_{av} in a Spinco no. 40 rotor to yield a pellet consisting essentially of rough microsomes.

To isolate detached ribosomes, such pellets were resuspended in TeaKM; a 10% solution of deoxycholate in water was added to a final concentration of 1%, and the mixture was layered over a 2-ml cushion of 2.0 M sucrose. Centrifugation for 24 h at 105,000 g_{av} in a Spinco no. 40 rotor yielded a pellet of detached ribosomes.

Free ribosomes were prepared from the 30,000 g supernate (see above) which was layered on a discontinuous sucrose gradient containing 2-ml layers each of 2.0 M and 1.75 M sucrose. Centrifugation for 24 h at 105,000 g_{av} in a Spinco no. 40 rotor yielded a pellet of free ribosomes.

Preparations of native small ribosomal subunits (S^N) from rabbit reticulocytes, of derived large ribosomal subunits (L^O) from rat liver free ribosomes by the puromycin-KCl procedure, and of pH 5 enzymes from a high speed supernate of Krebs ascites cells were as previously described (6, 13).

Isolation of Poly (A)-Containing RNA from MOPC 41 DL-1 Rough Microsomes

Pellets containing rough microsomes (3,000–5,000 A_{260} units) were resuspended in 60 ml 150 mM NaCl, 50 mM Tris-HCl pH 8.0, and 5 mM EDTA. A 10% solution of sodium dodecyl sulfate (SDS) was added to a final concentration of 1.5% and RNA was extracted from this mixture with phenol-chloroform-isoamyl alcohol (2) and fractionated on oligo (dT) cellulose as follows. RNA was resuspended in H_2O and subsequently adjusted to 400 mM NaCl, 50 mM Tris·HCl pH 7.5, and 0.2% SDS. The final concentration of RNA was 100–120 A_{260}units/ml. 6 ml of that solution was mixed at room temperature by gentle swirling with 4 ml of packed oligo (dT) cellulose which had been washed several times in 400 mM NaCl, 50 mM Tris·HCl pH 7.5, and 0.2% SDS. The cellulose was then sedimented at 1,000 g, washed twice with 100 mM NaCl, 50 mM Tris·HCl pH 7.5, 0.2% SDS, and transferred to a column. After washing with at least 10 bed volumes of a solution of 100 mM NaCl and 50 mM Tris·HCl pH 7.5, the poly (A)-containing RNA was eluted with a solution of 10 mM Tris·HCl pH 7.5, and 0.1% SDS. The poly (A)-containing RNA was twice precipitated with ethanol, then resuspended in double-distilled water to a concentration of ~10 A_{260}units/ml, and stored at −80°C.

Cell-Free Protein-Synthesizing Systems and Assays

The terms "readout" and "initiation" system have been adopted for the sake of brevity. In the readout system, previously started polypeptide chains are completed, whereas in the initiation system polypeptide chains are synthesized de novo.

INITIATION SYSTEM: The reaction mixture (250 μl) contained: 25 μmol of KCl, 5 μmol of HEPES·KOH (pH 7.3 at 20°C), 0.75 μmol of $MgCl_2$, 0.5 μmol of dithiothreitol (DTT), 0.25 μmol of ATP, 0.05 μmol of GTP, 1.5 μmol of creatine phosphate, a few crystals of creatine phosphokinase, 10 μCi of a reconstituted protein hydrolysate (algal profile) containing 15 ^{14}C-amino acids, 7.5 nmol each of the five amino acids not present in the algal hydrolysate (asparagine, cysteine, glutamine, methionine, and tryptophan) as well as S^N (0.4 A_{260} units), L^O (1.2 A_{260} units), 100 μl pH 5 enzymes and poly (A)-containing RNA (0.05 A_{260} units).

READOUT SYSTEM: The composition of this system was identical to that of the initiation system except that it contained either free or detached ribosomes or rough microsomes instead of S^N, L^O, and mRNA (in one case it also contained S^N).

Incubation in both systems was at 37°C. 10-μl aliquots (unless indicated otherwise in figure legends) were removed at indicated time intervals and spotted on 3M Whatman filter paper disks, which were processed according to Mans and Novelli (21). Radioactivity was determined in toluene-Liquifluor (New England Nuclear Corp., Boston, Mass.) in a Beckman LS 350 liquid scintillation counter at about 75% efficiency.

PROTEOLYSIS OF TRANSLATION PRODUCTS: 25-μl aliquots removed from the two systems described above after incubation (see figure legends), were cooled to 0–2°C in an ice bath, and each treated for 3 h at the same temperature with 3 μl of a solution containing trypsin and chymotrypsin (500 μg of each per ml). Proteolysis was terminated by the addition of 1 vol of 20% TCA, and the ensuing precipitate was prepared for SDS-polyacrylamide gel electrophoresis (PAGE) as described below.

ANALYSIS OF TRANSLATION PRODUCTS BY SDS-PAGE: 25-μl aliquots removed from the two systems either after completion of, or at various times during, incubation were cooled to 0–2°C in an ice bath and treated with an equal volume of ice-cold 20% TCA; after 1 h the ensuing precipitate was collected at 0–4°C by centrifugation in a swinging bucket rotor for 10 min at 2,000 g. The supernate was removed as completely as possible and the precipitate was dissolved by incubation for 20 min at 37°C in 30 μl of a solution containing 15% sucrose, bromophenol blue (serving both as a pH indicator for the sample and as a tracking dye for electrophoresis), 100 mM Tris base and 8 mM DTT (if the solution turned yellow [pH 3], Tris base was added in 1-μl aliquots to restore the blue color [pH 4.5 and higher]). Solubilization was completed by incubation in a boiling water bath for 2 min. After cooling to room temperature, 2 μl of a 0.5 M solution of α-iodoacetamide was added to each sample, and the mixture was incubated for 1 h at 37°C before a 25-μl aliquot was layered into a slot of a polyacrylamide slab gel.

Rabbit globin, porcine chymotrypsinogen, ovalbumin, and bovine albumin were treated in an identical manner and were used as standards for mol wt determinations.

The slab gel (1 mm thick) consisted of a 10–15% acrylamide gradient serving as a resolving gel and a 5% acrylamide stacking gel, both in SDS and buffers as described by Maizel (20). Electrophoresis was for 20 h and at constant current.

After electrophoresis, the slab gel was stained in a solution containing 0.2% Coomassie Brilliant Blue, 50% methanol, and 10% glacial acetic acid for 2 h and then destained in 50% methanol and 10% acetic acid. After destaining, the gel was soaked in the last solution with 5% glycerol added; this was helpful in preventing the gels from cracking during and after drying on Whatman 3 M paper.

AUTORADIOGRAPHY (AR) OF DRIED POLYACRYLAMIDE GELS AND DENSITOMETRIC ANALYSIS OF BANDS: Dried gels were exposed to medical X-ray film (Cronex 2D, du Pont de Nemours and Co., Inc., Wilmington, Del.), generally for a few days. The films were developed by conventional procedures.

The bands in the developed X-ray films were analyzed using a Joyce-Loebl densitometer (Joyce, Loebl and Co., Inc., Burlington, Mass.). The area under the resulting peaks was integrated and used as a quantitative measure for the radioactivity in the bands.

The validity of this quantitation procedure was verified by analyzing the autoradiographs derived from 10, 20, and 30-μl aliquots of a sample loaded into different slots; it was found that the area under each peak in the densitometry tracing was proportional to the amount loaded into the slot.

Preparation of Labeled Secretion Product from MOPC 41 and MOPC 41 DL-1

Freshly excised tumor freed of necrotic portions was sliced into small pieces using a razor blade. The tumor slices were washed several times in 150 mM NaCl, 20 mM Tris·HCl pH 7.4, and 5 mM $MgCl_2$, using centrifugation in a swinging bucket rotor at 500 *g* and at 4°C. Approximately 1 ml of packed tumor slices were resuspended in 3 ml of a medium containing minimal essential medium balanced salt solution for suspension cultures, vitamins, bicarbonate, and glucose as specified by Eagle (11) and supplemented with 50 μl (=50 μCi) of 15 ^{14}C-amino acids which were part of a reconstituted protein hydrolysate (see above). The resuspended slices were transferred to a 10-ml Erlenmeyer flask, gassed with 95% O_2-5% CO_2, and incubated for 5 h at 37°C. After incubation the tumor slices were centrifuged into a pellet at 1,000 *g*. The supernate was centrifuged for 1 h at 105,000 *g* in a Spinco no. 40 rotor. The resulting supernate contained secreted, radioactively labeled IgG light chains which were precipitated with 1 vol of ice-cold 20% TCA. The ensuing precipitate was prepared for SDS-PAGE as described above.

Electron microscopy

Pellets were fixed in 2% glutaraldehyde in TeaKM for 1 h at 0°C and postfixed in 2% OsO_4 in TeaKM for 1 h at 0°C. The pellets were stained with 0.5% uranyl acetate in acetate-Veronal buffer before dehydration and Epon embedding (12, 18). Sections were cut on a Porter-Blum MT2-B ultramicrotome (Dupont Instruments, Sorvall Operations, Newtown, Conn.) equipped with a diamond knife (Dupont Instruments, Wilmington, Del.). They were stained with uranyl acetate (38) and lead citrate (37) and viewed with a Siemens Elmiskop 101 at 80 kV.

Source of Materials

Oligo (dT) cellulose T-2 from Collaborative Research, Inc., Waltham, Mass. ATP, disodium salt; GTP, sodium salt; creatine phosphate, disodium salt; and creatine phosphokinase, salt-free powder from Sigma Chemical Co., St. Louis, Mo. α-iodoacetamide from Calbiochem, San Diego, Calif. DTT from R. S. A. Corp., Ardsley, N. Y. Coomassie Brilliant Blue and reconstituted protein hydrolysate (^{14}C), algal profile (1 mCi/1 ml) from Schwarz/Mann Div., Becton, Dickinson and Co., Orangeburg, N.Y., bovine pancreatic trypsin, 2× crystallized (185 U/mg), and bovine pancreatic α-chymotrypsin, 3× crystallized (49 U/mg), from Worthington Biochemical Corp., Freehold, N. J.

RESULTS

Cell Fractionation of MOPC 41 DL-1 Tumors

Among the characteristic ultrastructural features of murine myelomas are greatly dilated endoplastic reticulum (ER) cisternae and the occurrence of intracisternal A particles budding from the ER membranes. These features may have contributed to the unusually high density of rough microsomes, the bulk of which were found to band isopynically at 1.75 M sucrose. Some rough microsomes sedimented even through a 2.0 M sucrose cushion (conventionally used in the fractionation of liver cells (3) as a cutoff concentration for preventing sedimentation of rough microsomes) and therefore were present as contaminants in the free ribosome fraction (see Fig. 2). Rough microsomes were further fractionated by detergent treatment to isolate a fraction referred to as detached ribosomes. Sedimentation profiles of free as well as detached ribosomes in sucrose gradients are shown in Fig. 1. Many of the ribosomes, particularly in the case of detached ribosomes, were in the form of polysomes indicating that RNase action during cell fractionation was minimal. Characteristic for the profile of detached ribosomes (Fig. 1, BR) is the presence of significant amounts of large ribosomal subunits (designated L), whereas the profile of free ribosomes (Fig. 1, FR) is distinguished by the presence of a large amount of monosomes and a small but significant amount of S^N; large ribosomal subunits may be present but may not be resolved from the monosome peak. Furthermore, some material contained in the free-ribosome preparation had sedimented to the bottom of the sucrose gradient tube. This pellet was fixed, stained, sectioned, and inspected by electron microscopy. It showed both free ribosomes as well as rough microsomes (Fig. 2). A fraction collected from the sucrose gradient (Fig. 1, FR) comprising the monosome peak as well as the polysome region was sedimented by centrifugation and prepared for electron microscopy. It showed only free ribosomes without contamination by rough microsomes (electron micrograph not shown). Thus, for purification of free ribosomes, sucrose gradient centrifugation can be used to eliminate rough microsome contamination. Electron microscopy of the rough microsomal fraction showed the characteristic ribosome-studded vesicles (Fig. 3). Occasionally a few lysosomes were seen. Frequently the ER showed a thickening and there were also intracisternal particles characteristic for murine myelomas.

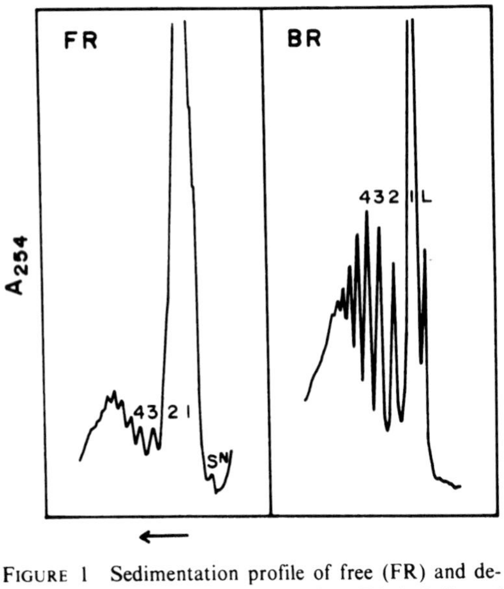

FIGURE 1 Sedimentation profile of free (FR) and detached (BR) ribosomes from MOPC 41 DL-1. Pellets of free and detached ribosomes were resuspended in ice-cold double-distilled water. 0.2-ml aliquots containing 3.0 A_{260} units were layered on 12.5 ml of 10–40% sucrose gradients in 100 mM KCl, 20 mM triethanolamine·HCl pH 7.4 at 20°C, and 3 mM $MgCl_2$. The gradients were centrifuged at 4°C in an SB 283 rotor of an IEC centrifuge for 100 min at 190,000 g_{av}. Fractionation of the sucrose gradients and recording of the optical density were as described (6). Arrow indicates direction of sedimentation. The native small ribosomal subunit peak is designated as S^N, the large ribosomal subunit peak as L, the mono-, di-, tri-, and tetrasome peaks, as 1, 2, 3, and 4, respectively.

In Vitro Translation of mRNA for the Light Chain of IgG

The crude mRNA prepared from rough microsomes as described under Materials and Methods was translated in a heterologous system, developed in this laboratory (13) (henceforth re-

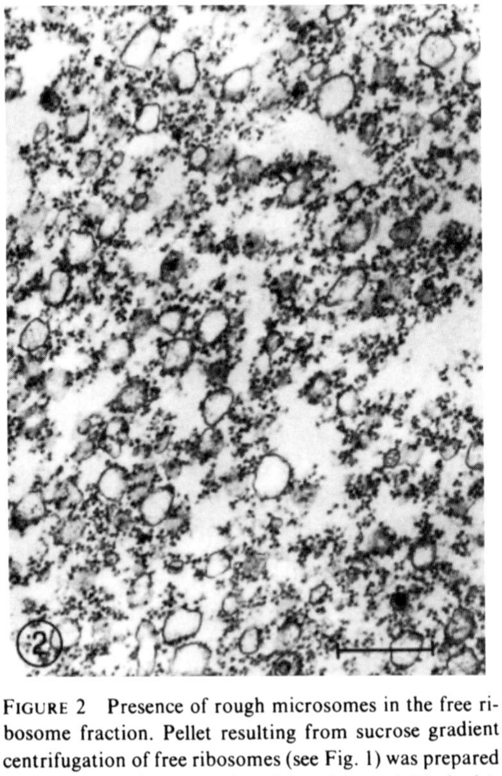

FIGURE 2 Presence of rough microsomes in the free ribosome fraction. Pellet resulting from sucrose gradient centrifugation of free ribosomes (see Fig. 1) was prepared for electron microscopy (see Materials and Methods). × 25,500. The bar denotes 0.5 μm.

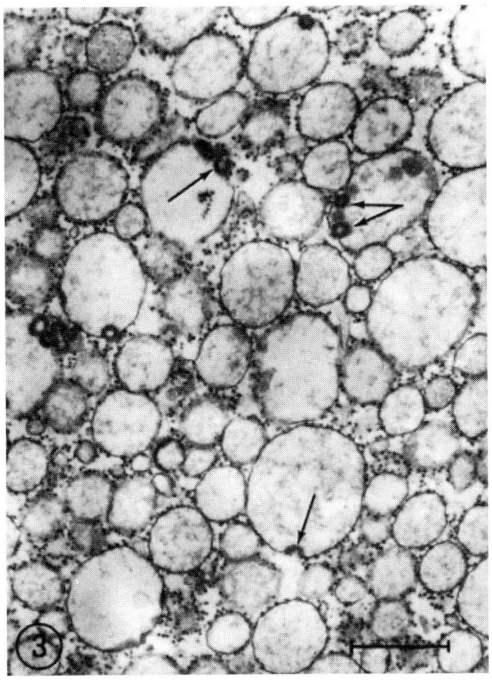

FIGURE 3 Rough microsome fraction was isolated and prepared for electron microscopy as described in Materials and Methods. Arrows point to local thickenings of the ER membrane and to intracisternal A particles. × 25,500. The bar denotes 0.5 μm.

ferred to as initiation system) consisting of S^N from rabbit reticulocytes (as a source of small ribosomal subunits as well as initiation factors), L^O prepared by the puromycin KCl procedure (6) from rat liver free ribosomes, and pH 5 enzymes from Krebs ascites cells. The time-course of polypeptide synthesis in this system in the presence or absence of mRNA is shown in Fig. 4. There was a more than twofold stimulation of polypeptide synthesis in the presence of mRNA. There was inhibition of polypeptide synthesis in the presence of aurintricarboxylic acid (ATA) at a concentration (10^{-4} M) which has been reported to inhibit initiation but not elongation in polypeptide synthesis (17). The extent of inhibition was similar in the absence (data not shown) and in the presence of mRNA.

Analysis of the products by SDS-PAGE and AR (Fig. 5) showed that a prominent radioactive band (slot B) was synthesized by the initiation system in the presence of the crude mRNA fraction (supposed to contain mostly light chain mRNA). This polypeptide has an estimated mol wt of 25,000 and is therefore larger in mol wt by ~4,000 than the secreted light chain of IgG of MOPC 41 DL-1 (slot S), which has a mol wt of 21,000. It was tentatively identified as the "precursor" of the light chain on the basis of work by other laboratories (15, 19, 23, 32, 34–36) showing that the primary translation product of mRNA for the light chain of IgG is longer than the secreted light chain.

In Vitro Translation of mRNA's Contained in Free Ribosomes, Bound Ribosomes, and Rough Microsomes

The time-course of polypeptide synthesis in a readout system, containing either free or detached ribosomes and pH 5 enzymes, is shown in Fig. 6 and is compared to the time-course of mRNA translation in an initiation system (see above). Because pH 5 enzymes contain only small amounts of initiation factors, polypeptide synthesis is essentially completed in the readout system after a 40-min incubation, whereas in the initiation system translation continues for more than 120 min, although at a slower initial rate.

Analysis of the products by SDS-PAGE and AR (Fig. 7) showed that two major products were synthesized by detached ribosomes. One of them (slot B–, upward pointing arrow) is a polypeptide of the same mol wt as the authentic secreted light

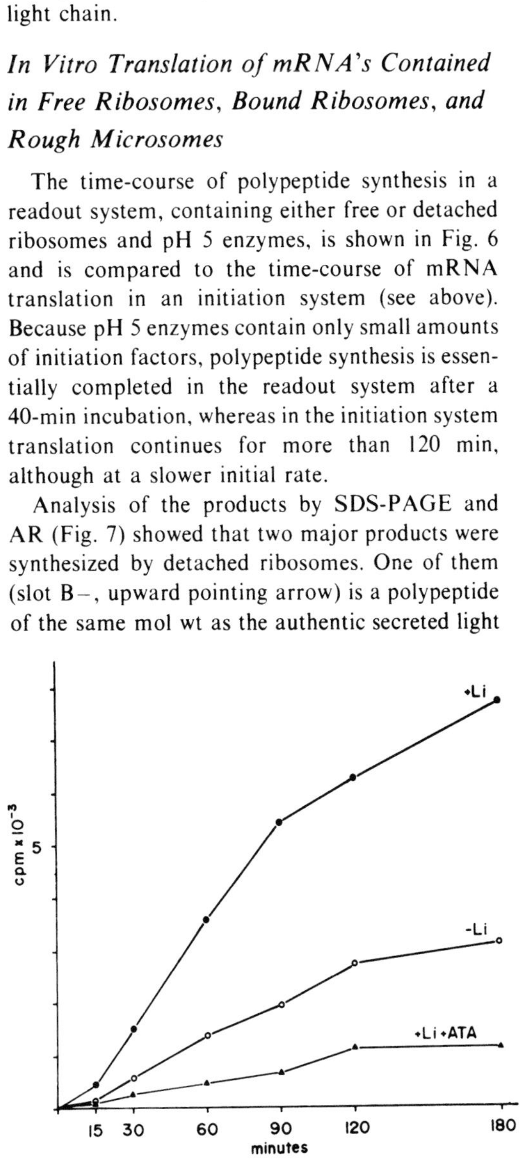

FIGURE 4 Time-course of polypeptide synthesis in an *initiation* system in the presence of mRNA for the light chain of IgG (+Li), in the presence of light chain mRNA and of 10^{-4} M ATA (+Li + ATA), and in the absence of mRNA (–Li). For details, see Materials and Methods.

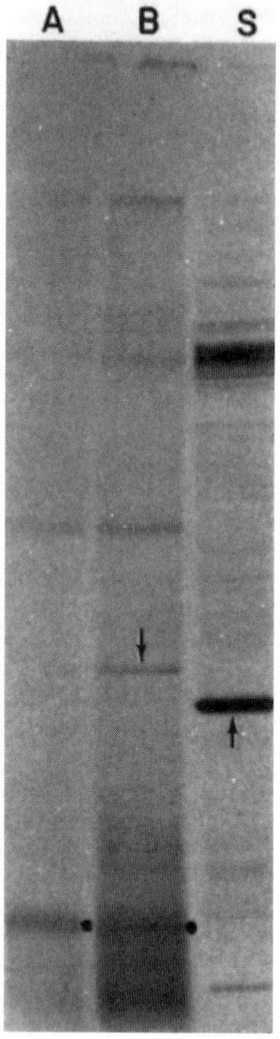

FIGURE 5 Analysis by SDS-PAGE and AR of an experiment of the type described in Fig. 4. Shown are the labeled products synthesized at the 180-min time point in the absence of mRNA (slot A) or in the presence of mRNA for the light chain of IgG (slot B). For comparison the labeled secreted light chain of IgG is shown in slot S (upward pointing arrow). Downward pointing arrow (slot B) indicates precursor of the light chain of IgG. Dots indicate the globin chains.

chain (shown in slot S); the other one (slot B−, downward pointing arrow) has the same mol wt as the precursor of the light chain synthesized in the initiation system (slot A). The products synthesized by crude *free* ribosomes (slot F−) also contained a band of a mol wt identical to that of the light chain of IgG. However, this band was shown to be due to the presence of rough microsomes in the crude free-ribosome fraction, since it was absent when purified free ribosomes collected from a sucrose gradient (see above) were tested in the same manner (data not shown).

It was shown recently that nascent polypeptides synthesized on free ribosomes from rat liver are sensitive to mild proteolysis, except for a fragment of ~40 amino acid residues on the carboxyl terminal which is thought to be protected because of its localization within the large ribosomal subunit (4). Products synthesized by rough microsomes of rat liver, on the other hand, were shown to be largely resistant to proteolytic attack since they are protected by the surrounding ER membrane (29). In agreement with these findings are the results shown in Fig. 7 slots B and F. The two major translation products of detached ribosomes were sensitive to mild proteolysis (slot B+); however, more than 60% (estimated from densitometric analysis) of the band in the position of the authentic light chain of IgG, apparently synthesized by rough microsomes present in the crude free-ribosome fraction, was resistant to proteolytic attack (slot F+; see also Fig. 14 slots A+ and A−).

The synthesis by detached ribosomes of two major polypeptides, one of them with the same mol

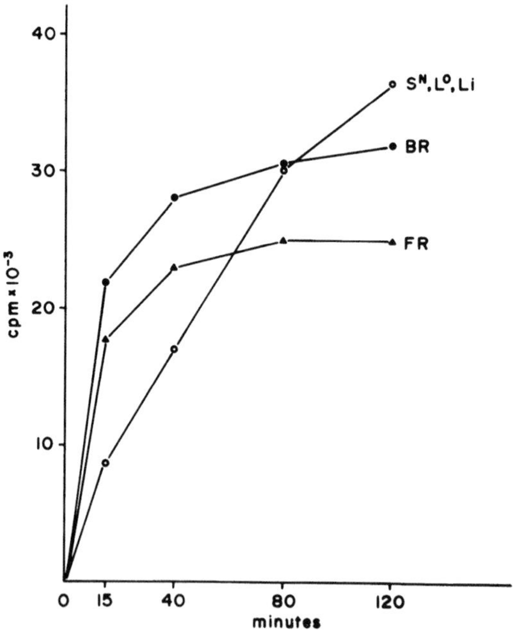

FIGURE 6 Time-course of polypeptide synthesis in a readout system containing 9.8 A_{260} units of crude free (FR) or 5.4 A_{260} units of detached (BR) ribosomes. For comparison the time-course of polypeptide synthesis in an initiation system containing mRNA for the light chain of IgG (S^N, L^O, Li) was included. In the latter case, 50-μl aliquots were counted, whereas in the former 10-μl aliquots were counted.

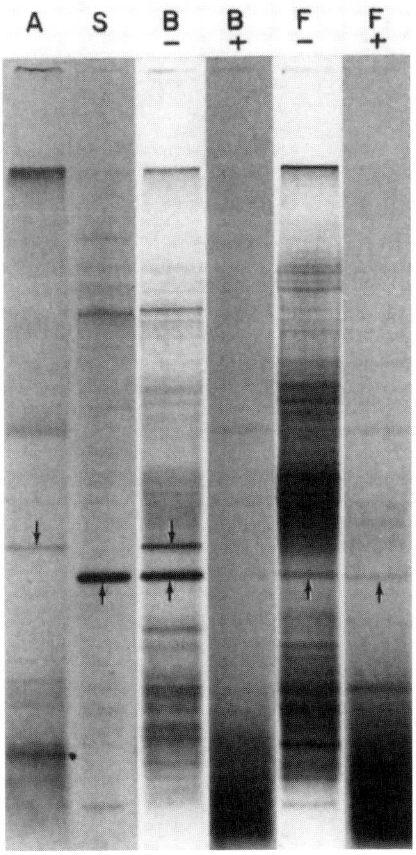

FIGURE 7 Analysis by SDS-PAGE and AR of products synthesized at the 120-min time point as described in Fig. 6 after incubation in the absence or presence of proteolytic enzymes. Shown are the labeled products synthesized in an initiation system in the presence of light chain mRNA (slot A), the labeled secreted light chain (slot S), and the products synthesized in a readout system containing detached ribosomes (slot B-) subsequently subjected to proteolysis (slot B+) or containing crude free ribosomes (slot F−), subsequently subjected to proteolysis (slot F+). Symbols used were as in Fig. 5.

wt as the precursor and the other with the same mol wt as the secreted light chain, suggested that isolated detached ribosomes are heterogeneous with respect to their content of processed and unprocessed nascent light chains. On the basis of the predictions made in the signal hypothesis, one could reason that those ribosomes located near the 5′ end of mRNA should contain unprocessed nascent light chains that still have their signal sequence, while those near the 3′ end should contain already proteolytically processed nascent light chains.

These assumptions were borne out by data obtained from a time-course experiment using detached ribosomes in a readout system; translation was stopped at various time points and the products were analyzed by SDS-PAGE and AR. To insure that there was no initiation in the readout system, a condition which was not met in the previous experiment, ATA was added in a concentration which has been reported (17) to inhibit initiation, but not elongation or release of nascent chains. That such conditions were achieved is demonstrated by the data shown in Fig. 4 and by the lack of inhibition observed in the readout system in the presence of ATA and detached ribosomes (data not shown). From the autoradiograph shown in Fig. 8 it is evident (slots 1 and 6) that at the earlier time points of readout, only processed chains were synthesized, apparently as a result of completion of chains by ribosomes near the 3′ end of mRNA. Only at later time points when ribosomes located further to the left on the mRNA have completed their readout were unprocessed chains synthesized (slots 9, 18, 25, and 50). The data of a quantitative analysis (see Materials and Methods) of this experiment are summarized in Fig. 9. It can be seen that synthesis of already processed chains was essentially completed after a 10-min incubation when there was only a barely detectable synthesis of unprocessed chains; the latter were synthesized only in the following 30 min. no significant synthesis of either chain was observed after a 50-min incubation (data not shown).

While in the preceding experiment initiation had to be ruled out in order to substantiate the conclusion that detached ribosomes contain both processed and unprocessed chains, the following experiment was performed under conditions in which initiation could take place. Such conditions were achieved by adding to the readout system, containing detached ribosomes, increasing amounts of S^N from rabbit reticulocytes as a source of initiation factors (13). Up to twofold stimulation in polypeptide synthesis was observed as a result of the addition of increasing amounts of S^N (Fig. 10). This stimulation could have been the result of the presence of small amounts of globin mRNA present in the S^N fraction. However, this was ruled out by product analysis using SDS-

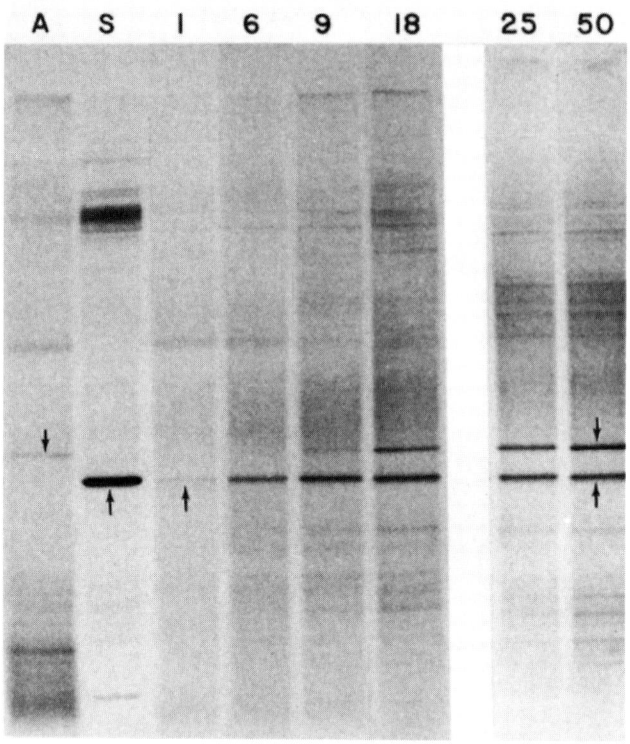

FIGURE 8 Analysis by SDS-PAGE and AR of the products synthesized by detached ribosomes (1.8 A_{260} units) during the course of readout in the presence of 10^{-4} M ATA. Readout was terminated at 1, 6, 9, 18, 25, and 50 min (slots 1, 6, 9, 18, 25, 50, respectively) by cooling 25-μl aliquots to 0°C and adding 25 μl 20% TCA. For comparison, the precursor of the light chain (slot A) synthesized in an initiation system (see Fig. 5) and the labeled secreted light chain (slot S) were included. Symbols used were as in Fig. 5. Slots 25 and 50 were from a separate slab gel.

PAGE and AR (Fig. 11). Only small amounts of globin were synthesized in response to increasing amounts of added S^N. Quantitative analysis (Fig. 12) of the radioactivity in the bands corresponding to the processed and unprocessed light chain of IgG revealed that the stimulation by increasing amounts of S^N can be accounted for by a proportional stimulation in the synthesis of unprocessed chains while there was no stimulation in the synthesis of processed chains.

Finally, readout experiments using isolated rough microsomes were performed. Fig. 13 shows the time-course of polypeptide synthesis in rough microsomes in the absence and presence of ATA, which produced only a slight inhibition of amino acid incorporation. Analysis of the products by SDS-PAGE and AR (Fig. 14) revealed the synthesis of a polypeptide corresponding to the mol wt of the processed light chain of IgG (slot A−). In contrast to the results obtained with detached ribosomes (see above), newly synthesized unprocessed chains were not detected. However, unprocessed light chains were synthesized when in vitro readout of rough microsomes took place in the presence of 10^{-4} M ATA (Fig. 14, slot B−).

The products synthesized in the readout experiment with rough microsomes were subjected to mild proteolysis and subsequently analyzed by SDS-PAGE and AR. As expected, the processed chains, presumably inside the microsomal vesicles, were largely protected from proteolytic attack. Densitometric analysis revealed that ~60% was resistant to proteolysis in agreement with the results shown in Fig. 7, slot F+. However, the

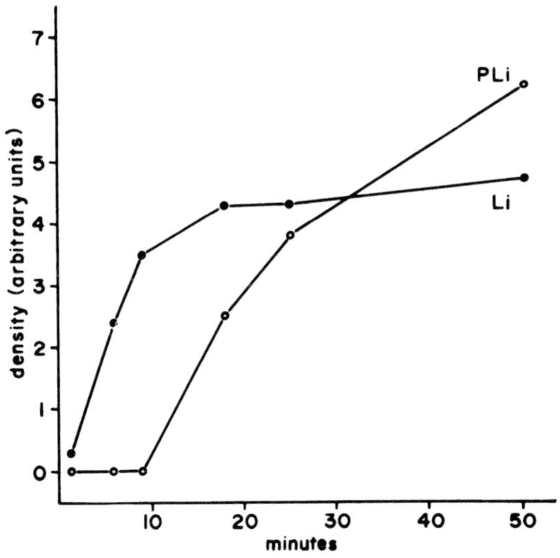

FIGURE 9 Quantitation by densitometry of the autoradiograph shown in Fig. 8. PLi and Li designate the unprocessed (downward pointing arrow in Fig. 8) and the processed light chains of IgG (upward pointing arrow in Fig. 8), respectively.

unprocessed chains synthesized on rough microsomes in the presence of ATA were degraded (Fig. 14, slot B+), supporting the interpretation that these chains were not segregated in the intravesicular space.

DISCUSSION

The results of cell fractionation reported here are in agreement with those of Cioli and Lennox (10). Although our cell fractionation was performed on solid MOPC 41 DL-1 tumors, a conventionally prepared free-ribosome fraction also was found to be contaminated by rough microsomes (see Fig. 2). This contamination was eliminated by isokinetic sucrose gradient centrifugation. Purified free ribosomes did not contain any detectable light chain mRNA activity, although it cannot be ruled out that some light chain mRNA was present and resulted in the synthesis of unprocessed light chains, which overlapped with the presence of other bands in the autoradiograph (see Fig. 7, slot F−).

Upon in vitro translation in an initiation system (see Materials and Methods) of a crude mRNA fraction containing the mRNA for the light chain of IgG, a product was synthesized which was larger by ~4,000 mol wt than the authentic secreted light chain. Similar results have been reported by several laboratories (15, 19, 23, 32, 34–36).

The majority of the data presented in this paper were obtained from in vitro translation of the light chain mRNA contained in the isolated rough microsome and detached ribosome fractions of MOPC 41 DL-1. It was shown here that both fractions contain unprocessed light chains (i.e., chains still containing the signal sequence) together with processed light chains, demonstrating that removal of the signal sequence in vivo takes place well before the nascent chain is completed. However, the difference between these two fractions is that in vitro only rough microsomes retain their ability for proteolytic removal of the signal sequence. Thus, in vitro completion of their nascent chains also results in concomitant proteolytic removal of the signal sequence from their unprocessed chains, and thus in the synthesis of only processed chains (see Fig. 14, slot A−). In contrast, detached ribosomes, having lost their ability

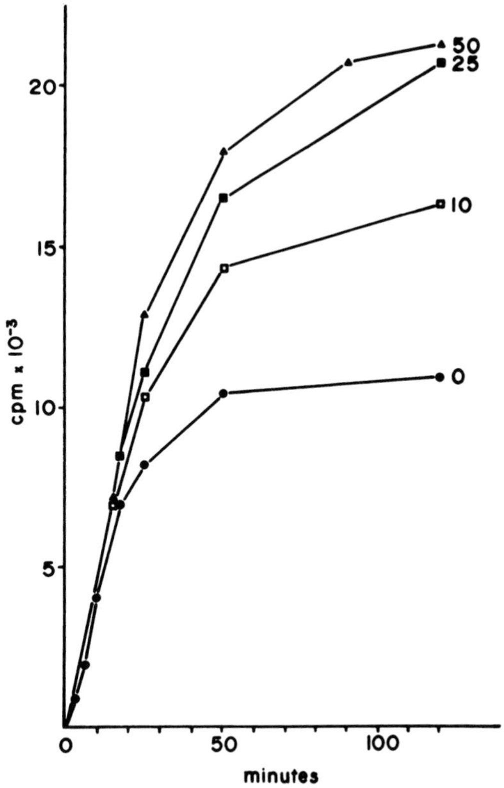

FIGURE 10 Time-course of polypeptide synthesis in a readout system containing detached ribosomes (1.8 A_{260} units) and S^N from reticulocytes as a source for initiation factors. Numbers next to curves designate the amount (in microliters) of S^N (10.5 A_{260} units/ml) present in the 250-μl assay.

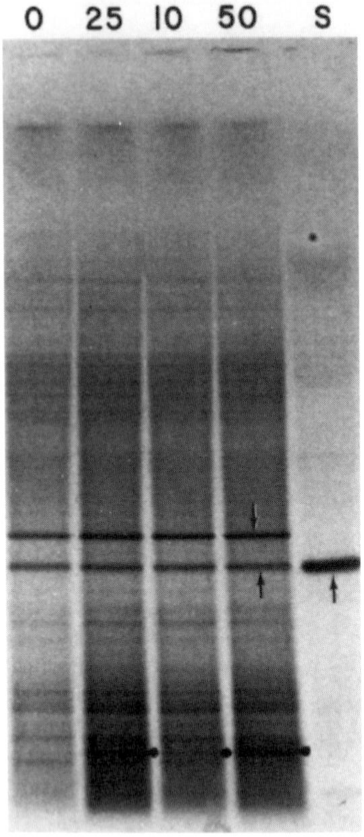

FIGURE 11 Analysis by SDS-PAGE and AR of the products synthesized at 120-min time point as described in Fig. 10. Slot numbers refer to microliters of S^N present in the readout system with detached ribosomes (see Fig. 10). For comparison, the labeled secreted light chain of IgG is included in slot S. Symbols used were as in Fig. 5

for in vitro processing, yielded both unprocessed as well as processed light chains upon completion of their nascent chains in vitro. This result can be rationalized by assuming that the processing activity is part of the membrane and was lost during the preparation of detached ribosomes by detergent solubilization of rough microsomes. Alternatively, the processing enzyme(s) may still be present in isolated detached ribosomes but in an inactivated form or may become inactivated rapidly during incubation in the readout system.

The data presented here also indicated that those polysomal ribosomes containing unprocessed chains are located on the mRNA to the left of those containing already processed chains (Figs. 8 and 9). Furthermore, the distribution of radioactivity between unprocessed and processed chains (more than one half in the former, see Fig. 9) indicated that probably more than one of the polysomal ribosomes contain unprocessed chains. This suggested that removal of the signal sequence occurs only after the ribosome has already translated a considerable portion of the mRNA. If removal of the signal sequence is an endoproteolytic event and if the processing activity is localized in the membrane *trans* rather than *cis* with respect to the ribosome-membrane junction, then the entire signal sequence would be required to have traversed the 70-Å distance which comprises the thickness of the ER membrane before processing could take place. Assuming that the signal sequence comprises ~20 amino acid residues, and adding to these ~19 and ~39 residues which comprise the portions of the nascent chain (3.6 Å per residue in the extended configuration) in the membrane and in the ribosome (4), respectively, then a total of ~78 amino acid residues have to be polymerized before processing can take place. Since the interribosomal distance on a ribosome-saturated polysome amounts to 90 nucleotides, or 30 codons, it is possible for the mRNA to accommodate two to three ribosomes

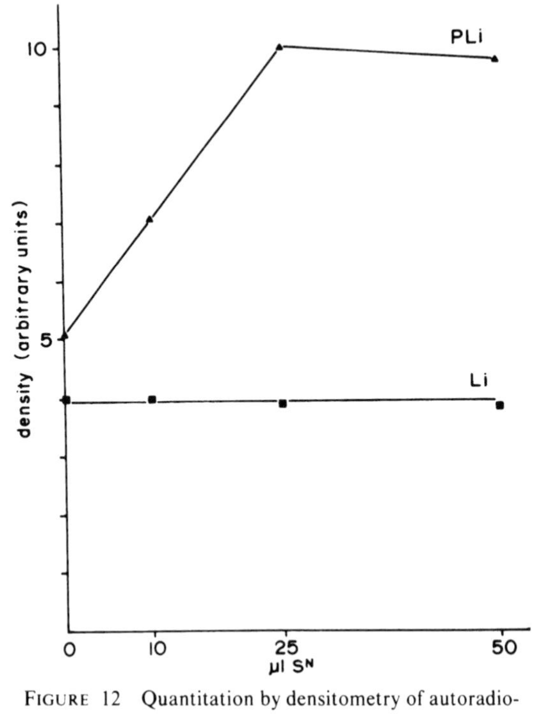

FIGURE 12 Quantitation by densitometry of autoradiograph shown in Fig. 11. PLi and Li designate the unprocessed and processed chains of IgG, respectively.

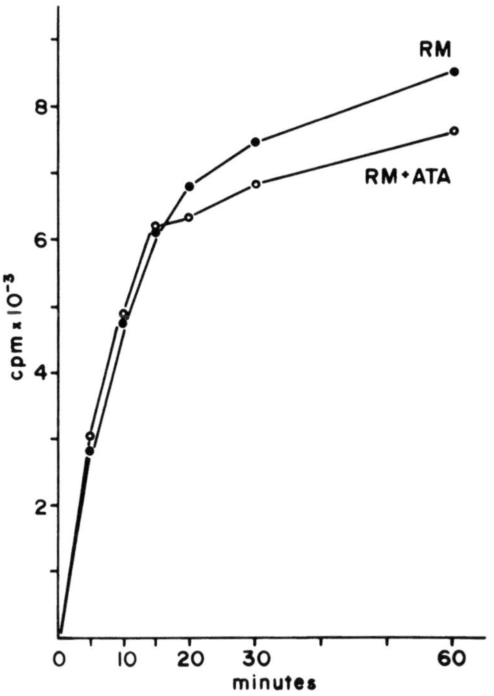

FIGURE 13 Time-course of polypeptide synthesis in the absence of ATA (RM) or presence (RM+ATA) of 10^{-4} M ATA in a readout system containing rough microsomes (4.3 A_{260} units).

still containing the signal terminal of their nascent chain, i.e., containing unprocessed chains.

The synthesis by detached ribosomes of unprocessed chains of the same mol wt as the translation product of light chain mRNA in a heterologous reconstituted system suggests that the latter is not an in vitro artifact. It could have been argued, otherwise, that translation of the light chain mRNA, which has been performed so far in all cases in heterologous systems, resulted in artifactual initiation at some point to the left of the initiation codon and therefore caused the synthesis of a larger precursor protein, while in vivo such a precursor protein would not have been synthesized.

By using ATA to inhibit initiation but not elongation and release of the nascent chain, it was clearly ruled out that the synthesis of unprocessed chains by detached ribosomes reflected in vitro initiation rather than completion of unprocessed chains. Conversely, the observed twofold stimulation of polypeptide synthesis by detached ribosomes, in a readout system which was supplemented by initiation factors, was shown to be entirely the result of increased synthesis of unprocessed chains, while the level of synthesis of processed chains was not affected. This result also demonstrated that initiation factors from rabbit reticulocytes which contain predominantly free ribosomes were able to perform in the translation of the light chain mRNA contained in bound ribosomes, supporting the contention that identical initiation factors are used for translation on free and bound ribosomes.

It should be noted that the presence of unprocessed chains in rough microsomes was detected only if completion of their nascent chains in vitro occurred in the presence of ATA. This result is of particular interest since Borgese et al. (8) have recently shown that ribosomes will not bind to stripped membranes in vitro in the presence of ATA. This result can therefore be rationalized as

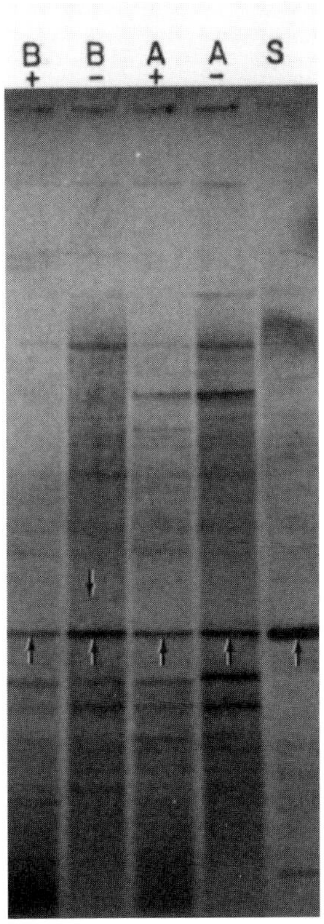

FIGURE 14 Analysis by SDS-PAGE and AR of the products synthesized at the 60-min time point as described in Fig. 13 and subsequently incubated in the absence or presence of proteolytic enzymes. Products synthesized in the absence or presence of 10^{-4} M ATA are shown in slots A and B, respectively. Products not subsequently incubated with proteolytic enzymes (see Materials and Methods) are shown in slots marked (−), those incubated in slots marked (+). For comparison the labeled secreted light chain of IgG is shown in slot S.

follows: Those ribosomes located farthest to the left on the mRNA in a rough microsome are not yet attached to the membrane. In the absence of ATA, attachment of these ribosomes as well as subsequent processing of their nascent chains can occur in vitro during chain completion, resulting in the synthesis of processed chains. In the presence of ATA, however, these ribosomes will not be able to attach. This in turn will deprive their nascent chains of access to the processing activity, but will not interfere with their completion in vitro. Alternatively, it is also possible that the synthesis of unprocessed chains in the presence of ATA does not result only from an interference in establishing the ribosome-membrane junction but is due to a direct inhibition of the processing activity by ATA.

Finally, it was demonstrated here that the microsomal membrane in the rough microsome fraction provides protection for the completed and released chain against proteolysis. Although protection of newly synthesized chains in microsomes (isolated from rat liver) has been demonstrated previously (29), it was assayed not by product analysis but by measuring the percentage of acid-insoluble radioactivity remaining after proteolysis. However, since this protection was observed to be significantly less than 100%, this type of assay could not distinguish between protection resulting from partial hydrolysis of all chains or resulting from a combination of complete resistance of some chains and complete hydrolysis of others. Using product analysis by SDS-PAGE and AR after proteolysis, it was established here (see also companion paper) that the protection of the majority of the product was complete in that there was no reduction in its mol wt. This constitutes important information since resistance to proteolysis (29) is at present the only rigorous assay for vectorially discharged chains. The original centrifugation assay (showing that newly made chains were sedimented with the membranes or were solubilized by detergent treatment) which was used to demonstrate vectorial discharge (26, 27) has subsequently been shown to be inadequate (8, 9, 31).

It should be emphasized that most of our conclusions remain to be confirmed by further characterization of the in vitro translation products, in particular of the band which we have assumed to be the unprocessed precursor of the light chain of IgG (on the basis of its mol wt). However, preliminary data[2] obtained by sequence analysis of 50 amino terminal residues of this putative precursor protein support our conclusions.

An Hypothesis for the Transfer of Proteins across Membranes

Most of the data which led to the formulation of an hypothesis for the transfer of proteins across membranes, referred to henceforth as the signal hypothesis, have been reviewed previously (5) and therefore will not be discussed here. Alternative hypotheses have been summarized previously (30) and will be omitted here. Instead, a more detailed version of the signal hypothesis than that presented previously (5), based on theoretical considerations as well as recent data from this and other laboratories, will be outlined.

As mentioned above, the essential feature of the signal hypothesis (illustrated in Fig. 15) is the occurrence of a unique sequence of codons, located immediately to the right of the initiation codon, which is present only in those mRNA's whose translation products are to be transferred across a membrane. No other mRNA's contain this unique sequence. Translation of the signal codons results in a unique sequence of amino acid residues on the amino terminal of the nascent chain. Emergence of this signal sequence of the nascent chain from within a space in the large ribosomal subunit triggers attachment of the ribosome to the membrane, thus providing the topological conditions for the transfer of the nascent chain across the membrane.

Following is an attempt to formulate this sequence of events in greater detail. It is suggested that translation of mRNA's containing signal codons begins on a free ribosome. Thus, initiation of translation of all mRNA's whether or not they contain signal codons proceeds by the same mechanism, eliminating the need for specialized ribosomes or initiation factors. Similarly, elongation will proceed on a free ribosome for both categories of mRNA's until anywhere from 10 to 40 amino acid residues of the nascent chain have emerged from the ribosome. Only at this point is the membrane able to distinguish between amino terminals of nascent chains as containing or not containing the unique signal sequence. If they lack the signal sequence, attachment of the ribosome to the membrane will not occur. If they contain the signal sequence, attachment may occur, but does not necessarily follow. It is conceivable, for example, that the availability of ribosome binding sites

[2] Devillers-Thiery, A., G. Blobel, and T. J. Kindt. Manuscript in preparation.

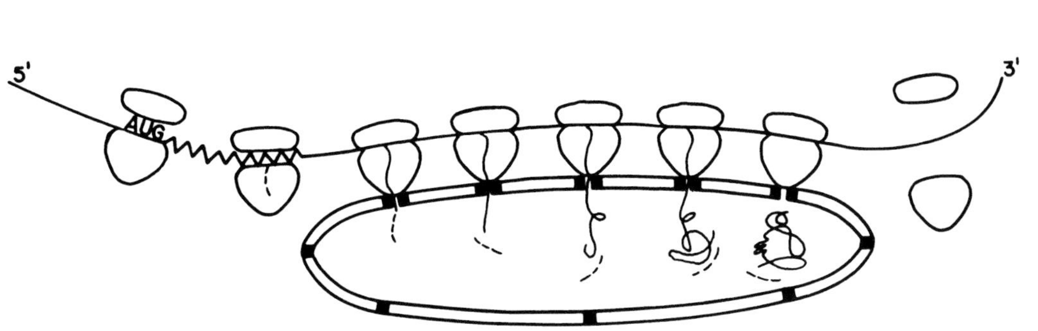

FIGURE 15 Illustration of the essential features of the signal hypothesis for the transfer of proteins across membranes. Signal codons after the initiation codon AUG are indicated by a zig-zag region in the mRNA. The signal sequence region of the nascent chain is indicated by a dashed line. Endoproteolytic removal of the signal sequence before chain completion is indicated by the presence of signal peptides (indicated by short dashed lines) within the intracisternal space. For details see text.

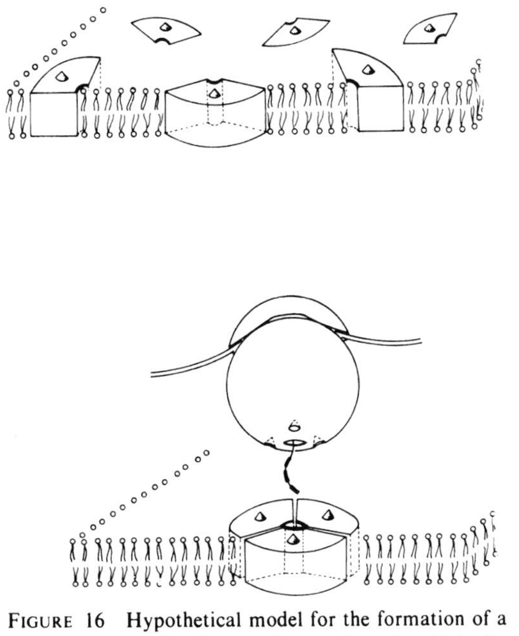

FIGURE 16 Hypothetical model for the formation of a transient tunnel in the membrane through which the nascent chain would be transferred. Specific regions of the signal sequence (indicated by line thickenings) of the nascent chain emerged from the large ribosomal subunit tunnel are recognized by one site each (indicated by a line thickening) on three membrane proteins. Recognition results in loose association of these proteins, subsequently "cross-linked" by interaction of sites on the large ribosomal subunit (indicated by notches around the tunnel exit on the large ribosomal subunit) with one site each on the three membrane proteins (indicated by cones). For details see text.

in the membrane may be limiting, allowing translation of signal sequence-containing peptides to continue to completion on nonattached ribosomes. In this event, chains containing the signal sequence would be released into the "soluble" compartment. These chains may be rapidly degraded, since they may be unable to assume their distinct native structure through the enzymatic modifications which are confined to the intracisternal space (e.g., proteolytic removal of the signal sequence or glycosylation). If, on the other hand, ribosome attachment occurs, translation will continue on a bound ribosome until the nascent chain is released and vectorially discharged. It is suggested that after discharge of the completed chain, the ribosome is detached from the membrane (evidence for the existence of a detachment factor will be presented elsewhere[3]). The released ribosome may again start translation of any mRNA, independent of whether the latter does or does not contain signal codons.

Ribosome attachment to, as well as detachment from, the membrane are likely to involve a complex sequence of events. A number of theoretical considerations form the basis for proposing the following model (illustrated in Fig. 16). The signal sequence of the nascent chain emerging from within a tunnel in the large ribosomal subunit may dissociate one or several proteins which have been found to be associated with the large ribosomal subunit of free ribosomes (7, 14, 22). Dissociation of these proteins may in turn uncover binding sites on the large ribosomal subunit. At the same time the emerging signal sequence also recruits two or more membrane receptor proteins and causes their loose association so as to form a tunnel in the membrane (see Fig. 16). This association is stabi-

[3] Blobel, G. Manuscript in preparation.

lized by each of these membrane receptor proteins interacting with the exposed sites on the large ribosomal subunit, with the latter playing the role of a cross-linking agent. Binding of the ribosome would link the tunnel in the large ribosomal subunit with the newly formed tunnel in the membrane in continuity with the transmembrane space. After release of the nascent chain into the transmembrane space, ribosome detachment from the membrane would eliminate the crosslinking effect of the ribosome on the membrane receptor proteins. The latter would be free again to diffuse as individual proteins in the plane of the membrane. As a result of their disaggregation, the tunnel would be eliminated. The tunnel, therefore, would not constitute a permanent structure in the membrane.

Recognition of the signal sequence by membrane receptor proteins may require precise synchronization with translation. Thus if, after emergence of the signal sequence from within the ribosome, translation were to continue without concomitant ribosome attachment, folding of the signal sequence with contiguous sequences of the nascent chain might effectively prevent subsequent ribosome membrane attachment. For this reason we postulate confluence of the large subunit and membrane tunnel, rather than other conceivable arrangements in order to provide topological conditions which would prevent the formation of secondary structures of the nascent chain at the ribosome membrane junction.

The model described above thus provides for a binding of the ribosome to the membrane which is functional, specific, and transient, that is limited in time as well as in space. Functional binding of the ribosome is coupled to tunnel formation in the membrane. By definition, then, nonfunctional binding does not involve tunnel formation and therefore does not provide the topology for the transfer of the nascent chain across the membrane. Functional binding also is specific in that it is limited only to those ribosomes which carry nascent chains containing the signal sequence. Finally, it is limited in time, in that it is linked to ongoing translation and it is limited in space, in that it occurs only on those membranes which contain ribosome binding sites and which possess sufficient fluidity to permit specific aggregation of these sites.

It should be noted that the signal hypothesis does not call for a direct attachment of mRNA to the membrane. In fact, it predicts that if initiation of protein synthesis were blocked, while elongation and release were to proceed, mRNA containing signal codons would be found in the soluble compartment of the cell rather than on rough microsomes. Cell fractionation then would recover these mRNA's in the free ribosome fraction, most likely in the form of mRNP's or bound to a free ribosome.

The original version of the signal hypothesis did not deal with the fate of the signal sequence. However, since the amino terminal sequence is different among the authentic secretory proteins, removal of the signal sequence before actual secretion was implied. It has been suggested (23) that the extra sequence in the amino terminal of the light chain precursor is removed by a membrane-associated activity. Data relevant to the proteolytic processing of the nascent chain are presented in this and in the following paper.

As already mentioned, the sequence of events suggested in the signal hypothesis may not be restricted to secretory proteins but may apply to the synthesis of all proteins which have to be transferred across a membrane. Thus, the mRNA's for lysosomal and peroxysomal proteins may contain identical signal codons as the mRNA's for secretory proteins, if the ribosome-ER junction is utilized for their transfer across the ER membrane. The same reasoning may apply to the synthesis of certain mitochondrial proteins which are synthesized in the cytoplasm. A junction of cytoplasmic ribosomes with the outer mitochondrial membrane has been described (16) and could be utilized for the transfer of these proteins across the outer mitochondrial membrane, presumably involving specific signal sequences and membrane recognition sites which are different from those of secretory proteins and the ER membrane, respectively. Finally, the synthesis of some membrane proteins may require transfer through the membrane before the protein could be inserted into the membrane. Depending on what particular membrane proteins would be required to be transferred for subsequent insertion, signal sequences and corresponding membrane recognition sites utilized for secretory and mitochondrial proteins could be involved. Other membrane proteins, in particular in those instances in which the site of synthesis is not separated from the site of insertion by the lipid bilayer, may be synthesized on free ribosomes.

These considerations make it evident that different signal sequences may exist. Moreover, the signal sequence for one group of proteins, e.g. for secretory proteins, may not be identical in all cases. Phylogenetic as well as ontogenetic variability may exist. Only after the sequence of a sufficient number of "signals" has been established will it be possible to recognize those sequence features which are essential for their postulated function, namely recognition by membrane binding sites. In addition, the signal sequence should contain information for its correct removal by the processing enzyme(s). One could therefore envision altered signal sequences which will not be recognized by membrane binding sites, with the result that transfer across the membrane could not take place. Likewise, an altered sequence may still be recognized by the membrane and result in transfer of the protein across the membrane but it may not serve as a substrate for the processing enzyme. The latter condition may be physiological for some proteins (for instance those proteins which need to retain the signal sequence for their proper function), pathological for others. The former condition, however, may be entirely pathological, since it would not lead to transfer across the membrane.

We thank Dr. G. Palade for his helpful comments and Mrs. N. Dwyer for preparing the illustrations for this and the following paper.

This investigation was supported by Grant Number CA 12413, awarded by the National Cancer Institute, DHEW.

Received for publication 30 June 1975, and in revised form 2 September 1975.

REFERENCES

1. Adelman, M. R., D. D. Sabatini, and G. Blobel. 1973. Ribosome-membrane interaction. Nondestructive disassembly of rat liver rough microsomes into ribosomal and membranous components. *J. Cell Biol.* **56**:206–229.
2. Aviv, H., and P. Leder. 1972. Purification of biologically active globin mRNA by chromatography on oligothymidylic acid-cellulose. *Proc. Natl. Acad. Sci. U. S. A.* **69**:1408–1412.
3. Blobel, G., and V. R. Potter. 1967. Studies on free and membrane-bound ribosomes in rat liver. I. Distribution as related to total cellular RNA. *J. Mol. Biol.* **26**:279–292.
4. Blobel, G., and D. D. Sabatini. 1970. Controlled proteolysis of nascent polypeptides in rat liver cell fractions. I. Location of the polypeptides within ribosomes. *J. Cell. Biol.* **45**:130–145.
5. Blobel, G., and D. D. Sabatini. 1971. Ribosome-membrane interaction in eukaryotic cells. *In* Biomembranes. L. A. Manson, editor. Plenum Publishing Corporation, New York. **2**:193–195.
6. Blobel, G., and D. D. Sabatini. 1971. Dissociation of mammalian polyribosomes into subunits by puromycin. *Proc. Natl. Acad. Sci. U. S. A.* **68**:390–394.
7. Borgese, N., G. Blobel, and D. D. Sabatini. 1973. In vitro exchange of ribosomal subunits between free and membrane-bound ribosomes. *J. Mol. Biol.* **74**:415–438.
8. Borgese, N., W. Mok, G. Kreibich, and D. D. Sabatini. 1974. Ribosomal-membrane interaction: in vitro binding of ribosomes to microsomal membranes. *J. Mol. Biol.* **88**:559–580.
9. Burke, G. T., and C. M. Redman. 1973. The distribution of radioactive peptides synthesized by polysomes and ribosomal subunits combined in vitro with microsomal membranes. *Biochim. Biophys. Acta.* **299**:312–324.
10. Cioli, D., and E. S. Lennox. 1973. Purification and characterization of nascent chains from immunoglobulin producing cells. *Biochemistry.* **12**:3203–3210.
11. Eagle, H. 1959. Amino acid metabolism in mammalian cell cultures. *Science (Wash. D.C.).* **130**:432–437.
12. Farquhar, M. G., and G. E. Palade. 1965. Cell junctions in amphibian skin. *J. Cell Biol.* **26**:263–291.
13. Freienstein, C., and G. Blobel. 1974. Use of eukaryotic native small ribosomal subunits for the translation of globin messenger RNA. *Proc. Natl. Acad. Sci. U. S. A.* **71**:3435–3439.
14. Fridlender, B. R., and F. O. Wettstein. 1970. Differences in the ribosomal protein of free and membrane bound polysomes of chick embryo cells. *Biochem. Biophys. Res. Commun.* **39**:247–253.
15. Green, M., P. N. Graves, T. Zehavi-Willner, J. McInnes, and S. Pestka. 1975. Cell-free translation of immunoglobulin messenger RNA from MOPC-315 plasmacytoma and MOPC-315 NR, a variant synthesizing only light chain. *Proc. Natl. Acad. Sci. U. S. A.* **72**:224–228.
16. Kellems, R. E., V. F. Allison, and R. A. Butow. 1975. Cytoplasmic type 80S ribosomes associated with yeast mitochondria. IV. Attachment of ribosomes to the outer membrane of isolated mitochondria. *J. Cell Biol.* **65**:1–14.
17. Lodish, H. F., D. Housman, and M. Jacobsen. 1971. Initiation of hemoglobin synthesis. Specific inhibition by antibiotics and bacteriophage ribonucleic acid. *Biochemistry.* **10**:2348–2356.
18. Luft, G. H. 1961. Improvements in epoxy embedding methods. *J. Biophys. Biochem. Cytol.* **9**:263–291.
19. Mach, B., C. Faust, and P. Vasalli. 1973. Purification of 14S messenger RNA of immunoglobulin

light chain that codes for a possible light-chain precursor. *Proc. Natl. Acad. Sci. U. S. A.* **70**:451–455.

20. Maizel, J. V. 1969. Acrylamide gel electrophoresis of proteins and nucleic acids. *In* Fundamental Techniques in Virology. K. Habel and N. P. Salzman, editors. Academic Press, Inc., New York. 334–362.
21. Mans, R. J., and G. D. Novelli. 1961. Measurement of the incorporation of radioactive amino acids into protein by a filter paper disk method. *Arch. Biochem. Biophys.* **94**:48–53.
22. McConkey, E. H., and E. J. Hauber. 1975. Evidence for heterogeneity of ribosomes within the HeLa cell. *J. Biol. Chem.* **250**:1311–1318.
23. Milstein, C., G. G. Brownlee, T. M. Harrison, and M. B. Mathews. 1972. A possible precursor of immunoglobulin light chains. *Nature New Biol.* **239**:117–120.
24. Palade, G. E. 1955. A small particulate component of the cytoplasm. *J. Biophys. Biochem. Cytol.* **1**:59–68.
25. Palade, G. E. 1958. *In* Microsomal Particles and Protein Synthesis. First Symposium of Biophysical Society. R. B. Roberts, editor. Pergamon Press, Inc., Elmsford, N.Y. 36.
26. Redman, C. M., and D. D. Sabatini. 1966. Vectorial discharge of peptides released by puromycin from attached ribosomes. *Proc. Natl. Acad. Sci. U. S. A.* **56**:608–615.
27. Redman, C. M., P. Siekevitz, and G. E. Palade. 1966. Synthesis and transfer of amylase in pigeon pancreatic microsomes. *J. Biol. Chem.* **241**:1150–1158.
28. Rolleston, F. S. 1974. Membrane-bound and free ribosomes. *Sub-Cell. Biochem.* **3**:91–117.
29. Sabatini, D. D., and G. Blobel. 1970. Controlled proteolysis of nascent polypeptides in rat liver cell fractions. II. Location of the polypeptides in rough microsomes. *J. Cell Biol.* **45**:146–157.
30. Sabatini, D. D., N. Borgese, M. Adelman, G. Kreibich, and G. Blobel. 1972. Studies on the membrane associated protein synthesis apparatus of eukaryotic cells. RNA Viruses and Ribosomes. Noord-Hollandsche Vitg. Mij., Amsterdam. 147–171.
31. Sauer, L. A., and G. N. Burrow. 1972. The submicrosomal distribution of radioactive proteins released by puromycin from the bound ribosomes of rat liver microsomes labeled in vitro. *Biochim. Biophys. Acta.* **277**:179–187.
32. Schechter, I. 1973. Biologically and chemically pure mRNA coding for mouse immunoglobulin L-chain prepared with the aid of antibodies and immobilized oligothymidine. *Proc. Natl. Acad. Sci. U. S. A.* **70**:2256–2260.
33. Schechter, I., D. J. McKean, R. Guyer, and W. Terry. 1974. Partial amino acid sequence of the precursor of immunoglobulin light chain programmed by messenger RNA in vitro. *Science (Wash. D.C.).* **188**:160–162.
34. Schmeckpeper, B. J., S. Cory, and J. M Adams. 1974. Translation of immunoglobulin mRNAs in a wheat germ cell-free system. *Mol. Biol. Rep.* **1**:355–363.
35. Swan, D., H. Aviv, and P. Leder. 1972. Purification and properties of biologically active messenger RNA for a myeloma light chain. *Proc. Natl. Acad. Sci. U. S. A.* **69**:1967–1971.
36. Tonegawa, S., and I. Baldi. 1973. Electrophoretically homogeneous myeloma light chain mRNA and its translation in vitro. *Biochem. Biophys. Res. Commun.* **51**:81–87.
37. Venable, J. H., and R. Coggeshall. 1965. A simplified lead citrate stain for use in electron microscopy. *J. Cell Biol.* **25**:407–408.
38. Watson, M. L. 1958. Staining of tissue sections for electron microscopy with heavy metals. *J. Biophys. Biochem. Cytol.* **4**:475–478.

Anderson R.G.W., Brown M.S., and **Goldstein J.L.** 1977. Role of the coated endocytic vesicle in the uptake of receptor-bound low density lipoprotein in human fibroblasts. *Cell* **10:** 351–364.

ENDOCYTOSIS OF AN EXTERNAL FLUID PHASE had been visualized by light microscopists for many years. During the 1960s, the electron microscope was used to characterize cells engaged in extensive endocytosis of specific proteins. The results suggested the hypothesis that cells could employ a pathway by which soluble, extracellular proteins bound to specialized regions on the plasma membrane and were internalized by a mechanism involving coated vesicles (1). The paper reproduced here used similar morphological methods to study the uptake by cultured cells of low-density lipoprotein (LDL), labeled either with radioisotope or ferritin. The structure of the coats on endocytic vesicles was examined in several labs (2, 3), and biochemical study identified clathrin as the principal protein subunit of this vesicle coat (4). Subsequent work in many labs has identified the machinery that provides specificity for the interaction between coat components and the relevant classes of transmembrane receptors (reviewed in 5). It was this paper by Anderson et al., however, that led to an understanding of a human disease state that was based on a failure to handle LDL properly. Mutation of key residues in the LDL-receptor resulted in the failure of proper lipid internalization (6). This and subsequent studies have deepened our understanding of receptor-mediated endocytosis in general and provided important insights into ways to study human disease.

1. Roth T.F. and Porter K.R. 1964. Yolk protein uptake in the oocyte of the mosquito *Aedes aegypti L. J. Cell Biol.* **20:** 313–332.
2. Crowther R.A., Finch J.T., and Pearse B.M.F. 1976. On the structure of coated vesicles. *J. Mol. Biol.* **103:** 785–798.
3. Heuser J. 1980. Three-dimensional visualization of coated vesicle formation in fibroblasts. *J. Cell Biol.* **84:** 560–583.
4. Pearse B.M.F. 1976. Clathrin: A unique protein associated with intracellular transfer of membrane by coated vesicles. *Proc. Natl. Acad. Sci.* **73:** 1255–1259.
5. Robinson M.S. 1994. The role of clathrin, adaptors and dynamin in endocytosis. *Curr. Opin. Cell Biol.* **6:** 538–544.
6. Davis C.G., Lehrman M.A., Russell D.W., Anderson R.G., Brown M.S., and Goldstein J.L. 1986. The J.D. mutation in familial hypercholesterolemia: Amino acid substitution in cytoplasmic domain impedes internalization of LDL receptors. *Cell* **45:** 15–24.

Cell, Vol. 10, 351-364, March 1977, Copyright © 1977 by MIT

Role of the Coated Endocytic Vesicle in the Uptake of Receptor-Bound Low Density Lipoprotein in Human Fibroblasts

Richard G. W. Anderson, Michael S. Brown, and Joseph L. Goldstein
Departments of Cell Biology and Internal Medicine
The University of Texas
Health Science Center at Dallas
5323 Harry Hines Boulevard
Dallas, Texas 75235

Summary

^{125}I-labeled and ferritin-labeled low density lipoprotein (LDL) were used as visual probes to study the surface distribution of LDL receptors and to examine the mechanism of the endocytosis of this lipoprotein in cultured human fibrobasts. Light microscopic autoradiograms of whole cells incubated with ^{125}I-LDL at 4°C showed that LDL receptors were widely but unevenly distributed over the cell surface. With the electron microscope, we determined that 60–70% of the ferritin-labeled LDL that bound to cells at 4°C was localized over short coated segments of the plasma membrane that accounted for no more than 2% of the total surface area. To study the internalization process, cells were first allowed to bind ferritin-labeled LDL at 4°C and were then warmed to 37°C. Within 10 min, nearly all the surface-bound LDL-ferritin was incorporated into coated endocytic vesicles that were formed by the invagination and pinching-off of the coated membrane regions that contained the receptor-bound LDL. With increasing time at 37°C, these coated vesicles were observed sequentially to migrate through the cytoplasm (1 min), to lose their cytoplasmic coat (2 min), and to fuse with either primary or secondary lysosomes (6 min). The current data indicate that the coated regions of plasma membrane are specialized structures of rapid turnover that function to carry receptor-bound LDL, and perhaps other receptor-bound molecules, into the cell.

Introduction

Biochemical and genetic experiments have established that normal human fibroblasts possess on their cell surfaces specific receptors for low density lipoprotein (LDL), the major cholesterol-carrying protein in human plasma (reviewed by Goldstein and Brown, 1976; Brown and Goldstein, 1976a). These receptors function to bind LDL with high affinity and to carry the lipoprotein particle into the cell, where its protein and cholesteryl ester components are degraded in lysosomes. The free cholesterol released from the lysosomal hydrolysis of LDL then becomes available for use by the cell for membrane synthesis and for the regulation of three important processes in cholesterol metabolism: first, it suppresses the activity of 3-hydroxy-3-methylglutaryl coenzyme A reductase, the rate-controling enzyme in cholesterol synthesis; second, it activates an acyl-CoA–cholesterol acyltransferase, a microsomal enzyme that attaches a long chain fatty acid to the incoming cholesterol so that it can be stored as esterified cholesterol; and third, it suppresses the synthesis of the LDL receptor, thus limiting the overall cellular uptake of LDL and preventing an overaccumulation of cholesterol by the cell (Brown, Dana, and Goldstein, 1974; Goldstein, Dana, and Brown, 1974; Brown and Goldstein, 1975).

Fibroblasts from patients with the homozygous form of familial hypercholesterolemia, which are genetically deficient in functional LDL receptors, lack the ability to take up LDL with high affinity (Goldstein and Brown, 1974; Goldstein et al., 1976a). Although these mutant cells passively ingest a small amount of LDL as a result of normal bulk fluid endocytosis, the efficiency of this receptor-independent uptake process is low compared with the efficiency of the receptor-mediated uptake process (Goldstein and Brown, 1974).

To understand better how LDL receptors function to promote the binding and adsorptive endocytosis of LDL, we labeled LDL with ferritin so that LDL binding sites could be visualized with the electron microscope (Anderson, Goldstein, and Brown, 1976). These studies showed that at 4°C, about 70% of the LDL receptors were concentrated on specialized coated indentations of the plasma membrane that occupied <2% of the cell surface. About 30 regions of coated membrane were seen per mm of surface length on both the normal and homozygous familial hypercholesterolemia fibroblasts, but only the normal cells bound LDL-ferritin at these coated regions.

Coated regions of membrane are so named because of the presence of a filamentous material covering the cytoplasmic surface of the plasma membrane (Roth and Porter, 1964; Fawcett, 1965). Several investigators have observed that coated regions of the plasma membrane are frequently indented to form so-called coated pits (Roth and Porter, 1964; Fawcett, 1965; Friend and Farquhar, 1967). In addition, several studies have shown that coated pits can transform into coated endocytic vesicles and that these vesicles constitute a pathway for the uptake of macromolecules by endocytosis (Roth and Porter, 1964; Fawcett, 1965; Friend and Farquhar, 1967; Bennett, 1969; Korn, 1975). In addition to their role in endocytosis, coated vesicles are believed to be involved in other functions within certain cells. For example, these vesicles are responsible for exocytosis of milk protein in the mammary gland (Franke et al., 1976) and for the delivery of lysosomal enzymes to secondary lysosomes in the vas deferens (Friend and Farquhar, 1967). Moreover, depending upon the cell type, these coated vesicles may be involved in either plasma membrane biogenesis (Franke and Herth, 1974; Franke et al., 1976) or membrane resorption (Heuser and Reese, 1973).

In view of our finding that LDL receptors appear to be preferentially located on coated regions of the plasma membrane, we have sought in the present studies to determine whether LDL enters fibroblasts through the transformation of a coated region into a coated pit and coated endocytic vesicle. In addition, we present light and electron microscopic data that further define the distribution of LDL receptors on the surface of human fibroblasts and the relationship of the receptor-mediated endocytosis of LDL to the endocytosis of other macromolecules.

Results

Distribution of ^{125}I-LDL Binding on Whole Cells

Figures 1A and 1B show autoradiograms of monolayers of normal fibroblasts that were exposed to a saturating level of ^{125}I-LDL for 2 hr at 4°C. The radioactivity showed an uneven distribution over the cell surface. In particular, there was a definite tendency for silver grains not to be seen directly over the nucleus (Figure 1A). On the other hand, there often was an area of high grain density in the region surrounding the nucleus. In many cells, the peripheral margins of the cell, including microextensions, were labeled (Figures 1A and 1B).

Although all the normal cells contained silver grains, the number of grains per cell was variable. Not infrequently, heavily labeled cells were observed adjacent to cells that contained only a few silver grains (Figure 1B). No consistent pattern of distribution of heavily labeled and lighly labeled cells could be discerned, nor was any characteristic cell morphology associated with either of the labeling patterns. When cells had been exposed to ^{125}I-LDL for only 10 min at 4°C (Figure 1C), the pattern of distribution of radioactivity was similar to that observed after the 2 hr incubations (Figure 1A). However, the absolute amount of radioactivity was higher after the longer incubation.

The distribution of radioactivity was also examined in normal cells that had been incubated in the presence of ^{125}I-LDL for 2 hr at 4°C and then warmed to 37°C for various times. When cells were warmed for 15 min, a time that is sufficient for the internalization of all receptor-bound ^{125}I-LDL (Brown and Goldstein, 1976b), many of the cells showed an enhanced concentration of silver grains over the body of the cell near the nucleus with a concomitant loss of grains over the peripheral margins. This shift in grain distribution was even more pronounced after the cells had been warmed for 60 min (Figure 2A). When the warming period was extended to 2 hr, a time sufficient to allow degradation of ^{125}I-LDL and excretion of the labeled ^{125}I-monoiodotyrosine (Goldstein and Brown, 1974; Brown and Goldstein, 1976b), virtually no silver grains were observed over any of the cells. As a control for these experiments, homozygous familial hypercholesterolemia fibroblasts were treated identically to the normal cells. As shown in Figure 2B, no silver grains were observed over any of these mutant cells under any of the conditions tested.

Binding and Endocytosis of Receptor-Bound LDL-Ferritin as Observed with the Electron Microscope

To study the ultrastructural dynamics of the internalization of receptor-bound LDL-ferritin, fibroblasts were allowed to bind the ferritin-labeled lipoprotein at 4°C, washed to remove unbound lipoprotein, and then warmed to 37°C for various periods of time before being fixed for electron microscopy. In cells that were fixed immediately following a 2 hr exposure to LDL-ferritin at 4°C, the distribution of ferritin cores was as previously described (Anderson et al., 1976). Most of the bound ferritin was located on the sides of indented, coated regions of membrane (Figure 3A) with occasional cores located on noncoated regions of membrane. It was noted that at 4°C, the labeled coated regions could assume a variety of shapes, including regions that displayed more than one indentation (Figure 3B).

As early as 1 min after the normal cells that had bound LDL-ferritin at 4°C were warmed to 37°C,

Figure 1. Autoradiograms of Normal Fibroblasts Exposed to ^{125}I-LDL for Various Times at 4°C

Monolayers of fibroblasts were incubated for the indicated time at 4°C with 6.5 μg protein per ml of ^{125}I-LDL, and autoradiograms were prepared as described in Experimental Procecures.

(A) After incubation for 2 hr, silver grains are localized over various regions of the cell. Some areas (circles) are free of silver grains, whereas other areas are labeled. Note that there are few silver grains over the nucleus (n), even though there are grains surrounding the periphery of this organelle. Magnification 1750×; bar = 3 μm.

(B) After incubation for 2 hr, heavily labeled cells and lightly labeled cells (ul) were present in the same culture, indicating that not every cell had the same amount of receptor activity. Magnification 1250×; bar = 5 μm.

(C) After incubation for 10 min, the distribution of silver grains is the same as after 2 hr, even though there are fewer grains per cell. Magnification 1750×; bar = 3 μm.

Coated Endocytic Vesicles and LDL Receptors
353

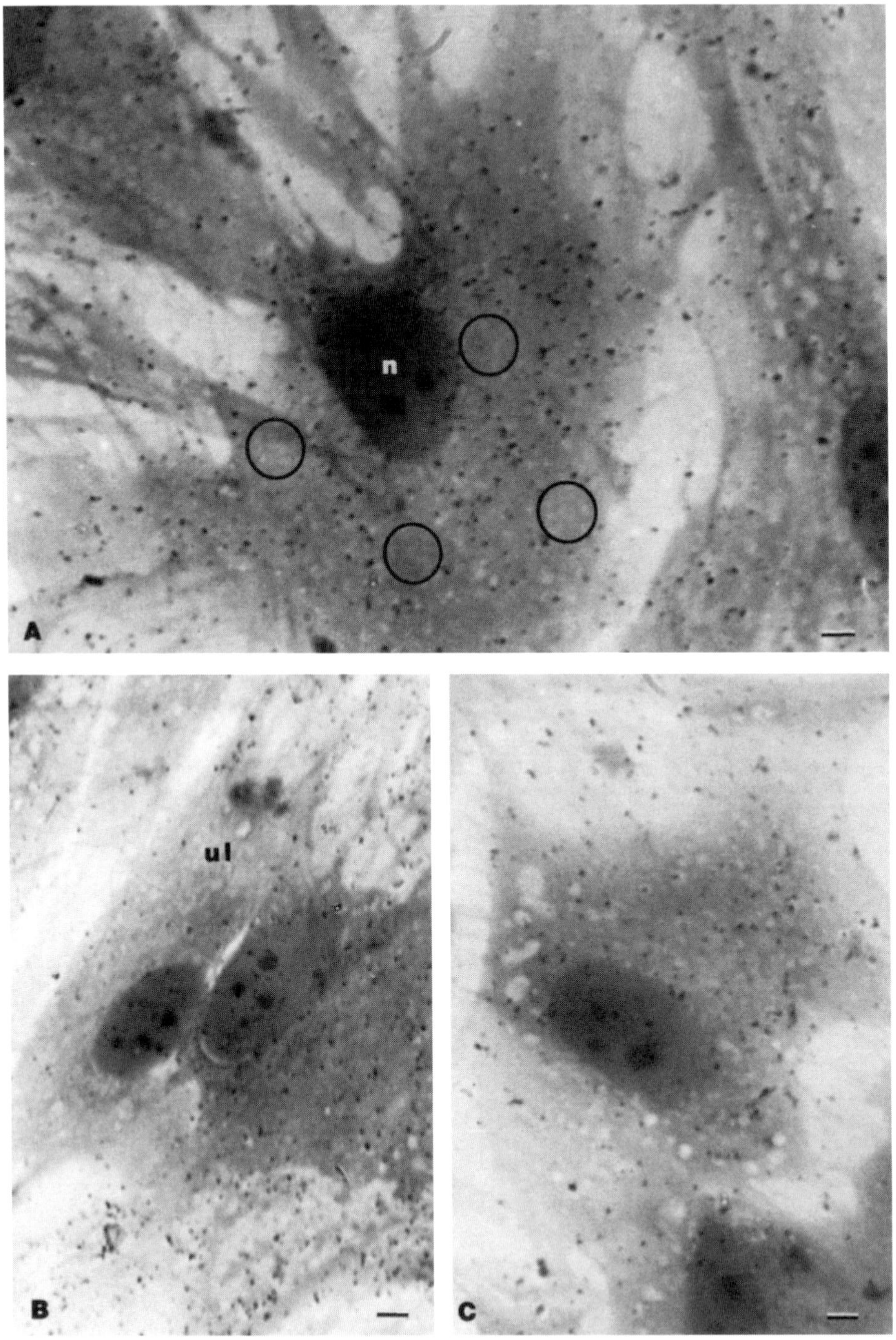

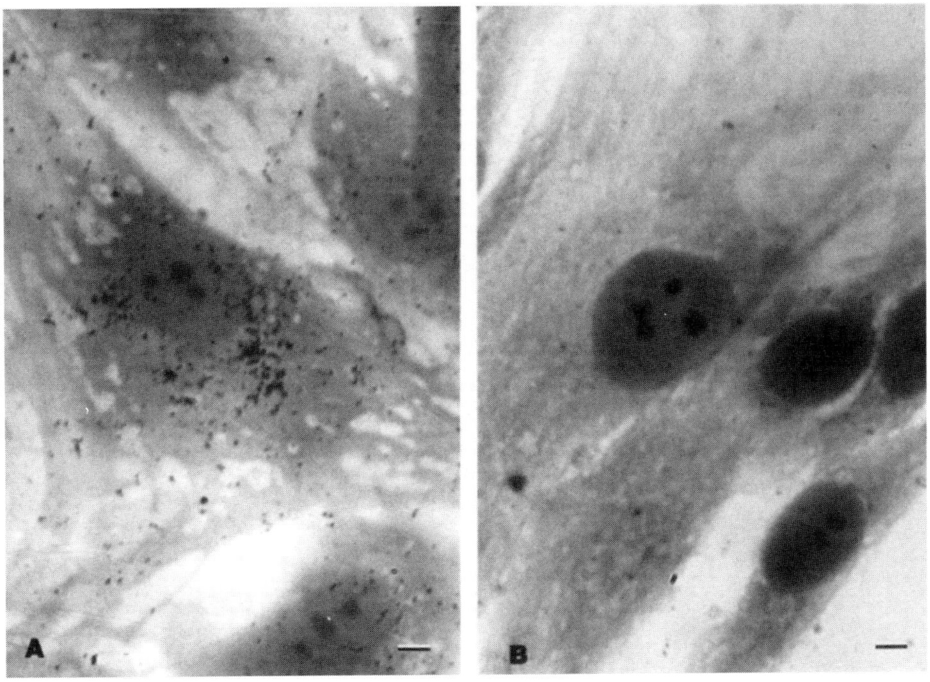

Figure 2. Autoradiograms of Normal and Homozygous Familial Hypercholesterolemia Fibroblasts Exposed to ^{125}I-LDL

(A) A normal cell that was incubated with 6.5 μg protein per ml of ^{125}I-LDL for 2 hr at 4°C, washed extensively, and then warmed to 37°C for 60 min as described in Experimental Procedures. Note the concentration of silver grains in the perinuclear region, probably within lysosomes. Magnification 1250×; bar = 5 μm.

(B) A mutant cell from a patient with the homozygous form of familial hypercholesterolemia that was incubated with 6.5 μg protein per ml of ^{125}I-LDL for 2 hr at 4°C under the same conditions and in the same experiment as the normal cells shown in Figures 1A and 1B. Note that in contrast to the normal cells, silver grains were not associated with the mutant cell. Magnification 1250×; bar = 5 μm.

some ferritin cores could be seen within the cell in coated endocytic vesicles (Figure 4C). In other areas of the same cell, however, LDL-ferritin could still be found on coated regions at the surface (Figure 4A). Between these two extremes were labeled, coated regions that had nearly completed the invagination step of endocytosis (Figure 4B). With further time at 37°C, more ferritin-containing vesicles were found in the cell and less ferritin was found on the cell surface.

The micrographs in Figures 4A-4E represent an attempt to reconstruct the apparent sequential steps in the formation of coated vesicles from coated pits in normal cells that were labeled with LDL-ferritin at 4°C and then warmed to 37°C. First, the coated pit becomes more indented, and the two portions of the membrane at the opening of the pit move closer together (Figures 4A and 4B). As the two portions of membane are about to fuse, LDL-ferritin attached to this region of membrane seems to be pushed either into the vesicle or out of the vesicle (Figure 4C). The result is that often after a vesicle is formed, a portion of the LDL-ferritin originally bound to the coated membrane remains on the cell surface (Figure 4C). After cells had been incubated for several minutes at 37°C, the coated vesicles, which were originally round or ovoid, began to lose their cytoplasmic coat of fuzzy material. Figure 4D shows a view of a ferritin-containing vesicle that is coated on one side and not the other. After the whole vesicle becomes uncoated, the resulting vesicles assume a variety of shapes (Figure 4E). Frequently, they appear to be collapsed (Figure 4F). In some sections, it appears as if small vesicles are budding off of a larger vesicle, but it is impossible to be certain that this image is not due to the plane of section through these irregularly shaped vesicles. In an extensive series of studies, the only detectable route of entry of LDL-ferritin into the cell was through the initial formation of coated vesicles.

Previous biochemical experiments have shown that internalized LDL eventually enters the lysosome (Goldstein, Brunschede, and Brown, 1975a; Goldstein et al., 1975a; Brown, Dana, and Goldstein, 1975). In the current ultrastructural studies, after the cells had been warmed to 37°C, ferritin-containing secondary lysosomes were frequently

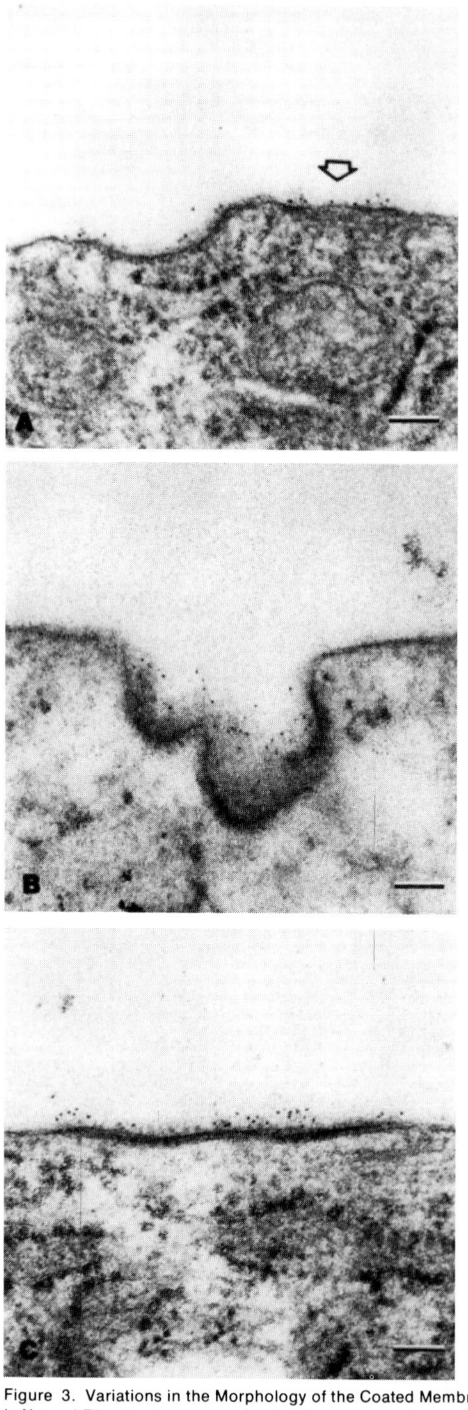

Figure 3. Variations in the Morphology of the Coated Membrane in Normal Fibroblasts Incubated with LDL-Ferritin at 4°C and 37°C

Monolayers of fibroblasts were incubated for 2 hr at the indicated temperature with 47.5 μg/ml of LDL-ferritin as described in Experimental Procedures. Bar = 1000 Å.

(A) A typical indented, coated region with LDL-ferritin bound at the sides of the indentation. Sometimes LDL-ferritin was found on the portion of membrane adjacent to the coated region (arrow). The cells were incubated at 4°C. Magnification 78,000×.

(B) A region of coated membrane that has two indentations. Presumably two coated vesicles will form from this segment of membrane. The cells were incubated at 4°C. Magnification 78,000×.

(C) A region of coated membrane that is not indented. This variation is observed most commonly in cells incubated at 37°C. Magnification 78,000×.

seen (Figure 4H). Although a series of static micrographs cannot unequivocally reveal the pathway from endocytic vesicle to lysosome, the micrographs in Figures 4E–4H are arranged to show what appears to be the probable sequence. After losing their cytoplasmic coats, the vesicles containing LDL-ferritin appear to increase in size. The amount of ferritin present in each vesicle appears to increase as the size of the vesicles increases (Figures 4F and 4G). At the same time, regions of increased electron density appear in the lumen of the vesicle, a finding that probably reflects the introduction of lysosomal enzymes. Finally, the vesicle increases in electron density and assumes the characteristics of a lysosome (Figure 4H). Considered together, the increase in size, electron density, and the quantity of ferritin that was observed sequentially in Figures 4E–4H are suggestive of the formation of a secondary lysosome that resulted from the fusion of primary lysosomes with several fused endocytic vesicles (reviewed by Korn, 1975).

Endocytosis of LDL-Ferritin at 37°C

Because of the possibility that the endocytosis of prebound LDL-ferritin might differ from the uptake process that occurs when LDL-ferritin is present continuously in the medium at 37°C, monolayers of normal fibroblasts were incubated with LDL-ferritin for various times at 37°C and then examined without ever being chilled. As with cells pretreated with LDL-ferritin at 4°C, the majority of LDL-ferritin was observed to bind at coated regions of plasma membrane. However, unlike in the 4°C pretreatment experiments, in the 37°C experiments many of the coated regions of membrane that bound LDL-ferritin were not indented (Figure 3C). Furthermore, at the higher temperature, relatively few coated endocytic vesicles were seen, even though ferritin-containing lysosomes were frequently present. The relative paucity of coated endocytic vesicles after continuous incubation at 37°C correlates with the relative paucity of coated membrane regions that were observed in cells that had bound LDL-ferritin at 4°C and were then warmed for periods > 4 min (see "Quantitative Analysis"). Considered together, the data indicate that the life span of a recognizable coated vesicle is relatively short at 37°C.

Cell
356

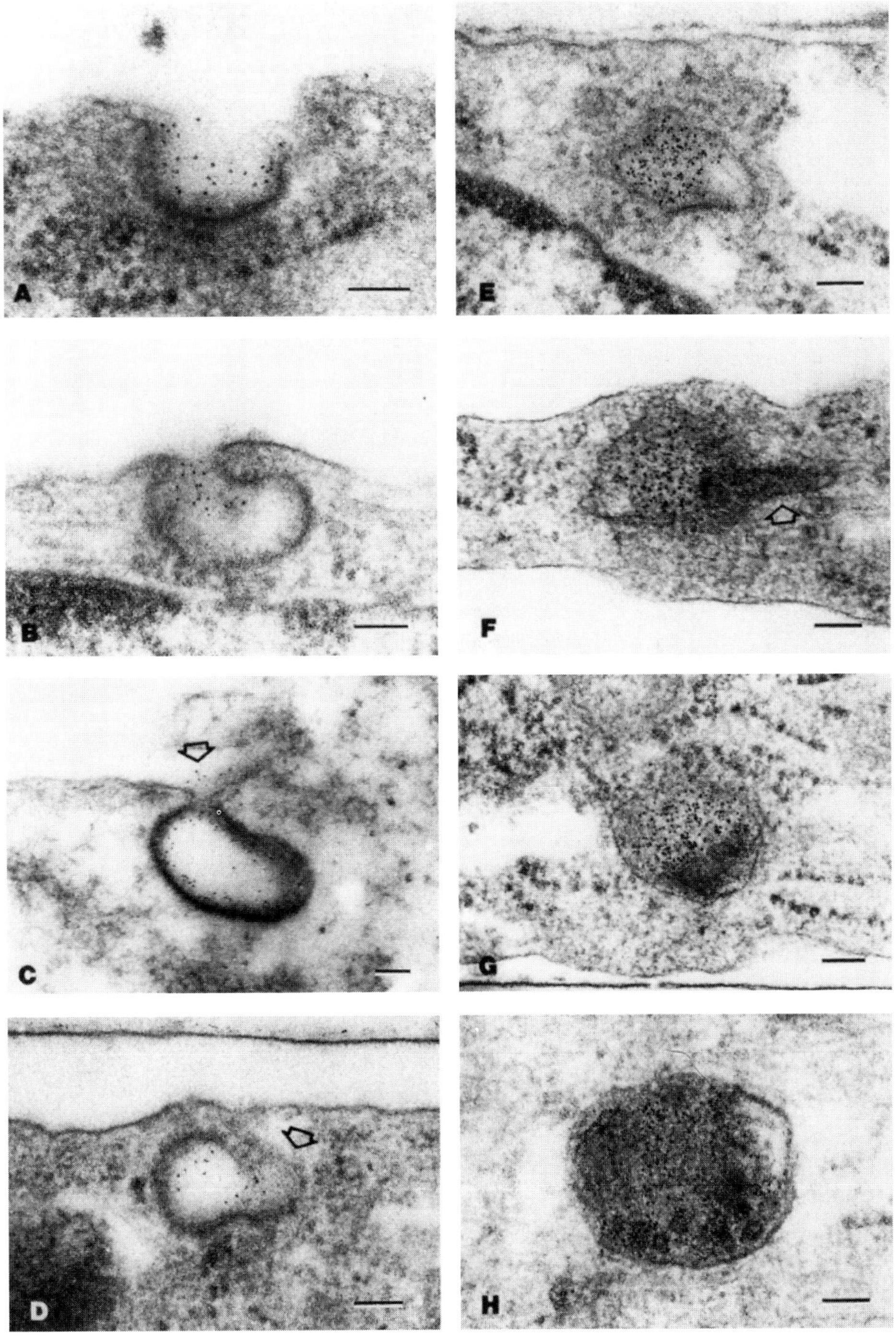

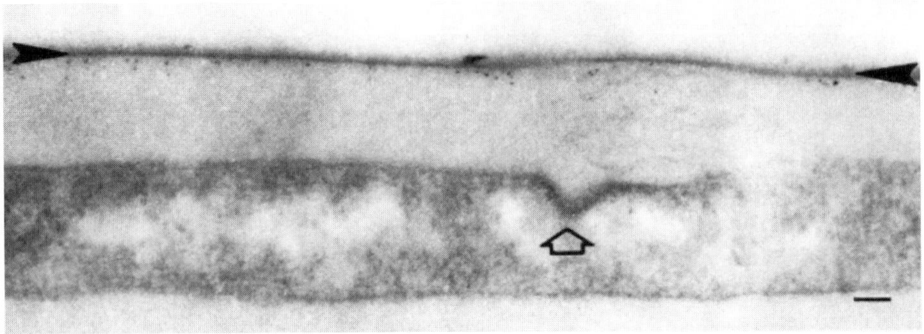

Figure 5. Unstained Electron Micrograph of a Fibroblast from a Familial Hypercholesterolemia Homozygote after Incubation with LDL-Ferritin

The fibroblast monolayer was incubated for 2 hr at 37°C with 100 μg/ml of LDL-ferritin as described in Experimental Procedures. Note that although LDL-ferritin is bound to the substratum of the dish (between solid arrow), no ferritin is bound to the cell surface, even in the indented, coated region of membrane (open arrow). Magnification 59,000×; bar = 1000 Å.

Incubation of Homozygous Familial Hypercholesterolemia Cells with LDL-Ferritin

Previous studies have shown that LDL-ferritin does not bind to the cell surface of homozygous familial hypercholesterolemia fibroblasts at 4°C (Anderson et al., 1976). In the current experiments, homozygous familial hypercholesterolemia fibroblasts were exposed to high concentrations of LDL-ferritin (100 μg/ml) at 37°C, and again no evidence was found for the binding of this complex to coated regions or to any other site on the cell surface. Rarely was ferritin found in secondary lysosomes in these cells. However, a considerable amount of ferritin was bound to the substratum to which the cells were attached (Figure 5). We believe that the substratum represents the site of the nonspecific or nonsaturable binding that has been observed biochemically when homozygous familial hypercholesterolemia fibroblasts are incubated with ^{125}I-LDL at 4°C and washed briefly with phosphate-buffered saline without albumin in the wash buffer (Stein et al., 1976).

Quantitative Analysis of the Endocytosis of Receptor-Bound LDL-Ferritin

To examine quantitatively the endocytosis of LDL-ferritin, monolayers of normal cells were treated with LDL-ferritin at 4°C for 2 hr, warmed to 37°C for various times, and prepared for electron microscopy. The distribution of ferritin cores was determined using previously described techniques (Anderson et al., 1976). Table 1 shows the data obtained after 0, 4, and 10 min of incubation at 37°C. When the cells were warmed, the total amount of LDL-ferritin on the cell surface decreased from 91 to 11 ferritin cores per mm of cell surface within 10 min. During the same interval, the percentage of coated regions that were labeled with LDL-ferritin declined from 60% to 11%. The decline in LDL-ferritin from the cell surface that occurred within 10 min of warming is believed to represent the internalization of LDL-ferritin, since extensive biochemical experiments with ^{125}I-LDL have shown that during the same interval, nearly all the receptor-bound ^{125}I-LDL is internalized and none disso-

Figure 4. Electron Micrographs Showing Representative Stages in the Endocytosis of LDL-Ferritin and Its Subsequent Appearance in the Lysosome

Normal fibroblasts were incubated with 47.5 μg/ml of LDL-ferritin for 2 hr at 4°C, washed extensively, and then warmed at 37°C for various times as described in Experimental Procedures. Bar = 1000 Å.

(A) A typical coated pit (time at 37°C, 1 min). Magnification 97,000×.

(B) A coated pit being transformed into an endocytic vesicle with LDL-ferritin included (time at 37°C, 1 min). Magnification 81,000×.

(C) Formation of a coated vesicle. As the plasma membrane begins to fuse to form the vesicle, some of the LDL-ferritin is excluded from the interior and is left on the surface of the cell (arrow) (time at 37°C, 1 min). Magnification 54,500×.

(D) A fully formed coated vesicle that appears to be losing its cytoplasmic coat on one side (arrow) (time at 37°C, 2 min). Magnification 75,000×.

(E) An endocytic vesicle that has completely lost its cytoplasmic coat. Note the irregular shape of this vesicle (time at 37°C, 2 min). Magnification 75,000×.

(F) An irregularly shaped endocytic vesicle that contains more LDL-ferritin than a typical coated vesicle and also has a region of increased electron density within the lumen (arrow) (time at 37°C, 6 min). Magnification 75,000×.

(G) An endocytic vesicle similar to (F) with more electron-dense material in the lumen (time at 37°C, 6 min). Magnification 69,000×.

(H) A secondary lysosome that contains LDL-ferritin (time at 37°C, 8 min). Magnification 75,000×.

Table 1. Quantitative Analysis of the Internalization of Receptor-Bound LDL-Ferritin in Normal Fibroblasts

Time at 37°C (Min)	Ferritin Cores Bound (Number per mm of Cell Surface) Coated Regions	Noncoated Regions	Total	% Bound Ferritin Associated with Coated Regions of Membrane	% Coated Regions That Were Labeled with Ferritin Cores	Number of Ferritin Cores per Ferritin-Labeled, Coated Region	Coated Membrane Regions (Number per mm of Cell Surface)
0	56 (60, 52)	35 (20, 50)	91 (80, 102)	63 (75, 51)	60 (75, 46)	3.7 (3.4, 4.0)	26 (26, 26)
4	41 (52, 31)	22 (23, 21)	63 (75, 52)	65 (69, 61)	38 (52, 25)	9.3 (6.5, 12)	12 (14, 11)
10	2 (4, 0)	9 (13, 5)	11 (17, 5)	18 (36, 0)	11 (22, 0)	1.5 (3.0, 0)	5 (9, 1)

Cell monolayers were prepared as described in Experimental Procedures. Each monolayer was chilled at 4°C for 30 min, after which it received 2 ml of ice-cold growth medium containing LDL-ferritin corresponding to 47.5 μg/ml of LDL-protein. Following a 2 hr incubation at 4°C, monolayers were washed at 4°C with phosphate-buffered saline and then either fixed immediately or warmed at 37°C for the specified time before fixation. Cells were processed for electron microscopy and quantitation as described in Experimental Procedures. Each value represents the average of two separate experiments; the data for each experiment are given in parentheses. In each experiment, 1 mm of surface length was quantified.

ciates intact from the cell surface (Brown and Goldstein, 1976b).

Of particular interest in these studies was the analysis of the change in distribution of the LDL-ferritin that was bound to regions other than coated regions. (In eight experiments in which 1 mm of cell surface was examined in each experiment, an average of 27% of the LDL-ferritin was found over such noncoated areas.) In the experiment shown in Table 1, initially a total of 91 ferritin cores was bound per mm of cell surface, of which 35 (37%) were bound at noncoated regions. After 4 min at 37°C, the total number of surface-bound ferritin cores had declined to 63 per mm. Of this total, 22 ferritin cores (35%) were still in noncoated regions. After 10 min at 37°C, nearly all the LDL-ferritin had disappeared from both the coated and noncoated regions. Part of the disappearance from noncoated regions may have been due to the migration of LDL-ferritin into coated regions, as suggested by the transient increase in the number of ferritin cores per labeled coated region after 4 min at 37°C.

The data in Table 1 also show that the number of coated regions declined after the cells were warmed from 4°C to 37°C. Because of this finding, we compared the number of coated regions in cells incubated at 4°C with cells that were incubated continuously at 37°C without ever being chilled. Based on nine different experiments in which at least 1 mm of cell surface was examined for each experiment, an average of 30.4 ± 5.5 (SEM) coated regions was observed per mm of cell surface at 4°C, whereas in four experiments at 37°C, an average of 13.7 ± 2.8 coated regions was observed. As noted above and in Figure 3C, the coated regions at 37°C tend not to be indented, and so they are less easily identified at the higher temperature. It is therefore, possible that the apparent decrease in the number of coated regions at 37°C is due to an underestimation based on their altered morphology.

Endocytosis of Macromolecules Other Than LDL-Ferritin

In addition to coated invaginations, cultured human fibroblasts exhibit smaller, flask-shaped invaginations that appear to carry out a type of micropinocytosis that does not involve coated membrane regions (Anderson et al., 1976). Even though these flask-shaped invaginations are numerous in fibroblasts, they never were observed to contain LDL-ferritin. To determine whether this restriction holds for other proteins, experiments were conducted to examine the uptake of three different macromolecules: native ferritin, ferritin-labeled immunoglobulin G, and horseradish peroxidase.

When native ferritin was incubated with fibroblasts at 4°C at concentrations up to 100 μg/ml, no ferritin was observed to bind to the cell surface or to enter the cell (Anderson et al., 1976). When ferritin was incubated with fibroblasts at 37°C at concentrations up to 2.5 mg/ml, ferritin was not seen bound to the cell surface or within any type of endocytic vesicle, even though some accumulation of ferritin in lysosomes took place. However, when higher concentrations of ferritin (10 mg/ml) were incubated with fibroblasts at 37°C, ferritin was observed to bind to the plasma membrane exclusively over coated regions (Figure 6A) and coated pits (Figure 6B). Moreover, under these conditions, ferritin could be seen in coated endocytic vesicles (Figure 6C) and in secondary lysosomes. No ferritin was ever observed in association with the smooth-surfaced, flask-shaped invaginations.

A similar experiment was conducted at 37°C using ferritin bound to immunoglobulin G. Although the concentration used (400 μg protein per ml) was high enough to permit binding of the complex to the substratum of the culture dish, no ferritin was bound to coated pits or to any other portion of the cell membrane. Scattered ferritin particles were observed in secondary lysosomes, but none could

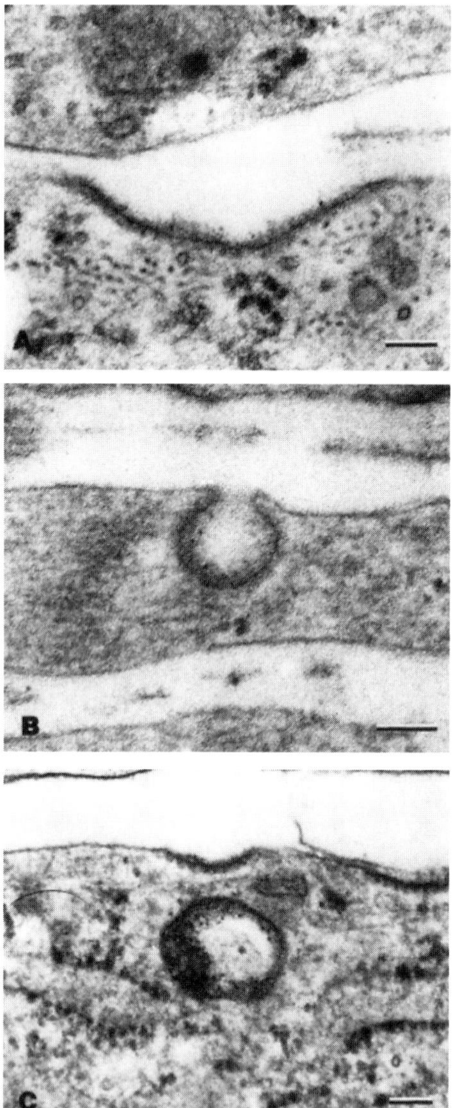

Figure 6. Binding and Endocytosis of Native Ferritin in Normal Fibroblasts

Monolayers of fibroblasts were incubated for 2 hr at 37°C with 10 mg/ml of native ferritin as described in Experimental Procedures. Bar = 1000 Å.

(A) The cell surface showing that native ferritin is exclusively located over coated regions of membrane. Magnification 81,000×.

(B) Formation of a coated endocytic vesicle that contains native ferritin. Magnification 90,500×.

(C) Endoctyic vesicle containing membrane-bound native ferritin. Magnification 62,500×.

be visualized within any type of endocytic vesicle.

Horseradish peroxidase has been used by several investigators to study bulk fluid endocytosis (Graham and Karnovksy, 1966; Friend and Farquhar, 1967; Steinman, Silver, and Cohn, 1974). We found that when human fibroblasts were incubated with horseradish peroxidase at a concentration of 5 mg protein per ml for 30 min at 37°C and then exposed to diaminobenzidine, reaction product was contained in several types of endocytic vesicles. Although the density of the reaction product obscured the characteristic fuzzy cytoplasmic coat of the coated pits and coated vesicles, it appeared that at least some of the peroxidase-containing vesicles were coated vesicles on the basis of their size, shape, and relationship to the cell surface (Figures 7A and 7B). In addition, horseradish peroxidase was observed to be contained within noncoated endocytic vesicles (Figure 7C). These noncoated endocytic vesicles were much smaller (average diameter 0.1 μm) than those coated endocytic vesicles that appeared to have lost their coat in the process of fusing with lysosomes (average diameter 0.35 μm, as shown in Figure 4E). In agreement with the findings of Steinman et al. (1974) in mouse L cells, we found no evidence for the binding of horseradish peroxidase to the cell surface of human fibroblasts. However, as mentioned above (Figure 7A), reaction product was occasionally observed in deeply invaginated, apparently coated membrane regions that were in the process of pinching off to form endocytic vesicles.

Discussion

The major conclusion of this study is that plasma LDL, which binds to the LDL receptor located primarily on coated regions of the cell surface membrane in human fibroblasts, is subsequently incorporated into coated endocytic vesicles that migrate through the cytoplasm and ultimately fuse with lysosomes. The internalization process is extremely efficient, since nearly all the receptor-bound LDL–ferritin enters the cell within 10 min at 37°C. Thus the kinetics of the ultrastructural internalization process for LDL–ferritin are identical to the kinetics obtained from biochemical studies using ^{125}I-labeled LDL, in which nearly all receptor-bound lipoprotein was found to enter the cell within 12 min (Goldstein et al., 1976a; Brown and Goldstein, 1976b). By these techniques, the coated region is identified as a region of membrane that contains a high concentration of LDL receptors and is specialized for carrying out the rapid adsorptive endocytosis of LDL.

In performing the current studies with LDL–ferritin, it was important to demonstrate that the complex was binding to the physiologic LDL receptor and not to some other molecule in the coated region. This consideration was especially important since in one of the earliest studies of coated mem-

Cell
360

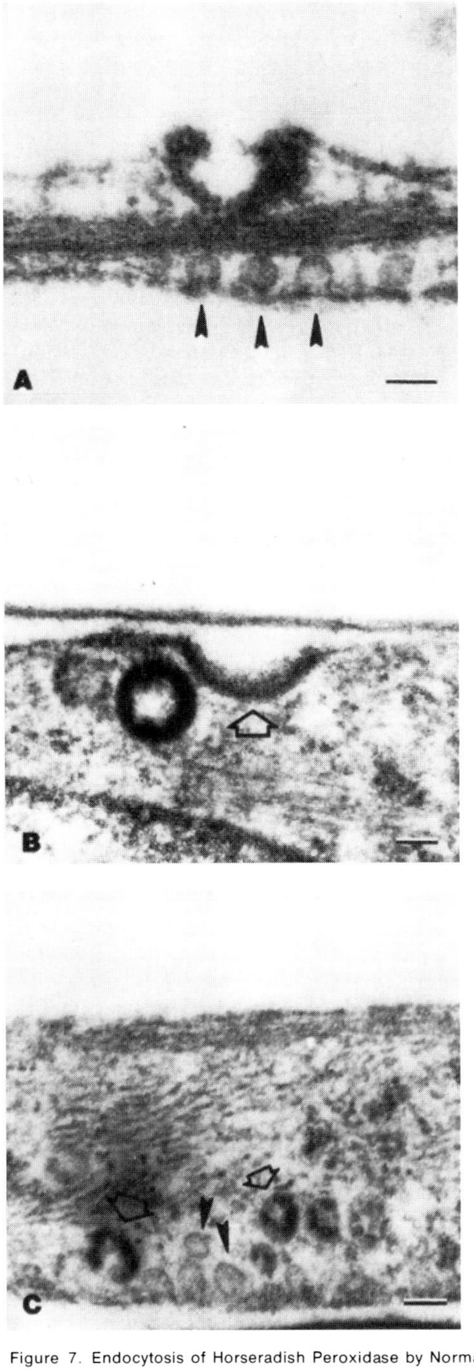

Figure 7. Endocytosis of Horseradish Peroxidase by Normal Fibroblasts

Monolayers of fibroblasts were incubated with 5 mg/ml of horseradish peroxidase for 60 min at 37°C as described in Experimental Procedures. Bar = 1000 Å.

(A) Formation of an apparent coated endocytic vesicle that contains reaction product. Although the characteristic fuzzy cytoplasmic coat cannot be distinguished, the morphology of this indented region corresponds to that of a coated pit. For comparison, the smaller vesicles (arrows) are typical examples of another type of noncoated, flask-shaped surface invagination. Magnification 72,000×.

(B) A formed endocytic vesicle that most probably is of the coated variety. Note that the adjacently located coated pit (arrow) that does not contain horseradish peroxidase is similar in size to the endocytic vesicle containing horseradish peroxidase. Magnification 62,500×.

(C) Groups of small endocytic vesicles containing horseradish peroxidase (open arrows). These vesicles when emtpy (solid arrows) do not appear to be coated. Magnification 62,500×.

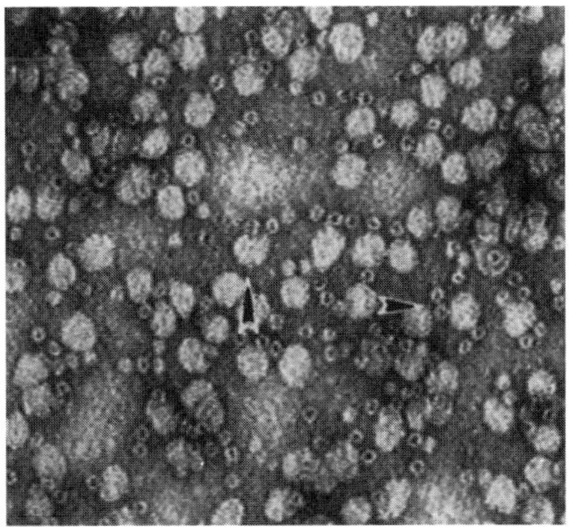

Figure 8. Electron Micrograph of a Negatively Stained Preparation of LDL–Ferritin

The LDL–ferritin was prepared and isolated as described in Experimental Procedures. On the average, two ferritins are found per LDL particle (arrows). Magnification 125,000×.

brane regions, Fawcett (1965) observed that endogenous ferritin was bound to coated regions on the membranes of guinea pig bone marrow erythroblasts (but not to the coated regions of hepatic Kuppfer cells). This finding led Fawcett to postulate that coated membrane regions in erythroblasts contained receptors that mediated the adsorptive endocytosis of endogenous ferritin. Furthermore, Lagunoff and Curran (1972) showed that rat macrophages ingest exogenously adminstered ferritin via coated endocytic vesicles. In the current studies, we too observed that ferritin, when incubated at high concentrations with human fibroblasts at 37°C, bound to the coated regions in a manner similar to that observed for LDL–ferritin.

The following observations establish, however, that under the conditions of the current study, the LDL–ferritin complex is binding only to the LDL receptor and not to a ferritin binding site. First, LDL–ferritin binding was never observed in cells from a familial hypercholesterolemia homozygote, which lack LDL receptors, even though these mutant cells bound free ferritin normally when it was added at high concentrations. Second, to observe significant binding of free ferritin in normal fibroblasts, it was necessary to add the probe at concentrations in the range of 10 mg/ml, which is about 50–100 fold higher than the concentration of ferritin used in the LDL-ferritin preparations. Third, binding of LDL-ferritin was reduced by competition with an excess of native LDL (Anderson et al., 1976). Fourth, the LDL-ferritin could be removed from the cell surface by heparin, an agent that specifically releases LDL from its receptor (Anderson et al., 1976; Goldstein et al., 1976a). Fifth, ferritin covalently coupled to another protein molecule (immunoglobulin G) did not bind to the cell surface of normal fibroblasts in the current studies when added to the cells at a concentration about 9 times higher than that used for LDL-ferritin. Thus although it seems probable that the coated regions may possess a low affinity binding site for ferritin, in the current studies it was possible to establish conclusively that this binding site was not responsible for the observed binding of the LDL-ferritin complex.

From the data in the current study, it appears probable that functional LDL receptors are normally concentrated over the coated regions, and that exposure of the cells to LDL is not required to induce movement of the LDL receptor into this specialized region of the cell surface. Thus most of the bound LDL-ferritin was found in coated regions whether the binding occurred at 4°C [where lateral movement of bound ligand is retarded (Frye and Edidin, 1970)] or at 37°C. Moreover, we have previously shown that the LDL-ferritin binding sites are located in coated regions even when the cells have been fixed with formaldehyde prior to their incubation with LDL (Anderson et al., 1976). The concentration of formaldehyde that was used was higher than that required to inhibit lateral mobility of surface-bound immunoglobulin on mouse lymphocytes (Abbas et al., 1975) and was also higher than that required to block completely the internalization of ^{125}I-LDL (Brown, Ho, and Goldstein, 1976). Finally, the ^{125}I-LDL autoradiograms indicated a similar distribution of surface-bound ^{125}I-LDL when the cells had been incubated at 4°C for as little as 10 min or as much as 2 hr (Figure 1). Considered together, these observations would make improbable ligand-induced patching or capping (de Petris and Raff, 1972) of receptor-bound LDL of the type reported for antigen cap formation that occurs when fibroblasts are exposed to bivalent antibodies to histocompatibility antigens (Edidin and Weiss, 1972). Consistent with this conclusion is the additional finding that the number and appearance of coated endocytic vesicles were similar in the familial hypercholesterolemia homozygote and normal cells when incubated with native LDL, even though the mutant cells failed to bind the lipoprotein. It therefore appears probable that the internalization of coated regions and formation of coated endocytic vesicles proceeds at the same rate whether or not LDL is bound to these regions.

Approximately 30% of the surface-bound LDL-ferritin was found scattered along the 98% of cell surface that was noncoated. The LDL-ferritin in noncoated regions appeared to be bound to LDL receptors, since it was not observed in the homozygous familial hypercholesterolemia cells. When the normal cells were warmed to 37°C, the amount of LDL-ferritin bound to noncoated regions declined in parallel with the LDL-ferritin that was bound to coated regions. We could not unequivocally determine whether the material bound at the noncoated regions was moving laterally into coated regions and then being internalized, or whether it was possibly dissociating from the cell surface. Throughout the entire study, however, we never observed LDL-ferritin entering the cell through noncoated endocytic vesicles.

In contrast to the restriction of LDL-ferritin uptake to coated endocytic vesicles, the uptake of horseradish peroxidase appeared to occur through both coated and noncoated endocytic vesicles. We believe that this difference is due largely to the preferential localization of LDL receptors on coated membrane regions. We cannot, however, exclude the possibility that part of this difference is due either to the difference in molecular weight of LDL-ferritin versus horseradish peroxidase, the larger LDL-ferritin particle being excluded from the small flask-shaped, noncoated vesicles, or to the greater sensitivity of the horeseradish peroxidase technique for visualization of endocytosis. Nevertheless, the current data do indicate that there is more than one route by which some extracellular macromolecules can enter human fibroblasts.

A useful tool in the current studies has been the technique of allowing the LDL-ferritin to bind to the receptor at 4°C and then following the fate of the bound complex when the cells are warmed to 37°C. Chilling the cells to 4°C appears to exert a preferential block on the actual pinching-off of the coated membrane regions so that a relative excess of indented, coated pits develops. When the cells are subsequently warmed to 37°C, the accumulated coated pits rapidly pinch off to form vesicles, producing a wave of endocytosis. On the other hand, when cells that have never been exposed to low temperatures are incubated continuously at 37°C, the pinching-off process and the initial migration of the coated vesicles into the cell interior occur rapidly, so that relatively few indented, coated membrane regions and coated endocytic vesicles are visible at any one instant in time.

Since the coated membrane regions that contain

functional LDL receptors appear to be internalized at least once every 10 min, a mechanism must exist for the rapid regeneration of these coated regions. Whether this occurs by the modification of uncoated plasma membrane or whether it involves the recycling to the cell surface of previously internalized coated pits remains to be resolved.

The coated pit-coated vesicle system has previously been shown to be involved in the adsorptive endocytosis of several proteins in other cell systems (Fawcett, 1965; Friend and Farquhar, 1967; Bennett, 1969; Korn, 1975). Of particular relevance to the current studies are the data of Roth and his co-workers, who have demonstrated that ovarian cells from both the mosquito and the chicken bind and take up huge amounts of lipoproteins by the coated pit-coated vesicle mechanism during the process of egg yolk formation (Roth and Porter, 1964; Yusko and Roth, 1976; Roth, Cutting, and Atlas, 1976). Moreover, in the chicken oocyte, the coated region has also been shown to contain specific receptors for immunoglobulins, another class of proteins that is transported in large amounts into the yolk (Roth et al., 1976).

In addition to their role in the selective transport of nutrient materials such as lipoproteins, the coated regions may also contain receptors for polypeptide hormone molecules that are destined to be taken up and degraded by cells. In a recent study, Carpenter and Cohen (1976) demonstrated that ^{125}I-labeled human epidermal growth factor binds to specific, high affinity receptor sites on the membranes of human fibroblasts at 4°C. When the cells were warmed to 37°C, the bound polypeptide hormone was internalized with extreme rapidity so that within 8 min, all the hormone was within the cell. Once internalized, the ^{125}I-labeled epidermal growth factor was degraded within lysosomes, and the ^{125}I-monoiodotyrosine was released into the culture medium. The kinetics of this receptor-mediated uptake process are essentially identical to those that we have observed for the receptor-mediated uptake of ^{125}I-LDL. This similarity suggests that the receptor for epidermal growth factor may be localized to the same coated regions of the plasma membrane that contain the LDL receptor, and hence the internalization process for both receptor-bound ligands may be identical. In this regard, it is also of interest that Terris and Steiner (1975) have reported that ^{125}I-insulin, which binds to a high affinity receptor on the plasma membrane of isolated rat hepatocytes, is subsequently internalized and degraded to its constituent amino acids with kinetics that are also similar to those for the uptake of LDL in human fibroblasts. On the basis of these data, it is tempting to postulate that the coated region-coated vesicle may represent a specific organelle that is adapted for the binding and rapid internalization of a variety of macromolecules that exert both nutritional and regulatory functions within the cell.

Experimental Procedures

Cells

Cultured fibroblasts were derived from skin biopsies obtained from a normal subject (D.S.) or a patient with the receptor-negative form of homozygous familial hypercholesterolemia (M.C.) (Goldstein et al., 1975c). Cells were grown in monolayer and used between the fifth and twentieth passage. Stock cultures were maintained in a humidified incubator (5% CO_2) at 37°C in 250 ml flasks containing 10 ml of growth medium consisting of Eagle's minimum essential medium (Gibco, Cat. No. F-11) supplemented with penicillin (100 units per ml); streptomycin (100 μg/ml); 20 mM Tris-HCl (pH 7.4); 24 mM $NaHCO_3$; 1% (v/v) nonessential amino acids; and 10% (v/v) fetal calf serum (Flow Laboratories). Unless otherwise noted, all experiments were performed in a similar format. On day 0, confluent monolayers of cells from stock flasks were dissociated with 0.05% trypsin-0.02% EDTA solution, and 1×10^5 cells were seeded into each 60 × 15 mm petri dish containing 3 ml of growth medium with 10% fetal calf serum. On day 3, the medium was replaced with 3 ml of fresh growth medium containing 10% fetal calf serum. On day 5, each monolayer was washed with 3 ml of phosphate-buffered saline, after which 2 ml of fresh medium containing 5% (v/v) human lipoprotein-deficient serum were added (final protein concentration 2.5 mg/ml). Experiments were initiated on day 7 after the cells had been incubated for 48 hr in the presence of lipoprotein-deficient serum.

Lipoproteins

Human LDL (density 1.019–1.063 g/ml) and lipoprotein-deficient serum (density >1.215 g/ml) were obtained from the plasma of healthy subjects and prepared by differential ultracentrifugation as previously described (Brown et al., 1974). ^{125}I-labeled LDL was prepared as previously described (Brown and Goldstein, 1974).

Coupling of Ferritin to LDL

Ferritin (Polysciences, Inc.) was activated with a 1200 fold molar excess of glutaraldehyde and coupled to LDL as previously described (Anderson et al., 1976). The LDL-ferritin complex was then purified by ultracentrifugation (Anderson et al., 1976), and samples were negatively stained with 1% sodium phosphotungstate to examine the number of ferritins bound to each LDL particle. Over 90% of the LDL particles were labeled with ferritin, and on the average there were two ferritins coupled to each LDL particle (Figure 8). The concentration of LDL-ferritin is expressed in terms of its content of LDL-protein, which was calculated on the basis of the measured cholesterol content of the complex (Anderson et al., 1976).

Binding of LDL-Ferritin to Fibroblasts

Monolayers of normal fibroblasts were cooled to 4°C for 30 min. The medium was removed and replaced with 2 ml of ice-cold growth medium containing 5% human lipoprotein-deficient serum and LDL-ferritin at a concentration corresponding to 47.5 μg/ml of LDL-protein. The monolayers were incubated at 4°C for 2 hr, after which each dish was washed 5 times at 4°C with ice-cold phosphate-buffered saline (3 ml per wash) and then switched to 2 ml of warm growth medium containing 5% lipoprotein-deficient serum. Samples of dishes were warmed to 37°C for the following time periods: 0, 1, 2, 4, 6, 8, and 10 min. At the end of each time period, the monolayers were fixed at room temperature for 30 min with 2% glutaraldehyde in 0.1 M sodium phosphate buffer (pH 7.3).

In other experiments, monolayers of either normal or homozygous familial hypercholesterolemia cells were incubated for various intervals at 37°C with growth medium containing 5% lipoprotein-deficient serum and 47.5 μg/ml of LDL–ferritin. At the end of the incubation, the cells were washed 5 times at 4°C with phosphate-buffered saline and fixed at room temperature with 2% glutaraldehyde in 0.1 M sodium phosphate buffer (pH 7.3).

Binding of ^{125}I-LDL to Fibroblasts

Normal and homozygous familial hypercholesterolemia fibroblasts were grown on glass coverslips. 1 × 10^5 cells were initially seeded on day 0 into each 60 petri dish that contained two glass coverslips (22 × 22 mm). The cells were grown exactly as described above. On day 7 after incubation for 48 hr in the presence of lipoprotein-deficient serum, the cells were chilled to 4°C for 30 min, then incubated at 4°C for 2 hr with growth medium containing 5% lipoprotein-deficient serum and 6.5 μg protein per ml of ^{125}I-LDL (235 cpm/ng protein). The medium was removed, and the cells were washed 6 times at 4°C with an albumin-containing buffer using the standard procedure previously described (Goldstein et al., 1976a). Some sets of coverslips were fixed immediately, while others received 2 ml of growth medium containing 5% lipoprotein-deficient serum and were warmed to 37°C in the petri dish for various times before fixation. The fixative was 6% paraformaldehyde in 0.1 M phosphate buffer (pH 7.3).

Autoradiography

Coverslips containing monolayers of fibroblasts that had been exposed to ^{125}I-LDL and fixed with paraformaldehyde as described above were washed with distilled water and air-dried. The coverslips were mounted on slides, cell side up, with Sanderaul (Bio/Medical Specialities, Santa Monica, California). The slides were dipped in a 1:2 dilution of Ilford L-4 liquid photographic emulsion (Polysciences, Inc.) in the darkroom and allowed to air-dry. The emulsion-coated slides were stored in light-tight boxes containing Drierite dessicant (Scientific Products) at 4°C for either 2 weeks or 4 weeks. At the end of these time periods, the slides were developed with D19-B developer, rinsed in distilled water, and fixed with Rapid-Fix (Eastman Kodak). The slides were stained with 1% toluidine blue and covered with a second coverslip. All autoradiograms were examined with an American Optical Microstar 10 and photographed with a Polaroid camera.

Exposure of Cells to Native Ferritin and Ferritin-Conjugated Antibody to IgG

Monolayers of normal fibroblasts were prepared and grown as described above. On day 7, the medium was replaced with growth medium containing 5% lipoprotein-deficient serum and either native ferritin at concentrations of 250 μg/ml, 2.5 mg/ml, and 10 mg/ml, or ferritin-labeled antibody (produced in goat) to rabbit IgG (Cappel Laboratories, Lot 8588) at a concentration of 400 μg protein per ml. The cells were exposed to these electron-dense probes for various periods of time at 4°C or 37°C, washed at 4°C 5 times with phosphate-buffered saline (3 ml per wash), and fixed for 30 min at room temperature with 2% glutaraldehyde in 0.1 M sodium phosphate buffer (pH 7.3).

Exposure of Cells to Horseradish Peroxidase

Monolayers of normal cells were prepared and grown as described above. On day 7, the cells were exposed at 37°C to growth medium containing 5% lipoprotein-deficient serum and 5 mg/ml of horseradish peroxidase (Sigma Chemical) for 30 min. The cells were washed 5 times at 4°C with phosphate-buffered saline and fixed for 5 min at room temperature with 2.5% glutaraldehyde in 0.1 M sodium cacodylate (pH 7.4). After removal of the fixative, the cells were washed with 40 mM sodium cacodylate (pH 7.3) and then reacted with diaminobenzidine plus H_2O_2 according to the method of Graham and Karnofsky (1966).

Electron Microscopy Procedures

Monolayers of cells were prepared for electron microscopy according to previously published procedures (Anderson et al., 1976). Fixed cells that were scraped from the culture dish and pelleted in microfuge tubes were post-fixed with 2% OsO_4 in 0.1 M sodium phosphate buffer (pH 7.3) for 2 hr, and then dehydrated in a graded series of alcohols and embedded in Araldite. Unless otherwise stated, all sections were cut in a Sorval MT-2B microtome, stained with uranyl acetate and lead citrate, and viewed with either a Philips 200 or a Philips 300 electron microscope.

Quantitative Electron Microscopy

The binding and endocytosis of LDL–ferritin were monitored using previously described quantitative procedures (Anderson and Brenner, 1971; Anderson et al., 1976).

Acknowledgments

Margaret Wintersole and Gloria Y. Burnschede provided excellent technical assistance. This work was supported by grants from the National Heart and Lung Institute and the National Institute of General Medical Sciences. M.S.B. is an established investigator of the American Heart Association. J.L.G. is a recipient of a research career development award from the National Institute of General Medical Sciences.

Received November 9, 1976

References

Abbas, A. K., Ault, K. A., Karnovsky, M. J., and Unanue, E. R. (1975). Non-random distribution of surface immunoglobulin on murine B lymphocytes. J. Immunol. *114,* 1197–1204.

Anderson, R. G. W., and Brenner, R. M. (1971). Accurate placement of ultrathin sections on grids: control by sol-gel phases of a gelatin flotation fluid. Stain Technol. *46,* 1–6.

Anderson, R. G. W., Goldstein, J. L., and Brown, M. S. (1976). Localization of low density lipoprotein receptors on plasma membrane of normal human fibroblasts and their absence in cells from a familial hypercholesterolemia homozygote. Proc. Nat. Acad. Sci. USA *73,* 2434–2438.

Bennett, H. S. (1969). The cell surface: movements and recombinations. In Handbook of Molecular Cytology, A. Lima-de Faria, ed. (Amsterdam: North-Holland), pp. 1295–1319.

Brown, M. S., and Goldstein, J. L. (1974). Familial hypercholesterolemia: defective binding of lipoproteins to cultured fibroblasts associated with impaired regulation of 3-hydroxy-3-methylglutaryl coenzyme A reductase activity. Proc. Nat. Acad. Sci. USA. *71,* 788–792.

Brown, M. S., and Goldstein, J. L. (1975). Regulation of the activity of the low density lipoprotein receptor in human fibroblasts. Cell *6,* 307–316.

Brown, M. S., and Goldstein, J. L. (1976a). Receptor-mediated control of cholesterol metabolism. Science *191,* 150–154.

Brown, M. S., and Goldstein, J. L. (1976b). Analysis of a mutant strain of human fibroblasts with a defect in the internalization of receptor-bound low density lipoprotein. Cell *9,* 663–674.

Brown, M. S., Dana, S. E., and Goldstein, J. L. (1974). Regulation of 3-hydroxy-3-methylglutaryl coenzyme A reductase activity in cultured human fibroblasts: comparison of cells from a normal subject and from a patient with homozygous familial hypercholesterolemia. J. Biol. Chem. *249,* 789–796.

Brown, M. S., Dana, S. E., and Goldstein, J. L. (1975). Receptor-dependent hydrolysis of cholesteryl esters contained in plasma low density lipoprotein. Proc. Nat. Acad. Sci. USA. *72,* 2925–2929.

Brown, M. S., Ho, Y. K., and Goldstein, J. L. (1976). The LDL pathway in human fibroblasts: relation between cell surface receptor binding and endocytosis of low density lipoprotein. Ann. N.Y. Acad. Sci. *275,* 244–257.

Carpenter, G., and Cohen, S. (1976). ^{125}I-labeled human epidermal growth factor: binding, internalization, and degradation in human fibroblasts. J. Cell Biol. *71*, 159–171.

de Petris, S., and Raff, M. C. (1972). Distribution of immunoglobulin on the surface of mouse lymphoid cells as determined by immunoferritin electron microscopy: antibody-induced, temperature-dependent redistribution and its implications for membrane structure. Eur. J. Immunol. *2*, 523–535.

Edidin, M., and Weiss, A. (1972). Antigen cap formation in cultured fibroblasts: a reflection of membrane fluidity and of cell mobility. Proc. Nat. Acad. Sci. USA. *9*, 2456–2459.

Fawcett, D. W. (1965). Surface specializations of absorbing cells. J. Histochem. Cytochem. *13*, 75–91.

Franke, W. W., and Herth, W. (1974). Morphological evidence for de novo formation of plasma membrane from coated vesicles in exponentially growing cultured plant cells. Exp. Cell Res. *89*, 447–451.

Franke, W. W., Lüder, M. R., Kartenbeck, J., Zerban, H., and Keenan, T. W. (1976). Involvement of vesicle coat material in casein secretion and surface regeneration. J. Cell Biol. *69*, 173–195.

Friend, D. S., and Farquhar, M. G. (1967). Functions of coated vesicles during protein absorption in the rat vas deferens. J. Cell Biol. *35*, 357–376.

Frye, L. D., and Edidin, M. (1970). The rapid intermixing of cell surface antigens after formation of mouse-human heterokaryons. J. Cell Sci. *7*, 319–335.

Goldstein, J. L., and Brown, M. S. (1974). Binding and degradation of low density lipoproteins by cultured human fibroblasts: comparison of cells from a normal subject and from a patient with homozygous familial hypercholesterolemia. J. Biol. Chem. *249*, 5153–5162.

Goldstein, J. L., and Brown, M. S. (1976). The LDL pathway in human fibroblasts: a receptor-mediated mechanism for the regulation of cholesterol metabolism. In Current Topics in Cellular Regulation, *11*, B. L. Horecker and E. R. Stadtman, eds. (New York: Academic Press), pp. 147–181.

Goldstein, J. L., Dana, S. E., and Brown, M. S. (1974). Esterification of low density lipoprotein cholesterol in human fibroblasts and its absence in homozygous familial hypercholesterolemia. Proc. Nat. Acad. Sci. USA *71*, 4288–4292.

Goldstein, J. L., Brunschede, G. Y., and Brown, M. S. (1975a). Inhibition of the proteolytic degradation of low density lipoprotein in human fibroblasts by chloroquine, concanavalin A, and Triton WR 1339. J. Biol. Chem. *250*, 7854–7862.

Goldstein, J. L., Dana, S. E., Faust, J. R., Beaudet, A. L., and Brown, M. S. (1975b). Role of lysosomal acid lipase in the metabolism of plasma low density lipoprotein: observations in cultured fibroblasts from a patient with cholesteryl ester storage disease. J. Biol. Chem. *250*, 8487–8495.

Goldstein, J. L., Dana, S. E., Brunschede, G. Y., and Brown, M. S. (1975c). Genetic heterogeneity in familial hypercholesterolemia: evidence for two different mutations affecting functions of low-density lipoprotein receptor. Proc. Nat. Acad. Sci. USA *72*, 1092–1096.

Goldstein, J. L., Basu, S. K., Brunschede, G. Y., and Brown, M. S. (1976a). Release of low density lipoprotein from its cell surface receptor by sulfated glycosaminoglycans. Cell *7*, 85–95.

Goldstein, J. L., Sobhani, M. K., Faust, J. R., and Brown, M. S. (1976b). Heterozygous familial hypercholesterolemia: failure of normal allele to compensate for mutant allele at a regulated genetic locus. Cell *9*, 195–203.

Graham, R. C., Jr., and Karnovsky, M. J. (1966). The early stages of absorption of injected horseradish peroxidase in the proximal tubules of mouse kidney: ultrastructural cytochemistry by a new technique. J. Histochem. Cytochem. *14*, 291–302.

Heuser, J. E., and Reese, T. S. (1973). Evidence for recycling of synaptic vesicle membrane during transmitter release at the frog neuromuscular junction. J. Cell Biol. *57*, 315–344.

Korn, E. D. (1975). Biochemistry of endocytosis. In MTP International Review of Science, *2*, C. F. Fox, ed. (London: Butterworths), pp. 1–26.

Lagunoff, D., and Curran, D. E. (1972). Role of bristle-coated membrane in the uptake of ferritin by rat macrophages. Exp. Cell Res. *75*, 337–346.

Roth, T. F., and Porter, K. R. (1964). Yolk protein uptake in the oocyte of the mosquite *Aedes Aegypti*, *L*. J. Cell Biol. *20*, 313–332.

Roth, T. F., Cutting, J. A., and Atlas, S. B. (1976). Protein transport: a selective membrane mechanism. J. Supramol. Struct. *4*, 527–548.

Stein, O., Weinstein, D. B., Stein, Y., and Steinberg, D. (1976). Binding, internalization, and degradation of low density lipoprotein by normal human fibroblasts and by fibroblasts from a case of homozygous familial hypercholesterolemia. Proc. Nat. Acad. Sci. USA *73*, 14–18.

Steinman, R., Silver, J. M., and Cohn, Z. A. (1974). Pinocytosis in fibroblasts: quantitative studies in vitro. J. Cell Biol. *63*, 949–969.

Terris, S., and Steiner, D. F. (1975). Binding and degradation of ^{125}I-insulin by rat hepatocytes. J. Biol. Chem. *250*, 8389–8398.

Yusko, S. C., and Roth, T. F. (1976). Binding to specific receptors on oocyte plasma membranes by serum phosvitin-lipovitellin. J. Supramol. Struct. *4*, 89–97.

Novick P., Field C., and Schekman R. 1980. Identification of 23 complementation groups required for post-translational events in the yeast secretory pathway. *Cell* **21:** 205–215.

Movements of recently synthesized proteins and lipids from the endoplasmic reticulum (ER) to the Golgi complex, and thence to other membrane systems of the cell, had been studied in mammalian cells for 20 years (Jamieson and Palade, Blobel and Dobberstein, and Balch et al., this volume), laying a groundwork for understanding many aspects of membrane traffic. However, a genetic analysis of the secretory pathway, as exemplified by the paper reprinted here, helped to reveal the complexity and sophistication of what goes on in cells. The genes identified in this paper by mutation were subsequently analyzed by the Schekman lab and also by former members of that group after they had established their own labs. For example, the identification of Sec4p as a small GTPase (1) initiated a major new emphasis in trafficking studies, focused on the molecules that regulate membrane behavior. Complementary studies were often carried out independently on different systems (2), and sometimes the greater ease of studying morphology or cell physiology in other systems allowed such studies to make important advances (3), but the deep power of the genetic approach led Schekman and his colleagues to a consistent series of successes, e.g. elucidating the role of Sec61p for protein translocation across ER membranes (4) and the involvement of other *SEC* genes in the formation of COPII-coated transport vesicles (5). While many labs have contributed significantly to our understanding of the organization of the yeast secretory machinery through the use of alternative approaches (e.g., 6), the genetic approach initiated here has been tremendously fruitful. It has now been extended by many colleagues to encompass the study of essentially all the compartments and membrane traffic pathways in the cell (e.g., 7).

---

1. Salminen A. and Novick P.J. 1987. A ras-like protein is required for a post-Golgi event in yeast secretion. *Cell* **49:** 527–538.
2. Melancon P., Glick B.S., Malhotra V., Weidman P.J., Serafini T., M.L. Gleason, L. Orci, and Rothman J.E. 1987. Involvement of GTP-binding "G" proteins in transport through the Golgi stack. *Cell* **51:** 1053–1062.
3. Donaldson J.G., Cassel D., Kahn R.A., and Klausner R.D. 1992. ADP-ribosylation factor, a small GTP-binding protein, is required for binding of the coatomer protein beta-COP to Golgi membranes. *Proc. Natl. Acad. Sci.* **89:** 6408–6412.
4. Deshaies R.J. and Schekman R. 1987. A yeast mutant defective at an early stage in import of secretory protein precursors into the endoplasmic reticulum. *J. Cell Biol.* **105:** 633–645.
5. Barlowe C., Orci L., Yeung T., Hosobuchi M., Hamamoto S., Salama N., Rexach M.F., Ravazzola M., Amherdt M., and Schekman R. 1994. COPII: A membrane coat formed by Sec proteins that drive vesicle budding from the endoplasmic reticulum. *Cell* **77:** 895–907.
6. Preuss D., Mulholland J., Franzusoff A., Segev N., and Botstein D. 1992. Characterization of the *Saccharomyces* Golgi complex through the cell cycle by immunoelectron microscopy. *Mol. Biol. Cell.* **3:** 789–803.
7. Bankaitis V.A., Johnson L.M., and Emr S.D. 1986. Isolation of yeast mutants defective in protein targeting to the vacuole. *Proc. Natl. Acad. Sci.* **83:** 9075–9079.

Cell, Vol. 21, 205–215, August 1980, Copyright © 1980 by MIT

Identification of 23 Complementation Groups Required for Post-translational Events in the Yeast Secretory Pathway

Peter Novick, Charles Field and Randy Schekman*
Department of Biochemistry
University of California, Berkeley
Berkeley, California 94720

Summary

Cells of a Saccharomyces cerevisiae mutant that is temperature-sensitive for secretion and cell surface growth become dense during incubation at the nonpermissive temperature (37°C). This property allows the selection of additional secretory mutants by sedimentation of mutagenized cells on a Ludox density gradient. Colonies derived from dense cells are screened for conditional growth and secretion of invertase and acid phosphatase. The *sec* mutant strains that accumulate an abnormally large intracellular pool of invertase at 37°C (188 mutant clones) fall into 23 complementation groups, and the distribution of mutant alleles suggests that more complementation groups could be found. Bud emergence and incorporation of a plasma membrane sulfate permease activity stop quickly after a shift to 37°C. Many of the mutants are thermoreversible; upon return to the permissive temperature (25°C) the accumulated invertase is secreted. Electron microscopy of *sec* mutant cells reveals, with one exception, the temperature-dependent accumulation of membrane-enclosed secretory organelles. We suggest that these structures represent intermediates in a pathway in which secretion and plasma membrane assembly are colinear.

Introduction

Studies of the secretory process in eucaryotic cells have focused on the molecular events associated with synthesis and processing of specific secretory and membrane proteins, and on the organelles which mediate passage from the endoplasmic reticulum to the cell surface. While much is known about the maturation of certain secretory and membrane polypeptides (such as insulin, Chan, Keim and Steiner, 1976; VSV glycoprotein, Katz et al., 1977), the molecular events associated with sorting, packaging, transport and exocytosis of the exported proteins remain obscure.

We have undertaken a study of the secretory apparatus in the yeast Saccharomyces cerevisiae. While yeast cells are not specialized for secretion as are, for example, the acinar cells of the pancreas (Palade, 1975), the ease of a combined genetic and biochemical approach allows the use of techniques which have been less feasible with traditional secretory systems. The use of conditional mutants has been crucial for the analysis of complex bacteriophage morphogenesis pathways, both in identifying intermediate structures and in providing biochemical assays for assembly steps (Wood and King, 1979). We believe that a similar approach may be useful in unraveling a eucaryotic morphogenesis pathway.

Yeast cell surface growth is restricted primarily to enlargement of the bud followed by cell division. Incorporation of new cell wall material, including secretion of the wall-bound enzymes invertase and acid phosphatase, is also restricted to the bud (Tkacz and Lampen, 1972, 1973; Field and Schekman, 1980). Membrane-enclosed vesicles have been implicated in secretion and bud growth (Moor, 1967; Matile et al., 1971). Our recent report of a conditional mutant blocked in secretion and cell surface growth, which accumulates membrane-enclosed vesicles containing a secretory enzyme (Novick and Schekman, 1979), supports such a role for vesicles.

In this report we describe a technique for the enrichment of conditional secretory and cell surface growth mutants. We have identified a large number of complementation groups that are required for the movement of at least two secretory enzymes and one plasma membrane permease through a series of distinct membrane-enclosed organelles in a pathway that leads to the cell surface.

Results

Secretory mutants are defined as those strains which fail to export active invertase and acid phosphatase, but continue to synthesize protein under restrictive growth conditions. In a previous report (Novick and Schekman, 1979) we described a screening procedure that allowed the identification of two nonallelic secretory mutants (*sec*1-1, *sec*2-1) among a group of randomly selected temperature-sensitive yeast mutants. The first mutant (HMSF 1) stopped dividing and enlarging at the nonpermissive temperature (37°C), yet protein and phospholipid synthesis continued for at least 3 hr. This situation produced dense cells. Henry et al. (1977) showed that during inositol starvation of an auxotrophic strain, net cell surface growth stopped while cell mass increased. Starved cells could be resolved from normal cells on a Ludox density gradient. We have used the Ludox density gradient technique to select additional secretory mutants.

Density Enrichment
In the experiment shown in Figure 1, about 5×10^6 *sec*1-1 cells were mixed with 5×10^8 X2180 cells, and after 3 hr at 37°C the mixture was sedimented in a solution of Ludox. The resulting gradient was fractionated and the genotype of cells in diluted aliquots was determined. The 5% increase in density allowed *sec*1 cells to be separated completely from a 100 fold larger population of wild-type cells.

The density separation made feasible the isolation

* To whom correspondence should be addressed.

of a large number of additional secretory mutants. Mutagenized cultures were allowed to grow for several generations at 24°C and then transferred to 37°C for 3 hr. The cells were then sedimented in a Ludox gradient, and the densest 1–2% of the cells were pooled. Temperature-sensitive growth mutants were identified among the dense cells, and the secretion of acid phosphatase and invertase was measured by a modification of the previous procedure (Novick and Schekman, 1979). The density gradient procedure enriched for a variety of temperature-sensitive mutants and among them about 15% were secretion-defective (Table 1).

The procedure was used on three strains; NF1R and SF182-3B were mutagenized with ethyl methanesulfonate (EMS) and X2180-1A was treated with nitrous acid. A total of 485 secretion-defective mutants were isolated, among which three classes were found. Class A *sec* mutants (188 total) showed accumulation of invertase at the nonpermissive temperature. Class B *sec* mutants showed no accumulation of active invertase, although protein synthesis continued at a high rate at 37°C. Class C mutants did not secrete because protein synthesis was temperature-sensitive. This report will deal with the *sec*A mutants; analysis of the *sec*B mutants is in progress.

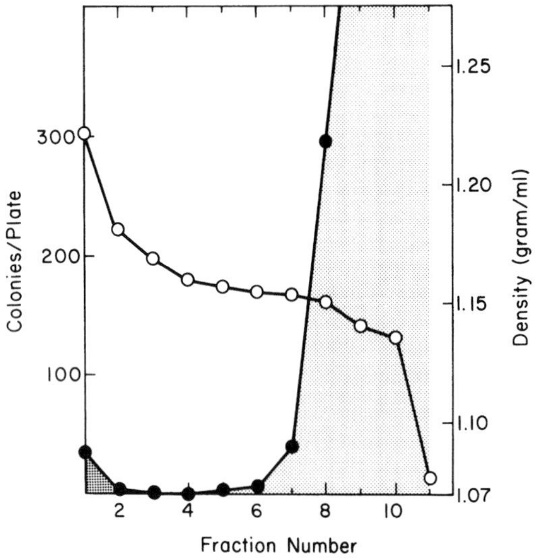

Figure 1. Density Gradient Separation of *sec*1-1 and X2180 Cells

SF150-5C and X2180-1A cells were grown in YPD medium at 25°C. After 3 hr at 37°C, the cells were sedimented, washed and resuspended in 1 ml of water. Cell aliquots were mixed (49.5 A_{600} units of X2180 and 0.5 A_{600} unit of SF150-5C) and sedimented on a Ludox gradient. Fractions were collected and diluted 10^4 fold, and 0.1 ml portions were spread on minimal medium (7 mM phosphate) agar plates. Colonies formed after 2 days at room temperature were stained for acid phosphatase activity (Hansche et al., 1978). Heavy stippling represents the phosphatase-constitutive (*pho*80) *sec*1 colonies (SF150-5C); light stippling represents phosphatase-repressed colonies (X2180).

The enrichment procedure was first performed on an *a* strain (NF1R) and then on an α strain (SF 182-3B). The *sec* mutants were arranged into complementation groups by standard genetic techniques. The enrichment was repeated with strain X2180-1A, and complementation analysis of the temperature-sensitive clones was performed with tester strains derived from the previously isolated *sec* mutants. New *sec* mutants that complemented all of the tester strains were crossed with X2180-1B, diploids were selected and sporulated, and the *sec* mutants were obtained in *a* and α mating type strains. Complementation analysis was performed again. By this procedure, 23 complementation groups were identified. EMS and nitrous acid produced a similar spectrum of mutant alleles (Table 2); *sec*2 and *sec*5 were the most common groups for each mutagen.

All the *sec* mutants were recessive in heterozygous diploids. Analysis of each complementation group was conducted with a representative allele chosen for optimum growth at 25°C, maximum inhibition of secretion at 37°C and maximum secretion of accumulated invertase upon return to 25°C. Each group showed 2:2 segregation of the temperature-sensitive phenotype which always coincided with the secretory defect.

Thermoreversible Accumulation of Invertase

External invertase synthesis is derepressed by a decreased supply of glucose, and the secreted enzyme remains in the yeast cell wall and can be assayed in a whole-cell suspension (Dodyk and Rothstein, 1964). Intracellular forms of the enzyme are measured in a spheroplast lysate, and this level does not rise significantly during derepression (Novick and Schekman, 1979).

The level of secreted invertase increased 13 fold when X2180 cells were transferred from YP + 5% glucose medium to YP + 0.1% glucose medium (Table 3). This increase was blocked to varying degrees in the *sec* mutants at 37°C. Derepression for 1 hr at 37°C produced a 4–18 fold increase in the intracellular level of invertase. Upon return to 25°C, in the

Table 1. Comparison of Screening Procedure with and without Density Enrichment

	Without Enrichment		With Enrichment	
Screening Stage	Colonies	%	Colonies	%
(1) Colonies tested	5,600	100	18,500	100
(2) TS mutants	291	5.2	2,830	15
(3) TS phosphatase secretion	63	1.1	980	5
(4) TS invertase secretion	16	.29	485	2.6
(5) TS invertase accumulation	2	.04	188	1.0

Table 2. Distribution of Mutants in the *sec*A Complementation Groups: EMS versus Nitrous Acid

	EMS		Nitrous Acid	
sec	Isolates	%	Isolates	%
1	8	11	4	3
2	28	39	41	35
3	3	4	0	0
4	7	10	2	2
5	10	14	16	14
6	3	4	3	3
7	1	1	3	3
8	6	8	4	3
9	3	4	4	3
10	1	1	2	2
11	1	1	11	9
12	1	1	3	3
13			4	3
14			4	3
15			2	2
16			2	2
17			1	1
18			2	2
19			1	1
20			1	1
21			1	1
22			4	3
23			1	1

presence of cycloheximide, all the mutants showed an increased level of secreted invertase. In most cases this represented the secretion of a large fraction of the accumulated enzyme. The mutants produced nearly normal levels of secreted and intracellular invertase when derepression was conducted at 25°C.

All 12 alleles of *sec*11 produced nearly normal levels of secreted invertase at 37°C. However, like the other mutants, *sec*11 accumulated about 7 fold more internal invertase than normal. Furthermore, 35% of the accumulated enzyme was secreted upon return to 25°C.

Other Defects

Secreted acid phosphatase first appeared in wild-type cells 1.5 hr after a transfer from YPD + 7 mM phosphate medium into phosphate-depleted YPD medium, and the rate of secretion was maximal from 2.5 to 5 hr after the shift. During this 2.5 hr period, the *sec* mutants secreted normally at 25°C, but produced at least 5 fold less external phosphatase at 37°C (Table 4).

Incorporation of a sulfate permease activity was used to assess the role of the *sec* gene products in plasma membrane assembly. In X2180, sulfate permease activity first appeared 30 min after cells were transferred from a minimal medium containing 1.5 mM methionine to a sulfate-free minimal medium. During a 2.5 hr period of derepression, most of the *sec* mutants produced normal levels of permease activity at 25°C, but showed significantly lower incorporation at 37°C (Table 4).

Bud emergence stopped quickly, as indicated by the nearly constant number of cells and buds, when the *sec* mutants were transferred from 25° to 37°C (Table 4). Cells arrested at all stages of the cell cycle, and no increase in cell size was noted. As expected from the enrichment procedure, all the *sec* mutants became denser than X2180 during a 3 hr incubation at 37°C (Table 4), although only a few of the strains became as dense as *sec*1. Certain other conditions, such as inhibition of protein synthesis or growth to stationary phase, caused X2180 cells to become dense at 37°C.

Mutants Accumulate Secretory Organelles

The reversible accumulation of invertase and the reduced incorporation of a membrane permease suggested that the *sec* mutants might accumulate an organelle of the secretory apparatus. This was confirmed for all but one of the mutants when thin sections were examined by electron microscopy. Wild-type cells grown at 37°C (Novick and Schekman, 1979, Figure 6A) or *sec* mutant cells grown at 25°C (Figure 2A) showed occasional enrichment of vesicles in the bud; short, thin tubules of endoplasmic reticulum (ER) were also seen apposed to the inner surface of the plasma membrane or in continuity with the nuclear membrane. Mutant cells incubated for 2 hr at 37°C showed several cytological aberrations. The groups were classified according to the organelle accumulated. The most common class, with 10 members (Table 5, Figure 2B), accumulated membrane-enclosed vesicles of 80–100 nm in diameter. These vesicles were not enriched in the bud.

A second class, with nine representatives, developed a more extensive network of ER than was seen in wild-type cells (Figure 3). The ER often lined the inner surface of the plasma membrane and wound through the cytoplasm where multiple connections with the nuclear membrane were visible (Figures 3A and 3B). The lumen of both the ER and the nuclear membrane was wider than the corresponding wild-type structure. Eucaryotic rough and smooth ER can be distinguished by the presence or absence of attached ribosomes. However, due to the high concentration of free ribosomes, it was not possible to identify specific associations with the yeast ER. While all nine members of this class showed extensive ER at 37°C, three complementation groups also produced small vesicles (~ 40 nm) which were often arranged in patches in the cytoplasm (Table 5, Figure 3C).

Table 3. Invertase Secretion and Accumulation by the *sec* Mutants

		Units/mg Dry Weight					
Strain	*sec* Group	External[a] (1 Hr 37°C)	Internal (1 Hr 37°C)	External (1 Hr 37°C → 3 Hr 25°C)	% Release[b]	External (1 Hr 25°C)	Internal (1 Hr 25°C)
X2180-1A		.38	.08	.33	0	.34	.14
HMSF 1	1-1	.02	.61	.28	43	.29	.15
HMSF 106	2-56	.03	.87	.36	38	.24	.18
HMSF 68	3-2	.02	.31	.05	9	.31	.22
HMSF 13	4-2	.05	.63	.13	11	.32	.30
HMSF 134	5-24	.03	.84	.08	6	.39	.17
HMSF 136	6-4	.03	.84	.46	52	.36	.14
HMSF 6	7-1	.04	.39	.10	16	.42	.29
HMSF 95	8-6	.03	.57	.07	7	.37	.22
HMSF 143	9-4	.09	1.05	.53	42	.20	.28
HMSF 147	10-2	.03	.68	.15	18	.31	.17
HMSF 154	11-7	.40	.53	.59	35	.56	.26
HMSF 162	12-4	.04	1.3	.90	64	.22	.11
HMSF 163	13-1	.19	.77	.64	58	.28	.14
HMSF 169	14-3	.07	.54	.32	46	.30	.12
HMSF 171	15-1	.17	.47	.33	34	.36	.19
HMSF 174	16-2	.06	1.50	.69	42	.45	.18
HMSF 175	17-1	.17	.70	.58	59	.29	.14
HMSF 176	18-1	.02	.97	.64	63	.36	.15
HMSF 178	19-1	.02	1.05	.49	45	.43	.21
HMSF 179	20-1	.07	.98	.68	63	.49	.24
HMSF 180	21-1	.29	.43	.48	44	.39	.18
HMSF 183	22-3	.04	.84	.58	64	.41	.18
HMSF 190	23-1	.03	.83	.57	65	.36	.15

[a] Cultures were grown overnight in YP + 5% glucose medium at 25°C to an A_{600} of 0.5–5.5. Cells (3 A_{600} units) were sedimented in a clinical centrifuge, resuspended in 3 ml of YP + 0.1% glucose medium and incubated at 37°C for 1 hr. An aliquot (1 ml) was then removed and added to a tube containing 1 mg of glucose and 0.1 mg of cycloheximide, and the mixture was incubated at 25°C for 3 hr. In a parallel experiment, 2 A_{600} units of the overnight culture were sedimented, and the cell pellet was resuspended in 2 ml of YP + 0.1% glucose medium and incubated at 25°C for 1 hr. At the end of each experiment samples were chilled on ice, centrifuged and resuspended in one half volume of 10 mM azide at 0°C.

[b] $\%\ \text{release} = \left(\frac{\text{Ext}_{37°C\to25°C} - \text{Ext}_{37°C}}{\text{Int}_{37°C}}\right) 100.$

A third class of mutant produced an organelle with no obvious counterpart in other eucaryotic cells. Because of its unique morphology, we call this structure a Berkeley body (Bb). The Bb, although varied in form, appeared to consist of two curved membranes with an enclosed electron-transparent lumen (Figure 4). In some sections the Bb was closed to form a toroid; in other sections it was open at one end to form a cup (Figure 4B). The toroid structure, with enclosed ribosomes and cytoplasm, may be an alternate view of the cup form; perpendicular planes of sectioning would give the image of one or the other. Two complementation groups (*sec*7, 14) made predominantly Bbs; two alleles of *sec*7, derived from different parent strains, were examined and each produced only Bbs. In addition to Bbs, *sec*14 also produced 80–100 nm vesicles (Figure 5A). Bbs were occasionally seen in sections of two complementation groups (*sec*2, 9) where 80–100 nm vesicles were the dominant structure.

Two exceptions to the major classes were observed: *sec*19 produced a mixture of the major organelles (Figure 5B), and *sec*11 did not build up any of the organelles.

Discussion

The results presented here show that in yeast, at least 23 gene products are required for the transport of secretory proteins from the site of synthesis to the cell

Table 4. Acid Phosphatase Secretion, Sulfate Permease Incorporation, Cell Division and Cell Density of the *sec* Mutants

Strain	*sec*	Acid Phosphatase[a] (Units/ml)			Sulfate Permease[b] (Units / mg Dry Weight)		Cell Number[c] (2 Hr 37°C / 0 Hr)	Cell Density[d] (g/ml)	
		2.5 Hr	5 Hr 37°C	5 Hr 25°C	25°C	37°C		25°C	37°C
X2180 1A		27	193	174	6.3	5.0	2.03	1.110	1.122
HMSF 1	1-1	27	28	147	7.4	.2	1.10	1.113	1.161
HMSF 106	2-56	50	48	170	5.8	.1	.92	1.109	1.141
HMSF 68	3-2	55	82	411	6.8	.3	1.10	1.109	1.142
HMSF 13	4-4	25	29	202	4.8	.2	1.07	1.116	1.146
HMSF 134	5-24	27	31	177	5.8	.4	1.04	1.110	1.161
HMSF 136	6-4	25	27	178	5.1	.1	1.20	1.111	1.159
HMSF 6	7-1	86	95	320	3.6	.03	.91	1.103	1.142
HMSF 16	8-1	44	81	223	4.5	.9	1.11	1.103	1.135
HMSF 143	9-4	30	30	177	6.1	.1	1.04	1.111	1.146
HMSF 147	10-2	25	47	70	7.2[e]	.03	.92	1.117	1.152
HMSF 154	11-7	16	15	107	6.1	2.0	1.48	1.117	1.142
HMSF 162	12-4	25	25	89	5.3[e]	.65	1.05	1.107	1.143
HMSF 163	13-1	22	18	154	4.2	.1	1.01	1.113	1.141
HMSF 169	14-3	25	26	121	5.3	1.5	.93	1.117	1.144
HMSF 171	15-1	23	41	190	6.7	1.7	1.15	1.117	1.159
HMSF 174	16-2	26	25	93	6.2	.02	.95	1.114	1.139
HMSF 175	17-1	32	30	200	6.1	.4	1.07	1.118	1.146
HMSF 176	18-1	27	25	202	5.6	.04	.93	1.117	1.155
HMSF 178	19-1	32	32	237	6.3	.3	.92	1.119	1.156
HMSF 179	20-1	23	21	136	1.5	.1	.94	1.116	1.141
HMSF 180	21-1	24	42	171	6.3	.2	1.08	1.113	1.142
HMSF 183	22-3	25	26	204	1.7	.1	1.09	1.115	1.151
HMSF 190	23-1	30	27	82	4.4	.1	1.02	1.114	1.145

[a] Cultures were grown overnight in YPD + 7 mm phosphate medium. Cells (4.5 A_{600} units) were centrifuged, resuspended in 3 ml of phosphate-depleted YPD medium and incubated at 25°C. After 2.5 hr, 1 ml aliquots were transferred to 37° and 0°C, and the rest were left at 25°C. Incubation was continued for 2.5 hr and the samples were chilled, centrifuged and resuspended in 1 ml of 10 mM azide at 0°C.

[b] Cultures were grown overnight in minimal medium + 1.5 mM methionine and 50 μM $(NH_4)_2SO_4$. Cells (1.2 A_{600} units) were centrifuged, washed once and resuspended in 1.2 ml of sulfate-free minimal medium. After 2.5 hr at 25° or 37°C, the tubes were chilled and 1 ml aliquots were removed for the permease assays; the rest were used for A_{600} determination.

[c] Cultures were grown overnight in YPD medium at 25°C. Cells (2 A_{600} units) were centrifuged and resuspended in 2 ml YPD medium; 0.5 ml was diluted with 10 mM azide at 0°C and the rest were incubated for 2 hr at 37°C. The ratio of the 2 hr/0 hr cell number is listed.

[d] Cultures were grown overnight in YPD medium. Cells (8 A_{600} units) were centrifuged and resuspended in 4 ml of YPD medium; 2 ml of each were incubated at 25° and 37°C. After 3 hr the cells were sedimented and resuspended in 0.5 ml of 10 mM azide at 0°C. X2180 cells grown to stationary phase at 37°C had a density of 1.131 g/ml; cells treated with 0.1 mg/ml of cycloheximide for 3 hr at 37°C had a density of 1.134 g/ml.

[e] HMSF 147 and 162 were auxotrophic and the sulfate permease experiment was carried out with prototrophic strains derived from crosses with X2180 (SF 226-1C, *sec*10; SF 292-2C, *sec*12).

surface. Thermosensitive defects in these gene products also block incorporation of a plasma membrane permease and stop bud growth. Taken together, these observations suggest that membrane growth and secretion are accomplished by parallel if not identical pathways. Furthermore, membrane-enclosed organelles accumulate in 22/23 of the mutants at 37° but not at 25°C. We propose that these structures are intermediates in the secretory pathway; their soluble contents are destined for secretion by exocytosis and their membranes will be incorporated into the plasma membrane by fusion.

In a previous report (Novick and Schekman, 1979), we described the detection of *sec*1-1 and *sec*2-1 in a collection of randomly selected temperature-sensitive mutants. No new *sec* mutants were found in a larger collection of temperature-sensitive strains, and therefore the density enrichment procedure was adopted. Although both of the original *sec* mutants become dense at 37°C and survive the enrichment, density

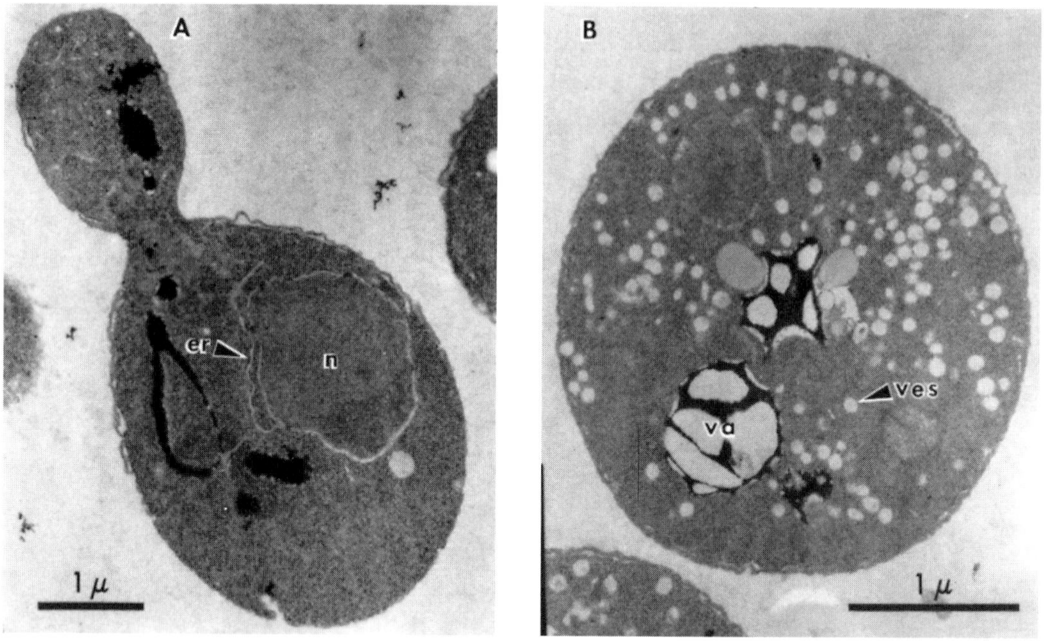

Figure 2. Thin Section Electron Micrographs of Cells Grown in YPD Medium
(A) HMSF 13 (*sec*4-2) grown at 25°C; (B) HMSF 171 (*sec*15-1) incubated at 37°C for 2 hr. Symbols: (n) nucleus; (va) vacuole; (er) endoplasmic reticulum; (ves) vesicles.

selection has the disadvantage that it eliminates mutants that die rapidly at 37°C. We assume that the density enrichment selects mutants that accumulate mass without a corresponding cell surface increase. The screening procedure, on the other hand, assumes only that secretion of cell wall mannoproteins is necessary for cell viability. For this reason, the density enrichment may eliminate mutants which fail to secrete but continue to expand, while the screening procedure will remove mutants which fail to enlarge but continue to secrete. Nevertheless, a large number of gene products are implicated in both aspects of cell surface growth.

It is likely that there are more than 23 *sec*A complementation groups required for the secretory process. Five of the groups reported here contain only one mutant allele, suggesting that groups exist for which no mutant has been found. Furthermore, the 23 *sec*A groups may represent fewer than 23 separate gene products; gene clusters coding for multifunctional proteins have been found in fungi and yeast. However, in such circumstances a single mutant frequently appears to fall into two otherwise distinct complementation groups (Fincham, 1977). No example of overlapping *sec* groups was found.

A trivial explanation for the large number of *sec* complementation groups is that mutant forms of various secreted proteins can act as inhibitors and block the passage of other cell surface molecules. Bassford and Beckwith (1979) and Bassford, Silhavy and Beck-

Table 5. Organelles Accumulated in the *sec* Strains

Strain (HMSF)	*sec*	Structure(s)
1	1-1	vesicles, Berkeley bodies
47	2-7	vesicles
3	3-1	vesicles
13	4-2	vesicles
81	5-8	vesicles
12	6-1	vesicles
6	7-1, -2	Berkeley bodies
93	8-4	vesicles
89	9-3	vesicles, Berkeley bodies
147	10-2	vesicles
154	11-7	
162	12-4	ER
163	13-1	ER
169	14-3	Berkeley bodies, vesicles
171	15-1	vesicles
174	16-2	ER
175	17-1	ER, small vesicles
176	18-1	ER, small vesicles
178	19-1	vesicles, Berkeley bodies, ER
179	20-1	ER
180	21-1	ER
183	22-3	ER, small vesicles
190	23-1	ER

Yeast Secretory Mutants
211

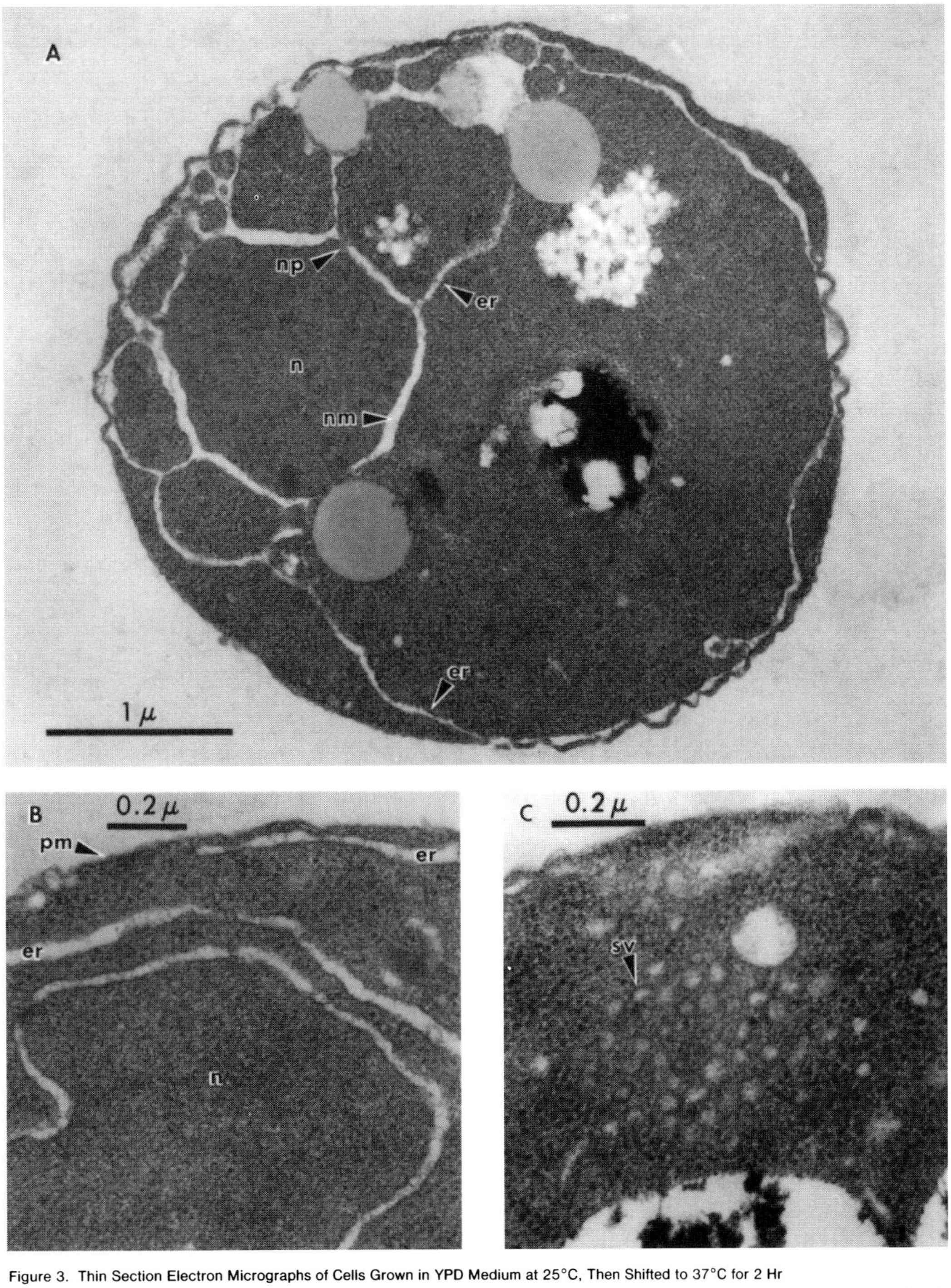

Figure 3. Thin Section Electron Micrographs of Cells Grown in YPD Medium at 25°C, Then Shifted to 37°C for 2 Hr
(A) HMSF 174 (*sec*16-2); (B) HMSF 190 (*sec*23-1); (C) HMSF 175 (*sec*17-1). Symbols are as in Figure 2 and (np) nuclear pore; (sv) small vesicle; (pm) plasma membrane; (nm) nuclear membrane.

Cell
212

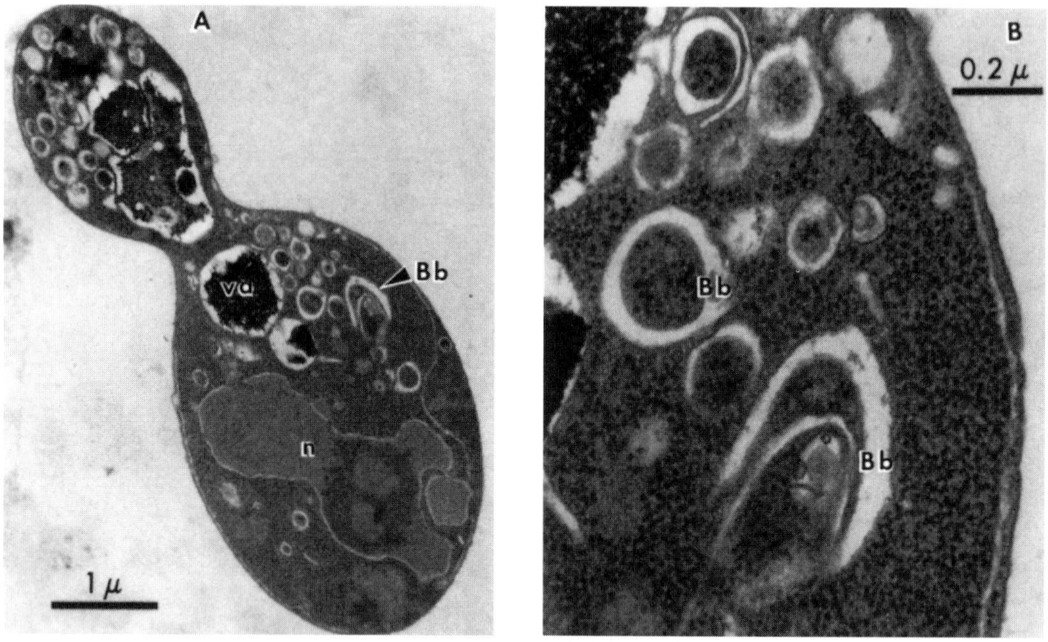

Figure 4. Thin Section Electron Micrograph of HMSF 6 (*sec*7-1) Grown in YPD Medium at 25°C, Then Shifted to 37°C for 2 Hr
(A) Low magnification; (B) a portion of the same cell at higher magnification. Symbols are as in Figure 2 and (Bb) Berkeley body.

with (1979) have shown that fusion of the E. coli *lac*Z (β-galactosidase) and *mal*E (periplasmic maltose binding protein) genes leads to the production of a hybrid protein which becomes stuck in the cytoplasmic membrane. The aberrant incorporation of this hybrid protein prevents the secretion of other proteins and the cells die. Such mutations are genetically dominant (T. Silhavy, personal communication); cell death resulting from the production of hybrid protein is not prevented by synthesis of a normal *mal*E gene product. In contrast, all the *sec* mutants are recessive and therefore the defective gene products are not likely to be secretion inhibitors.

Although 16 of the 23 representative mutant alleles are at least 5 fold reduced at 37°C in each of the four parameters of cell surface growth (Tables 3 and 4), several of the mutants are less restrictive (leaky). Some of the leaky strains (*sec*15, 17, 21) are in complementation groups for which only 1–2 mutant alleles are available. Among the groups with many members, some alleles are very restrictive at 37°C, while others are leaky. Certain strains are less restrictive for one enzyme marker than another. This may be due to the different growth conditions required for derepression of the marker enzymes. Thus while invertase appears in cells within 30 min of a transfer from 5 to 0.1% glucose, acid phosphatase production requires a 1.5 hr phosphate starvation. The *sec*11 group is an extreme example. Cells of all 12 mutant alleles secrete nearly normal amounts of invertase at 37°C, but secrete no acid phosphatase. Although no organelles accumulate in *sec*11 at 37°C, internal invertase levels rise 7 fold, and about one third of the accumulated invertase is secreted when cells are returned to 25°C. The *sec*11 gene product may be difficult to convert to a completely thermosensitive form, or it may not be required absolutely for the secretory process.

More important than the occasional leaky strain is the fact that in most *sec* mutants the block to secretion and membrane permease incorporation is coupled with the accumulation of secretory enzymes and membrane-enclosed organelles. van Rijn, Linnemans and Boer (1975) have shown by histochemical staining of wild-type cells that acid phosphatase is contained within ER, Golgi-like structures and vesicles. The acid phosphatase that accumulates in *sec*1 cells at 37°C is contained within vesicles (Novick and Schekman, 1979); the ER and Berkeley bodies (Bbs) produced in other mutants also contain this enzyme (B. Esmon, P. Novick and R. Schekman, manuscript in preparation). The various membrane-enclosed structures produced at 37°C probably represent functional intermediates in the secretory pathway, since most of the mutants secrete the accumulated invertase upon return to 25°C.

The rates of invertase synthesis and export may also be coupled. In many of the mutants, 2–4 fold more invertase accumulates in an internal pool at 37°C than is secreted by wild-type cells in a comparable period (Table 3).

Some of the mutants accumulate more than one

Yeast Secretory Mutants
213

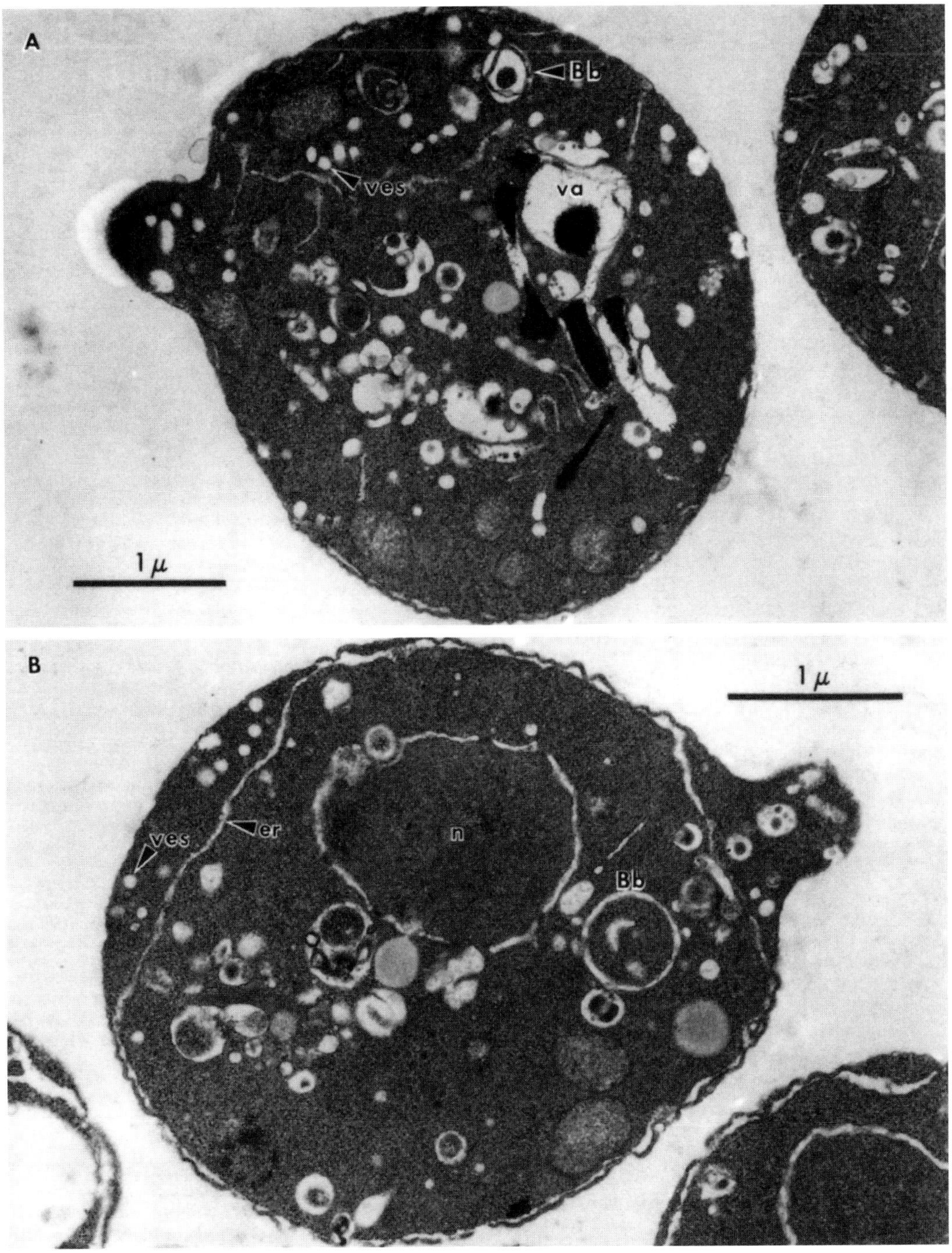

Figure 5. Thin Section Electron Micrographs of Cells Grown in YPD Medium at 25°C, Then Shifted to 37°C for 2 Hr
(A) HMSF 169 (*sec*14-3); (B) HMSF 178 (*sec*19-1). Symbols are as in Figures 2–4.

type of organelle. If a *sec* gene product acts at several stages in the pathway, a partial block might result in multiple structures. Alternatively, if the reactions which connect one intermediate to another are reversible, a block in the pathway could lead by mass action to the accumulation of an earlier intermediate. A third possibility is that unstable organelles may reversibly or irreversibly generate other structures.

The close resemblance of the structure in some *sec* mutants to ER suggests that the affected gene products are required at an early step in the pathway. The vesicle-accumulating mutants are probably defective in later steps. Bbs may be analogous to the Golgi apparatus; they may become distorted by the continued incorporation of new membrane and secretory material in the absence of discharge to a succeeding stage in the process.

An independent line of evidence supports these proposals. Analysis of the extent of glycosylation of invertase accumulated at 37°C in the *sec* strains indicates at least two stages in oligosaccharide assembly. The mutants that produce ER also accumulate a form of invertase that has only half as much carbohydrate as is found associated with the enzyme accumulated in the other *sec* mutants, or on the secreted enzyme (B. Esmon, P. Novick and R. Schekman, manuscript in preparation). The attachment of a core oligosaccharide to nascent mannoprotein chains in yeast is likely to occur by a dolichol-dependent reaction in the ER (Marriot and Tanner, 1979). This core accounts for about half of the carbohydrate found on secreted invertase (Lehle, Cohen and Ballou, 1979). The outer chain oligosaccharides are added by distinct enzymes (Raschke et al., 1973) in a dolichol-independent reaction (Parodi, 1979). This second stage of oligosaccharide assembly may occur after movement of glycoproteins from the ER to a Golgi-like organelle.

A pathway of ER → Bbs → vesicles → cell surface is indicated by cytological analysis of double *sec* mutant strains incubated at 37°C (P. Novick and R. Schekman, unpublished results). Further analysis should establish the order in which the *sec* gene products function and the relationship of this pathway to secretory and membrane protein maturation. The contribution of the secretory pathway to cell surface growth will be tested directly by the isolation and analysis of organelle and plasma membrane fractions from the *sec* mutant cells. By providing an enriched supply of intermediate organelles and by providing criteria for the authentic reconstruction of individual events in vitro, the *sec* mutants may aid in a biochemical dissection of the secretory pathway.

Experimental Procedures

Materials

S. cerevisiae isogenic haploid strains X2180-1A (*a*, gal2) and −1B (α, gal2) were from the yeast genetics stock center (Berkeley). NF1R, a spontaneous GAL^+ revertant, was derived from X2180-1A. HMSF 1 (*a*, *sec*1-1) was derived from X2180-1A (Novick and Schekman, 1979). Standard genetic techniques were used to construct SF182-3B (*a*, GAL^+) and SF150-5C (*a*, *sec*1-1, III *ACP*1-2, *pho*80-2). All other *sec* strains were derived from X2180-1A, NF1R or SF182-3B as described in the text.

YPD medium contained 1% Bacto-Yeast Extract, 2% Bacto-Peptone and 2% glucose; YP medium was the same with different levels of glucose. Phosphate-depleted YPD was prepared as described by Rubin (1973). Wickerham's minimal medium (Wickerham, 1946) was used with the following modifications: for phosphate-free medium, potassium chloride replaced potassium phosphate; for sulfate-free medium, chloride salts replaced all sulfate salts. Unless otherwise indicated, the carbon source was 2% glucose. Petri plates contained the indicated medium and 2% Difco agar. Liquid cultures were grown in flasks or tubes with agitation, and the experiments were initiated with exponentially growing cells from stock cultures at an A_{600} of 0.5–5. The absorbance of cell suspensions was measured in a 1 cm quartz cuvette at 600 nm in a Zeiss PM QII spectrophotometer; 1 A_{600} unit corresponds to 0.15 mg dry weight.

Other reagents were obtained as indicated: ethyl methanesulfonate, p-nitrophenolphosphate, glucose oxidase, O-dianisidine, peroxidase and cycloheximide were from Sigma; $H_2{}^{35}SO_4$ was from New England Nuclear; glutaraldehyde, osmium tetroxide and Spurr embedding medium were from Polysciences; Ludox AM was from Protex Wax (Oakland, California) and was purified as described by Price and Dowling (1977). Lyticase is a yeast lytic enzyme preparation useful in spheroplast formation (Scott and Schekman, 1980). Fraction II (30,000 U/mg; 1 unit will lyse 0.2 A_{600} of logarithmic phase S. cerevisiae in 30 min at 30°C) was used.

Isolation of Secretory (*sec*) Mutants

Stationary phase cultures were mutagenized with ethyl methanesulfonate as described (Novick and Schekman, 1979). For nitrous acid mutagenesis, stationary phase X2180-1A cells were collected on a nitrocellulose filter, washed twice with distilled water and resuspended in 5 ml of 0.5 M sodium acetate (pH 4.8) and 20 mg sodium nitrate. After 10 min at 30°C, 5 ml of 2.7% Na_2HPO_4 containing 1% yeast extract were added, and the cells were filtered and washed with water. The viability was 37%. The cells were then allowed to recover from mutagenesis by growth in YPD medium at 25°C for 16 hr.

In the largest enrichment experiment, 1.85 A_{600} of mutagenized cells were grown in 50 ml of YPD medium to a total A_{600} = 25.5. The culture was then incubated at 37°C for 3 hr and the cells were collected by filtration, washed and resuspended in 0.5 ml water. The cell sample was layered on 12.5 ml of a mixture containing Wickerham's salts (Wickerham, 1946) and 60% (v/v) of the purified stock Ludox suspension, in a Falcon 17 × 100 mm polypropylene tube. After centrifugation in a Sorval SS34 rotor (22,000 × g, 20 min, 4°C), the tube was punctured at the bottom and 1 ml fractions were collected. The A_{600} of the fractions was measured and corrected for the A_{600} of corresponding fractions from a cell-free gradient, and the densest 2% of the cells (1.5 A_{600} total) was diluted 2 fold with water. The cells were then centrifuged, the pellet was resuspended in 1 ml of water and diluted 400 fold, and 0.1 ml portions were spread on 200 YPD medium agar plates.

The plates were incubated at room temperature (20–25°C) for 2 days and colonies were replica-plated onto two YPD plates each; one replica was incubated at room temperature, the other at 37°C. After 26 hr the replicas were compared and temperature-sensitive colonies were picked from the master plate.

The temperature-sensitive clones were replica-plated onto phosphate-free minimal medium plates to derepress the synthesis of acid phosphatase, and after 10 hr at 24° or 37°C the replicas were stained for secreted acid phosphatase (Hansche, Beres and Lange, 1978).

The clones which showed temperature-sensitive secretion of phosphatase were screened for conditional secretion and internal accumulation of invertase. Cultures grown at 25°C in YP + 5% glucose medium were shifted to 37°C for 30 min, after which 2 A_{600} units of cells were sedimented for 1.5 min in a clinical centrifuge. The cell pellets were then resuspended in 2 ml of fresh YP + 0.1% glucose medium and cultures were incubated at 37°C for an additional 90 min. The cells were sedimented again and the pellets were resuspended in 10 mM NaN_3 at 0°C. Cell wall and internal invertase levels were measured as previously described (Novick and Schekman, 1979). Strains which secreted less invertase than the X2180 control were designated *sec* mutants; those that accumulated internal levels higher than the X2180 control were assigned to class A, and all others were put aside for further analysis.

Analytical Procedures

Cell number was determined with a hemocytometer; buds were counted as cells. Density was determined by sedimentation of 8 A_{600} units of cells on a Ludox gradient as described above. The region of highest cell concentration was estimated visually and a 0.5 ml sample was removed by puncturing the centrifuge tube with a syringe. Sample density was measured by weighing a 100 μl portion.

External (cell wall-bound) acid phosphatase was assayed at 37°C as described by van Rijn, Boer and Steyn-Parvé (1972); units of activity are nmole of p-nitrophenol released per min. External invertase was assayed at 37°C as described by Goldstein and Lampen (1975); units of activity are μmole of glucose released per min. Internal invertase was determined by assaying spheroplast lysates, prepared as previously described (Novick and Schekman, 1979). Sulfate permease was assayed at 37°C in 50 μM $(NH_4)_2SO_4$ as described by Breton and Surdin-Kerjan (1977); units of activity are nmole of SO_4^{2-} uptake per min. Radioactivity was measured in a Searle Delta 300 liquid scintillation counter.

Samples were prepared for electron microscopy by the procedure of Byers and Goetsch (1975).

Acknowledgments

We thank Susan Ferro and Frank Gadzhorn for help in the mutant screening process. We also thank Alice Taylor for her continued expert assistance with electron microscope techniques. This work was supported by grants from the NSF and the NIH.

The costs of publication of this article were defrayed in part by the payment of page charges. This article must therefore be hereby marked "*advertisement*" in accordance with 18 U.S.C. Section 1734 solely to indicate this fact.

Received April 22, 1980; revised May 27, 1980

References

Bassford, P. and Beckwith, J. (1979). *Escherichia coli* mutants accumulating the precursor of a secreted protein in the cytoplasm. Nature *277,* 538–541.

Bassford, P. J., Silhavy, T. J. and Beckwith, J. R. (1979). Use of gene fusion to study secretion of maltose-binding protein into *Escherichia coli* periplasm. J. Bacteriol. *139,* 19–31.

Breton, A. and Surdin-Kerjan, Y. (1977). Sulfate uptake in *Saccharomyces cerevisiae*: biochemical and genetic study. J. Bacteriol. *132,* 224–232.

Byers, B. and Goetsch, L. (1975). Behavior of spindles and spindle plaques in the cell cycle and conjugation of *Saccharomyces cerevisiae.* J. Bacteriol. *124,* 511–523.

Chan, S. J., Keim, P. and Steiner, D. F. (1976). Cell-free synthesis of rat preproinsulins: characterization and partial amino acid sequence determination. Proc. Nat. Acad. Sci. USA *73,* 1964–1968.

Dodyk, F. and Rothstein, A. (1964). Factors influencing the appearance of invertase in *Saccharomyces cerevisiae.* Arch. Biochem. Biophys. *104,* 478–486.

Field, C. and Schekman, R. (1980). Localized secretion of acid phosphatase reflects the pattern of cell surface growth in *Saccharomyces cerevisiae.* J. Cell Biol. *86,* in press.

Fincham, J. R. (1977). Allelic complementation reconsidered. Carlsberg Res. Commun. *42,* 421–430.

Goldstein, A. and Lampen, J. O. (1975). β-D-Fructofuranoside fructohydrolase from yeast. Meth. Enzymol. *42,* 504–511.

Hansche, P. E., Beres, V. and Lange, P. (1978). Gene duplication in *Saccharomyces cerevisiae.* Genetics *88,* 673–687.

Henry, S. A., Atkinson, K. D., Kolat, A. I. and Culbertson, M. R. (1977). Growth and metabolism of inositol-starved *Saccharomyces cerevisiae.* J. Bacteriol. *130,* 472–484.

Katz, F. N., Rothman, J. E. Knipe, D. M. and Lodish, H. F. (1977). Membrane assembly: synthesis and intracellular processing of the vesicular stomatitis viral glycoprotein. J. Supramol. Structure *7,* 353–370.

Lehle, L., Cohen, R. E., and Ballou, C. E. (1979). Carbohydrate structure of yeast invertase. J. Biol. Chem. *254,* 12209–12218.

Marriot, M. and Tanner, W. (1979). Localization of dolichyl phosphate- and pyrophosphate-dependent glycosyl transfer reactions in *Saccharomyces cerevisiae.* J. Bacteriol. *139,* 565–572.

Matile, P., Cortat, M., Wiemken, A. and Frey-Wyssling, A. (1971). Isolation of glucanase-containing particles from budding *Saccharomyces cerevisiae.* Proc. Nat. Acad. Sci. USA *68,* 636–640.

Moor, H. (1967). Endoplasmic reticulum as the initiator of bud formation in yeast (*Saccharomyces cerevisiae*). Arch. Mikrobiol. *57,* 135–146.

Novick, P. and Schekman, R. (1979). Secretion and cell surface growth are blocked in a temperature-sensitive mutant of *Saccharomyces cerevisiae.* Proc. Nat. Acad. Sci. USA *76,* 1858–1862.

Palade, G. (1975). Intracellular aspects of the process of protein synthesis. *Science 189,* 347–358.

Parodi, A. J. (1979). Biosynthesis of yeast mannoproteins: synthesis of mannan outer chain and of dolichol derivatives. J. Biol. Chem. *254,* 8343–8352.

Price, C. A. and Dowling, E. L. (1977). On the purification of the silica sol Ludox AM. Anal. Biochem. *82,* 243–245.

Raschke, W. C., Kern, K. A., Antolis, C. and Ballou, C. E. (1973). Genetic control of yeast mannan structure. J. Biol. Chem. *248,* 4660–4666.

Rubin, G. M. (1973). The nucleotide sequence of *Saccharomyces cerevisiae* 5.8S ribosomal ribonucleic acid. J. Biol. Chem. *248,* 3860–3875.

Scott, J. and Schekman, R. (1980). Lyticase: endoglucanase and protease activities that act together in yeast cell lysis. J. Bacteriol. *142,* 414–423.

Tkacz, J. S. and Lampen, J. O. (1972). Wall replication in *Saccharomyces* species: use of fluorescein-conjugated concanavalin A to reveal the site of mannan insertion. J. Gen. Microbiol. *72,* 243–247.

Tkacz, J. S. and Lampen, J. O. (1973). Surface distribution of invertase on growing *Saccharomyces* cells. J. Bacteriol. *113,* 1073–1075.

van Rijn, H. J. M., Boer, P. and Steyn-Parvé, E. P. (1972). Biosynthesis of acid phosphatase of baker's yeast. Factors influencing its production by protoplasts and characterization of the secreted enzyme. Biochim. Biophys. Acta *268,* 431–441.

van Rijn, H. J. M., Linnemans, W. A. M. and Boer, P. (1975). Localization of acid phosphatase in protoplasts from *Saccharomyces cerevisiae.* J. Bacteriol. *123,* 1144–1149.

Wickerham, L. J. (1946). A critical evaluation of the nitrogen assimilation tests commonly used in the classification of yeasts. J. Bacteriol. *52,* 293–301.

Wood, W. and King, J. (1979). Genetic control of complex bacteriophage assembly. In Comprehensive Virology, *13,* H. Frankel-Conrat and R. Wagner, eds. (New York and London: Plenum Press), pp. 581–624.

Balch W.E., Dunphy W.G., Braell W.A., and Rothman J.E. 1984. Reconstitution of the transport of protein between successive compartments of the Golgi measured by the coupled incorporation of N-acetylglucosamine. *Cell* **39**: 405–416.

THE ABILITY OF CELLS TO SORT AND TRANSPORT lipids and proteins produced in the endoplasmic reticulum to and through the Golgi complex to "post-Golgi" targets was established during the 1960s (Jamieson and Palade, this volume). Study of the mechanisms by which this transport was accomplished and regulated awaited the development of systems for in vitro analysis, so that aspects of the transfer process could be reconstituted and dissected (1). Experimental cell biology had traditionally examined cell function, either in vivo or following homogenization and cell fractionation to enrich a particular organelle for further study in vitro. Biochemical work had traditionally examined molecules or sets of molecules that accomplished a specific reaction or process in solution. The Rothman lab pioneered in a hybrid approach that used enriched fractions from cells compromised by mutation, and a corresponding fraction from wild-type cells that could complement the missing glycosylation activity in vitro. The approach was reminiscent of one used by students of bacteriophage assembly almost two decades earlier (2). With this system, Rothman and his colleagues described vesicular transfer of proteins from one Golgi compartment to another (3,4). Today there is much evidence that the authors of these papers misintreperted the direction of the transport they were seeing (5,6), but the paper reprinted here and its companions provided an assay that has allowed the identification of many essential components of the membrane traffic system.

---

1. Fries E. and Rothman J.E. 1980. Transport of vesicular stomatitis virus glycoprotein in a cell-free extract. *Proc. Natl. Acad. Sci.* **77:** 3879–3874.
2. Edgar R.S. and Wood W.B. 1966. Morphogenesis of bacteriophage T4 in extracts of mutant-infected cells. *Proc. Natl. Acad. Sci.* **55:** 495–505.
3. Braell W.A., Balch W.E., Dobbertin D.C., and Rothman J.E. 1984. The glycoprotein that is transported between successive compartments of the Golgi in a cell-free system resides in stacks of cisternae. *Cell* **39:** 511–524.
4. Balch W.E., Glick B.S., and Rothman J.E. 1984. Sequential intermediates in the pathway of intercompartmental transport in a cell-free system. *Cell* **39:** 525–536.
5. Letourneue F., Gaynor E.C., Hennecke S., Démollière S., Duden R., Emr S., and Cosson P. 1994. Coatomer is essential for retrieval of dilysine-tagged proteins to the endoplasmic reticulum. *Cell* **79:** 1199–1207.
6. Pelham H.R.B. 1994. About turn for the COPs? *Cell* **79:** 1125–1127.

Cell, Vol. 39, 405–416, December 1984 (Part 1), Copyright © 1984 by MIT 0092-8674/84/120405-12 $02.00/0

Reconstitution of the Transport of Protein between Successive Compartments of the Golgi Measured by the Coupled Incorporation of N-Acetylglucosamine

William E. Balch, William G. Dunphy,
William A. Braell, and James E. Rothman
Department of Biochemistry
Stanford University School of Medicine
Stanford, California 94305

Summary

Transport of the VSV-encoded glycoprotein (G protein) between successive compartments of the Golgi has been reconstituted in a cell-free system and is measured, in a rapid and sensitive new assay, by the coupled incorporation of ^{3}H-N-acetylglucosamine (GlcNAc). This glycosylation occurs when G protein is transported during mixed incubations from the "donor" compartment in Golgi from VSV-infected CHO clone 15B cells (missing a key Golgi GlcNAc transferase) to the next, successive "acceptor" compartment (containing the GlcNAc transferase) in Golgi from wild-type CHO cells. Golgi fractions used in this assay have been extensively purified, and account for all of the donor and acceptor activity in the cells. Together with several other lines of evidence, this indicates that the cell-free system is highly specific, measuring only transport between sequential compartments in the Golgi stack. Transport in vitro is almost as efficient as in the cell, and requires ATP and the cytosol fraction in addition to protein components on the cytoplasmic surface of the Golgi membranes.

Introduction

To understand the basis for the specificity of subcellular compartments, we must learn how cells transport proteins to their correct locations and keep them there (Rothman and Lenard, 1984). This extensive traffic in proteins and membranes is mediated by transport vesicles. We must find out how these vesicles bud off from membranes, taking away only a select set of proteins ("protein sorting"), and how each type of transport vesicle succeeds in fusing only with the membrane that envelops its designated target ("vesicle sorting"), ensuring accurate delivery of the desired proteins. Cell-free systems that faithfully reproduce these processes are needed to unravel this biochemistry. In this and two following papers (Braell et al., 1984; Balch et al., 1984), we describe and analyze a cell-free system that seems to reconstitute accurately intercompartmental transport as it occurs in the context of the Golgi stack. It seems likely that both the budding of transport vesicles (from one set of Golgi cisternae) and their select fusion (with the next set of cisternae) take place.

The Golgi is an organelle particularly well suited for a functional reconstitution because its stack can be isolated essentially intact (Fleischer, 1974) and because it is specialized for protein transport (Farquhar and Palade, 1981; Rothman, 1981; Tartakoff, 1982). Thus membrane components needed for transport should be especially concentrated in Golgi fractions. In addition, the actions of the series of glycosyltransferases present in the sequential cisternae of the stack offer a ready means to measure the transport of a protein between these successive compartments, making cumbersome physical separations unnecessary. Three distinct compartments, each consisting of a small number of cisternae, are distinguished within the stack by several, quite independent lines of evidence (Tartakoff, 1982, 1983; Rothman, 1981; Dunphy et al., 1981; Roth and Berger, 1982; Griffiths et al., 1982; Roth, 1983; Dunphy and Rothman, 1983; Goldberg and Kornfeld, 1983; Deutscher et al., 1983; Rothman et al., 1984a, 1984b). These compartments have been termed *cis, medial,* and *trans,* in the order of their encounter by transported glycoproteins (Griffiths et al., 1983). Galactose residues are added to the Asn-linked oligosaccharides of transported glycoproteins in the *trans* compartment; N-acetylglucosamine (GlcNAc) residues in an earlier compartment now known to consist of several *medial* cisternae (Dunphy et al., 1985). Fatty acids are attached covalently either just before or after entry into the Golgi, which occurs at the *cis* face (Bergmann and Singer, 1983).

Transport between the successive compartments of the Golgi stack is a vectorial process that is dissociative in nature. Thus, when transport vesicles are allowed to choose between targets in the Golgi population from which they had budded and those in a second, exogenous population of Golgi (introduced by cell fusion), the ensuing fusion occurs on an entirely random basis (Rothman et al., 1984a, 1984b). This implies that, even in the cell, transport vesicles are sorted on the basis of a biochemical specificity and not physical proximity. Preexisting cytoplasmic organization is not important for accurate transport between Golgi compartments.

In this and previous work (Fries and Rothman, 1980, 1981; Rothman and Fries, 1981; Dunphy et al., 1981; Rothman et al., 1984b) we have taken advantage of both the dissociative nature of transport in the Golgi and a mutant defective in a step in glycosylation (but not in transport) to design an assay that measures transfers between successive Golgi compartments in vitro. The mutant is clone 15B of CHO cells (Gottlieb et al., 1975; Tabas and Kornfeld, 1978), missing the Golgi enzyme GlcNAc transferase I. This enzyme is normally present in the *medial* cisternae of the stack (Dunphy et al., 1985), and is needed to initiate the steps resulting in the addition of peripheral GlcNAc, Gal, and sialic acid residues to form complex-type Asn-linked oligosaccharides in the Golgi (Hubbard and Ivatt, 1981; Schachter et al., 1983). We study the transport of the vesicular stomatitis virus (VSV)-encoded glycoprotein (G protein) because this single membrane protein replaces the entire population of newly synthesized proteins in the Golgi during a viral infection (Zilberstein et al., 1981). G protein is synthesized in the

This paper is dedicated to Professor Eugene P. Kennedy on the occasion of his 65th year.

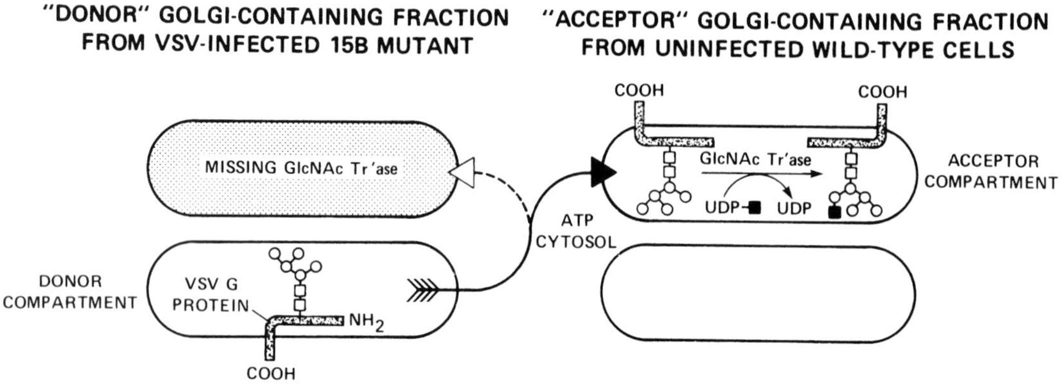

Figure 1. Assay for Transport of a Protein between Successive Golgi Compartments In Vitro, Based on a Transport-Coupled Glycosylation

A donor fraction containing Golgi membranes is prepared from VSV-infected clone 15B cells (a mutant cell line lacking UDP-GlcNAc glycosyltransferase I—GlcNAc Tr'ase). This fraction is incubated with an acceptor fraction also containing Golgi, but prepared from uninfected wild-type cells. Incorporation of ^{3}H-GlcNAc (from UDP-^{3}H-GlcNAc) into G protein occurs when G protein is transferred in a dissociative fashion between the two Golgi. Specifically, the assay measures transfers of G protein into the acceptor compartment in wild-type Golgi housing the GlcNAc transferase (corresponding to *medial* cisternae) from the immediately prior donor compartment (currently believed to be *cis* cisternae or vesicles derived from these) present in the 15B Golgi. This dissociative transfer (solid arrows) results from the random choice of target cisternae from among the two Golgi populations. Corresponding transfers should also occur within the 15B Golgi (dashed arrow) but are not recorded in the assay because of the lack of the GlcNAc transferase in 15B Golgi. Incorporation into G protein is measured directly by immunoprecipitation. The polypeptide of G protein will span the Golgi membrane before and after the transfer. The trimmed core oligosaccharide containing two N-acetylglucosamine (GlcNAc) (□) and 5 Man residues (O), is the immediate substrate for GlcNAc transferase I. It is not clear whether this trimming occurs upon arrival in the acceptor compartment or before departure from the prior donor compartment, but the process is diagrammed as if the latter were the case. ^{3}H-GlcNAc (■).

rough endoplasmic reticulum (ER), and it passes through the Golgi en route to the plasma membrane, from which it is incorporated into the envelope of budding virions.

For the assay, a Golgi-containing fraction is prepared from VSV-infected 15B cells. When this fraction is incubated, transport of G protein into the *medial* compartment may well occur, but of course GlcNAc cannot be added to G protein, since the needed enzyme is missing from the 15B cell Golgi. However, when a Golgi fraction from uninfected wild-type CHO cells is included (Figure 1), GlcNAc can be added to G protein following a dissociative form of this transfer, into the *medial* ("acceptor") compartment of the wild-type Golgi. Provided the in vivo specificity of membrane traffic and fusion is preserved in vitro, the incorporation of GlcNAc into G protein will be an exclusive measure of transfers of G protein between successive compartments in the Golgi (these compartments residing in different Golgi populations), even when these Golgi membranes are present as a minor fraction within a crude homogenate. In fact, the cell-free system seems to respect the same compartment boundaries that exist in the cell, exhibiting an appropriate specificity (Fries and Rothman, 1981; Dunphy et al., 1981; Rothman et al., 1984b). The G protein transported to the acceptor compartment thus originates within the immediately prior "donor" compartment present in the Golgi population from the infected 15B cells, currently believed to be the *cis* portion of the stack. Constructing an assay that measures the activities of donor and acceptor compartments in different Golgi fractions permits their separate biochemical manipulation prior to incubations. Since donor activities (e.g. the budding of vesicles) and acceptor activities (e.g. the fusion of vesicles) are certain to be distinct at a molecular level, this is an important advantage.

In this paper we report a much more rapid, quantitative, and flexible form of this assay, as well as the extensive purification of active and stable donor- and acceptor-containing Golgi fractions, opening the way to an analysis of the mechanisms of transfer. In the second paper of this series (Braell et al., 1984), we will report the use of this improved assay to localize the transferred G protein by electron microscopic autoradiography to morphologically intact Golgi stacks, derived from the wild-type and not from the 15B Golgi population. This finding implies that the transport in vitro is highly specific, and involves intact Golgi stacks. In the third paper (Balch et al., 1984), we exploit two-stage incubations and the selective inhibitory effects of N-ethylmaleimide to reveal two successive kinetic intermediates in the transport process. On the basis of electron microscopy, these seem to represent intermediate stages in the budding and fusion of transport vesicles.

Results

Transport-Coupled Incorporation of ^{3}H-GlcNAc into G Protein

In previous work we monitored the transport-coupled addition of GlcNAc to G protein indirectly, as judged by the ensuing resistance of the oligosaccharide to endoglycosidase H (Fries and Rothman, 1981). This procedure required that the polypeptide backbone of G be radioactively labeled in vivo, and it necessitated the use of SDS gels and autoradiography for analysis, taking several days. A more rapid, sensitive, and flexible assay was needed to

Reconstitution of Protein Transport
407

enable a further analysis and fractionation. We reasoned that by measuring the transport-coupled incorporation of ^{3}H-GlcNAc into G protein directly (from UDP-^{3}H-GlcNAc), these needs could be fulfilled (Figure 1). The donor fraction can be prepared from unlabeled VSV-infected 15B cells, and the ^{3}H-labeled G protein can be retained on filters as a rapidly formed immunoprecipitate and counted.

Postnuclear supernatants of VSV-infected CHO clone 15B and of uninfected wild-type cells were prepared as described in Experimental Procedures. These crude fractions were incubated together in the presence of ATP, an ATP regenerating system, and UDP-^{3}H-GlcNAc. Assays were terminated by the addition of detergent to solubilize the membranes, followed by anti-G protein antiserum. The immunoprecipitates (formed within 45 min at 37°C or overnight at 4°C) were collected on a Millipore filter, washed, and counted. As shown in Figure 2, a time-dependent incorporation of ^{3}H-GlcNAc into G protein was observed during the incubation. Only trace levels of ^{3}H were retained on the filter when the anti-G protein serum was replaced with a preimmune serum, implying that all of the ^{3}H retained had been in G protein. The time course of incorporation of ^{3}H-GlcNAc into G protein was very similar to that previously reported for the transport-coupled processing of G protein to the EndoH-resistant form in vitro (Fries and Rothman, 1981).

Analysis of a 60 min incubation using SDS-gel electrophoresis revealed that G protein is the major protein species labeled (Figure 2 inset, lane a), accounting for about 50% of the total radiolabel incorporated. The immunoprecipitate contained G protein as the only radiolabeled protein species (Figure 2, inset, lane b); G protein was quantitatively immunoprecipitated. No radiolabeled proteins were observed with preimmune serum (Figure 2 inset, lane c). The incorporation of ^{3}H-GlcNAc in incubations employing crude postnuclear supernatants exhibits all of the properties detailed in the following sections for incubations with sucrose-gradient-purified donor and acceptor Golgi fractions (data not shown). For example, no ^{3}H was incorporated into G protein when the postnuclear supernatant of uninfected 15B cells replaced the corresponding fraction from infected cells.

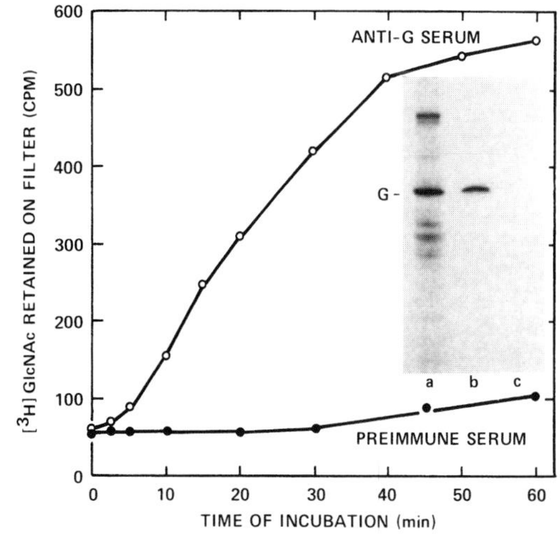

Figure 2. Incorporation of ^{3}H-GlcNAc into G Protein during Incubations of the PNS of VSV-Infected 15B Cells with the PNS of Uninfected Wild-Type Cells

Crude homogenates of VSV-infected clone 15B cells and wild-type cells were prepared as described in Experimental Procedures. The homogenates were centrifuged at 600 × g for 5 min, yielding a PNS fraction (about 15 mg/ml protein). Then, 35 μg of the PNS from the VSV-infected clone 15B cells and 70 μg of the PNS from uninfected wild-type cells were incubated for increasing times under standard assay conditions (see Experimental Procedures) except no cytosol was added. The incorporation of ^{3}H-GlcNAc into G protein was measured by collecting the immunoprecipitate on Millipore filters, using anti-G protein serum (O) or a preimmune serum (●), as in Experimental Procedures. For the inset autoradiograph, incubations were as described above for 60 min, followed by SDS gel electrophoresis. (Lane a) The incubation was terminated by the addition of 50 μl of gel sample buffer (GSB; 0.1 M Tris-HCl, pH 6.8; 2% sodium dodecylsulfate; 30 mM dithiothreitol; 10% glycerol; 1 μg/ml bromophenol blue), and boiled prior to SDS gel electrophoresis. (Lanes b and c) Incubations were terminated by addition of 50 μl detergent buffer (DB, see Experimental Procedures) and 15 μl of anti-G protein serum (lane b) or 15 μl preimmune serum (lane c). After an overnight incubation on ice, 25 μl of a 10% suspension of Staph A cells (Pansorbin, Calbiochem Corp., previously washed twice with washing buffer, (see Experimental Procedures) was added. After 1 hr on ice, the cells were pelleted in a microfuge in 15 sec, washed twice with 1 ml each of washing buffer, resuspended in 50 μl of water prior to addition of 50 μl GBS, and boiled. The cells were then pelleted for 3 min, and the supernatants were electrophoresed in a 10% polyacrylamide gel according to the method of Laemmli (1970). The dried gel was autoradiographed with fluorographic enhancement for 7 days.

Donor Activity Copurifies with the Golgi Membranes of VSV-Infected 15B Cells

Previous work using crude postnuclear supernatants (PNS) as donor and acceptor suggested that the assay employed measures only the reconstitution of a single stage within total ER–Golgi–plasma membrane transport pathway: the transfer of G protein between two successive locations in the Golgi complex (Fries and Rothman, 1981; Dunphy et al., 1981; Rothman et al., 1984b). The improved assay should measure the same process, except that the transport-coupled incorporation of GlcNAc is quantitated differently and more directly. To confirm this, and to offer novel lines of evidence for the compartmental specificity of the assay, it is necessary to show that the two assays measure the transfer of the same population of G protein molecules in Golgi, with identical properties and requirements. The following sections document these and related points.

Since the G protein should originate within the Golgi membranes of the homogenate of the VSV-infected 15B cells, fractionation of the homogenate on a density gradient should yield a Golgi-rich fraction containing all of the donor activity. The crude homogenate, prepared by breaking cells in the presence of 0.25 M sucrose and 10 mM Tris-HCl (pH 7.4), was adjusted to 1.4 M sucrose and 1 mM EDTA and loaded as a layer of a discontinuous sucrose gradient underneath layers of 1.2 M and 0.8 M sucrose (in 10 mM Tris-HCl, pH 7.4). After centrifugation, fractions were assayed for ER and Golgi marker enzymes, and for their donor activity; i.e., for their ability to provide G protein for transport-coupled glycosylation. As shown in Figure 3 (top), a single peak of donor activity (well separated from bulk protein) was observed at the 0.8/1.2 M sucrose interface (interface III). This activity copurified with the Golgi markers mannosidase I (the enzyme that trims high-mannose chains to the Man_5 intermediate; Hubbard and Ivatt, 1981) and galactosyltransferase (Figure 3, bottom). No donor activity was observed in the more dense fractions containing ER membranes, marked by the enzyme glucosidase I (Figure 3, bottom). The donor activity of the Golgi fractions was not significantly inhibited by the level of sucrose added when assaying the ER fractions from the bottom of the gradient (not shown). A minor peak of glucosidase I was detected in the fractions preceding the donor activity peak.

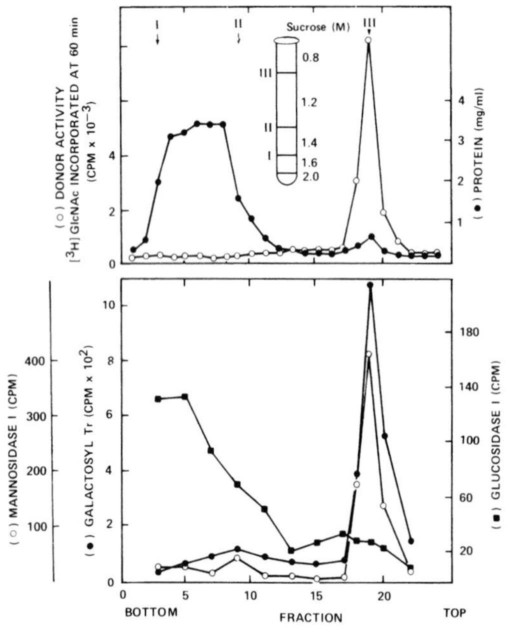

Figure 3. Copurification of Donor Activity with Golgi Markers following Sucrose Gradient Centrifugation of Crude Homogenates of VSV-Infected Clone 15B Cells

A crude homogenate prepared from VSV-infected clone 15B cells was fractionated using discontinuous sucrose density gradient centrifugation as described in Experimental Procedures, except that additional layers of sucrose below the homogenate were included. The gradient consisted of layers of 2.0 M sucrose (2 ml), 1.6 M sucrose (4 ml), the homogenate adjusted to 1.4 M sucrose (as described in Experimental Procedures, 8 ml), 1.2 M sucrose (12 ml), and 0.8 M sucrose (8 ml). All sucrose solutions were prepared with 10 mM Tris-HCl (pH 7.4). After centrifugation in the SW27 rotor for 2.5 hr at 25,000 rpm, fractions (about 1.3 ml) of the gradient were collected from the bottom. Then 2.5 μl of each fraction was assayed in the standard cocktail containing 5 μl of a gradient-purified membrane fraction prepared from wild-type cells (1 mg protein/ml) and 50 μg of gel-filtered CHO cytosol, using a 60 min incubation. Each gradient fraction was also assayed for the activity of mannosidase I, galactosyltransferase, and glucosidase I as described previously (Dunphy and Rothman, 1983). Protein was measured by the method of Lowry. (Top) Donor activity (O) and protein (●). (Bottom) Gal Transferase (●), Mannosidase I (O), and Glucosidase I (■). The arrows in the top panel labeled I, II, and III indicate the positions of the three significant interfaces, at 1.6/1.4 M (I), 1.4/1.2 M (II), and 1.2/0.8 M (III). The Golgi fraction is routinely harvested from interface III.

Table 1 shows that the pooled Golgi fraction (interface III) was about 40-fold purified over the PNS, as judged by the Golgi marker (mannosidase I) recovered in the greatest yield (82%). Donor activity was purified 20-fold, but this lower value is almost certainly an underestimate by about 2-fold since the recovery of donor activity (43%) was about half that of mannosidase I and no other peaks were found in the gradient. Galactosyltransferase was recovered (55%) and purified (25-fold) to about the same extent as donor activity. Less than 6% of glucosidase I activity was recovered in the active donor fraction. In general, different preparations yielded a 15 to 25 fold enrichment of donor activity, accompanied by a 40% to 70% yield.

An important aspect of this procedure is the collection of the Golgi fraction at an interface in a sucrose gradient, avoiding the formation of a pellet. Pelleting and washing steps, used in many purifications of Golgi membranes and needed for the physical separation of Golgi compartments (Dunphy et al., 1981), greatly reduce the yield of donor activity. The pooled Golgi fraction (interface III) will be referred to as the donor membrane fraction and is used routinely. It is frozen in aliquots in liquid nitrogen and is stable for months when stored at −80°C.

Acceptor Activity Copurifies with the Golgi Membranes of Wild-Type Cells

Using the density gradient centrifugation procedure described above, but now loading the homogenate of uninfected wild-type CHO cells, the major peak of acceptor activity was observed after 2.5 hr of centrifugation at the 0.8 M/1.2 M interface III (Figure 4, top), the same interface shown to contain the Golgi fraction active as donor (Figure 3, top). Acceptor activity cofractionated with GlcNAc transferase I, as well as another Golgi marker enzyme, galactosyltransferase (Figure 4, bottom). In addition, a broader and less purified peak of acceptor activity (typically representing 20% of the total recovered) was consistently observed just above interface II. This minor peak is no longer detected when the centrifugation is prolonged so as to allow a complete equilibrium to be reached (not shown). Most likely, the minor peak of the acceptor activity is present in smaller membranes that take longer to float to equilibrium at interface III. For routine purposes, the Golgi fraction was pooled from interface III after 2.5 hr of centrifugation. This fraction was well separated from ER membranes (marked by glucosidase I), and was about 10-fold purified over the PNS with respect to acceptor and Golgi marker activities, with yields of 30% to 40% (Table 2). Correcting for the yield, the actual degree of copurification of acceptor activity and Golgi markers in the Golgi fraction is likely to be about 30-fold, a value more consistent with that obtained for the donor Golgi fraction. In general, a 10 to 20 fold increase in specific acceptor activity accompanied by a yield of 25% to 50% was obtained for many independent preparations (data not shown), suggesting that the actual purification of Golgi

Reconstitution of Protein Transport
409

Table 1. Enrichment of Donor Activity and Golgi Marker Enzymes from VSV-Infected Clone 15B Cells upon Sucrose Density Gradient Centrifugation

Activity Assayed	Fraction Tested	Total Protein (mg)	Total Activity	Recovery of Activity (%)	Specific Activity	Enrichment of Activity (fold)
Donor[a]	PNS	67.5	4.2	[100]	62	[1]
	Golgi	1.5	1.8	43	1240	20
Mannosidase I[b]	PNS		8.0	[100]	0.12	[1]
	Golgi		6.6	82	4.46	37
Galactosyl transferase[c]	PNS		110	[100]	1.7	[1]
	Golgi		63	55	42.5	25
ER glucosidase I[d]	PNS		0.40	[100]	0.006	[1]
	Golgi		0.021	5.8	0.014	2.6

[a] A postnuclear supernatant (PNS) was prepared from the homogenate of VSV-infected 15B cells by centrifugation at 600 × g for 5 min. This PNS was fractionated using a discontinuous sucrose density gradient centrifugation as described in Experimental Procedures for homogenates. The Golgi fractions found at the 0.8/1.2 M sucrose interface (interface III, Figures 3 and 4) was collected. Assays of 1.5 μl of the PNS, or the Golgi fraction, were carried out in the standard cocktail for 60 min containing 5 μg of the acceptor Golgi membrane fraction, and 50 μg gel-filtered CHO cytosol. To compensate for the competitive effects of the endogenous UDP-GlcNAc pool found in the cytosol within the PNS fraction, 1.5 μl of cytosol from 15B cells (prepared from the 15B PNS and not gel-filtered) was added to each assay of donor Golgi fraction. Incorporation was linearly dependent upon the volume of fraction assayed in the range used, for both PNS and Golgi fractions. The total donor activity in the fraction was calculated based on incorporation of ^{3}H per unit volume assayed. The numbers shown represent cpm × 10^{-6}. Specific activity is reported as the total activity divided by the total protein in the fraction, and is shown as cpm/μg protein.

[b] The Golgi marker mannosidase I was assayed as described previously (Dunphy and Rothman, 1983). Activity is total units in the fraction. Specific activity reported in units/mg.

[c] The Golgi marker galactosyltransferase was assayed as described previously (Dunphy and Rothman, 1983). Activity is the total units (nmole/hr) of the fraction. Specific activity is reported as nmole/hr/mg protein.

[d] The endoplasmic reticulum (ER) marker glucosidase I was assayed as described previously (Dunphy and Rothman, 1983). Activity is the total units in the fraction. Specific activity is reported as units/mg.

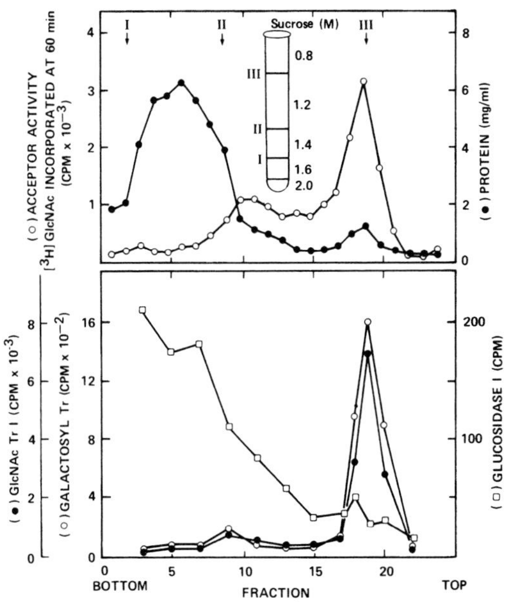

Figure 4. Copurification of Acceptor Activity with Golgi Markers following Sucrose Gradient Centrifugation of Crude Homogenates of Wild-Type Cells

The homogenate of uninfected wild-type cells was fractionated using discontinuous sucrose density gradient centrifugation as in Figure 3. Then, 2.5 μl of each fraction was assayed for 60 min in the standard cocktail containing 5 μl of the gradient-purified donor membrane fraction (1 mg/ml) prepared from VSV-infected clone 15B cells and 50 μg of gel-filtered cytosol (prepared from uninfected clone 15B cells). Each gradient fraction was also assayed for the activity of N-acetylglucosamine transferase I (GlcNAc TrI), galactosyltransferase (Gal Tr), and glucosidase I as described previously (Dunphy and Rothman, 1983). (Top) acceptor activity (O) and protein (●). (Bottom) Gal Tr (O), GlcNAc TrI (●), and Glucosidase I (□).

membranes from wild-type cells was typically about 40-fold over the PNS. This Golgi fraction is used routinely, and is referred to as the acceptor membrane fraction. Like the donor membrane fraction, acceptor is routinely frozen in aliquots in liquid nitrogen and stored at −80°C for up to several months.

When the gradient-purified donor or acceptor Golgi fractions were added back to the crude PNS, their respective activities were neither inhibited nor enhanced (data not shown). This reinforces the validity of the quantitative data in Tables 1 and 2. Both donor and acceptor activities were eliminated by trypsin, suggesting that cytoplasmically disposed proteins are required for their function (Balch and Rothman, submitted).

Summary of Requirements for the Transport-Coupled Incorporation of GlcNAc

Earlier work using crude PNS fractions as donor and acceptor revealed that transport-coupled oligosaccharide processing required ATP as well as a soluble, cytosolic fraction (Fries and Rothman, 1981). The same requirements should be evident in the new assay, as both donor and acceptor membrane fractions should be effectively freed of soluble proteins and ATP during their preparation on sucrose gradients. As shown in Table 3, mixtures of gradient-purified donor and acceptor fractions are inert unless both ATP and the cytosol fraction are provided.

Table 2. Enrichment of Acceptor Activity and Golgi Marker Enzymes from Uninfected Wild-Type CHO Cells upon Sucrose Density Gradient Centrifugation

Activity Assayed	Fraction Tested	Total Protein (mg)	Total Activity	Recovery of Activity (%)	Specific Activity	Enrichment of Activity (fold)
Acceptor[a]	PNS	72	5.6	[100]	77	[1]
	Golgi	2.6	1.8	32	685	8.9
Mannosidase I[b]	PNS		19.9	[100]	0.28	[1]
	Golgi		7.3	37	2.82	10.
Gal transferase[c]	PNS		256.0	[100]	3.6	[1]
	Golgi		75.0	29	28.9	8.1
GlcNAc transferase I[d]	PNS		164.0	[100]	2.3	[1]
	Golgi		70.0	42	26.8	12.
ER glucosidase I[e]	PNS		0.60	[100]	0.008	[1]
	Golgi		0.026	4.3	0.010	1.2

[a] The PNS of uninfected wild-type cells was prepared and also further fractionated using the sucrose density gradient, and the Golgi fraction (interface III) was harvested, exactly as in Table 1 except wild-type cells were used. The acceptor activity of the PNS and Golgi fractions were assayed in a manner precisely analogous to Table 1, employing 5 μg of donor Golgi fraction per 50 μl incubation, but adding 1.5 μl of non-gel-filtered cytosol from wild-type PNS to compensate for the endogenous UDP-GlcNAc pool. Total and specific activity are expressed as cpm × 10^{-6} and cpm/μg, as in Table 1.
[b] Activity is total units of the fraction. Specific activity is units/mg. See Table 1.
[c] Activity is total nmole/hr of the fraction. Specific activity is nmole/hr/mg. See Table 1.
[d] The Golgi marker GlcNAc transferase I was assayed as described previously (Dunphy and Rothman, 1983). Activity is the total units (nmole/hr) of the fraction.
[e] Activity is total units in the fraction. Specific activity is units/mg. See Table 1.

Replacement of the Golgi fraction from wild-type CHO cells with a comparable Golgi fraction prepared from uninfected clone 15B (which lacks GlcNAc transferase I) eliminated incorporation into G protein. This proves that all of the ^{3}H incorporated into G protein when wild-type membranes are provided as acceptor is via the in vivo enzymatic pathway, requiring GlcNAc transferase I. Addition of EDTA or omission of magnesium from the incubation cocktail eliminated incorporation.

A detailed analysis of conditions affecting the assay will be presented elsewhere (Balch and Rothman, submitted). Briefly, the apparent K_m for ATP is submicromolar; however, a regenerating system for ATP (creatine phosphate and creatine kinase) is needed to maintain ATP levels (Balch and Rothman, submitted). Inclusion of UTP helps protect ATP from hydrolysis. Transfer of G protein was optimal over a restricted range of close to physiological conditions including pH (6.9 to 7.2), temperature (35° to 39°C), salt (15 to 30 mM KCl), and osmolarity (0.15 to 0.3 M sucrose). Incorporation is specific for the in vivo substrate UDP-^{3}H-GlcNAc (K_m = 0.3 μM), which is stable throughout the time course of the incubation.

Table 3. Summary of Requirements of the Cell-Free System

Incubation Condition	^{3}H-GlcNAc Incorporated into G Protein[a]
1. Complete	[1]
2. − triphosphates and regenerating system	0.01
3. − cytosol	0.01
4. − VSV/15B donor fraction	0.01
5. − Wild-type Golgi fraction	0.03
6. − Wild-type Golgi fraction, + Golgi fraction from uninfected 15B	0.02
7. − Mg^{++}	0.02
8. + EDTA	0.03

Two and a half micrograms of the gradient-purified donor membrane fraction, 5 μg of the acceptor Golgi fraction, and 50 μg of gel-filtered CHO cytosol were incubated in the standard cocktail (see Experimental Procedures) containing ATP, UTP, CP, CPK, and UDP-^{3}H-GlcNAc for 60 min at 37°C for the complete incubation (1). (2–7) The indicated component(s) were selectively omitted from the assay. In 6, the acceptor Golgi fraction (from wild-type cells) was replaced with 5 μg of the gradient-purified Golgi fraction prepared from uninfected clone 15B cells. In 7, Mg acetate was omitted. In 8, Na_2EDTA was added to a complete assay at a final concentration of 5 mM.
[a] Values reported are normalized to the ^{3}H incorporated into G protein in the complete incubation (1).

The Glycosylated G Protein Is a Transmembrane Protein Sequestered in Sealed Golgi Membrane Vesicles

To test whether the transferred G protein is present in a sealed membrane compartment with the proper asymmetric membrane orientation, trypsin was added after a 60 min incubation. Analysis of the products of tryptic digestion by SDS-gel electrophoresis and autoradiography revealed that proteolysis in the absence of detergent quantitatively converted G protein to a fragment of slightly lower molecular weight without a loss of its ^{3}H-labeled oligosaccharide chains (Figure 5, lane 2 vs. lane 1). This shift in molecular weight is due to proteolytic cleavage within the small carboxy-terminal domain of G protein that is normally exposed on the cytoplasmic face of cellular membranes (Zilberstein et al., 1981). Only low molecular weight fragments were detected when trypsin was added in the presence of detergent (Figure 5, lane 3). In fact, 95% of radioactive GlcNAc incorporated into the oligosaccharides of G protein remained attached to immunoprecipitable polypeptide after this trypsin treatment. However, only 2% of the incorporated ^{3}H could be immunoprecipitated when the membranes were treated with trypsin in the presence

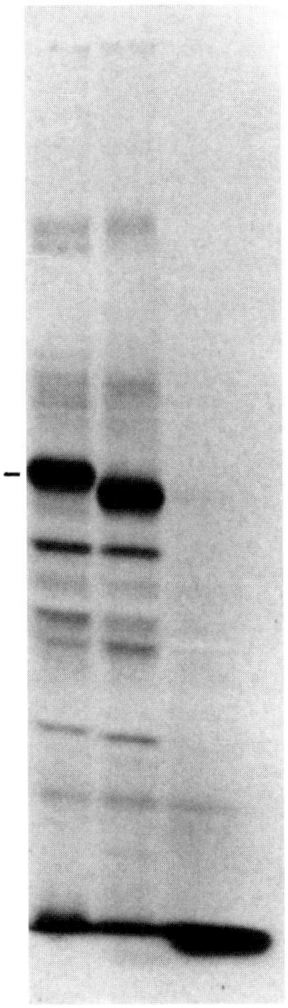

Figure 5. Transported G Protein Resides in Sealed Golgi Vesicles as a Transmembrane Protein with the Appropriate Orientation

A mixture of 2.5 μg of the donor membrane fraction, 5 μg of the acceptor Golgi, and 50 μg of gel-filtered cytosol was incubated for 60 min in the standard cocktail. Assays were terminated by the addition of 2.5 μl of 100 mM Na_2EDTA. Incubations shown in lanes 2 and 3 were treated with 5 μl of 1 mg/ml trypsin for 15 min at 37°C in the absence (lane 2), or presence (lane 3), of 0.1% Triton X-100 (added from a concentrated stock). Trypsin was inhibited by the addition of 5 μl of 1 mg/ml soybean trypsin inhibitor, and samples were mixed with an equal volume of gel sample buffer, boiled, electrophoresed in a 10% polyacrylamide gel, and autoradiographed, as in Figure 2.

of Triton X-100, added to disrupt the vesicles. These data establish that the transported G protein is retained in sealed Golgi vesicles with its carboxy-terminal domain on the outside and its oligosaccharide chains on the inside, as would be expected for membrane fission and fusion processes involving sealed compartments. This rules out the possibility that the GlcNAc transferase itself is artifactually released from wild-type Golgi membranes to act upon G protein in leaky or unsealed membranes from 15B cells.

We also found that addition of 0.1% Triton X-100 prior to initiation of the assay completely eliminated incorporation of ^{3}H-GlcNAc into G protein. GlcNAc transferase I is fully active in this concentration of Triton (Schachter et al., 1983). This detergent sensitivity distinguishes this transport-coupled glycosylation from an ordinary, uncoupled glycosyltransferase assay. The latter requires detergent to allow substrates and enzymes to mix; the former is profoundly inhibited because detergent dilutes G protein and the GlcNAc transferase with respect to each other. Evidently, these two proteins are concentrated together within the same vesicles as a result of specific transfer events, a prerequisite for efficient glycosylation.

The Donor Compartment Is Rapidly Depleted of G Protein When Protein Synthesis Is Inhibited In Vivo

Only G protein contained in the Golgi fraction of infected 15B cells is a substrate for transport-coupled glycosylation with GlcNAc (Figure 3). To test whether these G protein molecules within this fraction are in fact within the Golgi, we can investigate the time course with which the G protein population that comprises the substrate is depleted when protein synthesis is inhibited in vivo. Transport will continue normally in the absence of protein synthesis. So, when VSV-infected 15B cells are homogenized at various times after addition of cycloheximide, donor activity will progressively be reduced as the last molecules of G protein are transported out of the donor compartment. We found (Figure 6 inset, closed circles) that the donor activity of the whole PNS rapidly and completely disappeared after cycloheximide was added, with a half-time of about 10 min. At all times, the remaining donor activity banded with the Golgi at interface III in the sucrose gradient (Figure 6). As a control, a parallel experiment (Figure 6 inset, open circles) was carried out in which the wild-type cells, rather than the infected 15B cells, were treated with cycloheximide. No effect was observed on the acceptor activity of the PNS of wild-type cells.

These experiments show that it is a recently synthesized pool of G protein concentrated in the Golgi fraction of the 15B cell membranes that is subject to transport to receive GlcNAc. This pool is depleted with a half-time of ~10 min, a time course that is equal to that with which newly synthesized G protein enters the Golgi in CHO cells and far shorter than the time required to reach the cell surface (Fries and Rothman, 1981; Bergmann and Singer, 1983). This pool is not in the ER (Figure 3). Thus, independent kinetic evidence (Figure 6) and physical evidence (Figure 3) point to the conclusion that the transferred G protein originates in a Golgi compartment. These new results buttress the conclusion from earlier work that the assay defines a narrow segment of the transport pathway within the Golgi, and further imply that the new and old forms of the assay measure transport of the same molecules.

The Concentration of Cytosol Determines Both the Rate and Extent of Transport

The initial experiment with crude postnuclear supernatants to determine the feasibility of the new assay yielded a

Cell
412

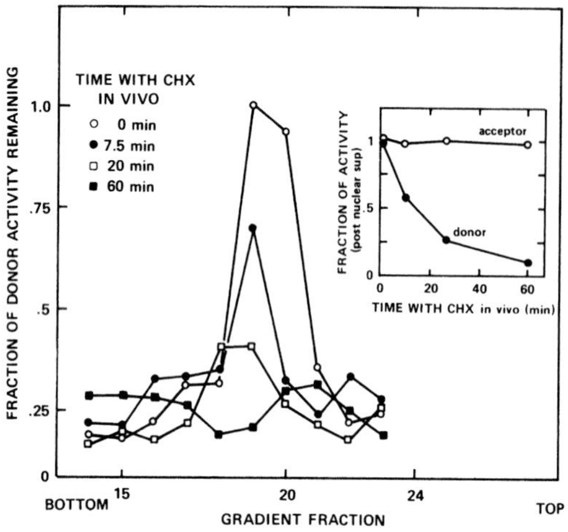

Figure 6. The Compartment in the Golgi Fraction That Donates G Protein In Vitro Is Rapidly Drained of G Protein In Vivo When Protein Synthesis Is Prevented

At 3.5 hr after infection, VSV-infected clone 15B cells were trypsinized into a suspension culture and incubated (starting at 4 hr after infection) for increasing periods of time with 100 μg/ml of cycloheximide (CHX) to inhibit G protein synthesis. At each time point, a sample of the cells was rapidly pelleted at 4°C, and a crude homogenate was prepared and fractionated using density gradient centrifugation as described in Figure 3. Fractions were collected from the bottom and assayed with 5 μg acceptor Golgi and 50 μg gel-filtered CHO cytosol (prepared from uninfected clone 15B cells). Only the top portion of each gradient, consisting of fractions 14–23 (which contained all the detectable donor activity), is shown. The incorporation of ^{3}H has been normalized to the largest value obtained (0 min CHX, fraction 19) for convenience. Inset: Total donor activity in the crude PNS fraction of the homogenates of VSV-infected 15B cells, as a function of time after CHX (closed circles), using untreated acceptor (as in Figure 2) for 60 min. A parallel control experiment was performed with CHX-treated wild-type cells (open circles) assayed using untreated donor to test the effect of CHX on acceptor activity. For this purpose, wild-type cells were treated in suspension with CHX (100 μg/ml) in vivo. At each time point, a crude PNS fraction was prepared. The values shown are the ^{3}H incorporated at a time point as a fraction of the untreated (i.e., no CHX) control.

complex time course for the incorporation of ^{3}H-GlcNAc into G protein (Figure 1). Similar kinetics were observed using the gradient-purified donor and acceptor Golgi fractions (Figure 7). The typical time course for incorporation in the presence of saturating amounts of cytosol (Figure 7, closed circles) shows a pronounced lag period of ~10 min after initiation of the incubation, followed by a phase of linear incorporation (extending from ~15 to 45 min), until eventually a plateau is reached by 60 min of incubation.

Both the observed rate of incorporation during the linear phase (Figure 7) and the level of the plateau attained (i.e. extent of incorporation measured at 60 min; see inset in Figure 7) are saturable functions of the concentration of cytosol, linear up to about 0.4 mg/ml cytosol protein. The lag period (indicated by extrapolation from the linear phase, dashed lines in Figure 7) is largely independent of the concentration of cytosol. The role of cytosol in the transfer of G protein between the donor and acceptor compart-

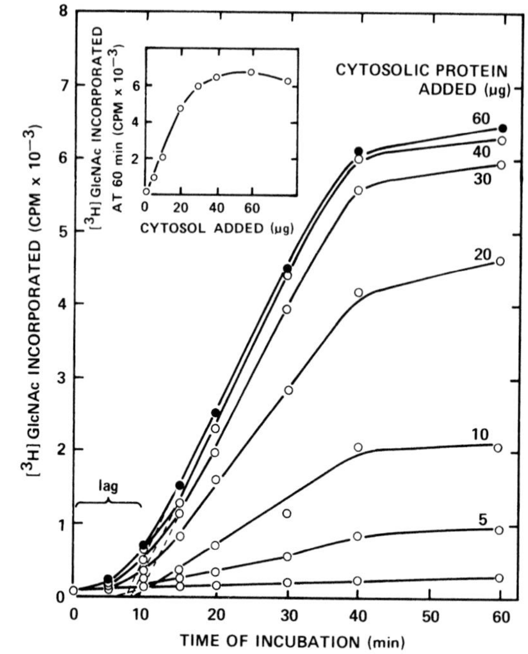

Figure 7. Effect of Cytosol Concentration on the Rate and Extent of Transport

Each incubation (50 μl) contained 2.5 μg donor membrane fraction, 5 μg of acceptor fraction, and gel-filtered CHO cytosol in the indicated amount in the standard assay cocktail. Incubations were stopped after various times at 37°C, and the ^{3}H-GlcNAc incorporated into G protein was determined. Inset: ^{3}H-GlcNAc incorporated into G protein at the plateau of incorporation as a function of the amount of cytosol protein added.

ments is considered in the third paper of this series (Balch et al., 1984).

Cytoplasmic component(s) active in the assay are found in a broad range of eucaryotic cell tissues, including 15B CHO cells, rat liver, and bovine brain, but not E. coli. The activity of the cytosol is sensitive to proteases and inhibited by N-ethylmaleimide, as will be detailed elsewhere (Balch and Rothman, submitted).

Discussion

We have developed an improved assay for measuring the transport of the VSV G protein between successive compartments of the Golgi via the coupled incorporation of GlcNAc. It is more direct than the former version of the assay (Fries and Rothman, 1981) and utilizes purified Golgi fractions, permitting the selective localization of the transferred G protein by electron microscopic autoradiography (Braell et al., 1984). Also, since radioactivity is incorporated only into transported molecules of G protein, there is a negligible background. This makes the new assay more sensitive and more quantitative, permitting a kinetic analysis that has revealed intermediates in the transfer process (Balch et al., 1984). Finally, the new procedure is rapid, so that it now takes less than a day to do what was previously a week's work. This kind of pace is essential to enable the fractionation of such a complex multicomponent system.

We believe that the new and old versions of the assay measure the same process, since their basic design is the same, as are those of their requirements and properties that can be compared. In the former assay the polypeptide chain of G protein had to be labeled in vivo before the donor membranes could be prepared. This enabled us to use pulse-chase experiments to pinpoint the source of the transferred G protein molecules within the Golgi and to measure the efficiency of transport (Fries and Rothman, 1981; Dunphy et al., 1981; Rothman et al., 1984b), which seems to be close to 100%—i.e., as efficient in the cell-free system as in vivo. A disadvantage of the new assay is that it measures only G protein molecules that have already arrived, making it more difficult to ascertain exactly which 15B membranes they came from and the absolute efficiency of their transport. However, the unique and extensive copurification of donor activity with the 15B Golgi membranes (Figure 3) and the prompt effect of cycloheximide upon the donor activity in the Golgi fraction (Figure 6) offer strong evidence that the new and old assays measure the transfer of the same, select population of G protein that has just entered the Golgi. A rough calculation of efficiency can be made based on the amount of ^{3}H-GlcNAc incorporated and the specific radioactivity of UDP-^{3}H-GlcNAc, taking the endogenous pool into account. The result (Balch and Rothman, submitted) is that about 15 ng of G protein, added in a total of 2.5 μg of 15B Golgi fraction, receives GlcNAc residues in vitro. This amount would represent roughly 25% of the total G protein content of this fraction, using the data of Quinn et al. (1984) for the content of Semliki Forest virus glycoproteins in Golgi membranes. Since the donor compartment from which transfer is measured is likely to consist of only one or a few cisternae of the total stack, the actual efficiency is likely to be several times higher.

Specificity of the Cell-Free System

A priori, the addition of GlcNAc to G protein in our assays could result from a nonspecific fusion of the Golgi membranes from wild-type cells with almost any membrane from 15B cells. This is because G protein in ER, Golgi, plasma membrane, and other membranes of the mutant is incompletely glycosylated, and thus potentially a substrate for the missing enzyme, GlcNAc transferase I. Specifically, G protein in the Golgi and all later compartments of 15B (plasma membranes, endosomes, and lysosomes, the latter two filling with progeny virions) will carry Man_5-containing oligosaccharides, the immediate substrates for GlcNAc transferase I (see Figure 1 and Hubbard and Ivatt, 1981). However, G protein in rough ER membranes carries Man_8 and Man_9 oligosaccharide chains (Zilberstein et al., 1981); thus prior action of Golgi mannosidase I to produce the Man_5 intermediate is needed before GlcNAc transferase I can act.

In spite of this multitude of subcellular membranes containing potential G protein substrates for glycosylation in vitro, the operative mechanism has proved highly specific for the Golgi membranes:

—Neither G protein present in the plasma membrane (or later derivatives, such as endosomes and lysosomes) nor G in the ER of 15B cells will receive GlcNAc in the in vitro system, even when the crudest membrane fractions are employed (Fries and Rothman, 1981; also, Figure 3). A nonspecific fusion of ER with Golgi membranes would presumably expose G protein to both Golgi mannosidase I and GlcNAc transferase I enzymes (60% of the mannosidase I activity of wild-type Golgi survives a 60 min incubation under standard assay conditions; data not shown).

—The donor activity distributes exclusively with Golgi membranes in sucrose gradients, and these are copurified some 40-fold (Figure 3). Enrichment is possible only when a select subset of subcellular membranes acts as donor.

—Only a freshly synthesized pool of G protein that has just entered the Golgi in pulse-chase experiments is subject to the glycosylation (Fries and Rothman, 1981; Dunphy et al., 1981; and Figure 7). These are the same molecules that can undergo dissociative transfers into *medial* cisternae in vivo (Rothman et al., 1984b).

—Once G protein undergoes this transfer within the 15B Golgi in vivo, it loses its ability to undergo what is apparently the same transport step in vitro (Rothman et al., 1984b). The cell-free system must therefore adhere to the compartment boundaries within the Golgi that exist in the cell.

There is also a high degree of specificity of the transport and/or fusion processes that occur in vitro for the Golgi membranes within the acceptor fraction from wild-type cells:

—The absolute efficiency with which G protein in the donor compartment goes on to receive GlcNAc in vitro is very high, perhaps approaching 100%, even when the crudest postnuclear supernatants of wild-type cells are used as acceptor. Since the Golgi membranes (which contain all of the GlcNAc transferase) are only a small fraction of the total membranes in such a crude fraction, nonspecific fusion with (or nonspecific transport to) the bulk membranes (which lack GlcNAc transferase) would result in a very low efficiency of glycosylation. A caveat is that massive, nonspecific fusions involving many organelles could occur such that each resultant giant vesicle would contain some GlcNAc transferase, permitting a high efficiency of glycosylation. This is ruled out by the electron microscope autoradiography experiments described in the next paper in the series (Braell et al., 1984) and also would be inconsistent with the high degree of specificity for compartment boundaries within the donor Golgi.

—The acceptor activity and GlcNAc transferase I activity copurify with similar yields. Such an enrichment implies a delivery process specific to the membranes of the Golgi fraction (Figure 4).

—Consistent with a high degree of specificity are the requirements for cytosol, ATP, and cytoplasmically exposed proteins on donor and acceptor membranes (Balch and Rothman, submitted)

Possible Mechanisms

It seems clear that G protein receives GlcNAc in vitro following a specific transport or fusion process involving membranes derived from successive Golgi compartments. That is, an authentic segment of the transport pathway between successive Golgi cisternal compartments has been reconstituted. But how much of the in vivo pathway exists in vitro and its exact correspondence to the Golgi structures that exist in the cell are not defined by the work described so far.

Several distinctions need to be made to help delineate the exact nature of the reconstitution and to characterize the mechanisms responsible. Does the G protein that is glycosylated in vitro start out in cisternae (or their remnants) or instead in transport vesicles that had already budded from cisternae at the time of homogenization? If G starts out in transport vesicles, then a specific fusion process has been reconstituted. If the glycosylated G protein starts out in Golgi cisternae, has the budding of a transport vesicle followed by its specific fusion been reconstituted? Or is there a *direct* but selective fusion between the appropriate cisternae of the 15B Golgi and those of the wild-type Golgi? The former possibility would represent a reconstitution of the entire process of vesicle-mediated intercompartmental transport; the latter something akin to a specific partial reaction. The experiments described in two subsequent papers (Braell et al., 1984; Balch et al., 1984) show that the bulk of Golgi membranes in both donor and acceptor fractions are in the form of stacks of cisternae, and suggest that G protein is transported between the stacks in the form of transport vesicles that form and fuse during the incubation.

Although the Golgi is particularly well suited for the study of protein transport, the principles embodied in the design of our assay system should be directly applicable to the reconstitution of other steps in secretion and in endocytosis. The general notion is to mix two homogenates, one of which contains the protein whose transport is monitored but lacks an enzyme that modifies the protein. The other homogenate contains the modifying enzyme in the compartment intended as target, but lacks the protein substrate. Constructions arranged by the use of wild-type cells and their mutants that have and lack the modifying enzyme can be used for reconstitution of steps in exocytic transport. Designs in which the enzyme is taken up by one cell population and the substrate by another cell population would be appropriate for reconstitution of endocytic transport in mixed homogenates.

Experimental Procedures

Materials

UDP-[6-^{3}H]GlcNAc (24 Ci/mmole) was from New England Nuclear; creatine phosphokinase (CPK, from rabbit muscle, Type I, 120 units/mg), UDP-GlcNAc, and GlcNAc-1-phosphate were from Sigma; [2, 8-^{3}H]-ATP (25 Ci/mmole) was from Amersham Corp. ATP and UTP were purchased from P-L Biochemicals.

Cells, Virus, and Antiserum

A wild-type line of Chinese hamster ovary (CHO) cells was maintained in suspension, and the CHO cell mutant, clone 15B (Gottlieb et al., 1975; obtained from S. Kornfeld, Washington University, St. Louis, Mo.), was grown in monolayers as described (Balch et al., 1983). Stock of VSV (Indiana serotype, San Juan isolate) was grown in monolayers of BHK cells as described (Balch et al., 1983). The resulting titer of the infection medium was typically about 10^9 plaque-forming units (pfu)/ml. Purified virions were used as a source from which G protein was purified to use as antigen in rabbits, following previously reported procedures (Balch et al., 1983). The activity of antiserum was followed by routinely using the in vitro transport assay.

Infection of 15B Cells to Yield Donor Fractions

Twenty plates (15 cm diameter) of densely confluent clone 15B cells (4–6 $\times 10^7$ cells per plate) were infected with 5–10 pfu of VSV per cell in serum-free growth medium (5 ml per plate) containing actinomycin D (5 μg/ml) and 20 mM HEPES-NaOH (pH 7.3). At 1 hr after the start of infection, 10 ml of complete growth medium was added to each plate. At 3.5 hr, cells were removed from the plates by trypsinization using the following procedure. The medium was aspirated, and each plate was rinsed with 10 ml of Tris-buffered saline (per liter: 0.4 g of KCl, 3.0 g of Tris base, 8.0 g NaCl, 0.1 g of $Na_2HPO_4 \cdot 12 H_2O$, adjusted to pH 7.4 with HCl), and then rinsed quickly with 5 ml of Tris-saline containing 0.05% trypsin and 0.02% Na_2EDTA. After 5 min at room temperature, cells in each plate were suspended in 2 ml of ice-cold complete medium by pipetting and pelleted (600 g for 5 min at 4°C). The pellet was washed once in homogenate buffer (HB; 0.25 M sucrose, 10 mM Tris-HCl, pH 7.4) and resuspended with HB to achieve a final volume equal to five times the volume of the cell pellet (i.e. a 20% suspension). Mild trypsinization does not account for the inability of G protein at the cell surface to be transferred, since cells can be harvested by scraping (without trypsin) and these two methods give similar results (data not shown).

Homogenization of Cells

For donor and acceptor fractions a crude homogenate was sometimes made from this suspension of VSV-infected clone 15B cells with 20 strokes of a very tight-fitting 7 ml Dounce homogenizer (Wheaton Co., Millville, NJ). Alternatively, we used a new device to break cells that made homogenization in 0.25 M sucrose much easier than with a Dounce homogenizer. Briefly, the cell suspension was forced repeatedly (via attached syringes) through a 0.5000 inch precision bore in a stainless steel block that contained a 0.4990 inch stainless steel ball (Industrial Tectonics Co., Ann Arbor, Mich.). Quantitative breakage of the cells required between 10 and 12 passes. The yield of activity and the properties of donor and acceptor fractions prepared with ball and Dounce homogenizers were very similar. However, the ball homogenizer had the advantage of permitting rapid and reproducible breakage of large volumes of cell suspension (20–30 ml) in less than 2 min. The ball homogenizer was used routinely. The details of construction of the homogenizer are given elsewhere (Balch and Rothman, submitted).

The crude homogenate obtained could be used at once or, more conveniently, frozen in liquid N_2 and stored at −80°C for later subcellular fractionation. Frozen crude homogenate was stable for several months. Immediately before subcellular fractionation, frozen homogenates should be thawed rapidly at 37°C and thereafter maintained on ice. This and later fractions rapidly lose activity at elevated temperatures.

For acceptor fractions, 2 liters of suspension of uninfected wild-type CHO cells were harvested at about 5 $\times 10^5$ cells/ml, washed, and suspended in HB as for the infected 15B cells, and homogenized in the same manner. The wild-type homogenate could be similarly frozen and stored.

Subcellular Fractionation: Preparation of Gradient-Purified Donor and Acceptor Golgi Membrane Fractions

The same procedure was used to prepare the Golgi fraction from VSV-infected clone 15B cells and from wild-type cells for use in routine assays. Six milliliter portions of the appropriate crude homogenate (in 0.25 M sucrose, 10 mM Tris-HCl, pH 7.4) were adjusted to 1.4 M sucrose by the addition of 6 ml of ice-cold 2.3 M sucrose containing 10 mM Tris-HCl (pH 7.4). Then, 1 mM Na_2EDTA was added from a 100 mM stock solution, vortexed vigorously to ensure uniform mixing, loaded into an SW27 tube, and overlaid with 14 ml of 1.2 M sucrose–10 mM Tris-HCl (pH 7.4), and then 8 ml of 0.8 M sucrose–10 mM Tris-HCl (pH 7.4). The gradients were centrifuged for 2.5 hr at 25,000 rpm (90,000 $\times$ g) in the SW27 rotor. The turbid band at the 0.8 M/1.2 M sucrose interface (interface III in Figures 3 and 4) was harvested in a minimum volume (~1.5 ml) by syringe puncture. This fraction was obtained at a protein concentration of 0.5 to 1.5 mg/ml protein, and is routinely used for the experiments reported, being referred to as either the donor or acceptor membrane fraction (when prepared from VSV-infected 15B cells or wild-type cells, respectively). The sucrose concentration in these fractions is always very close to 1.0 M. These fractions could be used immediately or, more conveniently, frozen in liquid N_2 in suitable aliquots and stored at −80°C. Frozen fractions are stable for several months and yield results identical to those obtained with freshly assayed fractions. The frozen membrane fractions should only be thawed shortly before the assay by a minimal exposure to 37°C, and maintained on ice prior to use. Protein was measured by the Lowry method.

Preparation of High Speed Supernatant Fractions (Cytosol)

Homogenate was prepared from uninfected wild-type CHO cells (or uninfected 15B cells) as described previously. This homogenate was centrifuged in an SW50.1 rotor for 60 min at 49,000 rpm. To remove inhibitory low molecular weight material (most likely the cytosolic pool of UDP-GlcNAc), 15 ml of the supernatant was filtered thru a Sephadex G-25 (Pharmacia Co.) column (2.5 × 50 cm) equilibrated with 25 mM Tris-HCl (pH 8.0)-50 mM KCl. The void volume fractions, containing all the excluded protein, were collected, pooled, and concentrated using a Amicon YM10 filter (Amicon Corp., Layton, Mass.) back to the volume originally loaded on the column. This gel-filtered cytosol could be routinely frozen in liquid N_2 in suitable aliquots and stored at −80°C. The cytosol retains full activity for several months under these conditions. Prior to use, frozen cytosol was thawed rapidly by a brief exposure to 37°C and maintained on ice.

Incubation Conditions to Achieve Transport In Vitro

In addition to donor membranes, acceptor membranes, and cytosol, standard incubations (50 μl) contained (final concentrations): 25 mM HEPES-KOH (pH 7.0), 25 mM KCl, 2.5 mM magnesium acetate (MgOAc), 50 μM ATP, 250 μM UTP, 2 mM creatine phosphate (CP), 7.3 IU/ml rabbit muscle creatine phosphokinase (CPK), and 0.4 μM (0.5 μCi) UDP-[^{3}H-GlcNAc]. This was prepared by combining 5 μl of a 10-times-concentrated buffer-salt stock solution (containing 250 mM HEPES-KOH, pH 7.0; 250 mM KCl; and 25 mM MgOAc) with 5 μl of a triphosphate and regenerating system stock solution (prepared daily by mixing 5 μl of CPK, 1600 IU/ml, stored at −80°C; 25 μl of 200 mM CP; 5 μl of 10 mM ATP, Na form, neutralized with NaOH; 10 μl of 100 mM UTP, Na form, neutralized; and 65 μl H_2O with 5 μl of UDP-^{3}H-GlcNAc, 100 μCi/ml, in H_2O, prepared daily by evaporation to dryness of an aliquot of an ethanolic stock of the sugar–nucleotide with a gentle stream of N_2, and then dissolving in H_2O, and 20 μl H_2O). To this mixture (on ice) was added, in the order listed, 5 μl of the gel-filtered cytosol (5–10 mg protein/ml), and 5 μl each of the donor and acceptor membrane fractions (0.5 to 1.5 mg protein/ml), which had been thawed as short a time as possible before use by a brief incubation (less than 15 sec) at 37°C and then chilled on ice. The assay was initiated by transfer to 37°C. Incubation was routinely for 60 min, and done in disposable glass tubes.

Quantitation of Transport by the Incorporation of ^{3}H-GlcNAc into G Protein

The incubation (50 μl) was terminated by transfer to ice, and the membranes were solubilized by addition of 50 μl of detergent buffer (DB) containing 50 mM Tris-HCl (pH 7.5), 250 mM NaCl, 1 mM Na_2EDTA, 1% Triton X-100, and 1% sodium cholate. Then, rabbit anti-G protein antiserum (or the same volume of preimmune serum as a control) was added. The amount of antiserum needed (generally 10–20 μl) to obtain maximal binding of ^{3}H-G protein was determined individually for each preparation of antiserum and of donor membrane. After 45 min at 37°C or an overnight incubation at 0–4°C, the immunoprecipitate was collected by its retention on Millipore filters (type HA, 0.45 μm; Millipore Corp., Bedford, Mass.). Each filter was first rinsed with 3 ml of a washing buffer (WB) (containing 50 mM Tris-HCl, pH 7.5; 250 mM NaCl; 5 mM Na_2EDTA; 1% Triton X-100). Before the filter can dry, the immunoprecipitated incubation mixture is diluted with 3 ml of WB and rapidly filtered. The tube is rinsed once with 3 ml WB, and this is filtered. The filter is then further washed in rapid succession with three portions of WB (3 ml each). It is important that incubations be filtered one at a time without any air drying of the filter between washes; otherwise there is a high nonspecific sticking of ^{3}H-sugar nucleotide to the filters. Filters are then dried under a heat lamp and counted in a scintillation cocktail.

Acknowledgments

This paper is dedicated to Professor Eugene P. Kennedy on the occasion of his 65th birthday. His former students thank him for his fine example.

We thank Ietje Kathman for her able technical assistance and Pamela Hilton for preparation of this manuscript. Also, we thank Dr. Suzanne Pfeffer for her critical comments on this paper. W. E. B. was supported by a National Institutes of Health postdoctoral fellowship; W. A. B. by a Jane Coffin Childs postdoctoral fellowship. This research was funded by National Institutes of Health grant AM 27044.

The costs of publication of this article were defrayed in part by the payment of page charges. This article must therefore be hereby marked "*advertisement*" in accordance with 18 U.S.C. Section 1734 solely to indicate this fact.

Received June 28, 1984; revised October 5, 1984

References

Balch, W. E., Fries, E., Dunphy, W., Urbani, L. J., and Rothman, J. E. (1983). Transport-coupled oligosaccharide processing in a cell-free system. Meth. Enzymol. *98*, 37–47.

Balch, W. E., Glick, B. S., and Rothman, J. E. (1984). Sequential intermediates in the pathway of intercompartmental transport of a cell free system. Cell *39*, in press.

Bergmann, J. E., and Singer, S. J. (1983). Immunoelectron microscopic studies of the intracellular transport of the membrane glycoprotein (G) of vesicular stomatitis virus in infected Chinese hamster ovary cells. J. Cell Biol. *97*, 1777–1787.

Braell, W. A., Balch, W. E., Dobbertin, D. C., and Rothman, J. E. (1984). The glycoprotein that is transported between successive compartments of the Golgi in a cell-free system resides in stacks of cisternae. Cell *39*, in press.

Deutscher, S. L., Creek, K. E., Merion, M., and Hirschberg, C. B. (1983). Subfractionation of the rat liver Golgi apparatus: separation of enzyme activities involved in the biosynthesis of the phosphomanosyl recognition marker in lysosomal enzymes. Proc. Nat. Acad. Sci. USA *80*, 3938–3942.

Dunphy, W. G., and Rothman, J. E. (1983). Compartmentation of Asn-linked oligosaccharide processing in the Golgi apparatus. J. Cell Biol. *97*, 270–275.

Dunphy, W. G., Fries, E., Urbani, L. J., and Rothman, J. E. (1981). Early and late functions associated with the Golgi apparatus reside in distinct compartments. Proc. Nat. Acad. Sci. USA *78*, 7453–7457.

Dunphy, W. G., Brands, R., and Rothman, J. E. (1985). Attachment of terminal *N*-acetylglucosamine to asparagine-linked oligosaccharides occurs in the central cisternae of the Golgi stack. Cell *40*, in press.

Farquhar, M. G., and Palade, G. E. (1981). The Golgi apparatus (complex)—(1954–1981)—from artifact to center stage. J. Cell Biol. *91*, 77s–103s.

Fleischer, B. (1974). Isolation and characterization of Golgi apparatus and membranes from rat liver. Meth. Enzymol. *31*, 180–191.

Fries, E., and Rothman, J. E. (1980). Transport of vesicular stomatitis virus glycoprotein in a cell-free extract. Proc. Nat. Acad. Sci. USA *77*, 3870–3874.

Fries, E., and Rothman, J. E. (1981). Transient activity of Golgi-like membranes as donors of vesicular stomatitis virus glycoprotein in vitro. J. Cell Biol. *90*, 697–704.

Goldberg, D. E., and Kornfeld, S. (1983). Evidence for extensive subcellular organization of asparagine-linked oligosaccharide processing and lysosomal enzyme phosphorylation. J. Biol. Chem. *258*, 3159–3165.

Gottlieb, C., Baenziger, J., and Kornfeld, S. (1975). Deficient uridine diphosphate-N-acetylglucosamine: glycoprotein N-acetylglucosaminyltransferase activity in a clone of Chinese hamster ovary cells with altered surface glycoproteins. J. Biol. Chem. *250*, 3303–3309.

Griffiths, G. R., Brands, R., Burke, B., Louvard, D., and Warren, G. (1982).

Viral membrane proteins acquire galactose in *trans* Golgi cisternae during intracellular transport. J. Cell Biol. *95,* 781–792.

Griffiths, G., Quinn, P., and Warren, G. (1983). Dissection of the Golgi complex: monensin inhibits the transport of viral membrane proteins from *medial* to *trans* Golgi cisternae in baby hamster kidney cells infected with Semliki Forest virus. J. Cell Biol. *96,* 835–850.

Hubbard, S. C., and Ivatt, R. J. (1981). Synthesis and processing of Asn-linked oligosaccharides. Ann. Rev. Biochem. *50,* 555–583.

Laemmli, U. K. (1970). Cleavage of structural proteins during the assembly of the head of bacteriophage T4. Nature *227,* 680–685.

Quinn, P., Griffiths, G., and Warren, G. (1984). Density of newly synthesized membrane proteins in intracellular membranes. II. Biochemical studies. J. Cell Biol. *98,* 2142–2147.

Roth, J. (1983). Application of lectin-gold complexes for electron microscopic localization of glycoconjugates on thin sections. J. Histochem. Cytochem. *31,* 987–999.

Roth, J., and Berger, E. G. (1982). Immunocytochemical localization of galactosyltransferase in HeLa cells: distribution with thiamine pyrophosphatase in *trans* Golgi cisternae. J. Cell Biol. *92,* 223–229.

Rothman, J. E. (1981). The Golgi apparatus: two organelles in tandem. Science *213,* 1212–1219.

Rothman, J. E., and Fries, E. (1981). Transport of newly synthesized vesicular stomatitis viral glycoprotein to purified Golgi membranes. J. Cell Biol. *89,* 162–168.

Rothman, J. E., and Lenard, J. (1984). Membrane traffic in animal cells, Trends Biochem. Sci. *9,* 176–178.

Rothman, J. E., Miller, R. L., and Urbani, L. J. (1984a). Intercompartmental transport in the Golgi is a dissociative process: facile transfer of membrane protein between two Golgi populations. J. Cell Biol. *99,* 260–271.

Rothman, J. E., Urbani, L. J., and Brands, R. (1984b). Transport of protein between cytoplasmic membranes of fused cells: Correspondence to processes reconstituted in a cell-free system. J. Cell Biol. *99,* 248–259.

Schachter, H., Narasimhan, S., Gleeson, P., and Vella, G. (1983). Glycosyltransferases in elongation of N-glycosidically linked oligosaccharides of the complex or N-acetyllactosamine type. Meth. Enzymol. *98,* 98–134.

Tabas, I., and Kornfeld, S. (1978). The synthesis of complex-type oligosaccharides. J. Biol. Chem. *253,* 7779–7786.

Tartakoff, A. M. (1982). Simplifying the complex Golgi. Trends Biochem. Sci. *7,* 174–176.

Tartakoff, A. M. (1983). The confined function model of the Golgi complex: center for ordered processing of biosynthetic products of the rough endoplasmic reticulum. Int. Rev. Cytol. *85,* 221–252.

Zilberstein, A., Snider, M. D., and Lodish, H. F. (1981). Synthesis and assembly of the vesicular stomatitis virus glycoprotein. Cold Spring Harbor Symp. Quant. Biol. *46,* 785–795.

Cytoskeleton

Tilney L.G. and **Porter K.R.** 1967. Studies on the microtubules in Heliozoa II. The effect of low temperature on these structures in the formation and maintenance of the axopodia. *J. Cell Biol.* **34:** 327–343.

MICROTUBULES WERE INITIALLY DISCOVERED in the axonemes of cilia and flagella (1), but after the introduction of glutaraldehyde as a fixative (2) they were recognized as ubiquitous in eukaryotic cells. While some groups thought of these slender pipes as cytoplasmic plumbing, Porter and his colleagues pursued the hypothesis that they were structural elements that contributed to cellular shape and organization. Several studies from Porter's group showed a correlation between the location of microtubules and the places where cells were changing shape (3), but the work by Tilney and Porter put the idea of cellular deformation by microtubule polymerization on a sound experimental footing. This paper and its companions showed that any of several physical or chemical treatments that dissolved microtubules led to the shortening of cytoplasmic protrusions. From these and related studies the role of microtubules in development and maintenance of cell form was established. Work in other labs led to the purification of tubulin, the protein subunit of microtubules (4), and the identification of conditions for polymerizing tubulin in vitro (5). Subsequent studies demonstrated the structural polarity of microtubules, both in vitro and in vivo (6), and showed the polar kinetics of microtubule growth (7). With the discovery of microtubule dynamic instability (Mitchison and Kirschner, this volume) and cytoplasmic motor enzymes (Vale et al, this volume), we have begun to get a solid, molecular understanding of the mechanisms by which microtubules contribute to cytoplasmic structure and cellular morphogenesis.

---

1. Fawcett D.W. and Porter K.R. 1954. A study of the fine structure of ciliated epithelia. *J. Morphol.* **94:** 221–281.
2. Sabatini D.D., Bensch K.G., and Barrnett R.J. 1962. Preservation of ultrastructure and enzymatic activity of aldehyde fixation. *J. Histochem. Cytochem.* **10:** 652–653.
3. Byers B. and Porter K.R. 1964. Oriented microtubules in elongating cells of the developing lens rudiment after induction. *Proc. Natl. Acad. Sci.* **52:** 1091–1099.
4. Borisy G.G. and Taylor E.W. 1967. The mechanism of action of colchicine. Colchicine binding to sea urchin eggs and the mitotic apparatus. *J. Cell Biol.* **34:** 535–548.
5. Weisenberg R.C. 1972. Microtubule formation in vitro in solutions containing low calcium concentrations. *Science* **177:** 1104–1105.
6. Heidemann S.R. and McIntosh J.R. 1980. Visualization of the structural polarity of microtubules. *Nature* **286:** 517–519.
7. Bergen L.G. and Borisy G.G. 1980. Head-to-tail polymerization of microtubules in vitro. Electron microscope analysis of seeded assembly. *J. Cell Biol.* **84:** 141–150.

STUDIES ON THE MICROTUBULES IN HELIOZOA

II. The Effect of Low Temperature on These Structures in the Formation and Maintenance of the Axopodia

LEWIS G. TILNEY and KEITH R. PORTER

From the Biological Laboratories, Harvard University, Cambridge, Massachusetts 02138

ABSTRACT

When specimens of *Actinosphaerium nucleofilum* are placed at 4°C, the axopodia retract and the birefringent core (axoneme) of each axopodium disappears. In fixed specimens, it has been shown that this structure consists of a highly patterned bundle of microtubules, each 220 A in diameter; during cold treatment these microtubules disappear and do not reform until the organisms are removed to room temperature. Within a few minutes after returning the specimens to room temperature, the axonemes reappear and the axopodia begin to reform reaching normal length 30–45 min later. In thin sections of cells fixed during the early stages of this recovery period, microtubules, organized in the pattern of the untreated specimens, are found in each reforming axopodium. Reforming axopodia without birefringent axonemes (and thus without microtubules) are never encountered. From these observations we conclude that the microtubules may be instrumental not only in the maintenance of the axopodia but also in their growth. Thus, if the microtubules are destroyed, the axopodia should retract and not reform until these tubular units are reassembled. During the cold treatment short segments of a 340-A tubule appeared; when the organisms were removed from the cold, these tubular segments disappeared. It seems probable that they are one of the disintegration products of the microtubules. A model is presented of our interpretation of how a 220-A microtubule transforms into a 340-A tubule and what this means in terms of the substructure of the untreated microtubules.

INTRODUCTION

The disposition of microtubules in cells and cell processes has led several investigators to suggest that these elements perform a skeletal role (Byers and Porter, 1964; Porter et al., 1964; Fawcett and Witebsky, 1964; Behnke, 1965; Tilney and Porter, 1965; Maser and Philpott, 1964; de-Thé, 1964; Pitelka, 1963; Roth, 1956) in the production and maintenance of linear cell extensions. Examples of these are found in cilia and flagella (Gibbons, 1961), in neurons (Gonatas and Robbins, 1965; Porter et al., 1964), in numerous structures in the protozoa (Tilney and Porter, 1964; Pitelka, 1963; Rudzinska, 1965), in cells in culture (Taylor, 1966), in certain types of sperm (Gordon, 1966, personal communication; Moses, 1966), or even in cells undergoing elongation such as during mitosis (de-Thé, 1964; Robbins and Gonatas, 1964; Roth et al., 1966) or in embryonic differentiation (Byers and Porter, 1964; Arnold, 1966).

It has further been noted that these elements are present in regions of the cytoplasm which can be characterized as gelled or regions of so called plasmagel (Bikle et al., 1966; Tilney et al., 1966). By virtue of this association, the microtubules might then be instrumental in the regulation of cytoplasmic movements within specific regions of the cytoplasm adjacent to these gelled areas. Such regions in which cytoplasmic flow occurs could then be defined as regions of plasmasol.

In order to investigate more rigorously the relationship of the microtubules to the production and maintenance of cell asymmetry, on the one hand, and to the regulation of motility, on the other hand, we undertook a group of experiments designed to cause the breakdown of the microtubules. We felt that through careful comparisons of the experimentally treated specimens with controls we could elucidate the role of the microtubules in the before-mentioned functions.

Toward this end, we have treated the protozoan, *Actinosphaerium nucleofilum*, with cold temperature (Tilney, 1965 *a* and *b*). It was reported a number of years ago (Inoué, 1952) that the birefringence of the mitotic spindle disappeared when dividing cells were placed in the cold. More recently, a number of investigators (Roth and Daniels, 1962; Ledbetter and Porter, 1963; Robbins and Gonatas, 1964; de-Thé, 1964; Roth, 1964; Manton, 1964; Harris, 1962; to mention only a few) have shown that each spindle fiber as resolved by the light microscope is, in reality, composed of a number of microtubules. Since each microtubule in the spindle appears morphologically similar to the microtubules in the cytoplasm of interphase cells, and since a reduction in birefringence of the isolated mitotic spindle results in a concomitant loss in the number of microtubules (Kane and Forer, 1965), it seemed reasonable to expect that cold temperature might bring about a destruction of cytoplasmic microtubules and that, in situations other than the mitotic spindle, temperature control might thus be used as an analytical tool to provide more information on the characteristics and function of these cytoplasmic elements.

Actinosphaerium nucleofilum seemed ideally suited for experiments of this kind. As is well known, its needle-like pseudopodia (axopodia), which can be up to 400 μ in length, contain within them a well defined system of microtubules, the so called axoneme (Kitching, 1964; Tilney and Porter, 1965). It was reasoned that, if the microtubules are instrumental in the maintenance of these slender protoplasmic extensions, then low temperature, which, as previously stated, should cause the breakdown of the microtubules, ought secondarily to cause retraction of the axopodia. Likewise, if the microtubules are important in the production of these linear cell extensions, then re-extension of the axopodia, when the Actinosphaerium is brought back to room temperature, should not occur in the absence of oriented microtubules.

It also proved worthwhile to examine the products of the microtubule disassembly since this cryopathology gave us some additional information about the composition of the microtubule through the behavior of these breakdown products.

MATERIALS AND METHODS

THE ORGANISMS: Cultures of Actinosphaerium were obtained from the Carolina Biological Supply Co.[1] and serially subcultured in a wheat medium (Looper, 1928). At intervals, ciliates were added as a food source.

LIGHT MICROSCOPIC PROCEDURES: For observation, cells were put either on a glass slide and covered by a cover slip supported on its edges, or in a shallow depression slide. Each preparation was allowed to equilibrate for a short period prior to cold treatment. Such preparations were viewed with an ordinary light microscope or with a Zeiss polarizing microscope equipped with a mercury arc lamp. With the latter, the 16x objective and a compensator setting of 3° proved optimal.

Living organisms were studied during the cold treatment by placing the microscope in a 4° cold room. Once the polarizing microscope had reached the temperature of the cold room, it could be used to study slides prepared as already described. To watch regrowth, we removed the microscope from the cold room and allowed it to equilibrate at room temperature. The slide was then taken from the cold room to room temperature and viewed immediately.

Photomicrographs were taken at intervals during and following cold treatment on 35-mm film. We subsequently developed the film in Diafine[2] to increase the effective film speed.

ELECTRON MICROSCOPIC PROCEDURES: Specimens were fixed after periods varying from 1 to 24 hr following the onset of cold treatment. The fixative was cold (4°C) 3% glutaraldehyde in 0.05 M phosphate buffer at pH 7.0 which contained 0.0015 M $CaCl_2$. The specimens were washed in 0.1 M phosphate buffer with 0.0015 M $CaCl_2$, postfixed in 1% OsO_4 in 0.1 M phosphate buffer and 0.0015 M $CaCl_2$, dehydrated rapidly, and embedded in Epon 812 (Luft, 1961).

Other specimens which had been cold treated for

[1] Burlington, North Carolina.

[2] Baumann Photo-chemical Corporation, Chicago.

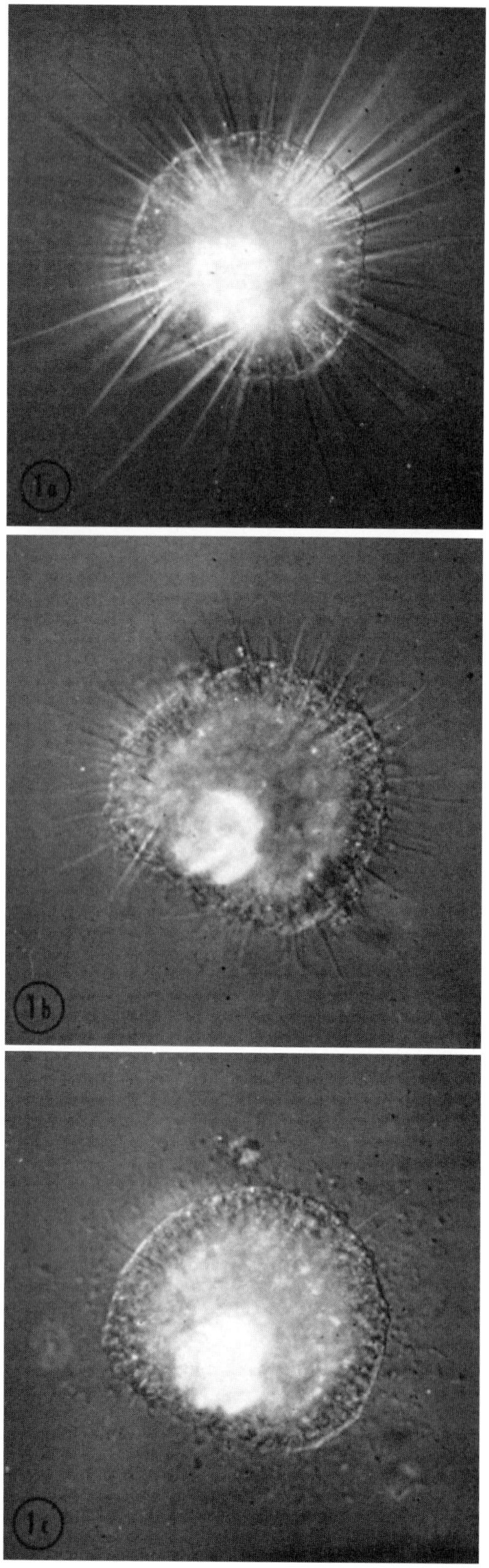

several hours were brought out of the cold room in small petri dishes which contained about 5 ml of fluid. These were examined with a dissecting microscope. When the axopodia first began to reform, about 10 min after removal from the cold, they were fixed in the identical fixative as that used on the cells in the cold the only difference being that the fixative was at room temperature. The remaining steps were carried out as already described.

Specimens were cut with a diamond knife on a Servall MT2 Porter-Blum ultramicrotome, stained with uranyl acetate (Watson, 1958) and lead citrate (Reynolds, 1963), and examined with the Siemens Elmiskop I electron microscope.

OBSERVATIONS

Normal Structure

The axoneme or birefringent core of each axopodium (in rare instances, a second axoneme parallel to the first is present within a single axopodium) extends without interruption from the tip of the axopodium to the medullary region of the cell body. The birefringence, which can be readily demonstrated in living specimens (Fig. 1 *a*), is positive, uniaxial, and has been shown to be form birefringence (McKinnon, 1909). The limiting membrane, ingested material such as rotifers (Fig. 1 *a*), and twinkling medullary particles of unknown nature are also birefringent.

When an axoneme is examined with the electron microscope, it is found to exist as a bundle of microtubules arranged parallel to one another and, in cross-sectional view, organized into an interlocking double coil. (For a more detailed description of the organization of the microtubules in the

FIGURE 1 These micrographs and the micrographs of Fig. 4 are all taken of the same organism with a Zeiss polarizing microscope. The large birefringent spot in the center of the organism is an ingested rotifer. × 170. *a*, 3 min at 4°C. Note the birefringent axonemes which extend from the medulla to the tip of each axopodium. *b*, 55 min at 4°C. The axopodia are reduced in length and have become beaded. Their short birefringent axonemes extend from the corticomedullary junction out into the axopodia. *c*, 2¼ hr at 4°C. Note the almost complete absence of axopodia. In four, possibly five, instances thin axonemes extend from the corticomedullary boundary out into short axopodia. In other instances, the birefringence of the axonemes has totally disappeared. A number of beads, presumably the short axopodia seen in Fig. 1 *b*, are seen surrounding the cell body. It is possible that they are connected together and to the cell body.

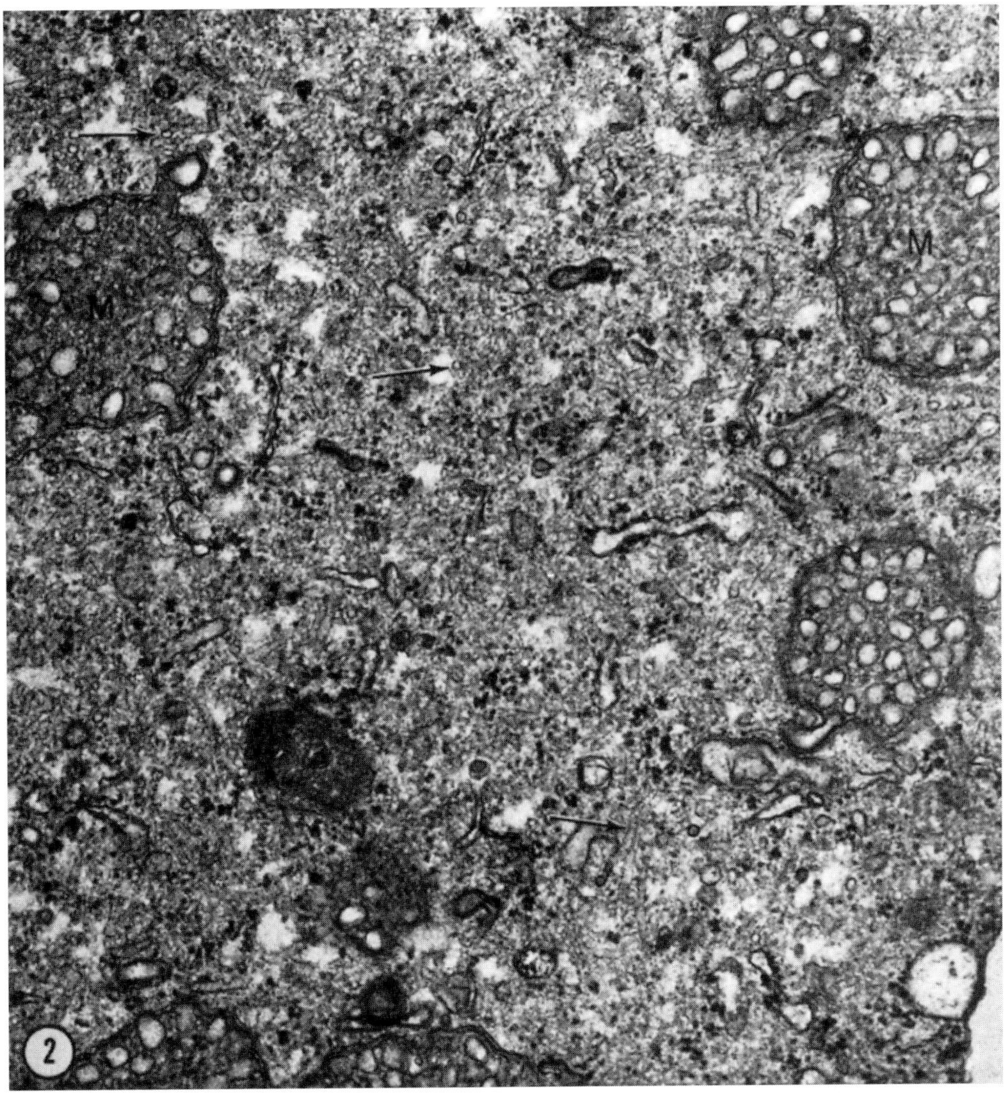

FIGURE 2 Micrograph taken through a portion of the cell body of an organism fixed after 2½ hr at 4°C. Mitochondria (*M*), short segments of rough-surfaced endoplasmic reticulum and profiles of tubules (see arrows) are apparent in this micrograph. The ground substance which we refer to in the text as "matrix substance" is very prominent in these cold-treated cells. × 43,000.

FIGURE 3 Micrographs taken of portions of cold-treated cells. *a*, Most prominent in this micrograph are tubular profiles measuring approximately 340 A in diameter. In some, a small dot can be seen in the center. × 82,000. *b*, Many of the 340-A tubules are present in longitudinal section in this micrograph. In the tubule designated by the arrows, one can make out diagonal striations across the wall of the tubule. These striations make approximately a 45° angle with the long axis of the tubule. × 110,000.

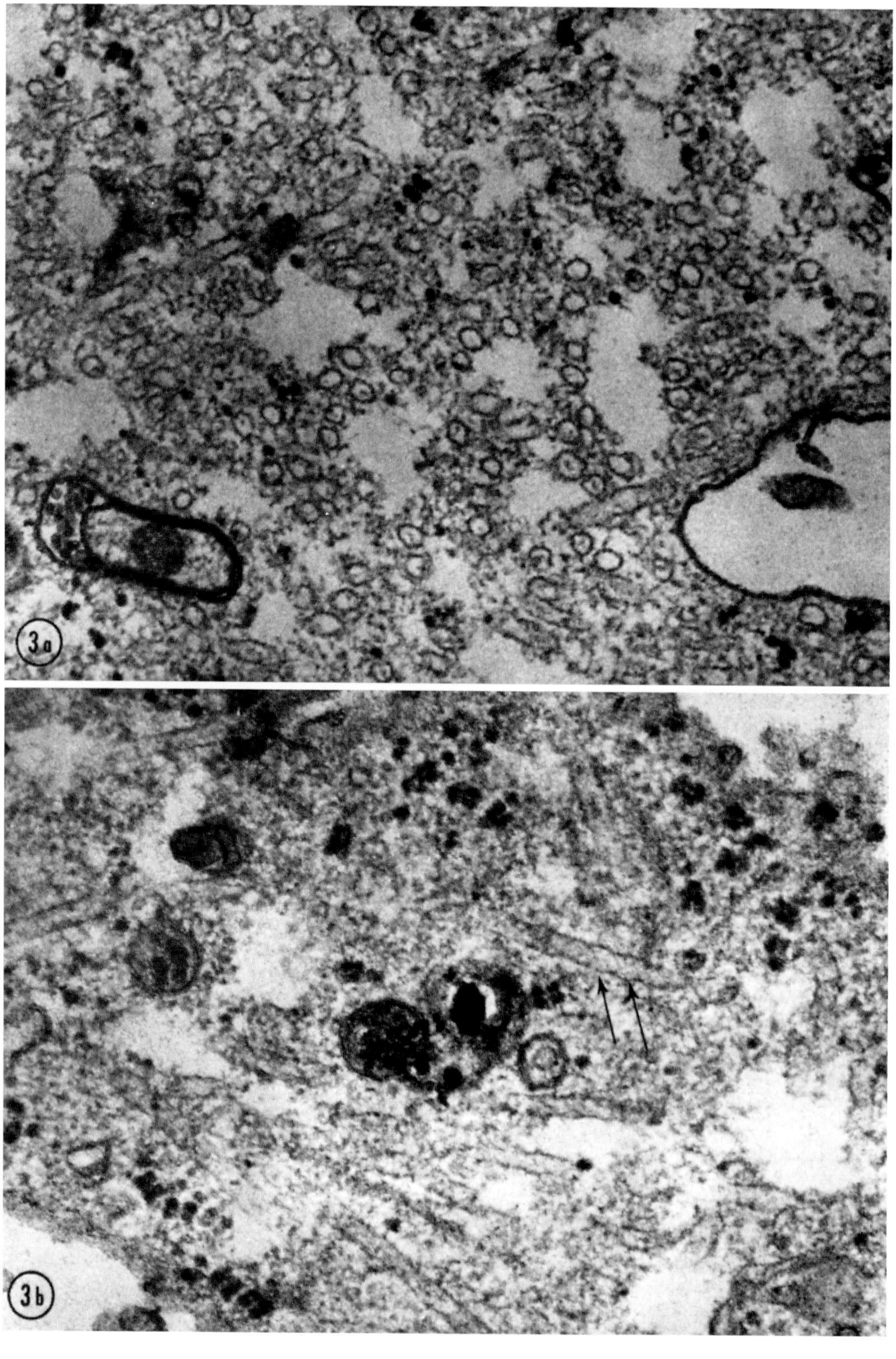
3a
3b

axoneme, see Tilney and Porter, 1965). Figs. 6 and 7, taken of recovering axopodia, depict this structure. Within each axopodium, dense granules, believed to function in food capture, and mitochondria are situated between the axoneme and the limiting plasma membrane. These cytoplasmic particles, in living organisms, move up and down the axopodia in the periaxonemal space.

Cold Treatment

1. LIGHT MICROSCOPIC OBSERVATION OF LIVING ANIMALS: A series of small beads appears along the length of most of the axopodia within a few minutes after preparations containing the organisms are placed at 4°C (Fig. 1 *b*). Several minutes later, the axopodia begin to shorten. In all but the final stage of axopodial retraction, the cytoplasm which once made up each axopodium withdraws into the cortical region of the cell body.

Alterations in the axoneme occur during the cold treatment as well. These can be studied most profitably in living organisms with the polarizing microscope. With this instrument, quantitative as well as qualitative measurements on the amount of birefringence in a single axoneme or in a portion of a single axoneme can be made with respect to the length of time the organisms are exposed to the cold. The accompanying Figs. 1 *a–c*, photomicrographs of the same organism, were taken after periods of 3 min, 1 hr, and 2¼ hr of cold treatment. As can be readily seen, the birefringence of the axoneme gradually disappears; each axoneme does not merely shorten in length but rather, while each decreases in length, the over-all birefringence of the whole axoneme diminishes. Thus, if we compare the birefringence of a specific region of the axoneme, for example that portion near the surface of the cell body, before axoneme shortening (Fig. 1 *a*) and during axoneme shortening (Fig. 1 *b*), we find that roughly the birefringence diminishes in this region in proportion to the amount of axoneme shortening.

After longer periods in the cold (Fig. 1 *c*) few axopodia remain, but in those that do persist axonemes can still be seen. A number of small beads lie just outside the cell body and persist in this position over long periods of cooling. Since these beads cannot account for all the cytoplasm in the axopodia, it would seem that some of the cytoplasm from the retracting axopodium has migrated into the cell body. It is probable that these beads are connected to each other and to the cell body by a plasma membrane and that during axopodial re-extension (Fig. 4) they again become a part of the axopodial cytoplasm.

With increased time in the cold, up to 1½ days, further changes in the axopodia and axonemes do not occur. An occasional short axopodium with a faintly birefringent axoneme similar to that depicted in Fig. 1 *c* remains. Redevelopment of straight rigid axopodia does not occur in the cold.

In contrast to the diminution in axoneme birefringence during cold treatment, the birefringence of the ingested material, the limiting membranes (this includes the plasma membrane on the surface of the cell body as well as the membranes which limit all the vacuoles), and twinkling medullary particles is unaltered.

Although no further changes take place in the axonemes and remaining axopodia with prolonged cold treatment, changes do take place in the size of the cortical vacuoles and the over-all density of the organisms. The cortical vacuoles increase in size and the optical density diminishes. We presume that these changes are related to the effect of low temperature on the contractile vacuole. This structure either completely fails to function or empties at a rate insufficient to counteract water inflow. Each organism gradually increases in diameter and eventually ruptures.

2. Electron Microscope Observations

a. CELLS FIXED AFTER 2¼ HR AT 4°C: Since few axopodia remain after 2¼ hr of cold treatment, it is not surprising that few were encountered in thin sections cut from several organisms. Present within some of these axopodia we found small numbers of microtubules (never more than 20); in other axopodia, which appeared larger in diameter, we found mitochondria, dense granules, small numbers of structures presumed to be excretion bodies (Tilney and Porter, 1965), and short segments of a randomly oriented, tubular component measuring 310–360 A in diameter. The tubular component was embedded in an amorphous material (Figs. 2 and 5). In no instance were both the 220- and the 340-A tubular profiles present in the same axopodium.

Since axopodia were rare, our attention focused on the cytoplasm of the cell body. As might be expected by the absence of birefringent axonemes (Fig. 1 *c*), organized arrays of microtubules were no longer present; in fact, the 220-A microtubules had completely disappeared from the cell body

(Fig. 2). Short segments of the larger, 310–360-A tubules, already described in some of the remaining axopodia, were commonly encountered scattered throughout the cell body. But in most instances, these units bore no specific orientation either to each other or to the free surface. A closer examination of their fine structure, in particular in sections cut parallel to their long axes, revealed that some of these tubular units possessed an irregular transverse striation. The pitch of the striae was not constant, but, in some preparations favorable for observation, it approximated a 45° angle relative to the central axis of the tubule (Fig. 3 *b*). The wall of each of the tubules was approximately 50 A thick, comparable to the wall thickness of the 220-A microtubules (Fig. 3 *a*). In the center of some of these elements there was a small electron-opaque dot about 75 A in diameter.

Certain alterations in the nature of the material which occupies the area between the formed elements of the cytoplasm also seem to occur in cold-treated specimens. This material in untreated cells is very inconspicuous and possesses a fluffy appearance. In cold-treated cells, on the other hand, the equivalent space is packed with a finely fibrous cotton-like material which is illustrated in Figs. 2,

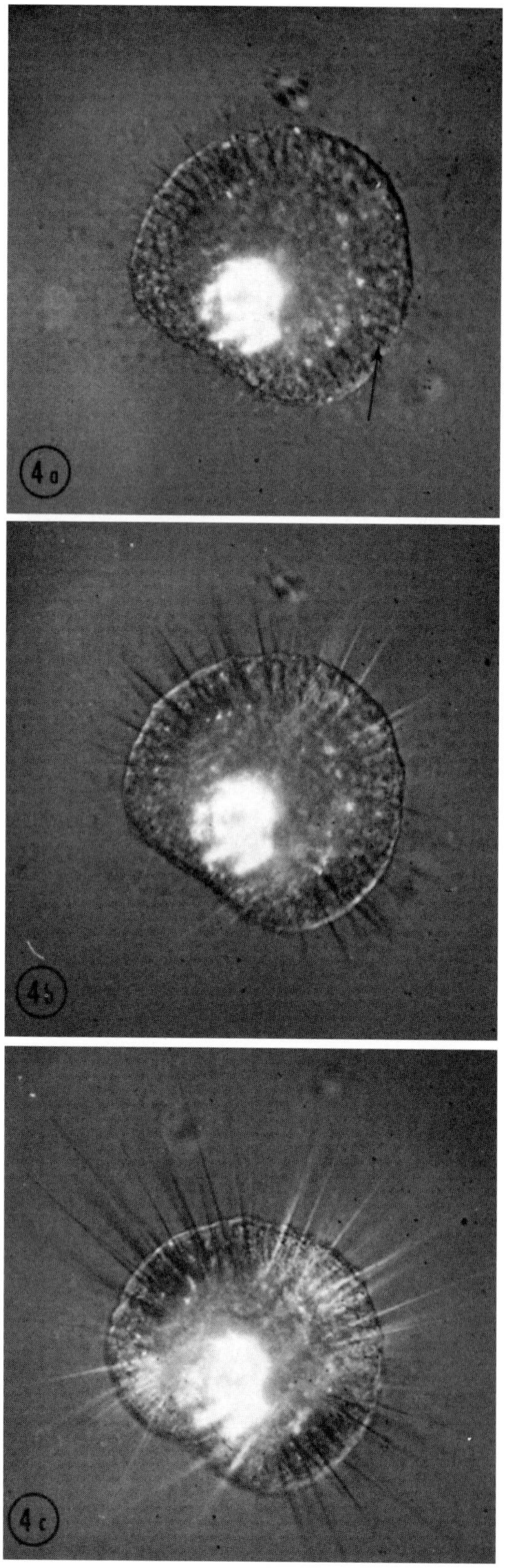

FIGURE 4 These micrographs are taken of the same organism illustrated in Fig. 1. In this series, the organism has been removed from the cold room and placed at room temperature; so these three micrographs depict recovery. As in Fig. 1, the highly birefringent spot in the center of the cell is due to an ingested rotifer. × 170. *a*, 3 min at room temperature. Even at this short time at room temperature, a few short axopodia had reformed; within each is a birefringent axoneme. Axonemes are beginning to form in the cell cortex (see arrow). *b*, 15 min at room temperature. Short axonemes now extend from the cell body out into the surrounding pond water. Within each is a strongly birefringent axoneme which extends from the tip of the axopodium into the medullary region. Through a comparison of Figs. 4 *b* and 1 *b* in which the axopodia are approximately the same length, it can be easily seen that not only are the reforming axopodia more rigid than the retracting ones but also the amount of birefringence is greater. *c*, 45 min at room temperature. By this time, the axopodia are of normal length. A comparison of Fig. 1 *a* with this figure reveals that axopodia, in both instances, seem to appear in approximately the same locations and in the same numbers.

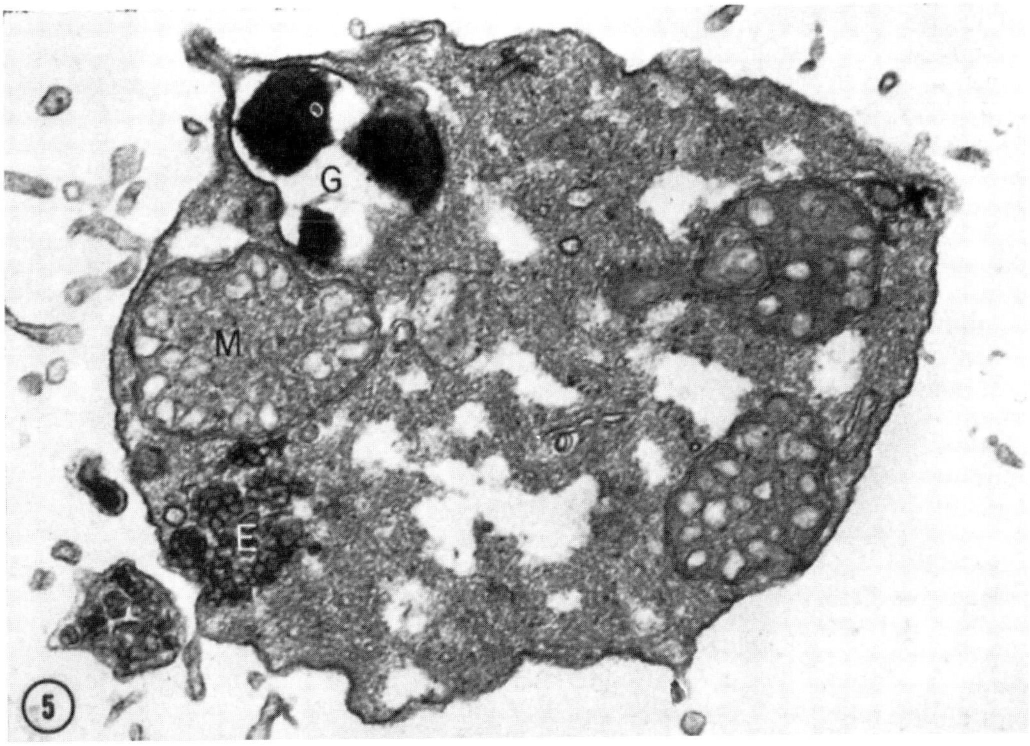

It is interesting to compare Fig. 5, a retracting axopodium, with Fig. 6, a reforming axopodium.

FIGURE 5 Section of an axopodium which remained after 2¼ hr of cold treatment. Beneath the limiting membrane are mitochondria (*M*), electron-opaque granules (*G*), excretion bodies (*E*), and a randomly arranged tubular component enmeshed in a dense matrix substance. Each tubule measures about 340-A in diameter. × 48,000.

3 *b*, 5, and 8. In the absence of a better name, we will refer to this material as "matrix substance." It is probable that it is derived in part from the disassembly of the axoneme and thus is related to the microtubules. The segments of 310–360-A tubules are commonly found in this matrix substance (Fig. 2).

b. CELLS FIXED AFTER 24 HR AT 4°C: The morphology of cells fixed after 24 hr at 4°C is comparable to that of cells fixed after shorter periods at 4°C. The frequency of the 310–360-A tubule segments has certainly not diminished by this stage. In certain micrographs these tubular units are oriented parallel to each other (similar to Fig. 3 *a*), but in most micrographs the tubules are randomly arranged within the matrix substance (Fig. 2). In one instance a short axopodium was found. This contained within it a few 220-A microtubules, but in no other instance were any 220-A microtubules seen at this stage. The matrix substance appears in the same quantity as is present after 2¼ hr of cold treatment.

3. *Recovery Following Low Temperature*

a. LIGHT MICROSCOPIC OBSERVATIONS OF LIVING ORGANISMS: Within 3–4 min after the preparations were removed from the cold, axopodia begin to reform; the few short axopodia which had remained during the cold treatment begin to elongate. Within each of these, the polarizing microscope reveals the presence of a birefringent axoneme (Figs. 4 *a* and *b*).

Birefringence attributable to an axoneme is first visible in the cortical region of the cell. These regions are indicated by the arrows in Fig. 4 *a* and appear to give rise to the axoneme depicted in Fig. 4 *b*. These reforming axonemes initially ex-

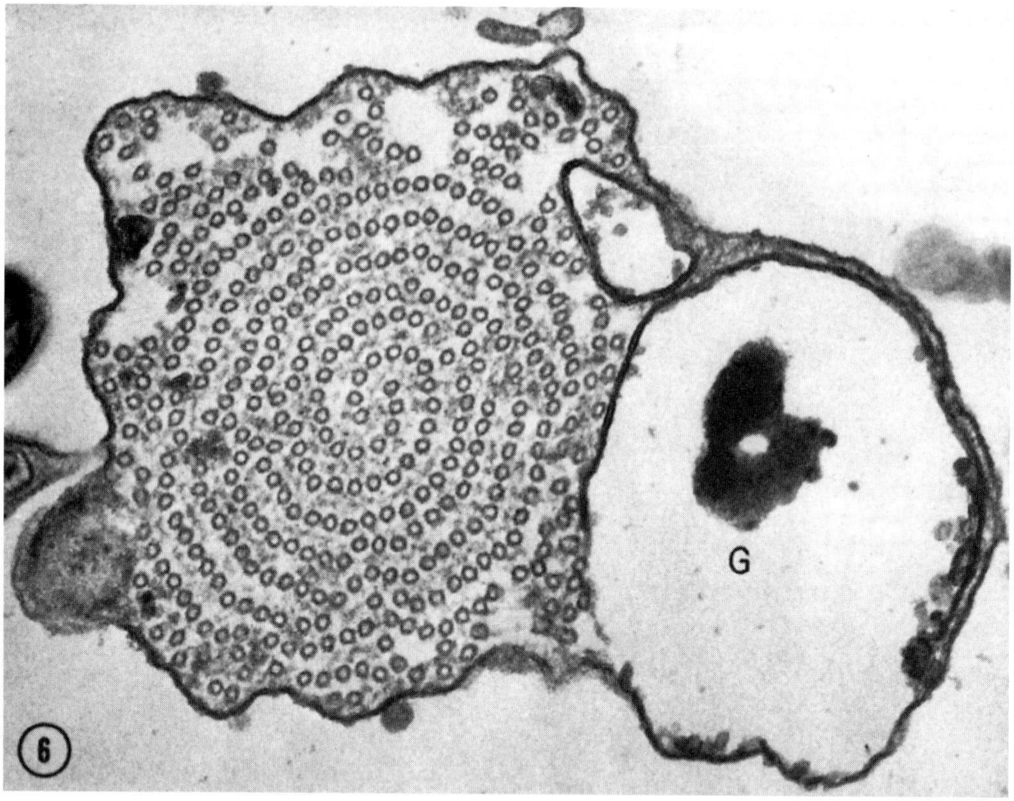

FIGURE 6 Transverse section through a recovering axopodium. Within the plasma membrane is a partially ruptured dense granule (*G*) and an axoneme consisting of a double coiled array of microtubules. Such an image is what one normally encounters in an untreated specimen. × 95,000.

tend in both directions. Once the basal end has made contact with the medullary region, it ceases to elongate basally; its only growth now will be in width. The apical end, on the other hand, continues to elongate as well as to grow in width. After making contact with the limiting membrane, it extends out into the surrounding medium. It is covered by the plasma membrane which becomes the axopodial membrane. As this reforming axopodium extends, presumably through the action of the axonemes, small granules move out into it and begin their characteristic movements. (For a more detailed description of these movements, see Tilney and Porter, 1965.) We have never observed a reforming axopodium which lacks a birefringent axoneme. It also appears that the residual beads present after full axopodial retraction (Figs. 1 *c* and 4 *a*) are incorporated into the reforming axopodia. (Compare Fig. 1 *c* with Figs. 4 *a* and *b*. Figures in the series 1 and 4 are from the same organism.)

It is interesting to compare the birefringence of reforming axopodia (Fig. 4 *b*) with the birefringence of retracting axopodia (Fig. 1 *b*). It is apparent by the comparison of these two figures (Figs. 1 *b* and 4 *b*) in which these cell extensions are approximately the same length, that the axonemes of the reforming axopodia are more strongly birefringent than the axonemes of the retracting axopodia. Of even greater significance is the fact that the surfaces of the reforming axopodia are smooth and relatively even while those of the retracting axopodia are beaded and clearly very irregular (Fig. 1 *b*).

By 35–40 min, the average length of the axopodium is within the normal range. Furthermore, organism movement has begun; a mechanism for

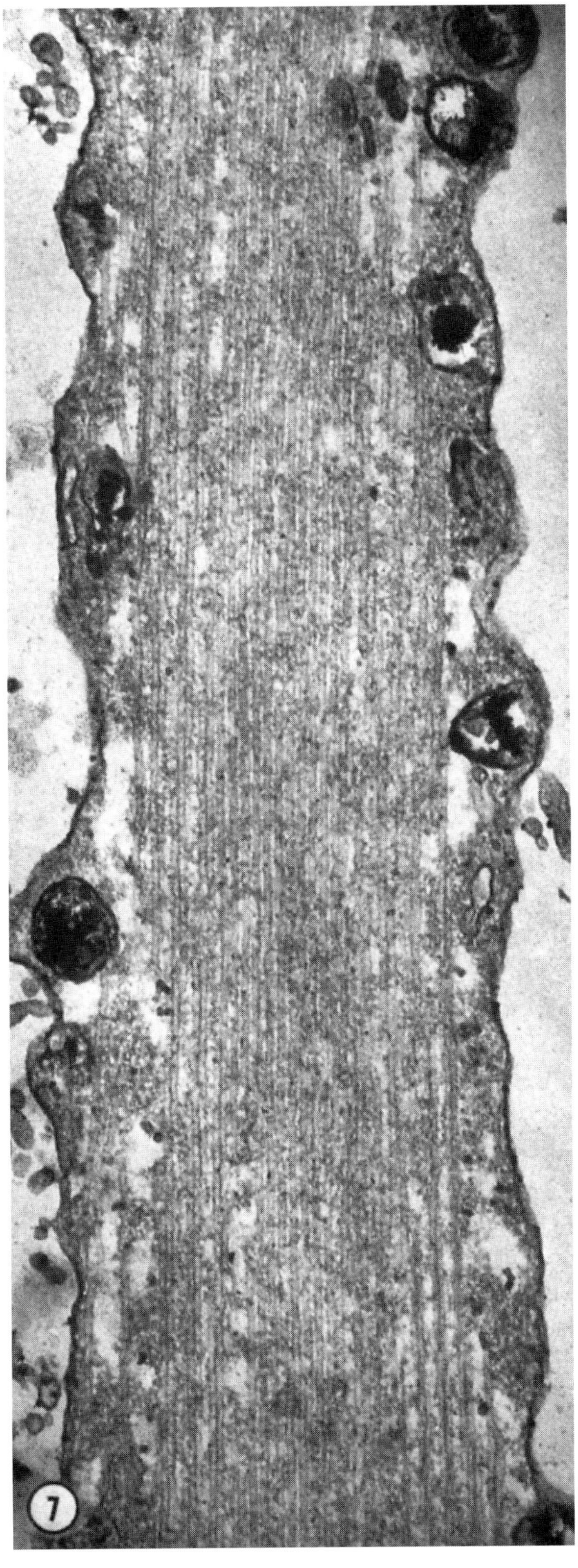

this movement, related to the axopodia, has been suggested (Tilney and Porter, 1965).

Through a careful comparison of Fig. 1 *a* with Fig. 4 *c* (micrographs of the same organism, which had been slightly compressed under a cover slip so that it could not change position), it is possible to show that the reforming axopodia appear in the same number and in approximately the same positions as before the cold treatment. These comparisons are technically difficult because a slight change in the focus will bring axopodia in a lower plane of focus into focus and those that were formerly in focus out of focus. When this is compensated for, one can match an axopodium before cold treatment to the same axopodium following cold treatment with reasonable certainty.

b. ELECTRON MICROSCOPIC OBSERVATIONS: Cells were fixed during the initial stages of axopodial reformation (about 10–15 min following removal from the cold). By this time, the axonemes were well developed and consisted of a bundle of 220-A microtubules (Figs. 6 and 7). These elements in cross-section were organized into their characteristic double-coiled configuration (Fig. 6) with an amorphous material present between the interlocking coils. (This material is present in untreated specimens as well.) In some cases irregularities or gaps in the double coil were seen, but, in all instances, microtubules were found in the reforming axopodia; never did we find a reforming axopodium without microtubules. Furthermore, these elements were invariably arranged parallel to the long axis of the process (Fig. 7). The diameter of the axoneme or the greatest number of microtubules present at the widest point, the axopodial base, was, in specimens fixed in the initial stage of recovery, seldom more than ½ that present in untreated specimens. (Absolute numbers of microtubules would be meaningless here, for there is considerable variation not only in the maximum number of microtubules in the axoneme in untreated organisms but also in the speed of recovery.) When recovery was completed, the number of microtubules was equivalent to that found in untreated specimens.

FIGURE 7 Longitudinal section through a recovering axopodium. Dense granules are present just peripheral to the axoneme. The latter is made up of microtubules arranged parallel to each other and to the long axis of the axopodium. × 56,000.

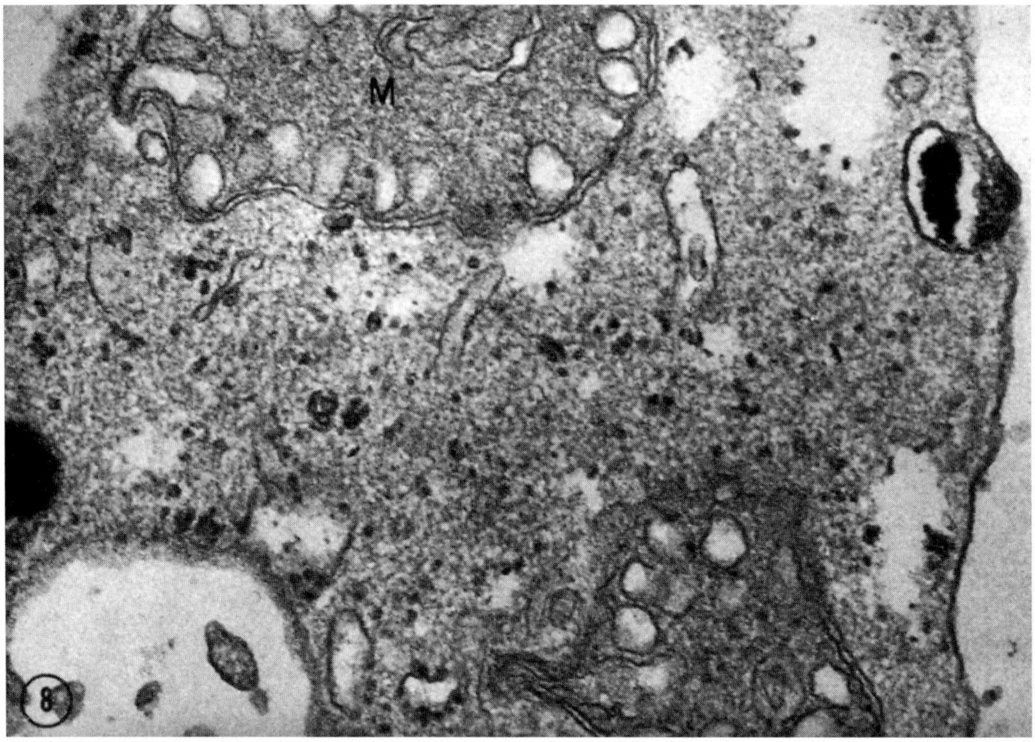

FIGURE 8 Portion of the cortex of the cell body from an organism in the early stages of recovery. Of particular interest here is the "matrix substance" in which ribosomes, mitochondria (*M*), and an occasional dense granule are embedded. This dense material is prominent in cells which have been fixed during cold treatment or in the early stages of recovery. We presume that it is derived from the breakdown of the axoneme. × 43,000.

Besides the microtubules which penetrate the medullary region of the cell body at the basal end of the axoneme, small numbers of microtubules, unassociated with one another, were also encountered in this zone of the cytoplasm. The 310–360-A tubular elements, on the other hand, were no longer present; the matrix substance was still abundant and found throughout the cytoplasm (Fig. 8).

DISCUSSION

The Role of the Microtubules in the Formation and Maintenance of the Axopodia

The observations presented above demonstrate that during the cold treatment the axonemes disappear and the axopodia withdraw so that after 2¼ hr at 4°C few axopodia and thus few axonemes remain. Thin sections of cells fixed at this time or after longer periods in the cold show that the microtubules which make up the axoneme and thus appear to relate to the birefringence of the axoneme in living material, disassemble and do not reassemble until the specimens are returned to room temperature.

Within a few minutes after returning the cells to 22°C (room temperature), the axonemes begin to reappear and the axopodia to reform. 30–45 min later, axopodia of normal length are encountered. In thin sections of axopodia from cells fixed during the early stages of this recovery period, microtubules are found in each reforming axopodium; they are organized into the semicrystalline array characteristic of untreated specimens and extend from the axopodial tip to the medullary region of the cell body. Reforming axopodia without axonemes are never encountered.

From these observations and the fact that the

limiting membrane alone is not known to support rigidly such long protoplasmic extensions in any other cell type, we conclude that the microtubules are intimately involved not only with the maintenance of the axopodia but also with their growth. Thus, if the microtubules are destroyed, these linear cell extensions ought to retract and not reform until these tubular units are reassembled. In work presented elsewhere, two other agents also antimitotic in nature, namely hydrostatic pressure (Tilney et al., 1966) and colchicine (Tilney, 1965 *a*, *b*), give similar results. In higher organisms as well, the microtubules appear to play a similar role (Tilney and Gibbins, 1966). It is, therefore, not unreasonable to suggest that these tubular elements may not only supply through their elongation the force necessary for axopodium regrowth but also may give form and rigidity to each newly extended axopodium.

The Relationship of the Microtubules to the Plasmagel

The work of Marsland and coworkers over the last 30 yr has demonstrated that protoplasmic gel structures undergo solational weakening if exposed to hydrostatic pressure or low temperature. In a recent study, Tilney et al. (1966) reported that the microtubular component which makes up the axonemes of Actinosphaerium disassembled when the cells were subjected to hydrostatic pressure and from this study suggested that the microtubules may be an important factor in the development of the plasmagel.

Porter (1966) has suggested that the microtubules may also act to shape and orient the plasmagel with which they are coexistent. Thus by organizing the surrounding matrix, the microtubules may define or enclose regions of plasmasol, regions in which cytoplasmic movement may occur. As pointed out in an earlier paper in this series (Tilney and Porter, 1965), the cytoplasmic movement of particles proceeds in linear paths or tracks similar to the tracks described in the movement of pigment granules in chromatophores (Bikle et al., 1966) or by the chromosomes in the mitotic spindle. One of the first reactions of the cells to cold temperature is a beading of the axopodia or a conversion of regions of plasmagel into nearly spherical solated volumes. One effect of this is that the tracts or paths through which the particles normally move no longer exist owing to the loss of the gelated region represented by the axoneme. This means, in our interpretations, that now the streaming particles, disintegration products of the microtubules, and material associated with the microtubules accumulate into local regions which we recognize in the light microscope as "beads."

Microtubules and Other Fibrous Proteins: Similarities in the Assembly and Disassembly

In this study, we have demonstrated that the polymerization of microtubules is favored by increasing the temperature, depolymerization by reducing the temperature. Inoué (1952; 1964) made quantitative measurements on the birefringence of the mitotic apparatus and found that an almost linear relationship existed between temperature and birefringence and that increasing the temperature resulted in an increase in birefringence. It is now well known that each spindle fiber contains a number of microtubules (de-Thé, 1964; Robbins and Gonatas, 1964; Roth et al., 1966; Harris, 1962); and recently Kane and Forer (1965) have demonstrated that a correlation exists between birefringence and microtubule number, both in isolated spindles (Kane and Forer, 1965) and after treatment with heavy water (Kane, 1966, personal communication). Thus, it appears that, at least in Actinosphaerium and in the mitotic apparatus, increasing the temperature favors polymerization of the microtubules. Furthermore, polymerization of the microtubules appears to involve an increase in volume (Tilney et al., 1966). Even though electron microscopy on mitotic spindles has not been carried out on cells fixed during the application of hydrostatic pressure, light microscope studies (Pease, 1941; Zimmerman and Marsland, 1964) have demonstrated that the mitotic apparatus solates under pressure so that it is probable that the microtubules in the mitotic spindle act as those in Actinosphaerium and depolymerize when the pressure is increased. Similar results have been observed on globular to filamentous transformation in F actin (Ikkai et al., 1966), tobacco mosaic virus protein polymerization (Lauffer, 1962), and in aggregation of protein and crystal formation in sickle cell anemia (Murayama, 1966). In all these cases as well as microtubule formation, polymerization appears to be an endothermic reaction involving an increase in volume.

In studies on the polymerization of the tobacco mosaic virus protein (Lauffer, 1962) and formation of the mitotic spindle (Inoué, 1964), it has been demonstrated that polymerization involves an increase in enthalpy and a decrease in free energy. In view of the endothermic nature of polymerization, one arrives at the rather unexpected conclusion that the entropy must increase during polymerization. This conclusion is opposed to one's intuitive understanding of thermodynamics, and it indicated to Lauffer that its explanation must be in the release of "something" during polymerization which is bound to the subunits of the polymer. In an elegant set of experiments he was able to show that this "something" was water.

Since the proteins of tobacco mosaic virus, of sickle cells, of the mitotic spindle, and of the axoneme are similar both in certain structural proportions and in behavioral characteristics (they all assemble into microtubules), it is not unreasonable to suggest that the mechanisms for polymerization of these proteins may be similar, that is, involving the release of water. Therefore, any change in the characteristics of water should affect polymerization or depolymerization, and one is not surprised that heavy water (D_2O) stabilizes the mitotic spindle (Marsland and Zimmerman, 1965) and appears to increase not only the birefringence (Inoué, 1964) but also the number of microtubules in the mitotic apparatus (Kane, 1966, personal communication). An increase in axoneme birefringence and number of microtubules in the axopodia was likewise observed when Actinosphaerium was subjected to heavy water (unpublished observations).

It may be useful to carry this analysis one step further and to suggest that each microtubule polymerizes by the addition at its end (or ends) of subunits, i.e., micelles which assemble one unit after another in the single position always available in a continuous coil. In the case of the tobacco mosaic virus (Casper, 1963), there appear to be 16½ subunits per turn while in the microtubules of plants and animals, where it is more difficult to work out the polymerization steps, the best evidence suggests that there are only 13 micelles per turn (Ledbetter and Porter, 1964; Gall, 1965). Further work on the polymerization of microtubules following in vitro isolation of the subunits should obviously be carried out.

Other proteins form a tubular component as well and possess some of the characteristics already mentioned. For example, the assembly of flagellin to form the bacterial flagella (Köffler, 1966; Kerridge et al., 1962) appears to be a similar case. (See Lauffer's reviews (1962, 1964) for other examples.)

The Significance of the 340-A Cold-Produced Tubules

The appearance of two types of structures in the cytoplasm during cold treatment, namely the 310–360-A tubular units and the amorphous material designated as matrix substance, and the simultaneous disappearance of the 220-A microtubules lead us to conclude that these new components are related to the breakdown of the axoneme. (This conclusion is not unreasonable since in untreated cells the microtubules make up a sizable proportion of the cytoplasm in the axopodium. We have calculated that the axonemes of Actinosphaerium comprise roughly 10% of the total bulk of the cytoplasm. These are, in turn, composed of microtubules and an amorphous material around the microtubules. Furthermore, in cold-treated cells changes in other cytoplasmic constituents do not occur.) One immediately wonders if the two types of tubules are related. Does the 220-A tubule transform into the 340-A tubular unit?

This question cannot be definitively answered until such a time as microtubules have been isolated in a pure fraction and made to disassemble and repolymerize in vitro. Nevertheless, all the facts, to be enumerated presently, point to the interpretation that the 220-A microtubules transform into the 340-A tubular units. A mechanism for this transformation is suggested, although it is recognized that other interpretations are possible.

It seems to us that there are five pertinent facts: 1) the 340-A tubules appear during the period of disappearance of the 220-A microtubules; 2) the 340-A tubule is 50% larger in diameter than the 220-A microtubule, yet is still a tubule; 3) no intermediate-sized tubular profiles are present, only the 220- and 340-A; 4) the wall thicknesses of both the 220- and the 340-A tubules are similar, about 50 A; 5) high resolution microscopy of the 340-A tubule reveals that there is a diagonal striation on the wall of the tubule. This banding makes an angle of approximately 45° with the long axis of the tubule and resembles the red stripes on a barber pole.

Before developing our interpretation of the above facts, we should mention several bits of information extracted from the literature which pertain to our interpretation. First, Pease (1963) and André and Thiéry (1963) have shown that the microtubules in sperm tails which had been dried on a grid and negatively stained tend to fray out at their tips; this reveals that the wall of each tubule is, in reality, composed of a series of 11 or more filaments about 40–50 A in diameter. Furthermore, the wall of microtubules in plant cells (Ledbetter and Porter, 1964), in neurons (Gall, 1965) and in dividing cells (Barnicot, 1966) has now been shown to be made of 9–13 filaments. (The number of filaments, although appearing to be variable, may be and probably is constant in the latter three cases. Two procedures were used here: thin sections and negatively stained preparations. In the latter, because the microtubules are dried down on the grid, it is difficult to reconstruct in three dimensions the number of subunits for, upon flattening of the tubule, superposition of the filaments will occur. We, therefore, put more confidence in the thin sections of Ledbetter and Porter and feel that in plant cells, in the mitotic apparatus, and in neurons, the wall is made up of 13 subunits.) Secondly, there appears to be a substructure to these filaments such that each filament can be considered to be composed of a series of globular subunits each about 40–50 A in length (Grimstone and Klug, 1966).

One interpretation of the above-mentioned data is that the 340-A tubule is formed by twisting the 220-A microtubule in such a way that adjacent

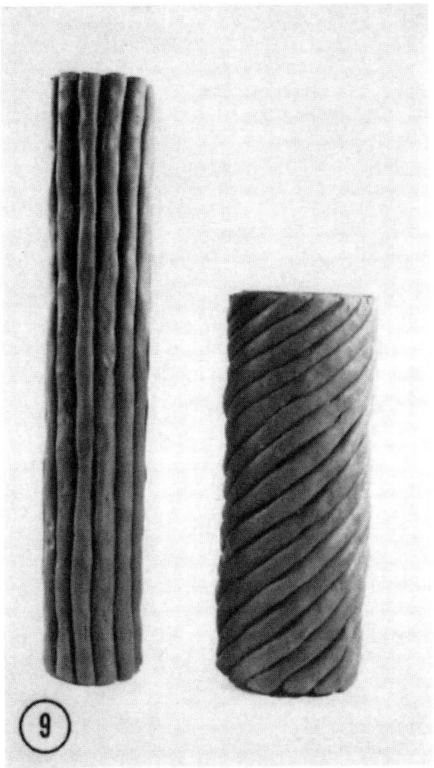

FIGURE 9 Plasticine model showing our conception of the transformation of the 220-A microtubule into the 340-A tubule produced during the cold treatment. The wall of each microtubule appears to be composed of filamentous units.

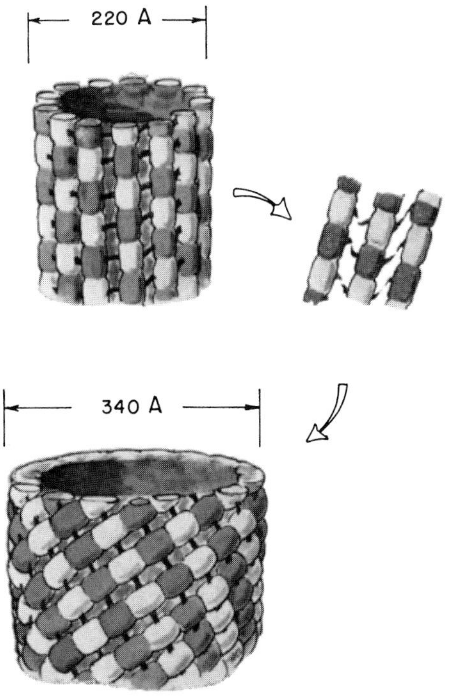

FIGURE 10 This drawing depicts our conception of how the 340-A tubule is formed. We suggest that the globular subunits which make up each filament which, in turn, aggregate to form the wall of the microtubule have discrete bonding sites along their surfaces. This is depicted in this drawing by the *Black* connecting bars. When the microtubules begin to transform into the 340-A tubular units, these bonds break and the filaments begin to twist and at the same time to slide past each other. Once the filaments have slid past one another by approximately 50-A, then the bonding sites on adjacent globular subunits along the length of the filaments are in register again so bonding can occur. It does. When this happens, the filaments lie at a 45° angle to the long axis of the tubular unit. Thus, each globular subunit has slid only one subunit past its neighbor on an adjacent filament.

filaments which make up the wall of the tubule slide relative to each other; at the same time, the diameter of the tubule, owing to the twisting motion, increases by 50%. It can, in fact, be demonstrated with models that if each of the filaments were to slide 50 A relative to the subunit on its left, then a 340-A tubule would be formed having a 50-A wall thickness and a banding at 45° to the long axis of the tubule. The 220- and 340-A tubules, as we envisage them, are illustrated in the clay model seen in Fig. 9.

There is one other fact so far unaccounted for by our interpretation, and that is that tubules intermediate in size between 220 and 340 A were not found. The explanation for this in terms of the model is illustrated in Fig. 10. We assume that there are discrete lateral bonding sites on the subunits such that there is only one position per subunit in which lateral bonding with an adjacent subunit can succeed. Thus, when the filaments slide past one another, the only place at which they can bond again is where the subunits on the filaments come in register. In this position, the filaments would lie at 45° to the long axis of the tubule and the diameter will have increased by exactly 50%.

Another possible interpretation which will fit all the data mentioned above is that the 220-A microtubule may break down in the cold into its component filaments, or into its subunits, and then reassemble as a 340-A tubule. We do not wish to give the impression that the 340-A tubular elements present in a cold-treated cell could account for the total 220-A microtubule population present prior to cold treatment. Rather, this tubular component may account for only a small number of the microtubules in an untreated or a fully recovered cell. The matrix substance may account for the remainder.

No 340-A tubular units were present in cells fixed during the initial stages of axopodial reformation. This indicates that these units either break down or reform into the 220-A microtubules or both during the early stages of recovery.

We wish to thank Mrs. Helen Lyman for her drawings in this paper and in the other papers in this series.

This study was supported by a grant to Dr. Keith R. Porter from the United States Public Health Service. Training Grant, 5TI GM-707.

Received for publication 20 January 1967.

BIBLIOGRAPHY

André, J., and J. P. Thiéry. 1963. Mise en evidence d'une sous structure fibrillaire dans les filaments axonematiques des flagelles. *J. Micr.* **2**:71.

Arnold, J. M. 1966. On the occurrence of microtubules in the developing lens of the squid, *Loligo pealii*. *J. Ultrastruct. Res.* **14**:534.

Barnicot, N. A. 1966. A note on the structure of spindle fibres. *J. Cell Sci.* **1**:217.

Behnke, O. 1965. Further studies on microtubules. A marginal bundle in human and rat thrombocytes. *J. Ultrastruct. Res.* **13**:469.

Bikle, D., L. G. Tilney, and K. R. Porter. 1966. Microtubules and pigment migration in the melanophores of *Fundulus heteroclitus* L. *Protoplasma.* **61**:322.

Byers, B., and K. R. Porter. 1964. Oriented microtubules in elongating cells of the developing lens rudiment after induction. *Proc. Natl. Acad. Sci.* **52**: 1091.

Caspar, D. L. D. 1963. Assembly and stability of the tobacco mosaic virus protein. *Advan. Protein Chem.* **18**:37.

Fawcett, D. W., and F. Witebsky. 1964. Observations on the ultrastructure of nucleated erythrocytes and thrombocytes with particular reference to the structural basis of their discoidal shape. *Z. Zellforsch.* **62**:785.

Gall, J. G. 1965. Fine structure of microtubules. *J. Cell Biol.* **27**:32A.

Gibbons, I. R. 1961. The relationship between the fine structure and direction of beat in gill cilia of a lamellibranch mollusc. *J. Biophys. Biochem. Cytol.* **11**:179.

Gonatas, N. K., and E. Robbins. 1965. The homology of spindle tubules and neuro-tubules in the chick embryo retina. *Protoplasma.* **59**:377.

Grimstone, A. V., and A. Klug. 1966. Observations on the substructure of flagellar fibres. *J. Cell. Sci.* **1**: 351.

Harris, P. 1962. Some structural and functional aspects of the mitotic apparatus in sea urchin embryos. *J. Cell Biol.* **14**:475.

Ikkai, T., T. Ooi, and H. Noguchi. 1966. Actin: volume change on transformation of G-form to F-form. *Science.* **152**:1756.

Inoué, S. 1952. Effect of temperature on the birefringence of the mitotic spindle. *Biol. Bull.* **103**:316.

Inoué, S. 1964. Organization and function of the mitotic spindle. *In* Primitive Motile Systems. R. D. Allen and N. Kamiya, editors. Academic Press Inc., New York. 549.

KANE, R. E., and A. FORER. 1965. The mitotic apparatus. Structural changes after isolation. *J. Cell Biol.* **25**(3):Part II, 31.

KERRIDGE, D., R. W. HORNE, and A. M. GLAUERT. 1962. Structural components of flagella from *Salmonella typhimurium*. *J. Mol. Biol.* **4**:227.

KITCHING, J. A. 1964. The axopods of the sun animalicule, *Actinophrys sol* (Heliozoa). *In* Primitive Motile Systems. R. D. Allen and N. Kamiya, editors. Academic Press Inc., New York. 445.

KÖFFLER, H. 1966. Formation of flagella-like filaments by self-assembly of flagellin molecules. *Proc. Roy. Micr. Soc.* **166**:68.

LAUFFER, M. A. 1962. Polymerization-depolymerization of tobacco mosaic virus protein. *In* The Molecular Basis of Neoplasm. University of Texas Press, Austin. 180.

LAUFFER, M. A. 1964. Protein-protein interaction: endothermic polymerization and biological processes. *In* Proteins and Their Reactions. H. W. Schultz and A. F. Anglemier, editors. Avi Publishing Company, Westport, Conn. Chapter 5.

LEDBETTER, M. C., and K. R. PORTER. 1963. A "microtubule" in plant cell fine structure. *J. Cell Biol.* **19**:239.

LEDBETTER, M. C., and K. R. PORTER, 1964. Morphology of microtubules of plant cells. *Science.* **144**: 872.

LOOPER, J. B. 1928. Observations on the food reactions of *Actinophrys sol. Biol. Bull.* **54**:485.

LUFT, J. H. 1961. Improvements in epoxy resin embedding methods. *J. Biophys. Biochem. Cytol.* **9**: 409.

MCKINNON, D. L. 1909. The optical properties of the contractile elements in Heliozoa. *J. Physiol.* **38**:254.

MANTON, I. 1964. Preliminary observations on spindle fibres at mitosis and meiosis in Equisitum. *J. Roy. Micr. Soc.* **83**:471.

MARSLAND, D. A., and A. M. ZIMMERMAN. 1965. Structural stabilization of the mitotic apparatus by heavy water in the cleaving eggs of *Arbacia punctulata. Exptl. Cell Res.* **38**:306.

MASER, M. D., and C. W. PHILPOTT. 1964. Marginal bands in nucleated erythrocytes. *Anat. Record.* **150**: 365.

MOSES, M. J. 1966. Chromosome reorganization and migration during spermateleosis in a Coccid (*Steatococcus tuberculatus*). *J. Cell Biol.* **31**:78A. (Abstr.)

MURAYAMA, M. 1966. Molecular mechanism of red cell "sickling." *Science.* **153**:145.

PEASE, D. C. 1941. Hydrostatic pressure effects upon the spindle figure and chromosome movement. I. Experiments on the first mitotic division of Urechis eggs. *J. Morphol.* **69**:405.

PEASE, D. C. 1963. The ultrastructure of flagellar fibrils. *J. Cell Biol.* **18**:313.

PITELKA, D. R. 1963. Electron-Microscopic Structure of Protozoa, The Macmillan Co., New York.

PORTER, K. R. 1966. Cytoplasmic microtubules and their function. *In* Principles of Biomolecular Organization. G. E. W. Wolstenholme and M. O'Connor, editors. CIBA Foundation Symposium. J. and A. Churchill, London. 308.

PORTER, K. R., M. C. LEDBETTER, and S. BADENHAUSEN. 1964. The microtubule in cell fine structure as a constant accompaniment of cytoplasmic movement. Electron Microscopy, 1964. Proceedings of 3rd European Regional Conference on Electron Microscopy. M. Titlbach, editor. Prague Czechoslavak Academy of Sciences. Volume B. 119.

REYNOLDS, E. S. 1963. The use of lead citrate at high pH as an electron-opaque stain in electron microscopy. *J. Cell Biol.* **17**:208.

ROBBINS, E., and N. K. GONATAS. 1964. The ultrastructure of a mammalian cell during the mitotic cycle. *J. Cell Biol.* **21**:429.

ROTH, L. E. 1956. An electron microscope study of the cytology of the protozoan *Euplotes patella*. *J. Biophys. Biochem. Cytol.* **3**:985.

ROTH, L. E. 1964. Motile systems with continuous filaments. *In* Primitive Motile Systems. R. D. Allen and N. Kamiya, editors. Academic Press Inc., New York. 527.

ROTH, L. E., and E. W. DANIELS. 1962. Electron microscopic studies of mitosis in Amebae. II. The giant ameba, *Pelomyxa carolinensis*. *J. Cell Biol.* **12**: 57.

ROTH, L. E., H. J. WILSON, and J. CHAKROBOITY. 1966. Anaphase structure in mitotic cells typified by spindle elongation. *J. Ultrastruct. Res.* **14**:460.

RUDZINSKA, M. A. 1965. The fine structure and function of the tentacle in *Tokophrya infusionum*. *J. Cell Biol.* **25**:459.

TAYLOR, A. C. 1966. Microtubules in the microspikes and cortical cytoplasm of isolated cells. *J. Cell Biol.* **28**:155.

DE-THÉ, G. 1964. Cytoplasmic microtubules in different animal cells. *J. Cell Biol.* **23**:265.

TILNEY, L. G. 1965 *a*. Microtubules in the asymmetric arms of Actinosphaerium and their response to cold, colchicine, and hydrostatic pressure. *Anat. Record.* **151**:426.

TILNEY, L. G. 1965 *b*. Microtubules in the Heliozoan *Actinosphaerium nucleofilum* and their relation to axopod formation and motion. *J. Cell Biol.* **27**: 107A. (Abstr.)

TILNEY, L. G., and J. R. GIBBINS. 1966. The relation of microtubules to form differentiation of primary mesenchyme cells in *Arbacia* embryos. *J. Cell Biol.* **31**:118A (Abstr.)

TILNEY, L. G., Y. HIRAMOTO, and D. MARSLAND. 1966. Studies on the microtubules in Heliozoa.

III. A pressure analysis of the role of these structures in the formation and maintenance of the axopodia of *Actinosphaerium nucleofilum* (Barrett). *J. Cell Biol.* **29**:77.

TILNEY, L. G., and K. R. PORTER. 1965. Studies on the microtubules in Heliozoa. I. Fine structure of *Actinosphaerium* with particular reference to axial rod structure. *Protoplasma.* **60**:317.

WATSON, M. L. 1958. Staining of tissue sections for electron microscopy with heavy metals. *J. Biophys. Biochem. Cytol.* **4**:475.

ZIMMERMAN, A. M., and D. MARSLAND. 1964. Cell division: Effects of pressure on the mitotic mechanisms of marine eggs (*Arbacia punctulata*). *Exptl. Cell Res.* **35**:293.

Summers K.E. and **Gibbons I.R.** 1971. Adenosine triphosphate-induced sliding of tubules in trypsin-treated flagella of sea-urchin sperm. *Proc. Natl. Acad. Sci.* **68:** 3092–3096.

THE BENDING OF CILIA AND FLAGELLA had been shown to correlate with a sliding of neighboring microtubules (1), suggesting that axonemes, like muscles, worked by a sliding filament mechanism (2). The Gibbons lab had isolated and characterized an ATPase, called dynein, that bound to axonemal microtubules and was plausibly the motor for their motions. This lab had also identified ATP-containing buffers that both optimized dynein's ATPase activity and reactivated demembranated cilia to swim in a very lifelike manner (3). The paper reproduced here used dark field optics to observe the behavior of demembranated cilia that had been digested briefly with trypsin under conditions that would normally promote axonemal bending. The sensitivity of the optics allowed the authors to watch and record the behavior of one or only a few microtubules as the axoneme disintegrated by a sliding of each microtubule over its neighbors, resulting in a many-fold elongation of the original structure. These observations put dynein-driven sliding of axonemal microtubules on a firm experimental footing. Subsequent work, by both biochemical and genetic approaches, demonstrated that dynein is the essential motor for axoneme motility (reviewed in 4), though we now know that dynein is actually a family of many related molecules, each composed of multiple polypeptides. The details of axoneme motility and morphogenesis are sufficiently complex that they are still the subjects of active study (e.g., 5–7). The paper reprinted here was a key advance, however, because it described the motile behavior of single microtubules, visualized with high contrast, high sensitivity optics. This microscopic approach anticipated the developments that led to the characterization of microtubule polymerization (Mitchison and Kirschner, this volume) and the discovery of kinesin (Vale et al., this volume).

---

1. Satir P. 1968. Studies on cilia. III. Further studies on the cilium tip and a "sliding filament" model of ciliary motility. *J. Cell Biol.* **39:** 77–94.
2. Brokaw C.J. 1971. Bend propagation by a sliding filament model for flagella. *J. Exp. Biol.* **55:** 289–304.
3. Gibbons I.R. 1965. Reactivation of glycerinated cilia from *Tetrahymena pyriformis*. *J. Cell Biol.* **1:** 400–402.
4. Luck D.J.L. 1984. Genetic and biochemical dissection of the eucaryotic flagellum. *J. Cell Biol.* **98:** 789–794.
5. Cole D.G., Diener D.R., Himelblau A.L., Beech P.L., Fuster J.C., and Rosenbaum J.L. 1998. *Chlamydomonas* kinesin-II-dependent intraflagellar transport (IFT): IFT particles contain proteins required for ciliary assembly in *Caenorhabditis elegans* sensory neurons. *J. Cell Biol.* **141:** 993–1008.
6. Pazour G.J., Dickert B.L., and Witman G.B. 1999. The DHC1b (DHC2) isoform of cytoplasmic dynein is required for flagellar assembly. *J. Cell Biol.* **144:** 473–481.
7. Porter M.E., Bower R., Knott J.A., Byrd P., and Dentler W. 1999. Cytoplasmic dynein heavy chain 1b is required for flagellar assembly in *Chlamydomonas*. *Mol. Biol. Cell.* **10:** 693–712.

Reprinted from
Proc. Nat. Acad. Sci. USA
Vol. 68, No. 12, pp. 3092–3096, December 1971

Adenosine Triphosphate-Induced Sliding of Tubules in Trypsin-Treated Flagella of Sea-Urchin Sperm

(motility/microtubule/cilia/sliding filament model/axonemes)

KEITH E. SUMMERS AND I. R. GIBBONS

Department of Biochemistry and Biophysics, and Pacific Biomedical Research Center, University of Hawaii, Honolulu, Hawaii 96822

Communicated by George E. Palade, October 12, 1971

ABSTRACT Axonemes isolated from the sperm of the sea urchin, *Tripneustes gratilla*, were briefly digested with trypsin. The digested axonemes retained their typical structure of a cylinder of nine doublet-tubules surrounding a pair of single tubules. The digestion modified the axonemes so that the subsequent addition of 0.1 mM ATP caused them to disintegrate actively into individual tubules and groups. The nucleotide specificity and divalent-cation requirements of this disintegration reaction paralleled those of flagellar motility, suggesting that the underlying mechanisms were closely related. Observations by dark-field microscopy showed that the disintegration resulted from active sliding between groups of the outer doublet-tubules, together with a tendency for the partially disintegrated axoneme to coil into a helix. Our evidence supports the hypothesis that the propagated bending waves of live-sperm tails are the result of ATP-induced shearing forces between outer tubules which, when resisted by the native structure, lead to localized sliding and generate an active bending moment.

Although much is known about the detailed fine structure of the flagellar axoneme (1–5), there is relatively little evidence concerning the functional role of this stiucture in the mechanism of flagellar motility. In this paper, we report preliminary observations on the active disintegration produced when ATP is added to axonemes whose structure has been modified by brief digestion with trypsin. By studying the movements of the flagellar tubules during this disintegration, we have been able to obtain information about the nature of the mechanical forces induced by ATP. Our results strongly support the "sliding filament" model of flagellar bending (6, 7).

Axonemes were isolated from sperm of the sea urchin, *Tripneustes gratilla*, by extraction and differential centrifugation in a solution containing 1% (w/v) Triton X-100, 0.1 M KCl, 5 mM $MgSO_4$, 1 mM ATP, 1 mM dithiothreitol, 0.5 mM EDTA, and 10 mM Tris-phosphate buffer, pH 7.0, (Gibbons and Fronk, manuscript in preparation). The resultant preparations contained about equal numbers of intact axonemes (50 μm long) and of shorter fragments formed by shearing. The centriole was visible, by dark-field microscopy, as a brighter granule at one end of many axonemes. A substantial proportion of the axonemes are capable of normal motility if diluted into reactivating solution at pH 8.0 (see above).

The suspension of axonemes was centrifuged and resuspended at a concentration of about 0.6 mg of protein/ml in Tris–Mg solution (2.5 mM $MgSO_4$, 0.1 mM dithiothreitol, 30 mM Tris·HCl, pH 7.8), at room temperature. Sufficient trypsin was then added to give a trypsin to axonemal protein ratio of about 1 to 1500. The course of the digestion was monitored by measurement of the turbidity of the suspension at 350 nm as a function of time. In routine preparations, further digestion was stopped by the addition of an excess of soybean trypsin-inhibitor after the turbidity had decreased to about 80% of its initial value. This suspension of digested axonemes could be stored at 0°C for up to 2 days, and constituted the starting material for most of our experiments. In some cases, the trypsin and inhibitor were removed by centrifugation of the digested axonemes and resuspension in fresh Tris–Mg solution.

Preliminary electron-microscopic examination of the digested preparations showed that the cylindrical structure of nine outer doublet tubules surrounding the two central-tubules remained largely intact in most axonemes (Fig. 1). Comparison with undigested preparations indicated that only slight structural changes resulted from digestion. The most apparent change was the disruption of the radial spokes near the point where they normally connect to the sheath surrounding the central tubules. Disruption of the nexin links that connect adjacent doublets (3, 5) was not directly apparent in the micrographs, but was revealed upon dialysis of digested axonemes against low concentrations of EDTA. The dynein arms on the doublets (3) appeared to be relatively resistant to the digestion. More detailed descriptions of the effects of digestion on the fine structure of the axoneme will be published elsewhere.

Although the trypsin digestion did not itself destroy the basic structure of the axoneme, it modified it in such a way that the structure became highly sensitive to ATP. Addition of a low concentration of ATP to the digested preparation caused a rapid disintegration of the axonemes into individual doublet-tubules and small groups (Fig. 2). The dynein arms and a portion of the spoke were still present on most of the separated doublets. The central tubules usually remained together as pairs, but they did not appear to be associated with the groups of doublets.

The specificity of the reaction responsible for this axonemal disintegration was examined by the use of the decrease in turbidity of the suspension, measured at 350 nm, as an assay of disintegration. With concentrations of ATP between 50 μM and 1 mM, complete disintegration of the axonemes occurred within 30 sec, and the decrease in turbidity amounted to about 50%. With lower concentrations of ATP, the decrease was smaller and slower to develop, amounting to about 25% after 3 min with 10 μM ATP. The reaction was highly

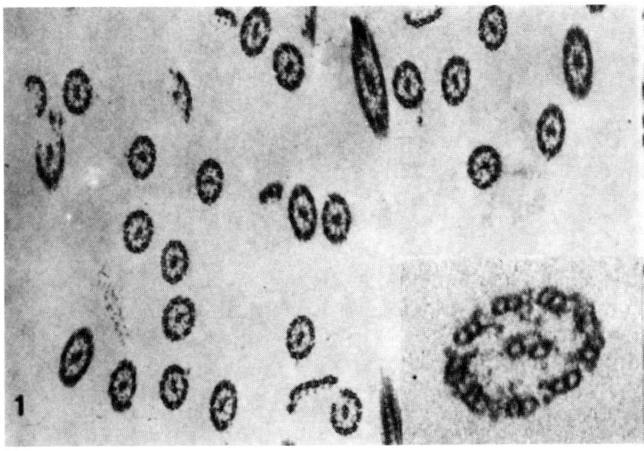

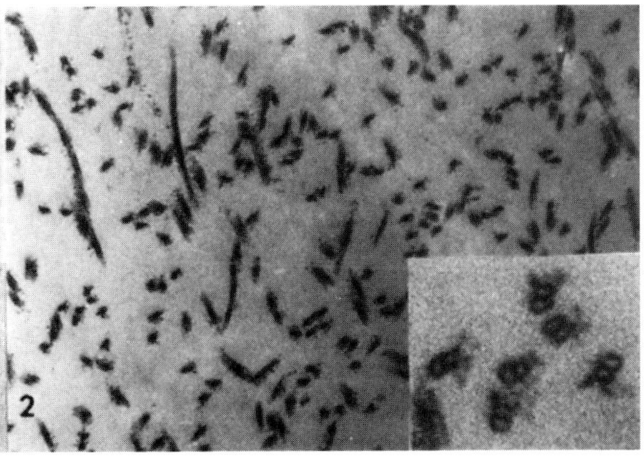

FIG. 1. Electron micrograph of trypsin-treated axonemes. The samples were fixed with glutaraldehyde and post-fixed with osmium tetroxide. After samples were embedded in epoxy resin, thin sections were cut, and then stained with uranyl acetate and lead citrate. ×27,000; insets ×109,000.

FIG. 2. The same preparation of trypsin-treated axonemes as in Fig. 1, but 0.5 mM ATP was added 3 min before fixation.

specific for ATP; other nucleotides, including GTP, CTP, and ITP, gave only a 3–5% decrease in turbidity, even when added at a concentration of 1 mM.

In addition to ATP, the presence of a divalent cation was required for the reaction to occur. When the digested axonemes were suspended in 0.1 M KCl, 0.5 mM EDTA, 30 mM Tris buffer, pH 7.8, the addition of 0.1 mM ATP gave no change in turbidity; subsequent addition of 2.5 mM $MgSO_4$ to the suspension gave a 40% decrease. $MnSO_4$ was about as effective as $MgSO_4$ in activating the response. Activation with 2.5 mM $CaCl_2$ gave a 27% decrease, but the response was slower than with $MgSO_4$ and took about 3 min to develop.

The manner in which the digested axonemes disintegrated was studied in more detail by dark-field microscopy. Under these conditions, structures probably as small as individual doublet-tubules were visible. When the doublets were in groups, they could not be resolved individually, but the approximate number in the group could be estimated from the relative intensity of its image. Photographs were taken at a magnification of ×210, with a 5–10 sec exposure. Because of Brownian movement, only structures attached to the coverglass could be photographed.

A suspension of the digested axonemes in Tris–Mg solution was placed in a trough 0.2 mm deep on a microscope slide and covered with a coverglass, except for a small opening at one end. A small volume of 0.5 mM ATP was placed into the opening, and evaporation and diffusion then drew the ATP slowly across the suspension. The time of disintegration of individual axonemes varied from a few seconds to several minutes, depending on such factors as the steepness of the ATP gradient and its rate of progression.

The overall disintegration of the axoneme appeared to result partly from active sliding movements between groups of tubules, and partly from the tendency for the partially disintegrated structure to coil into a helix. The sliding movements could be seen most clearly in short- to medium-length fragments of axoneme that initially were stuck to the coverglass along much of their length. In Fig. 3, the process begins with the protrusion of a group of tubules from the upper end of the axoneme. This protrusion gradually extends further and begins to coil around onto itself. Simultaneously with the growth of the protrusion, a region of reduced intensity extends upward from the lower end of axoneme, as a result of the smaller number of tubules remaining in this region. The lower end of the structure remains in a constant position on the coverglass. In the fourth micrograph of the series, two distinct steps of intensity are visible in the lower part of the structure, indicating the presence of two groups of tubules moving upward at different speeds. At the end of the disintegration, the protruded groups of tubules are partly outside the focal plane, but they appear to be at least as long as the initial axoneme and to have undergone further separation. One small group of the tubules has remained behind in the position of the original axoneme. The disintegration of a second, somewhat longer, axonemal fragment is shown in Fig. 5. In this case a large group of tubules is sliding toward the bottom-right of the micrograph. Although the free, forward end of the sliding group soon coils around out of the plane of focus, the progress of the group can be followed from the movement of its rear end as it moves downward, leaving behind it a stationary group of a few tubules attached to the coverglass. In the last micrograph of the series, the forward end of the sliding group has coiled completely around and come back into the focal plane, while the rear end has moved a total distance of about 25 μm, about equal to the length of the original axoneme. Further disintegration of the group into a tangle of mainly individual doublet-tubules occurred subsequently. No further movement was observed in the separated tubules.

In both of the above examples, the sliding groups of tubules left behind them a small stationary group of tubules attached to the coverglass in the same position as the original axoneme. It seems reasonable to suppose that these stationary tubules are the ones by which the axoneme had been attached to the coverglass, and that they formed the fixed surface relative to which the active sliding of the other tubules occurred. Where two groups of sliding tubules were observed in a given axoneme, they were most often moving in the same direction along the axoneme, with different speeds (Fig. 3). However, we have occasionally observed two groups of tubules moving in opposite directions relative to a third stationary group. The total length produced by sliding is sometimes greater than five times that of the original axonemal fragment.

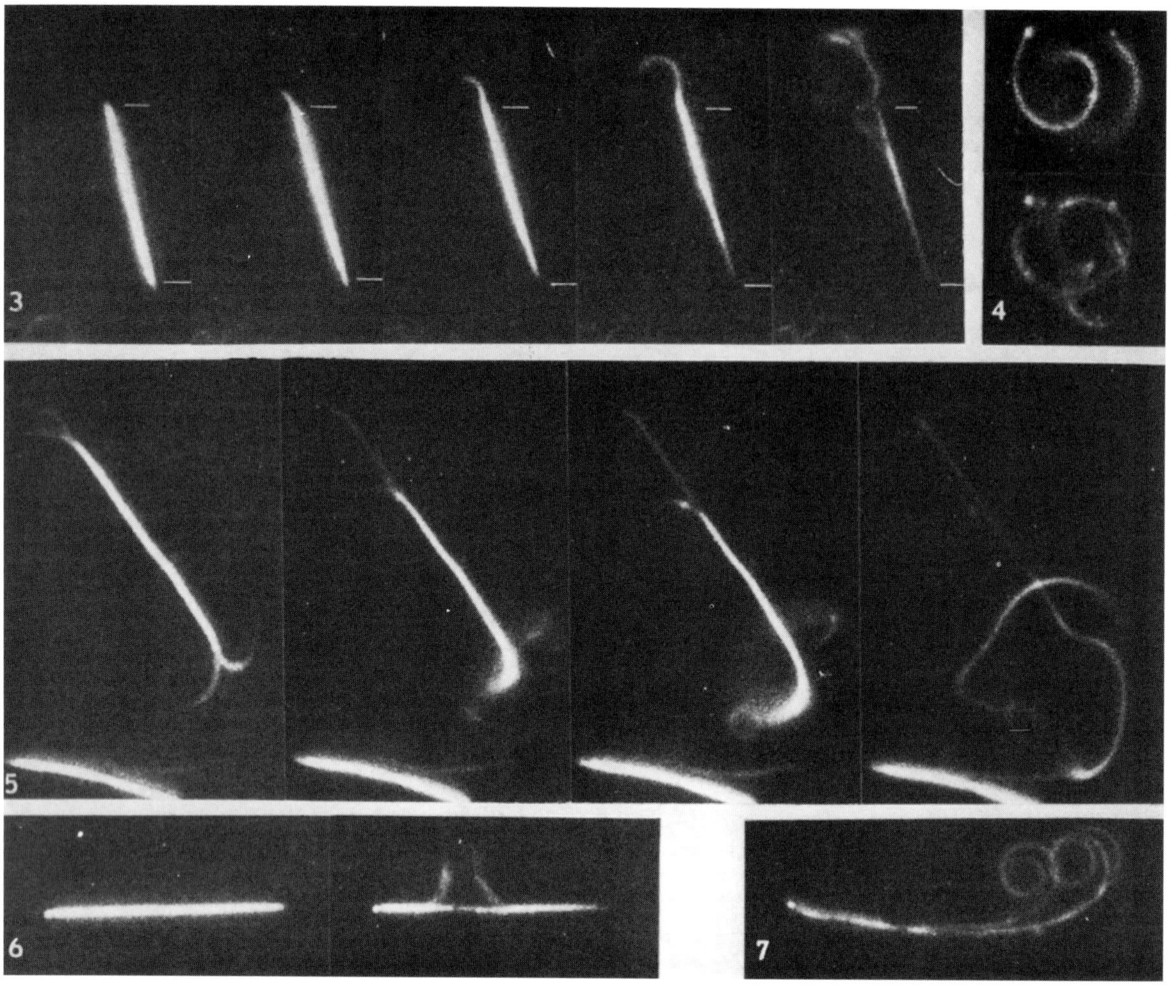

FIGS. 3–7. Dark-field light micrographs of trypsin-treated axonemes reacting to ATP. The successive micrographs in each series were taken at intervals of 10–30 sec. The *white lines* in Fig. 3 indicate the initial position of each end of this fragment. The centriole is visible in the *upper-left* corner of Fig. 4, and at the *left end* of the axonemes in Figs. 6 and 7. For further details see *text*. ×1730.

The disintegration of full-length axonemes and of fragments derived from the centriolar end is modified by the apparent presence of a block to sliding in the region of the centriole. With full-length axonemes that are attached to the coverglass, disintegration usually begins with the opening of a split about midway along their length. The group of tubules constituting one side of the split gradually billows out to form a large loop, while the group on the other side remains in about the original position on the coverglass (Fig. 6). As the size of the initial billow increases, the end of a moving group of tubules can be seen sliding up along the axoneme from the distal end toward the split, suggesting that the billow is a consequence of the force generated by active sliding in the region distal to the split. Further disintegration occurs subsequently, and the final result is a tangle of tubules, which is often still held together at the centriolar end. Short fragments of axoneme derived from the centriolar end are relatively resistant to disintegration, and they can often be found intact when full-length axonemes and distal fragments have disintegrated.

The disintegration of axonemes that are floating free in solution is difficult to study in detail, because they rotate irregularly through the focal plane as disintegration proceeds. In most cases, free axonemes first bend near the middle, and then coil into helices of rather uniform pitch (Fig. 4). The helices subsequently break down into balls of tangled tubules. No motion is seen once the separation into individual doublets is complete.

As mentioned above, partially disintegrated axonemes show a marked tendency to coil. This is illustrated by the helices formed in free-floating axonemes and by the coiling of the groups of tubules protruded from axonemes attached to the coverglass. Another expression of this tendency is seen in some axonemes, where disintegration occurs by means of one or more groups of tubules peeling back along the length of the axoneme, coiling up as they do so (Fig. 7). On occasion, we have observed what appear to be individual doublet-tubules peeling back in this way. In many axonemes coiling or peeling occurred simultaneously with sliding, giving complex patterns of disintegration. The relative predominance of sliding or coiling appeared to differ from one preparation to another, and it may be affected by the exact degree of digestion.

DISCUSSION

Our observations indicate that the principal factors responsible for the disintegration of the digested axonemes are the active sliding movements induced by ATP between groups of tubules, and the tendency for the partially disrupted structure to coil up. Although such other factors as length, attachment to the coverglass, presence or absence of the centriole, and exact degree of digestion exert a modifying influence on the detailed manner in which particular axonemes disintegrate, they are of less general significance.

The sliding movements that occur between groups of tubules indicate the presence of an active force induced by ATP, and are of particular interest because they seem likely to be closely related to the mechanism underlying normal motility. The force that tends to cause sliding might result either from the interaction of the outer doublets with the central sheath and tubules, or from the interaction of each outer doublet with its neighboring doublet. Interaction with sheath and tubules seems less probable, because the interaction would presumably have to occur through the spokes, and these structures appear to be the ones most damaged by the trypsin digestion. Moreover, it would be difficult to explain how the final length after sliding can be greater than three times the initial length. The second possibility, that the shearing forces derive from the interactions between adjacent doublets, appears more probable. In this case the interaction would occur through the dynein arms, and these appear to be relatively resistant to digestion. The close similarity between the specificity requirements of dynein ATPase (8) and those of the disintegration reaction provides support for believing that the dynein arms play an active mechanochemical role in inducing sliding.

We have no direct evidence concerning the detailed mechanism by which the dynein arms might induce sliding. One possibility would be for the binding and hydrolysis of ATP to cause a cyclic change in the angle of the arms, which, coordinated with the repeated making and breaking of their attachment to successive binding sites along the length of the B-tubule, would result in the arms "walking" one tubule along the other. Such a mechanism would be analogous to that thought to occur in muscle, where cyclic movements of the bridges on the myosin filaments cause sliding relative to the actin filaments (9).

If the interaction between tubules occurs through the arms, this implies that the arms on one doublet are capable of binding chemically to specific sites on the B-tubule of the adjacent doublet, and thus forming crossbridges between the two doublets. The gap of 5–10 nm usually observed in electron micrographs between the arms and the adjacent B-tubule could be a real separation and perhaps form part of the mechanism for coordination of the interactions between tubules (see below), although the possibility of this gap being an artefact resulting from the failure of glutaraldehyde fixation to preserve the interaction between the arm and B-tubule cannot be neglected.

Since the sliding movements between groups of tubules frequently continue for almost their whole length, a distance of 25 μm or more, the polarity of the structures generating the sliding must maintain a constant direction along the length of each doublet. Considerations of symmetry incline us to think that the direction of this polarity is probably the same on all nine doublets, but our observations on the patterns of sliding are not yet adequate to determine whether or not this is the case.

The coiling of partly disintegrated axonemes may be partly an indirect effect of the ATP-induced forces between tubules, but it seems unlikely that this is a complete explanation for this mechanism fails to account for the coiling of individual doublet-tubules. Further work is needed to elucidate the factors involved in this coiling. The coiling of the flagellar tubules is reminiscent of that of bacterial flagella (10), although the resemblance may be no more than coincidental.

The conditions necessary for obtaining the disintegration response closely match those necessary for obtaining propagated bending waves in undigested axonemes. In both cases, there is a high degree of specificity for ATP; other nucleotides are essentially inactive. The concentration of ATP that gives half-maximal disintegration (as judged by the decrease in turbidity) is about the same (10 μM) as the minimum concentration necessary for motility in undigested axonemes. The divalent-cation requirements are also similar, and can be satisfied by Mg^{++} or Mn^{++} or, less effectively, by Ca^{++} (ref. 11; Gibbons, B. H., manuscript in preparation). This similarity in conditions suggests that the active process underlying the ATP-induced disintegration of the trypsin-treated axonemes is closely related to that which produces bending waves in normal flagella.

Two general types of model have been proposed to explain the mechanism of flagellar bending. In the "local contraction" model, the bending of a short segment of flagellum is a direct result of the active contraction of longitudinal elements on one side of the flagellum within that segment (12–14), while in the "sliding filament" model, bending occurs as a result of active sliding between noncontractile longitudinal elements (1, 6, 7). The sliding filament model has been favored in recent theoretical work because it provides a better explanation for the fact that the propagation velocity of flagellar bending waves is constant, and does not change with the varying local viscous—resistive—moment they encounter (15–17). Brokaw has recently shown that the relationship between sliding and bending in such a model provides a possible mechanism for the generation of propagated bending waves (7).

Our demonstration that ATP causes active sliding between tubules in trypsin-treated flagella, together with Satir's earlier finding that the length of ciliary tubules remains constant during bending (6), provides strong experimental support for the general validity of the sliding filament model.

On this basis, the propagated bending waves of normal flagella can reasonably be explained as the result of ATP-induced shearing forces between adjacent doublet-tubules, which, when opposed by the elastic resistance of the native structure, lead to localized sliding and generate an active bending moment (7). We postulate that the trypsin treatment of the axonemes damages those structural components that are responsible for coordinating the shearing forces and for providing the elastic resistance that normally restrains sliding, with the result that unlimited sliding can occur, leading to complete disintegration of the axoneme. Since the spokes and the nexin links appear to be the components most susceptible to damage by trypsin, we would suggest that they may form part of the coordinating and resistive structures. The possibility that the trypsin modifies significantly the properties of the dynein itself must also be considered.

The coordination necessary to produce propagated bending waves implies that the active process that generates the forces between tubules must be sensitive to some local property of the wave, such as its curvature or the amount of shear in the structure. There is no direct evidence as to how active forces generated by the dynein arms, which are arranged in a basically helical pattern in the axoneme, could be coordinated to produce a planar bending wave. However, since the direction of bending is correlated with the plane of the central tubules (18, 19), it seems possible that the central tubules and sheath, together with the spokes that normally connect them to the outer doublets, are involved in the coordination process. The gap of 5–10 nm usually observed between the arms and the adjacent B-tubule could also form part of the coordination mechanism by preventing effective interaction of the arms with their corresponding sites on the B-tubule until the gap was closed by the distortions resulting from curvature or shear. Our evidence tentatively suggests that the active process can generate a force in only one direction between a given pair of tubules, but it is not easy to see how such a unidirectional force could give rise to almost symmetrical, planar waves, and the possibility of a reverse force being generated under appropriate conditions has to be considered. The differences reported in the structure and chemical properties of the two rows of dynein arms on each doublet (2, 20) suggest that the production of forces between tubules is complex. The largest bend angles normally observed in flagella are about 3 radians (ref. 21; Gibbons, I. R. manuscript in preparation), and bends of this magnitude indicate that sliding can amount to about 400 nm between tubules on opposite sides of the axoneme, with at least 100 nm between adjacent doublets, if the tubules are inextensible. This much sliding is considerably greater than can be accommodated by the spokes and nexin links, unless they have considerable ability to stretch or unless their links to the tubules are broken and remade. The further possibility that the tubules themselves undergo a significant amount of elastic compression and extension during extreme bending must also be considered. Although our observations provide strong support for the general basis of the "sliding filament" model, it seems likely that further evidence regarding the coordination and polarity of the active forces between tubules will be necessary before it will be possible to present a reasonably detailed hypothesis relating this theoretical model to the actual structure of the axoneme.

We thank Dr. C. J. Brokaw for sending us a copy of his manuscript before publication. This work has been supported in part by NIH grant GM15090, and NSF traineeship GZ-1986.

1. Afzelius, B. A., *J. Biophys. Biochem. Cytol.*, **5**, 269 (1959).
2. Allen, R. D., *J. Cell Biol.*, **37**, 825 (1968).
3. Gibbons, I. R., *Arch. Biol.*, **76**, 317 (1965).
4. Gibbons, I. R., and A. V. Grimstone, *J. Biophys. Biochem. Cytol.*, **7**, 697 (1960).
5. Stephens, R. E., *Biol. Bull.*, **139**, 438 Abs. (1970).
6. Satir, P., *J. Cell Biol.*, **39**, 77 (1968).
7. Brokaw, C. J., *J. Exp. Biol.*, in press (1971).
8. Gibbons, I. R., *J. Biol. Chem.*, **241**, 5590 (1966).
9. Huxley, H. E., and W. Brown, *J. Mol. Biol.*, **30**, 383 (1967).
10. Asakura, S., G. Eguchi, and T. Iino, *J. Mol. Biol.*, **16**, 302 (1966).
11. Brokaw, C. J., *Exp. Cell Res.*, **22**, 151 (1961).
12. Machin, K. E., *J. Exp. Biol.*, **35**, 796 (1958).
13. Brokaw, C. J., *Nature*, **209**, 161 (1966).
14. Lubliner, J., and J. J. Blum, *J. Theor. Biol.*, **31**, 1 (1971).
15. Sleigh, M. A., *Symp. Soc. Exp. Biol.*, **22**, 131 (1968).
16. Rikmenspoel, R., and M. A. Sleigh, *J. Theor. Biol.*, **28**, 81 (1970).
17. Brokaw, C. J., *J. Exp. Biol.*, **53**, 445 (1970).
18. Gibbons, I. R., *J. Biophys. Biochem. Cytol.*, **11**, 179 (1961).
19. Tamm, S. L., and G. A. Horridge, *Proc. Roy. Soc. Ser. B.*, **175**, 219 (1970).
20. Linck, R. W., *Biol. Bull.*, **139**, 429 Abs. (1970).
21. Brokaw, C. J., *J. Exp. Biol.*, **45**, 113 (1966).

Pollard T.D. and **Korn E.D.** 1973. *Acanthamoeba* myosin. I. Isolation from *Acanthamoeba castellanii* of an enzyme similar to muscle myosin. *J. Biol. Chem.* **248:** 4682–4690.

By the early 1970s, it was well established that actin was prevalent in essentially all eukaryotic cells (Lazarides and Weber, this volume), but research on myosin outside of muscle cells was less well developed. The work reproduced here took advantage of a micro-organism that was relatively easy to grow in large quantities to explore the possibility that myosin too could be found in non-muscle cells. The definition of myosin used was biochemical: an ATPase whose activity was stimulated by polymerized actin. By this criterion, a "myosin" was identified and purified, but its characteristics were surprising: it was too small and soluble to resemble the myosin that everyone knew at the time. For some years this "mini-myosin" appeared to be an orphan, and it was generally regarded as an enzyme peculiar to soil amebae. Similar enzymes, now called myosin 1, were later discovered in vertebrate tissues, e.g., the brush border at the apical pole of intestinal epithelial cells (1) and then in other micro-organisms (2). As the technology for cloning genes was applied to the study of myosin, a Pandora's box of diversity opened. It has become clear that myosin is really a superfamily, with members specialized for a variety of tasks that are still being identified and studied (3). In these many enzymes a basic mechanochemical motif, the myosin head, has been passed around to make a set of motors that can be specialized for a wide range of tasks, including cell movement, cytokinesis, vesicle traffic, and the maintenance of cell architecture. The paper reproduced here used straightforward biochemical methods to open up a field that will long have importance for students of cell organization and movement. Moreover, a companion to this paper (4) identified a cofactor protein that was required for actin filaments to activate the myosin's ATPase activity. This cofactor now has a homolog in p-21 activated protein kinase (PAK), an enzyme that participates in a variety of signal transduction pathways (5).

---

1. Collins J.H. and Borysenko C.W. 1984. The 110,000-dalton actin- and calmodulin-binding protein from intestinal brush border is a myosin-like ATPase. *J. Biol. Chem.* **259:** 14128–14135.
2. Jung G., Saxe C.L., Kimmel A.R., and Hammer J.A.D. 1989. *Dictyostelium discoideum* contains a gene encoding a myosin I heavy chain. *Proc Natl. Acad. Sci.* **86:** 6186–6190.
3. Mermall V., Post P.L., and Mooseker M.S. 1998. Unconventional myosins in cell movement, membrane traffic, and signal transduction. *Science* **279:** 527–533.
4. Pollard T.D. and Korn E.D. 1973. *Acanthamoeba* myosin. II. Interaction with actin and with a new cofactor protein required for actin activation of Mg 2^+ adenosine triphosphatase activity. *J. Biol. Chem.* **248:** 4691–4697.
5. Brzeska H., Szczepanowska J., Hoey J., and Korn E.D. 1996. The catalytic domain of *Acanthamoeba* myosin I heavy chain kinase. II. Expression of active catalytic domain and sequence homology to p21- activated kinase (PAK). *J. Biol. Chem.* **271:** 27056–27062.

THE JOURNAL OF BIOLOGICAL CHEMISTRY
Vol. 248, No. 13, Issue of July 10, pp. 4682-4690, 1973
Printed in U.S.A.

Acanthamoeba Myosin

I. ISOLATION FROM *ACANTHAMOEBA CASTELLANII* OF AN ENZYME SIMILAR TO MUSCLE MYOSIN*

(Received for publication, December 18, 1972)

THOMAS D. POLLARD‡ AND EDWARD D. KORN

From the National Heart and Lung Institute, Laboratory of Biochemistry, Section on Cellular Biochemistry and Ultrastructure, Bethesda, Maryland 20014

SUMMARY

An ATPase that accounts for about 0.3% of the total cell protein has been isolated in 90% purity from *Acanthamoeba castellanii*. The enzyme has been identified as a myosin-like ATPase by the following criteria. Maximal enzymatic activity occurs in the presence of EDTA and 0.5 M KCl, and is only 10 to 20% as high in the presence of Ca^{2+} and less than 1% in the presence of Mg^{2+}. The Mg^{2+} ATPase is activated by actin under physiological conditions and the enzyme binds to actin filaments in the absence but not in the presence of ATP. *Acanthamoeba* myosin shows some activity toward nucleoside triphosphates other than ATP but does not hydrolyze ADP or AMP. In contrast to myosins from other sources, *Acanthamoeba* myosin is soluble under physiological conditions, and its partition coefficient on gel filtration suggests a native molecular weight of about 180,000, which is smaller than other myosins. Several different experiments show that the isolated molecule is not formed by proteolytic cleavage of a larger native protein. The purified enzyme consists of three polypeptide chains with molecular weights of 140,000, 16,000, and 14,000. The amino composition is remarkable for the absence of cysteine.

This and the accompanying paper (4) are the first detailed reports on a myosin-like enzyme from the soil amoeba, *Acanthamoeba castellanii*. They form part of our continuing investigation of the molecular events generating the forces of ameboid movement. Although our ultimate goal is to understand movement at the cellular level, our immediate approach has been to purify and characterize the presumptive "motility" proteins because the mechanism of movement of an intact cell is clearly too complicated to understand without extensive biochemical and physical knowledge of the molecules involved.

Previous investigations in this laboratory have established the presence in *Acanthamoeba* of actin filaments (5), determined their general distribution within the cell (6, 7), provided evidence for the apparent association of some actin filaments with the plasma membrane (8, 9), and described in some detail the chemical, physical, and biological properties of the pure protein (5, 6). *Acanthamoeba* actin, although not identical to muscle actin, is very similar to it in amino acid composition and general sequence (5, 10), molecular weight (10), and ability to undergo polymerization to F-actin filaments to which muscle heavy meromyosin will bind in characteristic fashion (6) and which will activate the Mg^{2+}ATPase of muscle heavy meromyosin (5).

We now show that *Acanthamoeba* also contains an enzyme that shares many of the properties of muscle myosin. Both have ATPase activity that is partially activated by Ca^{2+}, maximally activated by K^+ in the presence of EDTA, and not activated by Mg^{2+}; muscle actin strongly activates the Mg^{2+} ATPase of both enzymes; and both enzymes bind reversibly to filaments of F-actin. We call the enzyme from the amoeba *Acanthamoeba* myosin, because it has these properties in common with muscle myosin. Other properties of the amoeba enzyme are unique among known myosins. *Acanthamoeba* myosin is smaller than myosins from other sources with a molecular weight of less than 200,000, it probably consists of one large and two small polypeptides, and a newly discovered "cofactor" protein is required for actin to activate its Mg^{2+} ATPase. In spite of these significant and interesting differences the properties of *Acanthamoeba* myosin are consistent with the proposal that it and *Acanthamoeba* actin act in concert with other proteins to generate force for cell motility by a mechanism that is closely related to that utilized for muscle contraction.

This paper describes the purification of *Acanthamoeba* myosin and some of the physical, chemical, and enzymatic properties of the purified protein. The accompanying paper (4) describes the interaction of *Acanthamoeba* myosin with muscle F-actin and the partial purification and characterization from *Acanthamoeba* of a cofactor protein required for the Mg^{2+} ATPase of the muscle actin-*Acanthamoeba* myosin complex.

EXPERIMENTAL PROCEDURE

Materials—Actin from the back and leg muscles of rabbits was purified free from troponin-tropomyosin using either gel filtration (11) or KCl washes (12). Imidazole, iodoacetamide, sodium dodecyl sulfate, ATP, ADP, ITP, and creatine phosphate were purchased from Sigma Chemical Company. Dithio-

* Some of the data in this paper have been published in preliminary reports (1–3).

‡ Present address, Department of Anatomy, Harvard Medical School, Boston, Massachusetts 02115.

threitol and GTP were purchased from Calbiochem Corp. AMP and CTP were products of P-L Biochemicals. Guanidine hydrochloride, ammonium sulfate, and urea were the ultrapure grade from Schwarz-Mann. Other chemicals were reagent grade, and deionized water was used throughout.

Culture Procedures—Acanthamoeba castellanii (Neff strain) was grown at 29° in 15-liter carboys as described (5) with the following modifications which result in healthier, faster growing cells. (*a*) Stock cultures were maintained in aerated cultures tubes (13). (*b*) Ten to twenty milliliters of stock culture were used to inoculate 1-liter Fernbach culture flasks, which were aerated by shaking continuously (14). (*c*) Single 4-day Fernbach cultures were used to inoculate the carboys. (*d*) The carboys were aerated with compressed air (not oxygen as before) and grew rapidly with a doubling time of about 15 hours. After 4 days the cells, which consisted of less than 2% cysts, were harvested by low speed centrifugation and were washed twice with 0.1 M NaCl. Average yield was about 100 g/15 liters.

ATPase Assay—Measurements were made at 29° (the optimal growth temperature of the cells) in 1.0 ml containing 0.5 M KCl, 2 mM ATP, 10 mM imidazole chloride, pH 7.0, and either 2 mM EDTA, 10 mM $CaCl_2$, or 10 mM $MgCl_2$. Activity under these conditions will be referred to as K^+-EDTA ATPase, Ca^{2+}-ATPase, or Mg^{2+} ATPase, respectively. Inorganic phosphate was measured by a modification of the method of Martin and Doty (15). The reaction was stopped and the protein precipitated by adding a mixture of 2 ml of 1:1 2-butanol-benzene with 0.5 ml of 3 N H_2SO_4 and 4% silicotungstic acid, and vortexing for about 1 s. Vortexing was repeated for 10 to 15 s after adding 0.2 ml of 10% ammonium molybdate. After the phases separated, 1.0 ml of the organic (upper) phase was added to 2.0 ml of 0.73 N H_2SO_4 in ethanol. The yellow phosphomolybdate complex was then reduced by adding 0.1 ml of a solution of stannous chloride prepared by diluting 1 ml of 10% stannous chloride in 12 N HCl with 25 ml of 1 N H_2SO_4. The blue color was read at 720 nm and characteristically gave an absorbance of 3.8 per μmole of phosphate.

When the samples to be assayed contained inorganic phosphate a radioactive assay was used. [γ-^{32}P]ATP-(New England Nuclear) was added in trace amounts to the nonradioactive 2 mM ATP. The reaction was stopped and the phosphomolybdate was extracted into 2 ml of isobutanol-benzene as usual. Then 0.2 ml of the organic phase was added to 15 ml of Aquasol (New England Nuclear), and radioactivity was measured in a Beckman LS-250 liquid scintillation spectrometer.

Protein Concentration—Where possible, protein concentration was estimated by the method of Lowry *et al.* (16) using bovine serum albumin as the standard. Alternatively, in very dilute samples of purified fractions freed of nucleotides, protein concentration was estimated from the absorbance at 280 nm, using $A_{280}^{1\%} = 6.5$. This approximate absorbancy was determined with highly purified (but not homogeneous) fractions of *Acanthamoeba* myosin by comparing the absorbance at 280 nm with the protein concentration estimated by the Lowry reaction (16).

Gel Electrophoresis—Proteins were reduced in a solution of 5 M guanidine hydrochloride, 50 mM DTT,[1] and 0.2 M NH_4HCO_3, pH 8.0, for 2 hours at 37° and alkylated for 15 min at 22° with 0.2 M recrystallized iodoacetamide at pH 8.0. The reaction was stopped by adding 2-mercaptoethanol to 2 M and the samples were then dialyzed against water and lyophilized. SDS-polyacrylamide gel electrophoresis was carried out according to Fairbanks *et al.* (17). Discontinuous polyacrylamide gel electrophoresis with 8 M urea in the gel and the sample was performed with the buffers described by Davis (18) but without a stacking gel. Instead, the sample was simply layered on the surface of the gel under the chamber buffer which contained no urea. The gels were stained with Coomassie blue according to Fairbanks *et al.* (17).

[1] The abbreviations used are: DTT, dithiothreitol; SDS, sodium dodecyl sulfate.

Isoelectric Focusing—Five per cent polyacrylamide gels with 8 M urea and pH 3 to 8 ampholytes (LKB) were prepared and run according to the procedure of Wrigley (19).

Densitometer Scans—Polyacrylamide gels stained with Coomassie blue were scanned at 550 nm using a gel-scanning accessory on a Beckman ACTA-III spectrophotometer.

Amino Acid Analysis—Reduced and alkylated proteins were hydrolyzed at 105° in 6 N HCl for 18 hours, and the amino acid composition was determined by the single column method (20) on a Beckman 121 amino acid analyzer. The system described by Kuehl and Adelstein (21) was used for the detection of methylated lysines and histidines.

Column Chromatography—Column chromatography and all other steps in the purification procedure were carried out at 4°. DEAE-cellulose (Whatman DE-52) was precycled with HCl and NaOH, and deaerated, and fines were removed according to the manufacturer's directions. Columns were packed using increasing pressures of nitrogen gas (22). Hydroxylapatite was made as described by Levin (23). There was considerable variation in the resolving power of three separate batches of hydroxylapatite made by this method.

As described under "Results," proteins were purified by selective adsorption to agarose at low ionic strength and elution with a salt gradient. For successful adsorption of the *Acanthamoeba* proteins, 8% agarose (Bio-Rad A-1.5m, 200 to 400 mesh) was suspended in 1.0 M KCl for 2 days before pouring into the column and equilibrating with starting buffer (Fig. 3). The capacity of the agarose was much less in one experiment in which the KCl wash was omitted.

When necessary, column flow was maintained with a Buchler peristaltic pump. Column effluents were monitored for absorbance at 280 nm with an ISCO monitor. Salt gradients were made with a Buchler Varigrad using Peterson's guidelines (22). Salt concentrations in column effluents were estimated by measurements of conductivity.

RESULTS

Identification of Myosin-like Activity in Amoeba Extracts—Cell extracts that contain F-actin and myosin usually decrease in viscosity upon the addition of ATP, which dissociates the complex of actin and myosin, and then return to the original viscosity as the ATP is hydrolyzed (24). Weihing and Korn (5) found this did not occur with extracts of *Acanthamoeba* known to contain F-actin. If myosin was present in those extracts, therefore, it must either have been in very low concentration or an atypical myosin. In either case, if a myosin-like enzyme was present it should be detectable by sensitive ATPase assays. We measured the ATPase activity of amoebae extracts in 0.5 M KCl because in that solvent muscle myosin has the property unique among ATPases of being inhibited by Mg^{2+} and activated maximally by EDTA and partially by Ca^{2+}. Homogenates of amoebae had considerable ATPase activity under all three assay conditions but it was readily shown that the Mg^{2+} ATPase and the K^+-EDTA ATPase were different enzymes. Most of the Mg^{2+}ATPase was localized in the mitochondria, microsomes,

TABLE I

Purification of Acanthamoeba myosin

Purification of *Acanthamoeba* myosin from 588 g of amoebae. Details of the procedures are given in the text, Table II, and Figures 3 to 5. SDS-gel electrophoretic analyses of the proteins at each step are shown in Figure 2. These results are representative of a typical preparation. Although there was some variation in the efficiency of each step, the final yield and specific activity of the purified enzyme was quite reproducible. In this experiment, the yield from the agarose column was lower and the yield from hydroxylapatite was higher than usual. Samples containing 2.0 and 4.0 μmoles per min of ATPase were removed for analysis at Steps 4 and 5, accounting for part of the apparent losses.

Step	Volume	Total protein	Total enzyme[a]		Specific activity
	ml	*mg*	*μmoles/min*	*%*	*μmoles/min/mg*
1. Homogenate	2,358	44,500	431	100	0.010
2. 125,000 × *g* supernatant	1,975	22,700	411	96	0.018
3. DEAE-peak	1,080	3,620	112	26	0.033
4. Ammonium sulfate (1.2 to 1.45 M fraction)	35	262	83	19	0.32
5. Agarose adsorption peak	109	8.2	20.6	4.8	2.5
6. Hydroxylapatite peak	50	3.3[b]	11.3	2.6	3.4

[a] K^+-EDTA ATPase was measured at 29° under standard conditions.

[b] Protein concentration was estimated from A_{280} assuming $A_{280}^{1\%} = 6.5$.

and plasma membranes and was sedimentable,[2] while essentially all of the K^+-EDTA ATPase was recovered in the 140,000 × *g* supernatant when pyrophosphate was included in the extraction medium (Table I).

When such high speed supernatants were fractionated by gel filtration in 0.5 M KCl on Sephadex G-200 or agarose (Fig. 1), the K^+-EDTA ATPase eluted as a single major peak with a partition coefficient corresponding to that of a globular protein of molecular weight between 150,000 and 200,000. Other peaks of ATPase activity were also observed on such columns: a consistent small peak of Ca^{2+}-ATPase (Fig. 1) immediately preceding the K^+-EDTA ATPase; and a variable peak of Ca^{2+}-ATPase of much higher molecular weight, usually in the void volume even on 4% agarose but having $K_D \cong 0.2$ in the experiment reproduced in Fig. 1. It seemed likely that these Ca^{2+}-ATPases were, at least in part, aggregated and partially denatured K^+-EDTA ATPase because they reappeared when the K^+-EDTA ATPase fraction isolated by gel filtration was rechromatographed following ammonium sulfate fractionation, isoelectric precipitation, or prolonged storage.

The K^+-EDTA ATPase peak from these gel filtration columns was identified as the sought for *Acanthamoeba* myosin by its ability to interact with purified muscle F-actin. At low ionic strength, muscle F-actin (1 mg per ml) activated the Mg^{2+}-ATPase 2- to 6-fold. Over 85% of the K^+-EDTA and Ca^{2+}-ATPase activities sedimented with the actin when the proteins from the gel filtration peak were mixed with muscle F-actin in 0.5 M KCl, 5 mM $MgCl_2$, 10 mM imidazole chloride, pH 7.0, and centrifuged at 140,000 × *g* for 90 min. In the presence of actin and 2 mM ATP, 90% of the ATPase activity remained in the supernatant and the actin pellet was enzymatically inactive.

[2] T. D. Pollard, R. Weihing, and E. D. Korn, unpublished results.

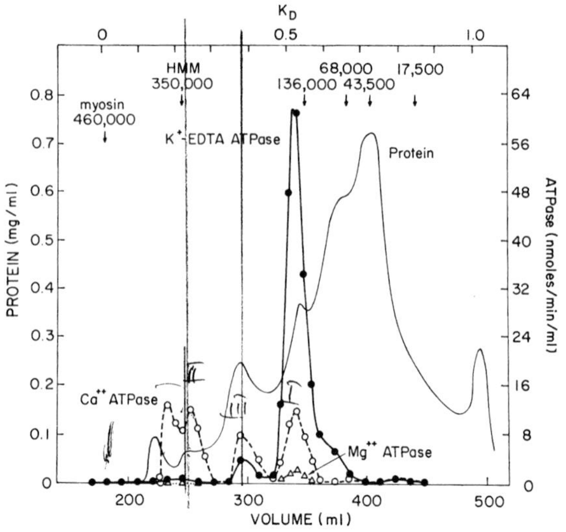

FIG. 1. Agarose gel filtration of *Acanthamoeba* extract. The extract was prepared by homogenizing cells in 2 volumes of 75 mM KCl, 12 mM sodium pyrophosphate, 5 mM DTT, and 30 mM imidazole chloride, pH 7.0, and centrifuging 75 min at 40,000 rpm in a Beckman 40 rotor. Then 7.5 ml of the clear supernatant was applied to a column (2.5 × 98 cm) of Bio-Rad A-1.5m (200 to 400 mesh) and the column was eluted by pumped descending flow at 12 ml per hour with 0.5 M KCl, 1 mM DTT, and 5 mM imidazole chloride, pH 7.0. Fractions of 6.3 ml were collected and analyzed for protein (——) and ATPase activity in the presence of K^+-EDTA (●), Ca^{2+} (○), or Mg^{2+} (△). Recoveries were 102% for K^+-EDTA ATPase and 90% for Ca^{2+}-ATPase. The partition coefficients (K_D) of a number of protein standards are given at the *top* of the graph: muscle myosin (460,000), muscle heavy meromyosin (350,000), bovine serum albumin dimer (136,000), bovine serum albumin monomer (68,000), ovalbumin (43,500), and myoglobin (17,500). The void volume ($K_D = 0$) and salt volume ($K_D = 1.0$) were determined by chromatographing blue dextran and ATP, respectively.

All available evidence suggests that these tests are specific for myosin. There is a report (25) of other enzymes binding to actin filaments, but this binding is apparently nonspecific because it occurs only at very low ionic strength.

Purification of Acanthamoeba Myosin—Standard techniques for precipitating myosin as a complex with actin which have proved useful in purifying myosin from other systems (24, 26–28) did not succeed. Although it was possible to precipitate most of the K^+-EDTA ATPase and *Acanthamoeba* actin at low ionic strength (0.05 M KCl) at pH 6.4 (but not at pH 6.8 or higher), most of the precipitated enzyme was irreversibly aggregated. Ammonium sulfate fractionation of crude extracts also caused irreversible aggregation. By the procedures outlined in Table I, however, it was possible to prepare highly purified *Acanthamoeba* myosin (Fig. 2).

For a typical large scale preparation, 500 to 1,000 g of amoebae were mixed with extraction solution (2 ml per g) containing 0.075 M KCl, 12 mM pyrophosphate, 5 mM DTT, and 30 mM imidazole chloride, pH 7.0. More than 95% of the cells were broken by rapid release from a Parr cell disruption bomb after equilibration for 5 min with N_2 at 300 p.s.i. The homogenate was then centrifuged in one or more Beckman Ti-15 zonal rotors (used as batch rotors) for 3 hours at 35,000 rpm (125,000 × *g*).

The pH of the 125,000 × *g* supernatant was adjusted to 8.0 with a saturated solution of Tris base and dialyzed against two changes of 10 volumes of 7.5 mM pyrophosphate, 0.5 mM DTT,

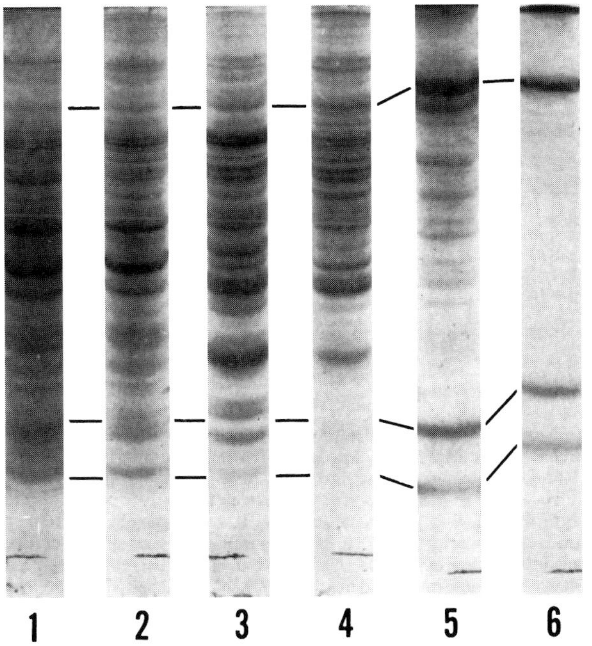

FIG. 2. Polyacrylamide gel electrophoresis in 1% SDS of the proteins at each step in the purification of *Acanthamoeba* myosin. The numbers of the gels correspond to the steps listed in Table 2. *Gels 1* to *4* are 7.5% acrylamide. *Gel 5* is 7.5% acrylamide but run at another time. *Gel 6* is 9% acrylamide. *Horizontal lines* mark the positions of the three *Acanthamoeba* myosin peptides in each gel.

and 10 mM Tris Cl, pH 8.0. Some material containing Ca^{2+}-ATPase activity usually precipitated and was removed by centrifugation. Without pyrophosphate up to 70% of the K^+-EDTA ATPase precipitated during dialysis and could not be solubilized. Several different approaches to DEAE-chromatography were tried but none was completely satisfactory. Continuous gradient elution (Fig. 3), which we usually used, separated the enzyme from nucleic acids and increased its specific activity, but recovery of enzymatic activity was poor. In preliminary small scale experiments the use of step gradient elution gave satisfactory purification and enzyme recoveries up to 70%.

Ammonium sulfate fractionation was carried out on the enzyme peak as eluted from the DEAE-column or after adjusting the eluate to a final concentration of 1 mM ATP, 1 mM $MgCl_2$, and 2 mM DTT. It was necessary to examine all of the ammonium sulfate fractions between 1.0 M and 1.6 M for ATPase activity (Table II) because of slight variations in the fractionation pattern from preparation to preparation.

The partially purified enzyme could be further purified up to 5-fold by gel filtration on agarose in 0.5 M KCl (see Table II in the accompanying paper (4)) but we achieved better results by adsorption of the enzyme to agarose at very low ionic strength. Elution of the column with an appropriate salt gradient gave a 10-fold purification of the enzyme (Fig. 4). If the agarose column was not washed with concentrated salt solutions after each run, its capacity was greatly reduced. Usually about 60% of the enzyme was recovered from the column, and it appeared from the relatively constant specific activity across the peak and from gel electrophoresis (Fig. 2) that the enzyme was approaching purity at this stage.

The final step in the purification of *Acanthamoeba* myosin was

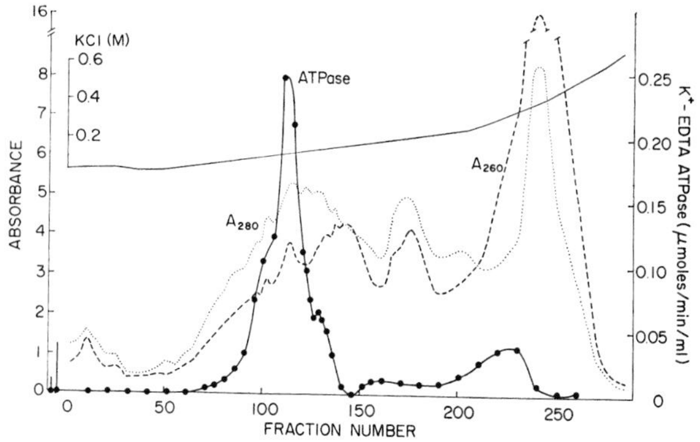

FIG. 3. Ion exchange chromatography on DEAE-cellulose: 1.7 liters of a 125,000 × g supernatant containing a total of 302 µmoles per min of ATPase activity was equilibrated with 7.5 mM sodium pyrophosphate, 0.5 mM DTT, and 10 mM Tris chloride, pH 8.0, by dialysis and then was applied to a column (5 × 80 cm) of Whatman DE-52 equilibrated with 10 mM KCl, 1 mM DTT, and 10 mM Tris chloride, pH 8.0. Two large fractions, I (400 ml) and II (1400 ml), whose A_{280} are shown by *vertical bars*, were collected during the application of the sample. The column was then eluted by a pumped flow of about 200 ml per hour with 4.5 liters of a linear 0.01 M to 0.25 M KCl gradient, followed by 2.5 liters of a linear 0.25 M to 0.8 M KCl gradient, all in 1 mM DTT and 10 mM Tris chloride, pH 8.0. Fractions of 25 ml were collected and analyzed for absorbance at 280 nm and 260 nm and K^+-EDTA ATPase. Recovery of ATPase was 53%. Fractions 108 to 118, which contained 28% of the recovered ATPase, were pooled for ammonium sulfate fractionation.

TABLE II

Ammonium sulfate fractionation

The fractionation was made at 4° by adding appropriate volumes of saturated ammonium sulfate containing 0.01 M EDTA, pH 7.0, to raise the ammonium sulfate concentration to the upper limit of a particular fraction. After standing for 15 min, the precipitate was pelleted at 10,000 × g for 10 min. Pellets were resuspended in 0.2 mM ATP, 1 mM DTT, and 2 mM Tris chloride, pH 7.6, and dialyzed overnight against the same buffer before assaying for K^+-EDTA ATPase under standard conditions.

Fraction assayed	Total protein	Total enzyme	Specific activity
	mg	*µmoles/min*	*µmoles/min/mg*
DEAE-peak (starting material)	1990	101	0.051
Ammonium sulfate fractions			
0 to 1.0 M	130	4	0.031
1.0 to 1.3 M	153	63	0.340
1.3 to 1.5 M	125	23.5	0.177
1.5 to 1.8 M	350	6	0.016
1.8 M supernatant	1230	4	0.001
Total of ammonium sulfate fractions	1988	100.5	

adsorption to a column of hydroxylapatite and elution with a gradient of phosphate buffer (Fig. 5). The recovery from these columns was generally quite good, 75 to 95%, in spite of the fact that the concentration of this labile enzyme was very low in the eluate.

Yield—Normally about 2 to 5% of the K^+-EDTA ATPase activity of the original homogenate was recovered in the final hydroxylapatite fractions as purified *Acanthamoeba* myosin. This amounted to about 0.5 to 1.0 mg of purified enzyme/100 g of cells. From the specific activity of the purified enzyme and

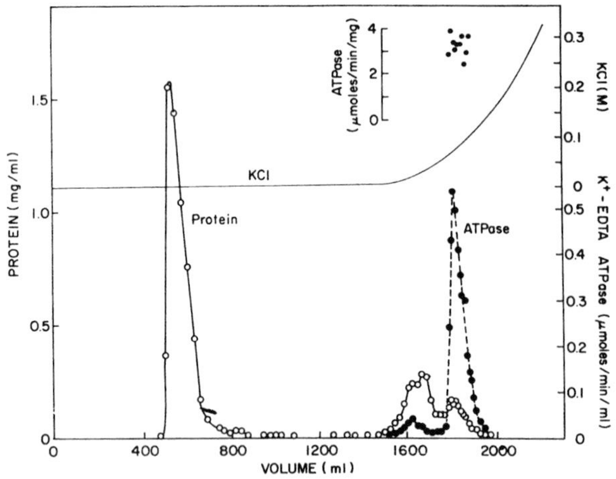

FIG. 4. Agarose adsorption of partially purified *Acanthamoeba* myosin. Proteins purified through the ammonium sulfate step were dialyzed against 0.2 mM ATP, 1 mM DTT, and 2 mM Tris chloride, pH 7.6, and then were applied (in 39.5 ml) to a column (5 × 75 cm) of 8% agarose (Bio-Rad A-1.5m (200 to 400 mesh)) equilibrated with the same buffer. After washing on the sample with 80 ml of starting buffer, the column was eluted with a concave KCl gradient in the same buffer made in a 9 × 100-ml chamber Varigrad with the following molar concentrations of KCl: 0, 0.025, 0.050, 0.075, 0.100, 0.200, 0.300, 0.400, 0.500. This gradient was followed by unbuffered 1.0 M KCl to wash the column. Downward flow of 32 ml per hour was maintained with 50 cm of hydrostatic pressure. Fractions were collected and measured for protein (○), and K^+-EDTA ATPase, expressed as μmoles per min per ml of column fraction (●). Fifty-nine per cent of the applied K^+-EDTA ATPase was recovered in the major *Acanthamoeba* myosin peak and the specific activity, shown in the *upper part* of the figure, increased about 10-fold.

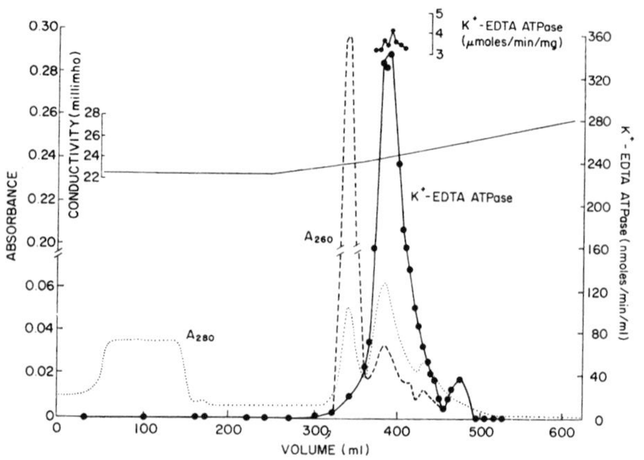

FIG. 5. Hydroxylapatite chromatography of the *Acanthamoeba* myosin peak from the agarose adsorption column (Fig. 4). The 100-ml sample, containing a total of 15.8 μmoles per min of ATPase, and a column (1.5 × 26 cm) of hydroxylapatite were separately equilibrated with 0.5 M KCl, 1 mM DTT, 1 mM phosphate, and 5 mM imidazole chloride, pH 7.0. After sample application, the column was eluted by pumped flow at 30 ml per hour with a concave phosphate gradient made with a 6 × 100-ml Varigrad chamber containing the following molar concentrations of potassium phosphate, pH 7.0, in the above buffer: 0, 0.01, 0.03, 0.05, 0.10, 0.30. The curves of absorbance are redrawn from the monitor record and measurements on individual fractions. The peak of material absorbing at 260 nm is ATP carried over from the agarose column with the sample. The peak of material which was not retarded by the column contained little or no protein. The K^+-EDTA ATPase is given both as activity per ml and as specific activity, with concentration of protein being estimated using $A^{1\%}_{280} = 6.5$. Recovery of ATPase was 94%.

assuming that all of the K^+-EDTA ATPase activity in the homogenate is *Acanthamoeba* myosin (which seems justified by the data in Table I and Fig. 1), about 0.3% of the cell's protein is *Acanthamoeba* myosin. In contrast, actin may account for about 10% of the amoeba's protein (3) so that there is a large molar excess of actin in the cell. The large amount of actin and the relatively small amount of myosin in the amoeba probably account for the difficulty in showing actomyosin-like decreases in viscosity when ATP was added to amoeba homogenates known to contain *Acanthamoeba* actin (5).

Stability—*Acanthamoeba* myosin was reasonably stable after purification through the ammonium sulfate or agarose adsorption steps, retaining up to 80% of its enzymatic activity for 1 week at 4° in 0.1 M KCl, 1 mM DTT, and 10 mM imidazole chloride, pH 7.0, at a concentration of about 0.2 mg per ml. In contrast, *Acanthamoeba* myosin purified through the hydroxylapatite step was difficult to store and lost as much as 10 to 25% of its enzymatic activity per day. Attempts to concentrate the hydroxylapatite fractions by ultrafiltration on Amicon membranes, by vacuum dialysis, or by ammonium sulfate precipitation resulted in losses of enzyme as large as 50%. Lyophilized protein could not be redissolved.

Enzymatic Activity—*Acanthamoeba* myosin catalyzed the hydrolysis of several nucleoside triphosphates but the highest rate was obtained with ATP which is presumably its natural substrate (Table III). The enzyme did not hydrolyze ADP or the several other nucleotides and related compounds that were tested, and ADP inhibited the hydrolysis of ATP (Table III). In experiments with [γ-^{32}P]ATP as the substrate it was shown that only the terminal phosphate was hydrolyzed since the inorganic phosphate released by hydrolysis had the same specific activity (counts per min per μmole) as the ATP substrate.

TABLE III

Substrate specificity of Acanthamoeba myosin

Substrate specificity of purified *Acanthamoeba* myosin (7 μg per ml) was measured in 0.5 M KCl, 2 mM EDTA, and 10 mM imidazole chloride, pH 7.0, at 29°. All rates were constant with time except for that with 2 mM ATP, which began to decline after about 0.3 μmoles of ATP were hydrolyzed, presumably due to inhibition by ADP.

Substrate	Rate of hydrolysis
	μmoles/min/mg protein
ATP (2 mM)	3.20
ITP (2 mM)	0.83
GTP (2 mM)	0.83
CTP (2 mM)	0.83
ADP (2 mM)	0.01
AMP (2 mM)	0
ATP (2 mM) + ADP (1 mM)	1.85
Pyrophosphate (2 mM)	0
Creatine phosphate (2 mM)	0

The effect of temperature on the K^+-EDTA ATPase activity is shown in Fig. 6. For unknown reasons, the activity remained constant at temperatures about 30° which is, perhaps coincidentally, the maximum temperature at which the cells will grow (29). Between 4° and 30° the energy of activation was 9 Cal per mole. The energy of activation of muscle myosin is 12 Cal per mole under these conditions (30).

The ATPase activity of *Acanthamoeba* myosin is affected by

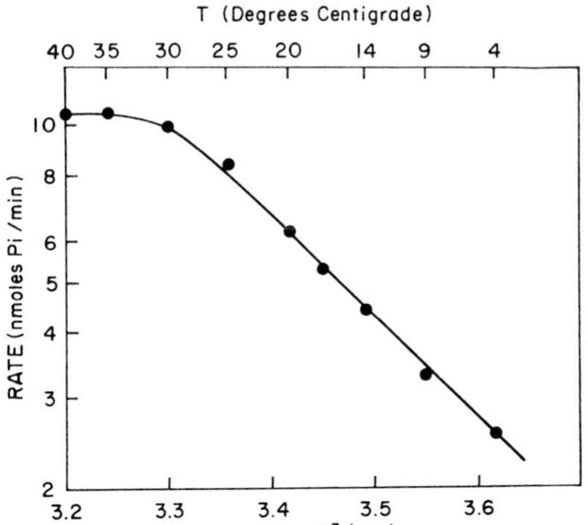

FIG. 6. Arrhenius plot of the temperature dependence of *Acanthamoeba* myosin K^+-EDTA ATPase. The energy of activation calculated from the linear part of the curve was 9 Cal per mole. Above 30° there was no increase in activity, although even at 40° the rate of hydrolysis was constant during the 7.5-min assay. Under standard conditions (29°), this preparation had a specific activity of 2.5 μmoles per min per mg, corresponding to an activity loss of about 30% during dialysis to remove the phosphate from the hydroxylapatite column.

TABLE IV

Effect of ions on ATPase activity of Acanthamoeba myosin

Acanthamoeba myosin was purified by step elution from DEAE-cellulose, ammonium sulfate fractionation, gel filtration, and hydroxylapatite chromatography. The resulting preparation had K^+-EDTA ATPase activity (0.5 M KCl-2 mM EDTA) of 2.2 μmoles per min per mg. Assays were carried out at 28° using 2 mM ATP and 10 mM imidazole chloride, pH 7.0, and the monovalent and divalent ions listed. All rates are expressed relative to the K^+-EDTA ATPase rate.

	2 mM EDTA	10 mM $CaCl_2$	10 mM $MgCl_2$
0.5 M KCl	1.00	0.12	0.01
0.05 M KCl	0.17	0.10	0.01
0.5 M NaCl	<0.01	0.10	

monovalent and divalent cations (Table IV) in very much the same way as muscle myosin (31, 32). In the presence of EDTA, ATPase activity was dependent on K^+, partial activity was retained in the presence of Ca^{2+}, and the enzyme was strongly inhibited by Mg^{2+}.

Criteria of Purity and Subunit Composition—The ultraviolet absorption spectrum of the purified enzyme had a maximum at 278 nm and a shoulder at 290 nm. The ration A_{280}:A_{260} was about 1.7, indicating little or no contamination by nucleic acid.

The high, constant specific activity (3 to 4 μmoles per min per mg of protein) across the peak eluted from hydroxylapatite (Fig. 5) indicated a high degree of purity, but better estimates of the purity of the enzyme were obtained by gel electrophoresis. Electrophoresis of purified *Acanthamoeba* myosin in the presence of either SDS (Fig. 2) or urea revealed only one slow moving polypeptide and two fast moving polypeptides. There were other minor bands but they were variable in amount and accounted for less than 10% of the stained proteins in most preparations. Isoelectric focusing in gels containing 8 M urea gave similar results of one major band (which concentrated at about

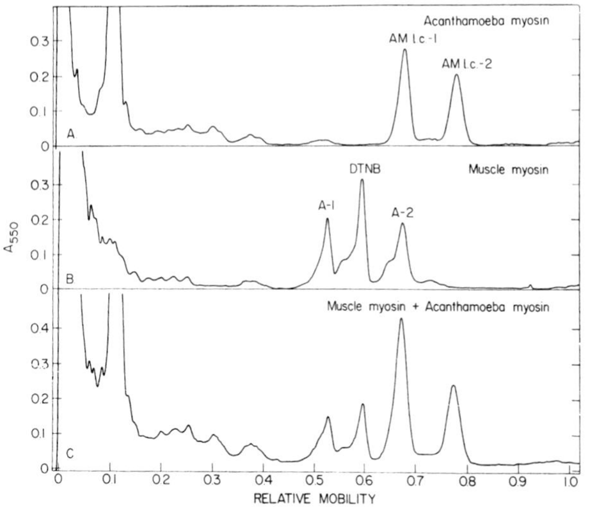

FIG. 7. Comparison of the light chains of *Acanthamoeba* myosin and muscle myosin by SDS gel electrophoresis. Polypeptides were separated by electrophoresis on 9% polyacrylamide gels containing 1% SDS, stained with Coomasie blue and densitometer scans obtained at 550 nm. *A*, purified *Acanthamoeba* myosin; *B*, muscle myosin; and *C*, *Acanthamoeba* myosin mixed with muscle myosin. For these gels the *Acanthamoeba* myosin consisted of equal portions of three separate preparations of *Acanthamoeba* myosin mixed together. The heavy chains are off scale at the *left*. The *Acanthamoeba* myosin light chains are labeled *AM lc-1* and *AM lc-2*. The muscle myosin light chains are labeled *A-1*, *DTNB*, and *A-2* according to the nomenclature of Weeds and Lowey (*33*). It is clear the *AM lc-1* and *A-2* are not separable electrophoretically.

pH 8) and two less intense bands (that concentrated at much lower pH levels).

The following observations suggest that all three polypeptides were, indeed, subunits of *Acanthamoeba* myosin. The three polypeptides copurified with the enzymatic activity as shown by gel electrophoresis in SDS (Fig. 2) and 8 M urea (not illustrated). Despite the approximately 10-fold difference in their molecular weight, the three polypeptides remained together after gel filtration on 8% agarose in 0.5 M KCl. The subunit composition was very reproducible. When equal portions of three different preparations of purified *Acanthamoeba* myosin were mixed and subjected to SDS-gel electrophoresis the same three bands were seen (Fig. 7*A*) in the same positions and relative concentrations as with the individual preparations (Fig. 2).

By analogy with the nomenclature employed for similar polypeptide subunits of muscle myosin the slowly migrating polypeptide will be referred to as the heavy chain and the rapidly migrating polypeptides as light chains, lc-1 and lc-2. In addition to their much lower molecular weight, lc-1 and lc-2 are much more acidic than the heavy peptide as shown by isoelectric focussing and by gel electrophoresis in urea.

Molecular Weight of Acanthamoeba Myosin—The heavy chains of five preparations of purified *Acanthamoeba* myosin had a reproducible molecular weight of 140,000 by SDS-gel electrophoresis. For unexplained reasons, the heavy chains of two earlier preparations had molecular weights of 115,000 and 120,000 (2). *Acanthamoeba* myosin lc-1 co-migrated with the smallest light chain (A-2) of muscle myosin (Fig. 7), which has a molecular weight of about 16,000 to 17,000 (33). The molecular weight of *Acanthamoeba* myosin lc-2 was about 14,000.

Native *Acanthamoeba* myosin had the partition coefficient of a globular protein of molecular weight between 150,000 and 200,000 as judged by gel filtration in 0.5 M KCl on 8% agarose (Bio-Rad A-1.5, Fig. 1), 6% agarose (Sephadex 6B), 4% agarose (Sephadex 4B), (3) (Fig. 4), Sephadex G-150 (4) (Fig. 4), and Sephadex G-200. Although the enzyme does adsorb to Bio-Rad agarose at low ionic strength, these estimates of molecular weight are probably reliable (*i.e.*, not influenced by adsorption) because such binding should be minimal in 0.5 M KCl and because the same results were found with Sephadex columns to which the enzyme does not adsorb even at low ionic strength. On this basis the native *Acanthamoeba* myosin would consist of a single heavy chain and one or more of the light chains. The native molecule must be relatively globular since an asymmetric molecule containing a polypeptide of 140,000 would have had much smaller partition coefficients than were observed.

Because the molecular weight of *Acanthamoeba* myosin was found to be considerably less than the molecular weight of muscle myosin (about 460,000 (43)), slime mold myosin (about 460,000 (35)) and platelet myosin (about 500,000 (36, 37)), the possibility was considered that the isolated *Acanthamoeba* "myosin" might have been a proteolytic fragment of a larger native molecule. The following observations make this unlikely. The yield of K^+-EDTA ATPase and its partition coefficient by gel filtration were unchanged when an amoeba homogenate was allowed to stand at 0° for 0, 2, and 4 hours before preparing the active supernatant by ultracentrifugation. Even very minor peaks of K^+-EDTA ATPase activity corresponding to a hypothetical larger native myosin were never seen on gel filtration of crude extracts (Fig. 1). There was no increase of ninhydrin-reactive material during the extraction and centrifugation of the amoeba homogenate indicating the absence of detectable proteolysis. Addition to the extraction solution of 5.4 mM diisopropyl fluorophosphate, and inhibitor of some proteases, did not affect the yield or partition coefficient of the *Acanthamoeba* myosin. SDS-gel electrophoresis of homogenates of *Acanthamoeba* incubated at room temperature for as long as 4 hours revealed no detectable loss of any high molecular weight polypeptide (9). When muscle myosin, which is highly susceptible to proteolytic cleavage (34), was incubated with an amoeba homogenate there was no decrease in muscle myosin ATPase activity nor was the molecule degraded into Subfragment 1 or heavy meromyosin as judged by gel filtration. Gel electrophoresis of the purified *Acanthamoeba* myosin in SDS or urea consistently showed only one heavy and two light chains in contrast to the heterogeneity of heavy meromyosin and Subfragment 1 which are produced by multiple cleavages of muscle myosin subjected to proteases (33, 38). From these observations it can be concluded that *Acanthamoeba* myosin *in situ* can have a higher molecular weight than the isolated enzyme only if the myosin were hydrolyzed immediately upon homogenization of the amoebae by an endogenous protease that is resistant to diisopropyl fluorophosphate and that does not attack muscle myosin or most of the other amoeba proteins.

Solubility of Acanthamoeba Myosin—In contrast to myosins from the cells of higher animals, the *Acanthamoeba* enzyme is soluble at low (physiological) ionic strength. It did not precipitate at pH 7.0 in 0.05 to 0.5 M KCl and had the same partition coefficient on Sepharose 6B in 0.1 M and 0.5 M KCl, indicative of no aggregation at low ionic strength. It was soluble in 0.2 mM ATP and 2 mM Tris chloride, pH 7.6, but insoluble in 0.05 M KCl below pH 6.5. No recognizable aggregates or thick filaments were observed in electron micrographs of negatively stained preparations of *Acanthamoeba* myosin in 0.05 to 0.1 M KCl, even with 10 mM $CaCl_2$ which stimulates filament formation by *Physarum* myosin (39).

TABLE V

Amino acid composition of two preparations of Acanthamoeba myosin

Amino acid	A	B
	moles/100,000 g	
Asp	108	99
Thr	38	38
Ser	50	51
Glu	114	108
Pro	58	61
Gly	102	108
Ala	70	74
Cys[a]	0	0
Val	46	47
Met	30	33
Ileu	44	45
Leu	75	74
Tyr	34	36
Phe	44	43
His	13	11
Lys	58	55
Arg	36	38
Total	920	921

[a] Analyzed as *S*-carboxymethylcysteine after reduction and alkylation. The apparent absence of cysteine was confirmed in two other preparations of *Acanthamoeba* myosin.

Amino Acid Composition—The amino acid analysis of purified *Acanthamoeba* myosin was remarkable because of the absence of cysteine in each of four different preparations tested (Table V). In all cases cysteine was analyzed as *S*-carboxymethylcysteine after reduction and alkylation of the protein. Hatano and Ohnuma (40) have reported the absence of cysteine in their preparations of *Physarum* myosin. Even apart from the apparent absence of cysteine the amino acid composition of *Acanthamoeba* myosin does not resemble the amino acid composition of muscle myosin or any of its proteolytic fragments that have been analyzed (34). No evidence was found for the presence of methylated lysines or histidine, which are constituents of some muscle myosins (21) and of *Acanthamoeba* actin (5), even when very large samples were applied to the amino acid analyzer.

DISCUSSION

The term "myosin" was originally used to describe a group of similar, but nonidentical, ATPases found in striated and smooth muscle cells. These muscle myosins share several characteristic properties that are thought to be central to the force-generating process (41): their ATPase activities are strongly inhibited by Mg^{2+} but, under physiological conditions, actin activates their Mg^{2+}ATPase activities; they bind to actin filaments forming complexes that are dissociated by Mg^{2+} and ATP; and, at physiological ionic strength, they aggregate into bipolar thick filaments which can cross-link actin filaments. Muscle myosins from different sources vary, however, in their specific enzymatic activities (42), in the number and size of their low molecular weight subunits (43, 44), in their content of methylated amino acids (45) and in the presence or absence of Ca^{2+}-sensitive regulatory activity (46).

Recently, myosins sharing many of these properties (Table VI) have been purified from the acellular slime mold *Physarum polycephalum* (28, 35, 39, 40), human platelets (36, 37), and

TABLE VI
Comparison of myosins from different cells

Source of myosin	Actin binding	ATPase activity[a]				Subunit molecular weights	Solubility in 0.05 M KCl	Filament formation	Reference
		Mg^{2+}	Mg^{2+} + actin	K^+-EDTA	Ca^{2+}				
Acanthamoeba	+	−	+++	++++	+	1 × 140,000 ? × 16,000 ? × 14,000	+	None known	This paper, 3, 4
Horse smooth muscle	+	−	+	+++	++		−	Bipolar thick filaments in dilute KCl, especially with Ca^{2+}	26, 47
Human platelet	+	−	+	++	++	2 × 200,000 ? × ∼20,000	−	Bipolar thick filaments in dilute KCl	36, 37
Mouse fibroblast	+	−	+	++	++	2 × 200,000 ? × ∼20,000	?	Small bipolar thick filaments in dilute KCl	48, 49
Physarum	+	−	±	−	+++	Native molecular weight 458,000	+	Bipolar thick filaments in Ca^{2+} or Mg^{2+}	35, 39, 40
Rabbit skeletal muscle	+	−	++++	++++	++	2 × 200,000 1 × 20,500 2 × 18,000 1 × 16,000	−	Bipolar thick filaments in dilute KCl	33, 34, 38

[a] Considerable variations in assay conditions make direct quantitative comparisons of ATPase activity meaningless, so we have used a semiquantitative − to ++++ scale to compare relative activities.

mouse fibroblasts (48, 49). These enzymes are presumed to participate in the development of force for movement in these nonmuscle cells (for a review, see Reference 50).

We have named the ATPase from *Acanthamoeba*, *Acanthamoeba* myosin because it possesses most of the characteristic properties of myosin, although it differs from other known myosins in its lower molecular weight, amino acid composition, and inability to form thick filaments. Naming this enzyme *Acanthamoeba* myosin broadens the meaning of the term myosin, but we find this preferable to coining a new and possibly confusing name for this protein which is likely to be the functional equivalent of myosin in the amoeba. Perhaps further investigation of other "primitive" organisms will reveal additional examples of myosins of this type.

Probably as a consequence of its low molecular weight and its globular shape, purified *Acanthamoeba* myosin seems unable to form filaments nor have myosin "thick filaments" been observed in several extensive investigations of the ultrastructure of the amoeba (6, 7). The absence of thick filaments makes it difficult to localize *Acanthamoeba* myosin within cells by electron microscopy but the fact that pyrophosphate, which dissociates myosin from actin, is required for complete extraction of the enzyme suggests that at least some of the enzyme may be bound to actin filaments *in situ*.

Despite its inability to form bipolar thick filaments, which are the mechanism by which other myosins make physical connections between actin filaments, *Acanthamoeba* myosin does cross-link F-actin (4). This, along with the reversible binding of *Acanthamoeba* myosin to actin filaments and the activation of the myosin Mg^{2+}ATPase by actin, which are thought to be directly related to the generation of force for movement, are considered in detail in the accompanying paper (4).

REFERENCES

1. POLLARD, T. D. (1971) *Fed. Proc.* **30,** Abstr. 1309
2. POLLARD, T. D., AND KORN, E. D. (1972) *Fed. Proc.* **31,** 502
3. POLLARD, T. D., AND KORN, E. D. (1973) *Cold Spring Harbor Symp. Quant. Biol.* **37,** 573–583
4. POLLARD, T. D., AND KORN, E. D. (1973) *J. Biol. Chem.* **248,** 4691–4697
5. WEIHING, R. R., AND KORN, E. D. (1971) *Biochemistry* **10,** 590
6. POLLARD, T. D., SHELTON, E., WEIHING, R. R., AND KORN, E. D. (1970) *J. Mol. Biol.* **50,** 91
7. BOWERS, B., AND KORN, E. D. (1968) *J. Cell Biol.* **39,** 95
8. POLLARD, T. D., AND KORN, E. D. (1973) *J. Biol. Chem.* **248,** 448–450
9. KORN, E. D., AND WRIGHT, P. L. (1973) *J. Biol. Chem.* **248,** 439–447
10. WEIHING, R. R., AND KORN, E. D. (1972) *Biochemistry* **11,** 1538
11. REES, M. K., AND YOUNG, M. (1967) *J. Biol. Chem.* **242,** 4449
12. SPUDICH, J. A., AND WATT, S. (1971) *J. Biol. Chem.* **246,** 4866
13. NEFF, R. J. (1969) *Symp. Soc. Exp. Biol.* **23,** 51
14. WEISMAN, R. A., AND KORN, E. D. (1966) *Biochim. Biophys. Acta* **116,** 229
15. MARTIN, J. B., AND DOTY, D. M. (1949) *Anal. Chem.* **21,** 965
16. LOWRY, O. H., ROSEBROUGH, N. J., FARR, A. L., AND RANDALL, R. J. (1951) *J. Biol. Chem.* **193,** 265
17. FAIRBANKS, G., STECK, T. L., AND WALLACH, D. F. H. (1971) *Biochemistry* **10,** 13
18. DAVIS, B. (1964) *Ann. N. Y. Acad. Sci.* **121,** 404
19. WRIGLEY, C. W. (1969) *Analytical Fractionation of Proteins According to Isoelectric Point by Gel Electrofocusing*, Shandon Press Ltd., London
20. GROSS, E. (1967) *Meth. Enzymol.* **25,** 238
21. KUEHL, W. M., AND ADELSTEIN, R. S. (1969) *Biochem. Biophys. Res. Commun.* **37,** 59
22. PETERSON, E. A. (1970) in *Laboratory Techniques in Biochemistry and Molecular Biology* (WORK, T. S., AND WORK, E., eds) p. 225, American Elsevier Publishing Co., New York
23. LEVIN, O. (1962) *Meth. Enzymol.* **5,** 27
24. SZENT-GYORGYI, A. (1951) *Chemistry of Muscle Contraction*, Academic Press, New York
25. ARNOLD, H., AND PETTE, D. (1970) *Eur. J. Biochem.* **15,** 360
26. YAMAGUCHI, M., MIYAZAWA, Y., AND SEKINE, T. (1970) *Biochim. Biophys. Acta* **216,** 411
27. BETTEX-GALLAND, M., AND LUSCHER, E. F. (1961) *Biochim. Biophys. Acta* **49,** 536
28. HATANO, S., AND TAZAWA, M. (1968) *Biochim. Biophys. Acta* **154,** 507

29. Neff, R. J., Neff, R. H., and Taylor, R. E. (1958) *Physiol. Zool.* **31,** 73
30. Sekine, T., and Kielley, W. W. (1964) *Biochim. Biophys. Acta* **81,** 336
31. Seidel, J. C. (1969) *Biochim. Biophys. Acta* **189,** 162
32. Seidel, J. C. (1969) *J. Biol. Chem.* **244,** 1142
33. Weeds, A. G., and Lowey, S. (1971) *J. Mol. Biol.* **61,** 701
34. Lowey, S., Slayter, H. S., Weeds, A. G., and Baker, H. (1969) *J. Mol. Biol.* **42,** 1
35. Adelman, M. R., and Taylor, E. W. (1969) *Biochemistry* **8,** 4976
36. Adelstein, R. S., Pollard, T. D., and Kuehl, W. M. (1971) *Proc. Nat. Acad. Sci. U. S. A.* **68,** 2703
37. Booyse, F. M., Hoveke, T. P., Zschocke, D., and Rafelson, M. E. Jr. (1971) *J. Biol. Chem.* **246,** 4291
38. Lowey, S. (1971) in *Subunits in Biological Systems* (Timasheff S. N., and Fasman, G. D., eds) p. 201, Marcel Dekker, Inc., New York
39. Nachmias, V. T. (1972) *Proc. Nat. Acad. Sci. U. S. A.* **69,** 2011
40. Hatano, S., and Ohnuma, J. (1970) *Biochim. Biophys. Acta* **205,** 110
41. Huxley, H. E. (1969) *Science* **164,** 1356
42. Barany, M. (1967) *J. Gen. Physiol.* **50,** 197
43. Lowey, S., and Risby, D. (1971) *Nature* **234,** 81
44. Sarkar, S., Sreter, F. A., and Gergely, J. (1971) *Proc. Nat. Acad. Sci. U. S. A.* **68,** 946
45. Kuehl, W. M., and Adelstein, R. S. (1970) *Biochem. Biophys. Res. Commun.* **39,** 956
46. Kendrick-Jones, J., Lehman, W., and Szent-Györgyi, A. (1970) *J. Mol. Biol.* **54,** 313
47. Schoenberg, C. (1969) *Tissue Cell* **1,** 83
48. Adelstein, R. S., Conti, M., Johnson, G., Pasten, I., and Pollard, T. D. (1972) *Proc. Nat. Acad. Sci. U. S. A.*, **69,** 3693–3697
49. Adelstein, R. S., and Conti, M. (1972) *Cold Spring Harbor Symp. Quant. Biol.*, **37,** 599–606
50. Pollard, T. D. (1972) in *Biology of Amoeba* (Jeon, K., ed) p. 291, Academic Press, Inc. New York

Lazarides E. and **Weber K.** 1974. Actin antibody: The specific visualization of actin filaments in non-muscle cells. *Proc. Natl. Acad. Sci.* **71**: 2268–2272.

By the early 1970s, electron microscopy had demonstrated the widespread distribution of slender microfilaments in eukaryotic cells. These filaments were recognized as actin polymers, by three criteria: their size and shape, their ability to bind heavy meromyosin (a proteolytic fragment of myosin that decorates polymerized actin in a polar and ATP-sensitive manner (1, 2)), and the discovery of actin as a prevalent cytoplasmic protein by standard biochemical methods (reviewed in 3). The paper reproduced here had a major impact on the development of our understanding of the cytoskeleton. It described comparatively straightforward methods for purifying abundant proteins and raise good antibodies to them. These methods came into widespread use for the study of both microfilaments and other polymers of the cytoskeleton, like microtubules (4) and intermediate filaments (reviewed in 5). It showed the efficacy of light microscopy and indirect immunofluorescence for localizing a particular cellular component, so long as the probes used were adequately specific. Much work on the structure of the cytoskeleton had previously been done by electron microscopy, taking advantage of the spatial resolution of this instrument to identify and characterize the structure under study. With good fluorescent probes, however, the power of the light microscope to show the localization of objects throughout the cell became obvious; a single image could do the work of many electron micrographs. The beauty of these fluorescent images, combined with the ease with which they could be obtained, opened a new era in research on cell structure. The methods have been applied widely, including for the characterization of alterations in the cytoskeleton that accompany human diseases (6).

---

1. Huxley H.E. 1963. Electron microscope studies on the structure of natural and synthetic protein filaments from striated muscle. *J. Mol. Biol.* **7:** 281–308.
2. Ishikawa H., Bischoff R., and Holtzer H. 1969. Formation of arrowhead complexes with heavy meromyosin in a variety of cell types. *J. Cell Biol.* **43:** 312–328.
3. Pollard T.D. and Weihing R.R. 1974. Actin and myosin and cell movement. *CRC Crit. Rev. Biochem.* **2:** 1–65.
4. Fuller G.M., Brinkley B.R., and Boughter J.M. 1975. Immunofluorescence of mitotic spindles by using monospecific antibody against bovine brain tubulin. *Science* **187:** 948–950.
5. Lazarides E. 1980. Intermediate filaments as mechanical integrators of cellular space. *Nature* **283:** 249–256.
6. Coulombe P.A., Hutton M.E., Vassar R., and Fuchs E. 1991. A function for keratins and a common thread among different types of epidermolysis bullosa simplex diseases. *J. Cell Biol.* **115:** 1661–1674.

Reprinted from

Proc. Nat. Acad. Sci. USA
Vol. 71, No. 6, pp. 2268–2272, June 1974

Actin Antibody: The Specific Visualization of Actin Filaments in Non-Muscle Cells

(immunofluorescence/microfilaments/sodium dodecyl sulfate gel electrophoresis)

ELIAS LAZARIDES AND KLAUS WEBER

Cold Spring Harbor Laboratory, Cold Spring Harbor, New York 11724

Communicated by J. D. Watson, March 11, 1974

ABSTRACT Actin purified from mouse fibroblasts by sodium dodecyl sulfate gel electrophoresis was used as antigen to obtain an antibody in rabbits. The elicited antibody was shown to be specific for actin as judged by immunodiffusion and complement fixation against partially purified mouse fibroblast actin and highly purified chicken muscle actin. The antibody was used in indirect immunofluorescence to demonstrate by fluorescence light microscopy the distribution and pattern of actin-containing filaments in a variety of cell types. Actin filaments were shown to span the cell length or to concentrate in "focal points" in patterns characteristic for each individual cell.

Eucaryotic cells contain three basic fibrous structures: filaments, microfilaments, and microtubules. These three structures are thought to be intimately involved in the maintenance of cell shape, in cell movement, and in other important cellular functions (1). Microfilaments are thought to contain actin. This assumption is based on the observation that these structures can be selectively decorated with heavy meromyosin, a specific proteolytic fragment of muscle myosin known to interact with muscle actin (2). Furthermore, actin is now now known to exist as a major protein component of a variety of non-muscle cellular types and in each case it has properties markedly similar to those of its muscle counterpart (3–7)*. The major protein of the microtubular system, tubulin, has been isolated and well characterized (9). The basic protein subunit of the filament structure, however, has not so far been identified. Presumptive muscle proteins like myosin (10–13) and tropomyosin (14) have been found in some non-muscle cells. However, their exact distribution within the cell, as well as their specific localization in one of these fibrous structures, is as yet undetermined.

We have developed a relatively simple way of selectively visualizing filamentous structures in the cell by using antibodies made against different structural proteins. The problem of purifying each antigen separately was circumvented by using sodium dodecyl sulfate (SDS) gel electrophoresis. The denatured proteins are antigenic, and the antibody obtained cross reacts with the native protein. Once specificity has been demonstrated, the antiserum obtained can be used in indirect immunofluorescence to visualize the structures in the cell with which the protein is associated.

In this paper we have used mouse fibroblast actin to test this approach. The actin, purified by SDS gel electrophoresis, was used to obtain an antibody in rabbits. The antiserum obtained was shown by immunodiffusion and complement fixation to be specific for actin. This antibody was then used in indirect immunofluorescence to show that microfilaments are polymers of actin. This technique also enabled us to demonstrate the complex network of actin filaments in a variety of cell types.

MATERIALS AND METHODS

Growth of Cells. Actin was isolated from the cell line SV101, a clone of mouse fibroblast 3T3 cells transformed by Simian virus 40. This transformed cell line was chosen because it grows to a higher saturation density than the parent 3T3 cell line (15). The cells were grown in roller bottles (Vitro Corp.) in Dulbecco's modified Eagle's medium containing 10% calf serum and 50 μg/ml of gentamycin. At confluency, the medium was removed and the cells were washed with phosphate-buffered saline (PBS). The cells were then scraped off the bottles, collected by low speed centrifugation, and stored at −70°.

Preparation of Actin. The cells were thawed and homogenized in 20 volumes of 95% ethanol. The precipitate was collected by low speed centrifugation, washed immediately with ether, and air dried. The yield from 10 bottles was approximately 1.2 g of ethanol-ether powder. The ethanol-ether powder was stirred at 4° in 0.01 M sodium phosphate buffer (pH 6.8), 10 mM $MgCl_2$, and 1 mM dithiothreitol (15 ml of buffer per g of powder) for 3–5 hr. The supernatant was made 30% in ammonium sulfate by adding 0.17 g of ammonium sulfate per ml of extract. After stirring for 30 min at 4°, the precipitate was collected by centrifugation and dissolved in and dialyzed against 0.01 M Tris·HCl (pH 7.5), 10^{-4} M $CaCl_2$, 1 mM dithiothreitol, and 10^{-4} M ATP. The actin could be further purified by a second precipitation at 30% ammonium sulfate saturation. Under these conditions, 1 g of ethanol-ether powder yields approximately 1 mg of actin.

Highly purified chicken muscle actin was a generous gift from Dr. Susan Lowey.

Antibody Preparation. The actin used as an antigen in rabbits was purified through SDS slab gel electrophoresis from the high speed supernatant (100,000 × *g*) of a mouse fibroblast cell homogenate (see *Results*). Approximately 400 μg of the antigen was injected in complete Freund's adjuvant and 2 weeks later the rabbits were boosted with an additional 400 μg. Blood was collected 6 weeks after the last injection and the serum was clarified by centrifugation at 10,000 rpm. The gamma globulin fraction was partially purified using precipitation with half saturated ammonium sulfate. It was

Abbreviations: SDS, sodium dodecyl sulfate; PBS, phosphate-buffered saline.

* The authors apologize for not referring to all the contributors in this field. The reader is referred to a recent detailed review of actin and myosin in non-muscle cells for complete references (8).

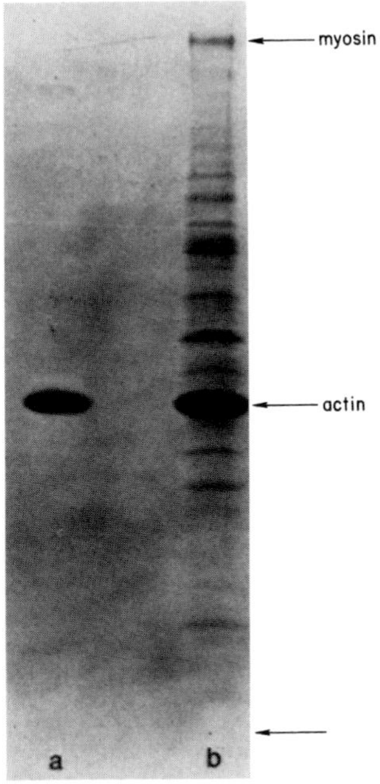

FIG. 1. SDS gel electrophoresis of actin. SDS polyacrylamide slab gel electrophoresis was performed in 12.5% slabs according to Studier (19). (*a*) Highly purified chicken muscle actin (10 μg). (*b*) A 30% ammonium sulfate cut of a low-salt extract from an ethanol powder of mouse fibroblasts (15 μg). A small amount of myosin copurifies with actin under the extraction conditions used (see *Materials and Methods*). The *arrow* at the bottom of the figure shows the dye front of the gel.

then dialyzed into 0.15 M NaCl, 0.01 M Tris·HCl (pH 7.5), and stored at −20° at a concentration of approximately 30 mg/ml.

Indirect Immunofluorescence. Cells were grown on glass coverslips in the appropriate medium (see figure legends). The coverslips were washed briefly in PBS to remove excess medium and fixed in PBS containing 3.5% formaldehyde for 20 min at room temperature. They were subsequently washed thoroughly in PBS, treated with absolute acetone at −10° for 7 min, and air dried. An appropriate dilution (1:20 in PBS) of the rabbit antibody was applied to the cells. After incubation in a humid atmosphere at 37° for 1 hr, the coverslips were washed 3 times in PBS and incubated for 1 hr with a 1:10 dilution of goat anti-rabbit globulins coupled to fluorescein made in PBS (Miles). The coverslips were washed 3 times in PBS and once in distilled water and mounted in Elvanol on a glass slide. Coverslips were viewed in a Zeiss PM III microscope with ultraviolet optics. Photographs were taken using Plus-X film (Kodak).

RESULTS

A protein purified from an ethanol powder of SV101 cells has several properties which identify it as mouse fibroblast actin: (*i*) it coelectrophoreses on SDS gels with highly purified chick muscle actin with a molecular weight of 45,000 (Fig. 1); (*ii*) it precipitates characteristically at a low ammonium sulfate saturation (12) (see below); (*iii*) it binds ATP on millipore filters; (*iv*) it is precipitable by 3 mM vinblastine sulfate

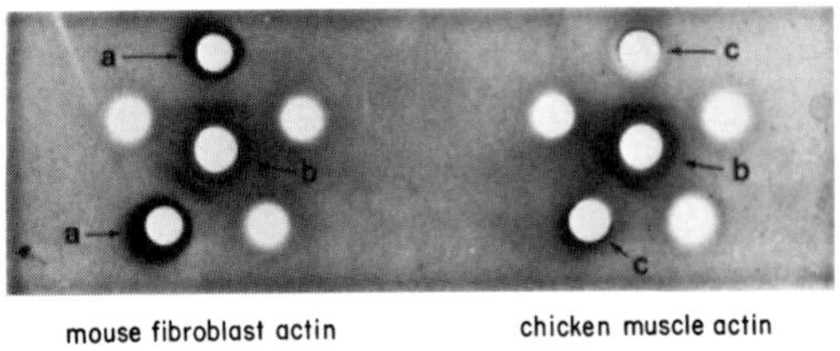

FIG. 2. Immunodiffusion of actin antiserum. (*a*) 4 μg (upper hole) and 2 μg (lower hole) of a 30% ammonium sulfate cut from mouse fibroblasts. The heavy precipitate around the holes is due to aggregated actin. (*b*) 20 μl of partially purified rabbit antiserum. (*c*) 5 μg (upper) and 2 μg (lower) of purified chicken muscle actin. Immunodiffusion was performed at 37° for 24 hr. The plates were washed for 24–36 hr at room temperature in 0.15 M NaCl, 0.01 M Tris·HCl (pH 7.5). They were subsequently stained in 0.25% Coomassie brilliant blue–50% methanol–7.5% acetic acid for 1 hr, destained in 7.5% methanol–7.5% acetic acid, and photographed.

(16, 17) and by 50 mM Mg^{+2} ions (18); and (*v*) it undergoes monomer to polymer transformations. Similar properties characterize skeletal muscle actin and the protein extensively studied and identified as actin-like (13, 16–18) in a variety of nonmuscle cellular types including cultured chick embryo fibroblasts (3) and human platelets (6). Mouse fibroblast actin purified as described in *Materials and Methods* is approximately 70% pure and shows one major polypeptide on SDS polyacrylamide gels and on polyacrylamide gels at pH 4.5 run in the presence of 8 M urea. Fig. 1*b* shows the purity of mouse fibroblast actin used in the experiments below. The gel is purposely overloaded to reveal all minor contaminants.

Approximately 10% of the total cellular actin precipitates at the 30% ammonium sulfate cut of a low-salt extract of an ethanol powder. Another 50% is recovered in the 30–60% ammonium sulfate cut. The remainder can be extracted as an actomyosin-like complex from the powder in the presence of 0.6 M NaCl. This differential fractionation appears to depend in part on the state of the polymerization of the actin under the extraction conditions used. In order to obtain sufficient actin to use as antigen, we therefore decided to purify this protein by SDS gel electrophoresis from the high-speed supernatant of homogenized SV101 cells.

SDS polyacrylamide gels separate proteins according to the molecular weights of their polypeptide chains (27) and therefore all actin will move with a uniform molecular weight on the gel regardless of its original state of polymerization. This method has the further advantage that SDS denatured proteins often make good antigens and allow one to use many different proteins as antigens from the same high-speed supernatant. The procedure is of general applicability and will be described in detail elsewhere (E.L. and K.W., manuscript in preparation).

The proteins moving as a major band at molecular weight 45,000 on SDS slab gels were recovered from the gels by elution. Approximately 800 μg of protein were obtained and this material was used as an antigen. Previous experiments involving ion exchange chromatography of the high-speed supernatant had shown that actin constituted more than 85% of the total protein moving in this molecular-weight range. The remaining 15% was shown to have different chromatographic properties than actin and showed a small number of minor species on polyacrylamide gels at pH 4.5 in 8 M urea.

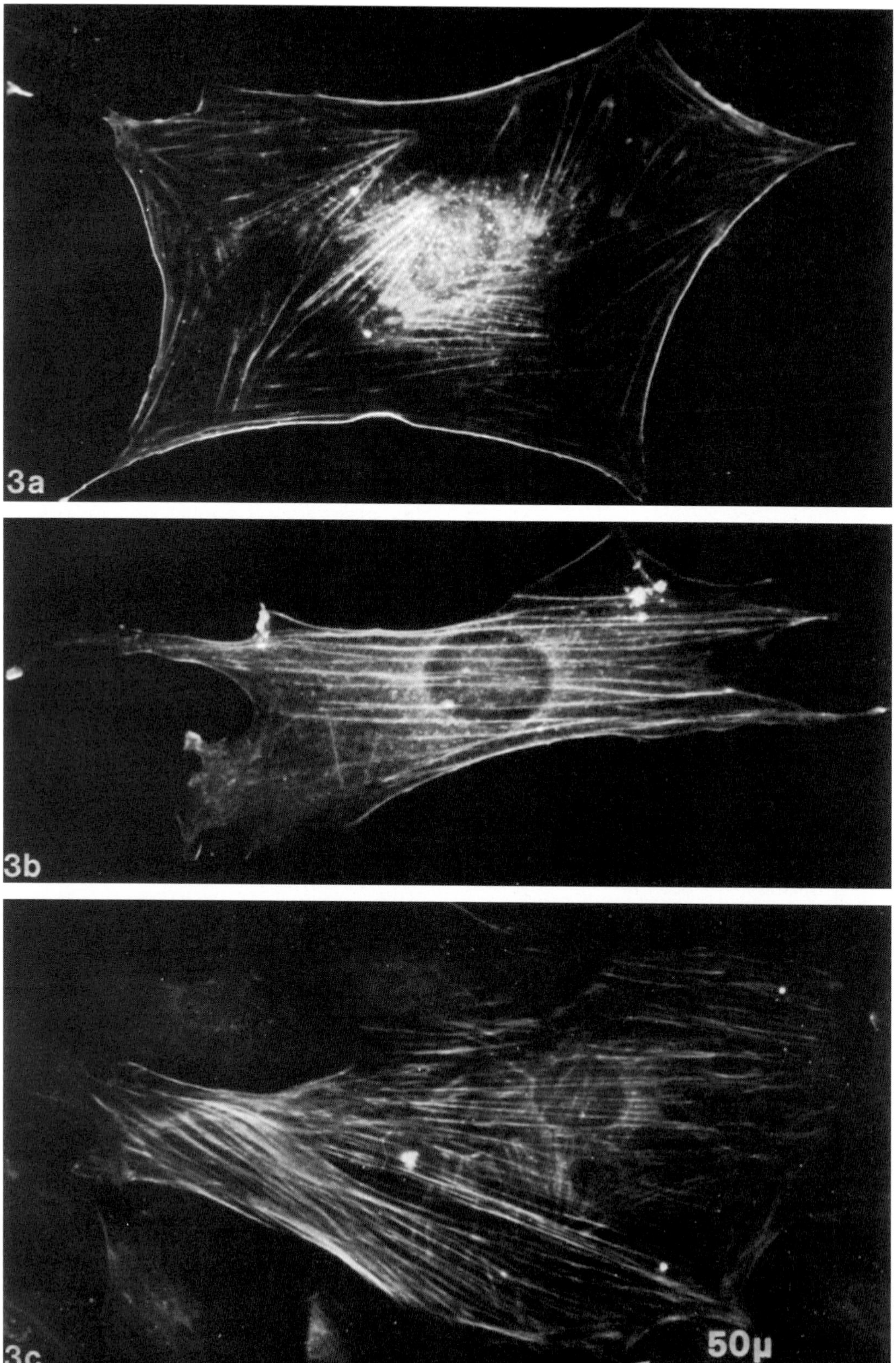

FIG. 3. *(Legend appears at bottom of the next page.)*

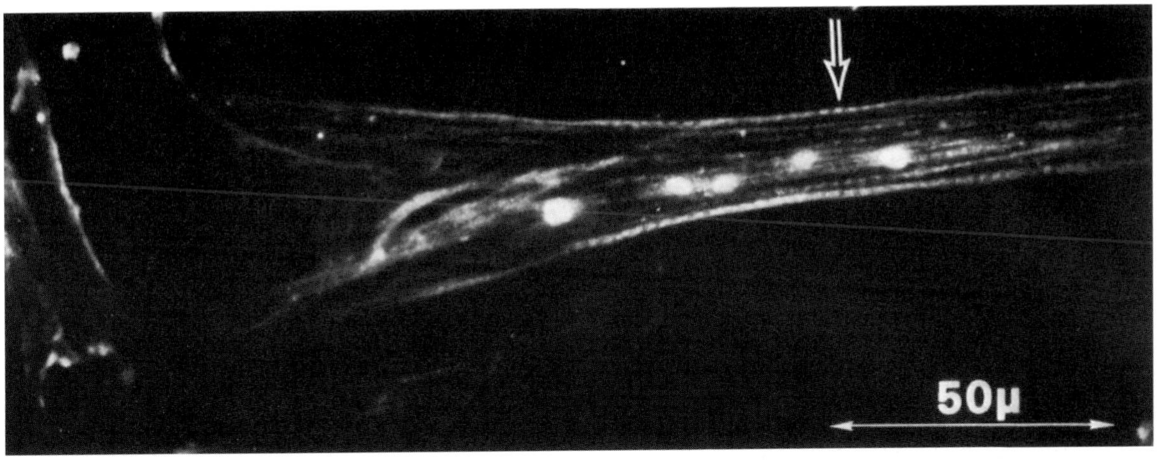

FIG. 4. Indirect immunofluorescence of actin antibody with primary chick embryo myoblasts. The *arrow* indicates the actin striations characteristic of myofibrils. The nuclear staining seen in this figure seems to be nonspecific. Purification of the antibody used in this experiment by 50% ammonium sulfate fractionation removes the nuclear fluorescence. The culture of primary chick embryo myoblasts was prepared for us by Dr. C. M. Chang using 5% chick embryo extract, 7% horse serum, and 3% fetal-calf serum in F-12 medium.

The antiserum is specific for actin as judged both by immunodiffusion and complement fixation, using the purified mouse fibroblast actin preparation (Fig. 1*b*) and the homogeneous preparation of chicken muscle actin (Fig. 1*a*). The results of the immunodiffusion assay are shown in Fig. 2. Both mouse fibroblast actin and chicken actin show a single precipitin line when diffused against the antiserum. In complement fixation the antibody shows complement fixing ability at a dilution of 1:150 both with the mouse fibroblast actin and the chicken actin. Complement fixation was performed by a modification of the micro method of Sever (20) (Osborn and Weber, in preparation).

Since the antibody appeared to be specific for actin, we attempted to use it for indirect immunofluorescence (21) in the hope of revealing the intracellular distribution of actin. Cells grown on coverslips were fixed in formaldehyde for optimal preservation of fibrous material and were then stained with antibody using the indirect immunofluorescent technique (see *Materials and Methods*). Fig. 3 shows the fluorescence pattern of the mouse fibroblast 3T3 cell line and a baby hamster kidney cell line (BHK) stained with the actin antibody. The fluorescent staining reveals a multitude of actin-containing filaments marking clearly the cell periphery and spanning the interior of the cell frequently parallel to each other. A multitude of patterns have been observed with varying degrees of complexity and each cell appears to have its own individual way of portraying its actin filamentous network. However, after observing hundreds of cells, two major patterns appear to prevail in the cell types tested so far. One is that shown in Fig. 3*b* and *c* where the fibers run parallel to each other along the long axis of the cell. The other is that shown in Fig. 3*a* where actin filaments seem to converge at what we have named "focal points." Control experiments with rabbit antiserum obtained before immunization reveal no fluorescent fibers. Furthermore, antibodies obtained against other cellular structural proteins reveal a very different fluorescent staining pattern.

As shown above (Fig. 2), the antibody cross reacts with chicken muscle actin. We therefore studied the staining pattern of cultured primary chick embryo myoblasts in early myogenesis (Fig. 4). Besides the usual actin-containing filaments, the fluorescence reveals the characteristic banding striations of actin in newly formed myofibrils.

DISCUSSION

It is known that actin is a major component of eucaryotic cells. It accounts for some 10% of the cells' proteins. Mouse fibroblast actin coelectrophoreses with chick muscle actin and has a molecular weight of 45,000. Not surprisingly, therefore, actin purified from SV101 fibroblasts shows the same properties as the protein extensively studied from a variety of different cell types.

SDS-denatured actin obtained by preparative gel electrophoresis of a cell extract after high speed centrifugation has proven to be a good antigen in rabbits. The antibody obtained reacts with native actin from SV101 cells and with highly purified chick muscle actin both by immunodiffusion and complement fixation. The antibody was directed against a specific class of proteins coelectrophoresing with actin and having a molecular weight of 45,000. While the possibility of obtaining antibodies against other minor proteins in the same molecular-weight range is likely, the possibility of obtaining antibodies to proteins with different polypeptide molecular weights is excluded. Thus the previous difficulty of obtaining an actin specific antibody due to the presence of contaminating tropomyosin is circumvented (22). Furthermore, actin is the only major component constituting more than 85% of the proteins migrating with a molecular weight of 45,000. We therefore believe that although the original antigen was not completely homogeneous, the small amounts of con-

FIG. 3 (*on preceding page*). Indirect immunofluorescence using actin antibody. (*a, c*) A sparse mouse fibroblast cell line (3T3). (*b*) A sparse hamster established cell line (BHK). Cells were grown on coverslips in Dulbecco's modified Eagle's medium containing 10% calf serum.

taminating proteins would not interfere with the final analysis.

Immunofluorescence clearly shows the presence of actin filamentous structures. Although limited by resolution, it has the advantage over electron microscopy in revealing the two dimensional mosaic of actin filaments of a whole cell. It also demonstrates that these fibers frequently span the whole length of the cell or converge to characteristic "focal points." Furthermore, the actin filament pattern observed with immunofluorescence corresponds to the microfilament pattern observed by electron microscopy. These latter structures exist as well organized bundles running in close association with and parallel to the plasma membrane, both in baby hamster kidney cells (23) and in 3T3 cells. In 3T3 cells, microfilament bundles are also seen frequently to converge together in patterns very similar to the "focal points" observed by immunofluorescence (unpublished observations with R. Goldman). Since the fluorescence is seen also in close association with the plasma membrane, the actin filaments observed with immunofluorescence correspond to the microfilaments observed by electron microscopy. This conclusion is further substantiated by the finding that only microfilaments are decorated with heavy meromyosin (2).

The technique of immunofluorescence is convenient and fast, and allows the screening of a large number of cells under a variety of experimental conditions. The availability of antibodies to other major structural proteins will enable us to use this technique to study the intracellular localization of these proteins. This experimental approach has been previously used successfully to localize myosin (24), the light chains of myosin (25), and troponin (26) in myofibrils.

The immunofluorescent demonstration that actin exists in filamentous structures gives us a tool to compare structural differences between normal and transformed cells during various stages of their cell cycle. We hope that the convenience of this technique will aid in answering major questions of cellular structure and movement.

We thank Dr. F. Miller at Stonybrook University for the use of his animal facilities and his help in the preparation of the antibodies. We thank Dr. R. Pollack for the use of his laboratory; Drs. R. Goldman and C.-M. Chang and Art Vogel for their help with the fluorescent microscopy; and Drs. M. Osborn, C. W. Anderson, R. Goldman, and R. Pollack for their comments on the manuscript. We also thank Dr. Susan Lowey for a generous gift of purified chicken muscle actin. This investigation was supported by the National Institutes of Health (Research Grant CA-13106 from the National Cancer Institute).

1. Goldman, R. D. & Knipe, D. M. (1972) *Cold Spring Harbor Symp. Quant. Biol.* **37**, 523–534.
2. Ishikawa, H., Bischoff, R. & Holtzer, H. (1969) *J. Cell Biol.* **43**, 312–328.
3. Yang, Y. & Perdue, J. G. (1972) *J. Biol. Chem.* **247**, 4503–4509.
4. Tilney, L. G. & Mooseker, M. (1971) *Proc. Nat. Acad. Sci. USA* **68**, 2611–2615.
5. Bettex-Galland, M. & Lüscher, E. F. (1965) *Advan. Protein Chem.* **20**, 1–35.
6. Probst, E. & Luscher, R. (1972) *Biochim. Biophys. Acta* **278**, 577–584.
7. Puskin, S. & Berl, S. (1972) *Biochim. Biophys. Acta* **276**, 695–709.
8. Pollard, T. D. & Weihing, R. R. (1973) "Cytoplasmic actin and myosin and cell movement," in *Critical Reviews in Biochemistry*, Vol. 2, pp. 1–65.
9. Weisenberg, R., Borisy, G. G. & Taylor, E. W. (1968) *Biochemistry* **7**, 4466–4478.
10. Adelstein, R. S., Pollard, T. D. & Kuehl, W. M. (1971) *Proc. Nat. Acad. Sci. USA* **68**, 2703–2707.
11. Adelstein, R. S., Conti, M. A., Johnson, G. S., Pastan, I. & Pollard, T. D. (1972) *Proc. Nat. Acad. Sci. USA* **69**, 3693–3697.
12. Stossel, T. P. & Pollard, T. D. (1973) *J. Biol. Chem.* **248**, 8288–8294.
13. Adelstein, R. S. & Conti, M. A. (1971) *Cold Spring Harbor Symp. Quant. Biol.* **37**, 598–605.
14. Cohen, I. & Cohen, C. (1972) *J. Mol. Biol.* **68**, 383–387.
15. Todaro, G. J., Green, H. & Goldberg, B. D. (1964) *Proc. Nat. Acad. Sci. USA* **51**, 66–73.
16. Bray, D. (1972) *Cold Spring Harbor Symp. Quant. Biol.* **37**, 567–571.
17. Wilson, L., Bryan, J., Ruby, A. & Mazia, D. (1970) *Proc. Nat. Acad. Sci. USA* **66**, 807–814.
18. Spudich, A. J. (1972) *Cold Spring Harbor Symp. Quant. Biol.* **37**, 585–593.
19. Studier, F. W. (1972) *Science* **176**, 367–376.
20. Sever, J. L. (1962) *J. Immunol.* **88**, 320–329.
21. Coons, A. H. & Kaplan, M. H. (1950) *J. Exp. Med.* **91**, 1–13.
22. Holtzer, H., Sanger, J. W., Ishikawa, H. & Strahs, K. (1972) *Cold Spring Harbor Symp. Quant. Biol.* **37**, 549–566.
23. Goldman, R. D. (1971) *J. Cell Biol.* **51**, 752–762.
24. Pepe, F. A. (1972) *Cold Spring Harbor Symp. Quant. Biol.* **37**, 97–108.
25. Lowey, S. & Holt, J. C. (1972) *Cold Spring Harbor Symp. Quant. Biol.* **37**, 19–28.
26. Perry, S. V., Cole, H. A., Head, J. F. & Wilson, F. J. (1972) *Cold Spring Harbor Symp. Quant. Biol.* **37**, 251–262.
27. Weber, K. and Osborn, M. (1969) *J. Biol. Chem.* **244**, 4406–4412.

Stossel T.P. and **Hartwig J.H.** 1976. Interactions of actin, myosin, and a new actin-binding protein of rabbit pulmonary macrophages. II. Role in cytoplasmic movement and phagocytosis. *J. Cell Biol.* **68:** 602–619.

ACTIN MICROFILAMENTS WERE KNOWN to bundle into "stress fibers" (Lazarides and Weber, this volume), and they had been found in the cell's cortex, where their presence was essential for the formation of a contractile ring, the machinery for animal cell cytokinesis (1). A drug that blocked cytokinesis (2) had a profound effect on the structure and activity of an interphase cell's cortex, suggesting that actin filament dynamics were essential for several activities of an animal cell's surface. Stossel and Hartwig identified pulmonary macrophages as a type of cell that could be collected in quantities sufficient for biochemical as well as cell biological study. In extracts of these motile and phagocytic cells they found not only actin and myosin but also the first "actin binding protein" (ABP). The interaction of ABP with microfilaments induced a gelation reminiscent of the structure at a cell's cortex (3). The paper reproduced here showed how this new protein interacted with filamentous actin to make gels that contract upon interaction with myosin; it also proposed ways in which these phenomena were related to the normal activities of the cortex in vivo, such as phagocytosis and cell motility. These observations defined a part of the machinery that a cell uses to regulate the structure and action of its periphery. Subsequent biochemical dissection of the actin regulatory system had identified more than 60 kinds of actin-binding proteins, including other filament cross-linking proteins, filament severing proteins (4) that promoted the solation of actin gels, proteins that cap a filament's end (5), proteins that bind actin monomers (6, 7), as well as others that initiate actin polymerization, (reviewed in 8). When some of these components are assembled in proper proportions, they can drive actin-dependent motile events in vitro that are strikingly similar to motions seen in vivo (9), suggesting that we are now well on our way to understanding the essential interactions between components of the actin cytoskeleton.

---

1. Schroeder T.E. 1972. The contractile ring. II. Determining its brief existence, volumetric changes, and vital role in cleaving *Arbacia* eggs. *J. Cell Biol.* **53:** 419–434.
2. Carter S.B. 1967. Effects of cytochalasins on mammalian cells. *Nature* **213:** 261–264.
3. Hartwig J.H. and Stossel T.P. 1975. Isolation and properties of actin, myosin, and a new actin-binding protein in rabbit alveolar macrophages. *J. Biol. Chem.* **250:** 5696–5705.
4. Yin H.L. and Stossel T.P. 1979. Control of cytoplasmic actin gel-sol transformation by gelsolin, a calcium-dependent regulatory protein. *Nature* **281:** 583–586.
5. Isenberg G.H., Aebi U., and Pollard T.D. 1980. An actin binding protein from *Acanthamoeba* regulates actin filament polymerization and interactions. *Nature* **288:** 455–459.
6. Carlsson L., Nystrom L.E., Sundkvisk I., Markey F., and Lindberg U. 1977. Actin polymerizability is influenced by profilin, a low molecular weight protein of nonmuscle cells. *J. Molec. Biol.* **115:** 465–483.
7. Tilney L.G., Bonder E.M., Coluccio L.M., and Mooseker M.S. 1983. Actin from *Thyone* sperm assembles on only one end of an actin filament: A behavior regulated by profilin. *J. Cell Biol.* **97:** 112–124.
8. Pollard T.D., Blanchoin L., and Mullins R.D. 2000. Biophysics of actin filament dynamics in nonmuscle cells. *Ann. Rev. Biophys. Biomol. Struct.* **29:** 545–576.
9. Loisel T.P., Boujemaa R., Pantaloni D., and Carlier M.F. 1999. Reconstitution of actin-based motility of *Listeria* and *Shigella* using pure proteins. *Nature* **401:** 613–616.

INTERACTIONS OF ACTIN, MYOSIN, AND A NEW ACTIN-BINDING PROTEIN OF RABBIT PULMONARY MACROPHAGES

II. Role in Cytoplasmic Movement and Phagocytosis

THOMAS P. STOSSEL and JOHN H. HARTWIG

From the Division of Hematology-Oncology, Children's Hospital Medical Center, The Sidney Farber Cancer Center, and the Department of Pediatrics, Harvard Medical School, Boston, Massachusetts 02115

ABSTRACT

Actin and myosin of rabbit pulmonary macrophages are influenced by two other proteins. A protein cofactor is required for the actin activation of macrophage myosin Mg^{2}ATPase activity, and a high molecular weight actin-binding protein aggregates actin filaments (Stossel T. P., and J. H. Hartwig. 1975. *J. Biol. Chem.* 250:5706–5711). When warmed in 0.34 M sucrose solution containing Mg^{2}-ATP and dithiothreitol, these four proteins interact cooperatively. Actin-binding protein in the presence of actin causes the actin to form a gel, which liquifies when cooled. The myosin contracts the gel into an aggregate, and the rate of aggregation is accelerated by the cofactor. Therefore, we believe that these four proteins also effect the temperature-dependent gelation and aggregation of crude sucrose extracts of pulmonary macrophages containing Mg^{2}-ATP and dithiothreitol. The gelled extracts are composed of tangled filaments.

Relative to homogenates of resting macrophages, the distribution of actin-binding protein in homogenates of phagocytizing macrophages is altered such that 2–6 times more actin-binding protein is soluble. Sucrose extracts of phagocytizing macrophages gel more rapidly than extracts of resting macrophages. Phagocytosis by pulmonary macrophages involves the formation of peripheral pseudopods containing filaments. The findings suggest that the actin-binding protein initiates a cooperative interaction of contractile proteins to generate cytoplasmic gelation, and that phagocytosis influences the behavior of the actin-binding protein.

Phagocytosis is a fundamental biological process. Early microscopists observing phagocytosis by macrophages theorized that consistency changes or gel-sol transformations of the peripheral cytoplasm might explain the flow of pseudopods around particles that characterizes engulfment (16). The appreciation in recent years that phagocytosis depends on energy metabolism (21), the formal demonstration that microfilaments are prominent in the cell periphery adjacent to particles being engulfed and in pseudopods embracing particles (27), and the unequivocal demonstration of contractile proteins in many nonmuscle cells (26) all support the idea that contractile proteins are involved in phagocytosis. The observation that actin polymerized subnormally in extracts of human polymorphonuclear leukocytes which ingested particles at markedly impaired rates provided further evidence for this idea (1).

As part of our endeavor to understand the mechanism by which phagocytic cells ingest particulate objects, we have been studying their contractile proteins and have purified actin and myosin from rabbit pulmonary macrophages. The myosin had low Mg^{2}-ATPase activity which was activated by F-actin at low ionic strength only in the presence of an additional protein cofactor which was partially purified from the macrophages. We discovered and purified a new high molecular weight protein which coprecipitated with actin when extracts of rabbit pulmonary macrophages, prepared with ice-cold 0.34 M sucrose-1 mM ethylenediaminetetraacetic acid (EDTA) at pH 7.0, were made 0.075–0.1 M in KCl and warmed to room temperature with stirring. In 0.1 M KCl, the purified high molecular weight protein-bound macrophage F-actin into dense arrays of interconnecting filaments which led to our naming it "macrophage actin-binding protein." We hypothesized that an interaction between macrophage actin, actin-binding protein, myosin, and cofactor might explain the assembly and movement of macrophage pseudopods which characterize ingestion (11, 33).

Kane reported that actin filaments formed a gel when extracts of sea urchin eggs prepared with glycerol and ethyleneglycolbis[β-aminoethyl ether]*N*,*N*′ tetraacetic acid (EGTA) were warmed and that the gelled actin was associated with other proteins, including a high molecular weight polypeptide similar in electrophoretic mobility in dodecyl sulfate to the macrophage actin-binding protein (14). Pollard subsequently discovered that if extracts of *Acanthamoeba castellanii* prepared with ice-cold 0.34 M sucrose containing 1 mM EGTA and 1 mM ATP were warmed to room temperature *without* stirring, the extracts gelled and subsequently underwent a rapid shrinkage resembling a contraction (22). In light of the possible relevance of cytoplasmic gelation in the phagocytic process, these phenomena appeared to provide a model for analyzing the mechanical interactions between macrophage contractile proteins and for determining whether phagocytosis influences these interactions.

We report here the results of our studies concerning the effects of phagocytosis on the gelation and contraction of extracts of rabbit pulmonary macrophages. We present evidence that this gelation and contraction is the result of a cooperative interaction between macrophage actin, actin-binding protein, cofactor, and myosin which correlates with enzymatic and binding interactions studied previously. The actin-binding protein, some further properties of which we describe, is the initiator of this interaction. Ingestion of particles increases the extractability of actin-binding protein from the macrophages which is the first example of a change in a contractile protein associated with phagocytosis. Some of this work has been presented in summary form (35).

MATERIALS AND METHODS

Preparation and Handling of Pulmonary Macrophages

Macrophages were obtained from the lungs of New Zealand albino rabbits by intratracheal lavage according to the technique of Myrvik et al. (20). The rabbits previously received complete Freund's adjuvant intravenously which increased cell yields (19). The macrophages were washed twice with 0.15 M NaCl at 4°C by centrifugation at 250 *g*-minute, and suspended at a final concentration of 5–10 mg cell protein/ml in Krebs-Ringer phosphate medium, pH 7.4.

The cells were then incubated for 20 min at 37°C with 0.2 vol of a suspension of *Escherichia coli* lipopolysaccharide-coated droplets of diisodecyl phthalate containing oil red O (31). These particles were treated with either heated (56°C, 30 min) or else fresh rabbit serum for 20 min at 37°C. Fresh but not heated serum opsonizes these particles by depositing a fragment of the third component of complement on their surfaces (32). The rate of ingestion of the particles by the macrophages was assayed spectrophotometrically (30). The macrophages were washed free of uningested particles twice with ice-cold 0.15 M NaCl with centrifugation at 250 *g*-min. The cells were then suspended in 15 vol of ice-cold distilled water, mixed by inversion, and packed by centrifugation at 250 *g*-min. The washed macrophages were suspended in 2 vol of an ice-cold homogenizing solution containing 0.34 M sucrose, 10 mM dithiothreitol, 5 mM ATP, 1 mM EDTA, 20 mM Tris-maleate buffer, pH 7.0 (hereafter called "sucrose solution"). ATP was omitted from the sucrose solution in some experiments. The cells were disrupted in this solution as reported previously (11). The homogenates were layered under an equal volume of 0.25 M sucrose, 10 mM Tris-maleate, pH 7.0 solution, and centrifuged at 100,000 *g*-h at 4°C. This step yielded the extract supernate, a pellet of membranes and organelles, and a pellicle of phagocytic vesicles which floated through the 0.25 M sucrose layer covering the extract supernate (34). The presence of this layer facilitated complete removal of these structures from the supernatant fraction.

Isolation of Macrophage Actin, Myosin, Cofactor, and Actin-Binding Protein

Macrophage actin was purified from the sucrose extracts by procedures previously reported (11). As observed by Pollard concerning *Acanthamoeba* extracts, warming of macrophage extract supernates from 4°C to 25°C for 1 h resulted in their gelation (discussed in more detail below). Centrifugation of the gels in round-bottom test tubes at 12,000 *g*-min (25°C) compressed them, yielding an opaque pellet. Myosin was isolated from the supernatant fluids remaining after centrifugation of the gelled extract supernates.[1] The supernates were cooled to ice-bath temperature, and an equal volume of ice-cold saturated ammonium sulfate containing 0.01 M EDTA, pH 7.0, was slowly added. The resulting precipitates were dialyzed in 0.6 M KI, 20 mM sodium thiosulfate, 5 mM ATP, 5 mM dithiothreitol, 20 mM Tris maleate, pH 7.0, and applied to 2 × 85-cm columns of Bio-Gel A15, 200–400 mesh (Bio-Rad Laboratories, Richmond, Calif.), with the upper sevenths equilibrated in the KI solution just defined. The rest of the columns were equilibrated and eluted with 0.6 M KCl, 5 mM dithiothreitol, 0.5 mM ATP, 10 mM Trismaleate solution by the method of Pollard et al. (25). Fractions with K^+- and EDTA-activated ATPase activity (myosin) and fractions which activated macrophage actomyosin Mg^{2+}-ATPase activity (cofactor) were pooled separately.

The pellets produced by centrifugation of gelled extract supernates were dissolved at a concentration of about 3 mg protein/ml in a solution of ice-cold 0.6 M KCl, 5 mM dithiothreitol, 20 mM Tris-maleate, pH 7.0. This solution was centrifuged for 3 h at 80,000 *g* at 4°C. Actin-binding protein was isolated from the supernatant fluid. All solutions were carefully kept at 0°–4°C. The supernatant fluids were concentrated with an Amicon pressure device (Amicon Corp., Scientific Sys. Div., Lexington, Mass.) with an XM-50 membrane and dialyzed against 0.6 M KI, 20 mM sodium thiosulfate, 5 mM dithiothreitol, 20 mM Tris-maleate solution, pH 7.0. The concentrates were applied to 2 × 85-cm columns of Bio-Gel A15, 200–400 mesh, equilibrated in and eluted with the above solution. Peaks with absorbance at 290 nm were pooled, concentrated as described above, and analyzed for the presence of a high molecular weight polypeptide (actin-binding protein) by polyacrylamide gel electrophoresis with sodium dodecyl sulfate. Rabbit skeletal muscle actin was prepared by the technique of Spudich and Watt (29), and rabbit skeletal muscle heavy meromyosin by the method of Lowey et al. (17).

[1] As discussed below, myosin becomes incorporated into the gel with time and contracts it. Salt accelerates the contraction rate. After 1 h of incubation at room temperture in the absence of added salt, most of the myosin in the extract was in the supernate rather than in the gel.

Analytical Procedures

Proteins were negatively stained with uranyl acetate as reported (11). Intact macrophages were fixed in 2.5% glutaraldehyde, 0.1 M sodium cacodylate, postfixed with Dalton's chrome osmium and 1% uranyl acetate, dehydrated with acetone, and embedded in Epon-Araldite. Thin sections were stained with lead citrate. These sections and negatively stained proteins were examined in a Philips 300 microscope with an accelerating voltage of 60 kV.

Assays of ATPase activity, protein concentration, polyacrylamide gel electrophoresis with sodium dodecyl sulfate, and quantitative densitometry of stained polyacrylamide gels were performed by means of methods cited previously (11). The amino acid composition of purified, reduced, and alkylated (5) actin-binding protein after 24-h acid hydrolysis was determined (AAA Laboratories, Seattle, Wash.). Changes in the turbidity of protein solutions or of supernatant extracts were continuously assayed using a recording spectrophotometer (Gilford Instrument Laboratories, Inc., Oberlin, Ohio) at 350 nm in 0.5-ml quartz cuvettes as described in an earlier report (36). The temperature of the solutions being analyzed was thermostatically controlled (Haake, Inc., Saddle Brook, N.J.) by circulating water through a jacketing device in the spectrophotometer.

RESULTS

The Contractile Proteins of Pulmonary Macrophages

The purity, subunit composition, and morphology of contractile proteins of macrophages were described previously (11) but are demonstrated in Figs. 1 and 2 for purposes of comparison with the morphology and protein composition of cytoplasmic extracts, gels, and contracted aggregates to be described. Macrophage actin migrated during electrophoresis as a single subunit of mol wt 42,000 on dodecyl sulfate polyacrylamide gels. Macrophage F-actin appeared as double helical filaments (Fig. 1). Macrophage myosin had heavy (mol wt 220,000) and light (mol wt 20,000 and 15,000) subunits, and in 0.1 M KCl at pH 7.0 formed bipolar filaments which were 300 nm long (Fig. 1). The actin-binding protein as originally purified co-migrated with one of the subunits of the erythrocyte protein spectrin during electrophoresis on dodecyl sulfate polyacrylamide gels. In 0.1 M KCl at pH 7.0, it appeared as beaded aggregates when negatively stained (11). When the protein was prepared by means of the modified procedure

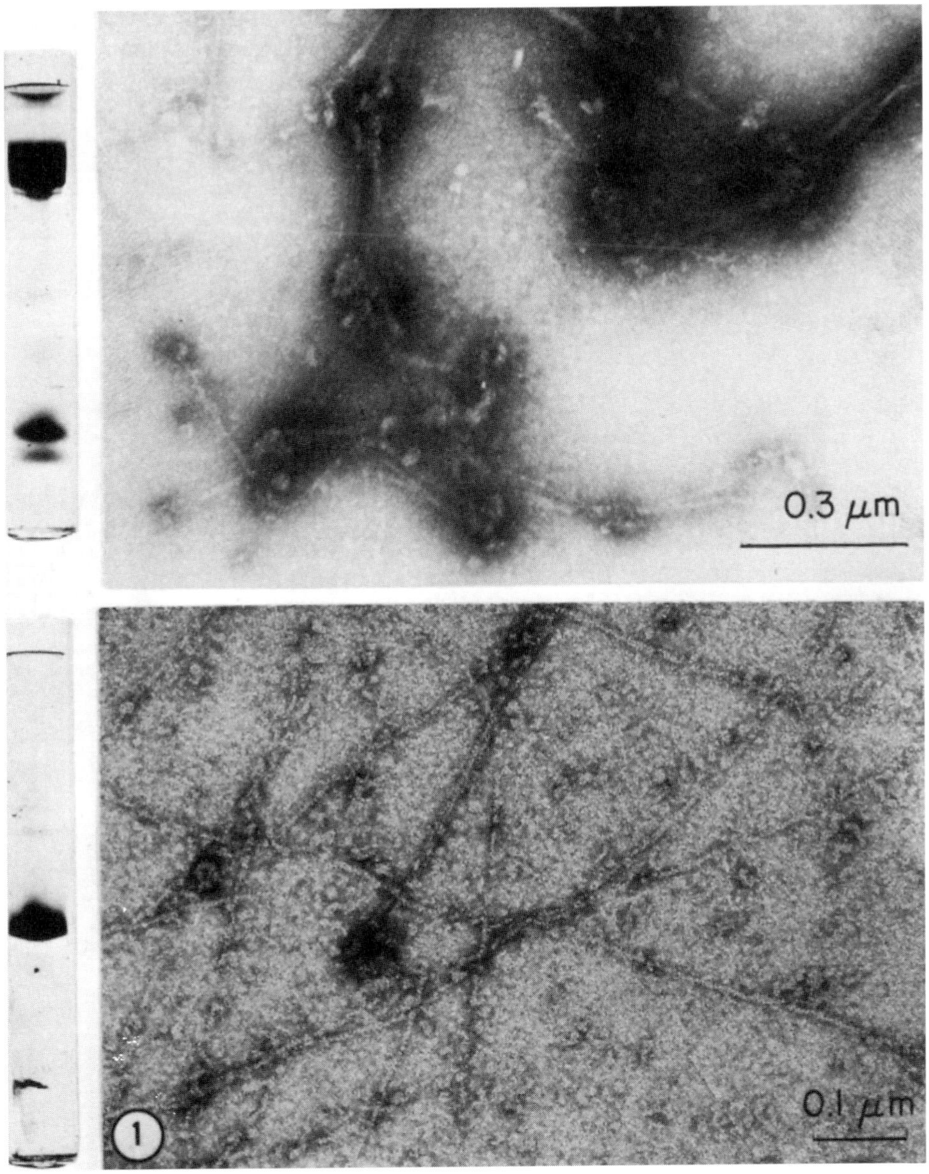

FIGURE 1 Morphology and subunit composition of the purified macrophage proteins, (top) myosin, and (bottom) actin. The reduced and denatured proteins are shown on Coomassie blue-stained dodecyl sulfate 5% polyacrylamide gels. Also shown is the morphology of macrophage myosin and actin negatively stained and photographed in the electron microscope.

described in this report, with careful attention paid to keeping the protein cold during purification, the purified protein migrated more slowly than the spectrin subunits during electrophoresis on polyacrylamide gels with dodecyl sulfate. Using other high molecular weight proteins for calibration, we now estimate that the actin-binding protein has a mol wt of approx. 270,000 (Fig. 2). We ascribe the faster mobility observed previously to proteolysis. Purified actin-binding protein warmed to 25°C for 20 min in 0.1 M KCl, 10 mM Tris-maleate buffer, pH 7.0, and negatively stained appeared as beaded hollow coils with diameters of 12 nm in the electron microscope (Fig. 2). The amino acid composition of macrophage actin-binding protein is indicated in Table I.

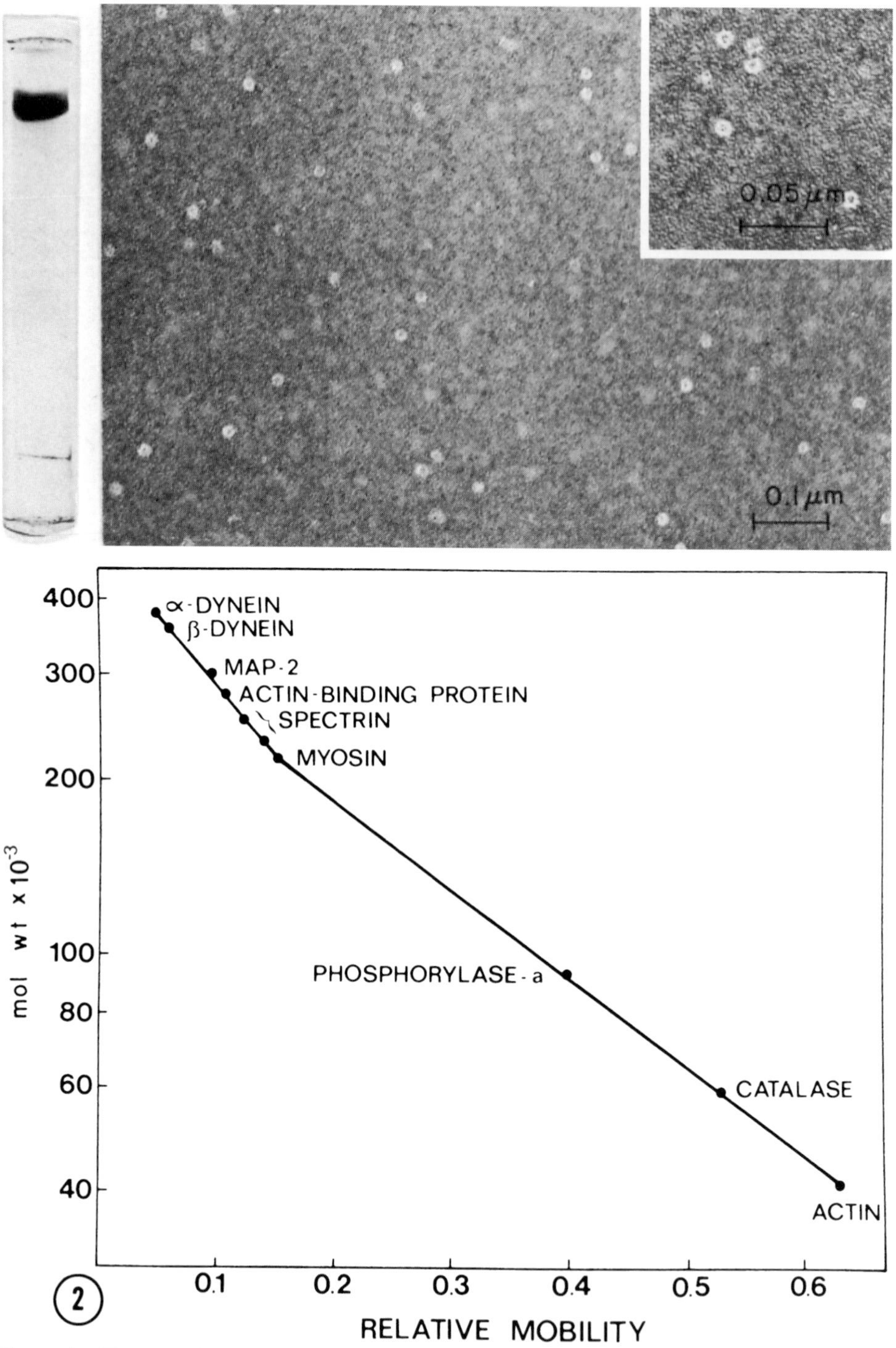

(2)

FIGURE 2 (Top) Morphology and subunit composition of macrophage actin-binding protein. A Coomassie Blue-stained dodecyl sulfate 5% polyacrylamide gel of purified macrophage actin-binding protein is shown. The micrographs show the morphology of purified macrophage actin-binding protein stained with uranyl acetate and viewed in the electron microscope. (Bottom) Comparison of the relative mobilities of reduced denatured proteins on dodecyl sulfate 5% polyacrylamide gells. Bands designated α-dynein, β-dynein and MAP-2 were observed in purified tubulin preparations (a gift of Dr. Michael Shelanski). The molecular weights of these proteins are those reported by Burns and Pollard (3) and Sloboda et al. (28). Rabbit erythrocyte spectrin and pulmonary macrophage actin-binding protein, myosin, and actin were purified as described by Hartwig and Stossel (11). Rabbit muscle phosphorylase-*a* and beef liver catalase were purchased from Sigma Chemical Co., St. Louis, Mo.

Morphology and Quantitation of Phagocytosis by Macrophages

Rabbit pulmonary macrophages observed with the phase-contrast microscope extended glassy pseudopods from the cell periphery when warmed to room temperature (Fig. 3). These pseudopods lacked the refractile bodies which were abundant in the interior cytoplasm. The pseudopods surrounded opsonized lipopolysaccharide-coated oil droplets and internalized them. Thin section electron micrographs of macrophages fixed while ingesting particles revealed that the pseudopods were composed of ill-defined arrays of randomly oriented interconnecting thin filaments and dense debris which excluded other recognizable intracellular organelles from the pseudopods. As shown in Fig. 4, pulmonary macrophages ingested opsonized particles rapidly but did not ingest unopsonized control particles.

Gelation and Contraction of Macrophage Extracts: Effect of Phagocytosis

Rabbit pulmonary macrophage extracts in 0.34 M sucrose containing dithiothreitol, ATP and EDTA (sucrose solution) gelled when warmed from ice-bath temperature to room temperature, and the gel was not disrupted by inversion (Fig. 5 *a*). Subsequently, strands formed a net in the gelled extract. This net then slowly pulled away from the vertical sides of the container. When the net was completely free of the sides of the vessel, it released from the bottom and underwent a rapid vertical contraction. This event produced a condensed aggregate which, because of trapped air bubbles, floated to the top of clear liquid squeezed out of the gel (Fig. 5 *b*).

Semiquantitative measurements of gelation were achieved by determining the time required for the supernatant extract to agglutinate and partially to immobilize latex particles on a microscope slide tilted back and forth, which gave an estimate of the strength of the gel. Viscometry was not suitable for quantifying gelation more precisely since the gel broke into chunks if disrupted. The optical density of extract supernates was assayed at a wavelength to maximize detection of light scattering, since light scattering has been used to monitor polymerization reactions (7). During warming and gelation, the turbidity of extract supernates increased (Fig. 6). Often an initial decrease in turbidity occurred which was clearly not due to condensation clearing from the cuvettes. If the gelled extracts (1 h, 25°C) were cooled, the gels liquified within 10 min (Fig. 6), and could be gelled again by warming. The total protein content of the liquified extract was not diminished by ultracentrifugation at 80,000 *g* for 3 h (4°C). During contraction of the gels, the net that formed in the gelled supernates was pulled out of the light path. Therefore, turbidity diminished markedly, and this diminution provided a quantitative assay for the final phase of contraction (hereafter called "final contraction").

Resting macrophage extracts prepared in sucrose solution gelled in 20–40 min. The final contraction of the extract gels assayed by clearing of the light path was very slow and required 3–12 h for completion. Both the gelation rate and the very gradual onset of final contraction were greatly increased by the addition of an excess of $MgCl_2$ (5 mM). The effect of $CaCl_2$ on the behavior of extract gels in the presence of excess $MgCl_2$ is shown in Fig. 7. $CaCl_2$ (either 1.3 mM or a buffered calcium solution 1 mM $CaCl_2$–0.6 mM EGTA) caused a flocculent precipitate to form immediately in the supernate, and gross gelation did not occur. Rapid gelation and final contraction occurred in either the absence of added $CaCl_2$ or in

TABLE I
Amino Acid Composition of Macrophage Actin-Binding Protein

Amino acid	Composition
	mol/100,000 g
Asp	76
Thr	54
Ser	59
Glu	99
Pro	62
Gly	103
Ala	64
Cys*	4
Val	74
Met	11
Ile	38
Leu	54
Tyr	27
Phe	28
His	19
Lys	52
Arg	36
Trp	10
Total	870

* Analyzed as *S*-carboxymethylcysteine after reduction and alkylation.

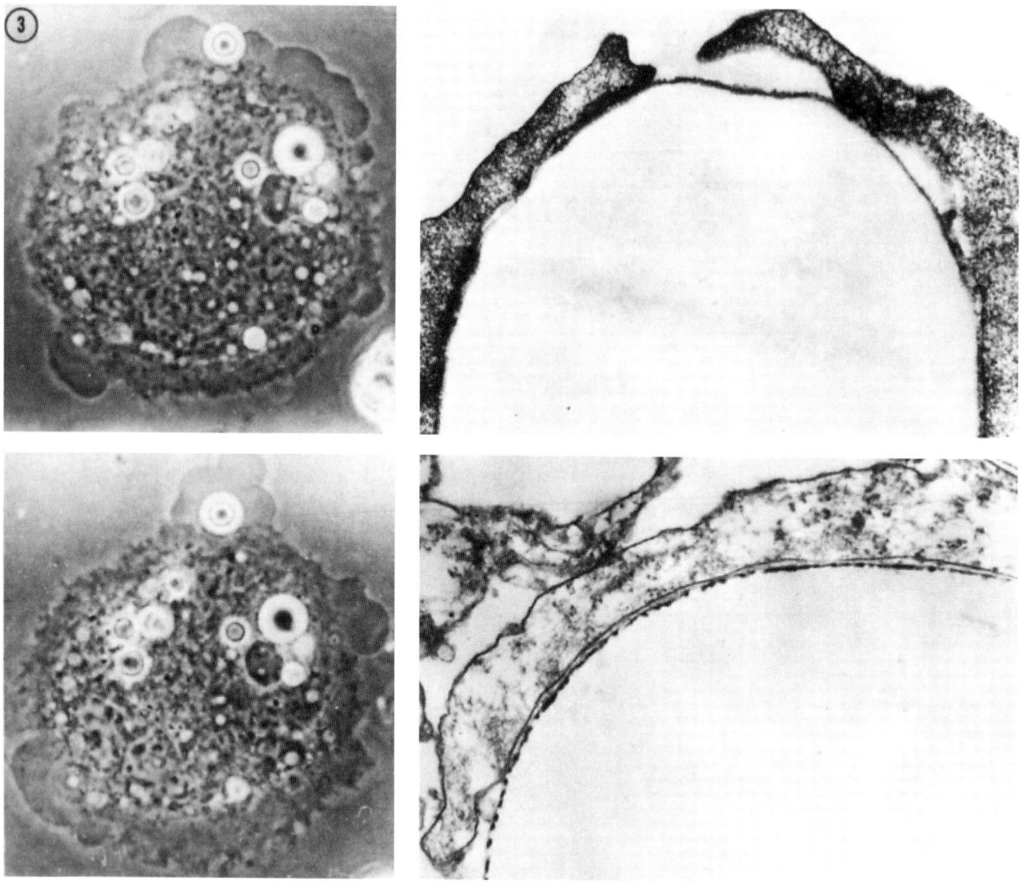

FIGURE 3 Morphology of rabbit pulmonary macrophages. (Left) Macrophages ingesting lipopolysaccharide-coated oil particles, phase contrast. × 1,000. (Right) Thin sections show macrophage pseudopods surrounding oil droplets. (Upper) × 42,500. (Lower) × 61,200.

the presence of 2.7 mM EGTA. Extracts prepared with sucrose solution minus ATP gelled but did not contract even after addition of $MgCl_2$.

Extracts of ingesting macrophages gelled more rapidly during warming than extracts of resting cells, as determined by the latex immobilization assay. Extracts from ingesting cells immobilized particles (gelled) in 10–20 min or half the time required for gelation of resting extracts, and the gels were stronger. Extracts of ingesting cells also scattered more light than extracts of resting cells during warming (Fig. 8, top). Extracts of ingesting cells containing $MgCl_2$ also gelled more rapidly than extracts of resting cells containing $MgCl_2$. On the other hand, the onset of the $MgCl_2$-stimulated final contraction was not accelerated in ingesting cells. In fact, resting cell extract gels began their final contraction sooner than ingesting cell extract gels (Fig. 8, bottom), possibly because more gel or a stronger gel had to be contracted.

Protein Composition of Intact Macrophages, Extract Supernates, and Gels: Effect of Phagocytosis

The concentration of proteins in macrophages, extract supernatants, and gels collected by low-speed centrifugation after 1 h of standing at 25°C was determined by quantitative densitometry of Coomassie Blue-stained dodecyl sulfate polyacrylamide gels after electrophoresis. The results are shown in Table II. Extracts prepared from cells that had ingested opsonized particles contained 58% more actin-binding protein than resting cells. The amounts of actin and myosin in the extract

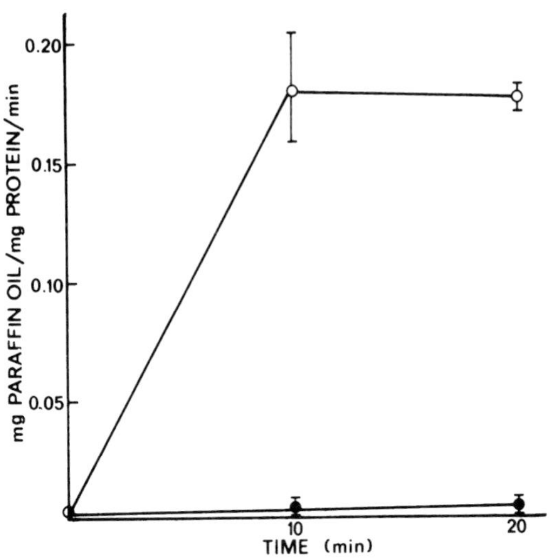

FIGURE 4 Ingestion of *E. coli* lipopolysaccharide-coated particles by rabbit pulmonary macrophages. Macrophages, 10 mg/ml cell protein, were incubated in 4 ml of Krebs-Ringer phosphate medium, pH 7.4, for 10 min at 37°C. At zero time, 1 ml of particles treated with fresh (O) or heat-inactivated (●) rabbit serum was added. Samples were removed from incubation flasks for washing and oil red O analysis at zero, 10, and 20 min.

supernates of resting and ingesting cells were not measurably different.

The quantities of actin, myosin, and actin-binding protein in whole cells did not differ whether the cells were resting or ingesting. Therefore, the increased amounts of actin-binding protein in extracts of ingesting macrophages relative to resting cells was the result of either redistribution, increased solubility, or increased extractability of the protein during phagocytosis.

An example of the protein composition of an extract gel (from resting cells) collected by low-speed centrifugation is shown in Fig. 9. Although there was some variation in the quantitative distribution of proteins in these gels, the major band was invariably actin. Actin-binding protein and myosin heavy chain were other prominent components. A polypeptide of approximately 90,000 mol wt, close to the molecular weight previously assigned to macrophage cofactor, was also present. Actin-binding protein and myosin were purified from these extract gels, which verified that the bands identified as these proteins on polyacrylamide gels in fact were these contractile proteins. Extract gels from ingesting cells contained 110% more actin-binding protein than extract gels from resting cells. The findings suggested that the increased rate of gelation of extracts of ingesting cells could be explained by the increased concentration of actin-binding protein in these extracts. This conclusion would mean that the acting-binding protein was responsible for the gelation of actin under these conditions. The validity of this idea was established by experiments with purified macrophage proteins.

Gelation and Contraction of Purified Macrophage and Muscle Proteins

During the warming of purified actin-binding protein from 0° to 25°C in 0.34 M sucrose solution, the turbidity of the solution increased (Fig. 10). Cooling reversed this turbidity and returned the absorbance to the initial basal measurement before warming. Purified rabbit skeletal muscle or macrophage G-actins alone in the 0.34 M sucrose solutions at the concentrations used did not gel or develop significant turbidity during warming. Purified actins plus purified macrophage myosin did not gel but slowly precipitated and settled such that turbidity decreased (Fig. 11). The addition of rabbit skeletal muscle or macrophage G-actin in sucrose solution at 0°C to actin-binding protein reconstructed the gelation phenomenon observed in the crude supernatant extracts. When the mixture of purified proteins was warmed to 25°C, the mixture gelled very rapidly (within minutes). But the increase in turbidity of the combined proteins with time was invariably less than that of actin-binding protein alone (Fig. 11). In repeated experiments the strength of the gel formed was roughly proportional to the quantity of actin-binding protein added. For example, at a weight ratio of actin-binding protein: rabbit muscle F-actin of 1:2 or 1:5, a solid gel that did not move at all on a tilted microscope slide formed within 15 min at room temperature in constant volumes of sucrose solution. At ratios of 1:10 and 1:20, the mixture was weakly mobile but fractured into chunks when perturbed. Actin alone (1 mg/ml) moved freely. A combination of purified actin, actin-binding protein, and myosin also gelled during warming; but marked increases in turbidity gradually occurred as observed in crude extract supernates. When the combination of actin, actin-binding protein, and myosin were left overnight at 25°C, final contraction occurred but only if myosin was present (Fig. 11). As reported previously, the presence of a macrophage cofactor increases markedly the Mg^{2+}-ATPase activity of macrophage actomyosin (Table III). Partially

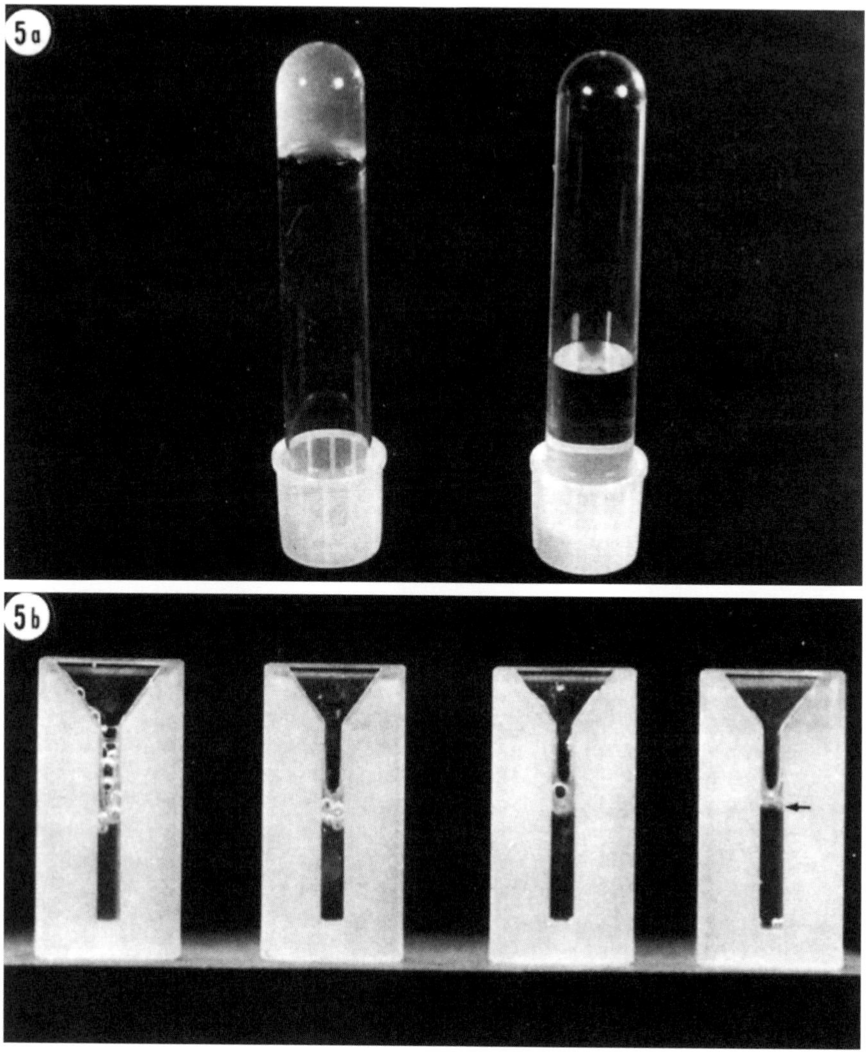

FIGURE 5 Gelation and contraction of rabbit pulmonary macrophage supernatant extracts in sucrose solution. (*a*) Macrophage supernate gelled by warming to 25°C for 60 min. Shown are inverted tubes of (left) gelled supernate and (right) water for comparison. (*b*) Stages in contraction of macrophage supernatant gels in cuvettes. From left to right, diffuse reticular filaments which progressively contract into the center of the cuvette, resulting in the final aggregate which has floated to the surface of the cuvette (arrow).

purified cofactor was added to purified actin, actin-binding protein, and myosin to determine whether this increase in enzymatic activity correlated with the mechanical events. The addition of the cofactor alone did not greatly increase the amount of turbidity generated or hasten the onset of final contraction. However, when cofactor was added in the presence of excess $MgCl_2$ (5 mM), turbidity increased rapidly and final contraction was accelerated (Fig. 12). The amount of turbidity formed was greater than that arising in the absence of cofactor plus Mg^{2+} (compare with Fig. 11). The effect of $CaCl_2$ on the behavior of the reconstituted system is shown in Fig. 13. Both the rate of increase in turbidity and the onset of final contraction occurred slightly sooner in EGTA than in the presence of free Ca^{2+}.

Analysis by polyacrylamide gel electrophoresis with dodecyl sulfate of the aggregate formed by final contraction of the gel formed from purified

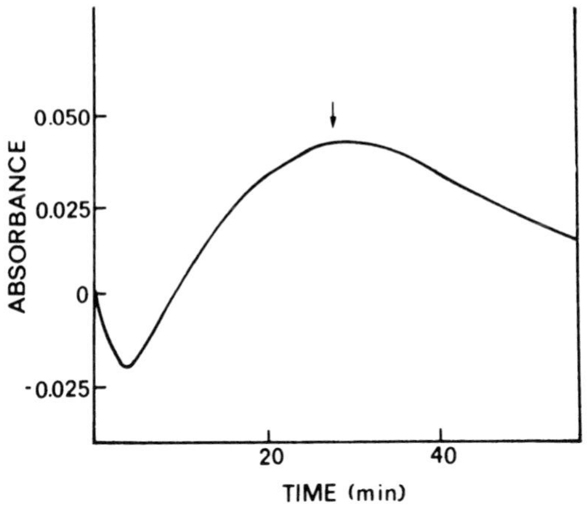

FIGURE 6 Effect of temperature on the turbidity (A_{350}) of a macrophage supernatant extract in sucrose solution. At zero time the extract was warmed to 25°C. The temperature of water circulating through the cuvette holder was lowered to 0°C at the time indicated by the arrow. Delay in cooling of the cuvette caused the fall in turbidity to occur more slowly than liquefaction of gelled extracts placed directly on ice.

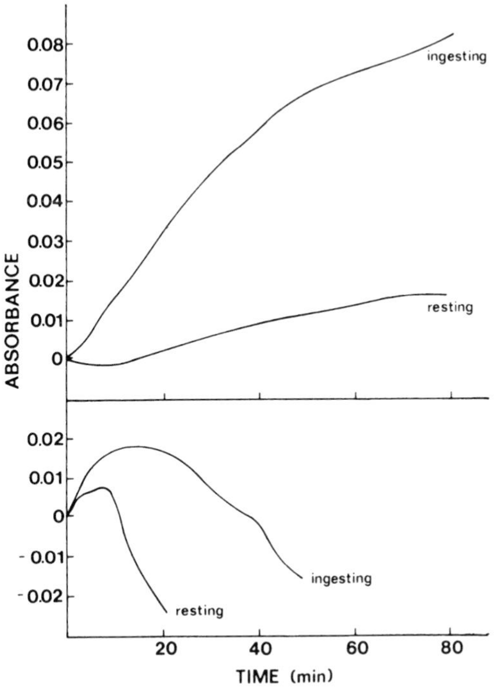

FIGURE 8 Temperature-dependent changes in turbidity of macrophage extracts prepared in sucrose solution from resting or ingesting cells. (Top) The ingesting extract was solidly gelled in 15 min; the resting in 30 min. The bottom panel shows turbidity of extracts to which 5 mM $MgCl_2$ was added. The abrupt decrease in absorbance was associated with final contraction of the gels. At zero time the extracts were warmed to 25°C. Total protein concentrations of ingesting and resting extracts were 9.1 mg/ml and 8.7 mg/ml, respectively.

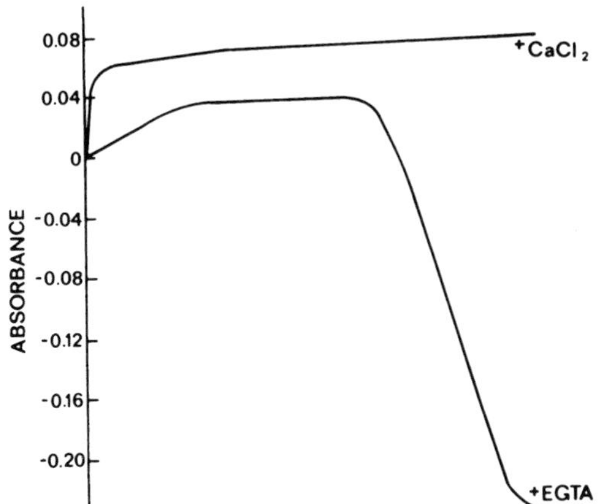

FIGURE 7 Effect of $CaCl_2$ or EGTA on the temperature-dependent changes in turbidity of macrophage supernate. At zero time, the supernates in sucrose solution containing added 5 mM $MgCl_2$ and either 1.3 mM $CaCl_2$ or 2.7 mM EGTA as indicated were warmed to 25°C. The abrupt decrease in absorbance was associated with contraction of the gel. Since protein was removed from the extract by the contraction, the absorbance was less than the starting optical density (explaining the negative readings).

proteins revealed five polypeptide bands, three of which corresponded to the macrophage proteins, actin, actin-binding protein, and myosin heavy chain, and two of which, with apparent mol wt of 155,000 and 90,000, have not been definitively identified (Fig. 14). These proteins were quantitatively removed from the supernatant fluid surrounding the aggregate. The 90,000 mol wt band was a polypeptide tentatively thought to represent macrophage cofactor (33).

The Morphology of Contracting Gels

Macrophage extracts were allowed to gel for 120 min and collected by low-speed centrifugation. The gelatinous pellet was gently disrupted at 25°C with 100 vol of 0.1 M KCl, 10 mM Tris-maleate, pH 7.0, in a Dounce homogenizer (Kontes Glass Co.,

TABLE II

Comparison of Actin-Binding Protein, Myosin and Actin Concentrations between Resting and Ingesting Macrophages, Extracts, and Gelled Extracts Collected by Centrifugation

	Resting*	Ingesting	P‡
	%	%	
Whole cells			
Actin-binding protein	1.82 ± 0.48	1.72 ± 0.45	NS
Myosin	2.06 ± 0.19	1.91 ± 0.67	NS
Actin	10.38 ± 2.01	12.58 ± 1.90	NS
Supernate			
Actin-binding protein	1.72 ± 1.04	2.72 ± 1.00	<0.01
Myosin	4.28 ± 0.68	4.69 ± 1.05	NS
Actin	12.56 ± 2.01	14.81 ± 4.45	NS
Gel			
Actin-binding protein	2.26 ± 0.56	4.77 ± 1.54	<0.01
Myosin	3.47 ± 0.32	3.08 ± 0.62	NS
Actin	31.43 ± 2.60	32.45 ± 1.87	NS

* Numbers express percent ± standard deviation of the total stained protein by densitometry of 5% polyacrylamide gels with dodecyl sulfate where $n = 5$.
‡ P indicates the significance of the difference between resting and ingesting means.

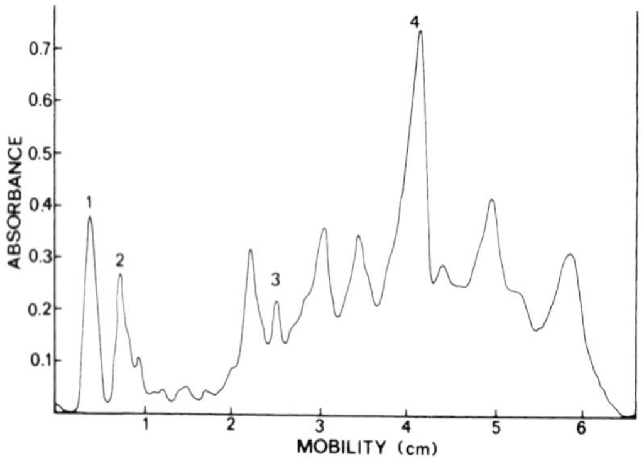

FIGURE 9 Densitometric scan of Coomassie Blue-stained dodecyl sulfate 5% polyacrylamide gel of a macrophage extracts gel collected by centrifugation after 60 min. The numbers label peaks corresponding to *1*, actin-binding protein; *2*, myosin heavy chain; *3*, cofactor; and *4*, actin.

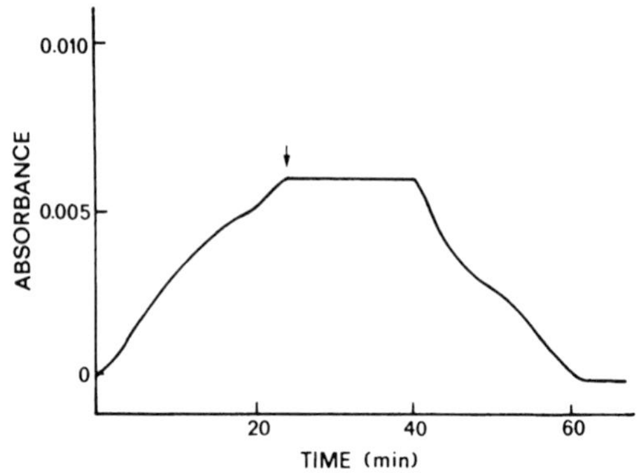

FIGURE 10 Effect of temperature on purified actin-binding protein. Purified actin-binding protein was dialyzed in sucrose solution and warmed to 25°C at zero time. The temperature of water circulating through the cuvette holder was lowered to 0°C at the time indicated by the arrow. Total protein concentration was 0.4 mg/ml.

Vineland, N.J.), and a sample was negatively stained and examined in the electron microscope (Fig. 15). The overall appearance was that of arrays of filaments and background debris (Fig. 15, top). Single and paired thin filaments (6–7 nm wide) were evident as were thicker (12–14 nm wide) filaments (middle). The pellet suspended in 0.1 M KCl solution was incubated with rabbit skeletal muscle heavy meromyosin, negatively stained, and examined (bottom). Most of the filaments observed had acquired arrowhead projections.

DISCUSSION

In a previous study, we showed that when cold sucrose extracts of pulmonary macrophages were warmed with stirring, a macrophage actin binding protein linked actin filaments into dense arrays of interconnecting cables (33). We have now demonstrated that cold sucrose extracts of pulmonary macrophages containing EDTA, dithiothreitol, and ATP gel and aggregate when warmed without stirring, and we have reproduced these phenomena with purified macrophage proteins. Identical results with extracts of human chronic myelogenous leukemic granulocytes (2) suggest that this may be a widespread cytological phenomenon. We have also shown that phagocytosis produces a change in the distribution of a macrophage actin-binding protein which secondarily results in enhanced gelation of macrophage extracts.

Gelation

Actin is the major component of macrophage gels, as it is of gels formed in extracts of other cells (14, 22). It has long been known that rabbit skeletal muscle F-actin is thixotropic (6), and recent studies on the state of rabbit skeletal muscle F-actin in solution have suggested that it exists as a weak gel (4). Whether rabbit macrophage F-actin differs from rabbit skeletal muscle F-actin with respect to the strength of a gel that it can form on its own (as has been suggested from work with *Acanthamoeba* actin [22]) is unknown. However, at the concentrations examined, purified rabbit macrophage or muscle G- or F-actin formed a gel (actually a rubbery solid which fractured when disrupted) only in the presence of purified actin-binding protein, and extracts of ingesting macrophages, containing more actin-binding protein than extracts of resting macrophages, gelled more rapidly than extracts of resting macrophages. Therefore, we were able to demonstrate that the actin-binding protein was responsible for the gelation of macrophage or muscle actin as it was for the precipitation of stirred actin filaments. While stirring does not prevent cross-linking of actin filaments by actin-binding protein, it must fracture the gel.

The macrophage actin-binding protein itself underwent a change during warming as evidenced by increases in turbidity. These alterations were reversed by cooling. The results were consistent with the actin-binding protein's initiation of gelation by undergoing a temperature-dependent aggregation, and the solubilization of crude macrophage extract gels by cooling supported this con-

TABLE III

Interaction of Macrophage Myosin, Cofactor, Actin-Binding Protein, and Rabbit Skeletal Muscle F-Actin

	Mg^{2+}-ATPase activity	
	nmol Pi/mg myosin protein/min	%
Complete system	380	100
−Actin-binding protein	387	102
−Myosin	12	3.1
−Cofactor	10	2.9
−Actin	9	2.3

The complete system contained, in 0.5 ml, 0.03 mg of macrophage myosin, 0.3 mg of rabbit skeletal muscle F-actin, 0.13 mg of crude cofactor, and 0.02 mg of actin-binding protein in 5 mM $MgCl_2$, 1.0 mM ATP, 40 mM KCl, and 10 mM Tris-maleate, pH 7.0. Incubations were carried out for 30 min at 37°C.

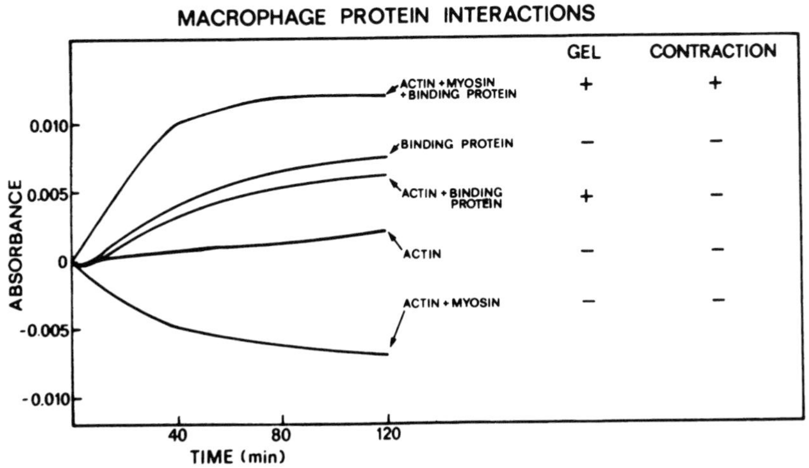

FIGURE 11 Effect of temperature on the turbidity of purified macrophage proteins in sucrose solution. At zero time the proteins in a total volume of 0.3 ml of sucrose solution were warmed to 25°C in cuvettes. Final protein concentrations were: actin, 2.0 mg/ml; actin-binding protein, 0.12 mg/ml; and myosin, 0.12 mg/ml. Gelation and/or final contraction occurred where indicated.

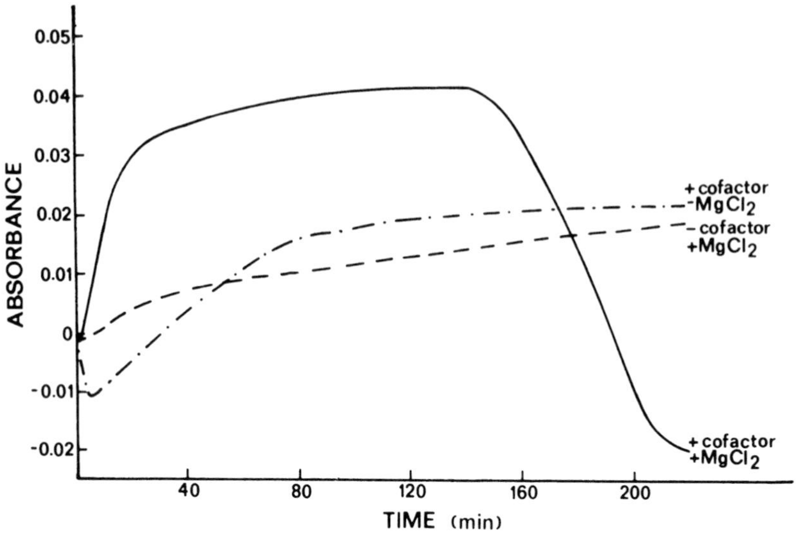

FIGURE 12 Effect of macrophage cofactor and $MgCl_2$ on the turbidity of actin-binding protein, myosin, and actin warmed to 25°C in sucrose solution. The rapid decrease in absorbance indicates final contraction. Total protein concentrations in a total volume of 0.3 ml of sucrose solution were: actin, 2.0 mg/ml; actin-binding protein, 0.12 mg/ml; myosin, 0.12 mg/ml; and cofactor, 0.5 mg/ml. The final $MgCl_2$ concentration was 5 mM.

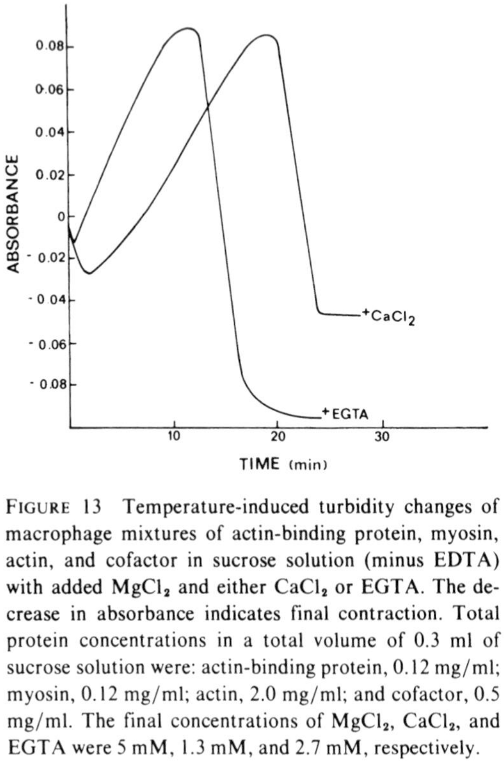

FIGURE 13 Temperature-induced turbidity changes of macrophage mixtures of actin-binding protein, myosin, actin, and cofactor in sucrose solution (minus EDTA) with added $MgCl_2$ and either $CaCl_2$ or EGTA. The decrease in absorbance indicates final contraction. Total protein concentrations in a total volume of 0.3 ml of sucrose solution were: actin-binding protein, 0.12 mg/ml; myosin, 0.12 mg/ml; actin, 2.0 mg/ml; and cofactor, 0.5 mg/ml. The final concentrations of $MgCl_2$, $CaCl_2$, and EGTA were 5 mM, 1.3 mM, and 2.7 mM, respectively.

cept. The increase in turbidity associated with warmed, purified actin-binding protein was diminished by the presence of actin undergoing gelation despite the presence of more protein in the system. This paradoxical finding suggested that the structure of actin-binding protein differed in the presence of actin. The optical phenomena could be explained if actin-binding protein, instead of aggregating, interacted with actin. Our earlier results indicated that 1 mol of actin-binding protein could bind as many as 100 G-actin monomers, although the maximal binding capacity of the actin-binding protein has not been formally determined (33). The actin-binding protein could promote filament assembly by binding actin monomers and bringing them into apposition. Sedimentation studies indicated that actin in sucrose extracts polymerized when warmed and depolymerized when cooled. Actin was sedimentable in extracts gelled by warming. Ultracentrifugation of gelled extracts, liquified by cooling, removed less than 10% of the actin and 5% of the total protein from the supernate. On the other hand, purified rabbit skeletal muscle actin polymerized by warming in sucrose remained completely sedimentable after cooling. While it is not clear at this time whether tempera-

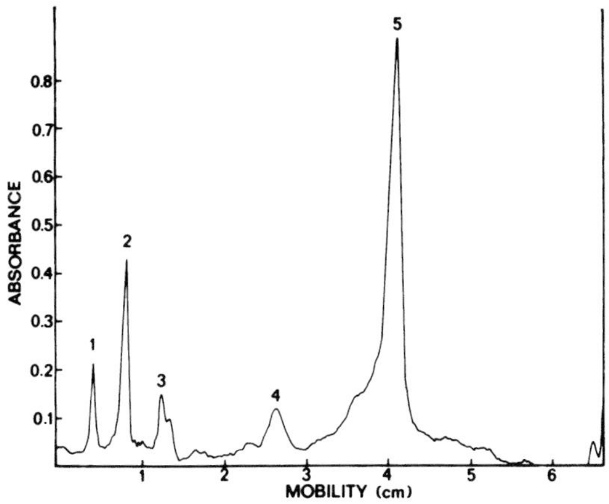

FIGURE 14 Densitometric scan of Coomassie Blue-stained dodecyl sulfate 5% polyacrylamide gel of the contracted aggregate made with purified macrophage proteins. The contracted aggregate formed when 0.03 mg of actin-binding protein, 0.03 mg of myosin, 0.13 mg of crude cofactor, and 0.5 mg of actin were warmed in 0.3 ml of sucrose solution containing added 5 mM $MgCl_2$ and 2.7 mM EGTA. The numbers label peaks corresponding to *1*, actin-binding protein; *2*, myosin heavy chain; *3*, 150,000 mol wt band; *4*, cofactor; and *5*, actin.

ture-dependent reversibility of actin assembly is a property of macrophage actin or is conferred by other proteins such as actin-binding protein, it appears that the actin polymers formed in macrophage extracts show unusual behavior relative to muscle F-actin alone.

The ability of the actin-binding protein to create a strong actin gel is unique but has counterparts in other muscle and nonmuscle proteins which bind to actin. Alpha-actinin cross-links actin filaments, but this process is more efficient in the cold than at high temperatures (8). Furthermore, this protein increases the Mg^{2+}-ATPase activity of actomyosin whereas the macrophage actin-binding protein has no influence on this activity. Tropomyosin alters the rheological properties of actin, but does not cause the type of temperature-dependent gelation observed here (13). The erythrocyte protein, spectrin, is associated with actin in the red cell (39) and has solubility and other properties which resemble somewhat those of the macrophage actin-binding protein (11, 18, 33). An erythrocyte protein isolated by sucrose density gradient centrifugation, possibly spectrin, was found to form ring structures of about 14-nm diam (9, 10), and a recent report showed negatively stained spectrin preparations to contain C-shaped structures (39). However, macrophage actin-binding protein and spectrin differ in several respects. The amino acid composition of macrophage actin-binding protein varies from that of spectrin as published (18). Reduced macrophage actin-binding protein yields one band on dodecyl sulfate polyacrylamide gels whereas spectrin yields two. The macrophage subunit has now been shown to be larger than either of the spectrin subunits. In one study, spectrin did not cause the gelation of rabbit skeletal muscle actin, although it influenced its viscosity (39). Nevertheless, the possibility remains that these two proteins are related in some way. Proteins with high molecular weight subunits and associated with actin have been subsequently discovered in other cell types. A protein purified from human leukemic granulocytes co-migrates, during electrophoresis on dodecyl sulfate polyacrylamide gels, with the macrophage protein, and it causes the gelation of human granulocyte or rabbit skeletal muscle actins (2). A protein in *Acanthamoeba* co-migrates, during electrophoresis on dodecyl sulfate polyacrylamide gels, with the human granulocyte and rabbit macrophage actin-binding proteins and is found in *Acanthamoeba* cytoplasmic actin gels (22). A protein with a high molecular weight subunit was also observed in cytoplasmic actin gels of sea urchin eggs (14) and of fibroblasts (40). These observations suggest that cytoplasmic actin-binding proteins may be widely distributed.[2]

A detailed study of the requirements and optimal conditions for gelation of macrophage extracts or of macrophage actin plus actin-binding protein has not yet been done. The gelation of *Acanthamoeba* extracts has been examined, and the general features were quire similar to those of the macrophage system presented here (22). The process in macrophages was temperature dependent and was accelerated by salt. $MgCl_2$ was an efficient accelerator. In crude macrophage extracts gelation occurred in the absence of added calcium ions and in the presence of EGTA. However, the

[2] Isolation of a high molecular weight subunit protein from chicken gizzard smooth muscle which cross reacts immunologically with non-muscle cells has recently been reported (Wang, K., J. F. Ash, and S. J. Singer. 1975. Filamin, a new high molecular-weight protein found in smooth muscle and non-muscle cells. *Proc. Natl. Acad. Sci. U.S.A.* **72**:4483–4486).

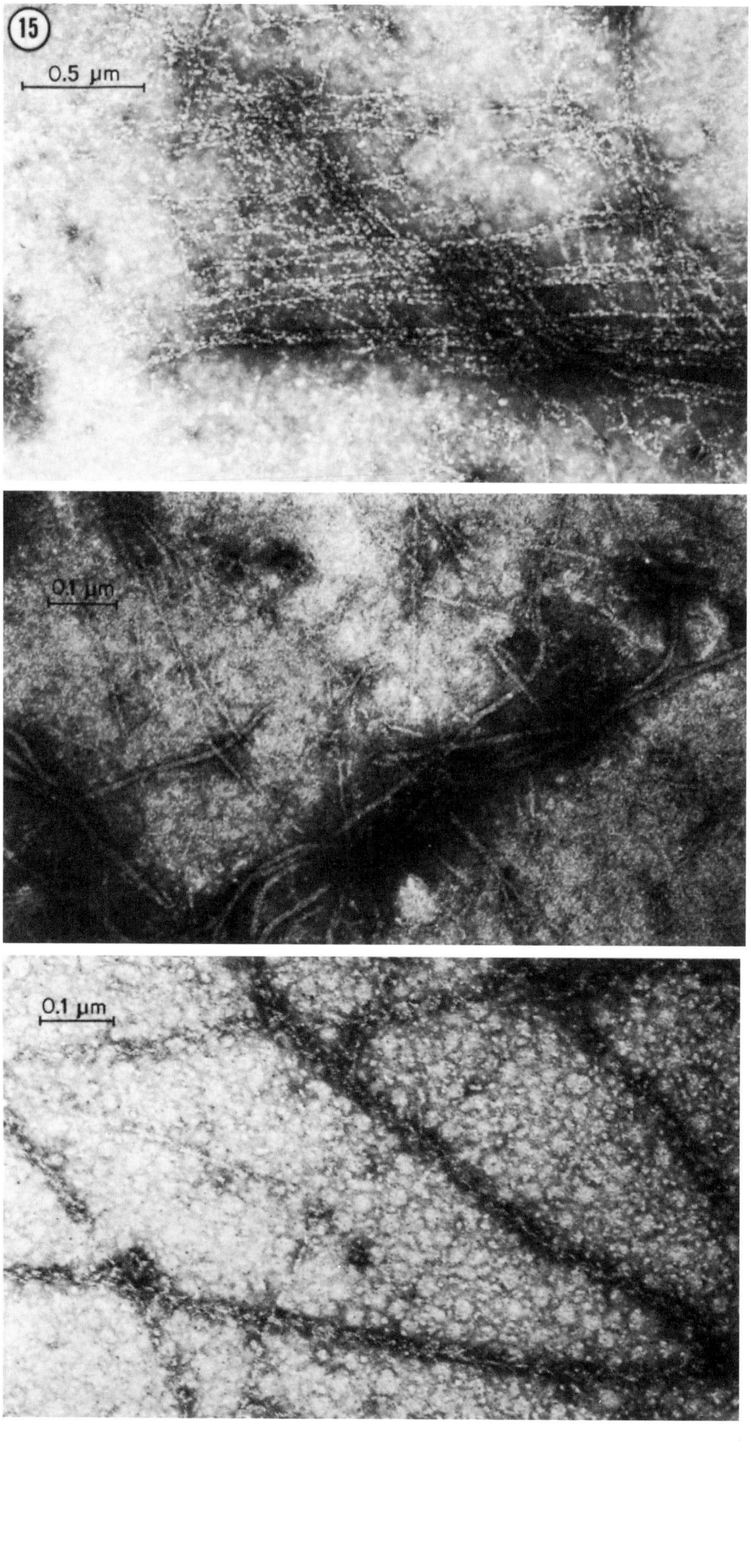
15
0.5 µm
0.1 µm
0.1 µm

actual free calcium content of these extracts even in the presence of EDTA is not known. The addition of various concentrations of calcium or calcium-EGTA caused marked precipitation of the extracts, and gelation did not occur. Gelation occurred when purified actin was combined with purified actin-binding protein in the presence of EGTA. In these reconstitution experiments, it is unlikely that free calcium was present. We conclude, therefore, that gelation does not require calcium, although it remains possible that calcium regulates gelation.

Various cytoplasmic extract preparations capable of undergoing movement have been studied by morphologic techniques in attempts to fathom the mechanisms of cytoplasmic motility (23, 24, 38, 41). Although some of these extracts were found to have temperature-dependent consistency changes (23), actual gelation has been described only recently, and only in extracts prepared with sucrose or glycerol (14, 22). Sucrose stabilizes actin, permitting it to polymerize after the removal of its bound divalent cations and nucleotides (15). The relevance of actin-binding protein-induced actin gelation in sucrose to cytoplasmic consistency changes in vivo remains to be proven. We speculate that high intracellular protein concentrations could have the same stabilizing effects as sucrose does in vitro.

Contraction

The properties of the aggregation of the macrophage-extract gel (final contraction) indicate that it is equivalent to the contraction or superprecipitation of skeletal muscle actomyosin and is a model of contractile force generation (37). The contraction of the crude macrophage extract gel requires added ATP and is accelerated by $MgCl_2$. Myosin is needed for aggregation of the actin-actin-binding protein gel, and the activity of the myosin is increased by macrophage cofactor and by $MgCl_2$. The process therefore parallels the Mg^{2+}-ATPase activity of macrophage actomyosin which is activated by a cofactor (33). The phenomenon differs from classic actomyosin superprecipitation with respect to the starting material, a solid actin-actin-binding protein gel rather than a thixotropic actomyosin gel, and with respect to the cofactor requirement. It is of interest that final contraction of the actin-actin-binding protein gel does occur, albeit very slowly, in the presence of myosin alone, which suggests that the low but finite Mg^{2+}-ATPase activity of macrophage actomyosin in the absence of cofactor is the transducer of the mechanical activity.

The precipitation of crude macrophage extracts in the presence of added calcium ions and their lack of contraction cannot be readily interpreted. The fact that the complete system of purified proteins contracts in EGTA solution suggests strongly that the reaction does not require calcium, but it leaves open the question as to whether the reaction can be regulated by calcium. Factors capable of conferring calcium regulation upon the system could easily have been lost or inactivated, although it was shown previously that the Mg^{2+}-ATPase activity of cofactor-activated macrophage actomyosin in relatively crude fractions was not influenced by calcium ions (33). It is notable that controlled reversible actin filament assembly, if it exists in cells, obviates the absolute need for a separate regulatory system. Contraction could be controlled merely by the availability of actin filaments. In essence, contraction could be controlled by gelation. The effect of myosin on mixtures of actin and actin-binding protein followed by warming reveals that myosin is responsible for the large changes in light scattering induced by warming of crude macrophage extracts. The measured turbidity therefore reflects the total cooperative interaction of the component proteins undergoing gelation and contraction. Presumably, the formation of the actin-actin-binding protein lattice provides binding sites for myosin. Turbidity equivalent to that observed in the crude extract occurs only when the myosin is activated by cofactor, $MgCl_2$, and ATP. Therefore, contraction must begin immediately in the complete system, and it culminates in the final aggregation of the lattice. Myosin filaments were not definitely iden-

FIGURE 15 Morphology of contracting macrophage extract gel warmed in sucrose solution, disrupted with 0.1 M KCl–10 mM Tris-maleate buffer, pH 7.0, and negatively stained with uranyl acetate. (Top) Tangled filaments with associated beaded debris; (Middle) Thick and thin filaments; (Bottom) Bare and decorated thin filaments after addition of rabbit skeletal muscle heavy meromyosin to the disrupted extract gel.

tified in contracting macrophage extract gels. As reported in a microscope study of superprecipitation of rabbit skeletal muscle actomyosin, when actin is present greatly in excess of myosin, myosin filaments are not easily identified (12).

Macrophage Contractile Proteins and Phagocytosis

The gelation and contraction phenomena described may be relevant to the mechanism of certain cytoplasmic motions including phagocytosis. The increased extractability of actin-binding protein following ingestion by macrophages is the first example of a change in a contractile protein in response to phagocytosis. Hypothetically, contact of an ingestible particle with the plasma membrane could effect release of some actin-binding protein from sites on the plasmalemma or other membranes, or else alter the nature of its membrane-association. The activated or released actin-binding protein would then promote the assembly of G-actin to F-actin and cross link F-actin filaments to form a gel. The existence of an agent other than myosin for cross-linking actin filaments allows myosin to move randomly oriented filaments without elaborate mechanical arrangements or without restrictions on the polarity of the actin filaments. In whatever direction actin filaments cross-linked by actin-binding protein are moved by myosin, the net effect will be an internal collapse of the filament network, i.e., centripetal contraction. The total volume of the contracting system in vitro remains the same during contraction, because liquid is squeezed out of the interstices of the gel. If this occurs in the intact cell, a simple and plausible system exists for creating localized compression and for transmitting force to other regions of the cell. In an elongated system, contraction would become complete first in the shortest dimension. This would explain why the gels in rectangular cuvettes first contracted in a horizontal direction. This would also explain how contraction of membrane-associated filaments could form long narrow pseudopods surrounding particles. At the end of the reaction, the actin filaments will tend to orient in parallel, which is what the electron micrographs have revealed in negatively stained gels and in thin sections of pseudopods of intact cells. If this is true, then the parallel filament bundles are the *result* and not the *cause* of cytoplasmic movement. If the randomly oriented filament networks are the generators of mechanical force, the parallel filament bundles must disassemble in order for movement to continue. The mechanisms for such reversibility, other than cooling, are now unknown.

We thank Ms. Ann Ballen and Ms. Antonia Labate for outstanding technical assistance and Dr. Thomas D. Pollard for sharing the results of his similar studies with *Acanthamoeba castellanii.*

This work was supported by U. S. Public Health Service grant HL-17742. Dr. Stossel is an established investigator for the American Heart Association.

Received for publication 23 June 1975, and in revised form 3 November 1975.

REFERENCES

1. Boxer, L. A., E. T. Hedley-Whyte, and T. P. Stossel. 1974. Neutrophil actin dysfunction and abnormal neutrophil behavior. *N. Engl. J. Med.* **293**:1093–1099.
2. Boxer, L. A., and T. P. Stossel. 1975. Isolation and interactions of contractile proteins from chronic myelogenous leukemia granulocytes (CMLG). *J. Cell Biol.* **67**(2, Pt. 2):40 a. (Abstr.).
3. Burns, R. G., and T. D. Pollard. 1974. A dynein-like protein from brain. *FEBS (Fed. Eur. Biochem. Soc.) Lett.* **40**:274–280.
4. Carlson, F. D., and A. B. Fraser. 1974. Dynamics of F-actin and F-actin complexes. *J. Mol. Biol.* **89**:273–281.
5. Craven, G. R., E. Steers, Jr., and C. B. Afinsen. 1965. Purification, composition, and molecular weight of the β-galactosidase of *Escherichia coli* K12. *J. Biol. Chem.* **240**:2468–2477.
6. Feuer, G., F. Molnar, E. Pettko, F. B. Straub. 1948. Studies on the composition and polymerisation of actin. *Hung. Acta Physiol.* **1**:150–163.
7. Gaskin, F., C. R. Cantor, and M. L. Shelanski. 1974. Turbidometric studies of the *in vitro* assembly and disassembly of porcine neurotubules. *J. Mol. Biol.* **89**:737–758.
8. Goll, D. E., A. Suzuki, J. Temple, and G. R. Holmes. 1972. Studies on purified α-actinin. I. Effect of temperature and tropomyosin on the α-actinin/F-actin interaction. *J. Mol. Biol.* **67**:469–488.
9. Harris, J. R. 1969. The isolation and purification of a macromolecular protein component from the human erythrocyte ghost. *Biochim. Biophys. Acta.* **188**:31–42.
10. Harris, J. R. 1971. Further studies on the proteins released from haemoglobin-free erythrocyte ghosts at low ionic strength. *Biochim. Biophys. Acta.* **229**:761–770.
11. Hartwig, J. H., and T. P. Stossel. 1975. Isolation and properties of actin, myosin and a new actin-binding protein in rabbit alveolar macrophages. *J.*

Biol. Chem. **250**:5699–5705.
12. Ikemoto, N., S. Kitagawa, and J. Gergely. 1966 Electron microscopic investigation of the interaction of actin and myosin. *Biochem. Z.* **345**:410–426.
13. Ishiwata, S. 1973. A study on the F-actin-tropomyosin-troponin complex. I. Gel-filament transformation. *Biochim. Biophys. Acta.* **303**:77–89.
14. Kane, R. E. 1975. Preparation and purification of polymerized actin from sea urchin egg extracts. *J. Cell Biol.* **66**:305–315.
15. Kasai, M., E. Nakano, and F. Oosawa. 1965. Polymerization of actin free from nucleotides and divalent cations. *Biochim. Biophys. Acta.* **94**:495–503.
16. Lewis, W. H. 1939. The role of a superficial plasmagel layer in changes of form, locomotion and division of cells in tissue cultures. *Arch. Exp. Zellforsch.* **23**:7–13.
17. Lowey, S., H. S. Slayter, A. G. Weeds, and H. Baker. 1969. Substructure of the myosin molecule. I. Subfragments of myosin by enzymatic degradation. *J. Mol. Biol.* **42**:1–29.
18. Marchesi, S. L., E. Steers, V. T. Marchesi, and T. W. Tillack. 1969. Physical and chemical properties of a protein isolated from red cell membranes. *Biochemistry.* **9**:50–57.
19. Moore, R. D., and M. D. Schoenberg. 1964. The response of histiocytes and macrophages in the lungs of rabbits injected with Freund's adjuvant. *Br. J. Exp. Pathol.* **45**:488–495.
20. Myrvik, Q. N., E. S. Leake, and B. Fariss. 1961. Studies on pulmonary alveolar macrophages from the normal rabbit: a technique to procure them in a high state of purity. *J. Immunol.* **86**:128–132.
21. Oren, R., A. E. Farnham, K. Saito, E. Milofsky, and M. L. Karnovsky. 1963. Metabolic patterns in three types of phagocytizing cells. *J. Cell Biol.* **17**:487–501.
22. Pollard, T. D. 1976. The role of actin in the temperature-dependent gelation and contraction of extracts of *Acanthamoeba*. *J. Cell Biol.* **68**:579–601.
23. Pollard, T. D., and S. Ito. 1970. Protoplasmic filaments of *Amoeba proteus*. The role of filaments in consistency changes and movement. *J. Cell Biol.* **46**:267–289.
24. Pollard, T. D., and E. D. Korn. 1971. Filaments of *Amoeba proteus*. II. Binding of heavy meromyosin by thin filaments in motile cytoplasmic extracts. *J. Cell Biol.* **48**: 216–219.
25. Pollard, T. D., S. M. Thomas, and R. Niederman. 1974. Human platelet myosin. I. Purification by a rapid method applicable to other nonmuscle cells. *Anal. Biochem.* **60**:216–266.
26. Pollard, T. D., and R. A. Weihing. 1974. Actin and myosin and cell movement. *CRC Crit. Rev. Biochem.* **2**:1–65.
27. Reaven, E. P., and S. G. Axline. 1973. Subplasmalemmal microfilaments and microtubules in resting and phagocytizing cultivated macrophages. *J. Cell Biol.* **59**:12–27.
28. Sloboda, R. D., S. A. Rudolph, J. L. Rosenbaum, and P. Greengard. 1975. Cyclic AMP-dependent endogenous phosphorylation of a microtubule-associated protein. *Proc. Natl. Acad. Sci. U. S. A.* **72**:177–181.
29. Spudich, J. A., and S. Watt. 1971. The regulation of rabbit skeletal muscle contraction. I. Biochemical studies of the interaction of the tropomyosin-troponin complex with actin and the proteolytic fragments of myosin. *J. Biol. Chem.* **245**:4866–4871.
30. Stossel, T. P. 1973. Quantitative studies of phagocytosis: kinetic effects of cations and of heat-labile opsonin. *J. Cell Biol.* **58**:346–356.
31. Stossel, T. P., and Z. A. Cohn. 1975. Phagocytosis. *In* Methods in Immunology. C. A. Williams and M. W. Chase, editors. The Williams & Wilkins Company, Baltimore, Md. **5**:261–295.
32. Stossel, T. P., R. J. Field, J. D. Gitlin, C. A. Alper, and F. S. Rosen. 1975. The opsonic fragment of the third component of human complement (C3). *J. Exp. Med.* **141**:1329–1347.
33. Stossel, T. P., and J. H. Hartwig. 1975. Interactions between actin, myosin and a new actin-binding protein of rabbit alveolar macrophages. Macrophage myosin Mg^{2+}-adenosine triphosphatase requires a cofactor for activation by actin. *J. Biol. Chem.* **250**:5706–5712.
34. Stossel, T. P., R. J. Mason, T. D. Pollard, and M. Vaughan. 1972. Isolation and properties of phagocytic vesicles. II. Alveolar macrophages. *J. Clin. Invest.* **51**:604–614.
35. Stossel, T. P., and S. H. Pincus. 1975. A new macrophage actin-binding protein: evidence for its role in endocytosis. *Clin. Res.* **23**:407 a. (Abstr.).
36. Stossel, T. P., and T. D. Pollard. 1973. Myosin in polymorphonuclear leukocytes. *J. Biol. Chem.* **248**:8288–8294.
37. Szent-Györgyi, A. 1947. Chemistry of Muscular Contraction. Academic Press, Inc., New York.
38. Taylor, D. L., J. S. Condeelis, P. L. Moore, and R. D. Allen. 1973. The contractile basis of amoeboid movement. I. The chemical control of motility in isolated cytoplasm. *J. Cell Biol.* **59**:378–394.
39. Tilney, L. G., and P. Detmers. 1975. Actin in erythrocyte ghosts and its association with spectrin. Evidence for a nonfilamentous form of these two molecules *in situ*. *J. Cell Biol.* **66**:508–520.
40. Weihing, R. R. 1975. Membrane association and polymerization of actin. Cell Motility. *Cold Spring Harbor Conf. on Cell Proliferation.* Vol. 3. In press.
41. Wolpert, L., C. M. Thompson, and C. H. O'Neill. 1964. Studies on the isolated membrane and cytoplasm of *Amoeba proteus* in relation to ameboid movement. *In* Primitive Motile systems in Cell Biology. R. D. Allen and N. Kamiya, editors. Academic Press, Inc., New York. 143–171.

Taylor D.L. and **Wang Y.-L.** 1980. Fluorescently labelled molecules as probes of the structure and function of living cells. *Nature* **284:** 405–410.

MANY STUDIES FROM THE 1960S AND 1970s described the structure and distribution of specific cell parts, generally using the static images available from fixed material. Work on live cells was hampered by the lack of probes with which to visualize particular cellular constituents. Whereas important studies had been accomplished by phase and polarization microscopy, such images revealed only the optical paths through different parts of the cell, not the protein composition of the structures seen, as could be visualized by immunofluorescence (Lazarides and Weber, this volume). It was the paper reproduced here that demonstrated the power of fluorescence optics to reveal the time-dependent distributions of proteins labeled non-destructively with appropriate fluorescent dyes in vivo. This paper also spelled out the controls required to minimize the likelihood of artifact. The methods described did not, however, become widely used, largely because the work required to do things right was so great. The discovery that the gene for a green fluorescent protein (GFP) could be expressed in a wide range of cells (1), even as a chimera with a gene of interest (2), has recently changed this situation quite dramatically. Now a wide range of proteins has being labeled by molecular biology and gene transformation; the distributions of many "GFP-tagged proteins" have been followed during cell function, growth, and reproduction (reviewed in 3). Tagging with GFP is even being used as a way to screen DNA libraries for gene products with interesting locations and behaviors. The ideas for how to follow proteins reliably in vivo were, however, all laid out in the work reprinted here. The need for tests to determine whether the labeled protein retained the function of its unlabeled counterpart was a particular focus of this work. The recent advances have facilitated the study of protein behavior in vivo by making it fast and easy enough to be really useful, but the controls identified in the earlier work are no less important today.

---

1. Chalfie M., Tu Y., Euskirchen G., Ward W.W., and Prasher D.C. 1994. Green fluorescent protein as a marker for gene expression. *Science* **263:** 802–805.
2. Wang S. and Hazelrigg T. 1994. Implications for bcd mRNA localization from spatial distribution of exu protein in *Drosophila* oogenesis. *Nature* **369:** 400–403.
3. Tsien R.Y. 1998. The green fluorescent protein. *Annu. Rev. Biochem.* **67:** 509–544.

Reprinted from Nature, Vol. 284, No. 5755, pp. 405-410, April 3 1980

Fluorescently labelled molecules as probes of the structure and function of living cells

D. Lansing Taylor & Yu-Li Wang
Department of Cell and Developmental Biology, Harvard University, 16 Divinity Avenue, Cambridge, Massachusetts 02138

A new approach to cell biology has been created by combining the techniques of micromanipulation of cells, fluorescence spectroscopy and low level light detection. These methods are now being used to study the spatial and temporal distribution, interaction and activity of specific molecules in living cells.

CELL functions such as locomotion, cell recognition, endocytosis, exocytosis, and cell division involve highly dynamic and transitory molecular interactions. Early cell physiologists studied these functions by probing living cells and developed sophisticated techniques for micromanipulation and microinjection[1-4]. A new approach bridging the gaps between cell physiology, biochemistry and ultrastructure is required to define the molecular mechanisms of complex cell functions. Such an approach would require a combination and coordination of (1) techniques for manipulating living cells, (2) techniques which provide signals from specific molecules with high sensitivity, and (3) techniques which detect weak signals from living cells. Recently, the methods of micromanipulation of cells[1-5], fluorescence spectroscopy[6-8], and low-level light detection[9-13] have been combined to provide such a new approach to cell biology. This new concept involves labelling purified molecules covalently with fluorescent probes, and incorporating the fluorescent conjugates into or onto living cells. Cells with associated fluorescent conjugates are then either viewed with an image intensifier coupled to a microscope or are investigated with a microspectrofluorometer. Therefore, the high sensitivity ($\sim 10^5$–10^6 molecules can be detected[14]), and versatility[6-8] of fluorescence techniques can be fully utilised in living cells to yield information at the molecular level.

This review is limited to applications of this approach in which purified molecules are labelled, administered and then studied at the light microscopical level. The technical aspects will be discussed in detail since this new approach requires careful application and interpretation. In addition, three categories of fluorescent conjugates will be discussed: (1) nonperturbing indicators of physiological processes, (2) biologically active agents and (3) functional cellular components (molecular cytochemistry)[15]. Some of the studies referenced could be included in more than one of these categories.

Experimental methods

The application of fluorescent conjugates to studies in single living cells demands the use of highly fluorescent fluorophores which absorb in the visible spectrum, so that adequate signals are obtained while minimising radiation damage and interference from autofluorescence[16]. Furthermore, the conjugates must be associated by stable covalent bonds and be devoid of non-covalently bound fluorophores. Recent advances in the preparation of fluorescent reagents have provided several compounds suitable for applications *in vivo*. Fluorochromes such as fluorescein[15-17], eosin[18], 7-chloro-4-nitrobenzo-2-oxadiazole (NBD)[19], and a series of long-wavelength rhodamine dyes[15,20-21] have been used successfully with living cells. In addition, various reactive derivatives of these fluorophores such as iodoacetamide[15], isothiocyanate[20] and sulphonyl chloride[13] can be obtained commercially. Many of the long-wavelength fluorophores used primarily by neurophysiologists should be useful when reactive derivatives become available[22].

Classical microinjection techniques are still the most direct methods for incorporating exogenous components into living cells[1,23]. Several different microinjection systems, using both hydraulic pressure[4,5,17] and compressed air[13,24,25], have been described in detail. Cells ranging in size from giant protozoans ($\sim$600 μm) to human fibroblasts ($\sim$15 μm) have been successfully microinjected. These techniques are time consuming and the numbers of cells which can be studied are limited. However, they have the advantages that only a very small volume of material is required for each experiment; components as large as

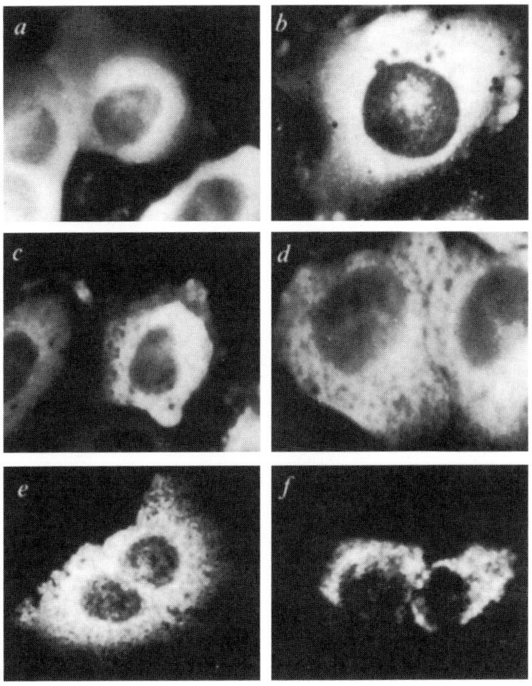

Fig. 1 Autophagy studied by injecting highly concentrated tetramethylrhodamine isothiocyanate labelled bovine serum albumin into HeLa cells. These fluorescence micrographs were taken of injected cells *a*, immediately; *b*, 30 min; *c*, 1 h; *d*, 90 min; *e*, 6 h thereafter. The labelled proteins were taken up into vesicles which ultimately fused with lysosomes (*f*). (From Stacey and Allfrey[45].)

2

organelles can be microinjected[4], mild solution conditions can be used to deliver the fluorescent conjugates[13,15,17,20], and injection volumes can be controlled.

During the past few years, new delivery methods using fusion techniques have been developed. The molecules are trapped inside carriers such as red blood cell ghosts[26] and liposomes[27,28] which are subsequently fused with target cells. While these methods allow the 'ultramicroinjection' of large populations of very small cells there are also several limitations: (1) relatively large volumes of fluorescent conjugates are required during the loading process, (2) the conjugates could become exposed to detergents, elevated temperatures, or organic solvents, (3) the proper entrapment and fusion steps must be carefully verified, and (4) the biological effects of the carriers must be controlled. Recently, fibroblasts have been successfully loaded with fluorescently labelled antibodies by red blood cell fusion techniques[29]. Continued improvements in fusion technology should make these approaches more generally useful in the future.

Cell perfusion techniques have been used to deliver small ions and molecules into plant cells[30]. A membrane permeation technique has now been reported which permits the incorporation of small molecules into living animal tissue culture cells[31]. The cells are permeabilised by short treatments with lysolecithin which make the cells leaky to exogenous molecules in the medium for a short period of time. This technique has advantages similar to those of fusion methods, yet does not require the use of carriers. Unfortunately, the application is limited to only very small molecules (molecular weight below 10,000) and to cells grown in monolayer culture. In addition, the cells remain viable for only a short period.

Recent developments in image intensification techniques have provided a more sensitive way of recording fluorescent images than classical photographic methods[10]. The use of standard photographic procedures requires long exposure times and intense illumination. Therefore, fluorescence photobleaching could be extensive, cell damage is possible, and dynamic processes are difficult or impossible to record. In contrast, TV image intensifiers coupled to fluorescence microscopes and video tape recorders allow the continuous recording of weakly fluorescent images in real time or time lapse without significant losses in resolution[10]. Some TV cameras also provide digital output for quantitative image analyses. When these intensifiers are coupled to optical prisms and multichannel analysers, they can also provide rapid microspectrofluorometric measurements[10,32].

Microspectrofluorometers have been developed for quantitative measurements of many fluorescence parameters with very low light intensities[33,34]. In particular, photomultipliers

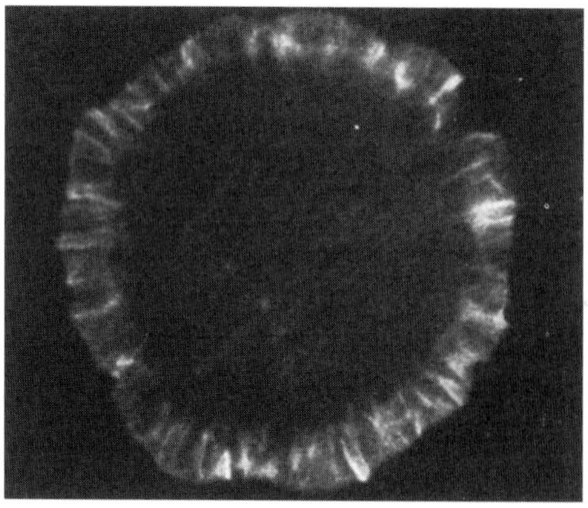

Fig. 2 Fluorescence photograph of a live chick embryo fibroblast that is lysolecithin permeabilised and labelled with NBD-phallicidin[19]. Most of the labelled actin in this spreading cell is located peripherally in the area of membrane ruffling. Some longitudinal stress fibres are also evident. (Micrograph courtesy of L. Barak.)

PURE ACTIN

Label with 5-iodoacetamidofluorescein

SELECT FUNCTIONAL AND OPTIMALLY LABELLED ACTIN

(*A*) Polymerisation–depolymerisation
(*B*) DEAE chromatography

CHARACTERISE LABELLED ACTIN

(*A*) Dye/protein
(*B*) Site of labelling
(*C*) Polymerisability
(*D*) Activation of myosin Mg^{2+} ATPase

DETERMINE FUNCTIONAL ACTIVITY *IN VITRO*

(*A*) Single cell models
(*B*) Bulk extracts

INCORPORATE LABELLED ACTIN INTO LIVING CELLS BY MICROINJECTION

Limit: 10% of endogenous actin concentration

DETERMINE FUNCTIONAL ACTIVITY *IN VIVO*

CORRELATE DISTRIBUTION OF ACTIN FLUORESCENCE WITH OVALBUMIN CONTROL

Fig. 3 Flow diagram of the steps involved in molecular cytochemistry. Actin is described as an example.

interfaced with single photon counting devices have provided instruments of high sensitivity[35]. These systems in combination with computer-controlled spectral scanning and data processing form the core of several sophisticated microspectrofluorometers which are capable of performing corrected spectral measurements on single living cells[35–38]. The use of lasers for excitation has also increased the versatility of microspectrofluorometers, because the state of polarisation, spot size, duration and intensity of the exciting light pulse can be controlled over large ranges.

Recently, fluorescence photobleaching recovery techniques have been introduced to measure the mobility of surface associated fluorophores[39–42]. These methods utilise a pulsed high-intensity laser microbeam to bleach a small area or volume of fluorescence, and the recovery of fluorescence is monitored with attenuated laser excitation. These measurements can yield diffusion constants of the mobile fraction, the percentage of fluorescent conjugates that are mobile, and information on the bulk directional flow of the conjugates. However, caution must be exercised in controlling the possible biological effects of photobleaching[43], and consideration must be given to the validity of the assumption that cell surfaces are flat and smooth. In addition, care must be taken to ensure that the membrane or surface markers are not internalised before making measurements. These potential problems have recently been addressed critically by Elson and Yguerabide[40].

Physical and chemical considerations

There are several possible physical and chemical problems which could give rise to experimental artefacts, and investigations using this new approach must be judged in part by the manner in which potential problems are considered. Factors which could affect the fluorescence intensity measured for the microscope include: (1) local accessible volume for the conjugate in the cell, (2) local environment of the conjugate, (3) optical properties of the microscope, and (4) local concentration of the conjugate.

Consideration of the local accessible volume is important for both cytoplasmic and membrane-associated components. For cytoplasmic components, the local accessible volume is controlled not only by the actual thickness of the cell, but also by the distribution of organelles which exclude the fluorescent conjugates. For membrane-associated components, the accessible volume is affected dramatically by the presence of membrane folds and microvillar structures. This problem can be identified and controlled by using a second fluorescent conjugate which would distribute uniformly within the accessible volume. Labelled bovine serum albumin, ovalbumin, or denatured proteins have been used in controls for cytoplasmic components[13,15,17,20] and fluorescent lipid probes[41] can be used in studies on membrane-associated factors. It is imperative that co-incorporation of the experimental conjugate and the control conjugate labelled with a second probe be accomplished in the same cell when the cell shape is irregular[13]. Adequate comparisons cannot be performed in separate cells when they have irregular geometries. However, separate cells can be used when the cell geometry is reasonably constant and simple[17].

The local environment around the fluorescent conjugate could also affect the fluorescence intensity by altering the quantum yield or fluorescence spectral properties. Local variations in ionic parameters such as *p*H and ionic strength as well as binding of other molecules to the conjugates could change the fluorescence parameters[6,7]. The sensitivities of fluorophores can either be utilised in characterising microenvironments or they must be controlled when other parameters are under investigation. Controls for the ionic environments can be performed by comparing the fluorescence of the experimental conjugates with the fluorescence of separate cells containing nonspecific molecules (such as ovalbumin) labelled with the same fluorophore[17], while controls for the effect of specific molecular interactions require the extrapolation of data from solution spectroscopic studies[16].

Knowledge of the optical properties of the microscope is important for interpreting fluorescence images and quantitating local fluorescence properties. The characterisation of the fluorescence image *in vivo* depends on the depth of focus of the microscope. A large depth of focus in relation to cell thickness

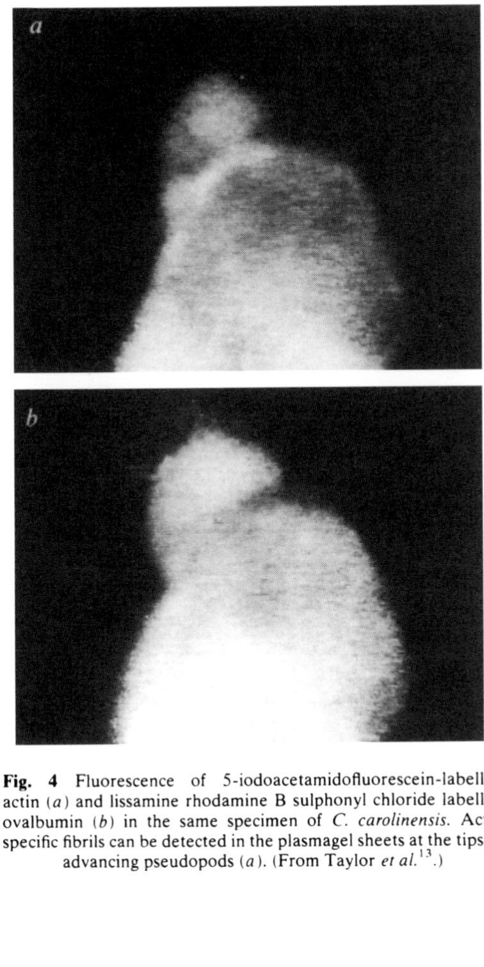

Fig. 4 Fluorescence of 5-iodoacetamidofluorescein-labelled actin (*a*) and lissamine rhodamine B sulphonyl chloride labelled ovalbumin (*b*) in the same specimen of *C. carolinensis*. Actin specific fibrils can be detected in the plasmagel sheets at the tips of advancing pseudopods (*a*). (From Taylor *et al.*[13].)

optimises the formation of an in focus image of the whole cell, while a small depth of focus yields an optical section of the cell. The latter condition would require several changes in focus to reconstruct the three-dimensional image of the cell. Quantitative measurements of fluorescence intensity are further affected by the extent of selecting light from specific planes of the specimen. This latter problem has been solved by using a combination of laser illumination and diaphragms placed in the image plane[39].

The local concentration of the fluorescent conjugates also affects the fluorescence intensity measured or visualised in different parts of cells. The distribution of the conjugates can be determined by applying a combination of the controls used for accessible volume and environmental sensitivity.

Nonperturbing indicators of physiological processes

Nonperturbing fluorescent conjugates can be used as indicators of mobility on, within, or between cells. Fluorescence indicators of normal cell-surface mobility have been studied for many years. In an early study Jeon and Bell[44] used fluorescently labelled antibodies to the cell surface of *Amoeba proteus* and a basic protein derived from a papain preparation to label the cell surface. Results based on a double labelling technique suggested that part of the cell surface could move independently of the lipid portion of the membrane.

In an elegant study, rhodamine-labelled peptides of different sizes have been microinjected into living cells coupled by gap junctions to probe the permeability and exclusion limit of the gap junction channels[45,46]. Molecules of molecular weight up to 1,200 pass through the channels of *Chironomus* salivary gland cells which indicates that the channels are ~1.0–1.4 nm in diameter. The selectivities of the various channels have been further characterised by varying the total charge of the labelled peptides[45]. For example, several mammalian cell channels can discriminate between 1–3 negative charges on the peptide backbones. The larger electronegativity inhibits passage. In addition, the permeability of multiple components has been directly compared by using mixtures of conjugates prepared with different probes[45,46]. A control for peptide degradation has also been reported.

Stacey and Allfrey used a similar technique and microinjected a wide variety of rhodamine conjugated proteins into living HeLa cells[47] to study the process of autophagy (Fig. 1). The segregation of microinjected proteins into autophagic vacuoles exhibited a high degree of selectivity with higher molecular weight proteins turning over faster than low molecular weight proteins. One protein (haemoglobin) never became autophagocytosed. The results with fluorescently conjugated proteins were verified with immunofluorescent and autoradiographic techniques.

More sophisticated applications of nonperturbing fluorescent conjugates involve the use of environmentally sensitive fluorophores and spectral analyses to probe intracellular environments. Fluorescein-labelled ovalbumin has been used to measure the cytoplasmic *p*H of single cells[48] based on the observation that the excitation spectra of fluorescein is highly *p*H dependent[49]. The *p*H is measured by determining the ratio of fluorescence emission intensity when the cells are excited at two different wavelengths. The effects of local pathlength are normalised since the ratio of intensities is determined. This fluorescence technique of measuring *p*H is less perturbative and permits better spatial resolution than standard microelectrode methods.

When applying fluorescent conjugates as indicators, the biological effects of the conjugates and the experimental procedures should be carefully examined to make sure that they are actually nonperturbing. The possible degradation of the conjugates inside the cells should also be considered, especially when the conjugates are used as size indicators[46,47]. This latter problem could be checked more critically in the future by isolating the labelled proteins from the cells after incorporation

4

Fig. 5 Tetramethylrhodamine isothiocyanate labelled α-actinin incorporated into a living gerbil fibroma cell. (From Feramisco[20].)

by fusion techniques. Furthermore, when spectral characteristics are used as indicators of local environments such as *p*H, the possible effects of other environmental factors must be ruled out. In all cases, calibration must be performed *in situ*[48].

Biologically active agents

Biologically active agents which can stimulate, alter or block cell functions include lectins, antibodies, drugs and toxins. These agents are used in probing cell structures and functions. Many of these agents can be fluorescently labelled, permitting detailed investigations on the distribution and mobility of binding sites. The relationship between distribution and biological effects of the labelled agents can also be studied.

The plant lectin, concanavalin A (Con A), has been used as a model ligand in investigations on receptor–ligand complexes[50–57]. Fluorescently labelled Con A has facilitated studies on the mobility of Con A receptors, the ultimate fate of the ligand and the relationship between receptor mobility and Con A stimulated cellular events[51–57]. It must be emphasised that Con A has many complex effects on cells[57] and it is difficult to define the complete molecular sequences of events. Similar but more specific experiments have also been performed with antibodies prepared against various cell-surface receptors[56–58]. These cell-surface associated fluorescent conjugates have been used to measure the mobility of membrane components using several different approaches. The simplest method is the observation of large cells such as muscle fibres[59] in the fluorescence microscope. Small labelled spots on the cells can be detected when spread over relatively large distances. A second approach involves fusing cells containing different surface fluorescent conjugates and following the redistribution of the labelled conjugates in the microscope[60]. Poo and colleagues[61] have also studied the accumulation of labelled surface receptors in a uniform electric field applied across the surfaces of cells. Finally, several investigators have used the photobleaching techniques discussed above.

Some drugs can also be fluorescently labelled while maintaining activity. The purpose of labelling the drugs is to study the correlation between the cellular effects and the localisation within the cell. Examples are the tubulin-binding drug colchicine[62], and the actin-binding cyclic peptides phallicidin[19] and phalloidin[63]. At the proper concentration, colchicine can cause microtubule depolymerisation and phallicidin and phalloidin can cause actin polymerisation and stabilisation (see ref. 64 for references). Fluorescent conjugates of these drugs have been used to localise pools of tubulin[62] and F-actin[19,63] respectively in fixed or extracted cells (Fig. 2). However, it is not yet clear whether these drugs have simply identified pools of tubulin and F-actin or have also altered the normal organisation of the target proteins.

Functional cellular components (molecular cytochemistry)

Molecular cytochemistry[15] has been defined as the reincorporation of functional cellular components into or onto living cells following purification, fluorescent labelling and assaying function *in vitro*. The experimental protocol is shown in Fig. 3 using actin as a model. The feasibility of molecular cytochemistry has been initially demonstrated using actin labelled with a non-destructive probe, 5-iodoacetamido-fluorescein (IAF). The labelled actin has been proven to be functional in the purified form, in cell-free extracts, and in single-cell models[13,15–17]. Microinjection of this fluorescent

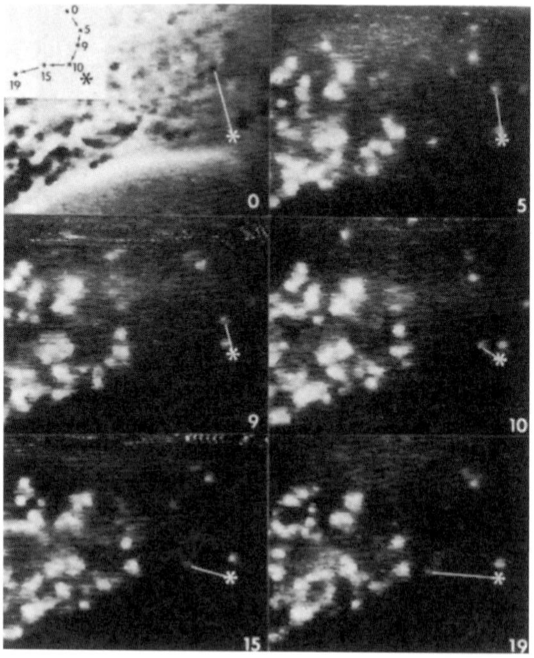

Fig. 6 Saltatory motion of fluorescent vesicles in 3T3 cells 24 h after incubation with rhodamine-labelled α_2-macroglobulin. A time lapse video tape recording at a 9 : 1 time lapse ratio was made of a cell with vigorous intracytoplasmic vesicle saltatory motion. The numbers in the lower right corner of each single frame image represent the real time in seconds after the beginning of the sequence. A phase image is presented at zero time followed by fluorescence images. The asterisk represents an arbitrary non-moving reference point. The inset summarises the motion of this single fluorescent vesicle. (From Willingham and Pastan[12].)

5

actin analogue into living cells has provided a direct way of visualising actin-containing structures and following the changes in actin distribution during cellular processes such as fertilisation of eggs[17], cytokinesis[17], amoeboid movement[13,15], pinocytosis[13], and Con A capping[13]. Extensive controls *in vitro* and *in vivo* have made the interpretation of the fluorescent images possible[13,15–17] (Fig. 4). This approach has recently been applied to other cells[20,21] and other contractile proteins[20] (Fig. 5).

The use of fluorescent conjugates specific for receptors in the cell surface has permitted the characterisation of the kinetics of the distribution and the ultimate fate of many ligands. These investigations have included studies on α_2-macroglobulin[12,65,66], epidermal growth factor[66–68], insulin[66–69], a chemotactic peptide[70], low-density lipoproteins[71], and acetylcholine receptors[71–74]. All of these fluorescent ligands initially label the cell surface uniformly, but at least some of the ligand–receptor complexes aggregate into patches within a few minutes. Most of the fluorescent conjugates eventually become internalised[69,75]. The role of internalisation has not yet been defined in detail. The redistribution of some of these ligand–receptor complexes in the plane of the membrane has been quantitated using the fluorescence photobleaching techniques.

Proper use of molecular cytochemistry demands the application of various biological controls in addition to the analysis of physical considerations discussed above. The functional activity of the conjugates must be carefully demonstrated. It requires that: (1) the labelling reaction does not abolish the normal biochemical activity of the substrate, (2) the conjugates have access to the intracellular domains where the function is performed, and (3) the conjugates are not rapidly degraded *in vivo*. The biochemical activities can be readily tested using *in vitro* assays, cell-free extracts or lysed cell models[13,16]. The use of site-specific probes would provide functionally homogeneous conjugates and would yield unequivocal results with *in vitro* assays[16]. The studies with labelled actin have yielded the most definitive biological controls[13,15–17] to date.

The function inside living cells is much more difficult to assay. For IAF-labelled actin mentioned above, well characterised cellular responses such as cortical wound healing, can be used to test the functionality *in situ*[13]. Furthermore, a comparison of the fluorescent images of biologically active conjugates with nonfunctional conjugate controls would also indicate the formation of structures related to biological functions[13,15,17]. Note that the absence of fluorescence from a cellular domain, such as the nucleus[13,17], does not necessarily imply that the endogenous component is not present. Negative results could be due to physical exclusion of the conjugate, slow turnover time for the incorporation into specific structures, as well as the absence of a cellular component from a particular region. Therefore, negative results must also be interpreted with caution.

Future prospects

This new approach to cell biology is likely to develop extensively both technically and in its range of biological applications. At the technical level, improvements in the hardware and software applied to intensified image recording and spectral analysis should permit sophisticated manipulations of the experimental data. Measurements of fluorescence polarisation[14,76,77], resonance energy transfer[60,78–80] and fluorescence lifetime could be performed and factors such as accessible volume could be corrected automatically. Detailed molecular information including microviscosity, local polarity, rotational freedom and formation of supramolecular structures could be determined.

The application of this new technique is expected to extend to many different areas of cell biology, through the use of fluorescent conjugates of nonperturbing indicators, biologically active agents and functional cellular components. Components binding specific ions or ligands could be labelled with environmentally sensitive probes and used as indicators of intracellular environments. One such possibility would be the calcium binding protein calmodulin[81]. Some drug studies previously relying on radioactive derivatives could be performed with fluorescent conjugates which would allow the direct observation of uptake, distribution and turnover. Furthermore, the new techniques of blocking cell functions with specific antibodies[82,83] or modified cellular components[84] can be combined with the present approaches yielding a more powerful technique. The most dramatic advances will probably be seen in the area of molecular cytochemistry using functional conjugates of proteins, lipids, nucleic acids, carbohydrates and even whole organelles. Careful use of molecular cytochemistry is expected to bring new insights into areas as diverse as cell motility, virus–cell interactions, nuclear–cytoplasmic interactins, cell–cell interactions in tissues, axonal transport, gene expression, and assembly of organelles at both morphological and molecular levels. A direct connection between cell physiology, biochemistry and ultrastructure would then become a reality.

We thank J. Heiple, V. Fowler, B. Luna, L. Tanasugarn, L. Simons, E. Haas and M. Rizzo for helpful discussions, and L. Stryer and J. Yguerabide for reading the manuscript. Valuable collaboration with B. Ware and F. Lanni on fluorescence photobleaching is also acknowledged. Some of the research reported here was supported by NSF grant PCM-7822499 and NIH grant AM 18111.

1. Chambers, R. & Chambers, E. L. *Explorations into the Nature of the Living Cell* (Harvard University Press, Cambridge, Massachusetts, 1961).
2. Lorch, I. J. in *The Biology of Amoeba* (ed. Jeon, K. W.) 1–36 (Academic, New York, 1973).
3. Jeon, K. W., Lorch, I. J., & Danielli, J. F. *Science* **167**, 1626–1627 (1976).
4. Hiramoto, Y. *Expl Cell Res.* **27**, 416–426 (1962).
5. Nichols, K. M. & Rikmenspoel, R. *J. Cell Sci.* **29**, 233–247 (1978).
6. Guilbault, G. G. *Practical Fluorescence. Theory, Methods and Techniques* (Dekker, New York, 1973).
7. Fluorescence Spectroscopy (eds Pesce, A. J., Rosen, C. G. & Pasby, T. L.) (Dekker, New York, 1971).
8. Thaer, A. A. & Sernetz, M. (eds) *Fluorescence Techniques in Cell Biology* (Springer, Berlin, 1973).
9. Reynolds, G. T. *Quant. Rev. Biophys.* **5**, 295–347 (1972).
10. Reynolds, G. T. & Taylor, D. L. *BioScience* (in the press).
11. Sedlacek, H. H., Gundlach, H. & Ax, W. *Behring, Inst. Mitt* **59**, 64–70 (1976).
12. Willingham, M. C. & Pastan, I. *Cell* **13**, 501–507 (1978).
13. Taylor, D. L., Wang, Yu-Li & Heiple, J. *J. Cell Biol.* (submitted).
14. Sengbusch, G. V. & Thaer, A. in *Fluorescence Techniques in Cell Biology* (eds Thaer, A. A. & Sernetz, M.) 31–39 (Springer, Berlin, 1973).
15. Taylor, D. L. & Wang, Y.-L. *Proc. natn. Acad. Sci. U.S.A.* **75**, 857–861 (1978).
16. Wang, Yu-Li & Taylor, D. L. *J. Histochem. Cytochem.* (submitted).
17. Wang, Y.-L. & Taylor, D. L. *J. Cell Biol.* **82**, 672–679 (1979).
18. Cherry, R. J., Cognoli, A., Oppliger, M., Schneider, G. & Semenza, G. *Biochemistry* **15**, 3653–3656 (1976).
19. Barak, L. S., Yocum, R. R., Nothnagel, E. A. & Webb, W. W. *Proc. natn. Acad. Sci. U.S.A.* (in the press).
20. Feramisco, J. R. *Proc. natn. Acad. Sci. U.S.A.* **76**, 3967–3971 (1979).
21. Kreis, T. E., Winterhalter, K. H. & Birchmeier, W. *Proc. natn. Acad. Sci. U.S.A.* **76**, 3814–3818 (1979).
22. Waggoner, A. S. *A. Rev. Biophys. Bioengng* **8**, 47–68 (1979).
23. Diacumakos, E. G. *Meth. Cell Biol.* **7**, 288–311 (1978).
24. Blinks, J. R. *et al. Meth. Enzym.* **58**, 292–328 (1978).
25. Rose, B. & Lowenstein, W. R. *J. Membrane Biol.* **28**, 87–199 (1976).
26. Schlegel, R. A. & Richsteiner, M. C. *Meth. Cell Biol.* **20**, 341–354 (1978).
27. Tyrrell, D. A., Heath, T. D., Colley, C. M. & Ryman, B. E. *Biochim. biophys. Acta* **457**, 259–302 (1976).
28. Gregoriadis, G. *New Engl. J. Med.* **295**, 704–770 (1976).
29. Yamaizumi, M., Uchida, T., Mekada, E. & Okada, Y. *Cell* **18**, 1009–1014 (1979).
30. Williamson, R. E. *J. Cell Sci.* **17**, 655–668 (1975).
31. Miller, M. R., Castellot, J. J. Jr & Pardee, A. B. *Expl Cell Res.* **120**, 421–425 (1979).
32. Kohen, E., Kohen, C. & Thorell, B. *Expl Cell Res.* **81**, 477–482 (1973).
33. West, S. S. in *Physical Techniques in Biological Research* Vol. 3C (ed. Pollister, A. W.) 253–321 (Academic, New York, 1969).
34. Ploem, J. S., Starke, J. A., De, Bonnet, J. & Wasmund, H. *J. Histochem. Cytochem.* **22**, 668–677 (1974).
35. Malmstadt, H. V., Franklin, M. L. & Horlick, G. *Analyt. Chem.* **44**, 63A–76A (1972).
36. Wreford, N. G. M. & Schafield, G. G. *J. Microsc.* **103**, 127–130 (1974).
37. Rost, F. W. D. & Pearse, A. G. E. *J. Microsc.* **94**, 93–105 (1971).
38. Cova, S., Prenna, G. & Mazzini, G. *Histochem. J.* 279–299 (1974).
39. Axelrod, D., Koppel, D. E., Schlessinger, J., Elson, E. & Webb, W. W. *Biophys. J.* **16**, 1055–1060 (1976).
40. Elson, H. & Yguerabide, J. *J. supramolec. Struct.* **12** (in the press).
41. Smith, B. A. & McConnell, H. M. *Proc. natn. Acad. Sci. U.S.A.* **75**, 2759–2763 (1978).
42. Koppel, D. E. *Biophys. J.* **28**, 281–292 (1979).
43. Sheetz, M. P. & Koppel, D. E. *Proc. natn. Acad. Sci. U.S.A.* **76**, 3314–3317 (1979).
44. Jeon, K. W. & Bell, L. G. E. *Expl Cell Res.* **33**, 531–539 (1964).
45. Flagg-Newton, J., Simpson, I. & Loewenstein, W. R. *Science* **205**, 404–407 (1979).
46. Simpson, I., Rose, B. & Loewenstein, W. R. *Science* **195**, 294–296 (1977).
47. Stacey, D. W. & Allfrey, V. G. *J. Cell Biol.* **75**, 807–817 (1977).
48. Heiple, J. & Taylor, D. L. *J. Cell Biol.* (submitted).
49. Ohkuma, S. & Poole, B. *Proc. natn. Acad. Sci. U.S.A.* **75**, 3327–3331 (1978).
50. Bittiger, H. & Schnebli, H. P. (eds) *Concanavalin A as a Tool* (Wiley, New York, 1976).
51. Schlessinger, J. *et al. Proc. natn. Acad. Sci. U.S.A.* **73**, 2407–2413 (1976).
52. Yahara, I. & Edelman, G. M. *Expl Cell Res.* **81**, 143–155 (1973).
53. Ryan, G., Borysenko, J. Z. & Karnovsky, M. J. *J. Cell Biol* **62**, 351–370 (1974).
54. Oliver, J. M., Ukena, T. E. & Berlin, R. D. *Proc. natn. Acad. Sci. U.S.A.* **71**, 394–398 (1974).
55. Condeelis, J. R. *J. Cell Biol.* **80**, 751–758 (1979).
56. Taylor, R. B., Duffus, P. H., Ross, M. C. & DePetris, S. *Nature new Biol.* **233**, 225–229 (1971).

6

57. Schreiner, G. F. & Unanue, E. R. *Adv. Immun.* **24,** 38–165 (1976).
58. Woda, B. A., Yguerabide, J. & Feldman, J. D. *J. Immun.* **123,** 2161–2167 (1979).
49. Edidin, M. & Fambrough, D. *J. Cell Biol.* **57,** 27–53 (1973).
60. Keller, P., Person, S. & Snipes, W. *J. Cell Sci.* **28,** 167–177 (1977).
61. Poo, M-m, Poo, W. J. H. & Lam, J. W. *J. Cell Biol.* **76,** 483–501 (1978).
62. Clark, J. & Garland, D. *J. Cell Biol.* **76,** 619–627 (1978).
63. Wulf, E., Deboben, A. Bautz, F. A., Faulstich, H. & Wieland, T. *Proc. natn. Acad. Sci. U.S.A.* **76,** 4498 (1979).
64. Taylor, D. L. & Condeelis, J. S. *Int. Rev. Cytol.* **56,** 57–144 (1979).
65. Pastan, I., Webb, W. W. & Elson, E. L. *Proc. natn. Acad. Sci. U.S.A.* **74,** 2909–2913 (1977).
66. Schlessinger, J., Schechter, Y., Willingham, M. & Pastan, I. *Proc. natn. Acad. Sci. U.S.A.* **75,** 2659–2663 (1978).
67. Scheckter, Y., Hernaez, L., Schlessinger, J. & Cuatrecasas, P. *Nature* **278,** 835–838 (1979).
68. Haigler, H., Ash, J. F., Singer, S. J. & Cohen, S. *Proc. natn. Acad. Sci. U.S.A.* **75,** 3317–3321 (1978).
69. Maxfield, F. R., Schlessinger, J., Schechter, Y., Pastan, I. & Willingham, M. C. *Cell* **14,** 805–810 (1978).
70. Niedel, J. E., Kahane, I. & Cuatrecasas, P. *Science* **205,** 1412–1414 (1979).
71. Krieger, M. *et al. J. supramolec. Struct.* **10,** 467–478 (1979).
72. Axelrod, D. *et al. Proc. natn. Acad. Sci. U.S.A.* **73,** 4594–4598 (1976).
73. Anderson, M. J. & Cohen, M. W. *J. Physiol., Lond.* **237,** 385–400 (1974).
74. Block, R. J. *J. Cell Biol.* **82,** 626–643 (1979).
75. Silverstein, S. C., Steinman, R. M. & Cohn, Z. A. *A. Rev. Biochem.* **46,** 669–722 (1977).
76. Axelrod, D. *Biophys. J.* **26,** 557–574 (1979).
77. Nihei, T., Mendelson, R. A. & Botts, *J. Biophys. J.* **14,** 236–242 (1974).
78. Stryer, L. *A. Rev. Biochem.* **47,** 819–846 (1978).
79. Wu, C. W. & Stryer, L. *Proc. natn. Acad. Sci. U.S.A.* **69,** 1104–1108 (1972).
80. Fernandez, S. M. & Berlin, R. D. *Nature* **264,** 411–415 (1976).
81. Potter, J. D. *et al.* in *Calcium Binding Proteins and Calcium Function* (eds Wasserman, R. H. *et al.*) 239–249 (North-Holland, New York, 1977).
82. Mabuchi, I. & Okuno, M. *J. Cell Biol.* **74,** 251–263 (1977).
83. Rungger, D., Rungger-Brändle, E., Chaponnier, C. & Gabbiani, G. *Nature* **282,** 320–321 (1979).
84. Meeusen, R. L. & Cande, W. Z. *J. Cell Biol.* **82,** 57–65 (1979).

Mitchison T. and **Kirschner M.** 1984. Dynamic instability of microtubule growth. *Nature* **312:** 237–242.

THE LABILITY OF CYTOPLASMIC and mitotic microtubules was established even at their discovery, thanks to extensive work on spindle birefringence, which suggested that spindle fibers were highly dynamic (Inoué and Sato, this volume). This lability was further documented by direct studies of microtubule disassembly in vitro (1) and of microtubule turnover in both mitotic and interphase cells (2). The mechanism for such rapid microtubule turnover was, however, obscure. Biochemical evidence suggested that microtubules exchanged tubulin subunits at their ends only, yet the speed of tubulin exchange between microtubules and the pool of soluble subunits (half-times <1min, even for polymers many micrometers long) defied understanding by conventional ideas about protein polymerization and depolymerization (3). In the paper reproduced here the authors presented evidence that microtubules display a novel kind of polymerization steady state. Mitchison and Kirschner used immunofluorescence microscopy to visualize individual polymers and to show that each microtubule could exist in either of two states: growing or shrinking. A sample of microtubules at polymerization steady state comprised a mixture of these populations, with more microtubules growing than shrinking because growth was the slower process. The authors identified the hydrolysis of GTP bound to polymerizable tubulin as the key regulatory factor in the transition between these states, and they recognized that a microtubule, which is composed largely of GDP-bound tubulin, is intrinsically unstable. It is able to grow only because caps of GTP-bound subunits stabilize its ends; when these are lost, either by dissociation or hydrolysis, the instability of the polymer becomes evident through its rapid disassembly. This "dynamic instability" was soon visualized directly by dark field light microscopy (4), revealing that the two ends of each microtubule behaved differently, and that some microtubule-associated proteins had a profound effect on the frequencies of transition between shrinking and growing states. Similar behavior was then described in living cells (5), even at mitosis, so the ideas of dynamic instability are now being incorporated into models for spindle behavior and the mechanisms for chromosome movement (6).

---

1. Carlier M.F., Hill T.L., and Chen Y. 1984. Interference of GTP hydrolysis in the mechanism of microtubule assembly: an experimental study. *Proc. Natl. Acad. Sci.* **81:** 771–775.
2. Saxton W.M., Stemple D.L., Leslie R.J., Salmon E.D., Zavortink M., and McIntosh J.R. 1984. Tubulin dynamics in cultured mammalian cells. *J. Cell Biol.* **99:** 2175–2186.
3. Oosawa F. and Kasai M. 1962. A theory of linear and helical aggregation of macromolecules. *J. Mol. Biol.* **4:** 10–21.
4. Horio T. and Hotani H. 1986. Visualization of the dynamic instability of individual microtubules by dark-field microscopy. *Nature* **321:** 605–607.
5. Cassimeris L., Pryer N.K., and Salmon E.D. 1988. Real-time observations of microtubule dynamic instability in living cells. *J. Cell Biol.* **107:** 2223–2231.
6. Rieder C.L. and Salmon E.D. 1994. Motile kinetochores and polar ejection forces dictate chromosome position on the vertebrate mitotic spindle. *J. Cell Biol.* **124:** 223–233.

Reprinted from Nature, Vol. 312, No. 5991, pp. 237-242, 15 November 1984

Dynamic instability of microtubule growth

Tim Mitchison & Marc Kirschner

Department of Biochemistry and Biophysics, University of California at San Francisco, San Francisco, California 94143, USA

We report here that microtubules in vitro *coexist in growing and shrinking populations which interconvert rather infrequently. This dynamic instability is a general property of microtubules and may be fundamental in explaining cellular microtubule organization.*

MICROTUBULES are structural filaments in the cytoplasm which are spatially organized and extremely dynamic[1,2]. Recently, considerable effort has been directed towards understanding what produces and stabilizes specific arrangements of microtubules in cells and by what means microtubules can completely reorganize their spatial distribution. In the accompanying paper[3], we suggest that microtubules nucleated by centrosomes can grow transiently at tubulin concentrations below those at which free microtubules are stable, and that nucleated microtubules coexist as shrinking and growing populations which rarely interconvert. This behaviour is clear only when individual microtubules rather than bulk populations are studied. Here we generalize the results from microtubules nucleated by centrosomes to free microtubules. We examine the detailed kinetics of microtubule assembly to try to account for these unusual dynamic properties.

Microtubule dilution

The crucial experiment demonstrating unusual dynamics in the preceding paper was that in which the microtubule number and length distributions were measured after centrosomes were regrown initially at a high tubulin concentration, then diluted[3]. The conclusion we drew from that experiment was that some microtubules continued to grow at the same time as others were lost by depolymerization from their distal ends. It seemed likely that this was a general property of microtubules. We therefore describe here a similar experiment with free microtubules (Fig. 1). Microtubules were first made by spontaneous polymerization in an assembly-promoting buffer. These were then used as seeds and diluted extensively into a tubulin solution well above the steady-state concentration and allowed to elongate for 4 min. The seeds, which were initially 1-2 μm long, elongated to form a sharp distribution with a mean length of 18.3 μm (Fig. 1*b*). This actively growing population was either fixed immediately or first diluted with warm buffer to just above (15 μM) or below (7.5 μM) the steady-state concentration. To assess both the length and number concentration, fixed microtubules were quantitatively sedimented onto grids for electron microscopy. This procedure gave a highly reproducible number concentration (Fig. 1) and when a known polymer mass was used it gave the expected mean length (see, for example, Fig. 4).

The result of this experiment was very similar to that found using centrosome nucleated microtubules[3], that is, below the steady-state concentration microtubules were found to both grow and shrink (Fig. 1). Above the steady-state concentration, the number of microtubules remained approximately constant (Fig. 1*a*) and their length increased from 18.3 to 40.2 μm in

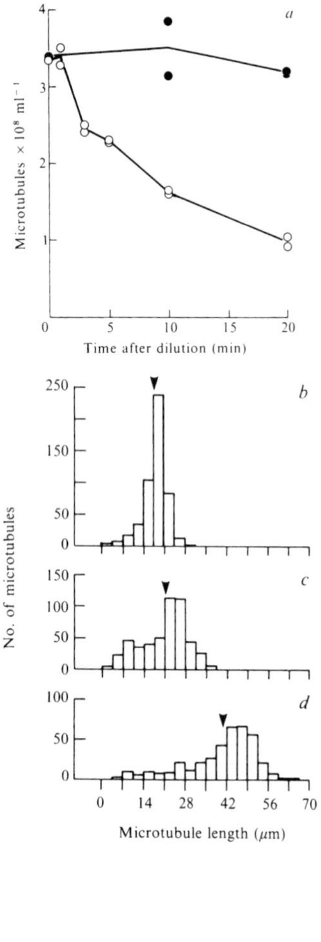

Fig. 1 Dilution after seeded assembly. *a*, For number concentration, the number of microtubule ends per field was averaged for 16 fields, divided by two and scaled to final number concentration using the geometry of the rotor and the dilution factor. Duplicate grids were made at each time point. Final tubulin concentration: closed circles 15 μM, open circles 7.5 μM. Microtubule lengths were determined by digitizing as in Fig. 6 of ref. 3. *b–d*, Length histograms. *b*, Fixed before dilution (500 measured, mean 18.3 μm, s.d. = 4.0 μm); *c*, 10 min at 7.5 μM (500 measured, mean 21.5 μm, s.d. = 8.6 μm). The mean is significantly increased ($P < 0.0001$) with respect to the starting population. *d*, 10 min at 15 μM (250 measured, mean 40.2 μm, s.d. = 12.0 μm). The arrow points to the mean of each distribution.

Methods: Preparation of tubulin, measurement of protein concentration, buffers used and generation of microtubule seeds were as described in Fig. 5 of ref. 3. 1/100 volume of seeds was added to prewarmed tubulin, 25 μM in PB. After 4 min at 37 °C, 10-μl aliquots of the growing microtubules were added to 100 μl of 1% glutaraldehyde in PB′ (PB without GTP) at 26 °C or 1 ml of prewarmed tubulin at 15 μM (above steady-state concentration) or 7.5 μM (below steady-state concentration) in PB, and gently mixed. Incubation was continued at 37 °C, and aliquots were fixed at the indicated time intervals. After 3 min in the fixative, the microtubules were diluted with cold PB′, and sedimented onto polylysine-coated 150-mesh Parlodion grids in the airfuge EM 90 rotor (Beckman) at 90,000*g* for 15 min. Grids were previously polylysine-coated by immersion in 1 μg ml⁻¹ polylysine for 10 min, then dried by aspiration. Grids were dried and shadowed as described in Fig. 4 of ref. 3. Random fields were photographed at a final magnification of ×1,500.

2

10 min, retaining a fairly sharp distribution (Fig. 1*d*). Below the steady-state concentration, however, the number concentration decreased with time (Fig. 1*a*), but the mean length still increased from 18.3 to 21.5 μm in 10 min (Fig. 1*c*). The rate of microtubule disappearance was similar to that of centrosomal microtubules at the same tubulin concentration, suggesting that a free minus end does not markedly decrease microtubule stability. This experiment demonstrates the coexistence of shrinking and growing populations with free microtubules and the transient growth of microtubules below the steady-state concentration, confirming that the centrosome data were demonstrating a general property of microtubules. The net polymer mass is clearly decreasing at the lower concentration, as the increase in mean length is more than offset by the decrease in number. At the higher concentration, the net polymer mass increases, confirming that the steady-state concentration lies between these concentrations (Fig. 1*a*). Since no renucleation occurs in this experiment, polymer mass will eventually decrease to zero at the lower concentration. Centrosomes present at the same concentration would, however, continuously renucleate microtubules[3].

Microtubule polymerization rates

In order to understand the dynamic processes occurring in the dilution experiment, we sought to determine the quantitative relationship between tubulin concentration and microtubule polymerization and depolymerization rates. A method giving data for plus and minus ends independently was chosen[4]. To determine polymerization rates, flagellar axonemes were mixed with prewarmed tubulin solutions, then fixed at various time intervals and the length of the microtubules polymerized onto each end of the axonemes was determined (the ends of the axoneme are distinguishable in the electron microscope[5]). To determine depolymerization rates, axonemes were first regrown with microtubules, then diluted (Fig. 2). Immunofluorescent visualization was used for higher tubulin concentrations (Fig. 2*a*) and electron microscopy for lower concentrations (Fig. 2*b*, *c*), with equivalent results. For the higher tubulin concentrations, the length increase is linear up to ~20 μm, at which length shear-induced breakage becomes difficult to avoid (Fig. 2*a*). At the lowest concentrations the mean length rapidly plateaued and only initial rates were used (Fig. 2*b*). The linear decrease of length with time (Fig. 2*c*) is consistent with endwise depolymerization.

Figure 3 plots the rate measurements as a function of tubulin concentration. The points on the positive portion of the data, which we refer to as the growing phase, show a linear relationship between growth rate and concentration. This is interpreted in terms of the simple first-order rate equation shown in the legend to Table 1. The values of k_T for the plus and minus end are quite similar to those derived for microtubule protein[4,6]. As in those studies, the fast-growing end corresponds to the end of the axoneme distal to the cell.

The data cannot be used to derive the *x* intercepts very accurately because they are so close to zero, and small changes in the slope produce large changes in their values. The data do not support the assertion that the intercepts for the plus and minus ends are significantly different. Thus, the growing phase of microtubules behaves like simple subunit addition to a polymer in equilibrium with its subunits, having the same extremely small critical concentrations for the two ends[7]. In any case, the *x* intercepts for both ends are much lower than the steady-state concentration measured on the same tubulin preparation[3], which was 14 μM. Thus axonemes, like centrosomes, can nucleate microtubules well below the steady-state concentration, and this can occur on both the plus and minus ends. This also confirms the data of Fig. 1, where microtubules with both plus and minus ends free continued to grow well below the steady-state concentration.

The shrinking phase of Fig. 3 confirms and extends earlier data[8] showing a large break in the bulk polymerization curve (measured by turbidity) where it crossed the *x* axis. For both the plus and minus ends, the observed depolymerization rate is 2–3 orders of magnitude greater than the extrapolated *y* intercept (k'_T) from the growing phase of the graph. Thus the off rates during depolymerization, which we call k'_D, are much larger than the off rates during polymerization, k'_T. We interpret the difference as being due to the existence of a GTP cap during polymerization, and hence the nomenclature of k'_T and k'_D where k'_T is the off rate of GTP tubulin from a GTP lattice, and k'_D is the off rate of GDP tubulin from a GDP lattice, in agreement with earlier studies[8].

It is important to distinguish the kind of data obtained by bulk measurements[8] from the measurement here of individual microtubules. The growing phase data of Fig. 2 is not an aggregate growth rate of both growing and shrinking microtubules because the axoneme template could not depolymerize and thus we only observed net growth. At most of the tubulin concentrations used, microtubules grew continuously during the time course, indicating that the transit of a growing microtubule into the shrinking phase is a relatively rare event. Only at 3 and 4 μM tubulin did phase transitions during the time course become significant. At these concentrations, microtubules depolymerized after transient growth, leading to plateauing of average length and a heterogeneous length distribution. Similarly, during the depolymerization experiments, the tubulin concentrations were sufficiently low that essentially all the microtubules shrank continuously. Thus the two arms of the plot in Fig. 2 cannot be connected. A single shrinking microtubule can transit from the growing phase to the shrinking phase over a wide range of concentrations (Fig. 1; ref. 3).

Table 1 Derived rate constants for the kinetics of microtubule assembly

	Plus end	Minus end	Units
Slope	0.135	0.042	$\mu m\ \mu M^{-1}\ min^{-1}$
k_T	3.82	1.22	$\times 10^6\ M^{-1}\ s^{-1}$
x intercept	0.1	0.9	μM
y intercept	−0.013	−0.039	$\mu m\ min^{-1}$
k'_T	0.37	1.1	s^{-1}
Depolymerization rate	−12.0	−7.5	$\mu m\ min^{-1}$
k'_D	340	212	s^{-1}

We assume 1,700 tubulin dimers per μm microtubule. The growth rate (J_T) is defined as $k_T c - k'_T d$, where *c* is the tubulin monomer concentration, k_T the second-order on-rate constant and k'_T the extrapolated off-rate constant during polymerization, calculated from *y* intercept of Fig. 3.

Steady-state dynamics

The dynamics at steady state are unclear, because at 14 μM tubulin (where the net polymer mass should be constant) the plus and minus ends should be growing at 1.88 and 0.55 μm min^{-1} respectively, or a total of 2.43 μm min^{-1} for free microtubules. We examined the nature of the steady state by measuring microtubule length and number concentration (Fig. 4). Microtubule polymerization was initiated by seeded assembly and followed by turbidity. After assembly to near steady state the microtubules were sheared. Following a transient decrease in turbidity, the microtubules rapidly attain a plateau in turbidity, corresponding to steady-state assembly (Fig. 4*a*). Sampling during the plateau of turbidity indicated that there is a steady increase in mean microtubule length, while the number concentration decreases steadily. Polymer concentration determined as the product of average length and number is constant (within an experimental error of 10%) and the mean polymer concentration (30 μM) is in good agreement with that expected from the bulk assembly data in our previous paper (Fig. 5; ref. 3).

The increase of mean length with time demonstrates that most microtubules are indeed growing at the steady-state concentration. Monomer for this growth must be supplied by the depolymerization of shrinking microtubules, and in the absence of renucleation this leads to the observed decline in number concentration. Thus the steady state can be interpreted as being due to the coexistence of two phases, with the majority of microtubules growing slowly balanced by the minority shrinking rapidly. The observed length fluctuations are much too large to be explained by random fluctuations of an equilibrium polymer. Using reasonable rate constants, a polymer 15 μm long would take a year to fluctuate to zero by fluctuations at equilibrium[9,10] and the addition of treadmilling would not give the observed rates[11]. End-to-end ligation of microtubules is also unlikely to explain the results, as experiments with pulses of biotin-labelled tubulin have shown only end addition as the mechanism of elongation (manuscript in preparation). The proportion in the two phases can be estimated from Fig. 4, because net growth must balance net disassembly at steady state. Unfortunately, the existence of different kinetics at the two ends complicates the analysis, but if most of the net growth occurs on plus ends at 1.9 $\mu m\ min^{-1}$ (the growth rate at 14 μM tubulin dimer) and disassembly occurs off both ends at an average of 9.7 $\mu m\ min^{-1}$, then there will be an average 5.2 growing microtubules for every shrinking one.

If transitions between growing and shrinking phases are very frequent, then little net microtubule elongation (or number loss) would be seen. The observed growth rate (Fig. 4) suggests that such transitions must be rare. A crude way to estimate the transition probabilities is to consider that for microtubules ~20 μm long, the half life for loss in number is ~20 min at the plateau in turbidity. Using the kinetic constants described above, such a microtubule would take ~1.7 min to depolymerize from the plus end, or ~2.7 from the minus end. If ~20% of the microtubules are depolymerizing at any one time, the half life should be ~10 min if they always depolymerize to completion once shrinking is initiated. The observed half life of 20 min may reflect that, on average, half the microtubules depolymerize to completion and half are recapped somewhere during the 20 μm of depolymerization, which represents 34,000 dissociation events. However, determination of the transition rates will require further experiments and detailed modelling.

We favour a model which explains the difference between growing and shrinking microtubules in terms of differences at their ends, in that growing microtubules have GTP-liganded caps whereas shrinking microtubules do not[8]. One observation which supports the idea that growing microtubules are stabilized by their GTP caps is shown in the absorbance traces in Fig. 4. When the seeded assembly mixture is sheared during the growth phase, there is a large transient decrease in the A_{350}, indicating considerable microtubule depolymerization. The extent of this drop, which can be up to one-third of the polymer present, depends on the extent of shearing. Merely drawing the solution into a Pasteur pipette has little effect on the A_{350}. Such a shear-induced depolymerization is not expected from simple theories of polymer polymerization[9]. We interpret this transient depolymerization as being due to the breakage of microtubules, which exposes GDP-liganded subunits at the new ends. These are unstable and start to depolymerize rapidly, a process which continues until new GTP caps can be established.

Microtubule assembly

A new model for microtubule assembly arising from the data presented here and other data[3,8,12] is presented in Fig. 5. The essence of the model is that two distinct phases of microtubules exist which are distinguished by the presence or absence of a GTP-liganded cap, and that the two phases interconvert infrequently. The ends without a cap are unstable at all monomer concentrations tested, that is, GTP subunits will not give net

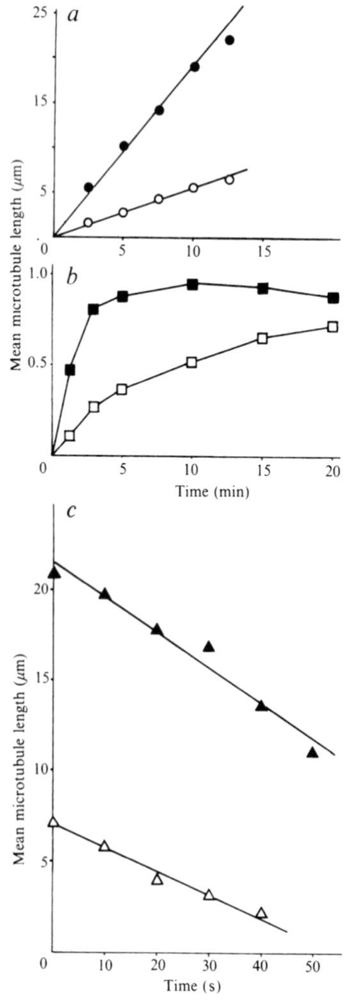

Fig. 2 Microtubule growth rate off axonemes. *a*, Polymerization at 15 μM tubulin (immunofluorescence). *b*, Polymerization at 3 μM tubulin (electron microscopy). *c*, Depolymerization at 2.5 μM tubulin (electron microscopy).

Methods: Axonemes were prepared by washing *Tetrahymena* cells in fresh growth medium, followed by 10 mM PIPES, 1 mM $MgCl_2$, 1 mM $CaCl_2$, *p*H 7.2 with KOH. Cells were resuspended at 25 °C in the same buffer and dibucaine was added to 0.5 mM[24]. After 5 min the cells shed their axonemes and rounded up. Cell bodies were removed by centrifuging twice at 2,000*g* for 5 min. The supernatant was made 5 mM in EDTA and 0.5% Triton X-100. Axonemes were pelleted at 25,000*g* for 20 min at 4 °C and resuspended in 50% (v/v) glycerol, 5 mM PIPES, 0.5 mM EDTA, 1 mM *β*-mercaptoethanol, *p*H 7 with KOH at a concentration of $\sim 10^{10}\ ml^{-1}$. They could be stored for months at −20 °C in this buffer. Solutions of tubulin in PB were prewarmed to 37 °C for 2 min and axonemes were added to a final concentration of $10^7\ ml^{-1}$. The mixture was incubated at 37 °C and aliquots were fixed as in Fig. 1 at five or six time points. The time course varied from 20 min at the lower to 5 min at the higher concentrations. Fixed regrown axonemes were diluted and sedimented onto polylysine-coated grids in the airfuge as in Fig. 1, or onto polylysine-coated glass coverslips (Fig. 1*b* of ref. 3) except that the spin was for 30 min. Regrown axonemes were then visualized by rotary shadowing and immunofluorescence respectively. For depolymerization experiments, axonemes were pregrown to ~20 μm on the plus and ~7.5 μm on the minus ends. They were then diluted and fixed at 10-s intervals and analysed by electron microscopy. Random axonemes were photographed at ×1,500 or ×3,000 in the electron microscope, or at ×250 in the light microscope and digitized directly from the negatives. At least 100 microtubules from each end were measured at each time point and the time courses were plotted. Rate of microtubule growth was determined by linear regression to the linear portion of the time course, which generally included all points with mean length less than 20 μm, with typical correlation coefficients of >0.95.

4

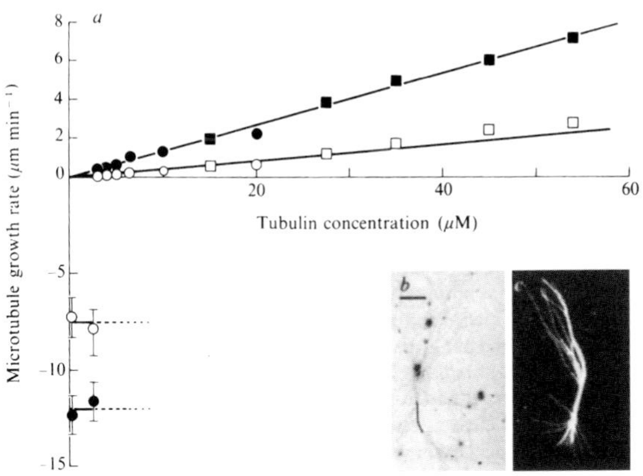

Fig. 3 ***a***, The growth rate of microtubules off axonemes obtained from data of the type in Fig. 2 plotted as a function of tubulin concentration. The closed symbols show data for the plus and the open for the minus ends. Data points plotted as squares were determined by immunofluorescence, and points plotted as circles by electron microscopy and shadowing. The line is a least-squares fit which minimizes relative deviations, that is $(di/yi)^2$ rather than $(di)^2$. All points have similar relative errors thus points at the lower tubulin concentration have smaller absolute errors, so minimizing relative deviation giv^es the most physically reasonable fit to the data. ***b***, Typical regrown axoneme visualized by rotary shadowing and electron microscopy. Scale bar, 4.5 μm. ***c***, Typical regrown axoneme visualized by immunofluorescence and printed to the same scale as ***b***.

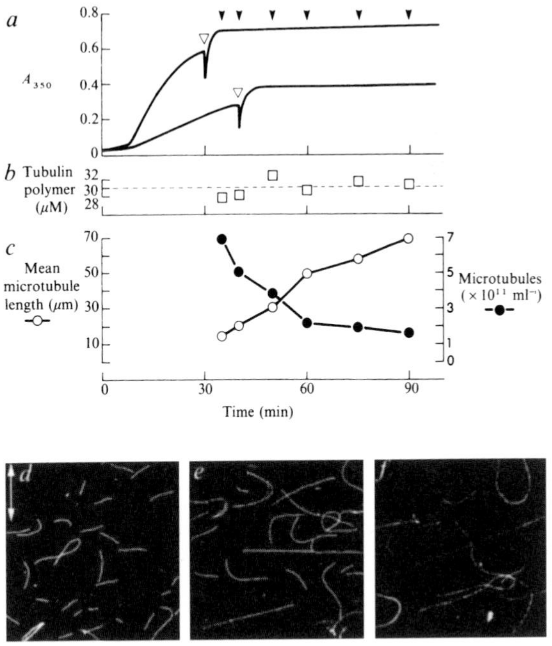

Fig. 4 Length redistribution at steady state. *a*, Turbidity record: upper trace, total tubulin = 59 μM; lower trace, total tubulin = 32 μM. Microtubules sampled at times shown by arrows. *b*, The product of mean length and number concentration, expressed as μM polymer, assuming 1,700 subunits μm^{-1}. *c*, Number concentration and mean length at the time point denoted by the arrows in *a*, upper trace. The same time scale is used for *a*, *b* and *c*; the points in *c* refer to the samples denoted by closed arrows in *a*. *d*, *e*, *f*, Part of typical immunofluorescence fields. Arrow = 30 μm. Time after shear: *d*, 5 min; *e*, 15 min; *f*, 60 min.

Methods: Microtubule seeds were prepared as in Fig. 5 of ref. 1. Tubulin in PB was added to a prewarmed cuvette in a Cary spectrophotometer and incubated throughout at 37 °C. At 8 min, 1/100 volume of seeds was added and the solution mixed. At the time indicated by the open arrow, the growing microtubules were sheared by passing the solution twice through a 22 gauge $1\frac{1}{2}$-inch needle, using a 1-ml syringe which had been prewarmed to 37 °C. From the solution in the upper trace, aliquots were removed for fixation at the times indicated by closed arrows. Aliquots (20 μl) were added to 2 ml of 1% glutaraldehyde in PB′ at 26 °C. After 3 min, the fixed microtubules were further diluted using cold PB′. Wide-mouthed pipette tips and very gentle mixing were used to minimize shear. 100-μl aliquots of diluted microtubules were spun onto 4.5-mm^2 polylysine-coated glass coverslips using the airfuge EM 90 rotor at 90,000*g* for 15 min. The microtubules were visualized by anti-tubulin immunofluorescence without methanol post-fixation. Microtubules were photographed at ×250 and ×100, and digitized for fluorescent axonemes as in Fig. 2. For the length data, 500 microtubules were measured at each time point. The higher magnification was used for the first three points, and the lower for the last three. For the number data, the number of microtubule ends per field at ×250 was averaged over 38 fields, divided by two and scaled using the magnification, the geometry of the rotor and the dilution factor.

addition to a GDP lattice in the concentration range studied. The GDP lattice thus depolymerizes rapidly, with off rates >100 times faster than the *y* intercept extrapolated from the growing phase kinetics (Table 1). The GTP-capped lattice, however, is stabilized, because the off rate of GTP subunits is slower by 2-3 orders of magnitude, so net addition of new GTP subunits occurs down to a very low tubulin concentration. However, the cap only exists because the hydrolysis rate lags slightly behind the polymerization reaction[13]. GTP hydrolysis is a first-order reaction initiated by the conformational change undergone by the subunit once incorporated into the polymer lattice. Thus, once a GTP subunit is incorporated, there is a fixed probability per unit time of hydrolysis occurring. Addition of GTP subunits, however, is a second-order reaction whose rate depends on monomer concentration. At high tubulin concentration there will always be a significant cap of GTP subunits at the growing tip of the polymer. In these circumstances nearly all microtubules will be in the growing phase. However, as the monomer concentration is decreased, the average cap size will decrease, statistical fluctuations in cap size will become important and the probability of GDP subunits becoming exposed on or near the polymer end increases. When this happens, the end subunits are rapidly lost into solution because their affinity for the GDP lattice is low. Once rapid depolymerization is initiated, it continuously exposes fresh GDP subunits at the polymer end and continues until the polymer disappears or a rare event recaps the microtubule and growth restarts. The events that cause a shrinking microtubule to become recapped are not clear. Depolymerization may terminate when encountering remaining GTP in the lattice, and the shrinking end could then become recapped by further addition of GTP-tubulin. The shrinking GDP end may also be capped by binding of GTP tubulin to the GDP end or by direct exchange of GDP for GTP on the terminal subunit.

It is clear, however, that phase transitions in either direction are rare compared with subunit addition in the growing phase or loss in the shrinking phase. Both phase transition probabilities should be quite sensitive to monomer concentration, and perhaps also to the free nucleotide concentrations. An explicit kinetic model incorporating these features has been developed recently[14]. Using plausible rate constants and Monte Carlo calculations, the essential features of persistence of growth and shrinking (with rare interconversions) have been demonstrated and a general model constructed[15].

An important question concerns the effect of other components of microtubules, such as microtubule-associated proteins (MAPs), on this dynamic behaviour. It has been suggested[8] that MAP-containing microtubules also have a GTP cap, as they

5

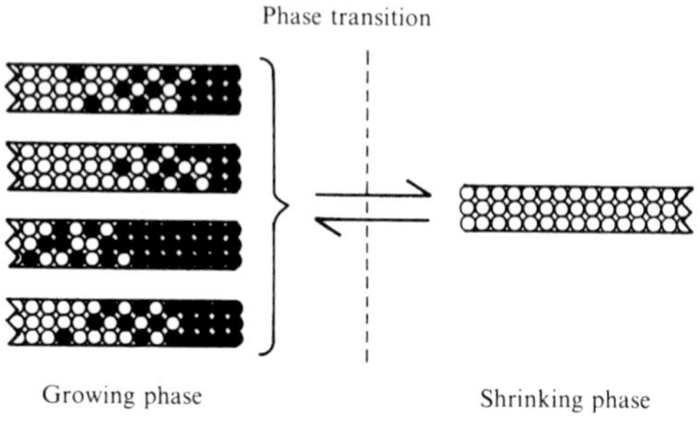

Fig. 5 Model for microtubule ends exchanging subunits with soluble dimer. The open circles represent tubulin liganded with GDP, and the closed circles represent GTP-tubulin. The growing phase of microtubules is stabilized by a GTP cap. This cap is of fluctuating length, but the average length increases with free tubulin concentration. The shrinking phase has lost this cap. Interconversion between the two phases is relatively rare.

also have a large discontinuity between growing and shrinking kinetics. Some experiments have demonstrated appreciable length fluctuations at steady state[11]. As MAPs bind to the microtubule lattice and stabilize it, they could slow the loss of GDP subunits and increase the probability that a shrinking microtubule would transit into the growing phase. This, in turn, would lower the steady-state concentration. The greater number of transitions at steady state induced by MAPs should tend to damp out the observed length fluctuations. Another factor which would tend to obscure length redistribution at steady state by offsetting the loss of microtubule number is some renucleation of microtubules, which may be strongly promoted by MAPs.

These considerations question the interpretation of isotope uptake experiments by microtubules at steady state as the result of a flux of subunits through the polymer or treadmilling[16,17]. The growing microtubule may behave instead like a simple equilibrium polymer, and extensive isotope uptake may occur through steady-state length redistribution. For microtubules with MAPs, length redistributions (and isotope uptake) would be expected to be less extensive, and in fact the measured isotope uptake is very small for tubulin-containing MAPs[16] compared with purer preparations of tubulin[18].

Such issues are important because of the question of the relevance of *in vitro* dynamic data with pure tubulin to the living cell. The fast depolymerization rate of microtubules as compared to the extrapolated values from the growing phase holds under various *in vitro* circumstances[8,11,19,20], and is likely to be true also *in vivo*. This may mean that the primary reason the cell has evolved GTP hydrolysis by tubulin is to have a polymer with built-in instability. Such a polymer could grow rapidly and be stabilized by end interactions, but it could also shrink rapidly, regulated by only small changes in conditions. The ability of microtubules to shrink rapidly may be of importance in reorienting the interphase microtubule array during locomotion or during morphological changes and in the extensive microtubule rearrangements of mitosis[2]. For example, using the off rates for the growing phase, a 20 μm microtubule would take more than 6 h to depolymerize at zero tubulin concentration, but using the off rates for the shrinking phase it would take about 1 min. The importance of the GTP cap in stabilizing growing microtubules raises the possibility that capping proteins may exist in cells and could be required for the long-term stabilization of a microtubule in a cell. Thus the centrosomes may be continually nucleating new microtubules, and only those ends which become capped or become stabilized in some other way such as by inhibition GTP hydrolysis may persist for long periods. A particular example of this could be the capture and capping of centrosomal microtubules by the kinetochore in prometaphase[21,22].

Various techniques have been used to study the dynamics of microtubules in living cells, and recently the powerful method of introducing fluorescently labelled tubulin molecules into living cells has been exploited. Recent results show that the mitotic spindle is much more dynamic than expected[23] and complete exchange of spindle microtubule and soluble subunits occurs within seconds. Interphase microtubules are considerably more stable but still exchange with soluble subunits within minutes. The dynamic behaviour induced in microtubules by the presence of fluctuating GTP caps provides the only satisfactory *in vitro* explanation for the rapid exchange of spindle microtubules.

We thank T. Hill, M.-P. Carlier and D. Pantaloni for helpful discussion, H. Martinez for the linear regression used in Fig. 2, and Cynthia Cunningham-Hernandez for help in preparing the manuscript. This work was supported by grants from the NIH and the ACS.

Received 27 April; accepted 6 September 1984.

1. Roberts, K. & Hyams, J. *Microtubules* (Academic, London, 1977).
2. McIntosh, J. R. *Mod. Cell Biol.* **2**, 115-142 (1983).
3. Mitchison, T. J. & Kirschner, M. *Nature* **312**, 232-237 (1984).
4. Bergen, L. G. & Borisy, G. G. *J. Cell Biol.* **84**, 141-150 (1980).
5. Allen, C. & Borisy, G. G. *J. molec. Biol.* **90**, 381-402 (1974).
6. Summers, K. & Kirschner, M. W. *J. Cell Biol.* **83**, 205-217 (1979).
7. Hill, T. L. & Kirschner, M. W. *Int. Rev. Cytol.* **84**, 185-234.
8. Carlier, M.-F., Hill, T. L. & Chen, Y.-D. *Proc. natn. Acad. Sci. U.S.A.* **81**, 771-775 (1984).
9. Oosawa, F. & Asakura, S. *Thermodynamics of the Polymerization of Protein* (Academic, New York, 1975).
10. Carlier, M.-F., Pantaloni, D. & Korn, E. D. *J. biol. Chem.* (in the press).
11. Kristofferson, D. & Purich, D. L. *Archs Biochem. Biophys.* **211**, 222-226 (1981).
12. Hill, T. L. & Carlier, M.-F. *Proc. natn. Acad. Sci. U.S.A.* **80**, 7234-7238 (1983).
13. Carlier, M.-F. & Pantaloni, D. *Biochemistry* **20**, 1918-1924 (1981).
14. Hill, T. L. & Chen, Y.-D. *Proc. natn. Acad. Sci. U.S.A.* (in the press).
15. Hill, T. *Proc. natn. Acad. Sci. U.S.A.* (in the press).
16. Margolis, R. L. & Wilson, L. *Cell* **13**, 1-8 (1978).
17. Farrell, K. W. & Jordan, M. A. *J. biol. Chem.* **257**, 3131-3138 (1982).
18. Cote, R. H. & Borisy, G. G. *J. molec. Biol.* **150**, 577-602 (1981).
19. Jameson, L. & Caplow, M. *J. biol. Chem.* **255**, 2284-2292 (1980).
20. Karr, T. L., Kristofferson, D. & Purich, D. L. *J. biol. Chem.* **255**, 8560-8655 (1980).
21. Pickett-Heaps, J., Tippit, D. H. & Porter, K. R. *Cell* **29**, 729-744 (1982).
22. Mitchison, T. J. & Kirschner, M. W. in *Molecular Biology of the Cytoskeleton* (eds Cleveland, D. W., Murphy, D. & Borisy, G. G.) (Cold Spring Harbor Laboratory, New York, in the press).
23. McIntosh, J. & Salmon, E. D. *J. Cell Biol.* (in the press).
24. Thompson, G. A., Baugh, C. C. & Walker, L. F. *J. Cell Biol.* **61**, 253-257 (1974).

Vale R.D., Reese T.S., and **Sheetz M.P.** 1985. Identification of a novel force-generating protein, kinesin, involved in microtubule-based motility. *Cell* **42:** 39–50.

MICROTUBULES HAD LONG BEEN KNOWN to be important for intracellular motility. Their role in mitotic chromosome movement was well established (Inoué and Sato, this volume), and their importance in axonal transport of vesicles and proteins was also widely appreciated (1). Microtubule-based movements in cilia and flagella were known to depend on the motor enzyme dynein, which could form mechanochemical cross-links between neighboring microtubules in the axoneme (Summers and Gibbons, this volume and 2). Other microtubule-dependent motions did not, however, appear to employ dynein. Several groups realized the value of cytoplasm isolated from squid giant axons for the study of intracellular transport in vitro. With recently improved microscopic methods, granules and even submicroscopic fibers could readily be seen to move in these preparations (3). Such methods, together with the observation that a non-hydrolyzable analog of ATP would freeze axoplasmic motion (4), led to the discovery of kinesin, a new microtubule-dependent motor enzyme. The paper reproduced here described the first isolation of this motor protein and opened a large field of inquiry. With the cloning of kinesin (5), it became possible to see the similarities between the primary structure of this motor enzyme and that of several proteins identified by genetics to be important for mitosis (e.g., 6). As a result, the paper by Vale et al. revealed the first view of what we now know to be an extensive superfamily of kinesin-like enzymes (7). These proteins are involved in a wide range of cellular motile functions; indeed, they contribute to the regulation of cell growth and form as well as to intracellular motions.

---

1. Fernandez H.L., Burton P.R., and Samson F.E. 1971. Axoplasmic transport in the crayfish nerve cord. The role of fibrillar constituents of neurons. *J. Cell Biol.* **51:** 176–192.
2. Gibbons B.H. and Gibbons I.R. 1973. The effect of partial extraction of dynein arms on the movement of reactivated sea-urchin sperm. *J. Cell Sci.* **13:** 337–357.
3. Brady S.T., Lasek R.J., and Allen R.D. 1985. Video microscopy of fast axonal transport in extruded axoplasm: A new model for study of molecular mechanisms. *Cell Motil.* **5:** 81–101.
4. Lasek R.J. and Brady S.T. 1985. Attachment of transported vesicles to microtubules in axoplasm is facilitated by AMP-PNP. *Nature* **316:** 645–647.
5. Yang J.T., Saxton W.M., and Goldstein L.S.B. 1988. Isolation and characterization of the gene encoding the heavy chain of *Drosophila* kinesin. *Proc. Nat Acad. Sci.* **85:** 1846–1886.
6. Enos A.P. and Morris N.R. 1990. Mutation of a gene that encodes a kinesin-like protein blocks nuclear division in *A. nidulans*. *Cell* **60:** 1019–1027.
7. Moore J.D. and Endow S.A. 1996. Kinesin proteins: A phylum of motors for microtubule-based motility. *Bioessays* **18:** 207–219.

Cell, Vol. 42, 39–50, August 1985, Copyright © 1985 by MIT

0092-8674/85/080039-12 $02.00/0

Identification of a Novel Force-Generating Protein, Kinesin, Involved in Microtubule-Based Motility

Ronald D. Vale,*†‡ Thomas S. Reese,* and Michael P. Sheetz,*†§
* Laboratory of Neurobiology,
National Institute of Neurological and
Communicative Disorders and Stroke
at the Marine Biological Laboratory
Woods Hole, Massachusetts 02543
† Department of Physiology
University of Connecticut Health Center
Farmington, Connecticut 06032
‡ Department of Neurobiology
Stanford University School of Medicine
Stanford, California 94305

Summary

Axoplasm from the squid giant axon contains a soluble protein translocator that induces movement of microtubules on glass, latex beads on microtubules, and axoplasmic organelles on microtubules. We now report the partial purification of a protein from squid giant axons and optic lobes that induces these microtubule-based movements and show that there is a homologous protein in bovine brain. The purification of the translocator protein depends primarily on its unusual property of forming a high affinity complex with microtubules in the presence of a nonhydrolyzable ATP analog, adenylyl imidodiphosphate. The protein, once released from microtubules with ATP, migrates on gel filtration columns with an apparent molecular weight of 600 kilodaltons and contains 110–120 and 60–70 kilodalton polypeptides. This protein is distinct in molecular weight and enzymatic behavior from myosin or dynein, which suggests that it belongs to a novel class of force-generating molecules, for which we propose the name kinesin.

Introduction

Most animal cells transport organelles within their cytoplasm (Schliwa, 1984). In neurons, bidirectional organelle transport within axons is thought to be the basis of fast axonal transport (Smith, 1980; Tsukita and Ishikawa, 1980). The squid giant axon, because of its large size, has been particularly useful for studying organelle movement (Allen et al., 1982; Brady et al., 1982). Axoplasm from the giant axon also can be dissociated, allowing bidirectional organelle movements to be visualized along individual cytoplasmic filaments with video-enhanced differential interference contrast microscopy (Vale et al., 1985a; Allen et al., 1985). These cytoplasmic filaments were identified as single microtubules by electron microscopy and immunofluorescence using antitubulin antibodies (Schnapp et al., 1985). Microtubules also support organelle movements in a variety of nonneuronal cells (Schliwa, 1984; Hayden et al., 1983).

Directed organelle movement within cells depends on ATP (Forman et al., 1984; Schliwa, 1984), and it has been generally assumed that movement is powered by an ATPase. Since all organelles move at the same velocity along isolated microtubules, the same type of molecular motor, or "translocator," may interact with a variety of organelles (Vale et al., 1985a). However, it is not known which proteins comprise the translocator. Dynein is the only molecule known to interact with microtubules to generate motive force (Gibbons, 1981), but attempts to identify dynein in neural tissue, where organelle movements are common, have been unsuccessful (Murphy et al., 1983).

Biochemical identification of the proteins involved in organelle movement requires an assay system for studying this phenomenon in vitro. Recently, we showed that isolated axoplasmic organelles move on purified squid optic lobe microtubules and that movement is enhanced by a soluble protein or proteins in axoplasm (Vale et al., 1985b). This experiment suggests that the translocator is a protein that binds reversibly to organelles and to microtubules. Organelle movement in the reconstituted system is similar to that in dissociated axoplasm except that movement along single microtubules in the reconstituted system is primarily unidirectional (Vale et al., 1985a; Schnapp et al., 1985).

The soluble fraction from axoplasm also induces movement of purified microtubules along a glass coverslip and movement of carboxylated latex beads along microtubules (Vale et al., 1985b). Native microtubules in extruded squid axoplasm also move along a glass coverslip (Allen et al., 1985). The soluble translocator from axoplasm that induces bead movement might also be responsible for moving inert beads after they are microinjected into axons or cultured cells (Adams and Bray, 1983; Beckerle, 1984). The velocities of microtubule and bead movements are the same (0.4 μm/sec), but both movements are 3 to 4 fold slower than organelle movements (Vale et al., 1985b). However, their similar sensitivities to vanadate suggest that organelle, bead, and microtubule movements may be driven by the same ATPase.

We describe the purification, from squid axoplasm and optic lobes, of a translocator protein that induces movement of microtubules on glass and movement of beads along microtubules. We also have identified a homologous protein in bovine brain. The characteristics of these proteins are distinct from myosin and dynein and appear to define a novel class of force-generating molecules.

Results

In Vitro Movement Assays

A high speed supernatant from axoplasm (S2 supernatant) promoted linear, ATP-dependent movement of puri-

§ To whom all correspondence should be addressed at his present address: Department of Cell Biology and Physiology, Washington University School of Medicine, St. Louis, Missouri 63110.

Cell
40

Table 1. Movements Supported by Squid and Bovine Translocators

Sample	Microtubule Movement along Glass	Microtubule Movement in Solution	Bead Movement	Organelle Movement[b]
Buffer	–	–	–	+
S2 axoplasmic supernatant[a]	0.44 ± 0.07 μm/sec	+	0.52 ± 0.06 μm/sec	+ + + to + + + +
Squid gel filtration	0.35 ± 0.06 μm/sec	+	0.34 ± 0.05 μm/sec	+ + to + + + +
Squid hydroxyapatite	0.37 ± 0.03 μm/sec	+	0.37 ± 0.06 μm/sec	+ + to + + + +
Bovine gel filtration	0.41 ± 0.05 μm/sec	+	0.59 ± 0.01 μm/sec	+ to + +

Microtubule, bead, and organelle movements promoted by S2 axoplasmic supernatant or purified squid or bovine translocator; movement occurred consistently in preparations for which rates are shown. Squid gel filtration peak fractions ($n = 7$; equivalent to lane 30 in Figure 4) and hydroxyapatite peak fractions ($n = 3$; equivalent to lane 40, 41 in Figure 5) were tested in motility buffer or 100 mM KCl, 50 mM Tris (pH 7.6), 5 mM $MgCl_2$, 0.5 mM EDTA, 2 mM ATP at final protein concentrations between 10 and 150 μg/ml. Bovine gel filtration peak fractions ($n = 3$; equivalent to lane 30 in Figure 7) were tested in KCl/Tris buffer at protein concentrations between 20 and 50 μg/ml. Organelle movement was assessed by viewing a 20 μm × 20 μm field of microtubules for 2.5–15 min and counting the number of different organelles that made directed movements along microtubules. The following rating system was employed: –, no movement in all preparations; +, 0–3 movements/min; + +, 3–7 movements/min; + + +, 7–15 movements/min; + + + +, 15–34 movements/min. The range of organelle movement in different preparations is reported. Microtubule, bead, or organelle movement assays are described in Experimental Procedures.
[a] Values from Vale et al. (1985b).
[b] The velocity of organelle movement in all samples was approximately 1.64 ± 0.24 μm/sec.

fied microtubules (freed of microtubule-associated proteins by 1 M NaCl extraction) at 0.4 μm/sec along a glass coverslip (Table 1). In the presence of S2 supernatant, microtubules in solution also moved relative to one another, to form a contracted aggregate of microtubules. Carboxylated latex beads treated with S2 supernatant also moved at 0.4 μm/sec along microtubules immobilized by poly-D-lysine treatment of the coverslip (Table 1). Microtubules or beads incubated with ATP in the absence of S2 supernatant did not exhibit directed movement. Furthermore, the number of organelles moving along microtubules was enhanced severalfold by S2 supernatant (Table 1; see also Vale et al., 1985b). The translocator is a protein whose activity is inhibited by vanadate (100 μM) and by a nonhydrolyzable ATP analog, adenylyl imidodiphosphate (AMP-PNP) (Vale et al., 1985b).

Because microtubule movement on glass and in solution is a sensitive and rapid assay of movement-inducing activity, we used this assay to purify the translocator protein in axoplasmic supernatant. Samples or fractions were scored positive for microtubule movement if movement was observed either on the glass coverslip or in solution.

Cosedimentation of the Translocator with Microtubules

Squid Axoplasm

Our previous studies suggested that the translocator does not bind tightly to microtubules (Vale et al., 1985b). Indeed, no microtubule movement occurred when microtubules were treated with S2 supernatant, centrifuged, washed, and resuspended in an ATP-containing buffer. However, it has been shown that AMP-PNP induces attachment of axoplasmic organelles to microtubules dissociated from squid axoplasm (Lasek and Brady, 1984). We thought that if this attachment of organelles were due to an increase in the affinity of a soluble translocator for microtubules, it might be possible to bind the translocator to microtubules with AMP-PNP to provide a means of affinity purification.

Addition of AMP-PNP to purified microtubules and axoplasmic supernatant (in the presence of 1 mM ATP) markedly increased the amount of a polypeptide (110 kilodaltons in SDS polyacrylamide gels) that copelleted with microtubules (Figure 1). This band often appeared as a doublet, possibly indicating that there are two forms of the protein that differ slightly in charge or molecular weight. This 110 kd polypeptide could be released from the microtubules with 0.1 M KCl and 5 mM ATP. When the released material was applied to new microtubules, they translocated along the coverslip and aggregated in solution (Figure 1). Two prominent proteins with molecular weights higher than 300 kd also associated with microtubules, but the amount bound was not influenced by AMP-PNP. Proteins released from microtubules treated with S2 supernatant in the absence of AMP-PNP, which included the 300 kd proteins but not the 110 kd protein, did not induce microtubule movement. The translocator also cosedimented with microtubules in the absence of AMP-PNP if the S2 supernatant was first depleted of ATP using hexokinase and glucose (not shown).

Squid Optic Lobes

A similar microtubule cosedimentation experiment was performed using a high speed supernatant (S3) from squid optic lobes in order to obtain larger quantities of translocator. When S3 supernatant from optic lobes was combined with microtubules and 5 mM AMP-PNP, a polypeptide of 110 kd copelleted with the microtubules. This prominent band was not seen in the absence of AMP-PNP treatment (Figure 2). Minor bands corresponding to molecular weights of 65, 70, and 80 kd were also more intense after AMP-PNP treatment. Two high molecular weight proteins (>300 kd) also associated with the microtubules (Figures 2 and 3). We refer to the higher and lower

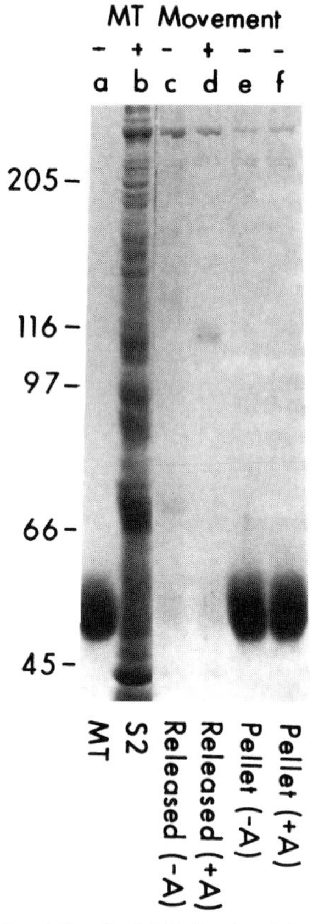

Figure 1. Cosedimentation of a Soluble Axoplasmic Translocator with Purified Microtubules in the Presence of AMP-PNP, and Its Subsequent Release by ATP

Axoplasmic supernatant (S2, lane b) was incubated with purified microtubules (lane a) at 100 µg/ml (with or without 10 mM AMP-PNP) for 10 min at 23°C, and then the microtubules were pelleted at 37,000 × g for 30 min at 4°C. Proteins released by ATP (see details below) from AMP-PNP-treated microtubules (lane d) show a prominent band at 110 kd and also induce microtubule movement (+). Neither this band nor movement (−) is associated with material released by ATP from control (AMP-untreated) microtubules (lane c), or with the microtubule pellets of AMP-PNP-treated (lane f) and untreated samples (lane e) after the ATP release.

Conditions for ATP release. After the original centrifugation of microtubules, the supernatant was discarded and microtubules were resuspended in 75 µl MTG buffer with or without 10 mM AMP-PNP, and then centrifuged as before. The microtubule pellet was washed with MTG buffer (50 µl), then resuspended in 30 µl MTG plus 5 mM ATP and 0.1 M KCl for 30 min at 23°C to release associated proteins. Microtubules were pelleted as before, the supernatants were collected (lanes c and d), and the microtubules were resuspended in 30 µl of MTG (lanes e and f). The positions of molecular weight standards are noted at left.

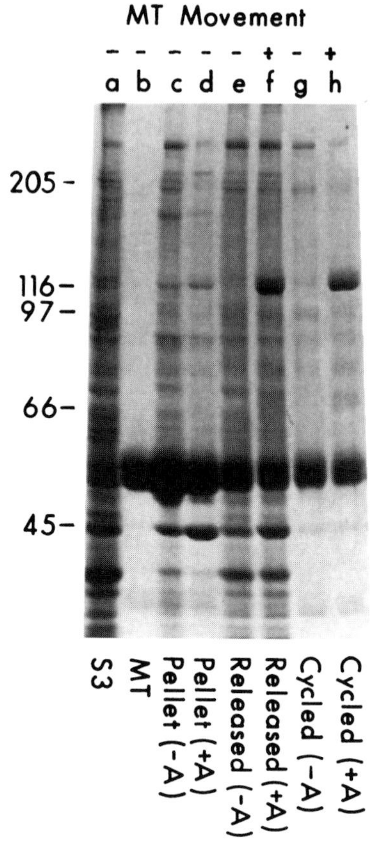

Figure 2. Cosedimentation of a Translocator from Squid Optic Lobes with Purified Microtubules in the Presence of AMP-PNP, and Its Subsequent Release by ATP

Soluble supernatant from squid optic lobes (S3, lane a) was incubated with 100 µg/ml of purified microtubules (lane b) in the presence or absence of 5 mM AMP-PNP as described in Experimental Procedures. Material released from microtubules by 5 mM ATP and 0.1 M KCl (45 min; 4°C) from AMP-PNP-treated samples (lane f) shows both a prominent band at 110 kd and movement-inducing activity; neither characteristic is associated with the supernatant of the untreated samples (lane e) or the final microtubule pellets resuspended in an equal volume of MTG with 2 mM ATP (lanes c and d) after ATP release.

The ATP-released translocator (lane f) was tested for rebinding to microtubules. The sample was diluted 5-fold in MTG (to reduce the concentration of ATP/KCl) and incubated with new microtubules (100 µg/ml) in the presence or absence of 5 mM AMP-PNP (15 min; 23°C). The microtubules were pelleted and washed, and proteins were released with 5 mM ATP and 0.1 M KCl for 45 min at 23°C. Material released from the AMP-PNP-treated sample contained the 110 kd polypeptide (lane h) and also induced movement (+); neither characteristic was prominent in the control (AMP-PNP-untreated) sample (lane g).

molecular weight species of the 300 kd polypeptides as HMW_1 and HMW_2, respectively. The amount of HMW_2 was somewhat variable, which may reflect its dissociation from microtubules during washing of the microtubule pellet (see below). Association of HMW_1 and HMW_2 with microtubules generally was unaffected by AMP-PNP, although in some experiments a small (2-fold) increase in the amount of HMW_1 was found in AMP-PNP-treated samples.

Unlike S2 supernatant from axoplasm, S3 supernatant from optic lobes did not induce microtubule movement, either because the concentration of translocator was too low or because inhibitory proteins were present. The AMP-PNP-dependent binding of the translocator to microtubules provided a method for concentrating its activity. Indeed, microtubules that were incubated with S3 optic lobe

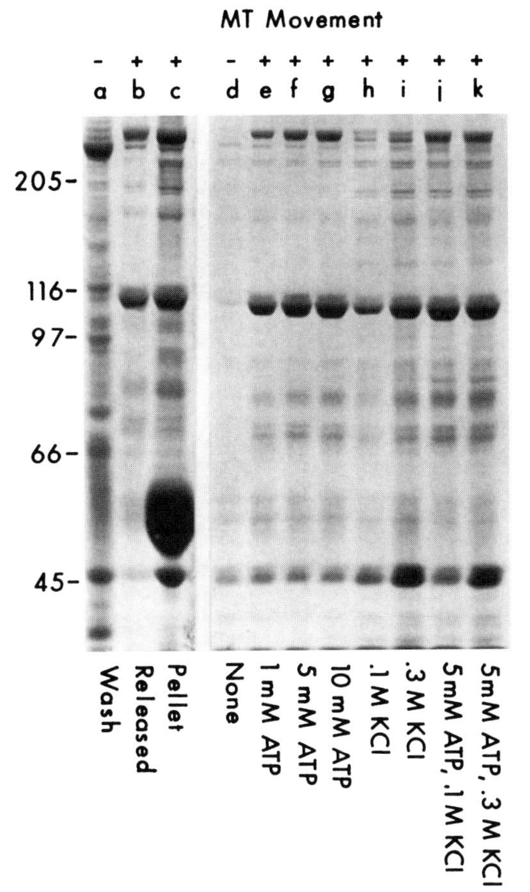

Figure 3. ATP- and KCl-Dependent Release of the Translocator from Microtubules

S3 supernatant was incubated with microtubules and AMP-PNP as described in Experimental Procedures. Many contaminating proteins (lane a), but not the translocator, were removed by resuspending the MT pellet in 2 ml of MTG buffer containing 10 mM AMP-PNP, centrifuging the microtubules, and removing the AMP-PNP-containing supernatant. The translocator could then be obtained by washing the pellet with MTG, releasing with 1 ml of MTG plus 10 mM ATP/0.1 M KCl (30 min; 23°C), centrifuging the microtubules, and collecting the supernatant that contains movement-inducing activity (lane b). The final microtubule pellet was then resuspended in 1 ml of MTG (lane c). To define further the conditions that release translocator from microtubules, S3 supernatant (8 ml) was incubated with microtubules and AMP-PNP, and the microtubule pellet was washed as described above. Microtubules were then aliquoted into eight 100 μl samples in MTG containing the indicated amounts of ATP and KCl and incubated for 30 min at 23°C (lanes d-k). The microtubules were pelleted, and 50 μl of each supernatant was run per lane. Lanes a-c are from a different experiment than lanes d-k.

supernatant and AMP-PNP, and then pelleted, washed, and resuspended in ATP-containing buffer moved along each other and on the glass coverslip. When microtubule-associated proteins, including the 110 kd polypeptide, were released with 5 mM ATP and 0.1 M KCl and then applied to new microtubules, these microtubules moved in a manner indistinguishable from the movement produced by S2 axoplasmic supernatant (Figure 2). Proteins released from microtubules not treated previously with AMP-PNP, which included HMW_1 and HMW_2 but not the 110 kd polypeptide, did not induce microtubule movement.

The released 110 kd polypeptide, if diluted to lower the concentration of ATP and KCl, could rebind to microtubules in the presence of AMP-PNP, and when rereleased, it still induced microtubule movement (Figure 2). This cycling of the translocator protein with microtubules also removed many of the minor protein bands. The 110 kd polypeptide did not pellet to a significant extent with microtubules when cycled in the absence of AMP-PNP, and the subsequently released supernatant induced little or no microtubule movement. The 65/70 kd doublet and 80 kd polypeptides also cycled selectively with microtubules in the presence of AMP-PNP.

HMW_2 and a variety of other proteins were released from microtubules by dilution in the absence of ATP or KCl, but the 110 kd polypeptide and HMW_1 remained attached (Figure 3, lane a). Hence, the purity of the 110 kd polypeptide could be increased without cycling by resuspending microtubules in buffer containing AMP-PNP to release HMW_2 and other proteins that dissociate upon dilution (Figure 3). When the 110 kd polypeptide was subsequently released with KCl or ATP, it comprised between 15% and 45% of the released protein, as determined by densitometry. The protein sample at this stage is referred to as MT-purified translocator. The major contaminants include the HMW proteins as well as tubulin, which was present in variable amounts because of some cold instability of the microtubules. Assuming 100% association of the translocator with microtubules in the presence of AMP-PNP, the concentration of translocator in the S3 supernatant is estimated to be approximately 5–10 μg/ml.

The 110 kd polypeptide dissociated from microtubules in either 0.1–0.3 M KCl or 1–10 mM ATP (Figure 3, lanes d–k). ATP was a more effective releasing agent than KCl, and combinations of ATP and KCl did not produce a greater effect than ATP alone. KCl (0.3 M) also released certain proteins that remained bound when only ATP was added. Between 40% and 90% of the 110 kd polypeptide was released by a single 30 min incubation at 23°C with 5 mM ATP and 0.1 M KCl, and the remainder was generally released by subsequent extraction under the same conditions. The HMW_1 protein was also released from microtubules by ATP, as were the 65/70 kd doublet and 80 kd polypeptides (Figure 3, lanes d–k).

Chromatographic Purification of the Translocator from Squid

Gel Filtration Chromatography

Supernatant containing the 110 kd and 65/70 kd doublet polypeptides released from microtubules promoted microtubule movement. To determine whether it is these polypeptides that induce movement, the MT-purified translocator was further purified by chromatography. The microtubule translocator eluted from a Bio-Gel A5m gel filtration column (Figure 4) as a single peak that corresponded precisely to the elution peak of the 110 kd poly-

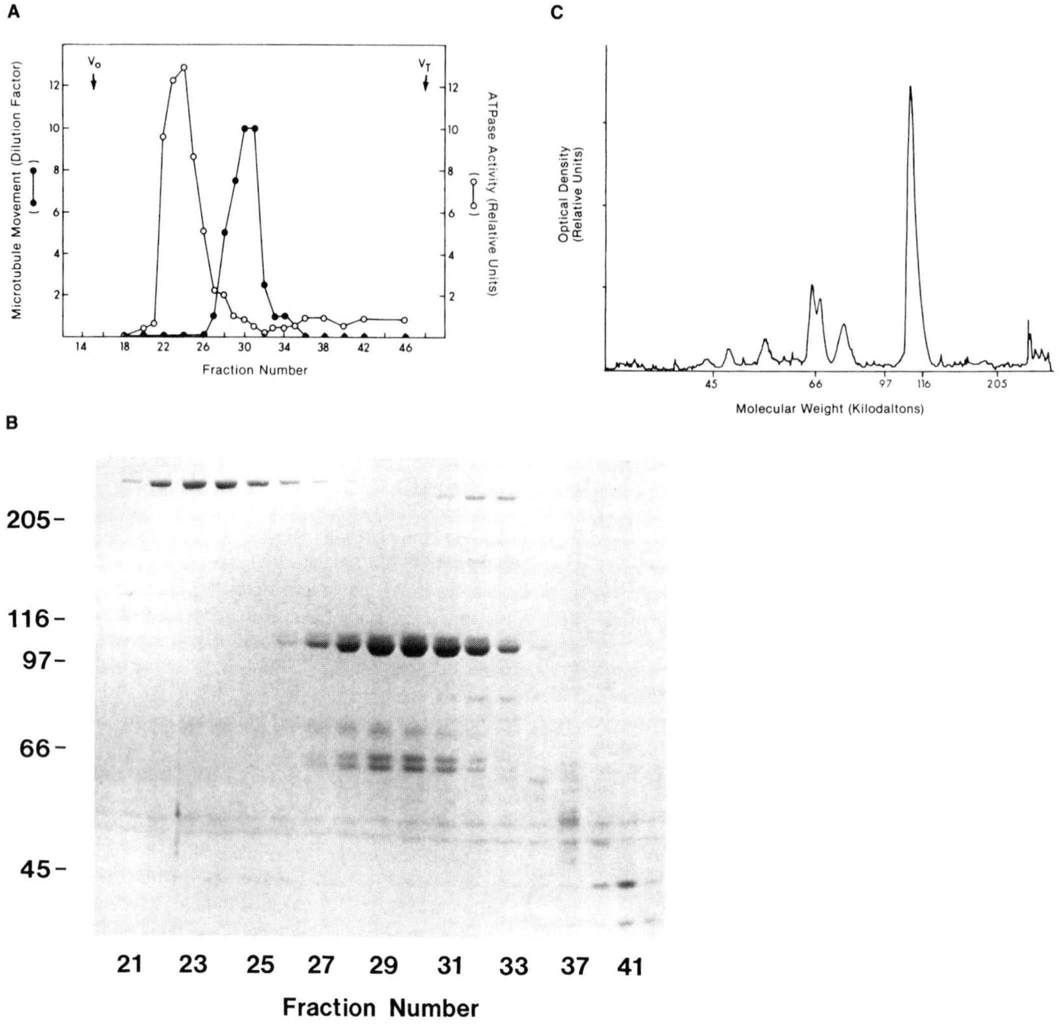

Figure 4. Gel Filtration Chromatography of Squid Translocator

(A) Microtubule-purified translocator was applied to a Bio-Gel A5m column, and the microtubule-movement-inducing and ATPase activities of the eluant fractions were determined. (B) The polypeptide compositions of fractions 21–45 were analyzed by SDS polyacrylamide gel electrophoresis (60 μl loaded per lane); fraction 30 contains approximately 65 μg/ml of protein. (C) A densitometer scan is shown of a lane from the fraction (30) with the greatest movement-inducing activity from another column. Chromatographic conditions: A 0.75 ml sample of microtubule-purified translocator (which moved microtubules at a 1:20 dilution) was applied to a 1 × 47 cm column equilibrated in 100 mM KCl, 50 mM Tris (pH 7.6), 5 mM $MgCl_2$, 0.5 mM EDTA, and 1 mM ATP. The column was run at 4°C at 5 cm/hr; 0.75 ml fractions were collected. Excluded (V_o) and included (V_T) volume fractions are noted at top. Microtubule movement was measured by serial dilution as described in Experimental Procedures. The tubulin doublet at 55 kd that streaks across the gel is not typical, but is an artifact of this particular gel.

peptide, as determined by SDS gel electrophoresis. The 110 kd polypeptide, which appeared as two bands on SDS polyacrylamide gels, eluted between the HMW_1 and HMW_2 proteins with a K_{av} of 0.45, approximately the elution peak of the globular protein thyroglobulin (molecular weight of 667 kd). Since the 110 kd polypeptide peak was clearly separated from the two HMW protein peaks, it is unlikely that the 110 kd polypeptide is in a complex with either of these proteins. On the other hand, the 65/70 kd doublet, which also demonstrated AMP-PNP-dependent binding to microtubules (see Figure 2), comigrated with the 110 kd peak, indicating that these polypeptides are tightly associated.

Densitometry scans of Coomassie-stained gels of the column fractions indicated a stoichiometric ratio of the 110 to 65/70 kd doublet polypeptides which ranged from 1.5:1 to 2.2:1 (measurements from five columns). The 65 and 70 kd polypeptides were present in approximately equal amounts. An 80 kd polypeptide also comigrated with the 110 kd polypeptide, but the molar ratio of 110 kd to 80 kd polypeptide (4:1 to 10:1) was less consistent between different preparations. Silver or nickel staining of 7.5% polyacrylamide gels of the column fractions did not reveal additional polypeptides that consistently copurified with the 110 kd polypeptide (not shown).

The major peak of ATPase activity eluting from the gel

Cell
44

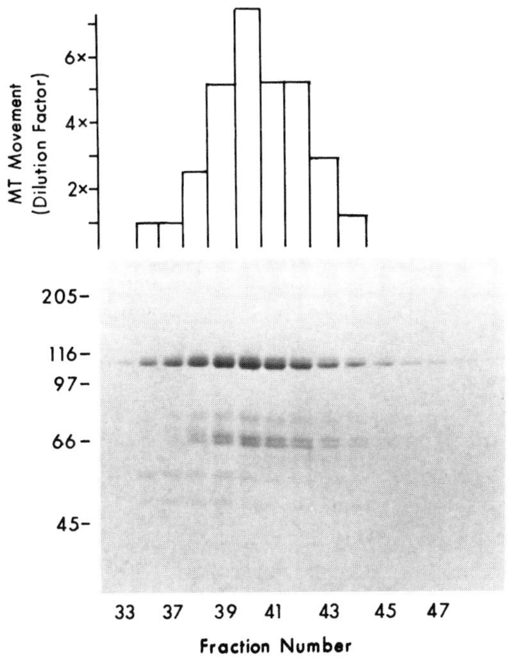

Figure 5. Hydroxyapatite Chromatography of Squid Translocator Purified by Gel Filtration

Samples corresponding to fractions 28–32 from the gel filtration column shown in Figure 4 were pooled and applied to a 1 × 6 cm hydroxyapatite column equilibrated with gel filtration buffer at pH 7.4. Proteins were eluted with a 0–0.3 M phosphate gradient, and 0.5 ml fractions were collected. Microtubule-movement-inducing activity, assayed by serial dilution, correlated with the presence of 110 and 65 kd polypeptides in eluant fractions. Fraction 40 contains approximately 35 μg/ml of protein.

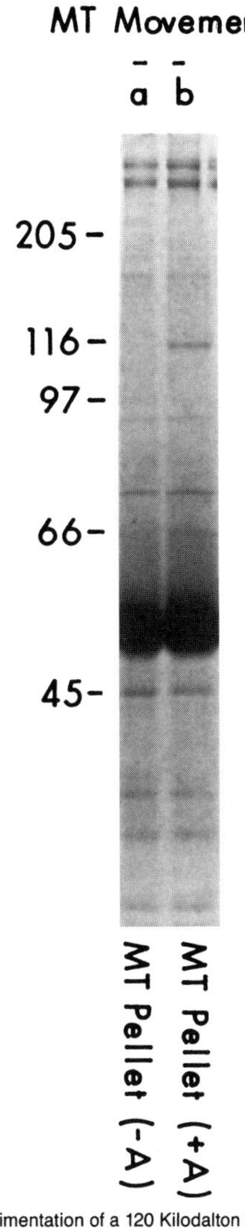

Figure 6. Cosedimentation of a 120 Kilodalton Bovine Brain Polypeptide with Squid Microtubules Depends upon AMP-PNP

Bovine brain S3 supernatant was incubated with squid optic lobe microtubules (100 μg/ml) in the absence (lane a) or presence (lane b) of 5 mM AMP-PNP for 15 min at 23°C. Microtubules were pelleted, washed once with PEM-taxol-GTP buffer, resuspended in 1/40 of the original volume with the same buffer containing 5 mM ATP and 0.1 M KCl, and then assayed for movement-inducing activity.

filtration column corresponded to the HMW_1 protein, while relatively little activity eluted with fractions that induced microtubule movement and contained the 110 kd polypeptide (Figure 4). The ATPase activities in the fractions corresponding to the peak elutions of HMW_1 and the 110 kd polypeptides were, respectively, 0.13 and 0.01 μmol/min/mg (in 100 mM KCl, 50 mM Tris, 5 mM $MgCl_2$, 0.5 mM EDTA, pH 7.6). In the 110 kd peak fractions, the ATPase was activated 1 to 2 fold by 0.25% Triton X-100 or 0.8 M KCl, but it was not significantly inhibited by 2.5 mM N-ethyl maleimide (NEM). The ATPase activity of the HMW_1 peak, on the other hand, was inhibited 80%–90% by 2.5 mM NEM when incubated 15 min at 23°C followed by 20 mM dithiothreitol (DTT). However, microtubule movement (as assayed by serial dilution) was not inhibited by treating the 110 kd peak fraction with 5 mM NEM for 30 min at 23°C followed by 20 mM DTT inactivation. Because microtubule movement occurred under conditions in which the HMW_1 ATPase was inhibited, this ATPase presumably is not involved in powering microtubule movement.

Translocator purified by gel filtration chromatography induced microtubule movement at total protein concentrations as low as 5 μg/ml and at ATP concentrations of 10–20 μM (at protein concentrations of 50 μg/ml). Microtubule movement also occurred at GTP concentrations of 0.5 mM; but CTP, TTP, and UTP did not support movement at concentrations of 5 mM. Microtubule movement induced by purified squid translocator was abolished by 100 μM vanadate and was only slightly inhibited by 25 μM vanadate. A similar dependence on vanadate concentration was observed with microtubule, bead, and organelle movements induced by S2 axoplasmic supernatant (Vale et al., 1985b).

Hydroxyapatite Chromatography

Fractions containing the 110 kd polypeptide from the Bio-Gel A5m column were pooled and applied to an hydroxyapatite column. The fraction that induced the most microtubule movement corresponded with the peak of the 110 kd polypeptide, which eluted at a phosphate concentration of approximately 0.18–0.2 M (Figure 5). The HMW proteins eluted earlier than the 110 kd polypeptide, and before the fractions that promoted microtubule movement. The 65/70 kd doublet and 80 kd polypeptides coeluted with the 110 kd polypeptide. Densitometry again revealed a stoichiometric ratio of the 110 kd polypeptides to the 65/70 kd doublet polypeptide that ranged from 1.5:1 to 2:1 (not shown). The extent of purification of the translocator after hydroxyapatite chromatography is at least 1000-fold, assuming a concentration of the translocator in S3 supernatant of 10 μg/ml (see above).

MT-purified translocator could be further purified by DEAE-Sephadex chromatography. Movement-inducing activity as well as the 110 kd and 65/70 kd doublet polypeptides bound to this column and were eluted together by 0.2 M KCl (not shown), also indicating that these polypeptides are involved in microtubule movement.

Purification of a Translocator from Bovine Brain

A protein homologous to the squid optic lobe translocator may be present in the vertebrate nervous system; therefore, we attempted to purify it by its affinity for microtubules in the presence of AMP-PNP. When S3 supernatant from bovine brain (which itself did not induce movement of microtubules) was incubated with squid microtubules, several proteins, including two of high molecular weight that might correspond to MAP_1 and MAP_2 (Vallee and Bloom, 1984), copelleted with the microtubules (Figure 6). If AMP-PNP was also added, another major polypeptide, of 120 kd, copelleted with the microtubules. This polypeptide attached with approximately equal affinity to squid or bovine microtubules in the presence of AMP-PNP (not shown). This and a minor polypeptide of approximately 62 kd were the only ones that demonstrated AMP-PNP-dependent binding to microtubules. However, microtubules associated with this protein did not move on glass or in solution (Figure 6). The 120 kd polypeptide could be released from the microtubules with 1–10 mM ATP, and when added to new microtubules, the released proteins still did not induce microtubule movement on glass or in solution.

When the MT-purified and ATP-released 120 kd polypeptide from bovine brain was applied to a Bio-Gel A5m column, the 120 kd polypeptide eluted in the same position as the 110 kd polypeptide from squid (Figure 7). The 120 kd polypeptide clearly separated from the high molecular weight polypeptides, but coeluted with the 62 kd polypeptide. The 120 kd and 62 kd polypeptides typically appeared as single bands rather than doublets. Densitometry indicated that the stoichiometric ratio of 120 kd polypeptide to 62 kd polypeptide was approximately 2:1 (measurements from three columns).

Unlike the material applied to the gel filtration column, fractions eluting from the column which contained the 120 kd polypeptide induced movement of squid and bovine microtubules in solution and along glass (Figure 7). Microtubule movement was detected at protein concentrations as low as 5 μg/ml. Fractions from the column, however, exhibited very little ATPase activity. Microtubule movement induced by fractions containing the bovine 120 kd polypeptide was inhibited by 100 μM vanadate but not by 5 mM NEM.

Movements Induced by Column-Purified Squid and Bovine Translocators

The S2 supernatant from axoplasm induces movements of microtubules on glass, microtubules in solution, latex beads on microtubules, and organelles on microtubules (Vale et al., 1985b; Table 1). The squid and bovine translocators were purified by microtubule affinity and column chromatography and assayed for their ability to move microtubules in solution and along glass as described above. The velocity of microtubule movement on glass was 0.3–0.5 μm/sec, whether induced by purified squid or bovine translocators or axoplasmic S2 supernatant (Table 1). Column-purified squid and bovine translocator also induced movement of carboxylated latex beads along microtubules at 0.3–0.5 μm/sec (Table 1). Preliminary studies suggest that bead movement along microtubules is unidirectional in the presence of purified translocator. The copurification of the factors inducing bead and microtubule movements suggests that they depend on activities of the same protein.

The squid translocator purified by gel filtration, or by gel filtration and hydroxyapatite chromatography, increased the number of organelles moving along purified microtubules severalfold compared to the number that moved in an ATP-containing buffer alone. Some purified protein preparations enhanced organelle movement to the same extent or to a greater extent than the S2 axoplasmic supernatant, although there was more variability in the amount of organelle movement between different preparations of purified squid translocator than between preparations of S2 axoplasmic supernatant. The purified bovine translocator, on the other hand, did not consistently increase the frequency of movement of organelles isolated from squid axoplasm. Organelle movement induced by S2 supernatant or purified translocator appears not to depend on nonspecific association of the translocator with the organelle surface, because trypsinized organelles, lipid vesicles, and vesicles, which are isolated with microtubules during microtubule purification, do not exhibit directed movement, even in the presence of active translocator (see also Vale et al., 1985b).

The velocity of organelle movement on purified microtubules was approximately 1.6 μm/sec in buffer alone, in S2 axoplasmic supernatant, or in the presence of column-purified 110 kd squid translocator (Table 1); this velocity is the same as that of organelles in dissociated axoplasm (Vale et al., 1985a, 1985b). Organelle movements in the presence of purified squid translocator or S2 supernatant were qualitatively indistinguishable.

Discussion

Purification of a Protein Involved in Microtubule-Based Movements

A translocator protein that induces movement of microtu-

Cell
46

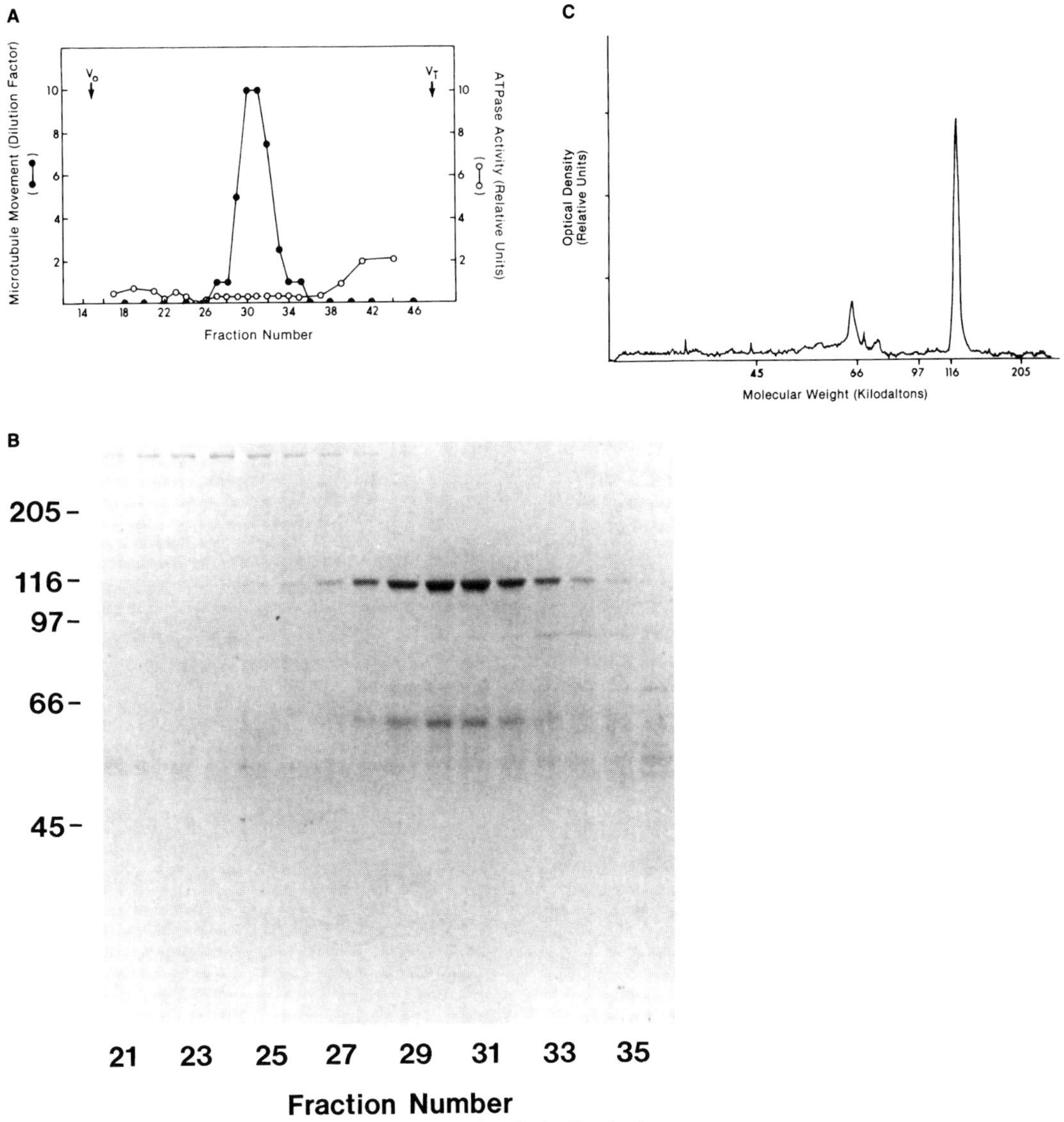

Figure 7. Gel Filtration Chromatography of the Microtubule-Purified Bovine Translocator

(A) Microtubule-purified bovine proteins (0.75 ml; purified as described in Experimental Procedures) were run on a Bio-Gel A5m column like that described in Figure 4, yielding the profile of ATPase and movement-inducing activities shown. (B) The polypeptide compositions of fractions 21–37 were analyzed by SDS polyacrylamide gel electrophoresis (40 μl per lane). (C) A densitometer scan of the Coomassie stained gel is shown of the lane corresponding to the fraction with greatest movement-inducing activity (30, which contains approximately 40–50 μg/ml of protein). Void (V_o) and excluded (V_T) volumes are indicated.

bules on a glass coverslip and in solution has been partially purified from squid axoplasm, squid optic lobes, and bovine brain. The principal step in purification involves cosedimentation of the translocator with microtubules in the presence of an ATP analog, adenylyl imidodiphosphate. Microtubule affinity has also been employed in the purification of other proteins such as MAP_1, tau, and dynein (Vallee and Bloom, 1984; Cleveland et al., 1977; Nasr and Satir, 1985).

Microtubule movement occurs only in the presence of a major polypeptide with a molecular weight of approximately 110–120 kd and with less abundant polypeptides with molecular weights of approximately 65, 70, and 80 kd. Binding of these polypeptides to microtubules is enhanced by AMP-PNP and is diminished by ATP, conditions that correlate with induction of microtubule movement. Moreover, the peak activity of the microtubule translocator coincides with the peak fractions of the 110 kd squid polypeptide in gel filtration, hydroxyapatite, and DEAE-Sephadex chromatography. We believe that the 110 kd squid and the 120 kd bovine polypeptides, although slightly different in molecular weight, are homologous proteins, because they share the ability to induce movement of microtubules and carboxylated latex beads, they bind to microtubules in the presence of AMP-PNP, they coelute in gel filtration columns, and they have similar sensitivities to NEM and vanadate.

The two high molecular weight microtubule-associated proteins are unlikely to be involved directly in microtubule movement. Neither protein coelutes from gel filtration and hydroxyapatite columns with the proteins that induce microtubule movement. Although HMW_1 is released from microtubules by ATP, its binding to them is typically not enhanced by AMP-PNP. Furthermore, the ATPase activity of HMW_1 is inhibited by NEM, while microtubule movement is not.

Several polypeptides from squid axons or optic lobes with molecular weights of 110, 80, 70, and 65 kd are consistently associated with microtubule movement. The 65/70 kd doublet polypeptides coelute with the 110 kd polypeptide in gel filtration and hydroxyapatite columns with a constant ratio of Coomassie blue staining intensity. A similar ratio of 120 kd to 62 kd polypeptides is observed with the bovine translocator. An 80 kd polypeptide also copurifies with the squid translocator, but typically not with the bovine translocator. This polypeptide, which was present in variable and generally less than stoichiometric amounts, could represent breakdown of the 110 kd polypeptide. In gel filtration columns, both the squid and bovine translocators elute with an apparent molecular weight of 600 kd. This result is consistent with the idea that both the bovine and squid translocators are complexes of the 110–120 kd and 62–70 kd polypeptides, but additional experiments are needed to establish the precise quaternary structure of this complex.

The AMP-PNP-dependent binding of the translocator to microtubules and its release by ATP indicates that it has a nucleotide binding site. The vanadate sensitivity of the translocator-induced microtubule movement, as well as the fact that ATP is a more effective substrate than GTP by an order of magnitude, suggests the presence of an ATPase. However, the fractions that induce the most movement have very little associated solution ATPase activity (ATP turnover rate of 0.1 sec^{-1}) compared to the ATPase activity of dynein (1.5–30 sec^{-1}; Bell et al., 1982). To obtain maximal ATPase hydrolysis of the translocator, a ternary complex consisting of translocator; microtubule; and bead, organelle, or glass surface may be necessary, though conditions for producing such complexes in high concentrations have not yet been found.

Kinesin: A Novel Class of Force-Generating Molecules

Dynein is the only protein known to bind to microtubules and perform work (Gibbons, 1981). Axonemal and cytoplasmic dyneins have been isolated from a variety of sources and have a number of common features that distinguish them as dyneins. By several structural and enzymatic criteria, however, the translocators isolated from squid and bovine neural tissue are not dyneins.

Dyneins are characterized by an ATPase subunit of high molecular weight (>300 kd) whereas the translocators described here contain no polypeptide with a molecular weight higher than 120 kd. Axonemal dyneins of sea urchin (Bell et al., 1979; Gibbons and Fronk, 1979), Tetrahymena (Porter and Johnson, 1983), and Chlamydomonas (Piperno and Luck, 1979) are quite large (18S–30S) and contain two or three high molecular weight ATPase subunits as well as several intermediate and low molecular weight subunits. The molecular weights of these dynein complexes have been estimated by scanning transmission electron microscopy to be between 1 million and 2 million daltons (Witman et al., 1983; Johnson and Wall, 1983). Lower molecular weight forms (10S–14S) have also been isolated, particularly by dissociating larger dynein complexes at low ionic strength, but they still contain a single high molecular weight ATPase (Gibbons and Fronk, 1979; Porter and Johnson, 1983; Pfister and Witman, 1984).

Cytoplasmic forms of dynein have been isolated (Pratt, 1980; Scholey et al., 1984; Asai and Wilson, 1985; Hisanga and Sakai, 1983; Hollenbeck et al., 1984). These proteins also contain a high molecular weight ATPase subunit that interacts with microtubules, but they have not yet been shown capable of generating force. Since the microtubule translocators isolated from neural tissue contain no 300 kd subunits, even in the presence of protease inhibitors, they appear to be structurally distinct from cytoplasmic and axonemal dyneins. The translocator also is distinct from high MW microtubule-associated proteins that have been characterized (Vallee and Bloom, 1984).

The translocator also differs from dynein with respect to its enzymatic properties. While AMP-PNP decreases the affinity of dynein for microtubules (Mitchell and Warner, 1981) and of myosin for actin (Greene and Eisenberg, 1980), it dramatically increases the affinity of the translocator for microtubules, even in the presence of equimolar concentrations of ATP. While NEM inhibits force generation by dynein (Mitchell and Warner, 1981; Cosson and Gibbons, 1978), as well as dynein's ability to bind to microtubules (Shimizu and Kimura, 1974), microtubule movement induced by squid or bovine translocators is largely unaffected by NEM. We conclude that the squid and bovine translocators represent a novel class of motility proteins that are structurally as well as enzymatically distinct from dynein, and we propose to call these translocators kinesin (from the Greek *kinein*, to move).

Possible Biological Roles for Kinesin

Kinesin appears to have a site for binding to a structure to be moved and another site capable of cyclic on and off interactions with microtubules to produce mechanical displacement (Vale et al., 1985b). If attached to an organelle or a bead, kinesin could act as a motor for organelle or bead translocation. The attachment to the organelle may be mediated by a protein on the organelle surface, since trypsinization of organelles blocks their movement (Vale et al., 1985b) whereas attachment to the bead would be through absorption to the anionic surface. Only those molecules in the correct orientation relative to the microtubule would generate forces resulting in unidirectional translocation, analogous to the manner in which myosin-coated beads move unidirectionally on Nitella actin cables (Sheetz and Spudich, 1983).

Although our results are consistent with the notion that

kinesin powers organelle translocation along microtubules, the evidence is circumstantial. Column-purified kinesin, like axoplasmic supernatant, can increase the number of organelles moving along purified microtubules. Kinesin also induces latex beads to move along microtubules, although at a lower velocity than the organelles. Furthermore, the motor powering organelle transport appears to form a stable attachment between vesicles and native axoplasmic microtubules in the presence of AMP-PNP (Lasek and Brady, 1984), and kinesin also forms a stable attachment to microtubules in the presence of AMP-PNP. Finally, organelles (Vale et al., 1985a) as well as kinesin can form rigor-like attachments with microtubules in the absence of ATP. On the basis of these findings, it seems likely that kinesin binds to organelles and moves them along microtubules, although there is no direct proof of an interaction of kinesin with organelles.

If kinesin serves as a motor for organelle transport, a particularly intriguing problem is how bidirectional movement of organelles on single microtubules is generated (Schnapp et al., 1985). Myosin (Sheetz and Spudich, 1983) and dynein (Gibbons, 1981) generate force in only one direction with respect to their filamentous substrates. Furthermore, movement of organelles, beads, and microtubules along purified microtubules in the presence of axoplasmic S2 supernatant (Vale et al., 1985b) or kinesin is primarily unidirectional, as is the movement of negatively charged beads injected into crab axons (Adams and Bray, 1983). Are there two different translocators that move organelles in opposite directions, or is kinesin modified in the axon to induce movement in both directions? Current data do not allow us to distinguish between these two alternatives.

In addition to organelle transport, there are other microtubule-based forms of intracellular motility for which the molecular motors are unknown. The most widely studied of such phenomena is the sliding of the microtubule in the spindle apparatus during mitosis. The ability of kinesin to move microtubules in solution suggests that it may have two microtubule-binding domains, at least one of which is capable of force generation. In a solution of microtubules, these forces would cause microtubule aggregation and contraction, as observed here and by Weisenberg and Ciani (1984). The same kinesin-induced microtubule–microtubule (or even microtubule–organelle) movements in an organized array of microtubules could generate microtubule sliding similar to that which occurs during mitosis. Similarly, kinesin could generate force between microtubules and the plasma or organelle membranes, which could account for some forms of cellular motility.

Experimental Procedures

Materials

Squid, Loligo pealeii, were obtained from the Marine Resources Department at the Marine Biological Laboratory at Woods Hole, Massachusetts. Giant axons were dissected from the squid as previously described (Vale et al., 1985a) and were frozen in liquid nitrogen. Freshly removed squid optic lobes were also frozen in liquid nitrogen. Fresh bovine brain was obtained from a slaughter house, placed on ice, and used within 1 hr of slaughtering. Carboxylated latex beads (0.15 μm diameter; 2.5% solid/solution) were obtained from Polyscience Inc. (Warrington, Pennsylvania) and glass coverslips (type 0) from Clay Adams (Parisippany, New Jersey). Taxol was a gift from Dr. Matthew Suffness at the National Cancer Institute. Bio-Gel A5m and hydroxyapatite were obtained from Bio-Rad Inc. (Richmond, New York). All other reagents were purchased from Sigma Chemical Co. (St. Louis, Missouri).

Preparation of Axoplasmic Supernatant and Organelles

S2 axoplasmic supernatant and organelles were prepared as previously described (Vale et al., 1985b) with the following modification in the sucrose gradient. A low speed (40,000 × g/min) supernatant (about 150 μl) from a homogenate of seven to ten axoplasms was incubated with 20 μM taxol and 1 mM GTP, and applied to a sucrose gradient in a 5 × 41 mm Ultra-Clear centrifuge tube (Beckman Inst., Palo Alto, California) consisting of 7.5% (125 μl), 15% (125 μl), and 35% (200 μl) sucrose layers. Sucrose solutions were made in motility buffer (175 mM potassium asparate, 65 mM taurine, 85 mM betaine, 25 glycine, 20 mM Hepes [pH 7.2], 6.4 mM $MgCl_2$, and 5 mM EGTA) containing 20 μM taxol and 1 mM GTP (this buffer is referred to as MTG buffer) and 2 mM ATP. The sucrose gradient was centrifuged at 135,000 × g for 70 min at 4°C in a SW50.1 rotor. The organelle band at the 15%–35% sucrose interface as well as 75–100 μl of the S2 supernatant above the gradient were collected.

Assays for Microtubule, Bead, and Organelle Movement

These assays (Vale et al., 1985b), as well as a complete account of the procedure for visualizing objects by video-enhanced differential interference contrast microscopy (B. Schnapp, submitted), will be described elsewhere. To measure microtubule movement, a 1 μl suspension of microtubules essentially free of microtubule-associated proteins (0.5–1 mg/ml solution; see below for preparation) was combined with 3–4 μl of a test sample on a glass coverslip. In some instances, a microtubule pellet was directly assayed for movement by placing an aliquot of the microtubules directly on a coverslip in an ATP-containing buffer. Microtubule movement on glass or microtubule movement relative to other microtubules in solution was visualized with the video microscope. A sample was scored positive for microtubule movement if movement was observed on glass or in solution. The presence or absence of movement in different parts of a preparation was so consistent that only a minute or two of observation was required to evaluate a sample. To determine the amount of movement-inducing activity of a sample, the maximum dilution that still supported microtubule movement was determined.

For determining movement of carboxylated latex beads, glass coverslips were treated overnight with a 1 mg/ml solution of poly-D-lysine, which prevented microtubules from moving along the glass; bead movements on microtubules could then be examined without the complication of simultaneous microtubule movement. A test sample (10 μl) was combined with beads (4 μl of a 200-fold dilution of a 2.5% solid stock solution in sample buffer) for 5 min at 4°C. Then 3–4 μl of this mixture was combined with 1 μl of microtubules on a poly-D-lysine-coated coverslip, and movement of beads along the microtubules was evaluated.

Organelle movement was assayed by combining 1 μl of microtubules with 3 μl of organelles (in motility buffer and 2 mM ATP) and 3 μl of test sample. The purified protein sample in motility buffer plus 2 mM ATP or in 100 mM KCl, 50 mM Tris (pH 7.6), 5 mM $MgCl_2$, 0.5 mM EDTA, 2 mM ATP yielded equivalent results with regard to organelle movement. High concentrations of phosphate, however, inhibited organelle movement, so fractions from the hydroxyapatite column were equilibrated with motility buffer using a Centricon filter (molecular weight cutoff of 30 kd; Amicon Corp., Massachusetts) precoated with bovine serum albumin. A video recording of a 20 × 20 μm field of microtubules was made over 2–5 min, and the number of different organelles moving per minute was determined. Velocities of microtubule, bead, and organelle movements were determined over distances of 3–10 μm.

ATPase Assay

ATPase activity was measured according to the procedure of Clarke and Spudich (1974). Protein samples (22.5 μl) in 0.1 M KCl, 50 mM Tris (pH 7.6), 5 mM $MgCl_2$, 0.5 mM EDTA were combined with 2.5 μl of 1–10 mM ATP containing ^{32}P-ATP (approximately 150,000 cpm) in 1.5 ml microfuge tubes at 23°C for 5 to 40 min.

Microtubule Affinity Purification of Squid Kinesin

Squid optic lobes (40 gm wet weight; either freshly dissected or previously frozen in liquid N_2) were homogenized in 1½–2 vol of motility buffer containing 1 mM ATP, 1 mM phenylmethylsulfonyl fluoride (PMSF), 10 μg/ml leupeptin, 10 μg/ml N-a-p-tosyl-L-arginine methyl ketone (TAME), 100 μg/ml soybean trypsin inhibitor, and 0.5 mM DTT with 40 strokes in a Dounce homogenizer. Homogenization and all subsequent steps were performed at 4°C unless specified. The homogenate was centrifuged for 30 min at 40,000 × g, and the supernatant was collected and recentrifuged at 150,000 × g for 60 min. The supernatant from this centrifugation (S2) was incubated with 20 μM taxol and 1 mM GTP for 25 min at 23°C to polymerize tubulin into microtubules (see Preparation of Microtubules).

The mixture was then layered over a 15%/50% (2–4 ml of each) sucrose gradient made in MTG buffer, and the gradient was centrifuged at 100,000 × g for 60 min in a SW27 rotor. The supernatant (S3, approximately 40 ml) was collected, taking care to avoid the first interface. S3 supernatant then was incubated with microtubules (100 μg/ml) in the presence of 5 mM AMP-PNP for 15 min at 23°C, and microtubules were sedimented at 40,000 × g for 30 min. The following conditions proved best for releasing the translocator in a relatively pure form, although different conditions were used in some experiments, as described in the figure legends. The microtubule pellet was resuspended in 1–2 ml of MTG containing 10 mM AMP-PNP for 15 min at 4°C to release proteins that dissociate from microtubules by dilution. Microtubules were pelleted at 37,000 × g for 30 min, and the pellet was washed once with MTG to remove AMP-PNP. The pellet was then resuspended in 1 ml of MTG containing either 10 mM ATP or 5 mM ATP and 0.1 M KCl for 30–40 min at 23°C. Microtubules were centrifuged as before, and the supernatant containing proteins that induced microtubule movement was removed (typically containing 1–4 mg/ml total protein) and, if not used that day, was stored in liquid nitrogen without significant loss of activity. The amount of the 110 kd polypeptide in the microtubule pellet and supernatant was assayed by polyacrylamide gel electrophoresis, and when necessary, the pellet was extracted again with 1 ml of MTG containing 10 mM ATP to retrieve additional translocator protein.

Microtubule Affinity Purification of Bovine Translocator

White matter (70 gm) was obtained from brains of freshly killed cows and homogenized at 4°C in a weight:volume ratio of 1:1 of 50 mM Pipes, 50 mM Hepes, 2 mM $MgCl_2$, 1 mM EDTA, 0.5 mM DTT, 1 mM PMSF, 10 μg/ml leupeptin, 10 μg/ml TAME, and 0.5 mM ATP (pH 7.0) with five bursts of 5 sec in a Polytron homogenizer. The homogenate was centrifuged at 25,000 × g for 30 min at 4°C, and the supernatant was collected and recentrifuged at 150,000 × g for 60 min at 4°C. GTP (1 mM) and taxol (10 μM) were added to the supernatant for 30 min at 25°C, and the polymerized microtubules were pelleted at 37,000 × g for 30 min at 20°C. The supernatant was removed and the microtubule pellet was further processed as described below. The supernatant (comparable to squid S3) was then incubated with squid or bovine microtubules (100 μg/ml) and AMP-PNP (5 mM) for 15 min at 23°C, and the microtubules were pelleted at 37,000 × g for 20 min at 20°C. The supernatant was discarded, and the pellet was resuspended in 5 ml of homogenization buffer containing 5 mM AMP-PNP without ATP for 15 min at 23°C. The microtubules were pelleted as described above, and the pellet was washed with 2 ml of homogenization buffer without AMP-PNP or ATP. The bovine translocator was released by incubation in 1–3 ml of homogenization buffer containing 10 mM ATP for 30–40 min at 23°C. Microtubules were pelleted as before, and the supernatant containing released bovine translocator was collected (1–2 mg/ml). Additional translocator was obtained in some instances by further extraction with 0.1 M KCl/10 mM ATP.

Preparation of Microtubules

Taxol-polymerized microtubules essentially free of microtubule-associated proteins were prepared from squid optic lobes either as previously described (Vale et al., 1985b) or in conjunction with the preparation of squid translocator. In the latter instance, S2 supernatant from squid optic lobes (see Microtubule Affinity Purification of Squid Kinesin) was incubated with 20 μM taxol and 1 mM GTP for 25 min at 23°C to polymerize soluble tubulin into microtubules. The mixture was layered over a 15%/50% (2–4 ml of each) sucrose gradient made in MTG buffer and centrifuged for 60 min at 100,000 × g in a SW27 rotor. The microtubule pellet was then washed once with 2 ml of 100 mM Pipes (pH 6.6), 5 mM EGTA, 1 mM $MgSO_4$ (PEM) containing 20 μM taxol and 1 mM GTP and resuspended in 4 ml of the same buffer. To extract microtubule-associated proteins, 2 ml of PEM-taxol-GTP containing 3 M NaCl was added to the microtubules, and the sample incubated at 23°C for 30 min. Microtubules were centrifuged at 37,000 × g for 30 min at 10°C, and the pellet was washed once with PEM-taxol-GTP. Microtubules were then resuspended in 2 ml of PEM-taxol-GTP at a concentration of 2–8 mg/ml and stored in liquid N_2. Microtubules obtained during the purification of translocator from bovine brain (see above) were further processed by the protocol of Regula et al. (1981). Bovine microtubules were also prepared according to Vallee (1982).

Polyacrylamide Gel Electrophoresis

Electrophoresis was performed in polyacrylamide gels (7.5%) under denaturing and reducing conditions according to the method of Laemmli (1970) and stained with Coomassie blue. Densitometry was performed using a Zenith Soft Laser Scanning densitometer equipped with a helium-neon laser.

Acknowledgments

We are grateful to Eric Steuer for his outstanding technical assistance. We thank Dr. Bruce Schnapp for his assistance and the use of the video microscope facility, and we also appreciate the efforts of Mary Wisgirda and Amy Sewell in collection of squid optic lobes and axons. This work was supported in part by grants from Hoechst-Roussell Pharmaceutical and National Institutes of Health GM33351 (to M. P. S.). M. P. S. is an Established Investigator of the American Heart Association.

The costs of publication of this article were defrayed in part by the payment of page charges. This article must therefore be hereby marked *"advertisement"* in accordance with 18 U.S.C. Section 1734 solely to indicate this fact.

Received April 9, 1985; revised May 14, 1985

References

Adams, R. J., and Bray, D. (1983). Rapid transport of foreign particles microinjected into crab axons. Nature *303*, 718–720.

Allen, R. D., Metuzals, J., Tasaki, I., Brady, S. T., and Gilbert, S. P. (1982). Fast axonal transport in squid giant axon. Science *218*, 1127–1128.

Allen, R. D., Weiss, D. G., Hayden, J. H., Brown, D. T., Fujiwake, H., and Simpson, M. (1985). Gliding movement of and bidirectional organelle transport along single native microtubules from squid axoplasm: evidence for an active role of microtubules in cytoplasmic transport. J. Cell Biol. *100*, 1736–1752.

Asai, D. J., and Wilson, L. (1985). A latent activity dynein-like cytoplasmic magnesium adenosine triphosphatase. J. Biol. Chem. *260*, 699–702.

Beckerle, M. C. (1984). Microinjected fluorescent polystyrene beads exhibit saltatory motion in tissue culture cells. J. Cell Biol. *98*, 2126–2132.

Bell, C. W., Fronk, E., and Gibbons, I. R. (1979). Polypeptide subunits of dynein 1 from sea urchin sperm flagella. J. Supramol. Struct. *11*, 311–317.

Bell, C. W., Fraser, C. L., Sale, W. S., Tang, W. Y., and Gibbons, I. R. (1982). Preparation and purification of dynein. Meth. Enzymol. *85*, 450–474.

Brady, S. T., Lasek, R. J., and Allen, R. D. (1982). Fast axonal transport in extruded axoplasm from squid giant axon. Science *218*, 1129–1131.

Clarke, M., and Spudich, J. A. (1974). Biochemical and structural studies of actomyosin-like proteins from non-muscle cells. J. Mol. Biol. *86*, 209–222.

Cleveland, D. W., Hwo, S.-Y., and Kirschner, M. W. (1977). Purification of tau, a microtubule-associated protein that induces assembly of

Cell
50

microtubules from purified tubulin. J. Mol. Biol. *116*, 207–225.

Cosson, M. P., and Gibbons, I. R. (1978). Properties of sea urchin sperm flagella in which the bending waves have been preserved by pretreatment with mono- and bi-functional maleimide derivatives. J. Cell Biol. *79*, 286a.

Forman, D. S., Brown, K. J., Promersberg, M. W., and Adelman, M. R. (1984). Nucleotide specificity for reactivation of organelle movement in permeabilized axons. Cell Motil. *4*, 121–128.

Gibbons, I. R. (1981). Cilia and flagella of eukaryotes. J. Cell Biol. *91*, 107–124.

Gibbons, I. R., and Fronk, E. (1979). A latent adenosine triphosphatase form of dynein 1 from sea urchin sperm flagella. J. Biol. Chem. *254*, 187–196.

Greene, L. E., and Eisenberg, E. (1980). Dissociation of the actin subfragment 1 complex by adenyl-5'imidodiphosphate, ADP and PPi. J. Biol. Chem. *225*, 543–548.

Hayden, J. H., Allen, R. D., and Goldman, R. D. (1983). Cytoplasmic transport in keratocytes: direct visualization of particle translocation along microtubules. Cell Motil. *3*, 1–19.

Hisanga, S., and Sakai, H. (1983). Cytoplasmic dynein of the sea urchin egg. II. Purification, characterization and interactions with microtubules and Ca-calmodulin. J. Biochem. *93*, 87–98.

Hollenbeck, P. J., Suprynowicz, F., and Cande, W. Z. (1984). Cytoplasmic dynein-like ATPase cross-links microtubules in an ATP-sensitive manner. J. Cell Biol. *99*, 1251–1258.

Johnson, K. A., and Wall, J. S. (1983). Structure and molecular weight of dynein ATPase. J. Cell Biol. *96*, 669–678.

Laemmli, U. (1970). Cleavage of structural proteins during the assembly of the bacteriophage T4. Nature *227*, 680–685.

Lasek, R. J., and Brady, S. T. (1984). Adenylyl imidodiphosphate (AMP-PNP), a non-hydrolyzable analogue of ATP produces a stable intermediate in the motility cycle of fast axonal transport. Biol. Bull. *167*, 503.

Mitchell, D. R., and Warner, F. D. (1981). BInding of dynein 21 S ATPase to microtubules. Effects of ionic conditions and substrate analogs. J. Biol. Chem. *256*, 12535–12544.

Murphy, D. B., Hiebsch, R. R., and Wallis, K. T. (1983). Identity and origin of the ATPase activity associated with neuronal microtubules. I. The ATPase activity is associated with membrane vesicles. J. Cell Biol. *96*, 1298–1305.

Nasr, T., and Satir, P. (1985). Alloaffinity chromatography: a general approach to affinity purification of dynein and dynein-like molecules. Anal. Biochem., in press.

Pfister, K. K., and Witman, G. B. (1984). Subfractionation of *Chlamydomonas* 18 S dynein into two unique subunits containing ATPase activity. J. Biol. Chem. *259*, 12072–12080.

Piperno, G., and Luck, D. J. L. (1979). Axonemal adenosine triphosphatases from flagella of *Chlamydomonas reinhardtii*. J. Biol. Chem. *254*, 3084–3090.

Porter, M. E., and Johnson, K. A. (1983). Characterization of the ATP-sensitive binding of *Tetrahymena* 30 S dynein to bovine brain microtubules. J. Biol. Chem. *258*, 6575–6581.

Pratt, M. M. (1980). The identification of a dynein ATPase in unfertilized sea urchin eggs. Dev. Biol. *74*, 364–378.

Regula, C. S., Pfeiffer, J. R., and Berlin, R. D. (1981). Microtubule assembly and disassembly at alkaline pH. J. Cell Biol. *89*, 45–53.

Schliwa, M. (1984). Mechanisms of intracellular organelle transport. In Cell Muscle Motility, Vol. 5, J. W. Shaw, editor. (New York: Plenum Publishing Co.), pp. 1–81.

Schnapp, B. J., Sheetz, M. P., Vale, R. D., and Reese, T. S. (1984). Filamentous actin is not a component of transport filaments isolated from squid axoplasm. J. Cell Biol. *99*, 351a.

Schnapp, B. J., Vale, R. D., Sheetz, M. P., and Reese, T. S. (1985). The structure of cytoplasmic filaments involved in organelle transport in the squid giant axon. Cell *40*, 455–462.

Scholey, J. M., Neighbors, B., McIntosh, J. R., and Salmon, E. D. (1984). Isolation of microtubules and a dynein-like MgATPase from unfertilized sea urchin eggs. J. Biol. Chem. *259*, 6516–6525.

Sheetz, M. P., and Spudich, J. A. (1983). Movement of myosin-coated structures on actin cables. Cell Motil. *3*, 484–485.

Shimizu, T., and Kimura, I. (1974). Effects of N-ethylmaleimide on dynein adenosine-triphosphatase activity and its recombining ability with outer fibers. J. Biochem. *76*, 1001–1008.

Smith, R. S. (1980). The short term accumulation of axonally transported organelles in the region of localized lesions of single myelinated axons. J. Neurocytol. *9*, 39–65.

Tsukita, S., and Ishikawa, H. (1980). The movement of membranous organelles in axons. Electron microscopic identification of anterogradely and retrogradely transported organelles. J. Cell Biol. *84*, 513–530.

Vale, R. D., Schnapp, B. J., Reese, T. S., and Sheetz, M. P. (1985a). Movement of organelles along filaments dissociated from axoplasm of the squid giant axon. Cell *40*, 449–454.

Vale, R. D., Schnapp, B. J., Reese, T. S., and Sheetz, M. P. (1985b). Organelle, bead and microtubule translocations promoted by soluble factors from the squid giant axon. Cell *40*, 559–569.

Vallee, R. B. (1982). A taxol-dependent procedure for the isolation of microtubules and microtubule-associated proteins (MAPs). J. Cell Biol. *92*, 435–442.

Vallee, R. B., and Bloom, G. S. (1984). High molecular weight microtubule-associated proteins. In Modern Cell Biology. P. H. Satir, series editor. (New York: Alan R. Liss), pp. 21–75.

Weisenberg, R. C., and Ciani, C. (1984). ATP-induced gelation-contraction of microtubules assembled *in vitro*. J. Cell Biol. *99*, 1527–1533.

Witman, G. B., Johnson, K. A., Pfister, K. K., and Wall, J. S. (1983). Fine structure and molecular weight of the outer arm dyneins of *Chlamydomonas*. J. Submicrosc. Cytol. *15*, 193–197.

Oakley B.R., Oakley C.E., Yoon Y., and Jung M.K. 1990. γ-Tubulin is a component of the spindle pole body that is essential for microtubule function in *Aspergillus nidulans*. *Cell* 61: 1289–1301.

SHORTLY AFTER THE CENTROSOME* WAS discovered in the late 19th century, its crucial roles in mitosis and cell motility were quickly recognized, primarily through the brilliant observations and speculations of Theodor Boveri. Later the electron microscope showed a close similarity in structure among microtubules, cilia, centrioles, and basal bodies: cilia consisted of nine microtubule doublets surrounding a central pair of tubules, whereas centrioles and basal bodies consisted of nine short microtubule triplets with no central tubules. Rabbit sera of unknown specificity were found that permitted immunostaining of centrioles and basal bodies (1), but without specific reagents little progress was made in identifying the functionally significant components of the centrosome. An important advance came with the discovery of γ-tubulin, a protein related to α- and β-tubulin of microtubules that was encoded by the *mip*A gene of the fungus *Aspergillus nidulans*. In the paper reproduced here Oakley et al. showed that γ-tubulin is required for proper microtubule function, particularly nuclear division. They also demonstrated by immunofluorescence that γ-tubulin is a component of the spindle pole bodies but not of the spindle itself, which contains α- and β-tubulin. Subsequently it was found that γ-tubulin occurs in centrosomes of *Xenopus, Drosophila*, and other organisms. Antibodies against γ-tubulin immunoprecipitate a complex, the γ-tubulin ring complex or γTuRC, which contains a well-defined set of centrosomal proteins (2). Purified γTuRC nucleates microtubule assembly and caps the minus ends of microtubules *in vitro*. Another component of the centrosome is a protein called pericentrin, which, as its name implies, is localized in the cytoplasmic region surrounding centrioles (3). Pericentrin interacts with both γ-tubulin and cytoplasmic dynein; how these components organize the mitotic spindle is currently a topic of active investigation (4, 5). Remarkable progress has also been made in elucidating the composition and structure of the spindle pole body in budding yeast (6, 7).

---

1. Connolly J. A. and Kalnins V. I. 1978. Visualization of centrioles and basal bodies by fluorescent staining with nonimmune rabbit sera. *J. Cell Biol.* **79:** 526–532.
2. Zheng Y., Wong M. L., Alberts B., and Mitchison T. 1995. Nucleation of microtubule assembly by a γ-tubulin-containing ring complex. *Nature* **378:** 578–583.
3. Doxsey S. J., Stein P., Evans L., Calarco P. D., and Kirschner M. 1994. Pericentrin, a highly conserved centrosome protein involved in microtubule organization. *Cell* **76:** 639–650.
4. Dictenberg J. B., Zimmerman W., Sparks C. A., Young A., Vidair C., Zheng Y., Carrington W., Fay F. S., and Doxsey S. J. 1998. Pericentrin and γ-tubulin form a protein complex and are organized into a novel lattice at the centrosome. *J. Cell Biol.* **141:** 163–174.
5. Purohit A., Tynan S. H., Vallee R., and Doxsey S. J. 1999. Direct interaction of pericentrin with cytoplasmic dynein light intermediate chain contributes to mitotic spindle organization. *J. Cell Biol.* **147:** 481–492.
6. Wigge P. A., Jensen O. N., Holmes S., Souès S., Mann M., and Kilmartin J. V. 1998. Analysis of the *Saccharomyces* spindle pole by matrix-assisted laser desorption/ionization (MALDI) mass spectrometry. *J. Cell Biol.* **141:** 967–977.
7. Adams I. R. and Kilmartin J. V. 1999. Localization of core spindle pole body (SPB) components during SPB duplication in *Saccharomyces cerevisiae*. *J. Cell Biol.* **145:** 809–823.

*In the classical cytological literature *centrosome* referred to the more or less spherical region of cytoplasm immediately surrounding the paired *centrioles*. In current usage the term centrosome includes both the centrioles and the surrounding cytoplasm. *Spindle pole body* refers to a smaller structure in fungal cells that plays the same role as the centrosome in higher eukaryotes.

Cell, Vol. 61, 1289–1301, June 29, 1990, Copyright © 1990 by Cell Press

γ-Tubulin Is a Component of the Spindle Pole Body That Is Essential for Microtubule Function in Aspergillus nidulans

Berl R. Oakley, C. Elizabeth Oakley, Yisang Yoon, and M. Katherine Jung
Department of Molecular Genetics
Ohio State University
Columbus, Ohio 43210

Summary

We have recently discovered that the *mipA* gene of A. nidulans encodes γ-tubulin, a new member of the tubulin superfamily. To determine the function of γ-tubulin in vivo, we have created a mutation in the *mipA* gene by integrative transformation, maintained the mutation in a heterokaryon, and determined the phenotype of the mutation in spores produced by the heterokaryon. The mutation is lethal and recessive. It strongly inhibits nuclear division, less strongly inhibits nuclear migration, and, as judged by immunofluorescence microscopy, causes a reduction in the number and length of cytoplasmic microtubules and virtually a complete absence of mitotic apparatus. We conclude that γ-tubulin is essential for microtubule function in general and nuclear division in particular. Immunofluorescence microscopy of wild-type hyphae with affinity-purified, γ-tubulin–specific antibodies reveals that γ-tubulin is a component of interphase and mitotic spindle pole bodies. We propose that γ-tubulin attaches microtubules to the spindle pole body, nucleates microtubule assembly, and establishes microtubule polarity in vivo.

Introduction

Microtubules are found in all eukaryotic cells and are essential for mitosis, meiosis, some forms of organellar movement, and other cytoskeletal functions. They are composed primarily of two similar proteins, α- and β-tubulin, which form a heterodimer that assembles into microtubules. The properties of microtubules are due in part, however, to proteins other than α- and β-tubulin. The best studied such proteins are microtubule-associated proteins. Microtubule-associated proteins coassemble with α- and β-tubulin, and some of them alter the assembly characteristics of microtubules in vitro and, presumably, in vivo (for review see Vallee et al., 1984). Recent work has revealed that a special class of microtubule-associated proteins (dynein, kinesin, and related proteins) is involved in microtubule-based motility (for review see Stebbings, 1988). In addition, there must be other proteins that are involved in the attachment of microtubules to kinetochores and that promote the assembly of microtubules at microtubule organizing centers such as the centrosome. The identification and characterization of proteins involved in microtubule function is important to understanding how microtubule assembly is regulated and how microtubules function.

Genetic and phenotypic analyses of three mutant alleles of the *mipA* (microtubule-interacting protein) locus of Aspergillus nidulans led Weil et al. (1986) to conclude that *mipA* is a nontubulin gene involved in microtubule function in vivo. The mutant *mipA* alleles were isolated as extragenic suppressors of *benA33*, a heat-sensitive (hs^-) β-tubulin mutation (Oakley and Morris, 1981; Oakley et al., 1985; Weil et al., 1986; Oakley et al., 1987a). The *benA* gene encodes the major β-tubulin expressed in hyphae (Sheir-Neiss et al., 1978), and this β-tubulin is essential for mitosis and nuclear migration (Oakley and Morris, 1980, 1981). The specificity of gene interactions between each of the mutant *mipA* alleles and several hs^- *benA* alleles demonstrated that the product of *mipA* interacts specifically, probably physically, with the β-tubulin encoded by *benA* (Weil et al., 1986). Unfortunately, each of the three mutant *mipA* alleles isolated is silent in combination with the wild-type *benA* allele, and, consequently, one cannot determine the role of the product of *mipA* in microtubule function without the complication of a *benA* mutation in the same strain (Weil et al., 1986).

We have recently cloned and sequenced the *mipA* gene and have found that the predicted product of the gene is a protein that is not α- or β-tubulin but is closely related to both proteins (Oakley and Oakley, 1989). This protein is clearly a member of the tubulin superfamily, and we have proposed that it be called γ-tubulin. Comparisons of the sequence of γ-tubulin with α- and β-tubulins from evolutionarily diverse organisms suggested strongly, moreover, that γ-tubulin did not diverge recently from an α- or β-tubulin but diverged from an ancestral tubulin at about the time of the α/β-tubulin divergence. The ancient origin of γ-tubulin, in turn, suggested that it is probably not restricted to A. nidulans and, in fact, γ-tubulin cDNAs from Drosophila melanogaster and Homo sapiens have recently been cloned and sequenced (Y. Zheng and B. R. O., unpublished data).

The existence of a new tubulin that plays some role in microtubule function was, to us, remarkable and unexpected and raised questions as to the function of γ-tubulin, the ubiquity of γ-tubulin, and the extent of the tubulin superfamily. Here we have begun to address the first question: the function of γ-tubulin in vivo.

Results

Creation of a Mutant *mipA* Allele by Gene Disruption

We have used the fact that transformation often occurs by homologous recombination in A. nidulans to create a mutation in the *mipA* gene. We first created plasmid pLO12 (Figure 1) by cloning an internal HaeIII fragment of the *mipA* gene into the unique SmaI site of plasmid pRG3 (Osmani et al., 1988). pRG3 contains the *pyr4* gene of Neurospora crassa cloned into the NdeI site of pUC19 (Yanisch-Perron et al., 1985). The *pyr4* gene complements *pyrG89* (Ballance et al., 1983), a mutation in the A. nidu-

Cell
1290

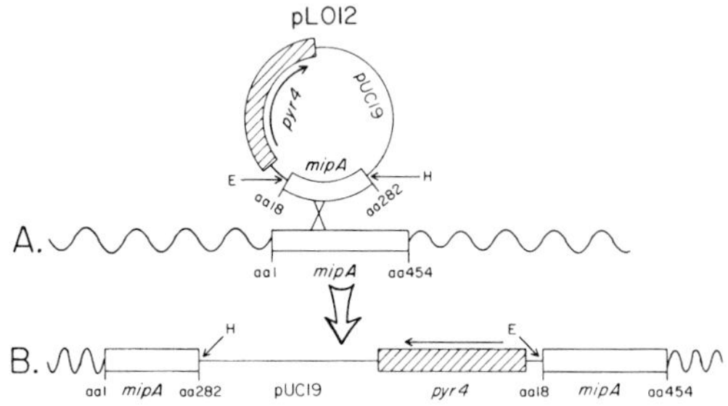

Figure 1. Creation of a Mutant *mipA* Allele by Gene Disruption

We constructed plasmid pLO12 by cloning a 1132 bp HaeIII fragment of *mipA* that encodes amino acids 18–282 into the unique SmaI site of plasmid pRG3. The EcoRI (E) and HindIII (H) sites of the pUC19 polylinker are designated to show the orientation of the fragment in the polylinker. Transformation of a strain that is wild type at the *mipA* locus should result in integration of pLO12 at *mipA* by homologous recombination as shown in (A). This integration results in the configuration shown in (B), in which there is a fragment containing amino acids 1–282 of the *mipA* gene controlled by the *mipA* promoter and a fragment containing the coding sequence for amino acids 18–454 of *mipA* without a promoter.

lans orotidine-5′-phosphate decarboxylase gene conferring pyrimidine auxotrophy (Palmer and Cove, 1975), but has sufficiently little homology to the *pyrG* gene that it does not direct integration at *pyrG*. The *mipA* HaeIII fragment is 1132 bp long and extends from a site in intron 2 to a site in exon 8 (see Oakley and Oakley, 1989). Amino acids 18–282 are encoded by the fragment. Integration of pLO12 by homologous recombination at the *mipA* locus (Figure 1A) would cause disruption of the *mipA* gene (Figure 1B), resulting in a fragment containing *mipA* amino acids 1–282 controlled by the *mipA* promoter and a frag-

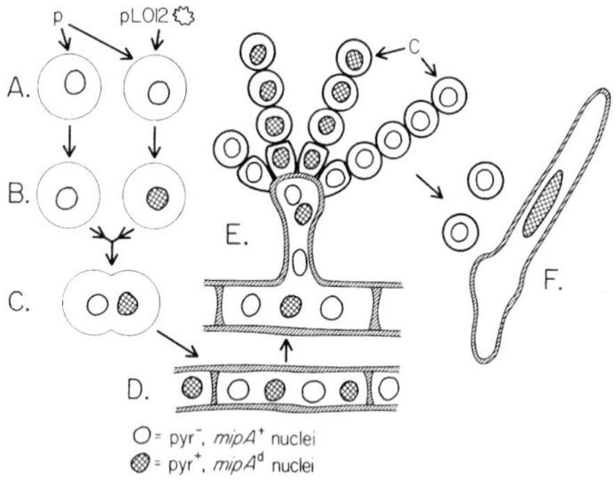

Figure 2. Maintenance of a Lethal *mipA* Gene Disruption in a Heterokaryon

In (A) protoplasts of A. nidulans strain G191 are transformed with pLO12. G191 carries *pyrG89* and is thus a pyrimidine auxotroph (pyr^-). Strains carrying *pyrG89* can not grow on media lacking uridine and uracil. In (B) one protoplast has been transformed with pLO12. This protoplast will be a pyrimidine prototroph (pyr^+) because of the complementation of *pyrG89* by *pyr4*, but integration of pLO12 at *mipA* results in the disruption of *mipA* ($mipA^d$). If the *mipA* gene is essential, this protoplast will not be viable. If transformed and untransformed protoplasts fuse (C), the resulting heterokaryon (D) will have functional *mipA* and *pyr4* genes and will be a viable pyrimidine prototroph. When the heterokaryon sporulates (E), the conidia (c), which are uninucleate, will have either $mipA^+$, pyr^- nuclei or $mipA^d$, pyr^+ nuclei. If the spores are germinated in selective medium (lacking uridine and uracil) (F), the spores with $mipA^+$, pyr^- nuclei will not germinate and the spores with $mipA^d$, pyr^+ nuclei will exhibit the phenotype caused by the disruption of *mipA*.

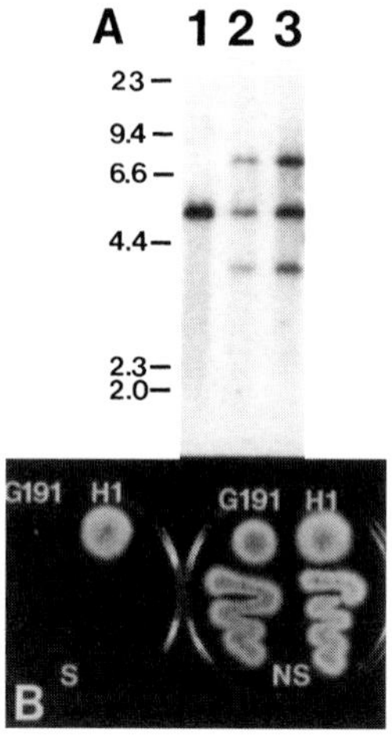

Figure 3. Identification of Heterokaryons Carrying *mipAd1*

(A) A Southern hybridization of BamHI digests of DNA of A. nidulans strain BRO2 (lane 1), which is an untransformed strain that is wild type at the *mipA* locus, and heterokaryons H1 (lane 2) and H2 (lane 3) created by the transformation of A. nidulans strain G191 with pLO12. Plasmid pLO6 (*mipA* cloned into pUC19) was used as the probe, and the sizes (in kilobases) and positions of fragments of a HindIII digest of bacteriophage lambda are shown at the left. As lane 1 shows, the *mipA* gene is on a 5.1 kb BamHI fragment. Plasmid pLO12 is 5.8 kb in length and carries a single BamHI site. Integration of pLO12 by homologous recombination at *mipA* creates fragments of 7.3 and 3.6 kb that hybridize to pLO6. In lanes 2 and 3 these fragments are present, as is the 5.1 kb fragment characteristic of the uninterrupted *mipA* locus. Thus, heterokaryons H1 and H2 carry both wild-type and disrupted *mipA* genes. Additional Southern hybridizations of pLO6 to EcoRI and PstI digests of BRO2, H1, and H2 DNA and of pLO12 to BamHI digests of G191, H1, and H2 DNA (data not shown) all indicate that H1 and H2 are heterokaryons carrying both disrupted and undisrupted *mipA* genes. Note: the DNA in lanes 2 and 3 were prepared by a miniprep procedure that leaves impurities that slightly retard the initial migration of DNA into agarose gels. The bands in lanes 2 and 3 are, thus, slightly retarded relative to lane 1.

(B) Incubation of hyphae and spores of G191 and H1 on selective medium (S) (lacking uridine and uracil) and nonselective medium (NS) (supplemented with uridine and uracil). Small squares of hyphal mats were cut out of colonies of G191 and H1 and placed on selective and nonselective media, and spores from G191 and H1 were streaked on to the same plates immediately below the hyphal squares. H1 was able to grow from hyphae on the selective medium because it carries both *mipAd1* and wild-type *mipA* nuclei. H1 conidia are uninucleate and thus carry either *mipAd1* or wild-type *mipA* nuclei but not both. Spores carrying the wild-type *mipA* gene lack the *pyr4* gene carried on pLO12 and cannot grow on the selective medium. Spores carrying *mipAd1* carry *pyr4* but still do not grow, indicating that *mipAd1* is lethal. The fact that H1 hyphae can grow on the selective medium shows that the lethality of *mipAd1* is recessive. As expected, on nonselective medium spores and hyphae of G191 and H1 grow. Note: in the center of each colony grown from hyphae is a region that might appear to be the hyphal inoculum. In fact, this raised region is typical of hyphal growth in circular colonies regardless of the form of inoculation. The actual hyphal inocula were all the size of the G191 inoculum on the selective medium.

ment containing *mipA* amino acids 18–454 without a promoter. If *mipA* is essential for growth or viability, one would expect such a mutation to be lethal. We have used the scheme developed by Osmani et al. (1988) to maintain and determine the phenotype of a *mipA* disruption as shown in Figure 2.

We transformed A. nidulans strain G191 (which is wild type at the *mipA* locus [$mipA^+$] but carries *pyrG89*) with pLO12 (Figure 2A). When protoplasts of A. nidulans are transformed, heterokaryons carrying transformed nuclei and untransformed nuclei often form (Figures 2B–2D). These heterokaryons can, in principle, form in three ways. First, as shown in Figure 2, two protoplasts, one with a transformed nucleus and one with an untransformed nucleus, can fuse (Figures 2B and 2C). Second, a plasmid can transform a multinucleate protoplast such that only one or some of the nuclei are transformed. Third, a nucleus in the G_2 phase of the cell cycle can be transformed with a plasmid such that integration occurs at one copy of *mipA* but not at both.

During transformation with pLO12, heterokaryons carrying both untransformed nuclei and nuclei carrying the *mipA* disruption should form. If the *mipA* disruption is a dominant lethal mutation, these heterokaryons will not be viable. If, however, the *mipA* disruption is not lethal, or is a recessive lethal mutation, the heterokaryon will be viable. In the heterokaryon, the transformed nuclei will provide the orotidine-5′-phosphate decarboxylase encoded by the *pyr4* gene (allowing the heterokaryon to grow on medium lacking uridine or uracil), and the untransformed nuclei will provide the product of the undisrupted *mipA* gene (γ-tubulin). When such heterokaryons sporulate (Figure 2E), the conidia (conidiospores) produced are uninucleate, and each conidium will contain either an untransformed nucleus or a transformed nucleus but not both. If the spores are incubated in selective medium (lacking uridine or uracil), the spores with untransformed nuclei will not germinate. The spores with transformed nuclei will be able to produce orotidine-5′-phosphate decarboxylase but will carry a disruption of the *mipA* gene. If *mipA* is not an essential gene, these spores should grow into colonies. If *mipA* is essential, however, the germlings from these spores should stop growing when the γ-tubulin carried over from the parental heterokaryon (if any) is exhausted. The phenotype exhibited by the growth-arrested germlings should reflect the function of γ-tubulin.

We tested spores from 18 colonies of G191 transformed with pLO12 for their ability to form colonies on selective medium. Spores from four transformants were unable to form colonies, suggesting that the transformants were balanced heterokaryons carrying untransformed nuclei and nuclei with lethal mutations. Three of these transformants were analyzed by Southern hybridizations using plasmid pLO6 as a probe. Plasmid pLO6 carries the *mipA* gene on a 2.8 kb HindIII–SacI fragment (Oakley and Oakley, 1989) cloned into the polylinker of pUC19. Two of the transformants had Southern hybridization patterns as predicted for a heterokaryon carrying both untransformed nuclei and nuclei with a single integration of pLO12 at the *mipA* locus (Figure 3A). We have designated these two

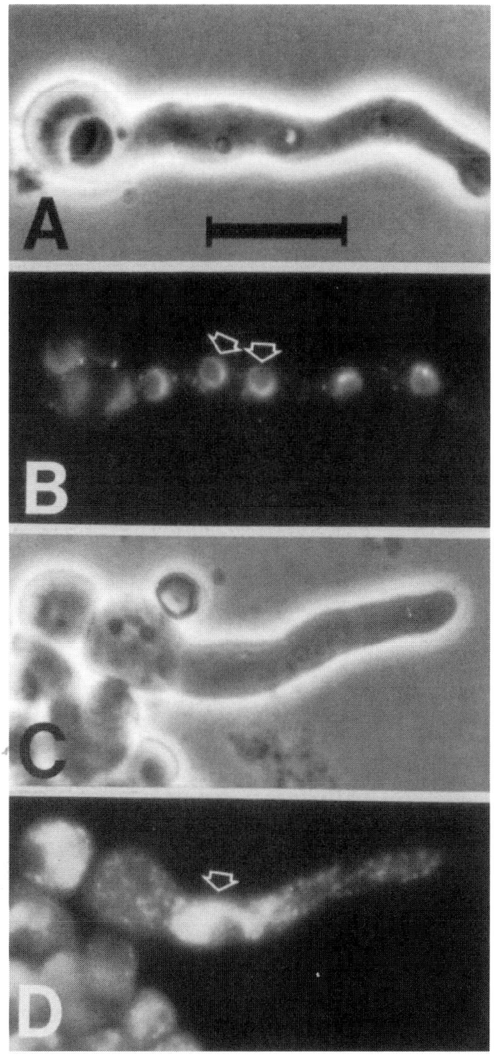

Figure 4. Inhibition of Nuclear Division by *mipAd1*

(A) A phase contrast micrograph of a germling of G191 incubated in nonselective medium.

(B) A fluorescence micrograph of the same field. The germling, which has been stained with the fluorescent DNA binding dye DAPI, contains eight nuclei, two of which are designated with arrows. The small points of DAPI fluorescence are mitochondrial genomes.

(C) A phase contrast micrograph of several spores from heterokaryon H1 incubated in selective medium. As expected, spores carrying $mipA^+$ have not germinated, but one, carrying *mipAd1*, has germinated.

(D) A fluorescence micrograph of the same field. DAPI staining reveals that the germling carrying *mipAd1* has a single large nucleus (arrow).

Nuclei and mitochondrial genomes in (D) were stained with a greater concentration of DAPI than those in (B) and are consequently brighter. Magnifications are all the same and the scale in (A) is 10 μm.

heterokaryons H1 and H2. Growth of H1 and spores of H1 on selective and nonselective media are shown in Figure 3B. We will call the mutant allele, caused by the integration of pLO12 at *mipA* as shown in Figure 1, *mipAd1* (for *mipA*

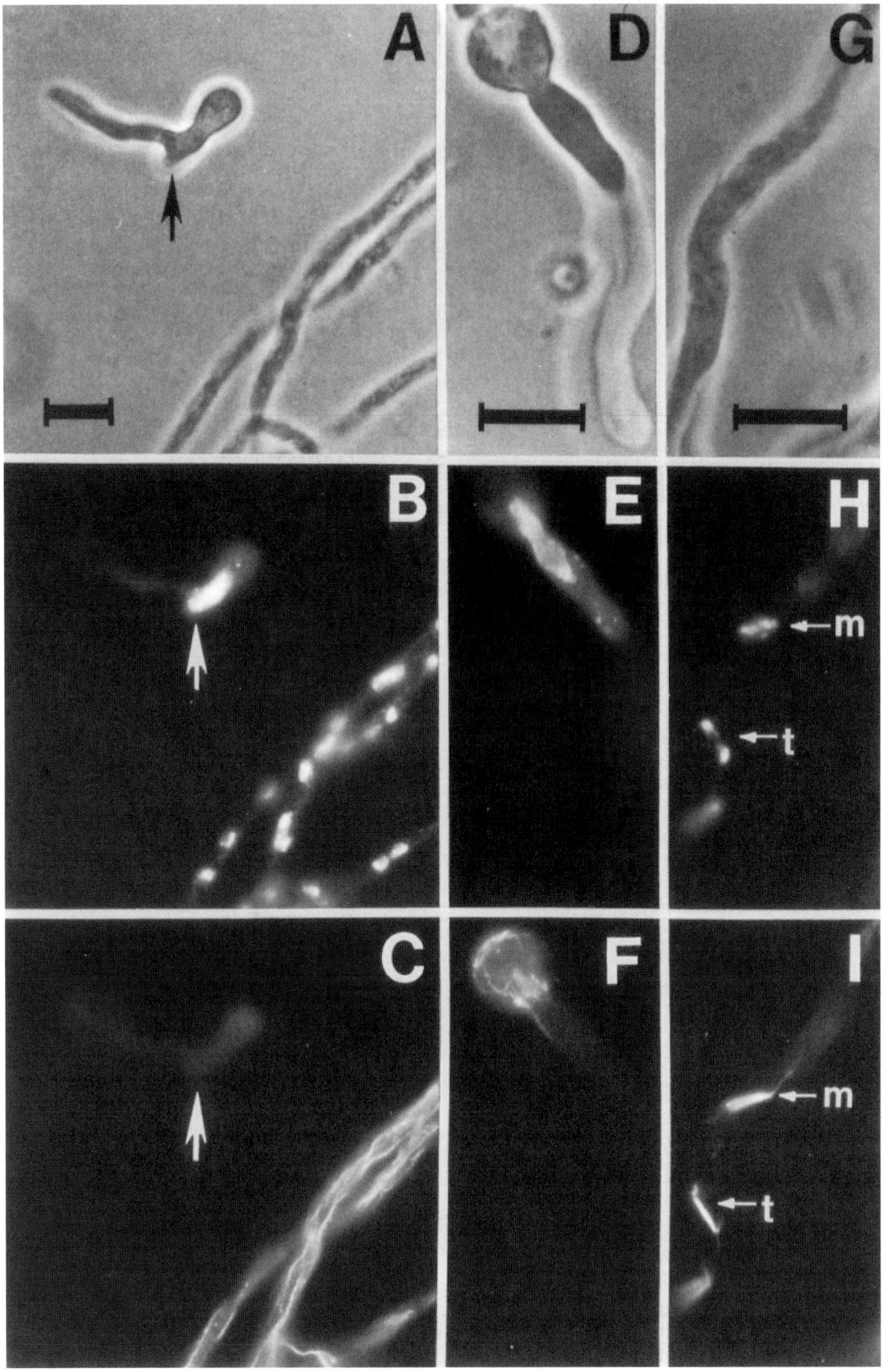
A
D
G
B
E
H
m
t
C
F
I
m
t

γ-Tubulin Function and Location
1293

disruptant 1). The third transformant had a restriction pattern indicating that multiple copies of pLO12 had integrated at a site away from *mipA* (results not shown). We presume that this transformant is a heterokaryon in which integration by nonhomologous recombination has disrupted an essential gene other than *mipA*.

A *mipA* Gene Disruption Blocks Nuclear Division

The fact that the heterokaryons containing untransformed and *mipAd1* nuclei are viable on selective medium, but the spores from these heterokaryons are not (Figure 3B), demonstrates that *mipAd1* is a recessive lethal mutation. To determine the growth-arrest phenotype caused by *mipAd1*, we germinated spores from heterokaryons H1 and H2 in selective medium. We first determined whether *mipAd1* blocked germination. Since *mipA*$^+$ spores carry *pyrG89* and do not carry *pyr4* (see Figures 1 and 2), they will not germinate on the selective medium we used. (This has been known for some time, but we verified that G191 spores do not germinate on our selective medium in the length of time we used for our experiments.) Any germlings seen must carry *pyr4* and *mipAd1*. The absence of any germlings when H1 and H2 spores are incubated in selective medium would indicate that *mipAd1* blocks germination. A substantial fraction of the H1 and H2 spores germinated (Figures 4 and 5), indicating that *mipAd1* does not block germination. However, there appeared to be a delay of germination of *mipAd1* spores relative to G191 spores incubated in nonselective medium. *mipAd1* germlings also grew more slowly than G191 germlings incubated in nonselective medium and never grew to more than a few hundred microns in length, even after long incubation times in which G191 germlings had grown to form a hyphal mat.

To determine the effects of *mipAd1* on nuclear division, we stained the germlings with DAPI (4′,6-diamidino-2-phenylindole), a fluorescent, DNA binding dye. Initial results with H1 and H2 were identical, and we have used H1 for all subsequent experiments. As Figure 4 and Table 1 show, nuclear division was dramatically inhibited in *mipAd1* germlings relative to G191 germlings incubated in nonselective medium. Most *mipAd1* germlings had a single nucleus, and in the few germlings that had more than one nucleus, the spacing between the nuclei was much greater than normal, indicating that nuclear division is inhibited relative to hyphal growth. Although we did not measure the volumes of the nuclei, it was obvious that nuclei in *mipAd1* germlings were much larger than nuclei in germlings of the untransformed control strain and (as

Table 1. Inhibition of Nuclear Division and Migration in *mipAd1* Germlings

	G191 (wild type)	*mipA* Disruptants
Number of nuclear divisions per germling[a]	2.6 ± 0.5	0.2 ± 0.4
Germlings exhibiting nuclear division[b]	100%	13.5%
Germling length (μm)[c]	28.5 ± 5.8	31.8 ± 9.3
Germling length (μm) per nucleus[d]	4.9 ± 1.4	29.1 ± 10.6
Germlings exhibiting nuclear migration[e]	100%	56%
Mitotic index[f]	4.3%	5.8%
Sample size	116	104

Conidia from heterokaryon H1 were incubated in selective medium for 9 hr, fixed, and stained with DAPI. As a control, untransformed G191 spores were incubated in nonselective medium for 6 hr before they were fixed and DAPI stained. The H1 conidia were incubated longer to allow the germlings to reach approximately the same length as the G191 germlings.

[a] Values represent the mean number of nuclear doublings per germling along with standard deviations. Thus, a value of 2 would be two doublings or four nuclei.

[b] Values represent the percentage of germlings with more than one nucleus.

[c] Values represent the mean germling length and standard deviation.

[d] Values represent the mean and standard deviation for the number of nuclei for each germling divided by the length of the germling.

[e] Values represent the percentage of germlings with at least one nucleus that has migrated completely out of the conidial swelling.

[f] Values represent the percentage of germlings with nuclei with condensed chromosomes. A. nidulans is coenocytic, and in germlings of the size examined all nuclei are in the same cytoplasm and enter mitosis together. Thus, in a germling, all nuclei are in mitosis or all are in interphase.

Figure 5. Patterns of Microtubules in Germlings Carrying *mipAd1* or the Wild-Type *mipA* Gene

(A) A phase contrast micrograph of spores of heterokaryon H1 germinated in nonselective medium. The arrow designates a germling carrying *mipAd1*. At the lower right of the panel are portions of four hyphae carrying *mipA*$^+$. The material has been stained with DAPI, and the microtubules have been stained using an antibody against α-tubulin.

(B) A fluorescence micrograph of the field in (A) taken through a DAPI filter set. The germling carrying *mipAd1* has a single large nucleus, while the hyphae carrying *mipA*$^+$ have many smaller, normal-sized nuclei.

(C) A fluorescence micrograph of the field in (A) taken through an FITC filter set to allow visualization of microtubules. No microtubules are visible in the germling carrying *mipAd1*, but numerous microtubules are visible in the hyphae carrying *mipA*$^+$. The shallow depth of focus makes the microtubules appear fragmented, but focusing through the hyphae revealed that many microtubules are more than 100 μm long. (A), (B), and (C) are the same magnification. The scale in (A) is 10 μm.

(D) Phase contrast image of a germling carrying *mipAd1*. (E) DAPI image of the field in (D) reveals that the germling has a single nucleus.

(F) An FITC image of the field in (D). A few microtubules are present in the region of the conidial swelling from which the germ tube extends. Focusing through the hypha revealed no additional microtubules. Magnifications in (D), (E), and (F) are the same; scale in (D) is 10 μm.

(G) Phase contrast image of nuclear division in a hypha of G191 grown in nonselective medium.

(H) DAPI image of the field in (G). Two nuclei are designated by arrows. One (m) is in medial nuclear division (there is no metaphase in A. nidulans), while the other (t) is in early telophase. The small, condensed A. nidulans chromosomes are visible in the medial nuclear division nucleus, and the chromosomes have separated into two masses in the telophase nucleus.

(I) FITC image of the field in (G). The mitotic spindles for the mitotic and telophase nuclei are designated. Cytoplasmic microtubules disassemble as nuclei enter mitosis, and only a few cytoplasmic microtubules, attached to the spindles, are visible. A small dark region corresponding to the SPB is visible at one end of the telophase nucleus. Magnifications in (G), (H), and (I) are the same. The scale in (G) is 10 μm.

Cell
1294

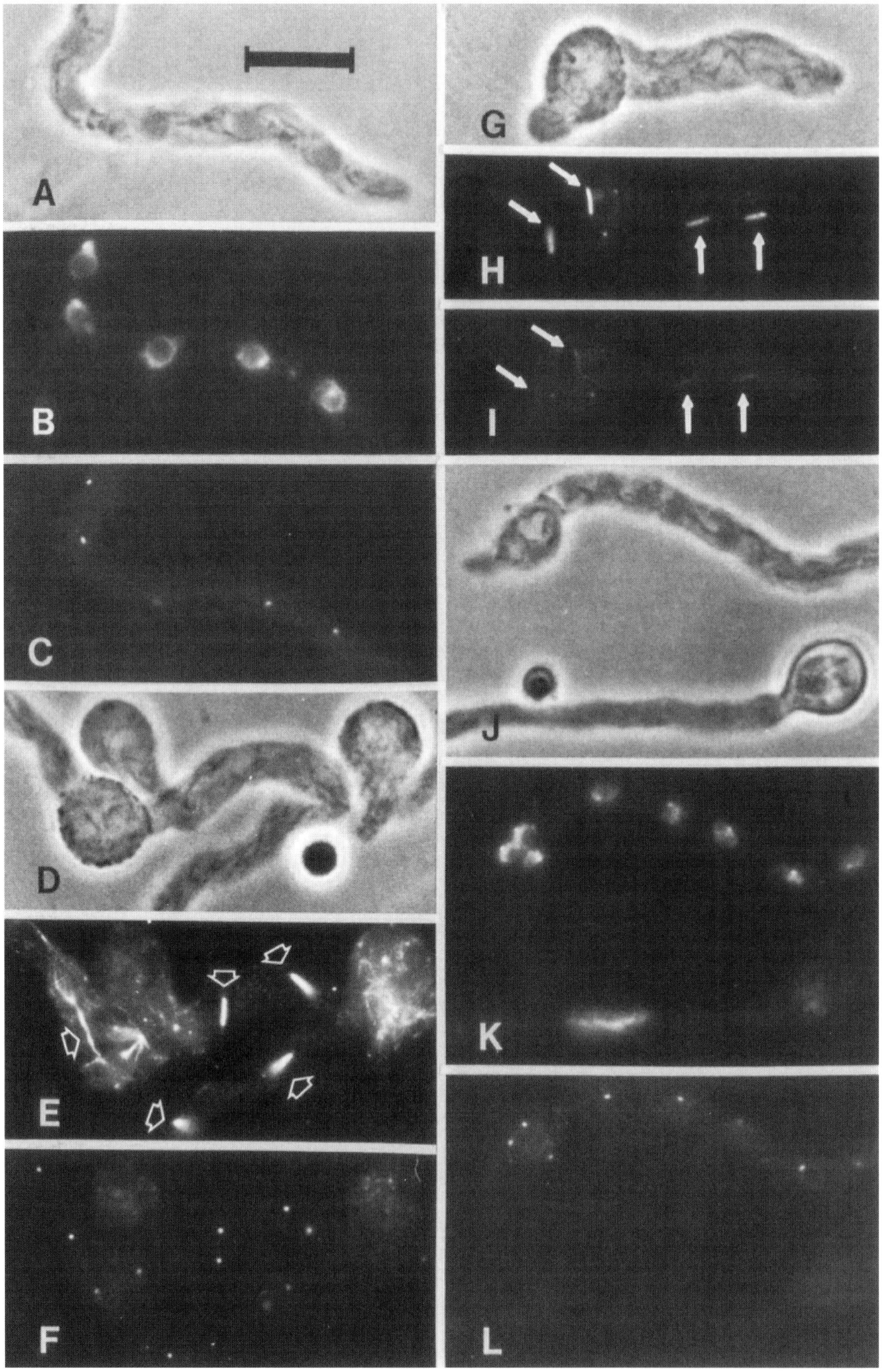

judged by DAPI staining) contained much more DNA (Figures 4 and 5). With prolonged incubation, nuclei in *mipAd1* germlings became even larger (results not shown). This indicates that *mipAd1* does not strongly inhibit DNA replication and this, coupled with the strong inhibition of nuclear division, results in large, presumably polyploid, nuclei.

Since Oakley and Morris (1980) have shown that the β-tubulin encoded by the *benA* gene (and by inference microtubules) is involved in nuclear migration in A. nidulans, we examined the effect of *mipAd1* on nuclear migration. As Table 1 shows, *mipAd1* partially inhibited nuclear migration.

Since nuclear division and nuclear migration involve microtubules and since genetic analyses of *mipA* mutations demonstrate that *mipA* plays some role in microtubule function (Weil et al., 1986), we were interested in determining the effects of *mipAd1* on the microtubule network. Osmani et al. (1988) have obtained good staining of cytoplasmic and mitotic apparatus (MA) microtubules using the monoclonal anti-α-tubulin antibody YOL1/34 (Kilmartin et al., 1982), and by using a slight modification of their procedure, we have obtained excellent staining of both classes of microtubules.

When conidia from heterokaryon H1 were incubated in selective medium so that only the *mipAd1*, *pyr4*$^+$ conidia germinated, microtubules were absent from most germlings, although a few cytoplasmic microtubules (Figure 5F) were present in a small fraction of the germlings (<10%). Networks of microtubules were visible, however, in swollen but ungerminated *mipA*$^+$, *pyrG89* conidia (data not shown). When conidia from heterokaryon H1 were incubated in nonselective medium, the *mipA*$^+$, *pyrG89* conidia germinated, grew normally, and contained normal arrays of cytoplasmic (Figure 5C) and mitotic (Figure 5I) microtubules. *mipAd1*, *pyr4*$^+$ conidia germinated but grew more slowly than the *mipA*$^+$, *pyrG89* conidia. As in medium lacking uridine, *mipAd1* germlings (Figures 5A–5C) lacked MA, and only a small fraction of the germlings had a few cytoplasmic microtubules.

To obtain good immunofluorescence staining of microtubules in A. nidulans, it is necessary to digest the hyphal wall with Novozym 234. Since Novozym 234 contains proteases, overdigestion causes microtubules to be lost. Old hyphae require longer digestion times than germlings to give optimal staining of microtubules, presumably owing to differences in wall thickness or composition. Since *mipAd1*, *pyr4*$^+$ germlings grow slowly, we thought it possible that their wall composition might be similar to that of old hyphae, and the apparent scarcity of microtubules might be due to incomplete digestion of the wall. Consequently, we varied times of digestion and concentrations of Novozym 234 in an effort to detect additional microtubules in *mipAd1*, *pyr4*$^+$ germlings. With the gentlest digestion conditions we used, no microtubules were visible, presumably because of insufficient wall digestion. As we increased enzyme amount and/or digestion times, a few cytoplasmic microtubules became visible in a small fraction (<10%) of the germlings. Additional increases in enzyme amount and/or digestion times gave similar results until, eventually, microtubule staining was lost, presumably owing to proteolysis of microtubules. In no case did more than a small fraction of the *mipAd1*, *pyr4*$^+$ germlings contain any cytoplasmic microtubules. Among several thousand *mipAd1*, *pyr4*$^+$ germlings examined, we found only two with MA. In both cases the germlings were very short and had apparently just germinated.

γ-Tubulin Is a Component of the Spindle Pole Body

To determine the intracellular location of γ-tubulin by immunofluorescence microscopy, we prepared γ-tubulin–specific antibodies. We elicited rabbit polyclonal antisera against a trpE–γ-tubulin fusion protein and affinity purified antibodies in two ways (see Experimental Procedures).

Figure 6. Localization of γ-Tubulin by Immunofluorescence Microscopy

(A–I) Photomicrographs of the diploid strain R21/R153.

(A) Phase contrast micrograph of a hypha. (B) Fluorescence micrograph of the same field showing nuclei as revealed by DAPI staining. (C) Fluorescence micrograph of the same field showing staining with γ-tubulin antibodies. The primary antibody was peptide-purified anti-γ-tubulin. The secondary antibody was TRITC-labeled goat anti-rabbit IgG. There is a single γ-tubulin spot associated with each nucleus.

(D–F) The material was treated with a mouse monoclonal antibody against α-tubulin (DM1A) and peptide-purified rabbit anti-γ-tubulin. FITC-labeled goat anti-mouse IgG and TRITC-labeled goat anti-rabbit IgG were used as secondary antibodies. (D) A phase contrast micrograph of germlings. (E) A fluorescence micrograph of the same field showing anti-α-tubulin staining. Several MA are visible (arrows). Other MA were partially or completely out of the plane of focus. The level of background staining is high owing to characteristics of the secondary antibody. (F) A fluorescence micrograph of the same field showing anti-γ-tubulin staining. γ-Tubulin–containing spots at the end of each MA are visible. Fainter spots are also visible, particularly in the conidial swelling at the upper right.

(G–I) This material has been treated with a mixture of peptide-purified anti-γ-tubulin and DM1A to which the synthetic peptide (7.5 μg/ml added 1 hr before use) that was used to purify the anti-γ-tubulin was added. FITC-labeled goat anti-rabbit IgG and TRITC-labeled goat anti-mouse IgG were used as secondary antibodies. (G) Phase contrast. (H) Anti-α-tubulin staining. MA are designated with arrows. (I) Anti-γ-tubulin staining. The secondary antibody combination was chosen so that MA are barely visible, owing to fluorescent "leak through" (the FITC filter set was less selective than the TRITC set). The MA are designated with arrows. The staining of spots at the ends of the MA by the anti-γ-tubulin antibody is eliminated by preadsorption. The staining of a few other, fainter spots in the cytoplasm was not eliminated by preadsorption.

(J–L) Anti-γ-tubulin staining of a germling carrying *mipAd1*. Anti-γ-tubulin affinity-purified using T7–γ-tubulin was the primary antibody, and TRITC-labeled goat anti-rat IgG was the secondary. The material was also stained with DAPI. (J) Phase contrast. (K) DAPI staining. The wild-type germling at the top has several nuclei. The germling at the bottom carries *mipAd1* and has a single nucleus. (L) Anti-γ-tubulin staining. Anti-γ-tubulin stained dots are visible in association with the nuclei of the wild-type germling but not in association with the nucleus in the germling carrying *mipAd1*. Because of the shallow depth of focus, a single micrograph cannot demonstrate the absence of stained dots in the germling carrying *mipAd1*, but by carefully focusing through the germling we were able to confirm that there were no such spots.

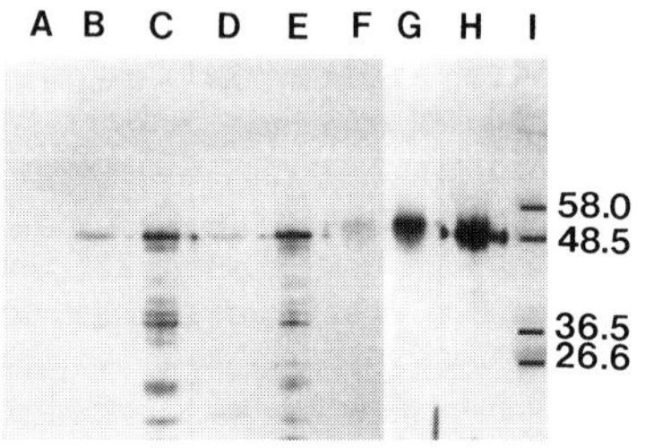

Figure 7. Determination of Specificity of Antibodies by Western Blotting

Lanes A, F, G, and H contain 0.15 μg of purified A. nidulans α- and β-tubulin. Lanes B and D contain 25 μg of protein from the wild-type A. nidulans strain, FGSC4. Lanes C and E contain 5.6 μg of protein from strain Ypma7, in which γ-tubulin overexpression was induced.

(A–C) Peptide-purified anti-γ-tubulin was used as the primary antibody. (A) Very faint staining of a band of the predicted size of γ-tubulin (50,000 daltons) may be visible. This staining is not due to binding of the antibody to α- or β-tubulin because if the sample is severely overloaded the antibody stains a thin band between the thick, overloaded α- and β-tubulin bands. We believe that a very small amount of γ-tubulin is copurifying with the α- and β-tubulin. (B) γ-Tubulin staining is visible as is staining of a faint lower band. The lower band is almost certainly a γ-tubulin breakdown product because it has the same mobility as a major breakdown product of overexpressed γ-tubulin. (C) Induction of γ-tubulin synthesis causes the expected increase in the intensity of the γ-tubulin band, revealing that the band is, indeed, γ-tubulin. Numerous lower bands appear, presumably due to proteolysis of the overexpressed γ-tubulin.

(D–F) The primary antibody was anti-γ-tubulin affinity-purified using T7–γ-tubulin. Results are similar to those in (A–C) except that this antibody recognizes a lower molecular weight band in the FGSC4 extract that is not present in the material in which γ-tubulin is overexpressed.

(G) Primary antibody DM1A (anti-α-tubulin).

(H) Primary antibody was an anti-β-tubulin antibody provided by Dr. Greg May.

First, we affinity purified antibodies on a column carrying a synthetic peptide corresponding to the C-terminal 13 amino acids of A. nidulans γ-tubulin. The amino acid sequence of this peptide is not found in any of the A. nidulans α- or β-tubulins, and antibodies that bind to this peptide should be specific for γ-tubulin. Second, we purified antibodies on a column carrying a γ-tubulin fusion protein expressed in Escherichia coli using a T7 vector system (Studier and Moffatt, 1986). Western blotting experiments (discussed in Experimental Procedures) revealed that antibodies affinity purified by each procedure bound to γ-tubulin. The peptide-purified antibodies appeared to be completely specific for γ-tubulin (Figure 7). The antibodies purified using the T7 fusion protein reacted with one low molecular weight band in addition to γ-tubulin and, on heavily overloaded blots of purified α- and β-tubulin, showed weak cross-reactivity with α-tubulin (Figure 7). The antibodies purified by each procedure produced identical results in immunofluorescence experiments, and we used the peptide-purified antibodies for all experiments unless otherwise indicated. We have used a diploid strain for most of our immunofluorescence experiments because nuclei and MA are larger in diploids than haploids.

In haploid and diploid hyphae that are wild type at the *mipA* locus, the γ-tubulin antibodies stained dots in the hyphae; in material stained with γ-tubulin antibodies and DAPI it was apparent that most of the dots were associated with the peripheries of nuclei (Figures 6A–6C and 6J–6L). It appeared, moreover, that there was one stained dot associated with the periphery of each interphase nucleus and two stained dots associated with each mitotic nucleus. We quantified the association of the stained dots with nuclei in 50 germlings chosen at random from three cover slips prepared in two separate experiments. The germlings contained 306 nuclei; 294 of these were interphase nuclei that had a single, stained spot associated with each of them. Nine were mitotic nuclei with two stained spots associated with each nucleus, and three were interphase nuclei without any stained spots. In view of the difficulties in making the cell walls permeable to antibodies, as previously discussed, we believe that the lack of stained spots with these three nuclei is due to the lack of antibody penetration.

The only organelle consistently associated with the nuclear envelope is the spindle pole body (SPB), which is, in fact, embedded in the nuclear envelope. In the G_1 phase of the cell cycle there is a single SPB associated with each nucleus. In haploid cells it is approximately 0.3 μm in diameter (see Figure 3 in Oakley and Morris, 1983) and thus would appear to be a small spot if observed by light microscopy. The SPB replicates approximately halfway through interphase (Oakley and Morris, 1983), and in G_2 there are two SPBs joined by a bridge (see Figure 2 in Oakley and Morris, 1983). Since the two G_2 SPBs are separated by much less than 0.1 μm, they would appear to be a single spot if observed by light microscopy. In mitosis, the microtubules of the MA assemble between the two SPBs (Oakley and Morris, 1983), and as they move apart the two SPBs would be distinguishable by light microscopy. The staining pattern we observed with the γ-tubulin antibodies is, thus, exactly as one would expect if γ-tubulin were localized at the SPBs.

To verify that the γ-tubulin antibodies were indeed staining SPBs, we double stained germlings with γ-tubulin antibodies and a mouse monoclonal antibody to α-tubulin (DM1A). We used secondary antibodies labeled with different fluorochromes to allow us to visualize, separately, microtubules and the spots stained with the γ-tubulin antibodies. We also stained with DAPI to visualize DNA. As before, the γ-tubulin antibodies brightly stained a single dot associated with each interphase nucleus. Cytoplasmic microtubules terminated at some but not all of these dots. Indeed, many nuclei were not associated with any microtubules. This result is consistent with the dots being SPBs because previous electron microscopy observations indicate that microtubules terminate at some but not all SPBs (B. R. O. and N. R. Morris, unpublished data). The most striking result of the double staining, however, is shown in Figures 6D–6F. At each end of each MA, we found a spot stained by the γ-tubulin antibodies. This result was true of all of the several hundred MA examined and was true of all mitotic stages, from the earliest stages of MA formation to the end of telophase. By switching between filter sets, we were able to determine that the brightly staining spots did not overlap the microtubule staining pattern but were at the ends of the microtubules. SPBs are at the ends of mitotic spindles; indeed, they are called SPBs because they are at the poles of mitotic spindles. Since the brightly stained spots are at the positions of the SPB at all mitotic stages, we conclude that the γ-tubulin antibodies are staining the SPB.

To verify that the SPB staining was due to the presence of γ-tubulin at the SPB, we preadsorbed a mixture of peptide-purified γ-tubulin antibodies and DM1A, a monoclonal α-tubulin antibody, with an excess of the synthetic γ-tubulin–specific peptide and used the preadsorbed mixture of antibodies to stain germlings. Although some non-SPB staining remained visible in the germlings (see below), the preadsorption appeared to eliminate SPB staining. To quantify the elimination of the SPB staining by preadsorption with the peptide, we examined 100 mitotic nuclei. As expected, the preadsorption did not affect staining by DM1A, and mitotic spindles were readily apparent. The preadsorption completely eliminated the staining of the SPB spots at the ends of the spindles by the γ-tubulin antibodies. We conclude that the SPB staining is due to γ-tubulin at the SPB.

In experiments with nonpreadsorbed antibodies, in addition to the SPB staining we noted a much fainter staining that colocalized with the microtubules of the MA (i.e., in a band between the two brightly stained SPB). This staining was sufficiently faint that it was difficult to reproduce photographically, but we detected it consistently, even with peptide-purified antibodies. Because the staining was so faint, it was difficult to determine if it was eliminated by preadsorption. We did not detect staining of cytoplasmic microtubules with γ-tubulin antibodies nor did we detect SPB staining with either of the α-tubulin antibodies we have used (DM1A and YOL1/34). The secondary antibodies produced only faint background fluorescence when used without primary antibodies.

We detected staining of some spots in the cytoplasm that stained more faintly than SPB and were not usually associated with nuclei (Figures 6F and 6I). These spots are thus not SPB. These spots were somewhat difficult to quantify because many of them were very faint. In addition, the number of these spots varied from experiment to experiment (compare, for example, Figures 6C and 6F). In the 50 germlings in which we quantified the SPB staining (see above), there were approximately 290 non-SPB spots. Preadsorption of peptide-purified antibodies with the γ-tubulin–specific synthetic peptide did not eliminate the staining of the non-SPB spots (Figure 6I). In 50 germlings stained with preadsorbed antibody and chosen at random, we counted 255 such spots. Given the variation in the number of these spots from experiment to experiment in germlings stained with nonpreadsorbed antibodies, we believe that the number in germlings stained with preadsorbed antibody is not significantly different from that in germlings stained with nonpreadsorbed antibodies. We conclude, thus, that the staining of most of these non-SPB spots is due to nonspecific binding of the antibodies, although we cannot rule out the possibility that some of them contain γ-tubulin.

To verify that *mipAd1* caused the expected loss of γ-tubulin, we germinated spores from heterokaryon H1 under nonselective conditions and double stained the germlings with γ-tubulin antibodies and DM1A (Figures 6J–6L). For these experiments we used antibodies affinity purified with the T7–γ-tubulin fusion protein rather than the peptide-purified antibodies. The peptide-purified antibodies specifically bind to the C-terminal end of γ-tubulin, while the fusion protein–purified antibodies should bind to all regions of γ-tubulin and thus detect any truncated γ-tubulin peptide produced in the disruptants. To quantify the results of these experiments, we examined 100 *mipAd1* and 100 *mipA*$^+$ germlings. There were stained SPB spots associated with nuclei in all 100 *mipA*$^+$ germlings. Among 100 *mipAd1* germlings, only two had any SPB staining. While the immunofluorescence of γ-tubulin in the disruptant germlings is subject to the same caveats that we have mentioned above for the staining of microtubules in the disruptants, the simplest explanation for this result is that *mipAd1* causes the expected loss of γ-tubulin.

Discussion

The *mipA* gene disruption, *mipAd1*, is a recessive lethal mutation that strongly inhibits nuclear division. γ-Tubulin, the product of the *mipA* gene, thus appears to be essential for growth in general and nuclear division in particular. A possible alternative interpretation is that *mipA* is not an essential gene but the disruption of *mipA* causes the production of a truncated protein, a poison peptide, that interferes with one or more essential cellular functions. Since the amino acid sequence of the N-terminal region of γ-tubulin is similar to that of α- and β-tubulin, a truncated γ-tubulin peptide might interact with the α/β-tubulin dimer and interfere with the assembly or function of microtubules. Ideally we would like to verify the loss of γ-tubulin in germlings carrying *mipAd1* and test for the existence of the putative poison peptide by Western blotting. These experiments are not feasible, however. The heterokaryon produces a mixture of spores, some carrying *mipAd1* and others carrying *mipA*$^+$, and there is, at present, no way of separating them. It is thus not possible to obtain a pure population of germlings carrying *mipAd1*. Identifying a putative poison peptide is further complicated by the fact that proteolytic cleavage products of γ-tubulin must always be present at some level. It would thus be difficult to determine whether a low molecular weight band on a Western blot was a normal proteolytic breakdown product of γ-tubulin or a poison peptide.

In the absence of Western blotting data, the fact that germlings carrying *mipAd1* show very little staining with γ-tubulin antibodies is a strong indication that *mipAd1* causes the expected loss of γ-tubulin. The fact, moreover, that *mipAd1* is recessive appears to rule out the existence of a poison peptide. If the disrupted gene produced a truncated peptide that interfered with microtubule assembly or function, the peptide should interfere with microtubule assembly or function in *mipAd1*/*mipA*$^+$ heterokaryons, such as H1, as well as germlings carrying only *mipAd1*. *mipAd1* should therefore be a dominant lethal mutation or a semidominant inhibitor of heterokaryon growth. Since *mipAd1* does not inhibit growth at all in *mipAd1/mipA*$^+$ heterokaryons, we conclude that it does not produce a truncated peptide that inhibits microtubule function. It follows that the inviability of germlings carrying *mipAd1* is due to the lack of functional γ-tubulin.

The most striking phenotype of *mipAd1* is the inhibition of nuclear division. Nuclear division is more completely inhibited by *mipAd1* than by several conditionally lethal α- and β-tubulin mutations of A. nidulans that have been characterized previously (Oakley and Morris, 1981; Oakley et al., 1985, 1987a), and the small amount of nuclear division that occurs is probably due to functional γ-tubulin carried over into the conidia from the parental heterokaryon. In germlings that have more than one nucleus, the nuclei are usually not identical in size, suggesting that chromosomes have not segregated equally into the daughter nuclei. We conclude that γ-tubulin is essential for nuclear division.

For nuclear division to occur, DNA replication, mitosis, and a number of other cell cycle events must be completed successfully. *mipAd1* could, in principle, block nuclear division by inhibiting mitosis or any of the other events of the cell cycle. *mipAd1* does not appear to block the cell cycle, however. If *mipAd1* blocked the cell cycle in G_1, S, or G_2, the mitotic index should be much lower than normal, and if it blocked in M, the mitotic index should be much higher than normal. The mitotic index in *mipAd1* germlings was near normal, however (Table 1). The fact that nuclei in *mipAd1* germlings are large and apparently contain more than the normal amount of DNA also seems to rule out the possibility that *mipAd1* blocks the DNA replication cycle. The simplest explanation of the data is that nuclei in *mipAd1* germlings go through the cell cycle and the chromosomes condense at M, but the nucleus does not divide before entering G_1.

The blockage of nuclear division without an increase in the mitotic index is surprising because antimicrotubule agents generally cause a cell cycle blockage in M as do conditionally lethal α- or β-tubulin mutations under restrictive conditions (Oakley and Morris, 1981; Toda et al., 1984; Hiraoka et al., 1984). (Note, however, that with *benA33*, a β-tubulin mutation of A. nidulans, the blockage is transient [Figure 2 in Oakley and Morris, 1981].) In Saccharomyces cerevisiae, in which chromosomal condensation is difficult to detect, conditionally lethal β-tubulin mutations cause a cell cycle arrest at a large budded stage (Huffaker et al., 1988), as do antimicrotubule agents (Pringle et al., 1986; Quinlan et al., 1980; Wood and Hartwell, 1982). Not all α-tubulin mutations of S. cerevisiae, however, arrest growth at the same point of the cell cycle (Schatz et al., 1988). Consequently, blockage in mitosis with consequent increase in mitotic index is not a necessary characteristic of all tubulin mutations. Antimicrotubule agents and the majority of tubulin mutations must directly or, more likely, indirectly, inhibit the activation of the cellular machinery that causes the M to G_1 transition, while *mipAd1* does not.

mipAd1 also inhibits migration of nuclei along hyphae. This inhibition is not complete but is similar to the inhibition obtained with heat-sensitive or cold-sensitive α- and β-tubulin mutations of A. nidulans (Oakley and Morris, 1981; Oakley et al., 1985, 1987a).

The genetic data of Weil et al. (1986) suggest that the product of *mipA*, which we have subsequently determined to be γ-tubulin, interacts specifically, probably physically, with the β-tubulin encoded by the *benA* gene. These data, coupled with the failure of mitosis and the inhibition of nuclear migration in *mipAd1* germlings, lead us to conclude that γ-tubulin is required for the functioning, and probably the assembly, of microtubules. While this conclusion may not be surprising, it is significant because identifying proteins required for microtubule function is central to understanding microtubule function, and very few proteins other than α- and β-tubulin are known to be necessary for microtubule function in vivo.

The immunofluorescence data we have obtained appear to show that *mipAd1* causes a substantial loss of cytoplasmic microtubules and a virtually complete loss of MA. While we cannot rule out the possibility that there are microtubules in *mipAd1* germlings that we did not detect, there were significant internal controls in our experiments in that there was excellent staining of MA microtubules and cytoplasmic microtubules in the G191 germlings on the same cover slips as the *mipAd1* germlings.

Immunofluorescence microscopy with γ-tubulin antibodies of germlings that are wild type at the *mipA* locus revealed that γ-tubulin is located at interphase and mitotic SPBs. The staining was bright and consistent, and staining of SPBs with the peptide-purified antibody was eliminated by preadsorption with the γ-tubulin–specific peptide. We conclude that γ-tubulin is located at the SPB. Light microscopy does not, however, afford the resolution necessary to determine whether γ-tubulin is a component of the SPB or is associated with the SPB.

We have found staining of some cytoplasmic spots that are not SPBs as well as faint staining that colocalized with the microtubules of the MA. Preadsorption of the peptide-purified antibody with the synthetic peptide used for purification of the antibodies did not eliminate the staining of the cytoplasmic spots. We conclude that few, if any, of these spots contain γ-tubulin. The mitotic spindle staining may indicate that there is a very small amount of γ-tubulin associated with (but not necessarily part of) the microtubules of the MA. Since the staining was very faint, it is possible that it may be due to weak, nonspecific binding of the γ-tubulin antibodies to the microtubules of the MA. We do not believe that the staining is due to weak cross-reactivity of γ-tubulin antibodies with α- or β-tubulin because we detected the staining with peptide-purified antibodies that do not cross-react with α- or β-tubulin (Figure 7). It is important to note, moreover, that the spindle staining is so faint that it is almost undetectable, and the predominant location of γ-tubulin is at the SPB.

The results reported here and the previous results of Weil et al. (1986) and Oakley and Oakley (1989) suggest a simple model for the function of γ-tubulin in vivo. To provide background for this model we briefly will discuss SPBs and related organelles. SPBs are members of a class of organelles called microtubule organizing centers (Pickett-Heaps, 1969). These organelles are found in cells of

γ-Tubulin Function and Location
1299

animals, fungi, algae, and protozoans and have a central role in the temporal and spatial regulation of microtubule assembly. They show a great deal of morphological diversity but share important characteristics. They nucleate the assembly of microtubules, i.e., microtubules assemble from microtubule organizing centers at concentrations of tubulin that will not support microtubule assembly in other regions of the cell. As a consequence, most of the microtubules in many types of cells are associated with, and attached to, microtubule organizing centers. In addition, centrosomes (animal microtubule organizing centers) have been shown to establish the polarity of microtubules (Heidemann et al., 1980). The establishing of microtubule polarity is, in turn, thought to be essential for mitosis and microtubule-mediated organellar translocation. As yet there is no evidence as to the mechanisms by which microtubule organizing centers nucleate microtubule assembly and establish microtubule polarity, but these mechanisms are fundamental to the functioning of microtubules in many, perhaps all, cells.

The genetic interactions between mutant *mipA* and *benA* alleles detailed by Weil et al. (1986) suggest a physical interaction between γ- and β-tubulin. This is not surprising because amino acid sequence similarities suggest that the structure of γ-tubulin may be similar to that of α- and β-tubulin. Thus, γ-tubulin might bind to β-tubulin at the same site that α-tubulin binds to β-tubulin. Since γ-tubulin is at the SPB and interacts with β-tubulin, we propose that it is bound tightly to the SPB and that it binds to the β-tubulin portion of the tubulin dimer at the ends of microtubules and attaches the microtubules to the SPB. Furthermore, since the disruption of the γ-tubulin gene causes the loss of MA and cytoplasmic microtubules, we propose that γ-tubulin nucleates the assembly of microtubules from the SPB. Aggregates of γ-tubulin could, for example, act as seeds for microtubule assembly as do small fragments of microtubules. An important implication of this model is that if γ-tubulin, which is at the SPB, binds specifically to β-tubulin, the β-tubulin portion of the dimer will be proximal to the SPB and the α-tubulin portion distal. This would establish the polarity of microtubules assembled from the SPB. Finally, posttranslational modification of γ-tubulin would be an attractive mechanism for the cell cycle–specific regulation of the assembly of microtubules.

Experimental Procedures

Strains and Media

A. nidulans strain G191 (*pabaA1*, *pyrG89*; *fwA1*, *uaY9*) was obtained from Dr. G. Turner (University of Bristol) via Dr. C. F. Roberts (University of Leicester). BRO2 (*yA2*; *benA33*) was constructed by B. R. O. in the laboratory of Dr. N. R. Morris (UMDNJ-Robert Wood Johnson Medical School). FGSC4 (Glasgow wild type) was obtained from the Fungal Genetics Stock Center (University of Kansas Medical Center, Kansas City, Kansas) via Dr. N. R. Morris. The diploid strain used for immunofluorescence experiments was constructed from strains R21 (*yA2*, *pabaA1*) and R153 (*wA3*, *pyroA4*), which were obtained from Dr. C. F. Roberts. Y12 (*pyrG89*, *pyroA4*, *yA2*) was created in our lab. E. coli strain JM109 (Yanisch-Perron et al., 1985) was used for transformation, propagation, and purification of plasmids. Media for selection against growth of strains carrying *pyrG89* were YG (5 g/l yeast extract, 20 g/l dextrose), YAG (YG with 15 g/l agar), or FYG, which is the same as YAG except that the agar is replaced by 25 g/l Pretested Burtonite 44c (TIC Gums). Nonselective solid medium for strains carrying *pyrG89* was YAG or FYG supplemented with 10 mM uridine and 1.0 mg/ml uracil. Nonselective liquid medium was YG supplemented with 10 mM uridine.

Disruption of *mipA*

Plasmid pLO12 was prepared by an alkaline lysis miniprep procedure (Maniatis et al., 1982) and purified additionally using Gene Clean (Bio 101 Inc.). G191 was then transformed with the DNA using the procedure of Oakley et al. (1987b). Transformants were selected on YAG osmotically balanced with 1.0 M sucrose.

Southern Hybridizations

Southern hybridizations were carried out as previously described (Oakley et al., 1987c). H1 and H2 DNAs were prepared by a miniprep procedure (Oakley et al., 1987c). BRO2 and G191 DNAs were prepared by the method of Oakley et al. (1987b) and purified over two cesium chloride gradients.

Fluorescence Microscopy

For observations of microtubules in heterokaryons carrying *mipAd1*, germlings were grown and stained with DAPI as previously described (Oakley et al., 1985) except that no agar was added to prevent clumping. DAPI concentrations were 30 ng/ml or 240 ng/ml. For immunofluorescence observations of microtubules in germlings of heterokaryon H1, specimens were prepared as follows. Conidia were harvested aseptically and suspended in sterile distilled water to a concentration of 1 × 10^8 per ml. Four microliters of the suspension was spread on a sterile cover slip and allowed to air dry. Cover slips were incubated in sterile petri dishes containing liquid medium at 32°C or, more often, 37°C. For fixation, the cover slips were transferred to 50 mM PIPES (pH 6.7), 25 mM EGTA, 5 mM $MgSO_4$, 5% dimethyl sulfoxide, 8% formaldehyde and incubated at room temperature for 45 min. The cover slips were then rinsed four times in PEM buffer (50 mM PIPES [pH 6.7], 25 mM EGTA, 5 mM $MgSO_4$). For wall digestion, the cover slips were transferred to a solution containing 50% v/v egg white, 5% w/v driselase (Sigma), 1% w/v Novozym 234 (Novo Industri), and 2.5 mM EGTA. Novozym was purified by ammonium sulfate precipitation before use. In attempting to detect more microtubules in *mipAd1* germlings, digestion times were varied from 30 min to 120 min at room temperature and Novozym was used at concentrations of 1% and 2% w/v. For digestion solutions containing 1% Novozym, digestion times of 60–90 min were optimal. Cover slips were rinsed four times in PEM and extracted in 100 mM PIPES (pH 6.7), 25 mM EGTA, and 0.1% v/v Nonidet P40 for 30 s. They were then rinsed four times in PEM before incubating for 45–90 min in primary antibody (YOL1/34 [Accurate] diluted in PEM). Cover slips were rinsed four times in PEM before incubation for 45–90 min in the dark in the secondary antibody solution (fluorescein isothiocyanate [FITC]-labeled, affinity-purified rabbit anti-rat IgG [Sigma] diluted in PEM). After rinsing in 60 mM Tris (pH 7.2), 200 mM NaCl four times, cover slips were stained in 700 ng/ml DAPI and washed four more times in 60 mM Tris (pH 7.2), 200 mM NaCl. They were then mounted in Citifluor (City University, London) before observation and photography using a Zeiss 1.30 NA, 100×, Neofluor objective. Photographs were taken with T-Max 400 film (Kodak) rated at ASA 1250 and developed in diafine (Acufine).

For staining with γ-tubulin antibodies, the same procedure was used with the following modifications. Cover slips were simply placed in petri dishes and medium inoculated with conidia (5 × 10^5/ml) was added. The conidia simply settled onto the cover slips and germinated. The germlings and ungerminated conidia adhered to the cover slips through incubation, fixation and staining. A batch of Novozym 234 (batch PPM 1961) was found to be low in protease activity and was used without ammonium sulfate precipitation. It was used at a final concentration of 2.5% with no driselase added. For secondary antibodies we used FITC- or tetramethylrhodamine B isothiocyanate (TRITC)-labeled polyclonal goat antibodies against rabbit IgG (Sigma). For double staining experiments we used DM1A, a mouse monoclonal antibody against α-tubulin (Sigma) along with our affinity-purified γ-tubulin antibodies. As secondary antibodies for DM1A, we used FITC- or TRITC-labeled goat polyclonal antibodies against mouse IgG (Sigma). To remove nonspecific binding, the secondary antibodies were preadsorbed for 1 hr or more on ice with an A. nidulans acetone powder prepared by the method of Harlow and Lane (1988). The extraction step was for 5 min. The cover slips were then washed once in PEM, and all subsequent washes were in 60 mM Tris (pH 7.2), 200 mM NaCl with 2% bovine serum albumin (BSA). DAPI staining was for 10 min at a DAPI concentration of 0.02–0.1 μg/ml in 60 mM Tris (pH 7.2), 200 mM NaCl with 1.6% BSA. Photographs were taken with T-Max 400 rated at ASA 800 or 1600 and developed in T-Max developer (Kodak).

Production and Purification of Antibodies

An A. nidulans γ-tubulin cDNA–E. coli trpE fusion polypeptide was created as follows. A cDNA encoding all but the three N-terminal amino acids of A. nidulans γ-tubulin was inserted into the trpE expression vector pATH11 to form plasmid pKJ20. pATH11 was modified from the vectors created by Spindler et al. (1984) and was generously given to us by Dr. Mary Crivellone from the lab of Dr. A. Tzagoloff at Columbia University. In this construction, sequences encoding the C-terminal 451 amino acids of A. nidulans γ-tubulin were fused in frame at the carboxyl end of sequences encoding 336 amino acids of the E. coli trpE protein. pKJ20 was transformed into bacterial strain RR1 (F^-, *hsdS20*[r_B^-, m_B^-], *ara-14*, *proA2*, *lacY1*, *galK2*, *rpsL20*[Sm^r], *xyl-5*, *sup-E44*, $lambda^-$), and expression of the fusion polypeptide was induced with indoleacrylic acid. Inclusion bodies containing the fusion polypeptide were harvested by centrifugation. The fusion polypeptide was purified by SDS–polyacrylamide gel electrophoresis and recovered by electroelution. Three New Zealand White rabbits were injected intradermally with 500 μg of the fusion polypeptide in complete Freund's adjuvant. Production of antibodies against γ-tubulin was monitored with ELISA assays using a synthetic peptide specific for A. nidulans γ-tubulin. All three rabbits produced antibodies against the peptide. The rabbits were boosted by intravenous injection of 50–100 μg of fusion polypeptide.

We affinity purified antibodies from one rabbit by a three stage procedure. In the first step, immune serum was passed over an affinity column with bacterial proteins as ligand to remove antibodies raised against the bacterial portion of the immunogen and against unrelated residual bacterial proteins. This column was prepared with total protein from RR1 bacteria overexpressing the *trpE* gene from pATH11. The total bacterial protein was coupled to 6-aminohexanoic acid–activated sepharose 4B (Sigma) in 0.1 M $NaHCO_3$ (pH 8.0), 0.5 M NaCl. Antibodies not retained by this column were carried through the next two steps.

In the second step, antibodies were isolated on an affinity column on which the ligand was a synthetic 14-amino-acid peptide (Peninsula Laboratories, Inc.), consisting of the 13 C-terminal amino acids of A. nidulans γ-tubulin (YLDPDAGKDEVGV) with a cysteine added at the N-terminus for ease of coupling. This sequence is not found in the A. nidulans α- or β-tubulin proteins. The peptide was coupled to 6-aminohexanoic acid–activated sepharose 4B in 0.1 M $NaHCO_3$ (pH 8.0), 0.5 M NaCl. Antibodies that did not bind to the bacterial protein column in step 1 were passed over the column several times. Unbound antibodies were retained for subsequent purification. The column was washed extensively with TBS (6 mM Tris, 166 mM NaCl [pH 7.5]) followed by TBS with 0.2% Triton X-100, followed by TBS. Bound antibodies were eluted in the opposite direction from sample application in 4.0 M $MgCl_2$, dialyzed against TBS, and concentrated by ultrafiltration.

Antibodies that did not bind to the peptide column in step 2 were affinity purified as follows. A cDNA encoding the C-terminal 451 amino acids of A. nidulans γ-tubulin was inserted in frame into the T7 expression vector pET-5a (Studier and Moffatt, 1986). The expressed fusion polypeptide contains 11 N-terminal amino acids of the bacteriophage T7 gene 10, 3 amino acids formed by linker sites, and the C-terminal 451 amino acids of A. nidulans γ-tubulin. The fusion polypeptide was expressed using the E. coli strain BL21(DE3) and procedures of Studier and Moffatt (1986). Inclusion bodies containing the fusion polypeptide were harvested by centrifugation, purified by SDS–polyacrylamide gel electrophoresis, and recovered by electroelution. Purified fusion polypeptide was coupled to an Affigel 15 column (Bio-Rad) in 50 mM MOPS (pH 7.5). After the serum was applied to the column, it was washed extensively with TBS, followed by TBS with 0.2% Triton X-100, followed by TBS. Bound antibodies were eluted with 4.0 M $MgCl_2$, dialyzed against TBS, and concentrated by ultrafiltration. Since the fusion polypeptide used as an antigen and the fusion polypeptide used for affinity purification share only A. nidulans γ-tubulin sequences, the eluate should contain only antibodies that bind to γ-tubulin. The specificity of the affinity-purified antibodies was evaluated by Western blotting (Figure 7).

Sample Preparation and Western Blotting

To demonstrate that the antibodies recognized γ-tubulin, a strain capable of induced overexpression of γ-tubulin was created as follows. A genomic AflII–HindIII fragment containing the wild-type *mipA* gene was cloned into plasmid pAL3 (Waring et al., 1989) creating plasmid pXMA29 (Y. Y. and B. R. O., unpublished data). In pXMA29 the entire γ-tubulin coding sequence is under control of the A. nidulans *alcA* (alcohol dehyrogenase) promoter. pXMA29 was transformed into A. nidulans strain Y12, creating strain Ypma7. For induction of γ-tubulin synthesis, Ypma7 was grown in YG for 15 hr at 37°C, washed with sterile distilled water, and incubated in inducing medium (minimal medium with 4 mM fructose and 50 mM butan-2-1) for 8 hr at 37°C. One hundred milligrams of wet mycelia was ground in 200 μl of extraction buffer (1.0 mM Tris-HCl [pH 6.8], 0.2% SDS, 2 mM phenylmethylsulfonyl fluoride) in a tissue homogenizer on ice. Extraction buffer (100 μl) was added, followed by 300 μl of 2× sample buffer (0.125 M Tris-HCl [pH 6.8], 4% SDS, 10% 2-mercaptoethanol, 20% glycerol, 0.004% bromophenol blue). The sample was mixed by vortexing, boiled for 5 min, cell wall debris was pelleted by centrifugation, and the supernatant was used for subsequent experiments. To prepare samples in which γ-tubulin synthesis was not induced, the wild-type strain FGSC4 was grown in YG without induction and extracted by the same procedure as Ypma7. A. nidulans α- and β-tubulin were purified through successive rounds of assembly and disassembly in taxol (Y. Y. and B. R. O., unpublished data).

We followed the Western blotting protocol recommended by Promega and used Promega reagents, except that blocking was for 1 hr, the primary antibody reactions were for 2 hr, and the secondary antibodies were alkaline phosphatase conjugated goat anti-rabbit or goat anti-mouse, which we adsorbed overnight with an A. nidulans acetone powder.

Acknowledgments

We would like to thank Robert Day, Dr. Patrick Dunne, and Dr. John Doonan for helping develop anti-tubulin immunofluorescence procedures; Jean McGowan for assistance in setting up cultures; Yixian Zheng for initial immunofluorescence observations of H1; and the Ohio State University College of Biological Sciences antibody facility for assistance in production of the antibodies. This work was supported by grant GM31837 from the National Institutes of Health.

The costs of publication of this article were defrayed in part by the payment of page charges. This article must therefore be hereby marked *"advertisement"* in accordance with 18 U.S.C. Section 1734 solely to indicate this fact.

Received December 19, 1989; revised April 9, 1990.

References

Ballance, D. J., Buxton, F. P., and Turner, G. (1983). Transformation of *Aspergillus nidulans* by the orotidine-5′-phosphate decarboxylase gene of *Neurospora crassa*. Biochem. Biophys. Res. Commun. *112*, 284–289.

Harlow, E., and Lane, D. (1988). Antibodies: A Laboratory Manual (Cold Spring Harbor, New York: Cold Spring Harbor Laboratory).

Heidemann, S. R., Zieve, G. W., and McIntosh, J. R. (1980). Evidence for microtubule subunit addition to the distal end of mitotic structures in vitro. J. Cell Biol. *87*, 152–159.

Hiraoka, Y., Toda, T., and Yanagida, M. (1984). The *NDA3* gene of fission yeast encodes β-tubulin: a cold-sensitive *nda3* mutation reversibly blocks spindle formation and chromosome movement in mitosis. Cell *39*, 349–358.

Huffaker, T. C., Thomas, J. H., and Botstein, D. (1988). Diverse effects of β-tubulin mutations on microtubule formation and function. J. Cell Biol. *106*, 1997–2010.

Kilmartin, J. V., Wright, B., and Milstein, C. (1982). Rat monoclonal antitubulin antibodies derived by using a new nonsecreting rat cell line. J. Cell Biol. *93*, 576–587.

Maniatis, T., Fritsch, E. F., and Sambrook, J. (1982). Molecular Cloning: A Laboratory Manual (Cold Spring Harbor, New York: Cold Spring Harbor Laboratory).

Oakley, B. R., and Morris, N. R. (1980). Nuclear movement is β-tubulin dependent in Aspergillus nidulans. Cell *19*, 255–262.

Oakley, B. R., and Morris, N. R. (1981). A β-tubulin mutation in A. nidulans that blocks microtubule function without blocking assembly. Cell *24*, 837–845.

Oakley, B. R., and Morris, N. R. (1983). A mutation in *Aspergillus nidulans* that blocks the transition from interphase to prophase. J. Cell Biol. *96*, 1155–1158.

Oakley, C. E., and Oakley, B. R. (1989). Identification of γ-tubulin, a new member of the tubulin superfamily encoded by *mipA* gene of *Aspergillus nidulans*. Nature *338*, 662–664.

Oakley, B. R., Oakley, C. E., Kniepkamp, K. S., and Rinehart, J. E. (1985). Isolation and characterization of cold-sensitive mutations at the *benA*, β-tubulin, locus of *Aspergillus nidulans*. Mol. Gen. Genet. *201*, 56–64.

Oakley, B. R., Oakley, C. E., and Rinehart, J. E. (1987a). Conditionally lethal *tubA* α-tubulin mutations in *Aspergillus nidulans*. Mol. Gen. Genet. *208*, 135–144.

Oakley, B. R., Rinehart, J. E., Mitchell, B. L., Oakley, C. E., Carmona, C., Gray, G. L., and May, G. S. (1987b). Cloning, mapping and molecular analysis of the *pyrG* (orotidine-5′-phosphate decarboxylase) gene of *Aspergillus nidulans*. Gene *61*, 385–399.

Oakley, C. E., Weil, C. F., Kretz, P. L., and Oakley, B. R. (1987c). Cloning of the *riboB* locus of *Aspergillus nidulans*. Gene *53*, 293–298.

Osmani, S. A., Engle, D. B., Doonan, J. H., and Morris, N. R. (1988). Spindle formation and chromatin condensation in cells blocked at interphase by mutation of a negative cell cycle control gene. Cell *52*, 241–251.

Palmer, L. M., and Cove, D. J. (1975). Pyrimidine biosynthesis in *Aspergillus nidulans*. Isolation and preliminary characterization of auxotrophic mutants. Mol. Gen. Genet. *138*, 243–255.

Pickett-Heaps, J. D. (1969). The evolution of the mitotic apparatus: an attempt at comparative ultrastructural cytology in dividing plant cells. Cytobios *3*, 257–280.

Pringle, J. R., Lillie, S. H., Adams, A. E. M., Jacobs, C. W., Haarer, B. K., Coleman, K. G., Robinson, J. S., Bloom, L., and Preston, R. A. (1986). Cellular morphogenesis in the yeast cell cycle. In Yeast Cell Biology, J. Hicks, ed. (New York: Alan R. Liss, Inc.), pp. 47–80.

Quinlan, R. A., Pogson, C. I., and Gull, K. (1980). The influence of the microtubule inhibitor, methyl benzimidazole-2-yl-carbamate (MBC) on nuclear division and the cell cycle in *Saccharomyces cerevisiae*. J. Cell Sci. *46*, 341–352.

Schatz, P. J., Solomon, F., and Botstein, D. (1988). Isolation and characterization of conditional-lethal mutations in the *TUB1* α-tubulin gene of the yeast *Saccharomyces cerevisiae*. Genetics *120*, 681–695.

Sheir-Neiss, G., Lai, M. H., and Morris, N. R. (1978). Identification of a gene for β-tubulin in Aspergillus nidulans. Cell *15*, 639–647.

Spindler, K. R., Rosser, D. S. E., and Berk, A. J. (1984). Analysis of adenovirus transforming proteins from early regions 1A and 1B with antisera to inducible fusion antigens produced in *Escherichia coli*. J. Virol. *49*, 132–141.

Stebbings, H. (1988). Cytoplasmic dynein graduates. Nature *336*, 14–15.

Studier, F. W., and Moffatt, B. A. (1986). Use of bacteriophage T7 RNA polymerase to direct selective high-level expression of cloned genes. J. Mol. Biol. *189*, 113–130.

Toda, T., Adachi, Y., Hiraoka, Y., and Yanagida, M. (1984). Identification of the pleiotropic cell division cycle gene *NDA2* as one of two different α-tubulin genes in Schizosaccharomyces pombe. Cell *37*, 233–242.

Vallee, R. B., Bloom, G. S., and Theurkauf, W. E. (1984). Microtubule-associated proteins: subunits of the cytomatrix. J. Cell Biol. *99*, 38s–44s.

Waring, R. B., May, G. S., and Morris, N. R. (1989). Characterization of an inducible expression system in *Aspergillus nidulans* using *alcA* and tubulin-coding genes. Gene *79*, 119–130.

Weil, C. F., Oakley, C. E., and Oakley, B. R. (1986). Isolation of *mip* (microtubule-interacting protein) mutations of *Aspergillus nidulans*. Mol. Cell. Biol. *6*, 2963–2968.

Wood, J. S., and Hartwell, L. H. (1982). A dependent pathway of gene functions leading to chromosome segregation in *Saccharomyces cerevisiae*. J. Cell Biol. *94*, 718–726.

Yanisch-Perron, C., Vieira, J., and Messing, J. (1985). Improved M13 phage cloning vectors and host strains: nucleotide sequences of the M13mp18 and pUC19 vectors. Gene *33*, 103–119.

Author Index